Control Keys

Ctrl A	Selects all objects
Ctrl B	Toggles snap mode
Ctrl C	Copies to the Clipboard
Ctrl D	Toggles dynamic UCS
Ctrl E	Cycles through isoplanes
Ctrl F	Toggles object snap mode
Ctrl G	Toggles grid display
Ctrl H	Toggles pick style
Ctrl I	Changes coordinate format
Ctrl K	Displays Hyperlinks dialog box
Ctrl L	Toggles ortho mode
Ctrl N	Opens new drawing
Ctrl O	Opens drawing files
Ctrl P	Displays Plot dialog box
Ctrl Q	Quits AutoCAD
Ctrl R	Cycles through viewports
Ctrl S	Saves drawing
Ctrl T	Toggles tablet mode
Ctrl U	Toggles polar tracking
Ctrl V	Pastes from the Clipboard
Ctrl X	Cuts to the Clipboard
Ctrl Y	Redoes last undo (MRedo)
Ctrl Z	Undoes last command

Ctrl 0	Toggles clean screen mode
Ctrl 1	Toggles Properties window
Ctrl 2	Toggles DesignCenter window
Ctrl 3	Toggles Tool Palettes window
Ctrl 4	Toggles SheetSet Manager
Ctrl 6	Toggles dbConnect Manager
Ctrl 7	Toggles Markup Set Manager
Ctrl 8	Toggles QuickCalc window
Ctrl 9	Toggles the command line

Ctrl Shift A	Toggles group mode
Ctrl Shift C	Copies with base point
Ctrl Shift H	Toggles display of open palettes
Ctrl Shift P	Toggles Quick Properties palette
Ctrl Shift S	Displays Save As dialog box
Ctrl Shift V	Pastes with insertion point

Function Keys

F1	Displays online help
F2	Toggles text & grap[illegible] windows
F3	T[illegible]
F4	
F5	
F6	
F7	
F8	[illegible]oggles ortho mode
F9	Toggles snap mode
F10	Toggles polar tracking
F11	Toggles object snap tracking
F12	Toggles dynamic input

Shift F1	Selects entire objects
Shift F2	Selects vertex subobjects
Shift F3	Selects edge subobjects
Shift F4	Selects face subobjects
Shift F5	Toggles solid history

Ctrl F4	Closes drawing
Ctrl F6	Switches to next drawing
Alt F4	Closes AutoCAD
Alt F8	Runs VBA
Alt F11	Opens VBA IDE

New in AutoCAD 2011

Other Keystrokes

Esc	Cancels commands and grips
Delete	Deletes selected objects
Enter	Executes and repeats commands
Shift	Ortho override
Ctrl	LockUI override; select faces, edges, or vertices

Shift A	Osnap override
Shift C	CENter override
Shift D	Snap/tracking override
Shift E	ENDpoint override
Shift M	MIDpoint override
Shift S	Osnap enforcement
Shift W	Toggles Navigation Wheel
Shift X	Polar override

THE ILLUSTRATED AUTOCAD® 2011 QUICK REFERENCE

Ralph Grabowski

Autodesk

Australia • Brazil • Japan • Korea • Mexico • Singapore • United Kingdom • United States

The Illustrated AutoCAD® 2011 Quick Reference
Ralph Grabowski

Vice President, Career and Professional Editorial: Dave Garza

Director of Learning Solutions: Sandy Clark

Acquisitions Editor: Stacy Masucci

Managing Editor: Larry Main

Senior Product Manager: John Fisher

Editorial Assistant: Andrea Timpano

Vice President, Career and Professional Marketing: Jennifer Baker

Marketing Director: Deborah Yarnell

Marketing Manager: Katie Hall

Associate Marketing Manager: Mark Pierro

Production Director: Wendy Troeger

Senior CPM: Angela Sheehan

Art Director: David Arsenault

Book Design and Typesetting: Ralph Grabowski

Library of Congress Control Number: 2010925287
ISBN-13: 978-1-1111-2516-5
ISBN-10: 1-1111-2516-3

Delmar Cengage Learning
5 Maxwell Drive
Clifton Park, NY 12065
USA

Cengage Learning products are represented in Canada by Nelson Education, Ltd.

For your course and learning solutions, visit **delmar.cengage.com**

Purchase any of our products at your local college store or at our preferred online store **www.ichapters.com**

Printed in the United States of America
1 2 3 4 5 XXX 13 12 11 10

About This Book

The Illustrated AutoCAD 2011 Quick Reference presents concise facts about all 679 commands found in AutoCAD 2011. Each command starts on its own page, and includes one or more of the following: command line options, dialog box and palette options, toolbar icons, shortcut menus, related commands, related system variables, and tips. The clearly formatted reference book is illustrated with hundreds of figures, as well as the following features:

- Variations of commands, such as the View and -View commands.
- Dozens of commands and system variables not documented by Autodesk.
- Quick start tutorials for new concepts, such as dimensional and geometric constraints.
- Icons, such as ☑ and ⊙, that indicate default settings of controls in dialog boxes and palettes.
- Over 100 definitions of acronyms and hard-to-understand terms.
- Nearly 1,000 context-sensitive tips.
- Names and descriptions of Express Tool commands listed in Appendix A.
- Obsolete commands that no longer work in AutoCAD listed in Appendix B.
- All system variables, including those not listed by the SetVar command, listed in Appendix C.
- Command alises listed in Appendix D.
- Commands not documented by Autodesk listed in Appendix E.

The name of each command is shown in mixed upper and lower case to help you understand names that are condensed. For example, VpClip is short for "ViewPort CLIP." Each command includes all alternative methods of command input:

- Alternate command spellings, such as Donut and Doughnut.
- Apostrophe (') prefix for transparent commands, such as 'Blipmode.
- Dash (-) prefix for command-line versions and plus (+) prefix for tabbed dialog boxes.
- Command aliases, such as L for the Line command.
- Menu bar and application menu selections, such as Draw ➪ Construction Line for the XLine command.
- Ribbon panel names, such as Home ➪ Modify for the Break command. Note that the ribbon's content varies according to the currently set workspace.
- Control-key combinations, such as Ctrl+E for the Isoplane toggle.
- Function keys, such as F1 for the Help command.
- Status bar and Quick Access toolbar buttons, such as MODEL for the MSpace command.

The version or release number indicates when the command first appeared in AutoCAD, such as Ver.1.0, Rel. 9, or 2011 — useful when working with older versions of AutoCAD. This 14th Edition is updated with AutoCAD 2011's revised user interface running on the Windows 7 operating system.

Special thanks to Stephen Dunning for his copy editing and to Bill Fane for his technical editing. *Soli Deo Gloria!*

Ralph Grabowski

Abbotsford, British Columbia, Canada

April 6, 2010

Table of Contents

Indicates commands new to AutoCAD 2011.

C

N

O

P

Q

R

S

T

U

V

W

X

Z

#

Appendices

About

Rel. 12 Displays the version number and other information about AutoCAD.

Command	*Aliases*	*Keyboard Shortcuts*	*Menu*	*Ribbon*
'about	...	...	**Help** ⇘**About**	...

Command: about

Displays dialog box:

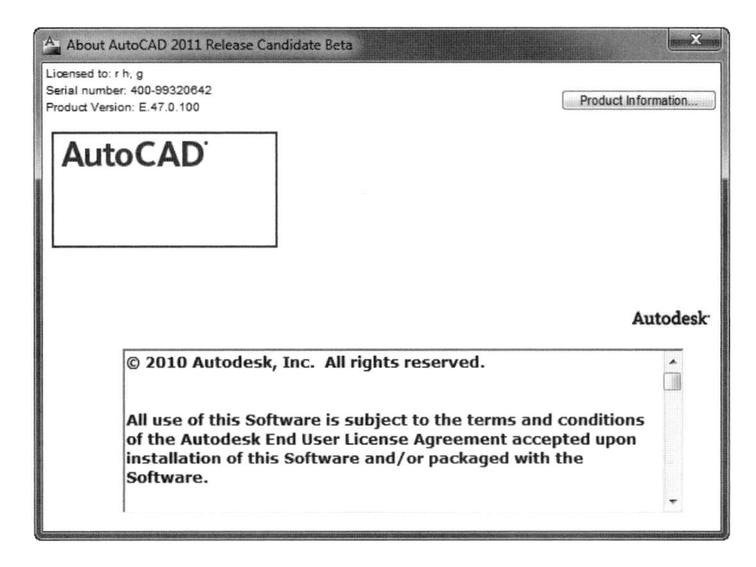

ABOUT AUTOCAD DIALOG BOX

X dismisses dialog box; alternatively, press ESC.

Product Information displays the Product Information dialog box.

PRODUCT INFORMATION DIALOG BOX

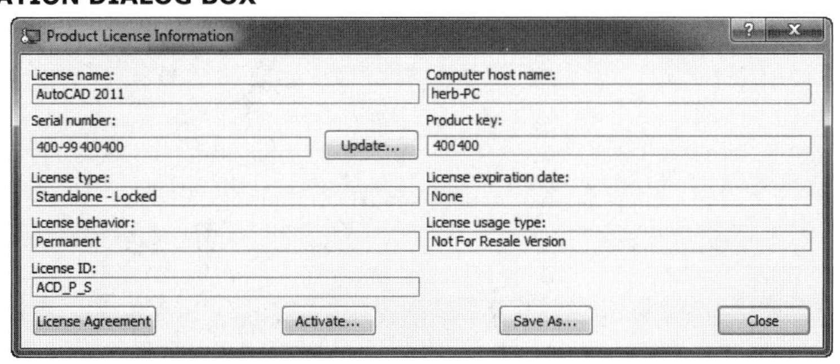

Update displays a dialog box for changing the software serial number.

License Agreement opens the computer's default word processor, and then displays the Autodesk Software License Agreement document (*license.rtf*).

Activate runs the Product Activation wizard.

Save As displays the Save As dialog box, which records the information displayed above as a *.txt* file.

Close returns to AutoCAD.

UPDATE DIALOG BOX

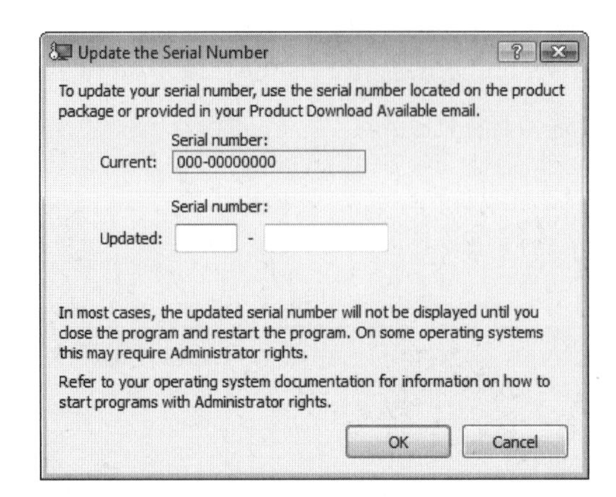

Updated provides space for a new serial number provided by Autodesk. For instance, 30-day trial versions of this software use the serial number 000-00000; changing to a paid version requires you to enter the new serial number here.

ACTIVATE (AUTODESK LICENSING) WINDOW

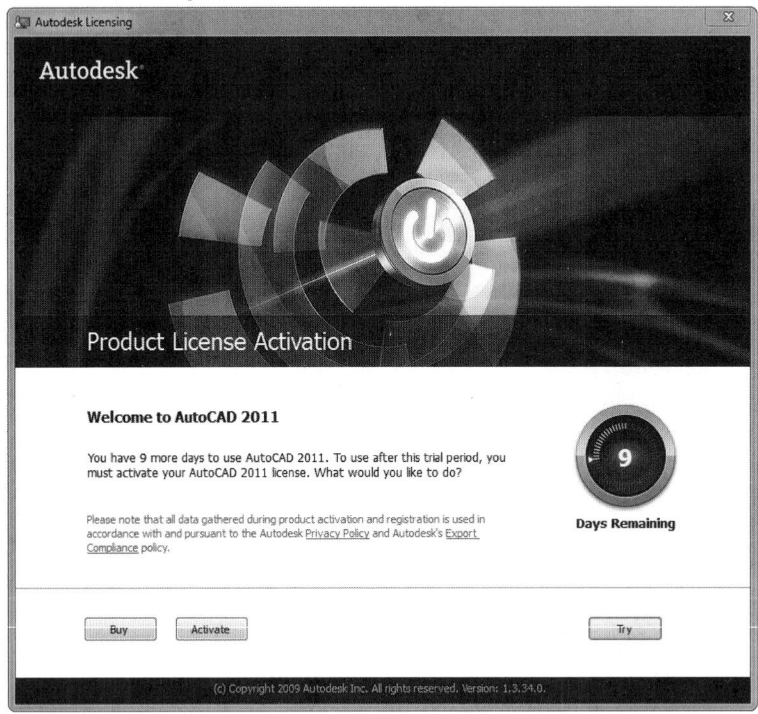

Run runs the software without a license.

Activate activates the license, registering AutoCAD with Autodesk.

RELATED COMMANDS

DwgProps reports information about the drawing.

Status displays information about the drawing and environment.

Time provides data about time.

Properties and **List** report information about selected objects.

RELATED SYSTEM VARIABLES

AcadVer reports the version number of AutoCAD, such as "18.0s."

Product reports the name of the software, such as "AutoCAD."

Platform reports the name of the operating system, such as "Microsoft Windows NT Version 6.1 (x86)" for Windows 7.

_PkSer displays the AutoCAD software serial number, such as "000-0000000."

_Server displays the network authorization code, such as "1."

_VerNum reports the internal build number, such as "D.48.0.300 (Unicode)."

AcisIn / AcisOut

Rel. 13 AcisIn imports *.sat* files into drawings to create 3D solids, 2D regions, and bodies; AcisOut exports 3D solids, 2D regions, and bodies in SAT format (up to v7).

Command	*Aliases*	*Keyboard Shortcuts*	*Menu*	*Ribbon*
acisin	...	...	**Insert** ↳**ACIS File**	**Insert** ↳**Import** ↳**ACIS File**
acisout	...	...	...	

Command: acisin

Displays Select ACIS File dialog box. Select a .sat file, and then click Open.

DIALOG BOX OPTIONS

Cancel dismisses the dialog box.

Open opens the selected *.sat* file.

AcisOut Command

Command: acisout
Select objects: *(Select one or more objects.)*
Select objects: *(Press Enter to end object selection.)*

Displays Create ACIS File dialog box. Enter a name for the .sat file, and then click Save.

DIALOG BOX OPTIONS

Cancel dismisses the dialog box.

Save saves the selected object(s) in a *.sat* file.

RELATED COMMANDS

AmeConvert converts AME v2.0 and v2.1 solid models and regions into solids.

3dPrint and **StlOut** export solid models as STL files.

RELATED SYSTEM VARIABLE

AcisOutVer specifies the ACIS version number for exporting models with the AcisOut command.

RELATED FILE

****.sat*** is the ASCII format of ACIS model files; short for "save as text."

TIPS

- AcisOut exports 3D solids, 2D regions, and bodies, which can be imported into other CAD systems.
- Other ACIS-based software can read *.sat* files.
- "ACIS" comes from the first names of the original developers, "Andy, Charles, and Ian's System." In Greek mythology, Acis was the lover of the goddess Galatea; when Acis was killed by the jealous Cyclops, Galatea turned the blood of Acis into a river.
- ACIS is the name of solids modeling technology from the Spatial Technologies division of Dassault Systemes, and is used by numerous 3D CAD packages. As of AutoCAD 2004, Autodesk uses its own ACIS-derived solids modeler, called ShapeManager.

ActBasepoint

2010 Sets base points in recorded action macros.

Command	*Aliases*	*Keyboard Shortcuts*	*Menu*	*Ribbon*
actbasepoint	...	...	...	**Manage** ↳**Action Recorder** ↳**Insert Base Point**

Command: actbasepoint
Specify a base point: *(Pick a point, or enter x,y coordinates.)*

COMMAND LINE OPTION

Specify a base point picks a x,y point in the drawing.

RELATED COMMANDS

ActRecord records macros.

ActManager manages *.actm* action macro files.

ActUserInput adds pauses for user input retroactively.

ActUserMessage adds dialog box messages to recorded action macros.

RELATED FILE

***.actm** stores action macros.

TIPS

- This command operates only while macros are being recorded.
- The new base point specified by this command affects the relative coordinates of commands that follow in the macro. When the macro is played back, users are asked to specify a base point. Effectively, this becomes a temporary UCS origin for the commands that follow.
- The base point's coordinates can be added, edited, and removed by the Action Recorder. See the ActStop command.
- *Caution!* ACTM files recorded with the ActBasepoint command cannot be played back in AutoCAD 2009.

ActManager

2010 Manages *.actm* action macro files.

Command	*Aliases*	*Keyboard Shortcuts*	*Menu*	*Ribbon*
'actmanager	**...**	**...**	**...**	**Manage** **⇘Action Recorder** **⇘Manage Action Macros**

Command: actmanager

Displays dialog box.

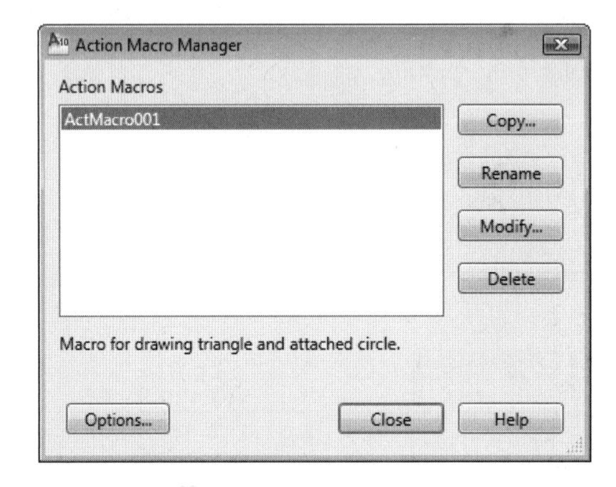

ACTION MACRO MANAGER DIALOG BOX

Select a macro to make button available.

Copy duplicates existing macros.

Rename renames the selected macro. Alternatively, click a macro name twice to rename it.

Modify edits the properties of the macro, not the macro itself (which are edited in the Action Tree); displays the Action Macro dialog box. See the ActRecord command.

Delete erases the selected macro by moving the related *.actm* file to the Recycle Bin (from where it can be recovered, if need be).

Options locates and modifies the path to the folder in which *.actm* files are stored; displays the Files tab of the Options dialog box. See the Options command.

RELATED COMMAND

ActRecord records macros.

RELATED SYSTEM VARIABLES

ActPath specifies path(s) to locate macros.

ActRecPath specifies the current path to store new macros.

RELATED FILE

***.actm** stores action macros in files.

ActRecord / ActStop

2009 Starts and stops recording macros; short for ACTtion RECORDer.

Command	Aliases	Keyboard Shortcuts	Menu	Ribbon
'actrecord	**arr**	...	**Tools** ↳**Action Recorder** ↳**Record**	**Manage** ↳**Action Recorder** ↳**Record**
actstop	**ars**	...	**Tools** ↳**Action Recorder** ↳**Stop**	**Manage** ↳**Action Recorder** ↳**Stop**
-actstop	**-ars**	...	...	...

Command: actrecord

A red "recording" dot appears next to the cursor; see figure.

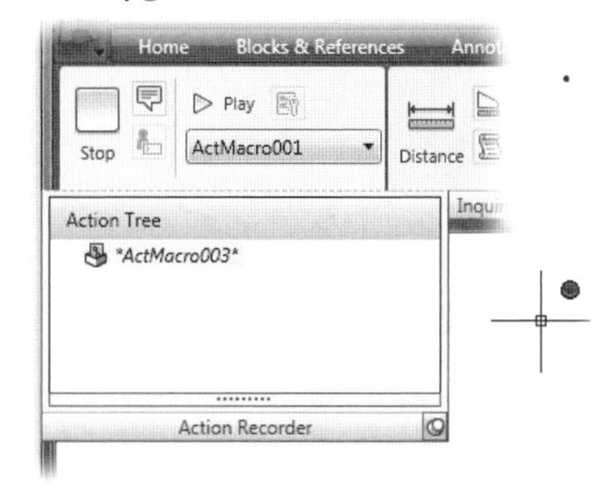

AutoCAD begins recording all user input until ActStop is entered.

If ActUi = 1, then the ActionTree also appears.

ACTSTOP Command

Command: actstop

Stops recording the macro, and then displays dialog box.

ACTION MACROS DIALOG BOX

Action Macro Command Name names the macro for later playback.

File Name reports the file name (user-assigned name + *.actm*).

Folder Path reports the folder in which the *.actm* file will be stored.

Description describes the macro, which is displayed in a tooltip.

Restore Pre-Playback View options

Determines whether the drawing view in effect before the macro was played back is restored during these two situations:

When Pausing for User Input the macro waits for input from the user:

☑ View restored.

☐ View is not restored.

Once Playback Completes the macro finishes playing back:

☑ View restored.

☐ View is not restored.

☑ **Check for Inconsistencies When Playback Begins** scans macros for inconsistencies between the current state of the drawing and its state when the macro was recorded.

SHORTCUT MENUS

Access this menu by right-clicking the name of a macro:

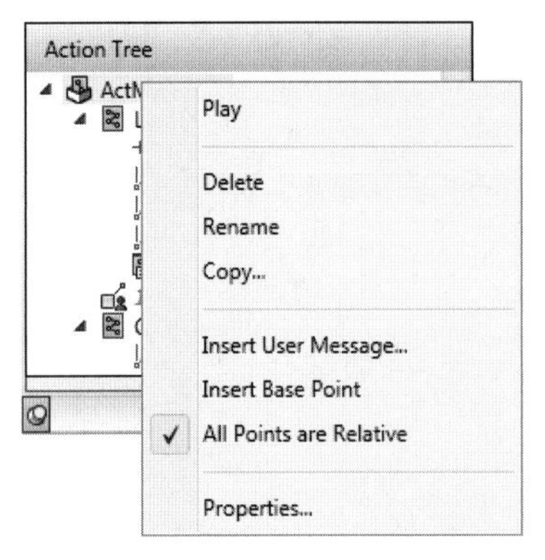

Play plays back the macro.

Delete erases the macro, and places the associated *.actm* file in the Recycling Bin folder (from which it can be retrieved).

Rename changes the name of the macro.

Copy copies the macro, and then prompts you for a new name.

Insert User Message prompts you to enter text for a dialog box.

Insert Base Point inserts a base point, which affects all relative coordinates that follow.

All Points are Relative makes all coordinates relative to each other; turning off this option changes all coordinates to absolute mode (relative to 0,0).

Properties displays the Action Macro dialog box.

Access this menu by right-clicking a command option.

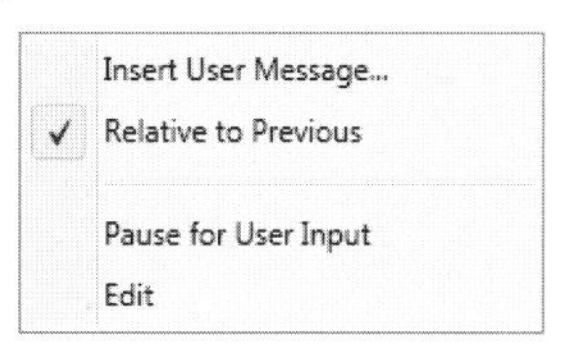

Insert User Message adds a prompt statement; opens the Insert User Message dialog box. See the ActUserInput command.

Relative to Previous makes the current coordinate relative to the previous one.

Request User Input forces the macro to pause until the user provides valid input, such as a coordinate pair.

Edit edits the selected option; commands cannot be edited, other than being deleted.

-ActStop Command

Command: -actstop

Enter action macro name <ActMacro001>: *(Enter a name.)*

Enter an option [Description/Setting/Exit] <Exit>: *(Enter an option.)*

COMMAND OPTIONS

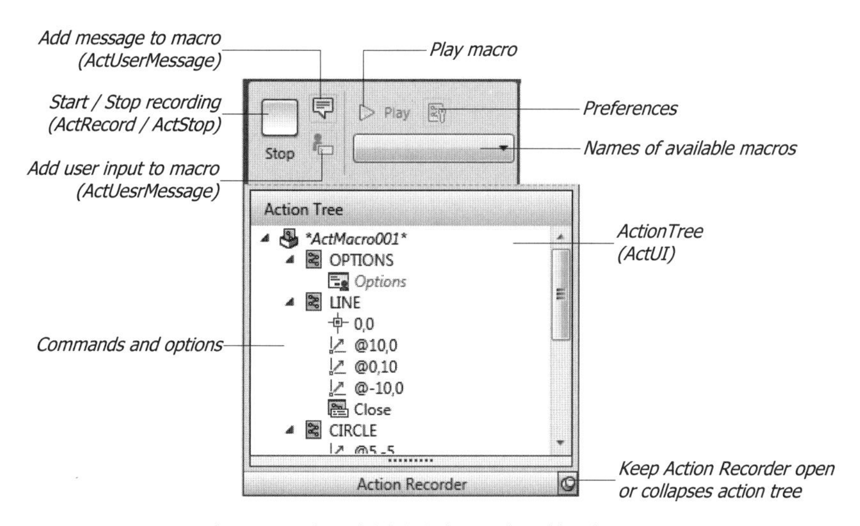

Enter action macro name names the macro, by which it is later played back.

Description prompts you to describe the macro.

Settings specifies how views are to be restored:

Restore the original view before a request for input [Yes/No] <Yes>: *(Enter Y or N.)*

Restore the original view after playback is complete [Yes/No] <Yes>: *(Enter Y or N.)*

Prompt during playback if inconsistencies are found [Yes/No] <Yes>: *(Enter Y or N.)*

Exit exits the command.

RELATED COMMANDS

ActManager manages *.actm* action macro files.

ActUserInput adds pauses for user input retroactively.

ActUserMessage adds dialog box messages to recorded action macros.

The following commands work only during the recording process:

ActBasepoint relocates the base point coordinates.

ActUserInput adds pauses for user input retroactively.

ActUserMessage adds dialog box messages to recorded action macros.

RELATED SYSTEM VARIABLES

ActPath specifies path(s) to locate macros.

ActRecorderState reports whether the Action Recorder is recording or not: (0) inactive; (1) actively recording a macro; (2) actively playing back a macro.

ActRecPath specifies the current path to store new macros.

ActUi determines when the Active Macro panel opens on the ribbon: (0) no change during recording or playback; (1) expands during playback; (2) expands during recording; (4) prompts for name and description when recording is completed.

RELATED FILE

***.actm** stores action macros in ASCII format.

TIPS

- To run the macro, enter its name at the 'Command:' prompt. For instance:

  Command: actmacro001

- The Action Tree lists the content of the macro.
- Point the ActPath system variable to folders on other computers on a network; this allows you to access macros stored on them.
- System variable ActUi changes macro preferences. Click the Preferences button on the ribbon's Tools panel for the related dialog box:

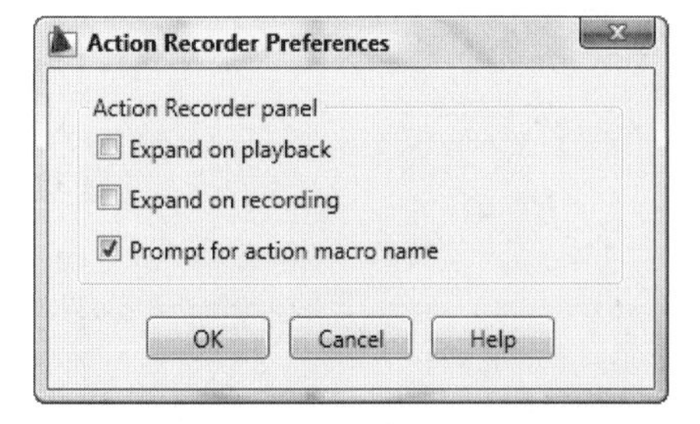

- It is more convenient to use the Action Recorder panel than to enter commands that start, stop, and edit macros.

ActUserInput

2009 Adds pauses for user input to recorded action macros.

Command	Aliases	Keyboard Shortcuts	Menu	Ribbon
'actuserinput	**aru**	**...**	**...**	**Manage ↳Action Recorder ↳Request User Input**

This command operates only during macro recording; enter the command transparently at a prompt, such as during the Line command, as shown below:

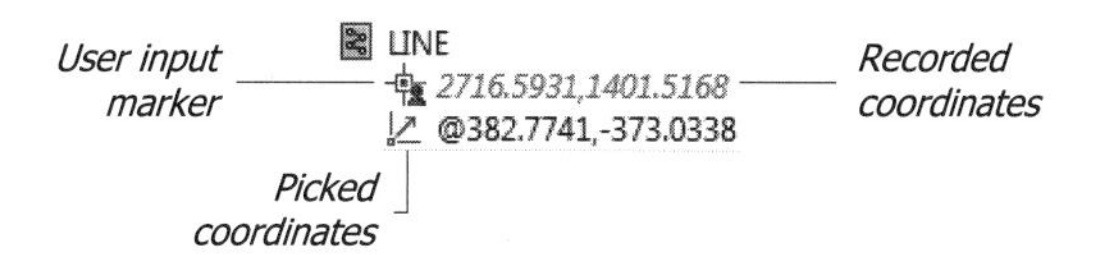

Command: actrecord
Command: line
Specify first point: 'actuserinput

Displays dialog box.

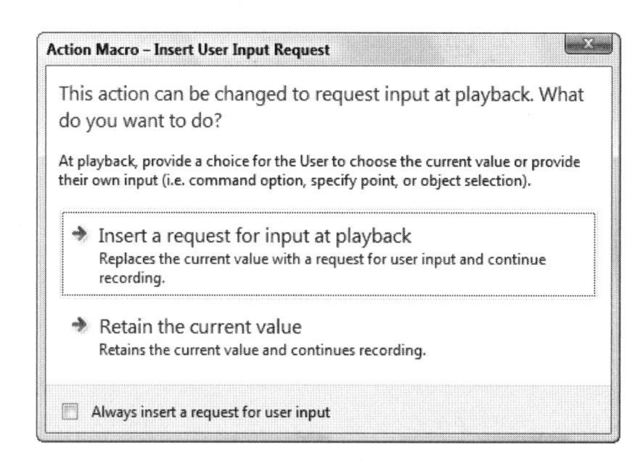

Click Insert a Request for Input at Playback. Notice that AutoCAD adds a User Input Marker in the Action recorder.

INSERT USER INPUT REQUEST DIALOG BOX

Insert a Request for Input at Playback pauses the macro during playback and waits for the user to respond.

Retain Current Value keeps the recorded value.

☐**Always Insert a Request for User Input** prevents this dialog box from appearing again.

INPUT REQUEST DIALOG BOX

When a running macro reaches a user-pause, it displays this next dialog box.

Provide Input pauses the macro and waits for the user to respond.

Uses the Recorded Value inputs the value recorded earlier during the macro recording.

Stop Playback halts the macro in mid-command.

☐**Always Pause for Input** prevents this dialog box from appearing again.

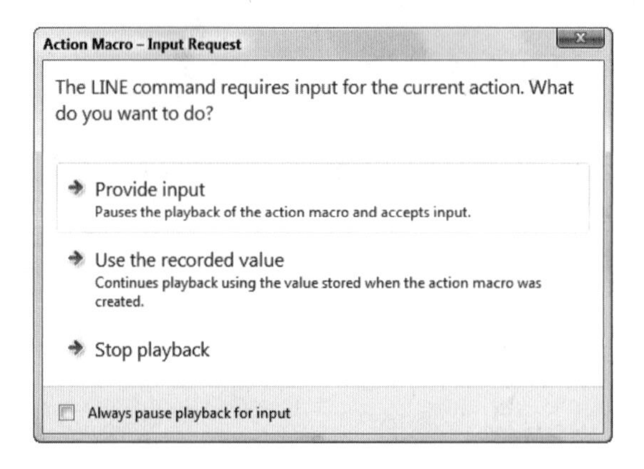

RELATED COMMANDS

ActRecord begins recording macros.

ActBasepoint relocates the base point coordinates.

ActUserMessage adds dialog box messages to recorded action macros.

TIPS

- After the macro has been recorded, you can add pauses by right-clicking coordinates and option names in the Action Tree:

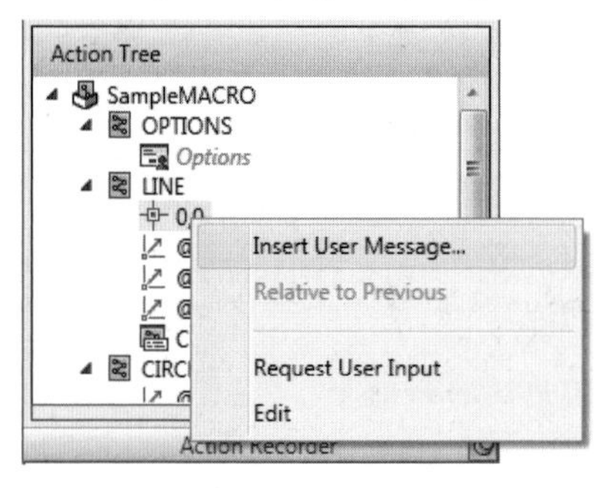

- This command cannot be used at the 'Command:' prompt; it can only be used transparently while a macro is being recorded. If you enter the command at an incorrect time, AutoCAD complains, "***ACTUSERINPUT command only allowed while recording an action macro."

ActUserMessage

2009 Adds dialog box messages to action macros.

Command	Aliases	Keyboard Shortcuts	Menu	Ribbon
'actusermessage	**arm**	...	...	**Manage** ↳**Action Recorder** ↳**Insert Message**
'-actusermessage	**-arm**	...	...	...

Command: actusermessage

Displays dialog box.

Enter a message, and then click OK.

Notice that AutoCAD adds a user message marker in the Action recorder.

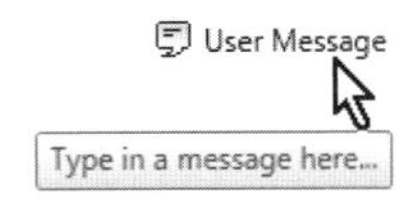

-ActUserMessage Command

Command: -actusermessage

Enter a message to display during playback <exit>: *(Enter a sentence of text, and then press Enter.)*

COMMAND LINE OPTIONS

Enter a message specifies the message to display by the running macro.

RELATED COMMANDS

ActRecord begins recording macros.

ActBasepoint relocates the base point coordinates.

ActUserInput add pauses for user input retroactively.

TIPS

- This command can only be used while a macro is being recorded. If you enter the command at an incorrect time, AutoCAD complains, "***ACTUSERMESSAGE command only allowed during recording an action macro."
- When a running macro reaches a user-pause, it displays this dialog box:

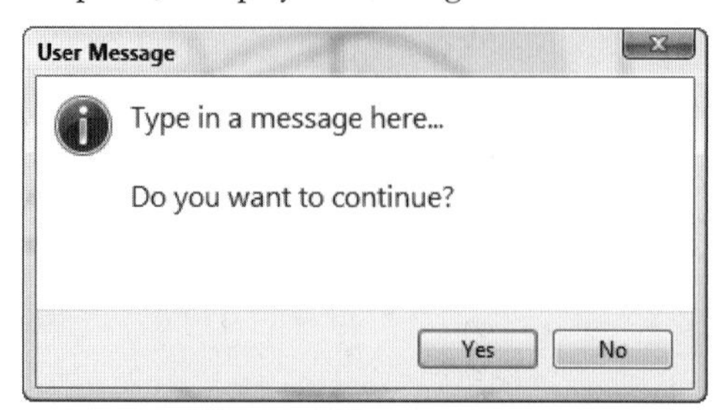

- After the macro has been recorded, you add dialog boxes by right-clicking the Action Tree:

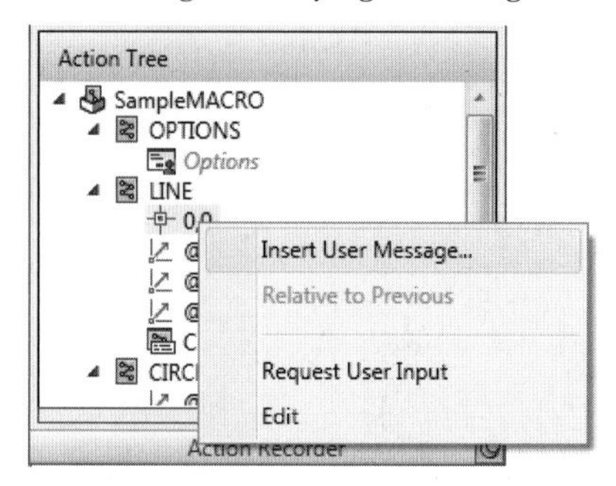

- Right-click a User Message marker to edit or delete it:

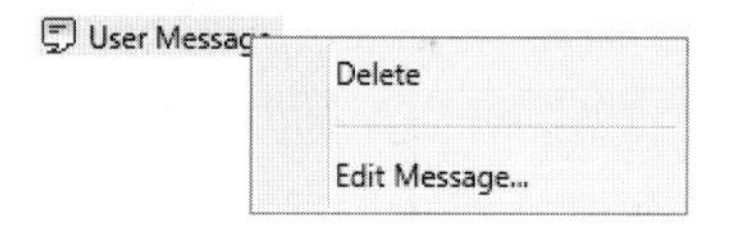

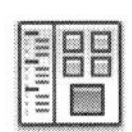

AdCenter / AdcClose

2000 Opens and closes the Design Center palette (short for Autocad Design CENTER).

Commands	Aliases	Keyboard Shortcuts	Menu	Ribbon
adcenter	**adc** **content** **dc, dcenter**	**Ctrl+2**	**Tools** **↳Palettes** **↳DesignCenter**	**Insert** **↳Content** **↳DesignCenter**
adcclose	**...**	**Ctrl+2**	**Tools** **↳Palettes** **↳DesignCenter**	**Insert** **↳Content** **↳DesignCenter**

Command: adcenter

Displays DesignCenter palette:

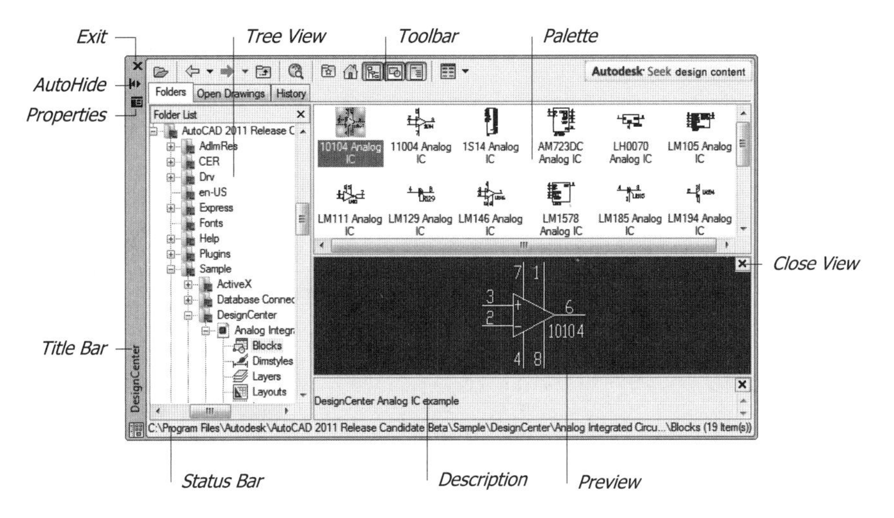

Command: adcclose

Closes the DesignCenter palette.

TABS

Folders displays the contents of the local computer, as well as of networked computers.

Open Drawings displays the contents of drawings currently open in AutoCAD.

History displays the drawings previously opened.

Palette toolbar

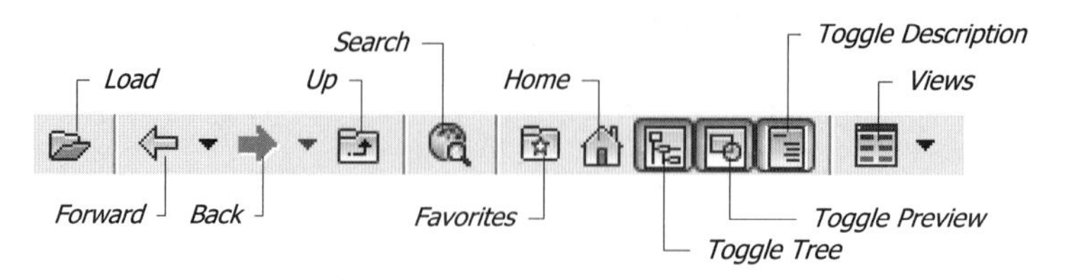

Load displays the Open dialog box to open the following vector and raster file types: *.dwg*, *.dws*, *.dwf*, *.dxf*, *.bil*, *.bmp*, *.cal*, *.cg4*, *.dib*, *.flc*, *.fli*, *.gif*, *.gp4*, *.ig4*, *.igs*, *.jpg*, *.jpeg*, *.mil*, *.pat*, *.pcx*, *.png*, *.rlc*, *.rle*, *.rst*, *.tga*, and *.tif*.

Back returns to the previous view.

Forward goes to the next view.

Up moves up one folder level.

Search opens the Search dialog box to search for AutoCAD files (on the computer) and for the following objects (in drawings): blocks, dimstyles, drawings, hatch patterns and *.pat* files, layers, layouts, linetypes, text styles, table styles, and xrefs.

Favorites displays the files in the *\documents and settings\<username>\favorites\autodesk* folder.

Home displays the files in the *\autocad 2011\sample\designcenter* folder.

Tree View Toggle hides and displays the Folders and Open Drawings tree views.

Preview toggles the display of the preview image of *.dwg* and raster files.

Description toggles the display of the description area.

View changes the display format of the palette area.

RELATED COMMAND

AdcNavigate specifies the initial path for DesignCenter.

RELATED SYSTEM VARIABLE

AdcState reports whether DesignCenter is open or not.

TIPS

- Use Design Center to track drawings and parts of drawings, such as block libraries.
- You can drag blocks, table styles, and other drawing parts from the Design Center palette into the drawing, and blocks onto Tool palette.
- Design Center allows you to share text and dimension styles between drawings.
- Switch the Design Center palette between floating and docked modes by right-clicking and selecting the option from the menu.
- Ctrl+2 opens and closes the Design Center palette each time you press the shortcut keystrokes.
- The Autodesk Seek design content button activates the Seek command for locating blocks and drawings on Autodesk's Web site. See the Seek command.

AdcNavigate / AdcCustomNavigate

2000 Specifies the initial path for Design Center to access content.

Command	*Aliases*	*Keyboard Shortcuts*	*Menu*	*Ribbon*
adcnavigate	...	...	...	...
adccustomnavigate				

Command: adcnavigate

Opens DesignCenter, if not already open.

Enter pathname <>: *(Enter a path, and then press Enter.)*

COMMAND LINE OPTION

Enter pathname specifies the path, such as *c:\program files\autocad2011\sample.*

ADCUSTOMNAVIGATE Command

An undocumented command.

Command: adccustomnavigate

Points to the folder containing custom objects, which are displayed in Design Center's Custom tab. The Custom tab is displayed only when custom applications are registered with AutoCAD, such as AutoCAD Architectural.

Enter pathname <>: *(Enter a path, and then press Enter.)*

RELATED COMMAND

AdCenter opens the Design Center palette.

TIPS

- AutoCAD uses the path specified by this command to locate content displayed by DesignCenter's Desktop option.
- You can enter paths to files, folders, or network locations:

Example of folder path:	*c:\program files\autocad 2011\sample*
Example of file path:	*c:\design center\welding.dwg*
Example of network path:	*\\downstairs\c\project*

AddSelected

2011 Creates new objects based on the object type and properties of the selected object.

Command	*Aliases*	*Keyboard Shortcuts*	*Menu*	*Ribbon*
addselected	...	...	...	...

Command: addselected
Select object: *(Choose an object.)*
AutoCAD launches the command needed to create the same object, and then assigns it the copied properties.

COMMAND LINE OPTION

Select object specifies the command to launch and properties to copy.

RELATED COMMANDS

SelectSimilar selects objects with similar properties.

MatchProp copies properties from one object and then applies them to other objects.

TIPS

- AutoCAD copies the properties — color, linetype, layer, and so on — of the selected object. For some objects, this command copies additional properties:

Object	*Additional Properties Copied*
Attribute definitions	Text style, height
Block references	Name
Dimensions	Dimension style, dimension scale
External references	Name
Gradients	Gradient name, color 1, color 2, angle, centered
Hatches	Pattern name, scale, rotation angle
Leaders	Dimension style, dimension scale
Mtext	Text style, height
Multileaders	Multileader style, scale
Tables	Table style
Text	Text style, height
Tolerances	Dimension style
Underlays	Name

- Typical command usage:

  Command: addselected
  Select object: *(Choose a line.)*
  _line Specify first point: *(Pick a point.)*
  Specify next point or [Undo]: *(etc.)*

- You access this command most easily by selecting an object, right clicking, and then choosing Add Selected from the shortcut menu.
- This command does not support 3D objects, including 3D meshes, surfaces, and 3D solids.

Adjust

2010 Changes the fade, contrast, brightness and/or monochrome settings for attached images and DWF, DWFx, PDF, and DGN underlays.

Command	Aliases	Keyboard Shortcuts	Menu	Ribbon
adjust	...	...	...	**Insert ↳Reference ↳Adjust**

Command: adjust
Select image or underlay: *(Select one or more images or underlays.)*
Select image or underlay: *(Select more objects, or press Enter to continue.)*

The next prompt varies, depending on the object selected.

COMMAND LINE OPTIONS

Select image or underlay selects the xref, image, or underlay to be modified. You can select more than one, but the objects must be of the same kind.

Image underlays
Enter image or underlay option [Fade/Contrast/Brightness] <Fade>: *(Enter an option.)*

Adjusts the fade, contrast, and brightness of images. See the ImageAdjust command.

DWF, DWFx, PDF, and DGN underlays
Enter image or underlay option [Fade/Contrast/Monochrome] <Fade>: *(Enter an option.)*

Adjusts the fade and contrast, and toggles between monochrome and colors of brightness of DWF, DWFx, PDF, and DGN underlays. See the DwfAdjust, PdfAdjust, and DgnAdjust commands.

RELATED COMMANDS

Attach attaches images, and DWF, DWFx, PDF, and DGN files as underlays.

Clip clips attached files.

DgnAdjust adjusts the fade, contrast, and monochrome of DGN attachments.

DwfAdjust adjusts the fade, contrast, and monochrome of DWF attachments.

ImageAdjust adjusts the fade, contrast, and brightness of raster image attachments.

PdfAdjust adjusts the fade, contrast, and monochrome of PDF attachments.

ULayers toggles visibility of layers in underlays.

TIPS

- To change the background color of underlays, use the Properties command's Adjust Colors for Background option.
- Monochrome mode changes color to shades of gray.

Ai_CircTan

Draws circles tangent to three points (an undocumented command).

Command	*Aliases*	*Keyboard Shortcuts*	*Menu*	*Ribbon*
'ai_circtan	...	...	...	...

Command: ai_circtan
Enter Tangent spec: *(Pick an object.)*
Enter second Tangent spec: *(Pick an object.)*
Enter third Tangent spec: *(Pick an object.)*

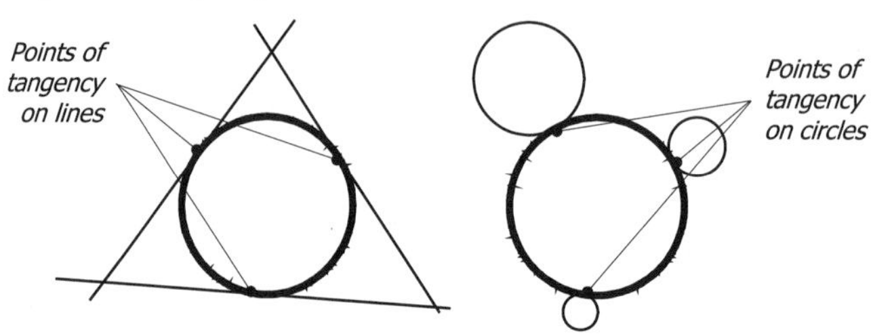

***Left**: Circle drawn tangent to three lines.*
***Right**: Circle drawn tangent to three circles.*

COMMAND LINE OPTION

Enter Tangent spec picks the objects to which the circle will be made tangent.

TIPS

- If a circle cannot be drawn between the three tangent points, AutoCAD complains, "Circle does not exist."
- This command is meant for use in toolbar and menu macros. It is an alternative to the Circle command's 3P option with the TANgent object snap.

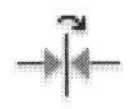

AiDimFlipArrow

Reverses the direction of selected arrowheads (undocumented command).

Command	Aliases	Keyboard Shortcuts	Menu	Ribbon
aidimfliparrow	...	...	...	... ...

Command: aidimfliparrow
Select objects: *(Select one or more arrowheads.)*
Select objects: *(Press Enter.)*

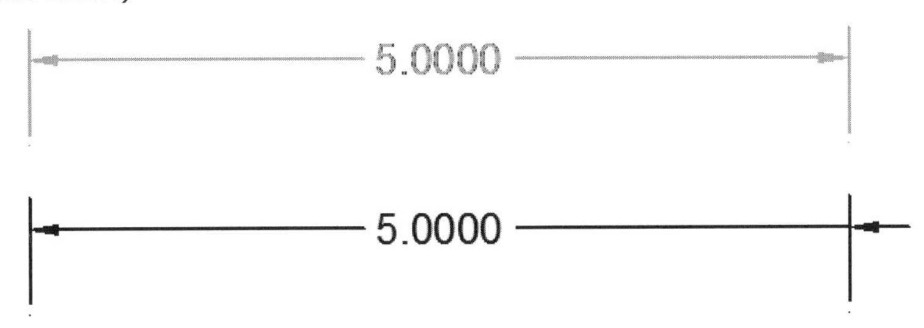

Top: *Arrowheads as drawn by AutoCAD's DimLinear command.*
Above: *Right arrowhead flipped by the AiDimFlipArrow command.*

COMMAND LINE OPTIONS

Select objects selects one or more arrowheads.

TIPS

- In some cases, AutoCAD places arrowheads differently from your preference. This command allows you retroactively to change the direction arrowhead's point.
- Selecting one arrowhead allows you to flip it independently of its companion arrowhead.
- This command has no effect on arrowheads in leaders.

AiDimPrec

Changes the displayed precision of existing dimensions (undocumented command).

Command	*Aliases*	*Keyboard Shortcuts*	*Menu*	*Ribbon*
aidimprec	...	...	...	...

Command: aidimprec
Enter option [0/1/2/3/4/5/6] <4>: *(Enter a digit.)*
Select objects: *(Select one or more dimensions; non-dimensions are ignored.)*
Select objects: *(Press Enter.)*

Before and after applying AiDimPrec = 1 to a decimal dimension.

Before and after applying AiDimPrec = 1 to a fractional dimension.

COMMAND LINE OPTIONS

Enter option specifies the precision (number of decimal places, or fractional equivalent); enter a number between 0 and 6.

Select objects selects one or more dimensions.

TIPS

- This command allows you retroactively to change the displayed precision of selected dimensions.
- Alternatively, you can select a dimension, right click, and then select Precision from the shortcut menu.
- Zero to six decimal places can be specified; fractional units are rounded to the nearest fraction:

  0 — Rounded to the nearest unit.
  1 — 1/2"
  2 — 1/4"
  3 — 1/8"
  4 — 1/16"
  5 — 1/32"
  6 — 1/64"
- *Caution!* Because AiDimPrec rounds off dimensions, it can display false values. The dimension line below measures 2.6875", but setting AiDimPrec to 0 rounds it to 3".

Applying AiDimPrec = 0 to a 2 $^{11}/_{16}$" dimension.

AiDimStyle

Saves and applies preset dimension styles (undocumented command).

Command	Aliases	Keyboard Shortcuts	Menu	Ribbon
aidimstyle	...	...	...	...

Command: aidimstyle
Enter option [1/2/3/4/5/6/Other/Save] <1>: *(Specify option; then press Enter.)*
Select objects: *(Select one or more dimensions.)*
Select objects: *(Press Enter.)*

COMMAND LINE OPTIONS

Enter option specifies a predefined dimension style, numbered 1 through 6.

Other applies a named dimension style to selected dimension(s).

Save saves the style of the selected dimension(s).

Select objects selects one or more dimensions.

Other option
Enter option [1/2/3/4/5/6/Other/Save] <1>: o

Displays dialog box after you select objects.

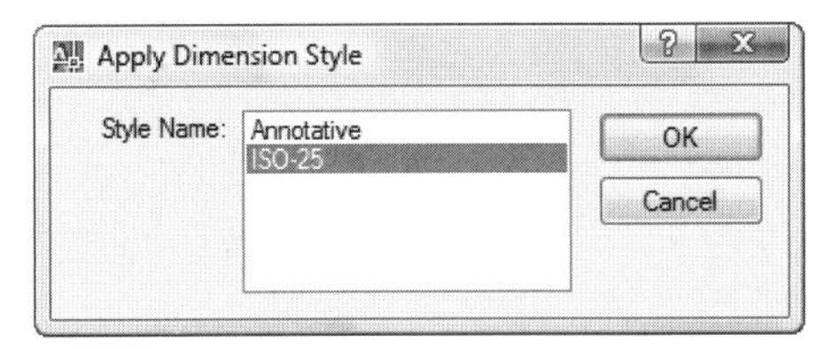

Style Name specifies the name of the dimension style to apply.

Save option
Enter option [1/2/3/4/5/6/Other/Save] <1>: s

Displays dialog box after you select exactly one dimension:

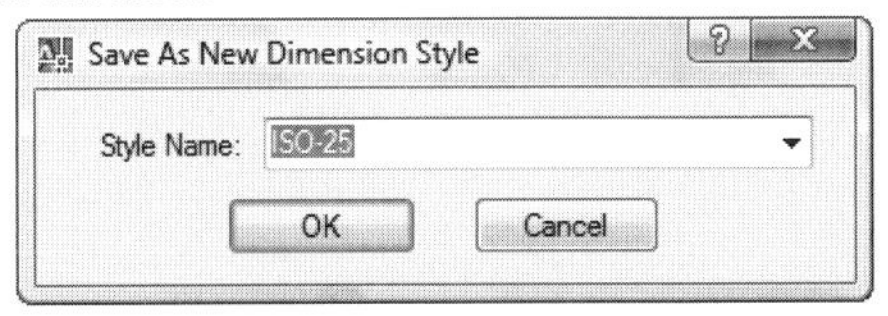

Style Name specifies the name by which to save the dimension style.

TIPS

- This command quickly applies and saves dimension styles: left-click a dimension to select it, right click, and then select Dim Style from the shortcut menu.

Ai_Dim_TextAbove/Center/Home

Moves dimension text relative to dimension lines (undocumented commands).

Commands	Aliases	Keyboard Shortcuts	Menu	Ribbon
ai_dim_textabove	...	...	...	...
ai_dim_textcenter				
ai_dim_texthome				

Command: ai_dim_textabove
Select objects: *(Select one or more dimensions.)*
Select objects: *(Press Enter.)*

Before and after applying Ai_Dim_TextAbove .

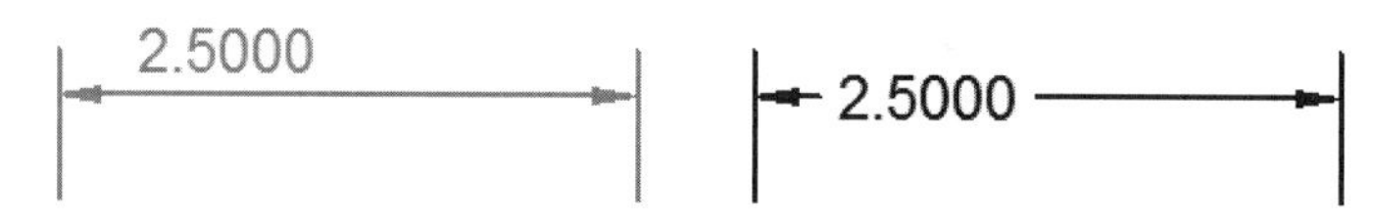

Before and after applying Ai_Dim_TextCenter.

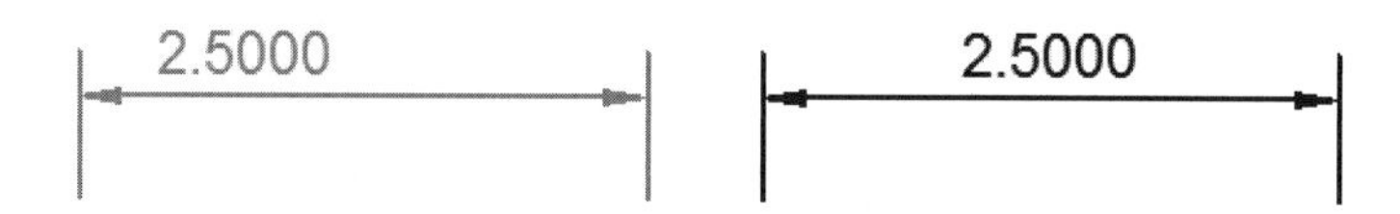

Before and after applying Ai_Dim_TextHome.

COMMAND LINE OPTION

Select objects selects one or more dimensions.

TIPS

- Select a dimension, right-click, and then select Dim Text Position from the menu:
  - **Ai_Dim_TextAbove** makes dimensions compliant with JIS dimensioning.
  - **Ai_Dim_TextCenter** centers text vertically on the dimension line, but not horizontally.
  - **Ai_Dim_TextHome** centers text horizontally on the dimension line, but not vertically.
- Use the DimTEdit command to align text to the left, center, or right on horizontal dimensions.

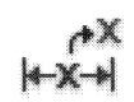

AiDimTextMove

Moves dimension text (undocumented command).

Command	Aliases	Keyboard Shortcuts	Menu	Ribbon
aidimtextmove	...	...	...	...

Command: aidimtextmove
Enter option [0/1/2] <2>: *(Enter an option, and then press Enter.)*
Select objects: *(Select one dimension.)*
Select objects: *(Press Enter.)*

Before and after applying AiDimTextMove = 0 to dimension text.

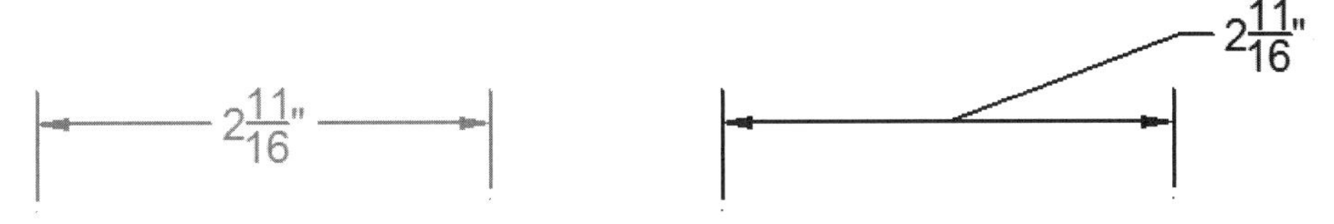

Before and after applying AiDimTextMove = 1 to dimension text.

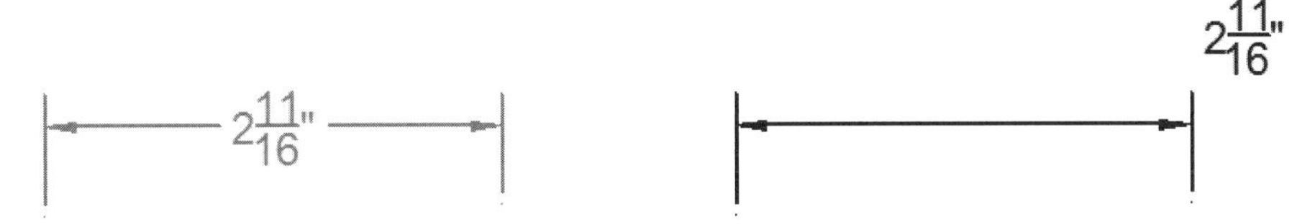

Before and after applying AiDimTextMove = 2 to dimension text.

COMMAND LINE OPTIONS

Enter option specifies the nature of text movement:

0 — Moves text with the dimension line.

1 — Adds a leader to the moved text.

2 — Moves text independently of dimension line and leader (default).

Select objects selects one or more dimensions.

TIPS

- This command allows you retroactively to change the position of dimension text. It is used by the right-click shortcut menu: select a dimension, right-click, and then select Dim Text Position from the shortcut menu.
- Although the command allows you to select more than one dimension, it operates on the first-selected dimension only.

Ai_Fms / Ai_PSpace

Ai_PSpace switches to layout mode; Ai_Fms does the same, and then goes to floating model space (short for Floating Model Space; both are undocumented commands).

Commands	Aliases	Keyboard Shortcuts	Menu	Ribbon
ai_fms	...	...	...	...
ai_pspace	...	...	...	...

Command: ai_pspace

Switches to the last active layout.

Command: ai_fms

Switches to the last active layout, and then to the first floating model viewport.

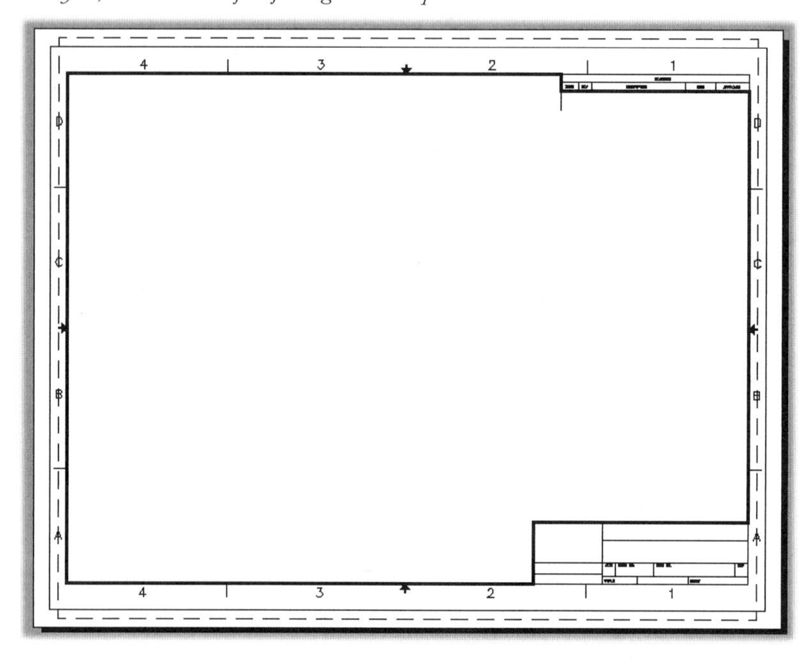

The heavy border indicates the currently-active floating viewport in model space.

COMMAND LINE OPTIONS

None.

TIPS

- The Ai_Fms command sets system variable Tilemode to 0, and then runs the MSpace command.
- These commands are meant for use with menu and toolbar macros.

Replaced Command

Ai_MOLC (make object's layer current) command was replaced by the LayMCur command (layer make current). See Lay... commands.

AiObjectScaleAdd / Remove

Adds (and removes) the current annotative scale to (and from) annotative objects (undocumented commands).

Commands	*Aliases*	*Keyboard Shortcuts*	*Menu bar*	*Ribbon*
aiobjectscaleadd	...	...	...	**Annotate** **↳Annotation Scaling** **↳Add Current Scale**
aiobjectscaleremove				**Annotate** **↳Annotation Scaling** **↳Remove Current Scale**

Command: aiobjectscaleadd
Select annotative objects: *(Pick one or more objects.)*
Select annotative objects: *(Press Enter to continue.)*
n objects updated to support annotation scale <*n*:*n*>.

AIOBJECTSCALEREMOVE Command

Command: aiobjectscaleremove
Select annotative objects: *(Pick one or more objects.)*
Select annotative objects: *(Press Enter to continue.)*
1 object scale removed.

COMMAND LINE OPTION

Select objects selects one or more objects; AutoCAD filters out non-annotative objects.

RELATED COMMANDS

ObjectScale lists all annotative objects associated with the selected object(s).

AnnoReset resets scale and location of selected annotative objects.

AnnoUpdate updates annotative objects to their styles.

Properties toggles the annotative property.

RELATED SYSTEM VARIABLE

CAnnoScale holds the name of the current annotation scale for the current viewport.

TIPS

- The AiObjectScaleAdd command adds the current annotative scale, as displayed by the Annotation Scale droplist.
- When the pickbox is over an annotative object, AutoCAD displays this icon:

  When the object has two or more scale factors, AutoCAD displays this icon:
- To see all the scale factors applied to an annotative object, use the ObjectScale command.

Align

Rel. 12 Moves, transforms, and rotates objects.

Command	*Aliases*	*Keyboard Shortcuts*	*Menu*	*Ribbon*
align	**al**	**...**	**Modify**	**Home**
			↳3D Operation	**↳Modify**
			↳Align	**↳Align**

Command: align
Select objects: *(Select one or more objects to be moved.)*
Select objects: *(Press Enter.)*
Specify first source point: *(Pick a point.)*
Specify first destination point: *(Pick a point.)*
Specify second source point: *(Pick a point.)*
Specify second destination point: *(Pick a point, or press Enter.)*

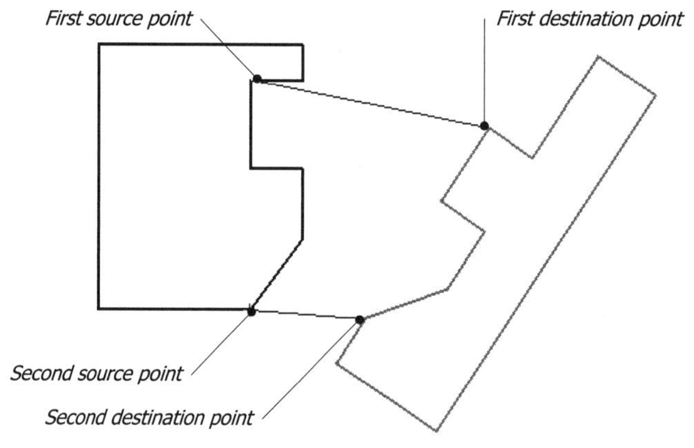

Specify third source point or <continue>: *(Pick a point, or press Enter.)*
Specify third destination point: *(Pick a point.)*

COMMAND LINE OPTIONS

First point moves object in 2D or 3D when one source and destination point are picked.

Second point moves, rotates, and scales object in 2D or 3D when only two source and destination points are picked.

Third point moves objects in 3D when three source and destination points are picked.

Continue option
Scale objects based on alignment points? [Yes/No] <N>: *(Type Y or N.)*

Scales and moves the objects.

TIPS

- Enter two pairs of points to define a 2D (or 3D) transformation, scaling, and rotation:

  First pair of points — Base point for alignment.
  Second pair of points — Rotation angle.
  Third pair — Planes aligned by source and destination points.

- Each pair of points defines a source and destination.

AllPlay

2009 Plays all show-motion views in sequence, without prompts (undocumented command).

Command	*Alias*	*Keyboard Shortcuts*	*Menu*	*Ribbon*
allplay	**aplay**	**...**	**...**	**...**

Command: allplay

COMMAND LINE OPTIONS

None.

RELATED COMMANDS

NavSMotion displays the show motion interface.

NewShot creates named views with show motion options.

SequencePlay plays all views in a category.

ViewGo displays the named view, along with the associated background setting, layer state, live section, UCS, and visual style.

ViewPlay plays the animation associated with a named view.

TIPS

- This command is useful for unattended playback of one or more show-motion animations, because these animations operate only inside AutoCAD; they cannot be exported as movie files.
- To control the playback of show-motion animations, it's best to use the NavSMotion command's user interface.

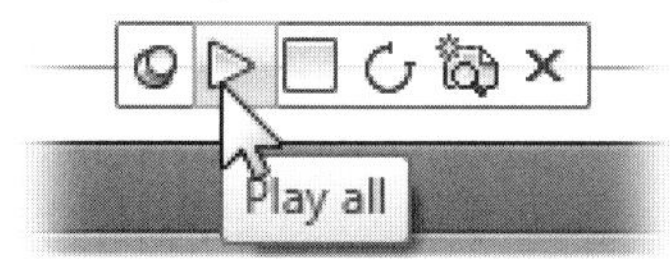

- Show-motion animations are created with views and the NewShot command.

AmeConvert

Rel. 13 Converts PADL solid models and regions created by AME v2.0 and v2.1 (AutoCAD Releases 11 and 12) to ShapeManager solid models.

Command	*Aliases*	*Keyboard Shortcuts*	*Menu*	*Ribbon*
ameconvert	...	...	...	...

Command: ameconvert
Select objects: *(Select one or more objects.)*
Processing Boolean operations.

COMMAND LINE OPTION

Select objects selects AME objects to convert; ignores non-AME objects, such as the ACIS solids produced by AutoCAD Release 13 through 2002, and ShapeManager solids from AutoCAD 2004-2011.

RELATED COMMAND

AcisIn imports ACIS models from a *.sat* file.

TIPS

- After conversion, the AME model remains in the drawing in the same location as the solid model. Erase, if necessary.
- AME holes may become blind holes (solid cylinders) in the solid model.
- AME fillets and chamfers may be placed higher or lower in the solid model.
- Once Release 12 PADL drawings are converted to AutoCAD 2011 solid models, they cannot be converted back to PADL format.
- This command ignores objects that are neither AME solids nor regions.
- Old AME models are stored in AutoCAD as anonymous block references.

DEFINITIONS

ACIS — solids modeling technology used by AutoCAD releases 13 through 2002.

AME — short for "Advanced Modeling Extension," the solids modeling module used by AutoCAD releases 10 through 12.

PADL — short for "Parts and Description Language," the solids modeling technology used by AutoCAD releases 10 through 12.

ShapeManager — the solids modeling technology used by AutoCAD releases 2004 through 2011.

AnalysisCurvature

2011 Displays color curvature gradients on 3D solid and surface models.

Command	*Alias*	*Menu*	*Ribbon*
analysiscurvature	**curvatureanalysis**	...	**Surface** ↳**Analysis** ↳**Curvature**

Command: analysiscurvature
Select solids, surfaces to analyze or [Turn off]: all
Select solids, surfaces to analyze or [Turn off]: *(Press Enter.)*

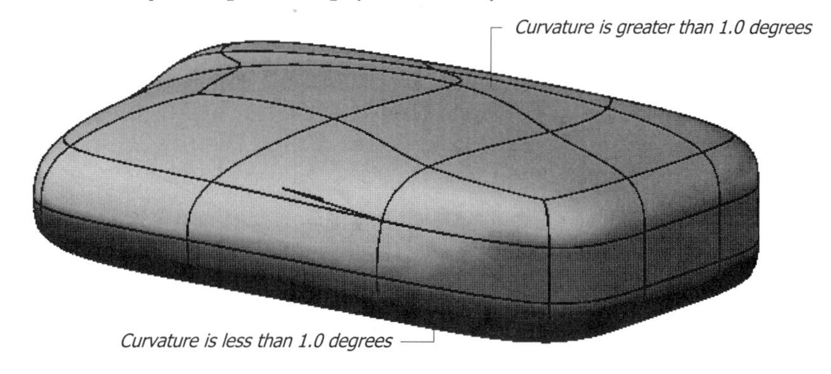

COMMAND LINE OPTIONS

Select solids, surfaces chooses the 3D objects to analyze.

Turn off turns off the analysis colors, returning the model to its original colors.

RELATED COMMANDS

AnalysisDraft displays color spacing gradients between parts and molds.

AnalysisOptions changes display options for curvature, draft, and zebra analyses; effects change in real time.

AnalysisZebra displays black and white surface continuity stripes on 3D models.

RELATED SYSTEM VARIABLES

VsaCurvatureHigh ("visual style analysis") specifies the value above which surface curvature is green; default = 1.

VsaCurvatureLow specifies the value below which surface curvature is blue; default = 1.

VsaCurvatureType determines the type of curvature analysis used by this command: Gaussian (default), mean, maximum, or minimum curvature.

TIPS

- Use this command to see Gaussian, minimum, maximum, and mean U and V surface curvatures of 3D models.
- The display is shown as a range of green to red to blue colors, which have the following meaning:

Color	*Meaning*
Green	Maximum curvature, or positive Gaussian values.
Red	Flat (mean curvature or zero Gaussian value).
Blue	Minimum curvature, or negative Gaussian values.

- This command works only with 3D solids and surfaces; it does not apply to 3D meshes or other objects.

AnalysisDraft

2011 Displays color spacing gradients between parts and molds.

Command	Alias	Menu	Ribbon
analysisdraft	**analysisdraft**	...	**Surface** ↳**Analysis** ↳**Draft**

Command: analysisdraft
Select solids, surfaces to analyze or [Turn off]: all
Select solids, surfaces to analyze or [Turn off]: *(Press Enter.)*

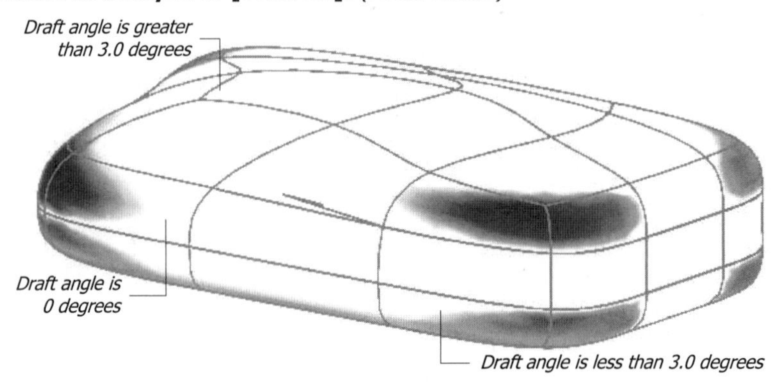

COMMAND LINE OPTIONS

Select solids, surfaces chooses the 3D objects to analyze.

Turn off turns off the analysis colors, returning the model to its original colors.

RELATED COMMANDS

AnalysisCurvature displays color curvature gradients on 3D solid and surface models.

AnalysisOptions changes display options for curvature, draft, and zebra analyses; effects change in real time.

AnalysisZebra displays black and white surface continuity stripes on 3D models.

RELATED SYSTEM VARIABLES

VsaDraftangleHigh specifies the draft angle above which portions of the model are displayed by green; default = 3 degrees.

VsaDraftangleLow specifies the draft angle below which portions of the model are displayed by blue; default = 3 degrees.

TIPS

- The maximum draft angle is shown as green, areas of minimum angle are blue, while zero angle areas are shown in red.
- This command works only with 3D solids and surfaces; it does not apply to 3D meshes or other objects.

AnalysisOptions

2011 Changes display options for curvature, draft, and zebra analyses.

Command	*Alias*	*Keyboard Shortcuts*	*Menu*	*Ribbon*
analysisoptions	...	...	...	**Surface** ↳**Analysis** ↳**Analysis Options**

Command: analysisoptions

Displays dialog box.

DIALOG BOX OPTIONS

Zebra Analysis tab

Select Object to Analyze button selects objects in the drawing.

Stripe Direction slider specifies the angle of cylindrical stripes; range is 0 to 90 degrees.

Type droplist selects the type of zebra stripes: Cylinder or Chrome Ball.

Size droplist selects the width of zebra stripes; range is Thinnest to Thickest.

Color 1 and **Color 2** droplists select stripe colors.

Clear Zebra Analysis button removes the stripes from the objects, returning them to their original color.

Curvature tab

Select Object to Analyze button selects objects in the drawing.

Display Style droplist specifies the display style: Gaussian, Mean, Max Radius, or Min Radius.

- Maximum curvature value displays green areas.
- Minimum curvature value displays blue areas.

Auto Range button assigns 80% of values between high and low ranges.

Max Range button determines the maximum and minimum ranges of all objects.

Clear Curvature Analysis button removes the curvature analysis display from the objects, returning them to their original color.

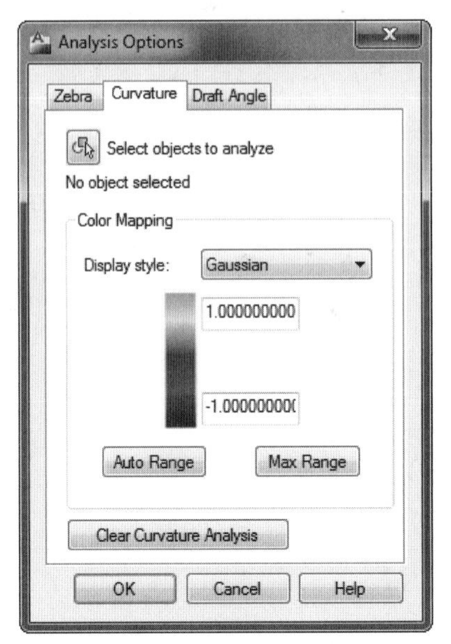

Draft Analysis tab

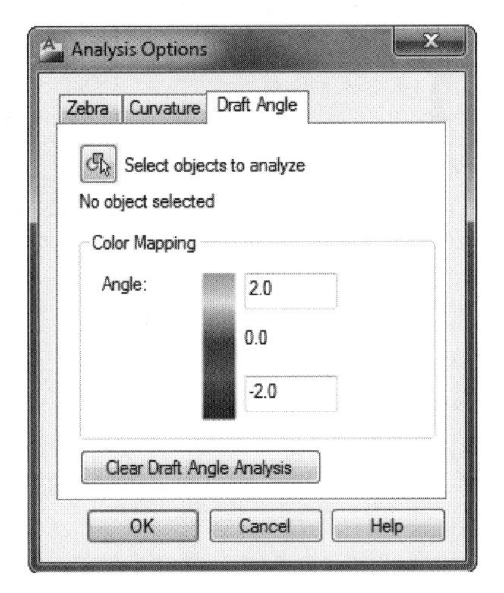

Select Object to Analyze button selects objects in the drawing.

Color Mapping Angle option controls the color of high and low draft angles.

Clear Draft Angle Analysis button removes the analysis colors from the objects, returning them to their original color.

RELATED COMMANDS

AnalysisCurvature displays color curvature gradients on 3D solid and surface models.

AnalysisDraft displays color spacing gradients between parts and molds.

AnalysisZebra displays black and white surface continuity stripes on 3D models.

RELATED SYSTEM VARIABLES

See system variables under the AnalysisCurvature, AnalysisDraft, and AnalysisZebra commands.

TIPS

- Analysis does not work when AutoCAD is using software graphics acceleration; to switch to hardware acceleration, click the Hardware Acceleration button on the status bar.
- This dialog box does not change the angle of chrome ball stripes; to do so, use the VsZebraDirection system variable.

AnalysisZebra

2011 Displays black and white surface continuity stripes on 3D models.

Command	Alias	Keyboard Shortcuts	Menu	Ribbon
analysiszebra	**zebra**	...	...	**Surface** ↳**Analysis** ↳**Zebra**

Command: analysiszebra
Select solids, surfaces to analyze or [Turn off]: all
Select solids, surfaces to analyze or [Turn off]: *(Press Enter.)*

COMMAND LINE OPTIONS

Select solids, surfaces chooses the 3D objects to analyze.

Turn off turns off the analysis colors, returning the model to its original colors.

RELATED COMMANDS

AnalysisCurvature displays color curvature gradients on 3D solid and surface models.

AnalysisDraft displays color spacing gradients between parts and molds.

AnalysisOptions changes display options for curvature, draft, and zebra analyses; effects change in real time.

RELATED SYSTEM VARIABLES

VsaZebraColor1 specifies the first color of zebra analysis stripes; default = white.

VsaZebraColor2 specifies the second color of zebra analysis stripes; default = black.

VsaZzebraDirection specifies the angle of zebra stripes; default = 90 degrees.

VsaZebraSize specifies the width of stripes; default = 45 degrees.

VsaZebraType determines the style of zebra analysis display: (0) chrome ball or (1; default) cylinder.

TIPS

- Zebra stripes illustrate the smoothness of curvatures: G0 (position), G1 (tangent), and G2 (curvature).
- This command works only with 3D solids and surfaces; it does not apply to 3D meshes or other objects.

AniPath

2007 Specifies animation paths along which cameras move; views along paths can be saved as movie files; short for "ANImation PATH."

Command	*Aliases*	*Keyboard Shortcuts*	*Menu*	*Ribbon*
anipath	...	...	**View** ↳**Motion Path Animation**	**Render** ↳**Animations** ↳**Animation Motion Path**

Command: anipath

Displays the Motion Path Animation dialog box.

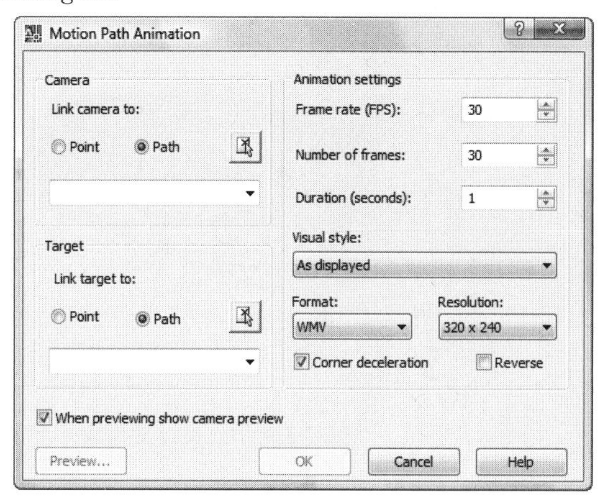

DIALOG BOX OPTIONS

Camera

⊙ **Point** places the camera in the drawing.

O **Path** guides the camera along an object in the drawing: a line, arc, elliptical arc, circle, polyline, spline, or 3D polyline.

Pick picks the camera point or selects the path in the drawing.

List lists previously selected points and paths.

Target

O **Point** points the camera at a target in the drawing.

⊙ **Path** moves the camera along a path defined by an object (line, circle, etc.) in the drawing.

Pick picks the target point or selects the path in the drawing.

List lists previously-selected target points and paths.

Animation Settings

Frame Rate (FPS) specifies the speed of the animation; ranges from 1 to 60 frames per second.

Number of Frames specifies the total number of frames to be captured for the animation.

Duration (seconds) specifies duration of animation in seconds; linked to Number of Frames.

Visual Style selects the preset visual style or rendering quality for the animation: as displayed, rendered, 3D hidden, 3D wireframe, conceptual, realistic, draft, low, medium, high, presentation.

Format specifies the animation file format: AVI, MOV, MPG, or WMV.

Resolution provides a list of resolutions, ranging from 160x120 to 1024x768.

☑ **Corner Deceleration** reduces the camera's speed around curves.

☐ **Reverse** reverses the animation direction.

☑ **When Previewing Show Camera Preview** displays the Animation Preview dialog box.

Preview previews the animation.

Animation Preview dialog box

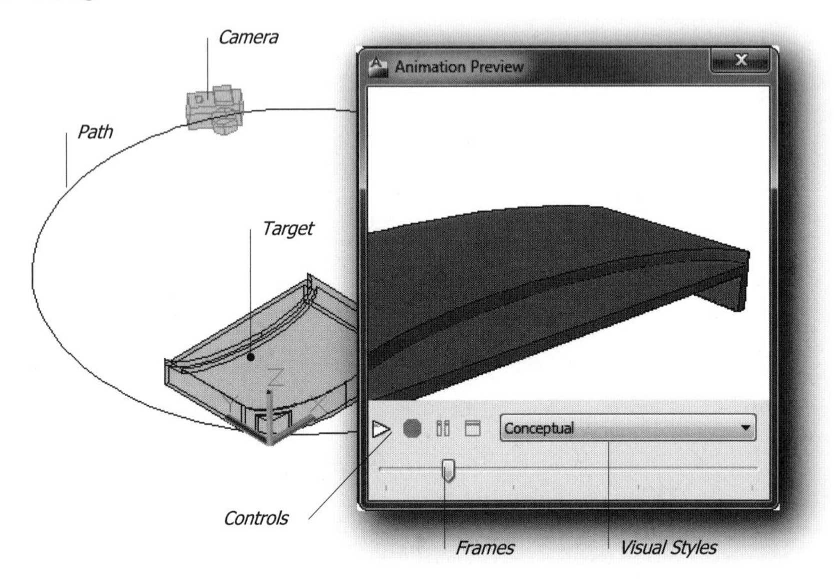

Animation Preview controls

▶ **Play** plays the animation once.

● **Record** records the animation, beginning with the current frame.

❚❚ **Pause** pauses the animation at the current frame; click Play to continue.

■ **Save** displays the Save As dialog box; saves the animation in AVI, MOV, MPG, or WMV formats.

Visual Styles selects the visual style for the animation.

Slider moves through the animation by frames.

RELATED COMMAND

VisualStyles creates and edits custom visual styles.

TIPS

- Cameras are created automatically when the motion path is specified.
- Deleting objects that define motion paths also deletes the path.
- Targets must be linked to paths when cameras are linked to points.
- Changing the frame rate determines the duration, and vice versa.
- MOV format requires QuickTime player; WMV format requires Media Player 9 or later.

AnnoReset / AnnoUpdate

2008 AnnoReset resets the location of all scale representations to the current one; AnnoUpdate updates objects when annotative styles change.

Commands	*Aliases*	*Menu*	*Ribbon*
annoreset	...	**Modify**	**Annotate**
		↳Annotative Object Scale	**↳Annotation Scaling**
		↳Synchronize Multiple-Scale Positions	**↳Synch Scale Positions**
annoupdate	...	...	...

Command: annoreset
Reset alternative scale representations to current position.
n **found,** *n* **was not annotative.**

ANNOUPDATE Command

Command: annoupdate
Select objects: *(Pick one or more objects.)*
Select objects: *(Press Enter.)*
n **found,** *n* **was updated.**

COMMAND LINE OPTION

Select objects selects annotative objects to be updated.

RELATED COMMANDS

ObjectScale assigns scale factors to annotative objects, and edits them.

AiObjectScaleAdd adds scale factors to annotative objects.

AiObjectScaleRemove removes scale factors from annotative objects.

RELATED SYSTEM VARIABLES

AnnoAllVisible toggles the display of annotative objects not at the current annotation scale.

AnnoAutoScale updates annotative objects when the annotation scale is changed.

AnnotativeDwg toggles whether the drawing acts like an annotative block when inserted into other drawings.

CAnnoScale reports the name of the current annotation scale for the current space; a separate scale is stored for model space and for each paperspace viewport.

CAnnoScaleValue reports the current value of the annotation scale.

TIPS

- Annotatively-scaled objects are displayed only when their assigned scale factor matches the model space's scale factor, whether in Model tab or in a layout tab's viewport. Unless, the technical editor notes, AnnoAllVisible is set to 1 (default), in which case all annotative objects appear in all viewports, regardless of scale — which seems to defeat the purpose of annotative objects.
- These objects can have annotative scale factors: text and mtext, attributes and blocks, leaders and multileaders, dimensions and tolerances, and hatch patterns. As well, linetype scaling can be made to match the model space scale factor using the MsLtScale system variable.
- See the ObjectScale command for more on annotative scaling.

Aperture

Ver. 1.3 Sets the size (in pixels) of the object snap target height, or box cursor.

Command	*Aliases*	*Keyboard Shortcuts*	*Menu*	*Ribbon*
'aperture	...	...	...	...

Command: aperture
Object snap target height (1-50 pixels) <10>: *(Enter a value.)*

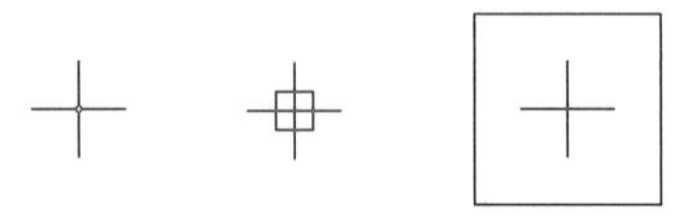

Aperture size = 1 (left), 10 (center), and 50 pixels (right).

COMMAND LINE OPTION

Height specifies the height of the object snap cursor's target; range is 1 to 50 pixels.

RELATED COMMAND

Options allows you to set the aperture size interactively (Drafting tab).

RELATED SYSTEM VARIABLES

ApBox toggles the display of the aperture box cursor (*undocumented*).

Aperture contains the current target height, in pixels:

1 — Minimum size.
10 — Default size, in pixels.
50 — Maximum size.

TIPS

- Besides the aperture cursor, AutoCAD has two other similar-looking cursors: the *osnap* cursor appears only during object snap selection; the *pick* cursor appears anytime AutoCAD expects object selection. The size of both cursors can be changed; to change the size of the pick cursor, use the Pickbox command.
- By default, the box cursor does not appear. Nevertheless, it determines how close you have to be to an object for AutoCAD to "snap" to it.
- To display the box cursor, use the Options command, select the Drafting tab, and then enable the Display AutoSnap Aperture box option. Alternatively, use the undocumented ApBox system variable.
- Use the Options command to change the size of the aperture visually: select the Drafting tab, and then move the Aperture Size slider.

AppLoad

Rel. 12 Creates a list of LISP, VBA, ARX, and other applications to load into AutoCAD (short for APPlication LOADer).

Command	*Alias*	*Keyboard Shortcuts*	*Menu*	*Ribbon*
appload	**ap**	**...**	**Tools** **↳Load Applications**	**Manage** **↳Applications** **↳Load Applications**

Command: appload

Displays dialog box.

LOAD/UNLOAD APPLICATIONS DIALOG BOX

Look in lists the names of drives and folders available to this computer.

File name specifies the name of the file to load.

Files of type displays a list of file types:

Filetype	***Meaning***
ARX	objectARX.
DVB	Visual Basic for Applications
DBX	objectDBX.
FAS	FASt load autolisp.
LSP	autoLiSP.
VLX	Visual Lisp eXecutable.

Load loads all or selected files into AutoCAD.

Loaded Applications displays the names of applications already loaded into AutoCAD.

History List displays the names of applications previously saved to this list.

☐ **Add to History** adds the file to the History List tab.

Unload unloads all or selected files out of AutoCAD.

Contents displays dialog box Startup Suite dialog box.

STARTUP SUITE DIALOG BOX

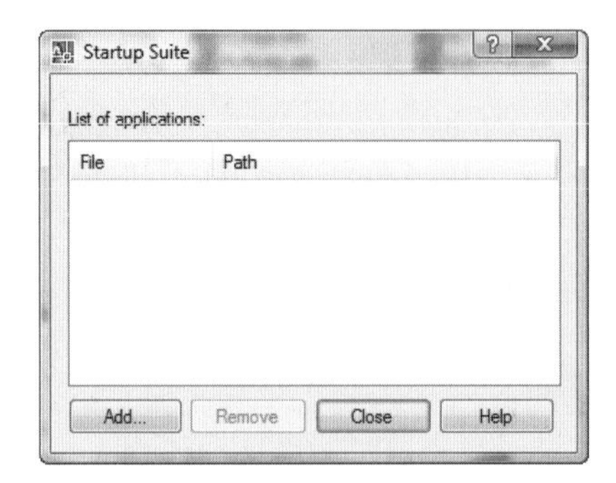

List of applications lists the file names and paths of applications to be loaded automatically each time AutoCAD starts. AutoLISP related files (*.lsp*, *.fas*, and *.vlx*) are loaded whenever a drawing is loaded or a new drawing is created. All others are loaded when AutoCAD starts.

Add displays the Add File to Startup Suite dialog box; allows you to select one or more application files.

Remove removes the application from the list.

Close returns to the Load/Unload Applications dialog box.

RELATED COMMAND

Arx lists ObjectARX programs currently loaded in AutoCAD.

RELATED AUTOLISP FUNCTIONS

(load) loads an AutoLISP program.

(autoload) predefines commands to load related AutoLISP programs.

STARTUP SWITCH

/ld loads an ARX or DBX application when AutoCAD starts up.

TIPS

- Use AppLoad when AutoCAD does not automatically load a command; alternatively, you can drag files from Windows Explorer into the Loaded Applications list.
- ARX, VBA, and DBX applications are loaded immediately; FAS, LSP, and VLX files are loaded after this dialog box closes.
- This command is limited to loading 50 applications at a time.
- Load the *asdkTPCtest.arx* application for undocumented Tablet PC support.
- The *acad2011doc.lsp* file establishes autoloader and other utility functions, and is loaded automatically each time a drawing is opened; the *acad2011.lsp* file is loaded only once per AutoCAD session. Use the AcadLspAsDoc system variable to control whether these files are loaded with AutoCAD.
- Support for VBA will be removed from AutoCAD 2012.

Arc

Ver. 1.0 Draws 2D arcs by eleven methods.

Command	Alias	Keyboard Shortcuts	Menu	Ribbon
arc	**a**	...	**Draw** ↳**Arc**	**Home** ↳**Draw** ↳**Arc**

Command: arc
Specify start point of arc or [CEnter]: *(Pick a point, or enter the CE option.)*
Specify second point of arc or [CEnter/ENd]: *(Pick a point, or enter an option.)*
Specify end point of arc: *(Pick a point.)*

COMMAND LINE OPTIONS

SSE (start, second, end) arc options

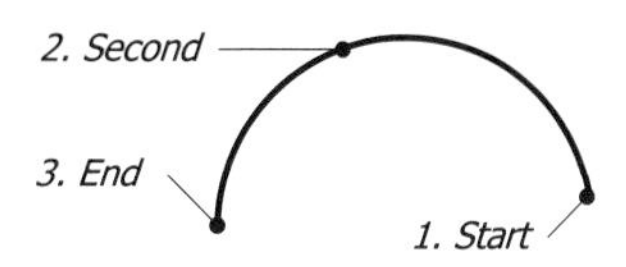

Start point indicates the start point of a three-point arc (point 1, above).

Second point indicates a second point anywhere along the arc (2).

Endpoint indicates the endpoint of the arc (3).

SCE (start, center, end), SCA (start, center, angle), and SCL (start, center, length) options

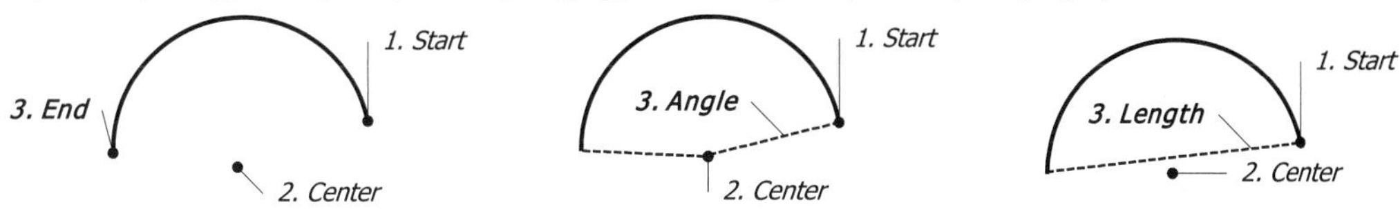

Start point indicates the start point of a two-point arc (1).

Center indicates the arc's center point (2).

Endpoint indicates the arc's end point (3).

Angle indicates the arc's included angle (3).

Length of chord indicates the length of the arc's chord (3).

SEA (start, end, angle), SED (start, end, direction), SER (start, end, radius), and SEC (start, end, center) options

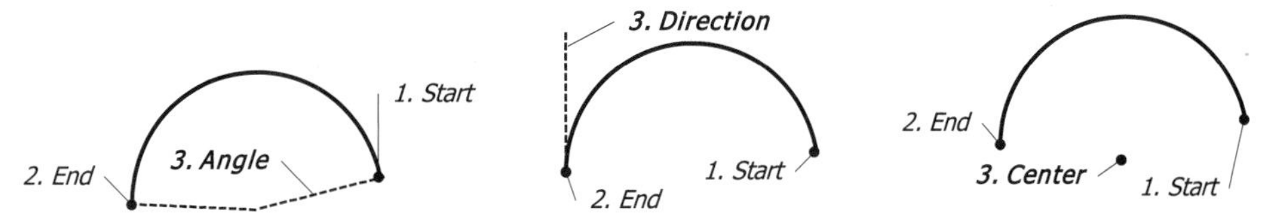

Start point indicates the start point of a two-point arc (1).

End indicates the arc's end point (2).

Angle indicates the arc's included angle (3).

Direction indicates the tangent direction from the arc's start point (3).

Center point indicates the arc's center point (3).

Radius indicates the arc's radius.

CSE (center, start, end), CSA (center, start, angle), and CSL (center, start, length) options

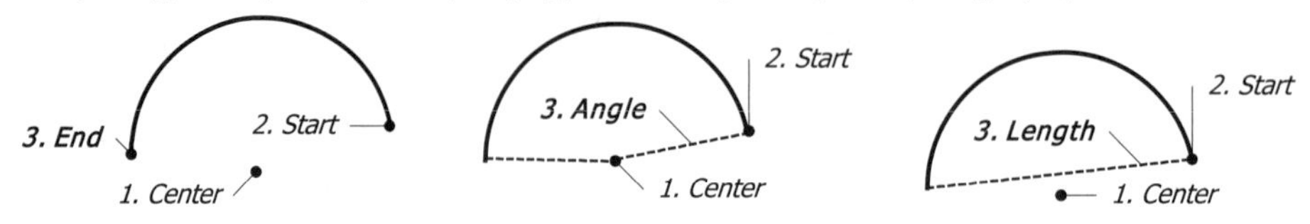

Center indicates the center point of a two-point arc (1).

Start point indicates the arc's start point (2).

Endpoint indicates the arc's endpoint (3).

Angle indicates the arc's included angle (3).

Length of chord indicates the length of the arc's chord (3).

Continued Arc option

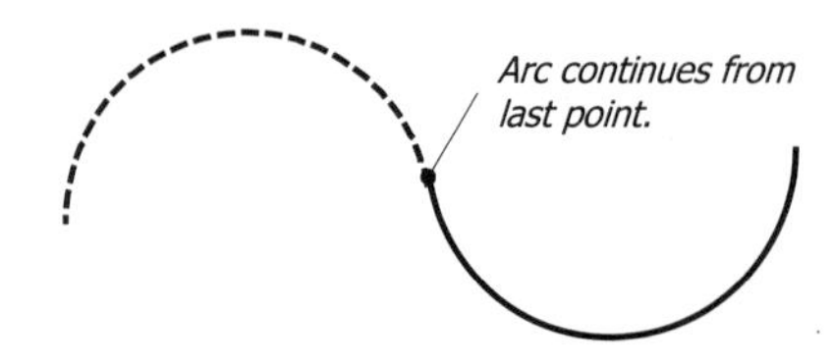

Enter continues the arc tangent from the endpoint of last-drawn line or arc.

RELATED COMMANDS

Circle draws an "arc" of 360 degrees.

DimArc dimensions the length of the arc.

DimCenter places center marks at the arc's center.

Ellipse draws elliptical arcs.

Pline draws connected polyline arcs.

ViewRes controls the roundness of arcs.

RELATED SYSTEM VARIABLES

LastAngle saves the included angle of the last-drawn arc (read-only).

WhipArc controls the smoothness in display of circles and arcs.; it does not beat them in to submission.

TIPS

- Arcs are drawn counterclockwise.
- In most cases, it is easier to draw a circle, and then use the Trim command to convert the circle into an arc.
- It can be easier to use the Fillet command to create arcs tangent to lines.
- To start an arc precisely tangent to the endpoint of the last line or arc, press Enter at the 'Specify start point of arc or [CEnter]:' prompt.
- You can drag the arc only during the last-entered option.
- Specifying an x,y,z-coordinate as the starting point of the arc draws the arc at the z-elevation.
- The components of AutoCAD arcs:

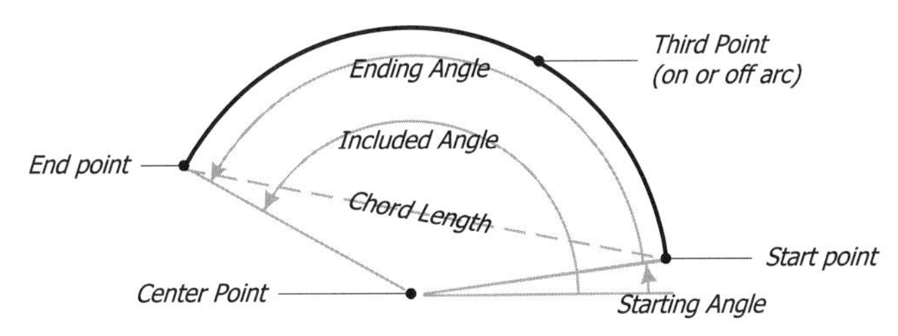

- AutoCAD 2006 adds three more stretch handles (shown as triangles) to arcs:

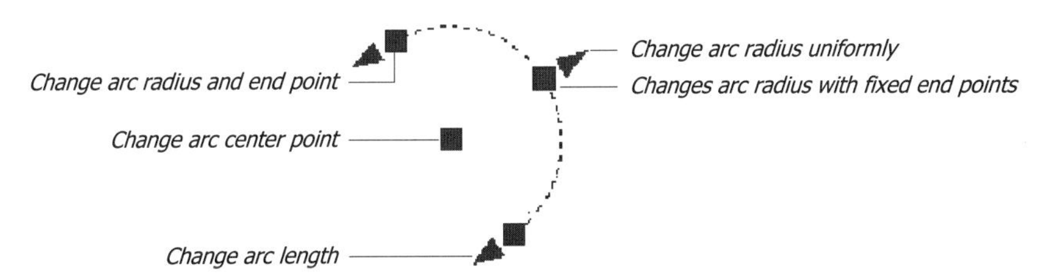

- When the chord length is positive, the minor arc is drawn counterclockwise from the start point; when negative, the major arc is drawn counterclockwise.

Archive

2005 Packages all files related to the current sheet set.

Commands	*Aliases*	*Keyboard Shortcuts*	*Menu Bar*	*Application Menu*	*Ribbon*
archive	...	...	...	**Publish ↳Archive**	...
-archive	...	...	...	...	

Command: archive

When no sheet set is open, AutoCAD complains, "No Sheet Set is Open," and terminates the command. (To open a sheet set, use the OpenSheetset command.)

When at least one sheet set is open, AutoCAD displays the following dialog box:

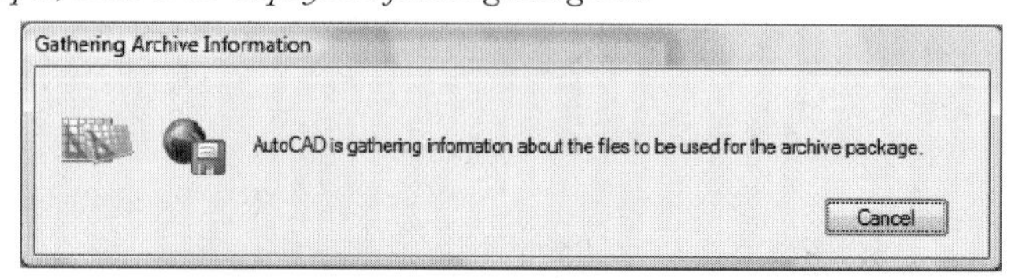

After a moment, AutoCAD displays the next dialog box.

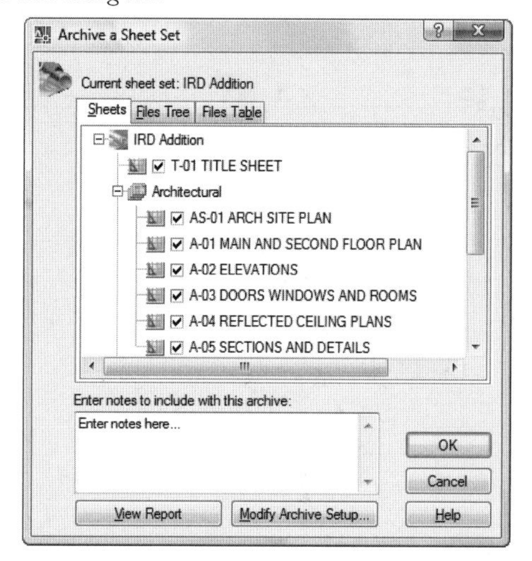

ARCHIVE A SHEET SET DIALOG BOX

Sheets tab displays sheets included with the archive.

Files Tree tab displays names of drawing and support files, grouped by category.

Files Table tab displays file names in alphabetical order.

Enter notes to include with this archive provides space for entering notes.

View Report displays the View Archive Report dialog box.

Modify Archive Setup displays the Modify Archive Setup dialog box.

Files Tree tab

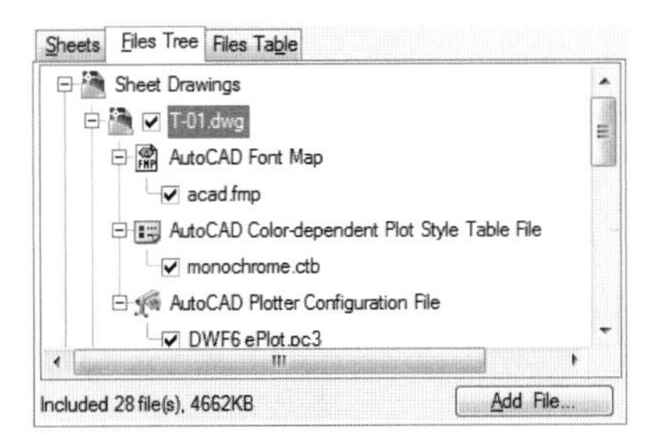

☑ File included in archive.

☐ File excluded from archive.

Add File displays the Add File to Archive dialog box, which adds files to the archive set.

Files Table tab

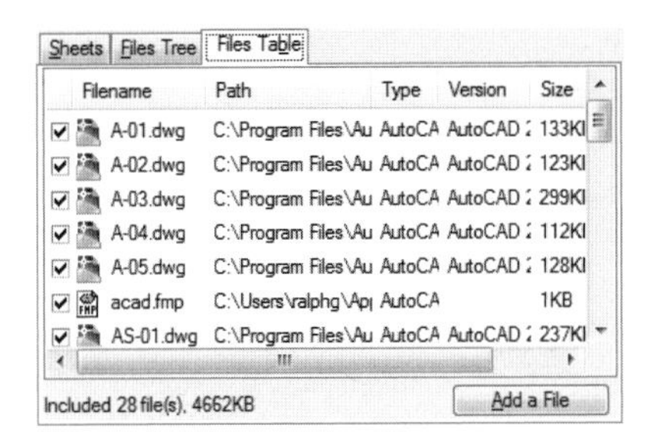

☑ File included in archive.

☐ File excluded from archive.

Add File displays the Add File to Archive dialog box, which adds files to the archive set.

VIEW REPORT DIALOG BOX

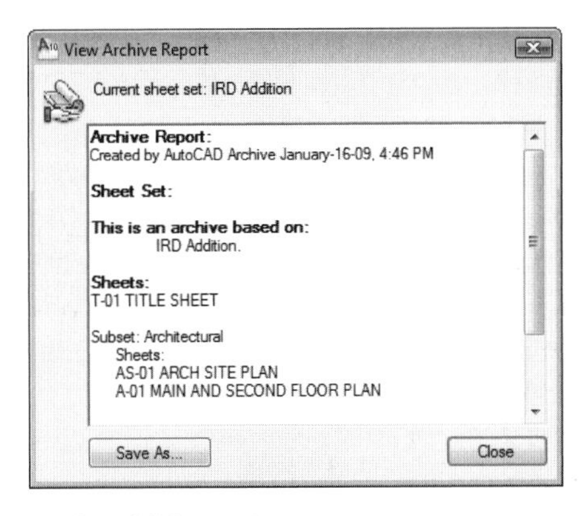

Save As saves the report in *.txt* format (plain ASCII text).

Close closes the dialog box, and then returns to the previous dialog box.

MODIFY ARCHIVE SETUP DIALOG BOX

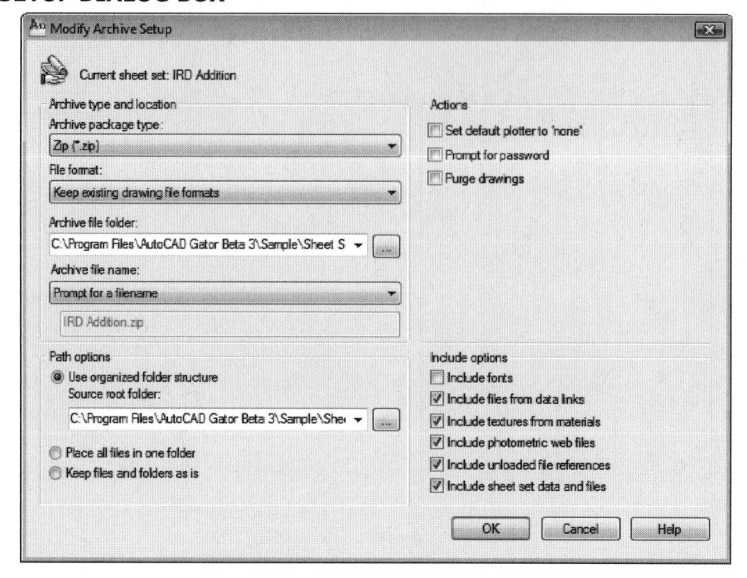

Archive Type and Location

Archive Package Type specifies how the files are packaged together:

- **Folder (set of files)** archives uncompressed files in new and existing folders.; best option when archiving to CD or FTP.
- **Self-Extracting Executable (*.exe)** archives files in a compressed, self-extracting file. Uncompress by double-clicking the file.
- **Zip (*.zip)** (*default*) archives files as a compressed ZIP file. Uncompress the file using PkZip or 7Zip; best option when sending by email.

File Format specifies the format of DWG files:

- **Keep existing drawing file formats** archives files in their native format.
- **AutoCAD 2010/LT 2010 Drawing Format** archives files in AutoCAD 2010 format.

Warning! Some objects and properties not found in earlier AutoCAD versions may be changed or lost when using older formats:

- **AutoCAD 2007/LT 2007 Drawing Format** archives files in AutoCAD 2007/8/9 formats.
- **AutoCAD 2004/LT 2004 Drawing Format** archives files in AutoCAD 2004/5/6 formats.
- **AutoCAD 2000/LT 2000 Drawing Format** archives files in AutoCAD 2000/i/1 formats.

Archive File Folder specifies the location in which to archive the files. When no location is specified, the archive is created in the same folder as the *.dst* file.

Archive File Name specifies how the archive file is named:

- **Prompt for a File Name** prompts the user for the file name.
- **Overwrite if Necessary** uses the specified file name, and overwrites the existing file of the same name.
- **Increment File Name if Necessary** uses the specified file name, and appends a digit to avoid overwriting the existing file of the same name.

Path Options

⊙**Use Organized Folder Structure** preserves the folder structure in the archive, but makes the changes listed below; allows you to specify the name of the folder tree. The option is unavailable when saving archives to the Internet. Autodesk reminds you of the following:

- Relative paths remain unchanged.
- Absolute paths inside the folder tree are converted to relative paths; absolute paths outside the folder tree are converted to "No Path," and are moved to the folder tree.
- *Fonts* folder created when font files are included in the archive.
- *PlotCFG* folder created when plotter configuration files are included.
- *SheetSets* folder created for sheet set support files; *.dst* sheet set data file is in root folder.

○ **Place All Files in One Folder** places all files in a single folder.

○ **Keep Files and Folders As Is** preserves the folder structure in the archive. The option is unavailable when saving archives to the Internet.

Actions

☐ **Set Default Plotter to 'None'** resets the plotter to None for all drawings in the archive.

☐ **Prompt for Password** displays a dialog box for specifying a password for the archive; not available when the Folder archive type is selected.

☐ **Purge Drawings** removes unused table entries from drawings; see Purge command.

Include Options

☐ **Include Fonts** includes all *.ttf* and *.shx* font files in the archive.

☑ **Include Files from Data Links** includes all database files.

☑ **Include Textures from Materials** includes all material definitions used by 3D models.

☑ **Include Photometric Web files** includes IES files that define web light distributions.

☑ **Include Unloaded File References** includes xrefs that are attached but not loaded.

☑ **Include Sheet Set Data File** includes the *.dst* sheet set data file with the archive.

-ARCHIVE Command

Command: -archive
Sheet Set name or [?] <Sheet Set>: *(Enter a name or ?.)*
Enter an option [Create archive package/Report only] <Create>: *(Type C or R.)*
Gathering files ...

Sheet Set name specifies the name of the sheet set to archive.

? lists the names of sheet sets open in the current drawing.

Create archive package creates the archive in ZIP format.

Report only displays the Save Report File As dialog box for saving the archive report as a text file; also prompts you for a Transmittal Note.

RELATED COMMANDS

eTransmit creates an archive of only the current drawing and its support files.

SheetSet creates and controls sheet sets.

TIPS

- This command works only when at least one sheet set is open in the current drawing.
- Archive files can become large when they hold many drawing files.
- The -Archive command is useful for creating scripts and macros that archive the same group of drawings and other files repeatedly.

Area

Ver. 1.0 Calculates the area and perimeter of objects.

Command	Alias	Keyboard Shortcuts	Menu	Ribbon
area	**aa**	...	**Tools** **↳Inquiry** **↳Area**	**Home** **↳Utilities** **↳Measure/Area**

Command: area

Specify first corner point or [Object/Add/Subtract]: *(Pick a point, or enter an option.)*

Specify next corner point or press ENTER for total: *(Pick a point, or press Enter.)*

Sample response:

Area = 1.8398, Perimeter = 6.5245

COMMAND LINE OPTIONS

First corner point specifies the first point to begin measurement.

Object selects the object to be measured.

Add switches to add-area mode.

Subtract switches to subtract-area mode.

Enter ends the area outline.

RELATED COMMANDS

Dist returns the distance between two points.

Id lists the x,y,z coordinates of a selected point.

List reports on the area and other properties of objects.

Properties reports the area of objects, as well as of single and multi-hatched areas.

MassProp returns surface area, and so on, of solid models.

MeasureGeom measures lengths, areas, angles, and volumes.

RELATED SYSTEM VARIABLES

Area contains the most recently-calculated area.

Perimeter contains the most recently-calculated perimeter.

MeasureGeom reports lengths, areas, angles, and volumes.

TIPS

- Before subtracting, you must use the Add option.
- At least three points must be picked to calculate an area; AutoCAD "closes the polygon" with a straight line before measuring the area.
- The Object option returns the following information:

  Circle, ellipse — area and circumference.
  Planar closed spline — area and circumference.
  Closed polyline, polygon — area and perimeter.
  Open objects — area and length.
  Region — net area of all objects in region.
  2D solid — area.

- Areas of wide polylines are measured along center lines; closed polylines must have only one closed area.

Array

Ver. 1.3 Creates 2D linear, rectangular, and polar arrays of objects.

Commands	*Aliases*	*Keyboard Shortcuts*	*Menu*	*Ribbon*
array	**ar**	**...**	**Modify** ↳**Array**	**Home** ↳**Modify** ↳**Array**
-array	**-ar**	**...**	**...**	**...**

Command: array

Displays dialog box.

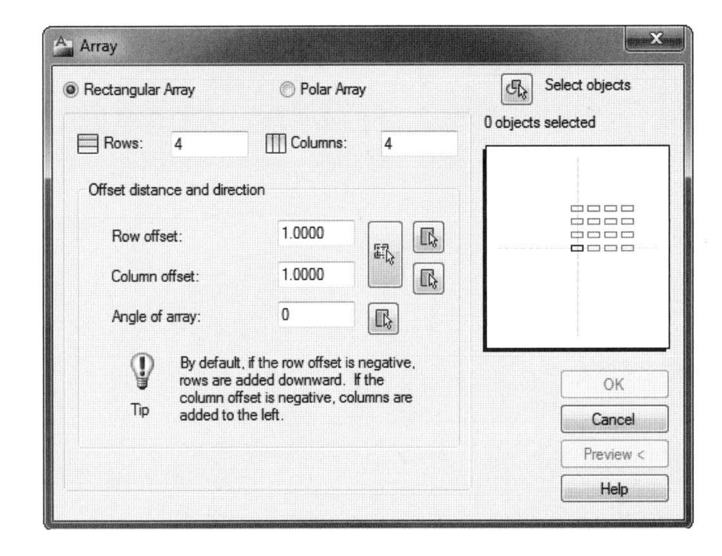

ARRAY DIALOG BOX

⊙ **Rectangular Array** displays the options for creating a rectangular array.

○ **Polar Array** displays the options for creating a circular array.

Select objects dismisses the dialog box temporarily, so that you can select the objects in the drawing:

Select objects: *(Select one or more objects.)*

Select objects: *(Press Enter.)*

Press Enter, or right-click to return to the dialog box.

Preview dismisses the dialog box temporarily, so that you can see what the array will look like.

Rectangular Array

Rows specifies the number of rows; minimum=1, maximum=32767.

Columns specifies the number of columns; minimum=1, maximum=32767.

Row offset specifies the distance between the center lines of the rows; use negative numbers to draw rows in the negative x-direction (to the left).

Column offset specifies the distance between the center lines of the columns; use negative numbers to draw rows in the negative y-direction (downward).

Angle of array specifies the angle of the array, which "tilts" the x and y axes of the array.

Polar Array

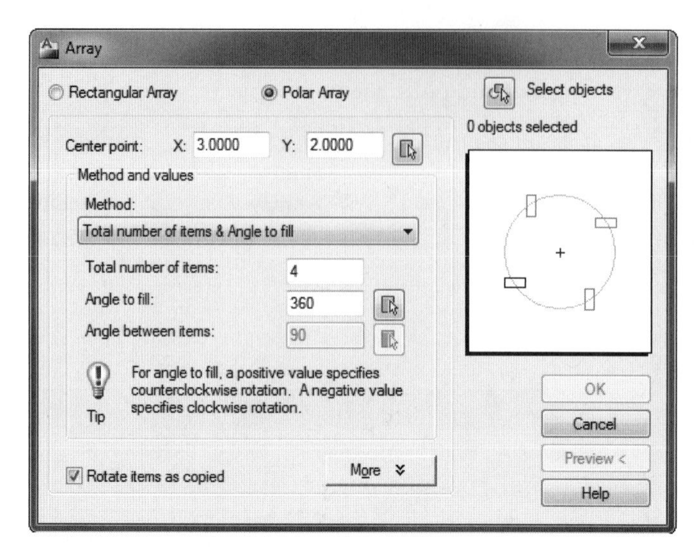

Center point specifies the center of the polar array.

Method specifies the method by which the array is constructed:

- Total number of items and angle to fill.
- Total number of items and angle between items.
- Angle to fill and angle between items.

Total number of items specifies the number of objects in the array; minimum=2.

Angle to fill specifies the angle of "arc" to construct the array; min=1 deg; max=360 deg.

Angle between items specifies the angle between each object in the array.

Rotate items as copied

☑ Objects are rotated so that they face the center of the array.

☐ Objects are not rotated.

More displays additional options for constructing a polar array.

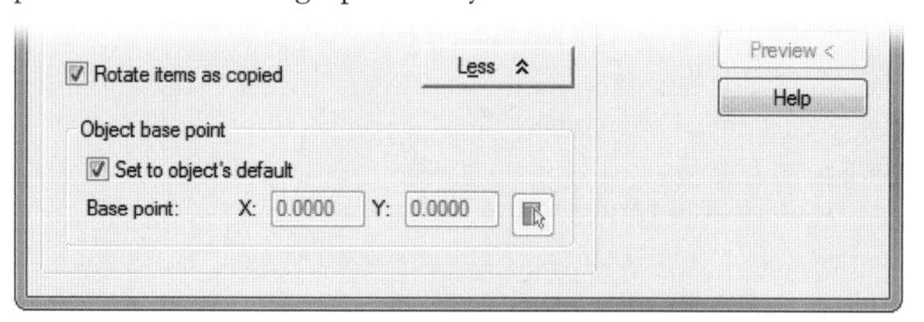

Object Base Point

Set to object's default:

☑ The base point of the object is used, as listed below.

- ☐ The base point is specified by the user; if objects are not rotated, select a base point to avoid unexpected results.

  Arc, circle, ellipse — center point.
  Polygon, rectangle — first vertex.
  Line, polyline, 3D polyline, ray, spline — start point.
  Donut — start point.
  Block, mtext, text — insertion point.
  Xline — midpoint.
  Region — grip point.

Base point specifies the x and y coordinates of the new base point.

-ARRAY Command

Command: -array
Select objects: *(Select one or more objects.)*
Select objects: *(Press Enter.)*
Enter the type of array [Rectangular/Polar] <R>: *(Type R or P.)*

Rectangular options
Enter the number of rows (---) <1>: *(Enter a value, or press Enter.)*
Enter the number of columns (|||) <1>: *(Enter a value, or press Enter.)*
Enter the distance between rows or specify unit cell (---): *(Enter a value.)*
Specify the distance between columns (|||): *(Enter a value.)*

Polar options
Specify center point of array or [Base]: *(Select one or more objects, or type B.)*
Enter the number of items in the array: *(Enter a value.)*
Specify the angle to fill (+=ccw, -=cw) <360>: *(Enter a value, or press Enter.)*
Rotate arrayed objects? [Yes/No] <Y>: *(Type Y or N.)*

COMMAND LINE OPTIONS

Rectangular creates a rectangular array from the selected object.

Polar creates a polar array from the selected object.

Base specifies the base point of objects and the center point of the array.

Center point specifies the center point of the array.

Rows specifies the number of horizontal rows.

Columns specifies the number of vertical columns.

Unit cell specifies the vertical and horizontal spacing between objects.

RELATED COMMANDS

3dArray creates a rectangular or polar array in 3D space.

Copy creates one or more copies of the selected object.

MInsert creates a rectangular block array of blocks.

RELATED SYSTEM VARIABLE

SnapAng determines the default angle of a rectangular array.

TIPS

- To create a rectangular array at an angle, use the Rotation option of the Snap command.
- Rectangular arrays are drawn upward in the positive x-direction, and to the right in the positive y-direction; to draw the array in the opposite directions, specify negative row and column distances.
- Polar arrays are drawn counterclockwise; to draw the array clockwise, specify a negative angle.

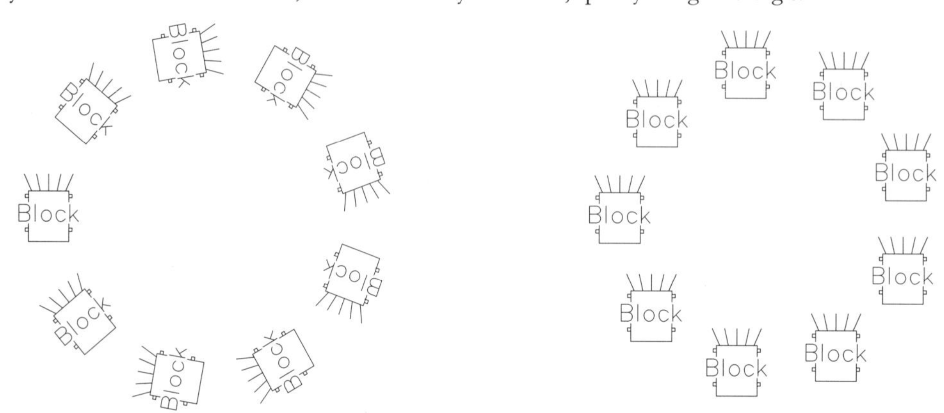

Nine-item polar arrays — rotated (left) and unrotated (right).

- For linear arrays, enter 1 for the number of rows or columns, or use Divide or Measure.
- As of AutoCAD 2009, you can use the mouse's roller wheel to zoom, pan, and orbit while previewing the array. Click Preview, and AutoCAD prompts at the command line:

  Pick or press Esc to return to dialog box or <right-click to accept array>:
  (Move the mouse, or enter an option.)

  Preview is available only with the Array command and not with the -Array command. Panning and orbiting may not work when the roller wheel has been redefined by the mouse driver.

Arx

Rel. 13 Loads and displays information about ARX programs (short for "Autocad Runtime eXtension").

Command	Aliases	Keyboard Shortcuts	Menu	Ribbon
arx	...	...	...	...

Command: arx
Enter an option [?/Load/Unload/Commands/Options]: *(Enter an option.)*

COMMAND LINE OPTIONS

? lists the names of currently loaded ARX programs.

Load loads the ARX program into AutoCAD.

Unload unloads the ARX program out of memory.

Commands lists the names of commands associated with each ARX program.

Options options

CLasses lists the class hierarchy for ARX objects.

Groups lists the names of objects entered into the "system registry."

Services lists the names of services entered in the ARX "service dictionary."

RELATED COMMAND

AppLoad loads LISP, VBA, ObjectDBX, and ARX programs via a dialog box.

RELATED AUTOLISP FUNCTIONS

(arx) lists currently-loaded ARX programs.

(arxload) loads an ARX application.

(autoarxload) predefines commands that load the ARX program.

(arxunload) unloads an ARX application.

RELATED FILE

*.*arx* are ARX program files.

TIPS

- Use the Load option to load external commands that do not seem to work.
- Use the Unload option of the Arx command to remove ARX programs from AutoCAD to free up memory.

Removed Commands

The following ASE (AutoCAD SQL Extension) commands were removed from AutoCAD 2000 and replaced by DbConnect: **AseAdmin**, **AseExport**, **AseLinks**, **AseRows**, **AseSelect**, and **AseSqlEd**.

Assist and **AssistClose** commands were removed from AutoCAD 2008, and replaced by the InfoCenter on the right end of the title bar.

Attach

2010 Attaches images and DWF, DWFx, PDF, and DGN files as underlays; attaches DWG files as externally referenced files.

Command	Aliases	Keyboard Shortcuts	Menu	Ribbon
attach	...	...	...	**Insert** ↳**Reference** ↳**Attach**
-attach				

Command: attach

Displays the Select Reference File dialog box.

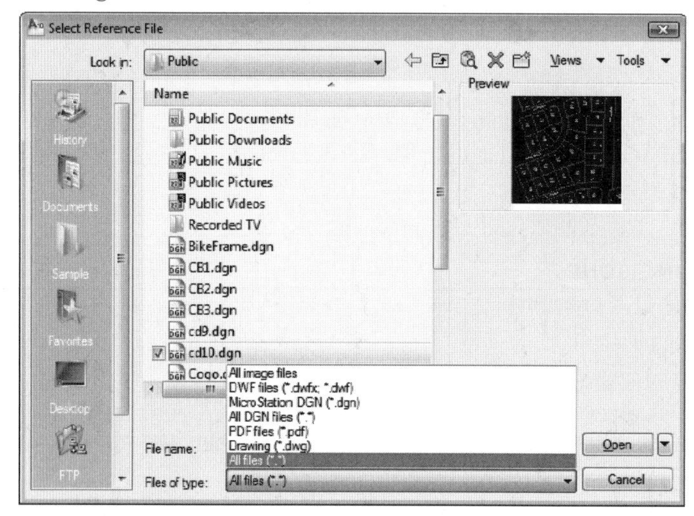

From the Files of Type droplist, choose a file type:

- *All image files.*
- *DWF files in .dwf or .dwfx format.*
- *MicroStation DGN files in V7 or V8 format.*
- *Other MicroStation files.*
- *PDF files*
- *Drawing files in .dwg format.*

Choose a file name, and then click Open.

The next prompt varies, depending on the object selected.

DWG Drawing files

Displays the Attach External Reference dialog box; see the XAttach command.

Image files

Displays the Attach Image dialog box; see the ImageAttach command.

DWF and DWFx files

Displays the Attach DWF Underlay dialog box; see the DwfAttach command.

MicroStation DGN files

Displays the Attach DGN Underlay dialog box; see the DgnAttach command.

Adobe PDF files

Displays the Attach PDF Underlay dialog box; see the PdfAttach command.

-ATTACH command

Command: -attach

Path to files to attach: *(Enter the drive, path, and file name; or press ~ to display the Select File dialog box.)*

The next prompt varies, depending on the object selected.

DWG Drawing files

Specify reference type [Attachment/Overlay] <Attachment>: *(Type A or O.)*

Attach Xref "*filename*": *filename.dwg*

"*filename*" loaded.

Specify insertion point or [Scale/X/Y/Z/Rotate/PScale/PX/PY/PZ/PRotate]: *(Pick a point, enter x, y coordinates, or enter an option.)*

Enter X scale factor, specify opposite corner, or [Corner/XYZ] <1>: *(Enter a scale factor, pick a point, or enter an option.)*

Enter Y scale factor <use X scale factor>: *(Enter a scale factor or press Enter.)*

Specify rotation angle <0>: *(Enter an angle or pick two points.)*

See the XAttach command.

Image files

Specify insertion point <0,0>: *(Pick a point, enter x,y coordinates.)*

Base image size: Width: 23.8, Height: 31.9, Inches

Specify scale factor or [Unit] <1>: *(Enter a scale factor, pick two points, or type U.)*

Specify rotation angle <0>: *(Enter an angle or pick two points.)*

See the ImageAttach command.

DWF and DWFx files

Enter name of sheet or [?] <*default*>: *(Enter the name of a sheet, or type ? for a list of sheet names.)*

Specify insertion point: *(Pick a point, enter x,y coordinates.)*

Base image size: Width: 35.5, Height: 23.0, Inches

Specify scale factor or [Unit] <1.0000>: *(Enter a scale factor, pick two points, or type U.)*

Specify rotation <0>: *(Enter an angle or pick two points.)*

See the DwfAttach command.

MicroStation DGN files
Enter name of model or [?] <*default*>: *(Enter the name of a model, or type ? for a list of model names.)*
Specify conversion units [Master/Sub] <Master>: *(Type M or S.)*
Specify insertion point: *(Pick a point, enter x,y coordinates.)*
Base image size: Width: 90.6, Height: 57.5, Inches
Specify scale factor or [Unit] <1.0000>: *(Enter a scale factor, pick two points, or type U.)*
Specify rotation <0>: *(Enter an angle or pick two points.)*

Adobe PDF files
Enter page number or [?] <1>: *(Enter the number of a page, or type ? for a list of page numbers.)*
Specify insertion point: *(Pick a point, enter x,y coordinates.)*
Base image size: Width: 8.5, Height: 11.0, Inches
Specify scale factor or [Unit] <1.0000>: *(Enter a scale factor, pick two points, or type U.)*
Specify rotation <0>: *(Enter an angle or pick two points.)*

AttachURL

Rel. 14 Attaches hyperlinks to objects and areas.

Command	*Aliases*	*Keyboard Shortcuts*	*Menu*	*Ribbon*
attachurl	...	...	...	...

Command: attachurl
Enter hyperlink insert option [Area/Object] <Object>: *(Type A or O.)*
Select objects: *(Select one or more objects.)*
Select objects: *(Press Enter.)*
Enter hyperlink <current drawing>: *(Enter an address.)*

COMMAND LINE OPTIONS

Area creates rectangular hyperlinks by specifying two corners of a rectangle.

Object attaches hyperlinks after you select one or more objects.

Enter hyperlink requires you to enter a valid hyperlink.

Area Options

First corner picks the first corner of the rectangle.

Other corner picks the second corner of the rectangle.

RELATED COMMANDS

Hyperlink displays a dialog box for adding a hyperlink to an object.

SelectUrl selects all objects with attached hyperlinks.

TIPS

- The hyperlinks placed in the drawing can link to *any* other file: another AutoCAD drawing, an office document, or a file located on the Internet.
- Autodesk recommends that you use the following URL (uniform resource locator) formats:

File Location	*Example URL*
Web Site	**http://***servername***/***pathname***/***filename***.dwg**
FTP Site	**ftp://***servername***/***pathname***/***filename***.dwg**
Local File	**file:///***drive***:/***pathname***/***filename***.dwg**
	or **file:////***localPC***/***pathname***/***filename***.dwg**
Network File	**file://***localhost***/***drive***:/***pathname***/***filename***.dwg**

- The URL (hyperlink) is stored as follows:

  One object — stored as xdata (extended entity data).
  Multiple objects — stored as xdata in each object.
  Area — stored as xdata in a rectangular object on layer URLLAYER.

- The Area option creates a layer named URLLAYER with the default color of red, and places a rectangle on this layer; do not delete the layer.

AttDef

ver. 2.0 Defines attribute modes and prompts (short for ATTribute DEFinition).

Commands	*Aliases*	*Keyboard Shortcuts*	*Menu*	*Ribbon*
attdef	**att**	...	**Draw**	**Insert**
	ddattdef		**⇘Block**	**⇘Attributes**
			⇘Define Attributes	**⇘Define Attributes**
-attdef	**-att**	...	...	...

Command: attdef

Displays dialog box.

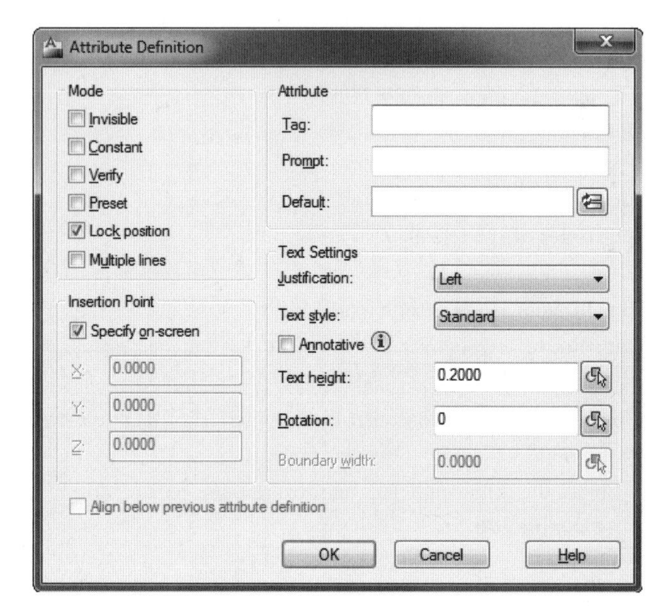

After you click OK, AutoCAD prompts:

Specify start point: *(Pick a point to locate the attribute text.)*

ATTRIBUTE DEFINITION DIALOG BOX

Mode options

☐ **Invisible** makes the attribute text invisible.

☐ **Constant** uses constant values for the attributes.

☐ **Verify** verifies the text after input.

☐ **Preset** presets the variable attribute text.

☑ **Lock Position** prevents attributes from moving;required for attributes in dynamic blocks.

☐ **Multiple Lines** allows multi-line attribute values,when on.

Insertion Point options

☑ **Specify On-screen** picks the insertion point with cursor.

X specifies the x coordinate insertion point.

Y specifies the y coordinate insertion point.

Z specifies the z coordinate insertion point.

Attribute options

Tag identifies the attribute.

Prompt prompts the user for input.

Default sets the default value for the attribute (formerly "Value).

Insert Field displays the Field dialog box; select a field, and then click OK. See the Field command.

Open Multiline Editor displays the multiline attribute editor; this option becomes available when the Multiple Lines option is turned on in the Modes area. See AttIPedit command.

Text Settings options

Justification sets the text justification.

Text Style selects a text style.

Annotative toggles annotative scaling for this attribute definition.

Text Height specifies the height of the text.

prompts 'Height:' at the command line; pick two points, or enter a value.

Rotation sets the rotation angle.

prompts 'Rotation angle:' at the command line; pick two points, or enter a value.

Boundary Width specifies the width of margins for multiline attribute values; available only when the Multiple Lines option is turned on.

prompts 'Multiline attribute width:' at the command line; pick two points, or enter a value.

☐ **Align below previous attribute definition** places the text automatically below the previous attribute.

-ATTDEF Command

Command: -attdef
Current attribute modes: Invisible=N Constant=N Verify=N Preset=N Annotative=N Multiple line=N
Enter an option to change [Invisible/Constant/Verify/Preset/Lock position/Annotative/Multiple lines] <done>: *(Enter an option.)*
Enter attribute tag: *(Enter text, and then press Enter.)*
Enter attribute prompt: *(Enter text, and then press Enter.)*
Enter default attribute value: *(Enter text, and then press Enter.)*
Specify start point of text or [Justify/Style]: *(Pick a point, or enter an option.)*
Specify height <0.200>: *(Enter a value.)*
Specify rotation angle of text <0>: *(Enter a value.)*

COMMAND LINE OPTIONS

Attribute mode selects the mode(s) for the attribute:

- **Invisible** toggles visibility of attribute text in drawing.
- **Constant** toggles fixed or variable value of attribute.

- **Verify** toggles confirmation prompt during input.
- **Preset** toggles automatic insertion of default values.
- **Lock position** toggles locking of the attribute value's position.
- **Annotative** toggles annotation scaling of attribute values.
- **Multiple lines** toggles multiline attribute values.

Start point indicates the start point of the attribute text.

Justify selects the justification mode for the attribute text.

Style selects the text style for the attribute text.

Height specifies the height of the attribute text; not displayed if the style specifies a height other than 0.

Rotation angle specifies the angle of the attribute text.

RELATED COMMANDS

AttDisp controls the visibility of attributes.

EAttEdit edits the values of attributes.

EAttExt extracts attributes and other data to tables or files on disc.

AttRedef redefines an attribute or block.

Block creates blocks with attributes.

Insert inserts blocks; if block has attributes, prompts for their values.

RELATED SYSTEM VARIABLES

AFlags holds the default value of modes in bit form.

AttMulti toggles the ability to create multiline attributes.

TIPS

- Constant attributes cannot be edited.
- Attribute tags cannot be *null* (have no value); attribute values may be null.
- You can enter any characters for the attribute tag, except spaces and exclamation marks. All characters are converted to uppercase.
- When you press Enter at 'Attribute Prompt,' AutoCAD uses the attribute *tag* as the prompt.

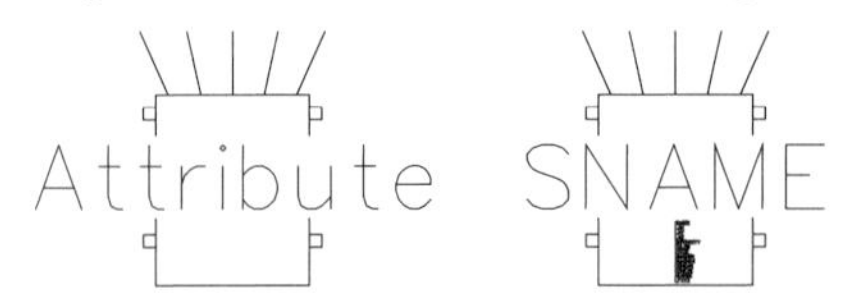

***Left**: Block with attribute value ("attribute").*
***Right**: Attribute tags ("sname" and so on).*

- AutoCAD does not prevent you from using the same tag over and over again. But during attribute extraction, you may have difficulty distinguishing between tags.

AttDisp

Ver. 2.0 Controls the display of all attributes in the drawing (short for ATTribute DISPlay).

Command	*Aliases*	*Keyboard Shortcuts*	*Menu*	*Ribbon*
'attdisp	...	...	**View** **↳Display** **↳Attribute Display**	**Home** **↳Block** **↳Retain Attribute Display**

Command: attdisp
Enter attribute visibility setting [Normal/ON/OFF] <Normal>: *(Enter an option.)*
Regenerating drawing.

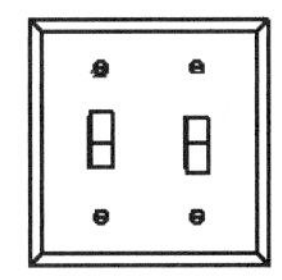

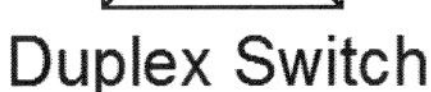

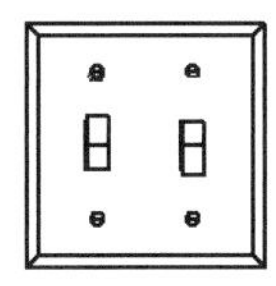

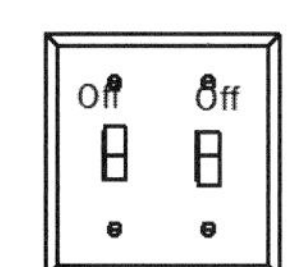

Duplex Switch

***Left to right:** Attribute displayed as Normal, Off, and On.*

COMMAND LINE OPTIONS

Normal displays attributes according to AttDef setting.

ON displays all attributes, regardless of AttDef setting.

OFF displays no attributes, regardless of AttDef setting.

RELATED COMMAND

AttDef defines new attributes, including their default visibility.

RELATED SYSTEM VARIABLE

AttMode holds the current setting of AttDisp:

- **0** — Off: displays no attributes.
- **1** — Normal: does not display invisible attributes.
- **2** — On: displays all attributes.

TIPS

- When RegenAuto is off, use Regen after AttDisp to change to attribute display.
- When you define invisible attributes, use AttDisp to view them.
- Use AttDisp to turn off the display of attributes, which increases display speed and reduces drawing clutter.

AttEdit

Ver. 2.0 Edits attributes in drawings (short for ATTribute EDIT).

Commands	*Aliases*	*Keyboard Shortcuts*	*Menu*	*Ribbon*
attedit	**ate** **ddatte**	...	**Modify** **↳Object** **↳Attribute** **↳Single**	...
-attedit	**-ate** **atte**	...	**Modify** **↳Object** **↳Attribute** **↳Global**	**Home** **↳Block** **↳Edit Attributes** **↳Multiple**

Command: attedit

Select block reference: *(Pick a block.)*

Displays dialog box.

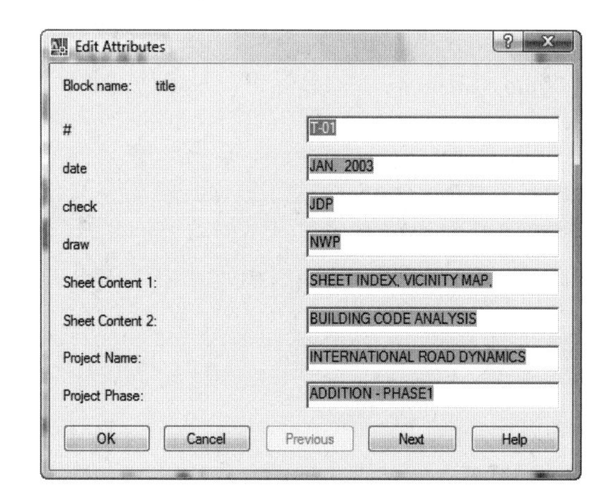

EDIT ATTRIBUTES DIALOG BOX

Block Name names the selected block.

Attribute-specific prompts allow you to change attribute values.

... **Open Multiline Editor** displays the multiline attribute editor; available with multiline attribute values only. See the AttIPedit command.

Previous displays the previous list of attributes, if any.

Next displays the next list of attributes, if any.

-ATTEDIT Command

Command: -attedit

One-at-time Attribute Editing options

Edit attributes one at a time? [Yes/No] <Y>: *(Type Y.)*

Enter block name specification <*>: *(Press Enter to edit all.)*

Enter attribute tag specification <*>: *(Press Enter to edit all.)*

Enter attribute value specification <*>: *(Press Enter to edit all.)*

Select Attributes: *(Select one or more attributes.)*

Select Attributes: *(Press Enter.)*

Enter an option [Value/Position/Height/Angle/Style/Layer/Color/Next] <N>: *(Enter an option, and then press Enter.)*

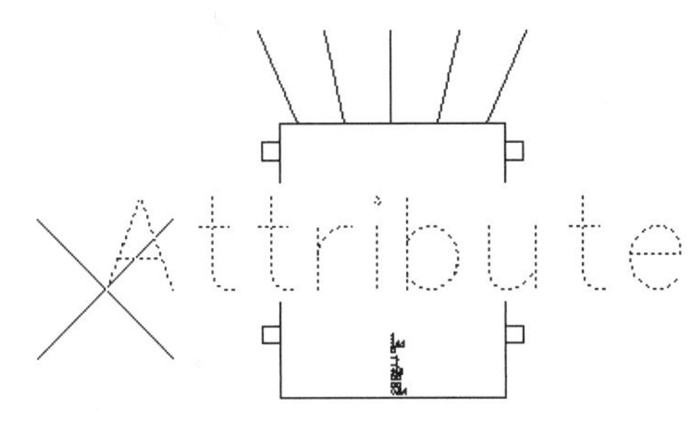

During single attribute editing, AttEdit marks the current attribute with an 'X.'

Global Attribute Editing options

Edit attributes one at a time? [Yes/No] <Y>: *(Type N.)*

Performing global editing of attribute values.

Edit only attributes visible on screen? [Yes/No] <Y>: *(Press Enter.)*

Enter block name specification <*>: *(Press Enter.)*

Enter attribute tag specification <*>: *(Press Enter.)*

Enter attribute value specification <*>: *(Press Enter.)*

Select Attributes: *(Select one or more attributes.)*

Select Attributes: *(Press Enter.)*

Enter string to change: *(Enter existing string, and then press Enter.)*

Enter new string: *(Enter new string, and then press Enter.)*

COMMAND LINE OPTIONS

Value changes or replaces the value of the attribute.

Position moves the text insertion point of the attribute.

Height changes the attribute text height.

Angle changes the attribute text angle.

Style changes the text style of the attribute text.

Layer moves the attribute to a different layer.

Color changes the color of the attribute text.

Next edits the next attribute.

RELATED SYSTEM VARIABLE

AttDia toggles use of AttEdit during the Insert command.

RELATED COMMANDS

AttDef defines an attribute's original value and parameter.

AttDisp toggles an attribute's visibility.

AttRedef redefines attributes and blocks.

EAttEdit edits all aspects of attributes.

AttIPedit edits attribute values in blocks.

Explode reduces attribute values to their tags.

TIPS

- Constant attributes cannot be edited with this command; only attributes parallel to the current UCS can be edited.
- The DdEdit command also displays this dialog box.
- Unlike other text input in AutoCAD, attribute values are case-sensitive.
- To edit null attribute values, use -AttEdit's global edit option, and enter \ (backslash) at the 'Enter attribute value specification' prompt.
- The wildcard characters **?** and ***** are interpreted literally at the 'Enter string to change' and 'Enter new string' prompts.
- To edit the different parts of an attribute, use the following commands:

  Attedit selects non-constant attribute *values* in one block.
  -AttEdit selects attribute *values* and *properties* (such as position, height, and style) in one block or in all attributes.

- When selecting attributes for global editing, you may pick the attributes, or use the following selection modes: Window, Last, Crossing, BOX, Fence, WPolygon, and CPolygon.
- You may use wildcards in the block name, tag, and value specifications:

#	matches any single numeric character.
@	matches any single alphabetic character.
.	matches any single non-alphabetic character.
*	matches any string.
?	matches any single character.
~	matches anything but the following pattern.
[ ]	matches any single character enclosed.
[~]	matches any single character not enclosed.
[-]	matches any single character in the enclosed range.
'	treats the next character as a non-wild-card character.

- AttEdit does not trim leading and trailing spaces from attribute values. Be sure to avoid entering them to prevent unexpected results.

AttExt

Ver. 2.0 Extracts attribute data from drawings to files on disk (short for ATTribute EXTract). The DataExtraction command is preferred for having more powerful abilities.

Commands	*Alias*	*Keyboard Shortcuts*	*Menu*	*Ribbon*
attext	**ddattext**	...	...	...
-attext	...	...	...	...

Command: attext

Displays dialog box.

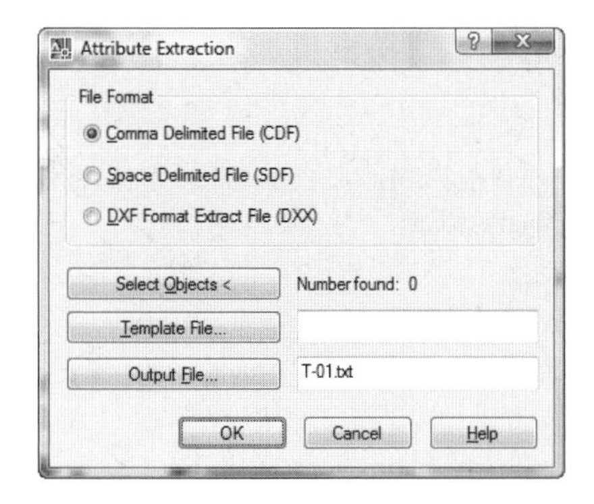

ATTRIBUTE EXTRACTION DIALOG BOX

File Format options

⊙ **Comma Delimited File (CDF)** creates a CDF text file, where commas separate fields.

○ **Space Delimited File (SDF)** creates an SDF text file, where spaces separate fields.

○ **DXF Format Extract File (DXX)** creates an ASCII DXF-format file.

Select Objects returns to the graphics screen to select attributes for export.

Template File specifies the name of the TXT template file for CDF and SDF files.

Output File specifies the name of the attribute output file, *.txt* for CDF and SDF formats, or *.dxx* for DXF format.

-ATTEXT Command

Command: -attext

Enter extraction type or enable object selection [Cdf/Sdf/Dxf/Objects] <C>: *(Enter an option.)*

Displays the Select Template File dialog box; select the template file.

Displays the Create Extract File dialog box.

COMMAND LINE OPTIONS

Cdf outputs attributes in comma-delimited format.

Sdf outputs attributes in space-delimited format.

Dxf outputs attributes in DXF format.

Objects selects objects from which to extract attributes.

RELATED COMMANDS

AttDef defines attributes.

DataExtraction provides a wizard for extracting attributes and other drawing data to tables and external data files. This is a more powerful and flexible command than AttExt.

RELATED FILES

.txt required extension for template file; extension for CDF and SDF files.

.dxx extension for DXF extraction files.

TIPS

- It is easier to use the DataExtraction command than this command for attribute extraction.
- **CDF** is short for "Comma Delimited File"; it has one record for each block reference; a comma separates each field; single quotation marks delimit text strings.
- **SDF** is short for "Space Delimited File"; it has one record for each block reference; fields have fixed width padded with spaces; string delimiters are not used.
- **DXF** is short for "Drawing Interchange File"; it contains only block reference, attribute, and end-of-sequence DXF objects; no template file is required.
- **CDF** files use the following conventions:

    Specified field widths are the maximum width.
    Positive number fields have a leading blank.
    Character fields are enclosed in **' '** (single quotation marks).
    Trailing blanks are deleted.
    Null strings are **''** (two single quotation marks).
    Use spaces; do not use tabs.
    Use the C:DELIM and C:QUOTE records to change the field and string delimiters to another character.

- Before you can specify the SDF or CDF option, you must create a template file.

AttIPedit

2008 Edits multiline attribute values (short for "ATTribute In Place EDITor").

Commands	*Alias*	*Keyboard Shortcuts*	*Menu*	*Ribbon*
attipedit	**ati**	...	...	...

Command: attipedit

Select attribute to edit: *(Pick an attribute value.)*

Select single-line attribute: displays simple in-place editor, like the Text command (no toolbar); right-click for editing options.

Select multi-line attribute: displays either a large or small toolbar, depending on the setting of AttIPe (default = 0):

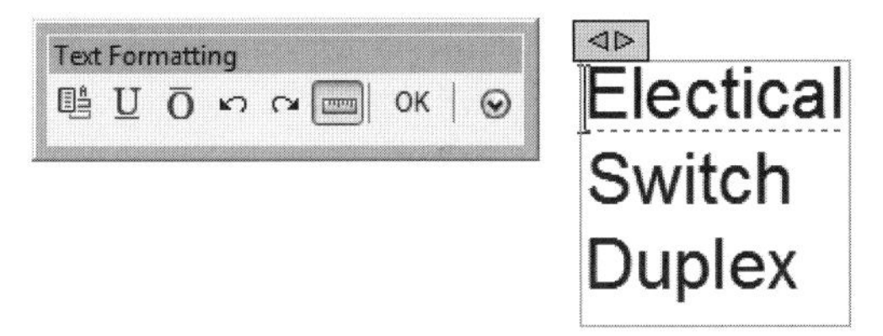

COMMAND LINE OPTION

Select attribute to edit selects attribute value to edit.

TOOLBAR AND SHORTCUT MENU OPTIONS

See the MText command.

RELATED COMMANDS

AttEdit edits all of a block's attributes through a dialog box.

AttDef defines new attributes.

EAttEdit edits all aspects of attributes.

RefEdit edits attributes in-place.

RELATED SYSTEM VARIABLES

AttIpe toggles the size of the toolbar; see tips.

AttMulti toggles the creation of multi-line attributes.

TIP

- This command works with text in attribute definitions (not yet part of a block insertion) and attributes in blocks.
- The AttIpe system variable toggles the size of the multi-line attribute editors. When set to 1, displays the MText-like toolbar. When 0, displays a condensed version of the toolbar.
- Double-clicking attribute values runs the Enhanced Attribute Editor dialog box (EAttEdit command).
- There are now more attribute editing commands than you can shake a stick at: AttEdit, -AttEdit, EAttEdit, BAttMan, RefEdit, DdEdit, Properties, and AttIPedit.

AttRedef

Rel. 13 Redefines blocks and attributes (short for ATTribute REDEFinition).

Command	*Alias*	*Keyboard Shortcuts*	*Menu*	*Ribbon*
attredef	**at**	...	...	...

Command: attredef
Name of Block you wish to redefine: *(Enter name of block.)*
Select objects for new Block...
Select objects: *(Select one or more objects.)*
Select objects: *(Press Enter.)*
Insertion base point of new block: *(Pick a point.)*

COMMAND LINE OPTIONS

Name of Block you wish to redefine specifies the name of the block to be redefined.

Select objects selects objects for the new block.

Insertion base point of new block picks the new insertion point.

RELATED COMMANDS

AttDef defines an attribute's original value and parameter.

AttDisp toggles an attribute's visibility.

EAttEdit edits the attribute's values.

Explode reduces attribute values to tags.

TIPS

- Existing attributes retain their values.
- Existing attributes not included in the new block are erased.
- New attributes added to an existing block take on default values.
- This command removes format and property changes applied with the AttEdit and EAttEdit commands, as well as extended data associated with the block.
- This command may negatively affect blocks and dynamic blocks provided by third-party applications.

AttSync

2002 Updates blocks with new attribute definitions (short for ATTribute SYNChronization).

Command	*Aliases*	*Keyboard Shortcuts*	*Menu*	*Ribbon*
attsync	...	...	...	**Home**
				↳**Block**
				↳**Synchronize Attributes**

Command: attsync
Enter an option [?/Name/Select] <Select>: *(Specify an option.)*
Select a block: *(Pick a block.)*
ATTSYNC block name? [Yes/No] <Yes>: *(Type Y or N.)*

COMMAND LINE OPTIONS

? lists the names of all blocks in the drawing.

Name names the block.

Select selects a single block with the cursor.

RELATED COMMANDS

AttDef defines an attribute.

BattMan edits the attributes in a block definition.

EAttEdit edits the attributes in block references.

TIPS

- This command is used together with other attributed-related commands, in the following order:
    1. AttDef and Block define attributes, and attach them to blocks.
    2. Insert inserts the block, and gives values to the attributes.
    3. EAttEdit changes the attributes in selected blocks, adding, deleting, or modifying the attribute definitions.
    4. BAttMan changes the attributes in the original blocks.
    5. AttSync updates the attributes to the new definition. (BAttMan also performs this task.)
- This command does not operate if the drawing lacks blocks with attributes. AutoCAD complains, "This drawing contains no attributed blocks."

Audit

Rel. 11 Examines drawing files for structural errors.

Commands	Aliases	Keyboard Shortcuts	Menu Bar	Application Menu	Ribbon
audit	...	...	**File** ↳**Drawing Utilities** ↳**Audit**	**Drawing Utilities** ↳**Audit**	...

Command: audit

Fix any errors detected? [Yes/No] <N>: *(Type Y or N.)*

Sample output shown in Text window.

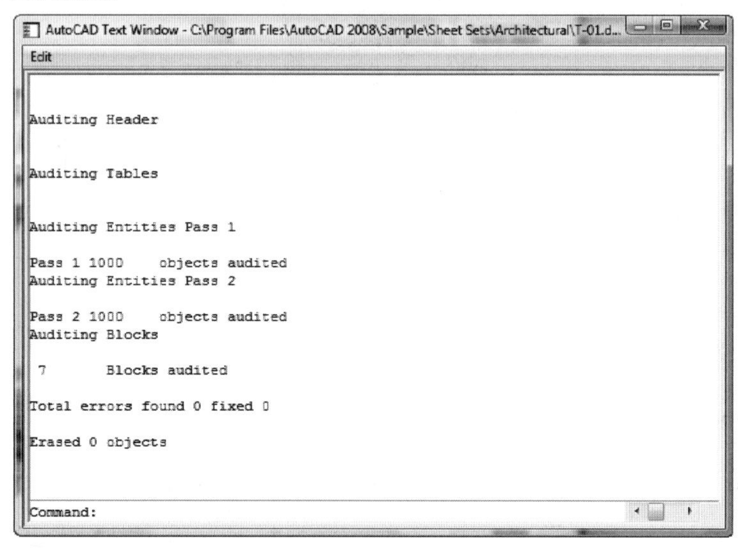

COMMAND LINE OPTIONS

N reports errors found in drawing files, but does not fix errors.

Y reports and fixes errors.

RELATED COMMANDS

DrawingRecovery displays a palette listing recovered drawings.

Save saves recovered drawings to disk.

Recover recovers damaged drawing files.

RELATED SYSTEM VARIABLE

AuditCtl creates *.adt* audit log files when set to 1.

RELATED FILE

.adt is the audit log file, which records the auditing process.

TIPS

- The Audit command is a diagnostic tool for validating and repairing the contents of *.dwg* files.
- Objects with errors are placed in the Previous selection set. Use an editing command, such as Copy, to view the objects.
- If Audit cannot fix a drawing file, try the Recover command.

AutoConstrain

2010 Adds geometric constraints to drawings automatically.

Command	Aliases	Keyboard Shortcuts	Menu	Ribbon
autoconstrain	...	...	**Parametric ↳Auto Constrain**	**Parametric ↳Geometric ↳Auto Constrain**

Command: autoconstrain
Select objects or [Settings] <Settings>: *(Select one or more objects, or type S.)*
Select objects or [Settings] <Settings>: *(Select more objects, or press Enter.)*

COMMAND LINE OPTIONS

Select objects chooses the objects to which to apply geometric constraints; press Ctrl+A to choose all objects in the drawing.

Settings specifies the default settings for this command; displays the Constraint Settings dialog box's AutoConstrain tab. See the ConstraintSettings command.

RELATED COMMANDS

ConstraintSettings specifies the initial conditions for creating dimensional and geometric constraints.

GeomConstraint applies geometric constraints manually.

DimConstraint applies dimensional constraints manually.

RELATED SYSTEM VARIABLES

ConstraintBarDisplay determines whether constraints are initially hidden or visible.

ConstraintBarMode determines which geometrical constraints are displayed on constraint bars.

ConstraintInfer toggles automatic inferences of constraints during drawing and editing *(new to AutoCAD 2011)*.

TIPS

- This command finds probable constraints between geometric objects in drawings. The objects must be within a tolerance distance, as specified by the Constraint Settings dialog box.
- Use the Constraint Settings dialog box to control which constraints are applied, and in which order of priority.
- See the DimConstraint, GeomConstraint, and Parameters commands for tutorials on using constraints.
- AutoCAD 2011 adds Equal to the list of constraints that can be applied automatically, as well as *inferred constraints* that apply constraints automatically as objects are drawn.

AutoPublish

2008 Automatically publishes the current drawing as a DWF file when saved or closed.

Commands	*Aliases*	*Keyboard Shortcuts*	*Menu*	*Ribbon*
autopublish	...	...	...	...

Command: autopublish

AutoPublish DWF or specify override [Location] <AutoPublish>: *(Press Enter, or type L.)*

COMMAND LINE OPTIONS

AutoPublish exports the drawing as a *.dwf* file immediately.

Location specifies the folder for published files; displays the Select A Folder for Generated Files dialog box.

DIALOG BOX OPTIONS

To access the options for this command: (1) enter the Options command, (2) choose the Plot and Publish tab, and then (3) click the Automatic DWF Publishing Settings button; displays dialog box.

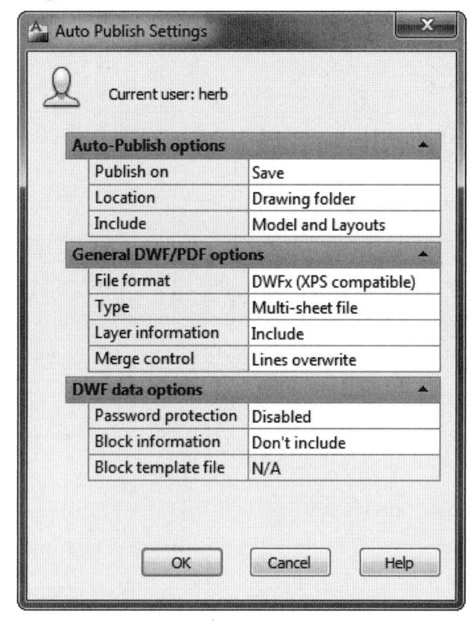

Auto-Publish Options

Publish On specifies when the drawing is exported in DWF format:

- **Save** exports the drawing when it is saved.
- **Close** exports the drawing when it is closed, whether saved or not.
- **Prompt on Save** asks if you want drawings exported in DWF format.
- **Prompt on Close** asks if you want the drawing exported in DWF format; this option works only if you agree to save changes to the drawing.

Location specifies the drive and folder in which the *.dwf* files are saved. Preset options include the drawing's folder, a folder named *dwf* below the drawing's folder, *my documents*, and any folder you choose.

Include determines whether the following are included in the *.dwf* file:

- **Model** exports only the Model tab.
- **Layouts** exports all layout tabs.
- **Model and Layouts** exports all tabs.

General DWF Options

DWF Format chooses the export format:

- DWF
- DWFx (XPS compatible)
- PDF

Type chooses between multiple single-sheet and single multisheet DWF files.

Layer Information toggles the inclusion of layers in the DWF file.

Merge Control determines how to handle crossing lines:

- **Lines Merge** colors of crossing objects blend together.
- **Lines Overwrite** top object overwrites objects underneath.

DWF Data Options

Password Protection provides an optional password for DWF files.

Block Information determines whether information about blocks and attributes is included.

Block Template File specifies the name of the BLK file to use in formatting block and attribute information:

- **Create** opens the Publish Block Template dialog box for creating new block templates.
- **Edit** opens the Select Block Template dialog box for selecting a BLK block template file.

RELATED COMMANDS

3dDwf exports 3D drawings in DWF format.

Archive creates multi-page DWF files from sheet sets.

Publish creates multi-page DWF files from one or more drawings.

Plot exports drawings in DWF format.

RELATED SYSTEM VARIABLES

AuttoPubFormat toggles the AutoPublish feature.

BackgroundPlot toggles whether plots are carried out in the background.

PublishCollate determines whether sheets are published one at a time, or all at once.

PublishHatch specifies how hatch patterns are treated when exported in DWF or DWFx format and then opened in Impression.

TIPS

- Use this command and its related system variable to have AutoCAD automatically generate *.dwf* files of the current drawing; this synchronizes DWG and DWF files, to ensure they are identical.
- This command does not turn on the AutoDwfPublish system variable, curiously enough.

Removed Commands

Background command was replaced by the Background option of the **View** command in AutoCAD 2007.

Base

Ver. 1.0 Changes the insertion point of the drawing, located by default at (0,0,0).

Command	*Aliases*	*Keyboard Shortcuts*	*Menu*	*Ribbon*
'base	...	...	**Draw** **↳Block** **↳Base**	**Insert** **↳Block** **↳Set Base Point**

Command: base
Enter base point <0.0,0.0,0.0>: *(Pick a point.)*

COMMAND LINE OPTION

Enter base point specifies the x,y,z coordinates of the new insertion point.

RELATED COMMANDS

Block specifies the insertion point of new blocks.

Insert inserts another drawing into the current drawing.

Xref references other drawings.

RELATED SYSTEM VARIABLE

InsBase contains the current setting of the drawing's insertion point.

TIPS

- Use this command to shift the insertion point of the current drawing.
- This command does not affect the current drawing. Instead, the relocated base point comes into effect when you insert or xref the drawing into another drawing.

Block Editor Commands

Commands used in the Block Editor, such as BAction and BActionTool, are found under the **BEdit** command.

BAttMan

2002 Edits all aspects of attributes in blocks; works with one block at a time (short for Block ATTribute MANager).

Command	Aliases	Shortcuts	Menu	Ribbon
battman	...	...	**Modify** ↳**Object** ↳**Attribute** ↳**Block Attribute Manager**	**Insert** ↳**Attributes** ↳**Manage**

Command: battman

When the drawing contains no blocks with attributes, displays error message: "This drawing contains no attributed blocks."

When drawing contains at least one block with attributes, displays dialog box:

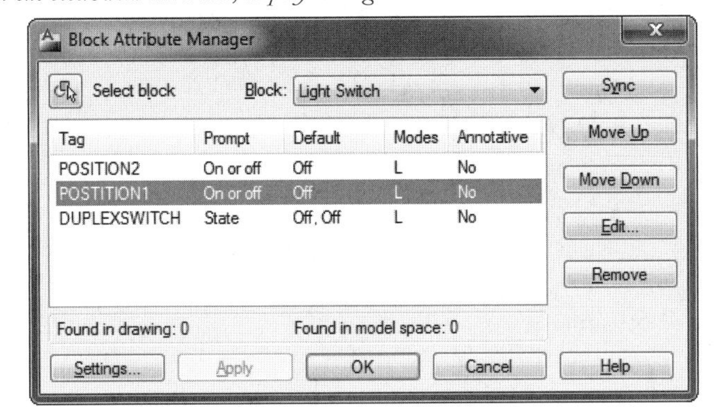

BLOCK ATTRIBUTE MANAGER DIALOG

 **Select block** hides the dialog box, and then prompts, 'Select a block:'.

Block lists the names of blocks in the drawing, and displays the name of the selected block.

Sync changes the attributes in block insertions to match the changes made here.

Move Up moves the selected attribute tag up the list; constant attributes cannot be moved.

Move Down moves the attribute tag down the list.

Edit displays the Edit Attribute dialog box; see the EAttEdit command.

Remove removes the attribute tag and related data from the block; it does not operate when the block contains a single attribute.

Settings displays the Settings dialog box.

Apply applies the changes to the block definition.

SETTINGS DIALOG BOX

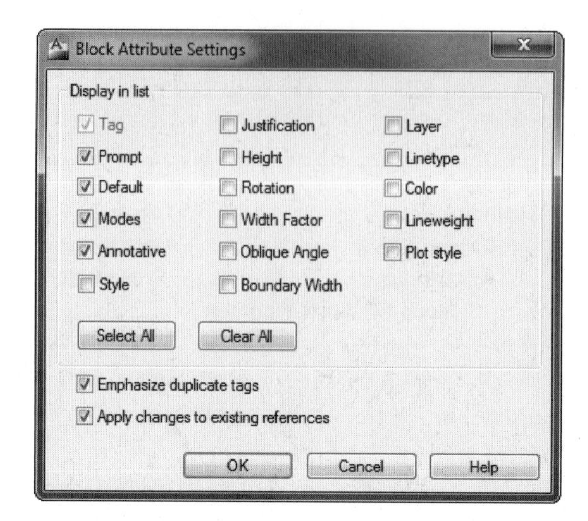

Display In List

☑ **Tag** toggles (on and off) the display of the column of tags.
☑ **Prompt** toggles display of the column of attribute prompts.
☑ **Default** toggles the display of the attribute's default value.
☑ **Modes** toggles the display of the attribute's modes: invisible, constant, verify, and/or preset.
☑ **Annotative** toggles annotative scaling of the attribute text.
☐ **Style** toggles the display of the attribute's text style name.

☐ **Justification** toggles the display of the attribute's text justification.
☐ **Height** toggles the display of the attribute's text height.
☐ **Rotation** toggles display of the attribute's text rotation angle.
☐ **Width Factor** toggles the display of the attribute's text width factor.
☐ **Oblique Angle** toggles the display of the attribute's text obliquing angle (slant).
☐ **Boundary Width** toggles the margin constraint of multi-line text.

☐ **Layer** toggles the display of the attribute's layer.
☐ **Linetype** toggles the display of the attribute's linetype.
☐ **Color** toggles the display of the attribute's color.
☐ **Lineweight** toggles the display of the attribute's lineweight.
☐ **Plot style** toggles the display of the attribute's plot style name (available only when plot styles are turned on).

Select All selects all display options.

Clear All clears all display options, except tag name.

Emphasize duplicate tags:

☑ Highlights duplicate attribute tags in red.
☐ Does not highlight duplicate tags.

Apply changes to existing references:

☑ applies changes to all block instances that reference this definition in the drawing.
☐ applies the new attribute definitions only to newly-inserted blocks.

RELATED COMMANDS

AttDef defines attributes.

Block binds attributes to symbols.

Insert inserts blocks, and then allows you to specify the attribute data.

TIPS

- Use this command to edit and remove attribute definitions, as well as to change the order in which attributes appear.
- The Sync option does not change the values you assigned to attributes.
- When an attribute has a mode of Constant, it cannot be moved up or down the list.
- Turning on all the options displays a lot of data. To see all the data columns, you can stretch the dialog box.

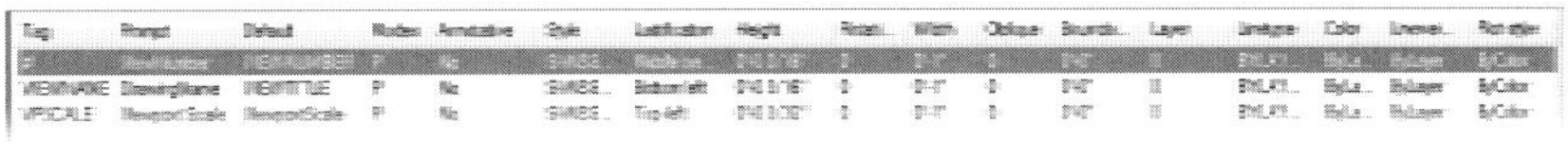

With the cursor, grab the edge of the dialog box to make it larger and smaller.

- An attribute definition cannot be changed to Constant via the Edit Attribute dialog box.
- The Remove option does not work when the block contains a single attribute.

BEdit

2006 Edits blocks and external references; assigns actions to blocks to create dynamic blocks (short for Block EDITor).

Command	Aliases	Keyboard Shortcuts	Menu	Ribbon
bedit	**be**	...	**Tools ↳Block Editor**	**Home ↳Block ↳Edit**
-bedit	...	...	...	...

Command: bedit

Displays dialog box.

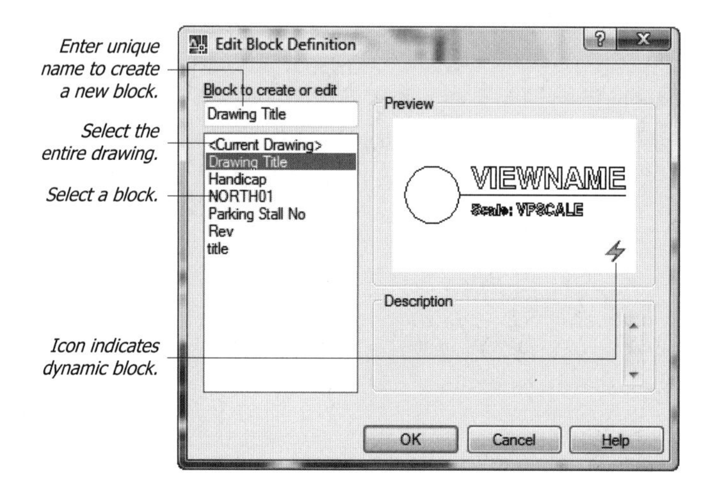

EDIT BLOCK DEFINITION DIALOG BOX

Block to create or edit

- To create a new block, enter a name.
- To edit an existing block, select its name from the list.
- To edit the drawing as a block, select <Current Drawing>.

OK enters the Block Editor.

Cancel returns to the drawing editor.

The following commands operate inside the Block Editor:

BAction	BActionBar	BActionSet	BActionTool
BAssociate	BAttOrder	BAuthorPalette	BAuthorPaletteClose
BClose	BCycleOrder	BCParameter	BConstruction
BGripSet	BLookupTable	BParameter	BSave
BSaveAs	BTable	BvState	BTestBlock
BvHide	BvShow	BwBlockAs	

Block Editor user interface

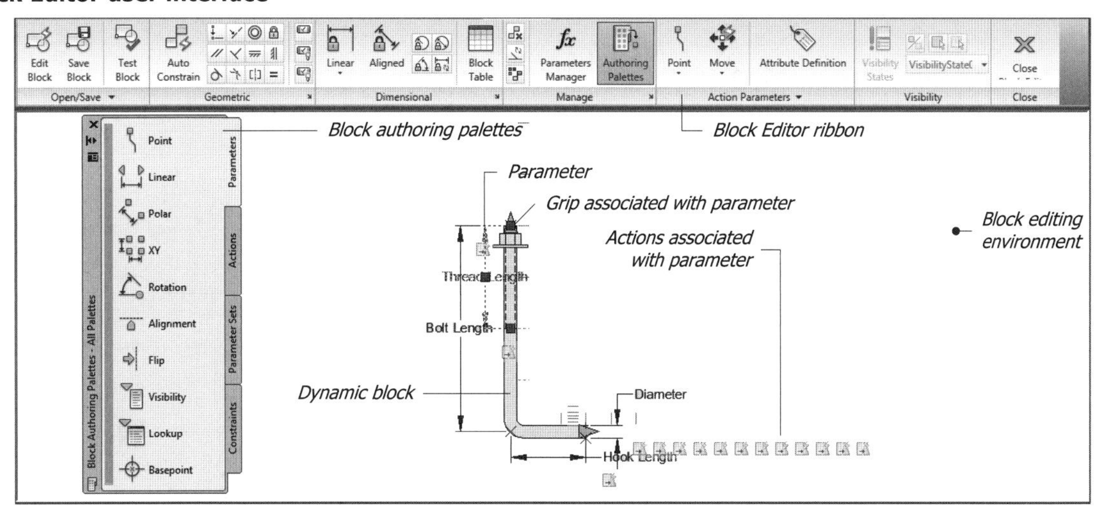

Block Editor: **BAction**

Adds actions to dynamic block definitions (alias: ac).

Command: baction
Select parameter: *(Select one parameter with which to associate an action.)*
Select action type [Array/Move/Scale/Stretch/Polar Stretch]: *(Enter an option.)*

Note: the actions available depend on the parameter selected. See list of actions and parameters later in this section.

Block Editor: **BActionBar**

Toggles the display of action bars.

Command: bactionbar
Select action: *(Choose an action.)*
Select parameters to show actions or [Showall/Hideall/RESet] <Show all>: *(Select one or more actions, or enter an option.)*

Select parameters displays action bars associated with parameters.

Showall displays all action bars.

Hideall hides all action bars.

RESet resets action bars to their original locations, and displays them all.

Block Editor: **BActionSet**

Respecifies selection set of objects associated with actions.

Select action: *(Select an action from the dynamic block definition.)*
Specify selection set for action object [New/Modify] <New>: *(Enter an option.)*

Modify modifies the existing selection set by adding and subtracting objects.

New creates a new selection set.

Block Authoring Palette: **BActionTool**

Defines changes to geometry of dynamic block references by changing custom properties.

Select action type [Array/Lookup/Flip/Move/Rotate/Scale/sTretch/Polar stretch]: *(Enter an option.)*
Select parameter: *(Select a parameter to associate with the action.)*
Specify selection set for action Select objects: *(Select one or more objects.)*

COMMAND OPTIONS

Array adds array actions to dynamic block definitions; selected objects are arrayed.

Flip adds flip actions to dynamic block definitions; objects are flipped about mirror line.

Lookup adds lookup actions to the dynamic block definitions; lookup actions display the shortcut menu with options.

Move adds move actions to dynamic block definitions; objects move in linear or polar directions.

Rotate adds rotate actions to dynamic block definitions; prompts "Specify action location or [Base type]."

Scale adds resize actions to dynamic blocks; objects resize in linear, polar, or x,y directions.

Stretch adds stretch actions to dynamic blocks; objects are stretched in point, linear, polar, or x,y directions.

Polar Stretch adds stretch actions to dynamic blocks; blocks are stretched by distance and angle.

Several actions use these options:

Base Type specifies the type of base point to use for the action:

- **Dependent** rotates dynamic block about the associated parameter's base point.
- **Independent** rotates dynamic block about a specified base point.

Multiplier changes the associated parameter value by the specified factor; prompts "Enter distance multiplier <1.0>."

Offset increases and decreases angles by the specified number of degrees; prompts "Enter angle offset <0>."

XY specifies distances in the x-, y-, and xy-directions from the parameter base point; prompts "Enter XY distance type [X/Y/XY]."

Block Editor: **BAssociate**

Associates orphaned actions with parameters.

Select action object: *(Select one action not associated with parameters.)*
Select parameter to associate with action: *(Select the parameter to associate.)*

Block Editor: **BAttOrder**

Controls the order in which attributes are listed.

Displays dialog box.

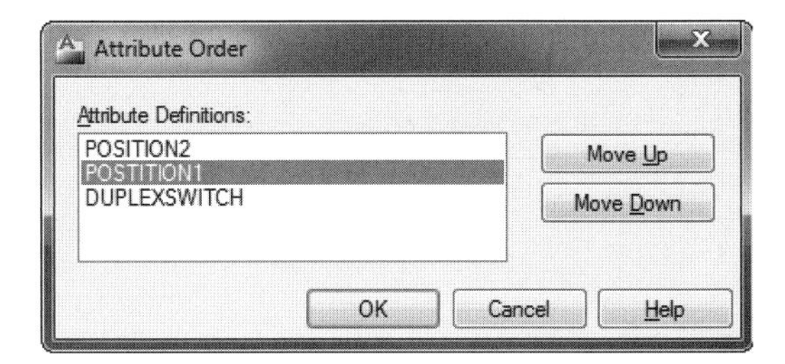

Move Up moves the selected attribute up the list.

Move Down moves the selected attribute down the list.

OK exits the dialog box, and returns to the Block Editor.

Cancel cancels changes, and returns to the Block Editor.

Block Editor: **BAuthorPalette** and **BAuthorPaletteClose**

Toggles the display of the Block Authoring Palettes window.

Block Editor: **BClose**

Closes the Block Editor environment, and returns to the drawing editor (*alias:* **bc**)

Block Editor: **BConstruction**

Converts objects into construction objects, which are seen only in the Block Editor.

Select objects or [Show all/Hide all]: *(Choose one or more objects, or enter an option.)*
Enter an option [Convert/Revert] <Convert>: *(Enter an option.)*

Select objects select objects to convert.

Show all shows all hidden construction objects.

Hide all hides all construction objects.

Convert converts objects into construction objects, which are displayed as gray dashed linetypes; their color, linetype, and layer cannot be changed.

Revert converts construction objects back into regular objects.

Block Editor: **BCParameter**

Converts dimensional constraints into parameter constraints; applies constraint parameters to objects; short for "block constraint parameter."

Select a dimensional constraint or [LInear/Horizontal/Vertical/Aligned/ANgular/Radius/Diameter] <last>: *(Select a dimensional constraint, or enter an option.)*

For details on options, see the DimConstraint command.

Block Editor: **BCycleOrder**

Changes the cycle order of grips in dynamic block references.

Displays dialog box.

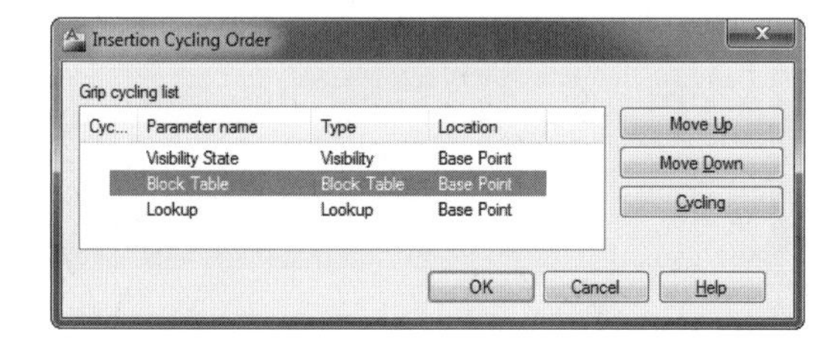

Move Up moves the selected parameter up the list.

Move Down moves the selected parameter down the list.

Cycling toggles cycling of the selected parameter.

Block Editor: **BGripSet**

Creates, resets, and deletes grips associated with parameters.

Select parameter: *(Select one parameter in the dynamic block definition.)*

Enter number of grip objects for parameter or reset position [0/1/2/4/Reposition]: *(Enter number of grips to display, or type R.)*

0 - 4 specifies the number of grips to display; number varies, depending on the parameter. Enter **0** to hide the display of grips.

Reposition repositions grips to their default positions (a.k.a. resets).

Grip	Name	Grip Manipulation	Associated Parameters
	Alignment	Aligns with objects within 2D planes.	Alignment
	Flip	Flips (mirrors).	Flip
	Linear	Moves along defined directions or axes.	Linear
	Lookup	Displays lists of items.	Visibility, Lookup
	Rotation	Rotates about axes.	Rotation
	Standard	Moves in any direction within 2D planes.	Base, Point, Polar, XY

Block Editor: **BLookupTable**

Creates lookup tables for dynamic block definitions.

If dynamic block definition contains a lookup action and at least one lookup parameter, displays dialog box.

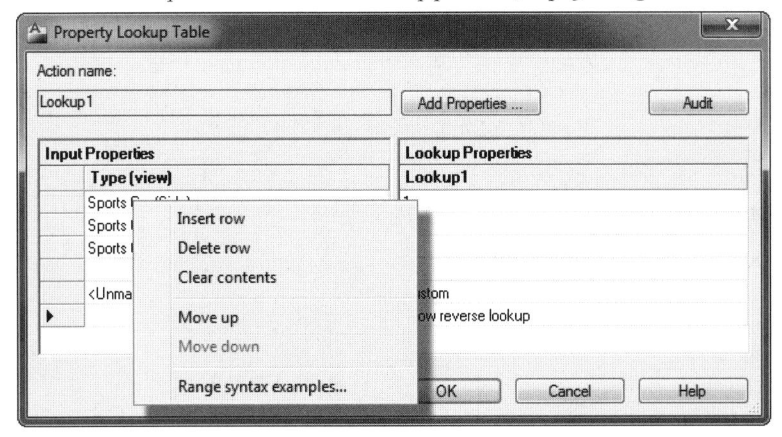

Add Properties displays the Add Parameter Properties dialog box for adding parameter properties to this table.

Audit checks data to ensure that rows are unique.

Right-click data or rows:

Insert Row adds a row above the selected row.

Delete Row erases the row and its data.

Clear Contents erases the data in the selected cell or row.

Move Up moves the row up.

Move Down moves the row down.

Range Syntax Examples displays help on writing ranges.

Right-click headings (columns):

Sort sorts data in the selected column.

Maximize All Headings makes each column wide enough to show the heading's wording.

Maximize All Data Cells makes each column wide enough to show the data's wording.

Size Columns Equally makes all columns the same width, fitted within the dialog box.

Delete Property Column erases the selected column.

Clear Contents erases the data in the selected cell or row.

Block Editor: **BParameter**

Adds parameters with grips to dynamic block definitions (*alias*: **param**).

Enter parameter type [Alignment/Base/pOint/Linear/Polar/XY/Rotation/Flip/Visibility/lookUp] <last>: *(Enter an option.)*

Select a parameter, and then associate an action with the parameter:

Parameter	Options	Action(s)	Comments
Alignment	Perpendicular	...	Rotates and aligns with other Base point objects automatically.
Base	Location	...	Defines base points for dynamic blocks.
Flip	Reflection Line	Flip	Flips objects about their reflection lines.
Linear	Start Point EndpointMove Chain Midpoint List Increment	Array Scale Stretch	Constrains grip movement to preset angles.
Lookup	Action Name Properties Audit Input Lookup	Lookup	Defines custom properties that evaluate values from lists and tables you define.
Point	Location Chain	Move Stretch	Defines x,y locations in drawings.
Polar	Base Point EndpointMove Chain List Increment	Array Polar Stretch Scale Stretch	Constrains grip movement to distances and angles.
Rotation	Base Point Radius Base Angle Default Angle Chain List Increment	Rotate	Defines angles.
Visibility	Display	...	Toggles the visibility states of blocks.
XY	Base Point EndpointMove Chain List Increment	Array Scale Stretch	Shows x,y distances from the base points of blocks.

Block Editor: **BSave** and **BSaveAs**

Save block definitions in the drawing, not to disk.

BSave saves the changes to the block definition. Changes are also saved when exiting the Block Editor (*alias*: **bs**)

BSaveAs saves the block definition by another name; displays dialog box.

Block Editor: **BTable**

Specifies groups of properties and their values; properties can be AutoCAD parameters, user-defined parameters, constraint parameters, and attributes.

When a table exists in the block, displays the Block Properties Table dialog box, detailed later.

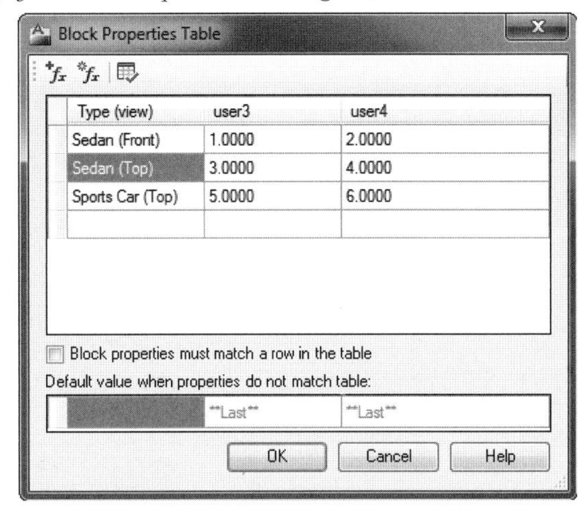

When a table does not exist yet, displays the following prompts:

Specify parameter location or [Palette]: *(Pick a point, or type P.)*

Enter number of grips [0/1] <1>: *(Type 0 or 1.)*

Specify parameter location picks the location of the parameter icon.

Palette prompts you, 'Display property in Properties palette? [Yes/No]:

- **Yes** displays block properties in Properties palette.
- **No** does not display block properties in palette.

Number of grips toggles the display of the accompanying grip:

- **0** Displays no grip.
- **1** Displays a triangular grip next to the table icon (as illustrated below).

BLOCK PROPERTIES TABLE DIALOG BOX

Add Properties adds parameters to the table; displays the Add Parameter Properties dialog box (detailed later).

New Properties creates and adds user-defined parameters to the table; displays the New Parameters dialog box.

Audit checks entries for errors; dialog boxes report errors.

(If you close the dialog box when errors exist, AutoCAD asks if you want to correct or ignore the errors.)

Grid Control Table reports the names and values of parameters entered into the table by the Add Properties and New Properties buttons. Click headings to change sort order; click cells to select values. Right-click headings and cells for shortcut menus.

Block Properties Must Match A Row In The Table determines how properties are modified:

☐ Properties in the grid control can be modified individually for a block reference.

☑ Properties must match the rows.

Default Value When Properties Do Not Match Table reports the default values of properties when they are changed without matching a row.

OK closes the dialog box, and places a table icon in the drawing at the location specified by the 'Specify parameter location' prompt.

 Double-click the icon to reopen this dialog box; double-click the grip to open the Properties palette.

The data in this dialog box are accessed by the BLookupTable command.

ADD PARAMETER PROPERTIES DIALOG BOX

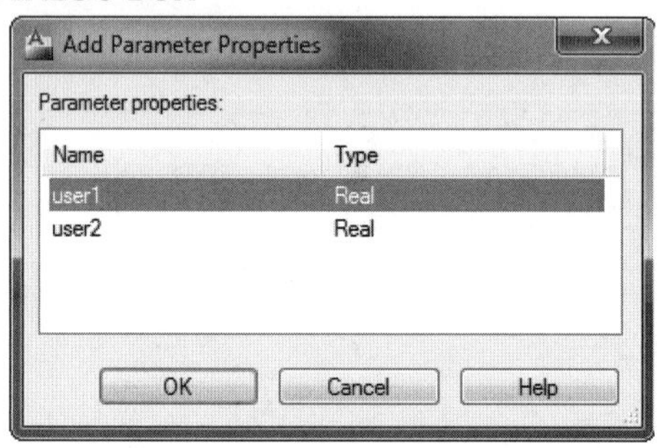

Name displays the names of parameters defined in the dynamic block. To add a parameter to the block properties table, choose one, and then click OK. Repeat. As you select parameters, they are removed from the list in this dialog box and added to the table in the Block Properties Table dialog box.

Type reports the type of parameter's type.

NEW PARAMETERS DIALOG BOX

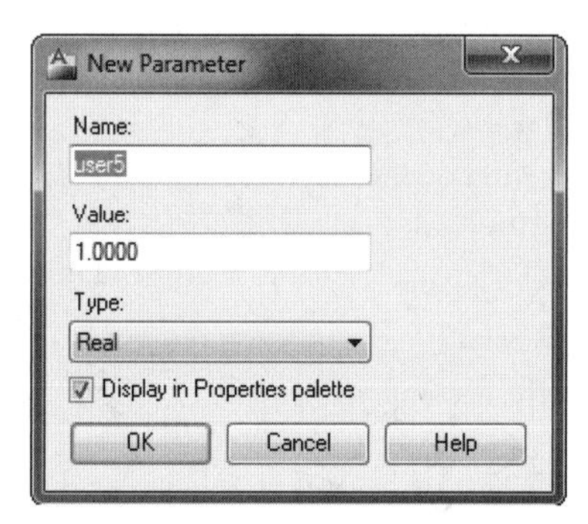

Name names the user-defined parameter.

Value specifies the value of the new parameter.

Type specifies the parameter's type; click the droplist to choose from real, distance, area, volume, angle, or string.

Display in Properties Palette toggles display of parameters in the Properties palette:

☑ Displays the user-defined parameter in the Properties palette.

☐ Does not display the parameter.

Block Editor: **BTestBlock**

Displays the dynamic block in a window that lets you test its actions; repeat the command to close the window.

This command has no options.

Note that the QSave, Save, and SaveAs commands save the block as a *.dwg* file when used inside the test window; after being saved, the block is opened as a drawing file in AutoCAD. The BTestBlock, BClose, and Close commands close the Test Block window, and return to the Block Editor.

Block Editor: **BvHide** and **BvShow**

Changes the visibility state of objects in dynamic block definitions.

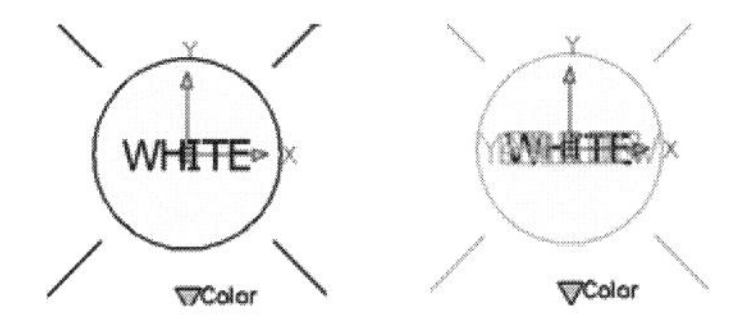

Left: *Visibility state set to Current (White).*
Right: *Visibility state set to All (White, Green, Red, Blue, Yellow, and Amber).*

Select objects to hide/show: *(Select objects to hide or show.)*
Hide/Show for current state or all visibility states [Current/All]: *(Enter an option.)*

Current sets the visibility of selected objects to the current visibility state.

All sets the visibility of selected objects to all visibility states, as defined by the block.

Block Editor: **BvState**

Creates, sets, renames, and deletes visibility states in dynamic blocks (*alias*: **vs**)

Displays dialog box.

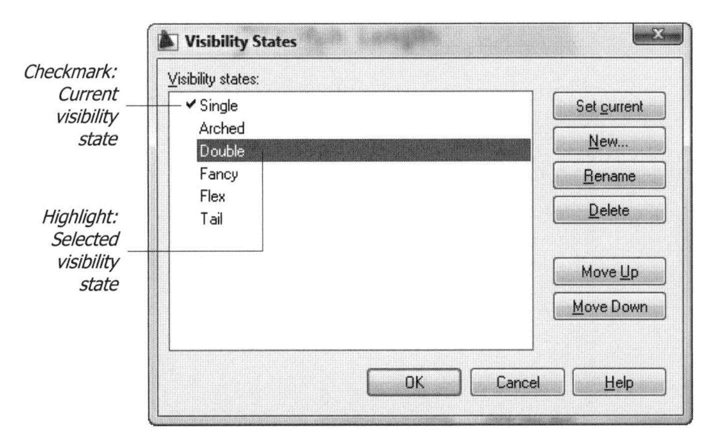

Set Current sets the selected visibility state as current in the Block Editor; does not affect the default visibility state of blocks in drawings.

New displays the New Visibility State dialog box (detailed below).

Rename changes the name of the selected visibility state.

Delete erases the selected visibility state.

Move Up moves the selected visibility state up the list.

Move Down moves the selected visibility state down the list.

NEW VISIBILITY STATE DIALOG BOX

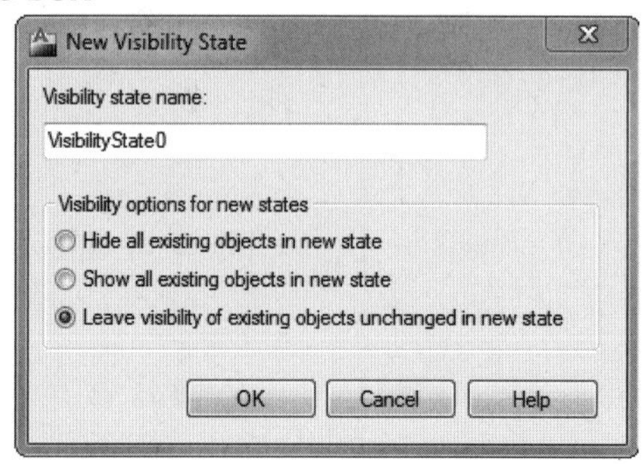

Visibility State Name names the new visibility state.

Visibility Options for New States:

○ **Hide All Existing Objects in New State** hides all objects in this visibility state.

○ **Show All Existing Objects in New State** shows all objects in this visibility state.

⊙ **Leave Visibility of Existing Objects Unchanged in New State** retains the visibility state of objects as the current visibility state.

Block Editor: **BwBlockAs**

Exports the block as a *.dwg* file on disk.

Displays dialog box.

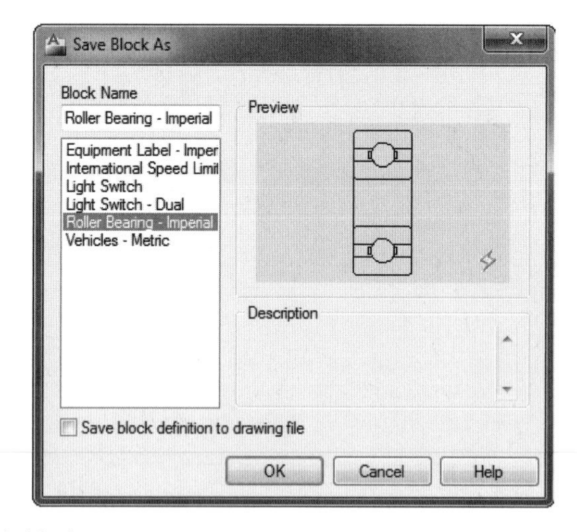

Block Name names the dynamic block.

Save block definition to drawing file determines if the block is also saved to disk:

☑ Block is saved as a *.dwg* file, like the WBlock command.

☐ Block is stored in the current drawing.

-BEDIT Command

Command: -bedit

Enter block name or [?]: *(Enter the name of a block or type ?.)*

Enter block name specifies the block to edit.

? lists the names of blocks in the drawing, if any.

RELATED COMMANDS

Block creates non-dynamic blocks.

Insert inserts blocks into drawings, including dynamic blocks.

RefEdit edits blocks and xrefs in-place.

RELATED SYSTEM VARIABLES

BlockEditLock toggles editing of dynamic blocks in the Block Editor.

BlockEditor reports whether the Block Editor is active.

The following system variables affect the Block Editor only:

BActionBarMode toggles the display of action bars related to selected parameters.

BActionColor specifies the text color of actions.

BDependencyHighlight toggles highlighting of objects dependent on selected authoring objects.

BGripObjColor specifies the color of custom grip object.

BGripObjSize specifies the size of grips.

BlockTestWindow reports whether the block test window is open.

BParameterColor specifies the color of block parameters.

BParameterFont specifies the SHX or TrueType font name for parameters.

BParameterSize specifies the size of parameter text and arrowheads.

BPTextHorizontal determines if parameter text is aligned or horizontal.

BTMarkDisplay toggles the display of parameter value set markers.

BVMode toggles the display of hidden objects between invisible or dimmed.

TIPS

- Selecting a block before entering the BEdit or -BEdit command immediately opens the Block Editor; alternatively, double-click a dynamic block to open it in the Block Editor.
- Entering names of nonexistent blocks creates new blocks in the Block Editor.
- Most of AutoCAD's drawing and editing commands are available in the Block Editor.
- Values in lookup tables are limited to 256 characters per cell, including distances and angles for points, linear, polar, XY, and rotation parameters; text string parameters; properties for flip; visibility parameter values; and architectural and mechanical units, such as 23'1/2". Invalid values are automatically reset to the last valid value.
- Commas are delimiters between values. Brackets [ ] specify inclusive ranges; parentheses ( ) specify exclusive ranges, as follows:

[8,31]	Any values between 8 and 31 inclusive.
(9,15)	Any values between 9 and 15, excluding 9 and 15.
[,25]	Any values up to and including 25.
(56,)	Any values larger than 56.

! This icon means that no action is associated with the parameter.

- When constraint bars are displayed, action bars are hidden, and vice versa.
- You can change the color, linetype, and layer of construction objects, but the changes will not show up until the construction objects revert to regular objects.
- Only one table can be placed by the BTable command in each dynamic block definition.
- Grips display tooltips describing their purpose, as illustrated below:

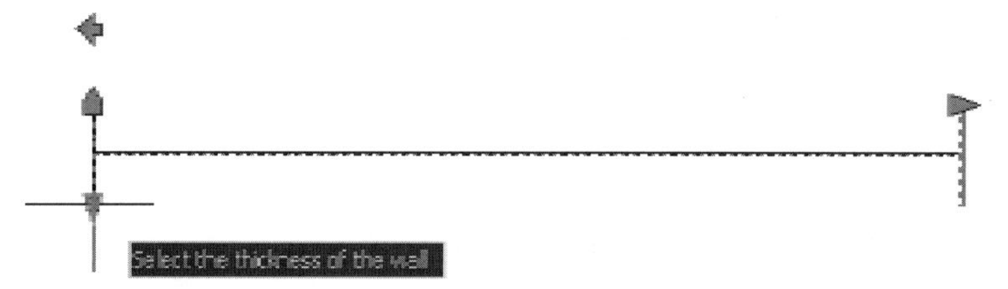

- If you cannot open the Block Editor, change the value of BlockEditLock to 0.
- The Block Edit's parameters are different from the drawing editor's parameters, but accomplish much the same thing.
- *New in AutoCAD 2011:*
  - The Parameters palette adds the Show and Order columns for toggling the display and determining the order in which constraints appear in the Properties palette.
  - Icons are now displayed for action parameters, attributes, and user parameters (as in the drawing editor version of the palette).

DEFINITIONS

Actions — determine how the geometry of dynamic blocks change when their custom properties are modified; actions are associated with parameters.

Action bars — indicate actions through small icons that associated with parameters.

Parameters — define custom properties for dynamic block references; actions must be associated with parameters.

Orphaned Actions — arise when parameters are removed from block definitions, orphaning their associated actions.

BESettings

2010 Sets colors, fonts, and other options for objects displayed by the Block Editor.

Command	*Aliases*	*Keyboard Shortcuts*	*Menu*	*Ribbon*
besettings	...	...	...	**Block Editor** ↳**Manage** ↳**Dialog box launcher**

Command: besettings

Displays dialog box.

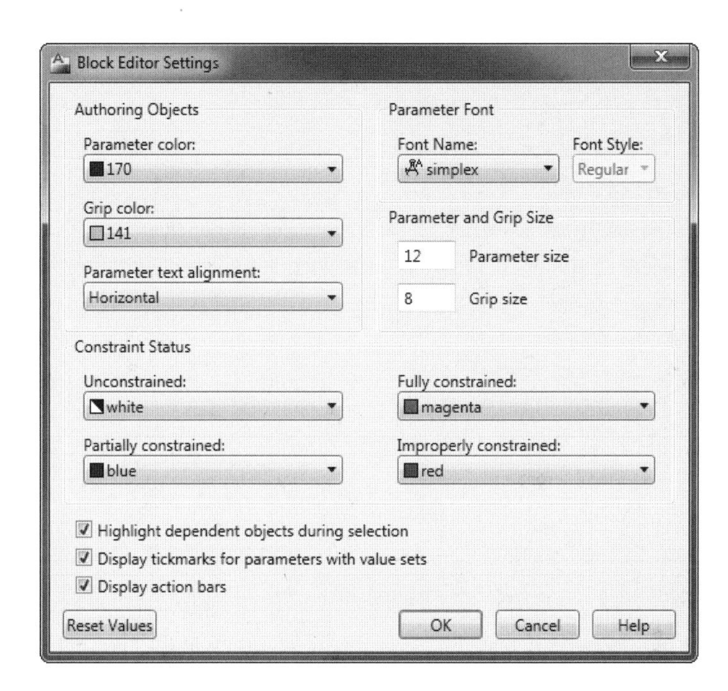

BLOCK EDITOR DIALOG BOX

Authoring Objects options

Parameter Color specifies the color of block parameters (stored in system variable *BParameterColor*); choose a color from the droplist, or else click Select Color for the Select Color dialog box.

Grip Color specifies the color of custom grip objects (*BGripObjColor*).

Parameter Text Alignment determines how parameter text is aligned (*BPTextHorizontal*):

- **Horizontal** forces the text always to be horizontal, no matter the angle of the parameter.
- **Aligned** aligns the text with the parameter.

Parameter Font options

Font Name specifies the font name; choose an SHX or TrueType name from the droplist (*BParameterFont*).

Font Style formats TrueType fonts; choose from regular, italic, bold, or bold-italic (does not apply to SHX fonts).

Parameter and Grip Size options

Parameter Size specifies the height of parameter text and size of arrowheads (*BParameterSize*).

Grip Size specifies the size of grip objects; measured in pixels (*BGripObjSize*).

Constraint Status options

Unconstrained specifies the color of unconstrained objects.

Partially Constrained specifies the color of partially constrained objects.

Fully Constrained specifies the color of fully constrained objects.

Improperly Constrained specifies the color of over constrained objects.

Highlight Dependent Objects During Selection highlights automatically objects dependent on the selected authoring object (*BDependencyHighlight*).

Display Tickmarks for Parameters With Value Sets toggles the display of parameter value set markers (*BTMarkDisplay*).

Display Action Bars toggles the display of action bars when parameters are selected (*BActionBarMode*).

Reset Values resets all settings to default values.

RELATED COMMAND

BEdit edits blocks.

RELATED SYSTEM VARIABLES

BActionBarMode toggles the display of action bars related to selected parameters.

BDependencyHighlight toggles highlighting of objects dependent on selected authoring objects.

BGripObjColor specifies the color of custom grip objects.

BGripObjSize specifies the size of grips.

BParameterColor specifies the color of block parameters.

BParameterFont specifies the SHX or TrueType font name for parameters.

BParameterSize specifies the size of parameter text and arrowheads.

BPTextHorizontal determines if parameter text is aligned or horizontal.

BTMarkDisplay toggles the display of parameter value set markers.

TIPS

- You can change the background color of the Block Editor through the Display tab of the Options dialog box: click the Colors button, and then choose Block Editor.

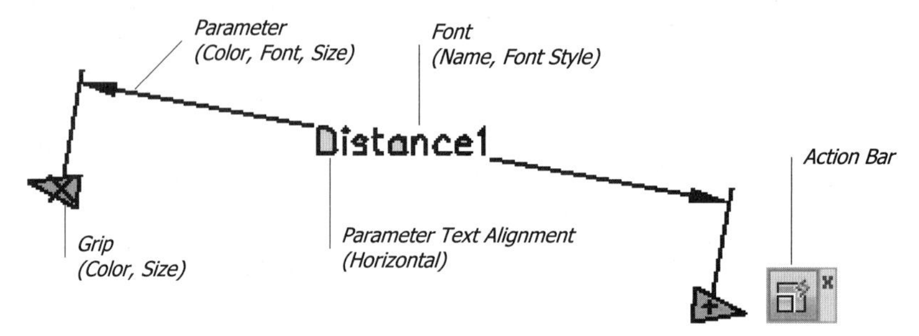

- When you use this dialog box inside the Block Editor, the changes are reflected immediately after exiting the dialog box.

Renamed Commands

BHatch and **-BHatch** commands are found under the Hatch and -Hatch commands.

Blipmode

Ver. 1.2 Turns the display of pick-point markers, known as "blips," on and off.

Command	*Aliases*	*Keyboard Shortcuts*	*Menu*	*Ribbon*
'blipmode	...	...	...	...

Command: blipmode
Enter mode [ON/OFF] <OFF>: on

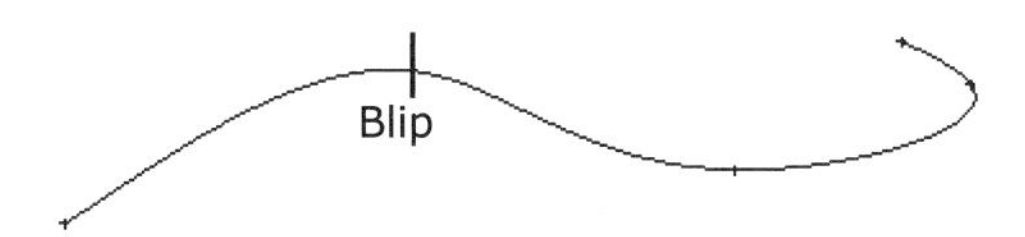

COMMAND LINE OPTIONS

ON turns on the display of pick-point markers.

OFF turns off the display of pick-point markers (default).

RELATED COMMANDS

Options allows blipmode toggling via a dialog box.

Redraw cleans blips off the screen.

RELATED SYSTEM VARIABLE

Blipmode contains the current blipmode setting.

TIPS

- You cannot change the size of the blip mark.
- Blip marks are erased by any command that redraws the view, such as Redraw, Regen, Zoom, and VPorts.

Block

Ver. 1.0 Defines a group of objects as a single named object; creates symbols.

Commands	*Aliases*	*Keyboard Shortcuts*	*Menu*	*Ribbon*
block	**b** **bmake** **bmod** **acadblockdialog**	...	**Draw** **↳Block** **↳Make**	**Insert** **↳Block** **↳Create**
-block	**-b**	...	...	...

Command: block

Displays dialog box.

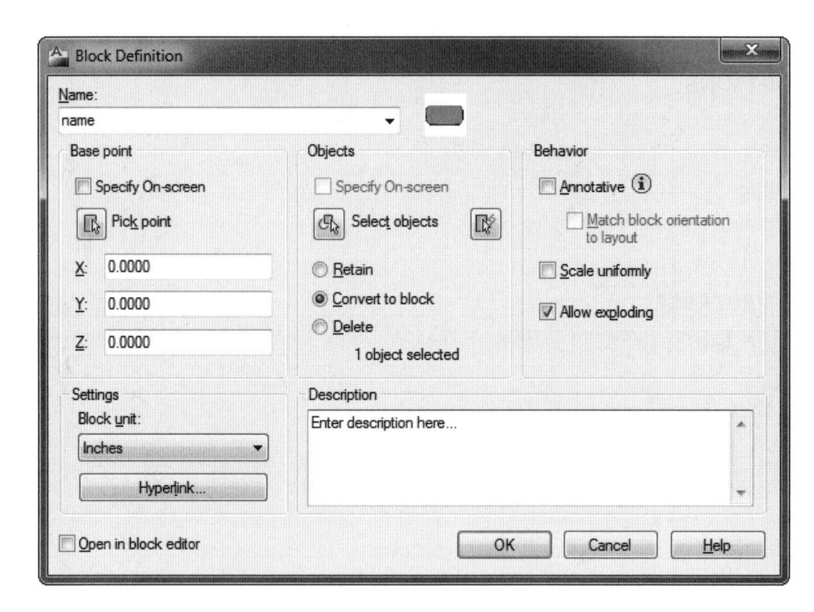

BLOCK DEFINITION DIALOG BOX

Name names the block (*maximum = 255 characters, including spaces, numbers, and other special characters, except* \ | / * ? : < *or* >).

Base Point

Specify On-screen prompts 'Specify insertion base point:' after you click OK.

Pick point dismisses the dialog box; AutoCAD prompts, 'Specify insertion base point:'. Pick a point that specifies the block's insertion point, usually the lower-left corner.

X, **Y**, and **Z** specify the x,y,z coordinates of the insertion point.

Objects

Specify On-screen prompts 'Select objects:' after you click OK.

Select objects dismisses the dialog box; AutoCAD prompts, 'Select objects:'. Select one or more objects that make up the block:

- ○ **Retain** leaves objects in place after the block is created.
- ⊙ **Convert to Block** erases objects making up the block, and replaces them with the block.

- ○ **Delete** erases the objects making up the block; the block is stored in drawing.

Quick Select displays the Quick Select dialog box; see the QSelect command.

Behavior

☑ **Annotative** toggles annotative scaling for this block.

- ☐ **Match Block Orientation to Layout** ensures the block rotates to match the layout.

☑ **Scale Uniformly** prevents different scale factors being applied to the x,y, and z directions.

☑ **Allow Exploding** allows the block to be exploded.

Settings

Insert Units selects the units for the block when dragged from the Design Center.

Hyperlink displays the Insert Hyperlink dialog box; see the Hyperlink command.

Description

Description describes the block.

Open in Block Editor opens the block in the Block Editor.

(AutoCAD 2008 removed the Preview Icon and X, Y, Z Scale options from this dialog box.)

-BLOCK Command

Command: -block
Enter block name or [?]: *(Enter a name, or type ?.)*
Specify insertion base point or [Annotative]: *(Pick a point, or type A.)*
Select objects: *(Select one or more objects.)*
Select objects: *(Press Enter.)*

COMMAND LINE OPTIONS

Block name allows you to name the block.

? lists the names of blocks stored in the drawing.

Insertion base point specifies the x,y coordinates of the block's insertion point.

Annotative applies annotative scaling to the block.

Select objects selects the objects and attributes that make up the block.

Annotative options
Create annotative block [Yes/No] <N>: *(Type Y.)*
Match orientation to layout in paper space viewports [Yes/No] <N>:

Create annotative block determines whether the block takes annotative scale factors.

Match orientation ensures the block rotates to match the layout.

RELATED COMMANDS

BEdit opens blocks in the Block Editor environment; makes dynamic blocks.

BlockIcon generates icons of blocks defined in earlier releases of AutoCAD.

Explode reduces blocks to their original objects.

Insert adds blocks or other drawings to the current drawing.

Oops returns objects to the screen after you create the block.

Purge removes unused blocks from drawings.

RefEdit edits blocks and xrefs in-place.

WBlock writes blocks as drawings to *.dwg* files on disk.

XRef displays another drawing in the current drawing.

RELATED FILES

All *.dwg* drawing files can be inserted as blocks.

RELATED SYSTEM VARIABLES

InsName default block name.

InsUnits drawing units for blocks dragged from the DesignCenter.

TIPS

- Blocks consist of these parts:

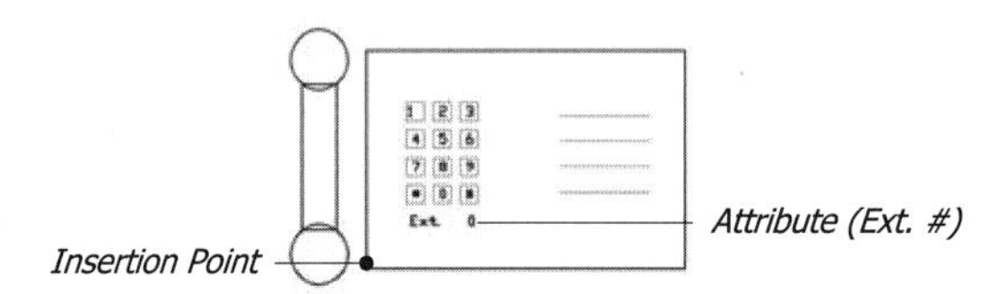

- The names of blocks have up to 255 alphanumeric characters, including $, -, and _.
- Use the INSertion object snap to select the insertion point of blocks.
- Objects within a block definition take on the properties specified by their layers, with one exception: if the objects were created on layer 0, then upon insertion they take on the properties specified by the host layer.
- AutoCAD has five types of blocks:

User blocks	Named blocks created by users.
Nested blocks	Blocks inside other blocks.
Unnamed blocks	Blocks created by AutoCAD.
Xrefs	Externally-referenced drawings.
Dependent blocks	Blocks in externally-referenced drawings.

- AutoCAD sometimes creates unnamed blocks, also called "anonymous blocks":

A*n	Groups.
D*n	Associative dimensions.
U*n	Created by AutoLISP or ARx applications.
X*n	Hatch patterns.

- AutoCAD automatically purges unreferenced anonymous blocks when drawings are first loaded.

BlockIcon

2000 Generates preview images for all blocks in drawings created with AutoCAD Release 14 or earlier.

Command	*Aliases*	*Keyboard Shortcuts*	*Menu*	*Ribbon*
blockicon	...	...	**File** ↳**Drawing Utilities** ↳**Update Block Icons**	...

Command: blockicon
Enter block names <*>: *(Enter a name, a wildcard pattern, or press Enter for all names.)*
n **blocks updated.**

COMMAND LINE OPTION

Enter block names specifies the blocks for which to create icons; press Enter to add icons to all blocks in the drawing.

RELATED COMMANDS

AdCenter displays the icons created by this command:

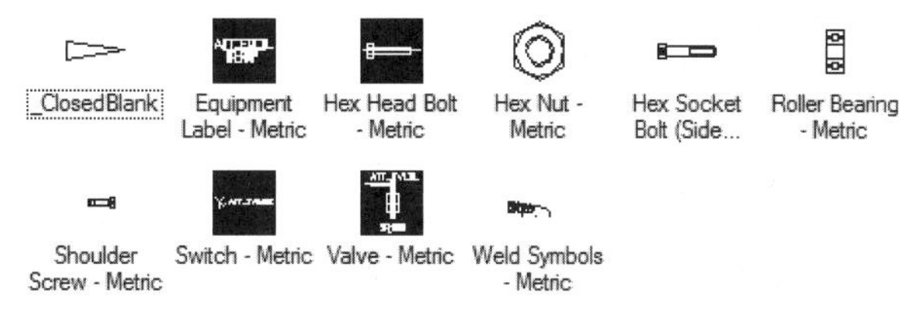

Block creates new blocks and their icons.

Insert inserts blocks; dialog box displays icons generated by this command.

TIP

- AutoCAD creates icons of blocks automatically; this command is necessary only for drawings older than AutoCAD 2000.

BmpOut

Rel. 13 Exports the current viewport as a raster image in BMP bitmap format.

Command	Aliases	Keyboard Shortcuts	Menu	Ribbon
bmpout	...	...	...	...

Command: bmpout

Displays Create BMP File dialog box. Enter a file name, and then click Save.

Select objects or <all objects and viewports>: *(Press Enter to select all objects, or select individual objects.)*

DIALOG BOX OPTION

Save saves drawings as BMP format raster files.

RELATED COMMANDS

JpgOut exports objects and viewports in JPEG format.

PngOut exports objects and viewports in PNG format.

TifOut exports objects and viewports in TIFF format.

WmfOut exports selected objects in WMF format.

RELATED WINDOWS COMMANDS

PRT SCR saves screen to the Clipboard.

ALT+PRT SCR saves the topmost window to the Clipboard.

TIPS

- The *.bmp* extension is short for "bitmap," a raster file standard for Windows.
- This command creates uncompressed *.bmp* files.

Boundary

Rel. 12 Creates boundaries as polylines or 2D regions.

Command	*Aliases*	*Keyboard Shortcuts*	*Menu*	*Ribbon*
boundary	**bo** **bpoly**	...	**Draw** ↳**Boundary**	**Home** ↳**Draw** ↳**Boundary**
-boundary	**-bo**	...	...	...

Command: boundary

Displays dialog box.

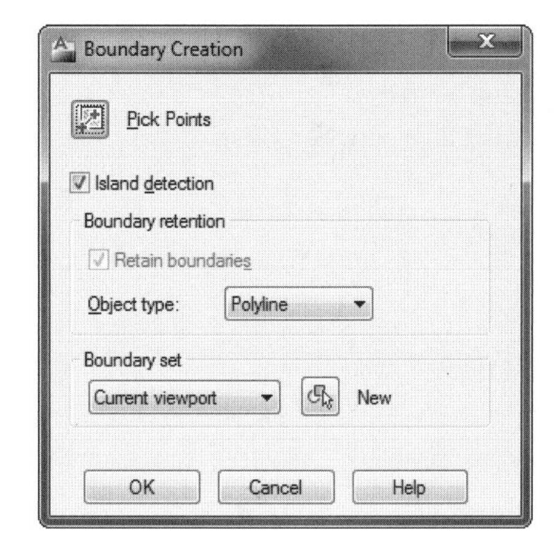

BOUNDARY CREATION DIALOG BOX

 **Pick Points** picks points inside closed areas.

Island detection: toggles island detection:

☑ Island detection is turned on.

☐ Island detection is turned off.

Boundary Retention

Retain Boundaries is grayed out (unavailable), because the point of this command is to create boundaries, not discard them!

Object Type constructs the boundary from one of the following:

- **Polylines** forms the boundary from a polyline.
- **Region objects** forms the boundary from a 2D region.

Boundary Set

Boundary Set defines how objects are analyzed for defining boundaries (default = current viewport).

New creates new boundary sets; AutoCAD dismisses the dialog box, and then prompts you to select objects.

-BOUNDARY Command

Command: -boundary
Specify internal point or [Advanced options]: *(Pick a point, or type A.)*

COMMAND LINE OPTIONS

Specify internal point creates a boundary based on the point you pick.

Advanced options options
Enter an option [Boundary set/Island detection/Object type]: *(Enter an option.)*

Boundary set defines the objects **-Boundary** analyzes when defining a boundary from a specified pick point. It chooses from a new set of objects, or from all objects visible in the current viewport.

Island detection:

- **On** uses objects within the outermost boundary as boundary objects.
- **Off** fills objects within the outermost boundary.

Object type specifies polyline or region as the boundary object.

RELATED COMMANDS

Hatch creates boundaries for hatches and fills.

PLine draws polylines.

PEdit edits polylines.

Region creates 2D regions from a collection of objects.

RELATED SYSTEM VARIABLE

HpBound specifies the default object used to create boundary:

0 — Draw as region.
1 — Draw as polyline (default).

TIPS

- Use this command to measure irregular areas:
  1. Apply the Boundary command to an irregular area.
  2. Use the Area command to find the area and perimeter of the boundary.
- Use Boundary together with the Offset command to help create *poching*, areas partially covered by hatching.

Box

Rel. 13 Draws a 3D box as a solid model.

Command	*Aliases*	*Keyboard Shortcuts*	*Menu*	*Ribbon*
box	...	...	**Draw** ↳**Modeling** ↳**Box**	**Home** ↳**Modeling** ↳**Box**

Command: box
Specify first corner or [Center]: *(Pick point 1, or type the C option.)*
Specify other corner or [Cube/Length]: *(Pick point 2, or enter an option.)*
Specify height or [2Point] <-1.2757>: *(Pick point 3, or type the 2 option.)*

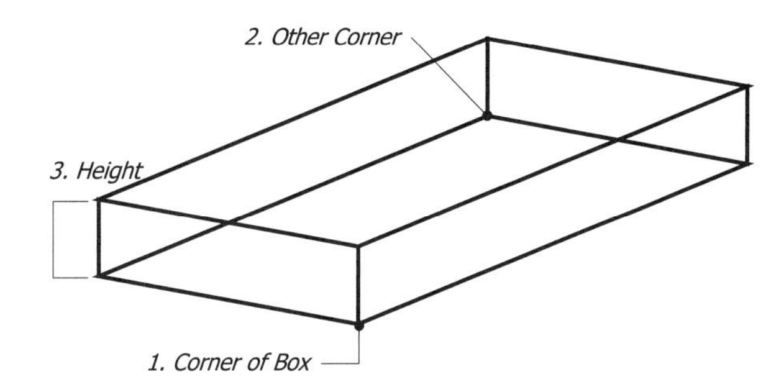

COMMAND LINE OPTIONS

First corner specifies one corner for the base of box.

Center draws the box about a center point.

Other corner specifies the second corner for the base of box.

Cube draws a cube box — all sides have the same length.

Length specifies the length along x axis, width along y axis, and height along z axis.

Height specifies the height of the box.

2point specifies height as the distance between two points.

RELATED COMMANDS

Ai_Box draws 3D surface model boxes.

Cone draws 3D solid cones.

Cylinder draws 3D solid tubes.

Sphere draws 3D solid balls.

Torus draws 3D solid donuts.

Wedge draws 3D solid wedges.

Mesh draws boxes as 3D mesh objects.

RELATED SYSTEM VARIABLES

DispSilh displays 3D objects as silhouettes after hidden-line removal and shading.

DragVs specifies the visual style during construction of 3D solid primitives.

TIP

- Once the box is placed in the drawing, you can edit it interactively. Select the box, and then notice the grips. Select a grip, which turns red, and then move it. Press Esc to exit direct editing.

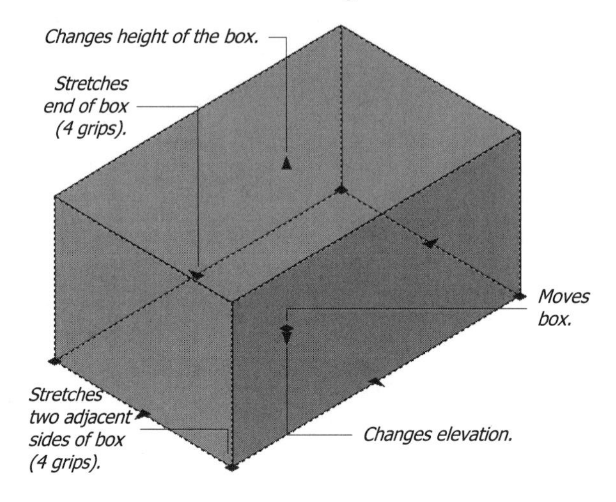

 Break

Ver. 1.4 Removes portions of objects.

Command	Alias	Keyboard Shortcuts	Menu	Ribbon
break	**br**	...	**Modify** ↳**Break**	**Home** ↳**Modify** ↳**Break**

Command: break
Select object: *(Select one object — point 1.)*
Specify second break point or [First point]: *(Pick point 2, or type F.)*

1. First Point 2. Second Point

Breaking a polyline at two points.

COMMAND LINE OPTIONS

Select object selects one object to break; the pick point becomes the first break point, unless the F option is used at the next prompt.

@ uses the first break point's coordinates for the second break point.

First Point options
Enter first point: *(Pick a point.)*
Enter second point: *(Pick a point.)*

First point specifies the first break point.

Second point specifies the second break point.

RELATED COMMANDS

Change changes the length of lines.

PEdit removes and relocates vertices of polylines.

Trim shortens the length of open objects.

TIPS

- Use this command to convert circles into arcs; pick the break points clockwise to keep the portion between pick points. Use the Join command to convert arcs to circles.
- This command can erase a portion of an object (as shown in the figure above) or remove the end of an open object.
- The second point does not need to be on the object; AutoCAD breaks the object at the point nearest to the pick point.
- The Break command works on the following objects: lines, arcs, circles, polylines, ellipses, rays, xlines, and splines, as well as objects made of polylines, such as donuts and polygons.

BRep

2007 Removes construction history from 3D solids and associativity from surfaces; short for "Boundary REPresentation."

Command	Aliases	Keyboard Shortcuts	Menu	Ribbon
brep	...	...	...	...

Command: brep
Select solids or surfaces: *(Select one or more 3D solid or surface models.)*
Select solids or surfaces: *(Press Enter to exit command.)*

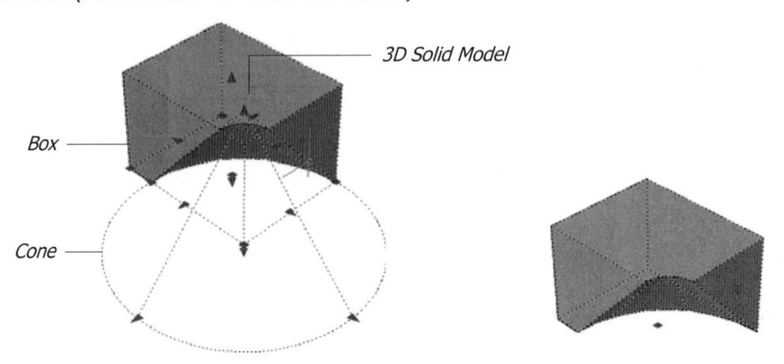

Left: *Before applying BRep, the model's original primitives can be selected.*
Right: *After applying BRep, the history of original primitives disappears.*

COMMAND LINE OPTION

Select solids or surfaces selects one or more objects from which to remove history or associativity.

RELATED SYSTEM VARIABLES

ShowHist toggles the preservation of history in solid models.

SolidHist toggles the display of primitive history entities.

SurfaceAssociativity toggles whether surfaces associate with their source objects *(new to AutoCAD 2011).*

TIP

- As of AutoCAD 2011, this command removes associativity from surfaces.

Browser / Browser2

Rel. 14 Prompts for a Web address, and then launches the Web browser.

Commands	*Aliases*	*Keyboard Shortcuts*	*Menu*	*Ribbon*
browser	...	...	...	...
browser2	...	...	...	...

Command: browser
Enter Web location (URL) <http://www.autodesk.com>: *(Enter a Web address.)*

Launches browser:

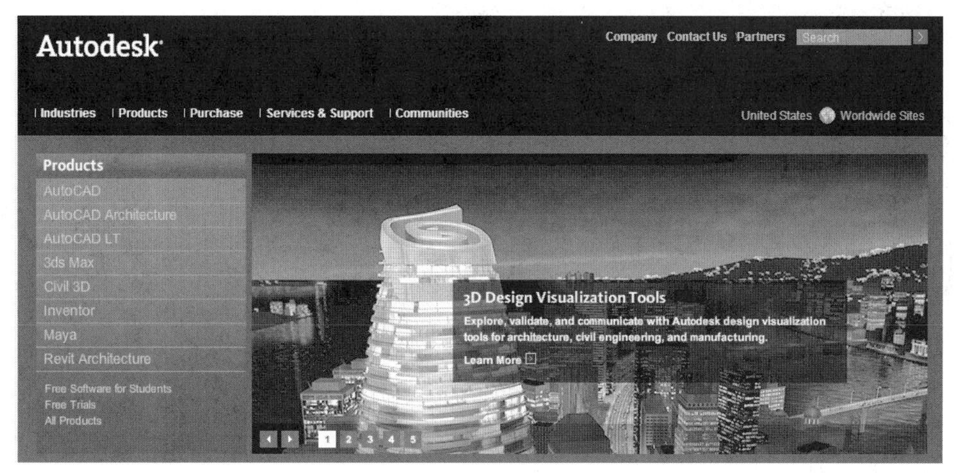

COMMAND LINE OPTION

Enter Web location specifies the URL; see the AttachURL command for information about URLs.

RELATED SYSTEM VARIABLE

InetLocation contains the name of the default URL.

TIPS

- *URL* is short for "uniform (or universal) resource locator," the universal file-naming system used on the Internet; also called a link or hyperlink. An example of a URL is http://www.autodeskpress.com, the Autodesk Press Web site.
- Many file dialog boxes also give you access to the Web browser; see the Open command.
- The undocumented Browser2 command also opens URLs.
- You can also use this command to access FTP sites and files on your computer.

Block Editor Commands

Commands used inside the Block Editor are found under the **BEdit** command.

Cal

Rel. 12 Calculates algebraic and vector geometry at the command line (*short for CALculator*).

Command	Aliases	Keyboard Shortcuts	Menu	Ribbon
'cal	...	...	...	...

Command: cal

>>Expression: *(Enter an expression, and then press Enter.)*

COMMAND LINE OPTIONS

( )	Grouping of expressions.
[ ]	Vector expressions.
+	Addition.
-	Subtraction.
*	Multiplication.
/	Division.
^	Exponentiation.
&	Vector product of vectors.
sin	Sine.
cos	Cosine.
tang	Tangent.
asin	Arc sine.
acos	Arc cosine.
atan	Arc tangent.
ln	Natural logarithm.
log	Logarithm.
exp	Natural exponent.
exp10	Exponent.
sqr	Square.
sqrt	Square root.
abs	Absolute value.
round	Round off.
trunc	Truncate.
cvunit	Units conversion using *acad.unt*.
w2u	WCS to UCS conversion.
u2w	UCS to WCS conversion.
r2d	Radians-to-degrees conversion.
d2r	Degrees-to-radians conversion.
pi	The value PI (3.14159).
xyof	x,y coordinates of a point.
xzof	x,z coordinates of a point.
yzof	y,z coordinates of a point.
xof	x coordinate of a point.
yof	y coordinate of a point.
zof	z coordinate of a point.
rxof	Real x coordinate of a point.
ryof	Real y coordinate of a point.
rzof	Real z coordinate of a point.
cur	x,y,z coordinates of a picked point.
rad	Radius of object.
pld	Point on line, distance from.
plt	Point on line, using parameter *t*.
rot	Rotated point through angle about origin.
ill	Intersection of two lines.
ilp	Intersection of line and plane.
dist	Distance between two points.
dpl	Distance between point and line.
dpp	Distance between point and plane.
ang	Angle between lines.
nor	Unit vector normal.
vec	Vector translation between two points.
vec1	Unit vector direction.
dee	Distance between two endpoints.
ille	Intersection of two lines defined by endpoints.
mee	Midpoint between two endpoints.
nee	Unit vector in the x,y-plane normal to two endpoints
vee	Vector from two endpoints.
vee1	Unit vector from two endpoints.
end	Endpoint object snap.
ins	Insertion point object snap.
int	Intersection object snap.
mid	Midpoint object snap.
cen	Center object snap.
nea	Nearest object snap.
nod	Node object snap.
qua	Quadrant object snap.
per	Perpendicular object snap.
tan	Tangent object snap.

ESC Exits Cal mode.

RELATED COMMAND

QuickCalc displays a graphical user interface for the Cal command.

RELATED SYSTEM VARIABLES

UserI1 — **UserI5** store integers.

UserR1 — **UserR5** store real numbers.

TIPS

- To use Cal, type expressions at the >> prompt.

  For example, to find the area of a circle (pi*r^2) with radius of 1.2 units, enter the following:

  Expression >> pi*(1.2^2)
  4.52389

- Cal recognizes these prefixes:

  ***** Scalar product of vectors.

  & Vector product of vectors.

- And the following suffixes:

  r Radian (degrees is the default).

  g Grad.

  ' Feet (unitless distance is the default).

  " Inches.

- Because 'Cal is a transparent command, it can perform calculations in the middle of other commands, and then return the value to that command.

  For example, to set the offset distance to the radius of a circle:

  Command: offset
  Specify offset distance or [Through] <Through>: 'cal
  >> Expression: rad
  >> Select circle, arc or polyline segment for RAD function: *(Pick circle.)* 2.0
  Select object to offset or <exit>: *(And so on.)*

- This command works with real numbers and integers. The smallest and largest integers are -32768 and 32767.
- The technical editor recommends: "For greater precision of PI, check out http://3.14159265358979323846264338327950288419716939937510582097494459.com."
- AutoCAD 2011 allows the use of expressions in some commands, such as in ChamferEdge, Extrude, and SurfExtend. At the 'Enter expression:' prompt, you can enter the following:
  - Basic arithmetic expressions, such as 1/2.54.
  - User variables created in the Parameters palette, such as Var2+0.5.
  - Dimensional constraint parameters, such as D1 and Dia2.

Camera

2000 Creates 3D perspective views.

Command	Aliases	Keyboard Shortcuts	Menu	Ribbon
camera	**cam**	**...**	**View** ↳**Create Camera**	**...**

Command: camera

Current camera settings: Height=0'-0" Lens Length=50.0 mm

Specify camera location: *(Pick a point, or enter coordinates.)*

Specify target location: *(Pick another point, or enter coordinates.)*

Enter an option [?/Name/LOcation/Height/Target/LEns/Clipping/View/eXit]<eXit>: *(Enter an option, or press Enter to exit the command.)*

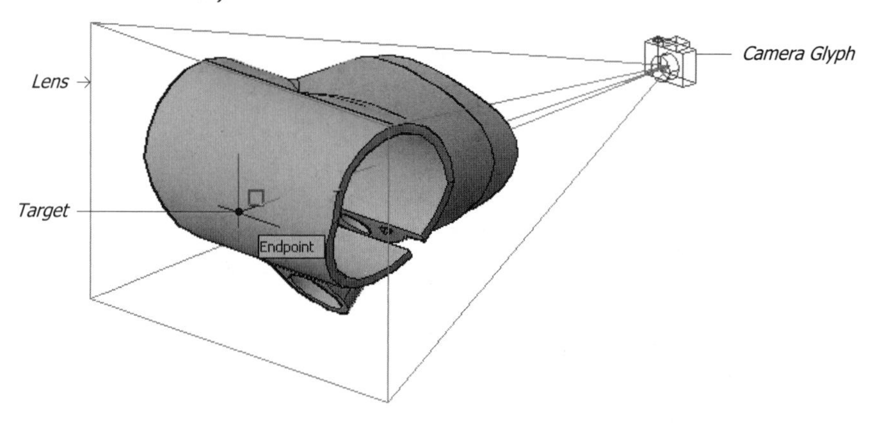

Turn on the camera glyph with the CameraDisplay system variable.

COMMAND LINE OPTIONS

Specify camera location specifies the x,y,z coordinates for the camera ("look from").

Specify target location specifies the x,y,z coordinates for the target ("look at").

? option

Enter camera name(s) to list <*> lists the names of cameras stored in the drawing, which are accessed through the View command.

```
Camera Name
-----------------
Camera1
Camera2
```

Name option

Enter name for new camera names the current camera-target position for later reuse.

LOcation option

Specify camera location locates the camera using x,y,z coordinates.

Height option

Specify camera height elevates the camera above the x,y-plane; saved in the CameraHeight system variable.

Target option

Specify target location locates the target point and determines the camera's direction.

LEns option

Specify lens length in mm specifies the angle of view; 50mm is normal; smaller numbers (such as 35) widen the view but also distort it, while larger numbers (such as 100) narrow the view and foreshorten it.

Clipping options

Enable front clipping plane? toggles the front clipping plane, which is used to cut off part of the view.

Specify front clipping plane offset from target plane specifies the location of the plane.

Enable back clipping plane? toggles the rear clipping plane.

Specify back clipping plane offset from target plane specifies the location of the plane.

View option

Switch to camera view? changes the viewpoint to that of the camera.

RELATED COMMANDS

DView has a CAmera option, which interactively sets a new 3D viewpoint.

View lists the names of saved camera views.

VPoint sets a 3D viewpoint through x,y,z coordinates or angles.

3dOrbit sets a new 3D viewpoint interactively.

RELATED SYSTEM VARIABLES

CameraDisplay toggles the display of the camera glyphs.

CameraHeight specifies the height of the camera above the x, y-plane.

TIPS

- After placing a camera, select its glyph to edit the viewpoint and to display the Camera Preview window. The window updates when grips modify the camera's specifications.

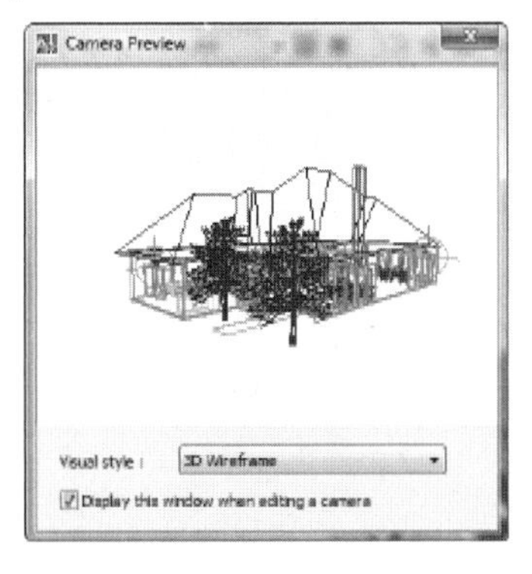

- Colors of the camera and field of view can be changed in the Options dialog box.
- Saved cameras are listed in the View command's dialog box.

- Grips interactively adjust the camera, target, and field of view. If clipping planes were specified, they can also be adjusted using grips.

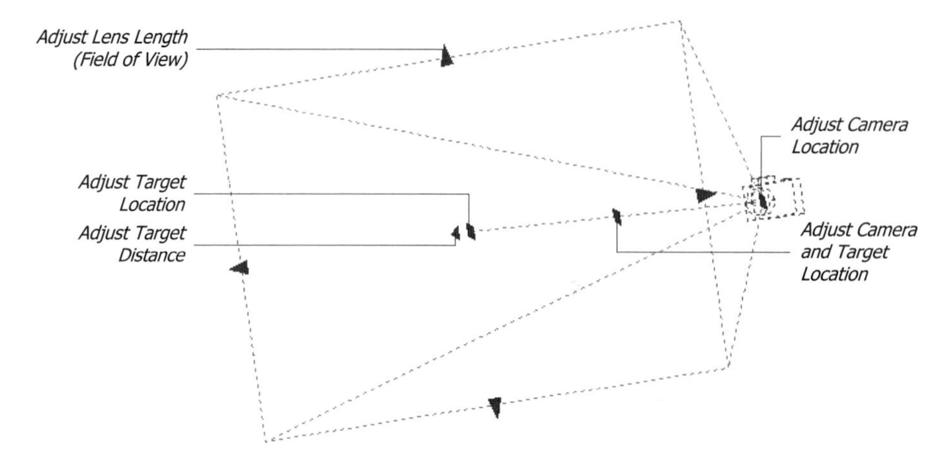

- After selecting a camera glyph, double-click it to display the Properties palette (at left, below):

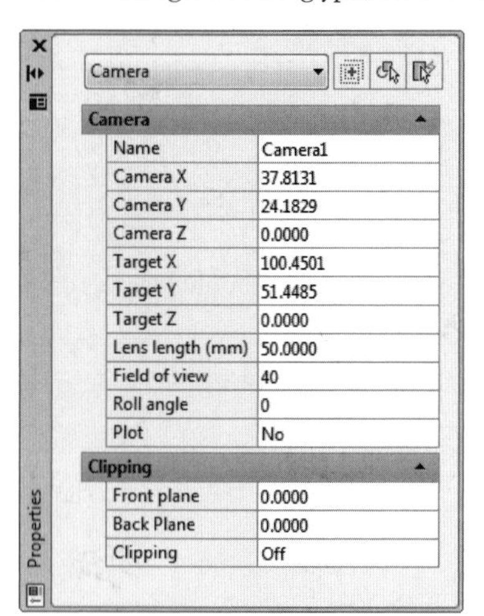

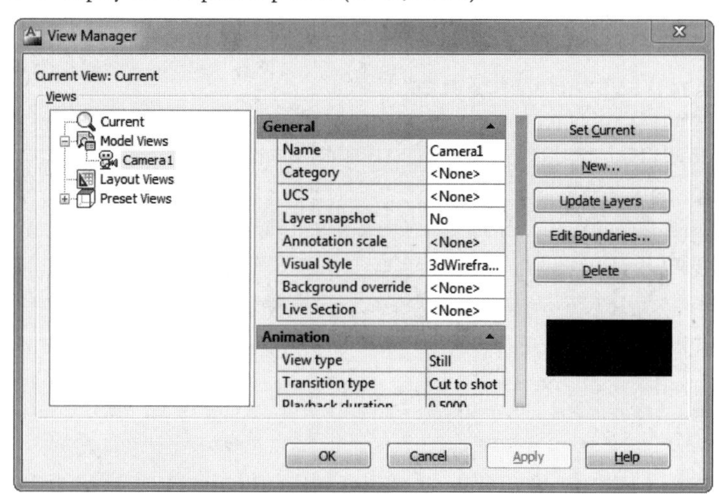

***Left**: Camera properties in the Properties palette.*
***Right**: Camera properties in the View Manager dialog box.*

- The View command also lists the names of cameras and their properties, but provides access to many more options.
- The 3D Navigation toolbar contains a droplist with the names of stored camera views. The Camera Adjustment toolbar has nothing to do with this command.
- Entering the Camera command automatically turns on perspective viewing mode.

Chamfer

Ver. 2.1 Bevels the intersection of two lines, all vertices of 2D polylines, and the faces of 3D solid models.

Command	*Alias*	*Keyboard Shortcuts*	*Menu*	*Ribbon*
chamfer	**cha**	...	**Modify** ↳**Chamfer**	**Home** ↳**Modify** ↳**Chamfer**

Command: chamfer
(TRIM mode) Current chamfer Dist1 = 0.0, Dist2 = 0.0
Select first line or [Undo/Polyline/Distance/Angle/Trim/mEthod/Multiple]: *(Pick an object, or enter an option.)*
Select second line: *(Pick an object.)*

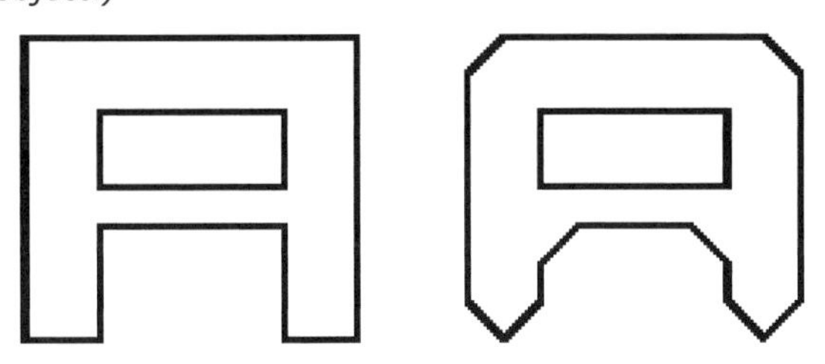

Left: *Original object drawn with polylines.*
Right: *Object chamfered with the Polyline option.*

COMMAND LINE OPTIONS

Select first line selects the first line, arc, face, or edge.

Select second line selects the second line, arc, face, or edge.

Undo reverses the last chamfer operation within the command.

Multiple allows more than one pair of lines to be chamfered.

Polyline options
Select 2D polyline: *(Pick a polyline.)*
n **lines were chamfered**

Select 2D polyline chamfers *all* segments of a 2D polyline; if the polyline is not closed with the Close option, the first and last segments are not chamfered.

Distance options
Specify first chamfer distance <0.5000>: *(Enter a value.)*
Specify second chamfer distance <0.5000>: *(Enter a value.)*

First distance specifies the chamfering distance along the line picked first.

Second distance specifies the chamfering distance along the line picked second.

Angle options
Specify chamfer length on the first line <1.0000>: *(Enter a distance.)*
Specify chamfer angle from the first line <0>: *(Enter an angle.)*

Chamfer length specifies the chamfering distance along the line picked first.

Chamfer angle specifies the chamfering angle from the line picked first.

Trim options

Enter Trim mode option [Trim/Notrim] <Trim>: *(Type T or N.)*

Trim trims or extends lines, edges, and faces after chamfer.

No trim does not trim lines, edges, and faces after chamfer.

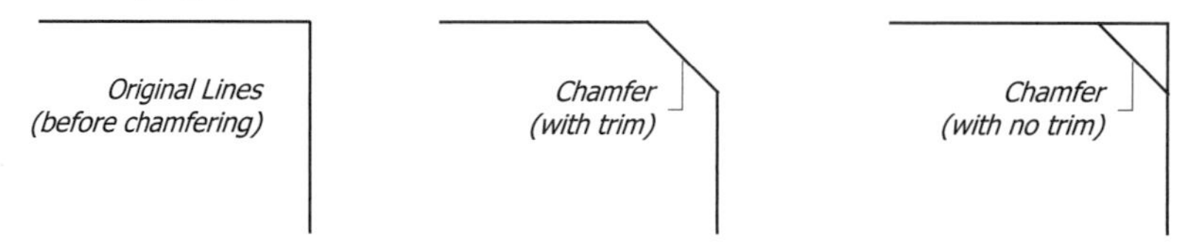

Intersecting lines before and after chamfering.

Method options

Enter trim method [Distance/Angle] <Angle>: *(Type D or A.)*

Distance determines chamfer by two specified distances.

Angle determines chamfer by the specified angle and distance.

Chamfering 3D Solids — Edge Mode Options

Command: chamfer
(TRIM mode) Current chamfer Dist1 = 0.0, Dist2 = 0.0
Polyline/.../<Select first line>: *(Pick a 3D solid model.)*
Select base surface: *(Pick surface edge on the solid model — see 1, below.)*
Next/<OK>: *(Type N, or select the next surface, or press OK to end selection.)*
Enter base surface distance <0.0>: *(Enter a value.)*
Enter other surface distance <0.0>: *(Enter a value.)*
Loop/<Select edge>: *(Select an edge — see 2, below.)*
Loop/<Select edge>: *(Press Enter.)*

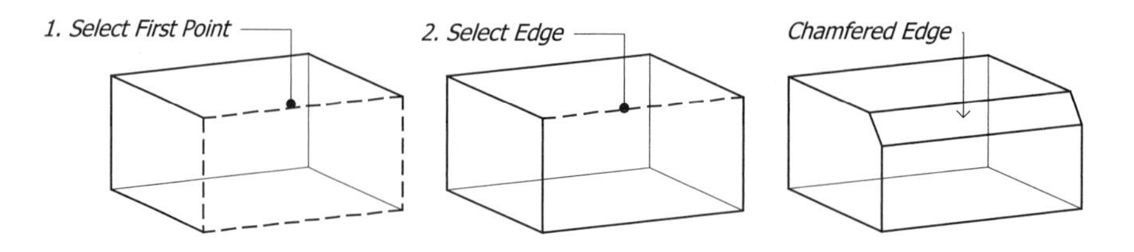

Chamfering 3D Solids — Loop Mode Options

Command: chamfer
(TRIM mode) Current chamfer Dist1 = 0.0, Dist2 = 0.0
Polyline/.../<Select first line>: *(Pick a 3D solid model.)*
Select base surface: *(Pick a surface edge — see 1, following.)*
Next/<OK>: *(Type N, or select the next surface, or OK to end selection.)*
Enter base surface distance <0.0>: *(Enter a value.)*
Enter other surface distance <0.0>: *(Enter a value.)*
Loop/<Select edge>: *(Type L.)*

Edge/<Select edge loop>: *(Pick a surface edge — see 2, below.)*
Edge/<Select edge loop>: *(Press Enter.)*

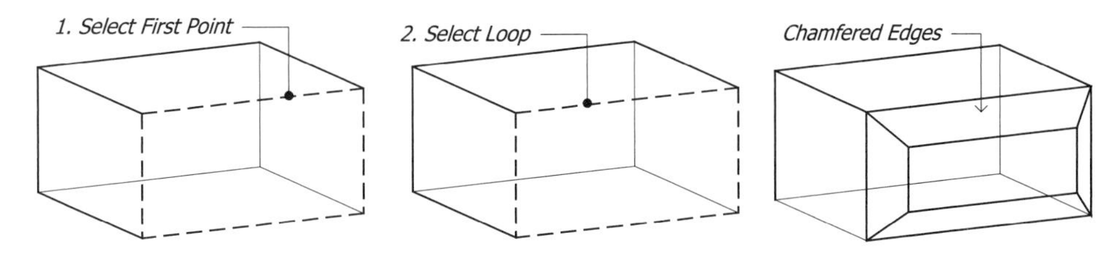

COMMAND LINE OPTIONS

Select first line selects the 3D solid.

Next selects the adjacent face; pressing Enter accepts the face.

Enter base surface distance specifies the first chamfer distance.

Enter other surface distance specifies the second chamfer distance.

Loop selects all edges of the face.

Select edge selects a single edge of the face.

RELATED COMMANDS

Fillet rounds the intersection with a radius.

SolidEdit edits the faces and edges of solids.

RELATED SYSTEM VARIABLES

ChamferA is the first chamfer distance.

ChamferB is the second chamfer distance.

ChamferC is the length of chamfer.

ChamferD is the chamfer angle.

ChamMode toggles chamfer measurement (0; default) by two distances or (1) by distance and angle.

TrimMode determines whether lines/edges are trimmed after chamfer.

TIPS

- When TrimMode is set to 1 and lines do not intersect, Chamfer extends or trims the lines to intersect before chamfering.
- When the two objects are not on the same layer, Chamfer places the chamfer line on the current layer.
- As of AutoCAD 2010, this command chamfers 3D mesh objects indirectly:
  1. Start the Chamfer command, and then select the 3D mesh objects.
  2. AutoCAD asks if you want to convert them into smooth or faceted 3D solids; choose Faceted.
  3. After conversion is complete, the Chamfer command continues; chamfer the object.
  4. Once chamfering is finished, use the MeshSmooth command to convert the 3D solids back to 3D mesh objects.
- AutoCAD 2011 introduces the ChamferEdge command for interactive beveling of 3D solid and surface edges.

ChamferEdge

2011 Bevels the edges of 3D solids and surfaces interactively.

Command	*Aliases*	*Keyboard Shortcuts*	*Menu*	*Ribbon*
chamferedge	...	...	**Modify**	**Solid**
			↳**Solid Editing**	↳**Solid Editing**
			↳**Chamfer Edge**	↳**Chamfer Edge**

Command: chamferedge
Distance1 = 1.0000, Distance2 = 1.0000
Select an edge or [Loop/Distance]: *(Pick an edge, or enter an option.)*
Select an edge belongs *[sic]* **to the same face or [Loop/Distance]:** *(Press Enter.)*

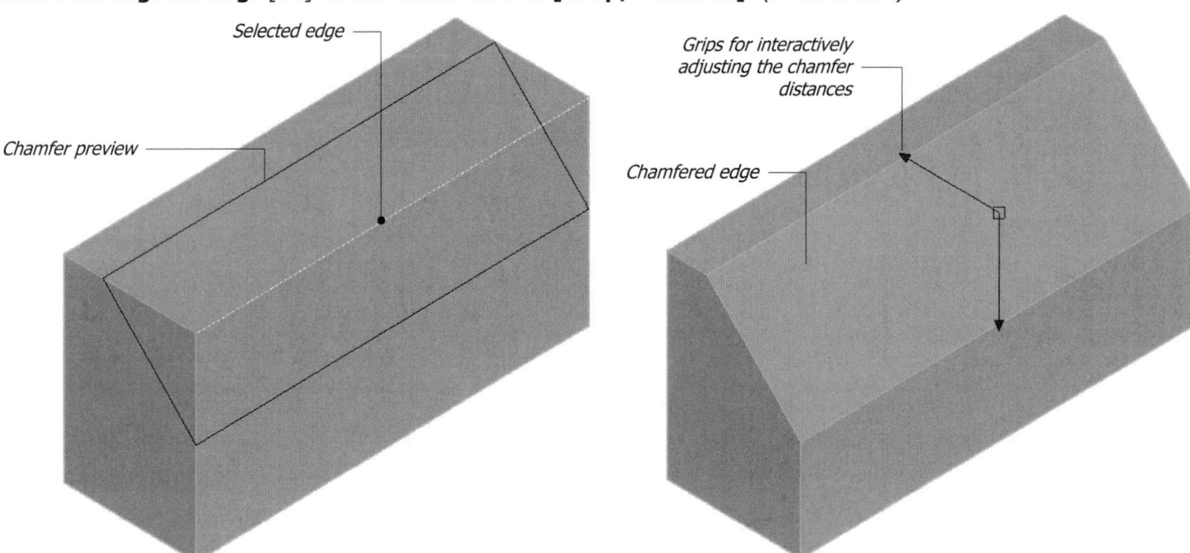

COMMAND LINE OPTIONS

Select an Edge selects the edge to chamfer.

Loop chamfers all edges on a single face; prompts you:

Select an edge loop or [Edge/Distance]: *(Choose an edge belonging to a face.)*

Edge returns to the edge selection prompt:

Select an edge or [Loop/Distance]: *(Choose an edge.)*

Distance specifies the chamfer distances:

Specify Distance1 or [Expression] <1.0>: *(Enter a distance, or type E.)*
Specify Distance2 or [Expression] <1.0>: *(Enter a distance, or type E.)*

Expression specifies the distance using formulas; see the Calc command for examples. Prompts you :

Enter expression: *(Enter a parametric formula, such as dia1*3.)*

TIPS

- You can select multiple edges at the 'Select an edge' prompt, but they must all be part of the same face.
- Instead of entering chamfer distances at the 'Distance' prompt, you can drag the chamfer's grips; after this command ends, you can no longer interactively edit the chamfer.

Change

Ver. 1.0 Modifies the color, elevation, layer, linetype, linetype scale, lineweight, plot style, material, annotative scaling, transparency, and thickness of most objects, as well as additional properties of lines, circles, blocks, text, and attributes.

Command	*Alias*	*Keyboard Shortcuts*	*Menu*	*Ribbon*
change	**-ch**	...	...	...

Command: change
Select objects: *(Pick one or more objects.)*
Select objects: *(Press Enter.)*
Specify change point or [Properties]: *(Pick a point, an object, or type P.)*
Enter property to change [Color/Elev/LAyer/LType/ltScale/LWeight/Thickness/TRansparency/Material/Plotstyle/Annotative]: *(Enter an option.)*

COMMAND LINE OPTIONS

Specify change point selects the object to change:

(pick a line) indicates the new length of lines.

(pick a circle) indicates the new radius of circles.

(pick a block) indicates the new insertion point or rotation angle of blocks.

(pick text) indicates the new location of text.

(pick an attribute) indicates the attribute's new text insertion point, text style, height, rotation angle, text, tag, prompt, or default value.

ENTER changes the insertion point, style, height, rotation angle, and text of text strings.

Properties options

Color changes the color of objects.

Elev changes the elevation of objects.

LAyer moves the object to different layers.

LType changes the linetype of objects.

ltScale changes the scale of linetypes.

LWeight changes the lineweight of objects.

Thickness changes the thickness of objects, except blocks and 3D solids.

Transparency changes the translucency of objects *(new to AutoCAD 2011)*.

Material changes the material assigned to objects.

Plotstyle changes the plot style of objects (available only when plot styles are turned on).

Annotative sets the annotative property for the objects.

RELATED COMMANDS

AttRedef changes blocks and attributes.

ChProp contains the properties portion of the Change command.

Color changes the current color setting.

Elev changes the working elevation and thickness.

LtScale changes the linetype scale.

Properties changes most aspects of all objects.

PlotStyle sets the plot style.

Materials assigns materials and textures to objects.

ObjectScale assigns annotative scale factors to objects.

RELATED SYSTEM VARIABLES

CAnnoScale contains the current annotative scale.

CeColor contains the current color setting.

CeLType contains the current linetype setting.

CeLWeight contains the current lineweight.

CeTransparency contains the level of translucency for newly created objects.

CircleRad contains the current circle radius.

CLayer contains the name of the current layer.

CMaterial contains the name of the default material.

CPlotstyle contains the name of the current plot style.

Elevation contains the current elevation setting.

LtScale contains the current linetype scale.

TextSize contains the current height of text.

TextStyle contains the current text style.

Thickness contains the current thickness setting.

TransparencyDisplay toggles the display of transparency.

TIPS

- The Change command cannot change the following properties:
  - The size of donuts; use grips to change them.
  - The radius and length of arcs; use grips to change them.
  - The length of polylines; use grips to change them.
  - The justification of text; use the JustifyText command.
- Use this command to change the endpoints of groups of lines to a common vertex:

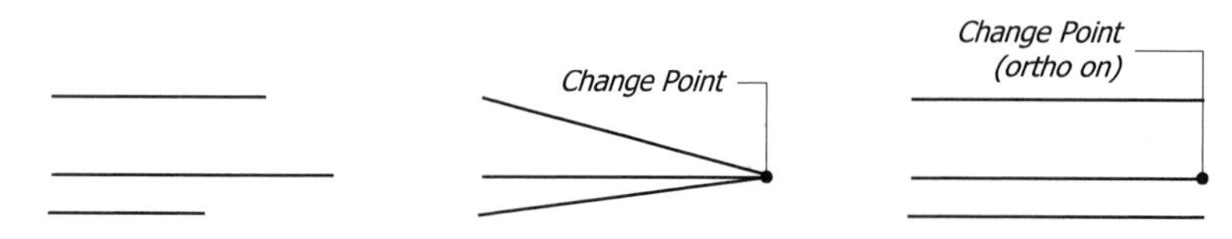

Left: *Original lines.*
Center: *Line endpoints changed with ortho mode turned off.*
Right: *Line endpoints changed with ortho on.*

- Turn on ortho mode to extend or trim a group of lines, without a cutting edge (unlike the Extend and Trim commands).
- The PlotStyle option is not displayed when plot styles are not turned on.
- This command is largely superseded by the Properties command.

CheckStandards

2001 Checks drawings for adherence to standards previously specified by the Standards command.

Command	Alias	Keyboard Shortcuts	Menu	Ribbon
'checkstandards	**chk**	...	**Tools ↳CAD Standards ↳Check**	**Manage ↳CAD Standards ↳Check**

Command: checkstandards

When the Standards command has not set up standards for the drawing, AutoCAD displays this error message:

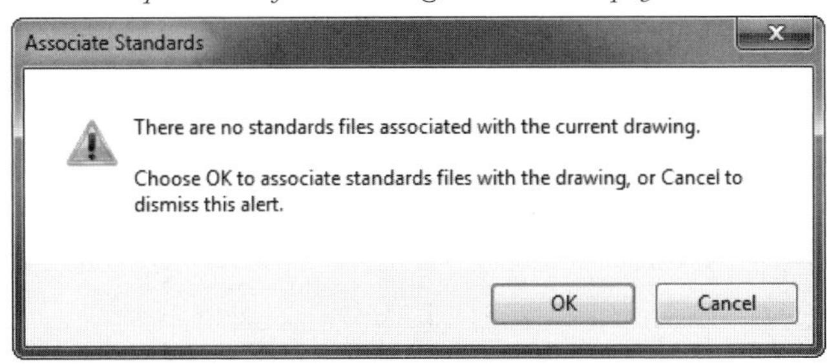

When standards have been set up for the drawing, this command checks the drawing against the CAD standards, and then displays this dialog box:

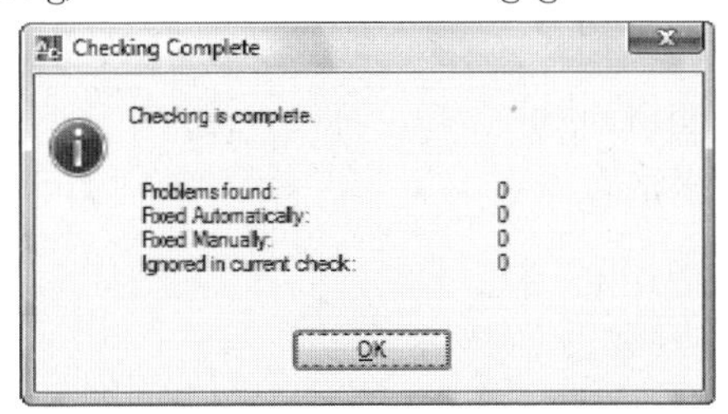

Click OK.

To change settings in the Configure Standards dialog box, click Settings.

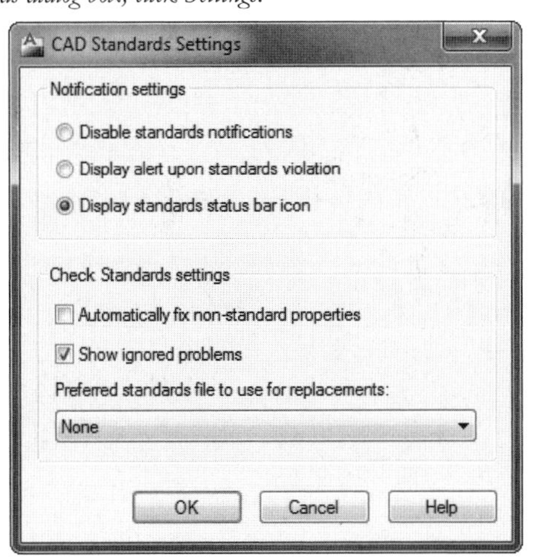

CAD STANDARDS SETTINGS DIALOG BOX OPTIONS

Notification Settings options

- ○ Disables standards notifications.
- ○ Displays alert upon standards violation, illustrated below.
- ⊙ Displays standards status bar icon.

Check Standards Settings options

Automatically fix non-standard properties:

- ☑ Fixes automatically properties not matching the CAD standard.
- ☐ Steps manually through properties not matching the CAD standard.

Show ignored problems:

- ☑ Displays problems marked as ignored.
- ☐ Does not display ignored problems.

Preferred standards file to use for replacements selects the default *.dws* file.

CHECK STANDARDS DIALOG BOX

When an object does not match the standards, AutoCAD displays this dialog box.

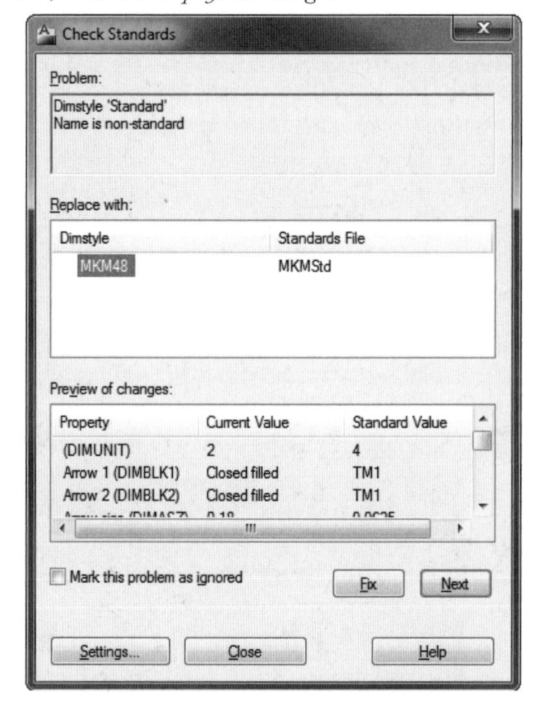

Problem describes properties in drawings that do not match the standard; this dialog box displays one problem at a time.

Replace With lists linetypes, text styles, etc. found in the *.dws* standards file.

Mark this problem as ignored:

- ☑ Ignores nonstandard properties, and marks them with the user's login name; some errors are always ignored by AutoCAD, such as settings for layer 0 and DefPoints.
- ☐ Does not ignore nonstandard properties.

Fix replaces the nonstandard property with the selected standard; the color check mark icon means fixes are available.

Next displays the next nonstandard property.

Settings displays the Check Standards Settings dialog box.

Close closes the dialog box; displays this warning dialog box if standards are not fully checked.

TRAY ICON

When drawings contain drawings standards, AutoCAD displays a book icon on the tray:

Click the icon to start the CheckStandards command. Right-clicking the icon displays the shortcut menu:

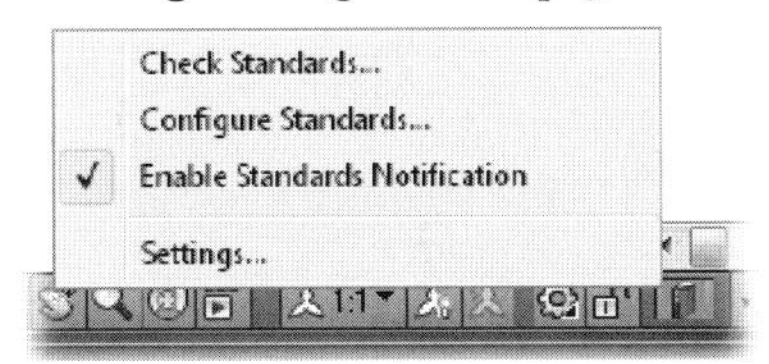

Check Standards runs the CheckStandards command.

Configure Standards runs the Standards command.

Enable Standards Notification toggles the display of the balloon.

Settings displays the CAD Standards Settings dialog box.

RELATED COMMANDS

Standards selects *.dws* standards files.

LayTrans translates layers between drawings.

RELATED SYSTEM VARIABLE

StandardsViolation determines whether alerts are displayed when a CAD standard is violated in the current drawing.

RELATED FILES

.dws* drawing standards file; stored in DWG format.

.chs* standard check file; stored in XML format.

TIPS

- While this dialog box is open, you can use the following shortcut keys:

F4	Fix problem.
F5	Next problem.

- The following balloon appears when standards are violated:

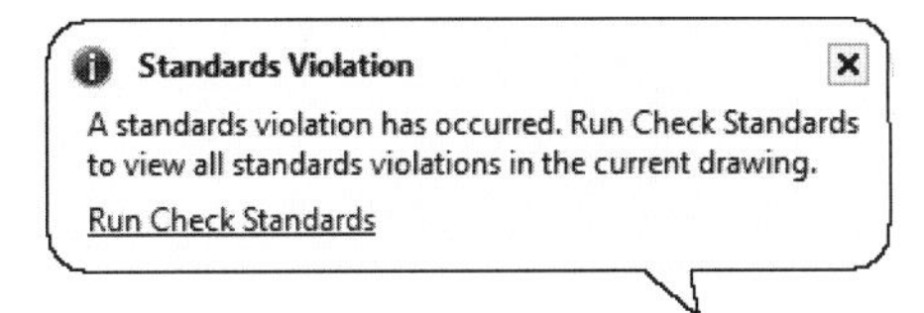

ChProp

Rel. 10 Modifies the color, layer, linetype, linetype scale, lineweight, plot style, annotative scale, transparency, thickness, material, plot style, and annotative scale factor of most objects (short for CHange PROPerties).

Command	*Aliases*	*Keyboard Shortcuts*	*Menu*	*Ribbon*
chprop	...	...	...	...

Command: chprop
Select objects: *(Select one or more objects.)*
Select objects: *(Press Enter.)*
Enter property to change [Color/LAyer/LType/ltScale/LWeight/Thickness/Plotstyle/TRansparency/Material/Annotative]: *(Enter an option.)*

COMMAND LINE OPTIONS

Color changes the color of objects.

LAyer moves objects to a different layer.

LType changes the linetype of objects.

ltScale changes the linetype scale.

LWeight changes the lineweight of objects.

Thickness changes the thickness of all objects, except blocks.

PlotStyle changes the plot style assigned to the objects (available only when plot styles are turned on).

Transparency changes the translucency of objects *(new to AutoCAD 2011).*

Material changes the material assigned to objects.

Annotative sets the annotative property for the objects.

RELATED COMMANDS

Change changes lines, circles, blocks, text and attributes.

Color changes the current color setting.

Elev changes the working elevation and thickness.

LtScale changes the linetype scale.

LWeight sets the lineweight options.

ObjectScale assigns annotative scale factors to objects.

Properties changes most aspects of all objects.

PlotStyle sets the plot style.

RELATED SYSTEM VARIABLES

See the Change command.

TIPS

- This command cannot change the elevation of objects, so use the Change or Properties command instead.
- The Plotstyle option is not displayed when plot styles are off.
- This command is largely superseded by the Properties command.

ChSpace

2007 Moves objects between paper and model space, scaling them appropriately (short for CHange SPACE).

Command	*Aliases*	*Keyboard Shortcuts*	*Menu*	*Ribbon*
chspace	...	...	**Modify** **↳Change Space**	**Home** **↳Modify** **↳Change Space**

Command: chspace

*In model tab, AutoCAD responds, "** Command not allowed in Model Tab **."*

In layout tabs, the command continues:

Select objects: *(Select one or more objects.)*

Select objects: *(Press Enter.)*

n **object(s) changed from PAPER space to MODEL space.**

Objects were scaled by a factor of 192.000000708072 to maintain visual appearance.

COMMAND LINE OPTION

Select objects selects the objects to be moved.

RELATED COMMANDS

SpaceTrans converts the height of text between model and paper space.

ExportLayout copies layouts to model space.

TIPS

- *Warning*! Objects are moved, not copied, which means they no longer appear in the other space.
- This command was formerly an Express Tool.

Circle

Ver. 1.0 Draws 2D circles by five different methods.

Command	*Alias*	*Keyboard Shortcuts*	*Menu*	*Ribbon*
circle	**c**	...	**Draw** **⇘Circle**	**Home** **⇘Draw** **⇘Circle**

Command: circle
Specify center point for circle or [3P/2P/Ttr (tan tan radius)]: *(Pick a point, or enter an option.)*
Specify radius of circle or [Diameter]: *(Pick a point, or type D.)*

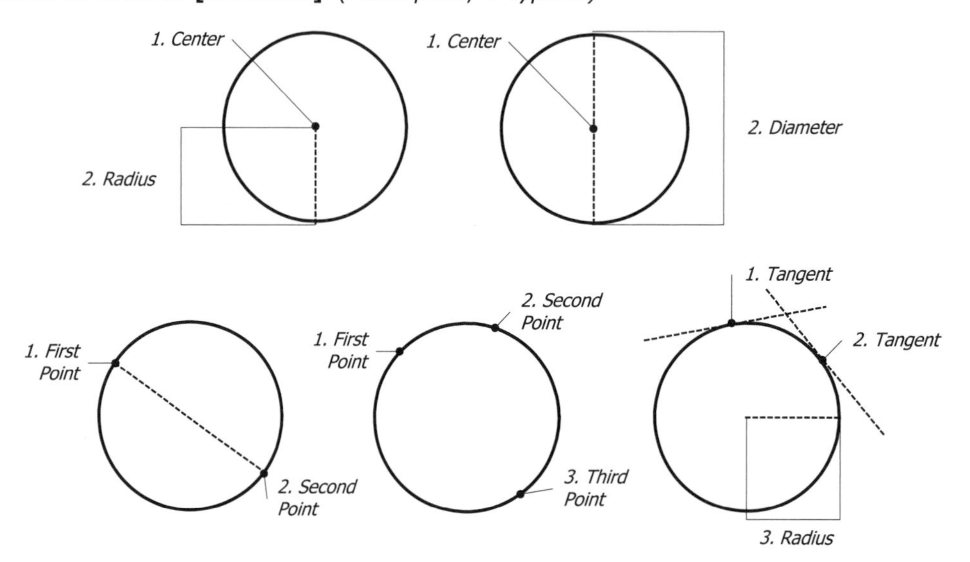

COMMAND LINE OPTIONS

Center and Radius or Diameter Options
Specify center point for circle or [3P/2P/Ttr (tan tan radius)]: *(Pick a point, or enter an option.)*
Specify radius of circle or [Diameter] <0.5>: *(Pick a point, or type D.)*
Specify diameter of circle <0.5>: *(Enter a value.)*

Center point indicates the circle's center point.

Radius indicates the circle's radius.

Diameter indicates the circle's diameter.

3P (three-point) Options
Specify first point on circle: *(Pick a point.)*
Specify second point on circle: *(Pick a point.)*
Specify third point on circle: *(Pick a point.)*

First point indicates the first point on the circle.

Second point indicates the second point on the circle.

Third point indicates the third point on the circle.

2P (two-point) Options
Specify first end point of circle's diameter: *(Pick a point.)*
Specify second end point of circle's diameter: *(Pick a point.)*

First end point indicates the first point on the circle.

Second end point indicates the second point on the circle.

TTR (tangent-tangent-radius) Options
Specify point on object for first tangent of circle: *(Pick a point.)*
Specify point on object for second tangent of circle: *(Pick a point.)*
Specify radius of circle <0.5>: *(Enter a radius.)*

First tangent indicates the first point of tangency.

Second tangent indicates the second point of tangency.

Radius indicates the first point of radius.

Tan Tan Tan *(menu-only option)*:

From the Draw menu, select Circle | Tan,Tan,Tan. AutoCAD prompts:

Command: _circle Specify center point for circle or [3P/2P/Ttr (tan tan radius)]: _3p Specify first point on circle: _tan to *(Pick an object.)*
Specify second point on circle: _tan to *(Pick an object.)*
Specify third point on circle: _tan to *(Pick an object.)*

AutoCAD draws a circle tangent to the three points, if possible.

RELATED COMMANDS

Ai_CircTan draws circles tangent to three objects.

Arc draws arcs.

Join turns arcs into circles.

Donut draws solid-filled circles or donuts.

Ellipse draws elliptical circles and arcs.

Sphere draws 3D solid balls.

Reverse reverses the direction the circle is drawn.

RELATED SYSTEM VARIABLE

CircleRad specifies the default circle radius.

TIPS

- The 3P (three-point) circle defines three points on the circle's circumference.
- When drawing TTR (tangent, tangent, radius) circles, AutoCAD draws the circles with tangent points closest to the pick points; note that more than one circle placement is possible.
- Using the TTR option automatically turns on the TANgent object snap.
- Giving circles thickness turns them into hollow cylinders.

ClassicImage

2007 Displays the Image Manager dialog box; replaced by the ExternalReferences command.

Command	Aliases	Keystroke Shortcuts	Menu	Ribbon
classicimage	...	...	...	...

Command: classicimage

Displays dialog box.

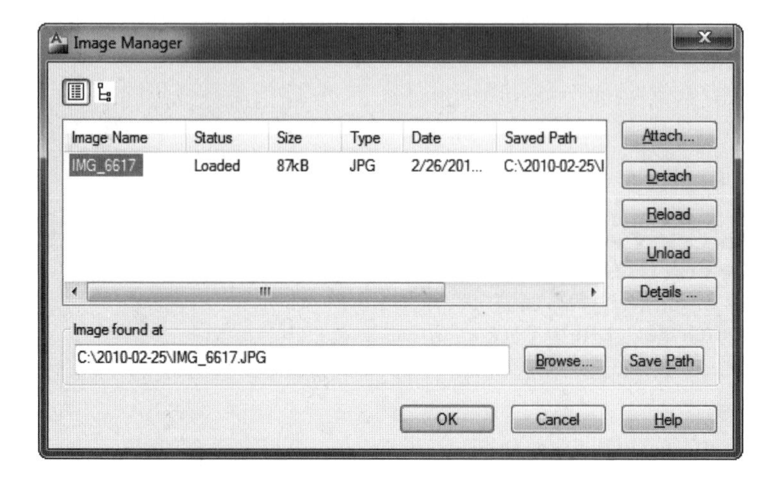

DIALOG BOX OPTIONS

Attach attaches an image as an underlay; displays the Attach Image dialog box; see the ImageAttach command.

Detach detaches images.

Reload reloads and displays the most-recently saved version of the image.

Unload unloads the image, but does not remove it permanently; rather it does not display the image.

Details displays information about the image.

Browse selects a path or file name; displays the Select New Path dialog box.

Save Path saves the path displayed by the Image Found At option.

RELATED COMMANDS

ExternalReferences replaces this command.

-Image is the command-line version.

TIP

- You may find this command's dialog box easier to use than the all-inclusive dialog box presented by the ExternalReferences command.
- Autodesk warns that this command may be removed in a future release.

ClassicLayer

2007 Controls the creation, status, and visibility of layers.

Command	Alias	Keyboard Shortcuts	Menu	Ribbon
'classiclayer	...	...	...	...

Command: classiclayer

Displays dialog box:

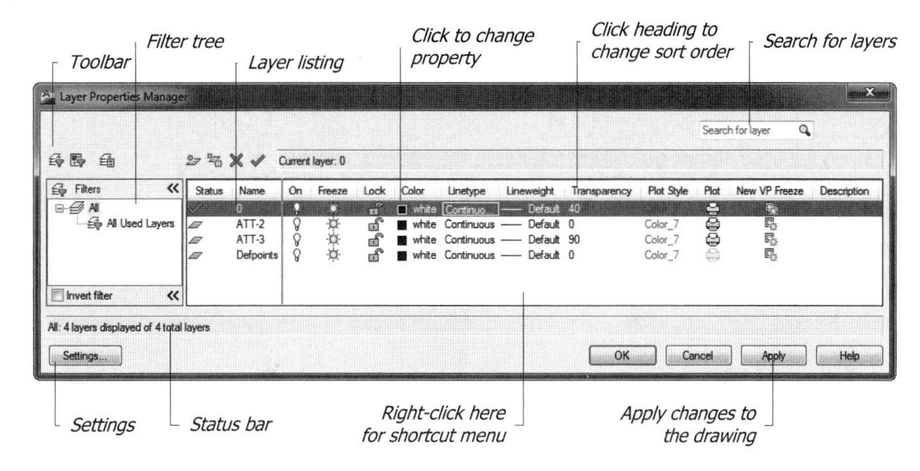

LAYER PROPERTIES MANAGER DIALOG BOX

This dialog box is nearly identical to the palette displayed by the Layer command. See the Layer command for details.

RELATED COMMAND

LayerPalette displays the Layer Properties Manager palette.

See the Layer command for related commands and system variables.

RELATED SYSTEM VARIABLE

LayerDlgMode toggles the purpose of the Layer command: (0) displays dialog box like the ClassicLayer command, or (1; default) displays palette like the LayerPalette command.

TIPS

- As of AutoCAD 2009, the Layer command was renamed ClassicLayer; Layer now displays the Layer Properties Manager palette.
- Autodesk warns that the ClassicLayer command will be removed in a future release of AutoCAD.
- You can open both the dialog box and palette. As you change properties in the dialog box, the palette updates immediately.
- AutoCAD 2011 adds support for transparency of layers in model space and in viewports.
- See the Layer command for details.

ClassicXref

2007 Displays the XRef Manager dialog box (undocumented command); replaced by the ExternalReferences command.

Command	*Aliases*	*Keystroke Shortcuts*	*Menu*	*Ribbon*
classicxref	...	...	...	...

Command: classicxref

Displays dialog box.

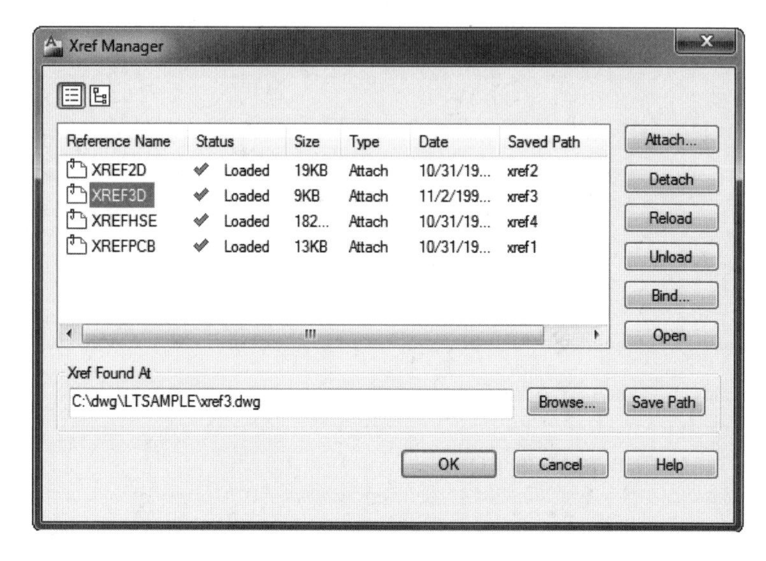

DIALOG BOX OPTIONS

Attach attaches a drawing as an xref (externally-referenced drawing); displays Attach Xref dialog box; see XAttach command.

Detach detaches xrefs.

Reload reloads and displays the most-recently saved version of the xref.

Unload unloads the xref; does not remove it permanently; rather it does not display the xref.

Bind binds named objects — blocks, dimension styles, layer names, linetypes, and text styles — to the current drawing; displays the Bind Xrefs dialog box; see the XBind command.

Open opens the selected xref in a new window for editing; see the XOpen command.

Xref Found At displays the path to the xref file.

Browse selects a path or file name; displays the Select New Path dialog box.

Save Path saves the path displayed by the Xref Found At option.

RELATED COMMANDS

ExternalReferences replaces this command.

-XRef is the command-line version.

TIPS

- You may find this command's dialog box easier to use than the all-inclusive dialog box presented by the ExternalReferences command.
- Autodesk indicates that this command will be removed in a future release.

CleanScreenOn / CleanScreenOff

2004 Maximizes the drawing area.

Command	Status Bar	Keyboard Shortcuts	Menu	Ribbon
cleanscreenon		**Ctrl+0** *(zero)*	**View** ↳**Clean Screen**	...
cleanscreenoff		**Ctrl+0**	**View** ↳**Clean Screen**	...

Command: cleanscreenon

AutoCAD turns off the title bar, toolbars, and window edges, and maximizes the AutoCAD window to the full size of your computer's screen:

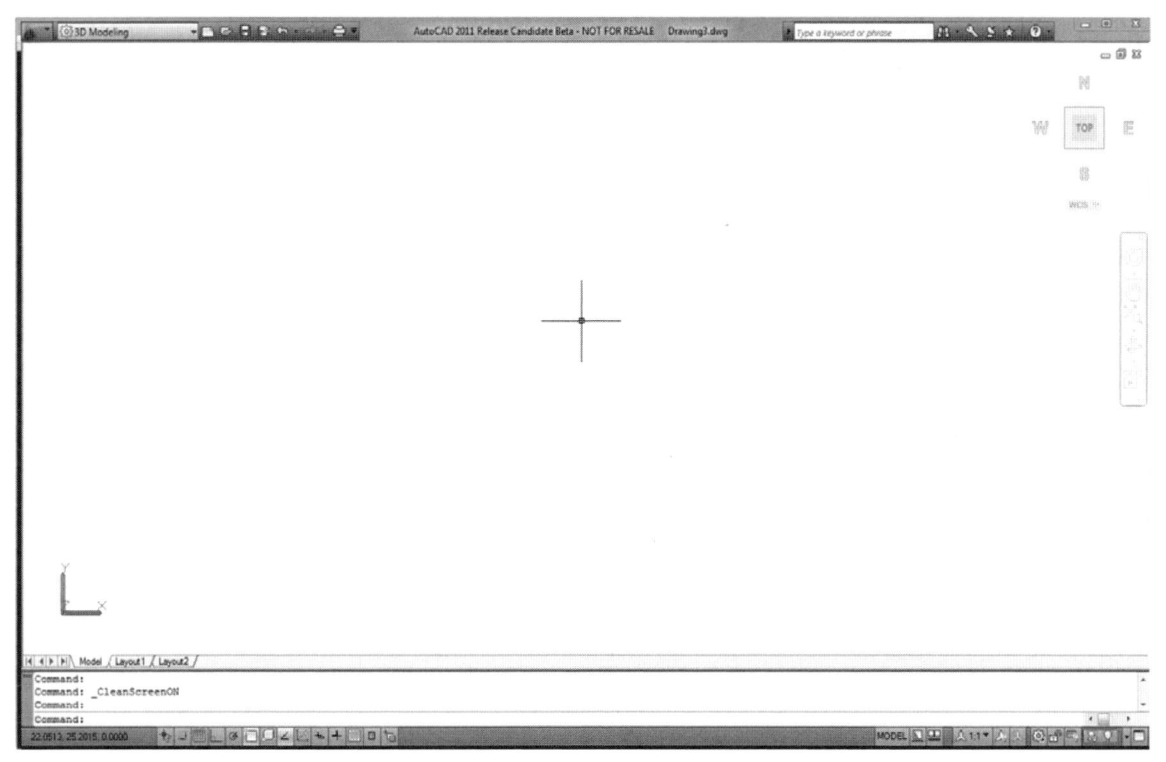

Command: cleanscreenoff

AutoCAD returns to normal.

TIPS

- Alternatively, you can click the Cleanscreen icon in the tray on the status bar.
- For an even larger drawing area, turn off the scroll bars and layout tabs through the Options | Display dialog box, and then drag the command prompt area into a window.
- To toggle this command, press Ctrl+0 (zero).
- The technical editor finds this command useful for capturing images, such as for illustrations in magazine articles.

Clip

2010 Clips attached files: images, external references, viewports, and underlays in PDF, DGN, DWF, or DWFx formats.

Command	*Aliases*	*Keyboard Shortcuts*	*Menu*	*Ribbon*
clip	...	...	...	**Insert** ↳**Reference** ↳**Clip**

Command: clip

Select object to clip: *(Select an image, a viewport, an xref, or a PDF, DGN, DWF, or DWFx underlay.)*

The prompts that follow vary according to the object selected.

COMMAND LINE OPTIONS

Select object to clip selects an object.

External References

Enter clipping option [ON/OFF/Clipdepth/Delete/generate Polyline/New boundary] <New>: *(Enter an option.)*

See the XClip command.

Images and Underlays

Enter image *(or DWF or DWFx or DGN or PDF)* clipping option [ON/OFF/Delete/New boundary] <New>: *(Enter an option.)*

See the ImageClip, DgnClip, DwfClip, and PdfClip commands.

Viewports

This option operates only in paper space:

Select clipping object or [Polygonal] <Polygonal>: *(Pick a viewport or type P.)*

See the VpClip command.

RELATED COMMANDS

Attach attaches images, DGN, DWF, DWFx, PDF, and DWG files.

VPorts creates viewports.

ImageClip clips attached raster images.

DgnClip clips DGN underlays.

DwfClip clips DWF underlays.

PdfClip clips PDF underlays.

VpClip clips viewport borders.

XClip clips externally-referenced drawings.

TIPS

- To select an image or underlay, you must pick its frame. If you cannot see the frame, turn it on with the Frame system variable.
- This command is a "universal" front-end for all underlay clipping commands.

Close / CloseAll

2000 Closes the current drawing, or all drawings; does not exit AutoCAD.

Command	*Aliases*	*Keyboard Shortcuts*	*Menu Bar*	*Application Menu*	*Ribbon*
close	...	**Ctrl+F4**	**File** ⇘**Close**	**Close** ⇘**Current Drawing**	...
closeall	...	...	**Window** ⇘**Close All**	**Close** ⇘**Close All Drawings**	

Command: close

Closes the current drawing.

When drawings are not saved since last change, displays dialog box.

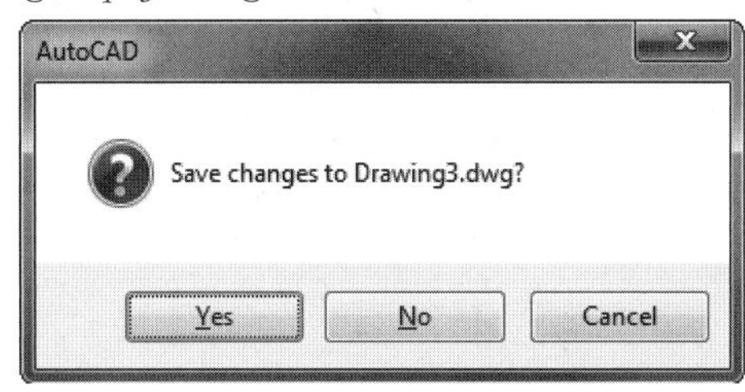

Command: closeall

Closes all open drawings.

AutoCAD DIALOG BOX

Yes displays the Save Drawing As dialog box, or saves the drawing if it has been previously saved.

No exits the drawing without saving it.

Cancel returns to the drawing.

RELATED COMMANDS

Quit exits AutoCAD.

Open opens additional drawings, each in its own window.

AutoPublish exports drawing as DWF, DWFx, or PDF files when the drawing is closed.

RELATED SYSTEM VARIABLES

DbMod reports whether the drawing has been modified since the last save.

AutomaticPub exports drawings in DWF/x and PDF formats, when set to 1.

TIPS

- As an alternative to this command, you can click the **X** button in the upper right corner of the drawing:

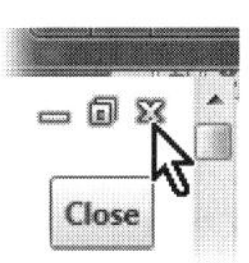

- If AutoPublish is turned on, then AutoCAD automatically publishes the drawing as a DWF or PDF file when you close the drawing. See the AutoPublish command.

Color

Ver. 2.5 Sets the new working color.

Command	Aliases	Keyboard Shortcuts	Menu	Ribbon
'color	**ddcolor**	...	**Format**	**Home**
	col		**↳Color**	**↳Properties**
	colour			**↳Object color**
'-color	...	...	...	...

Command: color

Displays dialog box.

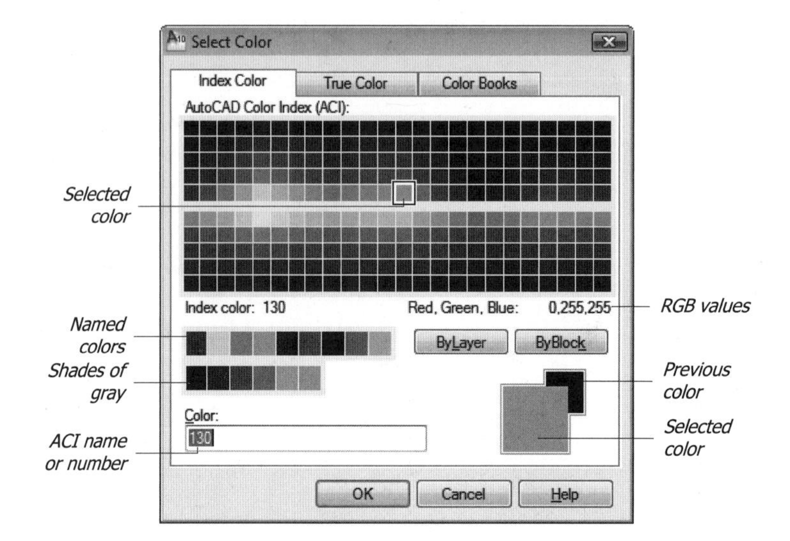

SELECT COLOR DIALOG BOX

Index Color tab

AutoCAD Color Index (ACI) selects one of AutoCAD's colors ("AutoCAD Color Index").

Bylayer sets the color to BYLAYER (color 256); color 257 is ByEntity in some situations.

Byblock sets the color to BYBLOCK.

Color sets the color by number or name.

True Color tab

Hue selects the hue (color), ranging from 0 (red) to 360 (violet).

Saturation selects the saturation (intensity of color), ranging from 0 (gray) to 100 (color).

Luminance selects the luminance (brightness of color), ranging from 0 (white) to 100 (black).

Color specifies the color number as hue, saturation, luminance.

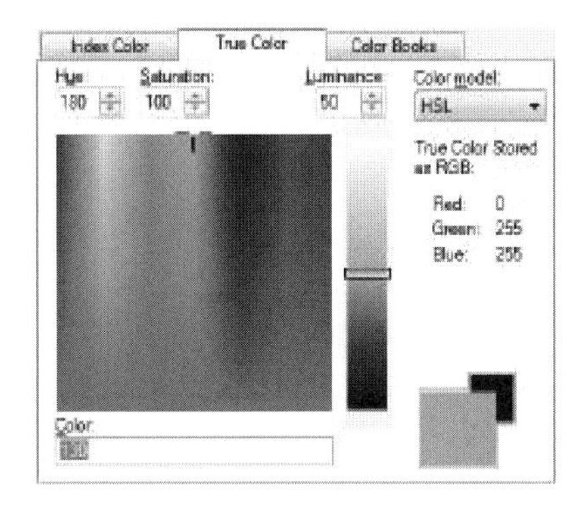

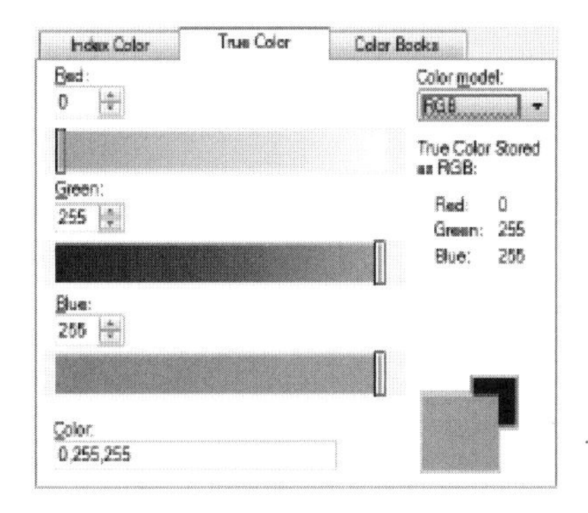

Left: *HSL specifies colors through hue saturation, and luminance.*
Right: *RGB specifies colors through red, green, and blue.*

Color model selects the type of color model:

- **HSL** is hue, saturation, luminance.
- **RGB** is red, green, blue..

Red selects the range of red from 0 to 255.

Green selects the range of green from 0 to 255.

Blue selects the range of blue from 0 to 255.

Color specifies the color number as red, green, blue.

Color Book tab

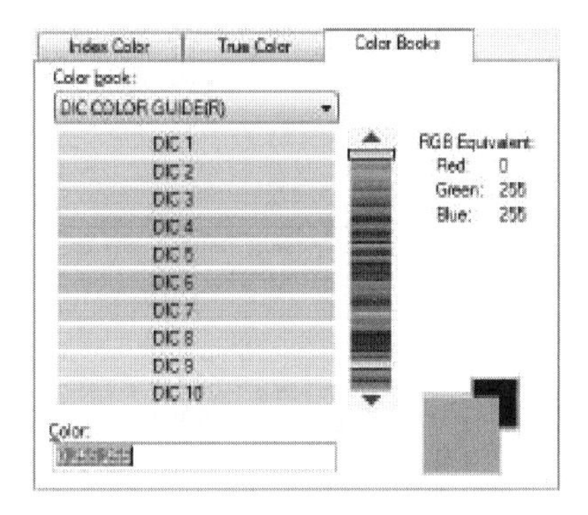

Color book selects a predefined collection of colors.

Color specifies the name of the Pantone, RAL, and DIC colors.

Historical notes: The **RAL** color system was designed in 1927 to standardize colors by limiting the number of color gradations, at first to just 30, but now to over 1,600. RAL (Reichs Ausschuß für Lieferbedingungen – German for "Imperial Committee for Supply Conditions") is administered by the German Institute for Quality Assurance and Labeling <www.ral.de>.

The **Pantone Color System** was designed in 1963 to specify color for graphic arts, textiles, and plastics, based on the assumption that people see colors differently. Designers typically work with Pantone's fan-format book of standardized colors <www.pantone.com>.

The **DIC Color Guide** is the Japanese standard for colors, developed by Dainippon Ink and Chemicals <www.dic.co.jp/eng/index.html>.

-COLOR Command

Command: -color

Enter default object color [Truecolor/COlorbook] <BYLAYER>: *(Enter a color number or name, or enter an option.)*

COMMAND LINE OPTIONS

BYLAYER sets the working color to the color of the current layer.

BYBLOCK sets the working color of inserted blocks.

color number sets the working color using number (1 through 255), name, or abbreviation:

Color Number	Color Name	Abbreviation	Comments
1	Red	R	
2	Yellow Y		
3	Green	G	
4	Cyan	C	
5	Blue	B	
6	Magenta	M	
7	White	W	Or black.
8 - 249			Additional colors.
250 - 255			Shades of gray.

Truecolor Options

Red, Green, Blue: *(Specify color values separated by commas.)*

Red, Green, Blue specifies color by red, green, and blue in the range from 0 to 255:

Red, Green, Blue: 255,128,0

COlorbook Options

Enter Color Book name: *(Specify a name, such as Pantone.)*

Enter color name: *(Enter a color name, such as 11-0103TC.)*

Enter Color Book name specifies the name of a color book.

Enter color name specifies the name of the color.

RELATED COMMANDS

ChProp changes the color via the command line in fewer keystrokes.

Properties changes the color of objects via a dialog box.

RELATED SYSTEM VARIABLE

CeColor is the current object color setting.

TIPS

- 'BYLAYER' means that objects take on the color assigned to that layer. When BYLAYER objects on layer 0 are part of a block definition, they take on the properties of the layer on which the block was inserted, or the layer to which the block was moved.
- 'BYBLOCK' objects in a block definition take on the color assigned to the block. This may be the color in effect at the time the block was inserted, or the color to which the block was changed.
- The Layer command's VP Color overrides allow different colors in each viewport.
- White objects display as black when the background color is white. Color "0" cannot be specified; AutoCAD uses it internally as the background color.
- The *colorwh.dwg* drawing has 255- and 16.7 million-color wheels.
- This command sets the color for anything created after the color is set. This is poor CAD practice; in almost all cases, the colors of objects should be set by placing them on layers with appropriate colors.

CommandLine / CommandLineHide

2006 Toggles the display of the command line palette.

Command	Alias	Keyboard Shortcuts	Menu	Ribbon
commandline	**cli**	**Ctrl+9**	**Tools** ↳**Command Line**	**View** ↳**Palettes** ↳**Command Line**
commandlinehide	**...**	**Ctrl+9**	**...**	**...**

Command: commandline

Displays the command line palette:

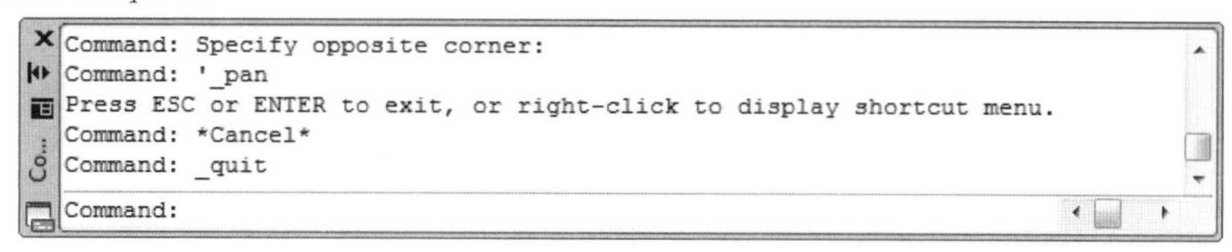

Command: commandlinehide

Hides the command line window.

RELATED SYSTEM VARIABLES

CmdInputHistory specifies the maximum number of previous commands remembered by AutoCAD.

CmdNames reports the name of the current command.

TIPS

- When the command line window is missing, you can still type commands: look for them in the drawing area. This works only when the DYN button is depressed on the status bar, and system variable DynMode is set to 1 or 3.
- AutoCAD displays the following dialog box to remind you how to get back to this palette:

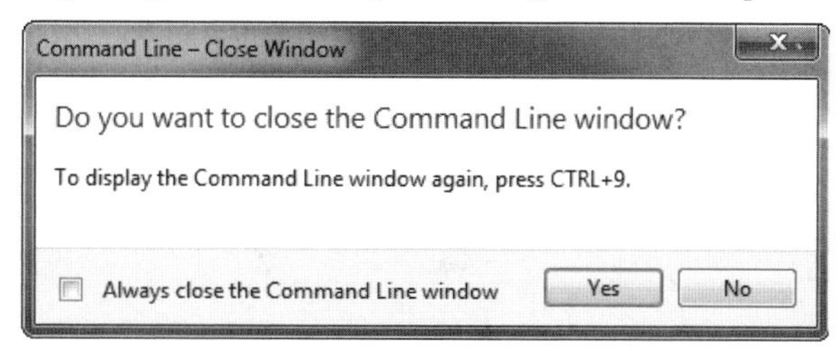

- The command line palette normally shows three lines of history; you can stretch the palette to show more or fewer lines.
- The palette can be made translucent: right-click the title bar, and then select Transparency from the shortcut menu. This does not work when AutoCAD is using hardware acceleration.
- Press F2 to display the text window, which displays the last 400 lines of command history.
- The command bar is a palette, which can be docked along any edge of the drawing area, or made to float — even on a second screen. Some users prefer to dock it at the top of the screen, particularly while teaching.

Compile

Rel. 12 Compiles *.shp* shape and font files and *.pfb* font files into *.shx* files.

Command	Aliases	Keyboard Shortcuts	Menu	Ribbon
compile	...	...	...	...

Command: compile

Displays Select Shape or Font File dialog box. Select a .shp or .pfb file, and then click Open.

DIALOG BOX OPTIONS

Open opens the *.shp* or *.pfb* file for compiling.

Cancel closes the dialog box without loading the file.

RELATED COMMANDS

Load loads compiled *.shx* shape files into the current drawing.

Style loads *.shx* and *.ttf* font files into the current drawing.

RELATED SYSTEM VARIABLE

ShpName is the current *.shp* file name.

RELATED FILES

**.shp* are AutoCAD font and shape source files.

**.shx* are AutoCAD compiled font and shape files.

**.pfb* are PostScript Type B font files.

TIPS

- As of AutoCAD Release 12, Style converts *.shp* font files on-the-fly; it is only necessary to use the Compile command to obtain *.shx* font files.
- Prior to Release 12, this command was an option on the number menu system that appeared when AutoCAD started up.
- As of Release 14, AutoCAD no longer directly supported PostScript font files. Instead, use the Compile command to convert *.pfb* files to *.shx* format.
- TrueType fonts are not compiled.

Cone

Rel. 11 Draws 3D solid cones with circular or elliptical bases.

Command	*Aliases*	*Keyboard Shortcuts*	*Menu*	*Ribbon*
cone	...	...	**Draw** ↳**Modeling** ↳**Cone**	**Home** ↳**Modeling** ↳**Cone**

Command: cone

Specify center point of base or [3P/2P/Ttr/Elliptical]: *(Pick center point 1.)*

Specify base radius or [Diameter] <2">: *(Specify radius 2, or type D.)*

Specify height or [2Point/Axis endpoint/Top radius] <2">: *(Specify height 3, or type A.)*

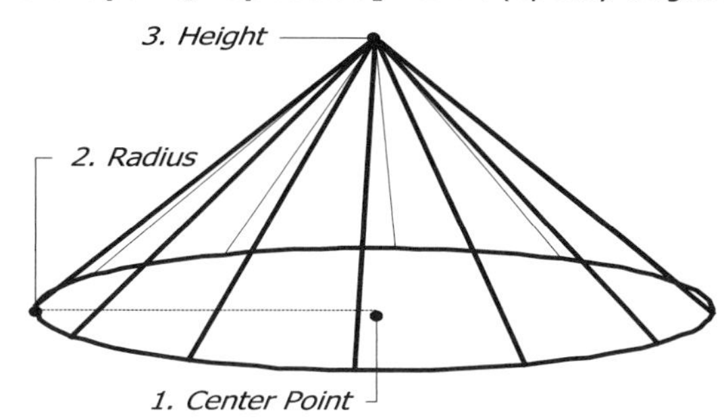

COMMAND LINE OPTIONS

Center point of base specifies the x,y,z coordinates of the center point of the cone's base.

3P picks three points that specify the base's circumference.

2P specifies two points on the base's circumference.

Ttr specifies two points of tangency (with other objects) and the radius.

Elliptical creates cone with an elliptical base.

Base radius specifies the cone's radius.

Diameter specifies the cone's diameter.

Height specifies the cone's height.

2Point picks two points that specify the z orientation of the cone.

Axis endpoint picks the other end of an axis formed from the center point.

Top radius creates a cone with a flat top.

RELATED COMMANDS

Ai_Cone draws 3D surface model cones.

Mesh draws 3D mesh cones.

RELATED SYSTEM VARIABLES

DragVS specifies the visual style during cone creation.

IsoLines specifies the number of isolines on solid surfaces.

TIPS

- You define the elliptical base in two ways: by the length of the major and minor axes, or by the center point and two radii.
- To draw a cone at an angle, use the Axis option.

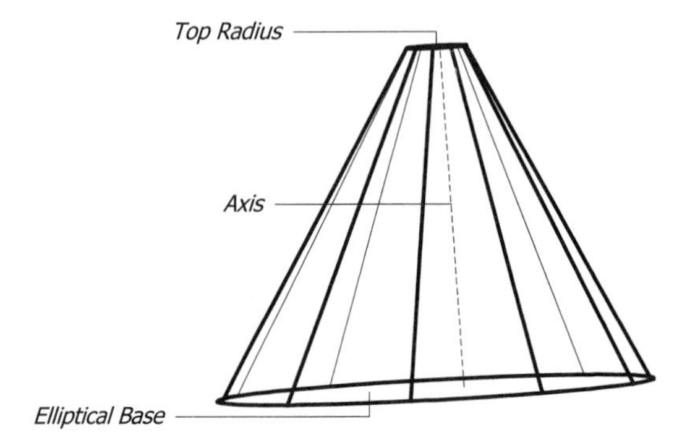

- Once the cone is placed in the drawing, you can edit it interactively. Select the cone, and then select a grip, which turns red, meaning it can be dragged. Press Esc to exit direct editing.

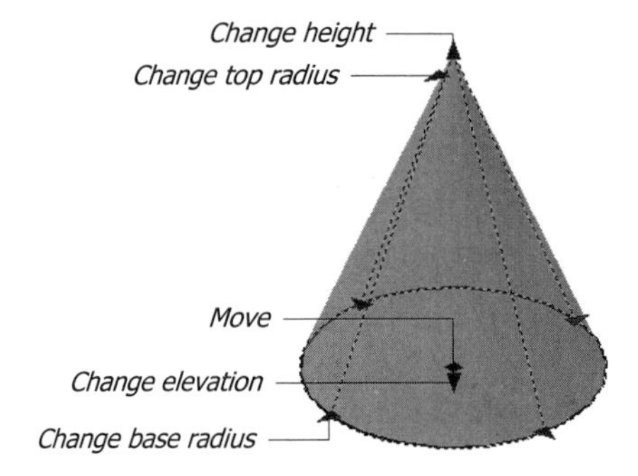

- Silhouette lines are displayed when IsoLines is set to 0 and DispSihl is set to 1. There is no need to hide or shade the cone to see them.

Changed Command

Config command now displays the Options dialog box; see the Options command.

ConstraintBar

2010 Toggles the display of constraint bars.

Command	Alias	Keyboard Shortcuts	Menu	Ribbon
constraintbar	**cbar**	...	**Parametric** ↳**Constraint Bars**	**Parametric** ↳**Geometric** ↳**Show All**

Command: constraintbar

Select objects to show constraints or [Showall/Hideall] <Showall>: *(Select one or more objects, or enter an option.)*

COMMAND LINE OPTIONS

Select objects selects objects, and then shows hidden constraints attached to them.

Showall shows all geometric constraint bars in the drawing.

Hideall hides all geometric constraint bars.

RELATED COMMANDS

AutoConstraint applies geometric constraints automatically.

ConstraintSettings specifies initial conditions for creating dimensional and geometric constraints.

DelConstraint erases constraints.

GeomConstraint applies geometric constraints manually.

DimConstraint applies dimensional constraints manually.

RELATED SYSTEM VARIABLES

ConstraintBarDisplay determines whether constraints are initially hidden or visible.

ConstraintBarMode toggles the display of constraint bars.

ConstraintInfer toggles the inference of constraints during drawing *(new to AutoCAD 2011)*.

TIPS

- *Constraint bars* are like toolbars that show constraint icons.
- Some or all constraint bars are displayed by AutoCAD; use this command's Showall option to display all of them.
- Selection preview controls the display of constraint bars. Place the cursor over a constraint bar, and AutoCAD highlights all geometry related to the constraint.

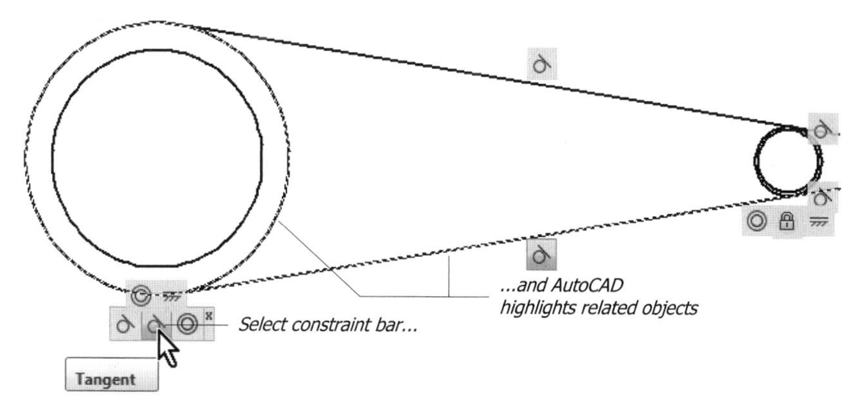

- Place the cursor over a constrained object, and AutoCAD highlights all constraint icons associated with the object.

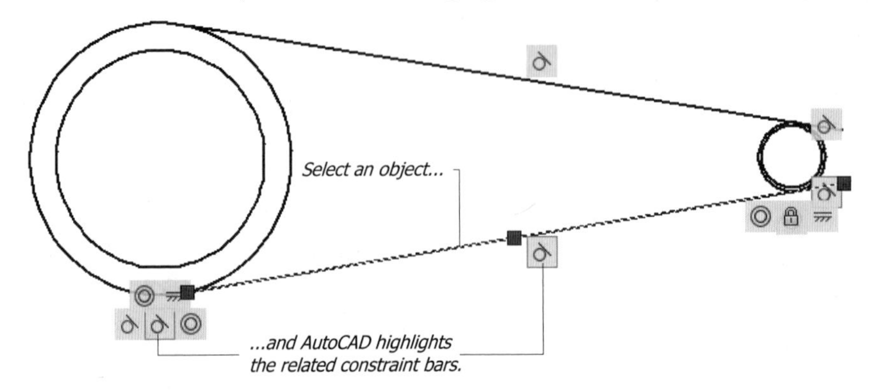

- Right-click a constraint bar for additional options:

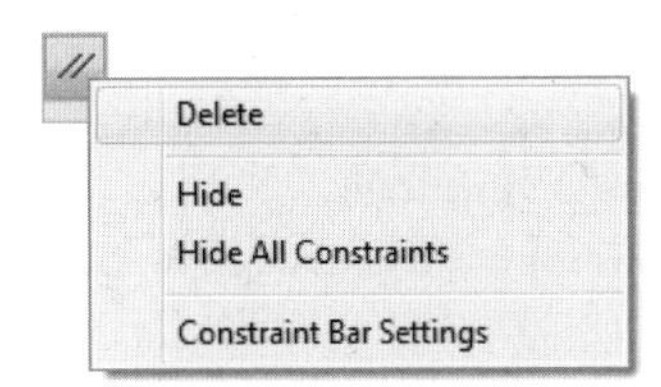

Delete erases geometric constraints from the selected object; see the DelConstraint command.

Hide hides the constraint bar for the selected object.

Hide All Constraints hides all constraint bars.

Constraint Bar Settings displays the Constraint Settings Dialog Box; see the ConstraintSetting command.

- Constraint bars contain these functions:

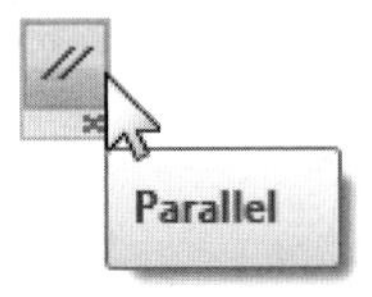

  - Highlight a bar, and its meaning appears in a tooltip.
  - Click the **x** to delete the constraint.

ConstraintSettings

2010 Specifies settings for constraints through a dialog box.

Command	Alias	Keyboard Shortcuts	Menu	Ribbon
constraintsettings	**csettings**	...	**Parametric** ↳**Constraint Settings**	**Parametric** ↳**Dimensional** ↳**Constraint Settings**

+constraintsettings

Command: constraintsettings

Displays a dialog box.

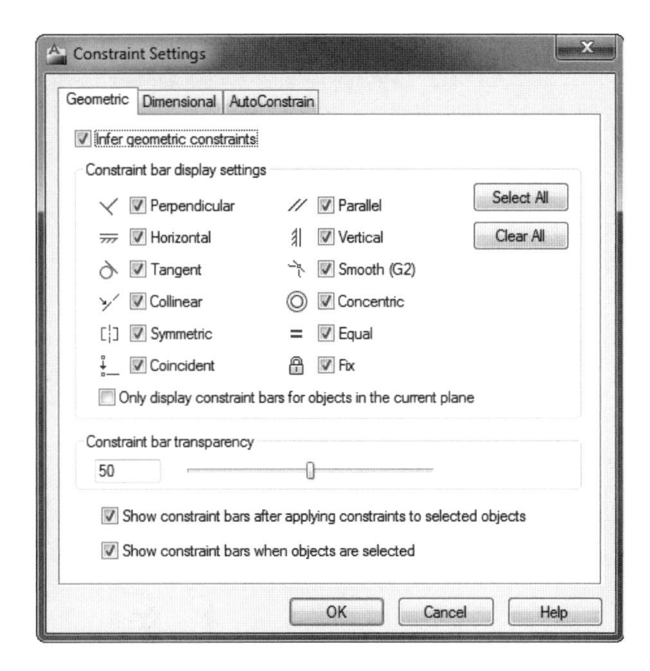

CONSTRAINT SETTINGS DIALOG BOX

Geometric tab

☑ **Infer Geometric Constraints** applies constraints when you draw and edit automatically *(stored in system variable ConstraintInfer; new to AutoCAD 2011).*

Constraint Bar Settings

☑ Toggles the display of each geometric constraint *(ConstraintBarMode).*

Select All turns on all geometric constraints.

Clear All turns off all geometric constraints.

☐ **Only Display Constraint Bars for Object in the Current Plane** displays constraint bars always (when off), and displays them only when in the current UCS (when on).

Constraint Bar Transparency

Constraint Bar Transparency changes the translucency of constraint bars. Enter a number, or drag the slider; range is from 10 (almost opaque) to 90 (almost transparent).

Show Constraint Bars After Applying Constraints to Selected Objects toggles the default visibility of constraint bars (*ConstraintBarDisplay*):

☐ Invisible.

☑ Visible.

Show Constraint Bars When Objects are Selected toggles the visibility of constraint bars during object selection *(new to AutoCAD 2011):*

☐ Invisible.

☑ Visible.

Dimensional tab

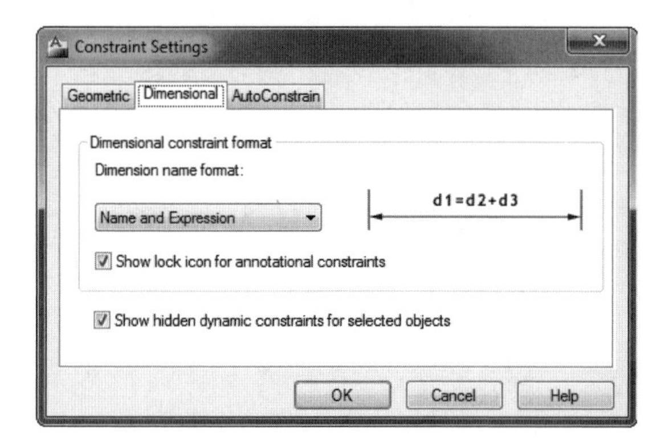

Dimensional Constraint Format

Dimension Name Format determines the format in which dimensional constraints are displayed (*ConstraintNameFormat*):

- **Name** displays the parameter's name only, such as "d1."
- **Value** displays the parameter's value only, such as "4.3210."
- **Name and Expression** displays the parameter's name and formula, such as "d2=d1/4."

Show Lock Icon for Annotational Constraints toggles the display of the lock icon (*DimConstraintIcon*):

☐ Not displayed.

☑ Displayed.

Show Hidden Dynamic Constraints for Selected Objects toggles the display of hidden constraints when objects are selected:

☐ Not displayed.

☑ Displayed.

(The Show All Dynamic Constraints option was removed in AutoCAD 2011, but its DynConstraintDisplay system variable remains.)

AutoConstrain tab

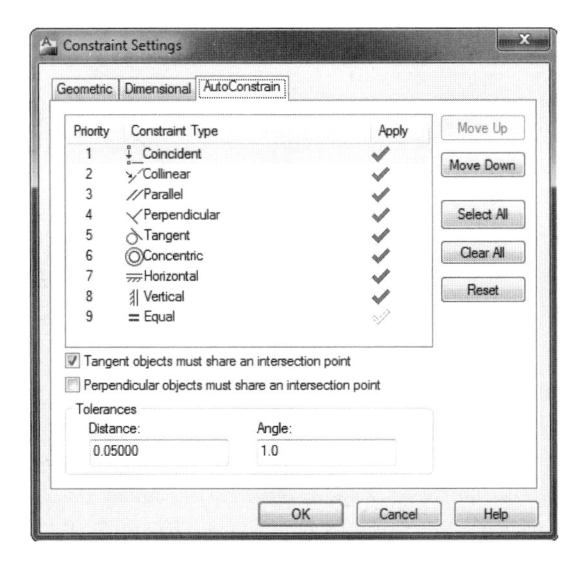

Priority specifies the order in which constraints are applied automatically with the AutoConstrain command; use the Move Up and Move Down buttons to change the order. *(Equal constraint is new to AutoCAD 2011.)*

Apply toggles individual geometric constraints.

Move Up moves the selected constraint up the priority list.

Move Down moves the selected constraint down the list.

Select All sets all geometric constraints as autoconstraints.

Clear All turns off autoconstraining.

Reset returns all AutoConstrain settings to default values.

Tangent Objects Must Lie on a Curve determines how two curves relate when applying tangent constraints:

☐ Tangent constraint is applied only when objects touch.

☑ Applied if objects would touch after being extended.

Perpendicular Objects Must Intersect determines how two lines relate for applying perpendicular constraints:

☐ Endpoint of one line must be within the distance tolerance of another line.

☑ Lines must intersect.

Tolerances

Distance specifies the distance within which constraints are applied automatically by the AutoConstrain command.

Angle specifies the angle within which constraints are applied automatically.

+CONSTRAINTSETTINGS command

Command: +constraintsettings
Tab index <0>: *(Enter an index number.)*

COMMAND LINE OPTIONS

Tab index specifies which tab of the dialog box to display: (**0**) Geometric tab, (**1**) Dimensional tab, (**2**) AutoConstrain tab.

RELATED COMMANDS

AutoConstrain applies geometric constraints automatically.

GeomConstraint applies geometric constraints manually.

DimConstraint applies dimensional constraints manually.

RELATED SYSTEM VARIABLES

ConstraintBarDisplay determines whether constraints are initially hidden or visible.

ConstraintBarMode determines which geometrical constraints are displayed by constraint bars.

ConstraintCursorDisplay *(undocumented)* toggles the display of the constraint icons that appear next to the cursor *(new to AutoCAD 2011).*

ConstraintInfer toggles the inference of geometric constraints as objects are drawn and edited *(new to AutoCAD 2011).*

ConstraintNameFormat specifies the format of text displayed by dimensional constraints.

DimConstraintIcon toggles the displays of the padlock icon in dimensional constraints.

DynConstraintBarDisplay toggles the display of dynamic constraints.

DynConstraintMode toggles the display of hidden dimensional constraints.

TIPS

- Geometric and dimensional constraints are illustrated below.

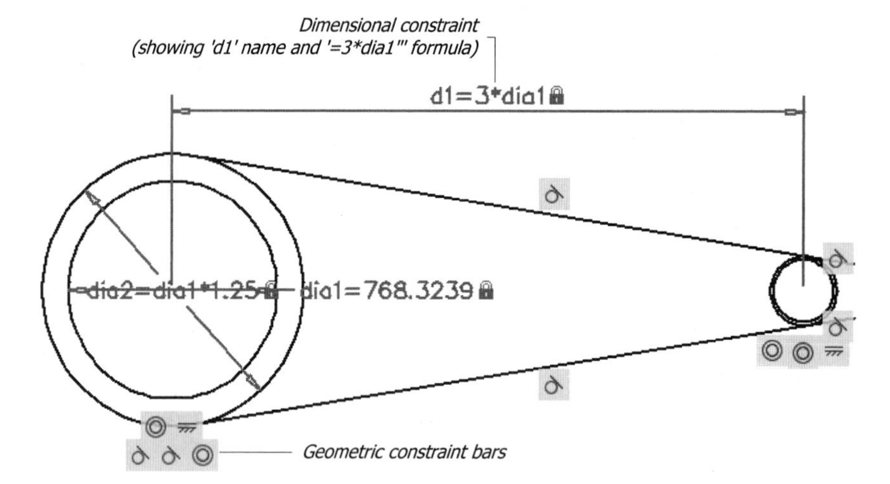

- Apply geometric constraints first.
- Apply a single Fix constraint to lock geometry to the UCS (or WCS).
- Accept default values for dimensional constraints, until all are applied.

Convert

Rel. 14 Converts 2D polylines and associative hatches (created in R13 and earlier) to an optimized "light-weight" format.

Command	*Aliases*	*Keyboard Shortcuts*	*Menu*	*Ribbon*
convert	...	...	...	...

Command: convert

Enter type of objects to convert [Hatch/Polyline/All] <All>: *(Enter an option.)*

Enter object selection preference [Select/All] <All>: *(Type S or A.)*

COMMAND LINE OPTIONS

Hatch converts associative hatch patterns from anonymous blocks to hatch objects; displays warning dialog box:

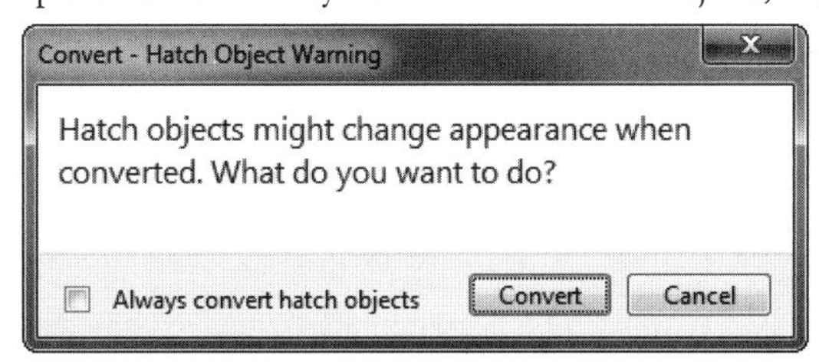

Yes converts hatch patterns to hatch objects.

No does not convert hatch patterns.

Polyline converts 2D polylines to Lwpolyline objects.

All converts all polylines and hatch patterns.

Select selects the hatch patterns and 2D polylines to convert.

RELATED COMMANDS

BHatch creates associative hatch patterns.

PLine draws 2D polylines.

RELATED SYSTEM VARIABLE

PLineType determines whether pre-Release 14 polylines are converted in AutoCAD.

0 — Not converted; PLine creates old-format polylines.

1 — Not converted; PLine creates lwpolylines.

2 — Converted; PLine creates lwpolylines (default).

TIPS

- When a Release 13 or earlier drawing is opened, AutoCAD automatically converts most (not all) 2D polylines to lwpolylines; hatch patterns are not automatically updated.
- Hatch patterns are automatically updated the first time the HatchEdit command is applied, or when their boundaries are changed.
- Polylines are not converted when they contain curve fit segments, splined segments, extended object data in their vertices, or 3D polylines.
- PLineType affects the following commands: Boundary (polylines), Donut, Ellipse (PEllipse = 1), PEdit (when converting lines and arcs), Polygon, and Sketch (SkPoly = 1).

ConvertCTB

2001 Converts a plot style file from CTB color-dependent format to STB named format (short for CONVERT Color TaBle).

Command	*Aliases*	*Keyboard Shortcuts*	*Menu*	*Ribbon*
convertctb	...	...	...	...

Command: convertctb

Displays Select File dialog box.

1. Select a .ctb file, and then click Open. AutoCAD displays the Create File dialog box.

2. Specify the name of an .stb file, and then click Save. When AutoCAD completes the conversion, it displays this dialog box:

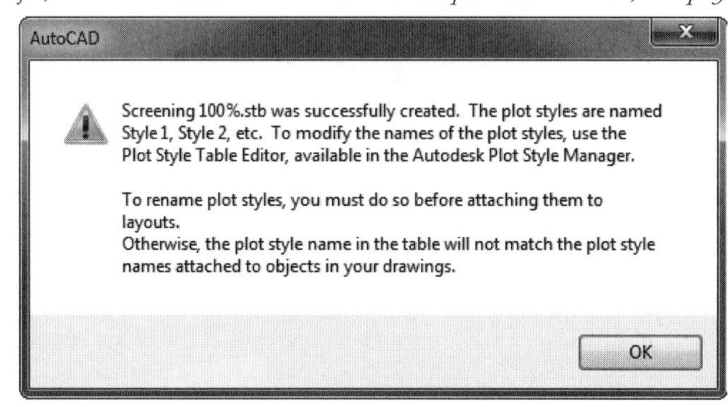

RELATED COMMANDS

ConvertPStyles converts a drawing between color-dependent and named plot styles.

PlotStyle sets the plot style for a drawing.

RELATED FILES

*.*ctb* are color-dependent plot style table files.

*.*stb* are named plot style tables.

TIPS

- "Color-dependent plot styles" are used by the *acad.dwt* template file, in which the color of the object controls the pen selection.
- "Named plot styles" is the alternative introduced with AutoCAD 2000, which allows plotter-specific information to be assigned to layers and objects.
- The technical editor recalls that named plot styles were the default in AutoCAD 2000, but Autodesk later changed the default to color-dependent styles.

ConvertOldLights / ConvertOldMaterials

2007 Converts lights and materials from drawings created in AutoCAD 2006 and earlier.

Command	*Aliases*	*Keyboard Shortcuts*	*Menu*	*Ribbon*
convertoldlights	...	...	...	...
convertoldmaterials		...	...	...

Command: convertoldlights

Lights defined in drawings created by AutoCAD 2006 and earlier are converted to the new format.

Command: convertoldmaterials

Materials defined in drawings created by AutoCAD 2006 and earlier are converted to the new format.

RELATED COMMANDS

LightList displays the Light List palette, which lists and edits lights in the current drawing.

Materials displays the Materials palette, which modifies materials.

RELATED SYSTEM VARIABLE

3dConversionMode determines how lights and materials are converted in drawings from AutoCAD 2007 and earlier.

TIP

- Autodesk warns that the conversion may not be correct in all cases. For example, the intensity of lights may need to be corrected, and the mapping of materials, adjusted.

ConvertPStyles

2001 Converts drawings to either color-dependent or named plot styles (short for Convert Plot STYLES).

Command	Aliases	Keyboard Shortcuts	Menu	Ribbon
convertpstyles	...	...	...	...

Command: convertpstyles

Displays warning dialog box.

When AutoCAD has completed the conversion, it displays the message, "Drawing converted from Named plot style mode to Color Dependent mode."

AUTOCAD DIALOG BOX

OK proceeds with the conversion.

Cancel prevents the conversion.

RELATED COMMANDS

ConvertCTB converts plot styles from the CTB color-dependent format to the STB style named format.

PlotStyle sets the plot style for a drawing.

TIPS

- "Color-dependent plot styles" (CTB) are used by the *acad.dwt* template file, in which the color of the object controls the pen selection.
- "Named plot styles" (STB) were introduced with AutoCAD 2000, which allows plotter-specific information to be assigned to layers and objects.
- When STB files were first introduced, users could only convert CTB drawings to STB, one way. Later, Autodesk added the ability to convert back.

Aliased Command

ConvToMesh command is also known as the MeshSmooth command.

ConvToNurbs

2011 Converts 3D solids and surfaces to NURBS surfaces.

Command	Aliases	Keyboard Shortcuts	Menu	Ribbon
convtonurbs	...	...	**Modify** ↳**Surface Editing** ↳**Convert to NURBS**	**Surface** ↳**Control Vertices** ↳**Convert to NURBS**

Command: convtonurbs
Select objects to convert: *(Select one or more objects.)*
Select objects to convert: *(Press Enter.)*

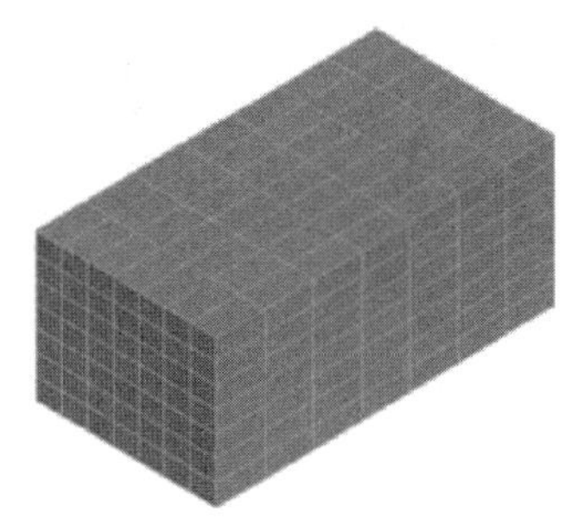

Left: *3D solid.*
Right: *3D NURBS surface.*

COMMAND LINE OPTIONS

Select objects to convert chooses the 3D solids and surfaces to convert to NURBS (nonuniform rational B-spline) surfaces.

RELATED COMMANDS

ConvToSolids converts meshes and closed 3D objects into 3D solids.

ConvToSurfaces converts meshes and open 3D objects into 3D surfaces.

MeshSmooth (ConvToMesh) converts 3D objects into mesh objects.

TIPS

- The two types of surfaces in AutoCAD are procedural surfaces (found in the current and earlier releases of AutoCAD) and NURBS surfaces (as of AutoCAD 2011).
- To convert 3D meshes to NURBS surfaces, first convert the meshes to solids or surfaces with the ConvToSolid or ConvToSurface commands.

ConvToSolid / ConvToSurface

2007 Converts closed objects to solids; converts open objects to surfaces.

Command	*Aliases*	*Keyboard Shortcuts*	*Menu*	*Ribbon*
convtosolid	**...**	**...**	**Modify** ↳**3D Editing** ↳**Convert to Solid**	**Home** ↳**Solid Editing** ↳**Convert to Solid**
convtosurface		**...**	**Modify** ↳**3D Editing** ↳**Convert to Surface**	**Home** ↳**Solid Editing** ↳**Convert to Surface**

Command: convtosolid
Select objects: *(Select one or more objects.)*
Select objects: *(Press Enter.)*

CONVTOSURFACE Command

Command: convtosurface
Select objects: *(Select one or more objects.)*
Select objects: *(Press Enter.)*

RELATED COMMANDS

Extrude extrudes 2D objects into 3D solids or surfaces.

Revolve rotates 2D objects into 3D solids or surfaces.

Thicken thickens 2D objects into 3D solids or surfaces.

Sweep rotates 2D objects into 3D solids or surfaces.

Loft converts 2D objects into lofted 3D solids or surfaces.

MeshSmooth converts 3D objects into mesh objects.

RELATED SYSTEM VARIABLE

DelObj determines whether to retain or erase objects used to create the surfaces and solids.

TIPS

- Use the Explode command to convert 3D solids with curved surfaces into surfaces.
- The ConvToSolid command works with these objects:

  Closed uniform-width polylines with thickness
  Closed zero-width polylines with thickness
  Circles with thickness
  3D mesh objects
  Closed 3D surfaces

- The ConvToSurface command works with these objects:

  Regions
  Arcs with thickness.
  Lines with thickness
  Planar 3D faces
  2D solids with thickness
  Mesh objects

Copy

Ver. 1.0 Makes one or copies of objects.

Command	*Aliases*	*Keyboard Shortcuts*	*Menu*	*Ribbon*
copy	**co**	**...**	**Modify**	**Home**
	cp		**↳Copy**	**↳Modify**
				↳Copy

Command: copy
Select objects: *(Select one or more objects — point 1.)*
Select objects: *(Press Enter.)*
Specify base point or [Displacement] <Displacement>: *(Pick point 2.)*
Specify second point of displacement or <use first point as displacement>: *(Point 3.)*
Specify second point or [Exit/Undo] <Exit>: *(Press Enter to end the command.)*

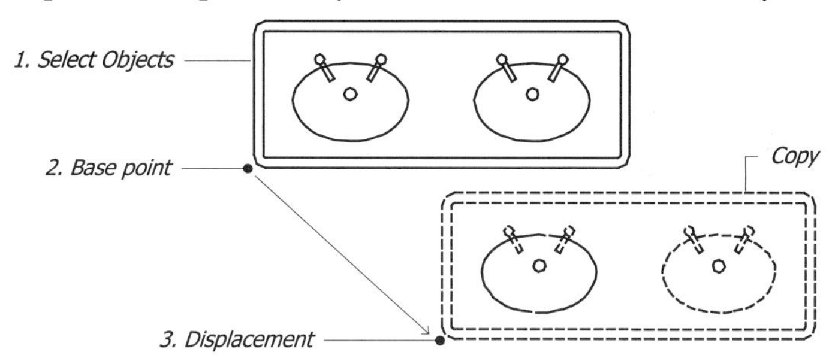

COMMAND LINE OPTIONS

Base point indicates the starting point; does not need to be on the object.

Second point indicates the point to which to copy.

Displacement prompts for two points to use as the displacement.

Undo undoes the last copy.

Exit exits the command.

RELATED COMMANDS

Array draws a rectangular or polar array of objects.

MInsert places an array of blocks.

Offset creates parallel copies of lines, polylines, circles, and arcs.

RELATED SYSTEM VARIABLE

CopyMode stops this command after making one copy.

TIPS

- Turn on ortho mode to copy objects in a precise horizontal and vertical direction, or Polar to achieve precise incremental angles; use OSnap to copy objects precisely to a geometric feature.
- Inserting a block multiple times is more efficient than placing multiple copies.
- Enter coordinates for the Basepoint, and then press Enter: objects move the relative distance specified by the coordinates.

CopyBase

2000 Copies selected objects to the Clipboard with a specified base point (short for COPY with BASEpoint).

Command	*Aliases*	*Keyboard Shortcut*	*Menu*	*Ribbon*
copybase	...	**Ctrl+Shift+C**	**Edit** ↳**Copy with Base Point**	...

Command: copybase
Specify base point: *(Pick a point.)*
Select objects: *(Select one or more objects.)*
Select objects: *(Press Enter.)*

COMMAND LINE OPTIONS

Specify base point specifies the base point.

Select objects selects the objects to copy to the Clipboard.

RELATED COMMANDS

CopyClip copies selected objects to the Clipboard with a base point equal to the lower-left extents of the selected objects.

PasteBlock pastes objects from the Clipboard into drawings as blocks.

PasteClip pastes objects from the Clipboard into drawings.

TIPS

- When PasteBlock pastes objects previously selected with the CopyBase command, AutoCAD prompts you 'Specify insertion point:', and then pastes the objects as a block with a name similar to A$C7E1B27BE.
- When you specify the All option at the 'Select objects:' prompt, CopyBase selects only objects visible in the current viewport.
- As of AutoCAD 2004, you can use the Ctrl+Shift+C shortcut for this command.
- For best results when pasting into another application, such as Word, the technical editor suggests using the WmfOut command to obtain scalable 2D drawings with lineweights.

CopyClip

Rel. 12 Copies selected objects from the drawing to the Clipboard (short for COPY to CLIPboard).

Command	*Aliases*	*Keyboard Shortcut*	*Menu*	*Ribbon*
copyclip	**...**	**Ctrl+C**	**Edit** **↳Copy**	**Home** **↳Clipboard** **↳Copy Clip**

Command: copyclip
Select objects: *(Select one or more objects.)*
Select objects: *(Press Enter.)*

COMMAND LINE OPTION

Select objects selects the objects to copy to the Clipboard.

RELATED COMMANDS

CopyBase copies objects to the Clipboard with a specified base point.

CopyHist copies Text window text to the Clipboard.

CopyLink copies the current viewport to the Clipboard.

CutClip cuts selected objects to the Clipboard.

PasteBlock pastes objects from the Clipboard into the drawing as a block.

PasteClip pastes objects from the Clipboard into the drawing.

RELATED WINDOWS COMMANDS

PrtScr copies the entire screen to the Clipboard

Alt+PrtScr copies the topmost window to the Clipboard.

TIPS

- Text is copied as AutoCAD objects to the Clipboard. Text copied from the command line is copied as plain text.
- To copy AutoCAD objects to another application, such as a word processor, Autodesk suggests using CopyClip, and then pasting as WMF format (through Edit | Paste Special) in the other document.
- When the All option is specified at the 'Select objects' prompt, CopyClip selects only objects visible in the current viewport and in the current space. (In paper space, for instance, this command copies only paper space objects, ignoring ones in model space.) The solution is to use the ExportLayout command.

CopyHist

Rel. 13 Copies all of the Text window text to the Clipboard (short for COPY HISTory).

Command	Aliases	Keyboard Shortcuts	Menu	Ribbon
copyhist	...	...	...	...

Command: copyhist

AutoCAD copies all text in the history window to the Clipboard.

COMMAND LINE OPTIONS

None.

RELATED COMMAND

CopyClip copies selected text from the drawing to the Clipboard.

RELATED WINDOWS COMMAND

ALT+PRT SCR copies the Text window to the Clipboard in graphics format.

TIPS

- To copy a selected portion of Text window text to the Clipboard, highlight the text first, and then select Copy from the Text window's Edit menu bar.

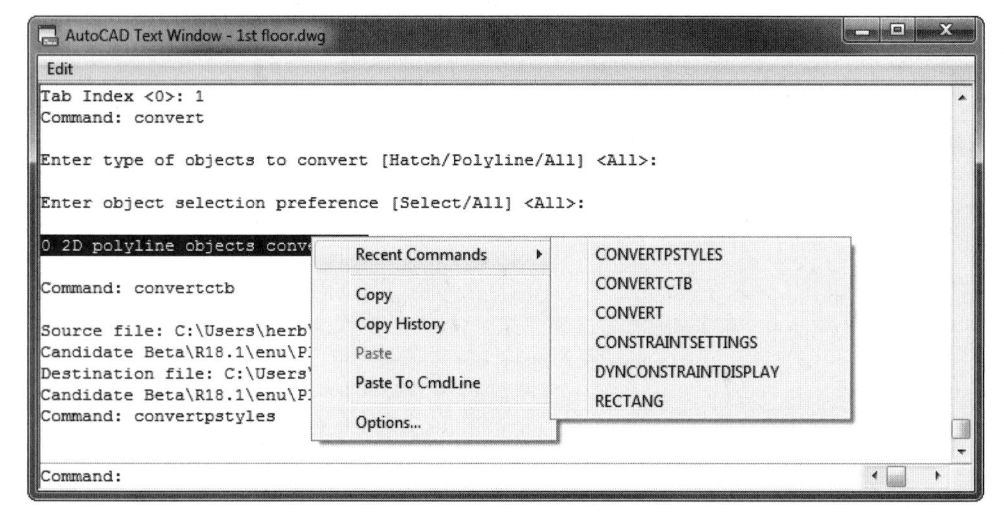

- To paste text to the command line, select Paste to Cmdline from the Edit menu. However, this only works when the Clipboard contains text — not graphics.
- As an alternative, right-click in the Text window to bring up the cursor menu.
- Pasting the history into Notepad helps creating script files.

CopyLink

Rel. 13 Copies the current viewport to the Clipboard; optionally allows you to link the drawing to AutoCAD.

Command	*Aliases*	*Keyboard Shortcut*	*Menu*	*Ribbon*
copylink	...	...	**Edit**	**Home**
			↳Copy Link	**↳Clipboard**
				↳Copy Link

Command: copylink

COMMAND LINE OPTIONS

None.

RELATED COMMANDS

CopyClip copies selected objects to the Clipboard.

CopyEmbed copies selected objects to the Clipboard.

CopyHist copies Text window text to the Clipboard.

CutClip cuts selected objects to the Clipboard.

PasteClip pastes objects from the Clipboard into the drawing.

RELATED WINDOWS COMMANDS

PRT SCR copies the entire screen to the Clipboard

ALT+PRT SCR copies the topmost window to the Clipboard.

TIPS

- In the other application, select Paste Special from the Edit menu to paste the AutoCAD image into the document; to link the drawing to AutoCAD, select the Paste Link option.
- AutoCAD does not let you link a drawing to itself.
- This command copies everything in the current viewport (if in model space) or the entire drawing (if in paper space).
- CopyEmbed is identical to CopyLink, except that CopyEmbed prompts you to select objects.
- If you select **Paste Special | AutoCAD Drawing** when pasting the drawing into a document, you can then double-click the drawing in the other application, which launches AutoCAD so that you can edit the drawing. *Caution*: this may cause AutoCAD to crash.

CopyToLayer

2007 Copies selected objects to the layer of another object; formerly an Express Tool.

Command	Aliases	Menu	Ribbon
copytolayer	...	**Format** **↳Layer Tools** **↳Copy Objects to New Layer**	**Home** **↳Layers** **↳Copy Objects to New Layer**
-copytolayer	...	...	...

Command: copytolayer
Select objects to copy: *(Pick one or more objects.)*
Select objects to copy: *(Press Enter.)*

Select object on destination layer or [Name] <Name>: *(Pick an object, or type N.)*
n **object(s) copied and placed on layer "***layername***".**
Specify base point or [Displacement/eXit] <eXit>: *(Enter an option or press Enter.)*

COMMAND LINE OPTIONS

Select objects to copy selects the objects to be copied.

Select object on destination layer specifies the layer through the selection of another object.

Name displays the Copy to Layer dialog box with names of all layers in the drawing.

Specify base point specifies the starting point of the copy.

Displacement specifies the displacement distance, from original to copy.

eXit exits the command.

Name dialog box

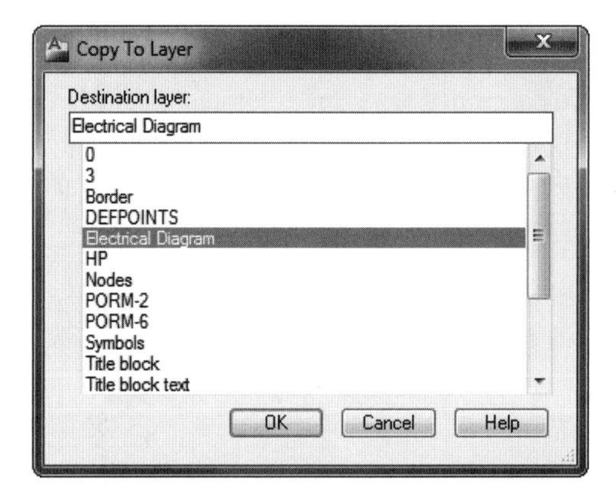

Select a layer name, and then click OK.

-COPYTOLAYER Command

Command: -copytolayer
Select objects to copy: *(Pick one or more objects.)*
Select objects to copy: *(Press Enter.)*

Specify the destination layer name or [?/= (select object)] <0>: *(Enter the name of a layer, or enter an option.)*
n **object(s) copied and placed on layer "***layername***".**
Specify base point or [Displacement/eXit] <eXit>: *(Enter an option or press Enter.)*
Specify second point of displacement or <use first point as displacement>:

COMMAND LINE OPTIONS

Select objects to copy selects the objects to be copied.

Select object on destination layer specifies the layer through the selection of another object.

? lists the names of layers in the drawing.

= prompts you to select an object whose layer should be used; AutoCAD prompts, "Select an object with the desired layer name."

Specify base point specifies the starting point of the copy.

Displacement specifies the displacement distance, from original to copy.

eXit exits the command.

RELATED COMMANDS

Copy copies objects onto the same layer.

Ai_Molc makes the layer of the selected object current.

ChSpace moves objects from paper to model space.

CUI

2006 Customizes many aspects of the user interface; replaces the Menu, Toolbar, and Customize commands (short for Customize User Interface).

Command	*Aliases*	*Keyboard Shortcuts*	*Menu*	*Ribbon*
cui	**toolbar**	**...**	**Tools**	**Manage**
	tbconfig		**↳Customize**	**↳Customization**
	to		**↳Interface**	**↳User Interface**

Command: cui

Displays the Customize tab of the dialog box:

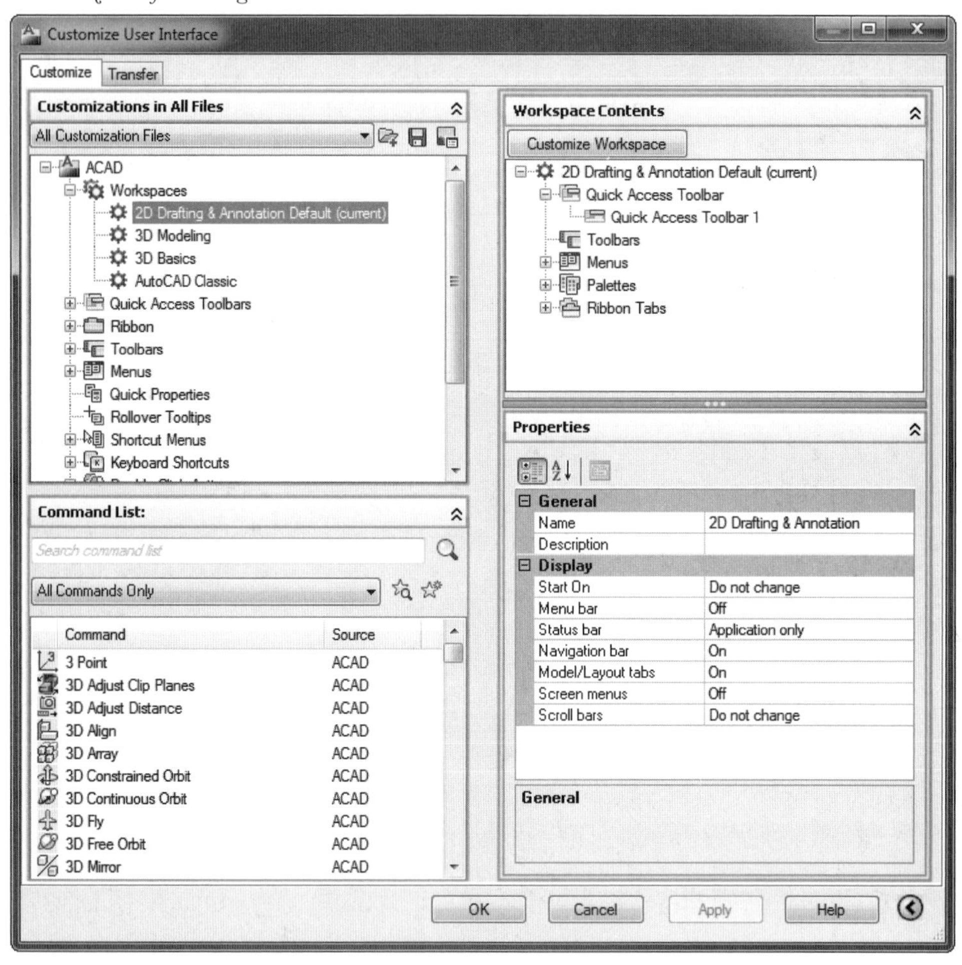

The dialog box consists of several panes; each pane is an area of the dialog box.

Customizations In All Files pane

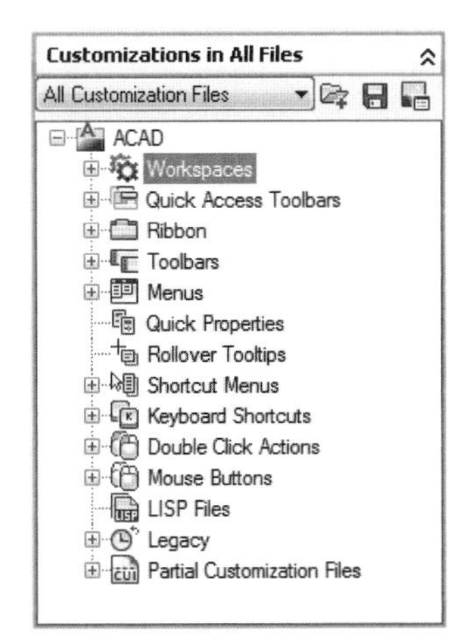

Right-click items to access the shortcut menus described later.

Categories lists categories of CUI files.

Load Partial Customization File displays the Open dialog box; see the CuiLoad command.

Save saves all current customization files.

Image Manager manages custom icons employed by the user interface; displays the Image Manager dialog box. (Not related to the Image command or raster image underlays.)

Command List pane

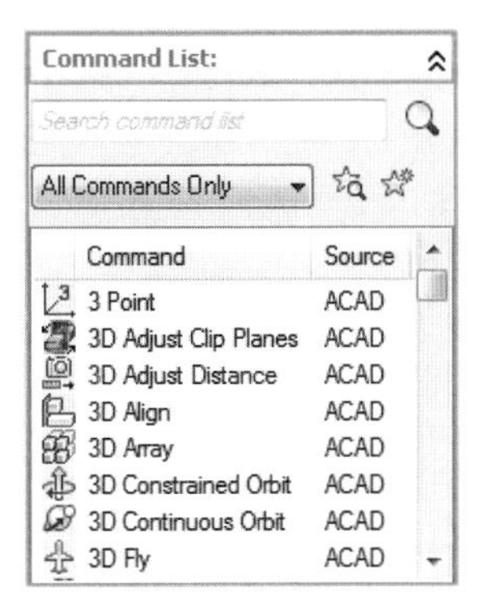

Right-click items to access shortcut menus.

Search Command List finds command descriptions by name.

Droplist lists categories of commands, such as All, File, and Seek.

Find Command or Text displays the Find and Replace dialog box.

Create a New Command creates new commands.

Properties pane

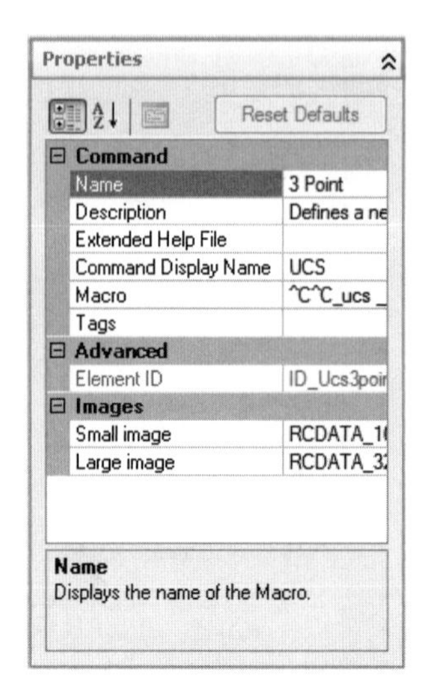

Content varies, depending on the item being edited:

Sorts properties according to categories.

Sorts properties alphabetically.

Toggles the display of the Tips box below the Properties grid.

Reset Defaults returns all properties to their default values.

Name specifies the name of the command or macro.

Description specifies the help text displayed on the status bar.

Macro executes the code when the command, menu item, toolbar button, or shortcut keystroke is invoked; click the **...** button to display the Long String Editor dialog box.

ElementID reports the internal identification assigned to the command by AutoCAD.

Large image shows the 24x24-pixel bitmap image displayed on toolbars and so on.

Small image shows the 16x16-pixel bitmap image displayed on toolbars and so on; click the **...** button to display the Select Image File dialog box for loading *.bmp* image files.

Button Image pane

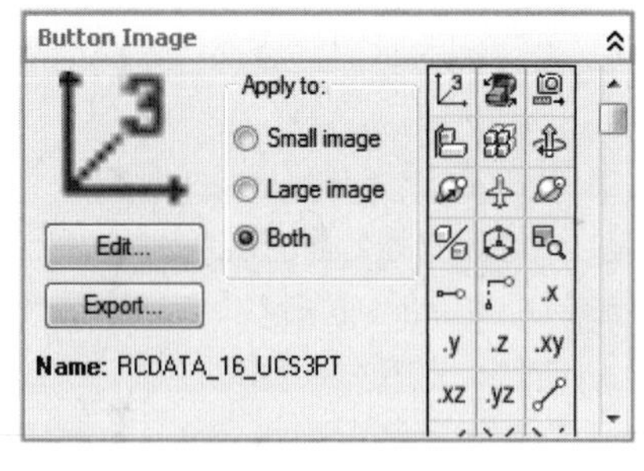

○ **Small** works with 16x16-pixel bitmap images.

○ **Large** works with 24x24-pixel bitmap images.

⊙ **Both** works with both sizes of images.

Edit displays the Button Editor dialog box.

Export exports the button as a *.bmp* (bitmap) file in the *C:\Users*login\ *AppData\Roaming\Autodesk\AutoCAD 2011 \ R18.1\enu\Support\Icons* folder.

SHORTCUT MENUS

Customizations In All Files pane

Right-click any item in the Customization In pane; content of shortcut menu varies with the item selected:

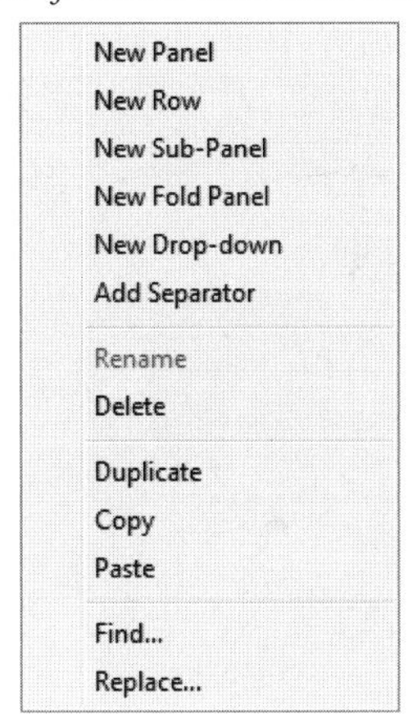

New adds an item specific to the section; examples include new panels, toolbars, and menus.

Add Separator adds separation lines between items.

Copy to Ribbon Panels copies item to ribbon.

Rename changes the name of the item.

Delete erases the item; AutoCAD displays a warning dialog box.

Duplicate copies the selected item.

Copy copies the item for reuse elsewhere in the dialog box.

Paste pastes an item copied from elsewhere into the CUI dialog box.

Find displays the Find tab of the Find and Replace dialog box.

Replace displays the Replace tab of the Find and Replace dialog box.

Command List pane

Right-click any item in the Command List pane:

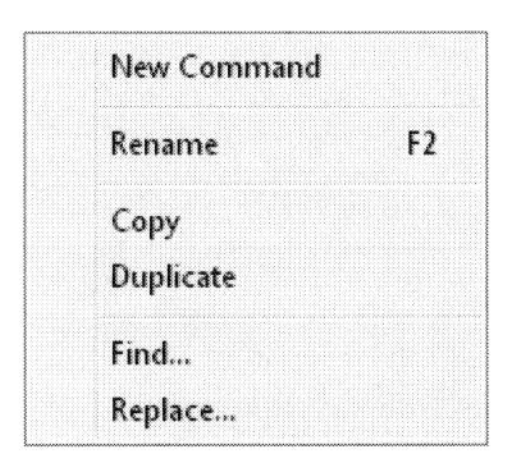

New Command adds new "commands" (macros, actually) to AutoCAD.

Rename renames the item in-place.

Copy copies item to the Clipboard; can be pasted elsewhere in CUI with the Paste command. Do not attempt to paste into other applications.

Duplicate copies the command.

Delete erases the item (available only for user-created commands).

BUTTON EDITOR DIALOG BOX

To access this dialog box: click the Button Image Edit button.

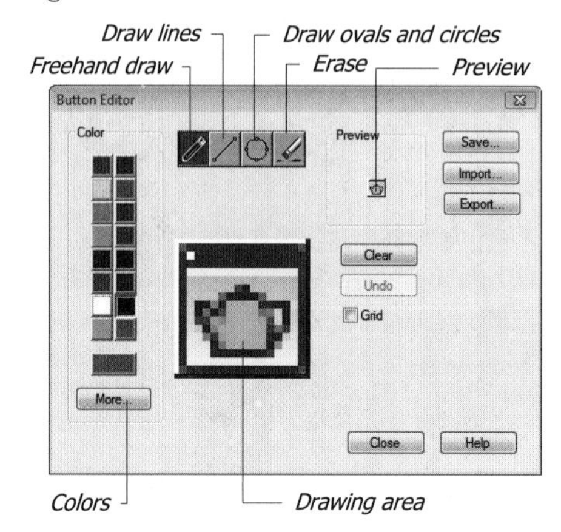

Save saves changes to the icon in AutoCAD's internal storage; displays the Save Image dialog box

Import imports *.bmp* files *(formerly the Open button).*

Export saves the icon as a *.bmp* file *(formerly the Save As button).*

Clear erases the icon.

Undo undoes the last operation.

Grid toggles grid lines in the icon drawing area.

More displays the Select Color dialog box.

IMAGE MANAGER DIALOG BOX

To access this dialog box, click the Image Manager icon in the Customizations in All Files pane:

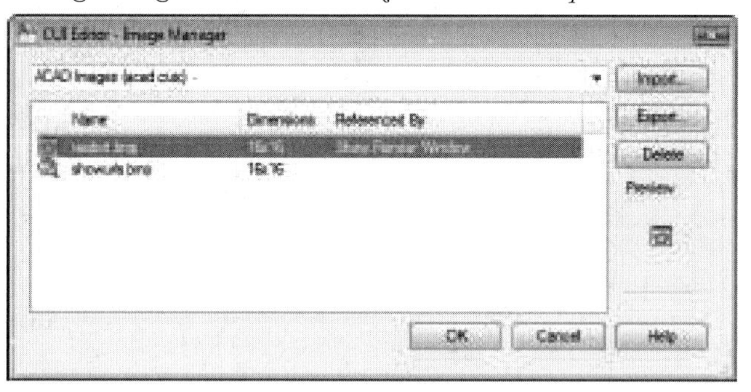

Name reports names of *.bmp* files.

Dimensions reports the size in pixels.

Referenced By reports the descriptive name of the command using the image.

Import opens *.bmp* image files.

Export saves the image as a *.bmp* file.

Delete removes the image from AutoCAD.

RELATED COMMANDS

CuiExport / CuiImport transfer settings between local *acad.cuix* and enterprise *.cuix* files.

CuiLoad / CuiUnload load and unload *.cuix* files; replaces the MenuLoad command.

Customize rearranges tool palettes.

QuickCui opens the CUI dialog box partially collapsed.

Tablet calibrates, configures, and toggles digitizing tablet menus.

Workspace creates, modifies, and saves workspaces.

RELATED SYSTEM VARIABLES

CuiState reports whether the Customize User Interface dialog box is open or not.

EnterpriseMenu specifies the path and name of the *.cui* file used for the enterprise.

LockUi prevents palettes and toolbars from moving; hold down Ctrl key to override the lock.

TempOverrides turns temporary override keys on and off.

Tooltips toggles the display of tooltips.

DEFINITIONS

CUI — customization user interface. It replaces the Menu, Toolbar, and Customize commands as of AutoCAD 2006.

Enterprise CUI — customization data shared by all users in an office; typically controlled by the CAD manager.

TIPS

- The LockUi system variable prevents the windows and toolbars from being moved; to override the latch, hold down the Ctrl key while dragging them.
- The file format changed in AutoCAD 2010 from *.cui* to *.cuix* in order to encapsulate bitmap images.
- AutoCAD 2011 adds the Navigation Bar item to the Workspace Contents pane.
- For more on the Transfer tab, see the CuiExport command.

CuiExport / CuiImport

2006 Transfers user interface customization data between *.cuix* files.

Command	Aliases	Keyboard Shortcuts	Menu	Ribbon
cuiexport	...	...	**Tools** ⇘**Customize** ⇘**Export Customizations**	...
cuiimport	...	...	**Tools** ⇘**Customize** ⇘**Import Customizations**	...

Command: cuiexport *or* cuiimport

Both commands display the same Transfer tab of the dialog box.

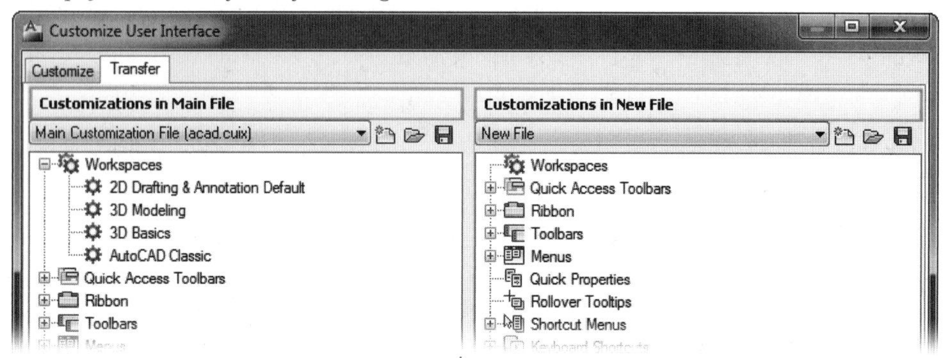

CUSTOMIZE USER INTERFACE DIALOG BOX

To import and export items, follow these steps:

1. Open a *.cui*, *.cuix* or *.mnu* file in one panel.
2. Open another *.cui*, *.cuix* or *.mnu* file in the second panel.
3. Drag items between the two panels.
4. Click the Save button.

Dialog Box Toolbar

Creates new customization files.

Opens customization and menu files; displays the Open dialog box.

Saves the current customization file; does not display a dialog box.

RELATED COMMANDS

Cui customizes elements of the user interface.

CuiLoad adds the content of *.cui*, *.cuix* and *.mnu* files to the user interface.

CuiLoad / CuiUnload

2006 Loads (and unloads) "partial" *.cuix* files.

Command	*Aliases*	*Keyboard Shortcuts*	*Menu*	*Ribbon*
cuiload	**menuload**	...	...	...
cuiunload	**menuunload**	...	...	...

Command: cuiload *or* cuiunload

Both commands display the same dialog box.

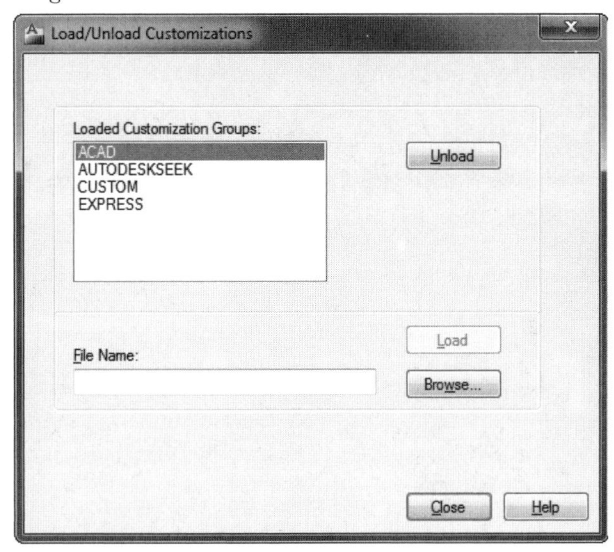

LOAD/UNLOAD CUSTOMIZATIONS DIALOG BOX

Loaded Customization Groups lists the names of loaded menu groups and files.

Unload unloads selected menu group.

Load loads the selected menu group file into AutoCAD.

File Name displays the name of the file.

Browse displays the Select Customization File dialog box; choose a *.cui*, *.cuix*, *.mnu*, or *.mns* file, and then click Open.

RELATED COMMANDS

CUI manipulates the user interface customization.

Tablet configures digitizing tablets for use with overlay menus.

RELATED FILE

.cuix* is the customization user interface file; stored in XML format.

TIPS

- The CUILoad command allows you to add *partial* user interfaces, without replacing the entire user interface structure.
- The *custom.cuix* file is meant for customizing menus independently of *acad.cuix*.
- As of AutoCAD 2006, these commands replace the MenuLoad and MenuUnload commands of earlier releases of AutoCAD.
- As of AutoCAD 2010, *.cuix* files replace *.cui* files.

CustomerInvolvementProgram

2008 Invites you to join the Customer Involvement Program (undocumented command; at 26 characters, the longest command name in AutoCAD).

Command	*Aliases*	*Menu*	*Ribbon*
customerinvolvementprogram	**...**	**Help** **⇘Customer Involvement Program**	**...**

Command: customerinvolvementprogram

Displays dialog box.

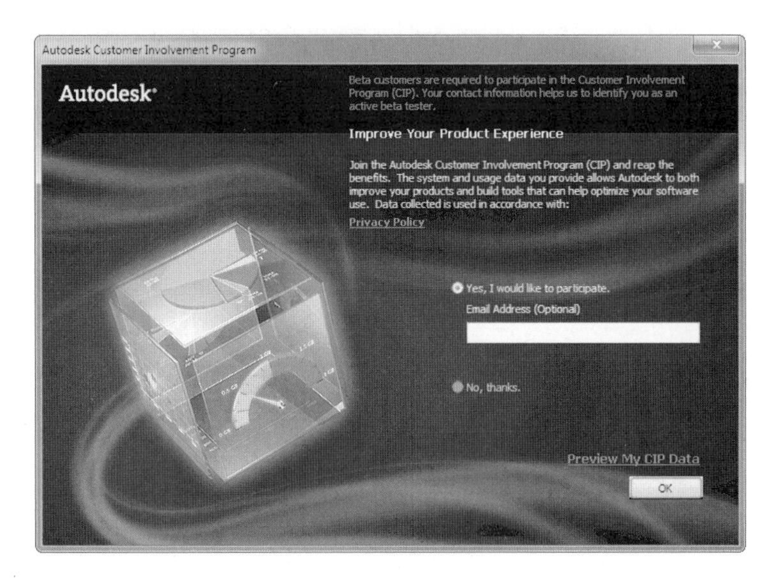

DIALOG BOX OPTIONS

○ **Yes, I would like to participate** contacts Autodesk via the Internet to let you join the program.

Email Address (Optional) provides Autodesk with your email contact address.

⊙ **No Thanks** allows you to keep the information to yourself.

OK dismisses the dialog box.

RELATED SYSTEM VARIABLE

CipMode reports the status of the Customer Involvement Program.

Customize

2000i Manages tool palettes, and creates palette groups.

Command	*Aliases*	*Keyboard Shortcuts*	*Menu*	*Ribbon*
customize	...	...	**Tools**	**Manage**
			↳Customize	**↳Customize**
			↳Tool Palettes	**↳Tool Palettes**

Command: customize

Displays dialog box.

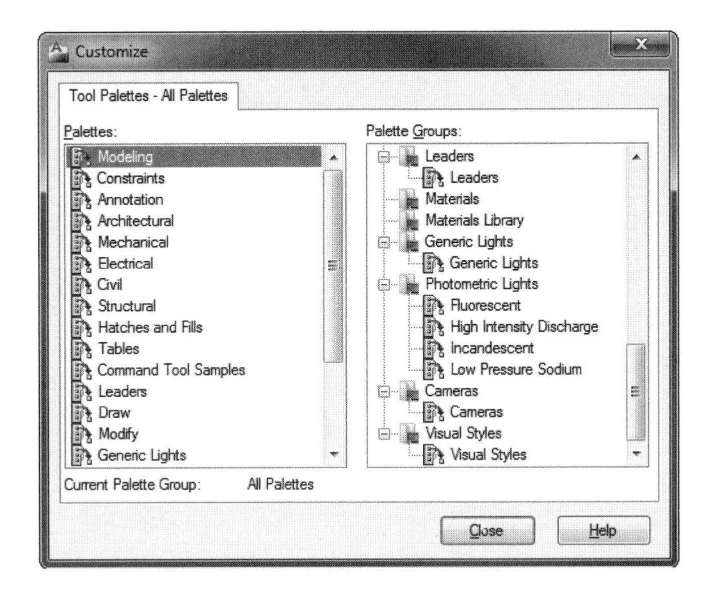

SHORTCUT MENUS

Palettes

Rename renames the selected palette.

New Palette creates a new, blank palette.

Delete removes the selected palette.

Import imports palettes (*.xtp* files).

Export exports the selected palette in an XML-like *.xtp* file.

Palette Groups

Accessing this menu by right-clicking a folder for the shortcut menu:

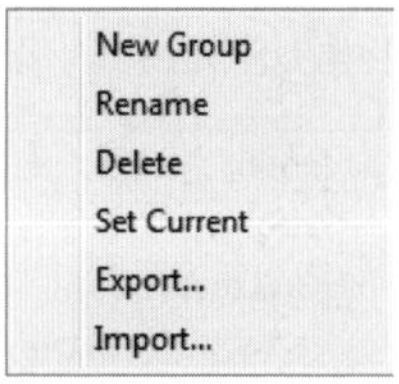

New Group
Rename
Delete
Set Current
Export...
Import...

New Group creates new groups and subgroups.

Rename renames the selected group.

Delete removes the group.

Set Current makes the palette group current.

RELATED COMMANDS

ToolPalette toggles the display of the Tool Palette.

TpNavigate displays the specified tool palette or palette group.

CutClip

Rel. 12 Cuts the selected objects from the drawing to the Clipboard (short for CUT to CLIPboard).

Command	*Aliases*	*Keyboard Shortcut*	*Menu*	*Ribbon*
cutclip	...	**Ctrl+X**	**Edit** ↳**Cut**	**Home** ↳**Utilities** ↳**Cut Clip**

Command: cutclip
Select objects: *(Select one or more objects.)*
Select objects: *(Press Enter.)*

COMMAND LINE OPTION

Select objects selects the objects to cut to the Clipboard.

RELATED COMMANDS

BmpOut exports selected objects in the current view to a *.bmp* file.

CopyClip copies selected objects to the Clipboard.

CopyHist copies the Text window text to the Clipboard.

CopyLink copies the current viewport to the Clipboard.

PasteClip pastes objects from the Clipboard into the drawing.

RELATED WINDOWS COMMANDS

PRT SCR copies the entire screen to the Clipboard.

ALT+PRT SCR copies the topmost window to the Clipboard.

TIPS

- When the All option is specified at the 'Select objects:' prompt in model space, CutClip cuts all objects in the current viewport, visible or not. In paper space, only paper space objects are cut.
- In the other application, use the **Edit | Paste** or **Edit | Paste Special** commands to paste the AutoCAD image into the document; the Paste Special command lets you specify the pasted format.
- You can use the Undo command (or Oops) to return the "cut" objects to the drawing.

CvAdd / CvRemove

2011 Adds and removes control vertices from NURBS splines and surfaces (short for Control Vertices).

Command	Aliases	Menu	Ribbon
cvadd	**insertcontrolpoint**	**Modify** ↳**NURBS Surface Editing** ↳**Add CV**	**Surface** ↳**Control Vertices** ↳**Add CV**
cvremove	**removecontrolpoint**	**Modify** ↳**NURBS Surface Editing** ↳**Remove CV**	**Surface** ↳**Control Vertices** ↳**Remove CV**

Command: cvadd
Select a NURBS surface or curve to add control vertices: *(Choose a surface or a spline.)*

Prompt sequence for NURBS surfaces:
Adding control vertices in U direction
Select point on surface or [insert Knots/Direction]: *(Pick a point, or enter an option.)*

Prompt sequence for NURBS splines (curves):
Select point on spline or [insert Edit point]: *(Pick a point, or type E.)*

COMMAND LINE OPTIONS

Select a NURBS surface or curve selects a surface or spline.

Options for surfaces:

Select point on surface adds a pair of control vertices to the edges of surface.

insert Knots adds a knot to the surface.

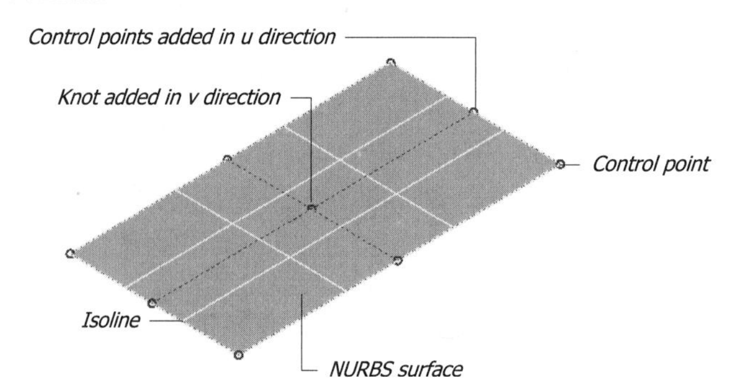

Direction switches the direction between u and v.

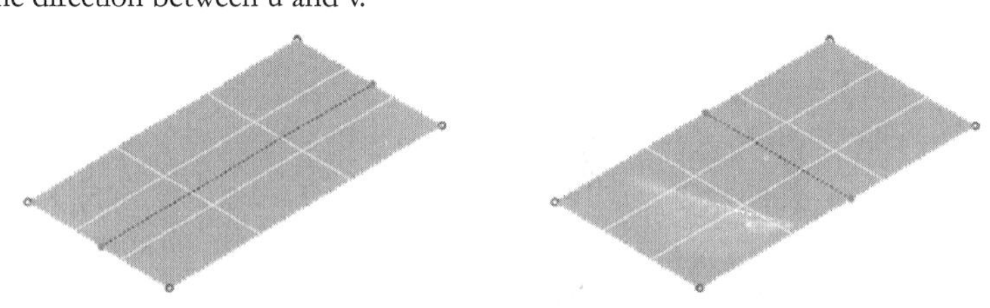

Left: *Placing a point along the u axis...*
Right: *...and along the v axis.*

Options for splines:

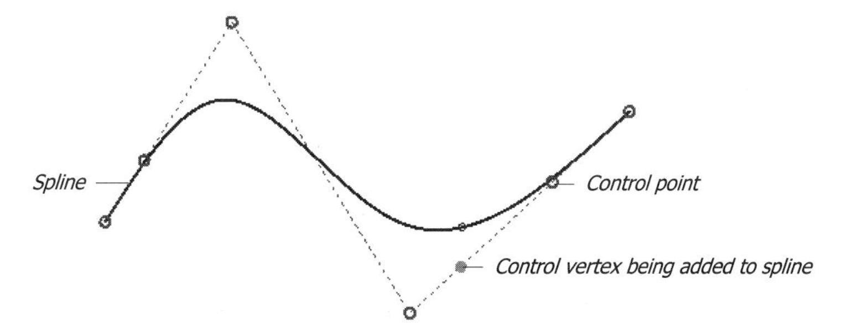

insert Edit point adds an edit point to the spline as shown by an orange dot; prompts you:

Select point on spline or [insert Control vertex]: *(Pick a point, or type C.)*

insert Control vertex adds a vertex point to the spline as shown by a brown circle; prompts you:

Select point on spline or [insert Edit point]: *(Pick a point, or type E.)*

 CVREMOVE Command

Command: cvremove

Select a NURBS surface or curve to remove control vertices: *(Choose a surface or a spline.)*

Prompts for surfaces:

Removing control vertices in U direction

Select point on surface or [Direction]: *(Pick a point on the surface, or type D.)*

Prompts for splines:

Select point on the curve: *(Pick a point on the spline.)*

RELATED COMMANDS

Spline draws NURBS splines when the SplMethod system variable is set to 1 (control vertices).

Commands such as Extrude, ConvToSurf, and ConvToNurbs create NURBS surfaces when the SurfaceModelingMode system variable is set to 1.

CvHide hides control vertices for all NURBS surfaces and curves.

CvRebuild rebuilds the shapes of NURBS surfaces and curves.

CvShow shows control vertices of selected NURBS surfaces and curves.

CvHide / CvShow

2011 Hides and shows control vertices on selected NURBS splines and surfaces.

Command	*Aliases*	*Menu*	*Ribbon*
cvhide	**pointoff**	**Modify** **↳NURBS Surface Editing** **↳Hide CV**	**Surface** **↳Control Vertices** **↳Hide CV**
cvshow	**pointon**	**Modify** **↳NURBS Surface Editing** **↳Show CV**	**Surface** **↳Control Vertices** **↳Show CV**

Command: cvhide

Hides all control vertices of all splines and surfaces.

CVSHOW Command

Command: cvshow

Select a NURBS surface or curve to display control vertices: *(Choose a surface or a spline.)*

Displays control vertices on the selected objects; shown as round circles and dotted lines.

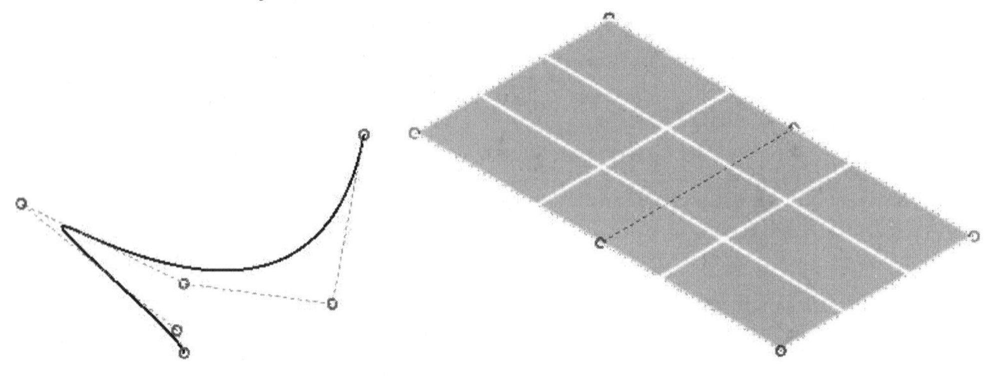

RELATED COMMANDS

Spline draws NURBS splines when the SplMethod system variable is set to 1 (control vertices).

Commands such as Extrude, ConvToSurf, and ConvToNurbs create NURBS surfaces when the SurfaceModelingMode system variables is set to 1.

CvAdd adds control vertices to NURBS surfaces and splines.

CvRebuild rebuilds the shapes of NURBS surfaces and curves.

CvRemove removes control vertices from NURBS surfaces and curves.

CvRebuild

2011 Hides and shows control vertices on selected NURBS splines and surfaces.

Command	*Aliases*	*Keyboard Shortcuts*	*Menu*	*Ribbon*
cvrebuild	**rebuild**		**Modify**	**Surface**
			↳Surface Editing	**↳Control Vertices**
			↳NURBS Surface Editing	**↳Rebuild CV**
			↳Rebuild Surface	

-cvrebuild

Command: cvrebuild

Select a NURBS surface or curve to rebuild: *(Pick a surface or spline.)*

Displays a different dialog box for surfaces and splines:

REBUILD SURFACE DIALOG BOX OPTIONS

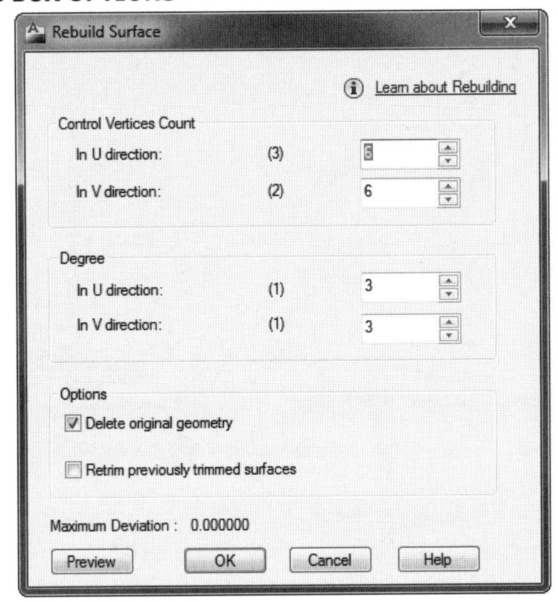

Control Vertices Count options

In U Direction specifies the number of control vertices in the v direction *(stored in the RebuildU system variable).*

In V Direction specifies the number of control vertices in the v direction (*RebuildU*).

Degree options

In U Direction specifies the degree in the u direction (*RebuildDegreeU*).

In V Direction specifies the degree in the v direction (*RebuildDegreeV*).

Options options

Delete Input Geometry toggles the removal of original surfaces following the rebuilding process (*RebuildOptions*).

Retrim Previously Trimmed Surface toggles the application of trimmed areas to rebuilt surfaces (*RebuildOptions*).

REBUILD CURVE DIALOG BOX OPTIONS

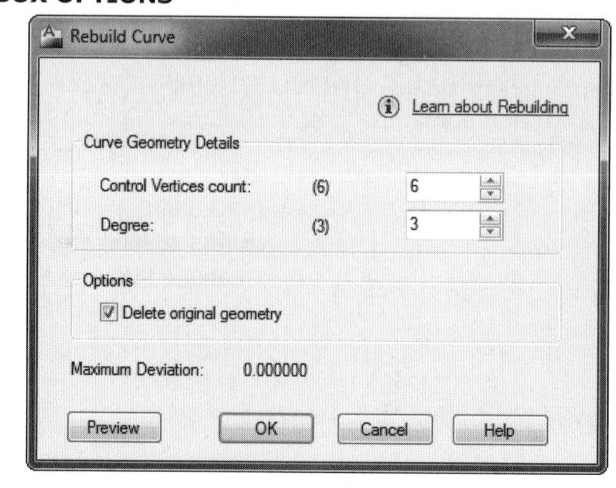

Curve Geometry Details options

Control Vertices Count specifies the number of control vertices along the spline *(stored in the Rebuild2dCv system variable).*

Degree specifies the number of control vertices per span (*Rebuild2dDegree*).

Options options

Delete Original Geometry toggles the removal of original splines following the rebuilding process (*RebuildOptions*).

RELATED COMMANDS

Spline draws NURBS splines when the SplMethod system variable is set to 1 (control vertices).

Commands such as Extrude, ConvToSurf, and ConvToNurbs create NURBS surfaces when the SurfaceModelingMode system variables is set to 1.

CvAdd adds control vertices to NURBS surfaces and splines.

CvRemove removes control vertices from NURBS surfaces and curves.

RELATED SYSTEM VARIABLES

Rebuild2dCv specifies the number of control vertices per spline; range is 2 to 32767; the technical editor notes that values above 25000 crash his computer.

Rebuild2dDegree specifies the degree for splines; range is 1 to 11.

Rebuild2dOption toggles what happens to original splines after rebuilding between (0) not deleted and (1; default) deleted.

RebuildDegreeU specifies the u-direction degree for NURBS surfaces; range is 2 to 11.

RebuildDegreeV specifies the v-direction degree for NURBS surfaces; range is 2 to 11.

RebuildOptions determines what happens to original surfaces and trimmed areas on rebuilt surfaces.

RebuildU specifies the default number of lines in the u direction for NURBS surfaces; range is 2 to 32767.

RebuildV specifies the default number of lines in the v direction for NURBS surfaces; range is 2 to 32767.

TIPS

- This command is useful for cleaning up surfaces and splines with too many control vertices.
- Surfaces become more complex with greater numbers of degrees; maximum is 11.
- The number in parentheses, such as (6), reports the current values of control vertices and degrees; the Maximum Deviation item reports the maximum deviation between the original and new surface or spline.
- Caution: rebuilding splines changes their shape; the shapes of surfaces are unaffected by rebuilds.

Cylinder

Rel. 12 Draws a 3D solid cylinder with a circular or elliptical cross section.

Command	*Aliases*	*Keyboard Shortcuts*	*Menu*	*Ribbon*
cylinder	**cyl**	**...**	**Draw** **⇘Modeling** **⇘Cylinder**	**Home** **⇘Modeling** **⇘Cylinder**

Command: cylinder

Specify center point of base or [3P/2P/Ttr/Elliptical]: *(Pick center point 1, or enter an option.)*

Specify base radius or [Diameter] <3>: *(Specify radius 2, or type D.)*

Specify height or [2Point/Axis endpoint] <1”>: *(Specify height 3, or type C.)*

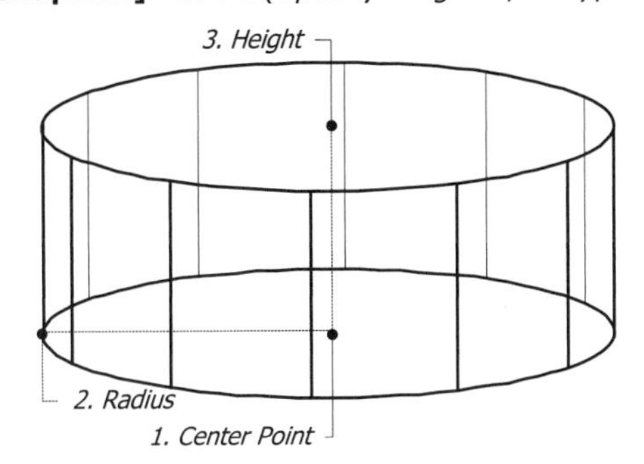

COMMAND OPTIONS

Center point of base specifies the x,y,z coordinates of the center point of the cylinder's base.

3P picks three points that specify the base's circumference.

2P specifies two points on the base's circumference.

Ttr specifies two points of tangency (with other objects) and the radius.

Elliptical creates elliptical cylinders.

Base radius specifies the cylinder's radius.

Diameter specifies the cylinder's diameter.

Height specifies the cylinder's height.

2Points picks two points that specify the z orientation of the cylinder.

Axis endpoint picks the other end of an axis formed from the center point.

RELATED COMMANDS

Ai_Cylinder creates cylinders made of 3D meshes.

Extrude creates cylinders and other extruded shapes with arbitrary cross sections and sloped walls.

Mesh creates cylinders as mesh objects.

RELATED SYSTEM VARIABLES

DragVs determines the visual style while creating 3D solid objects.

IsoLines determines the number of isolines on curved solids.

TIPS

- Once the cylinder is placed in the drawing, you can edit it interactively:
  1. Select the cylinder.
  2. Select one of the grips, which turns red.
  3. Drag the grip to change the cylinder's size or position.
  4. When done, press ESC to exit direct editing.

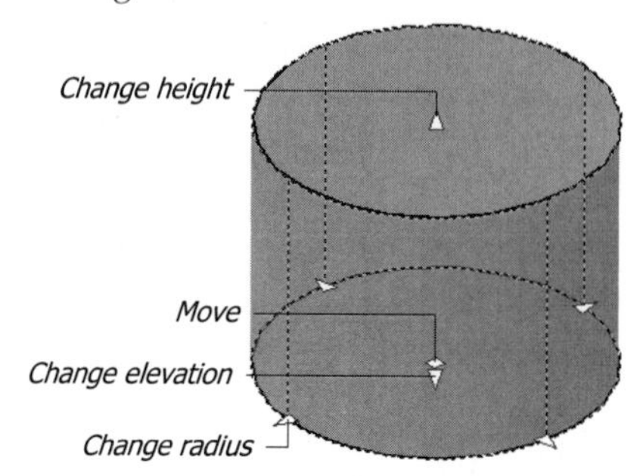

- The Axis endpoint option rotates the cylinder.

Removed Commands

Dashboard and **DashboardClose** commands are aliases for Ribbon and RibbonClose as of AutoCAD 2009.

DataExtraction

2008 Runs a wizard that extracts data from the drawing, and then places them in a table or in an external file; replaces functions of the EAttExt command.

Command	*Aliases*	*Keyboard Shortcuts*	*Menu*	*Ribbon*
dataextraction	**dx**	...	**Tools**	**Annotate**
	eattext		**↳Data Extraction**	**↳Tables**
				↳Extract Data
-dataextraction	...	...	...	...

Command: dataextraction

Displays wizard.

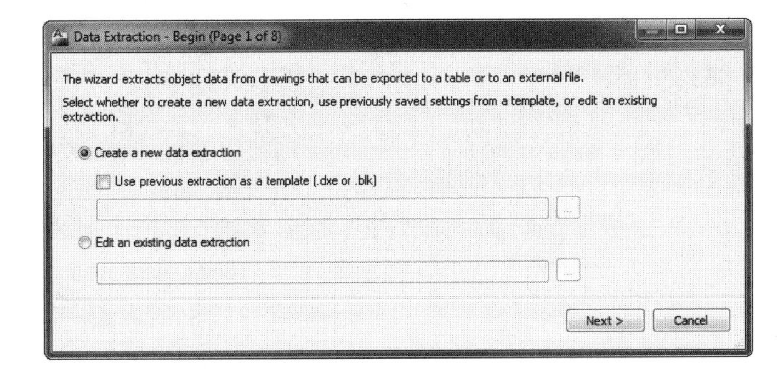

WIZARD OPTIONS

1. Begin page

⊙**Create a New Data Extraction** creates new templates for specifying which data to extract and how:

☐**Use Previous Extraction as a Template** reuses existing *.dxe* and *.blk* template files.

[...] **Open Template** opens dialog box for selecting data extraction template files (*.dxe* or *.blk*).

○**Edit an Existing Data Extraction** modifies an earlier template.

Next displays the Save Data Extraction As dialog box; enter a file name, and then click Save.

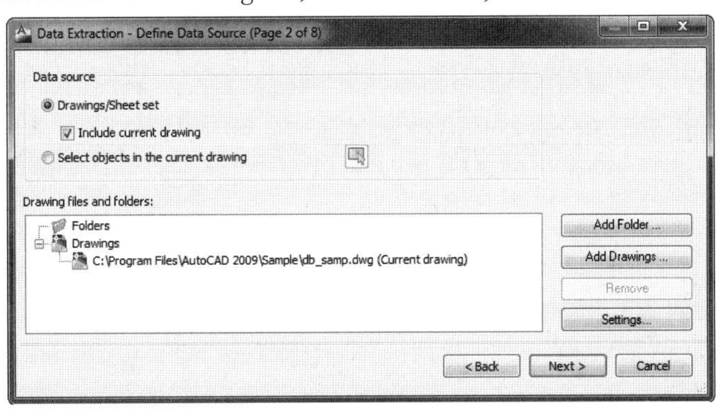

2. Define Data Source page

⊙**Drawings/Sheet Set** prompts you to select any drawing and folder for data extraction:

☑ **Include Current Drawing** includes the current drawing in the data extraction.

☐ Turn off to ignore data in the current drawing.

○ **Select Objects in the Current Drawing** selects only the current drawing for data extraction.

Select Objects selects specific objects in the current drawing.

Drawing Files and Folders options

Add Folder displays the Add Folder Options dialog box for selecting entire folders of drawing files to include in data extraction (detailed later).

Add Drawings displays the Select Files dialog box for selecting specific drawings files.

Remove removes the checked drawings or folders in the Drawing Files and Folders list from the data extraction.

Settings displays the Data Extraction - Additional Settings dialog box (detailed later).

3. Select Objects page

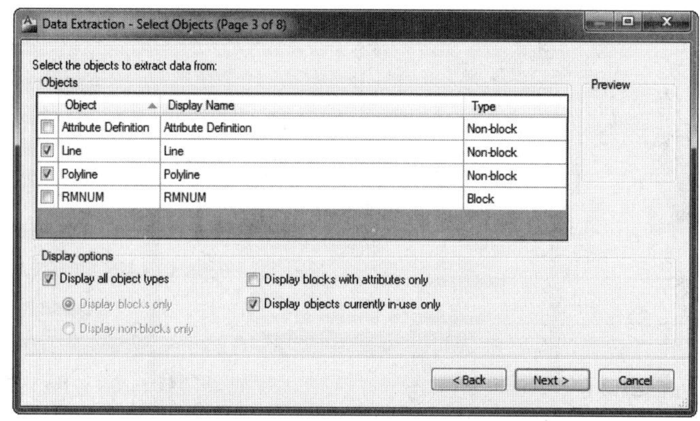

Objects options

Click a header to change the sort order.

Display Options options

☑ **Display All Object Types** lists all objects, both blocks and non-blocks.

⊙ **Display Blocks Only** lists only blocks and dynamic blocks; includes xrefs.

○ **Display Non-Blocks Only** lists all objects, except for blocks and xrefs.

☑ **Display Blocks with Attributes Only** lists names of blocks with attributes and dynamic blocks with actions and parameters.

☑ **Display Only Objects Used in Drawings** lists employed objects.

Shortcut menu

Check All selects all listed items.

Uncheck All deselects all listed items.

Invert Selection checks unchecked items and vice versa.

Edit Display Name allows you to change names in the Display Name column.

Check All
Uncheck All
Invert Selection
Edit Display Name

4. Select Properties page

Properties options

Click a header to change the sort order.

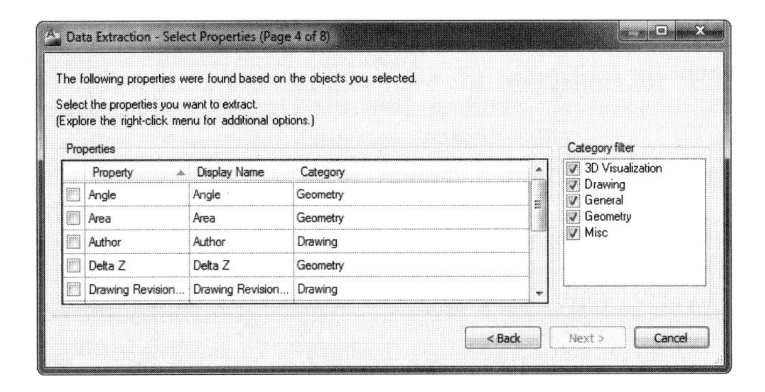

Category Filter options

☑Chooses subsets of object and drawing properties.

5. Refine Data page

Click a header to change the sort order.

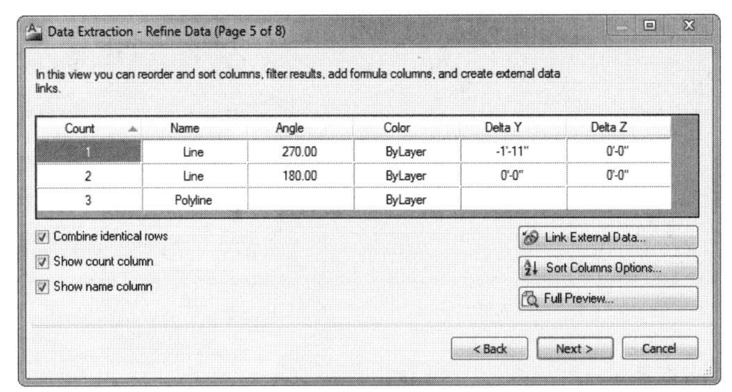

☑**Combine Identical Rows** combines rows with identical names.

☑**Show Count Column** toggles display of the Count column.

☑**Show Name Column** toggles the display of the Name column.

Link External Data displays the Link External Data dialog box; see DataLink command.

Sort Columns Options displays the Sort Columns dialog box for sorting data by more than one column (detailed later); click headers to sort by one column.

Full Preview displays a window showing the final output.

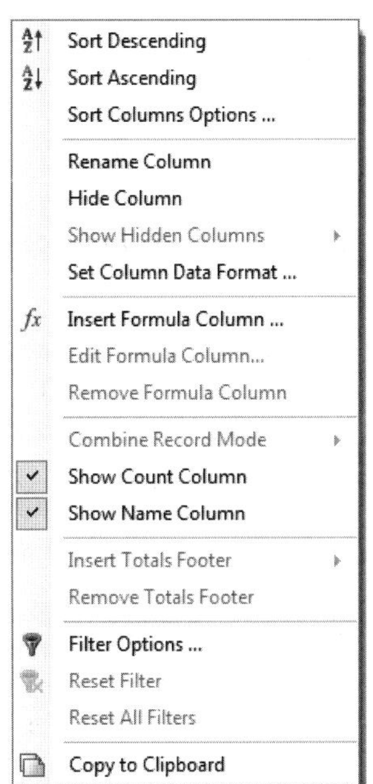

Right-click any row for this shortcut menu with many options:

Sort Descending sorts data in selected column by descending order.

Sort Ascending sorts data in selected column by ascending order.

Sort Column Options displays the Sort Columns dialog box (detailed later).

Rename Column changes the selected column's name.

Hide Column hides the selected column.

Show Hidden Columns shows a submenu of hidden column names.

Set Column Data Format displays the Set Cell Format dialog box (detailed later).

Insert Formula Column displays the Insert Formula Column dialog box (detailed later), and then inserts the formula in a new column to the right of the selected one.

Edit Formula Column displays the Edit Formula Column dialog box.

Remove Formula Column removes the selected formula column.

Combine Record Mode shows numeric data as separate values, or else combines identical numeric property rows into single rows.

Show Count Column toggles display of the Count column.

Show Name Column toggles display of the Name column.

Insert Totals Footer chooses from sum, max, min, or average for columns with numeric data; adds a row at the end of the table, inserting the result of the arithmetic in the cell.

Remove Totals Footer removes the Totals footer.

Filter Options displays the Filter Column dialog box (detailed later).

Reset Filter removes the filter of the selected column.

Reset All Filters resets filters of all columns.

Copy to Clipboard copies the whole enchilada to the Clipboard; you can paste the data into other applications, such as word processors and spreadsheets.

6. Choose Output page

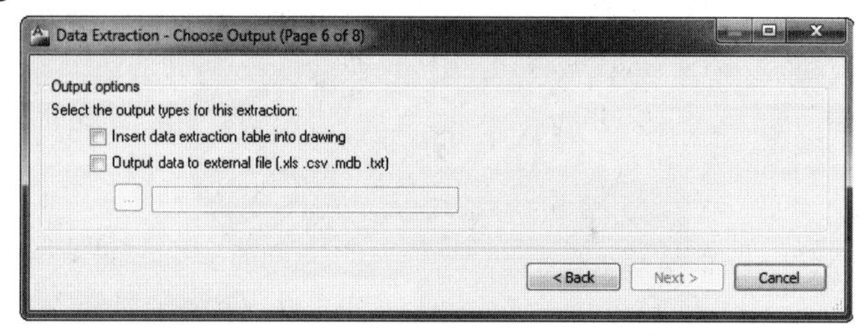

Output Options options

☐ **Insert Data Extraction Table into Drawing** inserts a table with the extracted data, at the end of this wizard automatically.

☐ **Output Data to External File** creates a data file in comma-separated (*.csv*), tab-separated (*.txt*), Access database (*.mdb*), or Excel spreadsheet (*.xls*) file format.

7. Table Style page

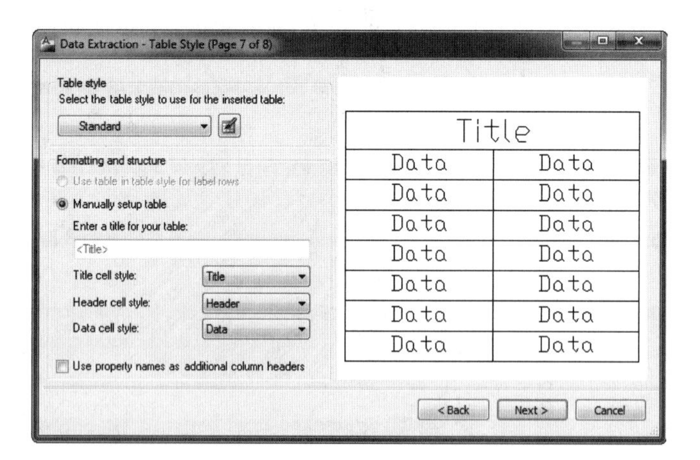

Table Style options

Table Style selects a named table style.

Formatting and Structure options

○ **Use Table in Table Style for Label Rows** (available when selected table style contains a template table) inserts a table with label cells in the top row and footer cells at the bottom.

⦿ **Manually Setup Table** enters a title and specifies the title, header, and data cell style.

Enter a Title for Your Table specifies the title for the table; this row is not overwritten when the table is updated.

Title Cell Style selects a cell style for the title row.

Header Cell Style selects a cell style for the header row.

Data Cell Style selects a cell style for the data rows.

☑ **Use Property Names as Additional Column Headers** uses the property name columns as the header row.

8. Finish page

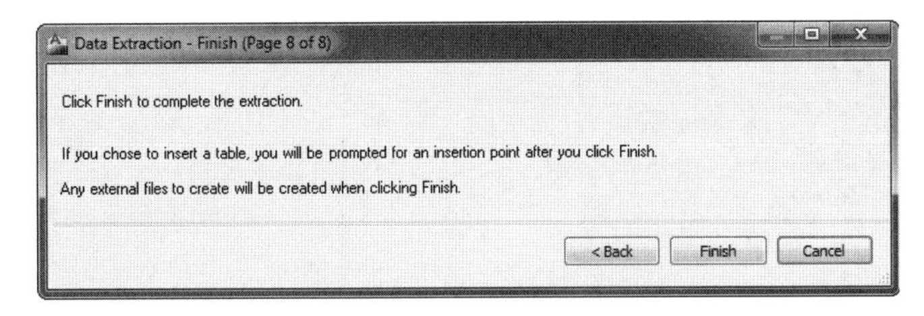

Finish inserts the table in the drawing, or exports the data to the specified file.

ADD FOLDER OPTIONS DIALOG BOX

Access this dialog box through Page 2's Add Folder button:

Folder reports the path to the folder.

Options options

☐ **Automatically Include New Drawings Added in this Folder to the Data Extraction** tracks when drawings are added to this folder, and includes them in the next data extraction process.

☑ **Include Subfolders** includes drawing files in subfolders.

☐ **Utilize a Wild-card Character to Select Drawings** reduces the number of drawings through the * (any number of characters) and ? (any single character) wildcards.

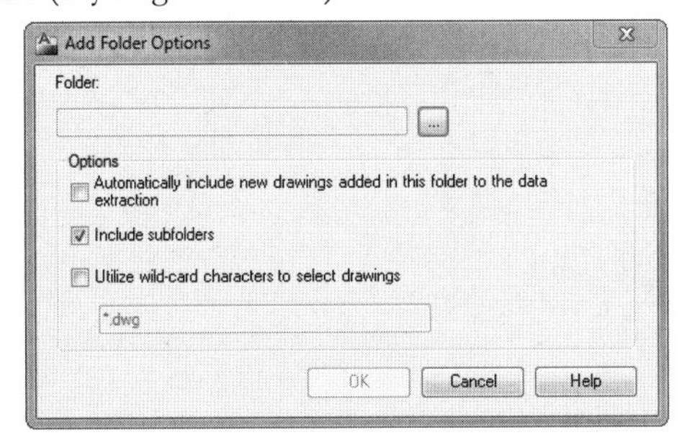

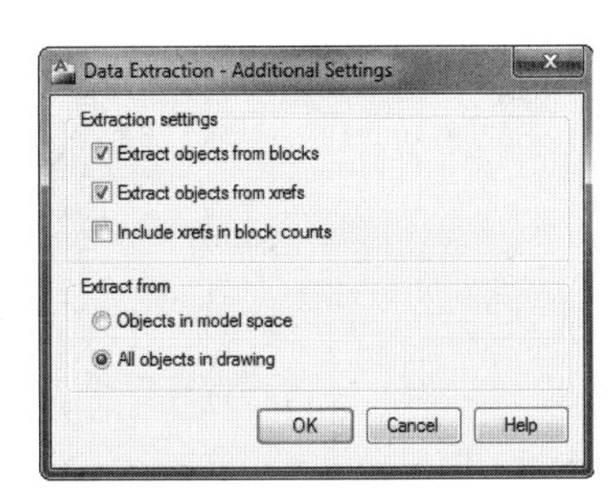

DATA EXTRACTION - ADDITIONAL SETTINGS DIALOG BOX

Access this dialog box through Page 2's Settings button:

Extraction Settings options

☑ **Extract Objects from Blocks** extracts data from objects inside of blocks.

☑ **Extract Objects from Xrefs** extracts data from objects and blocks in xrefs.

☑ **Include Xrefs in Block Counts** counts xrefs as blocks.

Extract From options

○ **Objects in Model Space** counts only objects in model space.

⊙ **All Objects in Drawing** counts objects in model and paper space.

SORT COLUMNS DIALOG BOX

Access this dialog box through Page 5's Sort Columns Options button:

Add adds more column names to the Column list.

Remove removes the selected column name.

Move Up/Down moves the selected column up (or down) the list; higher columns have sort preference over lower columns.

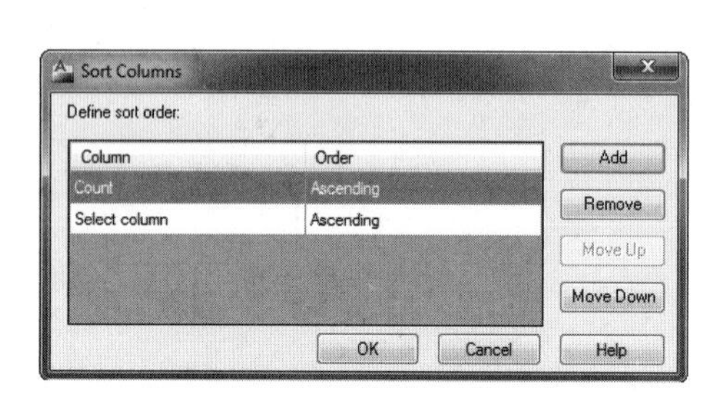

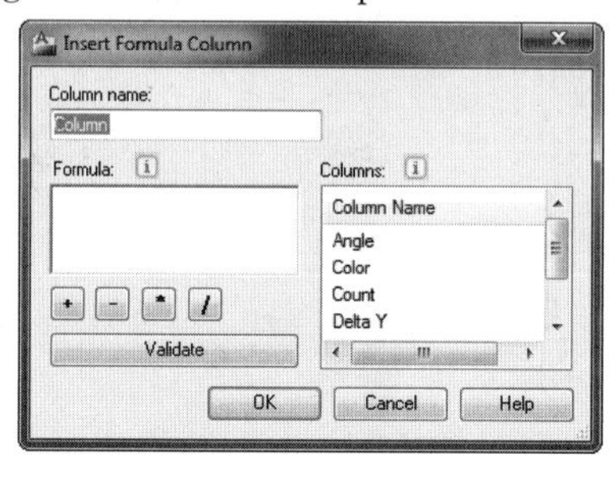

INSERT / EDIT FORMULA COLUMN DIALOG BOX

Access this dialog box by right-clicking a column in the Refine Data page, and then choosing Insert Formula Column:

Column Name specifies the name of the column whose content is to be edited.

Formula, Columns

*Drag an item from Columns into Formula, and then click a function button, such as + or *.*

+, -, *, / perform addition, subtraction, multiplication, or division on column values.

Validate checks that the formula will work.

FILTER COLUMN DIALOG BOX

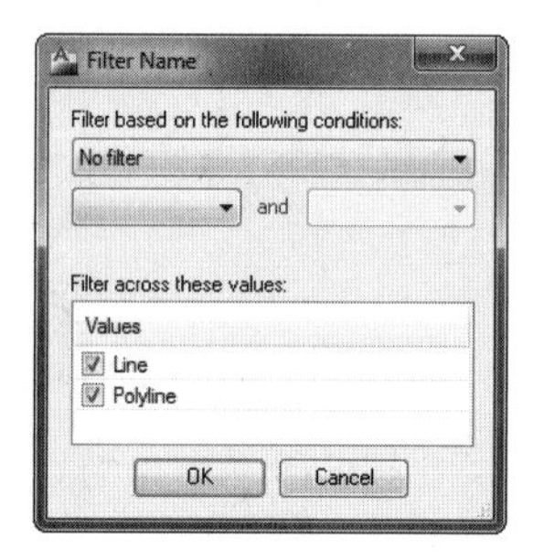

Access this dialog box by right-clicking a column in the Refine Data page, and then choosing Filter Column:

Filter Based on the Following Conditions selects amongst the following conditions:

Greater than	Greater than or equal to
Less than	Less than or equal to
Equal to	Not equal to
Between	Outside of

Filter Across These Values selects the values to filter:

☑ Include in filter.

☐ Include from filter.

-DATAEXTRACTION Command

Command: -dataextraction
Enter the template file path for the extraction type: *(Enter path and file name for the .blk attribute extraction template .dxe or data extraction file.)*
Enter the output filetype [Csv/Txt/Xls/Mdb] <Csv>: *(Enter an option.)*
Enter output filepath: *(Enter the path and file name.)*
Specify insertion point: *(Pick a point to locate the table in the drawing.)*

RELATED COMMANDS

AttDef extracts attributes only.

DataLink links tables with external data files.

dbConnect links external database files with objects.

RELATED SYSTEM VARIABLE

DxEval determines when AutoCAD should check for changes in extracted data tables.

TIPS

- This command replaces and extends the EAttExt command by outputting object properties and drawing information.
- One formula cannot be used for input to another.
- Excel and Access files are limited to a maximum of 255 columns.
- Although Autodesk's documentation states that blocks with non-uniform scale factors are not extracted, they are.
- AutoCAD warns you before inserting tables with more than 2,000 rows.

DataLink / DataLinkUpdate

2008 Links external spreadsheet files to tables in drawings.

Command	Aliases	Keyboard Shortcuts	Menu	Ribbon
datalink	**dl**	**...**	**Tools** **↳Data Links** **↳Data Link Manager**	**Insert** **↳Linking & Extracting** **↳Data Link**
datalinkupdate	**dlu**	**...**	**Tools** **↳Data Links** **↳Update Data Links**	**Annotate** **↳Tables** **↳Download from Source**

Command: datalink

Displays dialog box.

DATA LINK MANAGER DIALOG BOX

Links lists the links between the drawing and external spreadsheet files, if any.

Create a New A Excel Data Link displays the Enter Data Link Name dialog box, detailed later.

☑ **Preview** displays thumbnail images of spreadsheet cells.

ENTER DATA LINK NAME DIALOG BOX

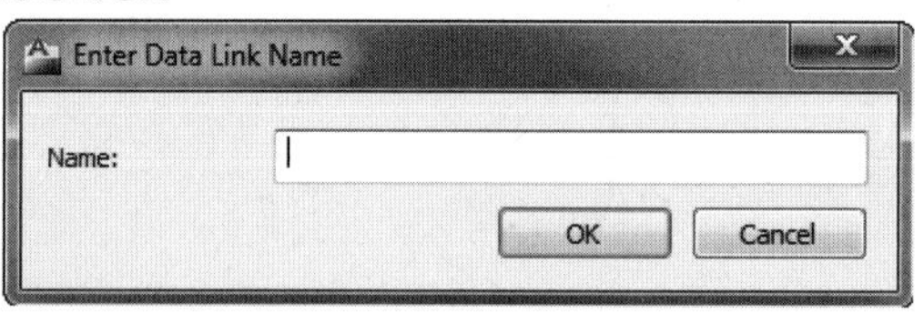

Name names the new data link.

OK displays the New Excel Data Link dialog box for selecting the spreadsheet file (illustrated below).

NEW EXCEL DATA LINK DIALOG BOX

File options

Browse For a File selects a previously entered file name.

... selects a spreadsheet *.xls* or *.xls*, or comma-delimited *.csv* file; the files don't need to be produced by Excel, despite the exclusive reference to Excel in this command.

Path Type specifies full, partial, or no path in an xref-like manner.

The Link Options and Preview items appear after the file is selected; see figure below.

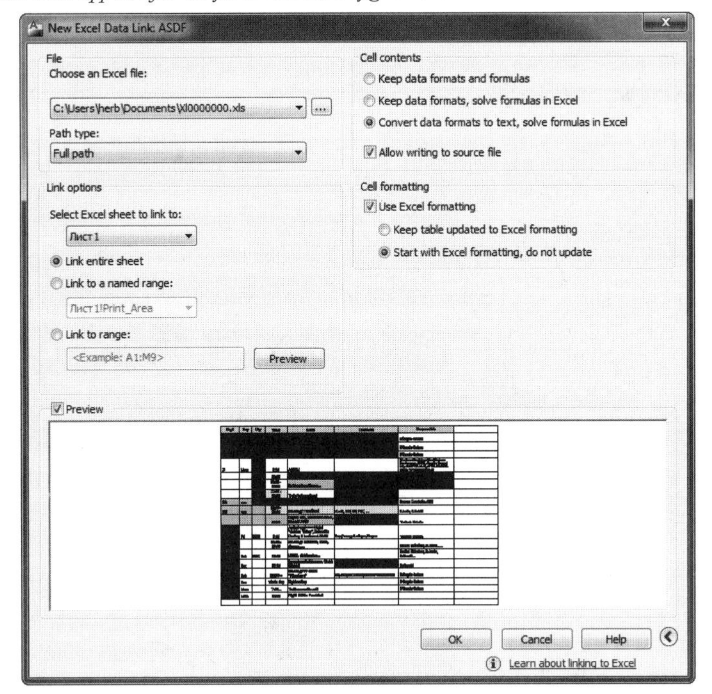

OK updates the dialog box to show the spreadsheet data:

Select the Excel Sheet to Link To shows the path and file name of the data file.

⊙ **Link Whole Sheet** uses all data in the file (for eventual linking with a table in the drawing).

○ **Link To a Named Range** uses a named range (subset) of data in the file.

○ **Link to Range** uses a range (subset) of data in the file, such as:

A3:E15	Part of the spreadsheet, from cell A3 to E15.
A:E	Entire columns, from A through E.
3:15	Entire rows, from 3 to 15.

☑ **Preview** updates the thumbnail image.

expands the dialog box to show more options:

☑ **Convert Data Types to Text** converts formulas to results, and displays them as text.

☐ **Retain Formulas** keeps formulas as formulas.

☑ **Allow Writing to Source File** permits the DataLinkUpdate command to update the external data file when the content of the table is changed in AutoCAD.

☑ **Use Excel Formatting** retains the formatting in the original data file, if any.

○ **Keep Table Updated to Excel Formatting** changes the formatting to reflect changes in the data file's formatting, after the DataLinkUpdate command has done its work.

⊙ **Start With Excel Formatting, Do Not Update** keeps the original formatting.

OK returns to the original Data Link Manager dialog box.

Click OK to exit the command. he data link has been created; it can now be used by the Table, DataExtraction, and -DataExtraction commands.

DATALINKUPDATE Command

Command: datalinkupdate
Select an option [Update data link/Write data link] <Update data link>: *(Enter an option.)*

COMMAND LINE OPTIONS

Update updates tables in AutoCAD, if the external sources have changed.

Write updates the external sources when the tables have changed.

TIPS

- There's a faster way to link tables in AutoCAD with external data files than using the DataLink and Table commands:

    1. In the spreadsheet program, copy a range of cells to the Clipboard.
    2. In AutoCAD, use the PasteSpec command to paste the cell as "AutoCAD Entities." Ensure that the Paste Link option is turned on.

- There is no need to use Excel; these commands work with the plain CSV (comma-separated value) text files output by many programs, as well as with the free spreadsheet software from OpenOffice (www.openoffice.org).
- When a drawing contains links to external data files, the following icon appears in the tray. (The same icon appears next to the cursor when it hovers over a linked table.)

- Tables linked to external files are locked by default, and cannot be edited. Use the table editing toolbar to unlock tables.
- When AutoCAD notices that the data in the tables are different from those in the external data files, it displays this warning balloon in the tray:

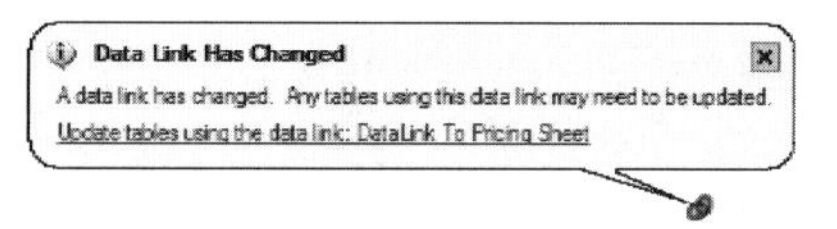

- "XLSX" refers to XML-format files output by Excel 2007.
- When linking to a password-protected data file, you may need to enter the password repeatedly.
- Data linking works both ways: edit values in the table in AutoCAD, and the spreadsheet updates, and vice versa.
- Caution: When you add rows or columns to the spreadsheet, AutoCAD does not do the same to the table. Instead, it deletes the existing table and generates a new one based on the changed spreadsheet file. The technical editor suggests this workaround: create two tables in AutoCAD. One reads and writes from the first's; the other is linked to the spreadsheet.

Dc...

2011 A group of commands that constrains objects dimensionally; these are direct versions of the DimConstraint command.

Command	Aliases	Keyboard Shortcuts	Menu	Ribbon
varies	...	...	...	**Parametric ↳Dimensional**

DCALIGNED Command

Constrains the distance between two points on objects; equivalent to the DimConstraint command's Aligned option.

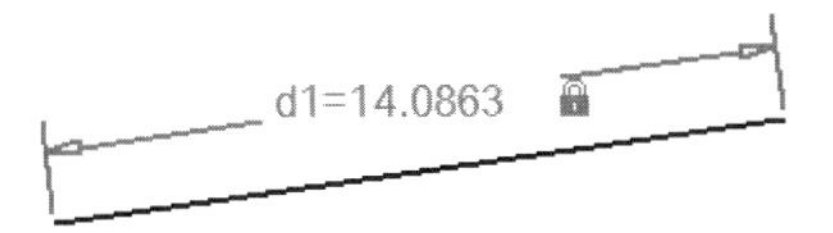

Command: dcaligned
Specify first constraint point or [Object/Point & line/2Lines] <Object>: *(Pick a point, or enter an option.)*
Specify second constraint point: *(Pick a point.)*
Specify dimension line location: *(Pick a point.)*
Dimension text = 16.5

COMMAND LINE OPTIONS

Specify first constraint point locates one end of the dimensional constraint.

Specify second constraint point locates the other end.

Specify dimension line location locates the dimension line.

Object applies the aligned constraint to an object: lines, polyline segments, arcs, and two points on an object.

Point & line applies the constraint between a point and a point on a line closest to the first point.

2Lines applies the constraint between two lines, making them parallel.

DCANGULAR Command

Constrains the angle between objects; equivalent to the DimConstraint command's ANgular option.

Command: dcangular
Select first line or arc or [3Point] <3Point>: *(Pick a point, or type 3.)*
Select second line: *(Select a point.)*

Specify dimension line location: *(Pick a point.)*
Dimension text = 90

COMMAND LINE OPTIONS

Select first line or arc selects a line, polyline segment, or an arc.

Select second line selects the second object.

3point selects three points in the drawing.

DCCONVERT Command

Converts associative dimensions to dimensional constraints; equivalent to the DimConstraint command's Convert option.

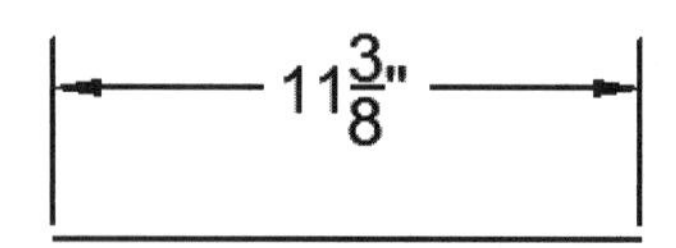

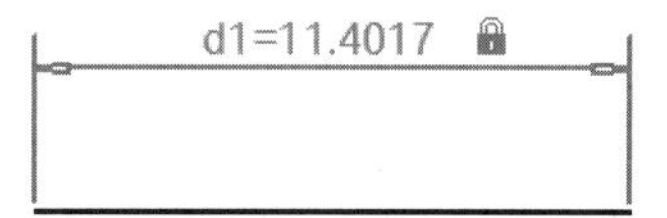

Left: *An associative dimension...*
Right: *...converted to a dimensional constraint.*

Command: dcconvert
Select associative dimensions to convert: *(Choose one or more dimensions.)*
Select associative dimensions to convert: *(Press Enter.)*
3 associative dimensions converted

COMMAND LINE OPTION

Select associative dimensions to convert chooses one or more associative dimensions to convert to dimensional constraints.

DCDIAMETER Command

Constrains the diameters of circular objects; equivalent to DimConstraint command's Diameter option.

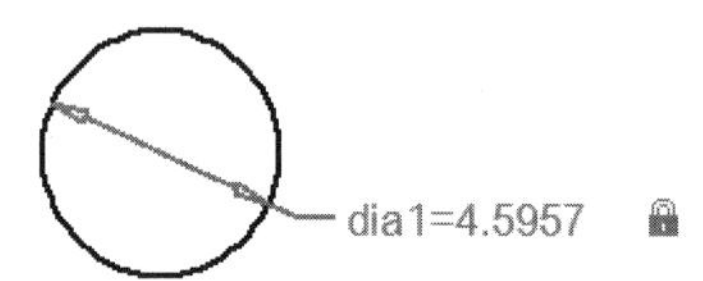

Command: dcdiameter
Select arc or circle: *(Choose a circle or an arc.)*
Dimension text = 6.5
Specify dimension line location: *(Pick a point.)*

COMMAND LINE OPTION

Select arc or circle chooses the arc, polyline arc, or circle to constrain.

DCFORM Command

Toggles the creation of dimensional constraints between dynamic and annotational; equivalent to the DimConstraint command's Form option.

Left: *Dynamic dimensional constraint.*
Right: *Annotational dimensional constraint.*

Command: dcform
Enter constraint form [Annotational/Dynamic] <Dynamic>: *(Type A or D.)*
Current settings: Constraint form = Dynamic
Enter a dimensional constraint option
[Linear/Horizontal/Vertical/Aligned/ANgular/Radius/Diameter/Form/Convert]
<Convert>: *(Carries on with the prompts from the DimConstraint command.)*

COMMAND LINE OPTION

Enter constraint form specifies the type of constraint, annotational or dynamic; see the DimConstraint command for differences between the two.

DCHORIZONTAL Command

Constrains the horizontal distance between objects; equivalent to the DimConstraint command's Horizontal option.

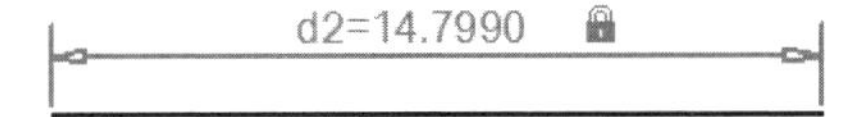

Command: dchorizontal
Specify first constraint point or [Object] <Object>: *(Pick a point, or press Enter to choose an object.)*
Select object: *(Select an object.)*
Specify dimension line location: *(Pick a point.)*
Dimension text = 16.5

COMMAND LINE OPTIONS

See DcAligned command.

DCLINEAR Command

Constrains objects horizontally, vertically, or rotationally; equivalent to the DimConstraint command's LInear option.

Command: dclinear
Specify first constraint point or [Object] <Object>: *(Enter an option, or press Enter to select an object.)*
Select object: *(Select an object.)*
Specify dimension line location: *(Pick a point.)*
Dimension text = 9.5

COMMAND LINE OPTIONS

See DcAligned command.

 **DCRADIUS Command**

Constrains the radii of curved objects; equivalent to the DimConstraint command's Radius option.

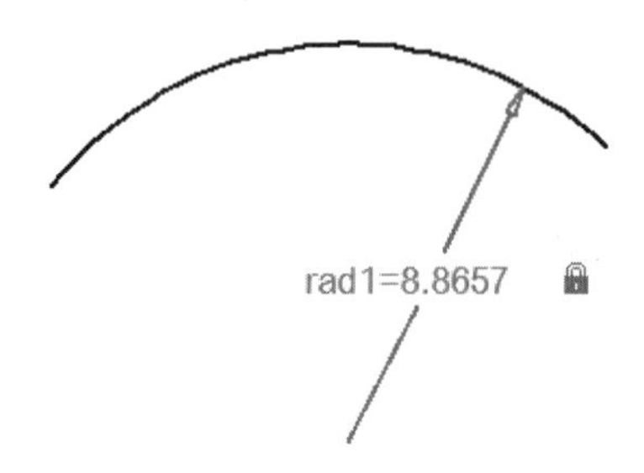

Command: dcradius
Select arc or circle: *(Choose a circle or arc.)*
Dimension text = 3.5
Specify dimension line location: *(Pick a point.)*

COMMAND LINE OPTIONS

See DcDiameter command.

 **DCVERTICAL Command**

Constrains the vertical distance between two objects or two points on an object; equivalent to the DimConstraint command's Vertical option.

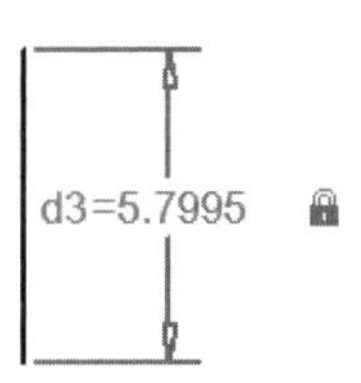

Command: dcvertical
Specify first constraint point or [Object] <Object>: *(Enter an option, or press Enter to select an object.)*
Select object: *(Select an object.)*
Specify dimension line location: *(Pick a point.)*
Dimension text = 9.5

COMMAND LINE OPTIONS

See DcAligned command.

RELATED COMMANDS

DimConstraint performs all the tasks of the above commands inside of a single command.

DelConstraint erases dimensional constraints.

BCParamaeters applies dimensional constraints with grips in the Block Editor.

ConstraintSettings specifies defaults for dimensional constraints.

Parameters creates links between dimensional constraints.

GeomConstraint applies geometric constraints to objects and geometry.

RELATED SYSTEM VARIABLES

DimConstraintDisplay toggles the display of dimensional constraints.

DimConstructionIcon toggles the display of the padlock icon.

DynConstraintMode toggles the display of hidden dimensional constraints during selection.

ConstraintNameFormat specifies the text format of dimensional constraints.

ConstraintRelax toggles the enforcement of constraints during editing.

ConstraintSolveMode toggles constraint behavior during editing.

TIPS

- See the DimConstraint command for details, tips, and tutorials.
- As of AutoCAD 2011, these commands replace the DimConstraint command on the ribbon.

DcDisplay

2011 Toggles the display of selected dynamic dimensional constraints.

Command	*Aliases*	*Keyboard Shortcuts*	*Menu*	*Ribbon*
dcdisplay	**...**	**...**	**Parametric** **↳Dynamic Dimensions** **↳Show/Hide**	**Parametric** **↳Dimensional** **↳Show/Hide**

Command: dcdisplay
Select objects: *(Select one or more dimensional constraints, or type S or H.)*
Select objects: *(Press Enter.)*
Enter an option [Show/Hide]<Show>: *(Type S or H.)*

d1=6.7608

Left: *Dimensional constraint shown...*
Right: *...and hidden.*

COMMAND LINE OPTIONS

Select objects selects the objects whose dimensional constraints will be shown or hidden.

Show displays the dimensional constraints associated with the selected objects.

Hide hides the constraints.

RELATED COMMANDS

DimConstraint adds dimensional constraints to objects.

HideObjects hides selected objects.

ConstraintBar toggles the display of geometric constraints.

RELATED SYSTEM VARIABLE

DynConstraintDisplay toggles the display of all dimensional constraints in the current drawing.

TIPS

- Use this command to declutter drawings of dimensional constraints.
- This command applies to dynamic constraints only; it has no effect on annotational constraints.
- Even when dimensional constraints are hidden by this command, they reappear when you selected the associated objects during other commands.
- This command has an undocumented option that is used by the ribbon to show and hide all dimensional constraints at once. At the 'Select objects' prompt, enter **s** (short for Showall) or **h** (short for Hideall), as follows:

  Command: dcdisplay
  Select objects: s
  Command:

- In AutoCAD 2010, this function was handled by the DynConstraintDisplay, which toggled the display of all dimensional constraints; in contrast, this command toggles the display of individual constraints.

DbConnect / DbClose

2000 Open and close the dbConnect Manager palette to connect objects with rows in external database tables (short for Data Base CONNECTion; undocumented commands).

Command	*Aliases*	*Keyboard Shortcuts*	*Menu*	*Ribbon*
'dbconnect	**dbc**	**Ctrl+6**	**Tools ↳Palettes ↳dbConnect**	...
dbcclose	...	**Ctrl+6**	**Tools ↳Palettes ↳dbConnect**	...

Command: dbconnect

Displays palette. (If a red x appears, the database is disconnected from the drawing.)

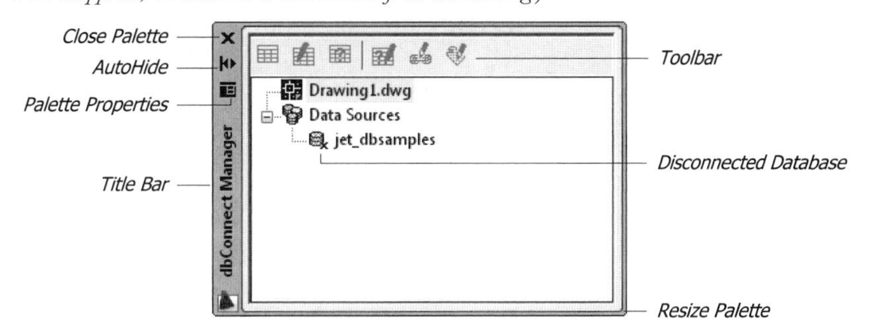

To connect the drawing with the database, right-click the database icon, and then select Connect.

The icons have the following meaning:

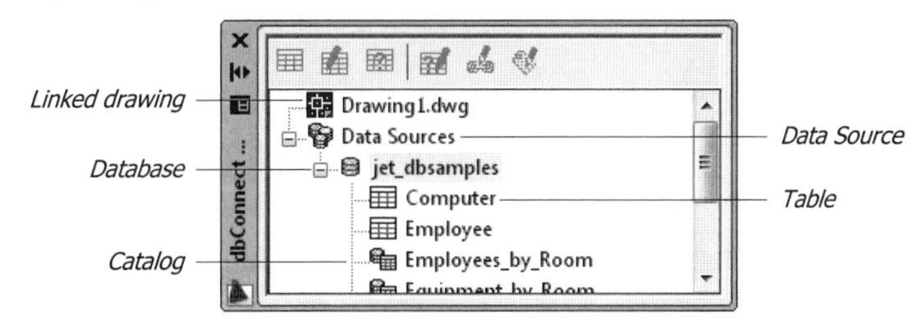

dbConnect Manager Toolbar

View Table opens an external database table in *read-only* mode; select a table, link template, or label template to make this button available.

Edit Table opens an external database table in *edit* mode; select a table, link template, or label template to make this button available.

Execute Query executes a query; select a previously-defined query to make this button available.

New Query displays the New Query dialog box when a table or link template is selected; displays the Query Editor when a query is selected.

New Link Template displays the New Link Template dialog box when a table is selected; displays the Link Template dialog box when a link template is selected; not available for link templates with links already defined in a drawing.

New Label Template displays the New Label Template dialog box when a table or link template is selected; displays the Label Template dialog box when a label template is selected.

VIEW TABLE AND EDIT TABLE WINDOWS

The View Table and Edit Table windows are identical, with the exception that View Table is read-only; hence all text in columns is grayed-out.

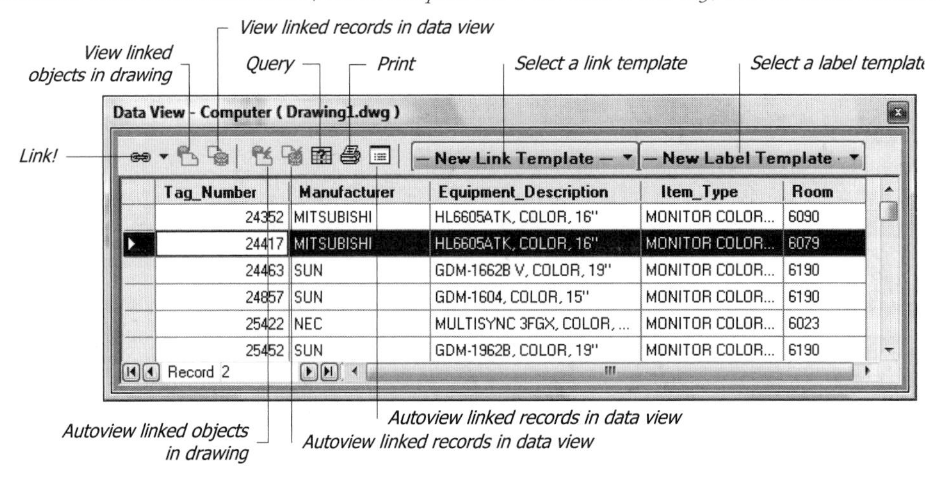

Link! creates a link or a label; click the droplist to select an option:

- **Create Links** turns on link creation mode.
- **Create Attached Labels** turns on the attached label creation mode.
- **Create Freestanding Labels** turns on freestanding label creation mode.

View Link Objects in Drawing highlights objects linked to selected records.

View Linked Records in Data View highlights records linked to selected objects in the drawing.

AutoView Linked Objects in Drawing highlights automatically objects that are linked to selected records.

AutoView Linked Records in Data View highlights records linked to selected objects in drawing.

Query displays the New Query dialog box.

Print Data View prints the data in this window.

Data View and Query Options displays the Data View and Query Options dialog box.

Select a Link Template lists the names of previously-defined link templates.

Select a Label Template lists the names of previously-defined label templates.

NEW QUERY DIALOG BOX

Access this dialog box by clicking the New Query button:

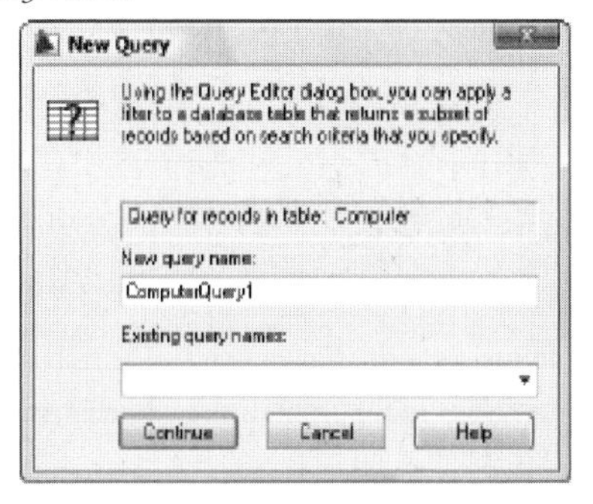

New Query Name names the query.

Existing Query Names uses an existing query.

Continue displays the Query Editor dialog box, illustrated below.

QUERY EDITOR DIALOG BOX

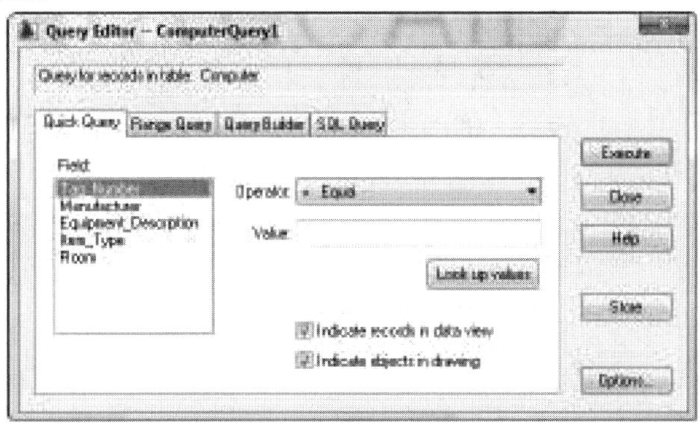

Field selects a field name.

Value specifies the value for which to search.

Execute executes the query.

Close closes the dialog box.

Store saves the settings.

Options displays the Data View and Query Options dialog box.

Range Query tab

Field lists the names of fields in the current table.

Value specifies the value tested by the operator.

Look Up Values displays the Column Values dialog box.

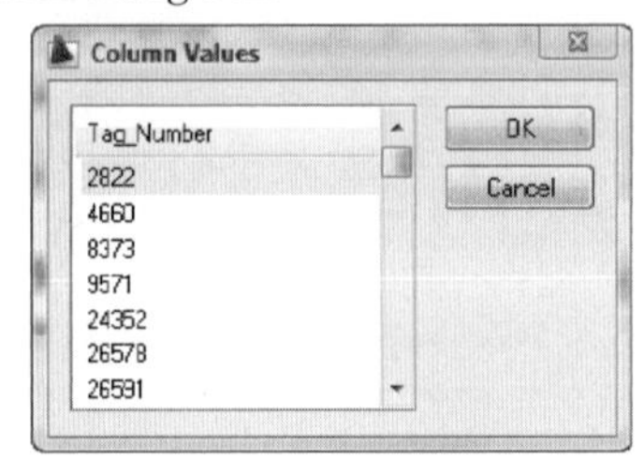

☑ **Indicate Records in Data View** highlights records that match the search criteria in the Data View window.

☑ **Indicate Objects in Drawing** highlights objects that match the search criteria in drawings.

Operator specifies a conditional operator:

Operator	Meaning
=	Equal — Exactly match the value (default).
<>	Not equal — Does not match the value.
>	Greater than — Greater than the value.
<	Less than — Less than the value.
>=	Greater than or equal — Greater than or equal to the value.
<=	Less than or equal — Less than or equal to the value.
Like	Contains the value; use the **%** wild-card character (equivalent to * in DOS).
In	Matches two values separated by a comma.
Is null	Does not have a value; used for locating records that are missing data.
Is not null	Has a value; used for excluding records that are missing data.

Query Builder tab

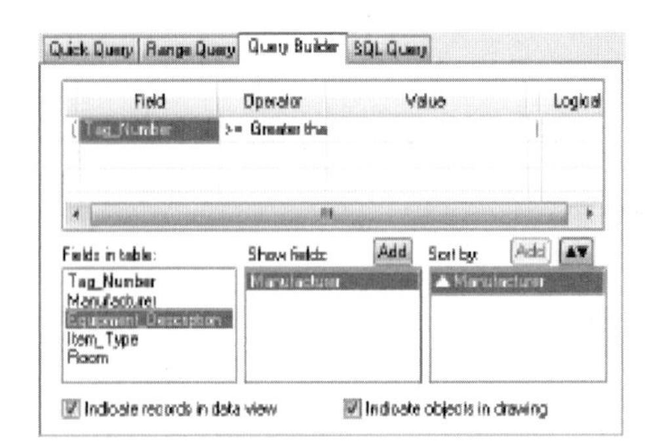

(specifies an opening parenthesis, which groups search criteria with parentheses; up to four sets can be nested.

) specifies a closing parenthesis.

Field specifies a field name; double-click the cell to display the list of fields in the current table.

Operator specifies a logical operator; double-click to display the list of operators.

Value specifies a value for the query; click **...** to display a list of current values.

Logical specifies an And or Or operator; click once to add And; click again to change to Or.

Fields in table displays the fields in the current table; when no fields are selected, the query displays all fields from the table; double-click a field to add it to the list.

Show fields specifies the fields displayed by the Data View window; drag the field out of the list to remove it.

Add adds a field from the Fields in Table list to the Show Fields list.

Sort By specifies the sort order: the first field is the primary sort; to change the sort order, drag the field to another location in the list; press DELETE to remove a field from the list.

Add adds a field from the Fields in Table list to the Sort By list (default = ascending).

⮝⮟ reverses sort order.

☑ **Indicate Records in Data View** highlights records that match the search criteria in the Data View window.

☑ **Indicate Objects in Drawing** highlights objects that match the search criteria in drawings.

SQL Query tab

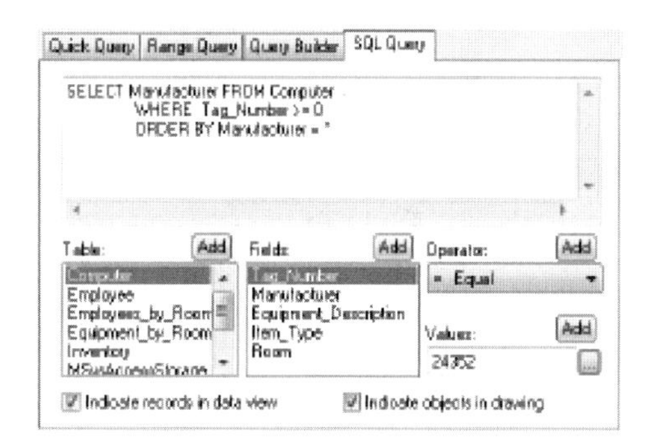

Table lists the names of all database tables available in the current data source.

Add adds the selected table to the SQL text editor.

Fields displays a list of field names in the selected database table.

Add adds the selected field to the SQL text editor.

Operator specifies the logical operator, which is added to the query (default = Equal).

Add adds the selected operator to the SQL text editor.

Values specifies a value for the selected field.

Add adds the value to the SQL text editor.

... lists available values for the field.

☑ **Indicate Records in Data View** highlights records that match the search criteria in the Data View window.

☑ **Indicate Objects in Drawing** highlights objects that match the search criteria in drawings.

DATA VIEW AND QUERY OPTIONS DIALOG BOX

Opened by the Query Editor dialog box.

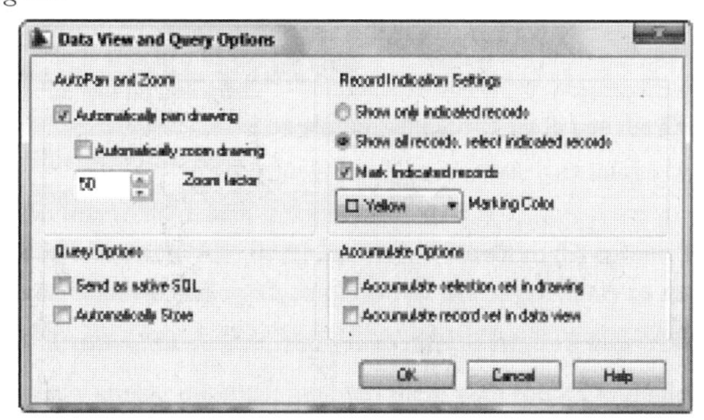

AutoPan and Zoom

☑**Automatically Pan Drawing** causes AutoCAD to pan the drawing automatically to display associated objects.

☐**Automatically Zoom Drawing** causes AutoCAD to zoom the drawing automatically to display associated objects.

Zoom Factor specifies the zoom factor as a percentage of the viewport area; range is 20 - 90; default = 50.

Query Options

Send as Native SQL makes queries to database tables:

☑ In the format of the source table.

☐ In SQL 92 format.

☐ **Automatically Store** stores queries automatically when they are executed (default = off).

Record Indication Settings:

⊙ **Show Only Indicated Records** displays the records associated with the current AutoCAD selections in the Data View window (default).

○ **Show All Records, Select Indicated Records** displays all records in the current database table.

☑ **Mark Indicated Records** colors linked records to differentiate them from unlinked records.

Marking Color specifies the marking color (default = yellow).

Accumulate Options

Accumulate Selection Set in Drawing

☑ Adds objects to the selection set as data view records are added.

☐ Replaces the selection set each time data view records are selected.

Accumulate Record Set in Data View

☑ Adds records to the selection set as drawing objects are selected.

☐ Replaces the selection set each time drawing objects are selected.

NEW LINK TEMPLATE DIALOG BOX

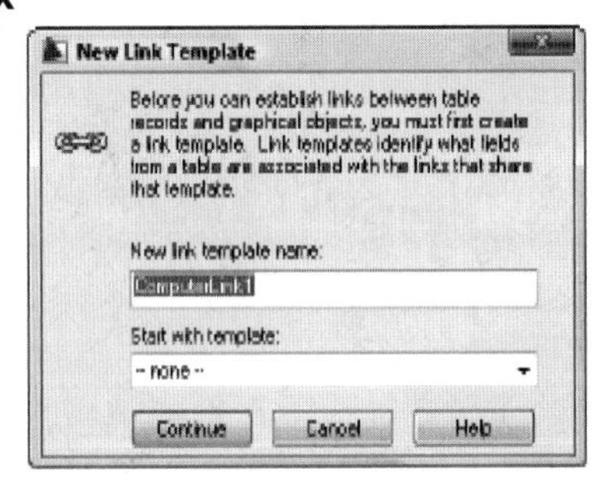

New link template name specifies the name of the link template.

Start with template reuses an existing template.

Continue displays the Link Template dialog box.

LINK TEMPLATE DIALOG BOX

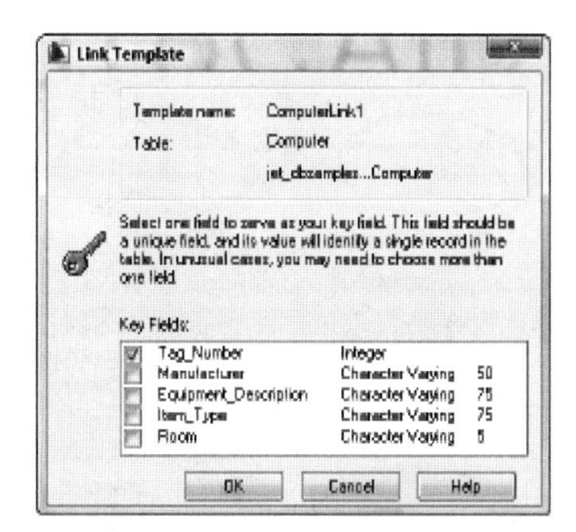

Key Fields selects one field name; you may select more than one field name, but AutoCAD warns you that too many key fields may slow performance.

NEW LABEL TEMPLATE DIALOG BOX

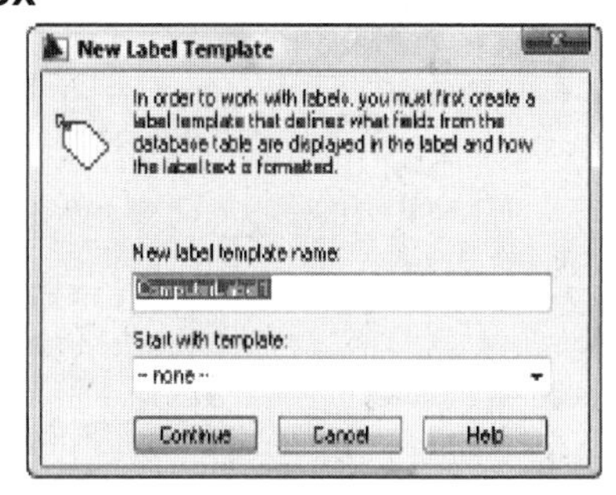

New Label Template Name names the label template.

Start with Template reuses an existing template.

Continue displays the Label Template dialog box.

LABEL TEMPLATE DIALOG BOX

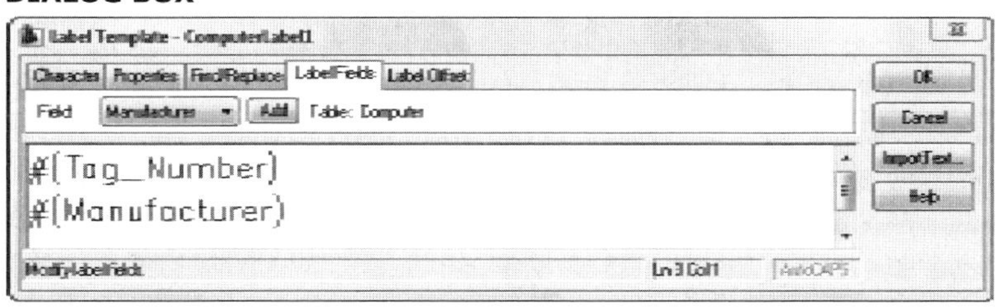

Field specifies the names of the fields.

Add adds the field to the label.

Label Offset displays the Label Offset tab.

Start specifies the justification of the label starting point.

Leader offset specifies the x, y distance between the label's starting point and the leader line.

Tip offset specifies the offset distance to the leader tip or label text.

See the MText command for more information.

RELATED COMMANDS

AttDef creates an attribute definition, akin to an internal database.

EAttExt exports attributes.

TIPS

- Query searches are case sensitive: "Computer" is not the same as "computer."
- OLE DB v2.0 must be installed before you use the dbConnect Manager.
- Leaders must have a length; to get rid of a leader, use a freestanding label.
- SQL is short for "structured query language," a standard method of querying databases.
- The properties of a link template can only be edited if it contains no links and if the drawing is fully (not partially) loaded.
- Before you can edit a record with an SQL Server table, you must define a *primary key*.
- This command cannot be used transparently (within another command) until it has been used at least once non-transparently.

Changed Command

DblClkEdit command changed to a system variable in AutoCAD 2009.

DbList

Ver. 1.0 Lists information on all objects in the drawing (short for Data Base LISTing).

Command	*Aliases*	*Keyboard Shortcuts*	*Menu*	*Ribbon*
dblist	...	...	...	...

Command: dblist

Sample listing:

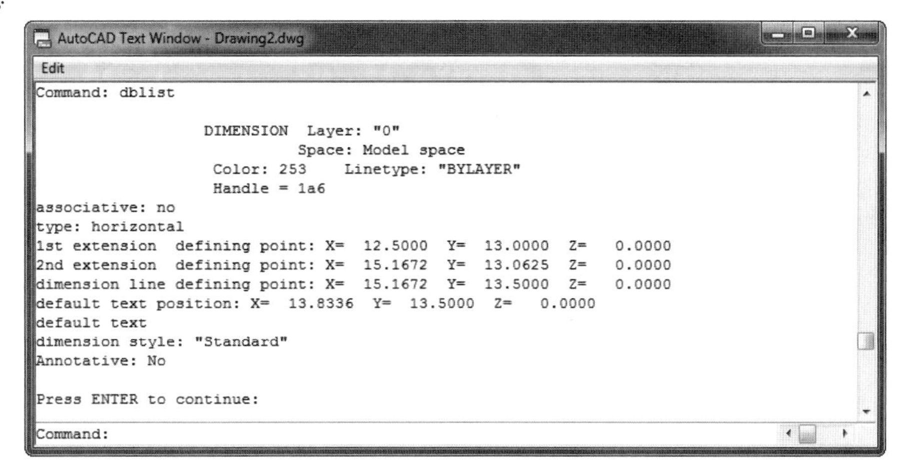

COMMAND LINE OPTIONS

ENTER continues display after pausing.

ESC cancels database listing.

RELATED COMMANDS

Area lists the area and perimeter of objects.

Dist lists the 3D distance and angle between two points.

Id lists the 3D coordinates of a point.

List lists information about selected objects in the drawing.

TIP

- This command is typically used to debug, and has little application for most users.

Renamed Commands

DdAttDef command was replaced by AttDef in AutoCAD 2000.

DdAttE command was replaced by AttEdit in AutoCAD 2000.

DdAttExt command was replaced by AttExt in AutoCAD 2000.

DdChProp command was replaced by Properties in AutoCAD 2000.

DdEdit command became an alias for the TextEdit command as of AutoCAD 2010.

DdGrips command was replaced by the Selection tab of the Options command in AutoCAD 2000.

DDim command was replaced by DimStyle in AutoCAD 2000.

DdInsert command was replaced by Insert in AutoCAD 2000.

DdModify command was replaced by Properties in AutoCAD 2000.

DdColor command was replaced by Color in AutoCAD 2000.

DdPtype

Rel. 12 Sets the style and size of points (short for Dynamic Dialog Point TYPE).

Command	*Aliases*	*Keyboard Shortcuts*	*Menu*	*Ribbon*
'ddptype	...	...	**Format ↳Point Style**	...

Command: ddptype

Displays dialog box:

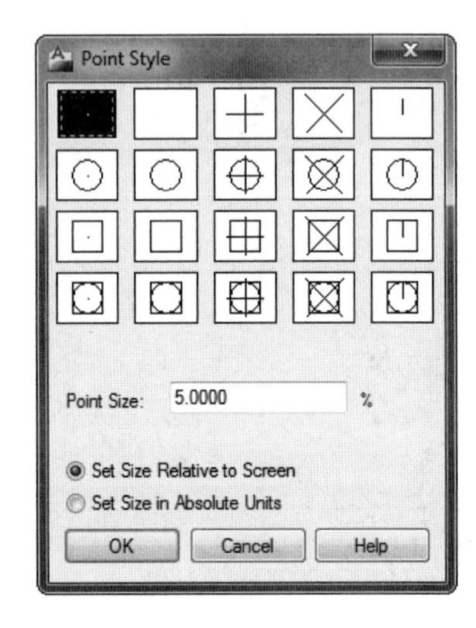

POINT STYLE DIALOG BOX

Point size sets the size in percentage or pixels.

⦿**Set Size Relative to Screen** sets the size as a percentage of the total viewport height.

○**Set Size in Absolute Units** sets the size in drawing units.

RELATED COMMANDS

Divide draws points along an object at equally-divided lengths.

Point draws points.

Measure draws points a measured distance along an object.

Regen displays the new point format with a regeneration.

RELATED SYSTEM VARIABLES

PdSize specifies the size of points:

0 — Point is 5% of viewport height (default).

positive — Absolute size in drawing units.

negative — Percentage of the viewport size.

PdMode determines the look of points.

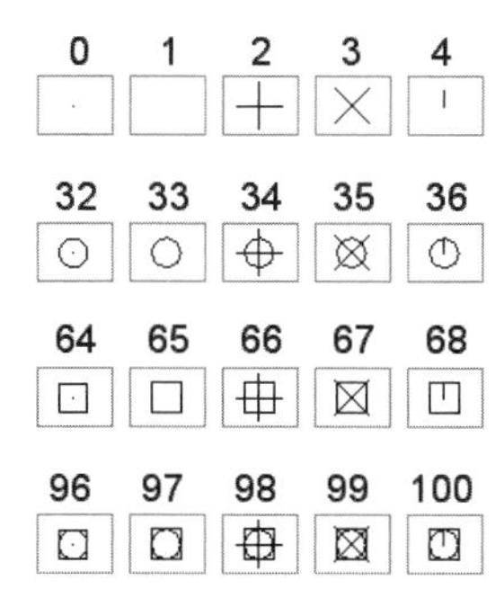

TIPS

- Points often cannot be seen in the drawing. To make them visible, change their mode and size.
- The two system variables listed above affect *all* points in the drawing.

Removed Commands

DdEModes command was removed from AutoCAD Release 14; it was replaced by the Object Properties toolbar and Object Properties panel on the Ribbon.

DdLModes command was removed from AutoCAD 2000; it was replaced by the Layer command.

DdLtype command was removed from AutoCAD 2000; it was replaced by the Linetype command.

DdRename command was removed from AutoCAD 2000; it was replaced by the Rename command.

DdRModes command was removed from AutoCAD 2000; it was replaced by the DSettings command.

DdSelect command was removed from AutoCAD 2000; it was replaced by the Selection tab of the Options command.

DdUcs and **DdUcsP** commands were removed from AutoCAD 2000; they were replaced by UcsMan.

DdUnits command was removed from AutoCAD 2000; it was replaced by the Units command.

DdView command was removed from AutoCAD 2000; it was replaced by the View command.

DdVPoint

Rel. 12 Changes the 3D viewpoint through a dialog box (short for Dynamic Dialog ViewPOINT).

Command	*Aliases*	*Keyboard Shortcuts*	*Menu*	*Ribbon*
ddvpoint	**vp**	...	**View** ↳**3D Views** ↳**Viewpoint Presets**	...

Command: ddvpoint

Displays dialog box.

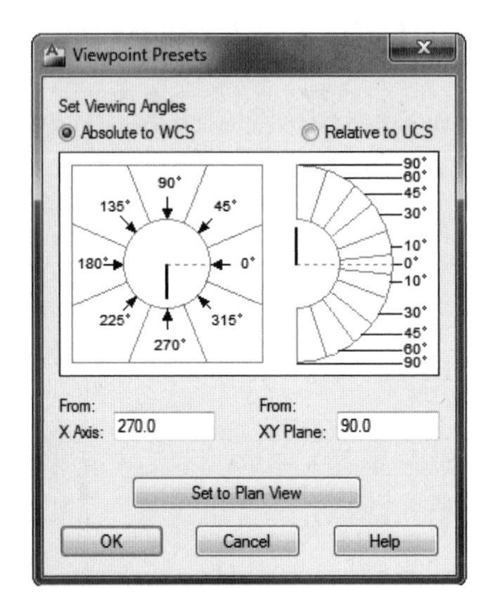

VIEWPOINT PRESETS DIALOG BOX

Set Viewing Angles options

⊙ **Absolute to WCS** sets the view direction relative to the WCS.

○ **Relative to UCS** sets the view direction relative to the current UCS.

From X Axis measures the view angle from the x axis.

From XY Plane measures the view angle from the x, y plane.

Set to Plan View changes the view to plan view in the specified UCS.

RELATED COMMANDS

NavVCube uses the viewcube to change 3D viewpoints.

VPoint adjusts the viewpoint from the command line.

3dOrbit changes the 3D viewpoint interactively.

RELATED SYSTEM VARIABLE

WorldView determines whether viewpoint coordinates are in WCS (1) or UCS (0).

TIPS

- After changing the viewpoint, AutoCAD performs an automatic zoom extents.
- In the image tile shown below, the black arm indicates the new angle.
- In the image tile, a second arm indicates the current angle.
- You can select an angle with your mouse:

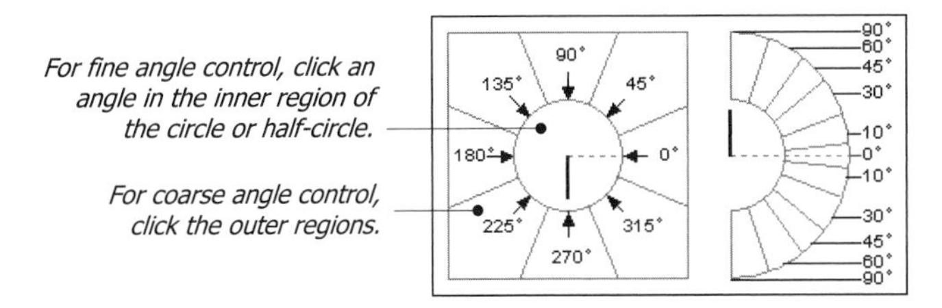

- WCS is short for "world coordinate system."
- UCS is short for "user-defined coordinate system."
- As an alternative, use the 3dOrbit command to set the 3D viewpoint.
- This command is not permitted in paper space (layout mode).

Delay

Ver. 1.0 Delays the next command, in milliseconds.

Command	*Aliases*	*Keyboard Shortcuts*	*Menu*	*Ribbon*
'delay	...	...	...	...

Command: delay
Delay time in milliseconds: *(Enter the number of milliseconds.)*

COMMAND LINE OPTION

Delay time specifies the number of milliseconds by which to delay the next command.

RELATED COMMAND

Script initiates scripts.

TIPS

- Use Delay to slow down the execution of a script file.
- The maximum delay is 32767, just over 32 seconds.

DelConstraint

2010 Erases constraints from drawings.

Command	Alias	Keyboard Shortcuts	Menu	Ribbon
delconstraint	**delcon**	...	**Parametric** ↳**Delete Constraints**	**Parametric** ↳**Manage** ↳**Delete Constraints**

Command: delconstraint
All constraints will be removed from selected objects...
Select objects: *(Choose one or more objects; enter All to remove all constraints.)*
Select objects: *(Choose more objects, or press Enter to exit the command.)*
n **constraints removed**

COMMAND LINE OPTION

Select objects chooses objects with constraints; these constraints will be erased from the drawing.

RELATED COMMAND

DimConstraint adds dimensional constraints to objects.

GeomConstraint adds geometric constraints to objects.

TIPS

- This command erases both visible and invisible constraints, and both dimensional and geometric constraints.
- Use the U or Undo command to reverse the effect of this command.

DetachURL

Rel. 14 Removes URLs from objects and areas.

Command	*Aliases*	*Keyboard Shortcuts*	*Menu*	*Ribbon*
detachurl	...	...	...	...

Command: detachurl
Select objects: *(Select one or more objects.)*
Select objects: *(Press Enter.)*

COMMAND LINE OPTION

Select objects selects the objects from which to remove URL(s).

RELATED COMMANDS

AttachUrl attaches a hyperlink to an object or an area.

Hyperlink attaches and removes hyperlinks via a dialog box.

SelectUrl selects all objects with attached hyperlinks.

TIPS

- When you select a hyperlinked area to detach, AutoCAD reports:

```
1. hyperlink ()
Remove, deleting the Area.
1 hyperlink deleted...
```

- When you select an object with no hyperlink attached, AutoCAD reports nothing.
- A URL (short for "uniform resource locator") is the universal file naming convention of the Internet; also called a link or hyperlink.

DgnAdjust

2008 Changes the look of MicroStation DGN underlays.

Command	*Aliases*	*Keyboard Shortcuts*	*Menu*	*Ribbon*
dgnadjust	...	...	...	...

Command: dgnadjust
Select DGN underlay: *(Select one or more DGN underlays.)*
Enter DGN underlay option [Fade/Contrast/Monochrome] <Fade>: *(Enter an option.)*

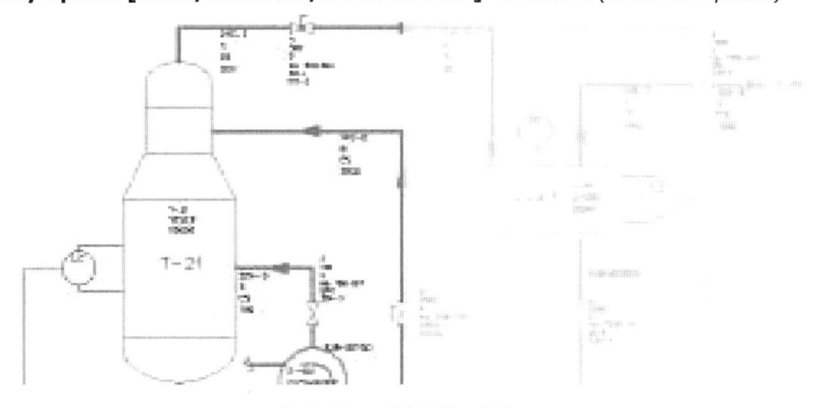

Left: *Normal DGN underlay.*
Right: *Underlay faded with DgnAdjust.*

COMMAND LINE OPTIONs

Select DGN underlay selects the underlay to adjust.

Enter fade value specifies the amount of fade:

0 — no fade.
25 — default value.
80 — maximum fade.

Enter contrast value specifies the change in contrast, ranging from 0 to 100.

Monochrome toggles monochrome colorization.

RELATED COMMANDS

Adjust adjusts fade, contrast, and monochrome of images, and DGN, DWF/x, and PDF files.

DgnAttach attaches *.dgn* files as underlays.

DgnClip clips DGN underlays.

DgnImport imports and translates *.dgn* files into AutoCAD drawing entities.

TIPS

- This command has no effect on the frame around the DGN underlay. Use the DgnFrame system variable to toggle its display.
- Use the Monochrome option to change all of the underlay's colors to black; use the Fade option to change the colors to gray.

DgnAttach

2008 Attaches MicroStation DGN files as underlays.

Command	Aliases	Keyboard Shortcuts	Menu	Ribbon
dgnattach	...	...	**Insert ↳DGN Underlay**	**Insert ↳Reference**
-dgnattach	...	...	...	...

Command: dgnattach

Displays Select DGN File dialog box.

Choose a .dgn file, and then click Open.

Displays dialog box.

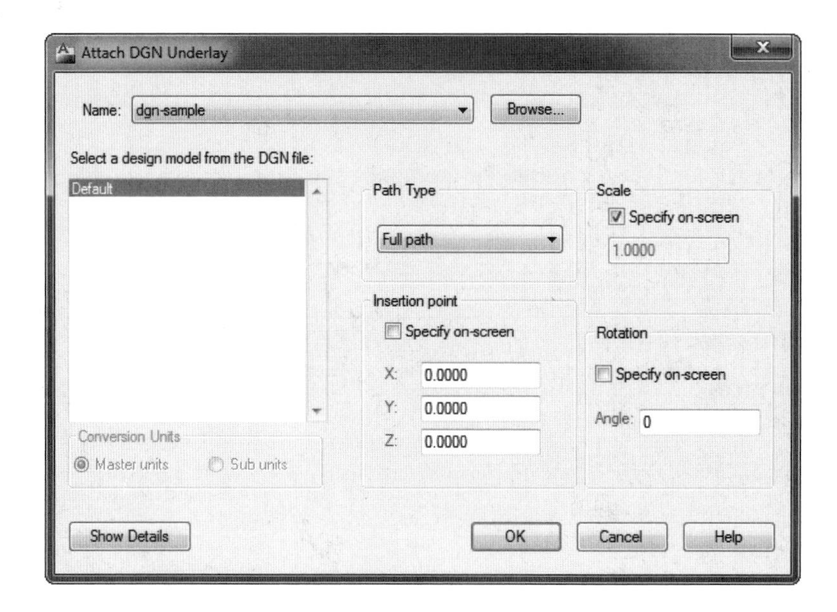

ATTACH DGN UNDERLAY DIALOG BOX

Name specifies the *.dgn* file to attach to the current drawing.

▼ Selects other *.dgn* files.

Browse displays the Select DGN File dialog box to select another *.dgn* file.

Select a design model from the DGN file lists the names of design models in the *.dgn* file; sheets are not listed.

Conversion Units specifies how AutoCAD's units should be matched:

- **Master Units** selects major units such as feet.
- **Sub Units** selects minor units such as inches.

Path Type specifies the type of path to the *.dgn* file:

- **Full Path** saves the full path name.
- **Relative Path** saves the relative path name.
- **No Path** saves the file name only.

Insertion Point specifies the *.dgn* underlay's insertion point (lower left corner); default =0,0,0.

Specify On-Screen:

☑ prompts you for the insertion point at the command line after you close the dialog box.

☐ enters the x,y,z coordinates in the dialog box.

Scale specifies the DGN underlay's scale factor; default = 1.0

Specify On-Screen:

☑ prompts you for the scale factor at the command line after you close the dialog box.

☐ enters the scale factor in the dialog box.

Angle specifies the DGN underlay's angle of rotation; default = 0 degrees.

Specify On-Screen:

☑ prompts you for the rotation angle at the command line after you close the dialog box.

☐ enters the angle in the dialog box.

-DGNATTACH Command

Command: -dgnattach
Path to DGN file to attach: *(Enter the path to folder.)*
Enter name of model or [?] <name of design model>: *(Enter the .dgn file name.)*
Specify conversion units [Master/Sub] <Master>: *(Enter an option.)*
Specify insertion point: *(Pick a point.)*
Specify scale factor or [Unit] <1.0000>: *(Enter a value, or type U.)*
Specify rotation <0>: *(Enter a value.)*

COMMAND LINE OPTIONS

Master / Sub specifies MicroStation units, such as feet for master units (MU) and inches for sub units (SU), or user-defined units, such as master = 1 and sub = 1000.

RELATED COMMANDS

Attach attaches images, DGN, DWF/x, and PDF formatted files.

DgnAdjust changes the look of attached *.dgn* files as underlays.

DgnClip clips DGN underlays.

DgnImport imports and translates *.dgn* files into AutoCAD drawing objects.

RELATED SYSTEM VARIABLES

DgnFrame toggles the display of the rectangular frame around DGN underlays.

DgnOsnap toggles whether osnap recognizes entities in the DGN underlay. The tooltip reporting the osnap mode prefixes "DGN:", as illustrated below:

TIPS

- This command displays 3D *.dgn* files in plan view.
- AutoCAD needs to know MicroStation units to convert them to its own units.
- Double-click the underlay's frame to display its properties; right-click it for options.
- To remove the underlay from the drawing, select it, and then press Del.
- Turn off the display by setting the Properties palette's Show Underlay option to off.

DgnClip

2008 Clips DGN underlays to isolate portions of the design file.

Command	Aliases	Keyboard Shortcuts	Menu	Ribbon
dgnclip	...	...	...	...

Command: dgnclip
Select DGN to clip: *(Select one underlay.)*
Enter DGN clipping option [ON/OFF/Delete/New boundary] <New boundary>: *(Enter an option.)*

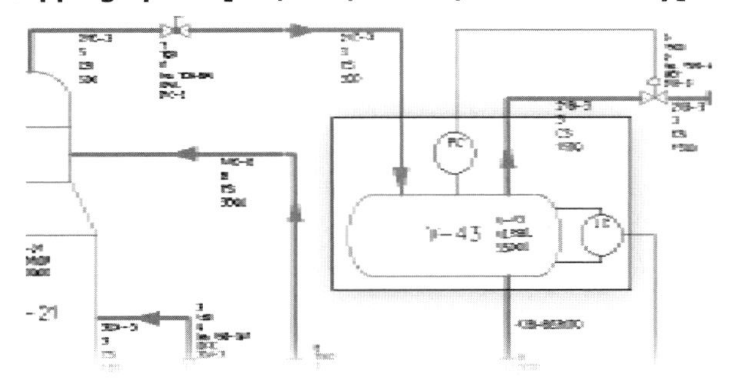

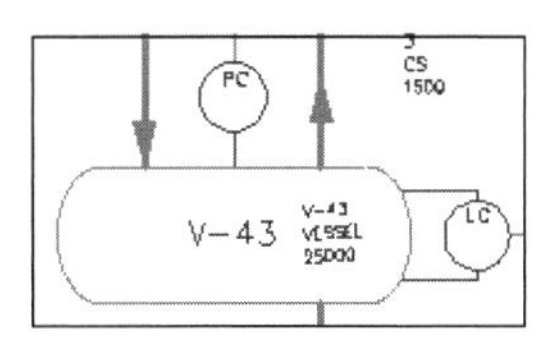

***Left:** Area being clipped.*
***Right:** Clipped area with frame.*

COMMAND LINE OPTIONS

ON turns on a previously-created clipping boundary.

OFF turns off the clipping boundary, so that it is hidden.

Delete removes the clipping boundary, restoring the DGN underlay.

New boundary creates a new clipping boundary:

- **Polygonal** creates a multi-sided boundary using prompts similar to the PLine command.
- **Rectangular** creates a rectangular boundary based on two pick points.

RELATED COMMANDS

Clip clips images, DGN, DWF/x, and PDF formatted files.

DgnAdjust adjusts the contrast, fade, and color of DGN underlays.

DgnAttach attaches MicroStation *.dgn* files as underlays.

RELATED SYSTEM VARIABLE

DgnFrame toggles the display of the frame around clipped DGN underlays.

TIPS

- After the clipping boundary is in place, you can use grips to adjust its size.
- This command can be started by selecting the DGN border, right-clicking, and then selecting DGN Clip from the shortcut menu.
- This command works only with DGN design files attached as underlays with the DgnAttach command; it does not work with design files imported with the DgnImport command.

DgnImport / DgnExport

2008 Imports 2D V7 and V8 *.dgn* files as AutoCAD objects; exports drawings as MicroStation elements.

Command	*Aliases*	*Keyboard Shortcuts*	*Menu Bar*	*Application Menu*	*Ribbon*
dgnimport	...	...	...	**Open** ↳**DGN**	**Insert** ↳**Import**
-dgnimport	...	...	...	...	...
dgnexport	...	...	...	**Export** ↳**DGN**	...
-dgnexport	...	...	...	...	...

Command: dgnexport

Displays Export DGN File dialog box. Name the .dgn file, choose V7 or V8 format, and then click Save.

Displays dialog box.

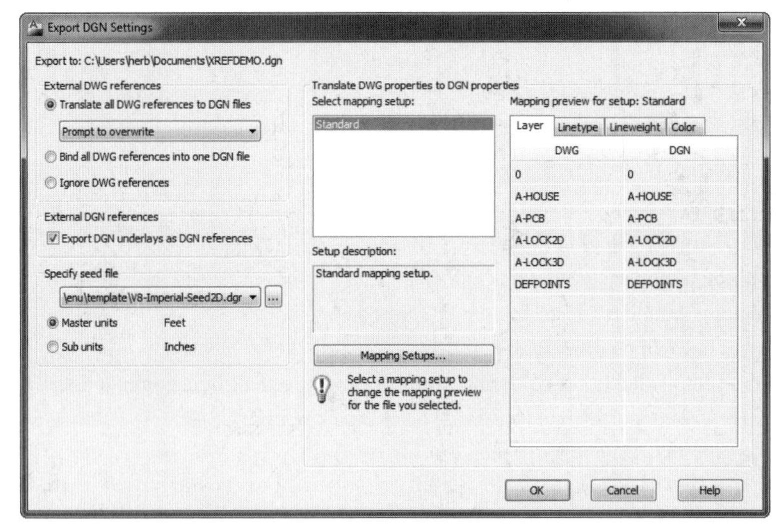

EXPORT DGN SETTINGS DIALOG BOX

External DWG References options

○ **Translate All DWG References to DGN Files** exports xrefs to *.dgn* files.

○ **Bind All DWG References into one DGN File** combines the parent *.dwg* and all xrefs into one *.dgn* file; xrefs are converted to cells (blocks).

⊙ **Ignore DWG References** leaves out xrefs.

External DGN References options

☑ **Export DGN Underlays** includes DGN underlays as *.dgn* references.

Specify Seed File options

[...] selects the seed (template) file for the exported *.dgn* file.

⊙ **Master Units** converts AutoCAD units to master units.

○ **Sub Units** converts AutoCAD units to sub units.

Translate DWG Properties to DGN Properties options

Select Mapping Setup lists the names of mapping setups.

Mapping Setups displays the DGN Mapping Setups dialog box; see the DgnMapping command.

Mapping Preview for Setup shows the mapped properties between the DWG and DGN file.

-DGNEXPORT Command

Command: -dgnexport
Enter DGN file format [V7/V8] <V8>: *(Enter an option.)*
Enter filename for DGN export <path>: *(Enter the path and file name.)*
Specify conversion units [Master/Sub] <Master>: *(Type M or S.)*
Specify mapping setup or [?] <Standard>: *(Enter a name or press Enter.)*
Specify seed file or [?] <path>: *(Enter a path and file name, or press Enter.)*
Translation successful.

The mapping setup is handled by the new DgnMapping command.

DGNIMPORT Command

Command: dgnimport

Displays Import DGN File dialog box. Choose a .dgn file, and then click Save.

Displays Import DGN Settings dialog box.

DIALOG BOX OPTIONS

See DgnExport command.

-DGNIMPORT Command

Command: -dgnimport
Enter filename for DGN import: *(Enter the path and file name.)*
Enter name of model or [?] <Default>: *(Enter a name or press Enter.)*
Specify conversion units [Master/Sub] <Master>: *(Type M or S.)*
Specify mapping setup or [?] <Standard>: *(Enter a name or press Enter.)*

RELATED COMMANDS

DgnAttach attaches MicroStation *.dgn* files as underlays.

DgnMapping controls the translation process.

RELATED SYSTEM VARIABLES

DgnImportMax limits the maximum number of MicroStation elements to import into AutoCAD. Default value is 1,000,000, while 0 means no limit.

DgnMappingPath reports the path to the folder that holds the *dgnsetups.ini* file (read-only).

TIPS

- The DgnImport command fails when the *.dgn* file does not include at least one design model. It does not import support files, such as *.cel, .s, .h, .cgm, .rdl, .d*, and *.dxf*.
- AutoCAD converts either MicroStation's master units or sub units to its own units.
- AutoCAD includes *.dgn* seed files for V7 and V8.

DgnLayers

2008 Toggles visibility of layers in DGN underlays.

Command	Aliases	Keyboard Shortcuts	Menu	Ribbon
dgnlayers	...	...	...	...

Command: dgnlayers
Select DGN underlay: *(Choose one underlay. right-click, and then choose DGN Layers.)*

Displays dialog box.

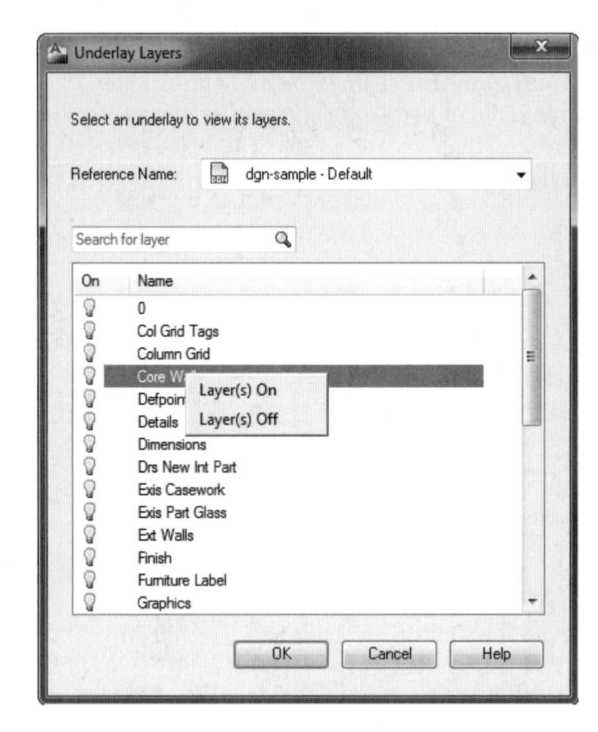

DGN LAYERS DIALOG BOX OPTIONS

Click the lightbulb icon to toggle layer visibility, or right-click for the shortcut menu.

Layer(s) On turns on the selected layers.

Layer(s) Off turns off the selected layers.

Apply changes the visibility of layers immediately, without leaving the dialog box.

RELATED COMMANDS

ULayers toggles visibility of layers in DGN, DWF/x, and PDF files.

DgnAttach attaches MicroStation *.dgn* design files as underlays.

DgnClip clips attached underlays.

TIPS

- This command works only for DGN files that are attached as underlays with the DgnAttach command.
- To toggle layers of DGN files imported as AutoCAD objects with the DgnImport command, use the Layers command.

DgnMapping

2008 Provides control over how MicroStation elements are translated into AutoCAD objects.

Command	*Aliases*	*Keyboard Shortcuts*	*Menu*	*Ribbon*
dgnmapping	...	...	...	...

Command: dgnmapping

Displays dialog box.

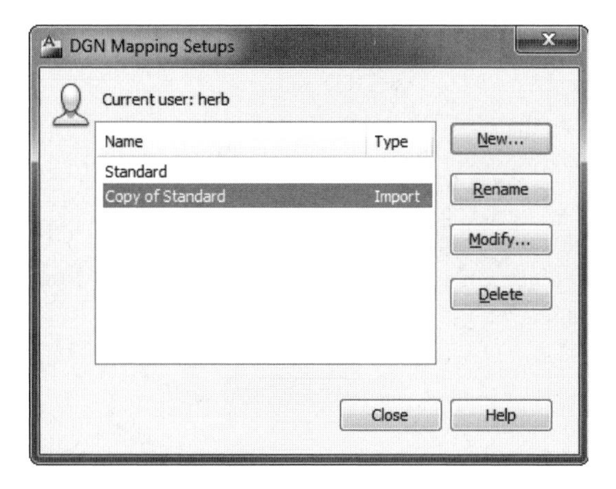

DIALOG BOX OPTIONS

New creates new mapping setups.

Rename renames the selected setup.

Modify displays the Modify DGN Mapping Setup dialog box.

Delete removes the selected setup from the list.

Modify DGN Mapping Setup dialog box

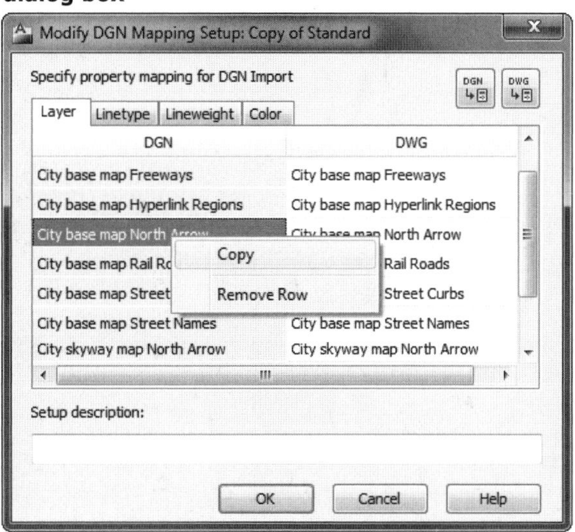

DWG gets layer and other information from another *.dwg* file.

DGN gets level and other information from another *.dgn* file.

To change a value in the DGN column, click an item and select it from the droplist:

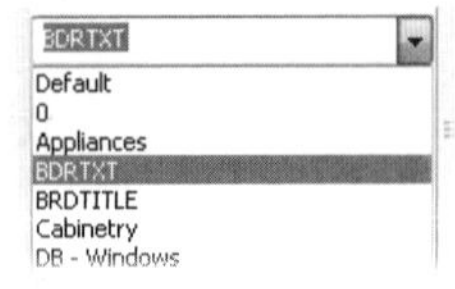

Setup Description describes the nature of this setup.

Right-click for shortcut menu:

Copy copies the layer name; it can be pasted on the DGN side.

Remove Row removes the layer pair.

The Linetype, Lineweight, and Color tabs are similar.

RELATED COMMANDS

DgnImport translates MicroStation *.dgn* design files into AutoCAD objects.

DgnExport translates AutoCAD objects into MicroStation elements in V7 or V8 format.

RELATED SYSTEM VARIABLES

DgnImportMax limits the number of MicroStation elements imported into AutoCAD.

DgnMappingPath points to the folder holding the *dgnsetups.ini* file.

TIPS

- This command has nothing to do with mapping or GIS. In this context, mapping refers to matching colors, linetypes, and so on.
- MicroStation contains an equivalent command for mapping AutoCAD objects to MicroStation elements.

Dim / Dim1

Ver. 2.1 Changes the prompt from 'Command' to 'Dim', allowing access to AutoCAD's original dimensioning mode (short for DIMensions).

Command	Alias	Keyboard Shortcuts	Menu	Ribbon
dim	...	...	...	...

Command: dim
Dim: *(Enter a dimension command from the list below.)*

COMMAND LINE OPTIONS

Aliases for the dimension commands are shown by UPPERCASE letters, such as **al** *for ALigned.*

ALigned draws linear dimensions aligned with objects; introduced with AutoCAD version 2.0; replaced by DimAligned.

ANgular draws angular dimensions that measure angles; introduced with AutoCAD v2.0; replaced by DimAngular.

Baseline continues dimensions from base points; introduced with AutoCAD v1.2; replaced by QDim and DimBaseline.

CEnter draws '+' center marks on circles' and arcs' centers; introduced with AutoCAD v. 2.0; replaced by DimCenter.

COntinue continues dimensions from previous dimensions' extension lines; introduced with AutoCAD v1.2; replaced by the QDim and DimContinue commands.

Diameter draws diameter dimensions on circles, arcs, and polyarcs; introduced with AutoCAD v. 2.0; replaced by the QDim and DimDiameter commands.

Exit returns to 'Command' prompt from 'Dim' prompt; introduced with AutoCAD v1.2.

HOMetext returns dimension text to its original position; introduced with AutoCAD v2.6; replaced by DimEdit command's Home option.

HORizontal draws horizontal dimensions; introduced with AutoCAD v1.2; replaced by DimLinear.

LEAder draws leaders; introduced with AutoCAD v2.0; replaced by the Leader and QLeader commands.

Newtext edits text in associative dimensions; introduced with AutoCAD v2.6; replaced by DimEdit's New option.

OBlique changes the angle of extension lines in associative dimensions; introduced with AutoCAD Release 11; replaced by the DimEdit command's Oblique option.

ORdinate draws x- and y-ordinate dimensions; introduced with AutoCAD R11; replaced by QDim and DimOrdinate.

OVerride overrides current dimension variables; introduced with AutoCAD R11; replaced by DimOverride.

RAdius draws radial dimensions on circles, arcs, and polyline arcs; introduced with AutoCAD v2.0; replaced by the QDim and DimRadius commands.

REDraw redraws the current viewport (same as 'Redraw).; introduced with AutoCAD v2.0.

REStore restores dimensions to the current dimension style; introduced with AutoCAD Release 11; replaced by the -DimStyle command's Restore option.

ROtated draws linear dimensions at any angle; introduced with AutoCAD v2.0; replaced by DimLinear.

SAve saves the current settings of dimension styles; introduced with AutoCAD Release 11; replaced by the -DimStyle command's Save option.

STAtus lists the current settings of dimension variables; introduced with AutoCAD v2.0; replaced by the -DimStyle command's Status option.

STYle defines styles for dimensions; introduced with AutoCAD v2.5; replaced by DimStyle.

TEdit changes the location and orientation of text in associative dimensions (Rel. 11); replaced by DimTEdit.

TRotate changes the rotation angle of text in associative dimensions; introduced with AutoCAD Release 11; replaced by the DimTEdit command's Rotate option.

Undo undoes the last dimension action; introduced with AutoCAD v2.0; replaced by the Undo command.

UPdate updates selected associative dimensions to the current dimvar setting; introduced with AutoCAD v2.6; replaced by the -DimStyle command's Apply option.

VAriables lists the values of variables associated with dimension styles, not dimvars; introduced with AutoCAD Release 11; replaced by-DimStyle command's Variables option.

VErtical draws vertical linear dimensions; introduced with AutoCAD v1.2; replaced by DimLinear.

RELATED DIM VARIABLES

Dim*xxx* specifies system variables for dimensions; see the DimStyle command.

DimAso determines whether dimensions are drawn associatively.

DimScale determines the dimension scale.

DIM1 Command

Displays the 'Dim' prompt for a single dimensioning command, and then returns to the 'Command' prompt (short for DIMension once).

Command: dim1
Dim: *(Enter a dimension command.)*

COMMAND LINE OPTIONS

Accepts all "original" dimension commands.

TIPS

- Most dimensions consist of the basic components illustrated below:

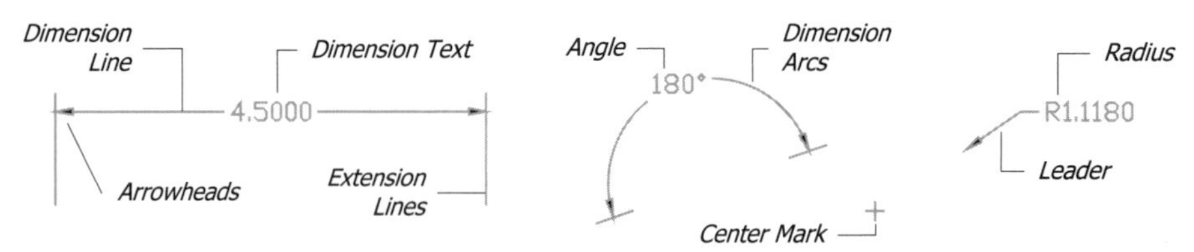

- The 'Dim' prompt dimension commands are included for compatibility with AutoCAD Release 12 and earlier. Only transparent commands and dimension commands work at the 'Dim' prompt. To use other commands, you must return to the 'Command' prompt by pressing Esc.
- *Defpoints* (short for "definition points") appear as small dots on layer DefPoints, and are used by earlier releases of AutoCAD to locate extension lines. When stretching dimensions, ensure defpoints are included; otherwise, dimensions are not updated automatically.
- The DefPoints layer *never* plots, making it useful for objects and notes that should never plot and for great practical jokes.
- As of AutoCAD 2002, defpoints are not used when DimAssoc = 2 (default); dimensions are attached directly to objects.
- Dimension text can have colored backgrounds set globally through the DimStyle command, or overridden with the Properties command.
- All the components of *associative dimensions* are treated as a single object; components of a nonassociative dimension are treated as individual objects.
- For additional control over dimensions, consult the following undocumented commands described elsewhere in this book: AiDimFlipArrow, AiDimPrec, AiDimStyle, Ai_Dim_TextAbove, Ai_Dim_TextCenter, Ai_Dim_TextHome, and AiDimTextMove.

DimAligned

Rel. 13 Draws linear dimensions aligned with objects.

Command	*Aliases*	*Keyboard Shortcuts*	*Menu*	*Ribbon*
dimaligned	**dal** **dimali**	**...**	**Dimension** **↳Aligned**	**Annotate** **↳Dimensions** **↳Aligned**

Command: dimaligned
Specify first extension line origin or <select object>: *(Pick a point, or press Enter to enable object selection, as shown by the following prompt.)*
Select object to dimension: *(Select an object.)*
Specify dimension line location or [Mtext/Text/Angle]: *(Pick a point, or enter an option.)*
Dimension text = *nnn*

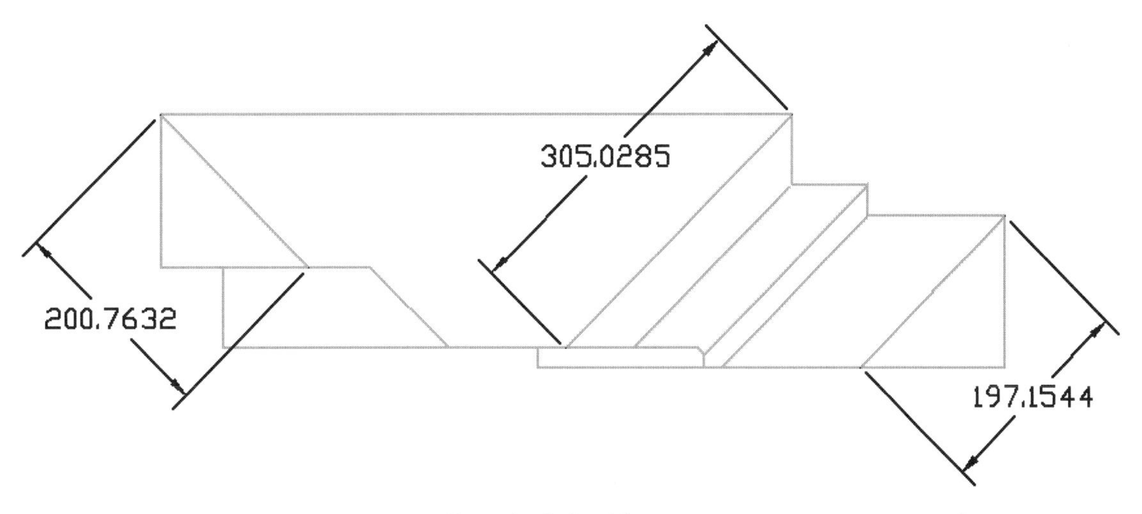

Examples of aligned dimensions.

COMMAND LINE OPTIONS

Specify first extension line origin picks a point for the origin of the first extension line.

Specify second extension line origin picks a point for the origin of the second extension line.

Select object selects an object to dimension, after you press Enter.

Select object to dimension picks a line, circle, arc, polyline, or explodable object; individual segments of polylines are dimensioned.

Specify dimension line location picks a point from which to locate the dimension line and text.

Mtext changes the wording of the dimension text.

Text changes the position of the dimension text.

Angle changes the angle of the dimension text.

RELATED DIM COMMAND

DimRotated draws rotated dimension lines with perpendicular extension lines.

DimAngular

Rel. 13 Draws dimensions that measure angles.

Command	Aliases	Keyboard Shortcuts	Menu	Ribbon
dimangular	**dan** **dimang**	...	**Dimension** **↳Angular**	**Annotate** **↳Dimensions** **↳Angular**

Command: dimangular
Select arc, circle, line, or <specify vertex>: *(Select an object, or pick a vertex.)*
Specify second angle endpoint: *(Pick a point; prompt varies, depending on object.)*
Specify dimension arc line location or [Mtext/Text/Angle/Quadrant]: *(Pick a point, or enter an option.)*
Dimension text = *nnn*

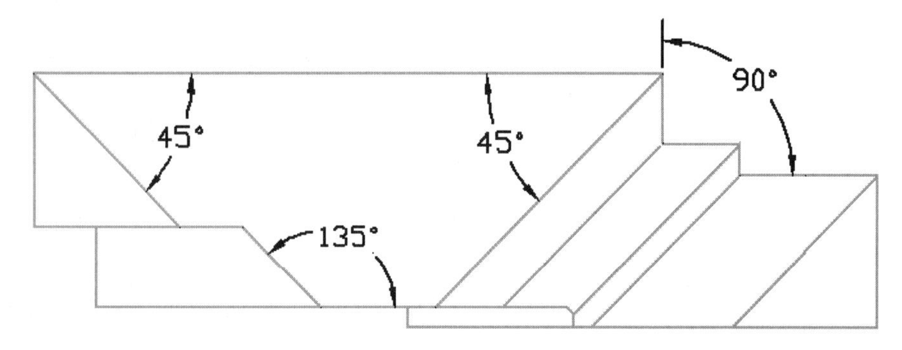

Examples of angular dimensions.

COMMAND LINE OPTIONS

Select arc measures the angle of the arc.

Circle prompts you to pick two points on the circle.

Line prompts you to pick two lines.

Specify vertex prompts you to pick points to make an angle.

Quadrant prompts you to select a point in a quadrant.

Specify dimension arc/line location specifies the location of the angular dimension.

Mtext changes the wording of the dimension text.

Text changes the position of the text.

Angle changes the angle of the dimension text.

RELATED DIM COMMANDS

DimArc dimensions the lengths of arcs.

DimCenter places a center mark at the center of an arc or circle.

DimRadius dimensions the radius of an arc or circle.

DimArc

2006 Dimensions the length along arcs and polyline arcs.

Command	Alias	Keyboard Shortcuts	Menu	Ribbon
dimarc	**dar**	...	**Dimension** ↳**Arc Length**	**Annotate** ↳**Dimensions** ↳**Arc Length**

Command: dimarc
Select arc or polyline arc segment: *(Select an arc.)*
Specify arc length dimension location, or [Mtext/Text/Angle/Partial]: *(Pick a point, or enter an option.)*
Dimension text = 15'-9 3/16"

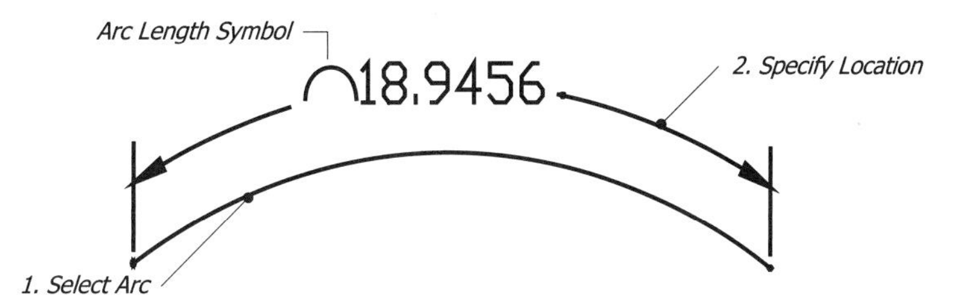

COMMAND LINE OPTIONS

Select arc measures the length of arcs.

Specify dimension arc length location specifies the location of the arc length dimension.

Mtext changes the wording of the dimension text.

Text changes the position of the text.

Angle changes the angle of the dimension text.

Partial prompts you to pick two points for a partial dimension:

Specify first point for arc length dimension: *(Pick a point.)*
Specify second point for arc length dimension: *(Pick another point.)*

RELATED SYSTEM VARIABLE

DimArcSym specifies the location of the dimension arc symbol.

TIPS

- Specify the placement of the arc length symbol in the Symbols and Arrows tab of the Dimension Style Manager.
- The arc symbol can be displayed above, or in front of, the dimension text.
- The extension lines can be *orthogonal* (angle less than 90 degrees) or *radial* (more than 90 degrees).
- The angle is either between 0 and 180 degrees, or between 180 and 360 degrees.

DimBaseline

Rel. 13 Draws linear dimensions based on previous starting points.

Command	Aliases	Keyboard Shortcuts	Menu	Ribbon
dimbaseline	**dba**	...	**Dimension**	**Annotate**
	dimbase		↳**Baseline**	↳**Dimensions**
				↳**Baseline**

Command: dimbaseline

Specify a second extension line origin or [Undo/Select] <Select>: *(Pick a point, or enter an option.)*

Dimension text = *nnn*

Specify a second extension line origin or [Undo/Select] <Select>: *(Pick a point, enter an option, or press Esc to exit command.)*

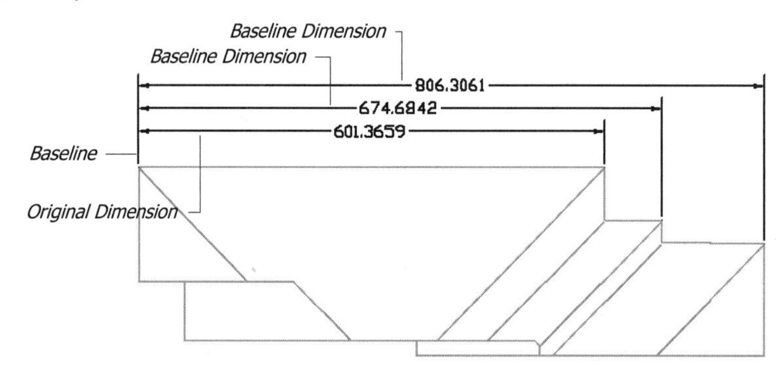

COMMAND LINE OPTIONS

Specify a second extension line origin positions the extension line of the next baseline dimension.

Select prompts you to select the base dimension.

Undo undoes the previous baseline dimension.

Esc exits the command.

RELATED DIM COMMANDS

Continue continues linear dimensioning from the last extension point.

QDim creates continuous or baseline dimensions quickly.

RELATED DIM VARIABLES

DimDli specifies the distance between baseline dimension lines.

DimSe1 suppresses the first extension line.

DimSe2 suppresses the second extension line.

DimBreak

2008 Breaks dimension and extension lines where they cross other objects.

Command	*Alias*	*Keyboard Shortcuts*	*Menu*	*Ribbon*
dimbreak	**dar**	**...**	**Dimension** ↳**Dimension Break**	**Annotate** ↳**Dimensions** ↳**Break**

Command: dimbreak
Select a dimension or [Multiple]: *(Select a dimension, or type M.)*
Select object to break dimension or [Auto/Restore/Manual] <Auto>: *(Select an object, or enter an option.)*
Select object to break dimension: *(Press Enter to exit the command.)*

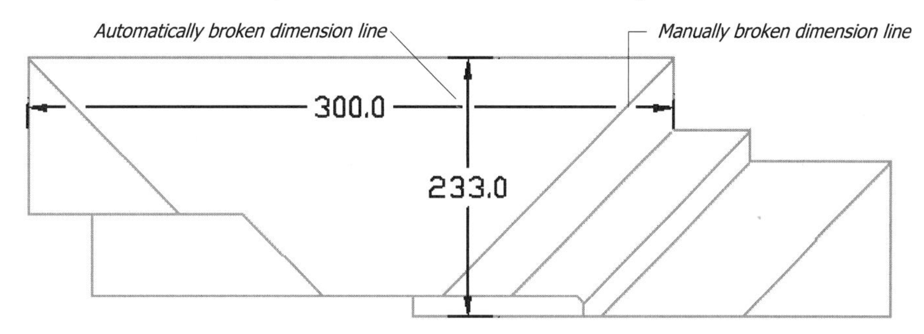

COMMAND LINE OPTIONS

Select a dimension selects a single dimension for breaking.

Multiple selects multiple dimensions to break.

Select object selects the object that crosses the dimension.

Auto breaks all dimension and extension lines that are crossed.

Restore removes dimension breaks.

Manual specifies two points to indicate the break location.

Multiple options

Select dimensions: *(Select one or more, or press Ctrl+A for all.)*
Select dimensions: *(Press Enter to end object selection.)*
Enter an option [Break/Restore]: *(Enter an option.)*

Break adds gaps, where necessary.

Restore removes all gaps, healing the dimension and extension lines.

TIPS

- Use the manual option to create breaks even where no objects cross.
- The gap is approximately 0.125 inches (3.75mm), and is affected by system variable DimScale.
- The gap is associated with the dimension: when the object changes size, so does the gap.
- There appears to be no system variable for specifying the gap, although it can be changed in DimStyle.

DimCenter

Rel. 13 Draws center marks and lines on arcs and circles.

Command	Alias	Keyboard Shortcuts	Menu	Ribbon
dimcenter	**dce**	...	**Dimensions** ↳**Center Mark**	**Annotate** ↳**Dimensions** ↳**Center Mark**

Command: dimcenter
Select arc or circle: *(Select an arc, circle, or explodable object.)*

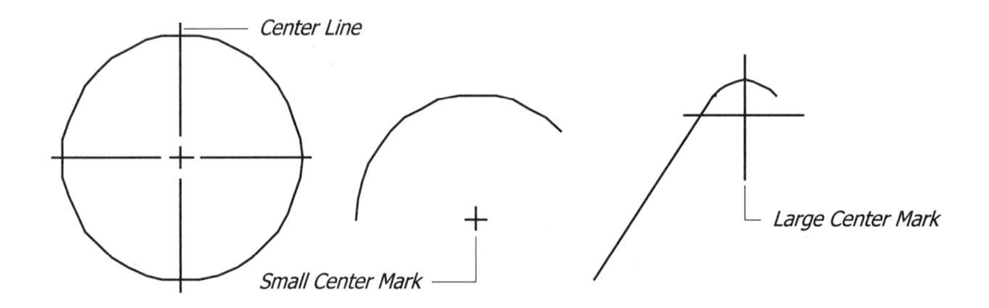

COMMAND LINE OPTION

Select arc or circle places the center mark at the center of the selected arc, circle, or polyarc.

RELATED DIM COMMANDS

DimAngular dimensions arcs and circles.

DimDiameter dimensions arcs and circles by diameter value.

DimRadius dimensions arcs and circles by radius value.

RELATED DIM VARIABLE

DimCen specifies the size and type of the center mark:

negative value — Draws center marks and lines.

0 — Does not draw center marks or center lines.

positive value — Draws center marks.

0.09 — Default value.

TIPS

- The center mark length defined by DimCen is from the center to one end of the mark; the center line length is the size of gap and extension beyond the circle or arc.

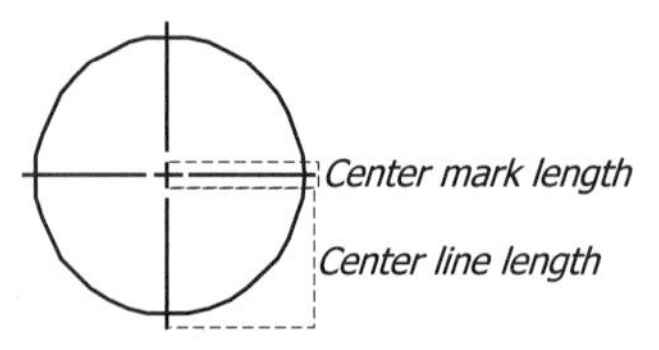

- Changing the center mark size for a dimension style does not update existing center lines and marks.

DimConstraint

2010 Adds dimensional constraints to objects and geometric features within objects.

Command	*Alias*	*Keyboard Shortcuts*	*Menu*	*Ribbon*
dimconstraint	**dcon**	...	**Parametric ↳Dimensional Constraints**	**Parametric ↳Dimensional**

Command: dimconstraint

Current settings: Constraint form = Dynamic

Select a dimension or [LInear/Horizontal/Vertical/ALigned/ANgular/Radial/Diameter/Form/Convert <Aligned>: *(Pick a dimension, or enter an option.)*

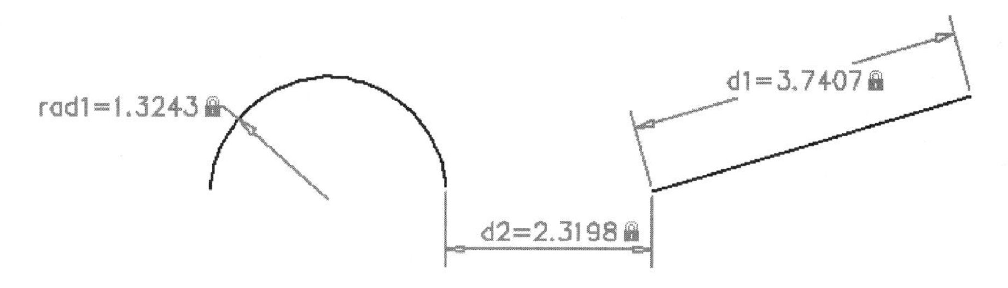

Left: *Radial dimensional constraint applied to an arc. (The padlock icon indicates dimensional constraints.)*
Center: *Horizontal dimensional constraint between arc and line.*
Right: *Aligned dimensional constraint applied to the line.*

COMMAND LINE OPTIONS

Select dimension selects the associative dimensions to be converted to a dimensional constraint.

See the **Dc...** *series of commands earlier in this book for details on each option:*

LInear applies linear dimensional constraints; can be horizontal, vertical, or aligned, depending on how the extension lines are dragged; see the DcLinear command for prompt sequence.

Horizontal applies horizontal dimensional constraints; see DcHorizontal.

Vertical applies vertical dimensional constraints; see DcVertical.

ALigned applies aligned dimensional constraints at an angle; see DcAligned.

ANgular applies angular dimensional constraints between two lines, on arcs, and on vertices; see DcAngular.

Radial applies radial dimensional constraints to arcs, circles, and polyarcs; see DcRadius.

Diameter applies diameter dimensional constraints to arcs, circles, and polyarcs; see DcDiameter.

Form toggles dimensional constraints between dynamic and annotational; see DcForm.

Convert converts associative dimensions into dimensional constraints; see DcConvert *(new to AutoCAD 2011).*

RELATED COMMANDS

Dc... replaces this command with a series of commands *(new to AutoCAD 2011).*

DcDisplay toggles the display of dynamic dimensional constraints.

DelConstraint erases dimensional constraints.

BCParamaeters applies dimensional constraints with grips in the Block Editor.

ConstraintSettings specifies defaults for dimensional constraints.

Parameters creates links between dimensional constraints.

GeomConstraint applies geometric constraints to objects and geometry.

RELATED SYSTEM VARIABLES

DimConstructionIcon toggles the display of the padlock icon.

DynConstraintDisplay toggles the global display of dimensional constraints.

DynConstraintMode toggles the display of hidden dimensional constraints during selection.

ConstraintNameFormat specifies the text format of dimensional constraints.

ConstraintRelax toggles the enforcement of constraints during editing.

ConstraintSolveMode toggles constraint behavior during editing.

TIPS

- Dimensional constraints can contain values like regular dimensions, as well as formulas and references to other dimensions. You can use the Parameters palette to define formulas; see the Parameters command.
- Use dimensional constraints to connect the sizes of objects:
    1. Apply a horizontal dimensional constraint to the line (d1), as illustrated below:

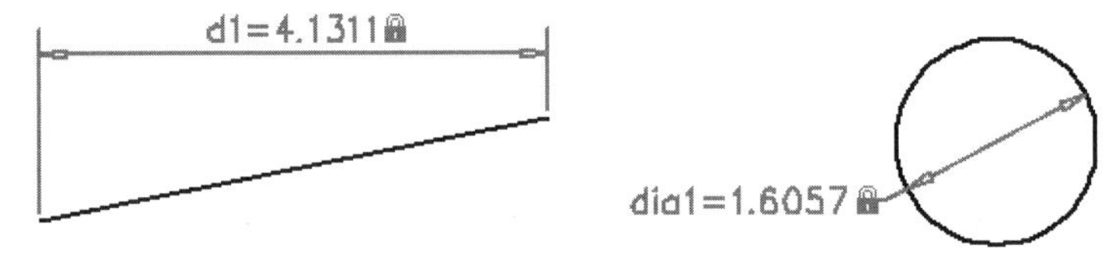

    2. Apply a diameter constraint to the circle (dia1), as illustrated above.
    3. Double-click the circle's text, and then edit it to reference the line:

       dia1 = d1 / 3

       This formula means that the diameter (dia1) of the circle is always 1/3 the length of the line (d1).

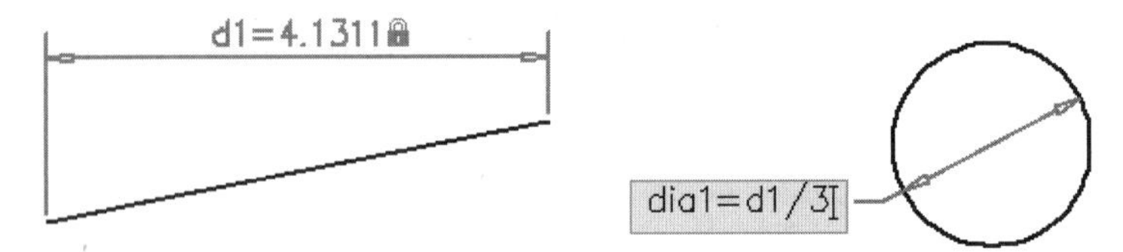

       When you now resize the line, the diameter of the circle changes to match: the line controls the circle.

    4. To resize the line or circle with dimensional constraints, (a) select the dimensional constraint, and then (b) drag the triangular handles.

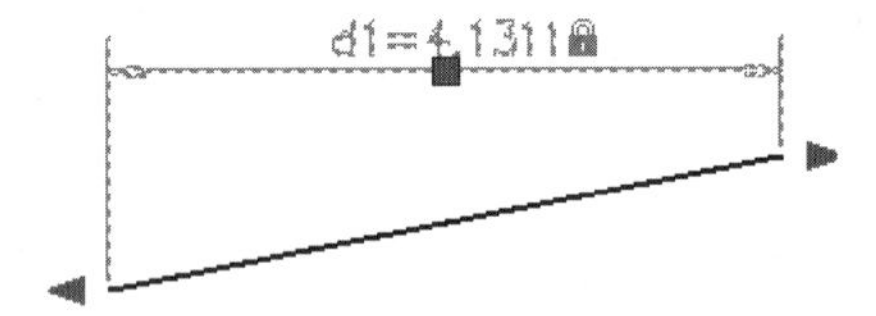

- To override a dimensional constraint, hold down the Ctrl key while dragging grips; AutoCAD makes copies of the object.
- Dimensional constraints can be applied to objects or to points on/within objects, as well as between points on two objects. For example, the distance between the center of the ellipse and circle is dimensionally constrained.

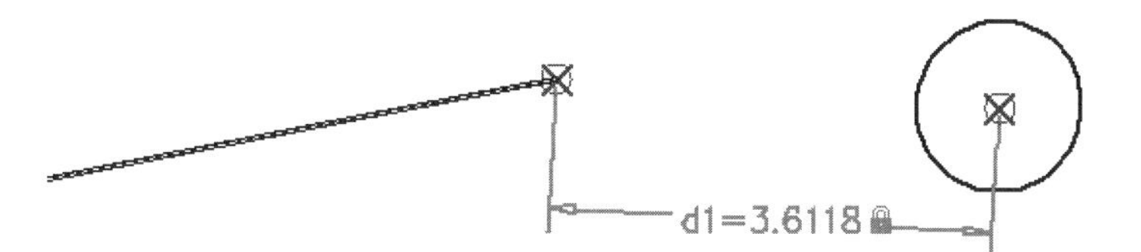

- There are three types of dimensional constraints:
    - **Dynamic** constraints do not show up in layouts, do not plot, have their own uneditable dimension style, and always appear the same size, no matter the zoom level; this is the default style of constraint.
    - **Annotational** constraints show up in layouts, are plotted, are affected by dimension styles, and change size with zoom levels. You cannot create these constraints; you must create dynamic constraints with the DimConstraint command, and then change them to annotational with the Properties command's Constraint Form property.
    - **Parametric** constraints are created in the Block Editor with the BCParameters command. These are like dynamic constraints, but also have user-definable grips for controlling the size of dynamic blocks.

If you intend to produce drawings that show dimensional constraints, then you should primarily use them in annotational form.

Quick Start Tutorial

Using Dimensional Constraints

You apply dimensional constraints just like regular dimensions, except that you need to keep in mind that dimensional constraints control the geometry. In this tutorial, dimensional constraints to force the length of a rectangle to be twice the diameter of a circle, and the circle to be at the center of the rectangle.

1. Create the simple drawing of a rectangle (Rectangle command) and circle (Circle), as shown below. The size of the rectangle and the circle's location within it are unimportant.

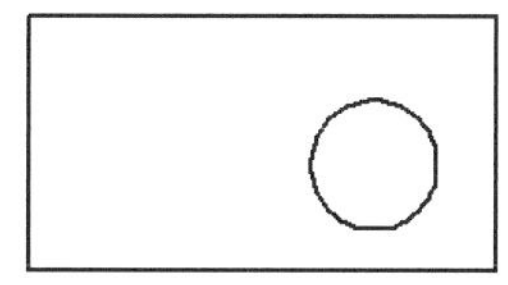

2. From the ribbon's Parametric tab, choose **Diameter**.

   Select arc or circle: *(Choose the circle.)*
   Dimension text = 290. Specify dimension line location: *(Pick a point.)*

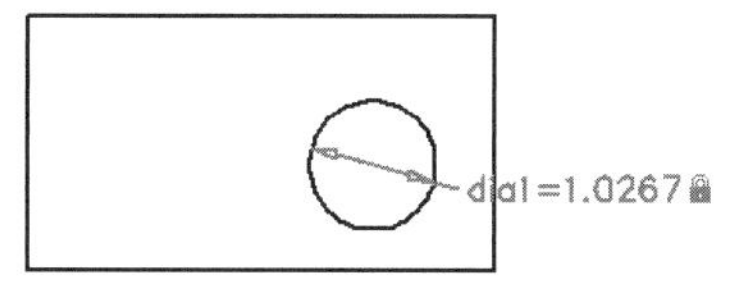

The padlock icon indicates that this dimension is a constraint. Notice that the dimension text is in the form of a formula: **dia1 = 290.5492.** *Dia1* means this is the first diameter constraint in the drawing.

3. Force the length of the sides of the rectangle to be twice the diameter of the circle. From the Parametric tab, choose **Horizontal**:

   Select first constraint point or [Object]: *(Pick an endpoint of the rectangle.)*

   Notice that AutoCAD highlights the endpoint with a red icon.

4. Choose the midpoint for the second point of the dimensional constraint:

   Select second constraint point: *(Pick the midpoint of the rectangle.)*

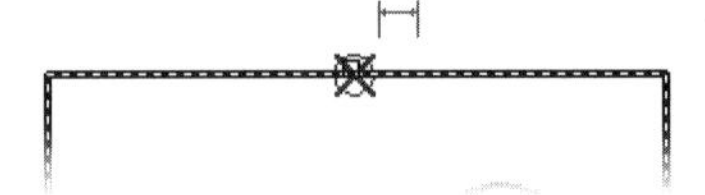

5. Locate the dimension line:

   Specify dimension line location: *(Pick a point.)*

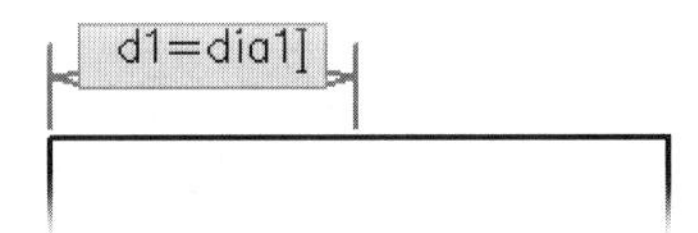

6. Edit the dimension text to read "d1 = dia1." This forces half of the line to equal the diameter of the circle. The rectangle changes shape immediately to meet the demand of the dimensional constraints.

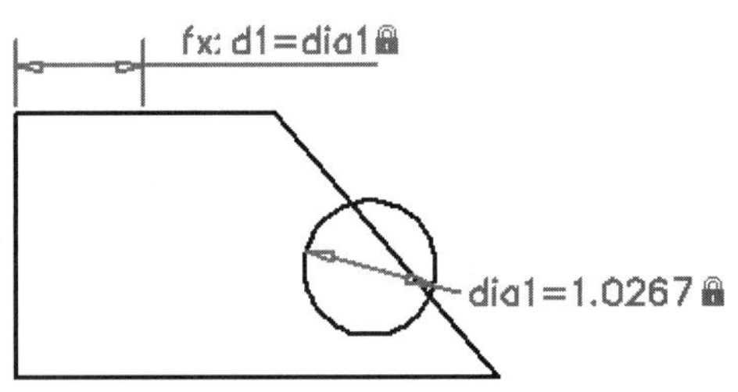

7. Click the diameter dimension. Notice the two triangle grips. Drag one of them. The circle not only changes diameter, but the polygon (no longer a rectangle) also changes size and shape.

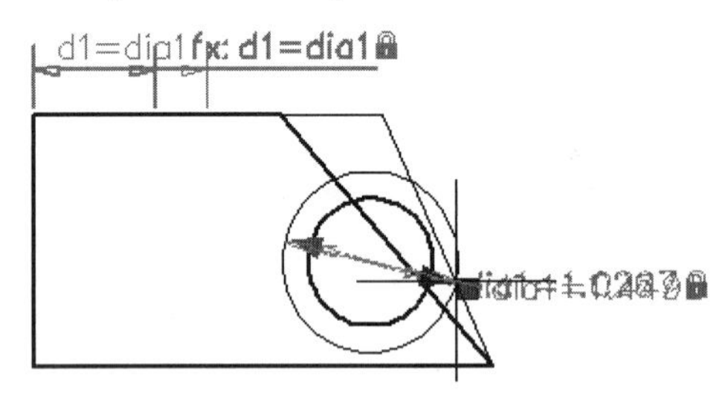

8. To force the circle to the center of the rectangle, add a horizontal and a vertical dimension, each of 0 units. Attach the dimensional constraints to the midpoints of lines and the center of the circle, as illustrated below:

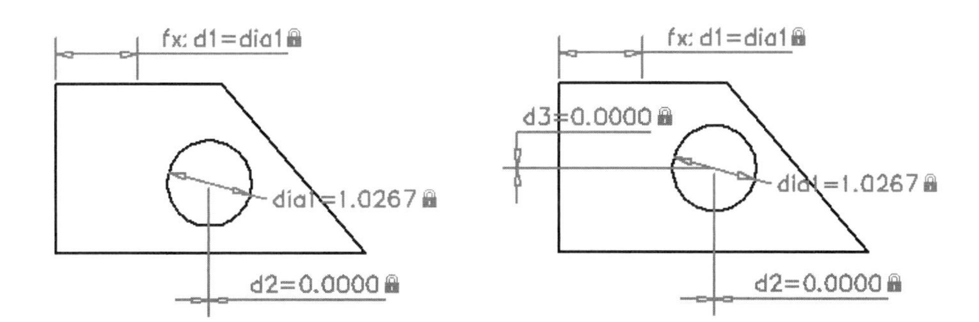

9. To force the polygon to stay rectangular, choose Horizontal from the Geometric panel, and then apply it to the horizontal lines of the polygon.

   Select an object or [2Points]: *(Pick a line.)*

   Repeat, applying Vertical geometric constraints to the vertical sides. As you apply the geometric constraints, notice that the polygon straightens out.

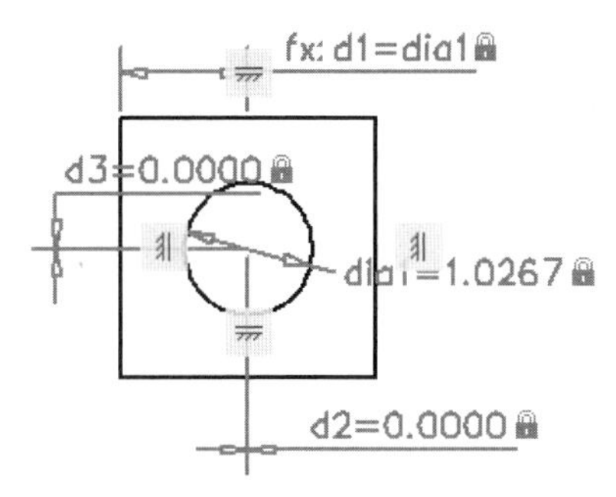

10. Drag the diameter constraint's arrow grips. As the circle and rectangle change size, notice that the rectangle remains rectangular.
11. Finish the job by adding Vertical dimensional constraints to keep the circle centered vertically. For fun, fill the rectangle with a gradient hatch pattern (Gradient command).

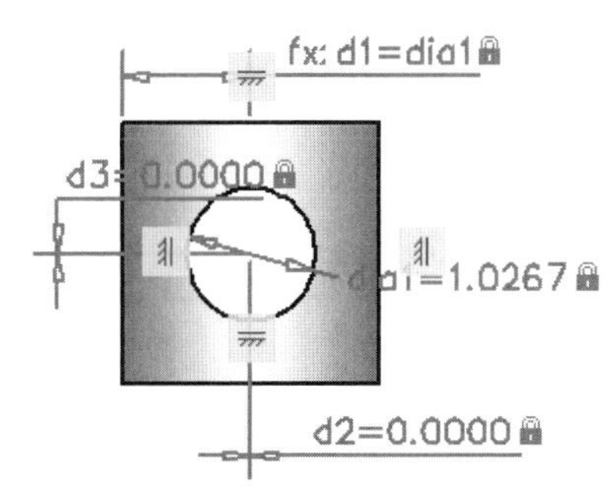

12. When you drag the diameter constraint's triangle grips, all three elements react: the circle, the square, and the gradient fill. The circle remains centered, the square remains square, and the gradient fill remains.

13. The dimensional constraints are a dynamic style; for plotting, you should change them to annotational style:

a. Start the Properties command, and then select all dimensional constraints.
b. In the **Constraint Form** droplist, choose **Annotational**.

Notice that the dimension text and lines turn black.

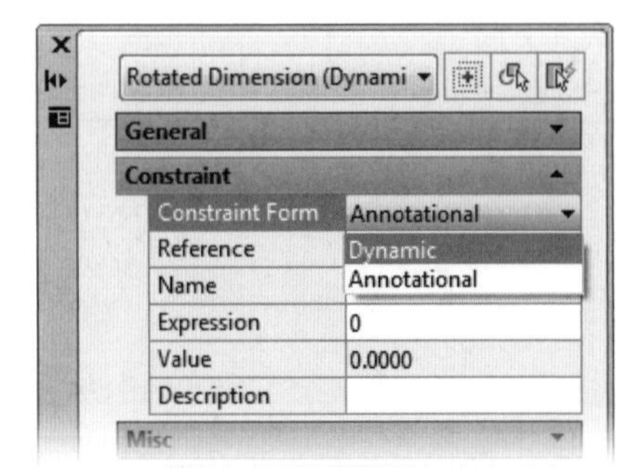

14. The annotational dimension still displays the dimension as a formula, instead of as a value.

a. Start the ConstraintSettings command, and then choose the Dimensional tab.
b. In the **Dimension Name Format** droplist, choose **Value**, and then click **OK**.

Notice that the dimensional constraints display their values only.

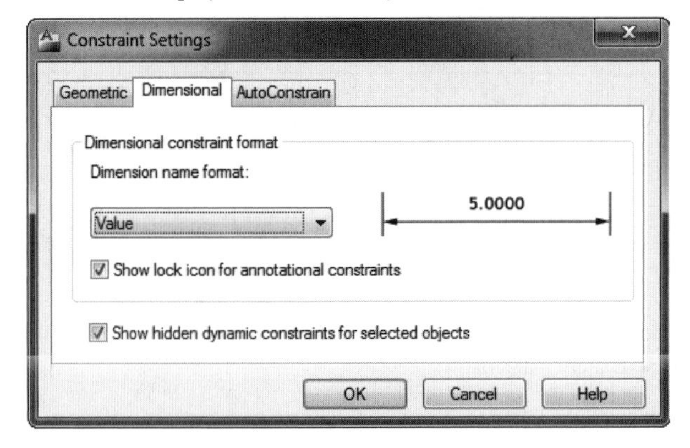

15. Plot the drawing with the Preview command to see that only annotational dimensions are plotted; dynamic dimensional constraints are not.

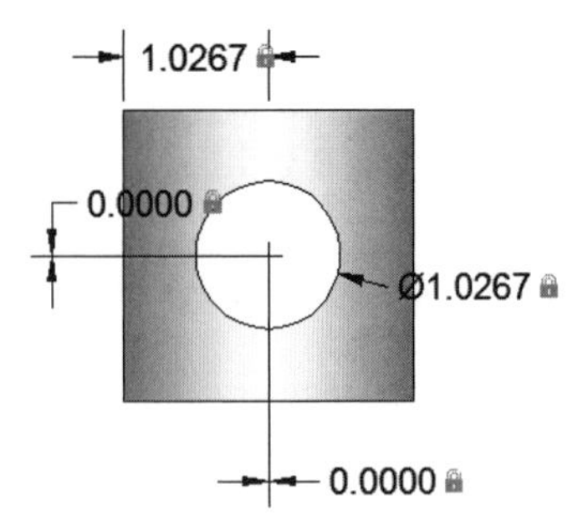

DimContinue

Rel. 13 Continues dimensioning from the second extension line of the previous dimension.

Command	*Aliases*	*Keyboard Shortcuts*	*Menu*	*Ribbon*
dimcontinue	**dco** **dimcont**	**...**	**Dimension** **↳Continue**	**Annotate** **↳Dimensions** **↳Continue**

Command: dimcontinue

Specify a second extension line origin or [Undo/Select] <Select>: *(Pick a point, enter an option, or press Enter to select a base dimension.)*

Dimension text = *nnn*

Specify a second extension line origin or [Undo/Select] <Select>: *(Pick a point, enter an option, or press Esc to exit command.)*

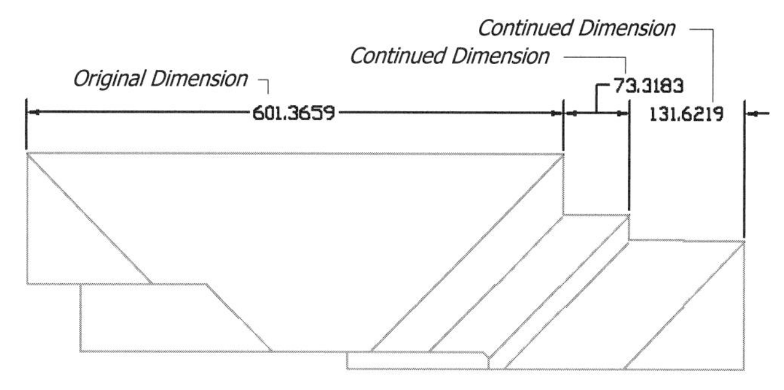

COMMAND LINE OPTIONS

Specify a second extension line origin positions the extension line of the next continued dimension.

Select prompts you to select the originating dimension.

Undo undoes the previous continued dimension.

Esc exits the command.

RELATED DIM COMMANDS

DimBaseline continues dimensioning from the first extension point.

QDim creates continuous or baseline dimensions quickly.

RELATED DIM VARIABLES

DimDli sets the distance between continuous dimension lines.

DimSe1 suppresses the first extension line.

DimSe2 suppresses the second extension line.

DimDiameter

Rel. 13 Draws diameter dimensions on arcs, circles, and polyline arcs.

Command	*Aliases*	*Keyboard Shortcuts*	*Menu*	*Ribbon*
dimdiameter	**ddi**	...	**Dimension**	**Annotate**
	dimdia		**↳Diameter**	**↳Dimensions**
				↳Diameter

Command: dimdiameter
Select arc or circle: *(Select an arc, circle, or explodable object.)*
Dimension text = *nnn*
Specify dimension line location or [Mtext/Text/Angle]: *(Pick point, or enter option.)*

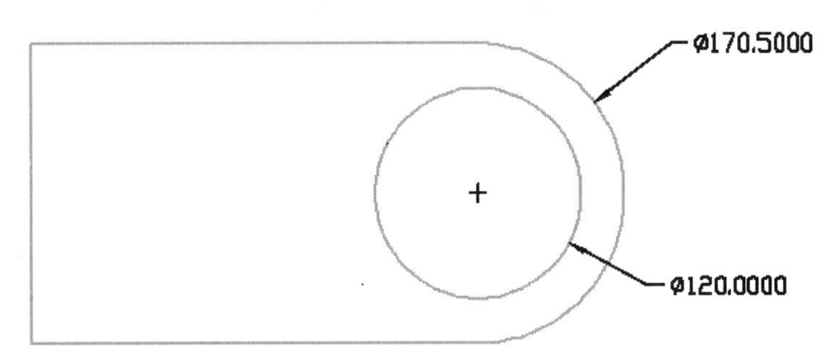

Examples of diameter dimensions.

COMMAND LINE OPTION

Select arc or circle selects an arc, circle, or polyarc.

Specify dimension line location specifies the location of the angular dimension.

Mtext changes the wording of the dimension text.

Text changes the position of the dimension text.

Angle changes the angle of the dimension text.

RELATED DIM COMMANDS

DimCenter marks the center point of arcs and circles.

DimRadius draws the radius dimension of arcs and circles.

TIPS

- When you drag the dimension line outside an arc, AutoCAD draws an extension to the arc:

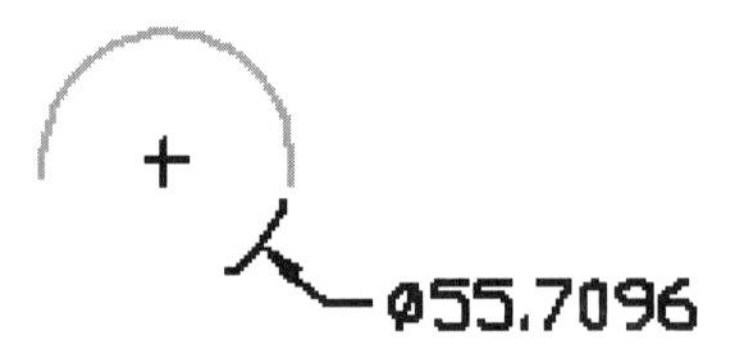

- The diameter symbol is automatically included. But if you need to add it manually, use the %%d code or the Unicode \U+2205.

DimDisassociate

2002 Converts associative dimensions to non-associative (the command that's the most difficult to spell correctly).

Command	Alias	Keyboard Shortcuts	Menu	Ribbon
dimdisassociate	**dda**	...	...	...

Command: dimdisassociate
Select dimensions to disassociate...
Select objects: *(Select one or more dimensions, or enter All to select all dimensions.)*
Select objects: *(Press Enter to end object selection.)*
nn **disassociated.**

COMMAND LINE OPTION

Select objects selects dimensions to convert to non-associative type.

RELATED DIM COMMANDS

DimReassociate converts dimensions from non-associative to associative.

DimRegen updates the location of associative dimensions.

RELATED SYSTEM VARIABLE

DimAssoc determines whether newly-created dimensions are associative:

0 — Dimension created exploded; all parts (such as dimension lines, arrow heads) are individual, ungrouped objects.

1 — Dimension created as a single object, but is not associative.

2 — Dimension created as a single object, and is associative (default).

TIPS

- As you select objects, this command ignores non-dimensions, dimensions on locked layers, and those not in the current space (model or paper).
- The command reports filtered and disassociated dimensions.
- The effects of this command can be reversed with the U and the DimReassociate commands.

DimEdit

Rel. 13 Applies editing changes to dimension text.

Command	*Aliases*	*Keyboard Shortcuts*	*Menu*	*Ribbon*
dimedit	**ded** **dimed**	...	**Dimension** ↳**Oblique**	...

Command: dimedit

Enter type of dimension editing [Home/New/Rotate/Oblique] <Home>: *(Enter an option, or press Enter for Home.)*

Select objects: *(Select one or more objects.)*

Select objects: *(Press Enter.)*

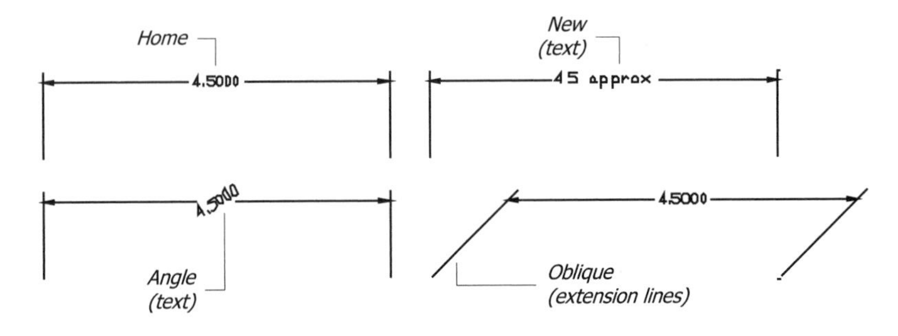

COMMAND LINE OPTIONS

Angle rotates the dimension text.

Home returns the dimension text to its original position.

Oblique rotates the extension lines.

New allows editing of the dimension text.

RELATED DIM COMMANDS

All.

RELATED DIM VARIABLES

Most.

TIPS

- When you enter dimension text with the DimEdit command's New option, AutoCAD recognizes <> as *metacharacters* representing existing text.
- Use the Oblique option to angle dimension lines by 30 degrees, suitable for isometric drawings; use the Style command to oblique text by the same angle. See the Isometric command.

DimInspect

2008 Creates and edits inspection-style text on dimensions.

Command	Aliases	Keyboard Shortcuts	Menu	Ribbon
diminspect	...	...	**Dimension ⇨Inspection**	**Annotate ⇨Dimensions ⇨Inspect**
-diminspect	...	...	...	...

Command: diminspect

Displays dialog box.

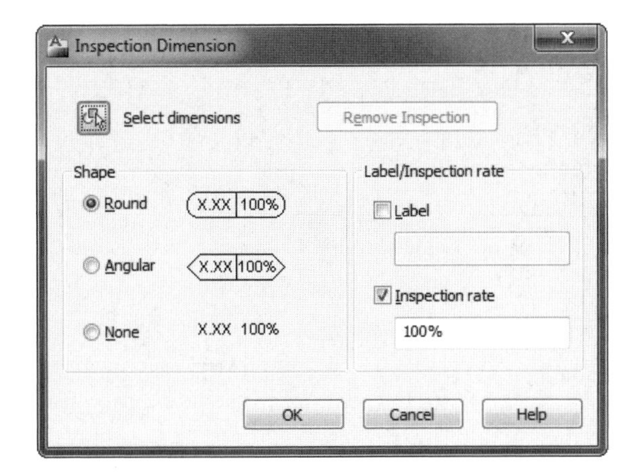

This command places the following inspection shapes:

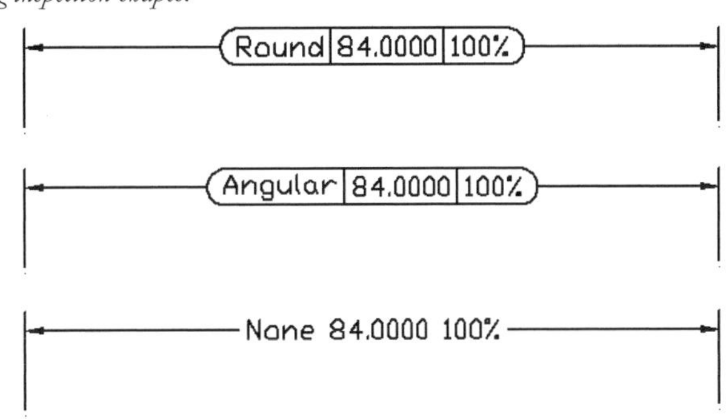

INSPECTION DIMENSION DIALOG BOX

Select Dimensions selects the dimensions for adding inspection codes.

Remove Inspection removes inspection text from selected dimensions.

Shape options

○ **Round** uses a rounded rectangle for the frame.

○ **Angular** uses a rectangle with angled ends.

⊙ **None** leaves out the frame.

Label/Inspection Rate options

☐ **Label** toggles the display of the label text.

☑ **Inspection Rate** toggles the percentage.

-DIMINSPECT Command

Command: -diminspect
Add inspection data or [Remove] <Add>: *(Enter an option.)*
Select dimensions: *(Select one or more dimensions.)*
Select dimensions: *(Press Enter to exit dimension selection.)*
Enter shape option [Round/Angular/None] <Round>: *(Enter an option.)*
Enter label data or <None>: *(Enter a value.)*
Enter inspection rate <100%>: *(Enter a value.)*

COMMAND LINE OPTIONS

Add adds inspection notation to the selected dimension.

Remove removes the inspection notations.

Round uses a rounded rectangle for the frame.

Angular uses a rectangle with angled ends.

None leaves out the frame.

Label data specifies the label text.

Inspection Rate specifies the percentage.

DEFINITION

Inspection Rate — specifies as a percentage how frequently parts should be inspected.

TIPS

- This command works with linear and radial dimensions.
- This command makes AutoCAD drawings more compatible with Inventor mechanical CAD software.
- You can switch dimensions between the three inspection styles by rerunning the DimInspect command, and then selecting another shape.
- This command accepts inspection rates greater than 100%.

DimJogged

2006 Places jogged radial dimensions; used for very large radii.

Command	Aliases	Keyboard Shortcuts	Menu	Ribbon
dimjogged	**jog**	...	**Dimension**	...
	djo		**⇘Jogged**	

Command: dimjogged
Select arc or circle: *(Pick an arc or a circle.)*
Specify center location override: *(Pick point 2.)*
Dimension text = 28.7205
Specify dimension line location or [Mtext/Text/Angle]: *(Pick point 3.)*
Specify jog location: *(Pick point 4.)*

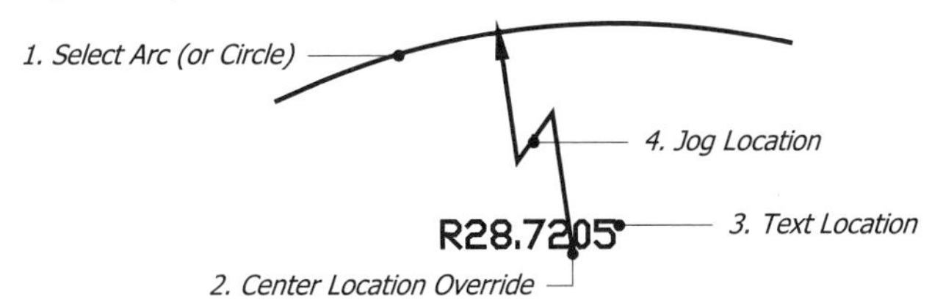

COMMAND LINE OPTIONS

Select arc or circle measures the radius of arcs and circles.

Specify center location override relocates the center from which to draw the dimension.

Specify dimension arc length location specifies the location of the arc length dimension.

Mtext changes the wording of the dimension text.

Text changes the position of the text.

Angle changes the angle of the dimension text.

Specify jog location positions the jog.

RELATED SYSTEM VARIABLE

DimJogAngle specifies the jog angle; default = 45 degrees.

TIPS

- The *transverse angle* of the jog can be set in the Dimension Style Manager; jogged dimensions can be edited using grips.
- This command is for arcs and circles whose centers are outside the drawing area. It is generally used with very shallow arcs.
- When you drag the dimension line outside an arc, AutoCAD draws an extension to the arc:

DimJogLine

2008 Adds jogged dimension lines to linear dimensions; used with very large objects.

Command	*Alias*	*Keyboard Shortcuts*	*Menu*	*Ribbon*
dimjogline	**djl**	**...**	**Dimension** ↳**Jogged Linear**	**Annotate** ↳**Dimensions** ↳**Jog Line**

Command: dimjogline
Select dimension to add jog or [Remove]: *(Pick a dimension, or type R.)*
Specify jog location (or press ENTER): *(Pick a point for the jog location, or press Enter for automatic placement.)*

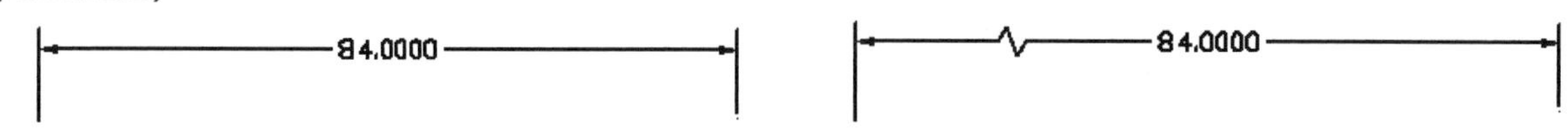

Left: *Linear dimension...*
Right: *...gets a jogged dimension line.*

COMMAND LINE OPTIONS

Select dimension selects the linear dimensions whose dimension lines should be jogged.

Remove removes jogs from selected dimensions.

Specify job location picks the location of the jog on the dimension lie.

Enter places jogs automatically, midway between the dimension text and the first extension line.

RELATED COMMANDS

All linear dimensioning commands, such as DimLinear and DimAligned.

RELATED SYSTEM VARIABLE

DimJogAng specifies the jog angle; default = 45 degrees.

TIPS

- You can move the location of the jog using grips editing.
- This command works with annotational dimensional constraints, but not with dynamic dimensional constraints.

DimLinear / DimHorizontal / DimRotated / DimVertical

Rel. 13 Drawi horizontal, rotated, and vertical dimensions.

Command	*Aliases*	*Keyboard Shortcuts*	*Menu*	*Ribbon*
dimlinear	**dli** **dimlin**	...	**Dimension** **↳Linear**	**Annotate** **↳Dimensions** **↳Linear**
dimhorizontal	...	...	...	...
dimrotated	...	...	...	...
dimvertical	...	...	...	...

Command: dimlinear
Specify first extension line origin or <select object>: *(Pick a point, or press Enter to select an object.)*
Specify second extension line origin: *(Pick a point.)*
Specify dimension line location or [Mtext/Text/Angle/Horizontal/Vertical/Rotated]: *(Pick a point, or enter an option.)*
Dimension text = *nnn*

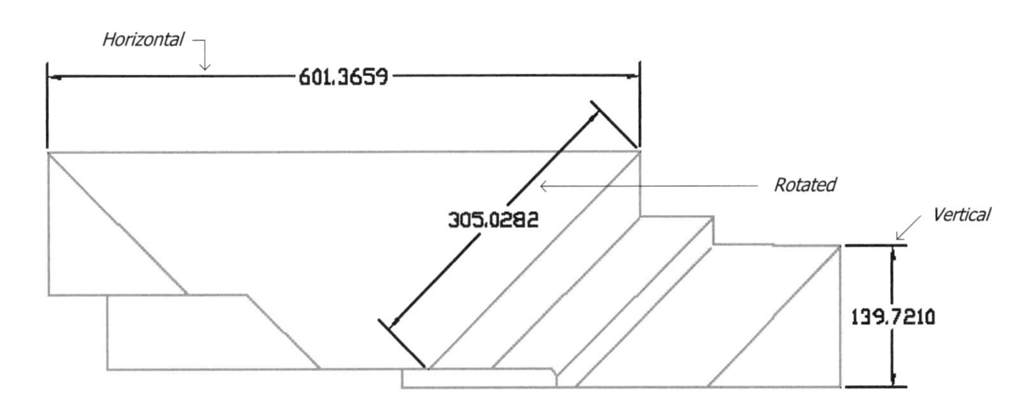

Examples of linear dimensions: horizontal, rotated (at 45 degrees), and vertical.

COMMAND LINE OPTIONS

Specify first extension line origin specifies the origin of the first extension line.

Select object dimensions a line, arc, or circle automatically.

Specify second extension line location specifies the location of the second extension line.

Specify dimension line location specifies the location of the dimension line.

Mtext displays the Text Formatting bar, which allows you to modify the dimension text; see the MText command.

Text option

Enter dimension text <*nnn*>: *(Enter dimension text, or press Enter to accept default value.)*

Enter dimension text prompts you to replace the dimension text on the command line.

Angle option
Specify angle of dimension text: *(Enter an angle.)*

Specify angle changes the angle of dimension text.

Horizontal option
Specify dimension line location or [Mtext/Text/Angle]: *(Pick a point or enter an option.)*

Horizontal forces dimension to be horizontal.

Vertical option
Specify dimension line location or [Mtext/Text/Angle]: *(Pick a point or enter an option.)*

Vertical forces dimension to be vertical.

Rotated option
Specify angle of dimension line <0>: *(Enter an angle.)*

Rotated rotates the dimension.

Select Object Options
Specify first extension line origin or <select object>: *(Press Enter.)*
Select object to dimension: *(Select one object.)*
Specify dimension line location or
[Mtext/Text/Angle/Vertical/Rotated]: *(Pick a point, or enter an option.)*
Dimension text = *nnn*

DIMHORIZONTAL Command

Draws horizontal dimensions (undocumented command; part of DimLinear).

Command: dimhorizontal
Specify first extension line origin or <select object>: *(Pick a point, or press Enter to select one object.)*
Specify second extension line origin: *(Pick a point.)*
Specify dimension line location or [Mtext/Text/Angle]: *(Pick a point, or select an option.)*
Dimension text = *nnn=*

COMMAND LINE OPTIONS

Specify first extension line origin specifies the first extension line's origin's location.

Select object dimensions a line, arc, or circle automatically.

Specify second extension line location specifies the second extension line's location.

Specify dimension line location specifies the dimension line's location.

Mtext displays the Text Formatting bar for modifying the dimension text; see the MText command.

Text prompts you to replace the dimension text on the command line: 'Enter dimension text <*nnn*>'.

Angle changes the angle of the dimension text: 'Specify angle of dimension text:'.

DIMROTATED Command

Draws rotated dimensions (undocumented command; part of DimLinear).

Command: dimrotated
Specify angle of dimension line <0>: *(Enter a rotation angle.)*
Specify first extension line origin or <select object>: *(Pick a point, or press Enter to select one object.)*
Specify second extension line origin: *(Pick a point.)*
Specify dimension line location or [Mtext/Text/Angle]: *(Pick a point, or select an option.)*
Dimension text = *nnn*

Figure: The aligned dimension is aligned with the two corner points. The rotated dimension is placed at 45 degrees.

COMMAND LINE OPTIONS

Specify angle of dimension line specifies the angle at which the dimension line is rotated.

Specify first extension line origin specifies the origin of the first extension line.

Select object dimensions a line, arc, or circle automatically.

Specify second extension line location specifies the location of the second extension line.

Specify dimension line location specifies the location of the dimension line.

Mtext displays the Text Formatting bar, which allows you to modify the dimension text; see the MText command.

Text prompts you to replace the dimension text on the command line: 'Enter dimension text <*nnn*>'.

Angle changes the angle of dimension text: 'Specify angle of dimension text:'.

DIMVERTICAL Command

Draws vertical dimensions (undocumented command; part of DimLinear).

Command: dimvertical
Specify first extension line origin or <select object>: *(Pick a point, or press Enter to select one object.)*
Specify second extension line origin: *(Pick a point.)*
Specify dimension line location or [Mtext/Text/Angle]: *(Pick a point, or select an option.)*
Dimension text = *nnn*

Figure: Examples of vertical dimensions.

COMMAND LINE OPTIONS

Specify first extension line origin specifies the origin of the first extension line.

Select object dimensions a line, arc, circle, or explodable object automatically.

Specify second extension line location specifies the location of the second extension line.

Specify dimension line location specifies the location of the dimension line.

Mtext displays the Multiline Text Editor dialog box, which allows you to modify the dimension text; see the MText command.

Text prompts you to replace the dimension text on the command line: 'Enter dimension text <*nnn*>'.

Angle changes the angle of the dimension text: 'Specify angle of dimension text:'.

RELATED DIM COMMANDS

DimAligned draws linear dimensions aligned with objects.

QDim dimensions objects quickly.

TIPS

- The DimRotated command draws dimensions at a specified angle, while the DimAligned command draws dimensions that align with the two pick points.

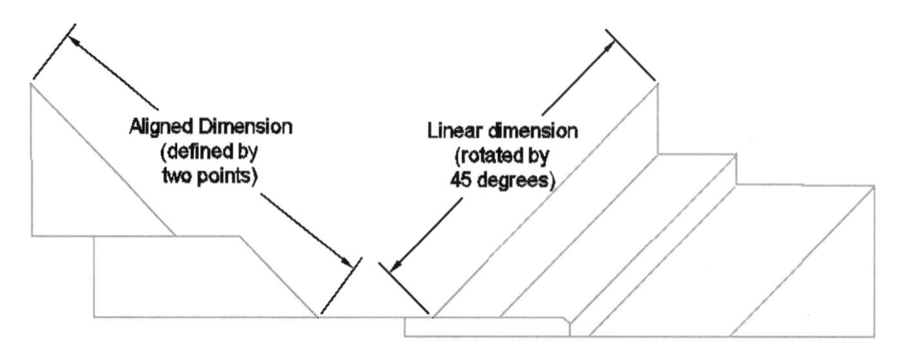

- You can specify the rotation angle of linear dimensions by picking two points at the Angle prompt.

DimOrdinate

Rel. 13 Draws x and y ordinate dimensions.

Command	Aliases	Keyboard Shortcuts	Menu	Ribbon
dimordinate	**dor**	...	**Dimension**	**Annotate**
	dimord		**↳Ordinate**	**↳Dimensions**
				↳Ordinate

Command: dimordinate
Specify feature location: *(Pick a point.)*
Specify leader endpoint or [Xdatum/Ydatum/Mtext/Text/Angle]: *(Pick a point, or enter an option.)*
Dimension text = *nnn*

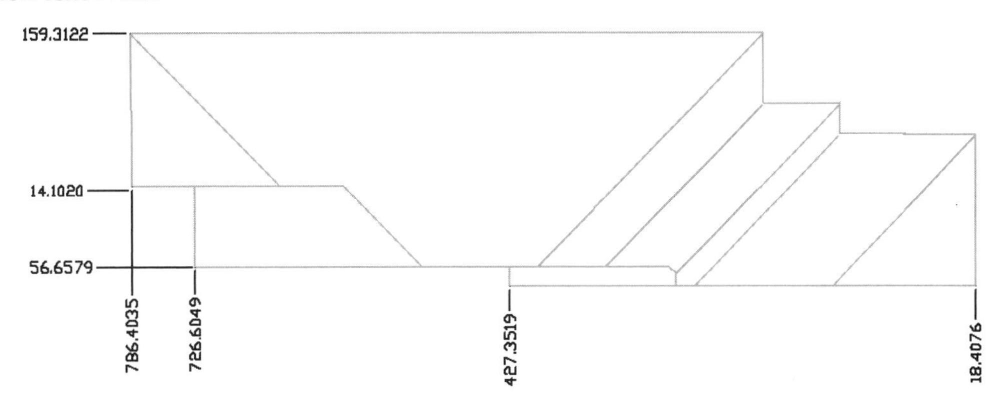

Examples of ordinate dimensions.

COMMAND LINE OPTIONS

Xdatum forces x ordinate dimension.

Ydatum forces y ordinate dimension.

Mtext displays the Text Formatting toolbar for modifying dimension text; see the MText command.

Text prompts you to replace dimension text on the command line: 'Enter dimension text <*nnn*>'.

Angle changes the angle of dimension text: 'Specify angle of dimension text:'.

RELATED DIM COMMANDS

Leader draws leader dimensions.

Tolerance draws geometric tolerances.

TIPS

- Strictly speaking, "ordinate" is the distance from the x-axis only, but AutoCAD and the rest of the drafting world use it to mean from both the x and y axes; distance from y axis is "abscissa."
- The 0,0 point is determined by the UCS origin. Use the UCS Origin command to relocate 0,0.
- Define the x=0 and y=0 ordinate dimensions, and then use DimBaseline for additional ones.

DimOverride

Rel. 13 Overrides the current dimension variables.

Command	*Aliases*	*Keyboard Shortcuts*	*Menu*	*Ribbon*
dimoverride	**dov**	...	**Dimension**	**Annotate**
	dimover		**↳Override**	**↳Dimensions**
				↳Override

Command: dimoverride

Enter dimension variable name to override or [Clear overrides]: *(Enter the name of a dimension variable, or type C.)*

Enter new value for dimension variable <*nnn*>: *(Enter a new value.)*

Select objects: *(Select one or more dimensions.)*

Select objects: *(Press Enter.)*

COMMAND LINE OPTIONS

Dimension variable to override requires you to enter the name of the dimension variable.

Clear removes the override.

New Value specifies the new value of the dimvar.

Select objects selects the dimension objects to change.

RELATED DIM COMMAND

DimStyle creates and modifies dimension styles.

RELATED DIM VARIABLES

All dimension variables.

DimRadius

Rel. 13 Draws radial dimensions on circles, arcs, and polyline arcs.

Command	*Aliases*	*Keyboard Shortcuts*	*Menu*	*Ribbon*
dimradius	**dra**	...	**Dimension**	**Annotate**
	dimrad		**↳Radius**	**↳Dimensions**
				↳Radius

Command: dimradius
Select arc or circle: *(Select an arc, circle, or explodable object.)*
Dimension text = *nnn*
Specify dimension line location or [Mtext/Text/Angle]: *(Pick a point, or enter an option.)*

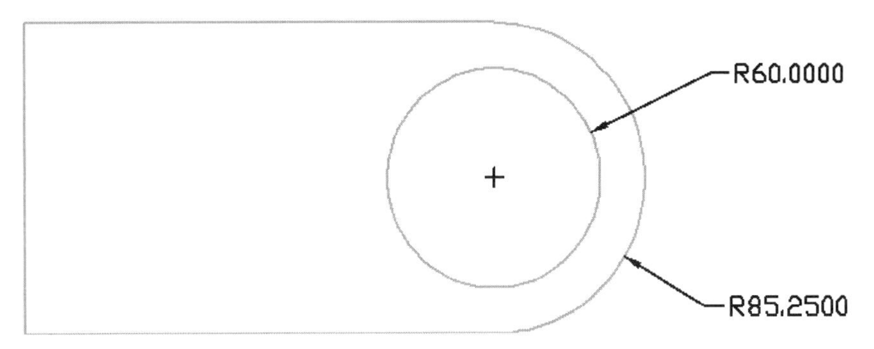

Examples of radial dimensions.

COMMAND LINE OPTIONS

Select arc or circle selects the arc, circle, or polyarc to dimension.

Specify dimension line location specifies the location of the angular dimension.

Mtext displays the Text Formatting bar, which allows you to modify the dimension text; see the MText command.

Text prompts you to replace the dimension text at the command line, 'Enter dimension text <*nnn*>'.

Angle changes the angle of dimension text, and prompts, 'Specify angle of dimension text:'.

RELATED DIM COMMANDS

DimCenter draws center marks on arcs and circles.

DimDiameter draws diameter dimensions on arcs and circles.

RELATED DIM VARIABLE

DimCen determines the size of the center mark.

TIP

- When you drag the dimension line outside an arc, AutoCAD draws an extension to the arc. See the DimDiameter command.

DimReassociate

2002 Reassociates dimensions with objects.

Command	Alias	Keyboard Shortcuts	Menu	Ribbon
dimreassociate	**dre**	...	**Dimension** ↳**Reassociate Dimensions**	**Annotate** ↳**Dimensions** ↳**Reassociate**

Command: dimreassociate

Select dimensions to reassociate...

Select objects: *(Select one or more dimensions, or enter All to select all dimensions.)*

nn **found.** *nn* **were on a locked layer.** *nn* **were not in current space.**

Select objects: *(Press Enter to end object selection.)*

The prompts that follow depend on the type of dimension selected.

Specify first extension line origin or [Select object]: *(Pick a point, or select an object.)*

Select arc or circle <next>: *(Press Enter to end object selection.)*

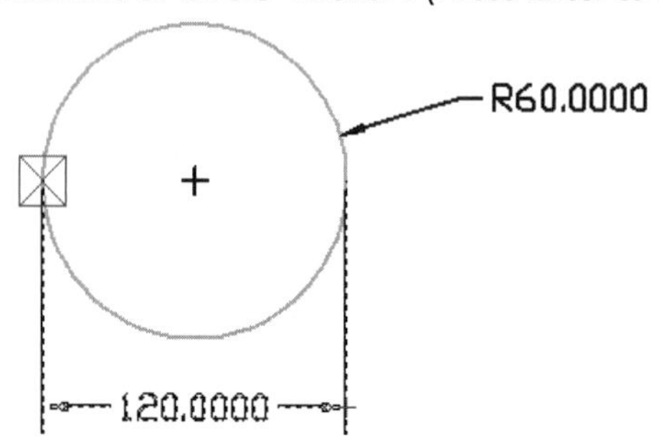

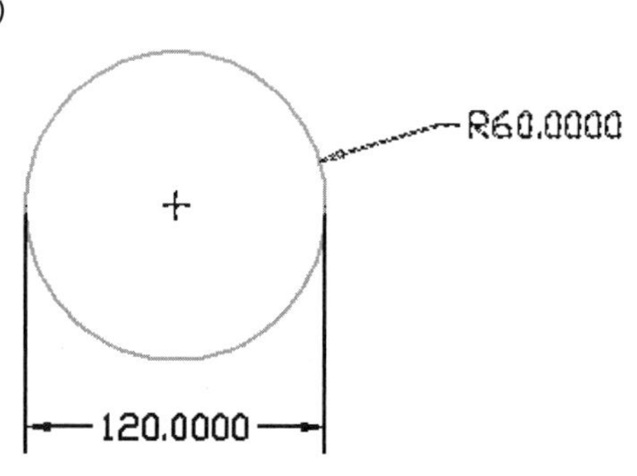

***Left:** Boxed X indicates that the line dimension is associated with the circle.*
***Right:** X indicates that the radius dimension is not associated.*

COMMAND LINE OPTIONS

Select objects selects the dimensions to reassociate.

Specify first extension line origin picks a point with which to associate the dimension.

Select arc or circle picks a circle or arc with which to associate the dimension.

Next goes to the next circle or arc.

RELATED DIM COMMANDS

DimDisassociate converts dimensions from associative to non-associative.

DimRegen updates the locations of associative dimensions.

TIPS

- As you select objects, this command ignores non-dimensions, dimensions on locked layers, and those not in the current space (model or paper).
- AutoCAD displays a boxed **X** to indicate the object with which the dimension is associated; an unboxed **X** indicates an unassociated dimension. The markers disappear when a wheelmouse performs a zoom or pan.

DimRegen

2002 Updates the locations of associative dimensions.

Command	*Alias*	*Keyboard Shortcuts*	*Menu*	*Ribbon*
dimregen	...	...	...	...

Command: dimregen

COMMAND LINE OPTIONS

None.

RELATED DIM COMMANDS

DimDisassociate converts dimensions from associative to non-associative.

DimReassociate converts dimensions from non-associative to associative.

TIP

- This command is meant for use after these conditions:
    - The drawing has been edited by a version of AutoCAD prior to 2004.
    - The drawing contains dimensions associated with an external reference, and the xref has been edited.
    - The drawing is in layout mode, model space is active, and a wheelmouse has been used to pan or zoom.

DimSpace

2008 Spaces parallel dimensions evenly.

Command	Aliases	Keyboard Shortcuts	Menu	Ribbon
dimspace	...	...	**Dimension ↳Space**	**Annotate ↳Dimensions ↳Adjust Space**

Command: dimspace
Select base dimension: *(Select one parallel linear or angular dimension.)*
Select dimensions to space: *(Select other dimensions to space equally.)*
Enter value or [Auto] <Auto>: *(Specify the spacing value, or type A.)*

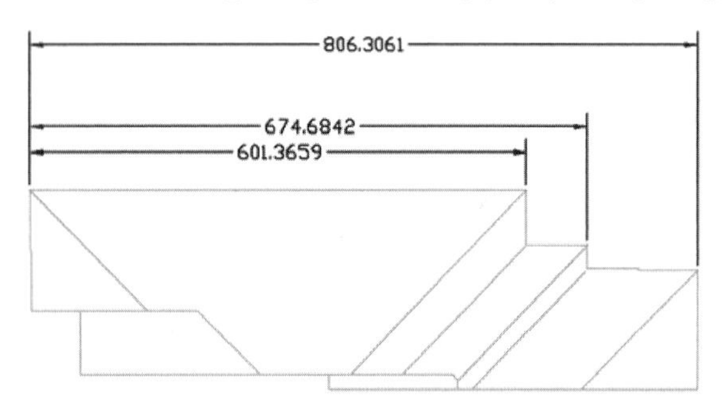

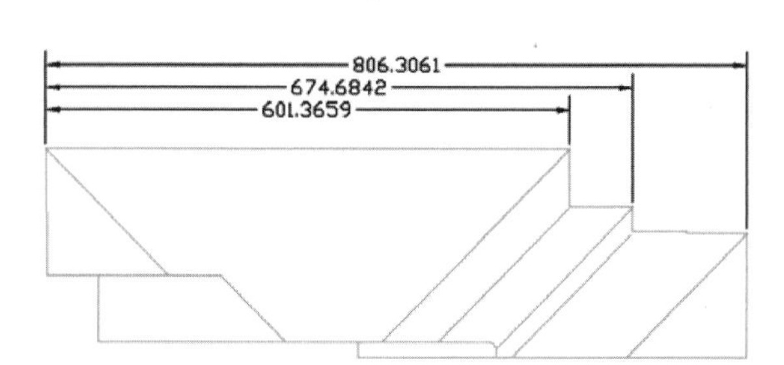

Left: *Dimensions before being spaced.*
Right: *After spacing to the base dimension.*

COMMAND LINE OPTIONS

Select base dimension specifies the dimension by which others are spaced.

Select dimensions to space specifies the dimensions to be relocated.

Enter value spaces the selected dimensions by the amount specified; entering 0 aligns them end to end.

Auto spaces the dimensions using 2x the height of the dimension text of the base dimension.

TIPS

- When dimension lines overlap, this command creates the spaced lines, but fails to move the dimensions into the spaces created — possibly a bug.

DimStyle

Rel. 13 Creates and edits dimension styles (short for DIMension STYLE).

Command	Aliases	Keyboard Shortcuts	Menu	Ribbon
'dimstyle	**d** **dst**	...	**Dimension** **↳Style**	**Annotate** **↳Dimensions** **↳Dimension Style**
	dimsty **ddim**	... ...	**Format** **↳Dimension Style**	...
'-dimstyle	...	...	**Dimension** **↳Update**	...

Command: dimstyle

Displays dialog box.

DIMENSION STYLE MANAGER DIALOG BOX

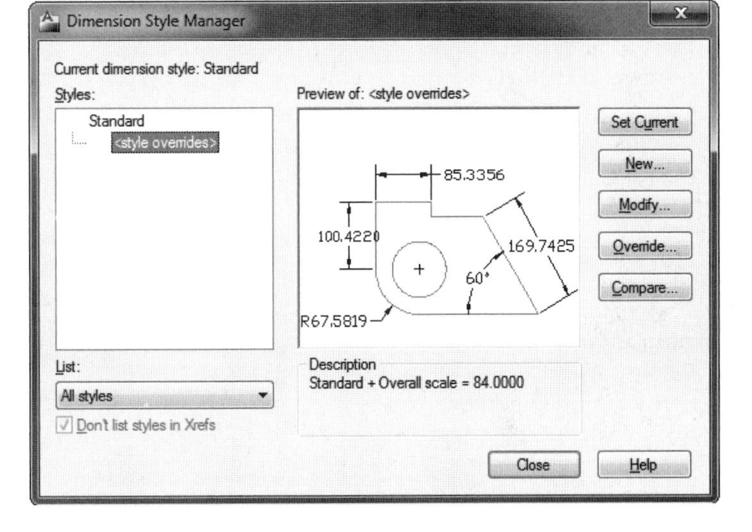

Styles lists the names of dimension styles in the drawing.

List modifies the style names listed under Styles:

- **All styles** lists all dimension style names stored in the current drawing (default).
- **Styles in use** lists only those dimstyles used by dimensions in the drawing.

Don't list styles in Xrefs

☑ Does not list dimension styles found in externally-referenced drawings.

☐ Lists xref dimension styles under Styles (default).

Set Current sets the selected style as the current dimension style.

New creates a new dimension style via Create New Dimension Style dialog box.

Modify modifies an existing dimension style; displays the Modify Dimension Style dialog box.

Override allows temporary changes to dimension styles; displays Override Dimension Style dialog box.

Compare lists the differences between dimension variables of two styles; displays the Compare Dimension Styles dialog box.

Shortcut Menu

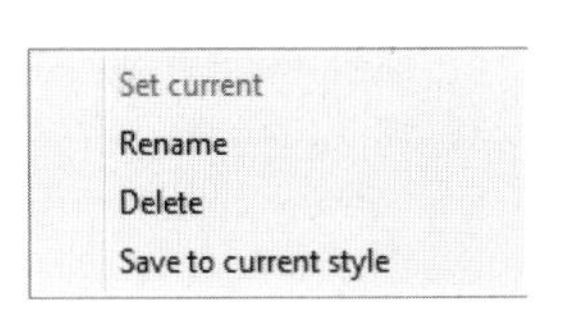

Access this menu by right-clicking a dimension style name under Styles.

Set Current sets the selected dimension style as the current style.

Rename renames dimension styles.

Delete erases selected dimension styles from the drawing; you cannot erase the Standard style and styles that are in use.

SELECT ANNOTATION SCALE DIALOG BOX

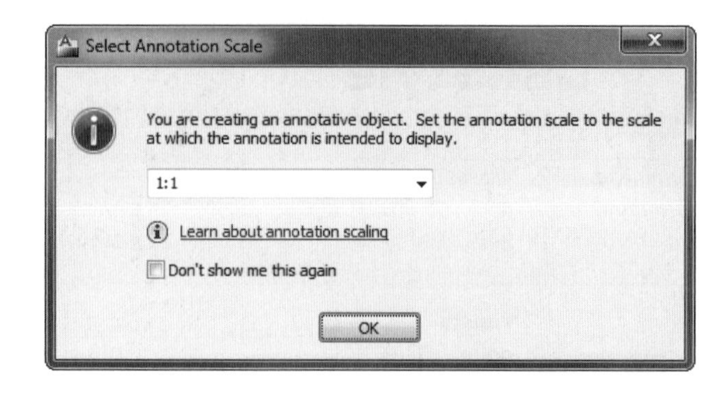

The first time a dimension is placed in a new drawing, and the style is annotative, AutoCAD displays this dialog box.

Set the Annotation Scale overrides the annotation scale specified by the dimension style.

☐ **Don't Show Me This Again** prevents the dialog box from appearing in the future.

CREATE NEW DIMENSION STYLE DIALOG BOX

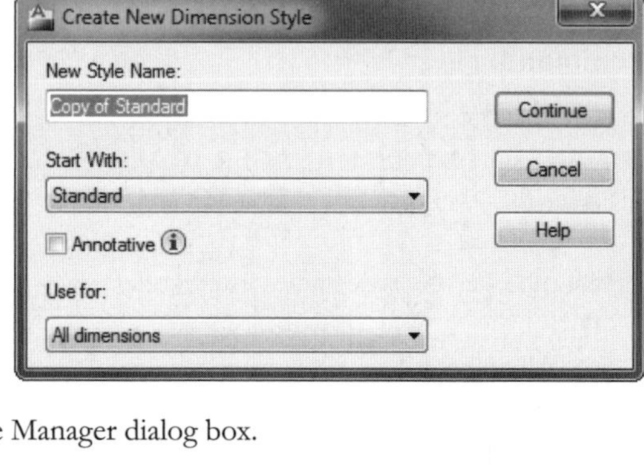

When creating new dimension styles, typically you copy an existing style, and then change it.

New Style Name names the new dimension style.

Start With lists the names of the current dimension style(s), which are used as the template for the new dimension style.

☑ **Annotative** makes the style annotative; see the Fit tab.

Use for creates a substyle that applies to a specific type of dimension type: linear, angular, radius, diameter, ordinate, leaders and tolerances.

Continue continues to the next dialog box, New Dimension Style.

Cancel dismisses this dialog box, and returns to the Dimension Style Manager dialog box.

Lines tab

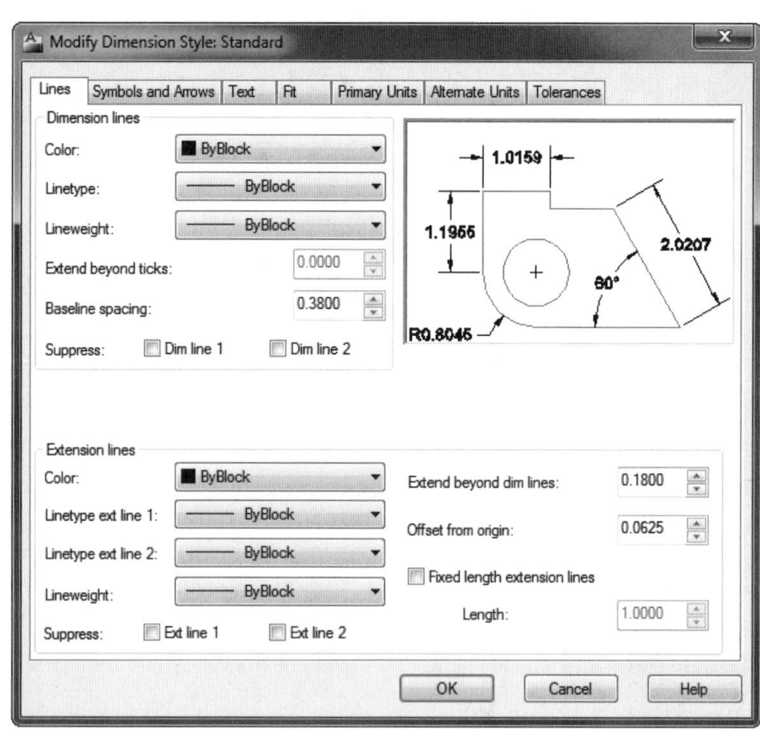

Sets the format of dimension lines and extension lines.

Dimension Lines options

Color specifies the color of the dimension line; select Other to display the Select Color dialog box (stored in dimension variable *DimClrD*).

Linetype specifies the linetype of the dimension line (*DimLtype*).

Lineweight specifies the lineweight of the dimension line (*DimLwD*).

Extend beyond ticks specifies the distance the dimension line extends beyond the extension line; used with oblique, architectural, tick, integral, and no arrowheads (*DimDlE*).

Baseline spacing specifies the spacing between the dimension lines of a baseline dimension (*DimDlI*); different than DimSpace.

Suppress suppresses dimension lines when outside the extension lines:

- ☐ **Dim line 1** for portion of dimension line left of text (*DimSD1*).
- ☐ **Dim line 2** for portion of dimension line right of text (*DimSD2*).

Extension Lines options

Color specifies the color of the extension line; select Other to display the Select Color dialog box (*DimClrE*).

Linetype ext 1 specifies the linetype of the first extension line (*DimLtEx1*).

Linetype ext 2 specifies the linetype of the second extension line (*DimLtEx2*).

Lineweight specifies the lineweight of the extension line (*DimLwE*).

Suppress suppresses extension lines:

- ☐ **Ext line 1** for left extension line (*DimSE1*).
- ☐ **Ext line 2** for right extension line (*DimSE2*).

Extend beyond dim lines specifies the distance the extension line extends beyond the dimension line; used with oblique, architectural, tick, integral, and no arrowheads (*DimExe*).

Offset from origin specifies the distance from the origin to the start of the extension lines (*DimExO*).

☐**Fixed length extension lines** specifies the length of extension lines (*DimFxlOn*):

- ☐ **Length** is 0.18 by default (*DimFXL*).

Symbols and Arrows tab

Sets the format of arrowheads, center marks, and arc lengths.

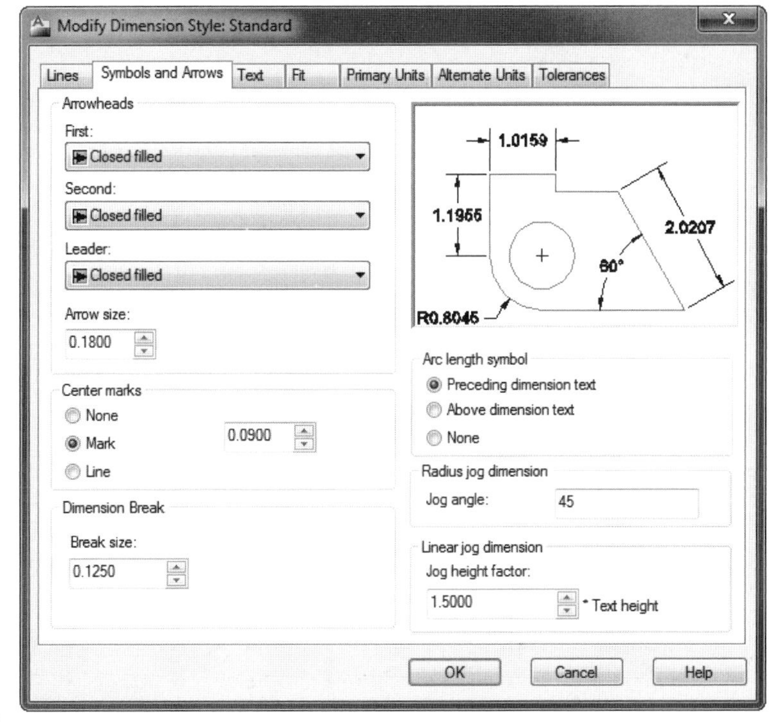

Arrowheads options

First specifies name of arrowhead for the first end of the dimension line (*DimBlk1*).

Second specifies arrowhead for the second end of the dimension line (*DimBlk2*); select User Arrow to display the Select Custom Arrow Block dialog box, illustrated on the next page.

Leader specifies the arrowhead for the leader; select User Arrow to display the Select Custom Arrow Block dialog box (*DimLdrBlk*).

Arrow Size specifies the size of arrowheads (*DimASz*).

Center Marks options

- ○ **None** places no center marks or center lines (*DimCen*=0).
- ⊙ **Mark** places center marks (*DimCen* > 0).
- ○ **Line** places center marks and center lines (*DimCen* < 0).

Size specifies size of center marks or center lines (*DimCen*; default = 0.09).

Dimension Break option

Break size specifies the gap used in "broken" dimension and extension lines. See the DimBreak command.

Arc Length Symbol options

⊙ **Preceding dimension text** places symbol in front of text (*DimArcSym*).

O **Above dimension text** places symbol above text.

O **None** places no symbol.

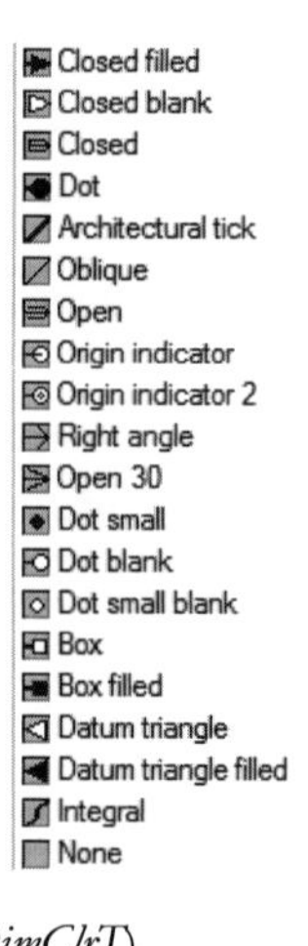

Radius Dimension Jog option

Jog Angle specifies the default jog angle for the DimJogged command (*DimJogAngle*).

Linear Jog Dimension option

Jog Height factor specifies the jog distance as a multiplier of the text height.

Text tab

Text Appearance options

Text style specifies the text style name for dimension text (*DimTxSty*).

[...] displays the Text Style dialog box; see the Style command.

Text color specifies the color of the dimension line; select Other to display the Select Color dialog box (*DimClrT*).

Fill color specifies the color of the background behind the text (*DimTFill* and *DimTFillClr*).

Text height specifies the height of the dimension text, when the height defined by the text style is 0 (*DimTxt*).

Fraction height scale scales fraction text height relative to dimension text; AutoCAD multiples this value by the text height (*DimTFac*).

☐ **Draw frame around text** draws a rectangle around dimension text; when on, dimension variable DimGap is set to a negative value (*DimGap*).

Text Placement options

Vertical specifies the vertical justification of dimension text relative to the dimension line:

- **Centered** centers dimension text in the dimension line (*DimTad* = 0).
- **Above** places text above the dimension line (*DimTad* = 1).
- **Outside** places text on the side of the dimension line farthest from the first defining point (*DimTad* = 2).
- **JIS** places text in conformity with JIS (*DimTad*= 3).

Horizontal specifies the horizontal justification of dimension text along the dimension and extension lines (*DimJust*; default = 0):

- **Centered** centers dimension text along the dimension line between the extension lines (*DimJust* = 0).
- **1st Extension Line** left-justifies the text with the first extension line (*DimJust* = 1).
- **2nd Extension Line** right-justifies the text with the second extension line (*DimJust* = 2).

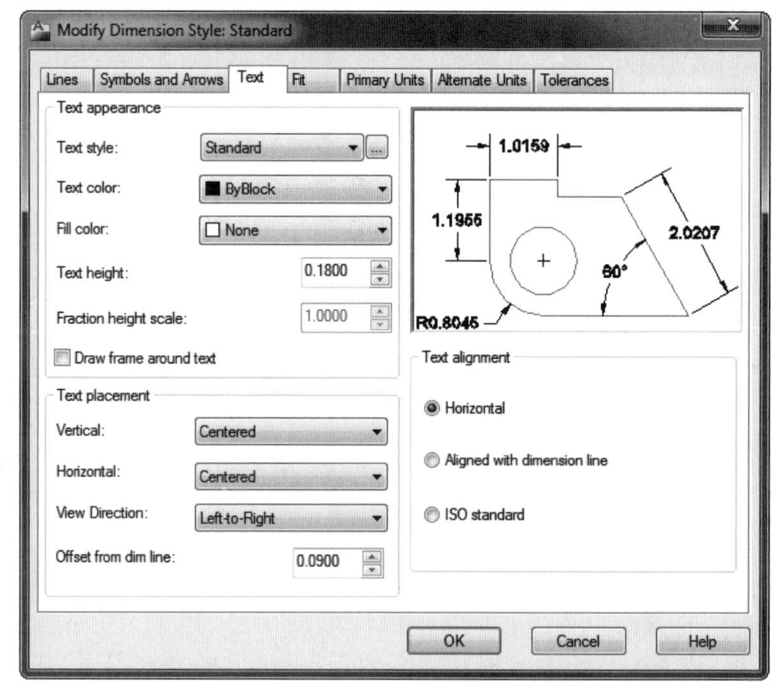

- **Over 1st Extension Line** places the text over the first extension line (*DimJust* = 3).
- **Over 2nd Extension Line** places the text over the second extension line (*DimJust* = 4).

Offset from dimension line specifies the text gap, the distance between the dimension text and the dimension line (*DimGap*).

Text Alignment options

Horizontal forces dimension text always to be horizontal (*DimTih* = on; *DimToh* = off).

Aligned with dimension line forces dimension text to be aligned with the dimension line (*DimTih* = off; *DimToh* = off).

ISO Standard forces text to be aligned with the dimension line when inside the extension lines; forces text to be horizontal when outside the extension lines (*DimTih* = off; *DimToh* = on).

Fit tab

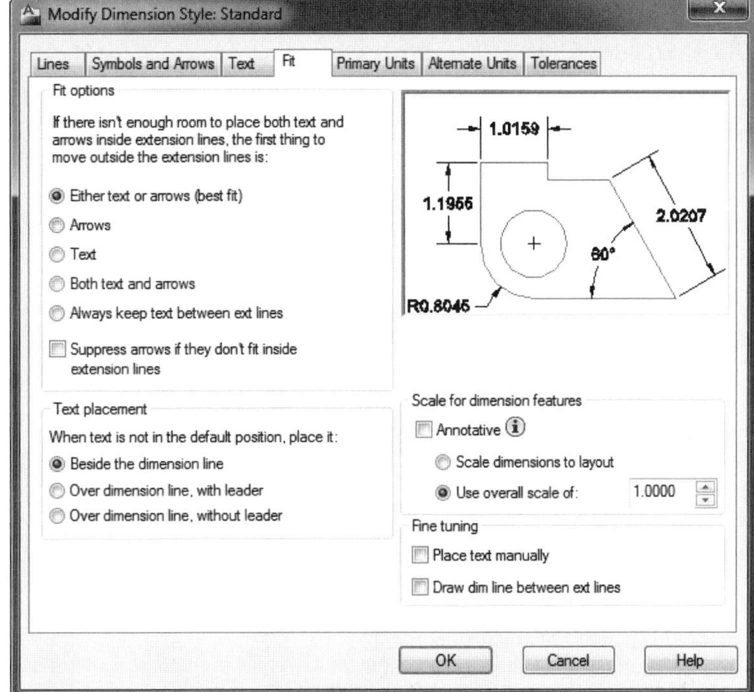

Fit options

If there isn't enough room to place both text and arrows inside extension lines, the first thing to move outside the extension lines is:

⊙ **Either the text or the arrows, whichever fits best** places dimension text and arrowheads between the extension lines when space is available; when space is not available for both, the text or the arrowheads are placed outside the extension lines, whichever fits best; if there is room for neither, both are placed outside the extension lines (*DimAtFit* = 3).

O **Arrows** places arrowheads between the extension lines when there is not enough room for arrowheads and dimension text (*DimAtFit* = 2).

O **Text** places text between extension lines when there is not enough room for arrowheads and dimension text (*DimAtFit* = 1).

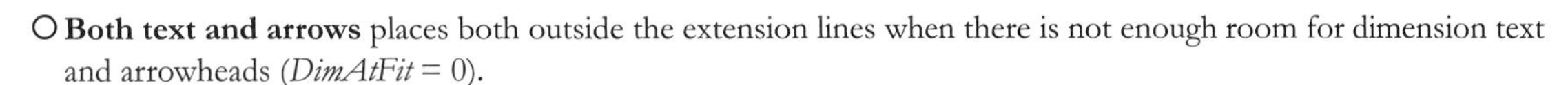

O **Both text and arrows** places both outside the extension lines when there is not enough room for dimension text and arrowheads (*DimAtFit* = 0).

O **Always keep text between ext lines** forces text between the extension lines (*DimTix*).

☐ **Suppress arrows if they don't fit inside the extension lines** suppresses arrowheads when there is not enough room between the extension lines.

Text Placement:

⊙ **Beside the dimension line** places dimension text beside the dimension line (*DimTMove* =0).

O **Over the dimension line, with a leader** draws a leader when dimension text is moved away from the dimension line (*DimTMove* = 1).

○ **Over the dimension line, without a leader** does not draw a leader when dimension text is moved away from the dimension line (*DimTMove* = 2).

Scale for Dimension Features

□ **Annotative** turns on the annotative property of dimensions (*DimAnno*).

○ **Scale dimension to layout** determines the scale factor of dimensions in layout mode; based on scale factor between current model space viewport and layout (*DimScale*).

⊙ **Use overall scale of** specifies the scale factor for all dimensions in the drawing; affects text and arrowhead sizes, distances, and spacing (*DimScale*).

Fine Tuning options

□ **Place text manually when dimensioning** places text at the position picked at the 'Dimension line location' prompt (*DimUpt*).

□ **Always draw dim line between ext lines** forces the dimension line between the extension lines (*DimTofl*).

Primary Units Tab

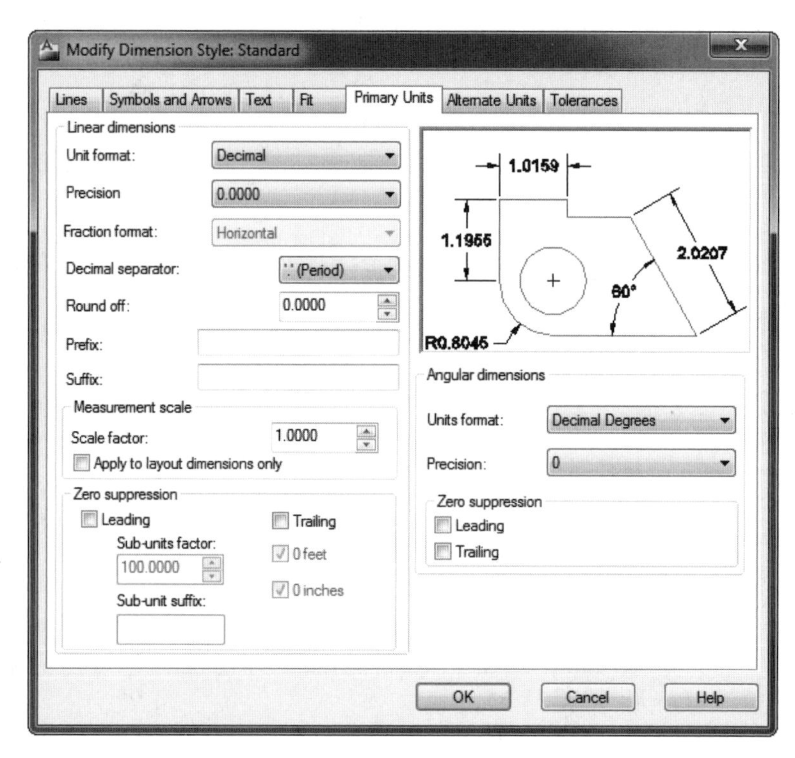

Linear Dimensions options

Unit Format specifies the linear units format; does not apply to angular dimensions (*DimLUnit*).

Precision specifies the number of decimal places (or fractional accuracy) for linear dimensions (*DimDec*).

Fraction Format specifies the stacking format of fractions (*DimFrac*):

- Horizontal stacked — $\frac{1}{2}$
- Diagonal stacked — 1/2

Decimal Separator specifies the separator for decimal formats (*DimDSep*).

Round Off specifies the format for rounding dimension values; does not apply to angular dimensions (*DimRnd*).

Prefix specifies a prefix for dimension text (*DimPost*); use these control codes to show special characters:

Control Code	Meaning
%%nnn	Character specified by ASCII number *nnn*.
%%o	Turns on and off overscoring.
%%u	Turns on and off underscoring.
%%d	Degrees symbol (°).
%%p	Plus/minus symbol (±).
%%c	Diameter symbol (Ø).
%%%	Percentage sign (%).

Suffix specifies a suffix for dimension text (*DimPost*); use the control codes listed above for special characters.

Measurement Scale

Scale factor specifies a scale factor for linear measurements, except for angular dimensions (*DimLFac*); for example, use this to change dimension values from imperial to metric.

☐ **Apply to layout dimensions only** applies the scale factor only to dimensions created in layout mode or paper space (stored as a negative value in *DimLFac*).

Zero Suppression options:

☐ **Leading** suppresses leading zeros in all decimal dimensions (*DimZin* = 4).

☐ **Trailing** suppresses trailing zeros in all decimal dimensions (*DimZin* = 8).

☑ **0 Feet** suppresses zero feet of feet-and-inches dimensions (*DimZin* = 0).

☑ **0 Inches** suppresses zero inches of feet-and-inches dimensions (*DimZin* = 2).

Angular Dimensions options

Units Format specifies the format of angular dimensions (*DimAUnit*).

Precision specifies the precision of angular dimensions (*DimADec*).

Zero Suppression options

Same as for linear dimensions (DimAZin).

Alternate Units Tab

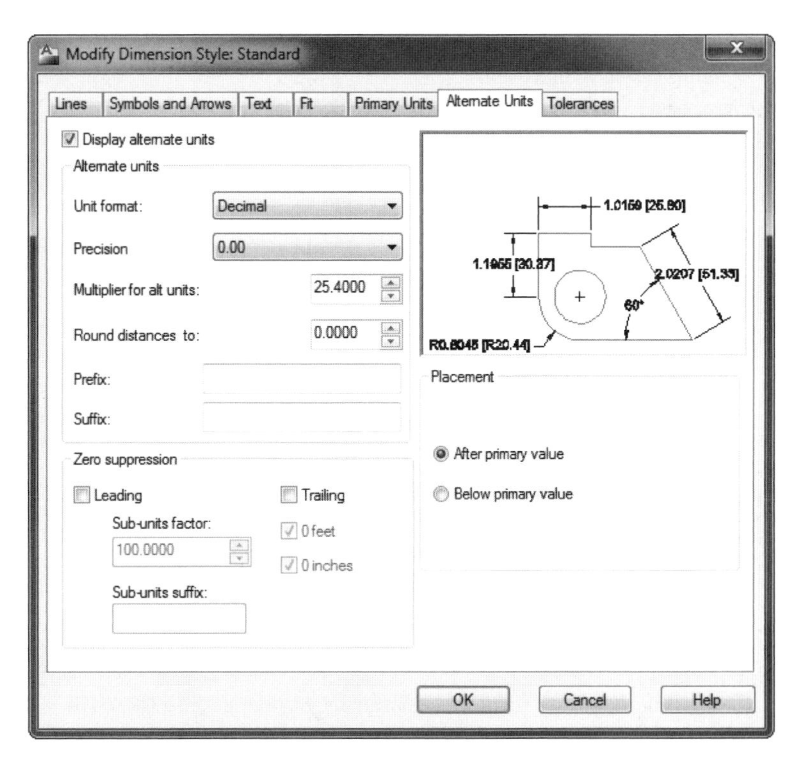

☐ **Display alternate units** adds alternate units to dimension text (*DimAlt*).

Alternate Units options

Unit Format specifies the alternate units format (*DimAltU*).

Precision specifies number of decimal places or fractional accuracy (*DimAltD*).

Multiplier for alt units specifies the conversion factor between primary and alternate units (*DimAltF*).

Round distances to specifies the format for rounding dimension values; does not apply to angular dimensions (*DimAltRnd*).

Prefix specifies a prefix for dimension text (*DimAPost*); you can use control codes to show special characters.

Suffix specifies a suffix for dimension text (*DimAPost*).

Zero Suppression options

☐ **Leading** suppresses leading zeros in all decimal dimensions (*DimAltZ* = 4).

☐ **Trailing** suppresses trailing zeros in all decimal dimensions (*DimAltZ* = 8).

☑ **0 Feet** suppresses zero feet of feet-and-inches dimensions (*DimAltZ* = 0).

☑ **0 Inches** suppresses zero inches of feet-and-inches dimensions (*DimAltZ* = 2).

Placement:

⊙ **After primary units** places alternate units behind the primary units (*DimAPost*).

○ **Below primary units** places alternate units below the primary units.

Tolerances tab

Specifies alignment of Deviation and Limits tolerances.

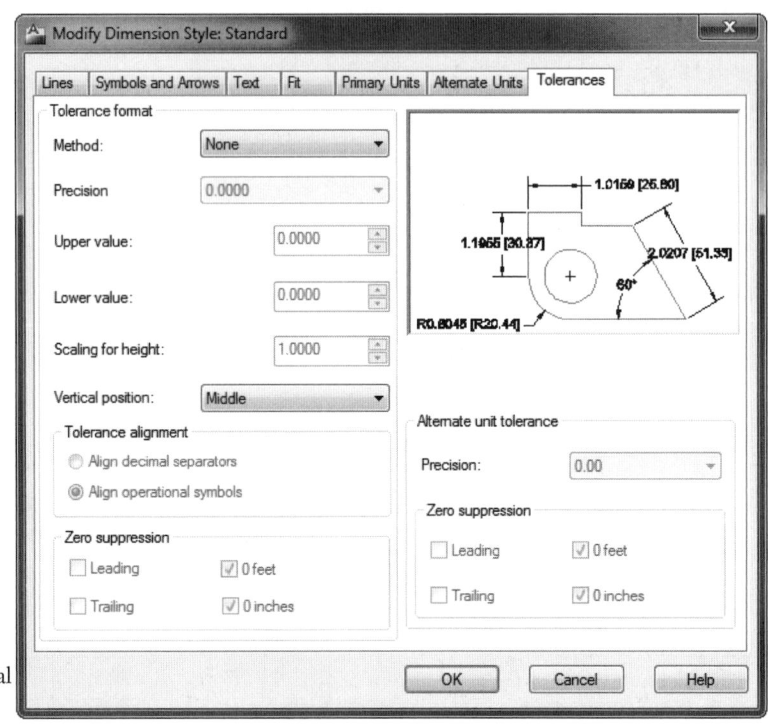

Tolerance Format options

Method specifies the tolerance format:

- **None** does not display tolerances (*DimTol*).
- **Symmetrical** places ± after the dimension (*DimTol*=0; *DimLim*=0).
- **Deviation** places + and – symbols (*DimTol*=1; *DimLim*=1).
- **Limits** places maximum over minimum value (*DimTol*=0; *DimLim*=1):

Maximum value = dimension value + upper value.

Minimum value = dimension value - lower value.

- **Basic** boxes the dimension text (*DimGap*=negative value).

Precision specifies the number of decimal places for tolerance values (*DimTDec*).

Upper value specifies the upper tolerance value (*DimTp*).

Lower value specifies the lower tolerance value (*DimTm*).

Scaling for height specifies the scale factor for tolerance text height (*DimTFac*).

Vertical position specifies the vertical text position for symmetrical and deviation tolerances:

Top aligns the tolerance text with the top of the dimension text (*DimTolJ*=2).
Middle aligns the tolerance text with the middle of the dimension text (*DimTolJ*=1).
Bottom aligns the tolerance text with bottom of dim text (*DimTolJ*=0).

Tolerance Alignment options

○ **Align Decimal Separators** lines up tolerances by their decimal points.

⊙ **Align Operator Symbols** lines up tolerances by their prefix symbols, + and - .

Zero Suppression options

☐ **Leading** suppresses leading zeros in all decimal dimensions (*DimTZin* = 4).

☐ **Trailing** suppresses trailing zeros in all decimal dimensions (*DimTZin* = 8).

☑ **0 Feet** suppresses zero feet of feet-and-inches dimensions (*DimTZin* = 0).

☑ **0 Inches** suppresses zero inches of feet-and-inches dimensions (*DimTZin* = 2).

Alternate Unit Tolerance options

Precision specifies the precision (number of decimal places) of tolerance text (*DimAltTd*).

Zero Suppression options: the same as for tolerance format (*DimAltTz*).

MODIFY DIMENSION STYLE DIALOG BOX

Dialog box is identical to the New Dimension Style dialog box.

OVERRIDE DIMENSION STYLE DIALOG BOX

Dialog box is identical to the New Dimension Style dialog box.

COMPARE DIMENSION STYLES DIALOG BOX

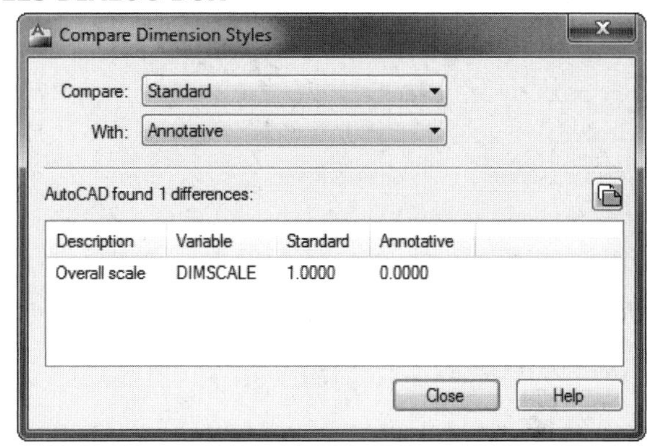

The list is blank when AutoCAD finds no differences. When With is set to <none> or the same style as Compare, AutoCAD displays all dimension variables.

Compare displays the name of one dimension style.

With displays the name of the second dimension style.

Copy to Clipboard copies the style comparison text to the Clipboard, which can be pasted in another Windows application.

Description describes the dimension variable.

Variable names the dimension variable.

Close closes the dialog box.

-DIMSTYLE Command

Command: -dimstyle
Current dimension style: Standard Annotative: No
Enter a dimension style option
[ANnotative/Save/Restore/STatus/Variables/Apply/?] <Restore>: *(Enter an option.)*
Current dimension style: Standard
Enter a dimension style name, [?] or <select dimension>: *(Enter an option.)*
Select dimension: *(Select a dimension in the drawing.)*
Current dimension style: Standard

COMMAND LINE OPTIONS

Annotative toggles the annotative property of dimension styles.

Save saves current dimension variable settings as a named dimension style.

Restore retrieves dimvar settings from a named dimension style.

STatus lists current setting of all dimension variables.

Variables lists the current setting for a specific dimension variable.

Apply updates selected dimension objects with current dimstyle settings.

? lists names of dimstyles stored in drawing.

INPUT OPTIONS

~dimvar (*tilde prefix*) lists the differences between current and selected dimstyles.

Enter lists the dimension variable settings for the selected dimension object.

RELATED DIMENSIONING COMMANDS

DDim changes dimension variable settings.

DimScale determines the scale of dimension text.

RELATED DIMENSION VARIABLES

All.

DimStyle contains the name of the current dimension style.

DimAnno toggles annotative scaling in dimensions.

TIPS

- At the 'Dim' prompt, the Style command sets the text style for the dimension text, and does *not* select a dimension style.
- Dimension styles cannot be stored to disk, only in drawings. However, you can import dimension styles from other drawings by using the XBind Dimstyle command, by dragging dimstyle names from Design Center, or by setting them up in template files.
- As of AutoCAD 2008, dimensions support annotative scale factors. This means that scale-dependent objects are displayed when their scale factor matches the scale of model space. This includes dimension text and symbols, arrowheads, leaders, and tolerances.
- When the annotative property is turned on, AutoCAD forces associativity in dimensions.
- All new drawings contain at least these two dimension sytles: Annotative and Standard (non-annotative).
- Dynamic dimensional constraints ignore dimension styles; annotative dimensional constraints follow dimstyles.

DimTEdit

Rel. 13 Changes the location and orientation of text in dimensions dynamically.

Command	Alias	Keyboard Shortcuts	Menu	Ribbon
dimtedit	**dimted**	...	**Dimension** ↳**Align Text**	**Annotate** ↳**Dimensions** ↳**Align Text**

Command: dimtedit

Select dimension: *(Select a dimension in the drawing.)*

Specify new location for dimension text or [Left/Right/Center/Home/Angle]: *(Pick a point, or enter an option.)*

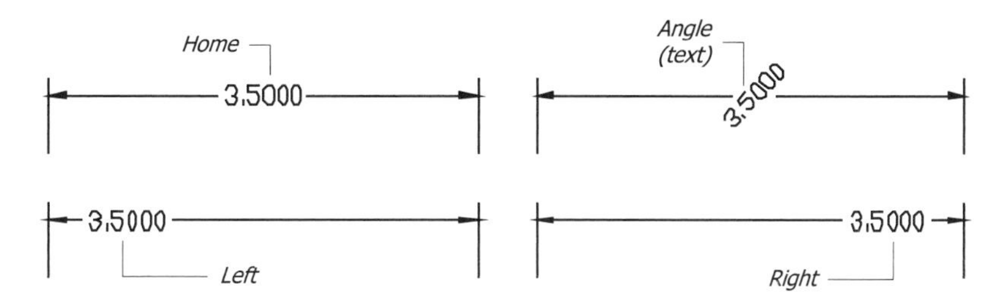

COMMAND LINE OPTIONS

Select dimension selects the dimension to edit.

Angle rotates the dimension text.

Center centers the text on the dimension line.

Home returns the dimension text to the original position.

Left moves the dimension text to the left.

Right moves the dimension text to the right.

RELATED DIM VARIABLES

DimSho specifies whether dimension text is updated dynamically while dragged.

DimTih specifies whether dimension text is drawn horizontally or aligned with the dimension line.

DimToh specifies whether dimension text is forced inside the dimension lines.

TIPS

- This command works only with dimensions created with DimAssoc = 1 or 2; use the DdEdit command to edit text in non-associative dimensions.
- An angle of 0 returns dimension text to its default orientation.

Dist

Ver. 1.0 Lists the 3D distances and angles between two points (short for DISTance).

Command	Alias	Keyboard Shortcuts	Menu	Ribbon
'dist	**di**	...	**Tools** ↳**Inquiry** ↳**Distance**	...

Command: dist
Specify first point: *(Pick a point.)*
Specify second point: *(Pick another point.)*

Sample result:
Specify first point: Specify second point:
Distance = 19, Angle in XY Plane = 22, Angle from XY Plane = 0
Delta X = 18, Delta Y = 7, Delta Z = 0

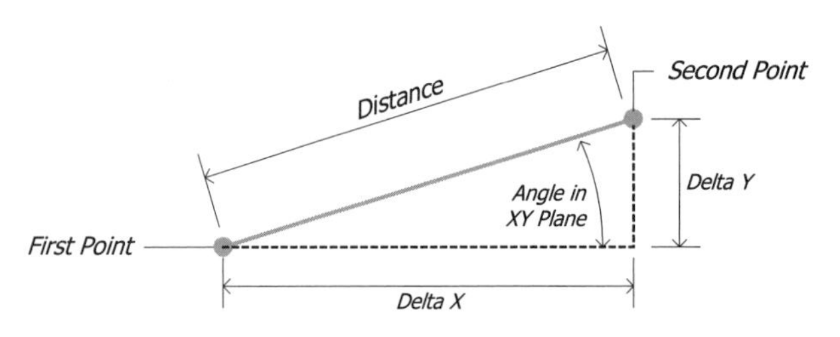

COMMAND LINE OPTIONS

Specify first point determines the start point of distance measurement.

Specify second point determines the endpoint.

RELATED COMMANDS

Area calculates the area and perimeter of objects.

Id lists the 3D coordinates of points.

MeasureGeom measures distances, radii, areas, angles, and volumes.

RELATED SYSTEM VARIABLE

Distance specifies the last calculated distance.

TIPS

- Use object snaps to measure precisely the distance between two geometric features.
- When the z-coordinate is left out, Dist uses the current elevation for z.
- Object snaps give 3D points and therefore the 3D distance.

Aliased Command

Details about the **DistantLight** command are found under the Light command.

Divide

Ver. 2.5 Places points or blocks at equal distances along objects.

Command	*Alias*	*Keyboard Shortcuts*	*Menu*	*Ribbon*
divide	**div**	**...**	**Draw**	**Home**
			↳Point	**↳Draw**
			↳Divide	**↳Divide**

Command: divide
Select object to divide: *(Select one object.)*
Enter the number of segments or [Block]: *(Enter a number, or enter B.)*

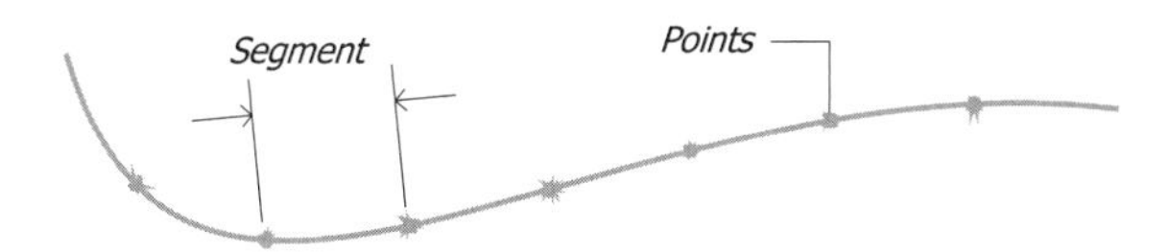

COMMAND LINE OPTIONS

Select object to divide selects a single open or closed object.

Enter the number of segments specifies the number of segments; must be a number between 2 and 32767.

Block Options

Enter name of block to insert: *(Enter the name of a block.)*
Align block with object? [Yes/No] <Y>: *(Type Y or N.)*
Enter the number of segments: *(Enter a number.)*

Enter name of block to insert specifies the block to insert. *Caution*! The block must be defined in the current drawing prior to starting this command.

Align block with object aligns the block's x axis with the object.

RELATED COMMANDS

Block creates the block to use with the Divide command.

DdPType controls the size and style of points.

Insert places a single block in the drawing.

MInsert places an array of blocks in the drawing.

Measure places points or blocks at measured distances.

TIPS

- The first dividing point on a closed polyline is its initial vertex; on circles, the first dividing point is in the 0-degree direction from the center.
- The points or blocks are placed in the Previous selection set, so that you can select them with the next 'Select Objects' prompt.
- Objects are unchanged by this command.
- Use the DdPType command to make points visible.

Donut

Ver. 2.5 Draws solid-filled circles as wide polylines consisting of a pair of arcs.

Command	*Aliases*	*Keyboard Shortcuts*	*Menu*	*Ribbon*
donut	**do** **doughnut**	...	**Draw** **↳Donut**	**Home** **↳Draw** **↳Donut**

Command: donut
Specify inside diameter of donut <0.5000>: *(Enter a value.)*
Specify outside diameter of donut <1.0000>: *(Enter a value.)*
Specify center of donut or <exit>: *(Pick a point.)*
Specify center of donut or <exit>: *(Press Enter to exit command.)*

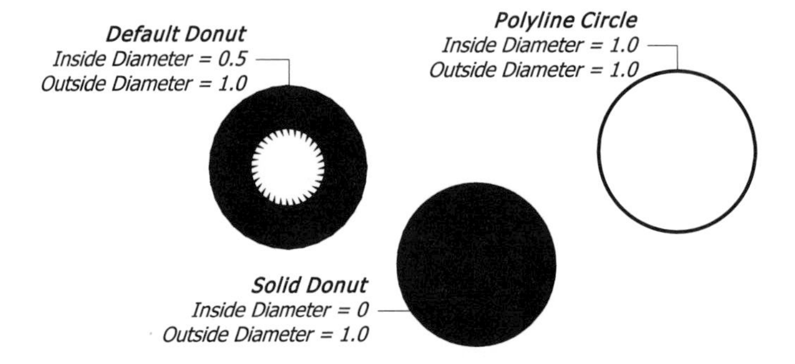

COMMAND LINE OPTIONS

Inside diameter specifies the inner diameter by entering a number or picking two points.

Outside diameter specifies the outer diameter.

Center of donut determines the donut's center point by specifying coordinates, or picking points.

Exit exits the command.

RELATED COMMANDS

Circle draws circles.

PEdit edits polylines.

Reverse reverses the vertex order of polylines.

RELATED SYSTEM VARIABLES

DonutId specifies the default internal diameter.

DonutOd specifies the default outside diameter.

Fill toggles the filling of donuts, as well as of wide polylines, hatches, and 2D solids.

TIPS

- This command repeats itself until canceled.
- Donuts are made of two polyline arcs.
- The technical editor reports that there is no matching Coffee command.

Dragmode

Ver. 2.0 Controls the display of objects during dragging operations.

Command	*Alias*	*Keyboard Shortcuts*	*Menu*	*Ribbon*
'dragmode	...	...	...	...

Command: dragmode
Enter new value [ON/OFF/Auto] <Auto>: *(Enter an option.)*

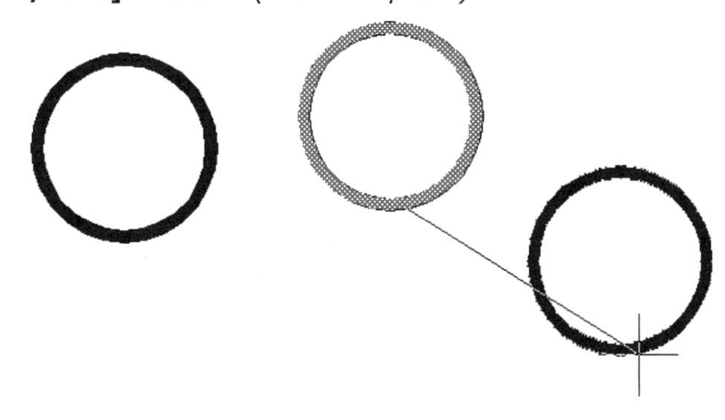

Highlight image (center) and drag image (right).

COMMAND LINE OPTIONS

ON enables dragging display only with the Drag option.

OFF turns off all dragging display.

Auto allows AutoCAD to determine when to display drag image.

COMMAND MODIFIER

Drag displays drag images when DragMode = on.

RELATED SYSTEM VARIABLE

DragMode is the current drag setting:

0 — No drag image.

1 — On if required.

2 — Automatic.

TIP

- Turn off DragMode and Highlight in very large drawings to speed up editing.

DrawingRecovery / DrawingRecoveryHide

2006 Displays and hides the Drawing Recovery Manager palette.

Commands	*Alias*	*Menu Bar*	*Ribbon*
drawingrecovery	**drm**	...	...
drawingrecoveryhide	...	...	...

Command: drawingrecovery

Displays palette (also displayed automatically following a software crash).

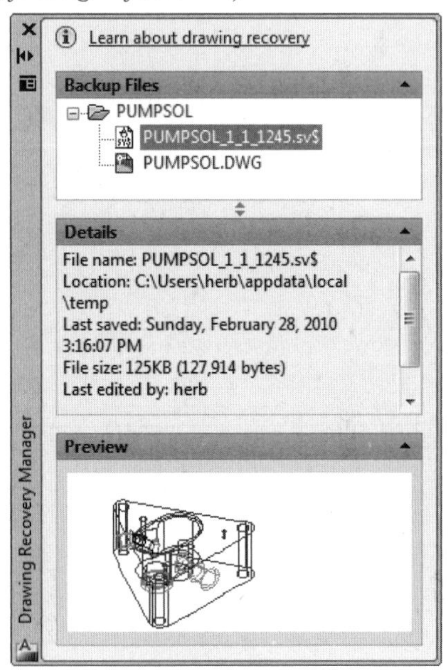

SHORTCUT MENU

Open opens the drawing in AutoCAD; see the Open command.

Properties displays the Properties dialog box; see the DwgProps command.

Command: drawingrecoveryhide

Closes the palette.

RELATED COMMANDS

Audit determines if drawing files have problems.

Recover recovers drawings at the command line.

RecoverAll recovers drawings and attached xrefs.

RELATED SYSTEM VARIABLES

DrState reports whether the Drawing Recovery palette is open.

RecoveryMode specifies when the Drawing Recovery palette appears, if at all.

RecoverAuto determines how you are notified about damaged files *(new to AutoCAD 2011)*.

DrawOrder

Rel. 14 Controls the display of overlapping objects.

Command	*Alias*	*Keyboard Shortcuts*	*Menu*	*Ribbon*
draworder	**dr**	...	**Tools**	**Home**
			↳Draw Order	**↳Modify**
				↳Bring to Front

Command: draworder
Select objects: *(Select one or more objects.)*
Select objects: *(Press Enter.)*
Enter object ordering option
[Above objects/Under objects/Front/Back] <Back>: *(Enter an option.)*

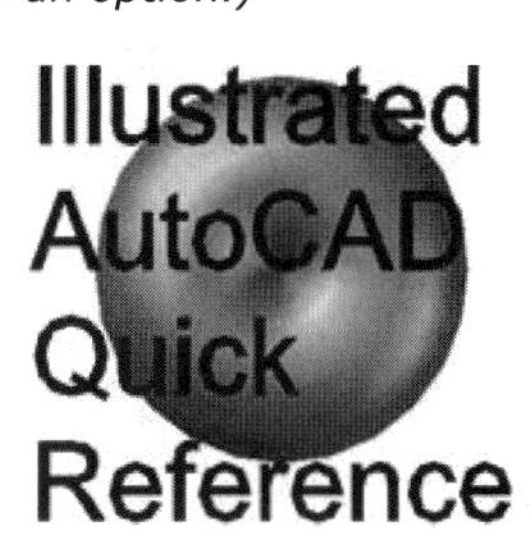

Left: *Text under solid.*
Right: *Text above solid.*

COMMAND LINE OPTIONS

Select objects selects the objects to be moved.

Above object forces selected objects to appear above the reference object.

Under object forces selected objects to appear below the reference object.

Front forces selected objects to the top of the display order.

Back forces selected objects to the bottom of the display order.

RELATED COMMANDS

Hatch displays hatching in front of or behind other objects.

TextToFront displays text and/or dimensions on top of overlapping objects.

RELATED SYSTEM VARIABLES

DrawOrderCtl controls draw order.

HpDrawOrder specifies display order of hatch patterns and fills; see the Hatch command.

TIPS

- When you pick more than one object for reordering, AutoCAD maintains the relative display order of the selected objects. The order in which you select objects has no effect on display order.
- When DrawOrderCtl is set to 3, editing operations may take longer.
- *Draw order inheritance* means that new objects created from objects with draw order are assigned the display order of the object selected first.

DSettings

2000 Controls the most common settings for drafting operations (short for Drafting SETTINGS).

Commands	Aliases	Keyboard Shortcuts	Menu	Ribbon
'dsettings	**ds** **se, ddrmodes** **os, osnap**	...	**Tools** **↳Drafting Settings**	...
'+dsettings	...	...	...	...

Command: dsettings

Displays tabbed dialog box.

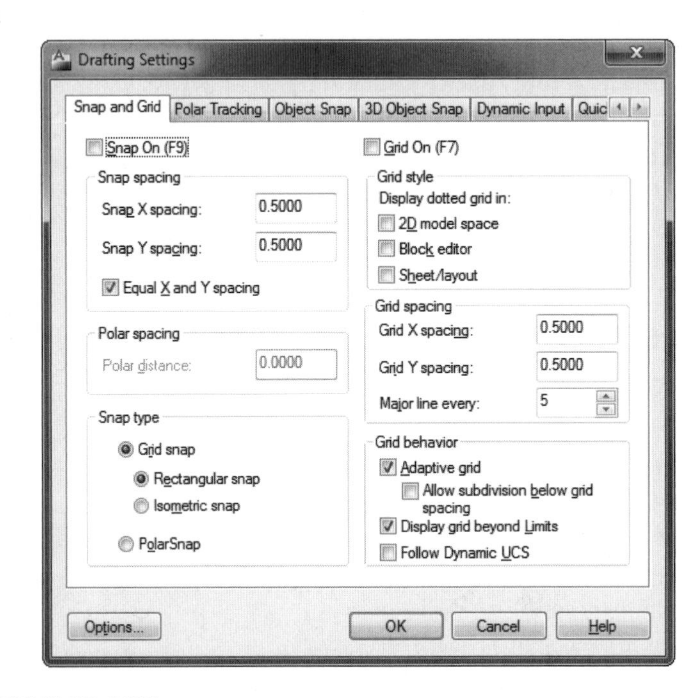

DRAFTING SETTINGS DIALOG BOX

Options displays the Options dialog box; does not work during transparent commands.

Snap and Grid tab

☐**Snap On (F9)** turns on and off snap mode *(stored in system variable SnapMode)*.

☐**Grid On (F7)** turns on and off the grid display (*GridMode*).

Snap Spacing options

Snap X Spacing specifies the snap spacing in the x direction (*SnapUnit*).

Snap Y Spacing specifies the snap spacing in the y direction (*SnapUnit*).

☑**Equal X and Y Spacing** sets the y spacing to that of x.

Polar Spacing options

Polar Distance specifies the snap distance, when Snap Type is set to Polar snap; when 0, the polar snap distance is set to the value of Snap X Spacing (*PolarDist*).

Snap Type options

⊙ **Grid Snap** specifies non-polar snap (*SnapType*).

⊙ **Rectangular Snap** specifies rectangular snap (*SnapStyl*).

○ **Isometric Snap** specifies isometric snap mode (*SnapStyl*).

○ **Polar Snap** specifies polar snap (*SnapType*).

SnapType	***SnapStyl***	***Meaning***
0 (off)	**0** (off)	Rectangular snap.
0 (off)	**1** (on)	Isometric snap.
1 (on)	**0** (off)	Polar snap.

Grid Style options *(new to AutoCAD 2011)*

Display dotted grid in displays dotted grid when the following options are off, lined grid otherwise.

☐ **2D model space** (*GridStyle* = 1).

☐ **Block editor** (*GridStyle* = 2).

☐ **Sheet/layout** (*GridStyle* = 4).

Grid Spacing options

Grid X spacing specifies spacing of grid in x direction; when 0, grid spacing = Snap X spacing (*GridUnit*).

Grid Y spacing specifies spacing of grid in y direction; when 0, grid spacing = Snap Y spacing (*GridUnit*).

Major Line Every specifies the number of minor grid lines between major grid lines (*GridMajor*).

Grid Behavior options

☑ **Adaptive Grid** changes the grid spacing, depending on the zoom level (*GridDisplay* = 2).

☐ **Allow Subdivision Below Grid Spacing** draws sub-grid lines (*GridDisplay* = 4).

☑ **Display Grid Beyond Limits** draws the grid beyond the extents set by the Limits command (*GridDisplay* = 1).

☐ **Follow Dynamic UCS** causes the grid to rotate with the dynamic UCS (*GridDisplay* = 8).

Polar Tracking tab

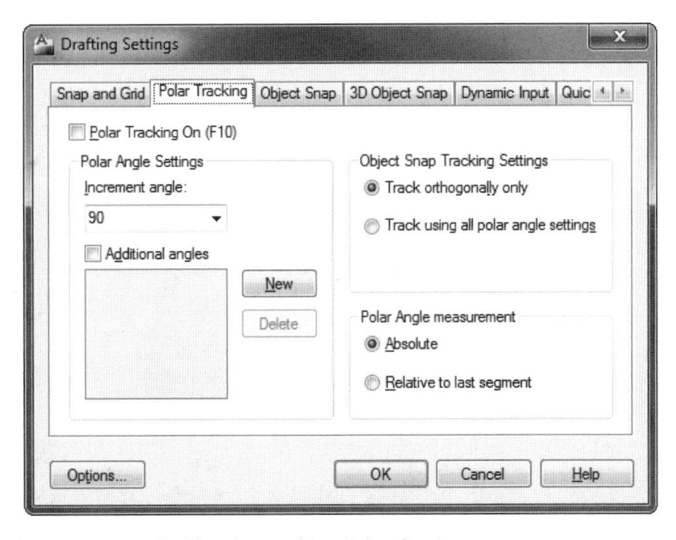

☐ **Polar Tracking On (F10)** turns on and off polar tracking (*AutoSnap*).

Polar Angle Settings options

Increment Angle specifies the increment angle displayed by the polar tracking alignment path; select a preset angle of 90, 60, 45, 30, 22.5, 18, 15, 10, or 5 degrees, or enter a value (*PolarAng*).

☑**Additional Angles** allows you to set additional polar tracking angles (*PolarMode*).

New adds up to ten polar tracking alignment angles (*PolarAddAng*)

Delete deletes added angles.

Object Snap Tracking Settings options

⊙**Track Orthogonally Only** displays orthogonal tracking paths when object snap tracking is on (*PolarMode*).

○**Track Using All Polar Angle Settings** tracks cursor along polar angle tracking path when object snap tracking is turned on (*PolarMode*).

Polar Angle Measurement options

⊙**Absolute** forces polar tracking angles along the current user coordinate system (UCS).

○**Relative To Last Segment** forces polar tracking angles on the last-created object.

Object Snap tab

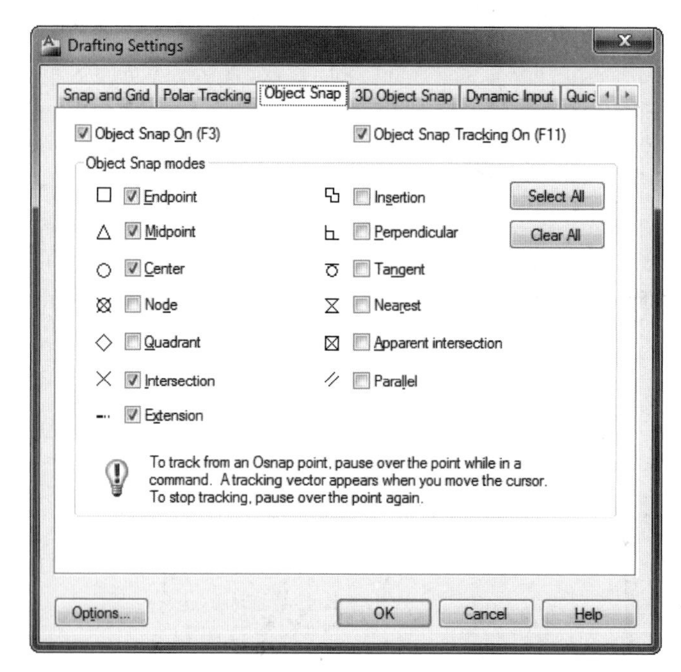

☐**Object Snap On (F3)** turns on and off running object snaps (*OsMode*).

☐**Object Snap Tracking On (F11)** toggles object snap tracking (*AutoSnap*).

Object Snap Modes options

☑**ENDpoint** snaps to the nearest endpoint of a line, multiline, polyline segment, ray, arc, and elliptical arc; and to the nearest corner of a trace, solid, and 3D face.

☐**MIDpoint** snaps to the midpoint of a line, multiline, polyline segment, solid, spline, xline, arc, ellipse, and elliptical arc.

☑**CENter** snaps to the center of an arc, circle, ellipse, and elliptical arc.

☐**NODe** snaps to a point.

☐**QUAdrant** snaps to a quadrant point (90 degrees) of an arc, circle, ellipse, and elliptical arc.

☑**INTersection** snaps to the intersection of a line, multiline, polyline, ray, spline, xline, arc, circle, ellipse, and elliptical arc; edges of regions; it does not snap to the edges and corners of 3D solids.

☑**EXTension** displays an extension line from the endpoint of objects; snaps to the point where two objects would intersect if they were infinitely extended; does not work with the edges and corners of 3D solids; automatically turns on intersection mode. (Do not turn on apparent intersection at the same time as extended intersection.)

☐**INSertion** snaps to the insertion point of text, block, attribute, or shape.

☐**PERpendicular** snaps to the perpendicular of a line, multiline, polyline, ray, solid, spline, xline, arc, circle, ellipse, and elliptical arc; snaps from a line, arc, circle, polyline, ray, xline, multiline, and 3D solid edge; in this case deferred perpendicular mode is automatically turned on.

☐**TANgent** snaps to the tangent of an arc, circle, ellipse, or elliptical arc; deferred tangent snap mode is automatically turned on when more than one tangent snap is required.

☐**NEArest** snaps to the nearest point on a line, multiline, point, polyline, spline, xline, arc, circle, ellipse, and elliptical arc.

☐**APParent intersection** snaps to the apparent intersection of two objects that do not actually intersect but appear to intersect in 3D space; works with a line, multiline, polyline, ray, spline, xline, arc, circle, ellipse, and elliptical arc; does not work on edges and corners of 3D solids.

☐ **PARallel** snaps to a parallel point when AutoCAD prompts for a second point.

Select All turns on all object snap modes.

Clear All turns off all object snap modes.

3D Object Snap tab *(new to AutoCAD 2011)*

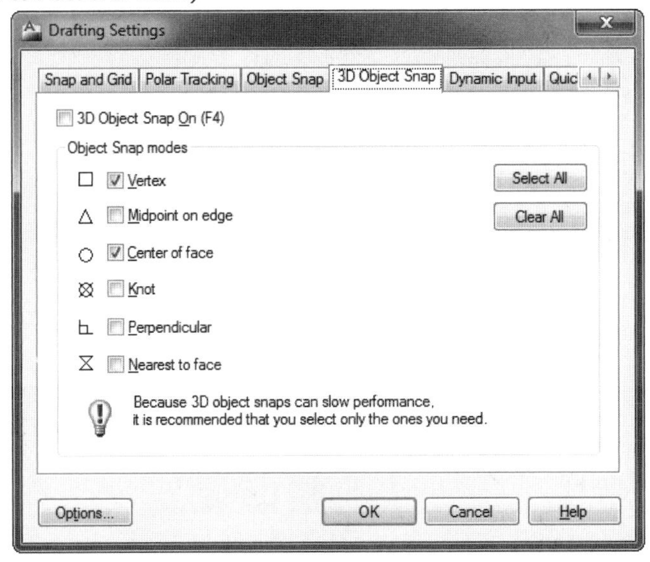

☐3D **Object Snap On (F4)** (ZNON) turns on and off 3D object snap modes *(3dOsMode = 1)*.

Object Snap Modes options

☑**Vertex** (ZVER) snaps to the nearest vertex *(3dOsMode = 2)*.

☐**Midpoint on Edge** (ZMID) snaps to midpoint of the nearest edge *(3dOsMode = 4)*.

☑**Center of Face** (ZCEN) snaps to the center of the nearest face (3d*OsMode = 8*).

☐**Knot** (ZKNO) snaps to the nearest knot on a spline (3d*OsMode = 16*).

☐**Perpendicular** (ZPER) snaps perpendicular to the nearest face (3d*OsMode = 32*).

☐**Nearest to Face** (ZNEA) snaps to the point nearest to a face (3d*OsMode = 64*).

Select All turns on all object snap modes.

Clear All turns off all object snap modes.

Dynamic Input tab

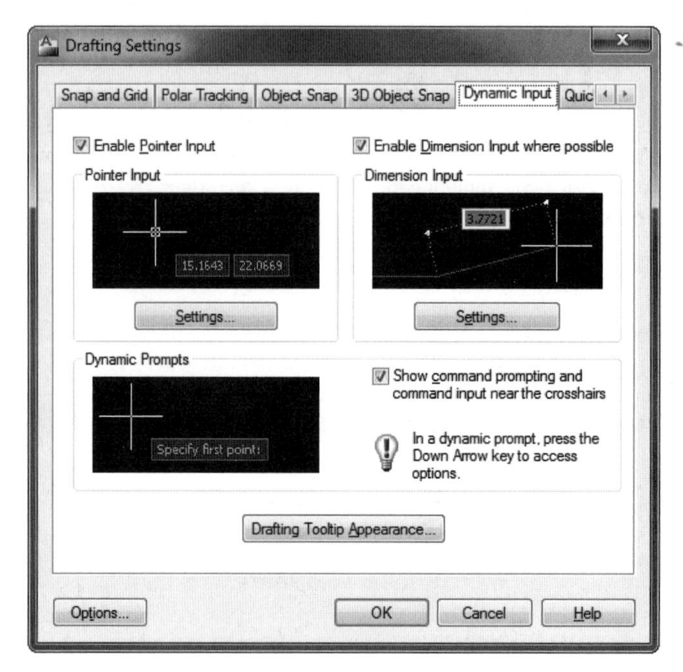

☑**Enable Pointer Input** toggles coordinate input at the drawing cursor, as illustrated below. Press the Tab key to switch between the x and y coordinates of the cursor position.

☑**Enable Dimension Input Where Possible** toggles the display of lengths and angles when drawing and editing. Press Tab to switch between distance and angle.

☑ **Show Command Prompting Near Crosshairs** switches the display of commands and prompts between the 'Command:' prompt and the drawing area. Press the Down Arrow key to display additional options.

Quick Properties tab

(Redesigned in AutoCAD 2011)

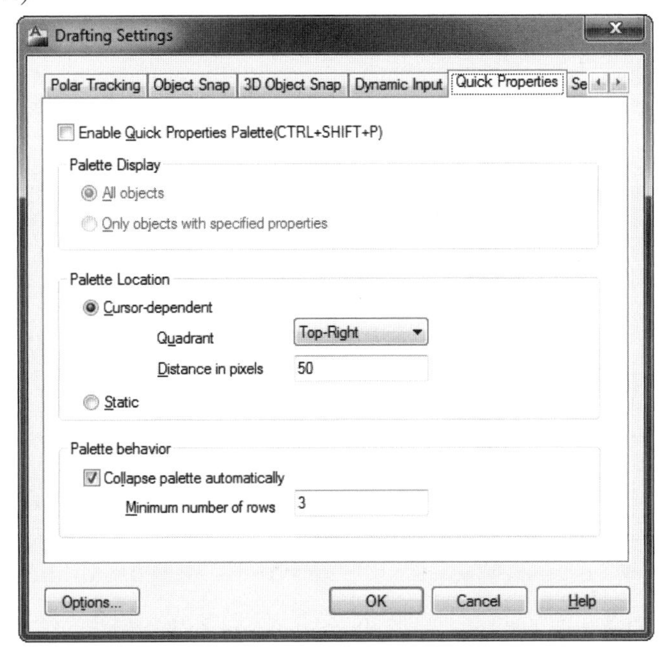

☑ Enable **Quick Properties Palette** displays the Quick Properties palette when an object is selected, depending on the following settings:

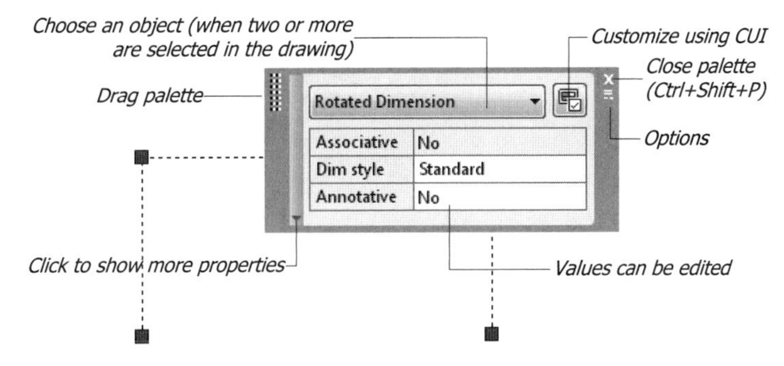

Palette Display options

⦿ **All objects** displays the palette when any object is selected.

○ **Only objects with specified properties** displays the palette only for objects defined by the Cui command.

Palette Location options

⦿ **Cursor-dependent** displays the palette relative to the cursor's position:

- **Quadrant** specifies the relative location, top-right, top-left, bottom-right, or bottom-left.
- **Distance in pixels** specifies the distance from the pick point; range is 0 and 400 pixels.

○ **Static** floats the Quick Properties panel in the same location until relocated manually; useful for locating on a second monitor or away from the editing area.

Palette Behavior options

☑**Collapse palette automatically** keeps the palette compressed until the cursor passes over it.

Minimum number of rows specifies the lines of properties to display in collapsed state; range is 1 to 30.

Selection Cycling tab *(new to AutoCAD 2011)*

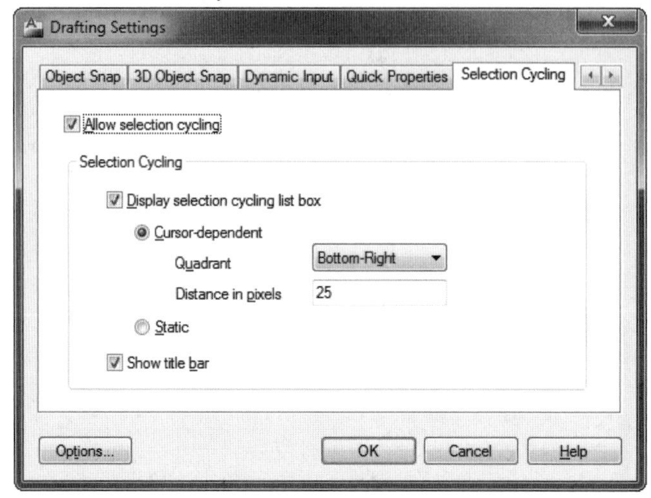

Allow selection cycling toggles the ability to choose among overlapping objects (*SelectionCycling*).

Selection Cycling options

☑ **Display selection cycling list box** toggles the display of the list box (cursor menu).

⊙**Cursor-dependent** locates the list box relative to the cursor:

Quadrant chooses from bottom right, top right, top left, and bottom left.

Distance in pixels specifies the distance between the cursor and the list box.

○ **Static** locates the list box in the same position; useful for placing on a second monitor.

☑ Show title bar toggles the title bar ("Selection"); when off, the list box is smaller.

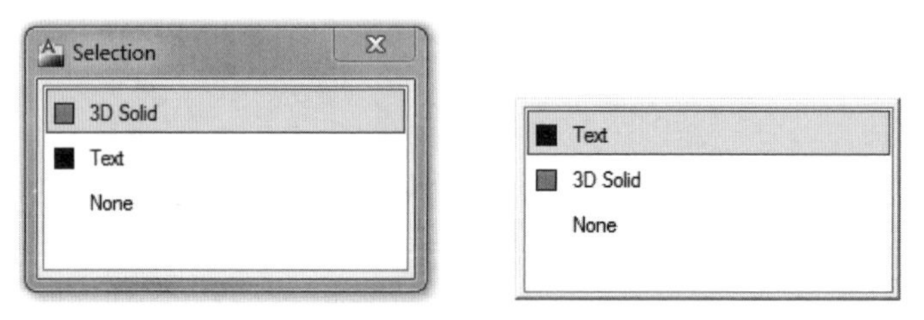

Left: *Selection cycling list box, with title bar...*
Right: *...and without.*

POINT INPUT SETTINGS DIALOG BOX

Access this dialog box from the Dynamic Input's Settings button:

Format options

For The Second or Next Points, Default To (*DynPiFormat* and *DynPiCoords*):

⊙**Polar Format** shows distance and angle; **,** returns to Cartesian format.

○**Cartesian Format** shows x and y distance; **<** returns to polar format.

⊙ **Relative Coordinates** shows coordinates relative to the last point; **#** returns to absolute format.

○ **Absolute Coordinates** shows coordinates relative to the origin; **@** returns to relative format. Direct distance entry is disabled when this option is turned on.

Visibility options

Show Coordinate Tooltips (*DynPiVis*):

○ **As Soon As I Type Coordinate Data** displays tooltips when you enter coordinates.

⊙ **When a Command Asks for a Point** displays tooltips when commands prompt for points.

○ **Always - Even When Not in a Command** displays tooltips always.

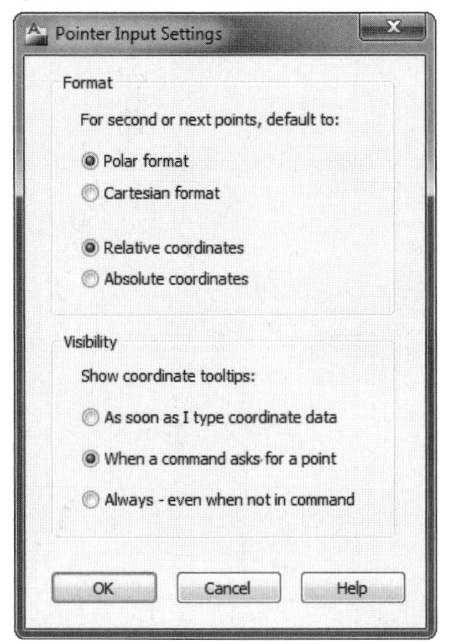

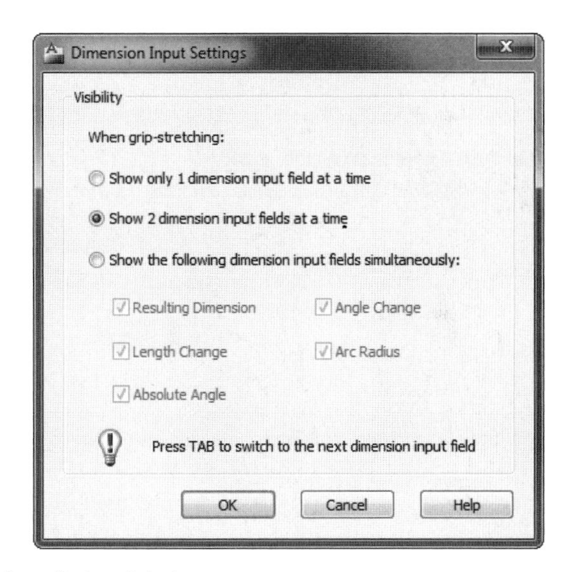

Left: *Pointer Input Settings dialog box.*
Right: *Dimension Input Settings dialog box.*

DIMENSION INPUT SETTINGS DIALOG BOX

Access this dialog box from the Dynamic Input's Settings button:

Visibility options

When Grip Stretching (*DynDiVis*):

○ **Show Only 1 Dimension Input Field at a Time** displays only the length dimension.

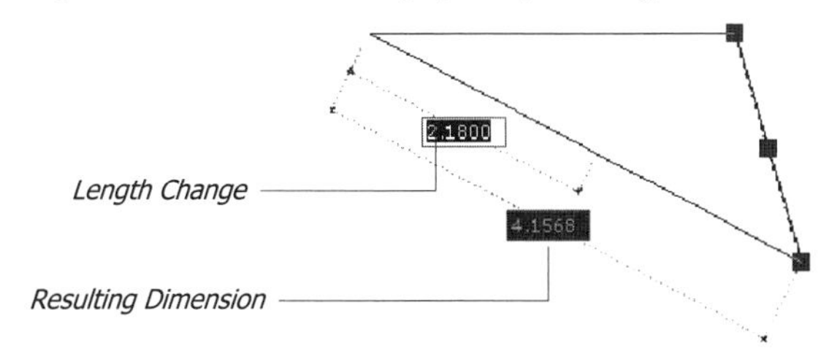

⊙ **Show 2 Dimension Input Fields at a Time** displays the length and resulting dimension.

○ **Show the Following Dimension Input Fields Simultaneously** (*DynDiGrip*):

☑ **Resulting Dimension** displays the length dimension tooltip.

☑ **Length Change** displays the change in length.

☑ **Absolute Angle** displays the angle dimension tooltip.

☑ **Angle Change** displays the change in the angle.

☑ **Arc Radius** displays the radius of arcs.

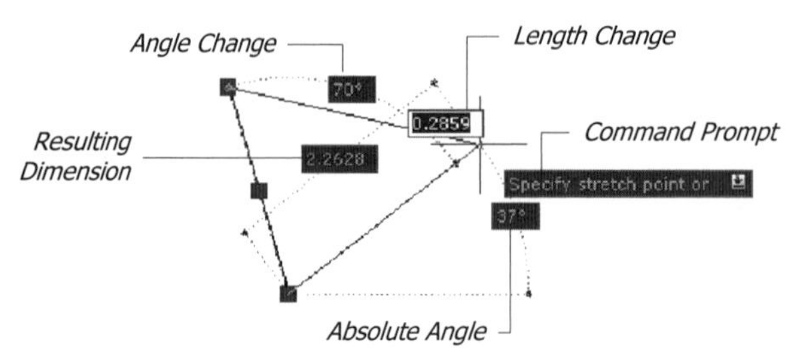

TOOLTIP APPEARANCE DIALOG BOX

Access this dialog box from the Dynamic Input's Drafting Tooltip Appearance button:

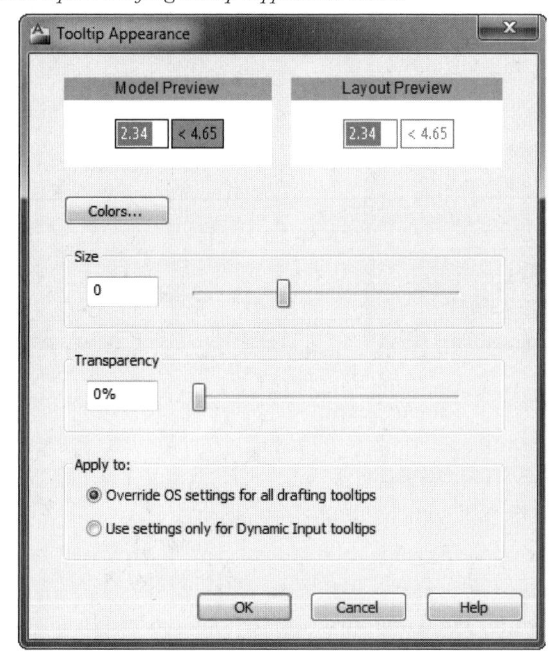

Color options

Model Color specifies tooltip colors in model space; displays the Select Color dialog box.

Layout Color specifies tooltip colors in model space.

Size specifies tooltip size; 0 = default size.

Transparency controls transparency of tooltips; 0% = opaque.

Apply To options

⊙ **Override OS Settings for All Drafting Tooltips** applies settings to all tooltips, overriding settings in the operating system.

○ **Use Settings Only for Dynamic Input Tooltips** applies settings to drafting tooltips only.

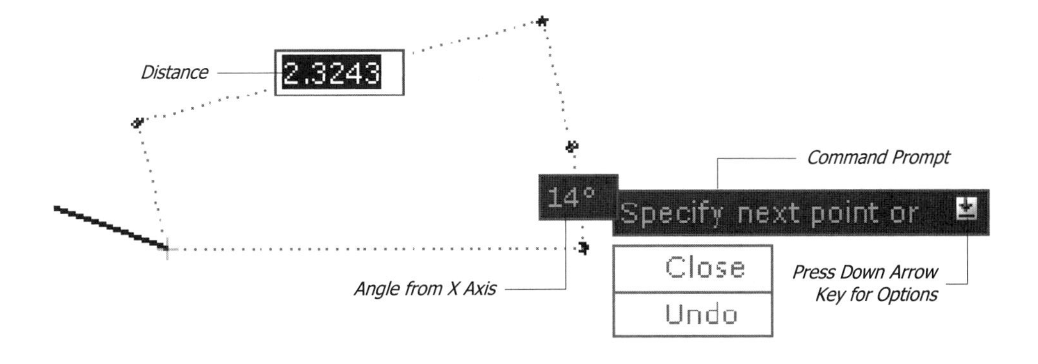

+DSETTINGS Command

Command: +dsettings
Tab Index <0>: *(Enter a digit.)*

COMMAND LINE OPTION

Tab Index displays the Drafting Settings dialog box with the associated tab:

0 — Displays Snap and Grid tab (*default*).

1 — Displays Polar Tracking tab.

2 — Displays Object Snap tab.

3 — Displays Dynamic Input tab.

4 — Displays Quick Properties tab.

5 — Displays 3D Object Snap tab *(new to AutoCAD 2011).*

6 — Displays Selection Cycling tab *(new to AutoCAD 2011).*

RELATED COMMANDS

Grid sets the grid spacing and toggles visibility.

Isoplane selects the working isometric plane.

Ortho toggles orthographic mode.

QuickProperties toggles the display of the QuickProperties palette.

Snap sets the snap spacing and isometric mode.

RELATED SYSTEM VARIABLES

AutoSnap controls AutoSnap, polar tracking, and object snap tracking.

CmdInputHistoryMax limits the maximum number of commands stored in history.

CrossingAreaColor specifies the color for crossing area selections.

DynDIGrip determines which dynamic dimensions are displayed during grips editing.

DynDIVis determines which dynamic dimensions are displayed during grips editing.

DynMode toggles dynamic input.

DynPICoords switches pointer input between relative and absolute coordinates.

DynPIFormat switches between polar and Cartesian format for coordinates.

DynPIVis toggles pointer input.

DynPrompt toggles prompts in dynamic input.

DynTooltips determines which tooltips are affected by appearance settings.

GridMode indicates the current grid visibility.

GridStyle determines the grid look (lines or dots) in 2D model, block editor, and paper space *(new to AutoCAD 2011)*.

GridUnit indicates the current grid spacing.

InputHistoryMode specifies where each command history is displayed.

MButtonPan toggles the middle mouse button between panning and *.cui* definition.

PolarAddAng specifies user-defined polar angles, separated by semicolons.

PolarAng specifies the increments of the polar angle.

PolarDist specifies the polar snap distance.

PreviewEffect specifies whether selection preview applies dashes and/or thickness to objects.

PreviewFilter excludes xrefs, tables, mtext, hatches, groups, and objects on locked layers.

SnapAng specifies the current snap rotation angle.

SnapBase sets the base point of the snap rotation angle.

OsMode holds the current object snap modes.

PolarMode holds the settings for polar and object snap tracking.

QpLocation specifies the location of the Quick Properties palette.

QpMode reports if the Quick Properties palette is open or closed.

SelectionArea toggles area selection coloring.

SelectionAreaOpacity specifies the transparency of the area selection color.

SelectionCycling toggles selection cycling *(new to AutoCAD 2011)*.

SelectionPreview determines when selection preview is active.

SnapIsoPair specifies the current isoplane.

SnapMode sets the current snap mode setting.

SnapStyl specifies the snap style setting.

SnapUnit sets the current snap spacing.

WindowAreaColor specifies the color for windowed area selections.

3dOsMode determines which 3D osnap modes are active *(new to AutoCAD 2011)*.

TIPS

- Use snap to set the cursor movement increment.
- Use the grid as a visual display to help you better gauge distances.
- Use object snaps to draw precisely to geometric features.
- m2p, a running object snap, snaps to the midpoint of two picked points.

DsViewer

Rel. 13 Displays bird's-eye view palette; provides real-time pan and zoom (short for DiSplay VIEWer).

Command	*Alias*	*Keyboard Shortcuts*	*Menu*	*Ribbon*
'dsviewer	**av**	**...**	**View** ↳**Aerial View**	**...**

Command: dsviewer

Displays the Aerial View palette.

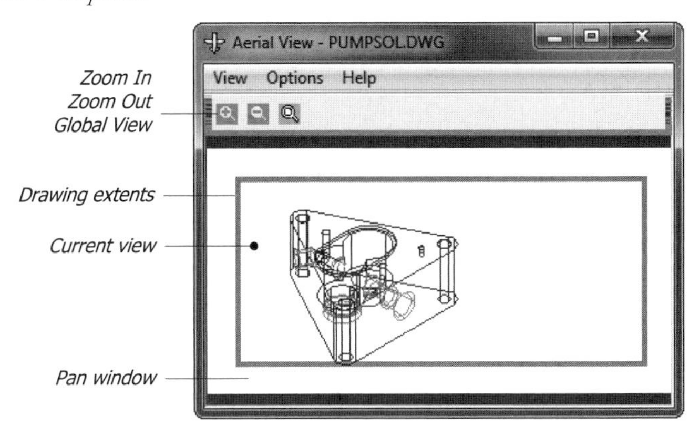

MENU BAR

View menu

Zoom In increases centered zoom by a factor of 2.

Zoom Out decreases centered zoom by a factor of 2.

Global displays entire drawing in Aerial View palette.

Options menu

Auto Viewport updates the Aerial View automatically with the current viewport.

Dynamic Update updates the Aerial View automatically with editing changes in the current viewport.

Realtime Zoom updates the drawing in real time as you zoom in the Aerial View palette.

RELATED COMMANDS

Pan moves the drawing view.

Zoom makes the view larger or smaller.

TIPS

- The purpose of the Aerial View is to let you see the entire drawing at all times, while panning and zooming.
- *Warning!* When in paper space, the Aerial View palette shows only paper space objects.
- To switch quickly between Pan (default) and Zoom modes, click on the Aerial View window; right-click or press Enter to lock the Aerial View palette and return cursor to editing window.
- This command works only in 2D visual modes.

Renamed Command

DText was renamed in AutoCAD 2000 as an alias of the Text command.

DView

Rel. 10 Zooms and pans 3D drawings dynamically, and turns on perspective mode for 3D drawings; superseded by the NavVCube command (short for Dynamic VIEW).

Command	*Alias*	*Keyboard Shortcuts*	*Menu*	*Ribbon*
dview	**dv**	...	...	...

Command: dview

Select objects or <use DVIEWBLOCK>: *(Select objects, or press Enter.)*

Enter option [CAmera/TArget/Distance/POints/PAn/Zoom/TWist/CLip/Hide/Off/Undo]: *(Enter an option.)*

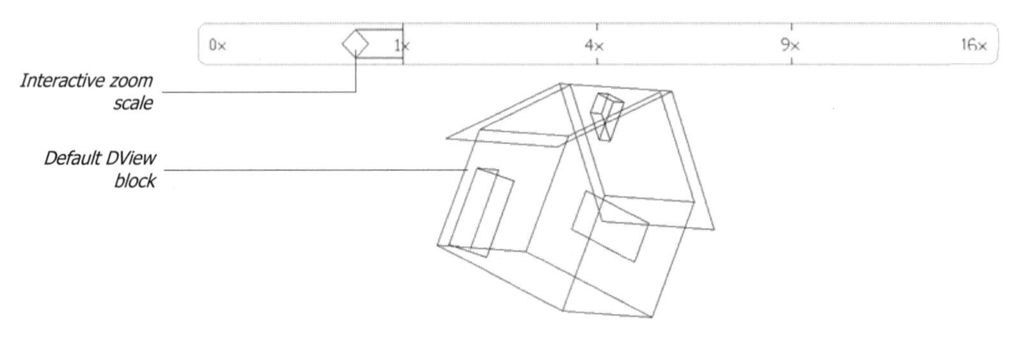

COMMAND LINE OPTIONS

CAmera indicates the camera angle relative to the target.

Toggle switches between input angles (*available after starting CAmera or TArget*).

TArget indicates the target angle relative to the camera.

Distance indicates the camera-to-target distance; turns on perspective mode.

POints indicates both the camera and target points.

PAn pans the view dynamically.

Zoom zooms the view dynamically.

TWist rotates the camera (the view).

CLip options

Enter clipping option [Back/Front/Off] <Off>:

Back clip options

Specify distance from target or [ON/OFF] <0.0>:

ON turns on the back clipping plane.

OFF turns off the back clipping plane.

Distance from target locates the back clipping plane.

Front Clip options

Specify distance from target or [set to Eye(camera)/ON/OFF] <1.0>:

Eye positions the front clipping plane at the camera.

Distance from target locates the front clipping plane.

Off turns off view clipping.

Hide removes hidden lines.

Off turns off the perspective view.

Undo undoes the most recent DView action.

eXit exits DView.

RELATED COMMANDS

NavVCube displays the viewcube for 3D navigation.

Hide removes hidden lines from non-perspective views.

Pan pans non-perspective views.

VPoint selects non-perspective viewpoints of 3D drawings.

Zoom zooms non-perspective views.

3dOrbit creates 3D views interactively, in parallel or perspective mode.

RELATED SYSTEM VARIABLES

BackZ specifies the back clipping plane offset.

FrontZ specifies the front clipping plane offset.

LensLength specifies the perspective view lens length, in millimeters.

Target specifies the UCS 3D coordinates of the target point.

ViewCtr specifies the 2D coordinates of current view center.

ViewDir specifies the WCS 3D coordinates of camera offset from target.

ViewMode specifies the perspective and clipping settings.

ViewSize specifies the height of view.

ViewTwist specifies the rotation angle of current view.

RELATED SYSTEM BLOCK

DViewBlock is the alternate viewing object displayed during DView.

TIPS

- The view direction is from the camera to target.
- Press Enter at the 'Select objects' prompt to display the DViewBlock house. You can replace the house block with your own by redefining the DViewBlock block.
- To view a 3D drawing in one-point perspective, use the Zoom option.
- Menus and transparent zoom and pan are not available during DView. Once the view is in perspective mode, you cannot use any command that requires pointing with the cursor.

Renamed Commands

DwfOut was made part of Plot with AutoCAD 2004; in AutoCAD 2010 it reappeared as ExportDWF.

DwfOutD was removed from AutoCAD 2000; it was combined with DwfOut.

DwfAdjust

2007 Adjusts the contrast, fade, and color of DWF and DWFx underlays.

Command	Alias	Keyboard Shortcuts	Menu	Ribbon
dwfadjust	...	...	...	**Insert** ↳**Reference** ↳**Adjust**

Command: dwfadjust
Select DWF underlay: *(Select objects.)*
Select DWF underlay: *(Press Enter.)*
Enter DWF Underlay option [Fade/Contrast/Monochrome] <Fade>: *(Enter option.)*

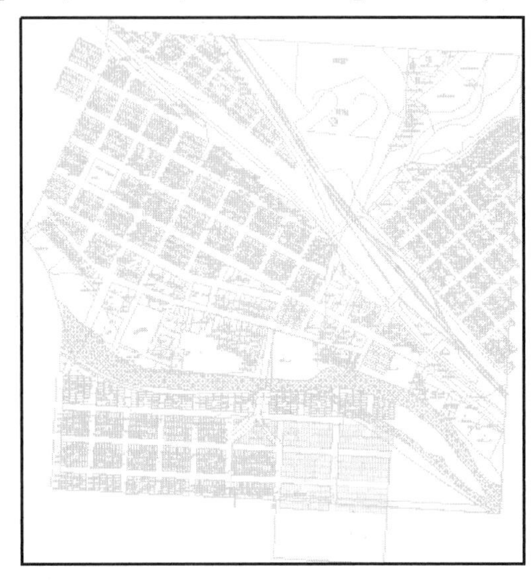

DWF underlay with fade set to maximum (90).

COMMAND LINE OPTIONS

Select DWF underlay selects the underlay to adjust.

Enter fade value specifies the amount of fade: (0) no fade, (25) default value, (90) maximum fade.

Enter contrast value specifies the change in contrast, ranging from 0 to 100.

Monochrome toggles monochrome colorization.

RELATED COMMANDS

Adjust adjusts images, as well as DWF, DWFx, DGN, and PDF underlays.

DwfAttach attaches *.dwf* files as underlays.

DwfClip clips DWF underlays.

DwfLayers toggles the display of layers.

RELATED SYSTEM VARIABLES

DwfFrame toggles display of the frame.

DwfAttach

2007 Attaches 2D *.dwf* or *.dwfx* files as xref-like underlays to the current drawing.

Commands	*Alias*	*Menu*	*Ribbon*
dwfattach	...	**Insert**	**Insert**
		↳DWF Underlay	**↳Reference**
			↳Attach
-dwfattach	...	...	...

Command: dwfattach

Displays the Select DWF File dialog box.

Select a file, and then click Open.

Displays the Attach DWF Underlay dialog box.

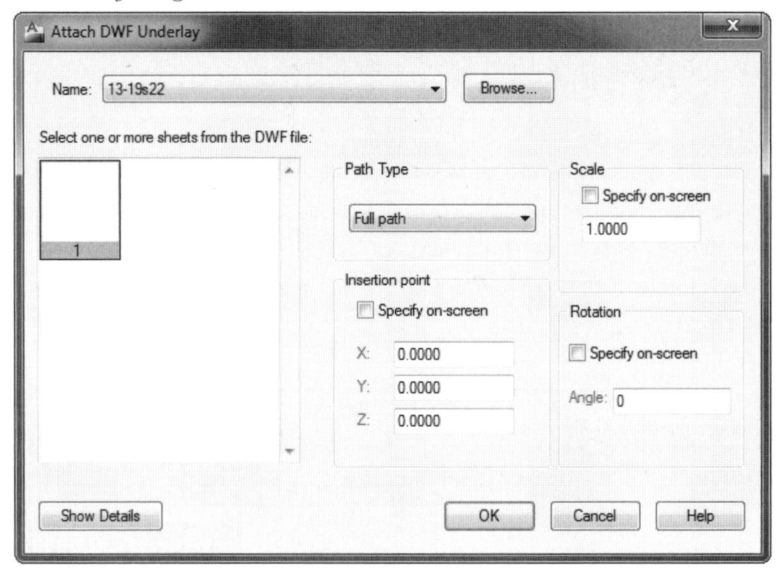

ATTACH DWF UNDERLAY DIALOG BOX

Name specifies the *.dwf* file to attach to the current drawing; click the droplist for other *.dwf* files. More than one *.dwf* file can be attached to drawings. (This droplist is disabled when the *.dwf* file is stored in Autodesk's Vault software.)

Browse displays the Select DWF File dialog box to select another *.dwf* file.

Select one or more sheets from the DWF file lists the names and preview images of sheets in the *.dwf* file; 3D *.dwf* sheets are not listed; one or more sheets can be attached at a time.

Path Type specifies the type of path to the *.dwf* file:

- **Full Path** saves the full path name.
- **Relative Path** saves the relative path.
- **No Path** saves the file name only.

Insertion Point specifies the *.dwf* underlay's insertion point (lower left corner); default =0,0,0.

Specify On-Screen:

☑ prompts you for the insertion point at the command line after closing the dialog box.

☐ enters the x,y,z coordinates in the dialog box.

Scale specifies the DWF underlay's scale factor; default = 1.0

Specify On-Screen:

☑ prompts you for the scale factor at the command line after closing the dialog box.

☐ enters the scale factor in the dialog box.

Angle specifies the DWF underlay's angle of rotation; default = 0 degrees.

Specify On-Screen:

☑ prompts you for the rotation angle at the command line after closing the dialog box.

☐ enters the angle in the dialog box.

RELATED COMMANDS

Adjust adjusts images, as well as DWF, DWFx, DGN, and PDF underlays.

Attach attaches images, as well as DWF, DWFx, DGN, and PDF files as underlays.

DwfAdjust adjusts the contrast, fade, and color of DWF underlays.

DwfClip clips DWF underlays.

DwfLayers toggles the display of layers.

ExportDWF and **ExportDWFx** export the current drawing as DWF files.

Markup places marked-up (red lined) *.dwf* files in the current drawing.

Plot creates single-sheet *.dwf* files of the current drawings.

Publish creates multi-sheet *.dwf* files of multiple drawings.

RELATED SYSTEM VARIABLE

DwfFrame toggles the display of the rectangular frame around DWF underlays.

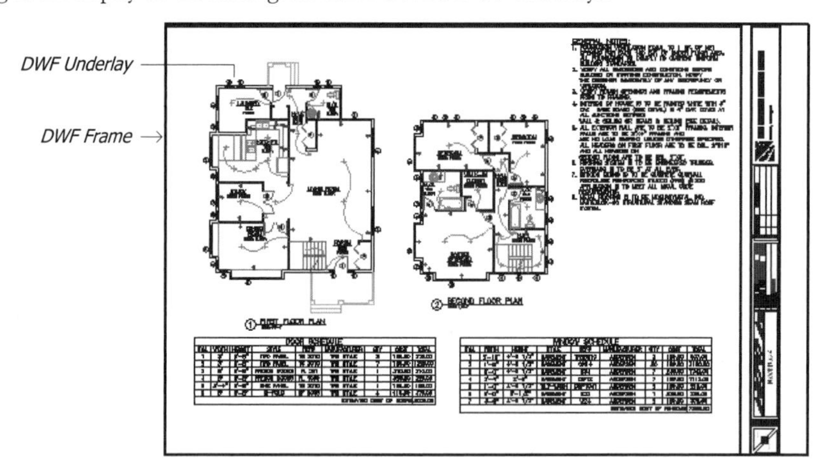

DwfOsnap toggles whether osnap recognizes the frame of DWF underlays.

TIPS

- This command cannot import 3D *.dwf* files. It attaches only DWF files with sheets.
- Double-click the underlay to display its properties.
- To remove the underlay from the drawing, select it, and then press Del.

DwfClip

2007 Clips DWF and DWFx underlays to isolate portions of the drawing.

Command	Alias	Keyboard Shortcuts	Menu	Ribbon
dwfclip	...	...	...	**Insert** ↳**Reference** ↳**Clip**

Command: dwfclip
Select DWF to clip: *(Select one underlay.)*
Enter DWF clipping option [ON/OFF/Delete/New boundary] <New boundary>: *(Enter an option.)*

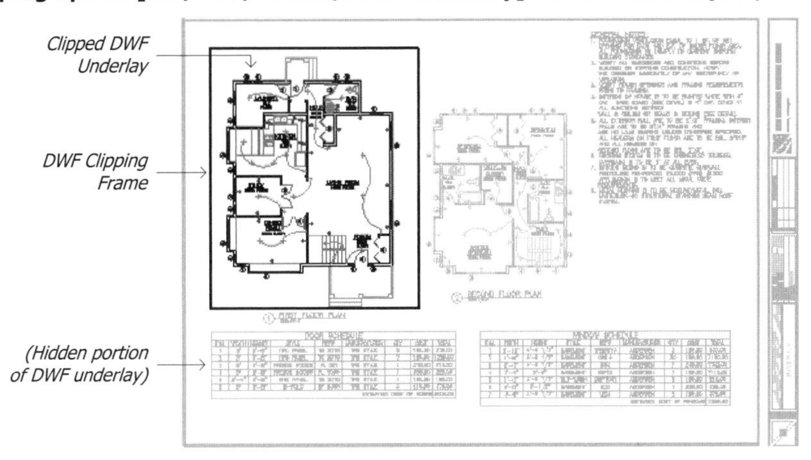

COMMAND LINE OPTIONS

ON turns on a previously-created clipping boundary.

OFF turns off the clipping boundary, so that it is hidden.

Delete removes the clipping boundary, restoring the DWF underlay.

New boundary creates a new clipping boundary:

- **Polygonal** creates a multi-sided boundary using prompts similar to the PLine command.
- **Rectangular** creates a rectangular boundary based on two pick points.

RELATED COMMANDS

Clip clips images, DWF/x, and DGN files.

DwfAdjust adjusts the contrast, fade, and color of DWF underlays.

DwfAttach attaches *.dwf* files as underlays.

DwfLayers toggles the display of layers.

RELATED SYSTEM VARIABLE

DwfFrame toggles the display of the frame around clipped DWF underlays.

TIPS

- After the clipping boundary is in place, you can use grips editing to adjust its size.
- This command can be started by selecting the DWF border, right-clicking, and then selecting DWF Clip from the shortcut menu.

DwfFormat

2009 Specifies the default DWF format.

Command	Alias	Keyboard Shortcuts	Menu	Ribbon
dwfformat	...	...	...	...

Command: dwfformat
Select default DWF format [Dwf/dwfX]: *(Type D or X.)*

COMMAND LINE OPTIONS

Dwf sets the default to DWF (compatible with AutoCAD 2008 and earlier).

dwfX sets the default to DWFx (compatible with Vista and IE).

RELATED COMMANDS

ExportDWF and **ExportDWFx** export drawings in DWF and DWFx formats.

Publish publishes sets of drawings and layouts in DWF and DWFx formats.

3dDwf exports 3D drawings in 3D DWF and 3D DWFx formats.

TIPS

- This command affects the default DWF format for the following commands:
  - Export.
  - Publish.
  - 3dDwf.
- Use this command to specify the default version of DWF, and regular or XPS-compatible DWFx.
- DWFx works only with Vista and 7 dialects of the Windows operating system and the IE Web browser, and only if DesignReview is installed.

DwfLayers

2008 Toggles the display of layers in DWF and DWFx underlays.

Command	*Alias*	*Keyboard Shortcuts*	*Menu*	*Ribbon*
dwflayers	...	...	...	...

Command: dwflayers
Select DWF underlay: *(Pick one underlay.)*

Displays the DWF Layers dialog box.

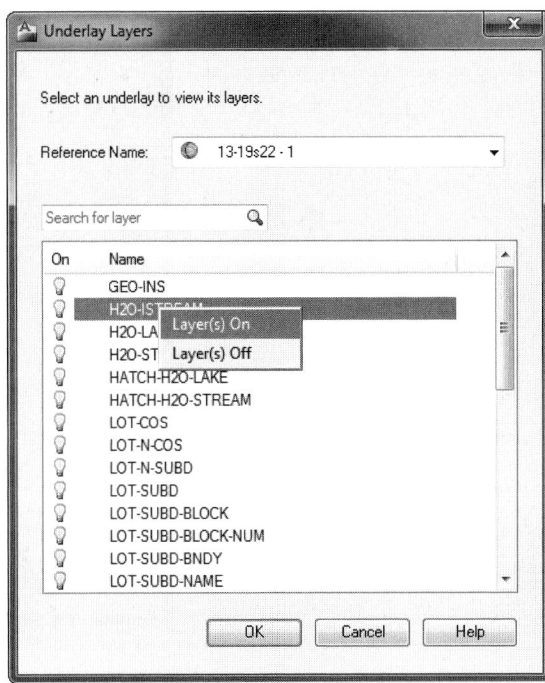

DWF LAYERS DIALOG BOX

Click the light bulb icon next to a layer name to turn the layer off or on.

Apply changes the visibility of layers without leaving the dialog box.

Shortcut Menu

Layer(s) On turns visibility of layers on.

Layer(s) Off turns visibility of layers off.

RELATED COMMANDS

DwfAttach attaches *.dwf* files as underlays.

ULayers toggles visibility of layers in PDF, DGN, and DWF/x underlays.

TIPS

- Use this command to reduce the amount of information displayed by the DWF. Alternatively, use the DwfAdjust command to fade the underlay.
- You can also access this command by right-clicking a DWF underlay, and then selecting DWF Layers from the shortcut menu.

DwgProps

2000 Records and reports information about drawings (short for DraWinG PROPertieS).

Command	*Alias*	*Keyboard Shortcuts*	*Menu Bar*	*Application Menu*	*Ribbon*
dwgprops	...	...	**File ↳Drawing Properties**	**Drawing Utilities ↳Drawing Properties**	...

Command: dwgprops

Displays tabbed dialog box; see below.

DRAWING PROPERTIES DIALOG BOX

Displays information about drawings obtained from the operating system.

General options

File Type indicates the type of file.

Location indicates the location of the file.

Size indicates the size of the file.

MS-DOS Name provides the MS-DOS file name truncated to eight characters with a three-letter extension.

Created indicates the date and time the file was first saved.

Modified indicates the date and time the file was last saved.

Accessed indicates the date and time the file was last opened.

Attributes options

Read-Only indicates the file cannot be edited or erased.

Archive indicates the file has been changed since it was last backed up.

Hidden indicates the file cannot be seen in file listings.

System indicates the file is a system file; DWG drawing files never have this attribute turned on.

PUMPSOL.DWG Properties
General Summary Statistics Custom
PUMPSOL.DWG
Type: AutoCAD Drawing
Location: C:\dwg\12SAMPLE\
Size: 122.00KB (125,181 bytes)
MS-DOS name: PUMPSOL.DWG
Created: Friday, May 08, 2009 12:48:28 PM
Modified: Tuesday, December 22, 1998 5:39:02 PM
Accessed: Friday, May 08, 2009 12:48:28 PM
Attributes: Read only Hidden Archive System
OK Cancel Help

Summary tab

Title specifies a title for the drawing; this is usually different from the file name.

Subject specifies a subject for the drawing.

Author specifies the name of the drafter of the drawing.

Keywords specifies keywords used by the operating system's Find or Search commands to locate drawings.

Comments contains comments on the drawing.

Hyperlink Base specifies the base address for relative links in the drawing, such as http://www .upfrontezine.com; may be an operating system path name, such as *c:\autocad 2011*, or a network drive name. *(Stored in system variable HyperlinkBase.)*

Statistics tab

Displays information about the drawing as obtained from the drawing.

Created indicates the date and time the drawing was first opened (*TdCreate*).

Modified indicates the date and time the drawing was last opened or modified (*TdUpdate*).

Last saved by indicates who last accessed the drawing (*LoginName*).

Revision number indicates the revision number; usually blank.

Total editing time indicates the total amount of time that the drawing has been open (*TdInDwg*).

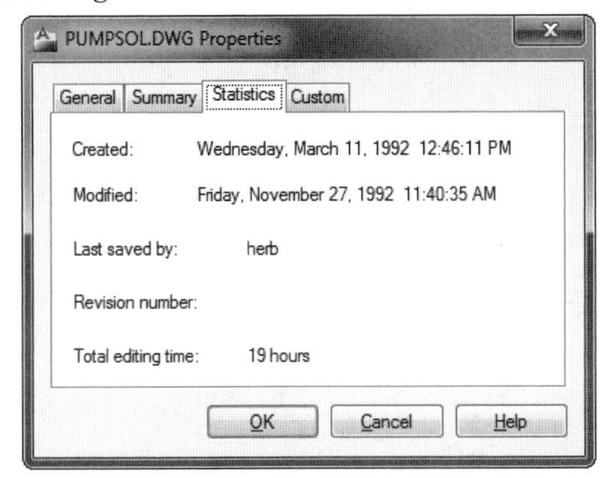

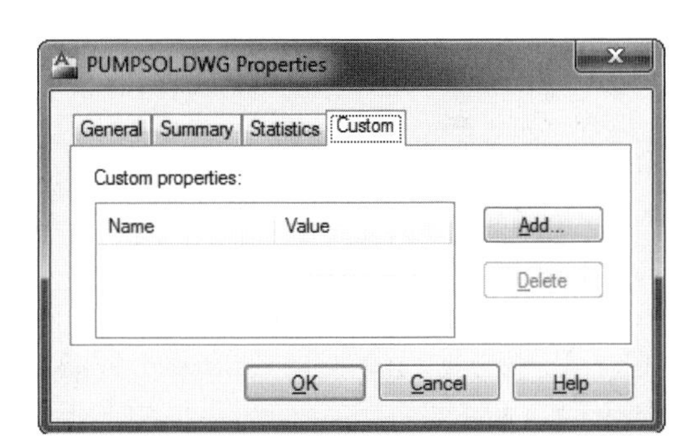

Left: *Statistics tab.*
Right: *Custom tab.*

Custom tab

Add displays the Add Custom Property dialog box; enter a name and a value. To edit values, click them under Value.

Delete removes the custom property.

RELATED COMMANDS

Properties lists information about objects in the drawing.

Status lists information about the drawing.

TIPS

- Use the Custom Properties tab to include project data in drawings.
- Use the Find button in the Design Center (or Search button in Windows Explorer) to search for drawings containing values stored in the Custom Properties tab, as well as for text in the drawing.
- You can use Google to search for text in *.dwg* files located on the Internet.
- Drawing properties can be exported to spreadsheets with the DataExtraction command.
- Windows Explorer can display drawing properties as column headings.

DxbIn

Ver. 2.1 Imports *.dxb* files into drawings (short for Drawing eXchange Binary INput).

Command	*Alias*	*Keyboard Shortcuts*	*Menu*	*Ribbon*
dxbin	...	...	**Insert** **↳Drawing Exchange Binary**	**Blocks & References** **↳Import** **↳DXB File**

Command: dxbin

Displays Select DXB File dialog box. Select a .dxb file, and then click Open.

DIALOG BOX OPTION

Open opens the *.dxb* file, and inserts it in the drawing.

RELATED COMMANDS

DxfIn reads DXF-format files.

Plot writes DXB-format files when configured for an ADI plotter.

TIPS

- To produce *.dxb* files, configure AutoCAD with the ADI plotter driver: after starting the PlotterManager command's Add-a-Plotter wizard, select "AutoCAD DXB File" as the manufacturer.
- This command was created for an early software product named CAD\camera, which attempted to convert raster scans into the DXB vector format. CAD\camera had the distinction of being Autodesk's first software release following the success of AutoCAD — and Autodesk's first failure following AutoCAD.

DxfIn / DxfOut

Ver. 2.0 Opens and exports *.dxf* files into and from drawings (short for Drawing interchange Format INput / OUTput; undocumented commands).

Commands	*Alias*	*Keyboard Shortcuts*	*Menu*	*Ribbon*
dxfin	...	...	**File** ↳**Open** ↳**DXF**	...
dxfout	...	...	**File** ↳**Save As** ↳**DXF**	...

Command: dxfin

Displays Select File dialog box. Select a .dxf file, and then click Open.

DXFOUT COMMAND

Command: dxfout

Displays Save Drawing As dialog box. Enter a file name, and then click Save.

DIALOG BOX OPTIONS

Save saves the drawing as a *.dxf* file.

Files of type creates a *.dxf* file compatible with these versions of AutoCAD:

- AutoCAD 2010 (compatible with 2011).
- AutoCAD 2007 (compatible with 2008 and 2009).
- AutoCAD 2004 (compatible with 2005 and 2006).
- AutoCAD 2000 (compatible with 2000, 2000i, and 2002, and AutoCAD LT).
- AutoCAD Release 12 and AutoCAD LT Release 2 (compatible with Release 13 and 14).

SAVEAS OPTIONS DIALOG BOX

From the Tool menu, select Options:

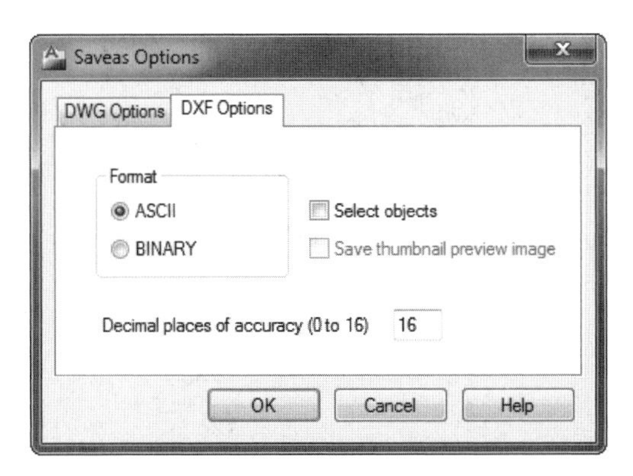

Format

⊙**ASCII** creates files in text format, readable by humans, and imported by most applications.

○**Binary** creates binary files with a smaller file size, but these cannot be read by all applications.

Additional

☐ **Select objects** selects objects to export, instead of the entire drawing.

☐ **Save thumbnail preview image** includes a preview image in the *.dxf* file.

Decimal places of accuracy (0 to 16) specifies the decimal places of accuracy.

RELATED COMMANDS

DxbIn reads DXB-format files.

AcisOut and **SatOut** save solid model objects in the drawing in ACIS-compatible SAT format.

SaveAs writes drawings in DWG and DXF formats.

TIPS

- Alternatively, you can use the Open command to open *.dxf* files.
- The *.dxf* file comes in two styles: *complete* and *partial*:

  Complete *.dxf* files contain all data required to reproduce a complete drawing.

  Partial *.dxf* files must be imported into existing drawings.

  To load a complete *.dxf* file, AutoCAD requires the current drawing to be empty. Partial *.dxf* files can be imported or inserted into any drawing, empty or not.
- Prior to AutoCAD 2006, DxfIn required a new drawing. (To create an empty drawing, use New with the Start from Scratch option.) As of AutoCAD 2006, *.dxf* files are opened in new drawings automatically.
- If you need to import the complete *.dxf* file into a non-empty drawing, use the Insert command, and insert the *.dxf* file with the Explode option turned on.
- Use the ASCII DXF format to exchange drawings with other CAD and graphics programs. Some applications, such as those for CNC (computer numerically controlled) machines, require 4 decimal places.
- Binary *.dxf* files are much smaller and are created much faster than ASCII binary files; few applications, however, read binary *.dxf* files.
- The AutoCAD Release 12 dialect of DXF is the most compatible with other applications.
- *Warning!* When saving a drawing in DXF format for earlier releases, AutoCAD erases or converts some objects into simpler objects.
- Autodesk documents the DXF format under Help > DXF Reference) and at at www.autodesk.com/dxf.

EAttEdit

2002 Edits attribute values and properties in a selected block (short for Enhanced ATTribute EDITor).

Command	*Alias*	*Keyboard Shortcuts*	*Menu*	*Ribbon*
eattedit	...	...	**Modify**	**Home**
			↳Object	**↳Block**
			↳Attribute	**↳Edit Single Attribute**
			↳Single	

Command: eattedit

If the drawing contains no blocks with attributes, AutoCAD complains, "This drawing contains no attributed blocks," and then the command exits.

When attributed blocks exist, the command continues:

Select a block: *(Select a single block.)*

Displays dialog box

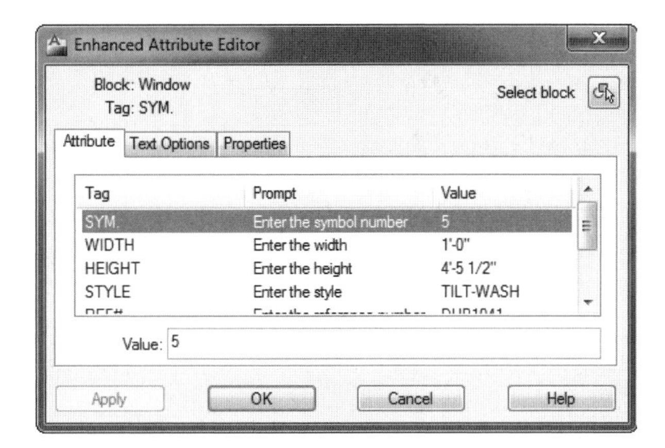

ENHANCED ATTRIBUTE EDITOR DIALOG BOX

Select block selects another block for attribute editing.

Apply applies the changes to the attributes.

Value modifies the value of the selected attribute; neither the tag nor the prompt can be modified by this command.

Text Options tab

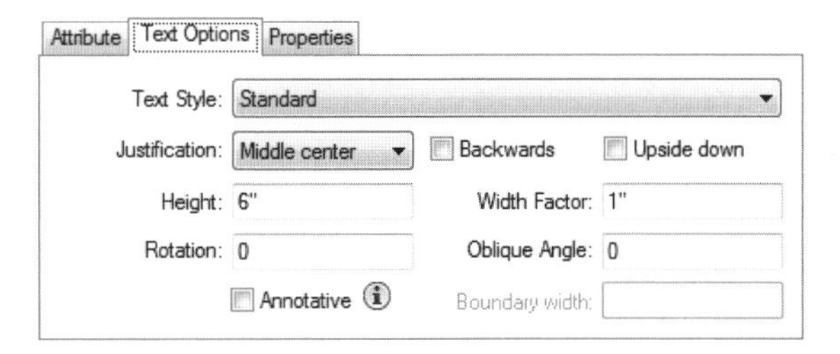

Text Style selects a text style name from the list; text styles are defined by the Style command; default is Standard.

Justification selects a justification mode from the list; default is left justification.

Height specifies the text height; can be changed only when height is set to 0.0 in the text style.

Rotation specifies the rotation angle of the attribute text; default = 0 degrees.

☐ **Backwards** displays the text backwards.

☐ **Upside Down** displays the text upside down.

Width Factor specifies the relative width of characters; default = 1.

Oblique Angle specifies the slant of characters; default = 0 degrees.

☐ **Annotative** toggles annotative scaling.

Boundary Width specifies margins for multi-line attributes.

Properties tab

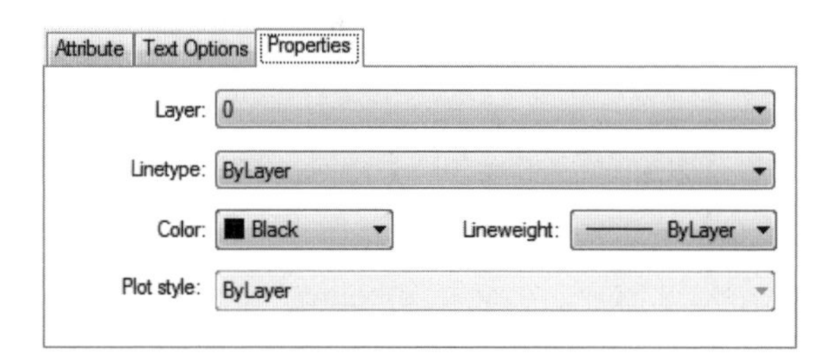

Layer selects a layer name from the list; layers are defined by the Layer command.

Linetype selects a linetype name from the list; linetypes are loaded into the drawing with the **Linetype** command.

Color selects a color from the list; to select from the full 255-color spectrum, select Other.

Lineweight selects a lineweight from the list; to display lineweights, click LWT on the status bar.

Plot style selects a plot style name from the list; available only if plot styles are enabled in the drawing.

RELATED COMMANDS

AttDef creates attribute definitions.

Block attaches attributes to objects.

BAttMan manages attributes.

EAttExt extracts attributes to a file.

TextEdit edits attributes.

TIPS

- This command edits only attribute values and properties; to edit all aspects of attributes, use BAttMan command.
- Blocks on locked layers cannot be edited.

Aliased Commands

EAttExt and **-EAttExt** commands are aliases for the DataExtraction and -DataExtraction commands as of AutoCAD 2008.

Edge

Rel. 12 Toggles the visibility of 3D faces.

Command	Alias	Keyboard Shortcuts	Menu	Ribbon
edge	...	...	...	...

Command: edge
Specify edge of 3dface to toggle visibility or [Display]: *(Pick an edge, or type D.)*
Enter selection method for display of hidden edges [Select/All] <All>: *(Type S or A.)*
Select objects: *(Select one or more objects.)*
Specify edge of 3dface to toggle visibility or [Display]: *(Press Esc to end the command.)*

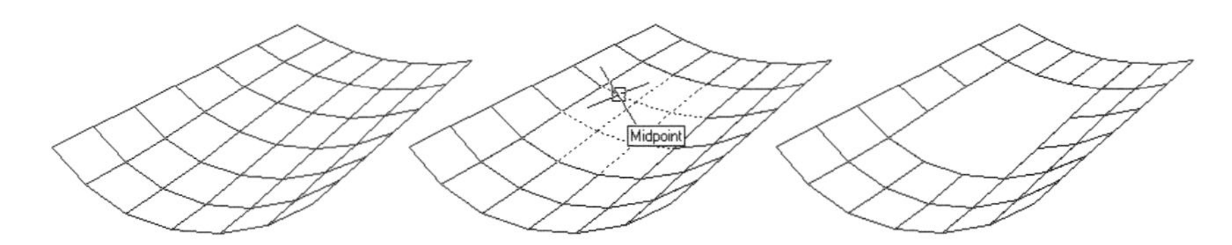

From left to right: *3D faces; edges selected with Edge; and invisible edges.*

COMMAND LINE OPTIONS

Specify edge selects the edge to make invisible.

Display Options

Select highlights invisible edges.

All selects all hidden edges, and regenerates them.

RELATED COMMAND

3dFace creates 3D faces.

RELATED SYSTEM VARIABLE

SplFrame toggles visibility of 3D face edges.

TIPS

- Make edges invisible to improve the appearance of 3D objects.
- Edge applies only to objects made of 3D faces; it does not work with polyface meshes or solid models.
- Use the Explode command to convert meshed objects into 3D faces.
- Re-execute Edge to display an edge that has been made invisible.
- The command repeats until you press Enter or Esc at the 'Specify edge of 3dface to toggle visibility or [Display]:' prompt.
- When you hide all edges of a 3D face, it can be difficult to unhide them. "No," advises the technical editor. "Just change to a visual style other than 3D wireframe."

EdgeSurf

Rel. 10 Draws 3D meshes as Coons patches between four boundaries (short for EDGE-defined SURFace); does not draw surfaces.

Command	*Alias*	*Keyboard Shortcuts*	*Menu*	*Ribbon*
edgesurf	...	...	**Draw** **↳Modeling** **↳Meshes** **↳Edge Mesh**	**Mesh Modeling** **↳Primitives** **↳Edge Surface**

Command: edgesurf

Current wire frame density: SURFTAB1=6 SURFTAB2=6

Select object 1 for surface edge: *(Pick an object.)*

Select object 2 for surface edge: *(Pick an object.)*

Select object 3 for surface edge: *(Pick an object.)*

Select object 4 for surface edge: *(Pick an object.)*

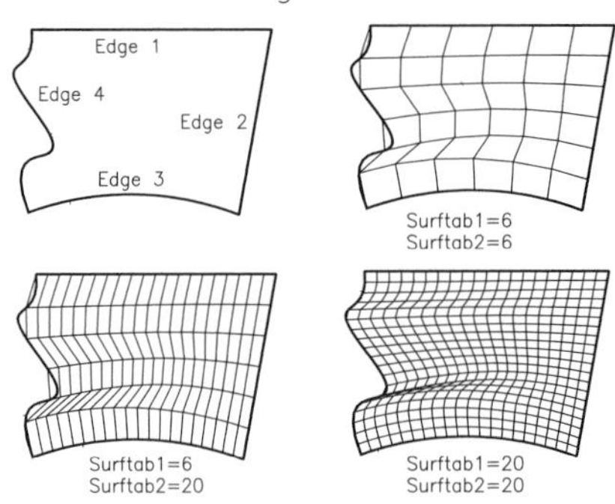

COMMAND LINE OPTION

Select object picks an edge.

RELATED COMMANDS

3dMesh creates 3D meshes by specifying every vertex.

3dFace creates 3D meshes of irregular vertices.

PEdit edits meshes created by EdgeSurf.

TabSurf creates tabulated 3D meshes.

RuleSurf creates ruled 3D meshes.

RevSurf creates 3D meshes of revolution.

RELATED SYSTEM VARIABLES

SurfTab1 stores the current *m*-density of meshing; the maximum mesh density is 32767.

SurfTab2 stores the current *n*-density of meshing; maximum is 32767.

MeshType determines whether this command creates polyface meshes or mesh objects.

TIP

- The boundary edges can be made from lines, arcs, and open 2D and 3D polylines; edges must meet at endpoints.

EditShot

2010 Edits the properties of show-motion animations.

Command	*Alias*	*Keyboard Shortcuts*	*Menu*	*Ribbon*
editshot	**eshot**	...	...	...

Command: editshot

Enter view name to edit: *(Enter the name of a view.)*

Displays the Shot Properties tab of the New View/Shot Properties dialog box.

COMMAND LINE OPTION

Enter view name to edit specifies the name of a view previously created by the View command.

DIALOG BOX OPTIONS

See NewShot command.

RELATED COMMANDS

NewShot creates named views with show motion options.

NavSMotion displays the show motion interface.

ViewPlay plays the animation associated with a named view.

TIP

- This command edits the properties of show-motion animations created by the NewShot command.

EditTableCell

2009 Performs all cell editing operations at the command prompt (undocumented command).

Command	*Alias*	*Keyboard Shortcuts*	*Menu*	*Ribbon*
edittablecell	...	...	...	...

Command: edittablecell
Select table: *(Select a table.)*
Cell or range in An:Bn format: *(Enter a cell name, such as A1, or a range of cells, such as A1:C3.)*
[Operation/Content/Format/Style/Datalink/Quit/ExecuteOneAction]: *(Enter an option.)*

COMMAND LINE OPTIONS

Operation inserts and deletes rows, columns, and cells; also inserts blocks, formulas, fields, and changes the look of borders; prompts you:

[BAck/insert ABove/insert BElow/delete ROw/insert LEft/insert RIght/delete COlumn/Merge/Unmerge/insert FOrmula/insert BLock/insert FIeld/BOrder/Quit]: *(Enter an option.)*

Content specifies the data type of cells (general, date, etc), locks and unlocks cells, and displays the Manage Cell Content dialog box; prompts you:

[BAck/Datatype/Lock/Unlock/Manage/Quit]: *(Enter an option.)*

Format changes the text alignment and background color of cells; prompts you:

[BAck/ALignment/BG color/Lock/Unlock/Quit]: *(Enter an option.)*

Style edits table and cell styles; also creates new table styles; prompts you:

[BAck/Edit table style/New table style/edit cell Style/Quit]: *(Enter an option.)*

Datalink creates new data links, and updates existing ones; prompts you:

[BAck/New/Update/Quit]: *(Enter an option.)*

Quit exits the command.

ExecuteOneAction exits the command after one option has been executed.

RELATED COMMAND

TablEdit edits tables.

TIPS

- This command prompts you to select a table and a cell range.
- This command is meant for use with macros and AutoLISP routines. It is more practical to double-click a cell, and then right-click for editing options.

Elev

Ver. 2.1 Sets elevation and thickness for creating extruded 3D objects (short for ELEVation).

Command	*Alias*	*Keyboard Shortcuts*	*Menu*	*Ribbon*
'elev	...	...	...	...

Command: elev

Specify new default elevation <0.0000>: *(Enter a value for elevation.)*

Specify new default thickness <0.0000>: *(Enter a value for thickness.)*

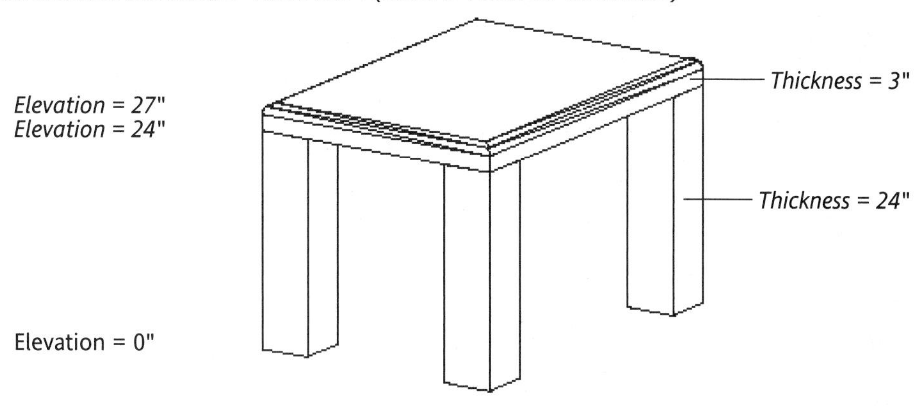

COMMAND LINE OPTIONS

Elevation changes the base elevation from z = 0.

Thickness extrudes new 2D objects in the z-direction.

RELATED COMMANDS

Change changes the thickness and z coordinate (elevation) of objects.

Move moves objects, including in the z direction.

Properties and **ChProp** change the thickness of objects.

RELATED SYSTEM VARIABLES

Elevation stores the current elevation setting.

Thickness stores the current thickness setting.

TIPS

- The current value of elevation is used whenever the z coordinate is not supplied.
- Thickness is measured up from the current elevation in the positive z-direction.
- The status bar reports the coordinates as x,y and elevation:

1'-0 1/2", -0'-2" , 0'-0"

Ellipse

Ver. 2.5 Draws ellipses by four different methods, and draws elliptical arcs and isometric circles.

Command	*Alias*	*Keyboard Shortcuts*	*Menu*	*Ribbon*
ellipse	**el**	**...**	**Draw** ↳**Ellipse**	**Home** ↳**Draw** ↳**Center**

Command: ellipse
Specify axis endpoint of ellipse or [Arc/Center]: *(Pick a point, or enter an option.)*
Specify other endpoint of axis: *(Pick a point.)*
Specify distance to other axis or [Rotation]: *(Pick a point, or type R.)*

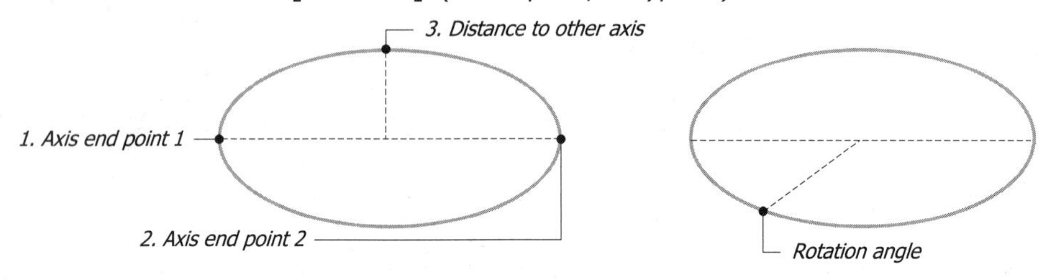

COMMAND LINE OPTIONS

Specify axis endpoint of ellipse indicates the first endpoint of the major axis.

Specify other endpoint of axis indicates the second endpoint of the major axis.

Specify distance to other axis indicates the half-distance of the minor axis.

Elliptical Arcs

Command: ellipse
Specify axis endpoint of elliptical arc or [Center]: *(Pick a point, or type C.)*
Specify other endpoint of axis: *(Pick a point.)*
Specify distance to other axis or [Rotation]: *(Pick a point, or type R.)*
Specify start angle or [Parameter]: *(Enter an angle or type P.)*
Specify end angle or [Parameter/Included angle]: *(Enter an angle, or enter an option.)*

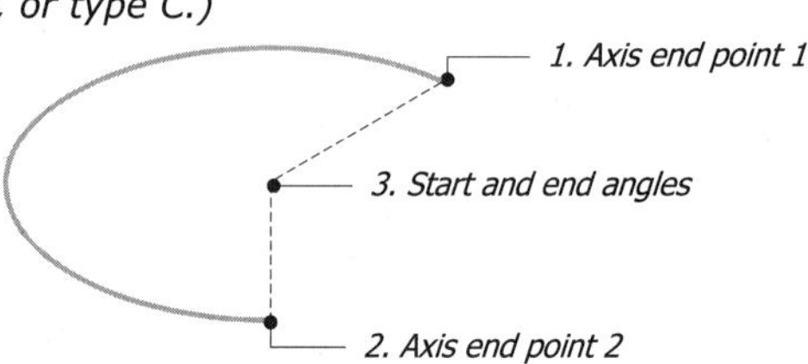

COMMAND LINE OPTIONS

Specify start angle indicates the starting angle of the elliptical arc.

Specify end angle indicates the ending angle of the elliptical arc.

Parameter indicates the starting angle of the elliptical arc; draws the arc with this formula:

p(u)=**c**+(**a***cos(u))+(**b***sin(u))

Parameter	*Meaning*
a	Major axis.
b	Minor axis.
c	Center of ellipse.

Included angle indicates an angle measured relative to the start angle, rather than to 0 degrees.

Center option

Specify center of ellipse: *(Pick a point.)*

Specify center of ellipse indicates the center point of the ellipse.

Rotation option

Specify rotation around major axis: *(Enter an angle.)*

Specify rotation around major axis indicates a rotation angle around the major axis:

0 degrees — minimum rotation: creates round ellipses, like circles.

89.4 degrees — maximum rotation: creates very thin ellipses.

Isometric Circles

This option appears only when system variable SnapStyl is set to 1 (isometric snap mode). Use F5 to switch among the three isometric drawing planes.

Command: ellipse

Specify axis endpoint of ellipse or [Arc/Center/Isocircle]: *(Type I.)*

Specify center of isocircle: *(Pick a point.)*

Specify radius of isocircle or [Diameter]: *(Pick a point, or type D.)*

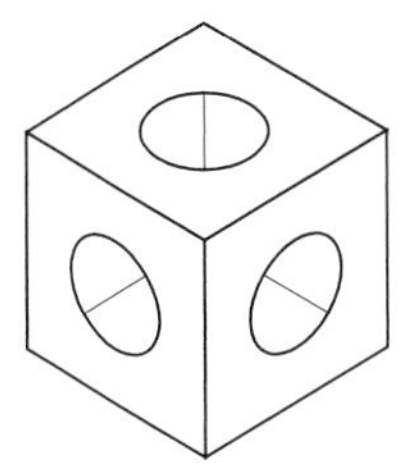

COMMAND LINE OPTIONS

Specify center indicates the center point of the isocircle.

Specify radius indicates the radius of the isocircle.

Diameter indicates the diameter of the isocircle.

RELATED COMMANDS

IsoPlane sets the current isometric plane.

PEdit edits ellipses (when drawn with a polyline).

Snap controls the setting of isometric mode.

RELATED SYSTEM VARIABLES

PEllipse determines how the ellipse is drawn.

SnapIsoPair sets the current isometric plane.

SnapStyl specifies regular or isometric drawing mode.

TIPS

- Previous to AutoCAD Release 13, Ellipse constructed the ellipse as a series of short polyline arcs. The PEllipse system variable controls how ellipses are drawn. When 0, true ellipses are drawn; when 1, a polyline approximation of an ellipse is drawn.
- When PEllipse = 1, the Arc option is not available.
- Use ellipses to draw circles in isometric mode. When Snap is set to isometric mode, Ellipse's Isocircle option projects a circle into the working isometric drawing plane. Press CTRL+E or F5 to switch between isoplanes.
- The Isocircle option only appears in the option prompt when Snap is set to isometric mode.

Removed Commands

End command was removed from AutoCAD Release 14, because users confused it for abbreviation for the ENDpoint object snap, unexpectedly ending their drawings ; it was replaced by Quit.

Erase

Ver. 1.0 Erases objects from drawings.

Commands	*Alias*	*Keyboard Shortcut*	*Menu*	*Ribbon*
erase	**e**	**Del**	**Modify** **↳Erase**	**Home** **↳Modify** **↳Erase**
...	...	...	**Edit** **↳Clear**	...

Command: erase
Select objects: *(Select one or more objects.)*
Select objects: *(Press Enter to end object selection.)*

COMMAND LINE OPTION

Select objects selects the objects to erase.

RELATED COMMANDS

Break erases a portion of a line, circle, arc, or polyline.

Fillet cleans up intersections by erasing selected segments.

Oops returns the most-recently erased objects to the drawing.

Trim cuts off the end of a line, arc, and other objects.

Undo returns the erased objects to the drawing.

TIPS

- The Erase L command erases the last-drawn item visible in the current viewport.
- *Warning!* The Erase All command erases all objects in the current space (model or layout), except on locked, frozen, and/or off layers.
- Oops brings back the most-recently erased objects; use U to bring back other erased objects.
- Objects on locked and frozen layers cannot be erased. To erase them, change the layers to unlocked and thawed.
- When two objects are connected by a constraint, this command erases the selected object(s) and related constraints.

eTransmit

2000i Collects drawings and related files into packages suitable for transmission by email or CD (short for Electronic TRANSMITtal).

Commands	Alias	Keyboard Shortcuts	Menu Bar	Application Menu	Ribbon
etransmit	...	...	**File** ↳**eTransmit**	**Send** ↳**eTransmit**	...
-etransmit	...	...	...	...	

Command: etransmit

If the drawing has not been saved since the last change, displays error dialog box.

OK saves the drawing, and then displays the Create Transmittal dialog box.

Cancel cancels the eTransmit command.

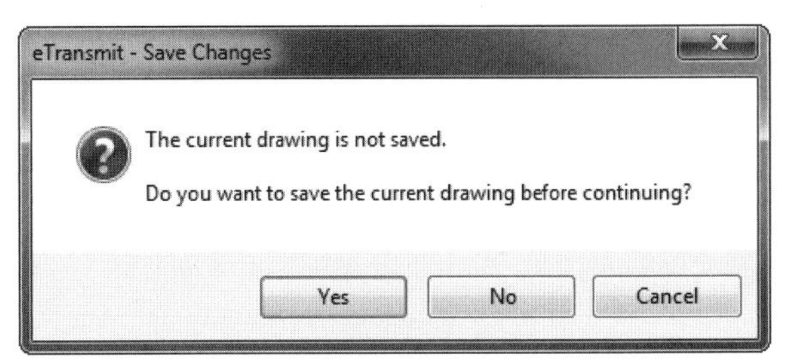

CREATE TRANSMITTAL DIALOG BOX

Files Tree tab

Sheets tab displays sheets included with transmittal.

Files Tree tab displays names of drawing and support files, grouped by category.

Files Table tab displays file names in alphabetical order.

Select a Transmittal Setup lists predefined setups.

Transmittal Setups displays the Transmittal Setup dialog box.

Enter Notes to Include With This Transmittal Package provides space for entering notes.

View Report displays the View Transmittal Report dialog box.

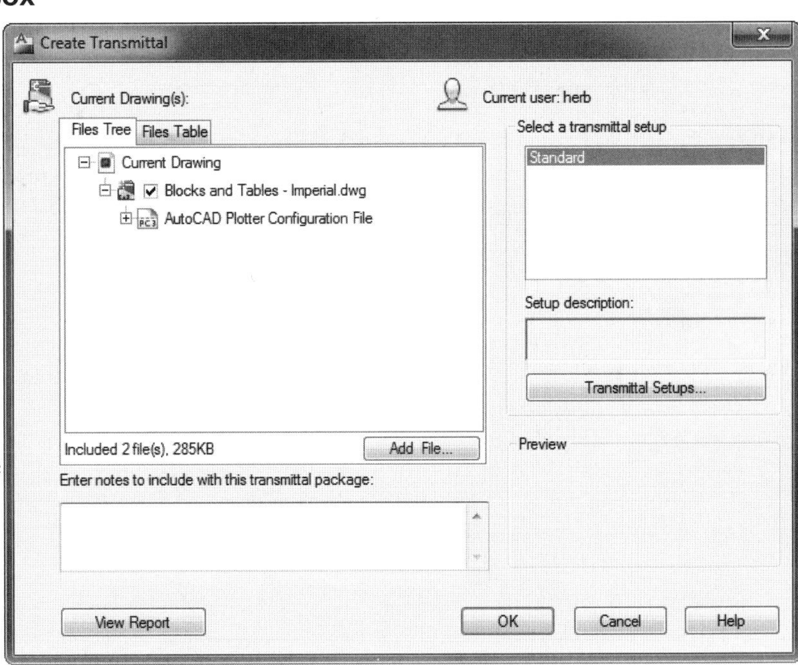

Files Table tab

☑ File included in transmittal.

☐ File excluded from transmittal.

Add File displays the Add File to Transmittal dialog box, to add files to the transmittal.

TRANSMITTAL SETUP DIALOG BOX

New creates new setups; displays New Transmittal Setup dialog box.

Rename renames setups, except for "Standard."

Modify changes setups; displays Modify Transmittal Setup dialog box.

Delete removes setups, except for "Standard."

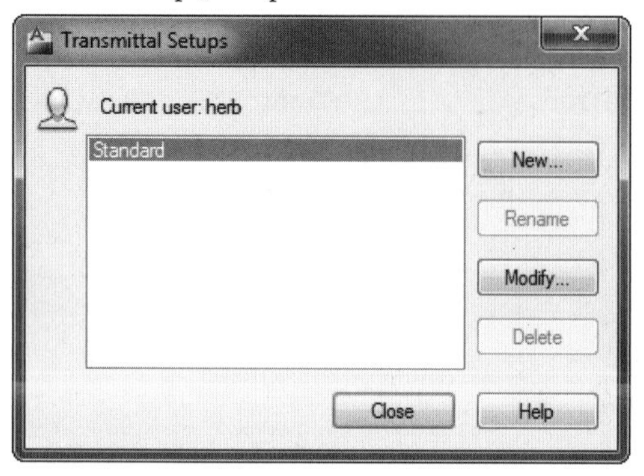

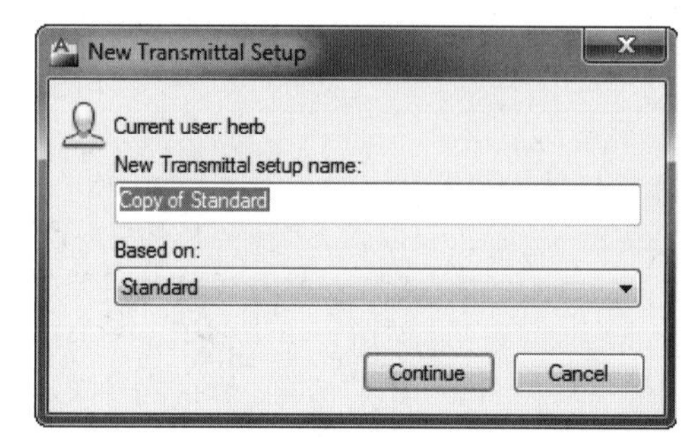

Left to right: *Transmittal Setup dialog box and New Transmittal Setup dialog box.*

NEW TRANSMITTAL SETUP DIALOG BOX

New Transmittal Setup Name specifies the name of the new setup.

Based on selects the setup to copy.

Continue displays the Modify Transmittal Setup dialog box.

MODIFY TRANSMITTAL SETUP DIALOG BOX

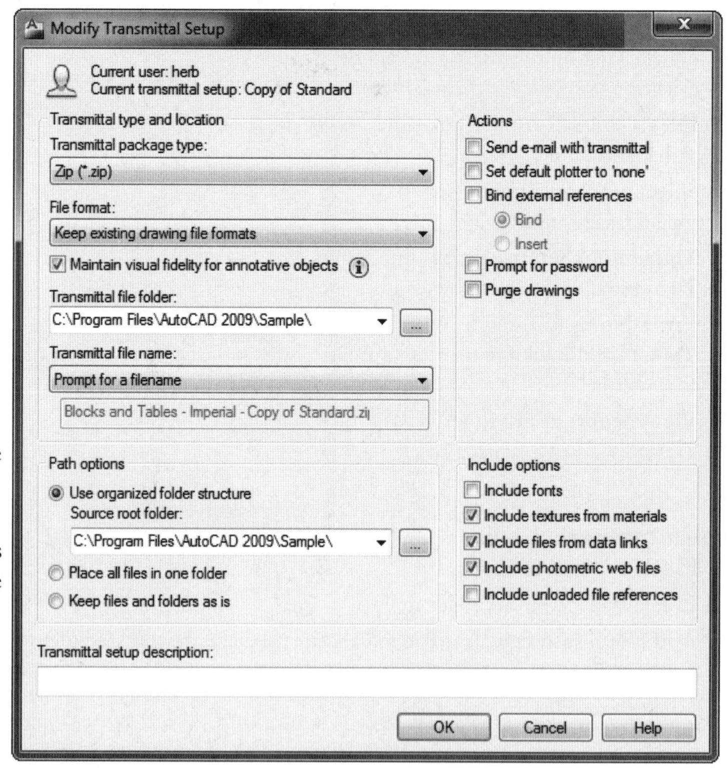

Transmittal Type and Location options

Transmittal Package Type:

- **Folder (set of files)** transmits uncompressed files in new and existing folders. Best option for transmitting files on CDs or by FTP.
- **Self-Extracting Executable (*.exe)** transmits files in a compressed, self-extracting file. Uncompress by double-clicking the file.
- **Zip (*.zip)** (*default*) transmits files as a compressed ZIP file. Uncompress the file using 7Zip or WinZip program. Best option when sending by email.

File Format (*Warning!* AutoCAD 2011 objects and properties not in DWG 2000-2009 may be changed or lost.)

- **Keep existing drawing file formats** transmits files in their version.

- **AutoCAD 2010/LT 2010 Drawing Format** transmits files in DWG 2010 format.
- **AutoCAD 2007/LT 2007 Drawing Format** transmits files in DWG 2007/8/9 format.
- **AutoCAD 2004/LT 2004 Drawing Format** transmits files in DWG 2004/5/6 format.
- **AutoCAD 2000/LT 2000 Drawing Format** transmits files in DWG 2000/i/1 format.

☑**Maintain Visual Fidelity for Annotative Objects** uses visual fidelity for annotative objects: each scale representation of annotative objects is saved to separate layers.

Transmittal File Folder specifies the location in which to collect the files. When no location is specified, the transmittal is created in the same folder as the *.dst* file.

Transmittal File Name:

- **Prompt for a File Name** prompts the user for the file name.
- **Overwrite if Necessary** uses the specified file name, and overwrites the existing file of the same name.
- **Increment File Name if Necessary** uses the specified file name, and appends a digit to avoid overwriting the existing file of the same name.

Path Options options

⊙**Use Organized Folder Structure** preserves the folder structure in the transmittal, but makes the changes listed below; it allows you to specify the name of the folder tree.

○**Place All Files in One Folder** places all files in a single folder.

○**Keep Files and Folders As Is** preserves the folder structure in the transmittal.

Actions options

☐**Send email with transmittal** opens the computer's default email software.

☐**Set Default Plotter to 'None'** resets the plotter to None for all drawings in the transmittal.

☐**Bind External References** merges all xrefs into the drawing:

⊙ **Bind** binds the xref.

○ **Insert** inserts the xrefs like a block.

☐**Prompt for Password** displays a dialog box for specifying a password; not available when the Folder archive type is selected.

☐**Purge Drawings** removes unused layers, linetypes, and so on, without prompting or warning.

Include Options options

☐**Include Fonts** includes all *.ttf* and *.shx* font files in the transmittal.

☑**Include textures from materials** includes material files.

☑**Include Files from Data Links** includes spreadsheet files linked to the drawing.

☑**Include Photometric Web Files** includes IEF (photometric web) files associated with web lights in the drawing.

☐**Include Unloaded File References** includes attached files that are in an unloaded state.

Transmittal Setup Description describes the setup.

-ETRANSMIT command

Command: -etransmit

Enter an option [Create transmittal package/Report only/CUrrent setup/CHoose setup/Sheet set] <Report only>: *(Enter an option.)*

COMMAND LINE OPTIONS

Create Transmittal Package creates transmittal packages from the current drawing and all support files; uses settings in the current transmittal setup.

Report Only displays the Save Report File As dialog box; enter a file name, and then click. **Save** saves the report in plain ASCII text format; does not create the transmittal package.

Current Setup lists the name of the current transmittal setup.

Choose Setup selects the transmittal setup.

Sheet Set specifies the sheet set and transmittal setup to use for the transmittal package; available only when a sheet set is open in the drawing.

RELATED COMMANDS

Archive creates archive sets of drawings, support files, and sheet sets.

Publish creates *.dwf* files with drawings and sheets.

PublishToWeb saves drawings as HTML files.

RELATED SYSTEM VARIABLE

SaveFidelity toggles visual fidelity for annotative objects.

TIPS

- The Password button is *not* available when you select the Folder option.
- Including a TrueType font (*.ttf*) is touchy, because sending a copy of the font might infringe on its copyright. All *.shx* and *.ttf* files included with AutoCAD may be transmitted. For this reason, the Include Fonts option is turned off by default. In addition, not including fonts saves some file space; recall that smaller files take less time to transmit via the Internet.
- "Self-extracting executable" means that the files are compressed into a single file with the *.exe* extension. The email recipient double-clicks the file to extract (uncompress) the files. The benefit is that recipients do not need to have a copy of 7Zip or WinZip on their computers; the drawbacks are that a virus could hide in the *.exe* file, and some e-mail programs prevent receipt of attachments with *.exe* extensions..
- Visual fidelity attempts to preserve data in drawings translated to earlier releases of AutoCAD, which don't support annotative scaling found in AutoCAD 2008 and later. When visual fidelity is turned on, annotative text appears on separated layers. This ensures the text is there, and can be toggled on and off by thawing and freezing the appropriate layers in each viewport.
- Files occasionally become corrupted when sent by email; the transmittal may need to be resent. If you continue to have problems, you may need to change a setting in your email software. Try changing the attachment encoding method from BinHex or Uuencode to MIME.

ExAcReload

2009 Reloads xrefs, and then reports whether they have changed (short for "EXternal AutoCad RELOAD"; undocumented command).

Commands	*Alias*	*Keyboard Shortcut*	*Menu*	*Ribbon*
exacreload	...	...	...	...

Command: exacreload
Reload Xref "xref1name": filename1.dwg
"xref1name" loaded: filename1.dwg

Reload Xref "xref2name": filename2.dwg
"xref1name" loaded: filename2.dwg has not changed.

COMMAND LINE OPTIONS

None.

RELATED COMMAND

ExternalReferences attaches *.dwg* files as externally-referenced drawings.

TIPS

- This command does nothing when no xrefs are attached, and does not report the changes made to the xref.
- This command is a keyboard substitute for the ExternalReferences command's Refresh button.

Explode

Ver. 2.5 Explodes polylines, blocks, associative dimensions, hatches, multilines, 3D solids, regions, bodies, and meshes into their constituent objects.

Command	*Alias*	*Keyboard Shortcuts*	*Menu*	*Ribbon*
explode	**x**	**...**	**Modify** ↳**Explode**	**Home** ↳**Modify** ↳**Explode**

Command: explode
Select objects: *(Select one or more objects.)*
Select objects: *(Press Enter to end object selection.)*

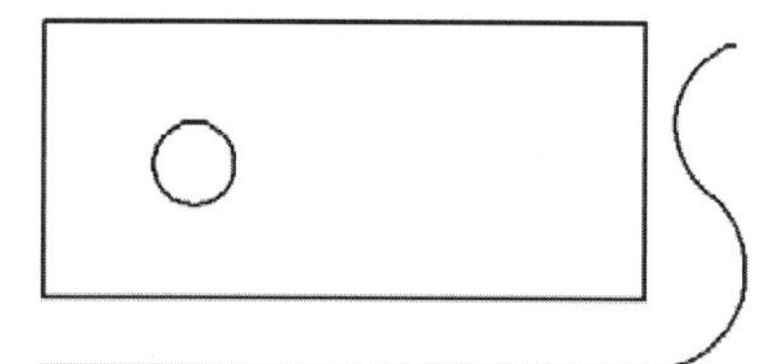

Left: *A variety of polylines.*
Right: *Polylines exploded into lines and arcs.*

COMMAND LINE OPTION

Select objects selects the objects to explode.

RELATED COMMANDS

Block recreates a block after an explode.

PEdit converts a line into a polyline.

Region converts 2D objects into a region.

Undo reverses the effects of this command.

Xplode controls the explosion process.

RELATED SYSTEM VARIABLE

ExplMode toggles whether non-uniformly scaled blocks can be exploded.

TIPS

- As of Release 13, AutoCAD can explode blocks inserted with unequal scale factors, mirrored blocks, and blocks created by the MInsert command.
- You cannot explode xrefs and dependent blocks (blocks from xref drawings), or single-line text. Use the TxtExp Express tool to explode text.
- Parts making up exploded blocks and associative dimensions of BYBLOCK color and linetype are displayed in color White (or Black when the background color is white) and Continuous linetype.
- While this command is not transparent, Xplode is.
- Flat faces of 3D solids become regions; curved faces become surfaces.
- Constraints are erased from exploded objects.

- The Explode command alters objects, as follows:

Objects	*Exploded Into*
Arcs in non-uniformly scaled blocks	Elliptical arcs
Associative dimensions	Lines, solids, and text
Blocks	Constituent parts
Circles in non-uniformly scaled blocks	Ellipses
Mtext and field text	Text
Multilines	Lines
Polygon meshes	3D faces
Polyface meshes	3D faces, lines, and points
3D meshes	3D faces
NURBS surfaces	Splines that show the surface outline but do no delineate bulges or other surface features *(new to AutoCAD 2011)*
Tables	Lines and text
2D polylines	Lines and arcs; width and tangency information lost
3D polylines	Lines
3D solids	Regions and surfaces
Regions	Lines, arcs, ellipses, and splines
Bodies	Single bodies, regions, and curves

- Resulting objects become the previous selection set.

Export

Rel. 13 Saves drawings in DWG and DXF and a few other formats.

Command	*Alias*	*Menu Bar*	*Application Menu*	*Ribbon*
export	**exp**	**File** ↳**Export**	**Export** ↳**Other Formats**	**Output** ↳**Export to DXF/PDF** ↳**Export**
-export	**-qpub**	...	...	...

Command: export

Displays dialog box.

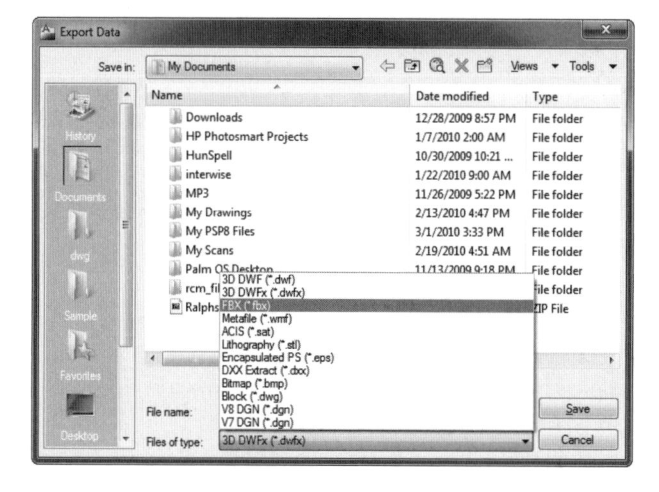

EXPORT DATA DIALOG BOX

Save in selects the folder (subdirectory) and drive into which to export the file.

Back returns to the previous folder (alternatively, press the *Alt+1* shortcut keystroke).

Up One Level moves up one level in the folder structure (*Alt+2*).

Search the Web displays a simple Web browser that accesses the Autodesk Web site (*Alt+4*).

Delete erases the selected file(s) or folder (*Del*).

Create New Folder creates a new folder (*Alt+5*).

Views displays files and folders in a list or with details.

Tools lists several additional commands, including the Options dialog box.

File name specifies the name of the file, or accepts the default.

Save as type selects the file format in which to save the drawing.

Cancel dismisses the dialog box, and returns to AutoCAD.

-EXPORT Command

Command: -export

Enter file format [Dwf/dwfX/Pdf] <dwfX> *(Enter an option.)*

Dwf, **dwfX**, or **Pdf** specifies the default file format for the ExportSettings command.

RELATED COMMANDS

AttExt exports attribute data in drawings in CDF, SDF, or DXF formats.

CopyClip exports objects from drawings to the Clipboard.

CopyHist exports text from the text screen to the Clipboard.

DgnExport exports drawings as MicroStation design files.

ExportSettings exports the current drawing in DWF, DWFx, or PDF format.

FbxExport exports 3D drawings as Motionbuilder Filmbox files for import into 3D Studio *(new to AutoCAD 2011)*.

Import imports several vector and raster formats.

LogFileOn saves the command line text as ASCII text in the *acad.log* file.

MassProp exports the mass property data as ASCII text in an *.mpr* file.

MSlide exports the current viewport as an *.sld* slide file.

Plot exports drawings in many vector and raster formats.

SaveAs saves drawings in AutoCAD's DWG format.

SaveImg exports renderings in TIFF, Targa, or BMP format.

3dDwf exports drawings in DWF or DWFx format.

RELATED SYSTEM VARIABLE

ExportePlotFormat specifies the default export format: (0) PDF, (1) DWF, or (2) DWFx.

TIPS

- The Export command exports the current drawing in the following formats:

Extension	*File Type*	*Related Command*
BMP	Device-independent bitmap	**BmpOut**
DGN	MicroStation V7/8 design file	**DgnExport**
DWF	3D Design Web Format	**3dDwf**
DWF	2D Design Web Format	**DwfOut** *(alias for Plot command)*
DWFx	3D XPS-compatible DWF	**3dDwf**
DWG	AutoCAD drawing file	**WBlock**
DXX	Attribute extract DXF file	**AttExt**
EPS	Encapsulated PostScript file	**PsOut**
FBX	Motionbuilder Filmbox	**FbxExport** *(new to AutoCAD 2011)*
SAT	ACIS solid object file	**AcisOut**
STL	Stereolithography	**StlOut**
WMF	Windows metafile	**WmfOut**

- This command acts as a "shell" by launching other AutoCAD commands that perform the actual export function.
- When drawings are exported in formats other than DWF or DXF, information is lost, such as layers and attributes.
- The formats supported by this command have changed over the years:
  - AutoCAD 2007 removed 3DS (3D Studio) and WMF formats, but added 3D DWF.
  - AutoCAD 2008 added the DGN V8 format.
  - AutoCAD 2009 added DWFx, 3D DWFx and DGN V7.
  - AutoCAD 2011 added FBX format.

Removed (and Restored) Command

ExpressTools (also known as "bonus CAD tools" in earlier versions of AutoCAD) were removed from AutoCAD 2002, but returned to AutoCAD 2004.

ExportDWF / ExportDWFx

2010 Exports the current drawing in DWF or DWFx format through a specially-designed file dialog box.

Commands	Aliases	Keyboard Shortcut	Application Menu	Ribbon
exportdwf	**qdwf**	...	**Export** **↳DWF**	**Output** **↳Export to DWF/PDF** **↳Export** **↳DWF**
exportdwfx	**qdwfx**	...	**Export** **↳DWFx**	**Output** **↳Export to DWF/PDF** **↳Export** **↳DWFx**

Command: exportdwf
Command: exportdwfx

Displays dialog box.

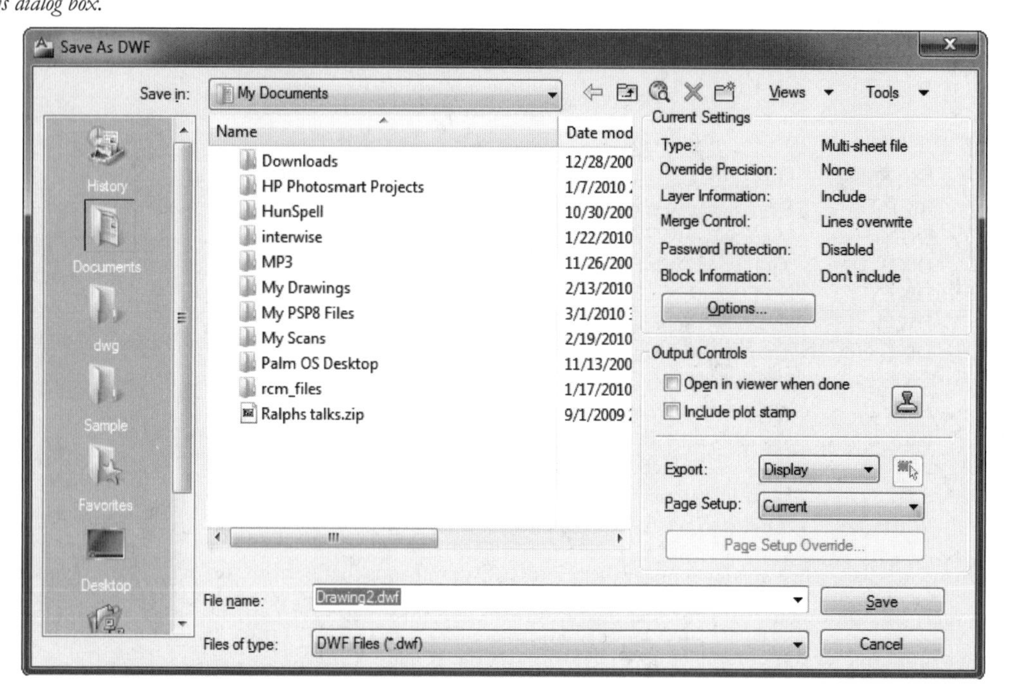

SAVE AS DWF DIALOG BOX

The dialog boxes are identical for both DWF and DWFx formats.

Current Settings options

Options specifies options for exporting the file; displays the Export to DWF/PDF Options dialog box. See the Publish command.

Output Controls options

☐ **Open in viewer when done** opens the DWF/x file in Design Review, if installed on your computer.

☐ **Include plot stamp** adds plot stamp data to the edges of the exported file.

Plot Stamp Settings specifies the plot stamp data to include; displays the Plot Stamp Settings dialog box. See the PlotStamp command.

Export specifies how much of the drawing to export:

- **Display** exports the view displayed by the current viewport.
- **Extents** exports the extents of the drawing (everything not frozen).
- **Window** exports a windowed area; dismisses the dialog box temporarily, and prompts you to pick two points.

Select Window allows you to redefine the windowed area to export.

Page Setup specifies the page setup to use:

- **Current** uses the current page setup; see the PageSetup command.
- **Override** changes the page setup by making the Page Setup Override button available.

Page Setup Override overrides the page setup; displays the Page Setup Override dialog box; see the ExportSettings command.

RELATED COMMANDS

Attach attaches DWF and DWFx files as underlays.

Export exports drawings in a variety of formats.

ExportPdf exports drawings in PDF format.

DgnExport exports drawings as MicroStation design files.

3dDwf exports drawings in 3D DWF format.

Plot exports drawings in a variety of raster and vector formats.

TIPS

- To export drawings in 3D format, use the 3dDwf command.
- DWF is short for "design Web format."

ExportLayout

2010 Exports layouts to *.dwg* files, and optionally imports them into model space.

Command	*Alias*	*Keyboard Shortcuts*	*Menu Bar*	*Application Menu*	*Ribbon*
exportlayout	...	...	...	**Save As** ↳**Export Layout as Drawing**	...

This command works only in layout mode.

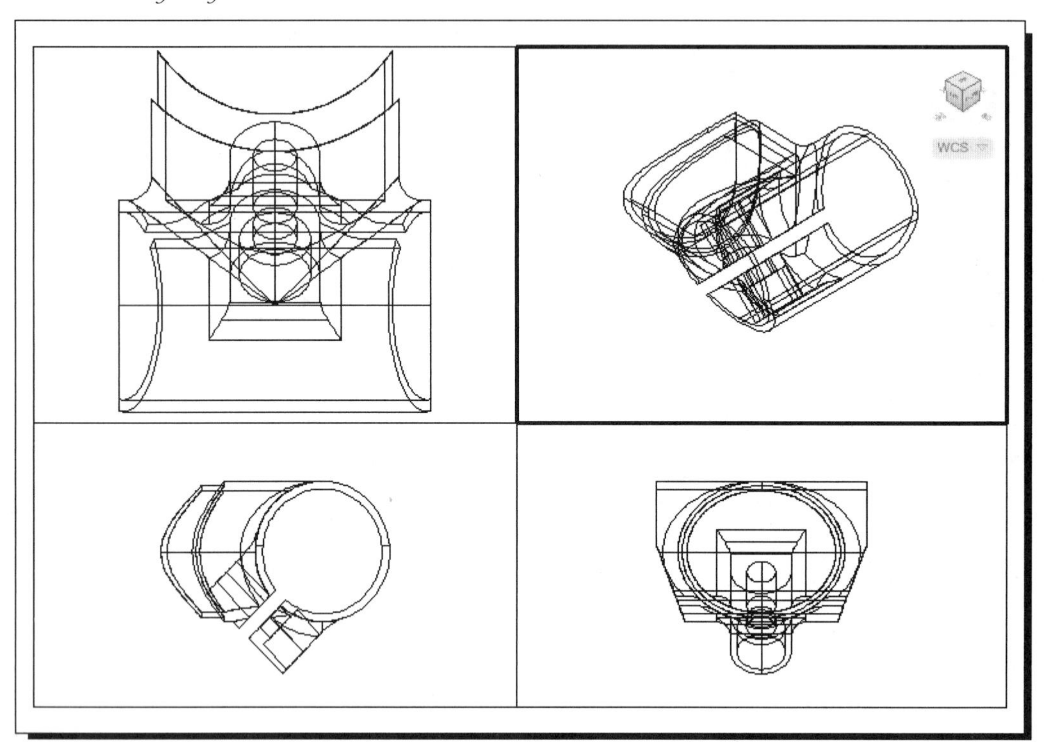

Command: exportlayout

Displays the Export Layout to Model Space Drawing dialog box. Specify the file name, or accept the default provided by AutoCAD. Click Save, and then wait...

When AutoCAD is done exporting, it displays dialog box:

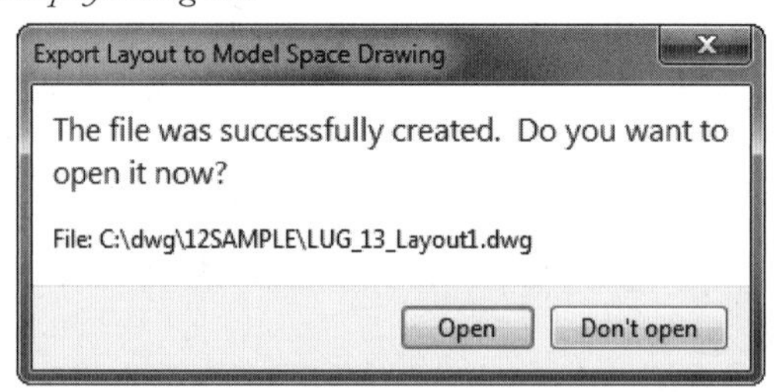

Click Open. AutoCAD displays the drawing in model space.

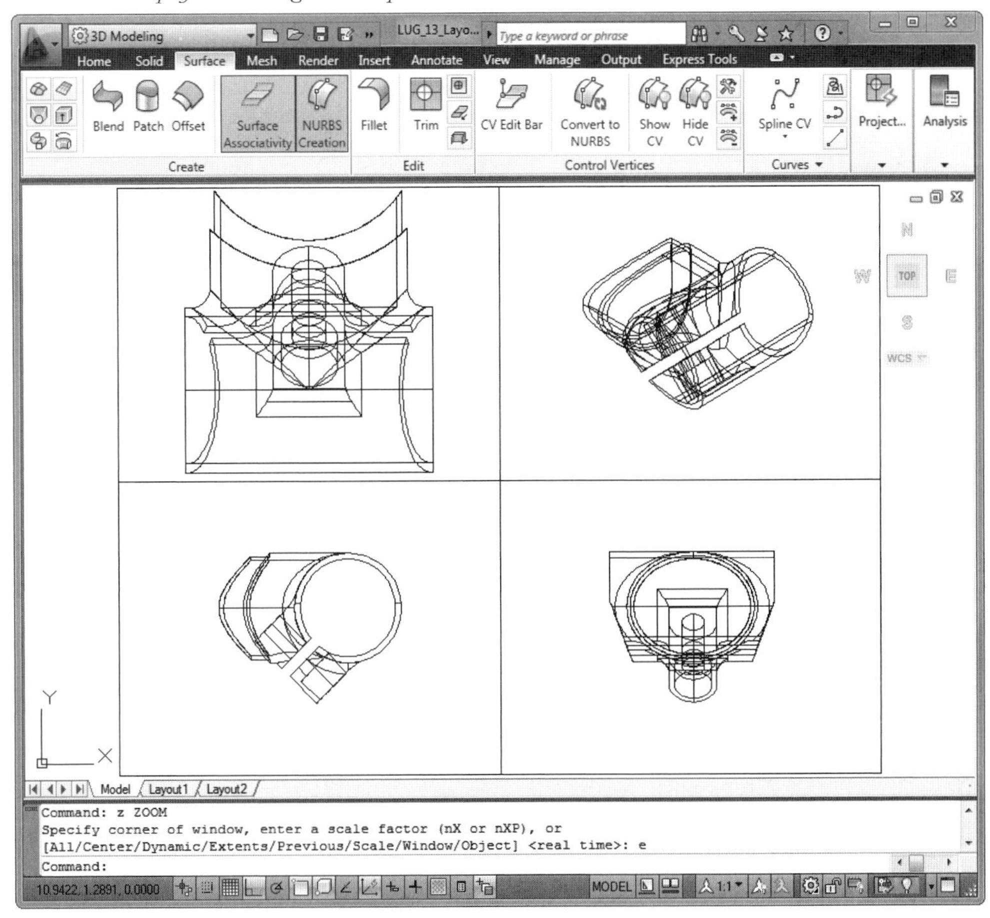

EXPORT LAYOUT TO MODEL SPACE DRAWING DIALOG BOX

Open opens the saved drawing in a new window.

Don't open doesn't open the saved drawing.

RELATED COMMANDS

Layout creates layouts.

TIPS

- Objects in layout space are exported at 1:1 scale; objects in model space are scaled by the viewport scale.
- All objects in model space are copied once per viewport at each viewport's scale factor. When this command is done, model space ends up looking like a layout.
- The viewport is mimicked by a rectangular polyline; 3D objects are still 3D.
- This command is a good way for producing single-space drawings for export (WMF, TIFF, etc) to another application's document.
- Portions of the drawing that extend beyond the viewport are sometimes not clipped.
- Extremely complex layouts sometimes export incorrectly.

ExportPDF

2010 Exports the current drawing in PDF format through a specially-designed file dialog box.

Commands	*Aliases*	*Keyboard Shortcut*	*Application Menu*	*Ribbon*
exportpdf	**qpdf**	**...**	**Export** **↳PDF**	**Output** **↳Export to DWF/PDF** **↳Export** **↳PDF**

Command: exportpdf

Displays dialog box.

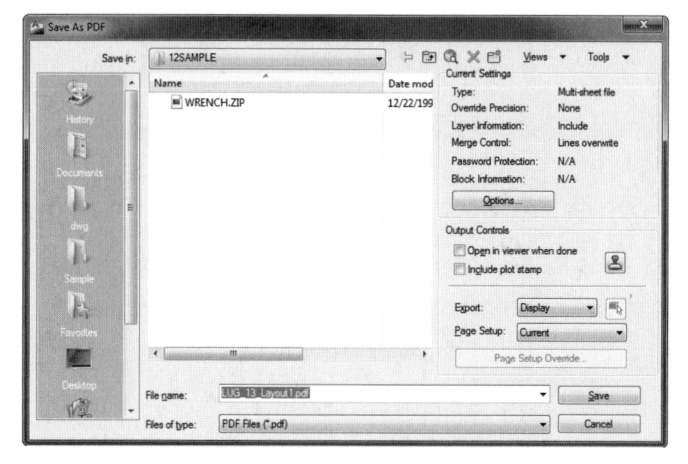

SAVE AS PDF OPTIONS

Current Settings options

Options specifies options for exporting the file; displays the Export to DWF/PDF Options dialog box. See the Publish command.

Output Controls options

☐**Open in viewer when done** opens the PDF file in Acrobat v7 or later, if installed on your computer.

☐**Include plot stamp** adds plot stamp data to the edges of the exported file.

Plot Stamp Settings specifies plot stamp data to include; displays the Plot Stamp Settings dialog box. See the PlotStamp command.

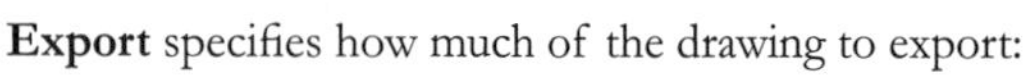

Export specifies how much of the drawing to export:

- **Display** exports the view displayed by the current viewport.
- **Extents** exports the extents of the drawing (everything not frozen).
- **Window** exports a windowed area; dismisses the dialog box temporarily, and prompts you to pick two points.

Select Window lets you reselect the windowed area to export.

Page Setup specifies the page setup to use:

- **Current** uses the current page setup; see PageSetup command.
- **Override** changes the page setup by making the Page Setup Override button available.

Page Setup Override overrides page setups; displays the Page Setup Override dialog box. See the ExportSettings command.

RELATED COMMANDS

PdfAttach attaches PDF files as underlays.

Export exports drawings in a variety of formats.

Plot exports drawings in a variety of raster and vector formats.

ExportSettings

2010 Specifies settings for exporting drawings in DWF and PDF formats; exports a windowed area of the drawing.

Command	*Alias*	*Shortcut Keystrokes*	*Menu*	*Ribbon*
exportsettings	...	...	...	**Output** **↳Export to DWF/PDF** **↳Export to DWF/PDF Settings**

Command: exportsettings
[Preview/Options/PAgesetup/Windowselect/Exportwindow]: *(Enter an option.)*

COMMAND LINE OPTIONS

Preview previews the drawing before exporting it; displays the Preview window. See the Preview command.

Options specifies the export options; displays the Export to DWF/PDF Options dialog box.

PAgesetup changes the page setup options; displays the Page Setup Override dialog box.

Windowselect prompts you to specify a windowed area of the drawing to export; displays the Specify Window Area dialog box.

Exportwindow exports the windowed area as specified by the Windowselect option; if a windowed area has not been specified, displays the Specify Window Area dialog box.

EXPORT TO DWF/PDF OPTIONS DIALOG BOX

Access this dialog box through the Options option.

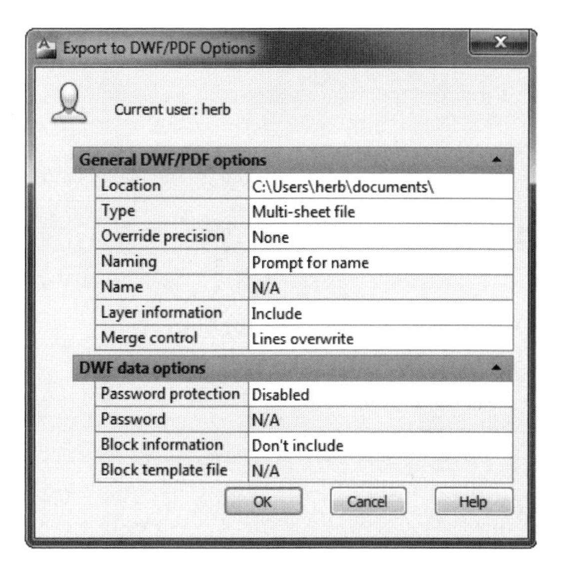

General DWF/PDF Options options

Location specifies the path to the folder in which to store files.

Type toggles between multi-sheet and single sheet files.

Override Precision chooses a preset precision:

- **None** does not override default precision.
- **For Manufacturing** sets precision to 0.001mm.
- **For Architecture** sets precision to 1/16".
- **For Civil Engineering** sets precision to 1/8".
- **Custom Preset** customizes preset precisions; opens the Precision Presets Manager dialog box, detailed later.

Naming determines how the file is named:

- **Prompt for Name** asks for the name in the file dialog box.
- **Specify Name** names the file in the Name field, below.

Name specifies the name of the file.

Layer Information toggles inclusion of layer names:

- **Include** includes layer names in exported files.
- **Don't Include** leaves out layer names; merges all graphics into one layer.

Merge Control determines how overlapping objects are handled:

- **Lines Merge** merges overlapping lines.
- **Lines Overwrite** overwrites overlapping lines.

DWF Data Options options

Password Protection toggles use of passwords to prevent unauthorized access to DWF files.

Password specifies the name of the password.

Block Information toggles the exportation of blocks.

Block Template File specifies the name of *.blk* or *.dxe* Block Template file.

PAGE SETUP OVERRIDE DIALOG BOX

Access this dialog box through the Options option.

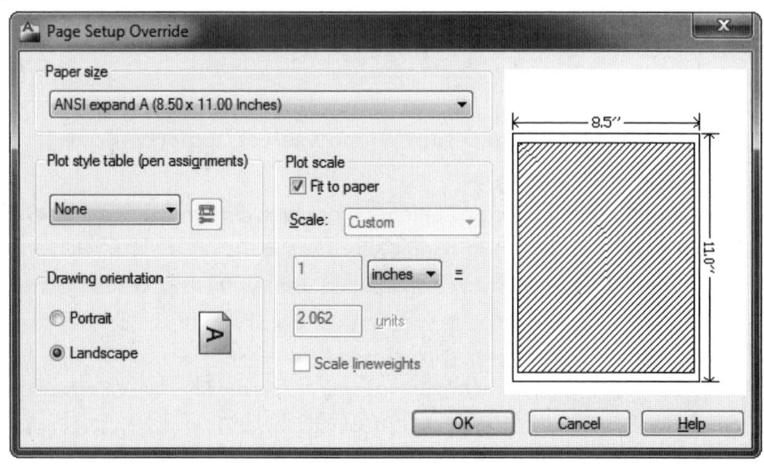

Paper Size specifies the size of media; choose a size from the droplist.

Plot Style Table options

Plot Style Table specifies the plot style to apply to the exported file. Choose a *.ctb* (color-based) or *.stb* (style-based) file from the droplist. Click New to create new plot styles; see the PlotStyle command.

Plot Style Table Editor edits the plot style; see the PlotStyle command.

Drawing Orientation options

○ **Portrait** plots the drawing normally.

⊙ **Landscape** rotates the plot by 90 degrees.

Plot Scale options

☑ **Fit to Paper** fits the drawing to the paper, ignoring the scale factor.

Scale specifies the scale factor.

Inches switches units between inches and millimeters.

☐ **Scale Lineweights** toggles the use of scaled lineweights.

SPECIFY WINDOW AREA DIALOG BOX

Access this dialog box through the Windowselect and Exportwindow options.

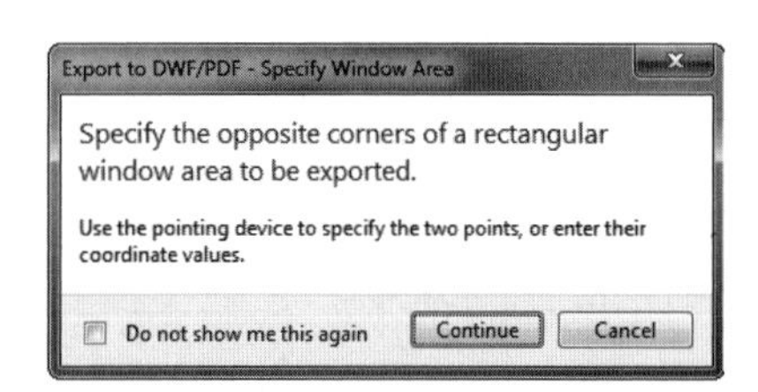

Continue dismisses this dialog box, and then AutoCAD prompts you in the command bar:

Specify first corner: *(Pick a point, or specify coordinates.)*
Specify opposite corner: *(Pick another point.)*

AutoCAD displays the Preview Window Area dialog box.

PREVIEW WINDOW AREA DIALOG BOX

Access this dialog box through the Exportwindow option.

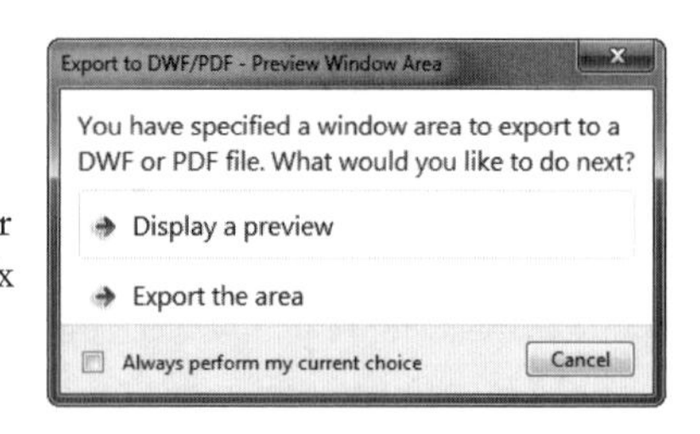

Display a Preview previews the windowed area of the drawing to be exported.

Export the Area exports the windowed area to file; displays the Save As PDF (or DWF or DWFx) dialog box. See the ExportPDF, ExportDWF, and ExportDWFx commands.

PRECISION PRESETS MANAGER DIALOG BOX

Access this dialog box through the Export to DWF/PDF Options dialog box's Override Precision option.

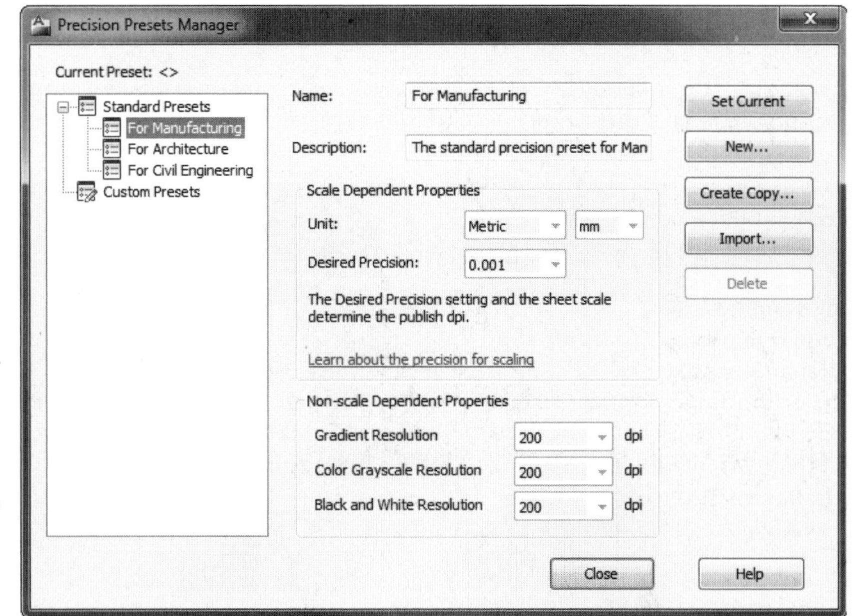

Current Preset specifies name of the current precision preset.

Name names the preset.

Description describes the purpose of the preset.

Scale Dependent Properties

Unit chooses metric or Imperial units and subunits (mm, feet, etc).

Desired Precision specifies the precision at which the drawing will be exported.

Non-scale Dependent Properties

A lower resolution setting reduces file size but makes gradients look posterized and images grainier.

Gradient Resolution specifies the resolution of gradient hatch areas.

Color Grayscale Resolution specifies the resolution of images that are color or grayscale.

Black and White Resolution specifies the resolution of monochrome images.

Set Current sets the selected preset as the working preset.

New creates new presets; displays the New Precision Preset dialog box, which prompts for a name and description.

Create Copy copies the selected preset; displays Copy Precision Preset dialog box.

Import imports *.dsd* Drawing Set Description files created by the Publish command.

Delete erases the preset.

NEW / COPY PRECISION PRESET DIALOG BOX

Access this dialog box through the Precision Presets Manager dialog box's New Precision Preset or Create Copy buttons.

Name specifies the name of the new preset.

Description describes the purpose of the preset; optional.

RELATED COMMANDS

ExportDWF, **ExportDWFx**, and **ExportPDF** export drawings in DWF, DWFx, and PDF formats.

Export exports drawings in several raster and vector formats.

-Export specifies the default file format, DWF, DWFx, or PDF.

BmpOut, **JpgOut**, and **TifOut** export drawings in raster formats.

3dDwf exports 3D drawings in 3D DWF format.

RELATED SYSTEM VARIABLES

ExportModelSpace specifies the plot area: (0) display, (1) extents, or (2) windowed area.

ExportPageSetup specifies whether to use (0) default or (1) overridden page setup values.

ExportPaperSpace specifies whether to publish (0) the current layout or (1) all layouts.

ExportePlotFormat specifies the default export format: (0) PDF, (1) DWF, or (2) DWFx.

-ExportToAutocad

2006 Converts AutoCAD Architecture custom objects to AutoCAD objects.

Command	*Aliases*	*Keyboard Shortcuts*	*Menu*	*Ribbon*
-exporttoautocad	**aectoacad** **exporttoautocad**	...	...	...

Command: -exporttoautocad
File format: 2007
Bind xrefs: Yes
Bind type: Insert
Filename prefix:
Filename suffix:
Export options [Format/Bind/bind Type/Maintain/Prefix/Suffix/?] <Enter for filename>: *(Enter an option.)*
Export drawing name <d:\path\filename.dwg>: *(Press Enter.)*

COMMAND OPTIONS

Format specifies the DWG version.

Bind determines whether xrefs are bound.

bind Type specifies the type of xref binding.

Maintain specifies how to deal with blocks in AutoCAD Architecture (formerly ADT) custom objects.

Prefix specifies a prefix for file names.

Suffix specifies a suffix for file names.

? lists the setting for each option.

Format options
Enter file format [r14/2000/2004/2007/2010] <2010>:

r14 saves drawing in Release 14 DWG format.

2000 saves drawing in 2000/2000i/2001 DWG formats.

2004 saves drawing in 2004/2005/2006 DWG formats.

2007 saves drawing in 2007/8/9 DWG formats.

2010 saves drawing in 2010 /2011 DWG format.

Bind option
Bind xrefs [Yes/No] <Yes>:

Yes binds externally-referenced files to the drawing.

No stores links to xref files.

bind Type option
Bind type [Bind/Insert] <Insert>:

Bind makes xrefs part of the drawing.

Insert inserts xrefs into the drawing as if they were blocks.

Maintain option
Maintain resolved properties [Yes/No] <Yes>:

Yes explodes blocks in AEC custom objects.

No keeps blocks as unexploded xrefs.

Prefix and Suffix options

Filename prefix <>:

Filename suffix <>:

Filename Prefix adds text as a prefix to the current drawing's file name.

Filename Suffix adds text as a suffix after the current drawing's file name.

RELATED SYSTEM VARIABLE

ProxyNotice controls the display of custom objects in drawings:

0 — Custom objects are not displayed.
1 — Custom objects are displayed as proxy objects.
2 — Custom objects are displayed by a bounding box.

TIPS

- Vertical applications based on AutoCAD have the ability to create *custom objects* which have special properties not otherwise found in AutoCAD. These objects are controlled by programming code written in ARX. Software such as AutoCAD Architecture and Mechanical Desktop use custom objects for intelligent doors, windows, and so on. When drawings made with these programs are opened in AutoCAD, custom objects are displayed as *proxy objects*, which cannot be fully edited by AutoCAD.
- This command explodes custom objects created in ADT so that they can be edited by AutoCAD, and then saved to file by another name in order to preserve the original drawing.
- Exploded custom objects lose their intelligence.
- The following commands are also available for working with AEC custom objects:

  • **-AecDwgUnits** sets the units for inserting AEC objects into AutoCAD drawings.

  • **AecDisplayManagerConfigsSelection** changes the display configuration.

  • **AecFileOpenMessage** toggles display of dialog box warning of opening drawings from older releases of ADT.

  • **AecFileSaveMessage** toggles display of warning message when saving to previous versions.

  • **AecObjectCopyMessage** toggles display of warning message when saving to newer versions.

  • **AecObjRelDump**, **AecObjrelDumpEnhancedReferences**, and **AecObjRelShowEnhancedReferences** perform AEC custom object database dumps to the text screen.

  • **AecObjRelShow** indicates which Architectural objects are related to each other.

  • **AecObjRelUpdate** updates Architectural objects.

  • **AecPostDxfinFix** fixes problems with imported Architectural drawings that cannot be repaired by AutoCAD's Audit command.

  • **AecSetXrefConfig** allows the same xref to be displayed several times.

  • **AecVersion** reports the version number of the Architectural object enabler.

Extend

Ver. 2.5 Extends the length of lines, rays, open polylines, arcs, and elliptical arcs to boundary objects.

Command	*Alias*	*Keyboard Shortcuts*	*Menu*	*Ribbon*
extend	**ex**	**...**	**Modify** **↳Extend**	**Home** **↳Modify** **↳Extend**

Command: extend
Current settings: Projection=UCS Edge=None
Select boundary edges ...
Select objects: *(Select one or more objects.)*
Select objects: *(Press Enter.)*
Select object to extend or shift-select to trim or [Project/Edge/Undo]: *(Select an object, or enter an option.)*
Select object to extend or shift-select to trim or [Project/Edge/Undo]: *(Press Enter to end the command.)*

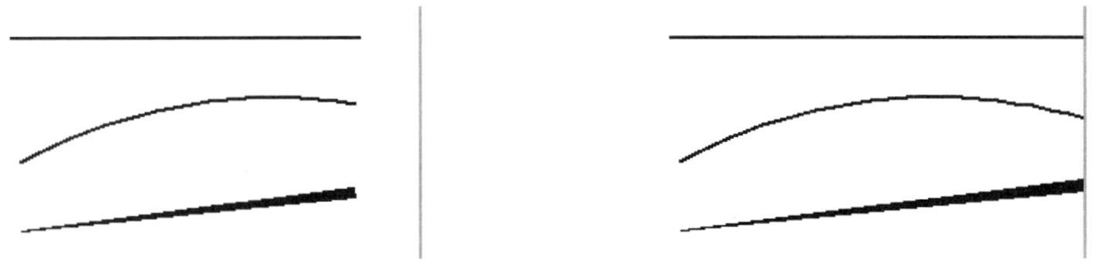

Left: *Line, arc, and variable-width polyline.*
Right: *Extended to gray line.*

COMMAND LINE OPTIONS

Select objects selects the objects to use for the extension boundary.

Select objects to extend selects the objects to be extended.

Shift-select to trim trims objects when you hold down the SHIFT key.

Undo undoes the most recent extend operation.

Project options
Enter a projection option [None/Ucs/View] <Ucs>: *(Enter an option.)*

None extends objects to boundary *(Release 12-compatible).*

Ucs extends objects in the x, y-plane of the current UCS.

View extends objects in the current view plane.

Edge options
Enter an implied edge extension mode [Extend/No extend] <No extend>: *(Enter an option.)*

Extend extends to implied boundary.

No extend extends only to actual boundary *(Release 12-compatible).*

RELATED COMMANDS

Change changes the length of lines.

Lengthen changes the length of open objects.

SolidEdit extends the face of a solid object.

Stretch stretches objects wider and narrower.

Trim reduces the length of lines, polylines and arcs.

RELATED SYSTEM VARIABLES

EdgeMode toggles boundary mode for the Extend and Trim commands.

ProjMode toggles projection mode for the Extend and Trim commands:

- **0** — None; Release 12 compatible.
- **1** — Current UCS (default).
- **2** — Current view plane.

TIPS

- The following objects (even when inside blocks) can be used as boundaries:

  2D polylines
  Lines
  3D polylines
  Rays
  Arcs
  Regions
  Circles
  Splines
  Ellipses
  Text
  Floating viewports
  Xlines
  Hatches

- When a wide polyline is the edge, Extend extends to the polyline's center line.
- Pick the object a second time to extend it to a second boundary line.
- Circles and other closed objects are valid edges: the object is extended in the direction nearest to the pick point.
- Extending a variable-width polyline widens it proportionately; extending a splined polyline adds a vertex.
- Objects can act as boundaries and be extendd.
- When selecting boundaries, press Enter first: this selects all visible objects as boundaries.

ExternalReferences / ExternalReferencesClose

2007 Displays the External References palette for centrally managing all externally-referenced drawings, images, and DGN, PDF, and DWF underlays.

Commands	*Aliases*	*Menu*	*Ribbon*
externalreferences	**er**	**Tools**	**Insert**
	image, im	**↳Palettes**	**↳External References**
	xref, xr	**↳External References**	
externalreferencesclose	**...**	**...**	**...**

Command: externalreferences

Opens the External References palette:

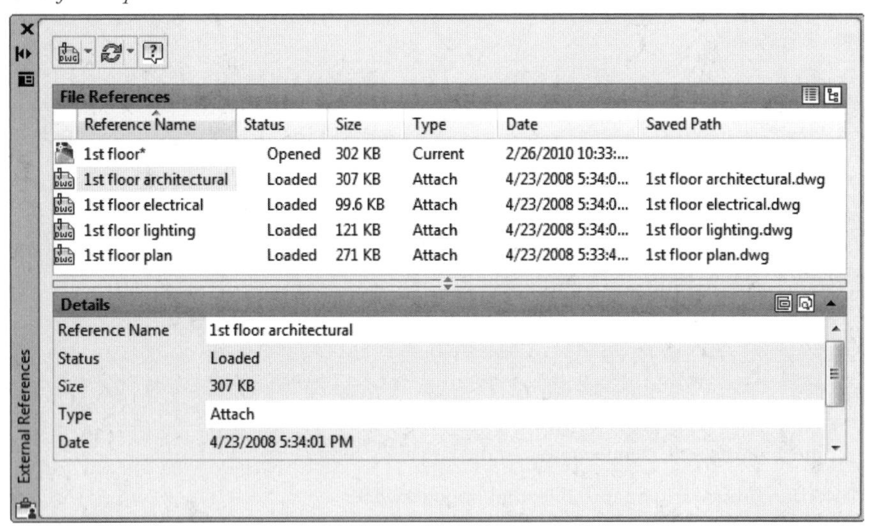

Command: externalreferencesclose

Closes the External References palette.

PALETTE TOOLBARS

Attach button

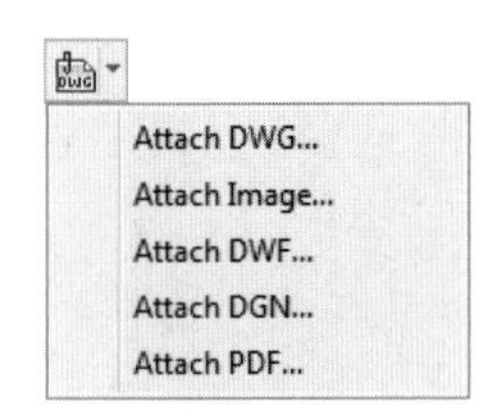

Attach DWG attaches *.dwg* files as external references; displays the Select Reference File dialog box, followed by the External Reference dialog box. See the XAttach command.

Attach Image attaches raster files as underlays; displays the Select Image File dialog box, followed by the Image dialog box. See the ImageAttach command.

Attach DWF attaches *.dwf* and *.dwfx* files as underlays; displays the Select DWF File dialog box, followed by the Attach DWF Underlay dialog box. See the DwfAttach command.

Attach DGN attaches MicroStation *.dgn* files as underlays; displays the Select DGN File dialog box, followed by the Attach DGN Underlay dialog box. See the DgnAttach command. (To import design files, use the DgnImport command.)

Attach PDF attaches *.pdf* files as underlays; displays the Select PDF File dialog box, followed by the Attach PDF Underlay dialog box. See the PDFAttach command.

Refresh button

Refresh refreshes the status of references.

Reload All References reloads references; useful when they have changed.

File References

 **List View** displays each external reference with information about its status, and so on.

 **Tree View** shows the connections between external references.

Right-click Menu

Right-click to display this shortcut menu:

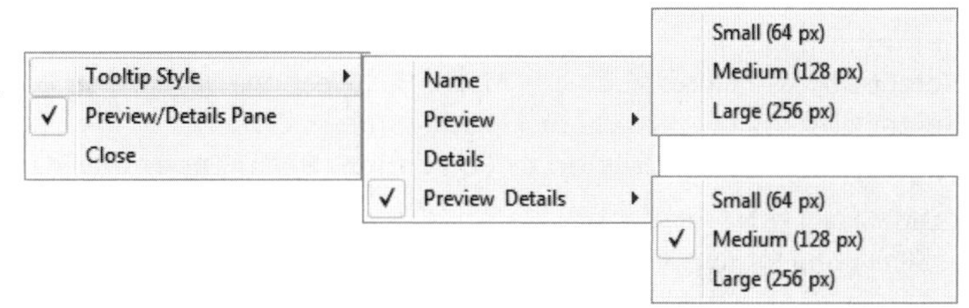

Tooltip Style specifies the size and content of tooltips.

Preview/Details Pane toggles the display of the preview pane.

Close closes the palette.

Details

 **Details** lists information about the selected external reference.

 **Preview** shows preview images of external references; no preview is available for old versions of *.dwg* files.

RELATED COMMANDS

Attach attaches image, *.dwf*, *.dwfx*, *.pdf*, and *.dgn* files as underlays.

DgnAttach attaches *.dgn* files.

DwfAttach attaches *.dwf* files.

ImageAttach attaches raster files.

PdfAttach attaches *.pdf* files.

XAttach attaches *.dwg* files.

RELATED SYSTEM VARIABLE

ErState reports whether this palette is open or closed.

TIPS

- The Status column displays brief reports that have the following meaning:

Status	*Meaning*
Loaded	File is attached to the drawing.
Unloaded	File will be unloaded.
Not Found	File cannot be found on search paths.
Unresolved	File cannot be read.
Orphaned	File is attached to an unresolved file.

- AutoCAD searches for externally-referenced files in (a) last known location, (b) current folder, and then (c) folders named by the search path. When it cannot find them, it offers you the opportunity to search for them.
- If Autodesk's Vault software is installed, the Attach from Vault option attaches files from the Vault.
- When xrefs are loaded, they initially appear dimmed; brighten them with the XDwgFadeCtl system variable.

Extrude

Rel. 11 Creates 3D solids and surfaces by extruding 2D objects; extrudes faces of 3D meshes..

Command	Alias	Keyboard Shortcuts	Menu	Ribbon
extrude	**ext**	...	**Draw**	**Home**
			↳Modeling	**↳3D Modeling**
			↳Extrude	**↳Extrude**

Command: extrude
Select objects to extrude or [MOde]: *(Select one or more objects, or type MO.)*
Select objects to extrude or [MOde]: *(Press Enter.)*
Specify height of extrusion or [Direction/Path/Taper angle/Expression]: *(Enter a height or an option.)*

COMMAND LINE OPTIONS

Select objects to extrude selects the 2D objects to extrude, called the "profile."

MOde specifies the type of 3D object to create *(new to AutoCAD 2011)*; prompts you:

Closed profiles creation mode [SOlid/SUrface] <Solid>: *(Type SO or SU.)*

- **SOlid** option creates 3D solids.
- **SUrface** option creates 3D surfaces.

Specify height of extrusion specifies the extrusion height. As of AutoCAD 2007, the extrusion dynamically changes height as you move the cursor.

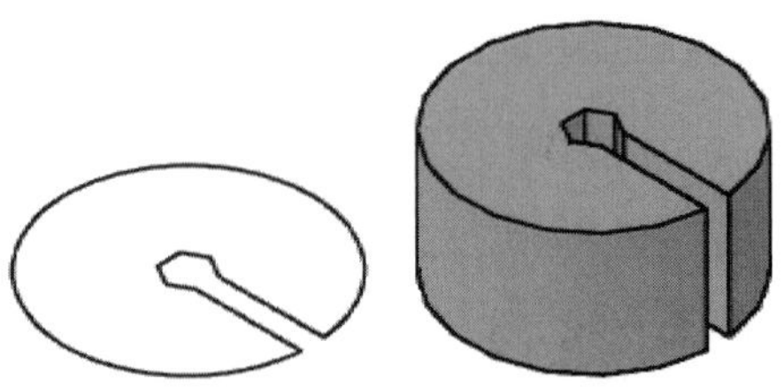

Left: *Original object (the profile).*
Right: *Extrusion by height.*

Direction specifies the extrusion height, direction, and angle by picking two points.

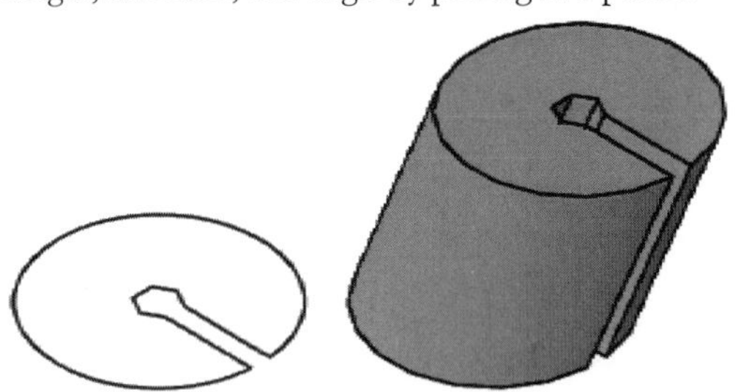

Extrusion by direction (picking two points).

Path extrudes the profile along another object, called the "path."

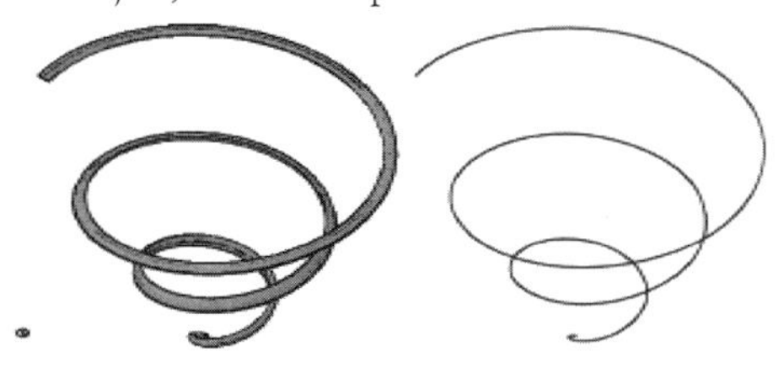

Left: *Original object (the profile) and the extrusion by path.*
Right: *The path object (a helix).*

Taper angle extrudes the profile at an angle; ranges from -90 to +90 degrees.

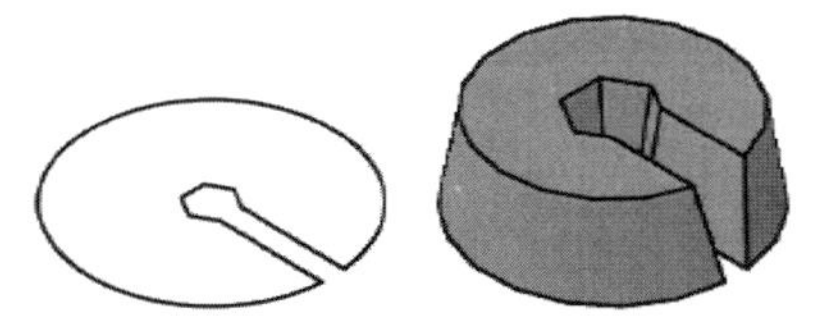

Extrusion by 10-degree taper (specifying an angle for the walls).
Notice that the outside tapers in, while the inside tapers out.

Expression determines the extrusion distance by a formula; prompts you:

Expression: *(Enter a mathematical or parametric expression; see the Calc command for examples.)*

RELATED COMMANDS

Revolve creates 3D solids by revolving 2D objects.

Elev gives thickness to non-solid objects to extrude them.

Thicken creates 3D solids by thickening surfaces.

TIPS

- Prior to AutoCAD 2007, this command worked only with closed objects, creating 3D solids. It now works with open and closed objects: open ones, such as arcs, create 3D surface; while closed ones, such as circles, create 3D solids.
- When extruding an object along a path, AutoCAD relocates the path to the object's centroid.
- This command extrudes the following objects: lines, traces, arcs, elliptical arcs, ellipses, circles, 2D polylines, 2D splines, 2D solids, regions, planar 3D faces, planar surfaces, and planar faces on solids. To select the face of a solid object, hold down the Ctrl key when picking.
- This command can use the following objects as extrusion paths: lines, circles, arcs, ellipses, elliptical arcs, 2D and polylines, 2D and 3D splines, helixes, solid edges, and edges of surfaces.
- You *cannot* extrude polylines with less than 3, or more than 500, vertices; similarly, you cannot extrude crossing or self-intersecting polylines.
- Objects within a block cannot be extruded; use the Explode command first.
- The taper angle must be between 0 (default) and +/-90 degrees; positive angles taper in from the base; negative angles taper out. This command does not work if the combination of angle and height makes the object's extrusion walls intersect.
- You can interactively extrude 3D mesh objects: (1) Ctrl+click a face, and then (2) drag the red dot.
- Prior to AutoCAD 2011, this command automatically created solids from closed objects and surfaces from open ones.

FbxExport / FbxImport

2011 Exports the current drawing in FBX format; imports FBX files.

Commands	*Aliases*	*Keyboard Shortcut*	*Application Menu*	*Ribbon*
fbxexport	...	...	**Export** ↳**FBX**	**Output** ↳**Export to DWF/PDF**
-fbxexport				
fbximport				
-fbximport				

Command: fbxexport

Displays the FBX Export dialog box; enter a file name, and then click Save.

Displays FBX Export Options dialog box:

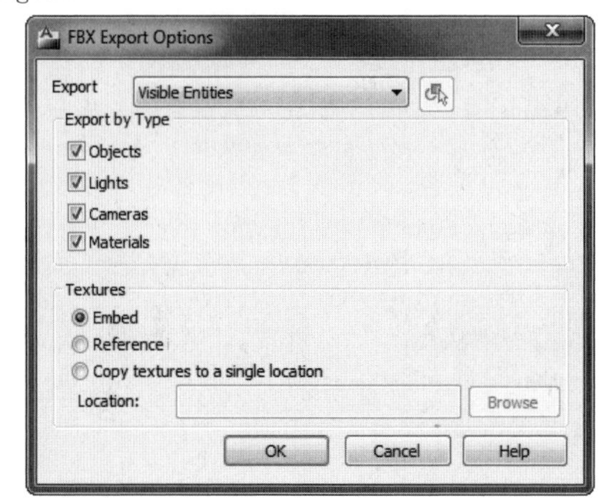

FBX EXPORT OPTIONS DIALOG BOX

Export specifies what to export, Visible Entities or Selected Entities.

Export by Type options

☑ **Objects** exports 3D objects and 2D objects with thickness.

☑ **Lights** exports point, spot, distant, photometric, sun, and sky lights.

☑ **Cameras** exports cameras and camera views.

☑ **Materials** exports materials that are applied to the objects being exported.

Textures options

⦿ **Embed** embeds textures in the exported FBX file.

○ **Reference** records the path to the textures' folder.

○ **Copy textures to a single location** copies the texture file to the specified location.

Location specifies the folder in which to place the copied textures files.

Browse displays the Browse for Folder dialog box.

-FBXEXPORT Command

See FbxExport command for the meaning of options:

Command: -fbxexport
Enter what should be exported? [Selected/Visible] <Visible>: *(Type S or V.)*
Current export type settings: Objects=Yes, Lights=Yes, Cameras=Yes, Materials=Yes
Specify types to export [Settings, All types] <Current Settings>: *(Enter an option.)*
Enter path and name of file to export<C:\filename.fbx>: *(Press Enter.)*

Settings options
Export objects? [Yes/No] <Yes>: *(Type Y or N.)*
Export lights? [Yes/No] <Yes>: *(Type Y or N.)*
Export cameras? [Yes/No] <Yes>: *(Type Y or N.)*
Export materials? [Yes/No] <Yes>: *(Type Y or N.)*
Following types will be exported: Objects, Lights, Cameras, Materials
Enter how to handle textures [Embed/Reference originals/Copy originals] <Embed>: *(Enter an option.)*

FBXIMPORT Command

Command: fbximport

Displays the FBX Import dialog box; enter a file name, and then click Open.

Displays FBX Import Options dialog box:

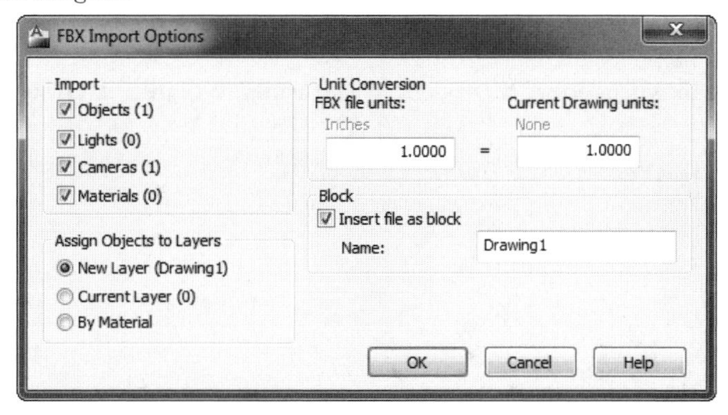

DIALOG BOX OPTIONS

Import options

The numbers in parentheses indicate the number of items in each category to be imported.

☑ **Objects** imports objects from the FBX file.

☑ **Lights** imports point, spot, distant, photometric, sun, and sky lights.

☑ **Cameras** imports cameras.

☑ **Materials** imports materials.

Assign Objects To Layers options

⊙ **New Layer** assigns imported objects to a new layer.

○ **Current Layer** assigns imported objects to the current layer.

○ **By Material** creates a new layer for every material name; objects are placed on layers according to their material.

Unit Conversion options

⊙ **FBX File Units** reports the units in the FBX file.

○ **Current Drawing Units** reports the units in the drawing.

Block options

☑ **Insert File As Block** inserts the FBX file as a block definition table. To see the imported objects, use the Insert command to place the block.

Name names the block definition.

-FBXIMPORT Command

See FbxImport command for the meaning of options:

Command: -fbximport
Enter path and name of file to import<C:\Drawing1.fbx>: *(Enter path and file name.)*
Current import type settings: Objects=Yes, Lights=Yes, Cameras=Yes, Materials=Yes
Specify types to import [Settings, All types] <Current Settings>: *(Type S or A.)*
Enter layer assignment to [New layer/Current layer/by Material] <New layer>: *(Enter an option.)*
Enter unit conversion (FBX file: Inches, Current Drawing: Inches) <1:1>: *(Enter a conversion factor.)*
Insert file as block? [Yes/No] <Yes>: *(Type Y or N.)*
Name the block: *(Enter a name for the block.)*
Import of C:\Drawing1.fbx succeeded. 1 Objects, 0 Materials, 0 Lights, 1 Cameras

Settings options:
Import objects? [Yes/No] <Yes>: *(Type Y or N.)*
Import lights? [Yes/No] <Yes>: *(Type Y or N.)*
Import cameras? [Yes/No] <Yes>: *(Type Y or N.)*

RELATED COMMAND

Export exports drawings in FBX and other formats.

RELATED SYSTEM VARIABLE

FbxImportLog toggles the creation of log files that report the status of imported *.fbx* files.

TIPS

- The purpose of this command is to exchange 3D drawings between AutoCAD and 3D Studio (and similar software).
- The current camera is exported from AutoCAD as Camera0.
- Data from FBX files are imported into AutoCAD as 3D mesh objects.
- The LightingUnits system variable determines whether lights are imported as regular (0) or photometric (1 or 2) lights.
- Settings in AutoCAD determine properties of plot glyphs, glyph displays, targets, and render shadow details.
- To import cameras into AutoCAD accurately, ensure that they are created as target cameras in 3D Studio.

Field

2005 Places automatically updatable field text in drawings.

Command	*Alias*	*Keyboard Shortcuts*	*Menu*	*Ribbon*
field	...	...	**Insert**	**Insert**
			↳Field	**↳Data**
				↳Field

Command: field

Displays dialog box, the content of which varies depending on the field name selected.

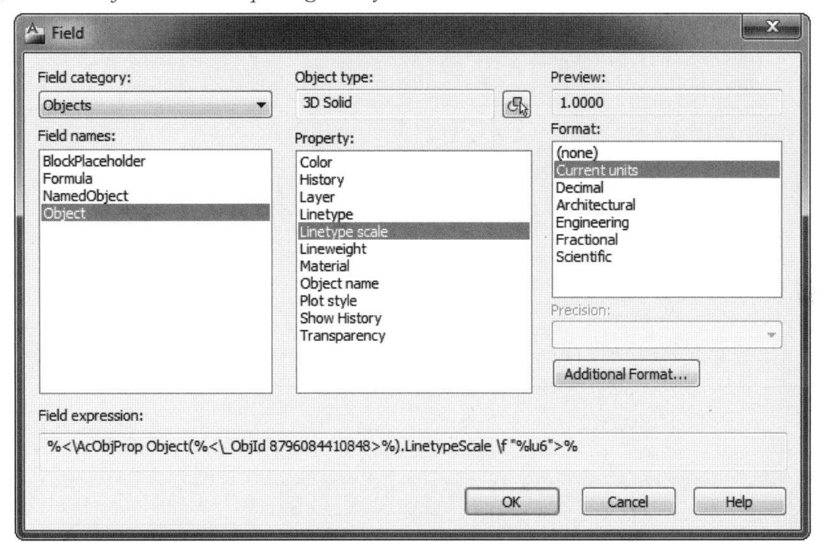

Select a field name and its options. Click OK. AutoCAD prompts you to place the field as mtext:

MTEXT Current text style: "Standard" Text height: 0.2000

Specify start point or [Height/Justify]: *(Pick a point, or enter an option.)*

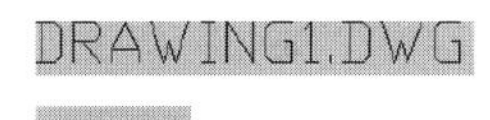

If the field has no value, AutoCAD inserts dashes as placeholders.

FIELD DIALOG BOX

Field category lists groups of fields.

Field names lists the names of fields that can be placed in drawings.

Format lists optional formats available.

Field Expression illustrates field codes inserted in the drawing; cannot be edited directly.

Additional options appear, depending on the field name selected.

RELATED COMMANDS

UpdateField forces selected fields to update their values.

Find finds field text in drawings.

MText places field text in drawings.

RELATED SYSTEM VARIABLES

FieldDisplay toggles the gray background to fields (default = 1, on).

FieldEval determines when fields are updated (default = 31, all options on).

TIPS

- When FieldEval=31 (all on), all fields update whenever you use the Plot, Save, and other *trigger* commands.
- To edit fields: (1) double-click a field; (2) in the mtext editor, right-click the field text; (3) from the shortcut menu, select Edit Field. To convert fields to mtext, select Convert Field to Text.
- The gray background helps identify field text in drawings.
- Field text has an annotative property when created with annotative styles or changed with the Properties or MText command.
- AutoCAD supports the following field names:

Field Category	Field Names	Comments
Date & Time	Create Date	Date drawing created.
	Date	Current date and time.
	Plot Date	Date last plotted.
	Save Date	Date last saved.
Document	Author	Data stored in the Drawing Properties dialog Filename box;
	Comments	see DwgProps command.
	Filesize	
	HyperlinkBase	
	Keywords	
	LastSavedBy	
	Subject	
	Title	
Linked	Hyperlink	Links to other drawings and files.
Objects	Block	Placeholder; can be used in Block Editor only.
	NamedObject	Blocks, layers, linetypes, etc.
	Object	Object selected from drawing.
	Formula	For tables only: Average, Sum, Count, Cell, and Formula.
Other	Diesel Expression	Allows use of Diesel macros.
	LISP Variable	All AutoLISP and VLISP variable names.
	System Variable	Access to all system variables.
Plot	DeviceName	Name of the plotter.
	Login	Login name.
	PageSetup	Page setup name.
	PaperSize	Size of paper.
	PlotDate	Same as under Date & Time.
	PlotOrientation	Orientation of the plot.
	PlotScale	Scale of the plot.
	PlotStyleTable	Name of the plot style table.
Sheetset	SheetSet	Information about sheet sets.
	SheetSetPlaceholder	Placeholder for future sheet sets.
	SheetView	CurrentSheetSetSubset
	CurrentSheetNumber	CurrentSheetSet
	CurrentSheetNumberandTitle	CurrentSheetSubset
	CurrentSheetTitle	CurrentSheetCategory
	CurrentSheetIssuePurpose	CurrentSheetRevisionDate
	CurrentSheetRevisionNumber	CurrentSheetCategory
	CurrentSheetCustom	CurrentSheetDescription
	Current SheetSetCustom	CurrentSheetSetDescription
	CurrentSheetProjectMilestone	CurrentSheetSetProjectNumber
	CurrentSheetProjectName	CurrentSheetSetProjectPhase

Removed Command

Files command was removed from Release 14. In its place, use Windows Explorer.

Fill

Ver. 1.4 Toggles whether hatches and wide objects — traces, multilines, solids, and polylines — are displayed and plotted with fills or as outlines.

Command	*Alias*	*Keyboard Shortcuts*	*Menu*	*Ribbon*
'fill	...	...	...	...

Command: fill
Enter mode [ON/OFF] <ON>: *(Type* **ON** *or* **OFF***.)*

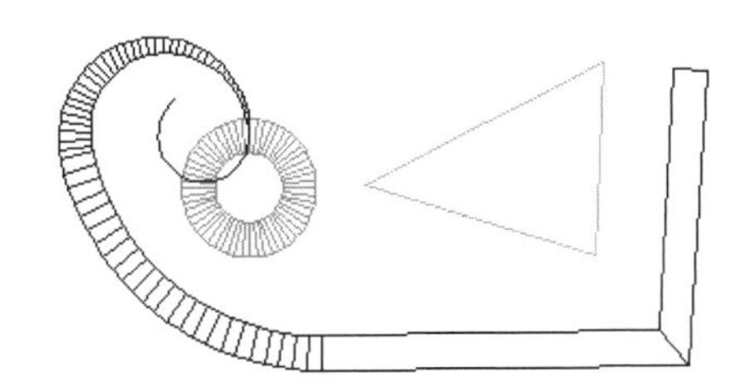

Left: *Fill on.*
Right: *Fill off.*

COMMAND LINE OPTIONS

ON turns on fill after the next regeneration.

OFF turns off fill after the next regeneration.

RELATED COMMAND

Regen changes the display to reflect the current fill or no-fill status.

RELATED SYSTEM VARIABLE

FillMode holds the current setting of fill status.

TIPS

- The state of fill (or no fill) does not come into effect until the next regeneration:

  Command: regen
  Regenerating model.

- Traces, 2D solids, and polylines are only filled in plan view, regardless of the setting of Fill. When viewed in 3D, these objects lose their fill.
- Fill affects objects derived from polylines, including donuts, polygons, rectangles, and ellipses, when created with PEllipse = 1.
- Fill does *not* affect TrueType fonts, which have their own system variable — TextFill — which toggles their fill/no-fill status for plots only.
- Fill does not toggle in rendered mode.

Fillet

Ver. 1.4 Joins intersecting lines, polylines, arcs, circles, and faces of 3D solids with a radius.

Command	*Alias*	*Keyboard Shortcuts*	*Menu*	*Ribbon*
fillet	**f**	...	**Modify** ↳ **Fillet**	**Home** ↳ **Modify** ↳ **Fillet**

Command: fillet
Current settings: Mode = TRIM, Radius = 0.0
Select first object or [Undo/Polyline/Radius/Trim/mUltiple]: *(Select an object, or enter an option.)*
Select second object or shift-select to apply corner: *(Select another object.)*

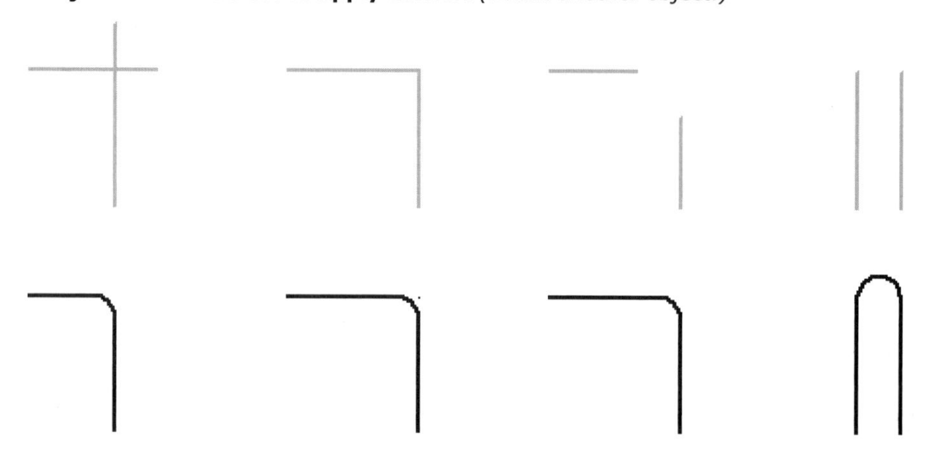

Lines before (top) and after (bottom) applying the Fillet command:
1. Crossing lines. 2. Touching lines. 3. Non-intersecting lines. 4. Parallel lines.

COMMAND LINE OPTIONS

Select first object selects the first object to be filleted.

Select second object selects the second object to be filleted.

Shift select to apply corner sets the radius to 0 temporarily.

Undo undoes the last fillet operation without leaving the command.

mUltiple prompts you to select additional pairs of objects to fillet.

Polyline option
Select 2D polyline: *(Pick a polyline.)*

Select 2D polyline fillets all vertices of a 2D polyline; 3D polylines cannot be filleted.

Radius option
Specify fillet radius <0.0>: *(Enter a value.)*

Specify fillet radius specifies the filleting radius.

Trim option
Enter Trim mode option [Trim/No trim] <Trim>: *(Type T or N.)*

Trim trims objects when filleted.

No trim does not trim objects.

Filleting 3D solids
Select first object or [Undo/Polyline/Radius/Trim/Multiple]: *(Pick a 3D solid.)*
Enter fillet radius or [Expression] <0.2000>: *(Specify the radius, or type E.)*
Select an edge or [Chain/Radius]: *(Pick an edge, or enter an option.)*
Select an edge or [Chain/Radius]: *(Press Enter.)*
n **edge(s) selected for fillet.**

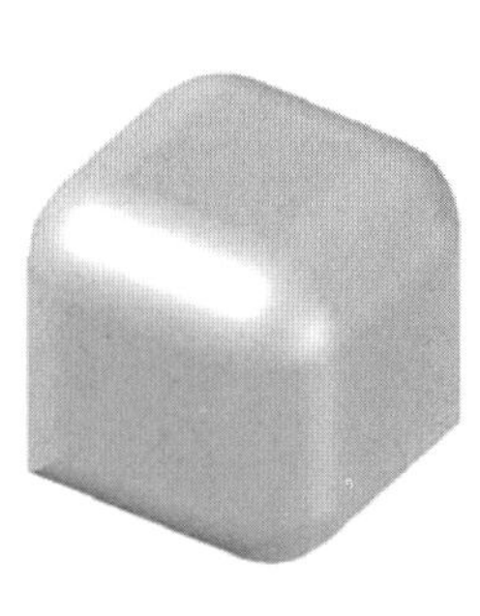

3D box before and after filleting.

Expression option *(new to AutoCAD 2011)*
Enter expression: *(Enter a math or parametric equation; see Calc command for examples.)*

Edge option
Select an edge or [Chain/Radius]: *(Pick an edge.)*

Select edge selects a single edge.

Chain option
Select an edge or [Chain/Radius]: *(Type C.)*
Select an edge chain or [Edge/Radius]: *(Pick an edge.)*

Select an edge chain selects multiple edges.

Radius option
Select an edge or [Chain/Radius]: *(Type R.)*
Enter fillet radius <1.0000>: *(Enter a value.)*

Enter fillet radius specifies the fillet radius.

RELATED COMMANDS

FilletEdge fillets 3D solids interactively *(new to AutoCAD 2011)*.

SurfFillet fillets surfaces *(new to AutoCAD 2011)*.

Chamfer bevels intersecting lines or polyline vertices.

SolidEdit edits 3D solid models.

RELATED SYSTEM VARIABLES

FilletRad specifies the current filleting radius.

TrimMode toggles whether objects are trimmed.

TIPS

- Pick the end of the object you want filleted; the other end will remain untouched.
- The lines, arcs, or circles need not touch.
- As a faster alternative to the Extend and Trim commands, use the Fillet command with a radius of zero.
- If the lines to be filleted are on two different layers, the fillet is drawn on the current layer.
- The fillet radius must be smaller than the length of the lines. For example, if the lines to be filleted are 1.0m long, the fillet radius must be less than 1.0m.
- Use the Close option of the PLine command to ensure a polyline is filleted at all vertices.

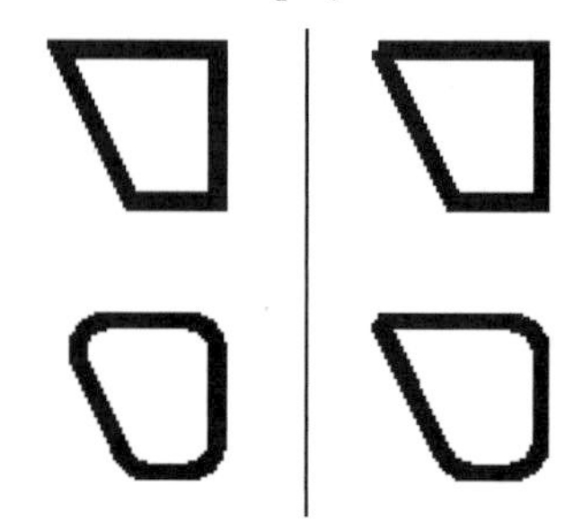

Left: *Polyline closed with Close option.*
Right: *Polyline not closed with Close option.*

- Filleting polyline segments from different polylines joins them into a single polyline.
- Filleting a line or an arc with a polyline joins it to the polyline.
- You can apply internal and external fillets in the same operation, after the two objects are unioned with the Union command (illustrated below).

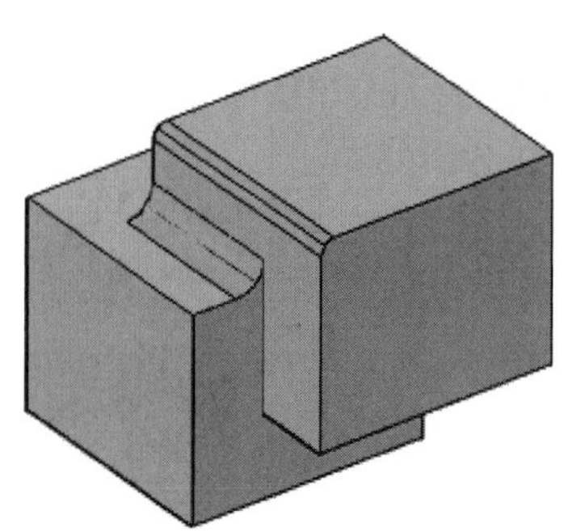

- Filleting a pair of circles does not trim them.
- As of AutoCAD Release 13, the Fillet command fillets a pair of parallel lines; the radius of the fillet is automatically determined as half the distance between the lines.
- As of AutoCAD 2010, this command can fillet two 3D mesh objects, but in an indirect manner:
    1. Convert the meshes into faceted solids with the ConvToSolid command.
    2. Perform the fillet operations with the Fillet command.
    3. Convert the solids back into 3D meshes object with the MeshSmooth command.
    Important! Ensure that the level of smoothness is set to 0 with the MeshSmoothLess command, otherwise the results will look incorrect.

FilletEdge

2011 Fillets the edges of 3D solid objects interactively.

Command	Aliases	Keyboard Shortcuts	Menu	Ribbon
filletedge	...	...	**Modify** ↳**Solid Editing** ↳**Fillet Edge**	**Solid** ↳**Solid Editing** ↳**Fillet Edge**

Command: filletedge

Radius = 1.0000
Select an edge or [Chain/Radius]: *(Select an edge, or enter an option.)*

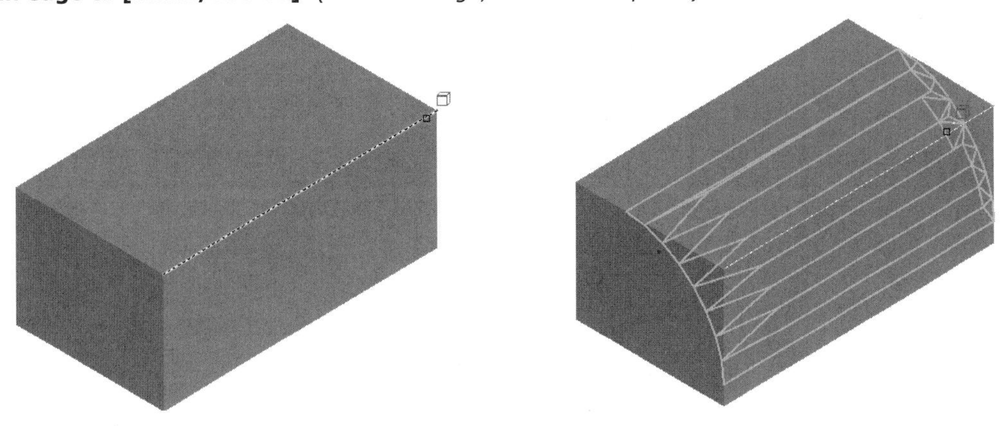

Left: *Select an edge and...*
Right: *...the fillet is immediately previewed.*

Select an edge or [Chain/Radius]: *(Press Enter to continue.)*
n **edge(s) selected for fillet.**
Press Enter to accept the fillet or [Radius]: *(Press Enter, type R, or drag the blue triangle grip to adjust the radius interactively.)*

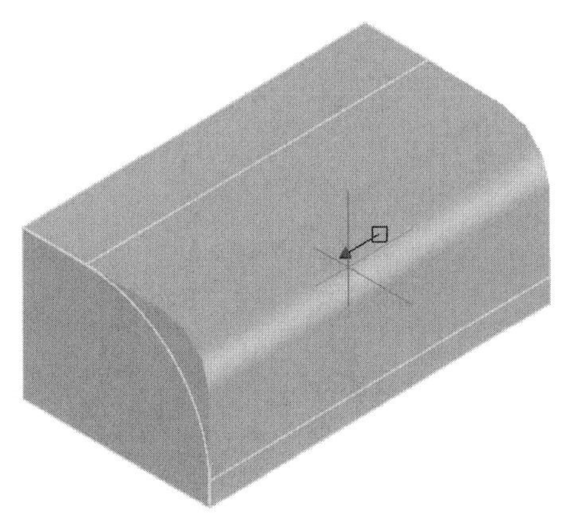

Dragging the red triangle grip to adjust the radius

As you drag the red triangle grip, AutoCAD prompts you:

>>Specify radius or [eXit]: *(Drag grip, enter a radius, or type X.)*

COMMAND LINE OPTIONS

Select an Edge selects one edge at a time.

Radius specifies the radius of the fillet.

Chain selects more than one edge at a time.

Press Enter to accept radius ends the command.

RELATED COMMANDS

Fillet rounds intersecting lines, solids, and polyline vertices.

SurfFillet fillets surfaces *(new to AutoCAD 2011).*

ChamferEdge chamfers 3D solids interactively *(new to AutoCAD 2011).*

SolidEdit edits 3D solid models.

RELATED SYSTEM VARIABLE

FilletRad3d reports the current fillet radius used by the FilletEdge and SurfFillet commands.

TIPS

- After you exit this command, you can no longer interactively adjust the fillet radius.
- The technical editor reports that the Chain option seems unnecessary: "I get the same result, even when I simply select multiple edges."

Filter

Rel. 12 Creates filter lists that can be applied when creating selection sets.

Command	*Alias*	*Keyboard Shortcuts*	*Menu*	*Ribbon*
'filter	**fi**	...	...	...

Command: filter

Displays dialog box.

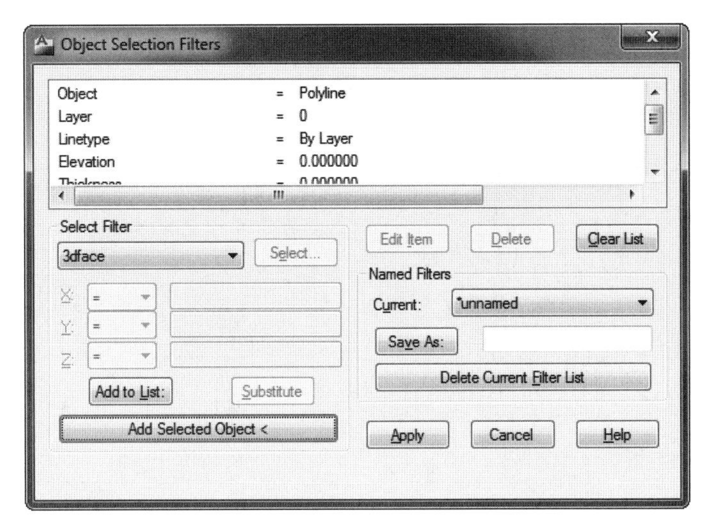

After you click Apply, AutoCAD continues in the command area:

Applying filter to selection.
Select objects: *(Select one or more objects.)*
Select objects: *(Press Enter to end object selection.)*
Exiting filtered selection.

AutoCAD highlights filtered objects with grips.

OBJECT SELECTION FILTERS DIALOG BOX

Select Filter options

Select displays all items of the specified type in the drawing.

X, Y, Z specifies the object's coordinates.

Add to List adds the current select-filter option to the filter list.

Substitute replaces a highlighted filter with the selected filter.

Add Selected Object selects the object to be added from the drawing.

Edit Item edits the highlighted filter item.

Delete deletes the highlighted filter item.

Clear list clears the entire filter list.

Named Filter options

Current selects the named filter from the list.

Save As saves the filter list with a name and the *.nfl* extension.

Delete Current Filter List deletes the named filter.

Apply closes the dialog box, and applies the filter operation.

COMMAND LINE OPTION

Select options selects the objects to be filtered; use the All option to select all non-frozen objects in the drawing.

RELATED COMMANDS

Any AutoCAD command with a 'Select objects' prompt.

QSelect creates a selection set quickly via a dialog box.

Select creates a selection set via the command line.

RELATED FILE

***.nfl** is a named filter list.

TIPS

- The selection set created by Filter is accessed via the P (previous) selection option.
- Alternatively, 'Filter is used transparently at the 'Select objects:' prompt.
- Filter uses the following grouping operators:

**Begin OR	*with*	**End OR
**Begin AND	*with*	**End AND
**Begin XOR	*with*	**End XOR
**Begin NOT	*with*	**End NOT

- Filter uses the following relational operators:

Operator	Meaning
<	Less than.
<=	Less than or equal to.
=	Equal to.
!=	Not equal to.
>	Greater than.
>=	Greater than or equal to.
*	All values.

- Save selection sets by name to an *.nfl* (short for *named filter*) file on disk for use in other drawings or editing sessions.
- The filter function is normally accessed through the QSelect command's dialog box or button on the Properties palette.

Find

2000 Finds, and optionally replaces, text in drawings.

Command	*Alias*	*Keyboard Shortcuts*	*Menu*	*Ribbon*
find	...	...	**Edit**	**Annotate**
			↳Find	**↳Text**
				↳Find Text

Command: find

Displays collapsed dialog box.

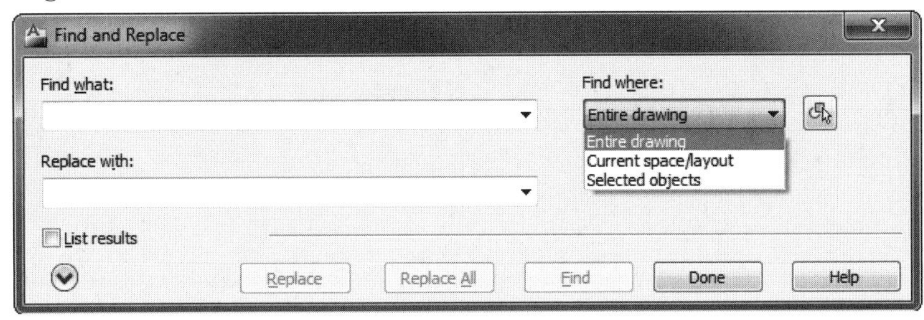

FIND AND REPLACE DIALOG BOX

Find text string specifies the text to find; enter text or click the down arrow to select one of the most recent lines of text searched for.

Replace with specifies the text to be replaced (only if replacing found text); enter text or click the down arrow to select one of the most recent lines of text searched for.

Find where:

- **Current Selection** searches the current selection set for text; click the Select objects button to create a selection set, if necessary.
- **Entire Drawing** searches the entire drawing.

Select objects allows you to select objects in the drawing; press Enter to return to dialog box.

Replace replaces a single found text with the text entered in the Replace with field.

Replace All replaces all instances of the found text.

Find finds the text entered in the Find text string field.

List Results expands dialog box to show found text:

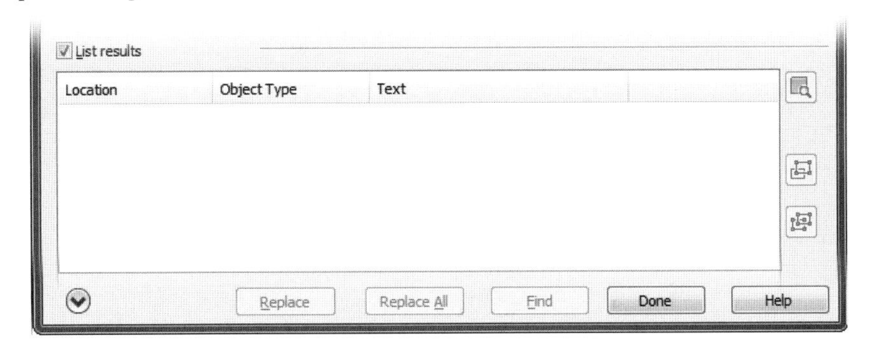

List Results options

zooms to the text in the drawing.

creates a selection set of the selected text.

creates a selection set from all found text.

expands the dialog box to show Search Options and Text Types:

Search Options options

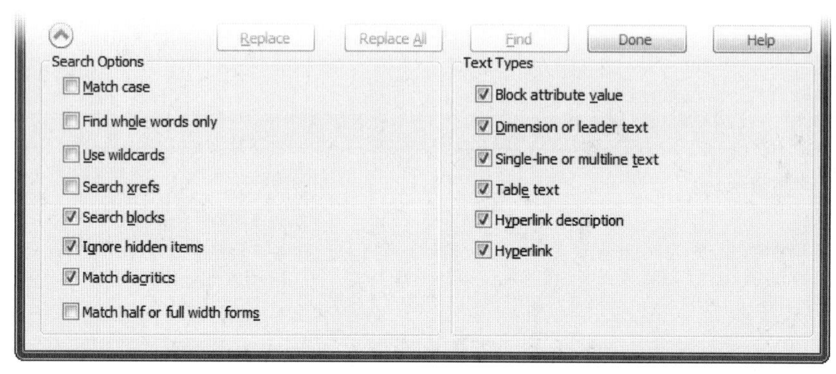

☐ **Match case** finds text that exactly matches the uppercase and lowercase pattern; for example, when searching for "Quick Reference," AutoCAD would find "Quick Reference" but not "quick reference."

☐ **Find whole words only** finds text that exactly matches whole words; for example, when searching for "Quick Reference," AutoCAD would find "Quick Reference" but not "Quickly Reference."

☐ **Use Wildcards** uses **?** for any single character and ***** for any number of characters.

☐ **Search Xrefs** extends the search to attached xrefs, which can prolong the search time.

☑ **Search Block**s extends the search to text inside blocks.

☑ **Ignore Hidden Items** does not search text on frozen or off layers, in invisible attributes, or text with visibility states in dynamic blocks.

☑ **Match Diacritics** searches for text with accent marks.

☐ **Match Half/Full Width Forms** applies to East Asian languages.

Text Types

☑ **Block Attribute Value** finds text in attributes.

☑ **Dimension Annotation Text** finds text in dimensions.

☑ **Text (Mtext, DText, Text)** finds text in paragraph text placed by the MText command, single-line text placed by the Text command, and field text placed by the Field command.

☑ **Table Text** finds text in tables.

☑ **Hyperlink Description** finds text in the description of a hyperlink.

☑ **Hyperlink** finds text in a hyperlink.

RELATED COMMANDS

DdEdit edits text.

Properties edits selected text.

TIPS

- To find database links, use the dbConnect command.
- The QSelect command places text in a selection set.

FlatShot

2004 Creates 2D blocks flattened from 3D models in the current viewport; short for "FLATten snapSHOT"; formerly an Express Tool.

Command	Alias	Keyboard Shortcuts	Menu	Ribbon
flatshot	**fshot**	...	...	**Home** ↳**Section** ↳**Flatshot**

Command: flatshot

Displays dialog box.

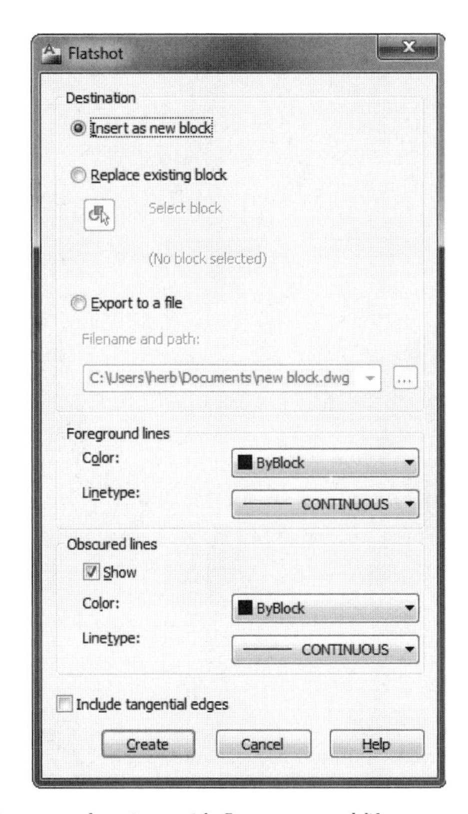

Choose options and then click Create. The command continues with -Insert command-like prompts:

Units: Unitless Conversion: 1.0000

Specify insertion point or [Basepoint/Scale/X/Y/Z/Rotate]: *(Pick a point, or enter an option.)*

Enter X scale factor, specify opposite corner, or [Corner/XYZ] <1>: *(Press Enter, or enter an option.)*

Enter Y scale factor <use X scale factor>: *(Press Enter, or enter a scale factor.)*

Specify rotation angle <0>: *(Press Enter, or enter a rotation angle.)*

DIALOG BOX OPTIONS

Destination options

Insert As New Block inserts the flattened model as a block.

Replace Existing Block replaces the existing block, if it already exists.

Select Block selects the block to replace; press Enter to return to the dialog box.

Export to a File saves the flattened model to a *.dwg* file on disk.

Foreground Lines options

Color specifies the color of non-hidden (non-obscured) lines; click Select Color to select other colors.

Linetype specifies the linetype of non-hidden lines; click Select Linetype to load additional linetypes into the drawing.

Obscured Lines options

Show toggles the display of hidden-lines.

Color specifies the color of hidden (obscured) lines.

Linetype specifies the linetype of hidden lines.

Create closes the dialog box, and then creates the flattened view; the original model is retained.

RELATED COMMANDS

ExportLayout exports layouts to model space, flattening 3D models.

SolView creates 2D views of 3D models.

SectionPlane slices 2D sections from 3D models

TIPS

- This command works with solids and surfaces only; it does not work with other kinds of 3D models, such as those made of faces and thickened 2D objects.

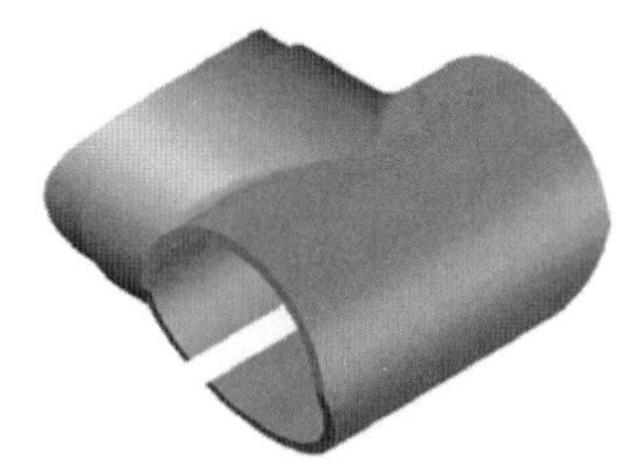
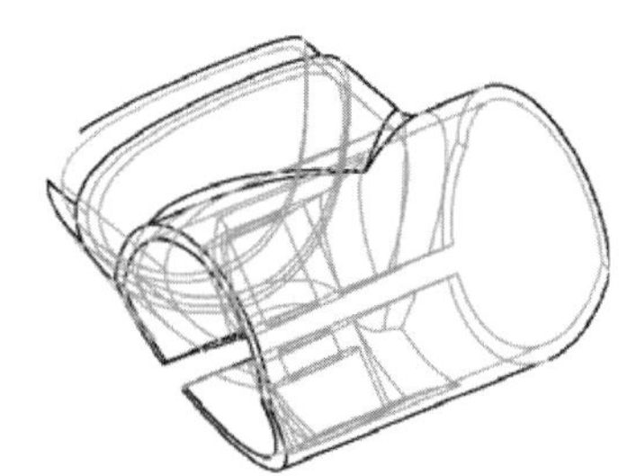

Left: Original 3D model.
***Right:** 2D block created by the FlatShot command.*

- The command creates a block from the 3D geometry; the block can be exploded with the Explode command for further editing.
- This command flattens the current screen view onto the current UCS. For the best result, set the UCS to View before using this command. This ensures the flattened view looks like the screen view. Finally, insert the new block into the WCS.
- As of AutoCAD 2010, this command flattens 3D mesh objects; the process may be quite slow with even simple drawings, as AutoCAD coverts the mesh faces to ShapeManager faces.

Relocated and Removed Commands

FreePoint, **FreeSpot**, and **FreeWeb** commands are found in the Light command reference.

Fog command was removed from AutoCAD 2007; it is now an alias for the RenderEnvironment command.

GifIn command was removed from Release 14. In its place, use Image.

Gc...

2011 A group of commands that geometrically constrain objects; more direct versions of the GeomConstraint command.

Command	*Aliases*	*Keyboard Shortcuts*	*Menu*	*Ribbon*
varies	...	...	...	**Parametric** ↳**Geometric**

See GeomConstraint command for tutorials and additional information on geometric constraints.

GCCOINCIDENT Command

Constrains selected objects to points: lines, polyline segments and arcs, circles, arcs, axes of ellipses, splines, and two constraint points.

Left: *Two separate lines.*
Right: *GcCoincident forcing end point of second line (gray) to touch the first line (black).'*

Command: gccoincident
Select first point or [Object/Autoconstrain] <Object>: *(Pick a point on an object, or enter an option.)*
Select second point or [Object] <Object>: *(Pick an object, or type O.)*

COMMAND LINE OPTIONS

Select first point selects a point on an object.

Select second point selects a point on a second object.

Object selects an object, and prompts for a point or multiple points.

Autoconstrain applies coincident constraints to a group of selected objects automatically, if possible.

GCCOLLINEAR command

Constrains selected objects to lie along the same line: lines, polyline segments, ellipses, and multi-line text.

Left: *Two separate lines.*
Right: *GcCollinear forcing second line (gray) to lay in the same line as that of the first line (black).*

Command: gccollinear
Select first object or [Multiple]: *(Pick a object or type M.)*
Select second object: *(Pick another object.)*

COMMAND LINE OPTION

Select first object chooses the object that specifies the linear constraint.

Multiple chooses additional objects to be made colinear with the first.

Select second point chooses the object to be constrained linearly with the first.

GCCONCENTRIC Command

Constrains two arcs, circles, or ellipses to the same center point.

Left: *Two arcs.*
Right: *GcConcentric forcing the center point of the second arc (gray) to match that of the first arc (black).*

Command: gcconcentric
Select first object: *(Pick an object.)*
Select second object: *(Pick another object.)*

COMMAND LINE OPTIONS

Select first object selects the curve that defines the center point.

Select second object selects the object whose center point will match that of the first.

GCEQUAL Command

Constrains selected objects to the same size or radius: lines, polyline segments and arcs, circles, and arcs.

Left: *Two lines of different lengths.*
Right: *GcEqual forcing the second line (gray) to be the same length as the first (black).*

Command: gcequal
Select first object or [Multiple]: *(Pick a object or type M.)*
Select second object: *(Pick another object.)*

See GcCollinear command for options

GCFIX Command

Locks selected objects in one position.

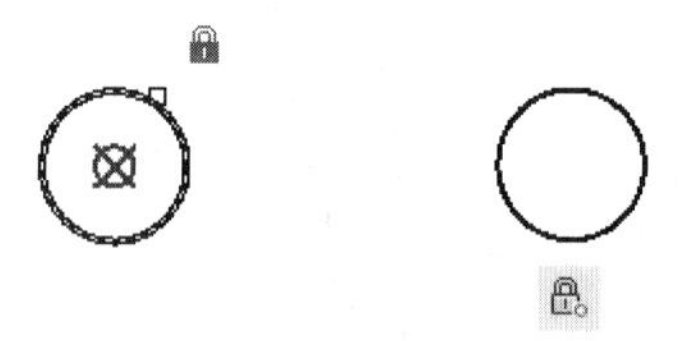

Left: *Circle selected...*
Right: *...and fixed into place by GcFix.*

Command: gcfix
Select point or [Object] <Object>: *(Pick a object or type O.)*

COMMAND LINE OPTIONS

Select first point selects a point on an object, which determines the fix location; the object can be rotated about the point.

Object selects an object, and then fixes the entire object into place; cannot be rotated.

GCHORIZONTAL Command

Constrains selected objects to be horizontal to the x axis of the current UCS: lines, polyline segments, ellipses, multiline text, and two valid points.

Left: *Line at an angle.*
Right: *GcHorizontal forcing the line to be horizontal.*

Command: gchorizontal
Select an object or [2Points] <2Points>: *(Pick an object or type 2P.)*

COMMAND LINE OPTIONS

Select an object selects an object to constrain.

2Points prompts you to pick two points on an object.

GCPARALLEL Command

Constrains selected objects to be parallel to each other: lines, polyline segments, ellipses, and multiline text.

Left: *Two lines.*
Right: *GcParallel forcing the second line (gray) to be parallel to the first (black).*

Command: gcparallel
Select first object: *(Pick an object.)*
Select second object: *(Pick another object.)*

See GcEqual command for options.

GCPERPEDICULAR Command

Constrains selected objects to be perpendicular to one another.

Command: gcperpedicular
Select first object *(Pick an object.)*
Select second object: *(Pick another object.)*

See GcEqual command for options.

Left: *Two lines.*
Right: *GcPerpedicular forcing the second line (gray) to be perpendicular to the first (black).*

GCSMOOTH Command

Constrains splines to be contiguous (and maintain G2 continuity) with other splines, lines, arcs, and polylines.

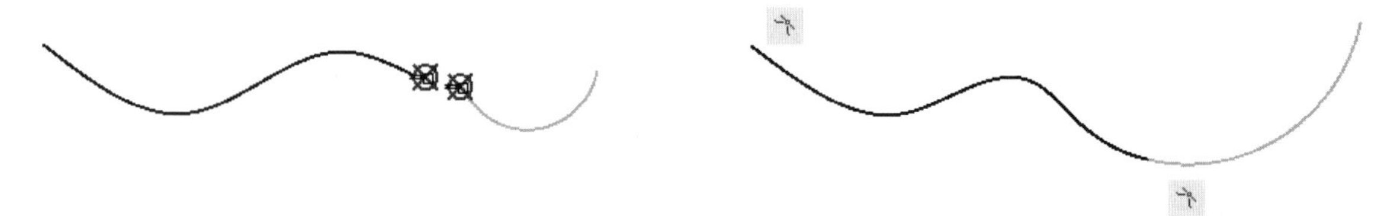

Left: *A spline and an arc.*
Right: *GcSmooth forcing the arc (gray) to smoothly connect to the spline (black).*

Command: gcsmooth
Select first spline curve:
Select second spline: *(Pick another object.)*

COMMAND LINE OPTIONS

Select first spline curve chooses the spline to act as the constraint condition.

Select second curve chooses the open object that is smoothed to match the first.

GCSYMMETRIC Command

Constrains lines, polyline segments, text, mtext, major and minor axes of ellipses and elliptical arcs to be symmetric about a mirror line.

Left: *Two lines on either side of a line of symmetry (gray).*
Right: *GcSymmetric forcing the second line to be symmetric to the first about the mirror line.*

Command: gcsymmetric
Select first object or [2Points] <2Points>: *(Pick an object or type 2P.)*
Select second object: *(Pick another object.)*
Select symmetry line: *(Pick a line, not two points.)*

COMMAND LINE OPTIONS

Select first object selects an object to constrain.

2Points prompts you to pick two points on an object.

Select second object chooses a second object.

Select symmetric line chooses the line that acts as the line of symmetry (mirror line); cannot pick two points.

GCTANGENT Command

Constrains two curves, or a line and a curve, to maintain tangency to each other, or their extensions.

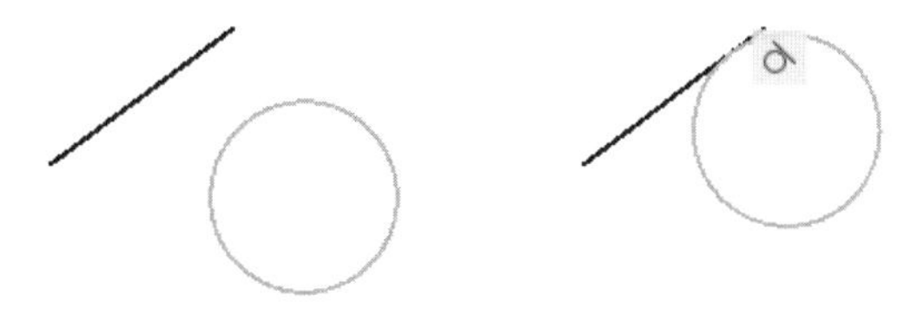

Left: *A line and a circle.*
Right: *GcTangent forcing the circle to be tangent to the line.*

Command: gctangent
Select first object: *(Pick an object.)*
Select second object: *(Pick another object.)*

See GcEqual command for options.

GCVERTICAL Command

Constrains selected objects to be vertical to the y axis in the current UCS.

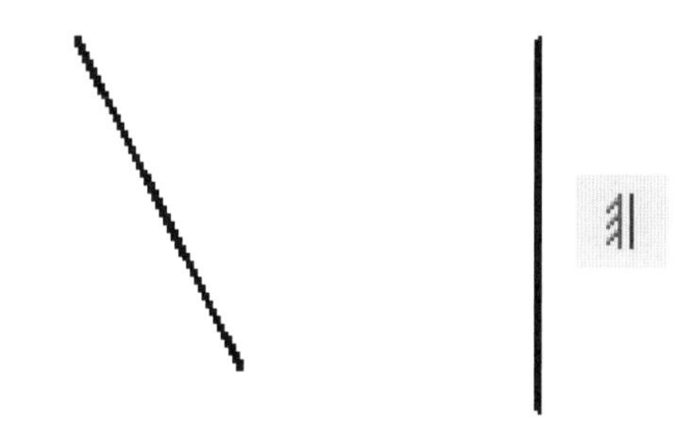

Left: *Line at an angle.*
Right: *GcVertical forces the line to be vertical.*

Command: gcvertical
Select an object or [2Points] <2Points>: *(Pick an object, or type 2P.)*

See GcHorizontal command for options.

RELATED COMMANDS

GeomConstraint applies all of these constraints (the master command).

DelConstraint removes constraints from selected objects.

ConstraintSettings specifies how geometric and dimensional constraints are applied.

RELATED SYSTEM VARIABLE

ConstraintBar toggles the display of constraint icons (a.k.a. "bars").

GeographicLocation

2007 Positions longitude-latitude in drawings; positions the sun (light) according to the date and location.

Command	Aliases	Menu	Ribbon
geographiclocation	**geo**	**Tools**	**Render**
	north	**⇘Geographic Location**	**⇘Sun & Location**
	northdir		**⇘Set Location**

Command: geographiclocation

Displays one of two dialog boxes:

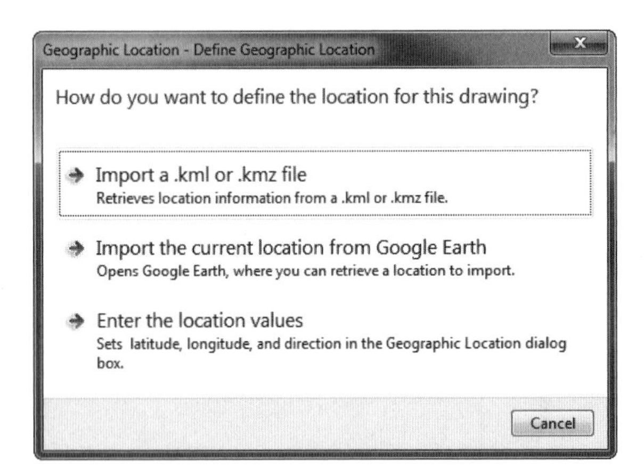

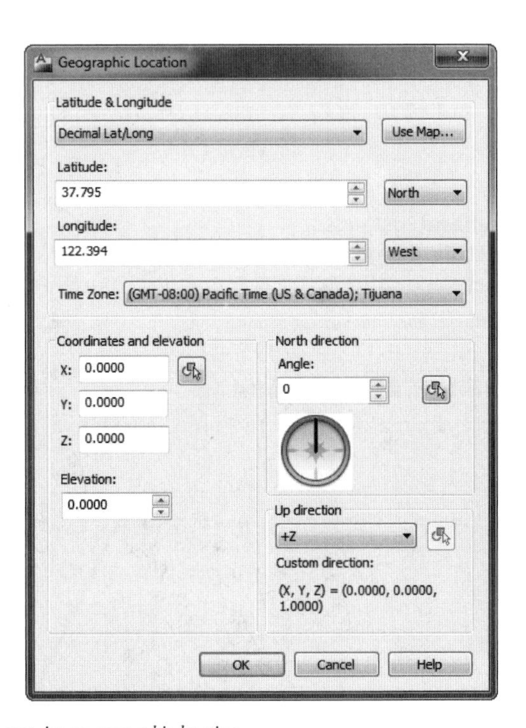

Left: *Dialog box displayed when the drawing contains no geographic location.*
Right: *Dialog box displayed when the drawing contains a geographic location.*

DEFINE GEOGRAPHIC LOCATION DIALOG BOX

When the drawing contains no geographic location, the location is set using this dialog box.

Import a Kml or Kmz File displays the Import A KML or KMZ File dialog box; gets the geographic location from the first coordinates stored in a Google Earth *.kml* or *.kmz* file.

Import the Current Location from Google Earth gets the location from Google Earth, if installed on your computer.

Enter the Location Values opens the Geographic Location dialog box.

LOCATION ALREADY EXITS DIALOG BOX

When the drawing contains a geographic location, the location is edited with this alternate dialog box.

Edit Current Geographic Location opens the Location Already Exits dialog box.

Redefine Geographic Location opens the Define Geographic Location dialog box.

Remove Geographic Location removes the latitude-longitude data and pushpin.

GEOGRAPHIC LOCATION DIALOG BOX

Access this dialog box through the Location Value button:

Latitude & Longitude options

Decimal Lat/Long chooses between decimal and degrees-minutes-seconds format.

Use Map selects location from the Location Picker dialog box.

Latitude sets the latitude between 0 and 90 degrees; when a city is selected from the map, its latitude is shown here; direction selects North or South of the equator.

Longitude sets the longitude from 0 to 180; direction selects East or West of the Prime Meridian.

Time Zone specifies the time zone.

Coordinates and Elevation options

X, Y, Z specify x, y, and z coordinates; alternatively, you can pick the position in the drawing.

Elevation provides an alternative method for specifying the z coordinate.

North Direction options

Angle specifies the direction of North; default is the positive y axis.

Up Direction specifies the up direction of the z axis; countries like Australia place South at the top of their maps.

LOCATION PICKER DIALOG BOX

Access this dialog box through the Use Map button.

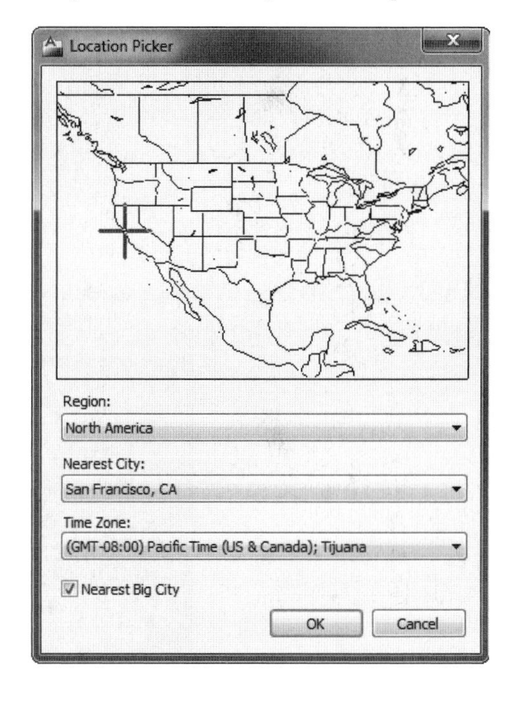

Map options

Click anywhere in the map to set the latitude and longitude; red cross shows the current location.

Region selects a region or the entire world.

Nearest City selects a city.

Time Zone specifies the time zone

☑ **Nearest Big City** displays the name of the nearest city to the latitude and longitude entered.

RELATED COMMAND

SunProperties opens the Sun Properties palette.

RELATED SYSTEM VARIABLES

GeoLatLongFormat specifies the format of latitude and longitude.

GeoMarkerVisibility toggles the visibility of geographic markers.

Latitude stores the current latitude.

Longitude stores the current longitude.

NorthDirection stores the angle from the positive y axis.

Timezone stores the current time zone.

RELATED FILES

KML — file format used by Google Earth; short for Keyhole Markup Language.

KMZ — compressed version of KML format; short for KM Zipped.

TIPS

- When a geographic location is added to the drawing, a pushpin icon indicates its location:

- To recall the position of the pushpin, pause the cursor over it until a tooltip appears:

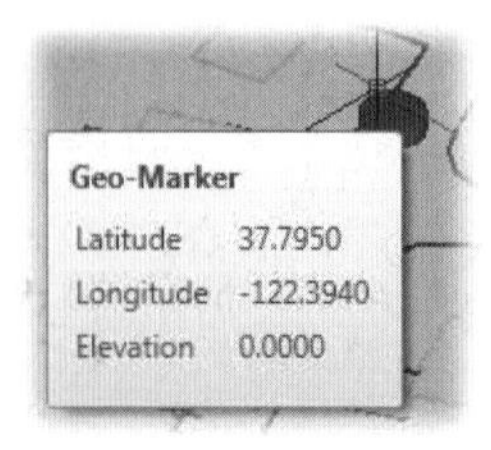

- The status bar can report x, y, z coordinates as longitude, latitude, and elevation. To change the display, right-click it, and then choose Geographic.

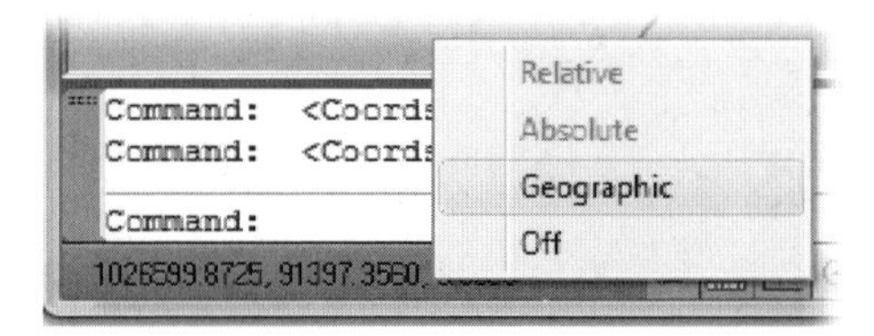

- When AutoCAD reads *.kml* or *.kmz* files, it reads the first set of coordinates it finds in the file. For a predictable result, have just one pair of coordinates in a file.
- When AutoCAD accesses Google Earth for coordinates, you get a more accurate result when it is zoomed in as far as possible. This command reads the coordinates at the center of the Google Earth screen.

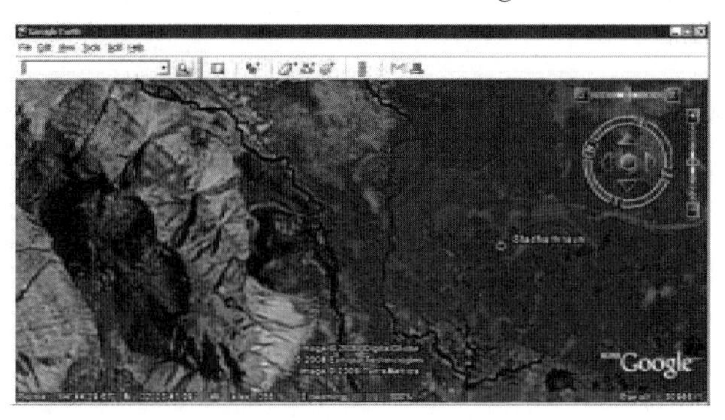

- This command uses names and coordinates of cities found in the *.map* files found in *\AutoCAD 2010\Support*, which are no longer user-editable.
- When the KML (or KMZ) file contains more than one location, AutoCAD uses the first one it comes across.

GeomConstraint

2010 Creates geometric relationships between objects.

Command	Alias	Keyboard Shortcuts	Menu	Ribbon
geomconstraint	**gcon**	...	**Parametric ↳Geometric**	**Parametric ↳Geometric**

Command: geomconstraint
Enter constraint type
[Horizontal/Vertical/Perpendicular/PArallel/Tangent/SMooth/Coincident/CONcentric/COLlinear/Symmetric/Equal/Fix]<Coincident>: *(Enter an option, or press Enter to exit the command.)*

After you enter an option, AutoCAD prompts you to pick one or more objects.

COMMAND LINE OPTIONS

Horizontal forces objects to be horizontal; prompts you:

Select an object or [2Points]: *(Choose one object, or type 2P.)*

Select an object chooses a line or a polyline segment.

2Points picks two points on the object; the object is forced horizontally along the two pick points. If an arc, then it is extended to make the two pick points horizontal.

Vertical forces objects to be vertical; displays the same prompt as Horizontal.

Perpendicular forces the first object to be perpendicular to a second; prompts you:

Select first object: *(Choose the object to be constrained.)*
Select second object: *(Choose the destination object.)*

Select first object chooses a line or a polyline segment to be made perpendicular.

Select second object picks the destination line or polyline segment; the first object is rotated to be perpendicular to the second.

PArallel forces one object to be parallel to a second; displays the same prompt as Perpendicular. Works with lines and polylines.

Tangent forces one straight or curved object to be tangent to a curved object; the objects need to touch each other. Displays the same prompt as Perpendicular. Works with lines, polylines and polyarcs, circles, arcs, and ellipses.

SMooth makes the connection between a spline and another open curve smooth; the endpoints become coincident. Technically, a G2 curvature continuous condition is applied. Prompts you:

Select first spline curve: *(Choose a spline.)*
Select second curve: *(Choose another spline, arc, or polyarc.)*

The first object selected must be a spline; the second object can be another spline, an arc, or a polyarc.

Coincident forces the endpoint of the second object to match a point of the first; prompts you:

Select first point or [Object/Autoconstrain] <Object>: *(Pick a constraint point, or enter an option.)*
Select second point or [Object] <Object>: *(Pick another constraint point, or type O.)*

Select first point specifies the end point of the first object.

Object selects an object, instead of a constraint point; works with lines, polylines and polyarcs, circles, arcs, ellipses, and splines. Prompts you:

Select object: *(Choose an object.)*
Select point or [Multiple]: *(Pick a point on the object, or type M.)*

Multiple allows you to select more than one additional point to coincide with the first.

Autoconstrain makes multiple objects constrained; see the AutoConstrain command.

Select second point specifies the end point of the second object, which will be made coincident to the first one.

CONcentric forces two curved objects to share the same center point; displays the same prompt as Perpendicular. Works with circles, arcs, polyarcs, and ellipses.

COLlinear forces one object to lie in the same line as a second; works with lines and polyline segments. Displays the same prompts as Coincident.

Symmetric forces two objects to have mirrorlike symmetry about an imaginary "symmetry" line; Prompts you:

Select first object or [2Points] <2Points>: *(Choose an object, or type 2P.)*
Select second object: *(Choose another object.)*
Select symmetry line: *(Pick an object.)*

Select first object chooses one object; works with lines, polylines and polyarcs, circles, arcs, and ellipses.

Select second object chooses the second object; the second object is scaled to be the same size as the first.

2Points prompts you to pick points on two objects; the objects maintain their size.

Select symmetry line chooses an object to use as the mirror (a.k.a. symmetry) line.

Equal forces one line or polyline to be the same length as a second, or forces one arc or circle to have the same radius as the second; prompts you:

Select first object or [Multiple]: *(Choose an object, or type M.)*
Select second object: *(Choose a second object.)*

Select first object chooses one object; works with lines, polylines and polyarcs, arcs, and circles.

Select second object chooses another object; becomes the same size as the first.

Multiple allows you to select additional objects; repeats the "Select object to make equal to the first" prompt until you press Enter.

Fix locks objects in position; prompts you:

Select point or [Object] <Object>: *(Pick a point on an object, or type O.)*

Select point specifies the point at which the object is locked; the object can be resized and rotated about the lock point.

Object locks the object in-place; it cannot be resized, rotated, or moved; works with all objects.

RELATED COMMANDS

DelConstraint removes constraints from selected objects.

Gc... commands applies each of these constraints individually; useful for applying the same constraint many times.

ConstraintSettings specifies how constraints are applied.

RELATED SYSTEM VARIABLES

ConstraintBar toggles the display of constraint icons (a.k.a. "bars").

ConstraintInfer toggles the inference of geometric constraints as objects are drawn and edited *(new to AutoCAD 2011)*.

TIPS

- Constraints can be applied to certain geometric features of many, but not all, objects.
- Constraints are like semipermanent object snaps.
- To remove constraints from objects, use the DelConstraint command.
- Toggle the display of constraint symbols with the ConstraintBarMode system variable.
- In Symmetry constraint, angles of lines are made symmetrical; center and radius of arcs and circles are made symmetric.
- Polylines and other objects are sometimes distorted to match the constraint conditions.
- Spline frames cannot be constrained.
- As of AutoCAD 2011, constraints are applied (inferred) automatically when ConstraintInfer is turned on.

Quick Start Tutorial

Applying Constraints

Geometric constraints can be applied manually with the GeomConstraint command, or automatically with the AutoConstrain command. Tutorial notes by technical editor, Bill Fane.

Geometric constraints are like "sticky" object snaps. For instance, the tangent osnap exists only for the split-second interval during which AutoCAD finds and then applies the tangency location; in contrast, geometric constraints force lines to remember that they are tangent to the arcs. Move one object, and the other moves along, as illustrated below:

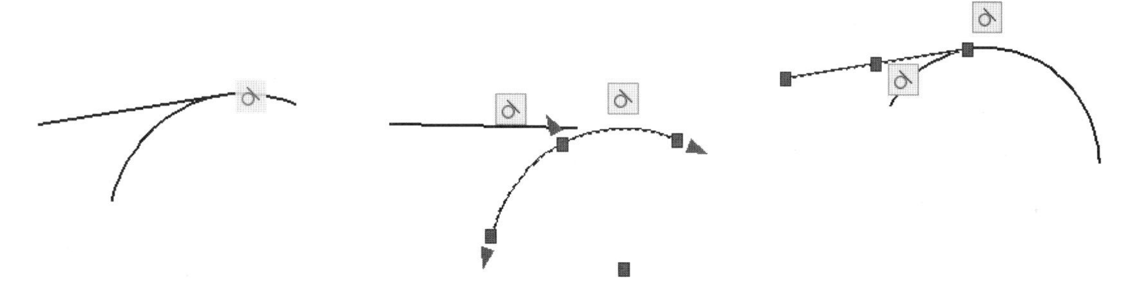

Left: *The line constrained tangentially to the arc.*
Center: *Move the arc, and the line moves with it. Even if the two no longer touch, they are still connected through the constraint.*
Right: *Move the line, and the arc moves with it.*

There is a geometric constraint that matches each of the object snap modes, such as Intersect and Perpendicular. As well, there are some extra ones, such as Equal, Symmetric, and Colinear.

(Objects do not need to touch for geometric constraints to work. A line in a front view can have Equal length to a line in the top view so that changing one line will change the other, and one end of the front-view line can always be Vertical to the end top-view line. The views stay in step.)

AutoCAD also provides dimensional constraints that define the sizes of geometric features. See the DimConstraint command. Geometric and dimensional constraints work together, and tare known as "parametrics." Sometimes, they can replace each other. For example, the Perpendicular constraint is the same as a dimensional constraint with a 90-degree angle.

AutoCAD does not allow you to add too many (or conflicting) constraints, known as "over constrained." Normally, you want drawings to be "fully constrained," which means there are no missing dimensions.

TUTORIAL

In this tutorial, you force the circle to remain centered in the keyhole shape with the Concentric constraint.

1. Draw the keyhole as a polyline.
2. Add the circle anywhere in the drawing.

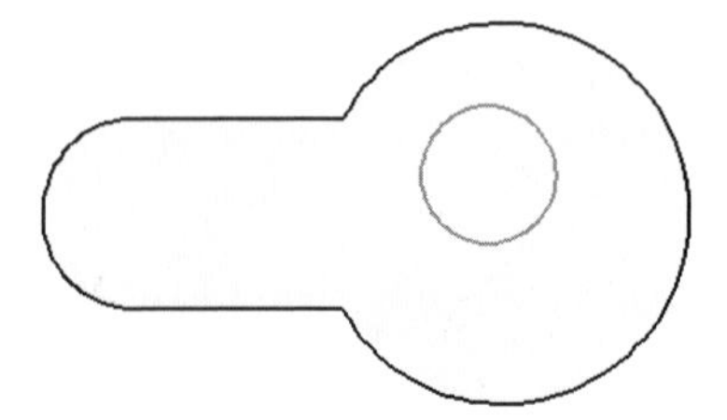

3. Start the GeomConstraint command, and then select the Concentric option:

   Command: geomconstraint
   Enter constraint type
   [Horizontal/.../CONcentric/.../Fix]<Coincident>: *(Type CON.)*

4. When AutoCAD asks you to select an object, choose the large polyarc at the right end of the polyline.

   Select first object: *(Choose the polyarc.)*

5. Choose the circle. Notice the pale blue concentric constraint icon that hovers near the cursor.

   Select second object: *(Choose the circle.)*

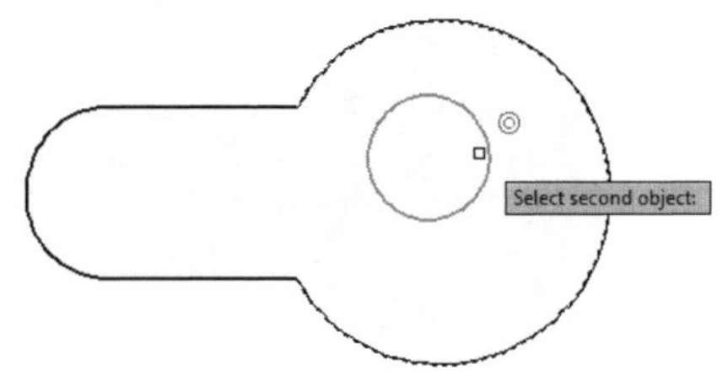

Notice that AutoCAD displays constraint bars, which indicate the type and location of the applied constraint.

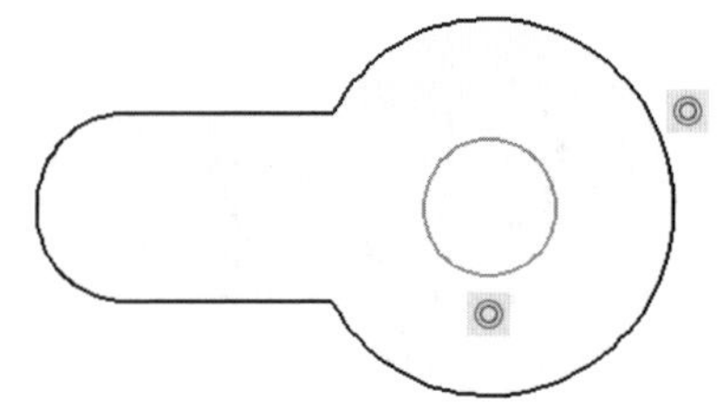

6. With the constraint in place, let's see what happens when the objects are moved. First, rotate the keyhole about its left end (base point centered on the small polyarc). Notice that the circle moves with the keyhole due to the concentric constraint.

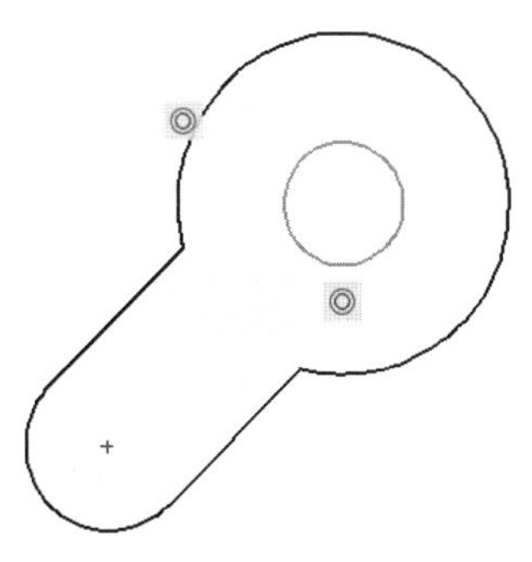

7. Drag the circle. Notice that the keyhole stretches itself to keep the large polyarc centered around the circle.

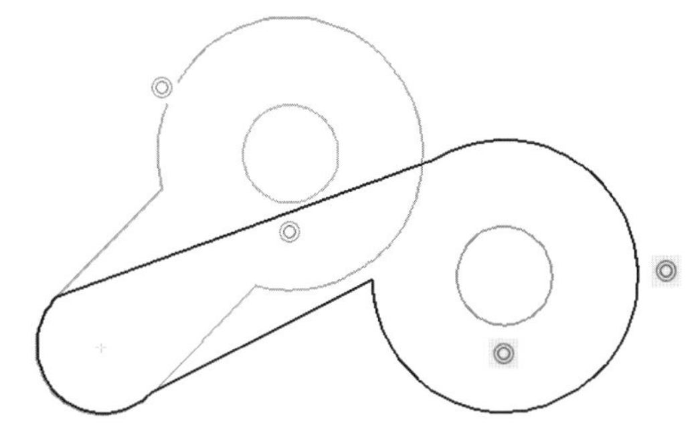

See the DimConstraint command for a tutorial on using dimensional constraints.

GotoUrl

2000 Goes to hyperlinks contained by objects.

Command	*Aliases*	*Keyboard Shortcuts*	*Menu*	*Ribbon*
'gotourl	...	...	...	...

Command: gotourl
Select objects: *(Select one or more objects.)*
Select objects: *(Press Enter to end object selection.)*
browser Enter Web location (URL) <http://www.autodesk.com>: http://www.upfrontezine.com

AutoCAD launches your computer's default Web browser, and attempts to access the URL.

COMMAND LINE OPTION

Select objects selects one or more objects containing a hyperlink.

RELATED COMMANDS.

Browser launches the Web browser.

Hyperlink attaches, edits, and removes hyperlinks from objects.

TIPS

- This command is meant for use by macros and menus.
- AutoCAD uses this command for the shortcut menu's Hyperlink | Open option.

Gradient

2006 Floods areas with gradient fills.

Command	Alias	Keyboard Shortcuts	Menu	Ribbon
gradient	**gd**	...	**Draw**	**Home**
			↳**Gradient**	↳**Draw**
				↳**Gradient**

Command: gradient

New command line prompt in AutoCAD 2011:

Pick internal point or [Select objects/seTtings]: *(Pick a point or enter an option.)*

COMMAND LINE OPTIONS

Pick internal point fills areas with the current gradient pattern in real time; change settings on the ribbon's Hatch Creation tab.

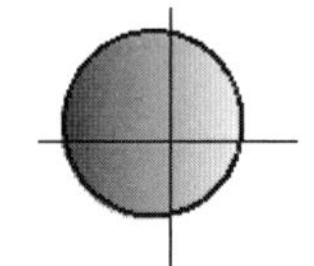

Select objects chooses objects.

seTtings displays the Hatch and Gradient dialog box.

HATCH AND GRADIENT DIALOG BOX

Color options

⦿ **One color** produces color-shade gradients.

○ **Two color** produces two-color gradients:

Orientation options

☑ **Centered** centers the gradient in the hatch area.

Angle rotates the gradient.

See the Hatch command for other options.

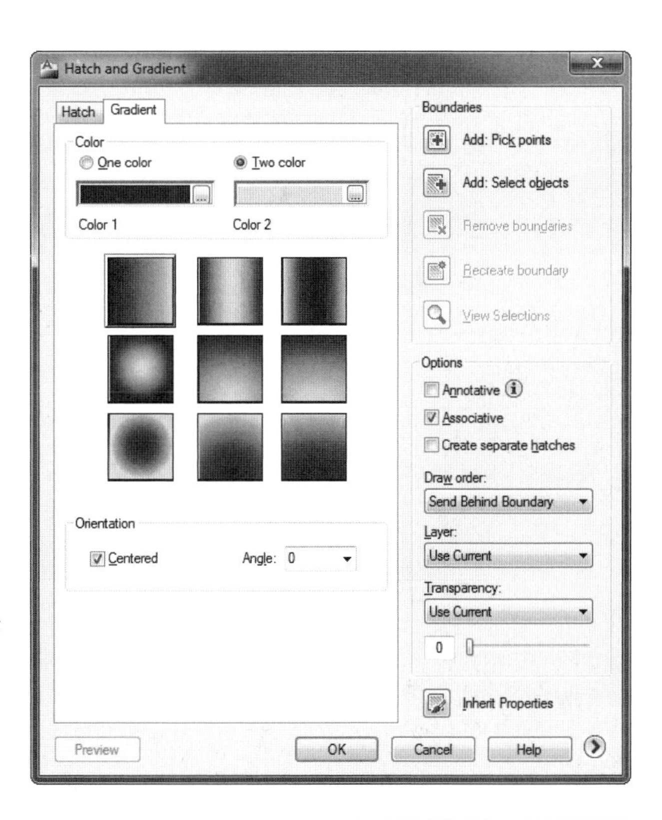

RELATED SYSTEM VARIABLES

GfAng specifies the angle of the gradient fill.

GfClr1 specifies first gradient fill color in RGB format.

GfClr2 specifies second gradient fill color in RGB format.

GfClrLum specifies luminescence of one-color gradient fills.

GfClrState specifies whether the gradient fill is one-color or two-color.

GfName specifies the gradient fill pattern: (1) linear, (2) cylindrical, (3) inverted cylindrical, (4) spherical, (5) inverted spherical, (6) hemispherical, (7) inverted hemispherical, (8) curved, and (9) inverted curved.

GfShift specifies whether the gradient fill is centered or shifted to the upper-left.

TIP

- The Gradient command simply displays the Gradient tab of the Hatch and Fill dialog box.

GraphScr

Ver. 2.1 Switches away from the text window to the graphics widow.

Command	*Alias*	*Keyboard Shortcuts*	*Menu*	*Ribbon*
'graphscr	**...**	**F2**	**...**	**...**

Command: graphscr

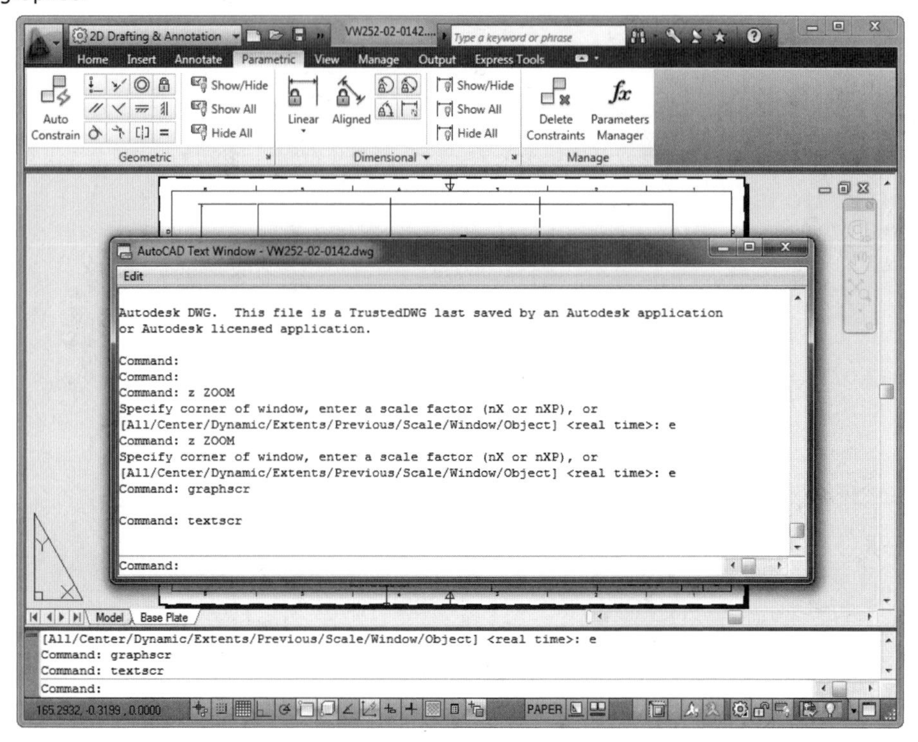

Behind: *Drawing window ("graphics screen:).*
In front: *Text window ("screen").*

RELATED COMMANDS

CopyHist copies text from the Text window to the Clipboard.

TextScr switches from the graphics window to the Text window.

RELATED SYSTEM VARIABLE

ScreenMode indicates whether the current screen is in text or graphics mode. Since the advent of Windows, it only holds the value of 3: dual screen, displaying both text and graphics.

TIP

- The Text window appears frozen when a dialog box is active. Click the dialog box's OK or Cancel button to regain access to the Text window.

Grid

Ver. 1.0 Displays a grid of non-plotting reference lines or dots.

Command	Aliases	Keyboard Shortcuts	Status Bar	Ribbon
'grid	...	**Ctrl+G F7**		...

Command: grid

Specify grid spacing(X) or [ON/OFF/Snap/Major/aDaptive/Limits/Follow/Aspect] <5'-0">: *(Enter a value, or enter an option.)*

When GridStyle is turned on, the grid displays as an array of dots in 2D model space, block editor, and/or layout tabs. Otherwise, the grid displays as perpendicular lines.

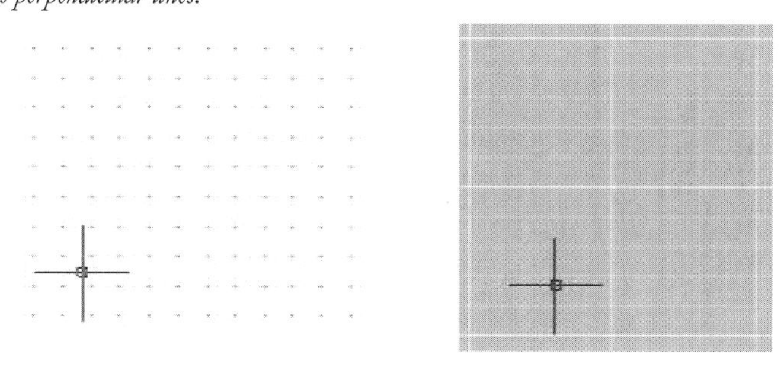

Left: *Grid of dots displayed when GridStyle is on.*
Right: *Grid of lines displayed by default.*

COMMAND LINE OPTIONS

Specify grid spacing(X) sets the x and y direction spacing; an **x** following the value sets the grid spacing to a multiple of the current snap setting, such as 2x.

ON turns on grid markings.

OFF turns off grid markings.

Snap makes the grid spacing the same as the snap spacing.

Major specifies the number of minor grid lines per major line; default = 5.

aDaptive displays fewer grid lines when zoomed out.

Limits determines whether the grid is limited to the rectangle defined by the Limits command.

Follow matches the grid plane to the dynamic UCS.

Aspect specifies different spacing for grid lines/dots in the x and y directions.

RELATED COMMANDS

DSettings sets the grid spacing and style via a dialog box.

Options sets the colors for grid lines.

Limits sets the limits of the grid in WCS.

Snap sets the snap spacing, which can equal the grid spacing.

RELATED SYSTEM VARIABLES

GridDisplay determines how the grid is displayed:

GridMajor specifies the number of minor grid lines per major lines.

GridMode toggles the grid in the current viewport.

GridStyle determines where the grid is displayed as dots *(new to AutoCAD 2011.)*

GridUnit specifies the current grid x, y-spacing.

LimCheck determines whether the grid is limited to the rectangle defined by the Limits command.

LimMin specifies the x, y coordinates of the lower-left corner of the grid display.

LimMax specifies the x, y coordinates of the upper-right corner of the grid display.

SnapStyl displays a normal or isometric grid.

TIPS

- The grid is most useful when set to the snap spacing, or to a multiple of the snap spacing.
- When the grid spacing is set to 0, it matches the snap spacing.
- The Options command's Display tab's Colors button sets the color of grid lines. The grid line along the x axis is colored red, while the y axis line is green.

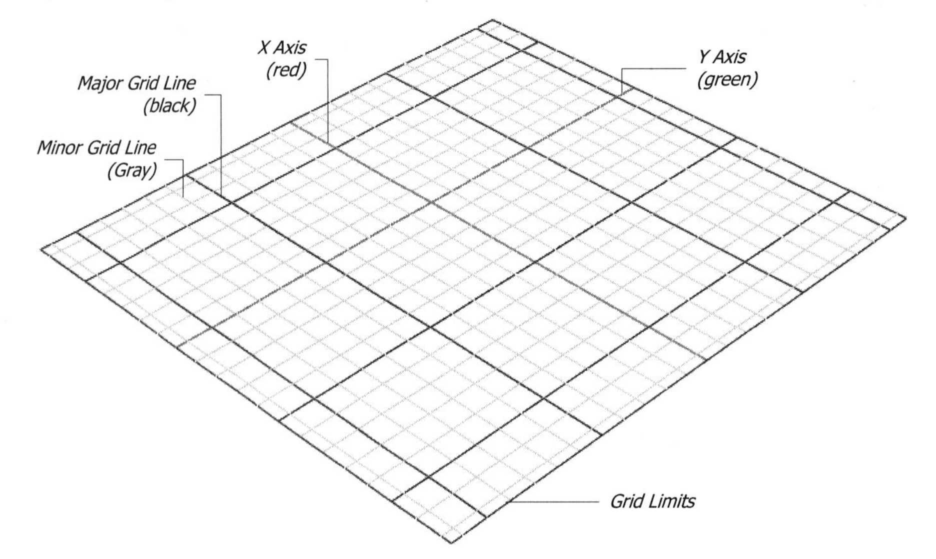

- The Snap command's Isometric option displays an isometric grid; when perspective mode is turned on, the grid is also displayed in perspective.
- You can set a different grid spacing in each viewport, and a different grid spacing in the x and y directions.
- Rotate the grid with the Snap command's Rotate option.
- Grid markings are not plotted; to create a plotted grid, use the Array command to place an array of points or lines.
- As of AutoCAD 2011, the grid displays as lines in all areas of AutoCAD by default; use the DSettings command's Snap and Grid tab's Grid Style options to determine if the grid displays as dots in 2D model space, the block editor, and/or layout tabs.

Group

Rel. 13 Creates named selection sets of objects.

Commands	*Aliases*	*Keyboard Shortcuts*	*Menu*	*Ribbon*
group	**g**	...	...	...
-group	**-g**	...	...	...

Command: group

Displays dialog box.

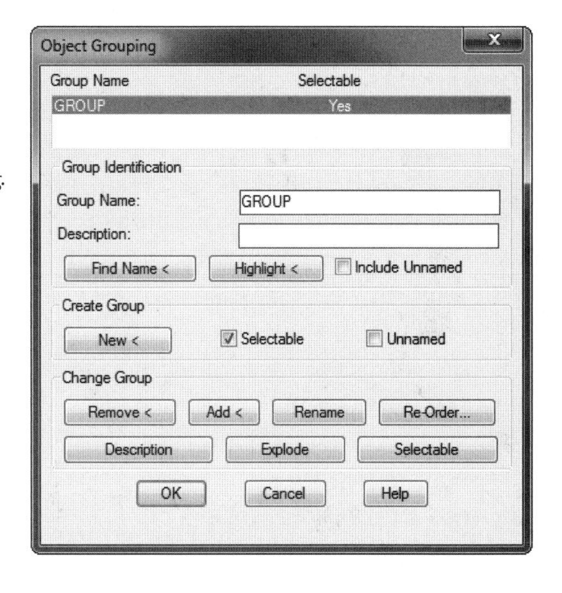

OBJECT GROUPING DIALOG BOX

Group Name lists the names of groups in the drawing.

Group Identification options

Group Name displays the name of the current group.

Description describes the group; may be up to 64 characters long.

Find Name lists the name(s) of group(s) that a selected object belongs to.

Highlight highlights the objects included in the current group.

☐**Include Unnamed** lists unnamed groups in the dialog box.

Create Group options

New selects objects for the new group.

☑**Selectable** toggles selectability: picking one object picks the entire group.

☐**Unnamed** creates an unnamed group; AutoCAD gives it the generic name ***A***n*, where *n* is a number that increases with each group.

Change Group options

Remove removes objects from the current group.

Add adds objects to the current group.

Rename renames the group.

Re-order changes the order of objects in the group; displays Order Group dialog box.

Description changes the description of the group.

Explode removes the group description; does not erase group members.

Selectable toggles selectability.

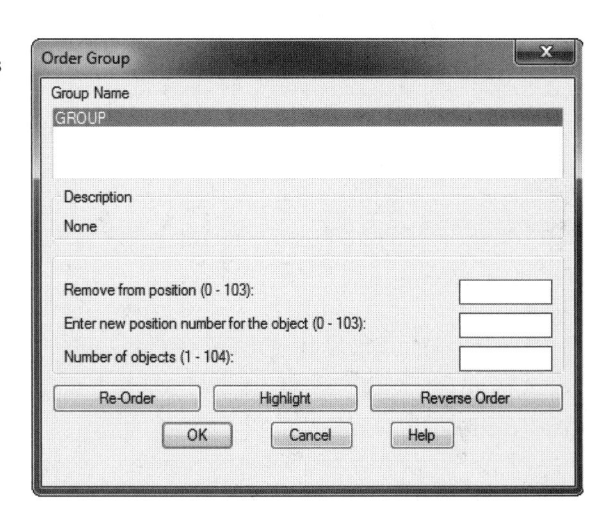

ORDER GROUP DIALOG BOX

Access this dialog box through the Re-Order button.

Group Name lists the names of groups in the current drawing.

Description describes the selected group.

Remove from position (0 - *n*) selects the object to move.

Replace at position (0 - *n*) moves the group name to a new position.

Number of objects (1 - *n*) lists the number of objects to reorder.

Re-Order applies the order changes.

Highlight highlights the objects in the current group.

Reverse Order reverses the order of the groups.

-GROUP Command

Command: -group
Enter a group option
[?/Order/Add/Remove/Explode/REName/Selectable/Create] <Create>: *(Enter an option.)*

COMMAND LINE OPTIONS

? lists the names and descriptions of currently-defined groups.

Order changes the order of objects within the group.

Add adds objects to the group.

Remove removes objects from the group.

Explode removes the group definition from the drawing.

REName renames the group.

Selectable toggles whether the group is selectable.

Create creates a newly-named group from the objects selected.

RELATED COMMANDS

Block creates named symbols from a group of objects.

Select creates selection sets.

RELATED SYSTEM VARIABLE

PickStyle toggles whether groups are selected by the usual selection process.

TIPS

- You can toggle groups on and off with the Ctrl+Shift+A shortcut keystroke.
- Use Ctrl+H to toggle between selecting the entire group and objects within the group.
- Consider a group as a named selection set; unlike a regular selection set, a group is not "lost" when the next group is created.
- Group descriptions can be up to 64 characters long.
- Anonymous groups are unnamed; AutoCAD refers to them as *A*n*.
- It is curious that such a useful command has disappeared from the menus, toolbars, and ribbon.

Changed Command

As of AutoCAD 2006, the **Hatch** command acts like the BHatch command, while the **-Hatch** command is like the -BHatch command.

Hatch

Ver. 1.4 Applies associative hatch pattern to objects within boundaries.

Commands	Aliases	Keyboard Shortcuts	Menu	Ribbon
hatch	**bh** **bhatch** **h**	...	**Draw** **↳Hatch**	**Home** **↳Draw** **↳Hatch**
-hatch	**-bhatch**	...	...	...

Command: hatch

New command line prompt in AutoCAD 2011:

Pick internal point or [Select objects/seTtings]: *(Pick a point or enter an option.)*

COMMAND LINE OPTIONS

Pick internal point fills areas with the current hatch pattern in real time; change settings on the ribbon's Hatch Creation tab.

Select objects chooses objects.

seTtings displays the Hatch and Gradient dialog box.

HATCH AND GRADIENT DIALOG BOX

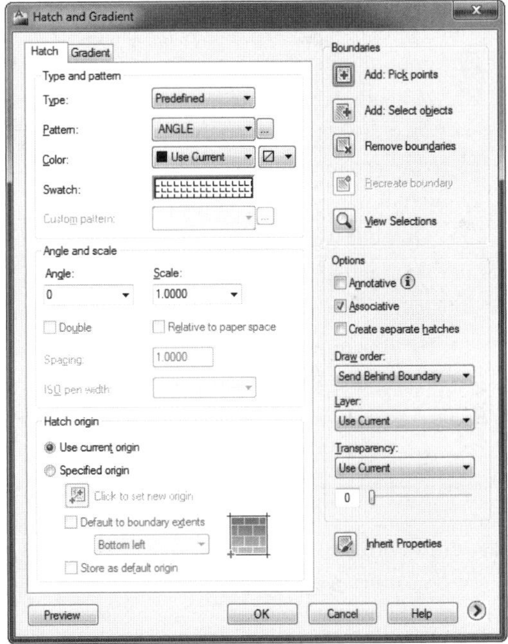

Type and Pattern options

Type selects the pattern type:

- **Predefined** uses hatches stored in *acad.pat* and *acadiso.pat* files.
- **User Defined** creates parallel-line hatches with user-defined spacing and angle.
- **Custom** uses hatches defined by *.pat* files on AutoCAD's search path.

Pattern selects hatch pattern.

[...] shows pattern type samples; displays the Hatch Pattern Palette dialog box, detailed later.

Color specifies the foreground and background colors *(new to AutoCAD 2011).*

Swatch displays a non-scaled preview of the hatch pattern; click to display the Hatch Pattern Palette dialog box.

Custom Pattern lists the names of custom patterns, if available.

Angle and Scale options

Angle specifies the hatch pattern rotation; default = 0 degrees.

Scale specifies the hatch pattern scale; default = 1.0.

☐**Double** applies the pattern a second time at 90 degrees to the first pattern; available when User-defined type is selected:

☐**Relative to Paper Space** scales the pattern using layout scale; available when hatch is applied in a layout viewport:

Spacing specifies the spacing between the lines of a user-defined hatch pattern.

ISO Pen Width scales patterns according to pen width; available for ISO patterns only.

Hatch Origin options

⊙ **Use Current Origin** uses x, y coordinates stored in the HpOrigin system variable.

○ **Specified Origin** specifies new hatch origins:

- **Click to Set New Origin** prompts "Specify origin point:" to locate new origin in drawing.
- ☑ **Default to Boundary Extents** places new origin at a corner or the center of rectangular *extents* of the hatch.
- ☐ **Store as Default Origin** stores the coordinates in the HpOrigin system variable.

Ver.Preview previews the hatch pattern.

Boundaries options

Add: Pick Points prompts, 'Pick internal point or [Select objects/remove Boundaries]:'. Pick a point inside an area; AutoCAD automatically detects the boundary to fill with the pattern.

Add: Select Objects prompts, 'Select objects or [picK internal point/remove Boundaries]:'. Pick one or more objects to fill with the hatch pattern.

Remove Boundaries prompts 'Select objects or [Add boundaries]:'. Pick one or more boundaries to be removed from the hatch pattern selection set.

Recreate Boundaries places polylines or regions around selected hatches.

View Selections views the pattern's selection set.

Options options

☐ **Annotative** toggles annotative scaling for hatch patterns.

☑ **Associative** creates hatches that update automatically when their boundaries or properties are modified; pattern is created as a hatch object.

☑ **Create separate hatches** makes multiple patterns independent; when off, multiple patters are created as a single hatch.

Draw Order:

- **Do not assign** places hatch normally.
- **Send to back** places hatch behind all other overlapping objects in the drawings.
- **Bring to front** places hatch in front of all other overlapping objects.
- **Send behind boundary** places hatch behind its boundary.
- **Bring in front of boundary** places hatch in front of its boundary.

Layer specifies whether to use the current or another layer.

Transparency specifies the level of translucency for the pattern *(new to AutoCAD 2011).*

Inherit Properties sets the hatch pattern parameters from an existing hatch pattern.

Gradient tab

See the Gradient command.

More Options

Islands options

☑ **Island detection** is turned on.

Island display style:

- ⊙ **Normal** turns hatching off and on each time it crosses a boundary; text is not hatched.
- ○ **Outer** hatches only the outermost areas; text is not hatched.
- ○ **Ignore** hatches everything within the boundary; text is hatched.

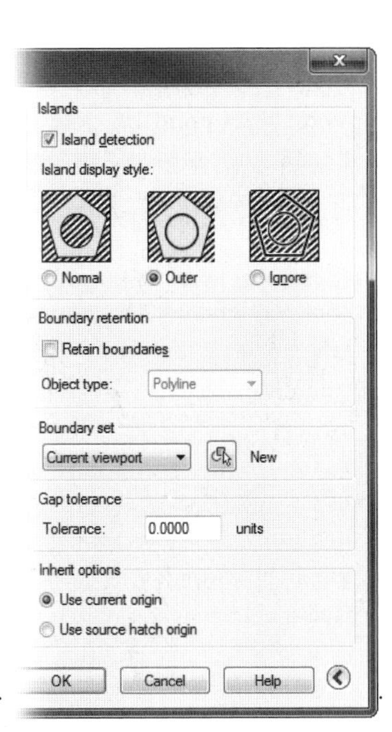

Boundary Retention options

Retain Boundaries:

☑ Keeps boundary polylines or regions (created during the boundary hatching process) are kept after the command finishes.

☐ Discards boundaries after hatching.

Object Type specifies the object from which to construct the boundary:

- **Polylines** form the boundary as a polyline.
- **Region objects** form the boundary as a 2D region.

Boundary Set options

Boundary Set defines how objects are analyzed for defining boundaries; not available when Select Objects is used to define the boundary.

 **New** creates new boundary sets; prompts you to select objects at the command line.

Gap Tolerance options

Tolerance specifies the maximum gap between boundary objects. (Objects don't need to touch to form a valid hatching boundary.) Range is 0 to 5000 units.

-HATCH Command

Command: -hatch
Current hatch pattern: ANSI31
Specify internal point or [Properties/Select objects/draW boundary/remove Boundaries/Advanced/ DRaw order/Origin/ANnotative/hatch COlor/LAyer/Transparency]: *(Pick a point, or specify an option.)*

COMMAND LINE OPTIONS

Specify internal point creates a boundary of the area surrounding the pick point.

Select objects selects one or more objects to fill with hatch pattern.

ANnotative does (or doesn't) apply the annotative property to the pattern.

hatch COlor specifies the foreground and background colors of the pattern *(new to AutoCAD 2011)*.

LAyer specifies the layer on which to place the pattern.

Transparency specifies the level of translucency for the pattern *(new to AutoCAD 2011)*.

Property options

Enter a pattern name or [?/Solid/User defined] <ANSI31>: *(Enter a name, or select an option.)*
Specify a scale for the pattern <1.0000>: *(Enter a scale factor.)*
Specify an angle for the pattern <0>: *(Enter an angle.)*

Enter a pattern name names the hatch pattern.

? lists the names of available hatch patterns.

Solid floods the area with a solid fill in the current color.

User defined creates a simple, user-defined hatch pattern.

Specify a scale specifies the hatch pattern angle (default = 0 degrees).

Specify an angle specifies the hatch pattern scale (default: = 1.0).

Draw Boundary options

Retain polyline boundary? [Yes/No] <N>: *(Type Y or N.)*
Specify start point: *(Pick point.)*

Specify next point or [Arc/Length/Undo]: *(Pick point, or enter an option.)*
Specify next point or [Arc/Close/Length/Undo]: *(Pick point or enter an option.)*
Specify start point for new boundary or <Accept>: *(Press Enter to end option.)*

Retain polyline boundary keeps the boundary created during the hatching process after -Hatch finishes.

Specify start point / next point prompts to draw a closed polyline.

Remove Boundaries options
Select objects or [Add boundaries]: *(Select one or more boundary objects.)*
Select objects or [Add boundaries/Undo]: *(Press Enter, or type U.)*

Select objects selects hatch boundaries to remove.

Undo adds the removed island.

Advanced options
Enter an option [Boundary set/Retain boundary/Island detection/Style/Associativity/Gap tolerance/separate Hatches]: *(Specify an option.)*

Boundary set defines the objects analyzed when a boundary is defined by a specified pick point.

Retain boundary keeps the boundary created by -Hatch.

Island detection uses objects within the outermost boundary for the boundary.

Style selects the pattern style: ignore, outer, or normal.

Associativity makes the pattern associative with its boundary.

Gap tolerance specifies the largest gap permitted in the boundary:

Specify a boundary gap tolerance value <0>: *(Enter a value between 0 and 5000.)*

Draw Order options
Enter draw order [do Not assign/send to Back/bring to Front/send beHind boundary/bring in front of bounDary] <send beHind boundary>: *(Specify an option.)*

do Not assign places hatches normally.

send to Back places hatches behind all other overlapping objects in the drawing.

bring to Front places hatches in front of all other overlapping objects.

send beHind boundary places hatches behind its boundary.

bring in front of bounDary places hatch in front of the boundary.

Origin options
[Use current origin/Set new origin/Default to boundary extents] <Use current origin>: *(Enter an option.)*

Use current origin uses the x, y coordinates stored in the HpOrigin system variable.

Set new origin prompts you to pick a point in the drawing as the new origin.

Default to boundary extents places the new origin at one of the four corners of the extents or at the center of the rectangular *extents* of the hatch.

RELATED COMMANDS

AdCenter places hatch patterns from other drawings.

Boundary traces polylines automatically around closed boundaries.

Convert converts Release 13 (and earlier) hatch patterns into Release 14-2011 format.

Gradient places gradient fills.

HatchEdit edits hatch patterns.

HatchToBack moves all hatches behind all other objects in the drawing in 2D wireframe mode *(new to AutoCAD 2011)*.

Properties changes the properties of hatch patterns.

ToolPalette stores and places selected hatch patterns and fill colors.

RELATED SYSTEM VARIABLES

DelObj toggles whether boundary is erased after hatch is placed.

FillMode determines whether hatch patterns are displayed.

HatchBoundSet (undocumented) reports the contents of the hatch boundary set.

HatchCreation (undocumented) reports whether the Hatch command is active.

HatchType (undocumented) specifies the type of hatch pattern.

HpAng specifies the current hatch pattern angle.

HpAnnotative determines whether new hatch patterns take on annotative scale factors.

HpAssoc determines whether or not hatches are associative.

HpBackgroundColor specifies the background fill color for hatch patterns.

HpBound specifies the hatch boundary object: polyline or region.

HpBoundRetain toggles whether boundary objects are created for new hatches.

HpColor specifies the default color for new hatch patterns.

HpDlgMode determines when dialog boxes are displayed by the Hatch, Gradient, and HatchEdit commands.

HpDouble specifies single or double hatching.

HpDrawOrder controls the display order of the hatch pattern relative to other overlapping objects.

HpGapTol reports the current gap tolerance.

HpInherit determines whether the MatchProp command copies hatch origins from the source objects.

HpIslandDetection determines how islands are hatched.

HpIslandDetectionMode toggles the type of island detection (controls HpIslandDetection).

HpLastPattern (undocumented) reports the name of the last hatch pattern used.

HpLayer specifies the name of the layer to use for new hatches and fills.

HpMaxlines specifies the maximum number of hatch lines to draw.

HpName names the current hatch pattern.

HpObjWarning specifies maximum number of hatch objects that can be selected before AutoCAD sounds the alert.

HpOrigin specifies the x, y coordinates for the origin of new hatches.

HpOriginMode determines the default hatch origin point.

HpOriginStoreAsDefault (undocumented) specifies the whether the user-defined hatch origin is stored as the default value.

HpQuickPreview toggles the display of hatch previews when the cursor is inside boundaries.

HpRelativePs (undocumented) toggles whether hatch pattern scaling is relative to the paper space scale factor.

HpScale specifies the current hatch pattern scale factor.

HpSeparate determines whether one hatch object or a separate one is created from several closed boundaries.

HpSpace specifies the current hatch pattern spacing factor.

HpTransparency sets the default translucency percentage for new hatches; has no effect on existing patterns.

OsnapHatch determines whether object snap snaps to hatch patterns.

PickStyle controls the selection of hatch patterns.

SnapBase specifies the starting coordinates of hatch pattern.

RELATED FILES

acad.pat contains the ANSI and other hatch pattern definitions.

acadiso.pat contains the ISO hatch pattern definitions.

TIPS

- This command first generates a boundary, and then hatches the inside area. Use the Boundary command to create only the boundary.
- As of AutoCAD 2011, the Hatch, HatchEdit, and Gradient commands display the Hatch Creation tab on the ribbon.

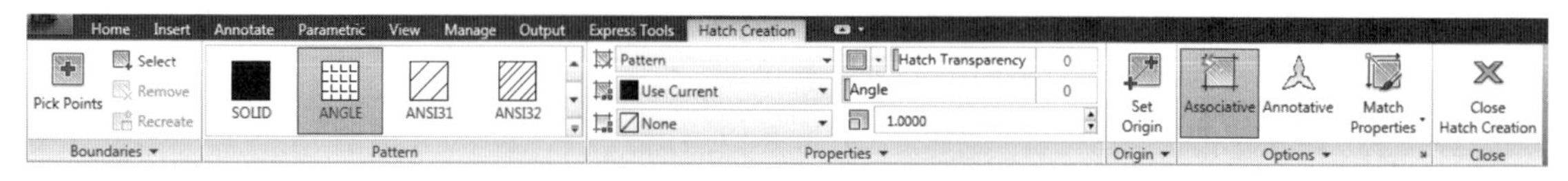

- Hatch stores hatching parameters in the pattern's extended object data.
- Bring the pattern to the front (through the Draw Order option) for easier editing of the hatch.
- ANSI patterns are all defined as 45° lines. To keep at 45°, the angle should be specified as 0.
- The Trim command trims hatches.
- When you pick a non-hatchable area, AutoCAD displays a dialog box with helpful tips:

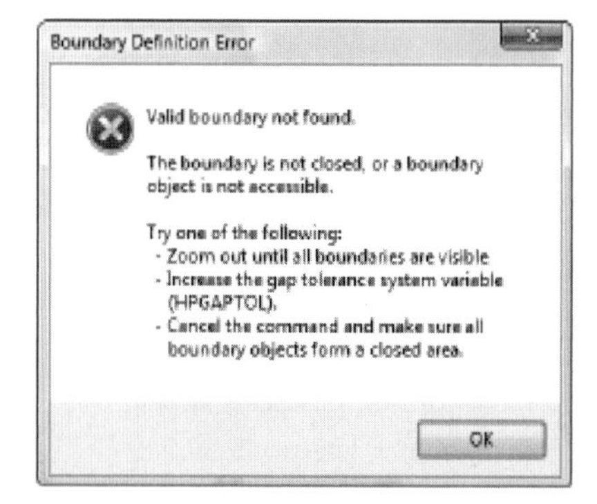

- Set the OsnapHatch system variable to 0 (default) to avoid object snapping to hatch and fill patterns.
- As of AutoCAD 2010, non-associative hatches can be edited with grips; as of 2011, associative hatches can also be edited in this manner.
- Gaps in the hatching boundary are highlighted.
- The following figure illustrates the extents of a hatch object:

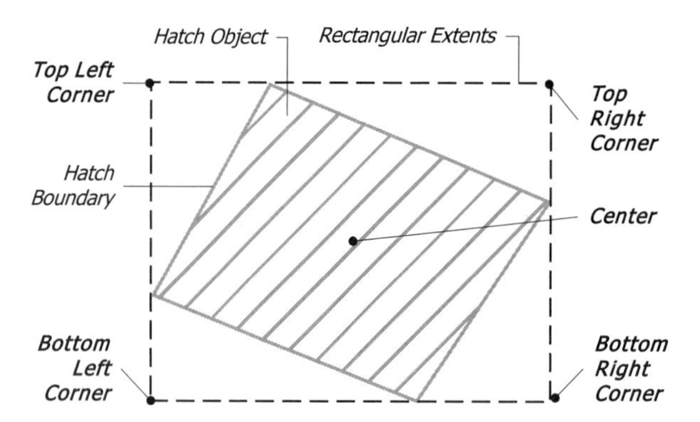

- Using constraints to resize a hatched object also resizes the hatch pattern

HatchEdit

Rel. 13 Edits associative hatch objects.

Commands	Alias	Keyboard Shortcuts	Menu	Ribbon
hatchedit	**he**	...	**Modify** ⇘**Object** ⇘**Hatch**	**Home** ⇘**Modify** ⇘**Edit Hatch**
-hatchedit	...	...	...	...

Command: hatchedit
Select hatch object: *(Select one hatch object.)*

Displays dialog box:

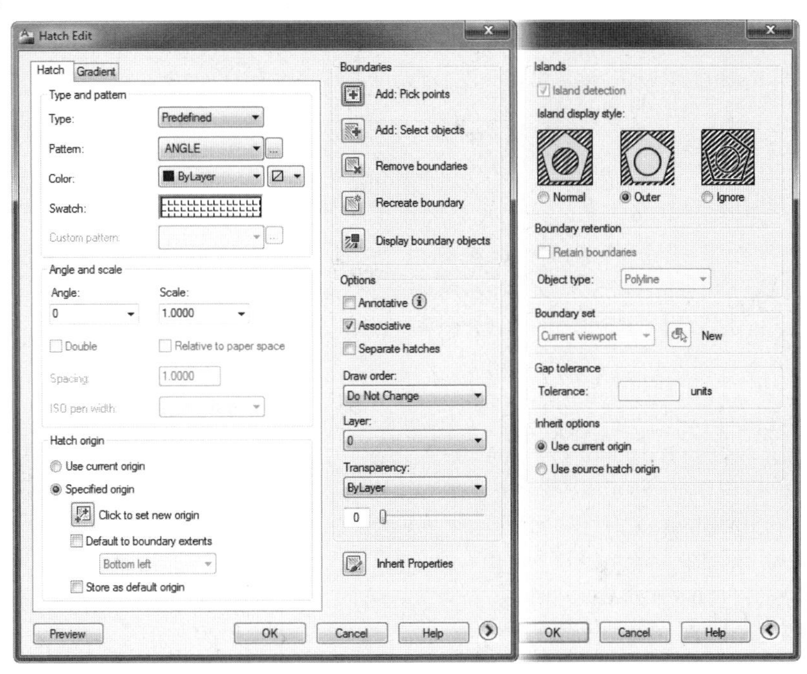

HATCH EDIT DIALOG BOX

See the Hatch command for options.

-HATCHEDIT Command

Command: -hatchedit
Select hatch object: *(Select a hatch or gradientc object.)*
Enter hatch option [DIsassociate/Style/Properties/DRaw order/ADd boundaries/Remove boundaries/recreate Boundary/ASsociate/separate Hatches/Origin/ANnotative/hatch COlor/LAyer/Transparency] <Properties>: *(Enter an option.)*

COMMAND LINE OPTIONS

See the Hatch command for options.

RELATED COMMANDS

Hatch applies associative hatch patterns.

Explode explodes a hatch pattern block into lines.

Gradient applies associative gradient fills to areas.

AddSelected creates new hatch patterns based on the properties of an existing one *(new to AutoCAD 2011)*.

MatchProp matches the properties of one hatch to other hatches.

Properties changes properties of hatch patterns.

TIPS

- HatchEdit works with associative and non-associative hatch objects.
- AutoCAD cannot change a non-associative hatch to associative; uncheck the Associative option to convert a non-associative hatch to associative. Also, you can use the Explode command to reduce hatches to lines.
- As an alternative to entering this command, double-click hatch patterns to display the Hatch Edit dialog box.
- To select a solid fill, click the outer edge of the hatch pattern, or use a crossing window selection on top of the solid fill.
- You can directly edit hatches using grips, as illustrated below *(new to AutoCAD 2011)*:

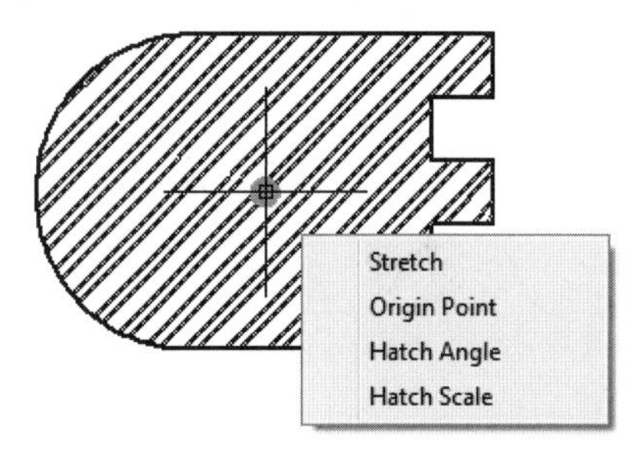

Click the blue dot grip for the context menu:

Stretch moves the hatch pattern out of its boundary.

Origin Point relocates the origin of the pattern, shifting it within its boundary. (There is no indicator for the origin.) this option uses the undocumented HatchSetOrigin command.

Hatch Angle rotates the pattern about its origin.

Hatch Scale changes the pattern's scale factor.

- Edit the hatch boundary by selecting it *(new to AutoCAD 2011)*:

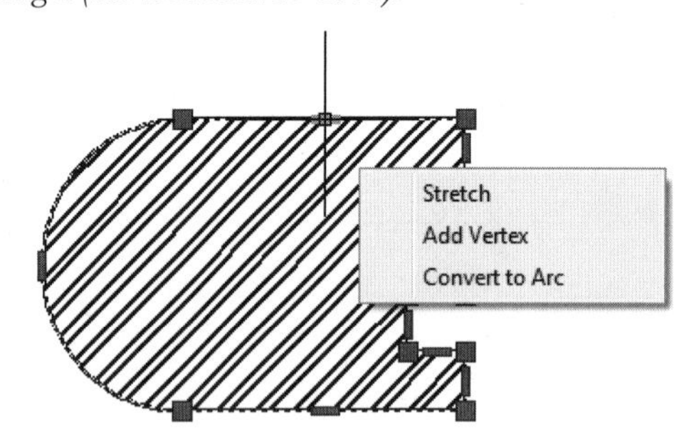

Click a grip on the boundary for the context menu:

Stretch stretches the boundary.

Add vertex adds a vertex at your next pick point.

Convert to arc (**line**) converts line segments to arcs, and vice versa.

Hatch...

2011 Utilities commands for hatch patterns and gradients.

Command	*Keyboard Shortcut*	*Menu*	*Ribbon*
hatchgenerateboundary	...	...	...
hatchsetboundary			
hatchsetorigin			
hatchtoback			

HATCHGENERATEBOUNDARY Command

Places non-associated polylines boundary around selected hatches and gradients.

Command: hatchgenerateboundary
Select hatch objects: *(Choose one or more hatches and/or gradients.)*
Select hatch objects: *(Press Enter.)*

COMMAND OPTION

Select hatch objects specifies the hatches and gradients about which to add the polyline boundary.

HATCHSETBOUNDARY Command

Trims or extends the areas of hatches and gradients to match different boundaries.

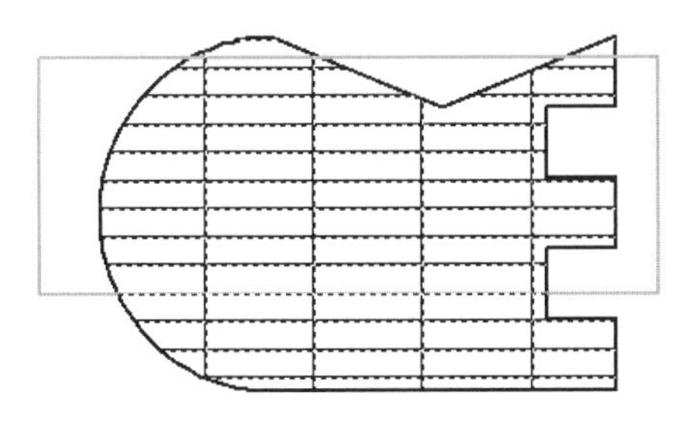

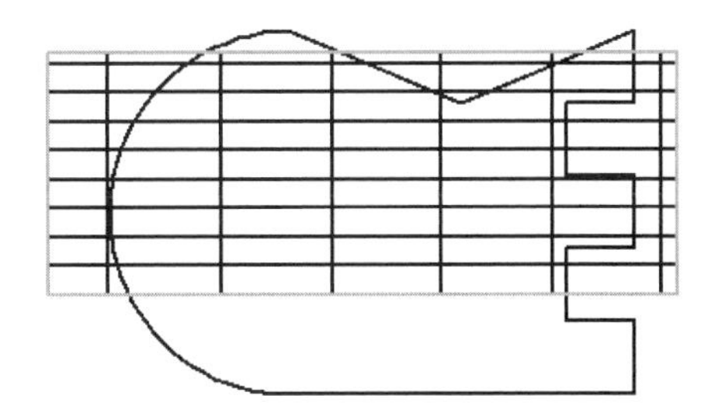

Left: *Hatch pattern in old boundary (black)...*
Right: *...and in new boundary (gray).*

Command: hatchsetboundary
Select hatch object: *(Choose one hatch and/or gradient.)*
Select objects to be used for the new boundary: *(Chose one or more objects.)*
Select objects to be used for the new boundary: *(Press Enter.)*
Erase selected linework? [Yes/No] <N>: *(Type Y or N.)*
Hatch boundary associativity removed.

COMMAND OPTIONS

Select hatch objects specifies a hatch or gradient whose boundary to modify.

Select objects to be used for the new boundary specifies the new boundary objects.

Erase selected linework determines whether the new boundary lines should be removed from the drawing.

HATCHSETORIGIN Command

Moves the origin of hatches and gradients.

Command: hatchsetorigin
Select hatch objects: *(Choose one or more hatches and/or gradients.)*
Select hatch objects: *(Press Enter.)*
Select new hatch origin: *(Pick a point or enter coordinates.)*

COMMAND OPTION

Select hatch objects specifies the hatches and gradients whose origin to move.

Select new hatch origin specifies the new origin point.

HATCHTOBACK Command

Changes the draw order of hatches and gradients to place them behind other objects in the drawing.

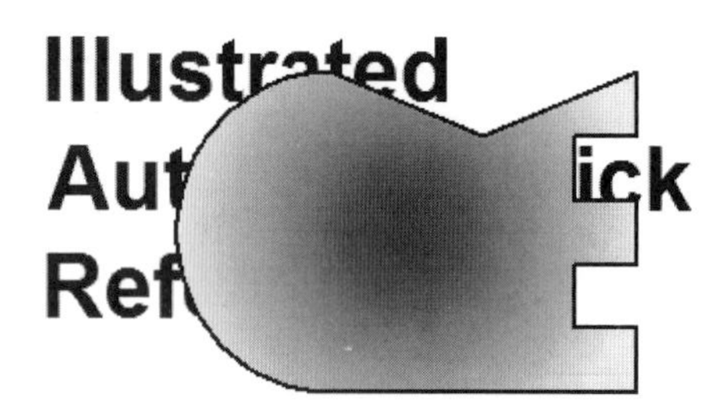

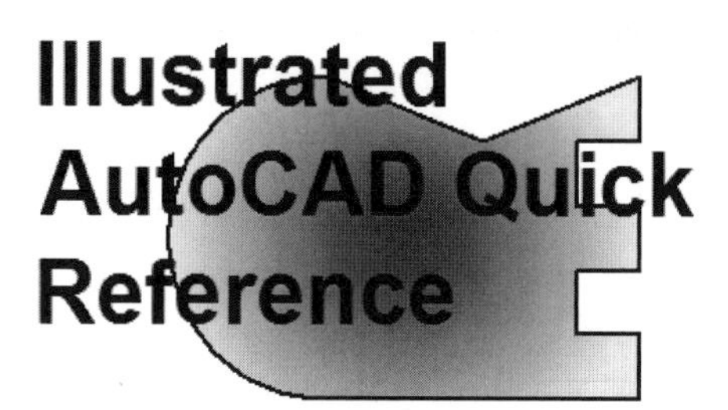

Left: *Gradient in front text.*
Right: *Gradient behind text.*

Command: hatchtoback
n **hatch object(s) sent to back.**

COMMAND OPTIONS

None.

RELATED COMMANDS

Hatch places hatch patterns and solid fills in drawings.

Gradient places gradient fills in drawings.

HatchEdit and **Properties** edit the properties of hatch patterns.

Boundary creates boundaries made of polylines or regions.

TextToFront changes the draw order of all text to place it in front of all other objects.

TIPS

- The boundary created by the HatchGenerateBoundary command is a non-associative polyline.
- The HatchSetBoundary command lets you expand and contract the extent of hatches and gradients when a floor plan or other view changes its components.
- The HatchSetOrigin command is useful for making hatches line up with other geometry, such as brick patterns in elevation views.
- The HatchToBack command also affects hatches and gradients on locked layers. This command does not work well in Realistic visual style and with wide polylines.

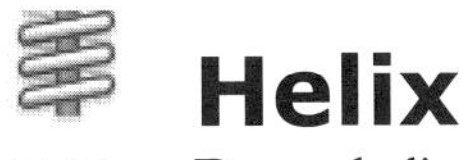

Helix

2007 Draws helixes.

Command	Aliases	Keyboard Shortcuts	Menu	Ribbon
helix	...	...	**Draw**	**Home**
			↳**Helix**	↳**Draw**
				↳**Helix**

Command: helix
Number of turns = 3.0000 Twist=CCW
Specify center point of base: *(Pick point 1.)*
Specify base radius or [Diameter] <1.0>: *(Enter radius 2, or type D.)*
Specify top radius or [Diameter] <1.5>: *(Enter radius 3, or type D.)*
Specify helix height or [Axis endpoint/Turns/turn Height/tWist] <1.0>: *(Enter height 4, or enter an option.)*

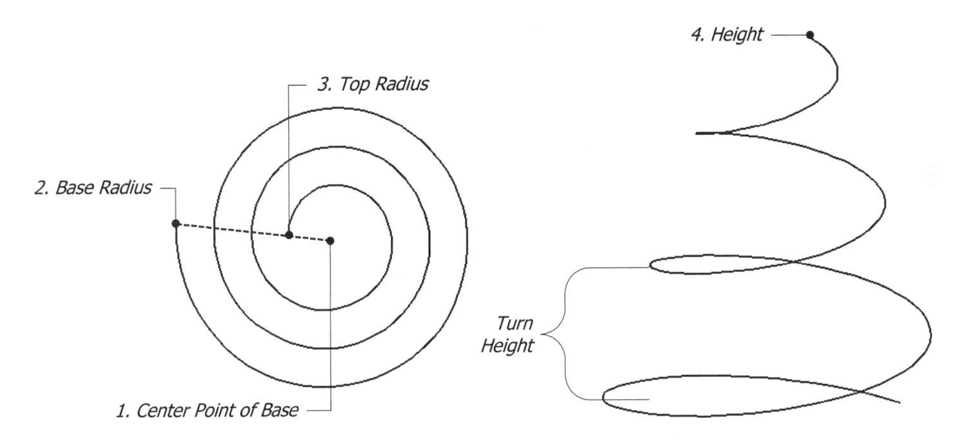

COMMAND OPTIONS

Center point of base specifies the coordinates of the helix's base; when z is not specified, the helix is drawn at the current elevation.

Base radius specifies the radius of the base.

Diameter specifies the diameter of the base.

Top radius specifies the radius of the top; value must be positive.

Helix height specifies the height of the helix; if negative, the helix is drawn downwards from its base.

Axis endpoint specifies a second point (from the base center point) to draw the helix at an angle.

Turns specifies the number of turns the helix makes; default = 3.

turn Height specifies the height of one turn.

tWist switches between drawing the helix clockwise and counterclockwise (default).

RELATED COMMAND

Reverse reverses the order of vertices in the helix.

TIP

- To draw a cylindrical spring, make the top and bottom radii the same.

Help

Ver. 1.0 Lists information for using AutoCAD's commands.

Command	Alias	Keyboard Shortcut	Menu	Ribbon
'help	**'?**	**F1**	**Help** ↳**Help**	**Help**

Command: help

Opens help in the computer's default browser or IE:

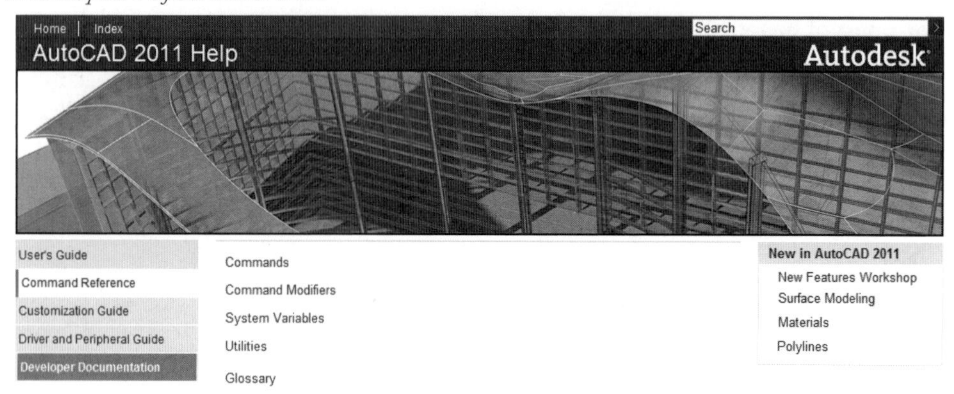

RELATED COMMANDS

All.

RELATED SYSTEM VARIABLE

HelpPrefix specifies the path to the help file; default = *C:\Program Files\Autodesk\Autocad 2011 \Help\Index.Html (new to AutoCAD 2011).*

RELATED FILE

index.html is AutoCAD 2011's primary help file.

TIPS

- Because 'Help, '?, and F1 are transparent commands, you can use them during another command to get help with the command's options.
- AutoCAD also provides help in the Infocenter on the title bar:

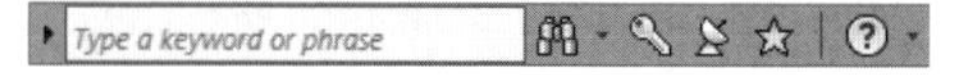

- As of AutoCAD 2011, Help is written in HTML and runs in your computer's browser.
- The Option command's System tab has several help options, including using online help from Autodesk's Web site when available.

Removed Commands

Hide command now operates only in 2D wireframe mode; as of AutoCAD 2007, it is an alias for VsCurrent in all other modes, such as 3D wireframe and Realistic.

HideObjects

2011 Turns off the display of selected objects.

Command	*Alias*	*Status Bar*	*Menu*	*Ribbon*
hideobjects	...		...	...

Command: hideobjects
Select objects: *(Choose one or more objects.)*
Select objects: *(Press Enter.)*

COMMAND OPTION

Select objects chooses the objects to be hidden from view.

RELATED COMMANDS

UnisolateObjects reveals all hidden objects *(new to AutoCAD 2011).*

IsolateObjects hides all objects, except those selected *(new to AutoCAD 2011).*

LayOn and **LayOff** turn on and off the layer associated with selected objects.

TIPS

- This command is handy for temporarily hiding objects until they are redisplayed with the UnisolateObjects command.
- Hiding selected objects can make it easier to see other objects that are nearby or underneath.
- This command can be accessed through the status bar: click the light bulb icon to display the shortcut menu:

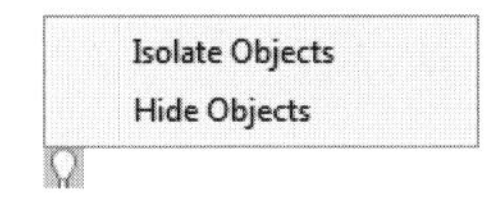

Light Bulb Color	*Meaning*
Yellow	No objects are hidden.
Red	At least one object is hidden or isolated.

- Use the LayOff command to turn off the display of all objects associated with the same layer.

HidePalettes

2009 Turns off all open palettes, including the command bar.

Command	*Alias*	*Keyboard Shortcut*	*Menu*	*Ribbon*
hidepalettes	**poff**	**Ctrl+Shift+H**	...	...

Command: hidepalettes

Closes all palettes; use ShowPalettes to restore them.

RELATED COMMANDS

ShowPalettes restores palettes that were closed.

CleanScreenOn maximizes the drawing area by reducing the number of user interface elements.

TIPS

- This command is handy for temporarily hiding all palettes, until they are reopened with the ShowPalettes command.
- Alternatively, press Ctrl+Shift+H to toggle the display of palettes.
- As an alternative to this command, consider turning on the Auto-Hide feature for palettes.

Replaced Command

HLSettings command was removed from AutoCAD 2007; it was replaced by the VisualStyles command.

HpConfig command was removed from AutoCAD 2000; it was replaced by the Plot command.

Hyperlink

2000 Attaches hyperlinks (URLs) to objects in drawings.

Commands	Aliases	Keyboard Shortcut	Menu	Ribbon
hyperlink	...	**Ctrl+K**	**Insert ↳Hyperlink**	**Block & References ↳Data ↳Hyperlink**
-hyperlink	...	...	...	...

Command: hyperlink
Select objects: *(Select one or more objects.)*
Select objects: *(Press Enter to end object selection.)*

Displays dialog box.

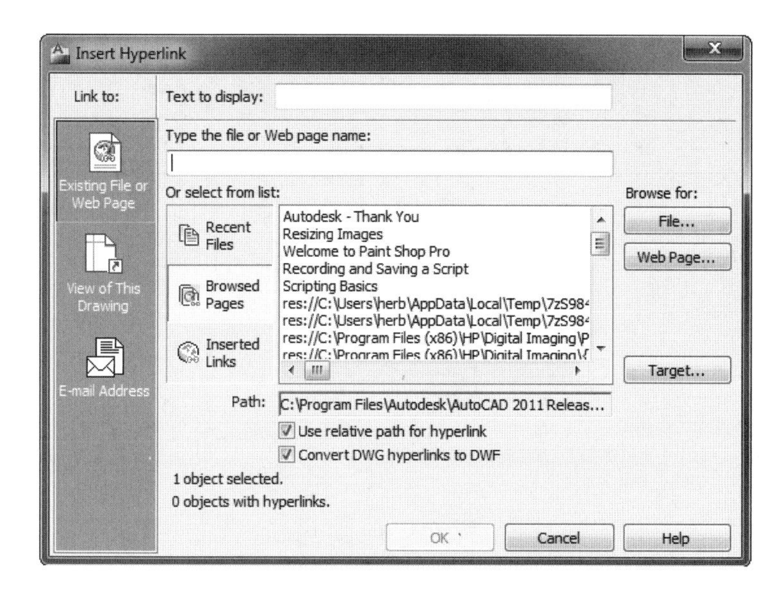

COMMAND LINE OPTION

Select object selects the objects to which to attach the hyperlink.

INSERT HYPERLINK DIALOG BOX

Existing File or Web Page page

Text to display describes the hyperlink; the description is displayed by the tooltip; when blank, the tooltip displays the URL.

Type the file or Web page name specifies the hyperlink (URL) to associate with the selected objects; the hyperlink may be any file on your computer, on any computer you can access on your local network, or on the Internet.

Or select from list:

- **Recent files** lists drawing files recently opened in AutoCAD.
- **Browsed pages** lists pages recently viewed with Web browsers and other software.
- **Inserted links** lists URLs recently entered in Web browsers.

File opens the Browse the Internet - Select Hyperlink dialog box.

Web Page starts up a simple Web browser.

Target specifies a location in the file, such as a target in an HTML file, a named view in AutoCAD, or a page in a spreadsheet document. Displays Select Place in Document dialog box.

Path displays full path and filename to the hyperlink; file name appears when Use Relative Path for Hyperlink is checked.

☑ **Use relative path for hyperlink** toggles use of the path for relative hyperlinks in the drawing; when "" (null), the drawing paths stored in AcadPrefix are used.

☑ **Convert DWG hyperlinks to DWF** preserves hyperlinks when drawings are exported in DWF format.

Remove link removes the hyperlink from the object; this button appears only if you select an object that already has a hyperlink.

View of This Drawing page

Text to Display specifies alternative wording of the hyperlink tooltip.

Select a View of This selects a layout, named view, or named plot.

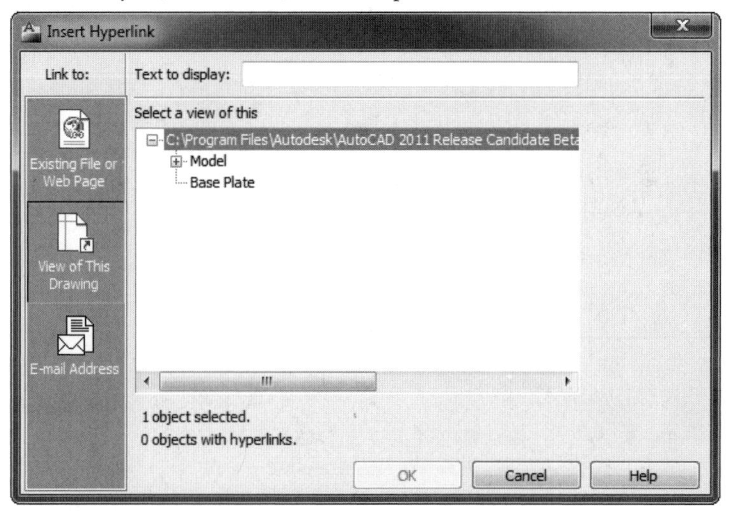

Email Address page

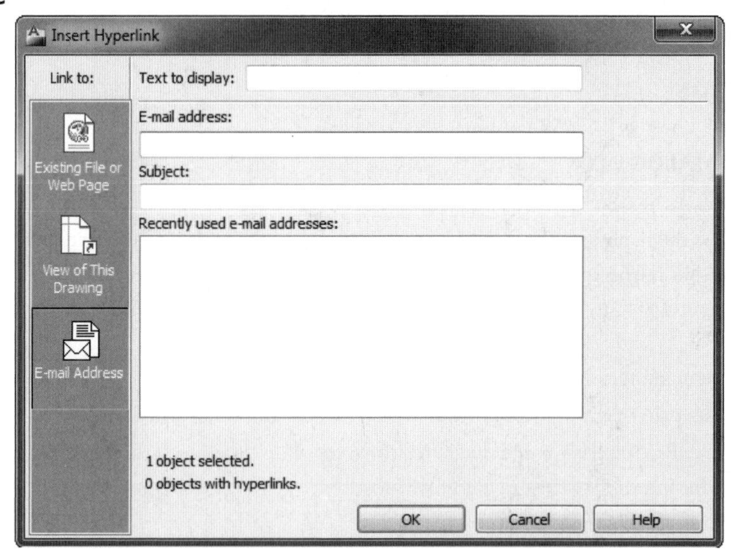

Text to Display specifies alternative wording of the hyperlink tooltip.

Email address specifies the email address; the *mailto:* prefix is added automatically.

Subject specifies the text that will be added to the Subject line.

Recently used e-mail addresses lists the email addresses recently entered.

-HYPERLINK Command

Command: -hyperlink
Enter an option [Remove/Insert] <Insert>: *(Enter an option.)*
Enter hyperlink insert option [Area/Object] <Object>: *(Enter an option.)*
Select objects: *(Select one or more objects.)*
Select objects: *(Press Enter to end object selection.)*
Enter hyperlink <current drawing>: *(Enter hyperlink address.)*
Enter named location <none>: *(Optional: Enter a bookmark.)*
Enter description <none>: *(Optional: Enter a description.)*

COMMAND LINE OPTIONS

Remove removes a hyperlink from selected objects or areas.

Insert adds a hyperlink to selected objects or areas.

Select objects selects the object to which the hyperlink will be added.

Enter hyperlink specifies the filename or hyperlink address.

Enter named location (*optional*) specifies a location within the file or hyperlink.

Enter description (*optional*) describes the hyperlink.

RELATED COMMANDS

HyperlinkOptions toggles the display of the hyperlink cursor, shortcut menu, and tooltip.

HyperlinkOpen opens hyperlinks (URLs) via the command line.

HyperlinkBack returns to the previous URL.

HyperlinkFwd moves forward to the next URL; works only when the HyperlinkBack command was used; otherwise AutoCAD complains, "** No hyperlink to navigate to **."

HyperlinkStop stops the display of the current hyperlink.

GoToUrl displays specific Web pages.

PasteAsHyperlink pastes hyperlinks to selected objects.

RELATED SYSTEM VARIABLE

HyperlinkBase specifies the path for relative hyperlinks in the drawing.

TIPS

- If the drawing has never been saved, AutoCAD is unable to determine the default *relative folder*. For this reason, the Hyperlink command prompts you to save the drawing.
- By using hyperlinks, you can create a *project document* consisting of drawings, contacts (word processing documents), project time lines, cost estimates (spreadsheet pages), and architectural renderings. To do so, create a "title page" of an AutoCAD drawing with hyperlinks to the other documents.
- To edit a hyperlink with Hyperlink, select object(s), make editing changes, and click OK.
- To remove a hyperlink with Hyperlink, select the object, and then click Remove Link.
- An object can have just one hyperlink attached to it; more than one object, however, can share the same hyperlink.
- Hyperlinks are also known as *URLs*, short for "uniform resource locator," a universal file naming system.

HyperlinkOpen / Back / Fwd / Stop

2000 Controls the display of hyperlinked pages (undocumented commands).

Commands	Aliases	Keyboard Shortcuts	Menu	Ribbon
'hyperlinkopen	...	...	...	...
'hyperlinkback	...	...	...	...
'hyperlinkfwd	...	...	...	...
'hyperlinkstop	...	...	...	...

Command: hyperlinkopen
Enter hyperlink <current drawing>: *(Enter a hyperlink option.)*
Enter named location <none>: *(Optional: Enter a bookmark.)*

Displays the specified Web page or file, if possible.

Command: hyperlinkback

Returns to the previous hyperlinked page.

Command: hyperlinkstop

Stops displaying the Web page.

Command: hyperlinkfwd

Hyperlinks to the next page; can be used only after the HyperLinkBack command.

COMMAND LINE OPTIONS

Enter hyperlink enters a URL (uniform resource locator) or a filename.

Enter named location enters a named view or other valid target.

RELATED COMMANDS

Hyperlink attaches a hyperlink to objects.

HyperlinkOptions specifies the options for hyperlinks.

GoToUrl displays a specific Web page.

RELATED TOOLBAR ICONS

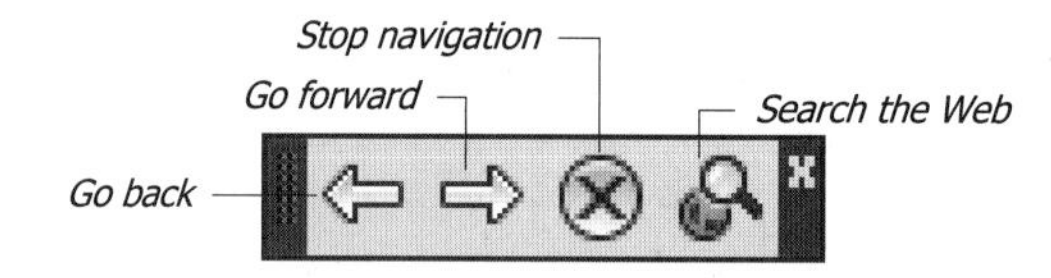

Go Back goes to the previous hyperlink; executes the HyperlinkBack command.

Go Forward goes to the next hyperlink; executes the HyperlinkFwd command.

Stop Navigation stops loading the current hyperlink file; executes the HyperlinkStop command.

Browse the Web displays the Web browser; executes the Browser command.

HyperlinkOptions

2000 Toggles the display of the hyperlink cursor, shortcut menu, and tooltip.

Command	*Aliases*	*Keyboard Shortcuts*	*Menu*	*Ribbon*
'hyperlinkoptions	...	...	...	...

Command: hyperlinkoptions
Display hyperlink cursor tooltip and shortcut menu? [Yes/No] <Yes>: *(Type Y or N.)*

COMMAND LINE OPTIONS

Display hyperlink cursor and shortcut menu toggles the display of the hyperlink cursor and shortcut menu.

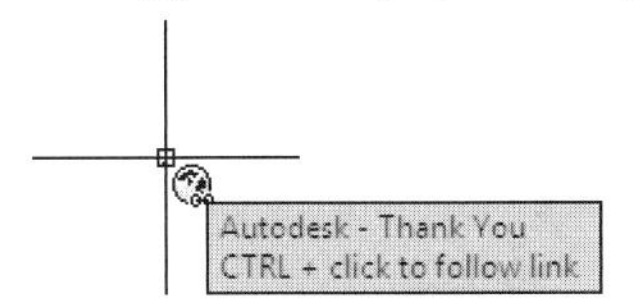

Display hyperlink tooltip toggles the display of the hyperlink tooltip.

SHORTCUT MENU

Select an object containing a hyperlink; then right-click to display the cursor menu.

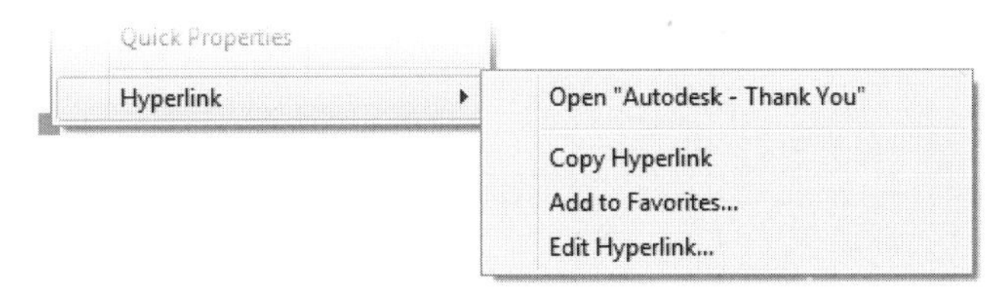

Hyperlink options

Open "url" launches the appropriate applications, and loads the file referenced by the URL.

Copy Hyperlink copies hyperlink data to the Clipboard; use the PasteAsHyperlink command to paste the hyperlink to selected objects.

Add to Favorites adds the hyperlink to a favorites list.

Edit Hyperlink displays the Edit Hyperlink dialog box; see the Hyperlink command.

RELATED COMMANDS

Hyperlink attaches a hyperlink to objects.

Options determines options for most other aspects of AutoCAD.

RELATED SYSTEM VARIABLE

HyperlinkBase specifies the path for relative hyperlinks in the drawing; when "" (null), the drawing paths stored in AcadPrefix are used.

TIP

- Answering "N" to this command makes hyperlinks unavailable.

Id

Ver. 1.0 Identifies the 3D coordinates of specified points (short for IDentify).

Command	*Aliases*	*Keyboard Shortcuts*	*Menu*	*Ribbon*
'id	...	...	**Tools** **↳Inquiry** **↳Id Point**	**Home** **↳Utilities** **↳Id Point**

Command: id
Specify point: *(Pick a point.)*

Sample output:
X = 1278.0018 Y = 1541.5993 Z = 0.0000

COMMAND LINE OPTION

Specify point picks a point.

RELATED COMMANDS

List lists information about picked objects.

Point draws points.

RELATED SYSTEM VARIABLE

LastPoint contains the 3D coordinates of the last picked point.

TIPS

- The Id command stores the picked point in the LastPoint system variable. Access that value by entering @ at the next prompt for a point value.
- Invoke the Id command to set the value of the LastPoint system variable, which can be used as relative coordinates in another command.
- If a 2D point is specified, the z coordinate displayed by Id is the current elevation setting; otherwise, the z-coordinate is that of the specified point.
- When you use Id with an object snap, the z coordinate is the object-snapped value.
- It is faster to enter "id" at the command line than to access it through menu or ribbons picks.

Replaced Command

Image was replaced by the ClassicImage and ExternalReferences commands.

-Image

Rel. 14 Controls the attachment of raster images at the command line.

Command	Alias	Keyboard Shortcuts	Menu	Ribbon
-image	**-im**	...	...	...

Command: -image

Enter image option [?/Detach/Path/Reload/Unload/Attach] <Attach>: *(Enter an option.)*

COMMAND LINE OPTIONS

? lists currently-attached image files.

Detach erases images from drawings.

Path lists the names of images in the drawing.

Reload reloads image files into drawings.

Unload removes images from memory without erasing them.

Attach displays the Attach Image File dialog box; see the ImageAttach command.

RELATED COMMANDS

ExternalReferences replaces the dialog box version of the Image command.

ImageAdjust and **Adjust** control the brightness, contrast, and fading of images.

ImageAttach and **Attach** attach images to the current drawing.

ImageClip and **Clip** create clipping boundaries around images.

ImageFrame and **Frame** toggle the display of image frames.

ImageQuality toggles between draft and high-quality mode.

RELATED SYSTEM VARIABLE

ImageHlt toggles whether the entire image is highlighted.

TIPS

- The -Image command loads files with the following extensions: *.bil*, *.bmp*, *.cal*, *.cg4*, *.dib*, *.flc*, *.fli*, *.flx*, *.gif*, *.gp4*, *.igs*, *.jpg*, *.mil*, *.pct*, *.pcx*, *.png*, *.rlc*, *.rle*, *.rst*, *.tga*, and *.tif*. Note that *.igs* is not IGES, but a raster format used for mapping. Autodesk's RasterDesign software permits viewing of additional raster formats, such as *.sid* and *.ecw*. This command handles raster images of these color depths:
  - **Bitonal** — black and white (monochrome).
  - **8-bit gray** — 256 shades of gray.
  - **8-bit color** — 256 colors.
  - **24-bit color** — 16.7 million colors.
- AutoCAD can display one or more images in any viewport. There is no theoretical limit to the number and size of images.
- Images do not display in shaded visual modes.
- The DrawOrder command can place images "behind" other objects.
- Images are always xrefs; the source image file must be present when the drawing is opened.
- The Image command was replaced by the ClassicImage and ExternalReferences commands in AutoCAD 2007.
- Images are always attached with a 1:1 aspect ratio. To change this, turn the image into a block, and then insert the block with unequal x and y scale factors. IntelliCAD and other CAD programs allow unequal aspect ratios; their drawings opened in AutoCAD honor the ratio.

ImageAdjust

Rel. 14 Controls brightness, contrast, and fade of attached raster images through a dialog box.

Commands	Alias	Keyboard Shortcuts	Menu	Ribbon
imageadjust	**iad**	...	**Modify** ↳**Object** ↳**Image** ↳**Adjust**	...
-imageadjust	**adjust**	...	...	...

Command: imageadjust
Select image(s): *(Select one or more image objects.)*
Select image(s): *(Press Enter to end object selection.)*

Displays dialog box.

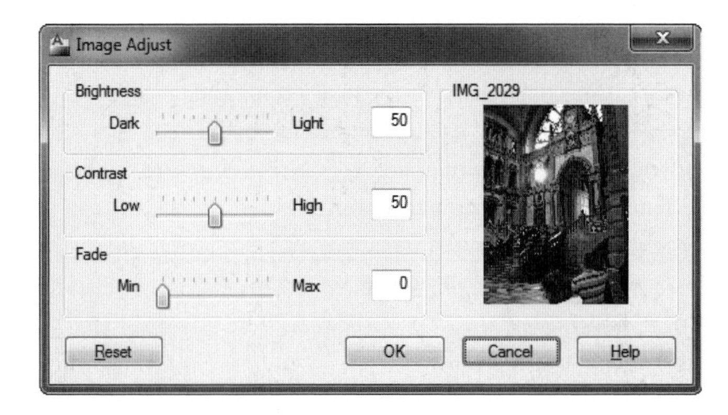

IMAGE ADJUST DIALOG BOX

Brightness options

Dark reduces the brightness of the image as values approach 0.

Light increases the brightness of the image as values approach 100.

Contrast options

Low reduces the image contrast as values approach 0.

High increases the image contrast as values approach 100.

Fade options

Min reduces the image fade as values approach 0.

Max increases the image fade as values approach 100.

Reset resets the image to its original parameters; default values are:
- **Brightness** — 50.
- **Contrast** — 50.
- **Fade** — 0.

-IMAGEADJUST Command

Command: -imageadjust

Select image(s): *(Select one or more images.)*

Select image(s): *(Press Enter to end object selection.)*

Enter image option [Contrast/Fade/Brightness] <Brightness>: *(Enter an option.)*

COMMAND LINE OPTIONS

Contrast option

Enter contrast value (0-100) <50>: *(Enter a value.)*

Enter contract value adjusts the contrast between 0% contrast and 100%; default = 50.

Fade option

Enter fade value (0-100) <0>: *(Enter a value.)*

Enter fade option adjusts the fading between 0% faded and 100%; default = 0.

Brightness option

Enter brightness value (0-100) <50>: *(Enter a value.)*

Enter brightness value adjusts the brightness between 0% brightness and 100%; default = 50.

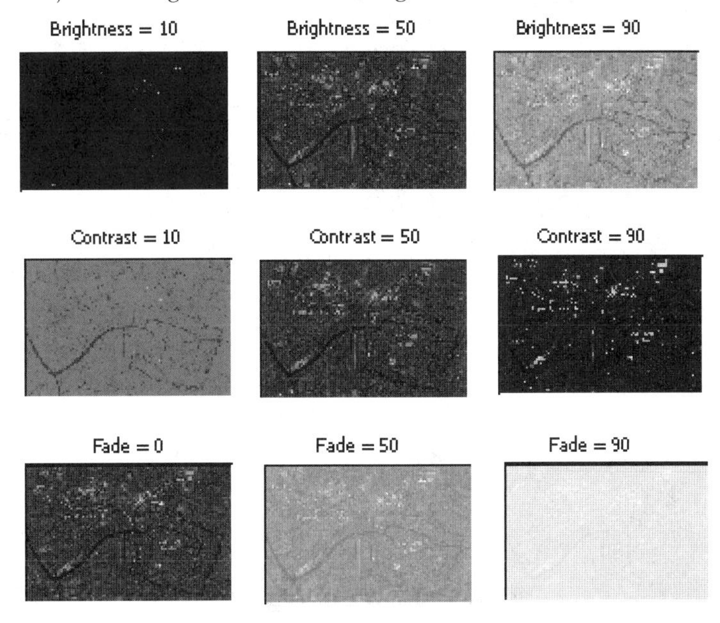

RELATED COMMANDS

Adjust adjusts brightness, contrast, and fading of images, xrefs, DWF/x, and PDF files, and xrefs through the command line.

ExternalReferences loads raster image files in drawings.

ImageAttach and **Attach** attach images to the current drawing.

ImageClip and **Clip** create clipping boundaries on images.

ImageFrame and **Frame** toggle the display of image frames.

ImageQuality toggles display between draft and high-quality mode.

ImageAttach

Rel. 14 Selects raster files to attach to drawings.

Command	Alias	Keyboard Shortcuts	Menu	Ribbon
imageattach	**iat**	...	**Insert** ↳**Raster Image**	**Image** ↳**Reference** ↳**Attach**

Command: imageattach

Displays file dialog box. Select an image file, and then click Open.

Displays dialog box.

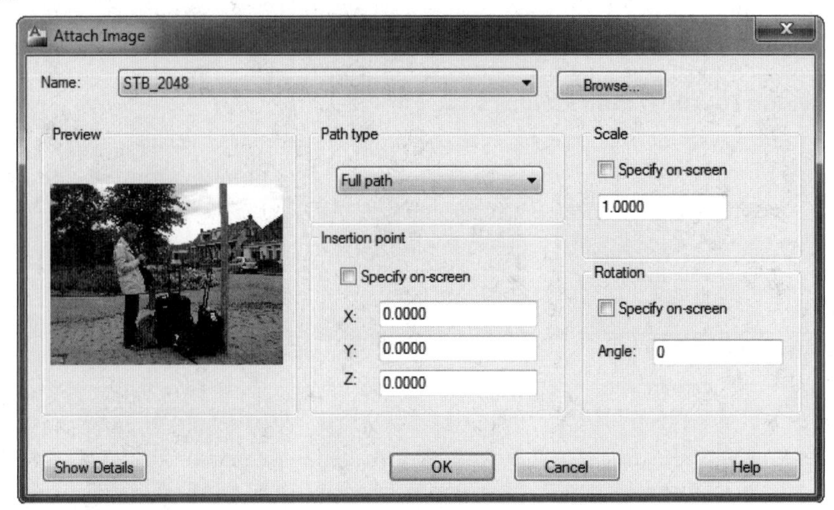

ATTACH IMAGE DIALOG BOX

Name selects names from a list of previously-attached images.

Browse selects files; displays Select Image File dialog box.

Path Type options

Full Path saves the path to the image file.

Relative Path saves the path relative to the current drawing file.

No Path does not save the path.

Insertion Point options

☑**Specify On-Screen** specifies the insertion point of the image in the drawing, after the dialog box is dismissed.

X, Y, Z specify the x, y, and z coordinates of the lower-left corner of the image.

Scale options

☑**Specify on-screen** specifies the scale of the image (relative to the lower-left corner) in the drawing after the dialog box is dismissed.

Scale specifies the scale of the image; a positive value enlarges the image, while a negative value reduces it.

Rotation options

☐ **Specify on-screen** specifies the rotation angle of the image about the lower-left corner in the drawing, after the dialog box is dismissed.

Angle specifies the angle to rotate the image; positive angles rotate the image counterclockwise.

Details expands the dialog box to display information about the image:

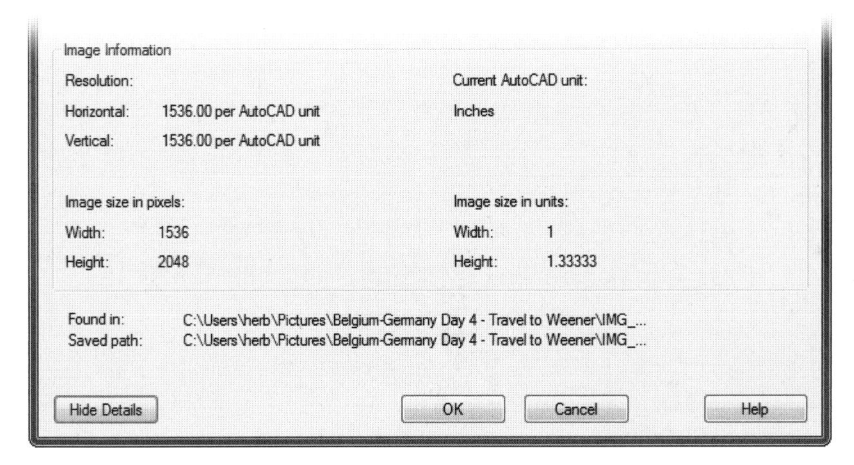

RELATED COMMAND

ExternalReferences controls the loading of raster image files in drawings.

Attach attaches images, xrefs, PDF files, and DWF/x files.

RELATED SYSTEM VARIABLE

InsUnits specifies the drawing units for the inserted image.

TIPS

- For a command-line version of the ImageAttach command, use the -Image command's Attach option.
- This dialog box no longer selects units from the Current AutoCAD Unit list box; as of AutoCAD 2000, use the InsUnits system variable.
- You can specify different x and y scales if you cheat using these instructions provided by the technical editor:
    1. Insert the image.
    2. Turn the image into a block.
    3. Insert the block using different x and y scale factors.

ImageClip

Rel. 14 Clips raster images.

Command	Aliases	Keyboard Shortcuts	Menu	Ribbon
imageclip	**icl**	**...**	**Modify** **↳Clip** **↳Image**	**Insert** **↳Reference** **↳Clip**

Command: imageclip
Select image to clip: *(Select one image object.)*
Enter image clipping option [ON/OFF/Delete/New boundary] <New>: *(Enter an option.)*

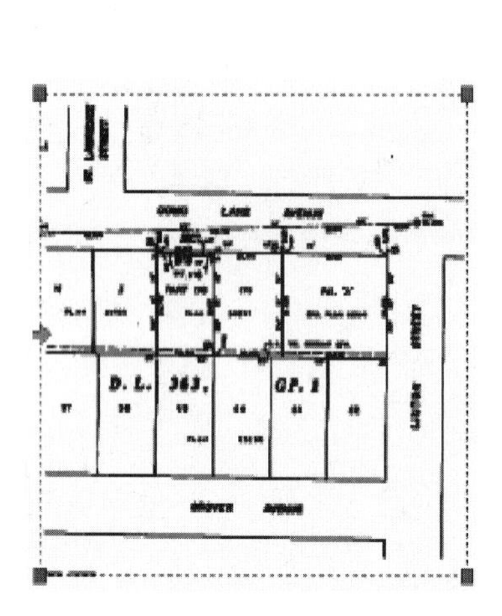

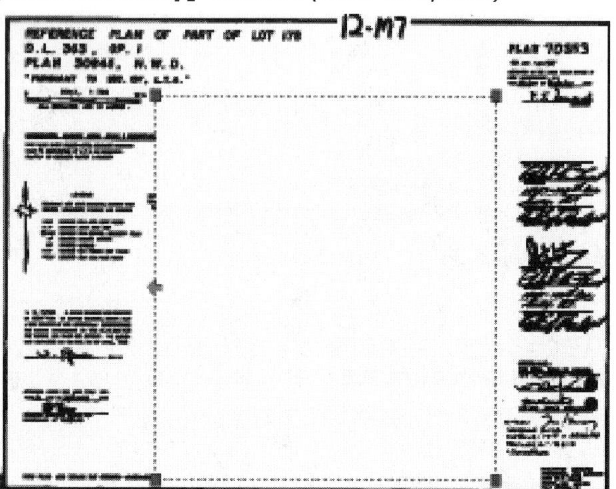

Left: A scanned image clipped with a rectangular boundary.
Right: The clipping boundary inverted.

COMMAND LINE OPTIONS

New boundary creates a new clipping path from a polyline, or as a polygonal or rectangular shape.

Select image selects one image to clip.

ON turns on the previous clipping boundary.

OFF turns off the clipping boundary.

Delete erases the clipping boundary.

New Boundary options
Outside mode - Objects outside boundary will be hidden.
Specify clipping boundary or select invert option:
Enter clipping type [Select polyline/Polygonal/Rectangular/Invert clip] <Rectangular>: *(Enter an option.)*

Select polyline creates a clipping path from an existing polyline.

Polygonal creates a polygonal clipping path.

Rectangular creates a rectangular clipping boundary.

Invert clip inverts the clipped region.

Select Polyline options

Select polyline: *(Select a polyline.)*

Select polyline chooses a single polyline made of straight segments: no arcs, splines, or crossovers. The polyline can be open or closed.

Polygonal options

Specify first point: *(Pick a point.)*
Specify next point or [Undo]: *(Pick a point, or type U.)*
Specify next point or [Undo]: *(Pick a point, or type U.)*
Specify next point or [Close/Undo]: *(Pick a point, or enter an option.)*
Specify next point or [Close/Undo]: *(Enter C to close the polygon.)*

Specify first point specifies the start of the first segment of the polygonal clipping path.

Specify next point specifies the next vertex.

Undo undoes the last vertex.

Close closes the polygon clipping path.

Rectangular options

Specify first corner point: *(Pick a point.)*
Specify opposite corner point: *(Pick a point.)*

Specify first corner point specifies one corner of the rectangular clip.

Specify opposite corner point specifies the second corner.

When you select an image with a clipped boundary, AutoCAD prompts:

Delete old boundary? [No/Yes] <Yes>: *(Type N or Y.)*

Delete old boundary?

Yes removes the previously-applied clipping path.

No exits the command.

Invert Clip options

When you enter I for the Invert Clip option, AutoCAD prompts you:

Inside mode - Objects inside boundary will be hidden.

AutoCAD then repeats the 'Enter clipping type [Select polyline/Polygonal/Rectangular/Invert clip]' prompt.

RELATED COMMANDS

Attach attaches images and other files to the current drawing.

Clip clips attached files.

ExternalReferences controls the loading of raster image files in the drawing.

ImageAdjust controls the brightness, contrast, and fading of the image.

ImageFrame toggles the display of the image's frame.

ImageQuality toggles the display between draft and high-quality mode.

TIPS

- You can use object snap modes on the image's frame, but not on the image itself.
- To clip a hole in the image, use the Invert Clip option. To invert a clipping boundary, click the arrow-shaped grip.
- You can edit the clipping boundary by editing its grips.

Changed Commands

ImageFrame command is now a system variable.

ImageQuality

Rel. 14 Toggles the quality of raster images.

Command	Aliases	Keyboard Shortcuts	Menu	Ribbon
imagequality	...	...	**Modify** ↳**Object** ↳**Image** ↳**Quality**	...

Command: imagequality

Enter image quality setting [High/Draft] <High>: *(Type H or D.)*

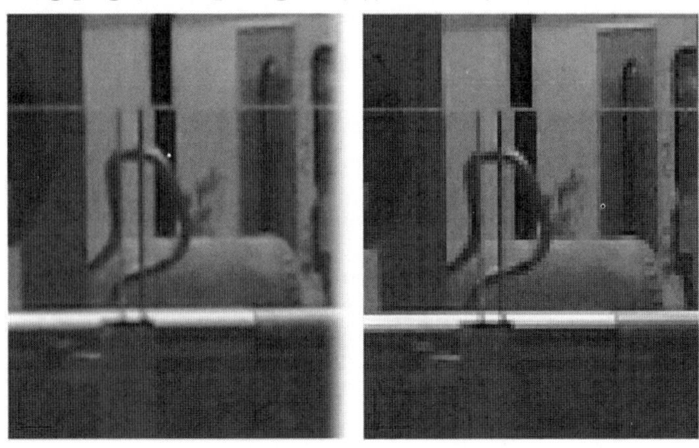

Left: *High quality looking smoother.*
Right: *Draft quality looking sharper.*

COMMAND LINE OPTIONS

High displays images at a higher quality.

Draft displays images at a lower quality.

RELATED COMMANDS

ExternalReferences controls the loading of raster image files in the drawing.

ImageAdjust controls the brightness, contrast, and fading of the image.

ImageAttach attaches an image in the current drawing.

ImageClip creates a clipping boundary on the image.

ImageFrame toggles the display of the image's frame.

TIPS

- High quality displays the image more slowly; draft quality displays the image more quickly.
- I find that *draft* quality looks better (crisper) than high quality (blurred), but the technical editor finds the quality varies with the type of image, depending on whether it is made of lines or shades, or is inserted by scale rather than by image file resolution.
- This command affects the display only; AutoCAD always plots images in high quality.

Import

Rel. 13 Imports vector files into drawings.

Command	Aliases	Keyboard Shortcuts	Menu	Ribbon
import	**imp**	...	...	**Blocks & References** **↳Import** **↳Import**

Command: import

Displays dialog box.

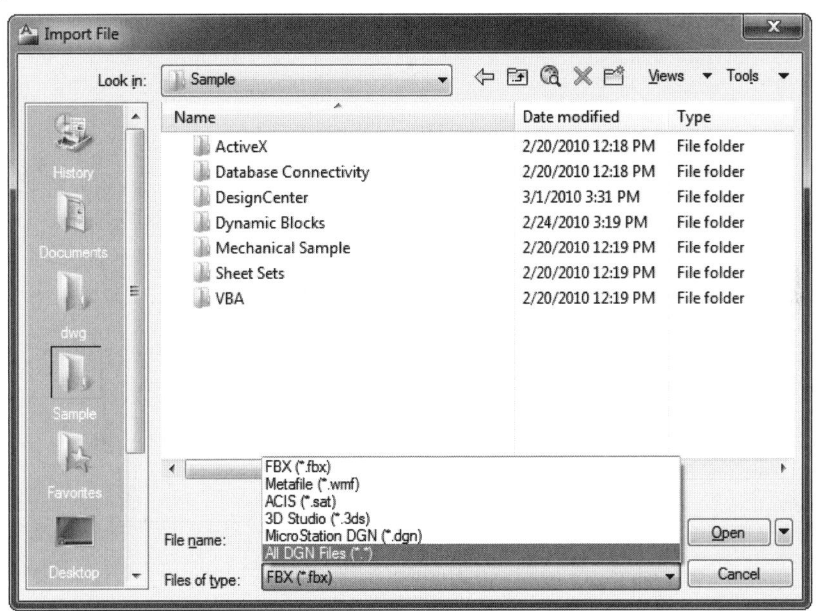

IMPORT FILE DIALOG BOX

Look in selects the folder (subdirectory) and drive from which to import the file.

File name specifies the name of the file, or accepts the default.

File of type selects the file format in which to import the file:

- Metafile (*.wmf)
- ACIS (*.sat)
- 3D Studio (*.3ds)
- FBX (*.fbx) from 3D Studio *(new to AutoCAD 2011)*
- MicroStation DGN (*.dgn) for V7 and V8
- All DGN Files (*.*), such as *.sht* sheet files

Open imports the file.

Cancel dismisses the dialog box, and returns to AutoCAD.

RELATED COMMANDS

AppLoad loads AutoLISP, VBA, and ObjectARX routines.

DgnImport imports MicroStation V7 and V8 DGN design files.

DxbIn imports a DXB file.

ExternalReferences displays *.dwg*, image, and *.dwf* files in the current drawing.

Export exports the drawing in several vector and raster formats.

FbxImport imports MotionBuilder Filmbox FBX files from 3D Studio *(new to AutoCAD 2011)*.

Load imports SHX shape objects.

Insert places another drawing in the current drawing as a block.

InsertObj places an OLE object in the drawing via the Clipboard.

MatEditorOpen imports rendering material definitions *(new to AutoCAD 2011)*.

MenuLoad loads customization and menu files into AutoCAD.

Open opens AutoCAD (any version) *.dwg* and *.dxf* files.

PasteClip pastes objects from the Clipboard.

PasteSpec pastes or links objects from the Clipboard.

Replay displays renderings in TIFF, Targa, or GIF formats.

VSlide displays *.sld* slide files.

XBind imports named objects from another *.dwg* file.

TIPS

- The Import command acts as a "shell" command; it launches other AutoCAD commands that perform the actual import function. Other options may be available with those commands, such as insertion point and scale.

Format	*Meaning*	*Related Command*
Metafile	Windows metafile WMF	**WmfIn**
ACIS	ASCII SAT	**AcisIn**
3D Studio	3D Studio	**3dsIn**
DGN	MicroStation V7 and V8	**DgnImport**
FBX	MotionBuilder Filmbox	**FbxImport**

- To import *.dxf* files, use the Insert command.
- AutoCAD no longer imports PostScript and EPS files; use the PDFAttach command to attach PDF files as underlays.

Removed Commands

Impression command was removed from AutoCAD 2011; it was not replaced.

Imprint

2007 Imprints objects on the faces of 3D solids; updated from the SolidEdit command's Body/Imprint option.

Command	Aliases	Keyboard Shortcuts	Menu	Ribbon
imprint	...	...	**Modify** **⇘Solid Editing** **⇘Imprint Edges**	**Home** **⇘Solid Editing** **⇘Imprint**

Command: imprint

Select a 3D solid: *(Select one 3D solid.)*

Select an object to imprint: *(Select another 2D, 3D, surface, or solid object.)*

Delete the source object [Yes/No] <N>: *(Type Y or N.)*

Select an object to imprint: *(Press Enter to exit the command.)*

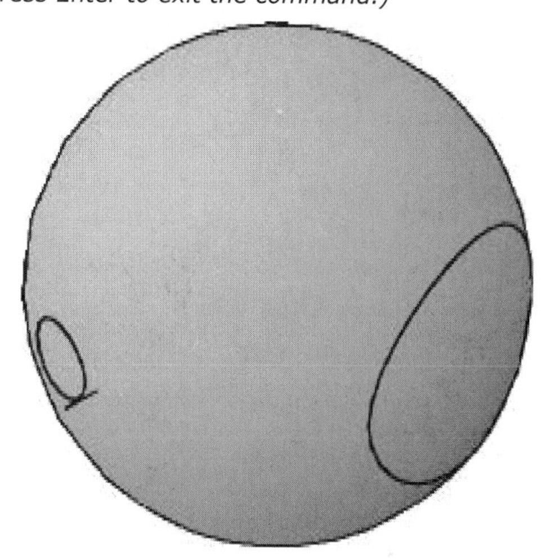

Two circles resulting from the imprinting a cone on the sphere.

COMMAND OPTIONS

Select a 3D solid selects the 3D solid to be imprinted.

Select an object selects the object that will do the imprinting; can be an arc, circle, line, 2D polyline, ellipse, spline, region, and 3D polyline, or 3D solid. Also called the "source" object.

Delete the source object toggles deletion of the source object.

RELATED COMMANDS

SolidEdit performs the same function with its Body/Imprint option.

PressPull extrudes and indents faces created by this command.

TIP

This command creates new faces that can be used by the PressPull command to extrude and indent the faces.

Removed Commands

INetCfg command was removed from AutoCAD 2000; it was replaced by Windows' Internet configuration.

INetHelp command was removed from AutoCAD 2000; it was replaced by AutoCAD's standard online help.

Insert

Ver. 1.0 Inserts previously-defined blocks into drawings.

Commands	*Aliases*	*Keyboard Shortcuts*	*Menu*	*Ribbon*
insert	**i**	...	**Insert**	**Home**
	inserturl		**↳Block**	**↳Block**
				↳Insert
-insert	**-i**	...	...	....

Command: insert

Displays dialog box.

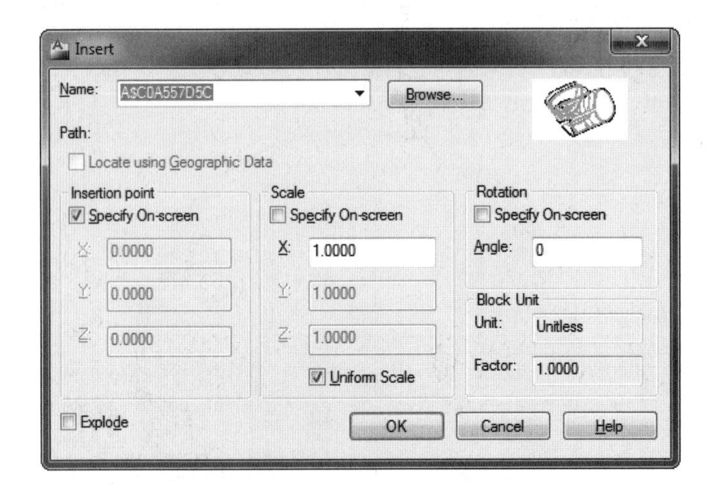

INSERT DIALOG BOX

Name selects the name from a list of previously-inserted blocks.

Browse displays the Select Drawing File dialog box; select a file in DWG or DXF format.

Insertion Point options

☑**Specify On-Screen** specifies the insertion point in the drawing, after you close dialog box.

X, Y, Z specifies the x, y, and z coordinates of the insertion point (at the base point, usually the lower-left corner of the block).

Scale options

☑**Specify on-screen** specifies the scale of the block (relative to the lower-left corner) after dialog box is dismissed.

X, Y, Z specifies the x, y, and z scales of the block; positive values enlarge the block, while negative values reduce it.

☐ **Uniform Scale** forces the y and z scale factors the same as the x scale factor.

Rotation options

☐ **Specify on-screen** specifies the rotation angle of the block (lower-left corner), after the dialog box is dismissed.

Angle specifies the angle to rotate the block; positive angles rotate the block counterclockwise.

Unit Block options

Unit specifies the units for the block based on the value stored in the InsUnit variable.

Factor reports the scale factor based on the InsUnit variable.

☐ **Explode** explodes the block upon insertion.

-INSERT Command

Command: -insert
Enter block name or [?]: *(Enter name, or ?.)*
Specify insertion point or [Basepoint/Scale/X/Y/Z/Rotate]: *(Pick a point, or enter an option.)*
Enter X scale factor, specify opposite corner, or [Corner/XYZ] <1>: *(Enter a scale factor, or enter an option.)*
Enter Y scale factor <use X scale factor>: *(Enter a scale factor, or press Enter.)*
Specify rotation angle <0>: *(Enter a rotation angle.)*

COMMAND LINE OPTIONS

Block name specifies the name of the block to be inserted.

? lists the names of blocks stored in the drawing.

Specify insertion point specifies the lower-left corner of the block's insertion point.

Basepoint places the block temporarily to allow you to specify a new base point. AutoCAD prompts:

Specify base point: *(Pick a point.)*

Scale specifies the scale for the x, y, and z factors equally. AutoCAD prompts:

Specify scale factor for XYZ axes: *(Enter a scale factor.)*

X indicates the x scale factor.

Y indicates the y scale factor.

Z indicates the z scale factor.

Rotate specifies the rotation angle.

The following options are hidden:

PScale supplies a predefined x, y, and z scale factor.

PX supplies a predefined x scale factor.

PY supplies a predefined y scale factor.

PZ supplies a predefined z scale factor.

PRotate supplies a predefined rotation angle.

Corner indicates the x and y scale factors by pointing on the screen.

XYZ displays the x, y, and z scale submenu.

INPUT OPTIONS

In response to the 'Block Name' prompt, you can enter:

~ displays a dialog box of drawings stored on disk:

Block name: ~

* inserts block exploded:

Block name: **filename*

= redefines existing block with a new block:

Block name: *oldname=newname*

In response to the 'Insertion point' prompt, you can enter:

Scale — specifies the x, y, and z scale factors.
PScale — presets the x, y, and z scale factors.
XScale — specifies the x scale factor.
PxScale — presets the x scale factor.
YScale — specifies the y scale factor.
PyScale — presets the y scale factor.

ZScale — specifies the z scale factor.
PzScale — presets the z scale factor.
Rotate — specifies the rotation angle.
PRotate — presets the rotation angle.

RELATED COMMANDS

Block creates a block out of a group of objects.

Explode reduces inserted blocks to their constituent objects.

ExternalReferences displays drawings stored on disk in the drawing.

MInsert inserts a block as a blocked rectangular array.

Rename renames blocks.

WBlock writes blocks to disk.

RELATED SYSTEM VARIABLES

ExplMode toggles whether non-uniformly scaled blocks can be exploded.

InsBase specifies the name of the most-recently inserted block.

InsUnits specifies the drawing units for the inserted block.

TIPS

- You can insert any other AutoCAD drawing into the current drawing.
- A *preset* scale factor or rotation means the dragged image is shown at that scale, but you can enter a new scale when inserting.
- Drawings are normally inserted as a block; prefix the file name with an * (*asterisk*) to insert the drawing as separate objects.
- Redefine all blocks of the same name in the current drawing by adding the = (*equal*) suffix after its name at the 'Block name' prompt.
- Insert a mirrored block by supplying a negative x or y scale factor, such as:

  X scale factor: -1

- AutoCAD converts negative z scale factors to absolute values, which makes them positive.
- As of AutoCAD Release 13, you can explode a mirrored block and a block inserted with different scale factors, when the system variable ExplMode is turned on.
- As of AutoCAD 2004, the Insert command no longer imports designXML files.
- As of AutoCAD 2008, the preview image shows the annotation icon for blocks with annotative scaling. Blocks with annotative scaling cannot have non-uniform scaling.
- When inserting blocks with attached xrefs, the xrefs are retained.

InsertObj

Rel. 13 Places OLE objects as linked or embedded objects in drawings (short for INSERT OBJect).

Command	Alias	Keyboard Shortcuts	Menu	Ribbon
insertobj	**io**	**...**	**Insert** **↳OLE Object**	**Blocks & References** **↳Data** **↳OLE Object**

Command: insertobj

Displays dialog box.

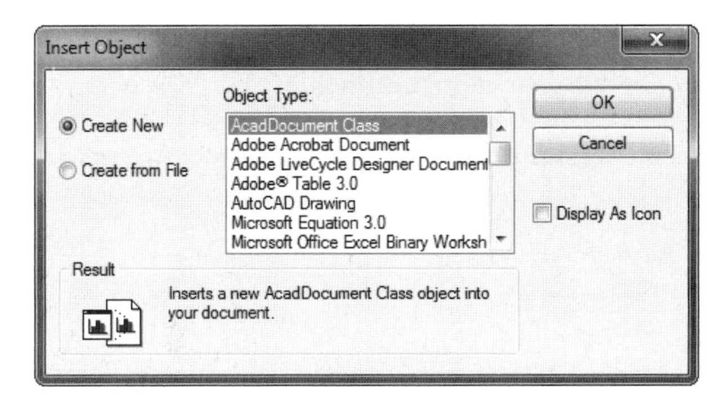

INSERT OBJECT DIALOG BOX

⊙**Create New** creates new objects in other applications, and then embeds them in the drawing.

○**Create from File** selects a file to embed in or a link to the current drawing.

Object Type selects an object type from the list; the related application automatically launches if you select the Create New option.

☑**Display As Icon** displays the object as an icon, rather than as itself.

Change Icon selects another icon.

RELATED COMMANDS

OleLinks controls the OLE links.

PasteSpec places an object from the Clipboard in the drawing as a linked object.

RELATED SYSTEM VARIABLES

MsOleScale determines the scale of OLE objects placed in model space.

OleHide toggles the display of OLE objects.

OleQuality determines the plot quality of OLE objects.

OleStartup specifies whether the source applications of embedded OLE objects load for plotting.

RELATED WINDOWS COMMANDS

Edit | Copy copies an object to the Clipboard for use in other Windows applications.

File | Update updates an OLE object from another application.

Renamed Command

InsertUrl was removed from AutoCAD 2000; replaced by the Insert command's Browse | Search the Web option.

Interfere

Rel. 11 Determines the interference of two or more 3D solid objects; optionally creates a 3D solid body of the common volumes.

Commands	*Alias*	*Menu*	*Ribbon*
interfere	**inf**	**Modify** ⇘**3D Operations** ⇘**Interference Checking**	**Home** ⇘**Solid Editing** ⇘**Interference**
-interfere	...	...	...

Command: interfere

Select first set of objects or [Nested selection/Settings]: *(Select one or more solid objects, or enter an option.)*

Select second set of objects or [Nested selection/checK first set] <checK>: *(Press Enter to check, or enter an option.)*

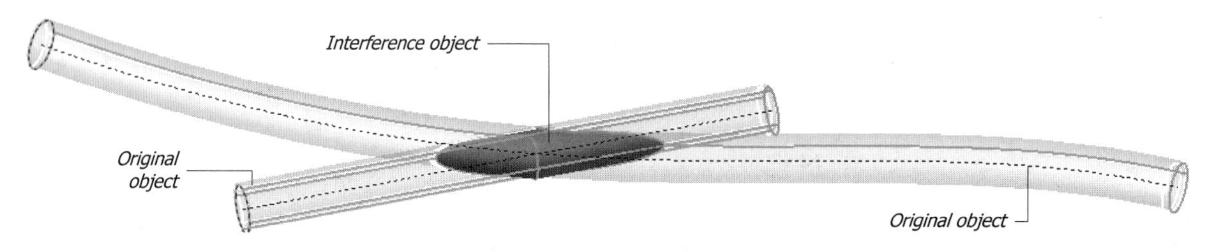

Interfering solid shaded at center.

COMMAND LINE OPTIONS

Select first set of objects checks solids in a single selection set for interference with one another.

Nested selection selects solids nested in blocks and xrefs.

Settings displays the Interference Settings dialog box.

checK displays the Interference Checking dialog box.

INTERFERENCE SETTINGS DIALOG BOX

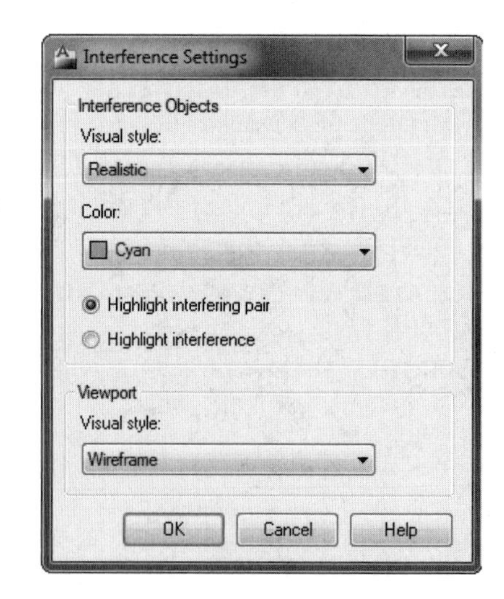

Interference Objects options

Visual Style specifies the visual style of interference objects.

Color specifies the color of interference objects.

⊙**Highlight Interfering Pair** highlights the pair of solids that interfere with each other.

○**Highlight Interference** highlights the interference object(s) created from the interfering pair.

Viewport options

Visual Style specifies the visual style while checking for interference.

INTERFERENCE CHECKING DIALOG BOX

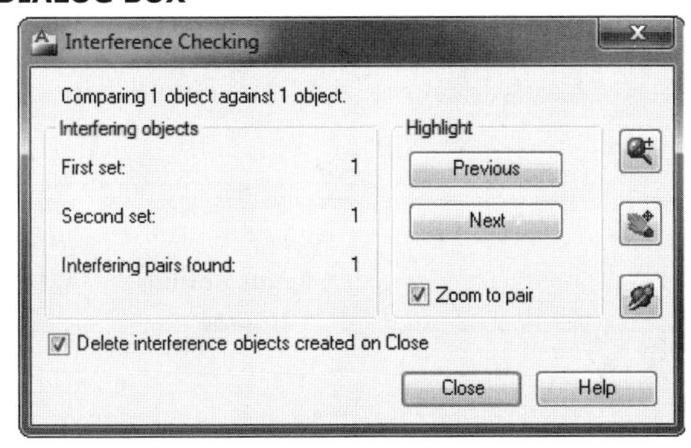

Highlight options

Previous and **Next** highlight the previous and next interference objects.

☑ **Zoom to Pair** zooms to interference object(s) when you click Previous and Next.

☑ **Delete Interference Objects Created on Close** deletes the interference object(s) after you close the dialog box.

Zoom closes the dialog box temporarily, and then starts the Zoom command.

Pan dismisses the dialog box temporarily, and then starts the Pan command.

3D Orbit dismisses the dialog box temporarily, and then starts the 3dOrbit command.

-INTERFERE Command

Command: -interfere

Select first set of objects or [Nested selection]: *(Select one or more solid objects, or enter an option.)*

Select second set of objects or [Nested selection/checK first set] <checK>: *(Press Enter to check, or enter an option.)*

Comparing 2 objects with each other.

Interfering objects (first set): 2

(second set): 0

Interfering pairs: 1

Create interference objects? [Yes/No] <No>: *(Type Y or N.)*

RELATED COMMANDS

Intersect creates a new volume from the intersection of two volumes.

Section creates a 2D region from a 3D solid.

Slice slices a 3D solid with a plane.

RELATED SYSTEM VARIABLES

InterfereColor specifies the color of interference objects.

InterfereObjVs specifies the visual style of interference objects.

InterfereVpVs specifies the visual style for the current viewport during interference checking.

Intersect

Rel. 11 Creates 3D solids, 2D regions, or surfaces through the Boolean intersection of two or more solids or regions, or surfaces.

Command	Alias	Keyboard Shortcuts	Menu	Ribbon
intersect	**in**	**...**	**Modify ↳Solids Editing ↳Intersect**	**Home ↳Solid Editing ↳Intersect**

Command: intersect
Select objects: *(Select one or more solid objects.)*
Select objects: *(Press Enter to end object selection.)*

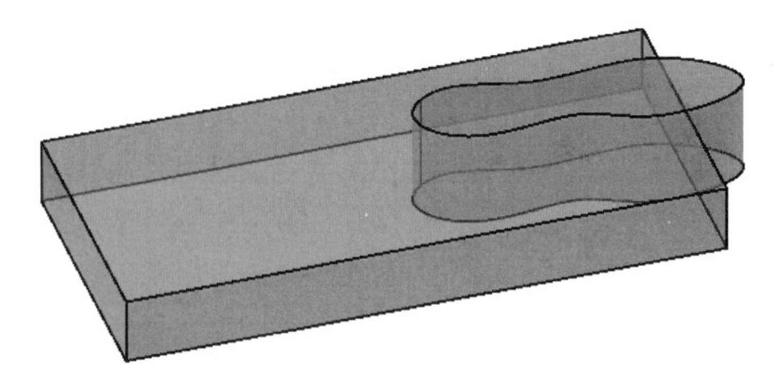

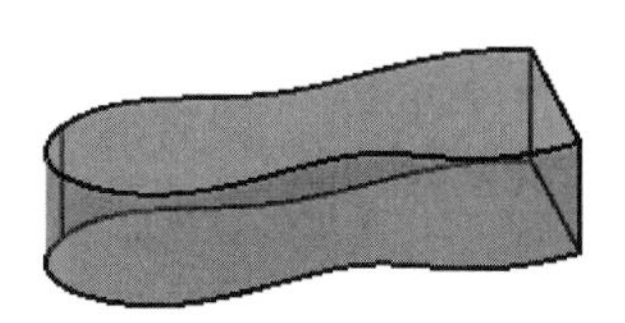

Left: *Two intersecting solids.*
Right: *The net result of applying the Intersect command.*

COMMAND LINE OPTION

Select objects selects two or more objects to intersect.

RELATED COMMANDS

Interfere creates a new volume from the interference of two or more volumes.

Subtract subtracts one 3D solid from another.

Union joins 3D solids into a single body.

RELATED SYSTEM VARIABLES

ShowHist toggles display of history in solids.

SolidHist toggles the retention of history in solids.

TIPS

- You can use this command on a mix of 2D regions, 3D solids, and 3D surfaces.
- When the selection set is a mixture of objects, this command sorts them into subsets, and then applies the intersection action among objects in these subsets:
    - All 3D solids and 3D surfaces.
    - All coplanar 2D regions.
- Here is the difference between the Interference and Intersect commands: **Intersect** *erases* all of the 3D solid parts that do not intersect, while **Interfere** *creates a new object* from the intersection; it does not erase the original objects.

IsolateObjects / UnisolateObjects

Turns off and on the display of unselected objects.

Command	Alias	Status Bar	Menu	Ribbon
isolateobjects	**isolate**		...	...
unisolateobjects	**unhide** **unisolate**		...	...

Command: isolateobjects
Select objects: *(Choose one or more objects.)*
Select objects: *(Press Enter.)*

AutoCAD hides all objects that were not selected.

COMMAND OPTION

Select objects chooses the objects not to be hidden from view.

UNISOLATEOBJECTS Command

Command: unisolateobjects

AutoCAD displays all hidden objects.

COMMAND OPTIONS

None.

RELATED COMMANDS

HideObjects turns off the display of selected objects.

LayOn and **LayOff** turn on and off the layer associated with selected objects.

LayIso and **LayUniso** isolate and unisolate layers associated with selected objects.

TIPS

- This command is handy for temporarily hiding objects until they are redisplayed with the UnisolateObjects command.
- The IsolateObjects command performs the opposite function of the HideObjects command:

Command	Function
IsolateObjects	Hides all objects, except selected ones.
HideObjects	Hides all selected objects.

- This command can be accessed through the status bar: click the light bulb icon to display the shortcut menu:

Isolate Additional Objects
Hide Objects
End Object Isolation

Light Bulb Color	Meaning
Yellow	No objects are hidden or isolated.
Red	At least one object is hidden or isolated.

- Use the LayIso command to turn off the display of all objects not associated with the same layer.

Isoplane

Ver. 2.0 Changes the crosshair orientation and grid pattern among the three isometric drawing planes.

Command	*Aliases*	*Keyboard Shortcuts*	*Menu*	*Ribbon*
'isoplane	...	**Ctrl+E F5**	...	...

Command: isoplane
Enter isometric plane setting [Left/Top/Right] <Top>: *(Enter an option, or press Enter.)*

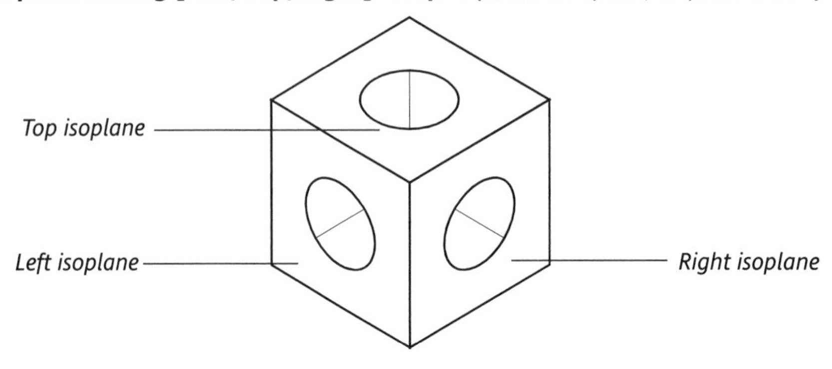

COMMAND LINE OPTIONS

Left switches to the left isometric plane.

Top switches to the top isometric plane.

Right switches to the right isometric plane.

Enter switches to the next isometric plane in the following order: left, top, right.

RELATED COMMANDS

Options displays a dialog box for setting isometric mode and planes.

Ellipse draws isocircles.

Snap turns on isometric drawing mode.

RELATED SYSTEM VARIABLES

SnapIsoPair contains the current isometric plane.

GridMode toggles grid visibility.

GridUnit specifies the current grid x, y spacing.

LimMin specifies the x, y coordinates of the lower-left corner of the grid display.

LimMax holds the x, y coordinates of the upper-right corner of the grid display.

SnapStyl specifies a normal or isometric grid:

- **0** — Normal (default).
- **1** — Isometric grid.

Renamed Command

JogSection command was renamed SectionPlaneJog as of AutoCAD 2010.

Join

2006 Joins open objects to make one object or closed objects.

Command	*Aliases*	*Keyboard Shortcuts*	*Menu*	*Ribbon*
join	**j**	**...**	**Modify** ↳**Join**	**Home** ↳**Modify** ↳**Join**

Command: join
Select source object: *(Select an object.)*
Select *objects* **to join to source or [cLose]:** *(Select other objects, or enter L to close an open object.)*

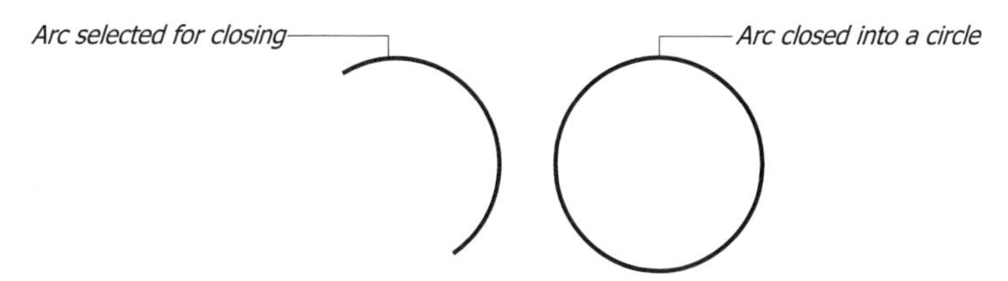

COMMAND LINE OPTIONS

Select source object selects objects to be joined or closed.

Select objects to join joins similar objects. The following objects can be joined, even when there is a gap between them:

- Colinear lines — joined into a single line.
- Arcs in the same imaginary circle — joined into a single arc.
- Elliptical arcs in the same imaginary ellipse — joined into a single elliptical arc.

These objects can only be joined when no gap exists between them:

- 3D and 2D polylines, lines, and arcs — joined into a single polyline.
- Splines — joined into a single spline.

cLose closes arcs and elliptical arcs:

- Arcs become circles.
- Elliptical arcs become ellipses.

RELATED COMMANDS

Break turns circles and ellipses into arcs.

PEdit joins lines, arcs, and polylines with its Join option.

TIP

- If arcs have been dimensioned with the DimArc command, the closed circles are disassociated from the dimension.
- As of AutoCAD 2011, this command joins 3D polylines with other open objects.
- The most complex object must be selected first. In order of decreasing complexity, these are 3D polylines, splines, 2D polylines, elliptical arcs, arcs, and lines.
- As of AutoCAD 2011, splines and 3D polylines no longer need to be coplanar, but must be colinear.

JpgOut

2004 Exports drawings in JPEG format.

Command	Aliases	Keyboard Shortcuts	Menu	Ribbon
jpgout	...	...	...	...

Command: jpgout

Displays Create Raster File dialog box. Specify a filename, and then click Save.

Select objects or <all objects and viewports>: *(Select objects, or press Enter to select all objects and viewports.)*

COMMAND LINE OPTIONS

Select objects selects specific objects.

All objects and viewports selects all objects and all viewports, whether in model space or in layout mode — basically doing a screen grab.

RELATED COMMANDS

BmpOut exports drawings in BMP (bitmap) format.

Image places raster images in the drawing.

Plot outputs drawings in JPEG and other raster formats.

PngOut exports drawings in PNG (portable network graphics) format.

TifOut exports drawings in TIFF (tagged image file format) format.

TIPS

- JPEG files are often used by digital cameras and Web pages.
- For more control over the resulting *.jpg* file, use the Plot command's PublishToWebJpg driver. For example, it can specify a high resolution, such as 2550x3300, whereas the JpgOut command is limited to the resolution of your computer's monitor (1280x1024, typically).
- When using the JpgOut command to save a rendering, the rendering effects of the VisualStyles command are preserved, but not those of the Render command.
- The drawback to saving drawings in JPEG format is that the image is less clear (due to artifacts) than in other formats; the advantage is that JPEG files are highly compressed.

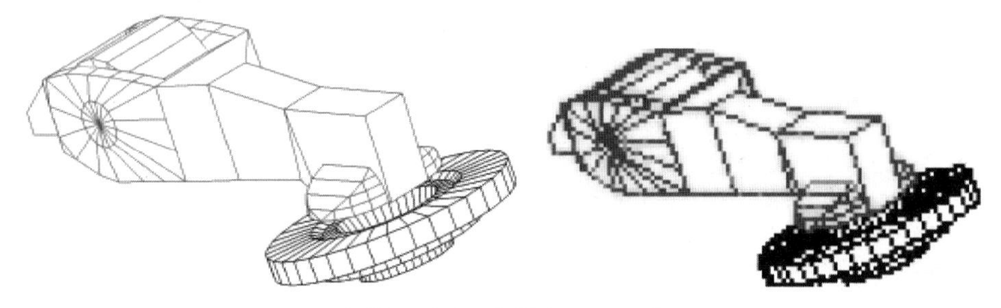

Left: *Original AutoCAD image zoomed-in.*
Right: *Enlarged JPEG image with artifacts.*

- This command provides no options for specifying the level of compression.
- JPEG is short for "joint photographic expert group."

JustifyText

2001 Changes the justification of text.

Command	*Aliases*	*Keyboard Shortcuts*	*Menu*	*Ribbon*
justifytext	...	...	**Modify** ↳**Object** ↳**Text** ↳**Justify**	**Annotate** ↳**Text** ↳**Justify**

Command: justifytext
Select objects: *(Select one or more text objects.)*
Select objects: *(Press Enter to end object selection.)*
Enter a justification option
[Left/Align/Fit/Center/Middle/Right/TL/TC/TR/ML/MC/MR/BL/BC/BR] <Left>: *(Enter an option, or press Enter.)*

COMMAND LINE OPTIONS

Select objects selects one or more text objects in the drawing.

Align aligns the text between two points with adjusted text height.

Fit fits the text between two points with fixed text height.

Center centers the text along the baseline.

Middle centers the text horizontally and vertically.

Right right-justifies the text.

TL justifies to top-left.

TC justifies to top-center.

TR justifies to top-right.

ML justifies to middle-left.

MC justifies to middle-center.

MR justifies to middle-right.

BL justifies to bottom-left.

BC justifies to bottom-center.

BR justifies to bottom-right.

RELATED COMMANDS

Text places text in the drawing.

TextEdit edits text.

ScaleText changes the size of text.

Style defines text styles.

Properties changes the justification, but moves text.

TIPS

- This command works with text, mtext, leader text, and attribute text.
- When the justification is changed, the text does not move to reflect the changed insertion point.

Lay...

2007 Manipulates layers; a group of commands formerly in Express Tools.

Command	*Aliases*	*Keyboard Shortcuts*	*Menu*	*Ribbon*
...	...	...	**Format**	**Home**
			↳Layer Tools	**↳Layers**

These commands perform focused operations on layers and objects. Many are paired, one command undoing the action of its mate.

LayOff turns off the layer of the selected object.
LayOn turns on all layers, except frozen layers.

LayIso turns off all layers except the ones holding selected objects; see LayIso command.
LayVpi isolates the selected object's layer in the current viewport.
LayWalk displays objects on selected layers, turning off all other layers.
LayUnIso turns on layers that were turned off with the LayIso command

LayFrz freezes the layers of the selected objects.
LayThw thaws all layers.

LayLck locks the layer of a selected object.
LayULk unlocks the layer of a selected object.

LayCur changes the layer of selected objects to the current layer.
LayMch changes the layers of selected objects to that of another selected object.
LayMCur makes the selected object's layer current (similar to Ai_Molc command).

CopyToLayer copies objects to another layer; see CopyToLayer command.
LayMrg moves objects to another layer.

LayDel erases all objects from the specified layer, and then purges the layer from the drawing.

LAYCUR Command

Changes the layer of selected objects to the current layer (short for "LAYer CURrent").

Command: laycur
Select objects to be changed to the current layer: *(Select one or more objects.)*
Select objects to be changed to the current layer: *(Press Enter.)*
One object changed to layer *layername* **(the current layer).**

LAYDEL and -LAYDEL Commands

Erase all objects from the specified layer, and then purge the layer from the drawing (short for "LAYer DELete"). The current layer cannot be deleted. If you make a mistake, use the U command to restore the purged layer name and its erased objects.

Command: laydel

Select object on layer to delete or [Name]: *(Select an object, or type N.)*

Selected layers: *layername*

Select object on layer to delete or [Name/Undo]:

********** WARNING **********

You are about to delete layer "*layername***" from this drawing.**

Do you wish to continue? [Yes/No] <No>: *(Type Y or N.)*

Deleting layer "*layername***".**

1 layer deleted.

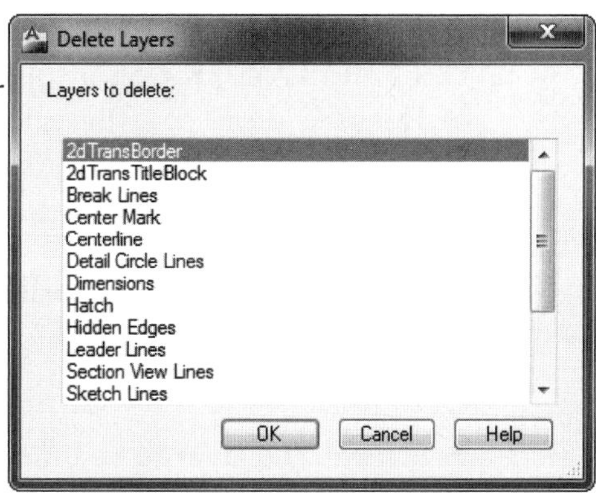

COMMAND OPTIONS

Displays the Delete Layers dialog box; select one or more layers to delete and purge.

Undo undoes the last layer delete.

LAYFRZ and LAYTHW Commands

LayFrz freezes the layers of the selected objects (short for "LAYer FReeZe") ; LayThw thaws all layers (short for "LAYer THaW").

Command: layfrz

Current settings: Viewports=Vpfreeze, Block nesting level=Block

Select an object on the layer to be frozen or [Settings/Undo]: *Select an object, or enter an option.)*

Layer "*layername***" has been frozen.**

Select an object on the layer to be frozen or [Settings/Undo]: *(Press Enter, or enter an option.)*

COMMAND OPTIONS

Settings offers these options:

Enter setting type for [Viewports/Block selection] selects options for viewports or blocks.

In paper space viewport use determines how to freeze layers in viewports:

- **Freeze** freezes the layers in all viewports.
- **Vpfreeze** freezes just the current viewport.

Enter Block Selection nesting lever determines how to freeze layers in blocks and xrefs:

- **Block** freezes the entire block's layer, but freezes the object's layer in xrefs.
- **Entity** freezes the object's layer.
- **None** freezes neither the block nor xref.

Undo undoes the last freeze action.

Command: laythw

All layers have been thawed.

LAYLCK and LAYULK Commands

LayLck locks the layer of the selected object (short for "LAYer LoCK"); LayULk unlocks the layer of a selected object (short for "LAYer UnLocK").

Command: laylck
Select an object on the layer to be locked: *(Select an object.)*
Layer "*layername*" has been locked.

Command: layulk
Select an object on the layer to be unlocked: *(Pick an object on a locked layer.)*
Layer "*layername*" has been unlocked.

LAYMCH and -LAYMCH Commands

LayMch and -LayMch change the layers of selected objects to that of a selected object (short for "LAYer MatCH").

Command: laymch
Select objects to be changed: *(Pick one or more objects.)*
Select objects: *(Press Enter.)*
Select object on destination layer or [Name]: *(Pick an object on another layer.)*
One object changed to layer "*layername*"

COMMAND OPTION

Name displays the Change to Layer dialog box; see LayDel command.

In the -LayMch command, the Name option displays the following prompt:

Enter layer name: *(Type the valid name of a layer.)*

LAYMCUR Command

Makes the selected object's layer current; replaces the Ai_Molc command (short for "LAYer Make CURrent").

Command: laycur
Select object whose layer will become current: *(Pick an object.)*
layername **is now the current layer.**

LAYMRG and -LAYMRG Commands

LayMrg and -LayMrg move objects to another layer (short for "LAYer MeRGe").

Command: laymrg
Select object on layer to merge or [Name]: *(Pick an object.)*

Selected layers: *layername*.
Select object on layer to merge or [Name/Undo]: *(Pick another object.)*
Selected layers: *layername, anothername*.
Select object on layer to merge or [Name/Undo]: *(Press Enter.)*
Select object on target layer or [Name]: *(Pick an object.)*

COMMAND OPTION

Name displays the Merge to Layer dialog box; see LayDel command.

In the -LayMrg command, the Name option displays the following prompt:

Enter layer name: *(Type the name of a layer.)*

LAYOFF and LAYON Commands

LayOff turns off the layer of the selected object; LayOn turns on all layers, except frozen ones.

Command: layoff
Current settings: Viewports=Vpfreeze, Block nesting level=Block
Select an object on the layer to be turned off or [Settings/Undo]: *(Select an object, or enter an option.)*
Layer "0" has been turned off.

See LayFrz for the meaning of the Setting and Undo options.

Command: layon
Warning: layer "*layername*" is frozen. Will not display until thawed.
All layers have been turned on.

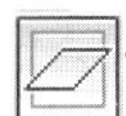

LayVpi

Isolates the selected object's layer in the current viewport (short for "LAYer ViewPort Isolate"). This command works only in paper space with two or more viewports.

Command: layvpi
Current settings: Viewports=Vpfreeze, Block nesting level=Block
Select objects on the layer to be isolated in viewport or [Settings/Undo]: *(Select one or more objects, or enter an option.)*
Layer *layername* **has been frozen in all viewports but the current one.**
Select an object on the layer to be isolated in viewport or [Settings/Undo]: *(Press Enter to exit the command.)*

See LayFrz for the meaning of the Setting and Undo options.

LayWalk

Displays objects on selected layers.

Command: laywalk

Displays dialog box.

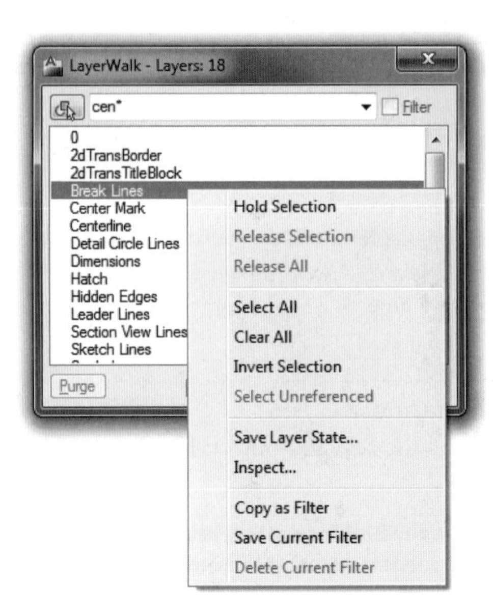

DIALOG BOX OPTIONS

Select objects selects one or more objects in the drawing, and then reports their layer name(s).

☐ **Filter** filters the layer list according to the rules of the selected filter.

Purge removes the selected layer, if it contains no objects.

☑ **Restore on Exit** restores layer display upon exiting the dialog box.

Close closes the dialog box and exits the command.

SHORTCUT MENU OPTIONS

Right-click the LayerWalk dialog box to access the following shortcut menu:

Hold Selection adds an asterisk (*) to the layer name, and always displays it.

Release Selection removes the asterisk from the selected layers.

Release All removes all asterisks, and then turns off the Always Show option.

Select All selects all layers, and displays them.

Clear All unselects all layers, hiding them.

Invert Selection selects unselected layers, and vice versa.

Select Unreferenced selects all layers with no objects; the Purge button removes them from the drawing.

Save Layer State saves the selected layers for use by the Layer States Manager.

Inspect displays the Inspect dialog box, which reports the total number of layers, the number of selected layers, and the total number of objects on the selected layers.

Layer / LayerClose

Ver. 1.0 Controls the creation, status, and visibility of layers.

Commands	*Aliases*	*Menu*	*Ribbon*
'layer	**la** **layerpalette** **ddlmodes**	**Format** **↳Layer**	**Home** **↳Layers** **↳Layer Properties**
layerclose	...	...	...
'-layer	**-la**	...	...
+layer	...	...	...

Command: layer

Displays palette:

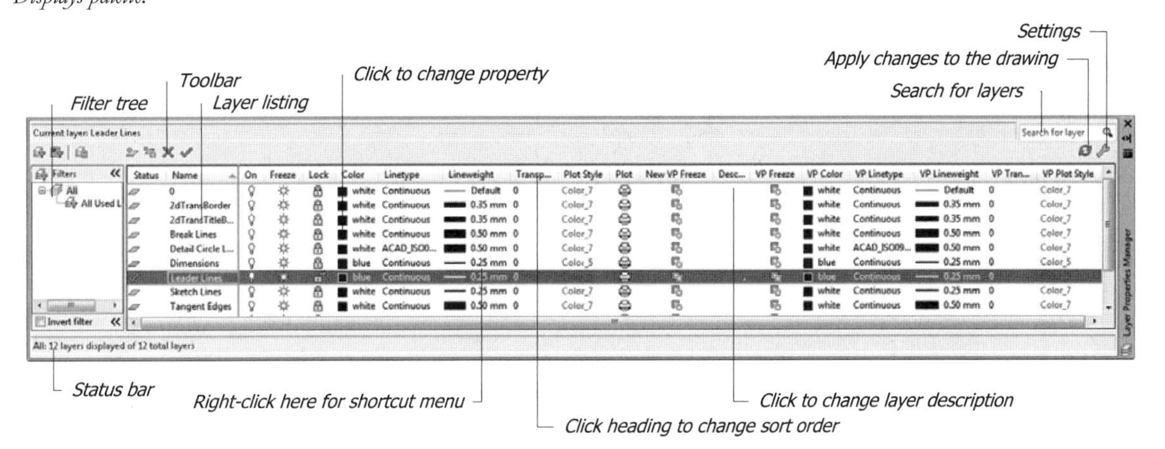

Command: layerclose

Closes the palette.

LAYER PROPERTIES MANAGER PALETTE

Click a header name to sort alphabetically (A-Z); click a second time for reverse-alphabetical sort (Z-A).

Status reports the layer status: current layer, empty layer, layer in use, or layer filter.

Name lists the names of layers in the current drawing.

On toggles layers between on and off.

Freeze toggles layers between thawed and frozen in all viewports.

Lock toggles layers between unlocked and locked.

Color specifies the color of objects on layers.

Linetype specifies the linetype for objects on layers.

Lineweight specifies the lineweight for objects on layers.

Transparency specifies the translucency of objects on layers *(new to AutoCAD 2011).*

Plot Style specifies the plot style for objects on layers; available only when plot styles are enabled in the drawing.

 **Plot** toggles layers between plot and no-plot.

Description describes the layer.

The following options appear only in layout mode (paper space):

 **Current VP Freeze** specifies whether the layer is frozen in the current viewport.

NewVP Freeze specifies whether the layer is frozen in new viewports.

VP Color overrides the color property for the current viewport.

VP Linetype overrides the linetype property for the current viewport.

VP Transparency overrides translucency for the current viewport *(new to AutoCAD 2011).*

VP Lineweight overrides the lineweight property for the current viewport.

VP Plot Style overrides the plot style for the current viewport.

☐ **Invert Filter** inverts the display of layer names; for example, when Show all used layers is selected, the Invert filter option causes all layers with no content to be displayed (ALT+I).

☑ **Indicate Layers in Use** displays icons next to layers that contain objects.

Apply applies the changes to the layers without exiting the dialog box.

OK exits the dialog box, and applies changes to the layers.

Cancel exits the dialog box, and leaves layers unchanged.

Palette toolbar

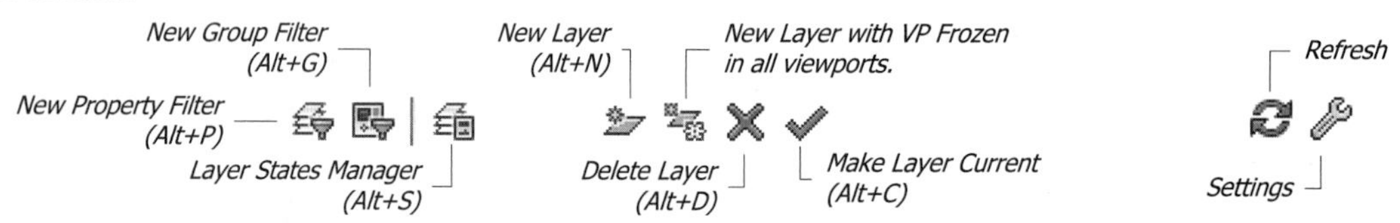

New Property Filter displays New Filter Properties dialog box.

New Group Filter adds an item to the group filter list, initially named "Group Filter 1."

Layer States Manager displays Layer States Manager dialog box; see LayerStates command.

New Layer adds layers to the drawing, initially named "Layer1."

New Layer VP Frozen in All Viewports freezes the new layer in all viewports.

Delete Layer purges the selected layer; some layers cannot be deleted: layers 0 and Defpoints, current layer, layers in xrefs, and layers with objects.

Make Current Layer sets the selected layer as current.

Refresh refreshes the list of layer names, properties, and statuses; useful when xrefs change (formerly Update).

Settings displays the Layer Settings dialog box.

Shortcut Menus

Access these menus by right-clicking almost any layer option.

Column Headers shortcut menu

Maximize Column changes the selected column width to display all column content.

Maximize All Columns changes the width of all columns to display content of all columns. To display complete headers, you must resize the column by dragging the header separators. You can resize the dialog box.

Filter Tree shortcut menu

Visibility ▸
Lock ▸
Viewport ▸
Isolate Group ▸
New Properties Filter...
New Group Filter
Convert to Group Filter
Rename
Delete

Depending on where you right-click in the filter tree area, some menu options may be grayed-out, indicating they are unavailable.

Visibility toggles the visibility of all layers in the selected filter or group:

- **On** displays, plots, and regenerates objects; includes objects during hidden-line removal.
- **Off** does not display or plot objects; does not include objects during hidden-line removal, and the drawing is not regenerated when the layer is turned on.
- **Thawed** reverses the Frozen option.
- **Frozen** does not display or plot objects; includes objects during hidden-line removal, and the drawing is regenerated when the layer is thawed.

Lock determines whether objects can be edited:

- **Lock** prevents objects from being edited; all other operations that don't involve editing are permitted, such as object snaps.
- **Unlock** allows objects to be edited.

Viewport specifies whether layers are frozen in layout mode (unavailable in model space):

- **Freeze** applies Current VP Freeze to all layers (in the filter or group) of the current viewport.
- **Thaw** turns off Current VP Freeze for layers in the filter or group.

Isolate Group turns off all layers not part of the filter or group; only layers that are part of the filter or group are visible in the drawing. In model space, this option applies to all layers; in layout mode, this option applies selectively, depending on the suboption selected:

- **All Viewports** freezes all layers not in the filter or group.
- **Active Viewport Only** freezes all layers (not in the filter or group) in the current viewport only.

New Properties Filter displays the Layer Filter Properties dialog box.

New Group Filter creates a new layer group filter; to add layers to groups:

1. Hold down the Ctrl key, and then select layer names in the layer list (right-hand pane).
2. Drag the layers onto the group filter name (left-hand pane).

Convert to Group Filter converts the selected filter to a group; the name does not change, but the icon changes to indicate a group filter.

Rename changes the name of the filter or group. As an alternative, click the name twice (slowly), and then enter a new name.

Delete erases the filter or group; does not erase the layers in the filter or group. The All, All Used Layers, and Xref filters cannot be erased.

Properties displays the Layer Filter Properties dialog box (available only when a filter is selected).

Select Layers adds and removes layers by selecting objects in the drawing (available only when a group is selected):

- **Add** allows you to select objects in the drawing; press Enter to return to the layer dialog box.
- **Replace** removes existing layers from the group, and adds the newly-selected layers.

Layer List shortcut menu

The content of this shortcut menu varies, depending on where you right-click:

Select Linetype/Color/Lineweight/Plotstyle opens the related dialog box.

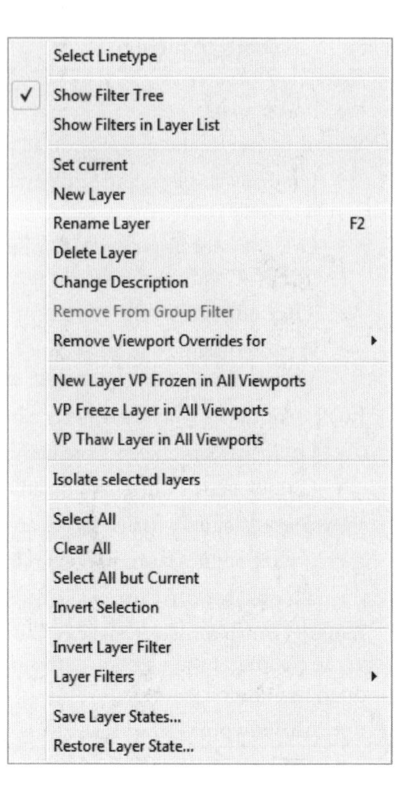

Show Filter Tree toggles filter tree view, left-hand pane of the palette.

Show Filters in Layer List toggles the display of filters in the layer list view (right-hand pane); when off (no check mark), only layers are shown.

Set Current sets the selected layer as current. Alternatively, press Alt+C or click the Set Current button on the dialog box's toolbar.

New Layer creates a new layer, naming it "Layer1" with the properties of the currently-selected layer. (Or, press Alt+N or click the New Layer button.)

Rename Layer renames the selected layer. (Or,. press F2.)

Delete Layer erases the selected layer(s) from the drawing; layers 0, Defpoints, the current layer, xref-dependent layers, and those containing objects cannot be erased. (Or, press Alt+D or click the Delete Layer toolbar button.)

Reconcile Layer adds the layer to the list of acceptable layers.

Change Description adds and changes layer description text. (Or, click the description twice, and then edit the text.)

Remove from Group Filter removes the selected layers from the selected group.

Remove Viewport Overrides for Linetype/Color/Lineweight/Plotstyle/ Selected Layer/All Layers returns overridden properties to ByLayer:

- **In Current Viewport Only** changes overridden layer properties to ByLayer in the current viewport only.
- **In All Viewports** changes overridden layer properties to ByLayer in all viewports.

New Layer VP Frozen in All Viewports creates a new layer, with its status frozen in all viewports.

VP Freeze Layer in All Viewports freezes selected layer(s) in all viewports.

VP Thaw Layer in All Viewports thaws selected layer(s) in all viewports.

Select All selects all layers; alternatively, press Ctrl+A.

Clear All unselects all layers; alternatively, click a single layer.

Select All but Current selects all layers, except the current layer.

Invert Selection unselects selected layers, and selects all other layers.

Invert Layer Filter displays all layers not in the selected filter; alternatively, press Alt+1.

Layer Filters displays a submenu listing the names of filters; alternatively, turn on the filter tree view (left-hand pane). Select a filter name to apply it to the layer list.

Save Layer States displays the New Layer State to Save dialog box; see LayerSate command.

Restore Layer State displays the Layer States Manager dialog box; alternatively, press Alt+S.

LAYER FILTERS PROPERTIES DIALOG BOX

Access this dialog box through the New Property Filter button.

Filter name names the filter.

Show Example displays on-line help.

Filter definition defines the parameters of the filter.

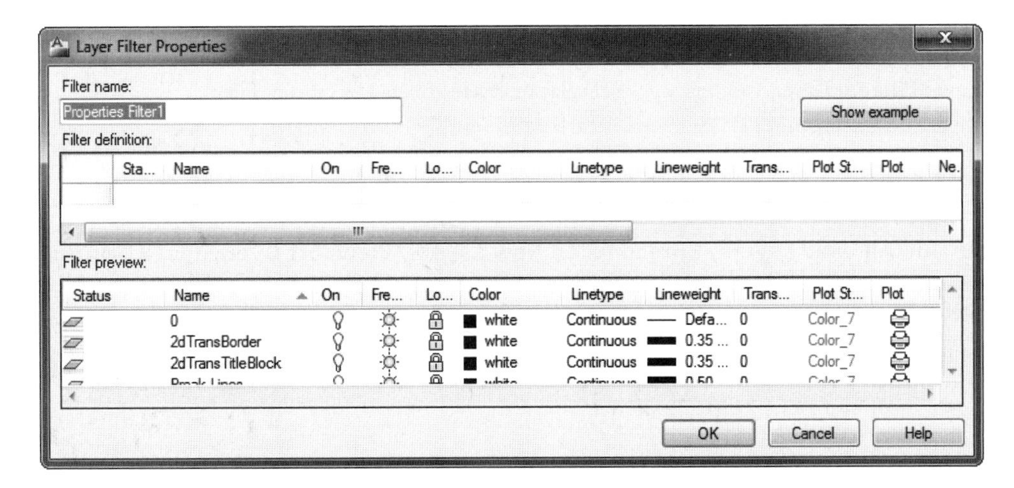

LAYER SETTINGS DIALOG BOX

Access this dialog box through the Settings button.

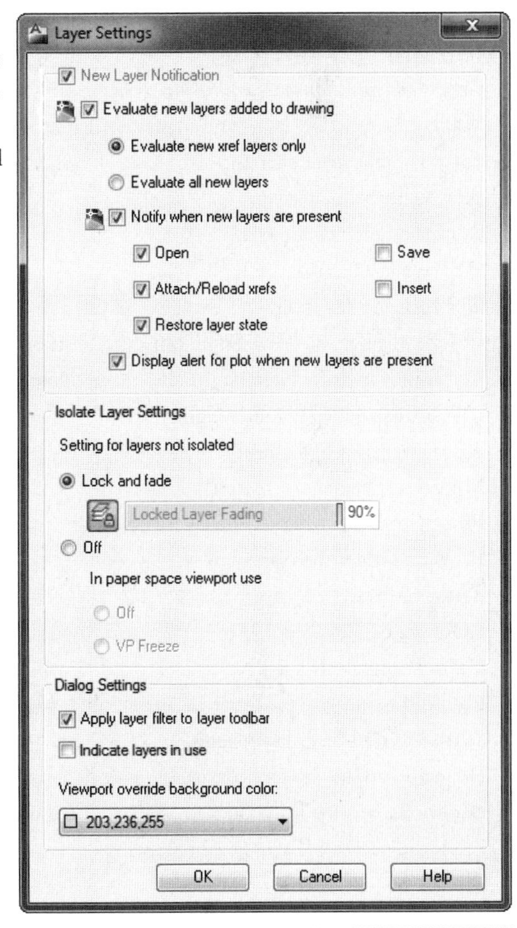

New Layer Notification Settings options

☑**Evaluate New Layers Added to Drawing** checks for new layers being added to drawings (a.k.a. "unreconciled layers") under these conditions:

- ⊙ **Evaluate New Xref Layers Only** checks for new layers added through attached xrefs.
- ○ **Evaluate All New Layers** checks for new layers added to drawings, including from xrefs.

☑**Notify when New Layers Are Present** toggles new layer notification under these conditions:

- ☑ **Open** when opening drawings.
- ☑ **Attach/Reload Xrefs** when attaching and reloading xrefs.
- ☑ **Restore Layer State** when restoring layer states.
- ☑ **Save** when saving drawings.
- ☐ **Insert** when inserting blocks.

☑**Display Alert for Plot When New Layers are Present** displays icon and balloon in the tray.

Isolate Layer Settings options

⊙ **Lock and Fade** specify isolation method of unselected layers:

- Toggle locking.
- Specify percentage of fading.

○ **Off** turns off unselected layers.

In paper space viewport use:

- ○ **Off** turns off unselected layers in paper space.
- ○ **VP Freeze** freezes unselected layers in paperspace.

Dialog Settings options

☑ **Apply Layer Filter to Layer Toolbar** toggles display of layers affected by current filter.

Viewport Override Background Color chooses the background color for viewport overrides.

+LAYER Command

Command: +layer
Specify filter [All/All Used Layers/Unreconciled New Layers/Viewport Overrides] <All>:*(Enter an option.)*

Specify Filter reports the names of layers associated with filters; an undocumented command.

-LAYER Command

Command: -layer
Current layer: "0"
Enter an option
[?/Make/Set/New/Rename/ON/OFF/Color/Ltype/LWeight/TRansparency/MATerial/Plot/Freeze/Thaw/LOck/Unlock/stAte/Description/rEconcile]: *(Enter an option.)*

COMMAND LINE OPTIONS

Groups and filters cannot be created by this command.

Make creates a new layer, and makes it current.

Set makes the layer current.

New creates a new layer.

Rename renames the layer.

OFF turns off the layer.

ON turns on the layer.

Color indicates the color for all objects drawn on the layer.

Ltype indicates the linetype for all objects drawn on the layer.

LWeight specifies the lineweight.

TRansparency specifies the translucency *(new to AutoCAD 2011)*.

Material specifies the material.

Plot determines whether the layer is plotted.

Freeze disables the display of the layer.

Thaw unfreezes the layer.

LOck locks the layer.

Unlock unlocks the layer.

PStyle specifies the plot style (available only when plot styles are attached to the drawing).

stAte sets and saves layer states.

Description describes the layer.

rEconcile sets the unreconciled property of unreconciled layers.

? lists the names of layers in the drawing.

stAte options

Enter an option [?/Save/Restore/Edit/Name/Delete/Import/EXport]: *(Enter an option).*

? lists the names of layer states in the drawing.

Save saves the layer state and properties by name; properties include on, frozen, lock, plot, newvpfreeze, color, linetype, lineweight, and plot style.

Restore restores a named state.

Edit changes the settings of named states.

Name renames named states.

Delete erases named states from the drawing.

Import opens layer states *.las* files.

Export saves a selected named state to a *.las* file.

RELATED COMMANDS

Lay... manipulates layers; a group of commands.

LayerP returns to the previous layer.

LayerState creates and restores layer states.

Change moves objects to different layers via the command line.

Properties moves objects to different layers via a dialog box.

LayTrans translates layer names.

Purge removes unused layers from drawings.

Rename renames layers.

SetByLayer forces VP override properties to ByLayer.

ULayers controls the display of layers in underlay objects, such as xrefs and PDFs.

View controls visibility of layers with named views.

VpLayer controls the visibility of layers in paper space viewports.

RELATED SYSTEM VARIABLES

CLayer contains the name of the current layer.

LayerDlgMode determines if the Layer command displays the dialog box or the palette.

LayerManagerState reports on the status of the Layer Properties Manager palette.

LayerFilterAlert removes extraneous filters.

LayerEval controls when AutoCAD checks the Unreconciled New Layer filter list for new layers.

LayerNotify specifies when alerts are displayed for unreconciled layers.

SetByLayerMode specifies which properties are changed by the SetByLayer command.

ShowLayerUsage toggles the layer usage icons.

VpLayerOverrides reports whether any layers in the current viewport have VP (viewport) property overrides.

VpLayerOverridesMode toggles whether viewport property overrides are displayed or plotted.

RELATED FILE

***.las** are layer state files, which use the DXF format.

TIPS

- A *frozen* layer cannot be seen or edited; a *locked* layer can be seen but not edited.
- Layer "Defpoints" never plots, no matter its setting.
- To create more than one new layer at a time in -Layer, use commas to separate layer names.
- For new layers to take on properties of an existing layer, select the layer before clicking New.
- If layer names appear to be missing, they may have been filtered from the list.
- When layers are added automatically, such as through xrefs, AutoCAD reports these as "unreconciled layers." A warning icon and balloon appear in the tray:

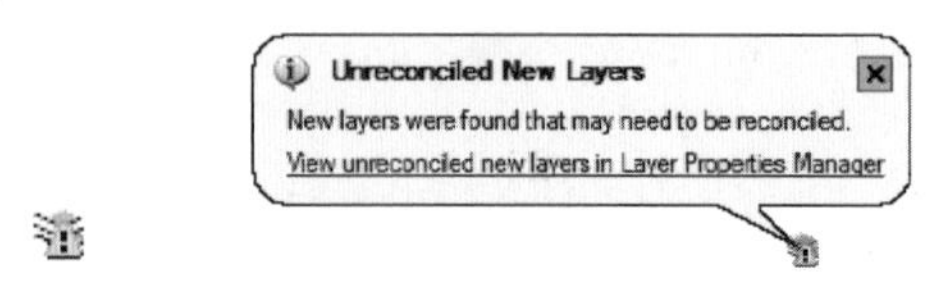

  - Click the icon to see the layer palette with all unreconciled layers listed.
  - Right-click the icon to control when you are warned about unreconciled layers.

- Turn off the Plot property for several random layers, and then drag the header until the Plot column is no longer shown. According to Lynn Allen, this is guaranteed to drive coworkers nuts!
- The layer manager in AutoCAD is incompatible with the layer manager found in earlier releases of Express Tools.
- As of AutoCAD 2009, the Layer Properties Manager is a palette, which means it can stay open all the time. Changes made to layers take effect immediately, so there is no Apply or OK button.
- As a palette, it can be docked, anchored, set to AutoHide, be partially transparent, and moved to a second monitor.
- Both Off and Freeze make objects invisible. AutoCAD, however, ignores objects on frozen layers, but not those of layers that are turned off. AutoCAD runs faster with frozen layers.

LayerP / LayerPMode

2002 Undoes changes made to layer settings (short for LAYER Previous).

Commands	*Aliases*	*Keyboard Shortcuts*	*Menu*	*Ribbon*
layerp	...	...	**Format** **↳Layer Tools** **↳Layer Previous**	**Home** **↳Layers** **↳Previous**
layerpmode	...	...	...	...

Command: layerp
Restored previous layer status

LAYERPMODE Command

Toggles layer-previous mode (short for LAYER Previous MODE).

Command: layerpmode
Enter LAYERP mode [ON/OFF] <ON>: *(Type ON or OFF.)*

COMMAND LINE OPTIONS

On turns on layer previous mode (default).

Off turns off layer previous mode.

RELATED COMMANDS

Layer creates and sets layers and modes.

LayerState creates and edits layer states.

Lay... manipulates layers; a group of commands.

View stores layer settings with named views.

RELATED SYSTEM VARIABLE

CLayer specifies the name of the current layer.

TIPS

- This command acts like an "undo" command for changes made to layers only, for example, the layer's color or lineweight.
- LayerPMode command must be turned on for the LayerP command to work.
- The LayerP command does not undo the renaming, deleting, purging, or creating of layers.
- When layer previous mode is on, AutoCAD tracks changes to layers.

LayerState

2008 Saves, restores, and manages named layer states.

Command	Aliases	Menu	Ribbon
layerstate	**las**	**Format**	**Home**
	lman	**↳Layer States Manager**	**↳Layers**
			↳Layer States Manager

Command: layerstate

Displays dialog box:

LAYER STATES MANAGER DIALOG BOX

Layer states lists the names of layer states that can be restored:

- To rename a layer state, click its name twice, and then edit the name.
- To edit the description, click the description twice, and then edit the text.
- To sort layer states, click the headers – Name, Space, and Description.

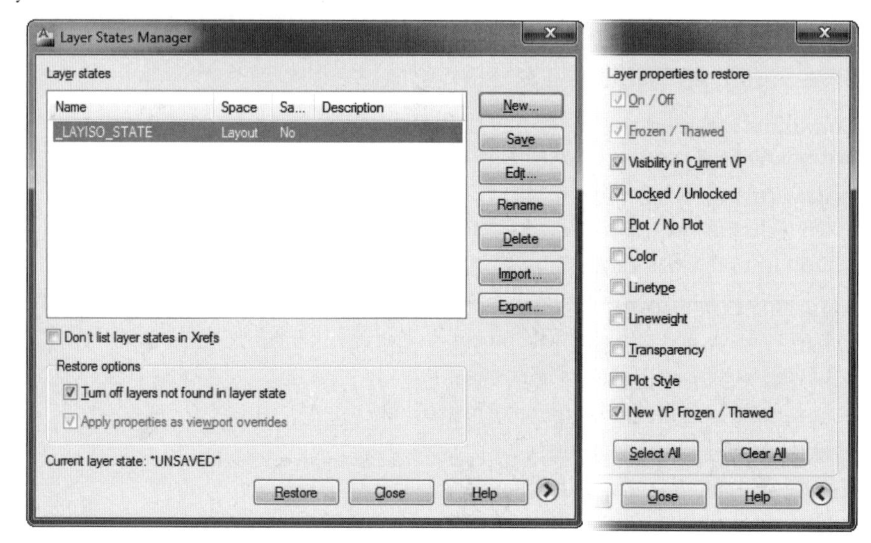

New displays the New Layer State to Save dialog box, detailed later.

Save saves the layer state settings.

Edit edits the selected layer state (detailed later).

Delete deletes the selected layer state.

Rename renames the selected layer state.

Import imports layer states from DWG drawings, DWS CAD standards, DWT templates, and LAS layer state files.

Export exports the selected layer state to LAS files.

☐ **Don't List Layer States in Xrefs** hides layer states stored in externally-referenced drawings.

Restore Options

☑**Turn off layers not found in layer state** turns off layers with unsaved settings.

☑**Apply Properties as Viewport Overrides** applies property overrides to the current viewport (available only when this command is started in a viewport.)

Restore closes the dialog box, and restores the selected layer state.

MORE RESTORE OPTIONS

Layer Properties to Restore toggles the properties to restore with the layers.

Select All turns on all properties.

Close All turns off all properties.

NEW LAYER STATE TO SAVE DIALOG BOX

New Layer State Name specifies the name of the layer state.

Description provides an optional description of the layer state.

OK displays the Layer States Manager dialog box, described earlier.

EDIT LAYER STATE DIALOG BOX

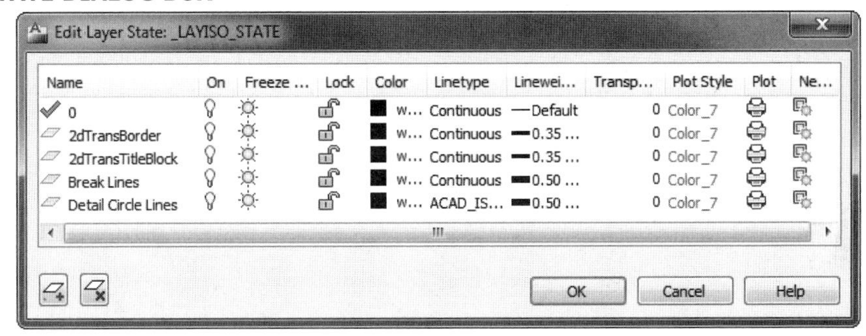

To change the property of a layer, click any icon. For example, click an On icon to turn off the layer.

To change the sort order, click headings. For example, click Color to sort by colors numbers.

To create a layer state, select one or more layer names, and then click OK. (Hold down the Ctrl key to select more than one layer.)

Add Layer to Layer State adds layers to the layer state; displays the Select Layers to Add to Layer State dialog box. This option does not work when all layers are included in the state.

Remove Layer From Layer State removes selected layers from the layer state.

OK returns to the Edit Layer State dialog box.

RELATED COMMANDS

Layer creates and sets layers and modes.

LayerP returns to the previous layer.

View stores layer settings with named views.

TIPS

- This command makes layer states more accessible by giving them their own command. You can still access this dialog box from the layers dialog box.
- Use LAS layer state files to share layer states among drawings.
- To control layer states at the command line, such as in macros, use the stAte option of the -Layer command.

LayIso / LayUniso

2007 Isolates or fades all layers except those holding selected objects (short for "LAYer ISOlate").

Command	*Aliases*	*Keyboard Shortcuts*	*Menu*	*Ribbon*
layiso	...	...	**Format** ↳**Layer Tools** ↳**Layer Isolate**	**Home** ↳**Layers** ↳**Isolate**

Command: layiso
Current setting: Lock layers, Fade = 90
Select objects on the layer(s) to be isolated or [Settings]: *(Select an object, or type S).*
Select objects on the layer(s) to be isolated or [Settings]: *(Press Enter.)*
Layer *layername* **has been isolated.**

Isolated layers appear normal; all other layers appear faded, or disappear from view:

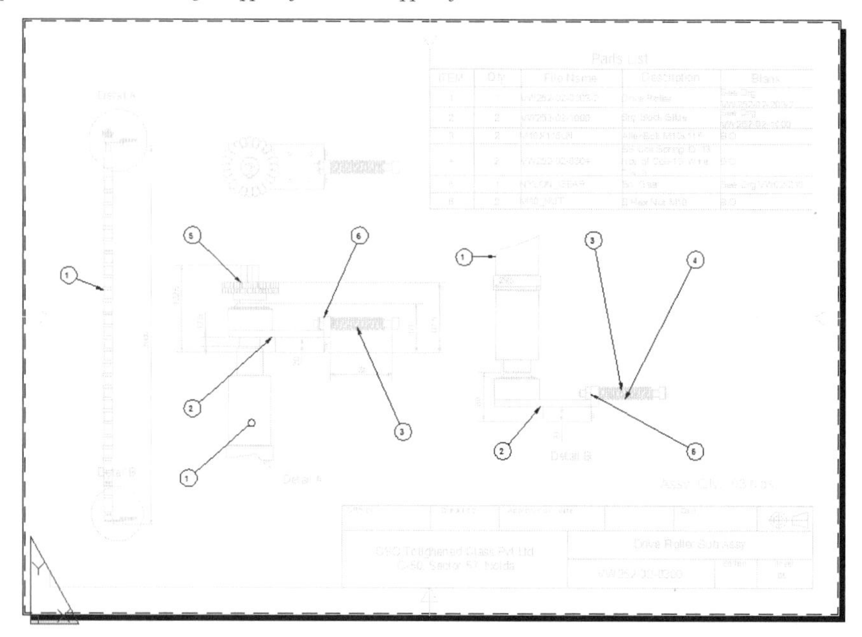

COMMAND OPTIONS

Select objects selects the objects whose layers will be isolated from all other layers.

Settings offers these options:

Enter setting for layers not isolated [Off/Lock and fade] <Lock>: *(Enter an option.)*

Off turns off all unselected layers; prompts you:

In paperspace viewport use [Vpfreeze/Off]: *(Enter V to freeze objects in just the selected viewport, or O to turn off layers in all viewports.)*

Lock and fade locks all unselected layers, and makes them look faded; prompts you:

Enter fade value (0-90): *(Enter a value between 0 for no fading and 90 for maximum fading.)*

-Layout

2000 Creates and deletes paper space layouts on the command line.

Command	Aliases	Keyboard Shortcuts	Menu	Ribbon
-layout	**layout** **lo**	...	**Insert** **↳Layout**	...

Command: -layout

Enter layout option [Copy/Delete/New/Template/Rename/SAveas/Set/?] <set>: *(Enter an option.)*

COMMAND LINE OPTIONS

Copy copies a layout to create a new layout.

Delete deletes a layout; the Model tab cannot be deleted.

New creates a new layout tab, automatically generating the name for the layout (default = first unused tab in the form Layout*n*), which you may override.

Template displays the Select File dialog box, which allows you to select a *.dwg* drawing or *.dwt* template file to use as a template for a new layout. If the file has layouts, it displays the Insert Layout(s) dialog box.

Rename renames a layout.

SAveas saves the layouts in a drawing template (*.dwt*) file. The last current layout is used as the default for the layout to save.

Set makes a layout current.

? lists the layouts in the drawing in a format similar to the following:

```
Active Layouts:
Layout: Layout1        Block name: *PAPER_SPACE.
Layout: Layout2        Block name: *Paper_Space5.
```

Shortcut Menu

Right-click any layout tab. (Layout tabs must be visible for the following options to work: If layout tabs are not visible, right-click the layout button on the status bar, and then select Display Layout and Model tabs.)

New layout
From template...
Delete
Rename
Move or Copy...
Select All Layouts
Activate Previous Layout
Page Setup Manager...
Plot...
Import Layout as Sheet...
Export Layout to Model...
Hide Layout and Model tabs
Model
Drive Roller (Brush)

New layout creates a new layout with the default name of "Layout*n*."

From template displays the Select File and Insert Layout dialog boxes.

Delete deletes the selected layout; displays a warning dialog box.

Rename asks if you want to rename the layout, or the layout and the sheet.

- **Rename Layout and Sheet** displays the Rename & Renumber dialog box; see SheetSet command.
- **Rename Layout Only** allows you to edit the name on the layout tab:

Move or Copy displays the Move or Copy dialog box. (Rather than use this dialog box, it is much easier to move tabs by dragging them, or to copy tabs by holding down the Ctrl key while dragging them.)

Select All Layouts selects all layouts.

Activate Previous Layout returns to the previously-accessed layout; useful when a drawing has many layouts.

Activate Model Tab returns to the Model tab.

Page Setup displays the Page Setup dialog box; see the PageSetup command.

Plot displays the Plot dialog box; see the Plot command.

Import Layout as Sheet turns the layout into a sheet; available only when the Sheet Set Manager is operating. Displays the Import Layouts as Sheets dialog box. A layout can be assigned only to one sheet; if needed, make copies of the layout for additional sheets.

Export Layout to Model turns the layout into a scaled model space object; see the ExportLayout command.

Hide Layout and Model tabs moves the tabs to the status bar.

MOVE OR COPY DIALOG BOX

Access this dialog box through the Move or Copy item on the shortcut menu.

Before layout lists the names of layouts; select one to appear before the current layout.

(move to end) moves the current layout to the end of layouts.

☐ **Create a copy** copies selected layouts.

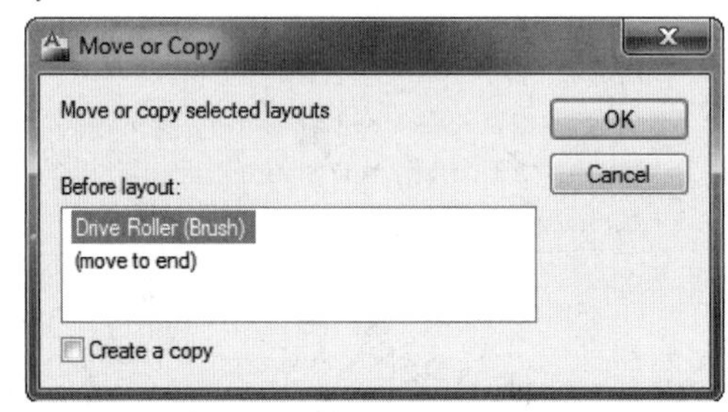

IMPORT LAYOUTS AS SHEETS DIALOG BOX

Access this dialog box through the Import Layout as Sheet item on the shortcut menu.

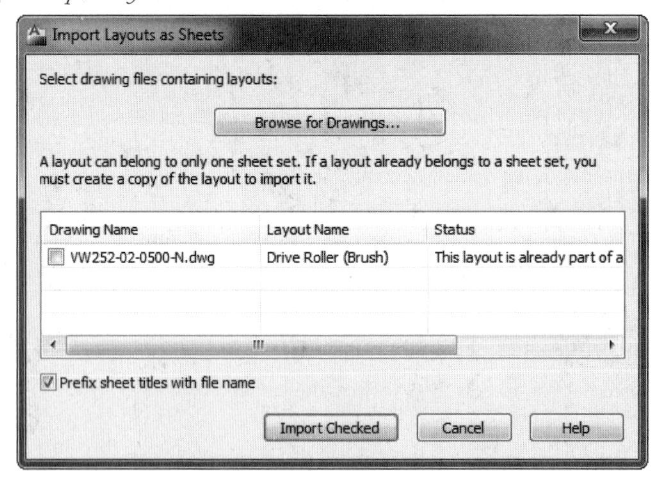

Browser for Drawings opens the Browse dialog box; select a drawing.

☐ *filename*.**dwg** selects drawings (when checked).

Prefix Sheet Titles with File Name identifies the source drawing for the sheets.

Import Checked imports the drawings with checked boxes.

RELATED COMMANDS

LayoutWizard creates and deletes paper space layouts via a wizard.

QvLayout turns on the thumbnail strip of layout views.

STARTUP SWITCH

/layout specifies the layout tab to show when AutoCAD starts up.

TIPS

- "Layout" is the name for paper space (as of AutoCAD 2000).
- Layout names can be up to 255 characters long, of which the first 31 characters are displayed in the tabs.
- Drawings can have up to 255 layouts each.
- Copy tabs by holding down the Ctrl key and dragging them; move tabs by dragging them. To copy or move more than one tab, hold down the Shift key to select them.
- Layout tabs can be moved between the drawing bar and status bar:
  - On the status bar, right-click the Layout button, and then select **Display Layout and Model tabs**:

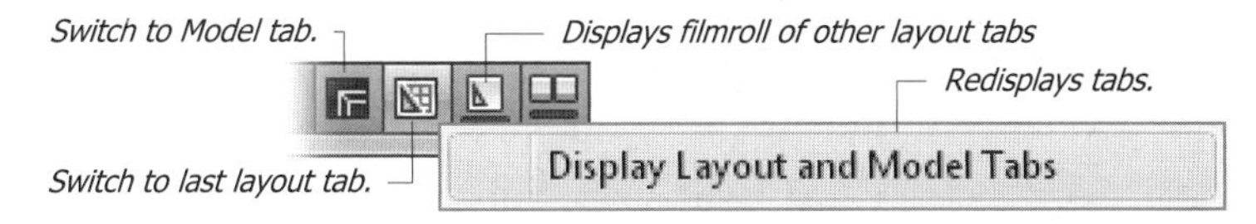

  - Right-click any layout tab, and then select **Hide Layout and Model tabs**:
- To switch between layouts, click tabs located below the drawing or on the status bar.

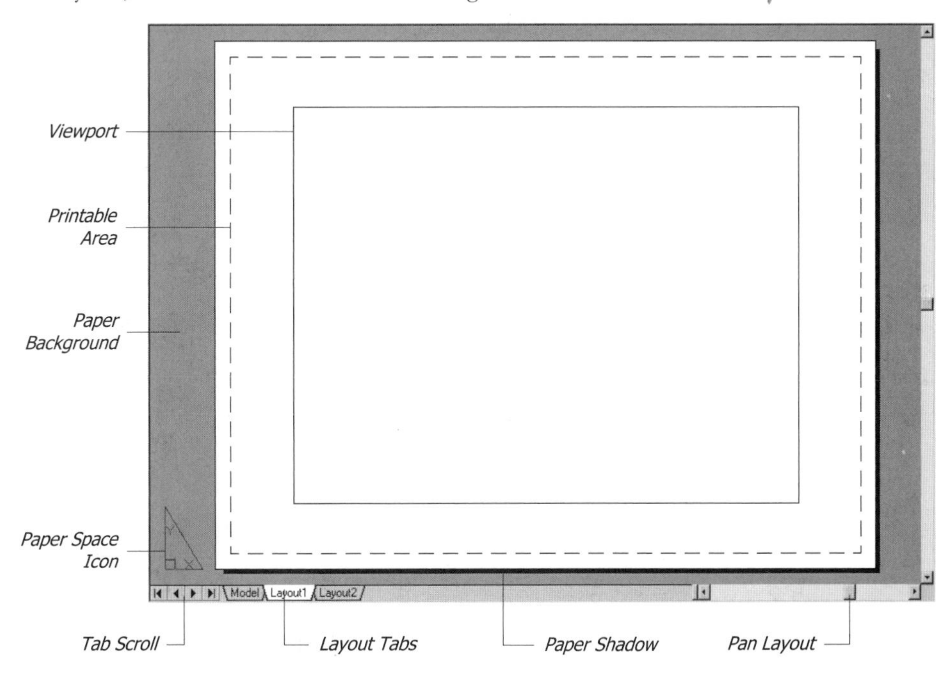

- The Model tab cannot be deleted, renamed, or copied.

LayoutWizard

2000 Creates and deletes paper space layouts via a wizard.

Command	*Aliases*	*Keyboard Shortcuts*	*Menu*	*Ribbon*
layoutwizard	...	...	**Insert** **↳Layout** **↳Create Layout Wizard**	...
			Tools **↳Wizards** **↳Create Layout**	

Command: layoutwizard

Displays dialog box.

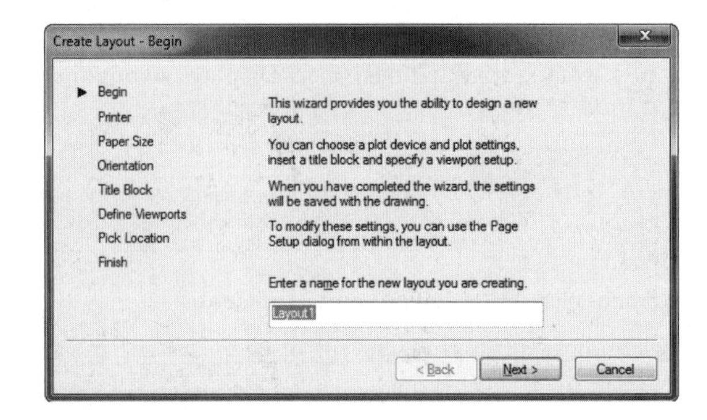

CREATE LAYOUT DIALOG BOX

Enter a name specifies the name for the layout.

Back displays the previous dialog box.

Next displays the next dialog box.

Cancel cancels the command.

Printer page

Select a configured plotter selects a printer or plotter to output the layout.

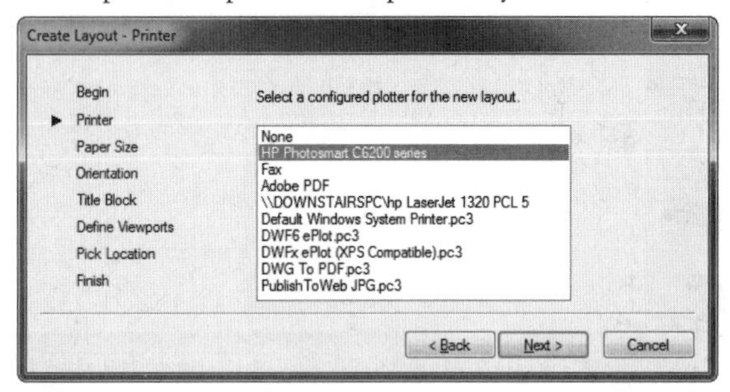

Paper Size page

Select a paper size... selects a size of paper supported by the output device.

Enter the paper units

⊙ **Millimeters** measures paper size in metric units.

○ **Inches** measures paper size in Imperial units.

○ **Pixels** measures paper size in dots per inch.

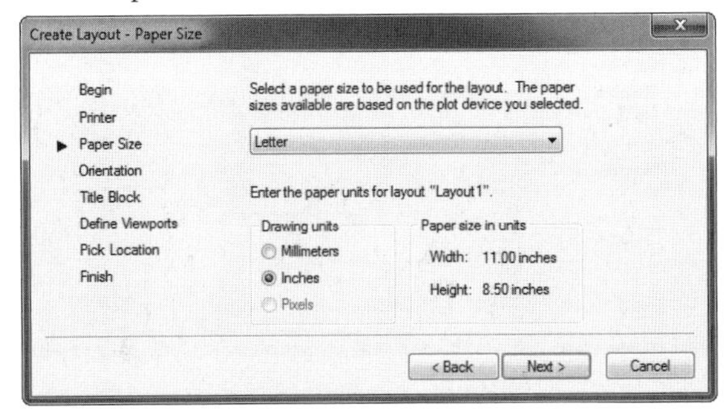

Orientation page

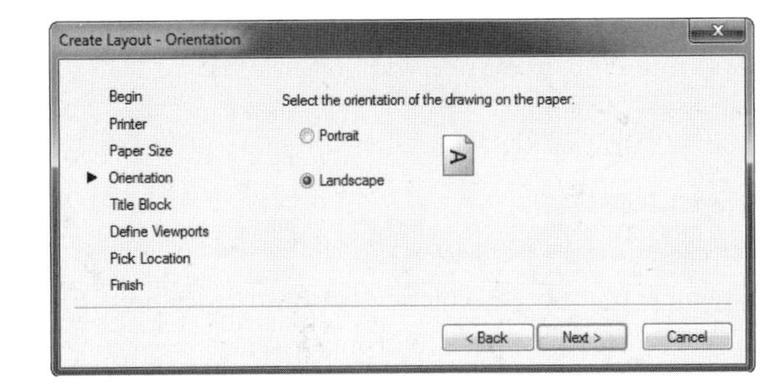

Select the orientation

- **Portrait** plots the drawing vertically.
- **Landscape** plots the drawing horizontally.

Title Block page

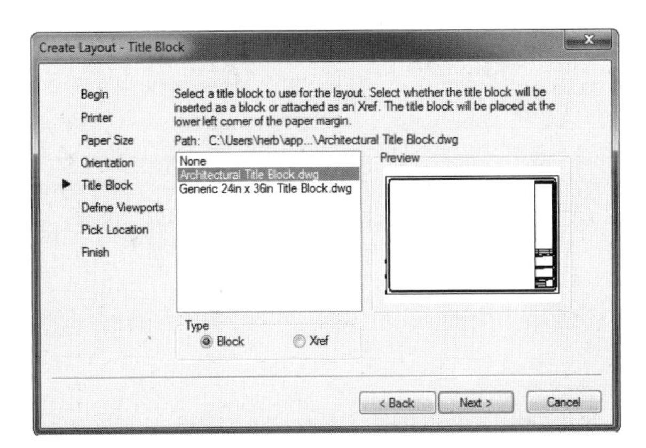

Select a title block... specifies a title border for the drawing as a block or an xref:

⊙ **Block** inserts the title border drawing.

○ **Xref** references the title border drawing.

Define Viewports page

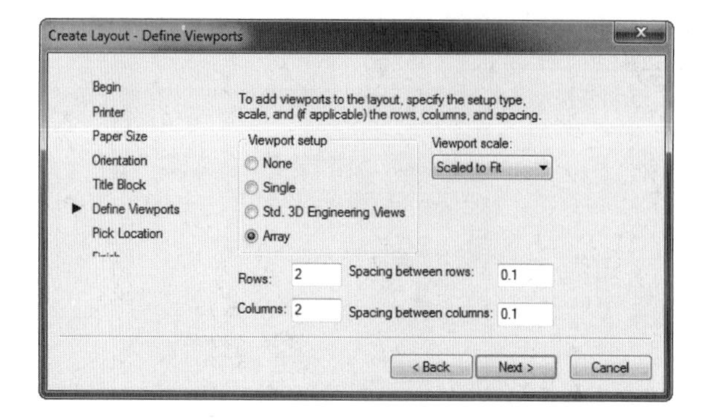

Viewport Setup

○ **None** creates no viewport.

⊙ **Single** creates a single paper space viewport.

○ **Std. 3D Engineering Views** creates top, front, side, and isometric views.

○ **Array** creates a rectangular array of viewports.

Viewport scale

- **Scaled to Fit** fits the model to the viewport.
- *mm:nn* specifies a scale factor, ranging from 100:1 to 1/128":1'0".

Rows and **Columns** specify the number of rows and columns for arrayed viewports.

Spacing between rows/columns specifies the vertical/horizontal distance between viewports.

Pick Location page

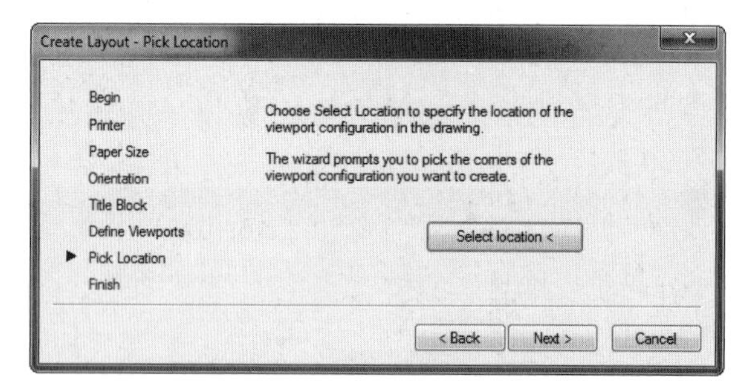

Select location specifies the corners of a rectangle holding the viewports. AutoCAD prompts you to pick a rectangle.

Finish page

Finish exits the dialog box and creates the layout.

RELATED COMMAND

Layout creates a layout on the command line.

RELATED SYSTEM VARIABLE

CTab contains the name of the current tab.

LayTrans

2002 Translates layer names (short for LAYer TRANSlation).

Command	*Aliases*	*Keyboard Shortcuts*	*Menu*	*Ribbon*
laytrans	...	...	**Tools** **↳CAD Standards** **↳Layer Translator**	**Manage** **↳Standards** **↳Layer Translator**

Command: laytrans

Displays dialog box.

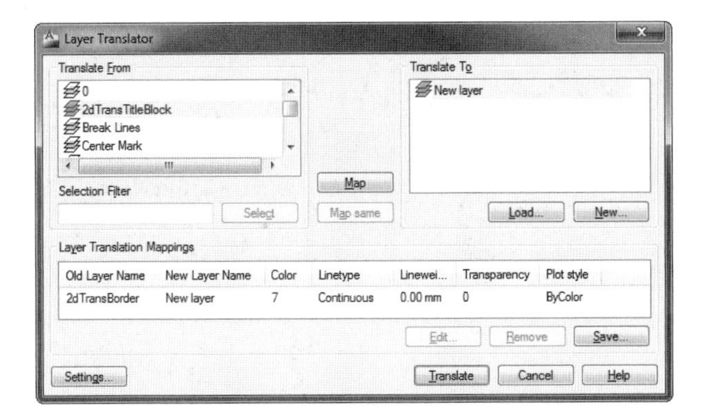

LAYER TRANSLATOR DIALOG BOX

Translate From

Translate From lists the names of layers in the current drawing; icons indicate whether the layer is being used (referenced):

Contains at least one object.

Contains no objects, and can be purged.

Selection Filter specifies a subset of layer names through wildcards.

Select highlights the layer names that match the selection filter.

Map maps the selected layer(s) in the Translate From column to the selected layer in the Translate To column.

Map same maps layers automatically with the same name.

Translate To

Translate To lists layer names in the drawing opened with the Load button.

Load accesses the layer names in another drawing via the Select Drawing File dialog box.

New creates a new layer via the New Layer dialog box.

Layer Translation Mappings options

Edit edits the linetype, color, lineweight, and plot style settings via the Edit Layer dialog box; identical to the New Layer dialog box.

Remove removes the selected layer from the list.

Save saves the matching table to a *.dws* (drawing standard) file.

Settings specifies translation options via the Settings dialog box.

Translate changes the names of layers, as specified by the Layer Translation Mappings list.

NEW LAYER DIALOG BOX

Access this dialog box through the New button. Identical to the Edit Layer dialog box.

Name specifies the name of the layer, up to 255 characters long.

Linetype selects a linetype from those available in the drawing.

Color selects a color; select Other for the Select Color dialog box.

Lineweight selects a lineweight.

Plot style selects a plot style from those available in the drawing; this option is not available if plot styles have not been enabled in the drawing.

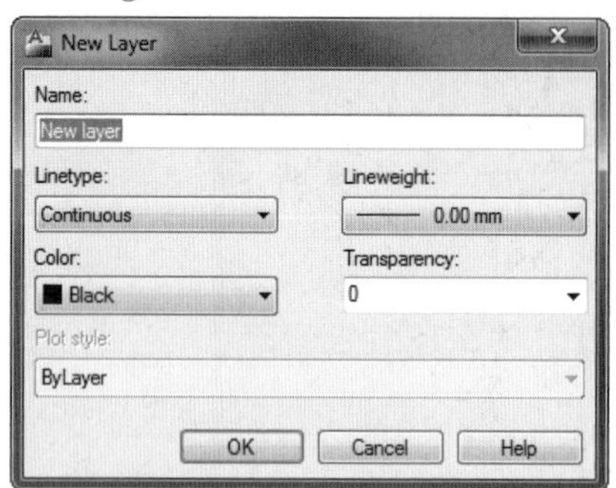

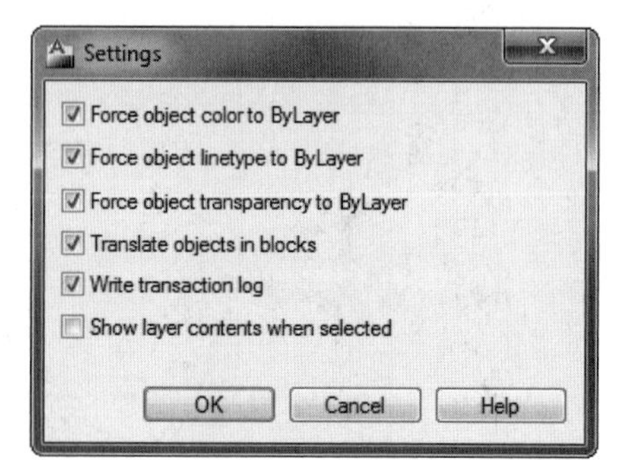

SETTINGS DIALOG BOX

Access this dialog box through the Settings button.

☑**Force object color to Bylayer** forces every translated layer to take on color Bylayer.

☑**Force object linetype to Bylayer** forces every translated layer to take on linetype Bylayer.

☑**Translate objects in blocks** forces objects in blocks to take on new layer assignments.

☑**Write transaction log** writes the results of the translation to a *.log* file, using the same filename as the drawing. When command is complete, AutoCAD reports:

```
Writing transaction log to filename.log.
```

☐**Show layer contents when selected** lists the names of selected layers in the Translate From list.

RELATED COMMANDS

Standards creates the standards for checking drawings.

CheckStandards checks the current drawing against a list of standards.

Layer creates and sets layers and modes.

RELATED FILES

.dws drawing standard file; saved in DWG format.

.log log file recording layer translation; saved in ASCII format.

TIPS

- You can purge unused layers (those prefixed by a white icon) within the Layer Translator dialog box:
  1. Right click any layer name in the Translate From list.
  2. Select Purge Layers. The layers are removed from the drawing.
- You can load layers from more than one drawing file; duplicate layer names are ignored. The first one wins.

Leader

Rel. 13 Draws leader lines with one or more lines of text.

Command	Aliases	Keyboard Shortcuts	Menu	Ribbon
leader	**lead**	...	...	...

Command: leader

Specify leader start point: *(Pick a starting point, such as at 1 illustrated below.)*

Specify next point: *(Pick the shoulder point — point 2.)*

Specify next point or [Annotation/Format/Undo] <Annotation>: *(Press Enter to start the annotation, or pick another point — point 3 — or enter an option.)*

Specify next point or [Annotation/Format/Undo] <Annotation>: *(Press Enter to specify the text — 4.)*

Enter first line of annotation text or <options>: *(Enter text — 5 — or press Enter for options.)*

Enter next line of annotation text: *(Press Enter.)*

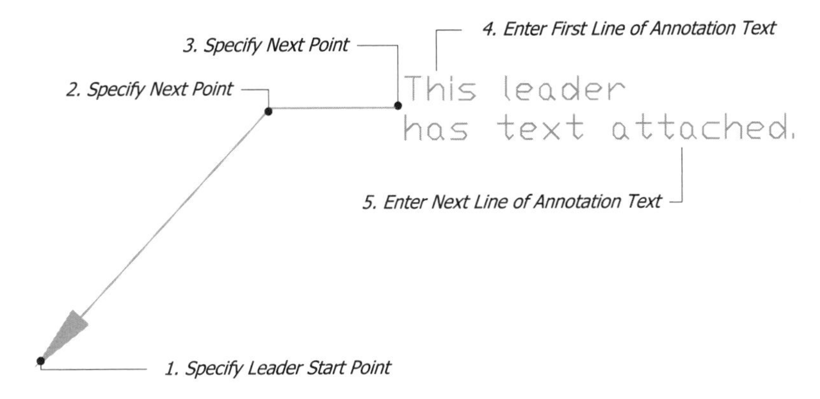

COMMAND LINE OPTIONS

Specify start point specifies the location of the arrowhead.

Specify next point positions the leader line's vertex.

Undo undoes the leader line to the previous vertex.

Format options

Enter leader format option [Spline/STraight/Arrow/None] <Exit>: *(Enter an option.)*

Spline draws the leader line as a NURBS (short for non uniform rational Bezier spline) curve; see the Spline command.

STraight draws a straight leader line (default).

Arrow draws the leader with an arrowhead (default).

None draws the leader with no arrowhead.

Annotation options

Enter first line of annotation text or <options>: *(Press Enter.)*

Enter an annotation option [Tolerance/Copy/Block/None/Mtext] <Mtext>: *(Enter an option.)*

Enter first line of annotation text specifies the leader text.

Tolerance places one or more tolerance symbols; see the Tolerance command.

Copy copies text from another part of the drawing.

Block places a block; see the -Insert command.

None specifies no annotation.

MText displays the Text Formatting toolbar; see the MText command.

RELATED DIMENSION VARIABLES

DimAsz specifies the size of the arrowhead and the hookline.

DimBlk specifies the type of arrowhead.

DimClrd specifies the color of the leader line and the arrowhead.

DimGap specifies the gap between hookline and annotation (gap between box and text).

DimScale specifies the overall scale of the leader.

RELATED COMMANDS

DimStyle defines styles for leaders.

MLeader draws multi-line leaders.

QLeader provides more options than Leader.

TIPS

- This command draws several types of leader:

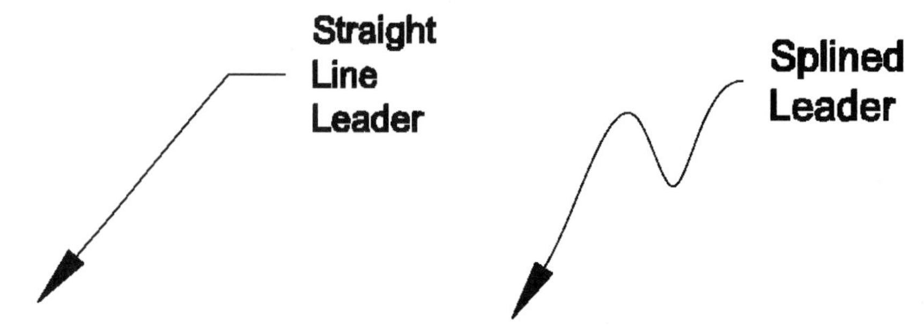

- The text in a leader is an mtext (multiline text) object.
- Use the \P metacharacter to create line breaks in leader text.
- Autodesk recommends using the MLeader command, which has a dialog box for settings and more options.
- As of AutoCAD 2008, leaders can take on the annotative property.
- For leaders with multiple arrowheads, use the MLeader command.

Lengthen

Rel. 13 Lengthens and shortens open objects by several methods.

Command	*Aliases*	*Keyboard Shortcuts*	*Menu*	*Ribbon*
lengthen	**len**	**...**	**Modify** **↳Lengthen**	**Home** **↳Modify** **↳Lengthen**

Command: lengthen
Select an object or [DElta/Percent/Total/DYnamic]: *(Select an open object.)*
Current length: ***n.nnnn***

COMMAND LINE OPTIONS

Select an object displays length; does not change the object.

DElta options
Enter delta length or [Angle] <0.0000>: *(Enter a value, or type A.)*
Specify second point: *(Pick a point.)*
Select an object to change or [Undo]: *(Select an open object, or type U.)*

Enter delta length changes the length by the specified amount.

Angle changes the angle by the specified value.

Undo undoes the most recent lengthening operation.

Percent options
Enter percentage length <100.0000>: *(Enter a value.)*
Select an object to change or [Undo]: *(Select an open object, or type U.)*

Enter percent length changes the length to a percentage of the current length.

Total options
Specify total length or [Angle] <1.0000)>: *(Enter a value, or type A.)*
Select an object to change or [Undo]: *(Select an open object, or type U.)*

Specify total length changes the length by an absolute value.

Angle changes the angle by the specified value.

DYnamic options
Select an object to change or [Undo]: *(Select an open object, or type U.)*
Specify new end point: *(Pick a point.)*

Specify new end point changes the length dynamically by dragging.

TIPS

- Lengthen command only works with open objects, such as lines, arcs, and polylines; it does not work with closed objects, such as circles, polygons, and regions.
- DElta changes lengths using these measurements: (1) distance from endpoint of selected object to pick point; or (2) incremental length measured from endpoint of angle.
- Angle option applies only to arcs.

Light

Rel. 12 Places point, spot, distant, target, and web lights in drawings for use by the Render command.

Command	*Aliases*	*Keyboard Shortcuts*	*Menu*	*Ribbon*
light	**pointlight**	...	**View**	**Visualize**
	distantlight		↳**Render**	↳**Lights**
	spotlight		↳**Light**	
	weblight			
	targetpoint			
	freepoint			
	freespot			
	freeweb			

Command: light

Enter light type [Point/Spot/Web/Targetpoint/Freespot/freeweB/Distant] <Freespot>: *(Enter an option.)*

COMMAND LINE OPTIONS

Point places point lights, which radiate light in all directions; you can also use the PointLight alias.

Spot places spot lights, which direct light at a target point; you can also use the SpotLight alias.

Web places web lights, which use IEF data to produce a precise illumination distribution; you can also use the WebLight alias.

Targetpoint places target point lights; you can also use the TargetPoint alias.

Freespot places spot lights without target points; you can also use the FreeSpot alias.

freeweB places web lights without target points; you can also use the FreeWeb alias.

Distant places distant lights, which do not attenuate, like the sun; you can also use the DistantLight alias.

Point light options

When system variable LightingUnits = 0 (default), displays the following prompts:

Specify source location <0,0,0>: *(Pick a point, or enter x, y, z coordinates.)*

Enter an option to change [Name/Intensity/Status/shadoW/Attenuation/Color/eXit] <eXit>: *(Press Enter, or enter an option.)*

Glyph placed by PointLight command. Point light glyphs can be relocated with their position grips.

Name names the light; if you do not give it a name, Autodesk assigns generic names, such as "Pointlight1."

Intensity specifies the light's brightness, ranging from 0 to the largest real number.

Status toggles the light on and off.

Shadow determines how the light casts shadows:

- **Off** — shadows are turned off.
- **Sharp** — shadows have sharp edges.
- **Soft** — shadows have soft edges; most realistic.

Attenuation specifies how the light diminishes or decays with distance:

- **None** — does not diminish.
- **Inverse Linear** — diminishes light intensity by the inverse of the linear distance.
- **Inverse Squared** — diminishes light intensity by the inverse of the square of the distance.
- **Attenuation Start Limit** — specifies where light starts (offset from the light's center); *Warning!* Attenuation limit options are not supported by OpenGL display drivers.
- **Attenuation End Limit** — specifies where light ends; no light is cast beyond this point.

Color specifies the light's color.

eXit exits the command and places the light glyph in the drawing.

The prompt changes when system variable LightingUnits = 1 or 2, as follows:

Enter an option to change [Name/Intensity factor/Status/Photometry/shadoW/Attenuation/filterColor/eXit] <eXit>: *(Enter an option.)*

Prompts that differ are listed here:

Intensity factor is the same as Intensity.

Photometry is a more accurate method of specifying the brightness and color of lights.

filterColor is the same as Color.

Photometric options

Enter a photometric option to change [Intensity/Color/eXit] <I>:
Enter intensity (Cd) or enter an option [Flux/Illuminance] <1500>:
Enter color name or enter an option [?/Kelvin] <D65>:

Intensity factor specifies brightness in terms of candelas (Cd), flux, or illuminance.

Color specifies color by name or by degrees Kelvin (absolute temperature).

Freepoint options

Options for Freepoint lights are identical to those of point lights, except that the target prompt is missing.

Spotlight and Targetpoint options

When system variable LightingUnits = 0, displays the following prompts:

Specify source location <0,0,0>: *(Pick a point.)*
Specify target location <0,0,-10>: *(Pick another point.)*
Enter an option to change [Name/Intensity/Status/Hotspot/Falloff/shadoW/Attenuation/Color/eXit] <eXit>: *(Press Enter, or enter an option.)*

Glyph placed by SpotLight command.

Name names the light; if you don't give it a name, generic names are assigned, such as "Spotlight1."

Intensity specifies the light's brightness, ranging from 0 to the largest real number.

Status toggles the light on and off.

Hotspot defines the angle of the brightest (inner) cone of light (a.k.a. beam angle); ranges from 0 to 160 degrees.

Falloff defines the angle of the full (outer) cone of light (field angle); range is 0 to 160 degrees.

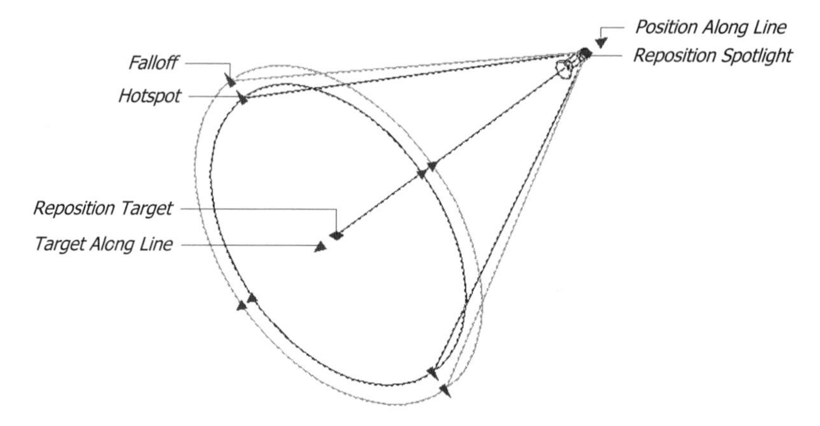

Spot and target lights can be edited using the grips shown above.

Shadow determines how the light casts shadows:

- **Off** — shadows are turned off.
- **Sharp** — shadows have sharp edges.
- **Soft** — shadows have soft edges; most realistic.

Attenuation specifies how the light diminishes or decays with distance:

- **None** — does not diminish.
- **Inverse Linear** — diminishes light intensity by the inverse of the linear distance.
- **Inverse Squared** — diminishes light intensity by the inverse of the square of the distance.
- **Attenuation Start Limit** — specifies where light starts (offset from the light's center).
- **Attenuation End Limit** — specifies where light ends; no light is cast beyond this point.

Color specifies the light's color.

eXit exits the command and places the light glyph in the drawing.

The prompts change when system variable LightingUnits = 1 or 2. See Pointlight options for details.

Freespot option

Options for Freespot lights are identical to those of spot lights, except that the target prompt is missing.

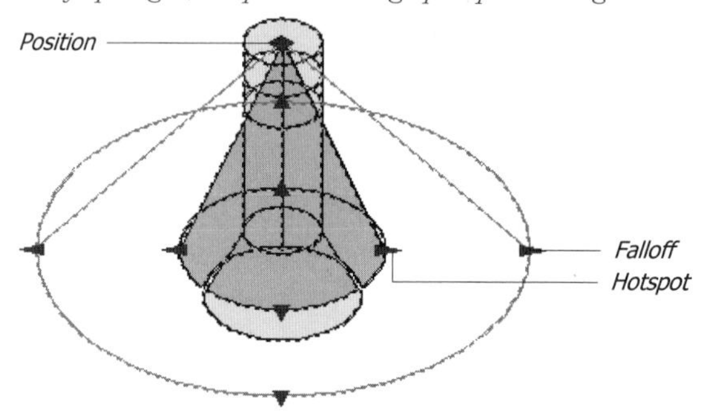

Free spot lights can be edited using grips.

Weblight options

This option is not available when system variable LightingUnits = 0. When LightingUnits = 1 or 2:

Specify source location <0,0,0>: *(Pick a point, or enter x,y,z coordinates.)*

Specify target location <0,0,0>: *(Pick a point, or enter x,y,z coordinates.)*

Enter an option to change [Name/Intensity factor/Status/Photometry/weB/filterColor/eXit] <eXit>: *(Press Enter, or enter an option.)*

Glyph placed by WebLight command. Web light glyphs can be relocated with their Position grips.

Options are the same as point lights, with the following exception:

weB specifies the IES file to use; found in the *C:\Autodesk\AutoCAD2011\x64\acad\Application Data\Autodesk\WebFiles* folder.

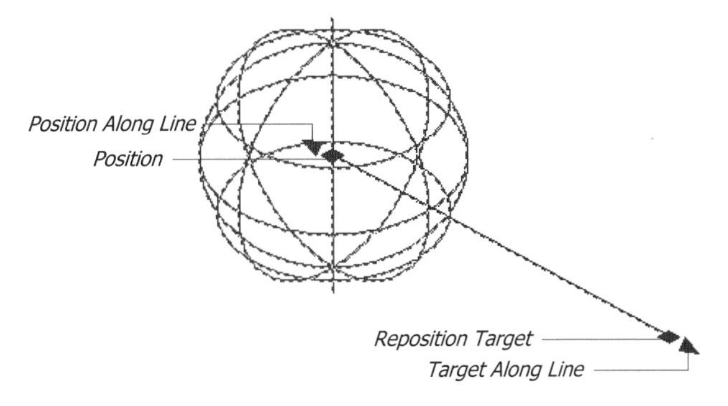

Web lights can be edited with grips.
***Above:** Without IES data.*
***Below:** With IES data showing the shape of the light's dispersion.*

Freeweb options

Options for Freeweb lights are identical to those of web lights, except that the target prompt is missing.

Distant light options

When system variable LightingUnits = 0, displays the following prompts:

Specify light direction FROM <0,0,0> or [Vector]: *(Pick a point, or type V.)*

Specify light direction TO <1,1,1>: *(Pick another point.)*

Enter an option to change [Name/Intensity/Status/shadoW/Color/eXit] <eXit>: *(Press Enter, or enter an option.)*

The prompt changes when system variable LightingUnits = 1 or 2. See Pointlight options, above.

From specifies the location of the distant light.

Vector prompts you to enter a vector direction, such as 0,-1,1.

To specifies the direction of the light.

Name names the light; Autodesk assigns generic names, such as "Pointlight1."

Intensity specifies the light's brightness, ranging from 0 to the largest real number.

Status toggles the light on and off.

shadoW determines how the light casts shadows:

- **Off** — shadows are turned off.
- **Sharp** — shadows have sharp edges.
- **Soft** — shadows have soft edges; most realistic.

Color specifies the light's color.

eXit exits the command and places the light in the drawing; distant lights show no glyphs.

RELATED COMMANDS

LightList displays a list of lights defined in the current drawing.

GeographicLocation determines the position of the sun based on longitude and latitude.

SunProperties specifies the properties of sunlight.

Ribbon provides additional controls for lights, sun position, and shadows.

Options changes the colors of light glyphs.

RELATED SYSTEM VARIABLES

LightGlyphDisplay toggles the display of light glyphs.

LightingUnits toggles photometric lighting.

ShadowPlaneLocation locates a ground plane onto which shadows can be cast.

3dConversionMode controls how material and light definitions are converted when a pre-AutoCAD 2008/9 drawing is opened.

TIPS

- This command works in model space only.
- Lights can be edited through the Properties palette, through grips editing, or through the Light List palette; see the LightList command.
- Turning on shadows slows down AutoCAD's display speed.

Shadows cast onto the shadow plane by the VsShadows system variable.

- As of AutoCAD 2007, light glyphs replace blocks to represent lights.
- When a drawing has no lights defined, AutoCAD assumes ambient light, which ensures every object in the scene has illumination; ambient light is an omnipresent light source.
- Use the ribbon's Render | Sun Status control to turn on the sun light.
- Use point lights to simulate light bulbs.
- New drawings have a bright light over your left shoulder and a dimmer one just over your right one.
- Light can be named using uppercase and lowercase letters, numbers, spaces, hyphens, and underscores to a maximum of 256 characters.
- Use the 3dConfig command's View Tune Log to identify the driver your computer's graphics board uses.

DEFINITIONS

Candela (cd) — international unit of luminous intensity.

Constant light — attenuation is 0; default intensity is 1.

Extents distance — distance from minimum lower-left coordinate to the maximum upper-right.

Falloff — angle of the full light cone; field angle ranges from 0 to 160 degrees.

Foot-candle (fc) — American unit of illuminance.

HLS color — colors by hue (color), lightness, and saturation (amount of gray).

Hotspot — brightest cone of light; beam angle ranges from 0 to 160 degrees.

IES — short for Illuminating Engineering Society.

Inverse linear light — light strength decreases to ½-strength two units of distance away, and ¼-strength four units away; default intensity is ½ extents distance.

Inverse square light — light strength decreases to ¼-strength two units away, and $^1/_8$-strength four units away; default intensity is ½ the square of the extents distance.

Kelvin (K) — absolute temperature, where 0K = absolute zero.

Lux (lx) — international unit of illuminance.

RGB color — three primary colors (red, green, blue) shaded from black to white.

Web lights — real-world light distribution of commercial lamps defined by IES.

LightList / LightListClose

2007 Edits the properties of lights placed in drawings.

Commands	Aliases	Keyboard Shortcuts	Menu	Ribbon
lightlist	...	...	**View** ↳**Render** ↳**Light** ↳**Light List**	**Render** ↳**Lights** ↳**Lights in Model**
lightlistclose	...	...	...	....

This command is for editing the properties of lights after they have been placed in drawings.

Command: lightlist

Displays palette (the sun light does not appear here; see the SunProperties command.)

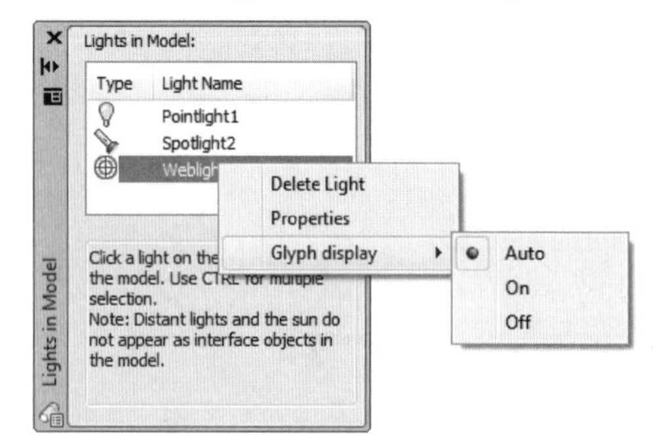

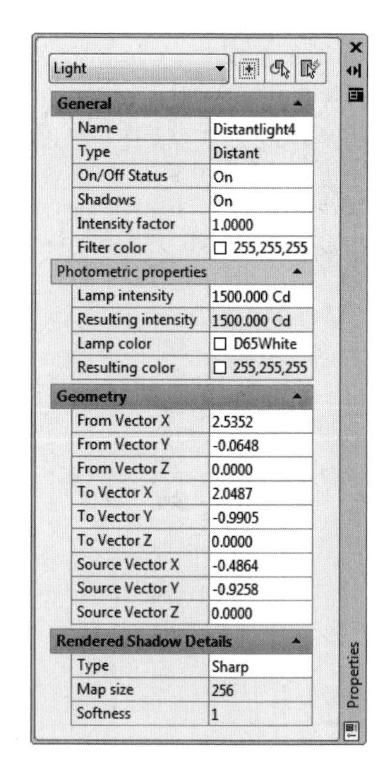

Command: lightlistclose

Closes the palette.

RIGHT-CLICK MENU

Right-click a light name to display the shortcut menu.

Delete Light erases the light from the drawing.

Properties displays the Properties palette.

Glyph Display toggles the display of light glyphs (symbols).

DISTANT LIGHT PROPERTIES

General properties

Name renames from its generic name, such as "Distantlight1."

On/Off Status toggles the light on and off.

Shadows toggles the display of shadows.

Intensity Factor specifies light's brightness; range 0 to the largest real number.

Filer Color specifies the light's filter color.

Photometric properties

These properties are displayed only when LightingUnits = 1 or 2:

Lamp Intensity specifies brightness in terms of candela (Cd), flux, or illuminance.

Lamp Color specifies color by name or by degrees Kelvin (absolute temperature).

Geometry properties

From Vector X, Y, Z specifies the x,y,z coordinates of the distant light.

To Vector X, Y, Z specifies the x,y,z coordinates of the light's target.

Source Vector X, Y, Z specifies x,y,z coordinates of light source based on altitude and azimuth.

Rendered Shadow Details properties

Type determines how the light casts shadows:

- **Off** — shadows are turned off.
- **Sharp** — shadows have sharp edges.
- **Soft** — shadows have soft edges; most realistic.

Map size specifies the size of the shadow map; range is 64x64 to 4096x4096.

Softness determines the softness of the shadow's edges.

PROPERTIES OF POINT AND TARGETPOINT LIGHTS

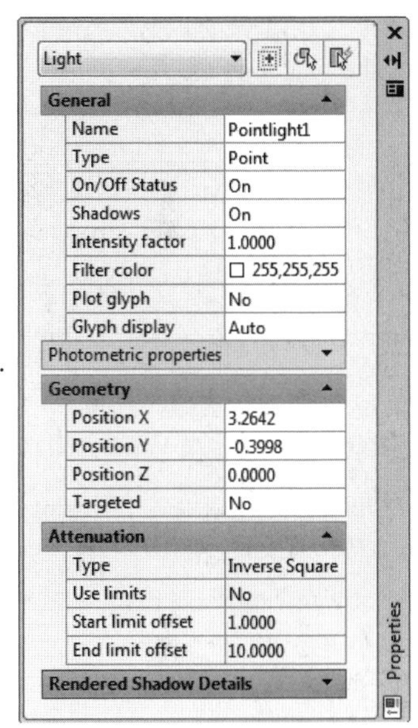

General Properties

See distant light properties.

Plot Glyph determines whether the light glyph is plotted.

Photometric properties

See distant light properties.

Geometry properties

Position X, Y, Z specifies the x,y,z coordinates of the light.

Target X, Y, Z specifies the x,y,z coordinates of the light's target (targetpoint light only).

Targeted toggles this light type between point and targetpoint.

Attenuation

Type specifies how the light diminishes or decays with distance:

- **None** — does not diminish.
- **Inverse Linear** — diminishes light intensity by the inverse of the linear distance.
- **Inverse Squared** — diminishes light intensity by the inverse of the square of the distance.

Use Limits toggles whether attenuation limits are employed:

Start Limit Offset — specifies where the light beam starts (offset from the light's center).

End Limit Offset — specifies where the light beam ends; no light is cast beyond this point.

Rendered Shadow Details properties

See distant light properties.

PROPERTIES OF SPOT AND FREESPOT LIGHTS

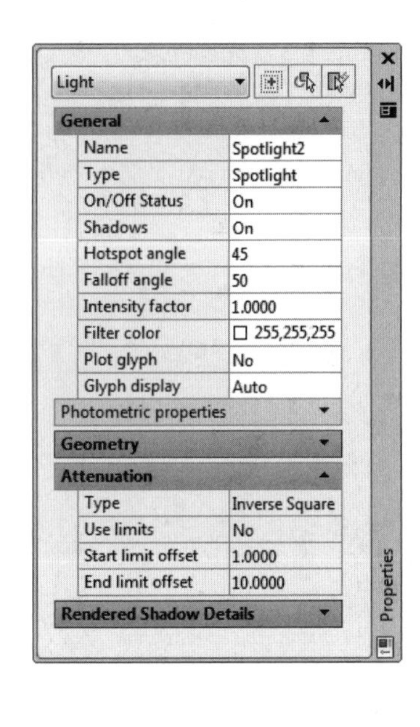

General Properties

See distant light properties.

Hotspot Angle defines the angle of the brighter, inner cone of light; from 0 to 160 degrees.

Falloff Angle defines the angle of the full, outer cone of light; ranges from 0 to 160 degrees.

Plot Glyph determines whether the light glyph is plotted.

Geometry & Photometric properties

See distant light properties.

Attenuation

Type specifies how the light diminishes or decays with distance:

- **None** — does not diminish.
- **Inverse Linear** — diminishes light intensity by the inverse of the linear distance.
- **Inverse Squared** — diminishes light intensity by the inverse of the square of the distance.

Use Limits toggles whether attenuation limits are employed:

Start Limit Offset — specifies where the light beam starts (offset from the light's center).

End Limit Offset — specifies where the light beam ends; no light is cast beyond this point.

Rendered Shadow Details properties

See distant light properties.

PROPERTIES OF WEB AND FREEWEB LIGHTS

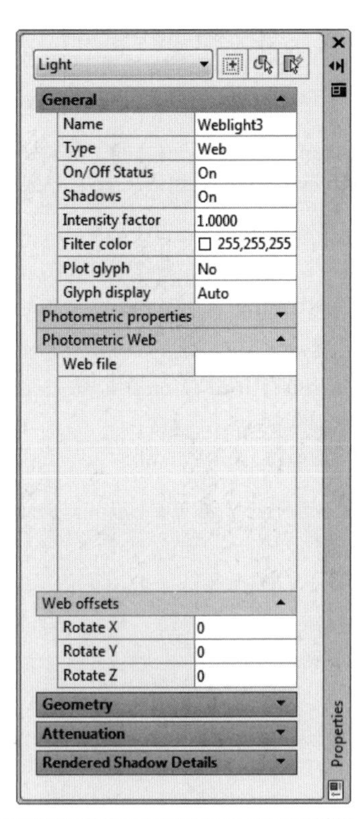

General Properties

See distant light properties.

Photometric properties

See point light properties.

Photometric Web properties

Web File selects the *.ies* file for specifying the light's 3D distribution of illumination.

Web Offsets properties

Rotate X, Y, Z rotates the web distribution orientation.

Geometry, Attenuation, and Rendered Shadow Details properties

See distant and spot light properties.

LAMP INTENSITY DIALOG BOX

Access this dialog box through Photometric Properties: choose Lamp Intensity, and then click the ▤ *button:*

⊙ **Intensity (Candela)** specifies candela of luminous intensity.

○ **Flux (Lumen)** specifies the rate of total energy leaving the lamp.

○ **Illuminance (Foot-candles)** specifies energy per area at the surface.

Distance specifies the distance; controlled by the InsUnits system variable.

Intensity Factor specifies a factor for the intensity of the lamp; can be used to darken or brighten lights.

Resulting Intensity reports the intensity multiplied by the factor.

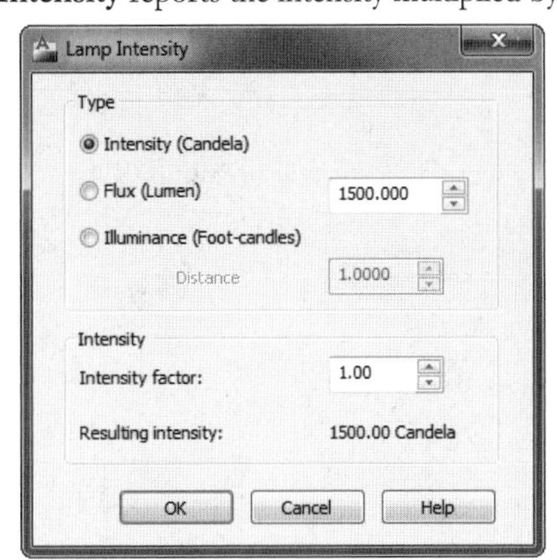

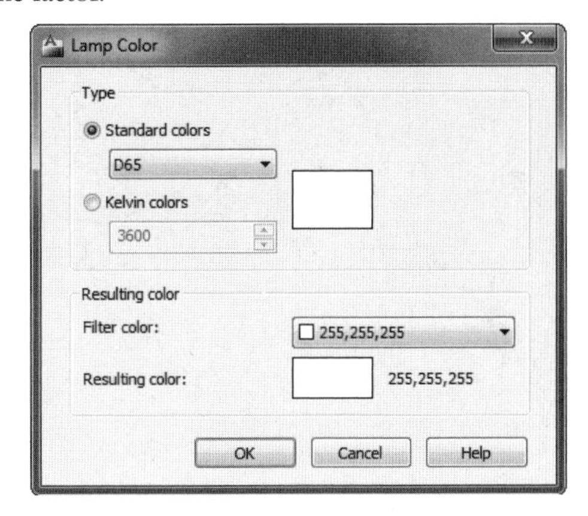

Left to right: *Lamp Intensity and Lamp Color dialog boxes*

LAMP COLOR DIALOG BOX

Access this dialog box through Photometric Properties: choose Lamp Color, and then click the ▤ *button:*

⊙ **Standard Colors** lists a catalog of standard colors (spectra), listed above.

○ **Kelvin Colors** specifies the color temperature of the light; range is 1000 to 20,000.

Filter Color specifies the color of the light's filter.

Resulting Color shows the effect of mixing colors of the light and of its filters.

RELATED COMMANDS

Light places regular and photometric lights in the current drawing; direct light commands are PointLight, SpotLight, WebLight, TargetPoint, FreeSpot, FreeWeb, Sun, and DistantLight.

Ribbon provides additional controls for lights.

RenderExposure changes the overall brightness of rendered scenes.

RELATED SYSTEM VARIABLES

LightGlyphDisplay toggles the display of light glyphs.

LightingUnits toggles between generic and photometric lighting.

LightsInBlocks toggles lights in blocks for rendering.

LinearBrightness determines the global brightness of generic lights.

LinearContrast determines the global contrast of generic lights.

LogExpBrightness determines the global brightness of photometric lighting.

LogExpContrast determines the global contrast of photometric lighting.

LogExpDaylight toggles global daylight of photometric lighting.

LogExpMidtones determines the global midtones of photometric lighting.

RenderUserLights toggles lights for renderings.

3dConversionMode controls how material and light definitions are converted when a pre-AutoCAD 2008 drawing is opened.

TIPS

- You can quickly choose a lamp color in the Photometric Properties section by clicking on Lamp Color:

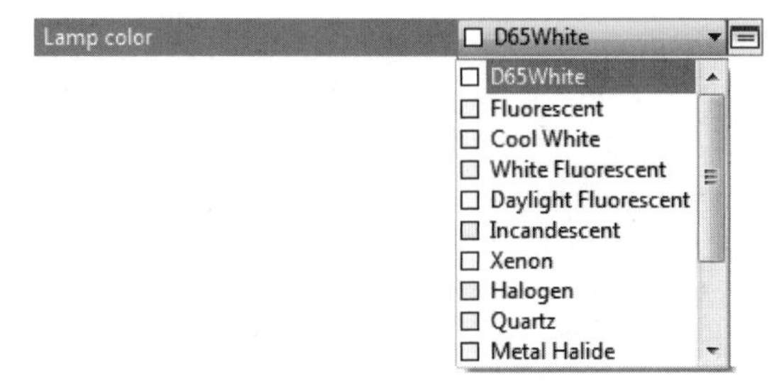

- Web lights use IES files, which are found in the *C:\Autodesk\AutoCAD2011\x64\acad\Application Data\Autodesk\WebFiles* folder.

Limits

Ver. 1.0 Defines the 2D limits in the WCS for the grid markings and in the Zoom All command; optionally prevents specifying points outside of limits.

Command	*Aliases*	*Keyboard Shortcuts*	*Menu*	*Ribbon*
'limits	...	...	**Format ↳Drawing Limits**	...

Command: limits

Reset Model *(or* **Paper***)* **space limits:**

Specify lower left corner or [ON/OFF] <0.0,0.0>: *(Pick a point, or type ON or OFF.)*

Specify upper right corner <12.0,9.0>: *(Pick another point.)*

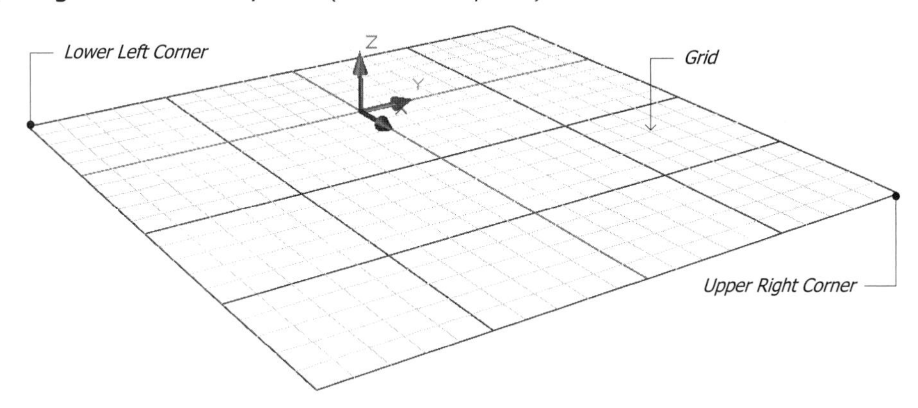

COMMAND LINE OPTIONS

OFF turns off limits checking.

ON turns on limits checking.

Enter retains limits values.

RELATED COMMANDS

Grid displays grid dots, which are bounded by limits.

Zoom displays the drawing's extents or limits (whichever is larger) with the All option.

RELATED SYSTEM VARIABLES

GridDisplay determines if the grid is constrained by the limits.

LimCheck toggles the limit's drawing check.

LimMin specifies the lower-right 2D coordinates of current limits.

LimMax specifies the upper-left 2D coordinates of current limits.

TIP

- As of AutoCAD 2007, the grid can ignore the limits set by this command; see the DSettings and Grid commands.

Line

Ver. 1.0 Draws straight 2D and 3D lines.

Command	Alias	Keyboard Shortcuts	Menu	Ribbon
line	**l**	**...**	**Draw** ↳**Line**	**Home** ↳**Draw** ↳**Line**

Command: line
Specify first point: *(Pick a starting point.)*
Specify next point or [Undo]: *(Pick another point, or type U.)*
Specify next point or [Undo]: *(Pick another point, or type U.)*
Specify next point or [Close/Undo]: *(Pick another point, or enter an option.)*
Specify next point or [Close/Undo]: *(Press Enter to end the command.)*

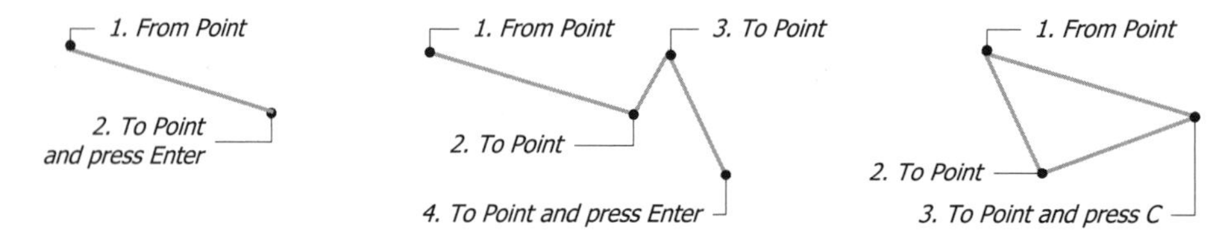

COMMAND LINE OPTIONS

Close closes the line from the current point to the starting point.

Undo undoes the last line segment drawn.

Enter continues the line from the last endpoint at the 'From point' prompt; terminates the Line command at the 'To point' prompt.

RELATED COMMANDS

MLine draws up to 16 parallel lines.

PLine draws polylines and polyline arcs.

Trace draws lines with width.

Ray creates semi-infinite construction lines.

Sketch draws continuous line segments, with up-down control.

XLine creates infinite construction lines.

Join and **PEdit** join lines and polylines.

RELATED SYSTEM VARIABLES

Lastpoint specifies the last-entered coordinate triple (x,y,z-coordinate).

Thickness determines the thickness of the line.

TIPS

- To draw 2D lines, enter x,y coordinate pairs; the z coordinate takes on the value of the Elevation system variable.
- To draw 3D lines, enter x,y,z coordinate triples.
- When system variable Thickness is not zero, the line has thickness in the z direction, which makes it a plane perpendicular to the current UCS.

Linetype

Ver. 2.0 Loads linetype definitions into the drawing, creates new linetypes, and sets the working linetype.

Commands	Aliases	Status Bar	Menu	Ribbon
'linetype	**lt** **ltype** **ddltype**		**Format** ↳**Linetype**	**Home** ↳**Properties** ↳**Linetypes**
'-linetype	**-lt** **-ltype**	...	...	...

Command: linetype

Displays dialog box.

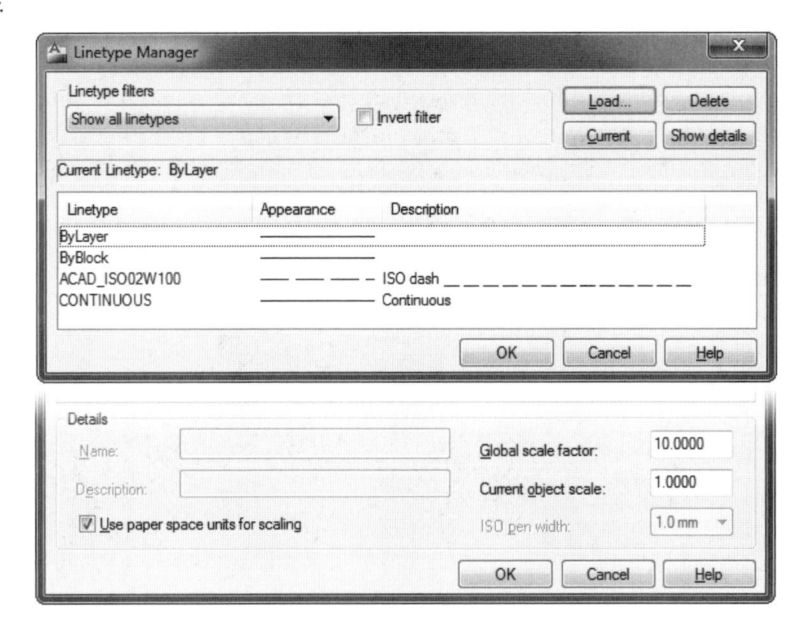

LINETYPE MANAGER DIALOG BOX

Linetype Filters options

- **Show all linetypes** displays all linetypes defined in the current drawing.
- **Show all used linetypes** displays all linetypes being used.
- **All xref dependent linetypes** displays linetypes in externally-referenced drawings.

☐ **Invert filter** inverts the display of layer names; for example, when Show all used linetypes is selected, the Invert filter option displays all linetypes not used in the drawing.

Loads displays the Load or Reload Linetypes dialog box.

Current sets the selected layer as current.

Delete purges the selected linetypes; some linetypes cannot be deleted, as described by the warning dialog box:

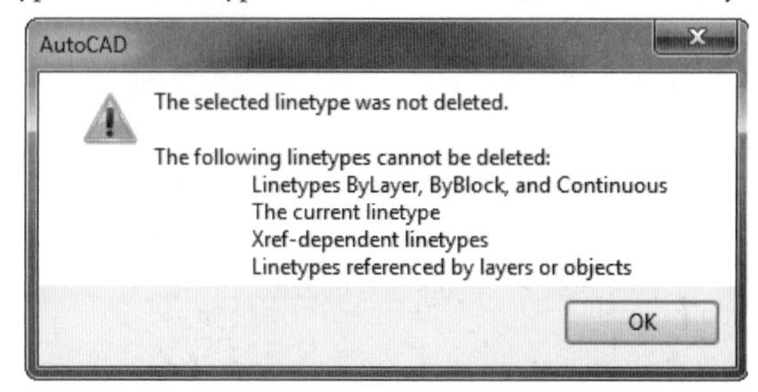

Show/Hide Details toggles the display of the Details portion of the Linetype Properties Manager dialog box.

Details options

Name names the selected linetype.

Description describes the linetype.

☑ **Use paper space units for scaling** specifies that paper space linetype scaling is used, even in model space.

Global scale factor specifies the scale factor for all linetypes in the drawing.

Current object scale specifies the individual object scale factor for all subsequently-drawn linetypes, multiplied by the global scale factor.

ISO pen width applies standard scale factors to ISO (international standards) linetypes.

LOAD OR RELOAD LINETYPES DIALOG BOX

File names the *.lin* linetype definition file.

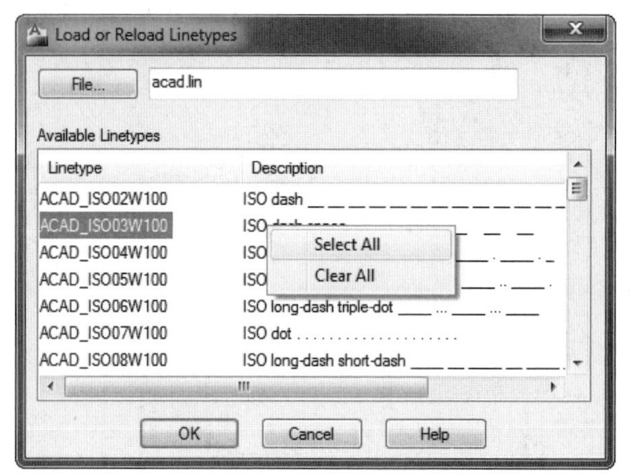

SHORTCUT MENU

Right-click any linetype name in the Linetype Manager dialog box:

Select All selects all linetypes.

Clear All selects no linetypes.

-LINETYPE Command

Command: -linetype
Enter an option [?/Create/Load/Set]: *(Enter an option.)*

COMMAND LINE OPTIONS

Create creates new user-defined linetypes.

Load loads linetypes from linetype definition (*.lin*) files.

Set sets the working linetype.

? lists the linetypes loaded into the drawing.

RELATED COMMANDS

Change and **Properties** change objects to different linetypes and linetype scales.

ChProp changes objects to a new linetype.

LtScale sets the scale of the linetype.

Rename changes the name of the linetype.

RELATED SYSTEM VARIABLES

CeLtype specifies the current linetype setting.

LtScale specifies the current linetype scale.

MsLtScale matches the linetype scale to the model view scale.

PsLtScale specifies the linetype scale relative to paper scale.

PlineGen controls how linetypes are generated for polylines.

TIPS

- The only linetypes defined initially in a new AutoCAD drawing are **Continuous** (unbroken lines), **Bylayer** (linetype specified by the layer), and **Byblock** (linetype specified by block definition).
- Linetypes must be loaded from *.lin* definition files before being used in a drawing. Better yet, add linetypes to template drawings so that they are loaded with every new drawing.
- It is faster to load all linetypes into the drawing, and later use the Purge command to remove unused linetypes.
- As of AutoCAD Release 13, objects can have independent linetype scales.
- To use MsLtScale correctly, set the value of other linetype scale system variables to 1:

  LtScale = 1 (or another size correct for plotting)
  CeLtScale = 1
  PsLtScale = 1
  MsLtScale = 1

- Changes in annotation scale appear with the next regeneration.

RELATED FILES

Standard linetypes are in the *acad.lin* and *acadiso.lin* files.

Linetype	Description
ACAD_ISO02W100	ISO dash
ACAD_ISO03W100	ISO dash space
ACAD_ISO04W100	ISO long-dash dot
ACAD_ISO05W100	ISO long-dash double-dot
ACAD_ISO06W100	ISO long-dash triple-dot
ACAD_ISO07W100	ISO dot
ACAD_ISO08W100	ISO long-dash short-dash
ACAD_ISO09W100	ISO long-dash double-short-dash
ACAD_ISO10W100	ISO dash dot
ACAD_ISO11W100	ISO double-dash dot
ACAD_ISO12W100	ISO dash double-dot
ACAD_ISO13W100	ISO double-dash double-dot
ACAD_ISO14W100	ISO dash triple-dot
ACAD_ISO15W100	ISO double-dash triple-dot
BATTING	Batting SSSSSSSSSSSSSSSSSSSSSSSSSSSSS
BORDER	Border
BORDER2	Border (.5x)
BORDERX2	Border (2x)
CENTER	Center
CENTER2	Center (.5x)
CENTERX2	Center (2x)
DASHDOT	Dash dot
DASHDOT2	Dash dot (.5x)
DASHDOTX2	Dash dot (2x)
DASHED	Dashed
DASHED2	Dashed (.5x)
DASHEDX2	Dashed (2x)
DIVIDE	Divide
DIVIDE2	Divide (.5x)
DIVIDEX2	Divide (2x)
DOT	Dot
DOT2	Dot (.5x)
DOTX2	Dot (2x)
FENCELINE1	Fenceline circle ---0----0----0----0----0---
FENCELINE2	Fenceline square ---[]----[]----[]----[]---
GAS_LINE	Gas line ---GAS---GAS---GAS---GAS---GAS---
HIDDEN	Hidden
HIDDEN2	Hidden (.5x)
HIDDENX2	Hidden (2x)
HOT_WATER_SUPPLY	Hot water supply ---- HW ---- HW ---- HW ----
PHANTOM	Phantom
PHANTOM2	Phantom (.5x)
PHANTOMX2	Phantom (2x)
TRACKS	Tracks
ZIGZAG	Zig zag

List

Ver. 1.0 Lists information about selected objects in the drawing.

Command	Aliases	Keyboard Shortcuts	Menu	Ribbon
list	**li**	...	**Tools**	**Home**
	ls		**↳Inquiry**	**↳Properties**
	showmat		**↳List**	**↳List**

Command: list
Select objects: *(Select one or more objects.)*
Select objects: *(Press Enter to end object selection.)*

Sample output for a block:

```
                  BLOCK REFERENCE  Layer: "0"
                            Space: Model space
                   Handle = 12b
        Block Name: "A$C0A557D5C"
                 at point, X=   3.5082  Y=  -1.1715  Z=   0.0000
     X scale factor:    1.0000
     Y scale factor:    1.0000
     rotation angle:   0
     Z scale factor:    1.0000
    Scale uniformly:   No
    Allow exploding:   Yes
```

COMMAND LINE OPTIONS

Enter continues the display.

Esc cancels the display.

F2 returns to graphics screen.

RELATED COMMANDS

Area calculates the area and perimeter of selected objects.

DbList lists information about *all* objects in the drawing.

MassProp calculates the properties of 2D regions and 3D solids.

TIPS

- This command lists the following information only under certain conditions:

  Color — when not set BYLAYER.
  Linetype — when not set BYLAYER.
  Thickness — when not 0.
  Elevation — when z coordinate is not 0.
  Extrusion direction — when z axis differs from current UCS.
  Transparency — when turned on.

- Object handles are described by hexadecimal numbers.
- Alternatives to this command include the Properties, Dist, and Id commands.

Removed Command

ListURL command was removed from AutoCAD 2000; it was replaced by **-**Hyperlink.

LiveSection

2007 Activates section planes to show the interior of 3D models.

Command	Aliases	Keyboard Shortcuts	Menu	Ribbon
livesection	...	...	...	**Home** **↳Section** **↳Live Section**

As an alternative, you can double-click a section plane to make it live.

Command: livesection

Select section object: *(Select one selection plane.)*

Selecting the live section activates its dynamic grips illustrated below.

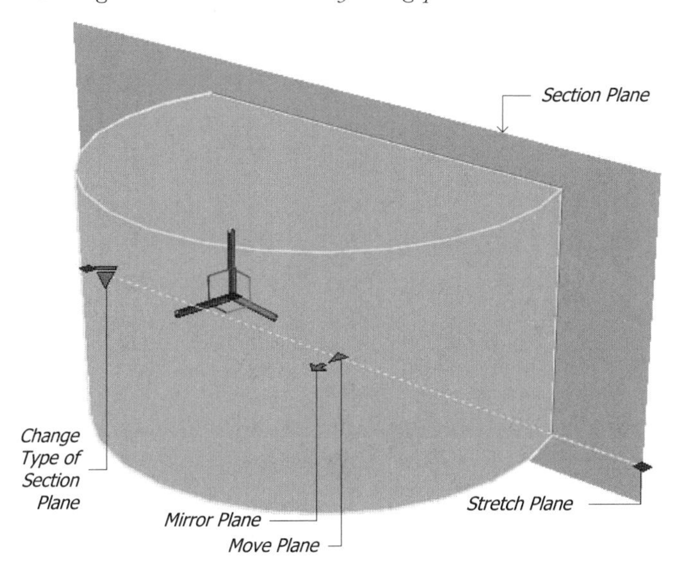

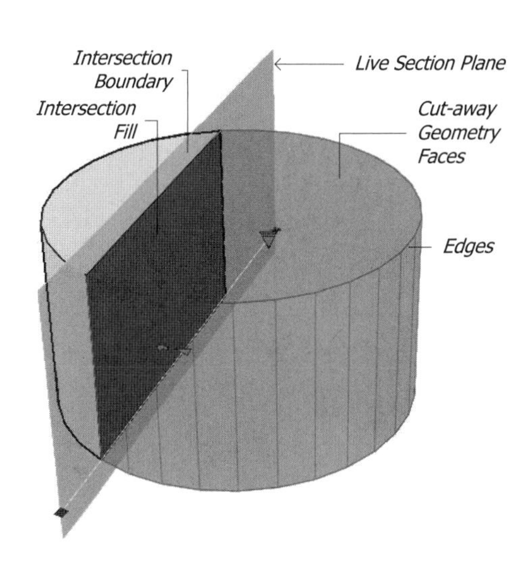

Editing live section planes.

COMMAND LINE OPTION

Select section object selects the section plane to make live.

SHORTCUT MENU OPTIONS

Right-click the live section plane for these additional options:

✓ Activate live sectioning
Show cut-away geometry
Live section settings...
Generate 2D/3D section...
Add jog to section

Activate live sectioning toggles live sectioning on and off.

Show cut-away geometry displays the hidden portions in another color.

Live section settings displays the Section Settings dialog box.

Generate 2D/3D section displays the Generate Section/Elevation dialog box, and then creates a second model of the 2D section or 3D elevation.

Add jog to section adds a 90-degree jog to the section plane; see the SectionPlaneJob command.

DYNAMIC GRIP MENU

Click the lookup grip:

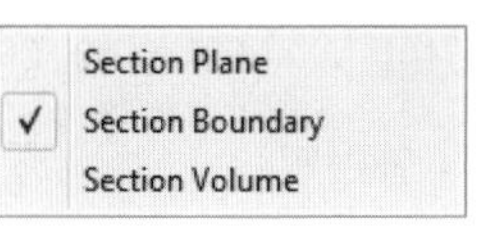

Section Plane displays the section line and cutting plane; the plane extends to "infinity" on all four sides.

Section Boundary displays a 2D box limiting the x,y extents of the cutting plane.

Section Volume displays a 3D box limiting the extent of the cutting plane in all directions.

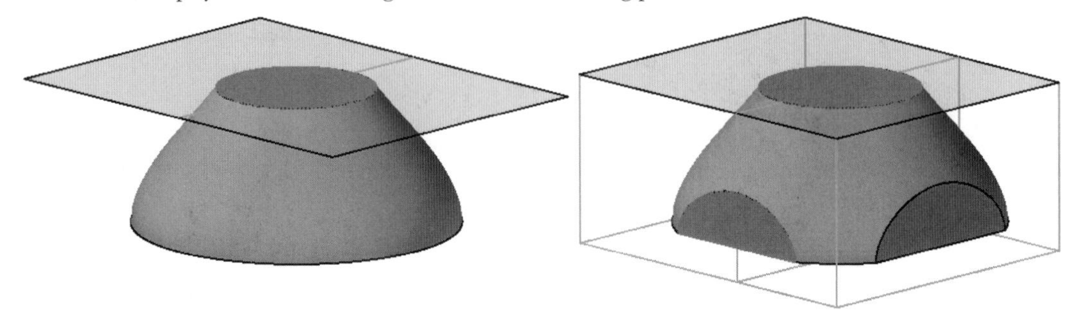

Left: *Section boundary.*
Right: *Volume boundary.*

SECTION SETTINGS DIALOG BOX

Access this dialog box by right-clicking a live section, and then choosing Live Section Settings from the shortcut menu.

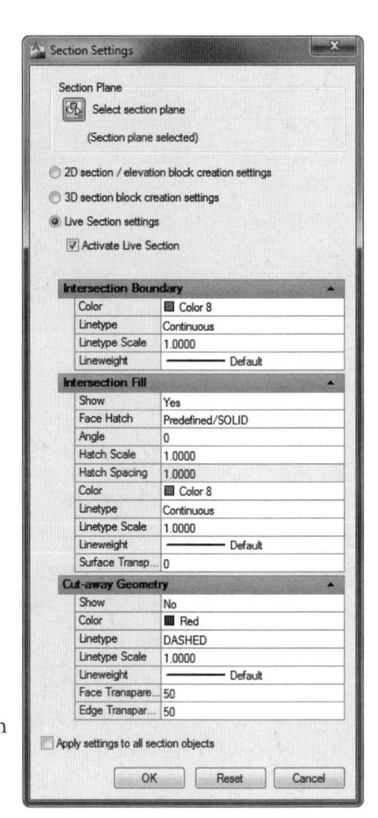

☑**Activate Live Section** turns on live sectioning for the selected section.

Intersection Boundary options

Color specifies the color of the intersection boundary.

Linetype specifies the linetype.

Linetype Scale specifies the linetype scale.

Lineweight specifies the lineweight.

Intersection Fill options

Show toggles the display of the intersection fill.

Face Hatch determines if the fill is solid, a hatch pattern, or none.

Angle specifies the angle of the hatch pattern.

Hatch Scale specifies the hatch pattern scale factor.

Hatch Spacing specifies the spacing between hatch lines.

Color specifies the color of the intersection fill.

Linetype specifies the linetype.

Linetype Scale specifies the linetype scale.

Lineweight specifies the lineweight.

Surface Transparency changes the translucency of the intersection fill; ranges from 0 to 100.

Cut-away Geometry options

Show toggles the display of cut-away geometry (removed by the live section plane).

Color specifies the color of the intersection fill.

Linetype specifies the linetype.

Linetype Scale specifies the linetype scale.

Lineweight specifies the lineweight.

Face Transparency changes the translucency of the cutaway faces.

Edge Transparency changes the translucency of the cutaway edges.

☑ **Apply Settings to All Section Objects** applies settings to future sections.

Reset returns settings to default values.

GENERATE SECTION/ELEVATION DIALOG BOX

Access this dialog box by right-clicking a live section, and then choosing Generate 2D/3D Section from the shortcut menu.

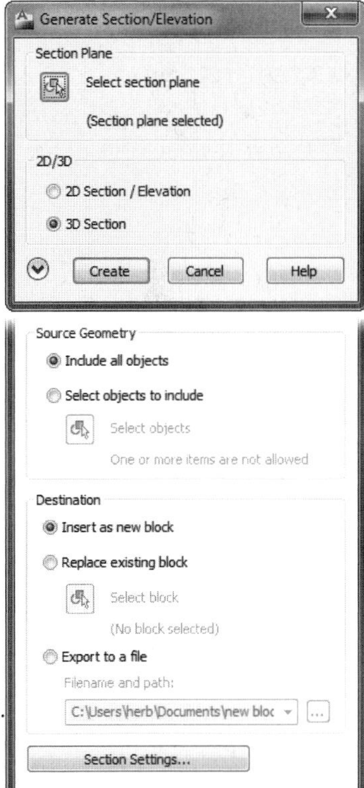

2D/3D options

⦿ **2D Section/Elevation** generates 2D sections (a.k.a. elevation).

○ **3D Section** generates 3D sections.

The following options are visible when the More button is clicked:

Source Geometry options

⦿ **Include All Objects** sections all 3D solids, surfaces, and regions in the drawing, blocks, and xrefs.

○ **Select Objects to Include** selects manually the 3D solids, surfaces, and regions to be sectioned.

Select Objects allows you to pick with a cursor the objects to section.

Destination options

⦿ **Insert as New Block** inserts the section as a block in the drawing.

○ **Replace Existing Block** replaces an existing section block.

Select Block selects the block to replace.

○ **Export to a File** saves the section as an AutoCAD drawing on disc.

Filename and Path specifies the *.dwg* file name and path by which the section is saved.

Section Settings displays the Section Settings dialog box.

Create creates the section.

RELATED AUTOCAD COMMANDS

SectionPlane places section planes in 3D objects.

SectionPlaneJog adds 90-degree jogs to section planes.

GenerateSection displays the Generate Section/Elevation dialog box.

TIPS

- Once live, section planes can be edited with dynamic grips.
- Live sectioning works on 3D objects and regions in model space only.
- Only one section plane can be live at a time.

Load

Ver. 1.0 Loads SHX-format shape files into drawings.

Command	Aliases	Keyboard Shortcuts	Menu	Ribbon
load	...	...	...	...

Command: load

Displays Load Shape File dialog box. Select an .shx file, and then click Open.

COMMAND LINE OPTIONS

None.

RELATED AUTOCAD COMMAND

Shape inserts shapes into the current drawing.

RELATED FILES

.shp* are source code files for shapes.

.shx* are compiled shape files.

In C:\Users\<login>\AppData\Roaming\Autodesk\AutoCAD 2011\R18.1\enu\Support folder:

gdt.shx and *gdt.shp* are geometric tolerance shapes used by the Tolerance command.

ltypeshp.shx and *ltypeshp.shp* are linetype shapes used by the Linetype command.

TIPS

- Shapes are more efficient than blocks, but are much harder to create.
- The Load command cannot load *.shx* files meant for fonts. For example, AutoCAD complains, "gdt.shx is a normal text font file, not a shape file."
- Do not confuse this command with the AutoLISP **(load)** function, which loads *.lsp* files.

Loft

2007 Creates 3D surface or solid lofts between two or more objects.

Command	*Aliases*	*Keyboard Shortcuts*	*Menu*	*Ribbon*
loft	...	...	**Draw** ↳**Modeling** ↳**Loft**	**Home** ↳**3D Modeling** ↳**Loft**

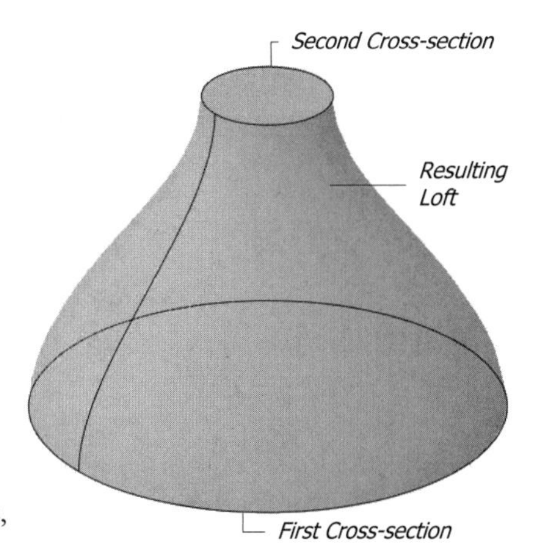

Command: loft

Current wire frame density: ISOLINES=4, Closed profiles creation mode = Solid

Select cross sections in lofting order or [POint/Join multiple edges/MOde]: *(Pick two or more cross-section objects or enter an option.)*

Select cross sections in lofting order or [POint/Join multiple edges/MOde]: *(Press Enter to continue.)*

Enter an option [Guides/Path/Cross sections only/Settings] <Cross sections only>: *(Press Enter, or enter an option.)*

AutoCAD creates an initial lofting, and then displays the Loft Settings dialog box.

Click OK. AutoCAD creates the final lofting.

COMMAND LINE OPTIONS

Select cross-sections selects the objects that define the lofting surface: lines, arcs, elliptical arcs, circles, ellipses, polylines, and points (for the first and last cross-sections only). As of AutoCAD 2011, you can also select subobjects: edges, faces, and vertices.

POint uses closed cross sections.

Join multiple edges prompts you to select multiple curves.

MOde switches between solids and surface creation.

Cross-sections only uses the cross-section objects selected earlier.

Path selects a single path object that guides the loft: a line, arc, elliptical arc, circle, ellipse, spline, helix, polyline, or 3D polyline.

Guides selects two or more guide-curve objects that define the loft; guides must start on the first cross-section, intersect each cross-section, and end at the last cross-section. Guide objects can be lines, arcs, elliptical arcs, 2D and 3D polylines, and splines.

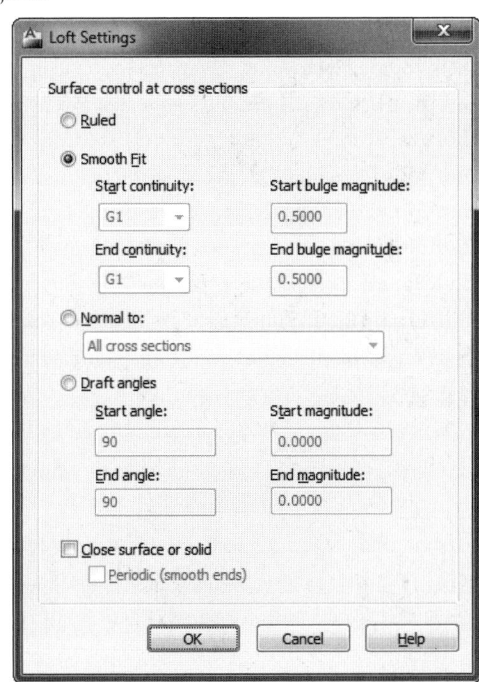

LOFT SETTINGS DIALOG BOX

⊙ **Ruled** creates smoothly-ruled lofts between cross-sections, but sharp-edged rules at cross sections.

○ **Smooth Fit** creates smoothly-ruled lofts between cross-sections, but sharp edges at the starting and ending cross sections.

○ **Normal to** controls the surface normal where the loft passes through the cross-sections:

- **All Cross Sections** makes surfaces normal to all cross-sections.

- **Start Cross Section** makes surfaces normal to the starting cross-section only.
- **End Cross Section** makes surfaces normal to the ending cross-section only.
- **Start and End Cross Sections** surface normal is normal to both starting and ending cross-sections.

O **Draft Angles** controls the draft angle and magnitude of the first and last cross-sections of the lofted solid or surface.

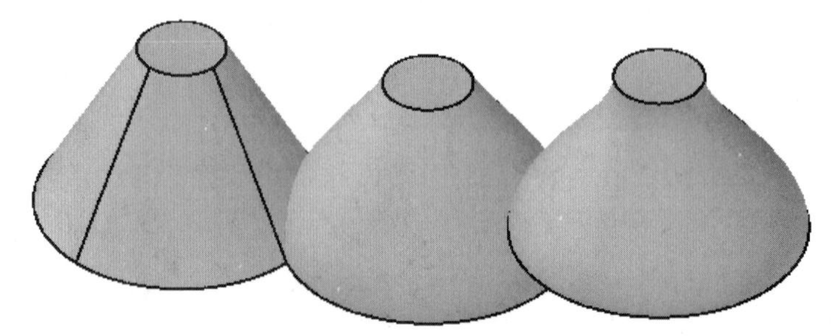

Left to right: *Normal, normal to start and end, normal to start.*

Start Angle specifies the draft angle of the starting cross-section.

Start Magnitude controls the distance between the starting cross-section in the direction of the draft angle but before the loft bends toward the next cross-section.

End Angle specifies the draft angle of the starting cross-section.

End Magnitude controls the distance between the ending cross-section in the direction of the draft angle but before the loft bends toward the previous cross-section.

☐ **Close Surface or Solid** (when off) creates open lofts; when on, closes the loft.

☑ **Preview Changes** shows changes to the loft as options are changed in this dialog box.

RELATED SYSTEM VARIABLES

DelObj toggles whether the cross-section geometry is erased after the loft is created.

LoftAng1 stores the default for the first lofting angle.

LoftAng2 stores the default for the last lofting angle.

LoftMag1 specifies the magnitude of the first lofting angle.

LoftMag2 specifies the magnitude of the last lofting angle.

LoftNormals specifies where lofting normals are placed.

LoftParam specifies the loft shape.

SurfU specifies the number of m-direction isolines on lofts.

SurfV specifies the number of n-direction isolines on lofts.

TIPS

- The cross-section curves must all be open or all closed; open ones create surfaces, while closed ones create solids.
- After the loft is created, its shape can be edited with grips, like spline curves. Click the triangle grip for additional options.
- Use this command to create solids from contour maps and boat hulls, among other complex shapes.
- As of AutoCAD 2011, you can also select subobjects for profiles and curves: edges, faces, and vertices.

LogFileOn / LogFileOff

Rel. 13 Turns on and off the command logging to *.log* files.

Command	*Aliases*	*Keyboard Shortcuts*	*Menu*	*Ribbon*
logfileon	...	...	...	...
logfileoff	...	...	...	...

Command: logfileon

AutoCAD begins recording command-line text to the log file.

Command: logfileoff

AutoCAD stops recording command-line text, and closes the log file.

COMMAND LINE OPTIONS

None.

RELATED AUTOCAD COMMAND

CopyHist copies all command text from the Text window to the Clipboard.

RELATED SYSTEM VARIABLES

LogFileName specifies the name of the log file.

LogFilePath specifies the path for the log files for all drawings in a session.

LogFileMode toggles whether text window is written to log file:

0 — text not written to file (default).

1 — text written to file.

TIPS

- AutoCAD places a dashed line at the end of each log file session.
- If log file recording is left on, it resumes when AutoCAD is next loaded, which can result in very large log files.
- The default log file name is the same as the drawing name, and is stored in folder *C:\Documents and Settings\username\Local Settings\Application Data\Autodesk\AutoCAD 2011\R18.1\enu*. You can give the file a different folder and name with the Options command's Files tab, or with system variables LogFileName and LogFilePath.
- *Historical note:* In some early versions of AutoCAD, Ctrl+Q meant "quick screen print," which output the current screen display to the printer. Ctrl+Q reappeared in AutoCAD Release 14 to record command text to a file. As of AutoCAD 2004, the Ctrl+Q shortcut quits AutoCAD, instead of toggling the log file — curious, given that there is already a keyboard shortcut, Alt+F4, that quits AutoCAD.

Removed Commands

LsEdit, **LsLib**, and **LsNew** commands were removed from AutoCAD 2007.

LtScale

Ver. 2.0 Sets the global scale factor of linetypes (short for Line Type SCALE).

Command	Aliases	Keyboard Shortcuts	Menu	Ribbon
'ltscale	**lts**	...	...	...

Command: ltscale
Enter new linetype scale factor <1.0000>: *(Enter a scale factor.)*
Regenerating drawing.

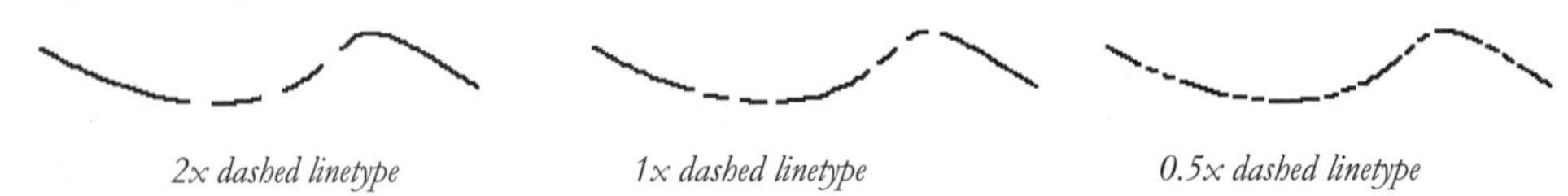

2x dashed linetype *1x dashed linetype* *0.5x dashed linetype*

COMMAND LINE OPTION

Enter new linetype scale factor changes the global scale factor of all linetypes in drawings.

RELATED COMMANDS

ChProp changes the linetype scale of one or more objects.

Properties changes the linetype scale of objects.

Linetype loads, creates, and sets the working linetype.

RELATED SYSTEM VARIABLES

LtScale contains the global linetype scale factor.

MsLtScale matches the linetype scale to the model space scale factor.

CeLtScale specifies the current object linetype scale factor relative to the global scale.

PlineGen controls how linetypes are generated for polylines.

PsLtScale specifies that the linetype scale is relative to paper space.

TIPS

- If the linetype scale is too large, the linetype appears solid.
- If the linetype scale is too small, the linetype appears as a solid line that redraws very slowly.
- The scale is not only set by the LtScale command, but also in the *acad.lin* file, which contains each linetype in three scales: normal, half-size, and double-size.
- To use MsLtScale correctly, set the value of these linetype scale system variables to 1:

  LtScale = 1 (or another size correct for plotting)
  CeLtScale = 1
  PsLtScale = 1
  MsLtScale = 1

  The annotative scaling does not come into effect until the next regeneration.
- You can change the linetype scaling of individual objects, which is then multiplied by the global scale factor specified by the LtScale command.

LWeight

2000 Sets the current lineweight (display width) of objects.

Commands	Aliases	Status Bar	Menu	Ribbon
'lweight	**lw**	**LWT**	**Format**	**Home**
	lineweight		**↳Lineweight**	**↳Properties**
				↳Lineweight
'-lweight	...	...	...	...

Command: lweight

Displays dialog box.

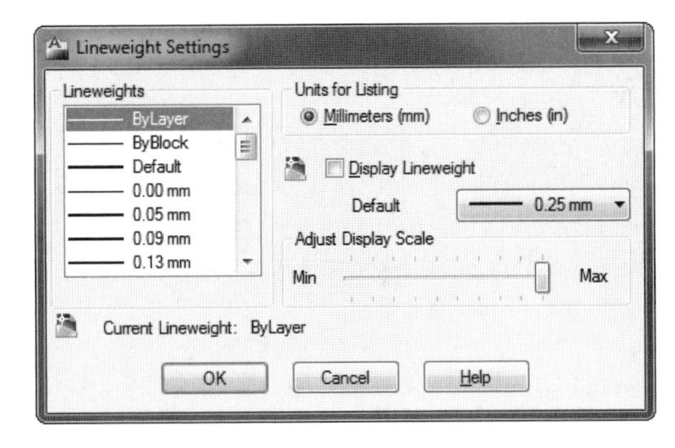

LINEWEIGHT SETTINGS DIALOG BOX

Lineweights lists lineweight values.

Units for Listing options

⊙ **Millimeters (mm)** specifies lineweight values in millimeters.

○ **Inches (in)** specifies lineweight values in inches.

☐ **Display Lineweight** toggles the display of lineweights; when checked, lineweights are displayed.

Default specifies the default lineweight for layers (default = 0.01" or 0.25 mm).

Adjust Display Scale controls the scale of lineweights in the Model tab, which displays lineweights in pixels. This setting should be adjusted to suit the resolution of the monitor.

SHORTCUT MENU

Right-click LWT on the status bar to display shortcut menu:

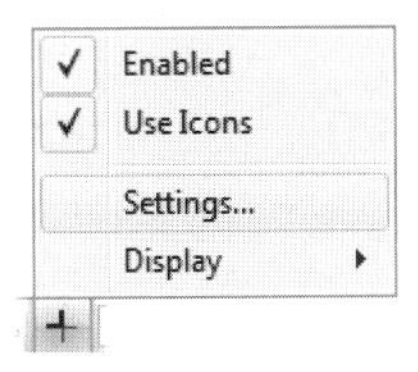

Enabled turns on lineweight display.

Settings displays Lineweight Settings dialog box.

-LWEIGHT Command

Command: -lweight

Enter default lineweight for new objects or [?]: *(Enter a value, or type ?.)*

ByLayer
ByBlock
Default
0.00 mm
0.05 mm
0.09 mm
0.13 mm
0.15 mm
0.18 mm
0.20 mm
0.25 mm
0.30 mm
0.35 mm
0.40 mm
0.50 mm
0.53 mm
0.60 mm
0.70 mm
0.80 mm
0.90 mm
1.00 mm
1.06 mm
1.20 mm
1.40 mm
1.58 mm
2.00 mm
2.11 mm

COMMAND LINE OPTIONS

Enter default lineweight specifies the current lineweight; valid values include Bylayer, Byblock, and Default.

? lists the valid values for lineweights:

```
ByLayer ByBlock Default
0.000" 0.002"  0.004"   0.005" 0.006"  0.007"
0.008" 0.010"  0.012"   0.014" 0.016"  0.020"
0.021" 0.024"  0.028"   0.031" 0.035"  0.039"
0.042" 0.047"  0.055"   0.062" 0.079"  0.083"
```

RELATED COMMAND

Properties changes lineweights of selected objects.

RELATED SYSTEM VARIABLES

LwDefault specifies the default linewidth; default = 0.01" or 0.25 mm.

LwDisplay toggles the display of lineweights in the drawing.

LwUnits determines whether the lineweight is measured in inches or millimeters.

CeLWeight specifies the lineweight for new objects.

TIPS

- To create custom lineweights for plotting, use the Plot Style Table Editor.
- A lineweight of 0 plots the lines at the thinnest width of which the plotter is capable, usually one pixel or one dot wide.
- Linewieghts are "hardwired" into AutoCAD, and cannot be customized.
- Lineweights can be set through layers, the Properties toolbar, and the ribbon's Home tab.
- The CeLWeight (short for current entity lineweight) system variable overrides the default set by the Layers command. The correct value of this system variable, therefore, is -1, ByLayer.

Replaced Command

MakePreview command was removed from AutoCAD Release 14; it was replaced by the RasterPreview system variable, which controls the creation of previews when drawings are saved.

Markup / MarkupClose

2005 Opens and closes the Markup Set Manager palette; opens DWF files.

Commands	*Alias*	*Keyboard Shortcuts*	*Menu*	*Ribbon*
markup	**msm**	**Ctrl+7**	**Tools** **↳Palettes** **↳Markup Set Manager**	**View** **↳Palettes** **↳Markup Set Manager**
markupclose	**...**	**Ctrl+7**	**...**	**...**

Command: markup

Displays palette.

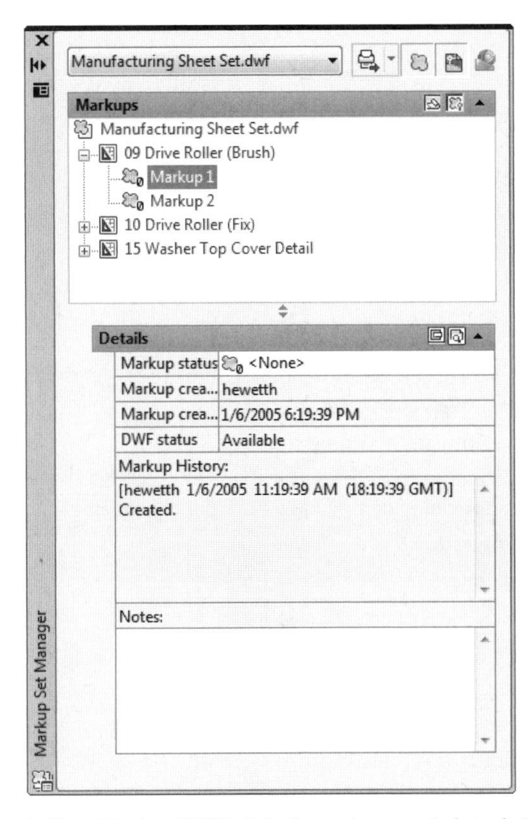

To open a markup set, select Open. In the Open Markup DWF dialog box, select a marked-up .dwf file, and then click Open.

(If the file contains no markup data, AutoCAD complains, 'Filename.dwf does not contain any markup data. Would you like to open this DWF file in the viewer?' Click Yes to open the file in the DWF Viewer; click No to cancel.)

Command: markupclose

Closes the palette.

PALETTE TOOLBAR

Republish All Sheets outputs sheets to printers or files; see the Publish command:

- **Republish All Sheets** republishes all sheets as a background job using Publish.
- **Republish Markup Sheets** republishes only sheets with markups.

View Redline Geometry toggles the display of markups (redlines).

View DWG Geometry toggles the display of AutoCAD geometry (objects).

View DWF Geometry toggles the display of DWG objects.

Markups buttons

Show All Sheets shows all sheets with markups.

Hide Non-Markup Sheets hides all sheets without markups.

▲ Collapses section.

Details buttons

Details displays details about the markup sheet.

Preview displays a bitmap image of the sheet, as illustrated below.

SHORTCUT MENU

Right-click a markup file:

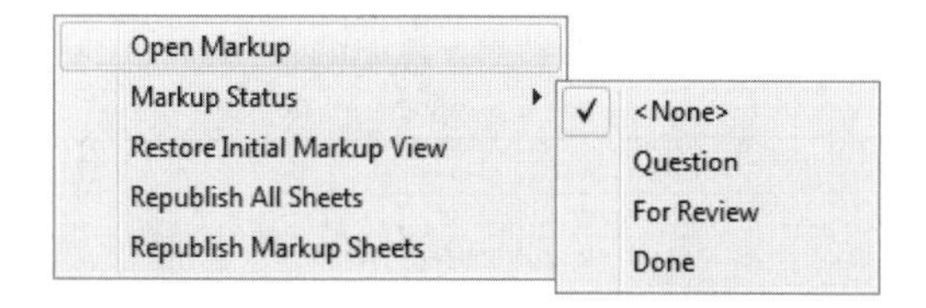

Open Markup opens the markup data in AutoCAD; alternatively, double-click the markup.

Markup Status displays a submenu for changing the status of a markup sheet:

- **<None>** indicates no change in status.
- **Question** indicates markup has questions to be answered.
- **For Review** indicates markup changes need to be reviewed.
- **Done** indicates markup is done.

Restore Initial Markup View resets all markup views to their original state.

Republish All Sheets republishes all sheets as a background job using Publish.

Republish Markup Sheets republishes only sheets with markups.

RELATED COMMANDS

OpenDwfMarkup opens *.dwf* files containing markup data.

DwfAttach attaches *.dwf* files as underlays.

RevCloud draws revision clouds in drawings.

RELATED SYSTEM VARIABLE

MsmState reports whether the Markup Set Manager palette is open.

TIPS

- This command displays markups (redlines) added to DWF files in DesignReview, included free with AutoCAD, or as a free download from www.autodesk.com/designreview.

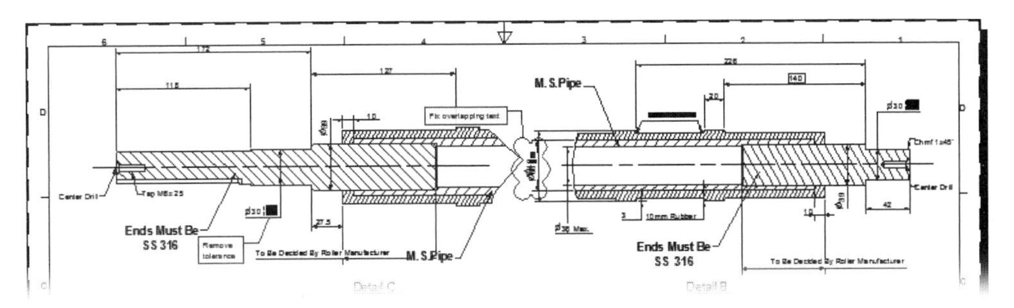

- MSM is short for "markup set manager." It works with DWG and DWF files from AutoCAD 2005 and later.
- This command only reads DWF files that have been marked up with Design Review (previously DWF Composer); it does not read unmarked up DWF files.
- When a DWF file is opened in AutoCAD, it cannot be edited; 3D DWF files cannot be marked up.
- As of AutoCAD 2009, this command also supports DWFx files, which are viewed by IE7/8 on the Windows Vista and 7 operating systems after Design Review is installed.
- To markup drawings inside AutoCAD, use the RevCloud and MLeader commands on a layer whose color is set to red.

MassProp

Rel.11 Reports the mass properties of 3D solid models, bodies, and 2D regions (short for MASS PROPerties).

Command	*Aliases*	*Menu*	*Ribbon*
massprop	...	**Tools** **⇘Inquiry** **⇘Region/Mass Properties**	...

Command: massprop
Select objects: *(Select one or more regions and/or solid model objects.)*
Select objects: *(Press Enter.)*
Write analysis to a file? [Yes/No] <N>: *(Type Y or N.)*

Example output of a solid sphere:

```
 ----------------   SOLIDS    ----------------

Mass:                      9.9378
Volume:                    9.9378
Bounding box:           X: 4.7662  --  7.1926
                        Y: 4.9973  --  8.1426
                        Z: -3.8483  --  1.6878
Centroid:               X: 5.8455
                        Y: 6.6391
                        Z: -0.2218
Moments of inertia:     X: 455.3659
                        Y: 354.4693
                        Z: 784.7290
Products of inertia:   XY: 384.7960
                       YZ: -10.0200
                       ZX: -14.3752
Radii of gyration:      X: 6.7692
                        Y: 5.9723
                        Z: 8.8862
Principal moments and X-Y-Z directions about centroid:
                        I: 17.1241 along [0.9515 0.3078 0.0000]
                        J: 16.6049 along [-0.2741 0.8472 -0.4552]
                        K: 4.6438 along [-0.1401 0.4331 0.8904]

Write analysis to a file? [Yes/No] <N>:
```

COMMAND LINE OPTIONS

Select objects selects the solid model objects (2D regions, 3D solids, and bodies) to analyze.

Write analysis to a file:

Yes writes mass property reports to *.mpr* files.

No does not write reports to file.

RELATED COMMANDS

Area calculates the area and perimeter of non-solid objects.

MeasureGeom reports distances, areas, radii, angles, and volumes.

Properties displays information about all objects.

RELATED FILE

*.*mpr* is the file to which MassProp writes its results (mass properties report).

TIPS

- This command can be used with 2D regions as well as 3D solids; it cannot be used with 3D surface models or 2D non-region objects.
- Distances are relative to the current UCS. For them to be meaningful, you should orient the UCS to a significant feature on the object.
- Mass properties are computed as if the selected regions were unioned and as if the selected solids were unioned.
- As of Release 13, AutoCAD's solid modeling no longer allows you to apply a material density to a solid model. All solids and bodies have a density of 1.
- AutoCAD only analyzes regions coplanar (lying in the same plane) with the first region selected.

DEFINITIONS

Bounding Box

– the lower-right and upper-left coordinates of a rectangle enclosing 2D regions.

– the x, y, z-coordinate triple of a 3D box enclosing 3D solids or bodies.

Centroid

– the x, y, z coordinates of the center of 2D regions.

– the center of mass of 3D solids and bodies.

Mass – equal to the volume, because density = 1; not calculated for regions.

Moment of Inertia

– for 2D regions = Area * Radius^2.

– for 3D bodies = Mass * Radius^2.

Perimeter – total length of inside and outside loops of 2D regions (not calculated for 3D solids).

Product of Inertia

– for 2D regions = Mass * Distance (centroid to y, z axis) * Distance (centroid to x, z axis).

– for 3D bodies = Mass * Distance (centroid to y, z axis) * Distance (centroid to x, z axis).

Radius of Gyration – for 2D regions and 3D solids = (MomentOfInertia / Mass)$^{1/2}$.

Volume – 3D space occupied by a 3D solid or body (not calculated for regions).

MatBrowserOpen / MatBrowserClose

2011 Matches the properties of table cells.

Command	Alias	Keyboard Shortcuts	Menu	Ribbon
matbrowseropen	**mat**	**...**	**View ↳Render ↳Materials Browser**	**Render ↳Materials ↳Materials Browser**
matbrowserclose				

Command: matbrowseropen

Displays the Materials Browser palette.

Command: matbrowserclose

Closes the Materials Browser palette.

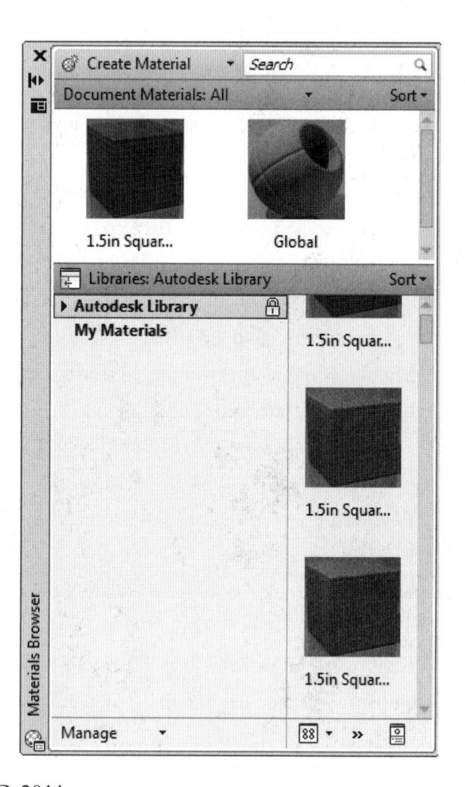

PALETTE OPTIONS

- **Create New Material** creates new materials based on existing ones; displays the Materials Editor palette. See MatEditorOpen command.
- **Search** searches for materials across all libraries.
- **Show/Hide Library Tree** toggles the display of the library pane.
- **Document Materials** filters the lists of the materials used by the current drawing: applies, selected, unused, and .
- Sort **Sort** determines how the library is displayed: by name, category, type, or material color.
- **Manage** opens, creates, removes, and renames libraries and categories.
- **Swatch Size** sets the size of material samples (swatches): list, grid, or text view.
- **Material Editor** accesses the Materials Editor palette.

RELATED COMMANDS

MatEditorOpen creates and edits materials.

MaterialAttach attaches materials to layers.

MaterialAssign assigns materials to objects.

MaterialMap adjusts the mapping of materials to objects.

ConvertOldMaterials converts materials from releases older than AutoCAD 2011.

RELATED SYSTEM VARIABLES

MatBrowserState reports on the state of the Materials Browser palette.

MaterialsPath reports the path to the material library folder; default is *C:\Program Files (x86)\Common Files\Autodesk Shared\Materials2011\assetlibrary_base.adsklib.*

3dConversionMode controls the conversion of old materials.

TIPS

- These commands replace the Materials and MaterialsClose commands as of AutoCAD 2011.
- The Autodesk Library lists materials found in Autodesk's master materials library, which are also used by other Autodesk applications. The My Materials library lists user-defined materials. Neither of these two can be renamed or removed.
- Materials appear when rendered, and only in Realistic and Xray visual styles.

MatchCell

2005 Matches the properties of table cells.

Command	*Aliases*	*Keyboard Shortcuts*	*Menu*	*Ribbon*
matchcell	...	...	...	...

Command: matchcell
Select source cell: *(Select a cell in a table.)*
Select destination cell: *(Select one or more cells.)*

COMMAND LINE OPTIONS

Select source cell gets property settings from the source cell.

Select destination cell passes property settings to the destination cells.

RELATED COMMANDS

Table creates new tables in drawings.

TableStyle defines table styles.

MatchProp copies properties between objects other than cells.

RELATED SYSTEM VARIABLE

CTable Style specifies the name of the current table style.

TIPS

- Use this command to copy formatting from one cell to another.
- This command can also be launched by right-clicking a selected cell, and then choosing Match Cell from the shortcut menu.
- Use the MatchProp command to copy properties from one table to another.

MatchProp

Rel.14 Matches the properties between selected objects (short for MATCH PROPerties).

Command	Aliases	Keyboard Shortcuts	Menu	Ribbon
'matchprop	**ma**	**...**	**Modify**	**Home**
	painter		**↳Match Properties**	**↳Clipboard**
				↳Match Properties

Command: matchprop

Select source object: *(Select a single object.)*

Current active settings: Color Layer Ltype Ltscale Lineweight Thickness PlotStyle Dim Text Hatch Polyline Viewport Table Material Shadow display Multileader

Select destination object(s) or [Settings]: *(Pick one or more objects, or type S.)*

Select destination object(s) or [Settings]: *(Press Enter to exit command.)*

COMMAND LINE OPTIONS

Select source object gets property settings from the source object.

Select destination object(s) passes property settings to the destination objects.

Settings displays dialog box.

PROPERTY SETTINGS DIALOG BOX

Access this dialog box thorugh the Settings option:

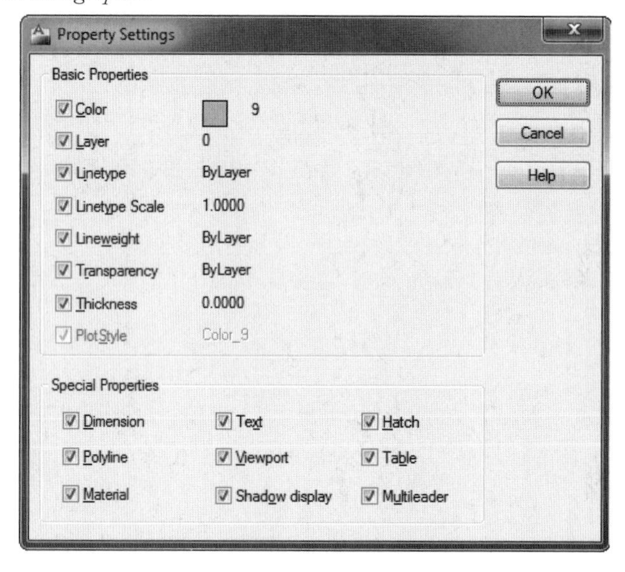

Basic Properties

☑**Color** specifies the color for destination objects; not available when OLE objects are selected.

☑**Layer** specifies the layer name for destination objects; not available when OLE objects are selected.

☑**Linetype** specifies the linetype for destination objects; not available when attributes, hatch patterns, mtext, OLE objects, points, or viewports are selected.

☑**Linetype Scale** specifies the linetype scale for destination objects; not available when attributes, hatch patterns, mtext, OLE objects, points, or viewports are selected.

☑**Lineweight** specifies the lineweight for destination objects.

☑**Transparency** specifies the translucency for destination objects *(new to AutoCAD 2011)*.

☑**Thickness** specifies the thickness for destination objects; available only for objects that can have thickness: arcs, attributes, circles, lines, mtext, points, 2D polylines, regions, text, and traces.

☑**Plot Style** specifies the plot style; not available when PStylePolicy = 1 (color-dependent plot style mode), or when OLE objects are selected.

Special Properties

☑**Dimension** copies the dimension styles of dimensions, leaders, and tolerance objects.

☑**Text** copies the text style of text and mtext objects.

☑**Hatch** copies the hatch pattern of hatched objects.

☑**Polyline** copies the width and linetype generation of polylines; curve fit, elevation, and variable width properties are not copied.

☑**Viewport** copies all properties of viewport objects, except clipping, UCS-per-viewport, freeze-thaw settings, and viewport scale.

☑**Table** copies style of table objects.

☑**Material** copies the material name.

☑**Shadow display** copies type of shadow casting.

☑**Multileader** copies the multileader styles.

RELATED COMMAND

Properties changes most properties of selected objects.

RELATED SYSTEM VARIABLE

PStylePolicy determines whether the PlotStyle option is available.

TIPS

- In other Windows applications, this command is known as Format Painter.
- Use this command to make pattern properties the same among multiple hatches.
- While this command is active, AutoCAD displays the paintbrush icon:

- Use the MatchCell command to copy properties from one table cell to another.

Removed Command

MatLib command was removed from AutoCAD 2007; it was replaced by the Materials group of the ToolPalette command.

MatEditorOpen / MatEditorClose

2011 Open and close the Materials Editor palette.

Command	*Aliases*	*Keyboard Shortcuts*	*Menu*	*Ribbon*
mateditoropen	...	...	**View** ↳**Render** ↳**Materials Editor**	**Render** ↳**Materials** ↳**Materials Editor**
mateditorclose				

Command: mateditoropen

Displays the Materials Editor palette.

Command: mateditorclose

Closes the Materials Editor palette.

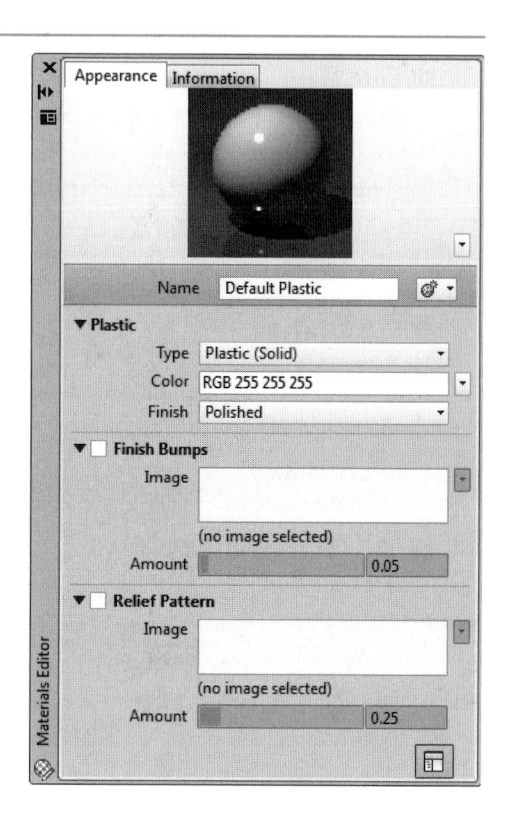

PALETTE OPTIONS

Appearance tab

Swatch Preview chooses the object and quality whose material to preview:

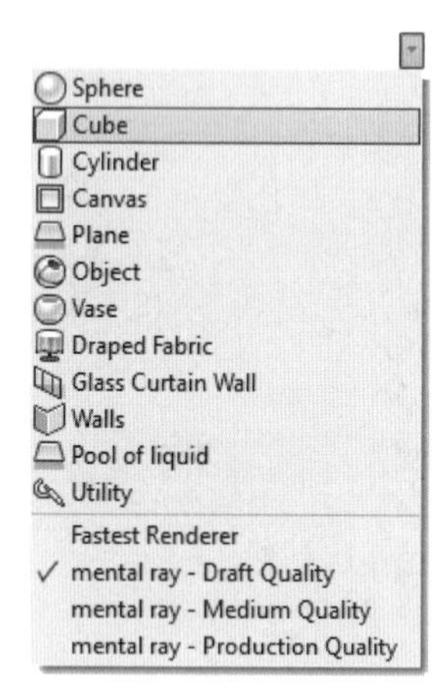

Left: *Swatch Preview droplist.*
Right: *Create droplist.*

Name names the material.

Create creates or duplicates the material. Selecting a material category from the droplist opens additional material editing palettes, as described below.

Show Materials Browser displays the Materials Browser; see the MatBrowserOpen command.

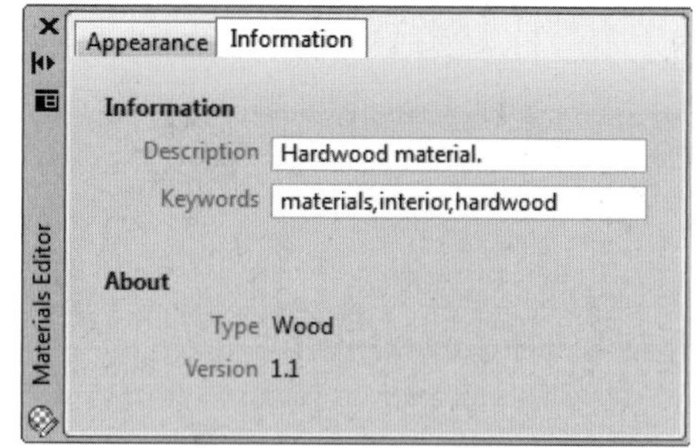

Information tab

Description describes the material.

Keywords tags the material and its appearance; used by the Material Browser palette's search function.

About reports the type, version number, and sometimes the location of the material.

ADDITIONAL PALETTES

Access these additional palettes through the Color droplist:

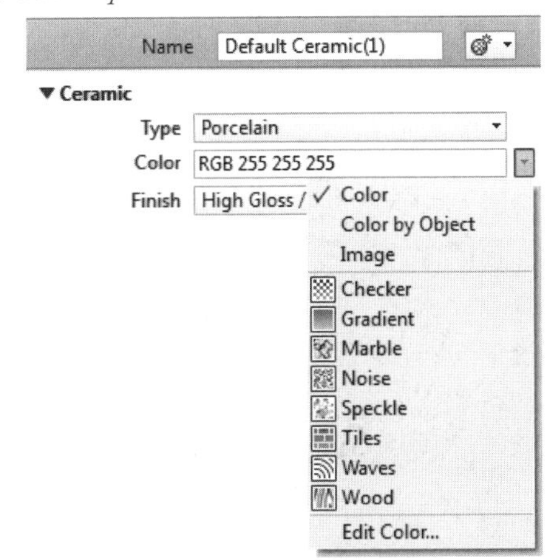

Examples of color palettes:

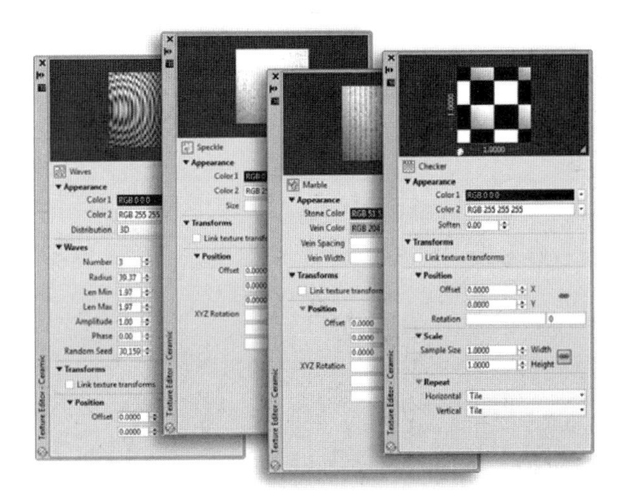

RELATED COMMANDS

MatBrowserOpen catalogs materials.

MaterialAttach attaches materials to layers.

MaterialMap adjusts the mapping of materials to objects.

RELATED SYSTEM VARIABLES

MatEditorState reports on the state of the Materials Editor palette.

MaterialsPath reports the path to the material library folder; default is *C:\Program Files (x86)\Common Files\Autodesk Shared\Materials2011\assetlibrary_base.adsklib.*

TIPS

- These commands replace the Materials and MaterialsClose commands as of AutoCAD 2011.
- Use the Materials Browser and Materials Editor palettes together, one for accessing materials, the other for editing them.

MaterialAttach / MaterialAssign

2007 Attaches rendering materials to layers; assigns them to objects (undocumented).

Commands	*Aliases*	*Keyboard Shortcuts*	*Menu*	*Ribbon*
materialattach	...	...	...	**Render** ⇘**Materials** ⇘**Attach by Layer**
materialassign	...	...	...	...

Command: materialattach

Before using this command, add materials to the drawing through the Materials group of the Tool palette. Displays dialog box:

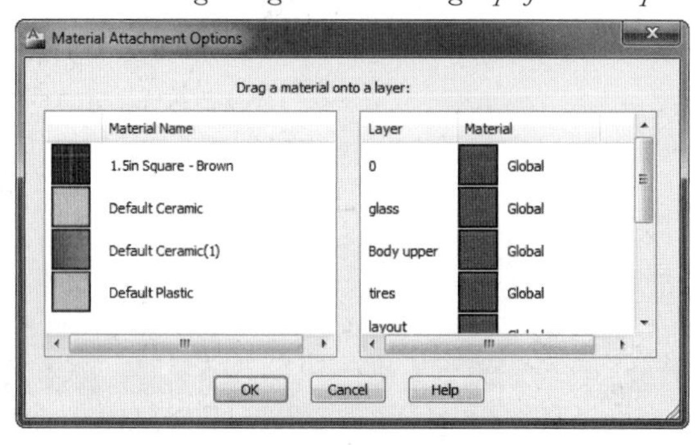

To attach a material to a layer, drag the material name onto the layer name.

MATERIAL ATTACHMENT DIALOG BOX

Material Name lists the names of materials defined in the drawing.

Layer lists the names of layers in the current drawing.

Material lists the material attached to the layer.

MATERIALASSIGN Command

Command: materialassign
Select objects: *(Select one or more objects.)*

Select objects assigns the current material to the selected objects (undocumented command).

RELATED COMMANDS

Materials creates and edits materials.

Render displays the materials applied to layers.

RELATED SYSTEM VARIABLE

CMaterial specifies the default material for newly-created objects.

TIPS

- Materials can be attached to objects by dragging them from the Materials Browser palette onto the objects.
- Materials can be matched between objects using the MatchProp command.
- Layers without a user-attached material are automatically assigned the Global material.
- Materials can be removed from drawings with the Purge command.

MaterialMap

2007 Maps materials to the faces of objects, and adjusts the mapping.

Command	*Alias*	*Keyboard Shortcuts*	*Menu*	*Ribbon*
materialmap	**setuv**	...	**View** **↳Render** **↳Mapping**	**Materials** **↳Material Mapping**

To see the effect of this command, change the visual style to Realistic with the VS alias.

Command: materialmap
Select an option [Box/Planar/Spherical/Cylindrical/copY mapping to/Reset mapping]<Box>: *(Enter an option.)*
Select faces or objects: *(Pick one or more objects.)*
Select faces or objects: *(Press Enter.)*
Accept the mapping or [Move/Rotate/reseT/sWitch mapping mode]: *(Press Enter, or enter an option.)*

COMMAND LINE OPTIONS

Planar maps the material once over the entire object, scaling it vertically and horizontally; meant for use on faces.

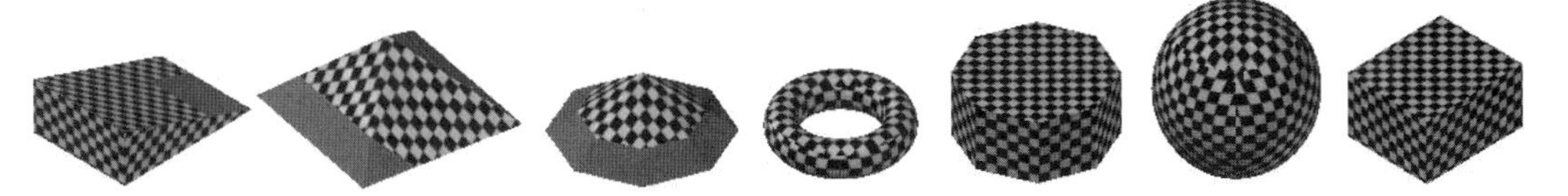

Box repeats the mapping on each face of objects; meant for rectangular objects; default mapping for new objects.

Cylindrical wraps the material horizontally, and scales the material vertically; meant for use on cylindrical objects.

Spherical wraps the material horizontally and vertically; meant for use on round objects.

copY mapping to copies the mapping style to other objects.

Reset mapping (and **reseT**) resets the u,v-mapping coordinates to their default values.

Select faces or objects selects the faces or objects to which the mapping style should be applied; hold down the Ctrl key to select faces.

Accept the mapping exits the command.

Move moves the material mapping interactively; see the Move3D command. Use the triangular grips to change the size of the map in the x, y, and z directions.

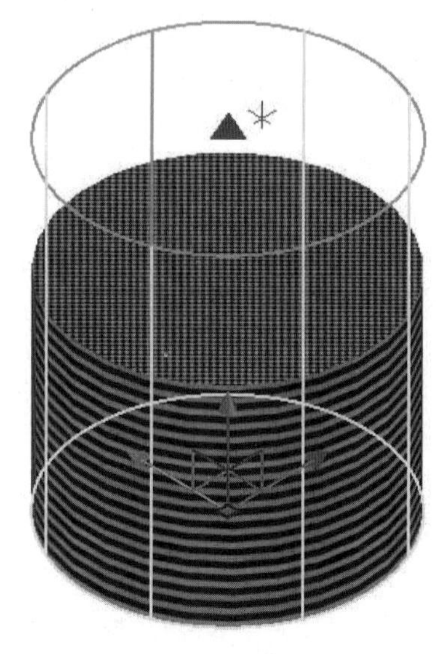

Moving the material map interactively.

Rotate rotates the material mapping interactively; see the Rotate3D command.

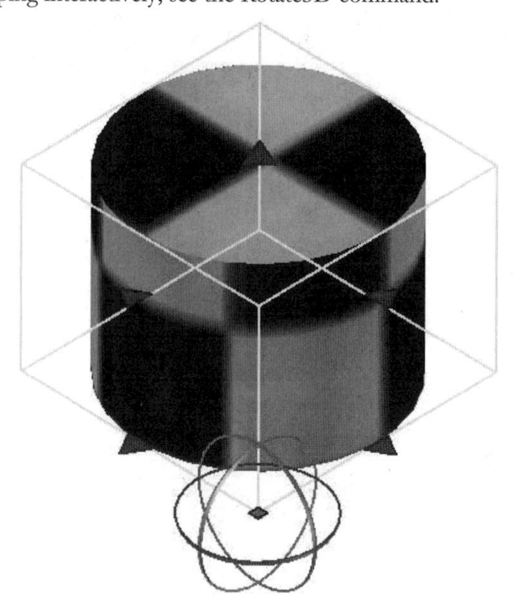

Rotating the material map interactively.

sWitch mapping mode switches to one of the other mapping modes: box, planar, spherical, or cylindrical.

RELATED COMMANDS

Materials creates and edits materials.

Render displays the materials applied to layers.

RELATED SYSTEM VARIABLE

CMaterial specifies the default material for newly-created objects; default = bylayer.

TIPS

- Materials can be mapped only to 3D solids, 3D surfaces, 3D meshes, faces, regions, and 2D objects with thickness.
- Materials affect only the appearance of objects and not their mass. All solid models have a uniform density of 1.

Renamed Commands

Materials and **MaterialsClose** are now hardwire aliases for the MatBrowserOpen and MatBrowserClose commands, as of AutoCAD 2011.

Measure

V. 2.5 Places points or blocks at constant intervals along lines, arcs, circles, and polylines.

Command	Aliases	Keyboard Shortcuts	Menu	Ribbon
measure	**me**	...	**Draw**	**Home**
			↳**Point**	↳**Draw**
			↳**Measure**	↳**Multiple Points**
				↳**Measure**

Command: measure

Select object to measure: *(Pick a single object.)*

Specify length of segment or [Block]: *(Enter a value, or type B.)*

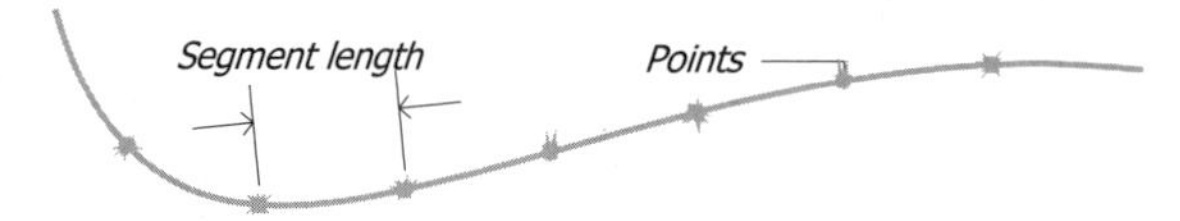

COMMAND LINE OPTIONS

Select object selects a single object for measurement.

Specify length of segment indicates the distance between markers.

Block Options

Enter name of block to insert: *(Enter name.)*

Align block with object? [Yes/No] <Y>: *(Type Y or N.)*

Specify length of segment: *(Enter a value.)*

Enter name of block indicates the name of the block to use as a marker; the block must already exist in the drawing.

Align block with object? aligns the block's x axis with the object.

RELATED COMMANDS

Block creates blocks that can be used with the Measure command.

Divide divides objects into a specific number of segments.

RELATED SYSTEM VARIABLES

PdMode controls the shape of a point.

PdSize controls the size of a point.

TIPS

- You must define the block before it can be used with this command.
- This command does not place points (or blocks) at the beginning or end of objects.
- The first point is placed at the distance measured from the endpoint closest to the pick point.

Removed Command

MeetNow command was removed from AutoCAD 2004.

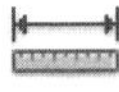

MeasureGeom

2010 Measures distances, areas, radii, angles, and volumes.

Command	Alias	Keyboard Shortcuts	Menu	Ribbon
measuregeom	**mea**	...	**Tools** ↳**Inquiry**	**Home** ↳**Utilities** ↳**Measure**

Command: measuregeom

Enter an option [Distance/Radius/Angle/Area/Volume] <Distance>: *(Enter an option.)*

As you pick points, AutoCAD reports the current measured distance as a dynamic dimension:

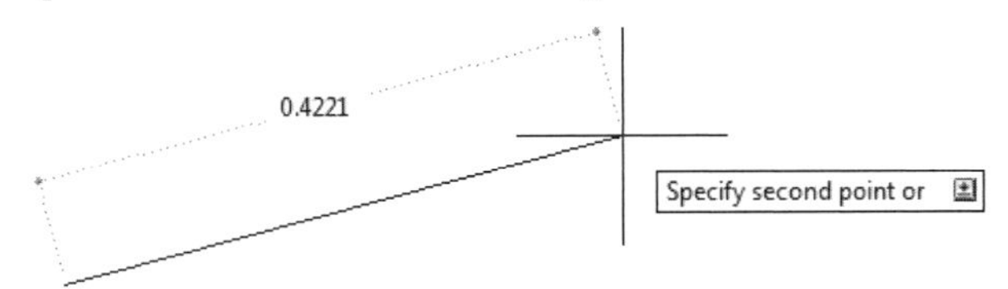

COMMAND LINE OPTIONS

Distance measures the distances between two or more points.

Radius measures the radius and diameter of arcs and circles.

Angle measures the angles of arcs, the angle on circles, the angle between two lines, and the angle between any three points making up a vertex.

Area measures the area and perimeter of objects and areas.

Volume measures the volumes of 3D objects, or of areas with height.

Distance options

Specify first point: *(Pick a point.)*

Specify second point or [Multiple Points]: *(Pick another point, or type M.)*

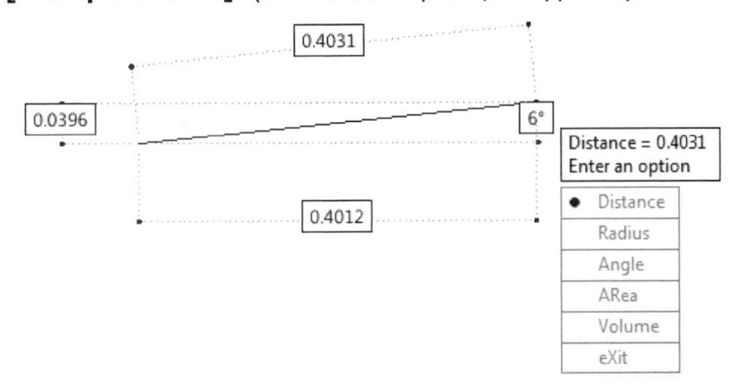

Specify first point specifies the starting point of the measured distance.

Specify second point specifies the ending point.

Multiple points continues prompting you for additional points, and reports the total of all segments:

Specify next point or [Arc/Length/Undo/Total] <Total>: *(Pick another point, or enter an option.)*

Arc measures arc-shaped distances with prompts similar to those of the Arc command.

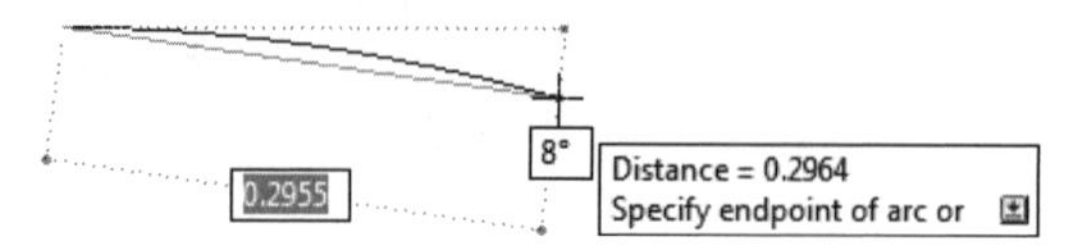

Length prompts you to "Specify length of line."

Undo "unmeasures" the last segment.

Total reports the total length, and ends the multi-point measurement.

Radius option

Select arc or circle: *(Pick an arc or a circle.)*

Select arc or circle reports the radius and diameter of the arc or circle in the command bar.

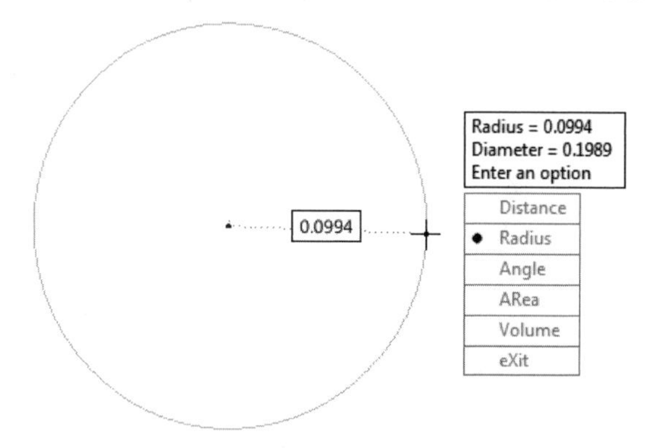

Angle options

Specify Arc, Circle, Line or <Vertex>: *(Pick an arc, a circle, line, or type V.)*

Specify arc reports the angle between the arc's two endpoints

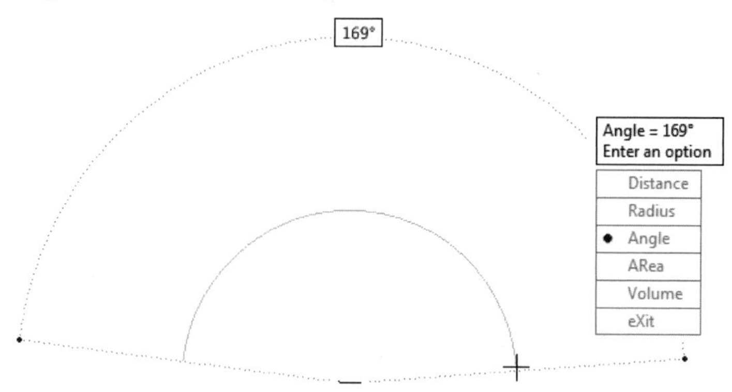

Circle reports the angle between two points you pick on the circle.

Line reports the angle between two lines; prompts you to select a second line.

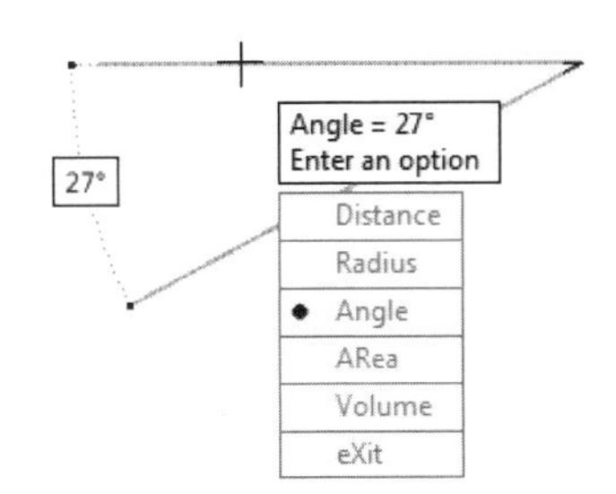

Vertex measures the angle between two points; prompts you:

Select angle vertex *(Pick a point that represents the vertex.)*

Specify first angle endpoint *(Pick a point that represents an endpoint.)*

Specify second angle endpoint: *(Pick another point that represents the other endpoint.)*

Area options

Specify first corner point or [Object/Add area/Subtract area/eXit] <Object>: *(Pick a point, or enter an option.)*

Specify next corner point or [Arc/Length/Undo]: *(Pick another point, or enter an option.)*

Specify next corner point or [Arc/Length/Undo/Total] <Total>: *(Enter another point, enter an option, or press Enter.)*

Added areas are shown in green; subtracted areas in red:

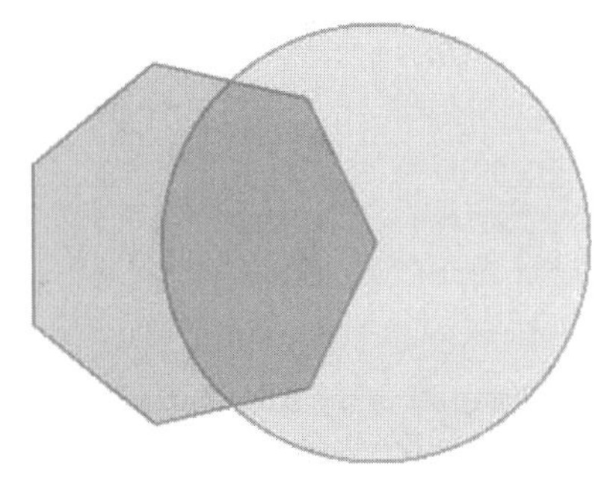

The area and perimeter are reported in the command bar:

Area = 13.20, Perimeter = 42.22

Total area = 0.80

Specify first corner point specifies the first point of an area.

Specify next corner point specifies additional points defining the corners of the area.

Object reports the area of a selected object.

Add area adds additional areas to the total; prompt changes to "ADD mode."

Subtract area removes areas from the total area; prompt changes to "SUBTRACT mode."

Total reports the total area and exits the command.

Undo undoes the last point.

eXit exits the command.

Volume options

Specify first corner point or [Object/Add volume/Subtract volume/eXit] <Object>: *(Pick a point, or enter an option.)*

Specify next corner point or [Arc/Length/Undo]: *(Pick another point, or enter an option.)*

Specify next corner point or [Arc/Length/Undo/Total] <Total>: *(Enter another point, enter an option, or press Enter.)*

The volume is reported in the command bar:

Total Volume = 0.0477

Specify first corner point specifies the first point of a volume.

Specify next corner point specifies additional points defining the corners of the volume.

Object reports the volume of a selected 3D object.

Add volume adds additional volumes to the total; prompt changes to "ADD mode."

Subtract volume removes volumes from the total area; prompt changes to "SUBTRACT mode."

Total reports the total volume and exits the command.

Undo undoes the last point.

eXit exits the command.

RELATED COMMANDS

Area calculates the area and perimeter of one or more areas.

Dist reports the distance and angle between two points.

ID reports the x, y, and z coordinates of picked points.

MassProp reports the mass properties of 3D solids.

Properties reports many geometric measurements of objects.

TIPS

- This command also measures surfaces, as of AutoCAD 2011.
- Measuring the area or volume of more than one object can be tricky. Here are the steps to take:

1. Start the MeasureGeom command, and then enter the ARea option:

    Enter an option [Distance/Radius/Angle/ARea/Volume] <Distance>: ar

2. Enter the Add area option, and then select the first object. (You have to add the first object to nothing before adding and subtracting others.)

    Specify first corner point or [Object/Add area/Subtract area/eXit] <Object>: a

3. Specify that you wish to add objects:

    Specify first corner point or [Object/Subtract area/eXit]: o

4. Even though the prompt says 'Select objects' in the plural, AutoCAD allows you to pick just one:

    (ADD mode) Select objects: *(Pick the first object.)*
    Area = 0.0213, Circumference = 0.5168
    Total area = 0.0213

4. Continue in Add mode to add the area of the next objects:

    (ADD mode) Select objects: *(Pick the next object.)*

5. To remove areas, switch to Remove mode by pressing Enter at the 'Select objects' prompt, and then enter s:

    (ADD mode) Select objects: *(Press Enter.)*
    Specify first corner point or [Object/Subtract area/eXit]: s

- When measuring the volume of more than one object, AutoCAD prompts for the height of each one.

Menu

V. 1.0 Loads *.cui*, *.cuix*, *.mns*, and *.mnu* menu files.

Command	*Aliases*	*Keyboard Shortcuts*	*Menu*	*Ribbon*
menu	...	...	...	...

Command: menu

Displays the Select Menu File dialog box.

Select a .cui, .cuix, .mns, or .mnu file, and then click Open.

COMMAND LINE OPTIONS

None.

RELATED COMMANDS

Cui controls most aspects of customization.

CuiLoad loads a partial menu file.

Tablet configures digitizing tablet for use with overlay menus.

RELATED SYSTEM VARIABLES

MenuName specifies the name of the currently-loaded menu file.

MenuCtl determines whether sides screen menu pages switch in parallel with commands entered at the keyboard.

MenuEcho suppresses menu echoing.

ScreenBoxes specifies the number of menu lines displayed on the side menu.

RELATED FILES

*.*cui* source file for menus, toolbars, and so on; stored in XML format.

*.*mns* source for obsolete (legacy) menu files; stored in ASCII format.

*.*mnu* source for obsolete (legacy) menu template files; stored in ASCII format.

TIPS

- AutoCAD automatically converts *.mns* and *.mns* files into *.cuix* files.
- The *.cuix* file defines the function of the screen menu, menu bar, cursor menu, icon menus, digitizing tablet menus, pointing device buttons, toolbars, help strings, and so on.
- AutoCAD 2010 changed the file format from *.cui* to *.cuix*, which added the abililty to hold bitmap images.
- AutoCAD includes the *custom.cuix* file for customizing menus independently of *acad.cuix*.

Replaced Commands

MenuLoad and **MenuUnload** commands became aliases for the CuiLoad and CuiUnload commands with AutoCAD 2006.

Mesh

2010 Creates 3D mesh primitives: boxes, cones, cylinders, pyramids, spheres, tori, and wedges.

Command	*Aliases*	*Keyboard Shortcuts*	*Menu*	*Ribbon*
mesh	...	...	**Draw** ↳**Modeling** ↳**Meshes** ↳**Primitives**	**Mesh Primitives** ↳**Box** *(etc)*

Command: mesh
Select primitive [Box/Cone/CYlinder/Pyramid/Sphere/Wedge/Torus/SEttings]: *(Enter an option.)*

COMMAND LINE OPTIONS

Box creates square and rectangular boxes.

Cone creates cones and truncated cones.

CYlinder creates round and elliptical cylinders.

Pyramid creates multisided pyramids and truncated pyramids.

Sphere creates spheres.

Wedge creates rectangular and cubic wedges.

Torus creates tori.

SEttings specifies the smoothness and tessellation values; range is 0 to 4.

Box options

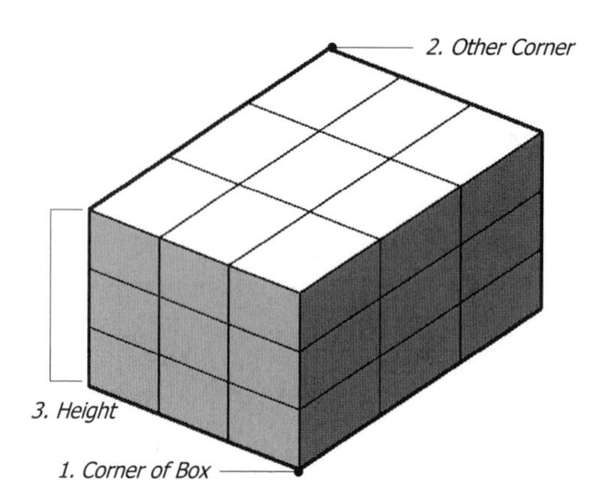

Specify first corner or [Center]: *(Pick point 1, or type the C option.)*
Specify other corner or [Cube/Length]: *(Pick point 2, or enter an option.)*
Specify height or [2Point] <1.0>: *(Pick point 3, or type the 2 option.)*

First corner specifies one corner for the base of the box.

Center draws the box about a center point.

Other corner specifies the second corner for the base of the box.

Cube draws a cubic box — all sides having the same length.

Length specifies the length along x axis, width along y axis, and height along z axis.

Height specifies the height of the box.

2point specifies height as the distance between two points.

Cone options

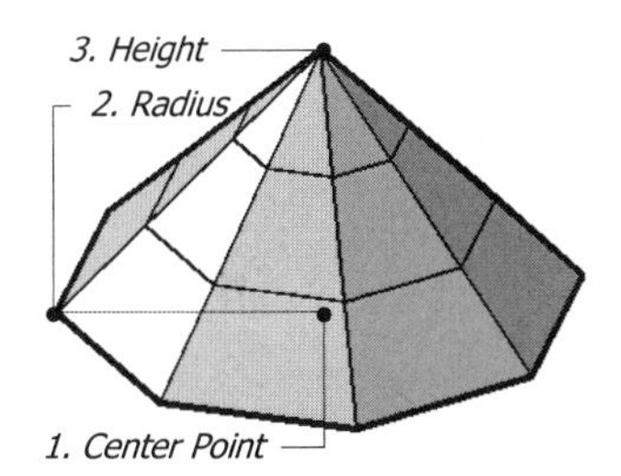

Specify center point of base or [3P/2P/Ttr/Elliptical]: *(Pick center point 1.)*
Specify base radius or [Diameter] <2">: *(Specify radius 2, or type D.)*
Specify height or [2Point/Axis endpoint/Top radius] <2">: *(Specify height 3, or type A.)*

Center point of base specifies the x, y, z coordinates of the center point of the cone's base.
3P picks three points that specify the base's circumference.
2P specifies two points on the base's circumference.
Ttr specifies two points of tangency (with other objects) and the radius.
Elliptical creates a cone with an elliptical base.

Base radius specifies the cone's radius.
Diameter specifies the cone's diameter.

Height specifies the cone's height.
2Point picks two points that specify the z orientation of the cone.
Axis endpoint picks the other end of an axis formed from the center point.
Top radius creates a cone with a flat top.

CYlinder options

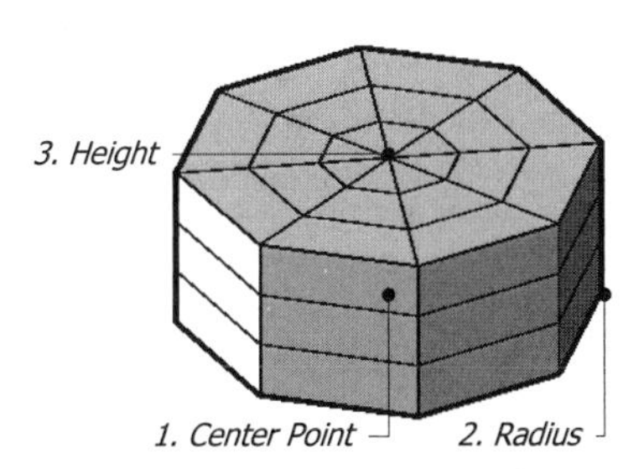

Specify center point of base or [3P/2P/Ttr/Elliptical]: *(Pick center point 1, or enter an option.)*
Specify base radius or [Diameter] <3>: *(Specify radius 2, or type D.)*
Specify height or [2Point/Axis endpoint] <1">: *(Specify height 3, or type C.)*

Center point of base specifies the x, y, z coordinates of the center point of the cylinder's base.
3P picks three points that specify the base's circumference.
2P specifies two points on the base's circumference.
Ttr specifies two points of tangency (with other objects) and the radius.
Elliptical creates elliptical cylinders.

Base radius specifies the cylinder's radius.

Diameter specifies the cylinder's diameter.

Height specifies the cylinder's height.

2Point picks two points that specify the z orientation of the cylinder.

Axis endpoint picks the other end of an axis formed from the center point.

Pyramid options

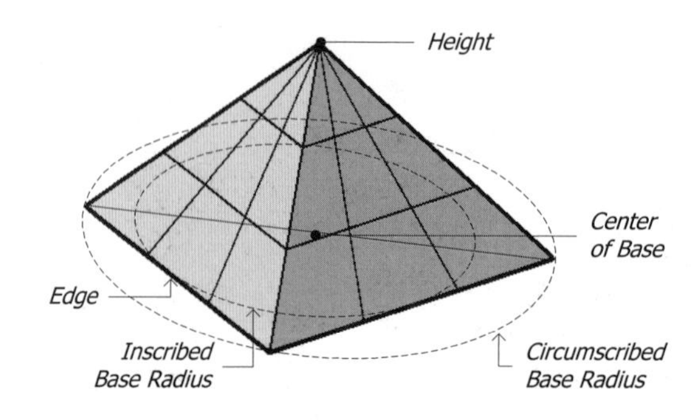

4 sides Circumscribed
Specify center point of base or [Edge/Sides]: *(Pick a point, or enter an option.)*
Specify base radius or [Inscribed]: *(Pick a point, or type I.)*
Specify height or [2Point/Axis endpoint/Top radius]: *(Pick a point, or enter an option.)*

The base is defined by a center point or edge length and the number of sides:

Specify center point of base locates the center of the pyramid's base by defining x, y, z coordinates or by picking a point in the drawing.

Edge specifies the size of the base by the length of one edge; the base is always made from equilateral sides.

Sides specifies the number of sides; the range is 3 to 32, and the default is 4.

When the base is drawn like a polygon (circumscribed or inscribed):

Specify base radius specifies the radius of a "circle" that circumscribes the base.

Inscribed specifies the radius of a "circle" inscribed within the base.

Specify height indicates the height by specifying x, y, z coordinates or by picking a point in the drawing.

2Point specifies the height by picking two points.

Axis endpoint specifies the location and angle of the top point; this allows the pyramid to be drawn at an angle; the top and bottom are perpendicular to the axis.

Top radius specifies the size of the flat top to the pyramid.

Sphere options

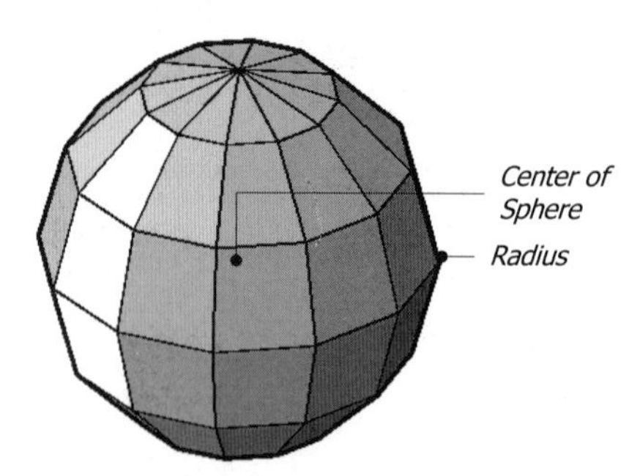

Specify center point or [3P/2P/Ttr]: *(Pick a point, or enter an option.)*
Specify radius or [Diameter] <1.0>: *(Enter a value, or type D.)*

Specify center point locates the center point of the sphere.

3P specifies three points on the outer circumference (equator) of the sphere.

2P specifies two points on the outer diameter of the sphere.

Ttr specifies two tangent points plus the radius of the outer circumference of the sphere.

Radius specifies the radius of the sphere.

Diameter specifies the diameter of the sphere.

Wedge options

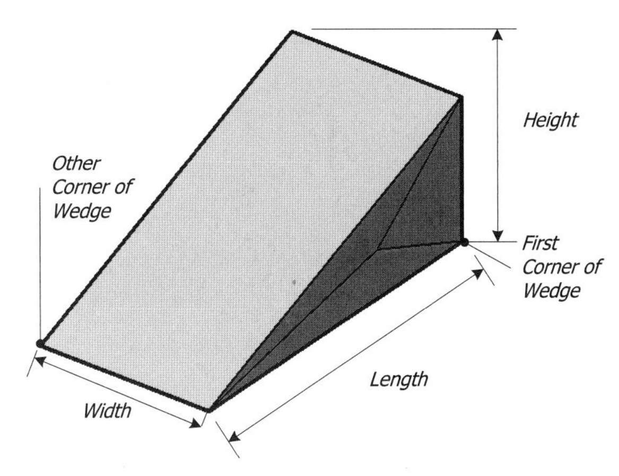

Specify first corner or [Center]: *(Pick a point, or type C.)*
Specify other corner or [Cube/Length]: *(Pick a point, or enter an option.)*
Specify height or [2Point]: *(Pick another point, or enter 2P.)*

First corner specifies the lower-left corner of the wedge.

Center draws the wedge's base about the center of the sloped face.

Cube draws cubic wedges.

Length specifies the length, width, and height of the wedge.

2Point picks two points to indicate the height.

Center options
Specify center: *(Pick a point.)*
Specify corner or [Cube/Length]: *(Pick a point, or enter an option.)*
Specify height or [2Point]: *(Pick a point.)*

Specify center of wedge indicates the center of the wedge's inclined face.

Specify opposite corner indicates the distance from the midpoint to one corner.

Cube options
Specify length: *(Specify the length.)*

Specify length indicates the length of all three sides.

Length options
Specify length: *(Specify the length.)*
Specify width: *(Specify the width.)*
Specify height: *(Specify the height.)*

Specify length indicates the length parallel to the x-axis.

Specify width indicates the width parallel to the y-axis.

Specify height indicates the height parallel to the z-axis.

Torus options

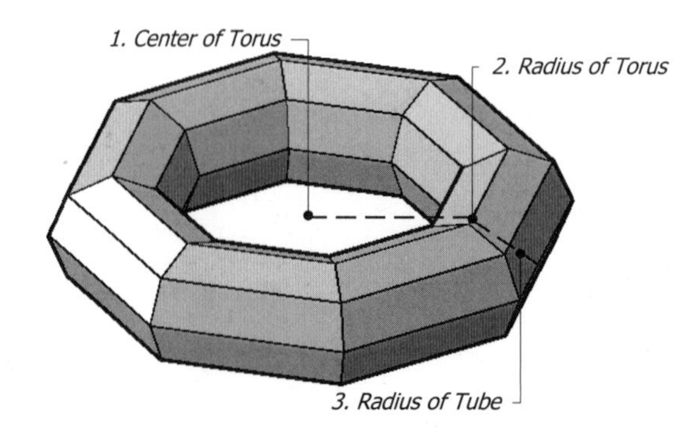

Specify center point or [3P/2P/Ttr]: *(Pick point 1.)*

Specify radius or [Diameter] <1.0>: *(Specify the radius, 2, or type D.)*

Specify tube radius or [2Point/Diameter]: *(Specify the radius, 3, or type D.)*

Center point indicates the center of the torus.

3P specifies three points on the outer circumference of the torus.

2P specifies two points on the outer diameter of the torus.

Ttr specifies two tangent points plus the radius of the outer circumference of the torus.

Specify radius indicates the radius of the torus.

Diameter indicates the diameter of the torus.

Tube radius indicates the radius of the tube.

2Point asks for two points to determine the diameter of the tube.

SEttings options

Specify level of smoothness or [Tessellation]: *(Enter a number, or type T.)*

Specify level of smoothness specifies smoothness between 0 (none) and 6.

Tessellation specifies the number of tessellation lines; displays the Mesh Primitives Options dialog box. See MeshPrimitivesOptions command.

RELATED COMMAND

MeshPrimitivesOptions specifies the default number of tessellation lines and smoothness for mesh primitives.

RELATED SYSTEM VARIABLES

DivMeshBoxHeight specifies the default number of subdivisions along the height of boxes.

DivMeshBoxLength specifies the default number of subdivisions along the length of boxes.

DivMeshBoxWidth specifies the default number of subdivisions along the width of boxes.

DivMeshConeAxis specifies the default number of subdivisions around the base of cones.

DivMeshConeBase specifies the default number of subdivisions on the base of cones.

DivMeshConeHeight specifies the default number of subdivisions between the base and the tip of cones.

DivMeshCylAxis specifies the default number of subdivisions around the trunk of cylinders.
DivMeshCylBase specifies the default number of subdivisions on the end caps of cylinders.
DivMeshCylHeight specifies the default number of subdivisions along the trunk of cylinders.

DivMeshPyrBase specifies the default number of subdivisions on the base of pyramids.
DivMeshPyrHeight specifies the default number of subdivisions between the base and the tip of pyramids.
DivMeshPyrLength specifies the default number of subdivisions along the base of pyramids.

DivMeshSphereAxis specifies the default number of subdivisions between the two polar ends of spheres.
DivMeshSphereHeight specifies the default number of subdivisions around spheres.

DivMeshTorusPath specifies the default number of subdivisions along paths swept by tori profiles.
DivMeshTorusSection specifies the default number of subdivisions around paths swept by tori profiles.

DivMeshWedgeBase specifies the default number of subdivisions on the base of wedges.
DivMeshWedgeHeight specifies the default number of subdivisions along the height of wedges.
DivMeshWedgeLength specifies the default number of subdivisions along the length of wedges.
DivMeshWedgeSlope specifies the default number of subdivisions along the slope of wedges.
DivMeshWedgeWidth specifies the default number of subdivisions along the width of wedges.

TIPS

- You can edit 3D mesh objects directly: hold down the **Ctrl** key as you choose a face, edge, or vertex, and then drag the sub-object to manipulate the 3D mesh surface, as illustrated below. You can select more than one face.

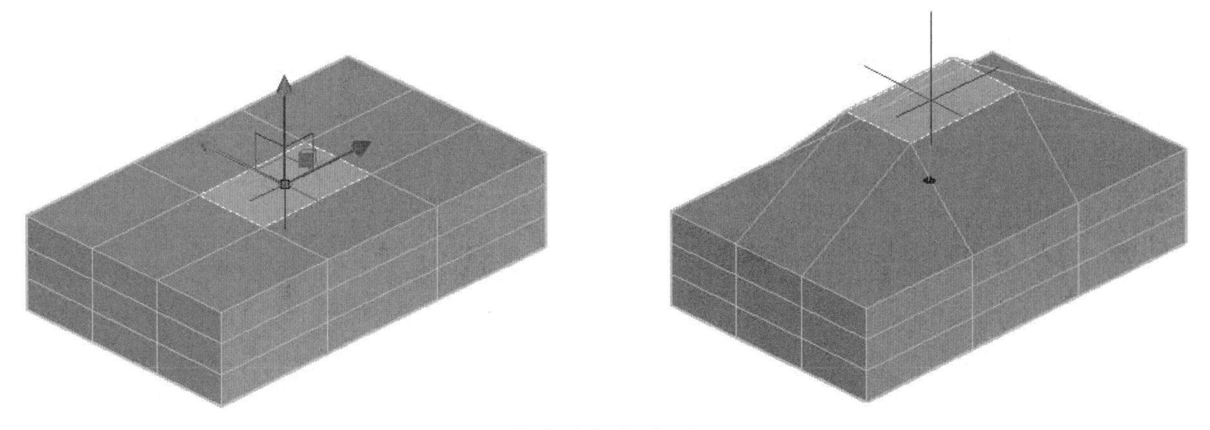

Left: *A face is selected.*
Right: *The face is dragged to create a new shape.*

- The Mesh command creates primitives such as boxes from 3D mesh objects, which were introduced with AutoCAD 2010. The Ai_Box command creates boxes made of polyfaces, while the Box command creates solid model boxes.
- 3D mesh objects created by this command are "watertight," because the surface encloses a hollow inside (water wouldn't be able to leak out). In contrast, the solid models created by the Box and other commands are solid inside.
- You can convert 3D mesh objects to solid models with the ConvToSolid command. Some editing command perform the conversion automatically, such as Fillet and Chamfer.

- If you find it difficult to select a specific subojbect, use the SubObjectSelectionMode system variable to filter the choices:

SubObjectSelectionMode	*Meaning*	*Icon*
0	No filter	
1	Vertices only	
2	Edges only	
3	Faces only	
4 *(applies to solids only)*	History only	

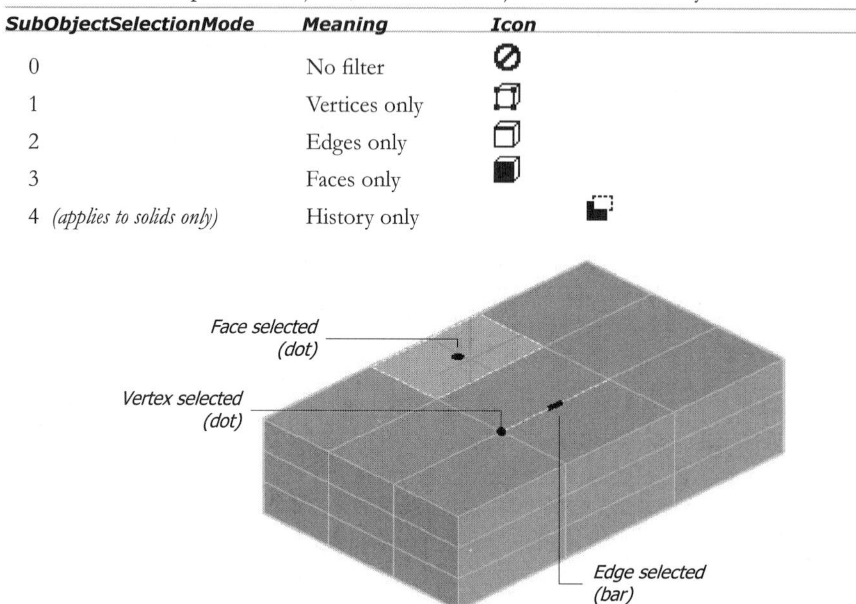

- The look of 3D mesh objects can be adjusted with the MeshCrease and MeshUncrease commands, MeshSmoothMore and Mesh SmoothLess, and MeshRefine commands.

MeshCap

2011 Covers open edges with a 3D mesh.

Command	Aliases	Keyboard Shortcuts	Menu	Ribbon
meshcap	...	...	**Modify** **↳Mesh Editing** **↳Close Hole**	**Mesh** **↳Mesh Edit** **↳Close Hole**

Command: meshcap
Select connecting mesh edges to create a new mesh face: *(Choose one or more edges of meshes.)*
Select connecting mesh edges to create a new mesh face: *(Press Enter.)*

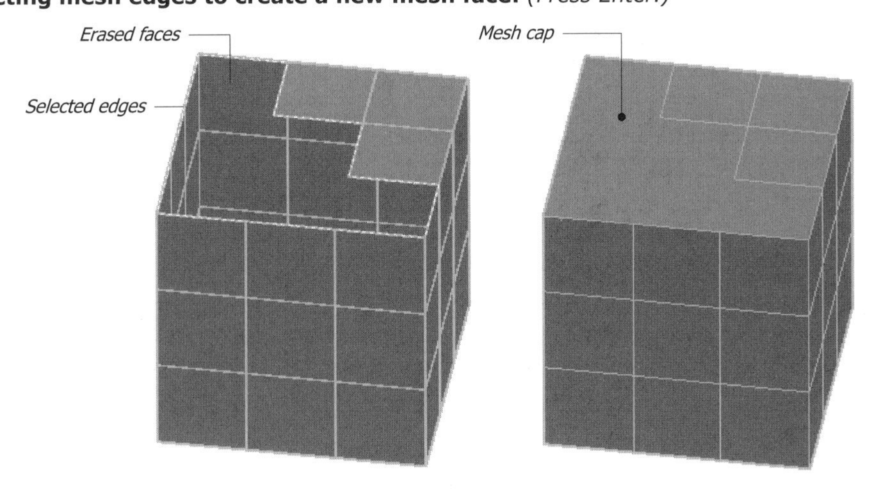

COMMAND LINE OPTION

Select connecting mesh edges chooses the edges to be capped; edges must connect. AutoCAD creates the cap out of a single multi-edged face.

RELATED COMMAND

SurfacePatch caps open surfaces.

TIPS

- This command can repair "leaky" meshes that did not translate properly from other programs.
- The Erase command removes faces from meshes, as well as edges and vertices (along with all adjacent faces). Use this command to replace the missing faces.

MeshCollapse

2011 Collapses faces and edges by merging vertices of 3D mesh objects to a single vertex.

Command	*Aliases*	*Keyboard Shortcuts*	*Menu*	*Ribbon*
meshcollapse	...	...	**Modify**	**Mesh**
			↳Mesh Editing	**↳Mesh Edit**
			↳Collapse Face or Edge	**↳Collapse Face or Edge**

Command: meshcollapse
Select mesh face or edge to collapse: *(Press Enter.)*

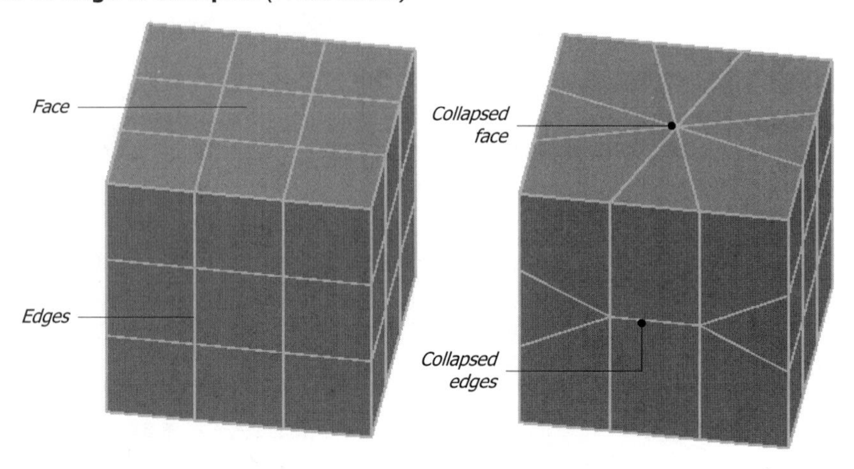

COMMAND LINE OPTION

Select mesh face or edge chooses the faces and edges to remove.

RELATED COMMANDS

MeshSplit splits one face into multiple faces.

MeshMerge merges two or more faces into a single face.

TIPS

- This command converges faces at the center of a selected edge or face. Adjacent faces adjust their shape to match the new topology.
- You get to select one edge or face, and then the collapse occurs.

MeshCrease / MeshUncrease

2010 Sharpens and removes edges of mesh sub-objects.

Command	Alias	Keyboard Shortcuts	Menu	Ribbon
meshcrease	**crease**	...	**Modify** ↳**Mesh Editing** ↳**Crease**	**Mesh** ↳**Mesh** ↳**Add Crease**
meshuncrease	**uncrease**	...	**Modify** ↳**Mesh Editing** ↳**Uncrease**	**Mesh Modeling** ↳**Mesh** ↳**Remove Crease**

Command: meshcrease
Select mesh subobjects to crease: *(Choose one or more faces on mesh objects.)*
Select mesh subobjects to crease: *(Choose more faces, or press Enter to continue.)*
Specify crease value [Always]: *(Enter a number, or type A.)*

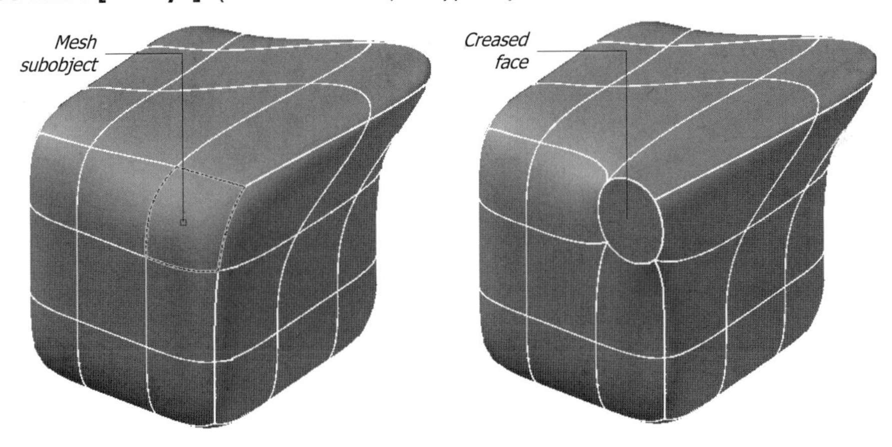

COMMAND LINE OPTIONS

Select mesh subobjects chooses the faces or edges to sharpen with a crease. Press Shift+click to remove a sub-object from the selection set.

Specify crease value specifies the crease value; enter a number between 0 and 6. (0 removes the crease.) The crease will disappear when its value is less than the mesh's smoothness value.

Always retains the crease, no matter the values of crease and smoothness.

MESHUNCREASE Command

Command: meshuncrease
Select crease to remove: *(Choose one or more creased faces on mesh objects.)*
Select crease to remove: *(Choose more faces, or press Enter to continue.)*

Select crease to remove chooses the faces or edges to uncrease.

TIP

- Enter A ("Always"), unless you have a reason for entering a crease number.

MeshExtrude

2011 Extrudes faces of 3D meshes by a height, in a direction, along a path, and with a taper angle.

Command	*Aliases*	*Keyboard Shortcuts*	*Menu*	*Ribbon*
meshextrude	...	...	**Modify** ↳**Mesh Editing** ↳**Extrude Face**	**Mesh** ↳**Mesh Edit** ↳**Extrude Face**

Command: meshextrude
Adjacent extruded faces set to: Join
Select mesh face(s) to extrude or [Setting]: *(Choose one or more faces, or type S.)*
Select mesh face(s) to extrude or [Setting]: *(Press Enter.)*
Specify height of extrusion or [Direction/Path/Taper angle] <1.0>: *(Enter an option.)*

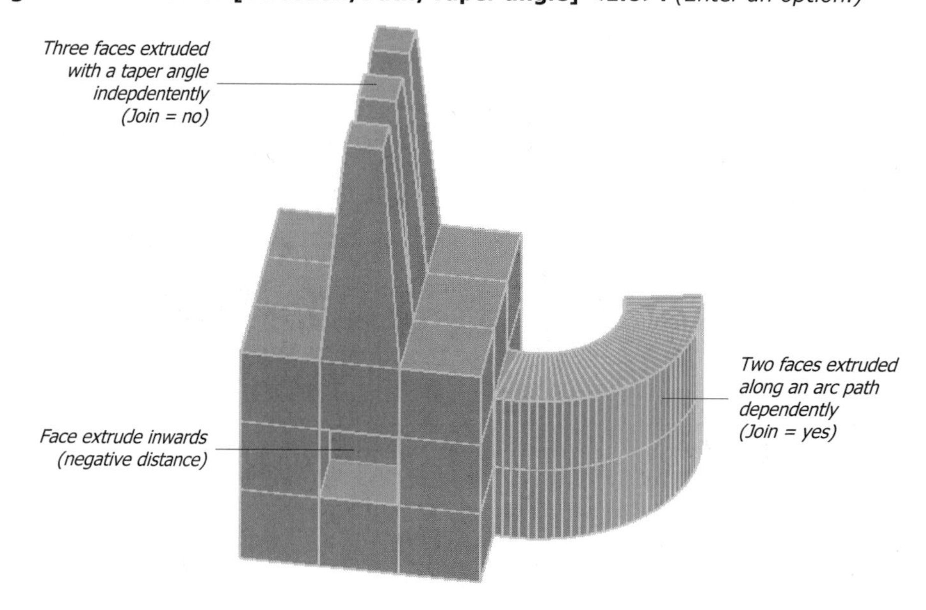

COMMAND LINE OPTIONS

Select mesh face(s) chooses the faces to extrude.

Setting determines whether multiple faces are treated as one or individually; prompts you:

Join adjacent mesh faces [Yes/No] <No>: *(Type Y or N.)*
Adjacent extruded faces set to: Unjoin

Height specifies the length of the extrusion; enter a negative value to push extrusion into the mesh.

Direction specifies the length and direction of the extrusion.

Path selects an open object that specifies the path of the extrusion, such as line, polyline, arc, or spline.

Taper angle specifies the angle at which the extrusion should taper; enter a negative angle to taper outward.

TIP

- It is not always apparent if adjacent faces are extruded when meshes have not been smoothed or have a taper applied.

MeshMerge

2011 Merges adjacent faces of 3D meshes into a single face.

Command	Aliases	Keyboard Shortcuts	Menu	Ribbon
meshmerge	...	...	**Modify** ↳**Mesh Editing** ↳**Merge Face**	**Mesh** ↳**Mesh Edit** ↳**Merge Face**

Command: meshmerge
Select adjacent mesh faces to merge: *(Choose one or more adjacent faces.)*
Select adjacent mesh faces to merge: *(Press Enter.)*

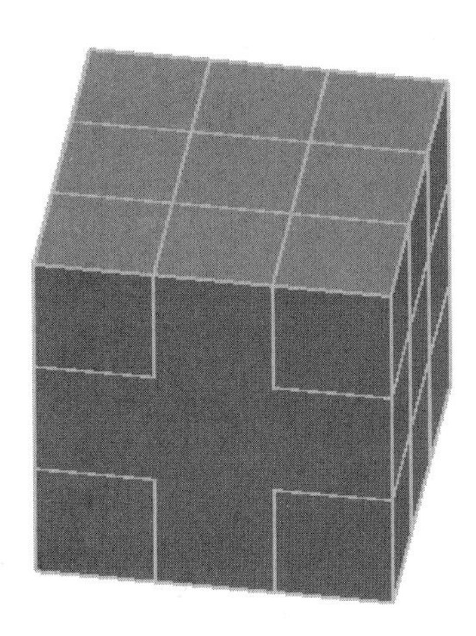

Left: *Selected, adjacent faces.*
Right: *The faces merged.*

COMMAND LINE OPTION

Select adjacent face chooses the faces to merge; the faces must be adjacent to one another. AutoCAD merges the selected faces into a single, multi-edged face.

RELATED COMMANDS

MeshCap fills in gaps between faces.

MeshCollapse reduces multiple faces to one.

TIP

- Autodesk recommends that selected faces be on the same plane; selecting faces around a corner may result in gaps.

MeshOptions

2010 Sets options for converting objects into 3D meshes.

Command	*Aliases*	*Keyboard Shortcuts*	*Menu*	*Ribbon*
meshoptions	...	...	...	**Home** ↳**Mesh** ↳**Mesh Tessellation Options**

Command: meshoptions

Displays dialog box.

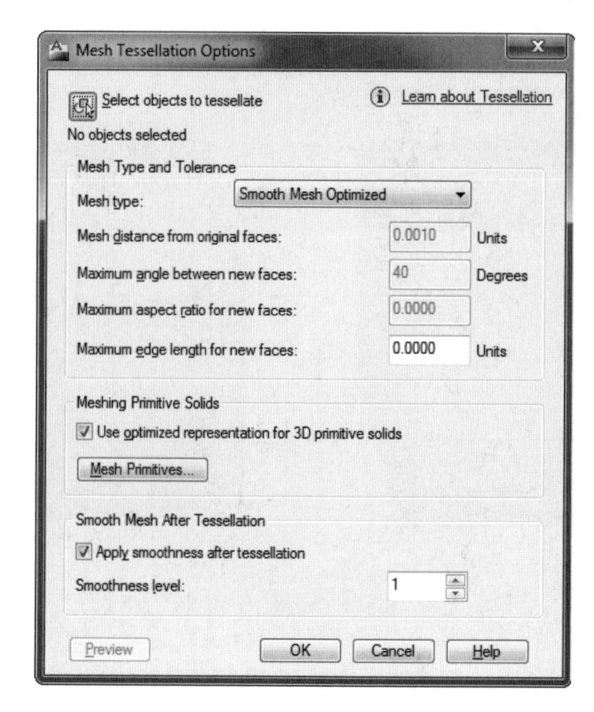

MESH TESSELLATION OPTIONS DIALOG BOX

Select Objects to Tessellate chooses objects in the drawing; the dialog box is temporarily dismissed. Choose 3D solids, 3D surfaces, 3D faces, polygon meshes, polyface meshes, regions, and closed polylines.

Mesh Type and Tolerance options

Mesh Type specifies the type of mesh into which the objects are converted:

- **Smooth Mesh Optimized** matches the mesh faces to the shape of the original objects.
- **Mostly Quads** uses quadrilateral faces as much as possible.
- **Triangles** uses triangular faces as much as possible.

Mesh Distance from Original Faces specifies the maximum deviation of mesh faces from the original object.

Maximum Angle Between New Faces specifies the surface normal's maximum angle between two adjacent faces.

Maximum Aspect Ratio for New Faces specifies the maximum height-width ratio of new mesh faces:

FaceTerGridRatio	*Meaning*
>1	Specifies the height over width ratio.
1	Forces height to equal the width; faces are square.
<1	Specifies the width over height ratio.
0	Ignores the aspect ratio; faces can have any aspect ratio.

Maximum Edge Length for New Faces specifies the maximum length of face edges; set to 0 to allow AutoCAD to determine the length.

Meshing Primitive Solids options

Use Optimized Representation for 3D Primitive Solids determines which settings are used for mesh object conversion:

☑ Use settings from the Mesh Primitive Options dialog box; see MeshPrimitiveOptions command.

☐ Use settings from this dialog box (Mesh Tessellation Options).

Mesh Primitives specifies the number of tessellation divisions for mesh primitives; displays the Mesh Primitive Options dialog box. See MeshPrimitiveOptions command.

Smooth Mesh After Tessellation options

Apply Smoothness After Tessellation determines if smoothness is applied:

☐ Applies smoothness.

☑ Does not apply smoothness.

Smoothness Level specifies the level of smoothness; range is 0 to 6, or the maximum your computer can handle, as determined by AutoCAD.

Preview shows the effect of the settings on the current drawing; dismisses the dialog box temporarily. Press Esc to return to the dialog box. This button is available only after you use the Select Objects to Tessellate button to select solids and other convertible objects.

RELATED COMMAND

MeshSmooth converts solids, regions, surfaces, and closed polylines into mesh objects.

RELATED SYSTEM VARIABLES

FaceTerDevNormal specifies maximum angle between surface normal and contiguous mesh faces.

FaceTerDevSurface specifies how closely converted mesh objects follow the shapes of the original solids and surfaces.

FaceTerGridRatio specifies maximum aspect ratio for mesh subdivisions of converted solids and surfaces.

FaceTerMaxEdgeLength specifies maximum edge length of converted solids and surfaces.

FaceTerMaxGrid specifies maximum number of u and v grid lines on converted solids and surfaces.

FaceTerMeshType specifies type of mesh created.

FaceTerMinUGrid specifies minimum number of u grid lines for converted solids and surfaces.

FaceTerMinVGrid specifies minimum number of v grid lines for converted solids and surfaces.

FaceTerPrimitiveMode toggles smoothness settings of converted objects between settings in the Mesh Tessellation Options and Mesh Primitive Options dialog boxes.

FaceTerSmoothLev specifies default smoothness level of converted objects.

TIPS

- Small values of FaceTerDevSurface give you more accurate conversions, but the large number of faces could slow down AutoCAD. To create accurate conversions of fillets, AutoCAD increases the number of faces in high curvature areas, while decreasing it in flatter areas.
- Setting system variable FaceTerGridRatio to 0 can create sliver-like faces, which are not desirable.
- Setting system variable FaceMaxEdgeLength to 0 lets AutoCAD use the size of the model to determine the size of mesh faces. If you find AutoCAD running slowly, increase the value of this system variable. Fewer faces mean less accurate conversion but better performance.

MeshPrimitiveOptions

2010 Sets options for creating 3D mesh objects.

Command	*Aliases*	*Keyboard Shortcuts*	*Menu*	*Ribbon*
meshprimitiveoptions	...	...	...	**Mesh Modeling** ↳**Primitives** ↳**Mesh Primitive Options**

Command: meshprimitiveoptions

Displays dialog box.

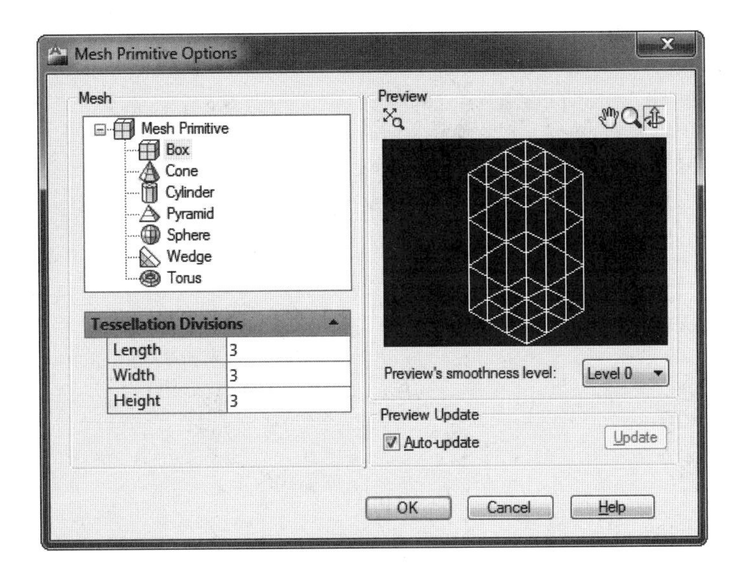

MESH PRIMITIVE OPTIONS DIALOG BOX

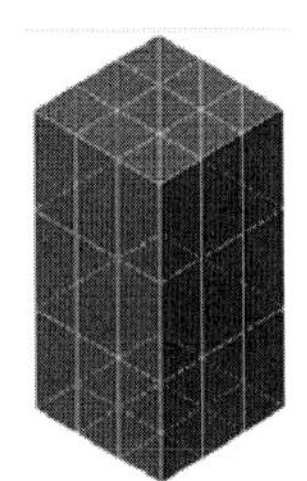

Box

Length specifies the default divisions along the length of boxes (x axis; default = 3).

Width specifies the default divisions along the width of boxes (y axis; 3).

Height specifies the default divisions along height of box es(z axis; 3).

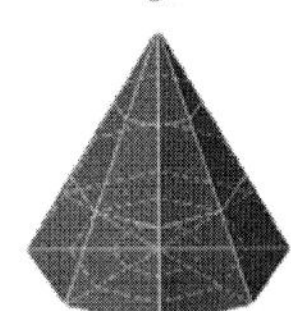

Cone

Axis specifies the default divisions around the perimeter of cones (default = 8).

Height specifies the default divisions between the base and top of cones (3).

Base specifies the default divisions between the perimeter and the center of the cones' base (3).

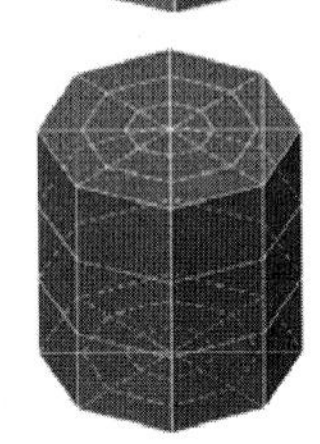

Cylinder

Axis specifies the default divisions around the perimeter of the cylinders' base (8).

Height specifies the default divisions between the base and top of cylinders (3).

Base specifies the default divisions between the perimeter and center of cylinders' base (3).

Pyramid

Length specifies the default divisions for the sides of pyramids' base (3).

Height specifies the default divisions between the base and the top of pyramids (3).

Base specifies the default divisions between the perimeter and center pyramids' base (3).

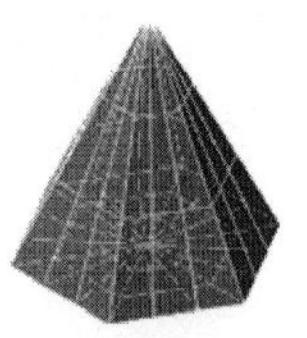

Sphere

Axis specifies the default number of radial divisions around the axis endpoint of spheres (12).

Height specifies the default divisions between axis endpoints of spheres (6).

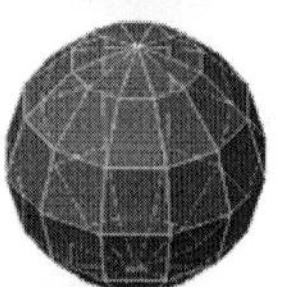

Wedge

Length specifies the default divisions along the length of wedges (x axis; 4).

Width specifies the default divisions along the width of wedges (y axis; 3).

Height specifies the default divisions along the height of wedges (z axis; 3).

Slope specifies the default divisions on the slope between the apex and the edge of the base (3).

Base specifies the default divisions between the midpoint of triangular sides of wedges (3).

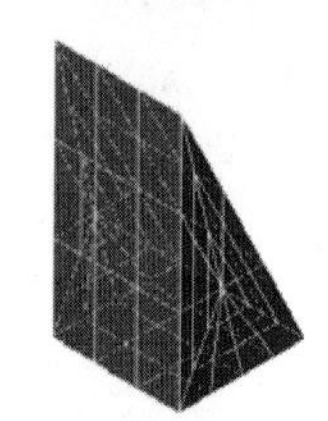

Torus

Radius specifies the default divisions around the circumference of tubes' profiles (8).

Sweep Path specifies the default divisions around the circumference of tubes' path (8).

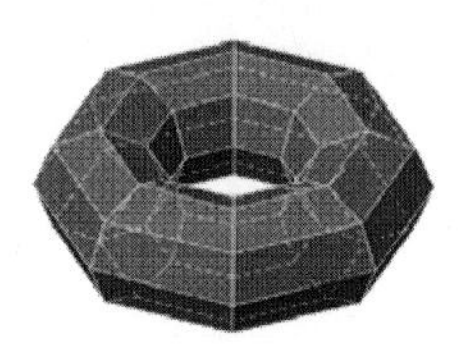

Preview options

Preview's Smoothness Level changes the smoothness level of the preview, but not of primitive meshes; range is from 0 (no smoothness) to 6 (maximum smoothness).

Auto-update updates the preview as options are changed.

Update updates the preview manually.

SHORTCUT MENU

Access this menu by right-clicking the preview area.

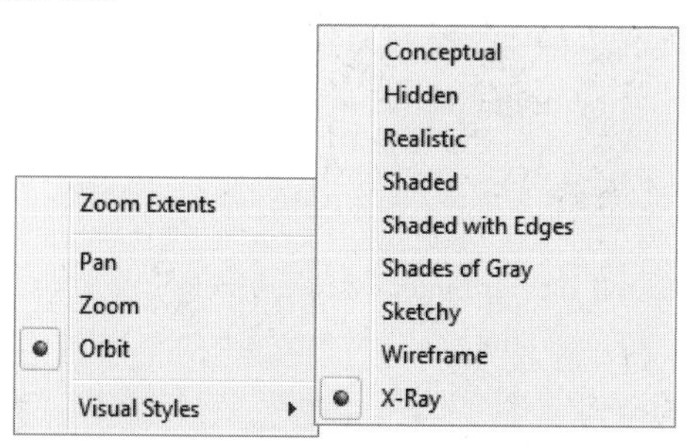

Zoom Extents zooms the sample image to fill the preview area.

Pan moves the sample image through clicking and dragging.

Zoom enlarges and reduces the sample image through clicking and dragging.

Orbit rotates the preview image in 3D space; see the 3dOrbit command.

Visual Styles determines the look of the sample image; see the VisualStyles command.

RELATED COMMANDS

MeshSmoothLess decreases the smoothness of mesh objects.

MeshSmoothMore increases the smoothness of mesh objects.

RELATED SYSTEM VARIABLES

DivMeshBoxHeight specifies the default number of subdivisions along the height of boxes.

DivMeshBoxLength specifies the default number of subdivisions along the length of boxes.

DivMeshBoxWidth specifies the default number of subdivisions along the width of boxes.

DivMeshConeAxis specifies the default number of subdivisions around the base of cones.

DivMeshConeBase specifies the default number of subdivisions on the base of cones.

DivMeshConeHeight specifies the default number of subdivisions between the base and the tip of cones.

DivMeshCylAxis specifies the default number of subdivisions around the trunk of cylinders.

DivMeshCylBase specifies the default number of subdivisions on the endcaps of cylinders.

DivMeshCylHeight specifies the default number of subdivisions along the trunk of cylinders.

DivMeshPyrBase specifies the default number of subdivisions on the base of pyramids.

DivMeshPyrHeight specifies the default number of subdivisions between the base and the tip of pyramids.

DivMeshPyrLength specifies the default number of subdivisions along the base of pyramids.

DivMeshSphereAxis specifies the default number of subdivisions between the two polar ends of spheres.

DivMeshSphereHeight specifies the default number of subdivisions around spheres.

DivMeshTorusPath specifies the default number of subdivisions along paths swept by tori profiles.

DivMeshTorusSection specifies the default number of subdivisions around paths swept by tori profiles.

DivMeshWedgeBase specifies the default number of subdivisions on the base of wedges.

DivMeshWedgeHeight specifies the default number of subdivisions along the height of wedges.

DivMeshWedgeLength specifies the default number of subdivisions along the length of wedges.

DivMeshWedgeSlope specifies the default number of subdivisions along the slope of wedges.

DivMeshWedgeWidth specifies the default number of subdivisions along the width of wedges.

TIP

- You can also create mesh objects using 3D solid, region, and surface drawing and editing commands, and then use the MeshSmooth command to convert them to meshes.

MeshRefine

2010 Increases the number of faces in mesh objects and sub-objects.

Command	Alias	Keyboard Shortcuts	Menu	Ribbon
meshrefine	**refine**	...	**Modify** ↳**Mesh Editing** ↳**Refine Mesh**	**Home** ↳**Mesh** ↳**Mesh Refine**

Command: meshrefine

Mesh object or face subobjects to refine: *(Pick a mesh object, or press Ctrl+pick to choose a subobject, such as a face.)*

Mesh object or face subobjects to refine: *(Press Enter.)*

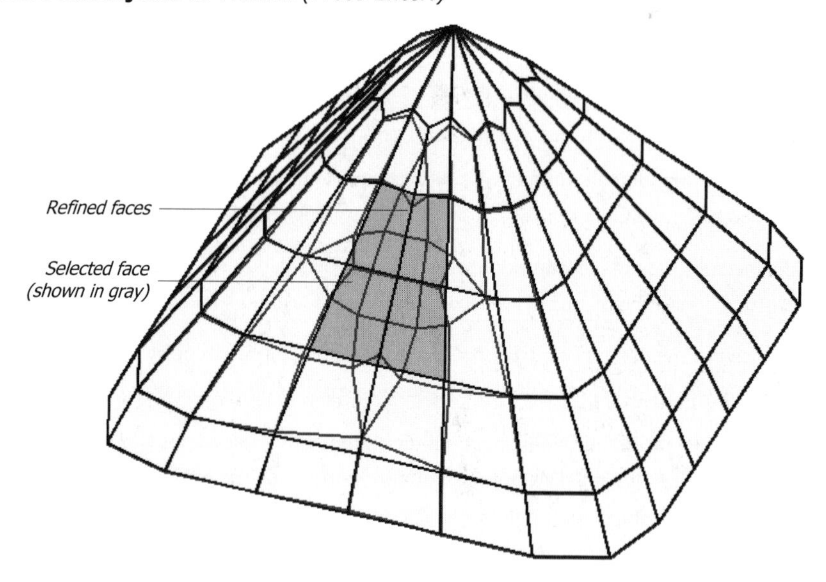

COMMAND LINE OPTIONS

Mesh object selects a single, entire mesh object.

Face subobject selects a face within the mesh object; to select the face, hold down the Ctrl key.

RELATED COMMAND

MeshSplit splits faces into two.

RELATED SYSTEM VARIABLE

SubObjSelectionMode determines which part of a sub-object is selected when you hold down the Ctrl key.

TIPS

- The purpose of this command is to increase the number of editable faces. This allows you to create more detailed model features.
- This command works only with mesh objects that have a smoothness level of 1 or higher. You may need to apply the MeshSmoothMore command first.
- Too much mesh detail can overwhelm your computer, so limit the areas in which mesh faces are refined.

MeshSmooth

2010 Converts objects into mesh objects.

Command	*Alias*	*Keyboard Shortcuts*	*Menu*	*Ribbon*
meshsmooth	**smooth**	...	**Draw**	**Home**
			↳Mesh Editing	**↳Mesh**
			↳Smooth Less	**↳Smooth Objects**

Command: meshsmooth
Select objects to convert: *(Choose one or more objects.)*
Select objects to convert: *(Press Enter to end object selection.)*

COMMAND LINE OPTION

Select objects to convert chooses the objects to convert to 3D mesh objects.

RELATED COMMAND

MeshOptions specifies how objects are converted to 3D mesh objects.

RELATED SYSTEM VARIABLES

FaceterDevNormal specifies maximum angle between surface normal and contiguous mesh faces.

FaceterDevSurface specifies how closely converted mesh objects follow the shape of the original solids and surfaces.

FaceterGridRatio specifies maximum aspect ratio for mesh subdivisions of converted solids and surfaces.

FaceterMaxEdgeLength specifies maximum edge length of converted solids and surfaces.

FaceterMaxGrid specifies maximum number of u and v grid lines on converted solids and surfaces.

FaceterMeshType specifies type of mesh created.

FaceterMinUGrid specifies minimum number of u grid lines for converted solids and surfaces.

FaceterMinVGrid specifies minimum number of v grid lines for converted solids and surfaces.

FaceterPrimitiveMode toggles smoothness settings of converted objects between settings in the Mesh Tessellation Options and Mesh Primitive Options dialog boxes.

FaceterSmoothLev specifies default smoothness level of converted objects.

TIPS

- This command converts the following objects into 3D mesh objects:
    - 3D solids
    - 3D surfaces
    - 3D faces
    - Polyface meshes
    - Polygon meshes
    - Regions (2D)
    - Surfaces and NURBS surfaces *(new to AutoCAD 2011)*.
    - Closed polylines (2D)
- The Mesh Tessellation Options dialog box specifies the parameters this command uses for converting objects; see the MeshOptions command.
- Mesh objects can be converted back to 3D solids with the ConvToSolid command and to surfaces with the ConvToSurface command.
- This command has nothing to do with smoothing mesh objects, despite its name.

MeshSmoothLess / MeshSmoothMore

2010 Decreases and increases the smoothness levels of 3D mesh objects.

Command	Alias	Keyboard Shortcuts	Menu	Ribbon
meshsmoothless	**less**	**...**	**Modify ↳Mesh Editing ↳Smooth Less**	**Home ↳Mesh ↳Smooth Less**
meshsmoothmore	**more**	**...**	**Modify ↳Mesh Editing ↳Smooth More**	**Home ↳Mesh ↳Smooth More**

Command: meshsmoothless
Select mesh objects to increase the smoothness level: *(Choose one or more mesh objects.)*
Select mesh objects to increase the smoothness level: *(Press Enter.)*

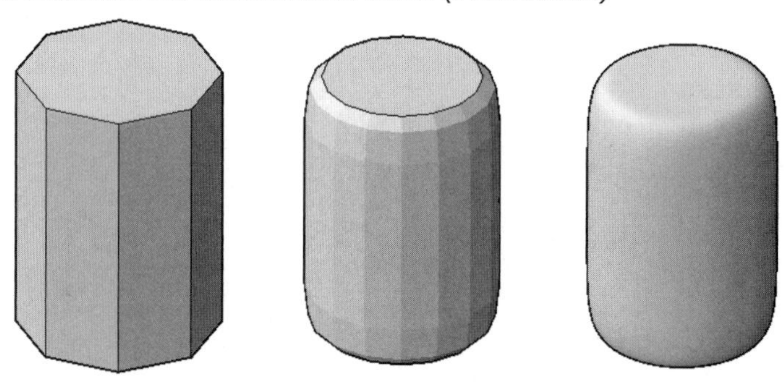

Left to right: *Smoothness settings of 0, 1, and 4.*

Command: meshsmoothmore
Select mesh objects to decrease the smoothness level: *(Choose one or more mesh objects.)*
Select mesh objects to decrease the smoothness level: *(Press Enter.)*

COMMAND LINE OPTION

Select mesh objects chooses the mesh objects to be modified; smoothness levels are decreased or increased by one level.

RELATED COMMAND

MeshRefine multiplies the number of faces in mesh sub-object areas.

RELATED SYSTEM VARIABLES

SmoothMeshGrid specifies maximum level of smoothness for underlying mesh facet grids displayed on mesh objects.

SmoothMeshMaxFace specifies maximum number of faces permitted for mesh objects.

SmoothMeshMaxLev specifies maximum smoothness level for mesh objects.

TIPS

- Mesh smoothness ranges from 0 (none) to 4.
- Maximum smoothness may tax your computer.
- These commands stop operating when smoothness reaches 0 or the maximum your computer can handle.
- Mesh objects that have been refined cannot change their smoothness level.

MeshSpin

2011 Rotates pairs of adjacent triangular faces on 3D meshes.

Command	*Aliases*	*Keyboard Shortcuts*	*Menu*	*Ribbon*
meshspin	...	...	**Modify** **↳Mesh Editing** **↳Spin Triangle Face**	**Mesh** **↳Mesh Edit** **↳Spin Triangle Face**

Use the MeshSplit command to change square faces into triangular ones.

Command: meshspin

Select first triangular mesh face to spin: *(Choose a triangular face.)*

Second adjacent triangular mesh face to spin: *(Choose an adjacent triangular face.)*

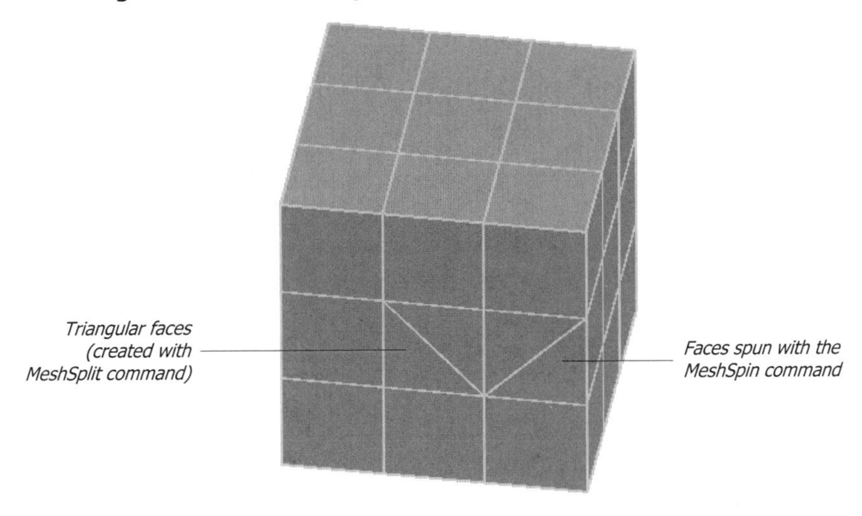

COMMAND LINE OPTIONS

First triangular mesh face chooses a triangular face.

Second adjacent triangular mesh face chooses the second triangular face, which must be adjacent to the first.

MeshSplit

2010 Splits single mesh faces into two.

Command	Alias	Keyboard Shortcuts	Menu	Ribbon
meshsplit	**split**	...	**Modify ↳Mesh Editing ↳Split Face**	**Mesh Modeling ↳Mesh Edit ↳Split Mesh Face**

Command: meshsplit
Select a mesh face to split: *(Pick a single face on a mesh object.)*
Specify first split point: *(Pick the starting point of the new mesh line.)*
Specify second split point: *(Pick the ending point of the new mesh line.)*

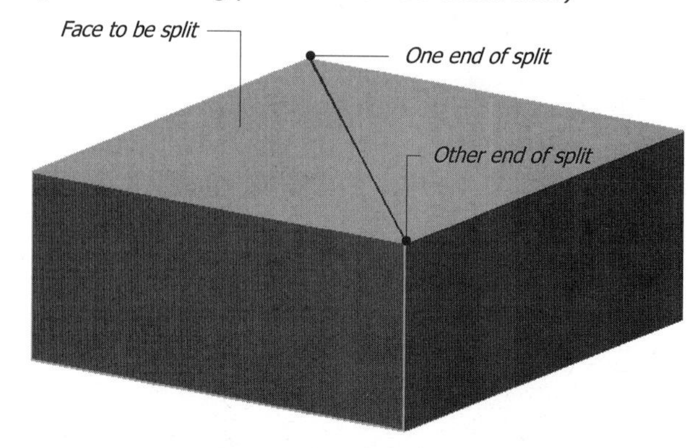

COMMAND LINE OPTIONS

Select a mesh face to split selects the face to mesh; works only with 3D mesh objects.

Specify first split point picks the first point of the split line.

Specify second split point picks the second point of the split line.

RELATED COMMAND

Mesh creates 3D mesh objects.

RELATED SYSTEM VARIABLE

SubObjSelectionMode determines which part of a sub-object is selected when you hold down the Ctrl key.

TIP

- After splitting a face, you can drag the resulting faces to create different shapes.

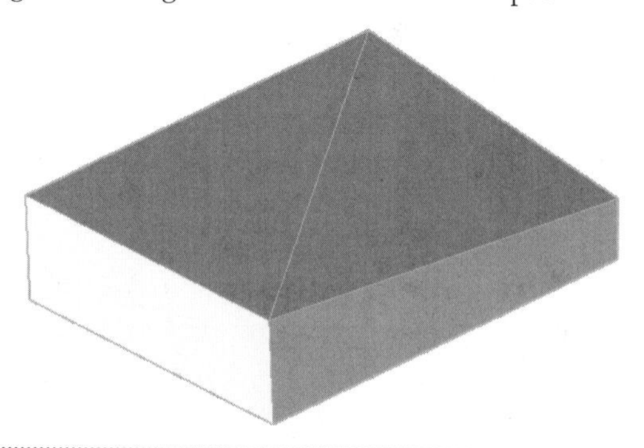

MigrateMaterials

2011 Converts materials from drawings older than AutoCAD 2011.

Command	*Aliases*	*Keyboard Shortcuts*	*Menu*	*Ribbon*
migratematerials	...	...	...	...

Command: migratematerials
n **materials migrated to the Materials Browser.**

COMMAND LINE OPTIONS

None.

RELATED COMMANDS

MatBrowserOpen catalogs materials.

MatEditorOpen edits materials.

RELATED SYSTEM VARIABLES

MaterialsPath reports the path to the material library folder; default is *C:\Program Files (x86)\Common Files\Autodesk Shared\Materials2011\assetlibrary_base.adsklib.*

3dConversionMode controls the conversion of old materials.

TIPS

- This command searches for old (a.k.a. "legacy") materials and converts them into the format used by AutoCAD 2011 and newer releases.
- Converted materials are added to the Materials Browser's library, and are stored in the *C:\Documents and Settings\login\Application Data\Autodesk\AutoCAD 2011\R18.1\enu\Support* folder.

MInsert

V. 2.5 Inserts an array of blocks as a single block (short for Multiple INSERT).

Command	Aliases	Keyboard Shortcuts	Menu	Ribbon
minsert	...	...	...	...

Command: minsert
Enter block name or [?]: *(Enter a name, type ?, or enter ~ to select a .dwg file.)*
Specify insertion point or [Scale/X/Y/Z/Rotate]: *(Pick a point, or enter an option.)*
Enter X scale factor, specify opposite corner, or [Corner/XYZ] <1>: *(Enter a value, pick a point, or enter an option.)*
Enter Y scale factor <use X scale factor>: *(Enter a value, or press Enter.)*
Specify rotation angle <0>: *(Enter a value, or press Enter.)*
Enter number of rows (---) <1>: *(Enter a value.)*
Enter number of columns (|||) <1>: *(Enter a value.)*
Enter distance between rows or specify unit cell (---): *(Enter a value.)*
Specify distance between columns (|||): *(Enter a value.)*

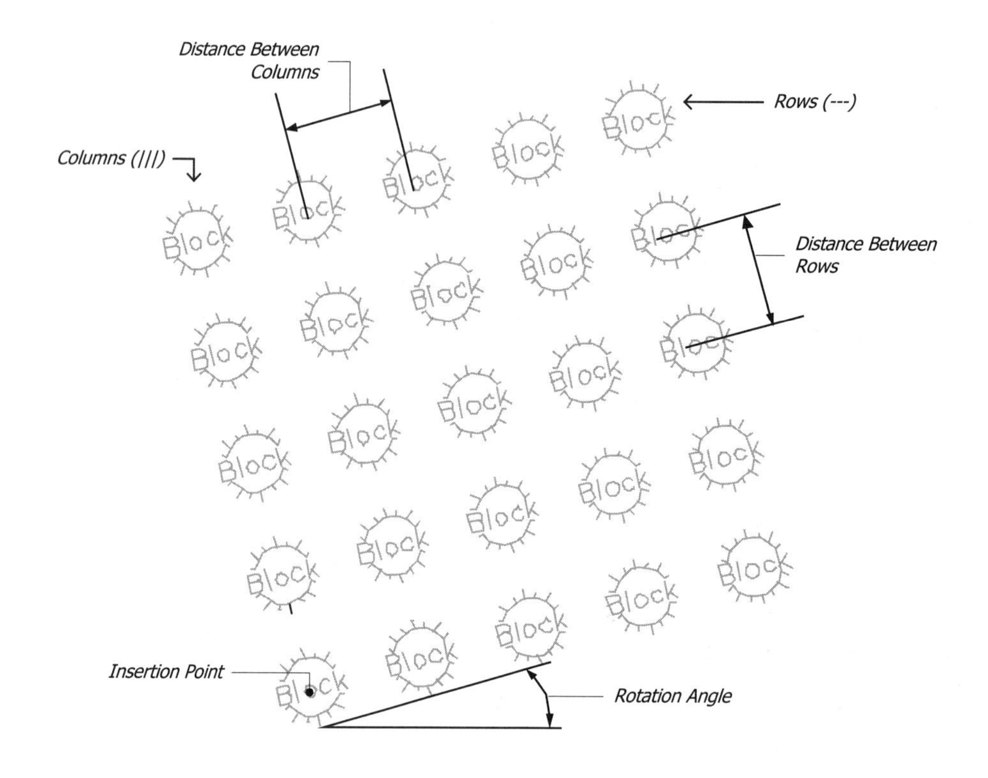

COMMAND LINE OPTIONS

Enter block name names the block to be inserted; the block must already exist in the drawing.

? lists the names of blocks stored in the drawing.

Specify insertion point specifies the x,y coordinates of the first block.

X scale factor indicates the x scale factor.

Specify opposite corner specifies a second point that indicates the x,y-scale factor.

Corner indicates the x and y scale factors by picking two points on the screen.

XYZ specifies x, y, and z scaling.

Specify rotation angle specifies the angle of the array.

Number of rows specifies the number of horizontal rows.

Number of columns specifies the number of vertical columns.

Distance between rows specifies the distance between rows.

Specify unit cell shows the cell distance by picking two points on the screen.

Distance between columns specifies the distance between columns.

RELATED COMMANDS

3dArray creates 3D rectangular and polar arrays.

Array creates 2D rectangular and polar arrays.

Block creates blocks.

Insert inserts single blocks.

TIPS

- The array placed by this command is a single block.
- The Explode command cannot explode blocks created by this command; as a workaround, use the Express | Modify | Flatten Objects command, which reduces the minsert block into individual blocks, which can be further exploded or edited.
- You may redefine blocks created by this command.

Mirror

V. 2.0 Creates a mirror copy of a group of objects in 2D space.

Command	Alias	Keyboard Shortcuts	Menu	Ribbon
mirror	**mi**	...	**Modify** ⇘**Mirror**	**Home** ⇘**Modify** ⇘**Mirror**

Command: mirror
Select objects: *(Pick one or more objects.)*
Select objects: *(Press Enter to end object selection.)*
Specify first point of mirror line: *(Pick a point.)*
Specify second point of mirror line: *(Pick another point.)*
Delete source objects? [Yes/No] <N>: *(Type Y or N.)*

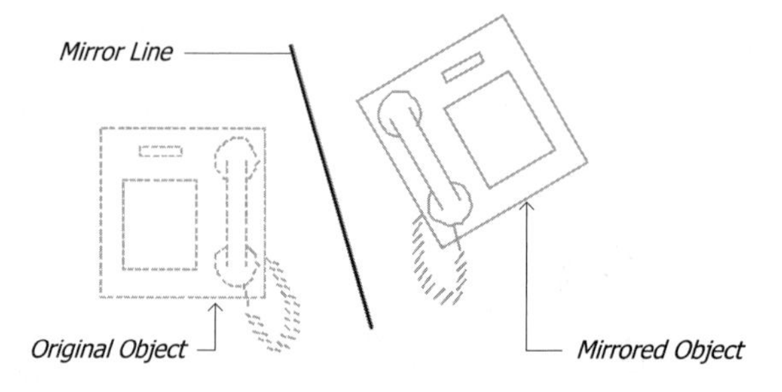

COMMAND LINE OPTIONS

Select objects selects the objects to mirror.

First point specifies the starting point of the mirror line.

Second point specifies the endpoint of the mirror line.

Delete source objects deletes selected objects.

RELATED COMMANDS

Copy creates non-mirrored copies of objects.

Mirror3d mirrors objects in 3D-space.

RELATED SYSTEM VARIABLE

MirrText determines whether text is mirrored by this command. Since AutoCAD 2005, this variable defaults to 0 (text is not mirrored).

TIPS

- This command cuts in half the drafting of symmetrical objects. For doubly-symmetrical objects, use Mirror twice.
- Although you can mirror a viewport in paper space, this does not mirror the model space objects inside the viewport.
- Turn on Ortho or Polar mode to ensure that the mirror is perfectly horizontal or vertical.
- The mirror line becomes a mirror plane in 3D; it is perpendicular to the xy-plane of the UCS containing the mirror line.

Mirror3D

Rel.11 Mirrors objects about a plane in 3D space.

Command	*Alias*	*Keyboard Shortcuts*	*Menu*	*Ribbon*
mirror3d	**3dmirror**	**...**	**Modify** **↳3D Operation** **↳3D Mirror**	**Home** **↳Modify** **↳3D Mirror**

Command: mirror3d
Select objects: *(Pick one or more objects.)*
Select objects: *(Press Enter to end object selection.)*
Specify first point of mirror plane (3 points) or
[Object/Last/Zaxis/View/XY/YZ/ZX/3points] <3points>: *(Pick a point, or enter an option.)*
Delete old objects? <N>: *(Type Y or N.)*

COMMAND LINE OPTIONS

Select objects selects the objects to be mirrored in space.

Specify first point specifies the first point of the mirror plane.

Object selects a circle, arc or 2D polyline segment as the mirror plane.

Last selects the last-picked mirror plane.

View specifies that the current view plane is the mirror plane.

XY specifies that the xy-plane is the mirror plane.

YZ specifies that the yz-plane is the mirror plane.

ZX specifies that the zx-plane is the mirror plane.

Zaxis defines the mirror plane by a point on the plane and the normal to the plane, i.e., the z axis.

3points defines three points on the mirror plane.

RELATED COMMANDS

Align translates and rotates objects in 2D planes and 3D space.

Mirror mirrors objects in 2D space.

Rotate3d rotates objects in 3D space.

RELATED SYSTEM VARIABLE

MirrText determines whether text is mirrored by the Mirror and Mirror3D commands:

0 — Text is not mirrored about the horizontal axis (*default*).

1 — Text is mirrored.

MLeader

2008 Draws multiline leaders.

Command	Alias	Keyboard Shortcuts	Menu	Ribbon
mleader	**mld**	...	**Dimension** ↳**Multileader**	**Annotate** ↳**Leader** ↳**Multileader**

Command: mleader

Specify leader arrowhead location or [leader Landing first/Content first/Options]: *(Pick a point, or enter an option.)*

Specify leader landing location: *(Pick a point.)*

Displays the Text Editor tab on the ribbon; enter text, and then click Close Text Editor.

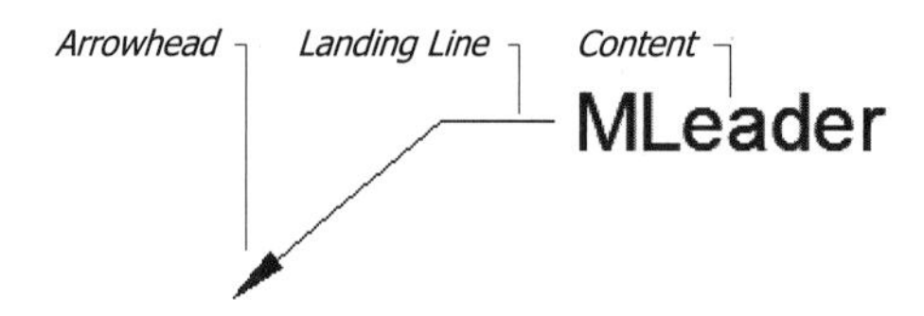

COMMAND LINE OPTIONS

The option you pick here affects all subsequent uses of this command, until changed:

Specify leader arrowhead location locates the start of the mleader.

leader Landing first draws the mleader starting with its landing line.

Content first draws the mleader's content first, such as its text or block

Options options

Enter an option [Leader type/leader lAnding/Content type/Maxpoints/First angle/Second angle/eXit options]: *(Enter an option.)*

Leader type specifies the type of leader line to use: straight, spline, or no leader line.

leader lAnding toggles use of the horizontal landing line.

Content type specifies the type of content: block or none.

Maxpoints specifies the maximum number of vertices.

First angle specifies the angle of the first point.

Second angle specifies the second angle.

eXit options returns to the earlier prompt line.

RELATED COMMANDS

MLeaderAlign aligns selected multileaders to a specified justification.

MLeaderCollect organizes selected multileaders (with blocks) as a group of blocks attached to a single leader.

MLeaderEdit edits mleaders, adding and removing leader lines.

MLeaderStyle displays the Multileader Style Manager dialog box for creating new styles and modifying existing ones.

Leader draws single-line leaders.

QLeader draws single-line leaders with additional options.

RELATED SYSTEM VARIABLE

CMleaderStyle reports the current multileader style.

TIPS

- Use the MLeaderEdit command to add leader lines to mleaders.

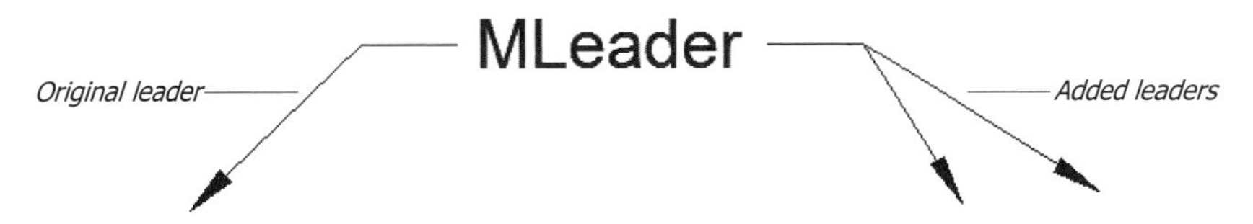

- Mleaders can be edited using grips. When you move the text or landing line, the other elements move with them, except for the arrowhead, which remains fixed in place.

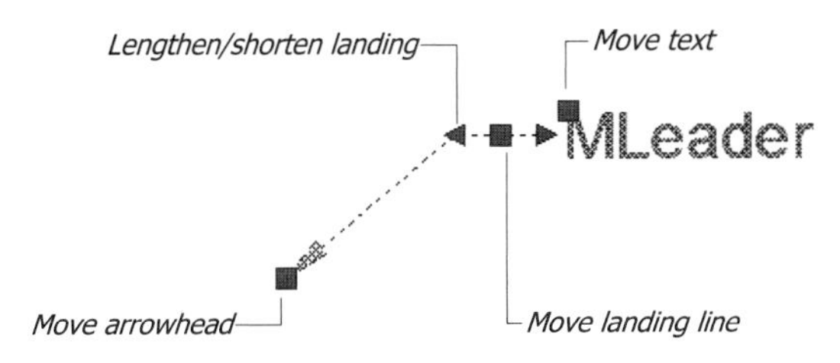

MLeaderAlign

2008 Aligns two or more multiline leaders by their text.

Command	*Alias*	*Keyboard Shortcuts*	*Menu*	*Ribbon*
mleaderalign	**mla**	...	...	**Annotate ↳Leaders ↳Align**

Command: mleaderalign
Select multileaders: *(Pick one or more multileaders.)*
Select multileader to align to or [Options]: *(Pick the reference multileader or type O.)*
Specify direction: *(Move cursor to locate alignment.)*

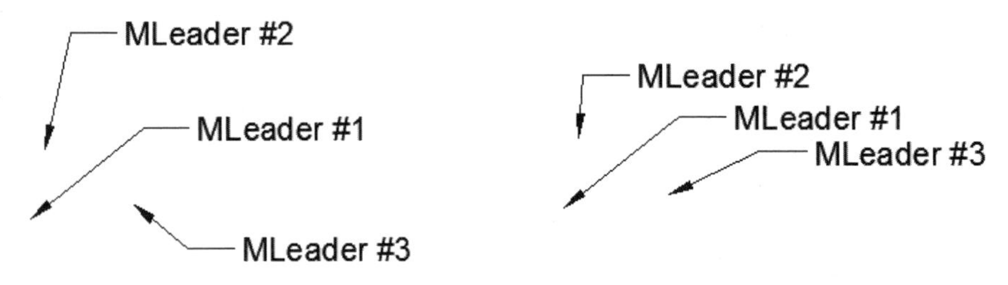

Leaders #2 and #3 aligned with leader #1.

COMMAND LINE OPTIONS

Select multileaders selects one or more multileaders to be aligned.

Select multileader to align to picks the reference mleader; the others will line up with this one.

Options displays further prompts.

Options options

Enter an option [Distribute/make leader segments Parallel/specify Spacing/Use current]: *(Enter an option.)*

Distribute spaces the mleaders evenly between two selected points.

Make Leader Segments Parallel places content so that each of the last leader segments in the selected multileaders is parallel.

Specify Spacing specifies spacing between the landing lines of the selected multileaders.

Use Current uses the current spacing between multileader content.

RELATED COMMANDS

MLeader draws multiline leaders.

MLeaderEdit edits mleaders, adding and removing leader lines.

MLeaderStyle displays the Multileader Style Manager dialog box for creating new styles and modifying existing ones.

TIP

- This command aligns leaders by their text or block; the arrowheads remain in place.

MLeaderCollect

2008 Collects the blocks of several multiline leaders into a single mleader with a row of blocks.

Command	*Alias*	*Keyboard Shortcuts*	*Menu*	*Ribbon*
mleadercollect	**mlc**	...	...	**Annotate** ↳**Leaders** ↳**Collect**

Command: mleadercollect
Select multileaders: *(Pick one or more multiline leaders.)*
Specify collected multileader location or [Vertical/Horizontal/Wrap]: *(Pick a point or enter an option.)*

COMMAND LINE OPTIONS

Specify collected multileader location picks the upper-left corner of the collected blocks.

Vertical collects the blocks into a vertical stack.

Horizontal collects the blocks into a horizontal stack.

Wrap specifies a width for wrapped collections; prompt, 'Specify wrap width or [Number]:'

Number specifies the maximum number of blocks per row.

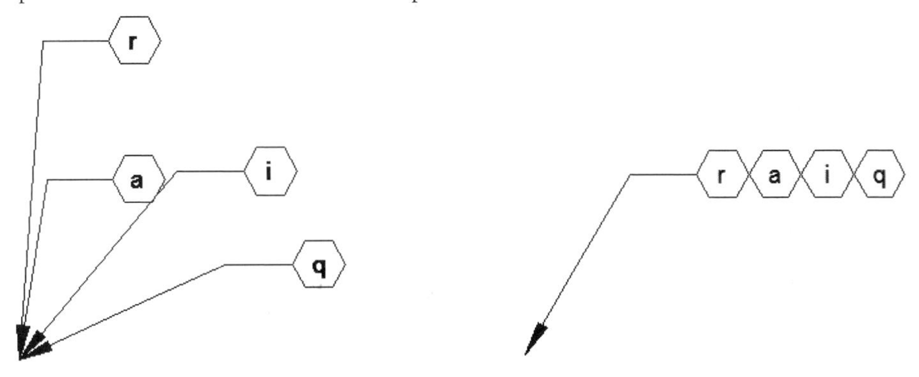

Left: *Several multiline leaders with block content.*
Right: *Collected into a single mleader.*

RELATED COMMANDS

MLeader draws multiline leaders.

MLeaderEdit edits mleaders, adding and removing leader lines.

MLeaderStyle displays the Multileader Style Manager dialog box for creating new styles and modifying existing ones.

TIPS

- This command collects leaders in the order in which you select them.

  This command works only with mleaders that have blocks for their content. If you select mleaders with text, AutoCAD complains, '1 was filtered out, 0 total'.

- This command is useful for combining multiple, drawing-crowding leaders into a single object.

MLeaderEdit

2008 Adds and removes leaders from multileader objects.

Command	Alias	Keyboard Shortcuts	Menu	Ribbon
mleaderedit	**mle**	...	...	**Annotate** ↳**Leaders** ↳**Add Leader**

Command: mleaderedit
Select a multileader: *(Select one mleader.)*
Select an option [Add leader/Remove leader]: *(Enter an option.)*
Specify leader arrowhead location: *(Pick a point.)*
Specify leader arrowhead location: *(Press Enter to exit command.)*

COMMAND LINE OPTIONS

Add Leader adds one leader line to the selected multileader; the new leader line is added on the left or right side, depending on where the cursor is:

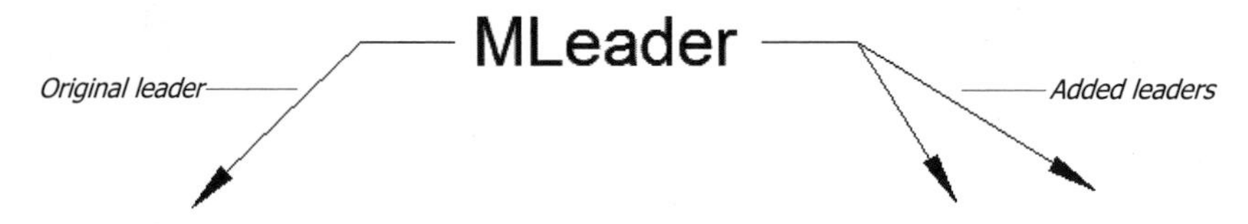

Moving cursor left or right of text, and then placing new leader lines.

Remove Leader removes one or more leader lines; prompts you, 'Specify leaders to remove:'.

RELATED COMMANDS

MLeader draws multiline leaders.

MLeaderAlign aligns selected multileaders to a specified justification.

MLeaderCollect organizes selected multileaders (with blocks) as a group of blocks attached to a single leader.

MLeaderStyle displays the Multileader Style Manager dialog box for creating new styles and modifying existing ones.

TIPS

- The Remove option removes all leader lines from an mleader; you can use the Add option to add leader lines to the remaining mleader text.
- Mleaders can be edited with grips and with the Properties palette.
- You can apply annotative scaling, colors, linetypes, lineweights, plot styles, and other properties to mleaders.

MLeaderStyle

2008 Creates and edits multiline leader styles.

Command	Alias	Keyboard Shortcuts	Menu	Ribbon
mleaderstyle	**mls**	...	**Format** ↳ **Multileader Style**	**Annotative** ↳**Leaders** ↳**Multileader Style**

Command: mleaderstyle

Displays dialog box.

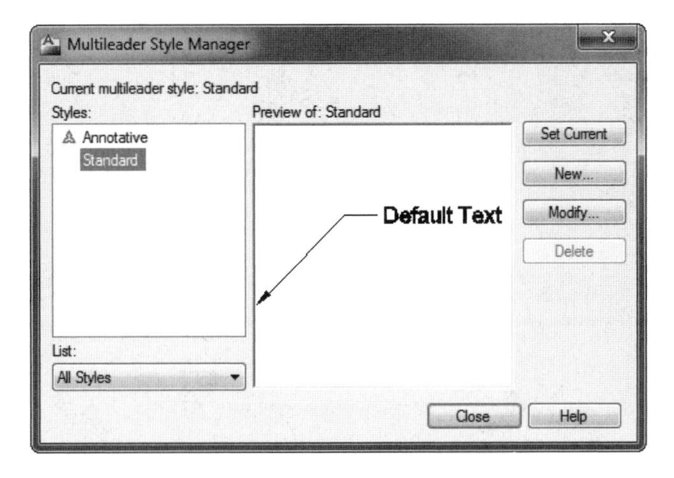

MULTILEADER STYLE MANAGER DIALOG BOX

Styles lists the names of multiline styles in the drawing.

List modifies the style names listed under Styles:

- **All styles** lists all multiline styles stored in the current drawing.
- **Styles in use** lists only those multiline styles used by the drawing.

Set Current sets selected style as the current multiline style.

New creates new styles; see Create New Multiline Style dialog box.

Modify modifies existing multiline styles; see Modify Multiline Style dialog box.

Delete erases the selected style, if it is not in use or is not named Standard.

CREATE NEW MULTILEADER STYLE DIALOG BOX

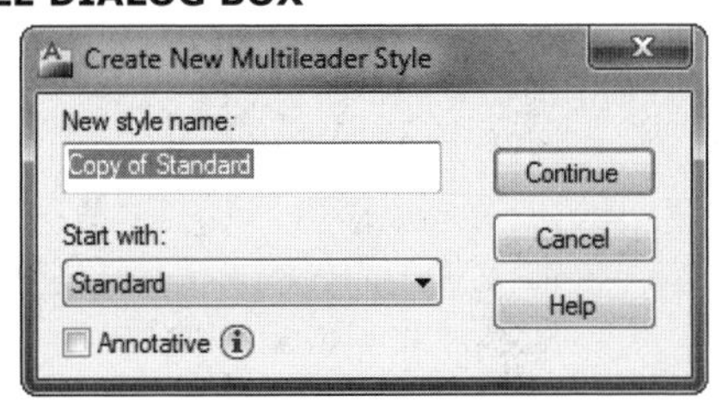

New Style Name names the new style.

Start With specifies an existing style as the template.

☐ **Annotative** toggles annotative scaling for multiline leaders.

Continue displays the Modify Multileader Style dialog box (detailed below).

MODIFY MULTILEADER STYLE DIALOG BOX

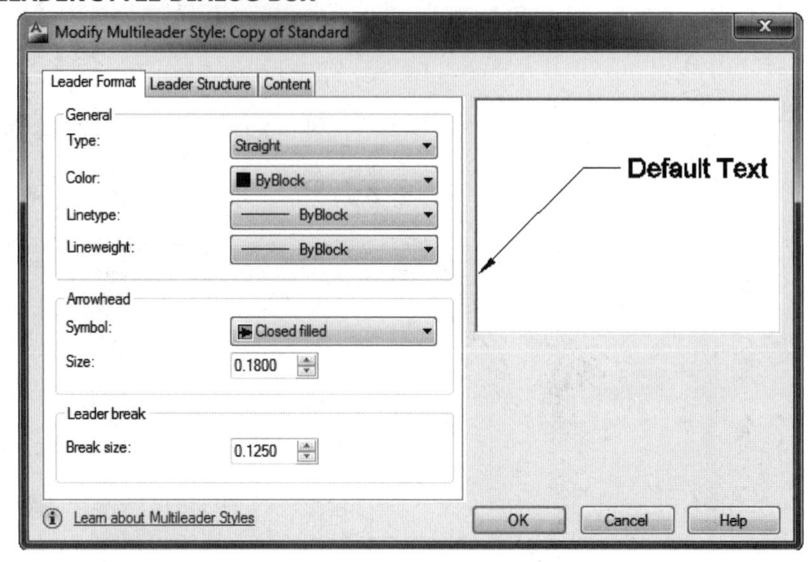

Leader Format tab

Type specifies the shape of the leader line: straight, spline, or none.

Color specifies the color of the leader line.

Linetype specifies the linetype of the leader line.

Lineweight specifies the lineweight of the leader line.

Symbol selects the arrowhead symbol.

Size specifies the size of arrowhead.

Break Size specifies the gap size when leader lines are broken by the DimBreak command.

Leader Structure tab

☑ **Maximum Leader Points** specifies a maximum number of vertices on the leader line.

☐ **First Segment Angle** specifies the increment angle of the first vertex (angle between first and second leader lines): 0, 15, 30, 45, 60 or 90 degrees.

☐ **Second Segment Angle** specifies the increment angle of a second vertex.

☑ **Automatically Include Landing** includes the horizontal landing line.

☑ **Set Landing Distance** specifies a fixed length for the landing line.

☐ **Annotative** toggles annotative scaling.

The following options are available only when Annotative is turned off:

○ **Scale Multileaders to Layout** scales the mleader based on the scale factor between model space and the current viewport.

⊙ **Specify Scale** specifies the scale factor.

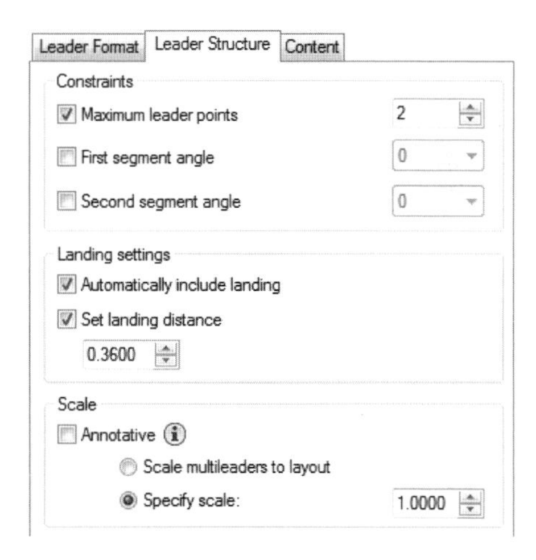

Content tab

Multileader Type chooses between text, blocks, or no content.

Content: MText options

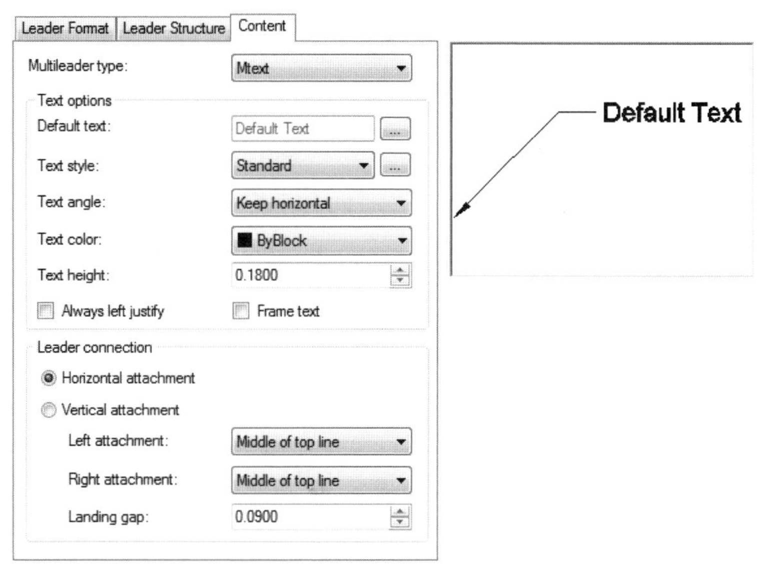

Default Text specifies default text for the mleader; click ... to access the mtext editor. See the MText command.

Text Style selects the text style; click ... to access the text style dialog box. See the Style command.

Text Angle specifies the angle of the multileader text:

- Keep horizontal
- As inserted
- Always right-reading

Text Color specifies the color of the multileader text. See the Color command.

Text Height specifies the height of the text.

☐ **Always Left Justify** forces text always to be left justified.

☐ **Frame Text** places a box around the text.

Left and **Right Attachment** specifies where the landing line is attached to the multileader text:

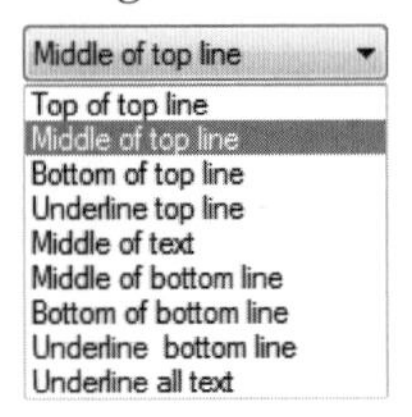

Attachment options for left and right content attachment to leader lines.

Landing Gap specifies the distance between the landing line and the text.

Content: Block options

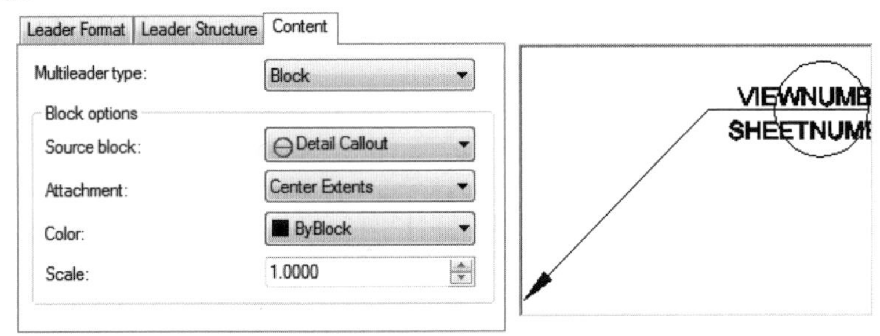

Source Block selects the predefined block used in place of multileader text:

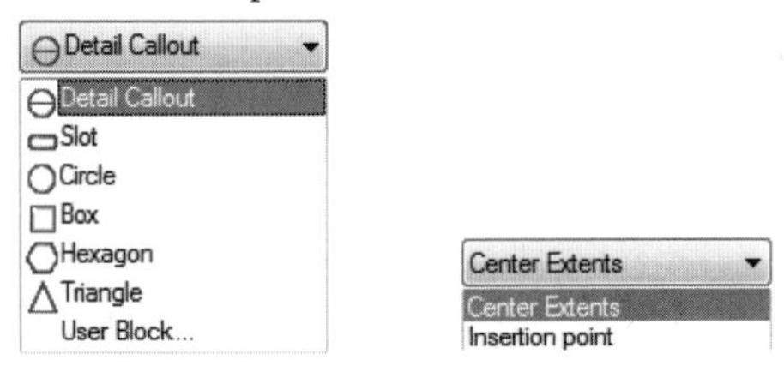

Left: *Predefined blocks available as leader content.*
Right: *Attachment options for blocks.*

Attachment specifies where the blocks are attached: extents, insertion point, or center point of the block.

Color specifies the color of the block content.

Content: None option

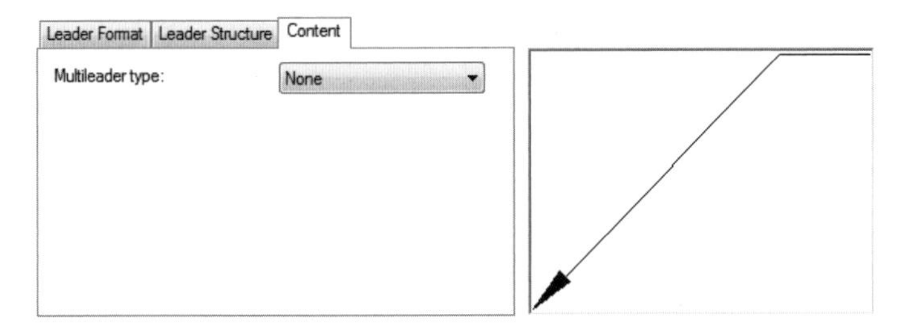

RELATED COMMANDS

MLeader draws multiline leaders.

MLeaderAlign aligns selected multileaders to a specified justification.

MLeaderCollect organizes selected multileaders (with blocks) as a group of blocks attached to a single leader.

MLeaderEdit edits mleaders, adding and removing leader lines.

RELATED SYSTEM VARIABLE

CMleaderStyle reports the current multileader style.

TIPS

- Set a multileader style before drawing mleaders.
- You can apply mleader styles from the the ribbon: select an mleader, and then choose a style name from one of the droplists.

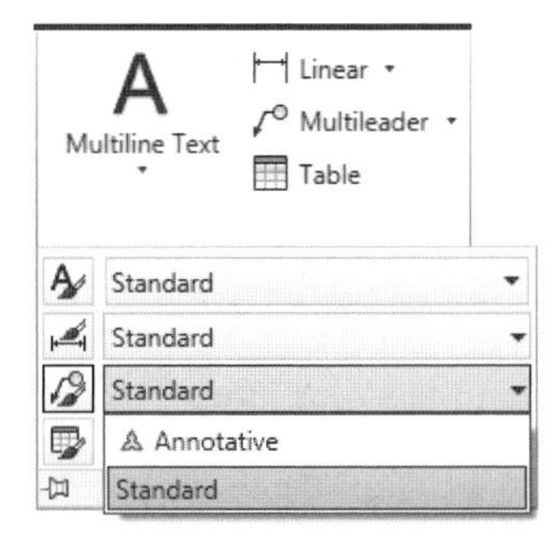

- Multileader styles can be overridden by the Properties palette.
- Changes to the multileader style retroactively update existing mleaders.

MlEdit

Rel.13 Edits multilines (*short for MultiLine EDITor*).

Commands	Aliases	Keyboard Shortcuts	Menu	Ribbon
mledit	...	...	**Modify** ↳**Object** ↳**Multiline**	...
-mledit	...	...	...	...

Command: mledit

Displays dialog box:

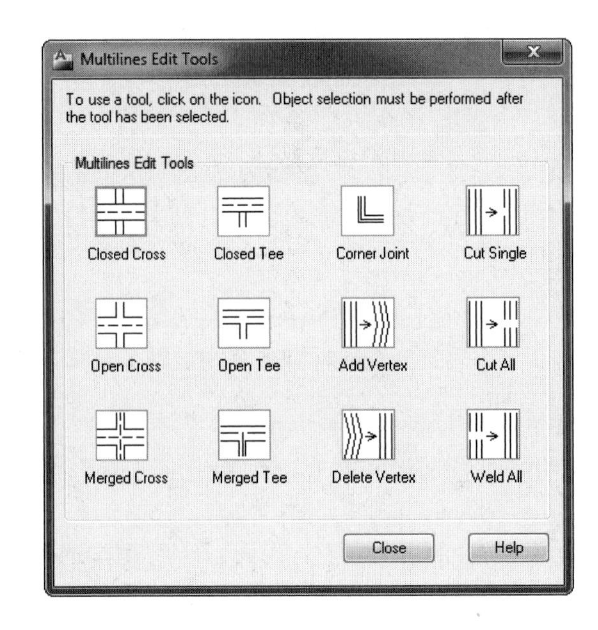

MULTILINE EDIT TOOLS DIALOG BOX

Closed Cross closes the intersection of two multilines.

Open Cross opens the intersection of two multilines.

Merged Cross merges a pair of multilines: opens exterior lines; closes interior lines.

Closed Tee closes T-intersections.

Open Tee opens T-intersections.

Merged Tee merges T-intersection by opening exterior lines and closing interior lines.

Corner Joint creates corner joints with pairs of intersecting multilines.

Add Vertex adds vertices (*joints*) to multiline segments.

Delete Vertex removes vertices from multiline segments.

Cut Single places gaps in a single line of multilines.

Cut All places gaps in all lines of multilines.

Weld All removes gaps from multilines.

-MLEDIT Command

Command: -mledit
Enter mline editing option [CC/OC/MC/CT/OT/MT/CJ/AV/DV/CS/CA/WA]: *(Enter an option.)*

COMMAND LINE OPTIONS

AV adds vertices.

DV deletes vertices.

CC closes crossings.

OC opens crossings.

MC merges crossings.

CT closes tees.

OT opens tees.

MT merges tees.

CJ creates corner joints.

CS cuts a single line.

CA cuts all lines.

WA welds all lines.

U undoes the most-recent multiline edit.

RELATED COMMANDS

MLine draws up to 16 parallel lines.

MlStyle defines the properties of a multiline.

RELATED SYSTEM VARIABLES

CMlJust specifies the current multiline justification mode.

CMlScale specifies the current multiline scale factor (default = 1.0).

CMlStyle specifies the current multiline style name (default = " ").

RELATED FILE

***.*mln** is the multiline style definition file.

TIPS

- Use the Cut All option to open up a gap before placing door and window symbols in a multiline wall.
- Use the Weld All option to close up a gap after removing the door or window symbol in a multiline.
- Use the Stretch command to move a door or window symbol in a multiline wall.
- When you open a gap in a multiline, AutoCAD does not cap the sides of the gap. You may need to add the end caps with the Line command.

MLine

Rel.13 Draws up to 16 parallel lines (short for Multiple LINE).

Command	Alias	Keyboard Shortcuts	Menu	Ribbon
mline	**ml**	...	**Draw ↳Multiline**	...

Command: mline
Current settings: Justification = Top, Scale = 1.00, Style = STANDARD
Specify start point or [Justification/Scale/STyle]: *(Pick a point, or enter an option.)*
Specify next point: *(Pick a point.)*
Specify next point or [Undo]: *(Pick a point, or type U.)*
Specify next point or [Close/Undo]: *(Pick a point, or else enter an option.)*

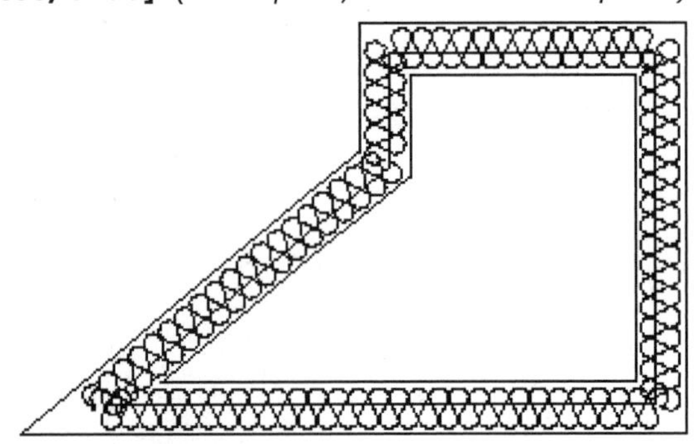

COMMAND LINE OPTIONS

Specify start point indicates the start of the multiline.

Specify next point indicates the next vertex.

Undo removes the most recently-added segment.

Close closes the multiline to its start point.

Justification options

Enter justification type [Top/Zero/Bottom] <top>: *(Enter an option.)*

Top draws the top line of the multiline at the cursor; the remainder of multiline is "below" the cursor.

Zero draws the center (*zero offset point*) of the multiline at the cursor.

Bottom draws the bottom of the multiline at the cursor; the remainder of the multiline is "above" the cursor.

Scale options

Enter mline scale <1.00>: *(Enter a value.)*

Enter mline scale specifies the scale of the width of the multiline; see Tips for examples.

STyle options

Enter mline style name or [?]: *(Enter style name, or type ?.)*

Enter mline style name specifies the name of the multiline style.

? lists the names of the multiline styles defined in the drawing.

RELATED COMMANDS

MlEdit edits multilines.

MlProp defines the properties of a multiline.

RELATED SYSTEM VARIABLES

CMlJust specifies the current multiline justification:

0 — Top (default).

1 — Middle.

2 — Bottom.

CMlScale specifies the current multiline scale factor (default = 1.0).

CMlStyle specifies the current multiline style name (default = " ").

RELATED FILE

**.mln* is the multiline style definition file.

TIPS

- Examples of scale factors:

Scale	*Meaning*
1.0	Specifies the default scale factor.
2.0	Draws multiline twice as wide.
0.5	Draws multiline half as wide.
-1.0	Flips multiline.
0	Collapses multiline to a single line.

- Multiline styles are stored in *.mln* files in DXF-like format.
- Use the MlEdit command to create (or close up) gaps for door and window symbols in multiline walls.
- Multiline definitions are stored in the *.mln* file, which describes multiline styles in a DXF-like format.
- You cannot change the properties of the Standard style.
- Once multilines are placed in drawings, you cannot change their properties or their style. Once drawn, that's it! The workaround is to erase, change styles, and then reapply the MLine command.

MlStyle

Rel.13 Defines the characteristics of multilines (short for MultiLine STYLE).

Command	Aliases	Keyboard Shortcuts	Menu	Ribbon
mlstyle	...	...	**Format ↳Multiline Style**	...

Command: mlstyle

Displays dialog box.

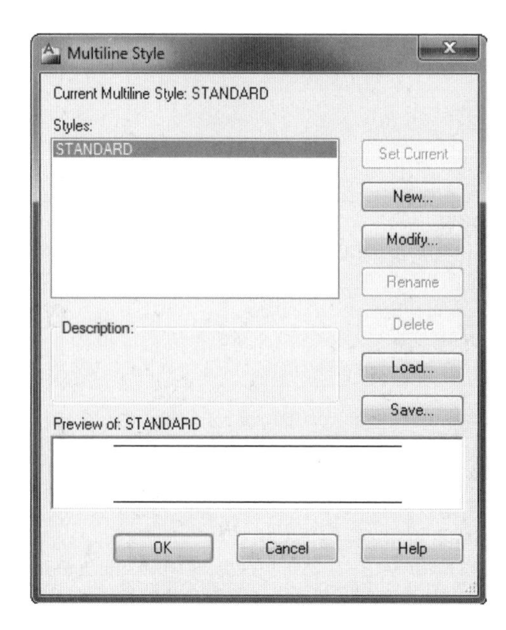

MUTLILINE STYLE DIALOG BOX

Set Current sets the selected multiline style name as the working style.

New creates new multiline styles; displays the Create New Multiline Style dialog box.

Modify changes the properties of existing multiline styles.

Rename changes the name of the selected multiline style.

Delete removes the multiline style from the Styles list.

Load loads styles from *.mln* files; displays the Load Multiline Styles dialog box.

Save saves a multiline style or renames a style; displays Save Multiline Style dialog box.

CREATE NEW MULTILINE STYLE DIALOG BOX

New Style Name specifies the name of the new multiline style. *Warning!* The name must contain no spaces.

Start With copies the multiline style from an existing style.

Continue displays the New Multiline Style dialog box.

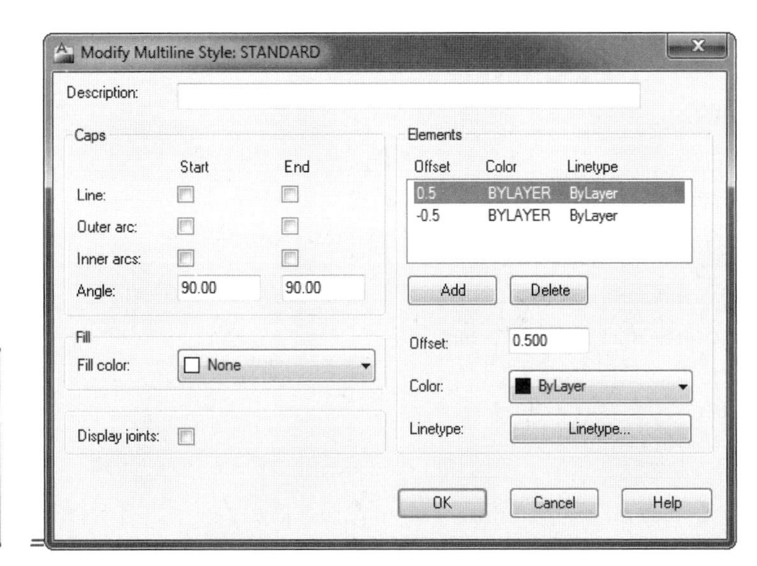

MODIFY MULTILINE STYLE DIALOG BOX

Caps options

☐**Line** draws a straight line start and/or end cap.

☐**Outer Arc** draws arcs to cap the outermost pair of lines.

☐**Inner Arcs** draws arcs to cap all inner pairs of lines.

Angle specifies the angle for straight line caps.

Fill options

Fill Color lists common colors, and displays the Select Color dialog box.

☐**Display Joints** toggles the display of joints (miters) at vertices; affects all multiline segments.

Elements options

Add adds an element (line).

Delete deletes an element.

Offset specifies the distance from origin to element; the origin is often the center line.

Color specifies the element color; displays Select Color dialog box.

Linetype specifies the element linetype; displays Select Linetype dialog box.

RELATED COMMANDS

MlEdit edits multilines.

MLine draws up to 16 parallel lines.

RELATED SYSTEM VARIABLES

CMlJust specifies the current multiline justification.

CMlScale specifies the current multiline scale factor.

CMlStyle specifies the current multiline style name.

TIP

- Once multilines are placed in drawings, you cannot change their properties or their style. Once drawn, that's it! The workaround is to erase, change styles, and then reapply the MLine command.

Model

2000 Switches to Model tab.

Command	Aliases	Status Bar	Menu	Ribbon
model	...	**PAPER**	...	...

Command: model

Switches to the model tab.

COMMAND LINE OPTIONS

None.

RELATED COMMANDS

Layout creates layouts.

MSpace switches to model space.

QVLayout displays a thumbnail strip of layouts.

RELATED SYSTEM VARIABLE

Tilemode switches between Model and Layout tabs.

TIPS

- This command automatically sets Tilemode to 1.
- As an alternative to this command, you can select the Model tab:

  ...or click the button or icon on the status bar :

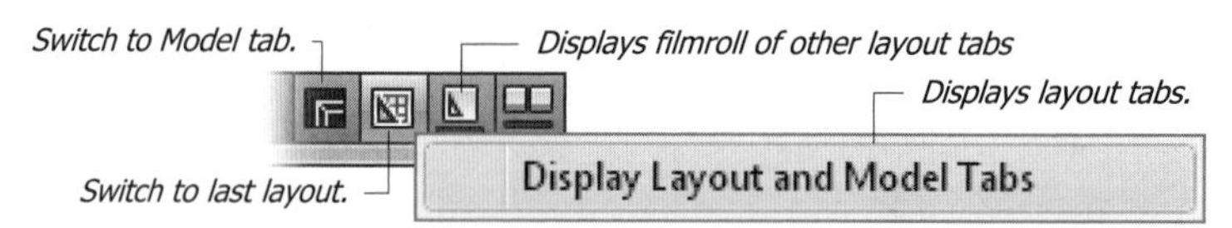

- The Model tab replaces the TILE button on the status bar of AutoCAD Release 13 and 14.
- *Historical notes:* The system variable is named "tile"mode, because model space can only display *tiled* viewports. (Paper space, or layout mode, can display overlapping viewports.) Turning off tiled-viewport mode switched AutoCAD to paper space, where viewports no longer must be tiled.

  Going back farther, it was a graphic board manufacturer, Control Systems, that first figured out how to make AutoCAD display four tiled viewports at once. Autodesk added the feature to AutoCAD Release 10.

  All of which leads to a question I cannot answer: Why can't viewports be floating in model space? Technical editor Bill Fane attempts an answer: "Because that's the way we've always done it!" Seriously, though... The idea is that the model resides in model space, and then is looked at through viewports in paper space. Because multi-view drawings can be created, scaled, and plotted more easily in paper space, there is no need for floating viewports in model space.

Move

V. 1.0 Moves one or more objects.

Command	*Alias*	*Keyboard Shortcuts*	*Menu*	*Ribbon*
move	**m**	...	**Modify** ↳**Move**	**Home** ↳**Modify** ↳**Move**

Command: move

Select objects: *(Select one or more objects.)*

Select objects: *(Press Enter to end object selection.)*

Specify base point or [Displacement] <Displacement>: *(Pick a point.)*

Specify second point of displacement or <use first point as displacement>: *(Pick a point, or press Enter.)*

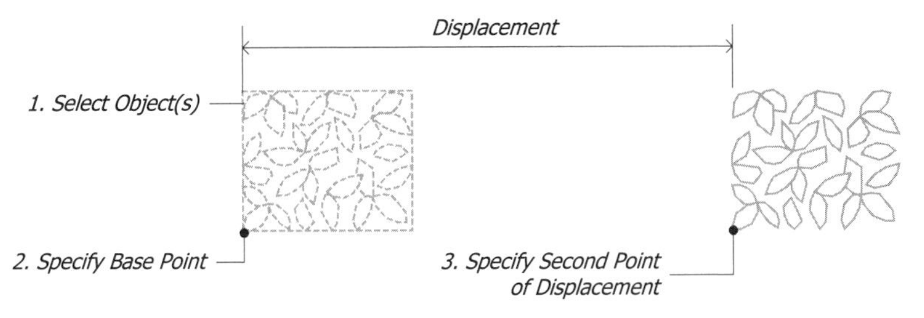

COMMAND LINE OPTIONS

Select objects selects the objects to copy.

Specify base point indicates the starting point for the move.

Displacement specifies relative x,y,z displacement when you press Enter at the next prompt.

Specify second point of displacement indicates the distance to move; you can use absolute (x,y), relative (@x,y), and polar (x<angle) coordinates.

RELATED COMMANDS

Copy copies the selected objects.

3dMove moves objects in 3D space.

Grips editing moves objects.

TIPS

- When you press Enter at the 'Specify Second Point of Displacement' prompt, AutoCAD uses the first point as the "displacement."

  For example, when you enter **4,3** at the 'Specify base point' prompt, and press Enter at the second prompt, AutoCAD moves the objects 4 units in the x direction and 3 units in the y.

- You can move objects without using the Move command: select the object, and then drag it (without grabbing a grip). After a brief pause, the object begins to move.

MRedo

2004 Reverses the effect of the Undo command (short for Multiple REDO).

Command	Aliases	Keyboard Shortcut	Menu	Ribbon
mredo	...	**Ctrl+Y**	...	...

Command: mredo
Enter number of actions or [All/Last]: *(Enter an option.)*

COMMAND LINE OPTIONS

Enter number of actions redoes the specified number of steps.

All redoes all commands undone.

Last redoes the last command.

RELATED COMMANDS

Redo redoes a single undo.

U undoes a single command.

Undo undoes one or more commands.

TIPS

- The MRedo button on the Quick Access toolbar lists the redoable actions:

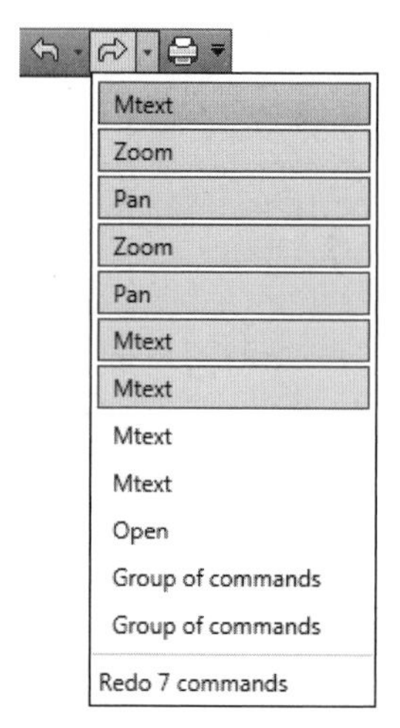

- This command allows you to redo several undoes, but does not allow you to skip over actions.

MSlide

Ver.2.0 Saves the current viewport as a *.sld* slide file on disk (short for Make SLIDE).

Command	*Aliases*	*Keyboard Shortcuts*	*Menu*	*Ribbon*
mslide	...	...	...	...

Command: mslide

Displays Create Slide File dialog box. Specify a file name, and then click Save.

COMMAND LINE OPTIONS

None.

RELATED COMMANDS

Save saves the current drawing as a DWG-format drawing file.

SaveImg saves the current view as a TIFF, Targa, or GIF-format raster file.

VSlide displays an SLD-format slide file in AutoCAD.

RELATED FILES

.sld files store slides created by this command.

.slb files store libraries of slides.

RELATED AUTODESK PROGRAM

SlideLib.exe compiles a group of slides into an SLB-format slide library file.

TIPS

- You view slides with the VSlide command.
- Viewing a slide was a predecessor to viewing raster and vector images inside AutoCAD.
- Slide files are used to create the images in the (rarely used) image dialog boxes.

MSpace

Rel.11 Switches the drawing from paper space to model space (short for Model SPACE).

Command	Alias	Keyboard Shortcuts	Menu	Ribbon
mspace	**ms**	...	...	...

Command: mspace

*In model tab, AutoCAD complains, '** Command not allowed in Model Tab **'.*

In model space of a layout tab, AutoCAD complains, 'Already in model space.'

In a layout tab (paper space), AutoCAD switches to model space in layout mode, and highlights a viewport:

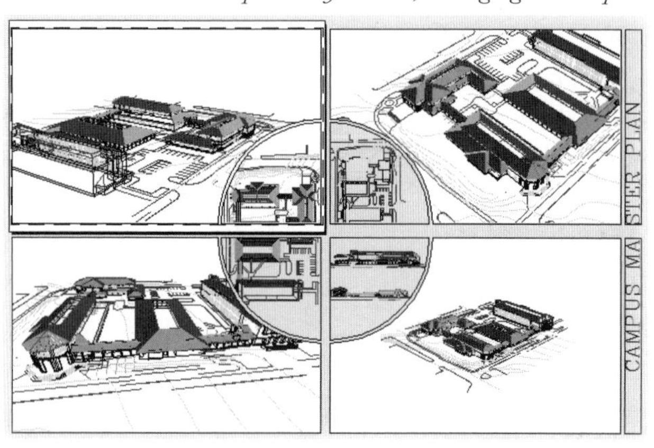

COMMAND LINE OPTIONS

None.

RELATED COMMANDS

PSpace switches from model space to paper space.

Model switches from layout mode to model mode.

Layout switches from model mode to layout mode.

QvLayout shows a thumbnail filmstrip of layouts.

RELATED SYSTEM VARIABLES

MaxActVp specifies the maximum number of viewports with visible objects; default=64.

TileMode specifies the current setting of tiled viewports.

TIPS

- To switch quickly between paper space and model space, click the Model and Paper icons on the status bar.
- AutoCAD clears the selection set when moving between paper space and model space.
- The model/layout *tabs* switch between model and layout spaces; the model/paper *button* switches between paper and model space within a layout.
- Double-click the viewport to switch from paper to model space; to return, double-click outside the viewport.

Aliased Command

MtEdit command is an alias of the TextEdit command as of AutoCAD 2010.

MText

Rel.13 Creates multiline, or paragraph, text objects that fit the width defined by the boundary box (short for Multline TEXT).

Commands	*Aliases*	*Keyboard Shortcuts*	*Menu*	*Ribbon*
mtext	**t**	...	**Draw**	**Home**
	mt		**↳Text**	**↳Annotation**
			↳Multiline Text	**↳Multiline Text**
-mtext	**-t**	...	...	...

Command: mtext

Current text style: "Standard" Text height: 0.20 Annotative: No

Specify first corner: *(Pick point 1.)*

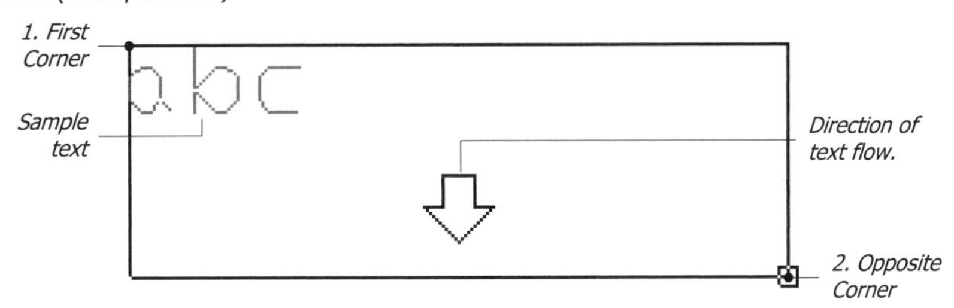

Specify opposite corner or [Height/Justify/Line spacing/Rotation/Style/Width/Columns]: *(Pick point 2, or enter an option.)*

(See the -MText command for command-line options.)

Displays text editing area and text formatting tabs. Shown below is a 2-column display.

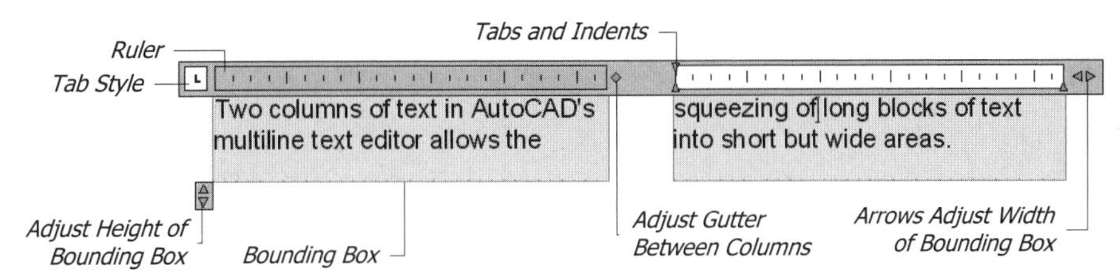

TEXT FORMATTING TAB

When MTextToolbar = 2, this command displays commands in the ribbon's Text Editor tab (redesigned in AutoCAD 2011); otherwise displays a floating toolbar:

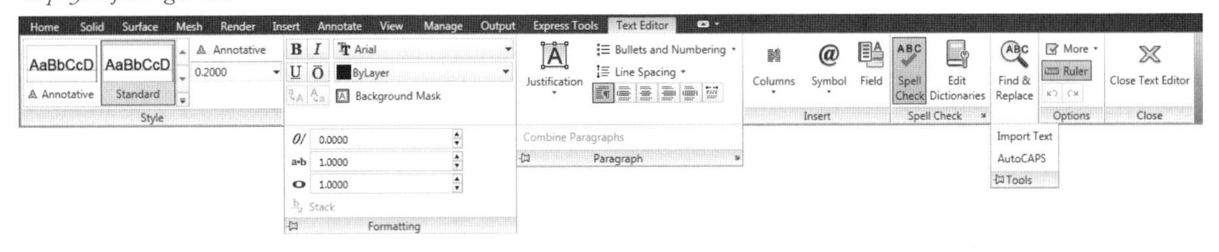

Style panel

Style selects a predefined text style; see the Style command.

Height specifies the height of the text in units (default = 0.2 units for Imperial drawings, 2.5 units for metric drawings).

Formatting panel

Bold boldfaces the text, if permitted by the font.

Italic *italicizes* the text, if permitted by the font.

Font selects a TrueType (*.ttf*) or AutoCAD (*.shx*) font (default=TXT).

Underline underlines the text.

Overline draws a line over the selected text, the inverse of underline.

Color selects color for text; choose Other Color to display the Select Color dialog box.

Upper/lowercase changes selected text to all uppercase or all lowercase.

Background Mask specifies the color for the background of the text.

Oblique Angle slants the text.

Tracking makes the spacing between characters tighter or looser.

Width Factor makes characters wider or narrower.

Stack Fraction stacks a pair of characters separated by slash.

Paragraph panel

Justification aligns the text within the bounding box.

Bullets applies bullet symbols, numbers (1., 2., 3., etc.), or letters (A., B., C., etc.) automatically to selected text.

Line Spacing places extra space between lines of text.

Alignment aligns the text horizontally against the left margin, centers it, aligns it to the right margin, forces all but the last line to fit between margins, or forces all lines in the paragraph to fit.

Combine Paragraphs combines selected text into a single paragraph.

Paragraph displays the Paragraph dialog box, detailed later.

Insert panel

Columns splits the bounding box into two or more static (fixed) or dynamic (as-needed) columns; choose Column Settings for the dialog box; see the Columns droplist later.

Symbol displays a menu of symbols; see the Symbol droplist later.

Field inserts field text from the Field dialog box; see the Field command.

Spell Check panel

Spell Check toggles real-time spell checking; unrecognized words are underlined.

Edit Dictionaries adds, removes, and edits words stored in the user-defined dictionary; see the Spell command.

Spell Check displays the spell checker dialog box; see the Spell command.

Tools panel

Find & Replace displays the Find and Replace dialog box.

Import Text imports text from ASCII (*.txt*) and RTF format files; displays the Open dialog box.

AutoCAPS uppercases initial letters of sentences as text is typed.

Options panel

More displays a menu of options; see the More droplist later.

Ruler toggles the display of the tab and indent ruler.

Undo undoes the last action.

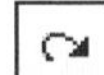

Redo undoes the last undo.

Close panel

Close Text Editor closes the tab, and exits the MText command.

DROPDOWN OPTIONS

Bullets and Numbering droplist

Off turns off bullets.

Numbered applies numbers, such as 1, I, and i.

Lettered applies uppercase or lowercase letters, such as A and a.

Bulleted applies bullets, such as •.

Start starts the number list from 1, or letter list from A, at the selected paragraph.

Continue forces the selected paragraph to take the next number/letter from the above paragraph.

Allow Auto-list applies bullets/lists as text is entered.

Use Tab Delimiter Only applies bullets/list only when Tab is pressed after a letter, number, or bullet character.

Allow Bullets and Lists allows mixing of bullets and list numbers/letters.

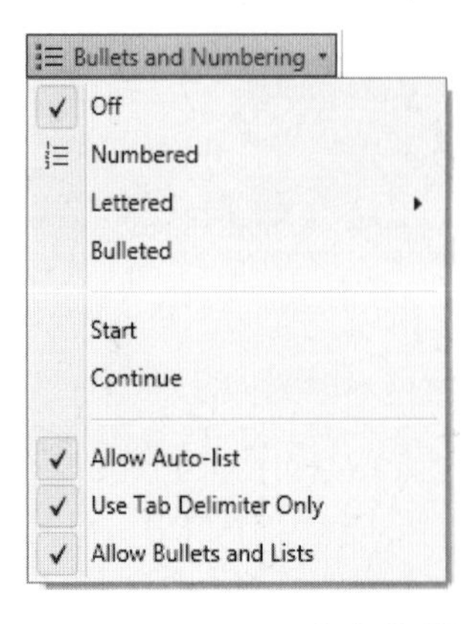

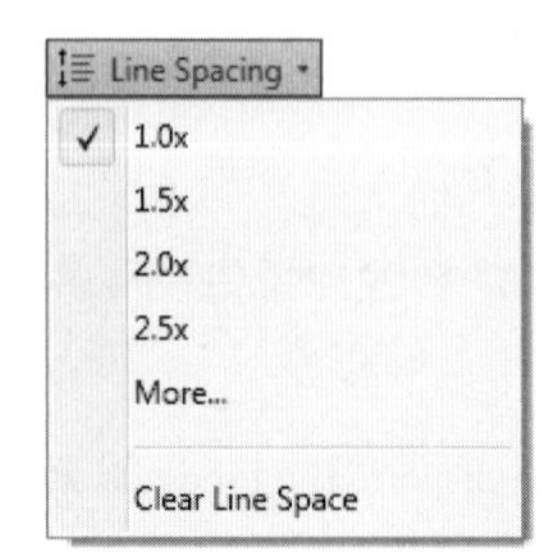

Left: *Bullets and Numbering droplist.*
Right: *Line Spacing droplist.*

Line Spacing droplist

1.0x to **2.5x** spaces lines by 1x (normal) to 2.5x.

More opens the Paragraph dialog box.

Clear Line Space returns the selected paragraph to its original spacing.

Columns dropdown

No Columns returns the bounding box to a single column.

Dynamic Columns creates columns as required by the amount of text:

- **Auto Height** stretches the column to fit the text, until you press Alt+Enter or enter sufficient text for AutoCAD to create another column automatically.
- **Manual Height** displays the Column Height marker at the bottom of the bounding box. You manually adjust the height; if it is too small for the text, AutoCAD adds another column.

Static Columns creates a specific number of columns, two or more:

- **2, 3, 4, 5, 6** splits the bounding box into 2, 3, or more columns.
- **More** displays the Column Settings dialog box.

Insert Column Break forces the following text to start at the top of the next column; alternatively, press Alt+Enter.

Column Settings displays the Column Settings dialog box.

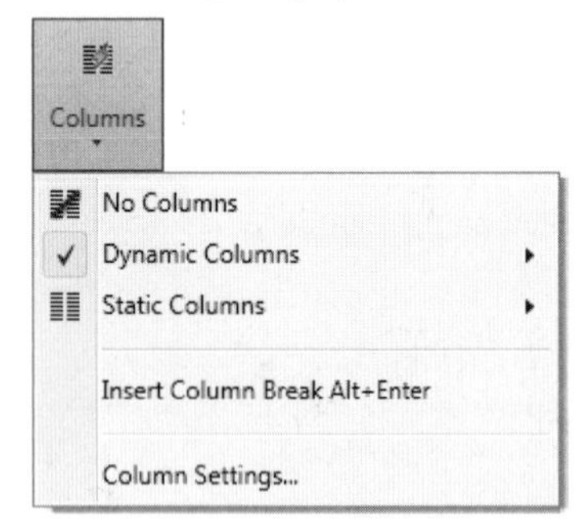

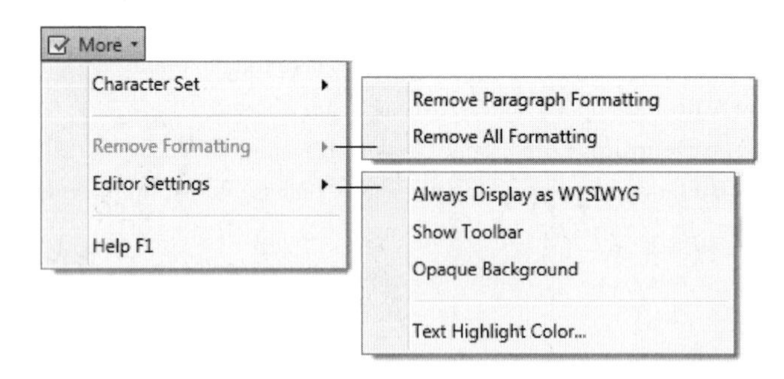

Left: *Columns droplist.*
Right: *More droplist.*

More droplist

Character Set selects alternate character sets, such as Western, Hebrew, and Thai.

Remove Formatting removes bold, italic, and underlined formatting from selected text.

Editor Settings sets the following options:

Always Display as WYSIWYG edits text in context.

Show Toolbar toggles the Text Formatting toolbar.

Opaque Background toggles opacity of the bounding box.

Text Highlight Color specifies the color of highlighted text; displays the Select Color dialog box.

Symbol droplist

Inserts the special characters illustrated at right.

Non-breaking Space inserts spaces that are not broken at the ends of lines; alternatively, press Ctrl+Shift+Space.

Other displays the Windows Character Map dialog box.

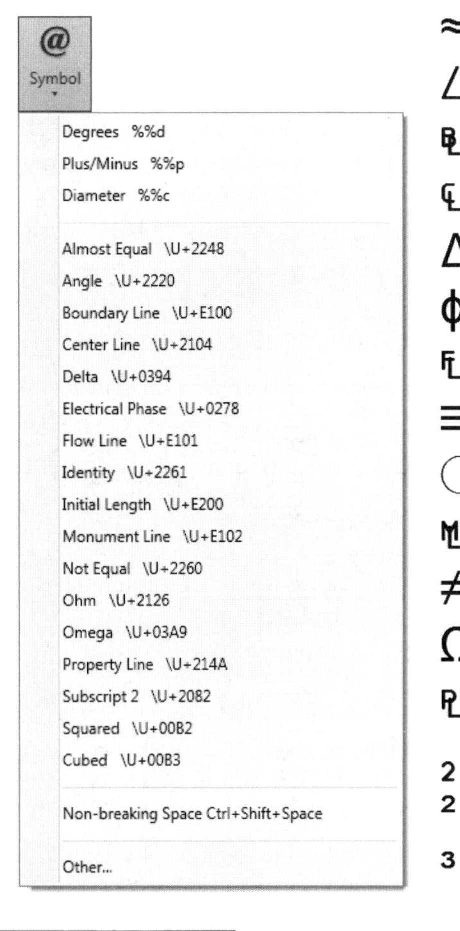

PARAGRAPH DIALOG BOX

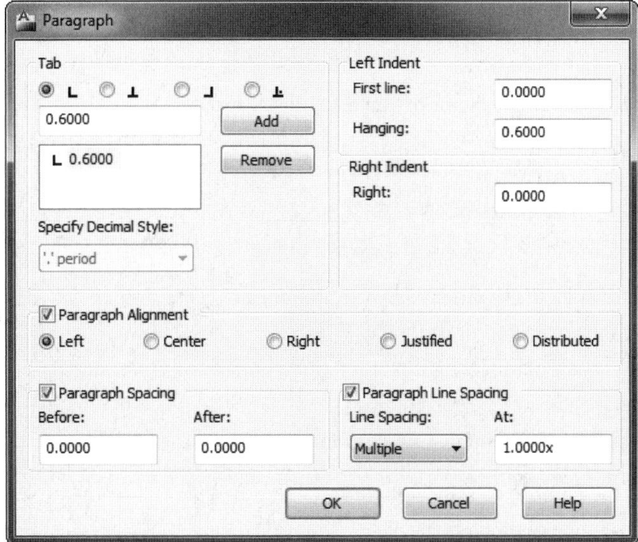

Tab options

⊙ ∟ **Left** aligns text to the right of the tab position.

○ ⊥ **Center** centers text about the tab position.

○ ⅃ **Right** aligns text to the left of the tab position.

○ ⊥ **Decimal** aligns text by the decimal marker.

Add adds tabs at the indicated distance.

Remove removes selected tabs.

Specify Decimal Style selects period, comma, or space.

Left Indent options

First Line specifies distance from left margin to the start of first line of text in a paragraph.

Hanging specifies distance from left margin to all other lines of text in the paragraph.

Right Indent options

Right specifies the distance from the right margin to the end of all text in the paragraph.

Paragraph Alignment options

⊙ **Left** aligns text against the left margin.

○ **Center** centers text between the two margins.

○ **Right** aligns text against the right margin.

○ **Justified** aligns text between both margins, except for the last line in the paragraph.

○ **Distributed** forces text to fit between margins, including the last line.

Paragraph Spacing options

☑ **Paragraph Spacing** toggles spacing between paragraphs.

Before and **After** specifies line spacing before (and after) the selected paragraph.

Paragraph Line Spacing options

☑ **Paragraph Line Spacing** toggles spacing between lines of text.

Line Spacing specifies the amount of space between lines:

- **Exactly** specifies the spacing in inches or other units.
- **At least** specifies the minimum spacing.
- **Multiple** specifies the spacing as a factor of the text height, such as 1x.

At specifies the distance in units or factors.

COLUMN SETTINGS DIALOG BOX

Access this dialog box through the Column droplist.

Column Type options

⊙ **Dynamic Columns** creates columns as required by the amount of text:

⊙ **Auto Height** stretches the column to fit the text.

○ **Manual Height** displays the Column Height marker at the bottom of the bounding box.

○ **Static Columns** creates a specific number of columns.

○ **No Columns** switches the bounding box to a single column.

Column Number specifies the number of static columns; range is 2 to 99.

Height specifies the height of the bounding box.

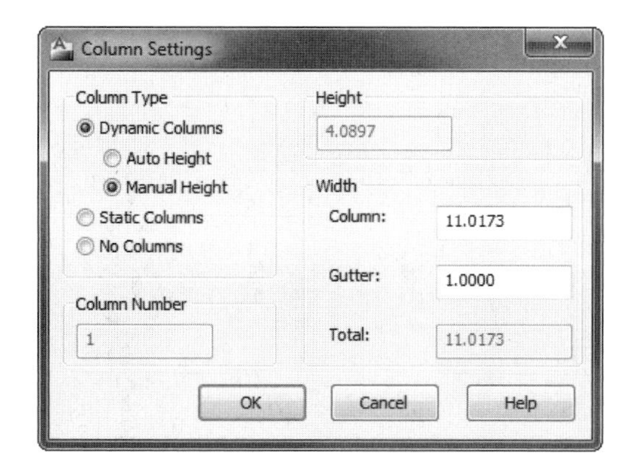

Width options

Column specifies the width of each column, margin to margin.

Gutter sets the distance between columns; the default is 5 times the text height.

Total reports the total width of the columns and gutter(s).

SET MTEXT WIDTH & HEIGHT SHORTCUT MENU

Access this menu by right-clicking the tab bar.

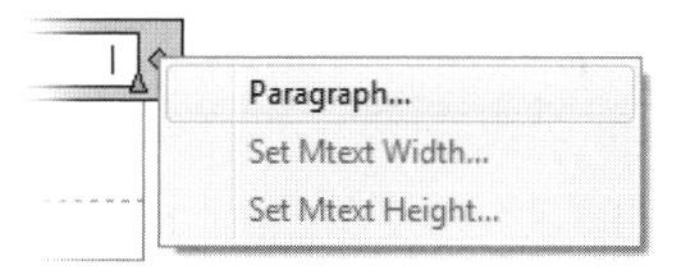

Paragraph displays the Paragraph dialog box.

Set MText Width changes the width of the mtext bounding box.

Set MText Height changes the height of the bounding box.

It is easier to change the boundary box's size by dragging its right and bottom borders.

BACKGROUND MASK DIALOG BOX

Access this dialog box through the Background Mask button.

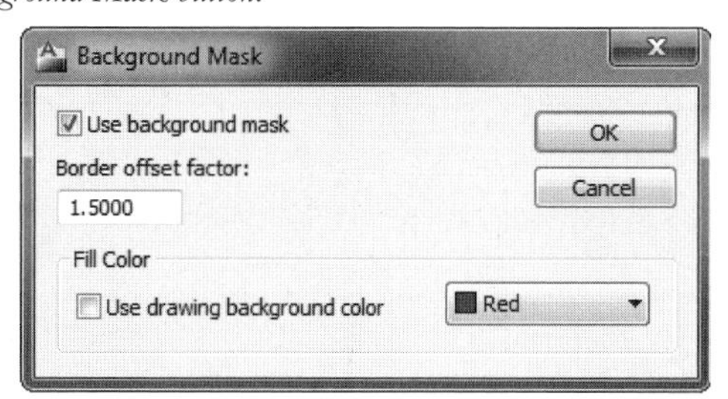

☑**Use background mask** toggles the display of the background mask. Note that the background mask applies to the full width of the mtext block, rather than the width of text.

Border offset factor determines the distance that the "margin" extends beyond the text. The *factor* is based on the text height: 1.0 means there is no offset; 1.5 means the offset distance is 1.5 times the text height. Maximum value = 5.0; minimum value = 1.0.

Fill Color option

Use drawing background color:

☑ Uses the background color (usually white or black).

☐ Uses the specified color; for more colors, choose Select Color.

FIND AND REPLACE DIALOG BOX

Access this dialog box through the Find and Replace button.

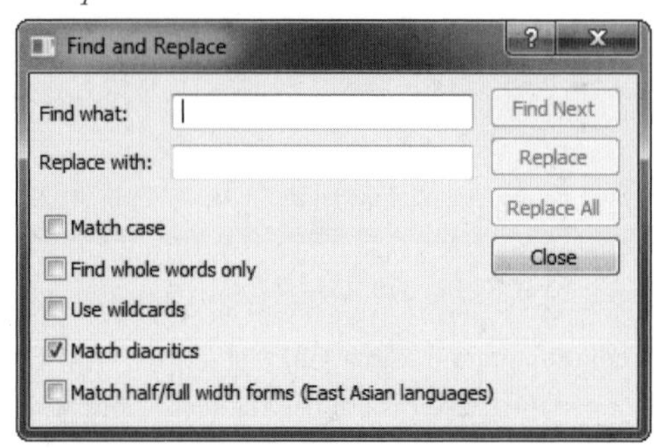

Find what specifies the text to search for. This dialog box searches only text within the bounding box; to search for text in the entire drawing, use the Find command.

Replace with specifies the replacement text; leave blank to search only.

Match case matches the case of the words.

Match whole word only

☑ Matches the entire word(s).

☐ Matches parts of the word(s).

Use Wildcards uses characters like ? to search for any character.

Match Diacritics searches for accents.

Match Half/full Width Forms searches for East Asian language forms.

Find Next finds the next occurrence of the word(s).

Replace replaces the found occurrence.

Replace All replaces all occurrences.

Cancel dismisses the dialog box.

-MTEXT Command

Command: -mtext
Current text style: "Standard" Text height: 0.20 Annotation: No
Specify first corner: *(Pick a point.)*
Specify opposite corner or [Height/Justify/Line spacing/Rotation/Style/Width/Columns]: *(Pick another point.)*
MText: *(Enter text.)*
MText: *(Press Enter to end the command.)*

COMMAND LINE OPTIONS

Height specifies the height of UPPERCASE text (*default = 0.2 units*).

Justify specifies a justification mode.

Line spacing specifies the distance between lines of text. AutoCAD prompts, 'Enter line spacing type [At least/Exactly] <At least>:" and then "Enter line spacing factor or distance <1x>:'.

Rotation specifies the rotation angle of the boundary box.

Style selects the text style for multiline text (default = STANDARD).

Width sets the width of the boundary box; a width of 0 eliminates the boundary box.

Columns specifies the number of text columns; AutoCAD prompts:

> **Enter column type [Dynamic/Static/No columns] <Dynamic>:** s
> **Specify total width: <16.0>:** *(Enter a number.)*
> **Specify number of columns: <2>:** *(Enter a number.)*
> **Specify gutter width: <1.0>:** *(Enter a number.)*
> **Specify column height: <2.0>:** *(Enter a number.)*

RELATED COMMANDS

TextEdit edits mtext; alternatively, double-click the mtext to edit.

Properties changes all aspects of mtext.

MtProp changes properties of multiline text.

PasteSpec pastes formatted text from the Clipboard into the drawing.

Spell checks spelling of words.

Style creates a named text style from a font file.

RELATED SYSTEM VARIABLES

CenterMt determines how the bounding box is stretched.

MTextEd names the external text editor for placing and editing multiline text.

MTextToolbar determines which toolbar is displayed during the MText command.

MTJigString specifies the sample text displayed when the bounding box is created; default = "abc"; maximum = 10 characters.

TIPS

- Use the MTextEd system variable to define a different text editor. To return to the older interface, enter the following system variables:

  MTextEd=oldeditor
  MTExtFixed=1

- The Import Text option is limited to ASCII (unformatted) and RTF (rich text format) text files.
- To import Word documents, copy the text to the Clipboard, and then press Ctrl+V in the mtext editor. Most, but not all, formatting is retained.
- To import formatted text, copy text from the word processor to the Clipboard, and then use AutoCAD's PasteSpec command.
- To link text in the drawing with a word processor, use the InsertObj command. When the word processor updates, the linked text is updated in the drawing.
- With some fonts, the mtext editor displays the diameter symbol as "%%c" and nonbreaking spaces as hollow rectangles, but these are displayed correctly in drawings.
- Stacked text can be created on either side of the following symbols:
  - **Carat** (^) stacks text as left-justified tolerance values.
  - **Forward slash** (/) stacks text as center-justified fractional-style values; the slash is converted to a horizontal bar.
  - **Pound sign** (#) stacks text with a tall diagonal bar.

  Use the stack tool a second time to unstack stacked text.
- As of AutoCAD 2009, this command checks spelling in real time.

Undocumented Command

MtProp changes the properties of multiline text (short for Multline Text PROPerties); displays the same text editor as does the MText command.

MView

Rel.11 Creates and manipulates overlapping viewports (short for Make VIEWports).

Command	Aliases	Keyboard Shortcuts	Menu	Ribbon
mview	**mv**	...	**View** ↳**Viewports**	...

Command: mview

*In Model tab, AutoCAD complains, '** Command not allowed in Model Tab **'.*

In a layout tab, AutoCAD prompts:

Specify corner of viewport or [ON/OFF/Fit/Shadeplot/Lock/Object/Polygonal/Restore/LAyer/2/3/4] <Fit>: *(Pick a point, or enter an option.)*

Specify opposite corner: *(Pick another point.)*

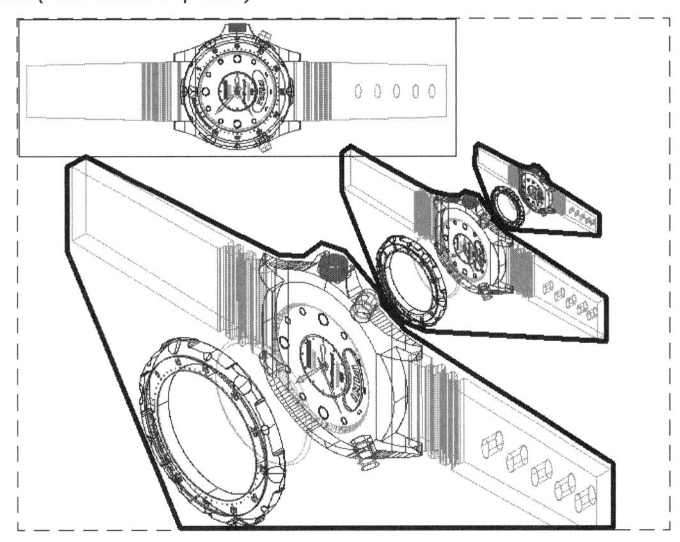

COMMAND LINE OPTIONS

Specify corner of viewport indicates the first point of a single viewport (default).

Fit creates a single viewport that fits the screen.

Shadeplot creates hidden-line or shaded views during plotting and printing.

Hideplot performs hidden-line removal during plotting and printing (*undocumented option*).

Lock locks the selected viewport's scale relative to model space.

Object converts a circle, closed polyline, ellipse, spline, or region into a viewport.

OFF turns off a viewport.

ON turns on a viewport.

Polygonal creates a multisided viewport of straight lines and arcs.

Restore restores a saved viewport configuration.

LAyer restores layer property overrides applied to viewports; properties are reset to ByLayer:

> **Reset viewport layer property overrides back to global properties [Y/N]?:** *(Type Y to reset changed properties to ByLayer.)*
>
> **Select viewports:** *(Select one or more.)*
>
> **Select viewports:** *(Press Enter to end viewport selection.)*

2 (Two Viewports) options

Enter viewport arrangement [Horizontal/Vertical] <Vertical>: *(Enter an option.)*

Specify first corner or [Fit] <Fit>: *(Pick a point, or enter an option.)*

Horizontal stacks two viewports.

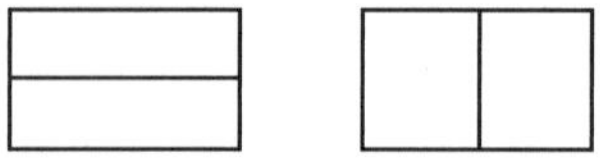

Vertical places two viewports side-by-side (default).

3 (Three Viewports) options

[Horizontal/Vertical/Above/Below/Left/Right]<Right>: *(Enter an option.)*

Specify first corner or [Fit] <Fit>: *(Pick a point, or enter an option.)*

Horizontal stacks the three viewports.

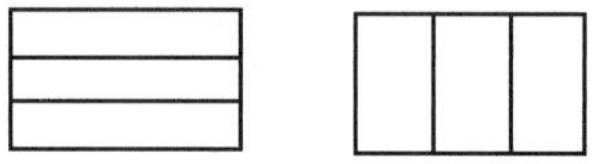

Vertical places three side-by-side viewports.

Above places two viewports above the third.

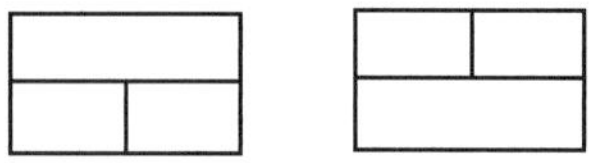

Below places two viewports below the third.

Left places two viewports to the left of the third.

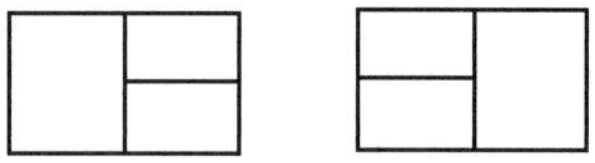

Right places two viewports to the right of the third (*default*).

4 (Four Viewports) options

Specify first corner or [Fit] <Fit>: *(Pick a point, or enter an option.)*

Fit creates four identical viewports that fit the viewport.

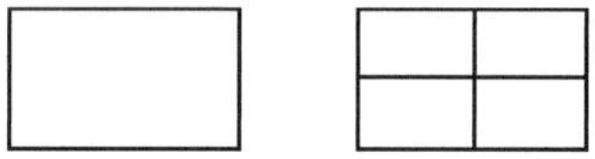

First Point indicates the area of the four viewports (default).

Shadeplot options

Shade plot? [As displayed/Wireframe/Hidden/Visual style/Rendered] <As displayed>: *(Enter an option.)*

As displayed plots the drawing as displayed.

Wireframe plots the drawing as a wireframe.

Hidden plots the drawing with hidden lines removed.

Visual style plots the drawing with specified visual style.

Rendered plots the drawing rendered.

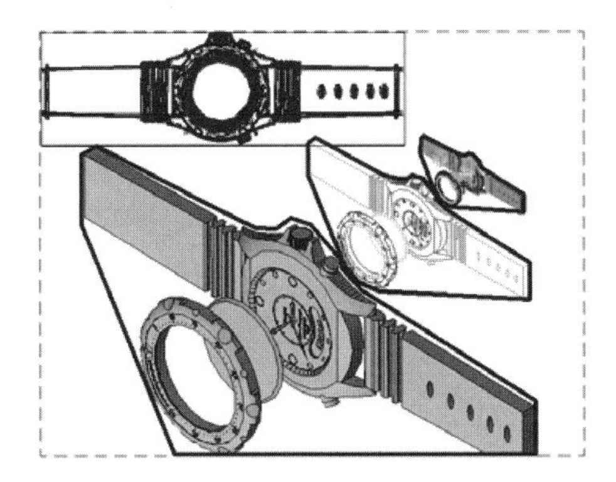

Viewports set to different visual styles.

Hideplot Options *(undocumented)*
Hidden line removal for plotting [ON/OFF]: on
Select objects: *(Select one or more viewports.)*

Hidden line removal for plotting toggles the removal of hidden lines for plotted output.

RELATED COMMANDS

Layout creates new layouts.

MSpace switches to model space.

PSpace switches to paper space before creating viewports.

RedrawAll redraws all viewports.

RegenAll regenerates all viewports.

VpLayer controls the visibility of layers in each viewport.

VPorts creates tiled viewports in model space.

Zoom zooms a viewport relative to paper space via the XP option.

RELATED SYSTEM VARIABLES

CvPort specifies the number of the current viewport.

MaxActVp controls the maximum number of active viewports; range is 2 to 64.

PsLtScale specifies linetype scaling in paper space.

TileMode controls the availability of overlapping viewports.

VpLayerOverride reports whether the current viewport has any overridden layers.

VpLayerOverrideMode toggles the plotting and display of overridden layers.

TIPS

- Drawings can have up to 32,767 viewports, but only 64 at most show content.
- Although MaxActVp limits the number of simultaneously-visible viewports, Plot plots all viewports.
- Snap, Grid, Hide, Shade, Ucs, and so on, can be set separately in each viewport.
- Press Ctrl+R to switch between viewports.
- Some of this command's options are also available from the Properties palette: select a viewport, right-click, and then select Properties from the shortcut menu.
- Click in an inactive viewport to make it active; the previously active viewport becomes inactive.
- Any shape of viewports can be locked.

MvSetup

Rel. 11 Sets up a drawing quickly, complete with a predrawn border. Optionally, sets up multiple viewports, sets the scale, and aligns views in each viewport (short for Model View SETUP).

Command	*Aliases*	*Keyboard Shortcuts*	*Menu*	*Ribbon*
'mvsetup	...	...	...	...

Command: mvsetup

When in model space:

Enable paper space? [No/Yes] <Y>: *(Type Y or N.)*

Command prompts in model tab (not paper space):

Enter units type [Scientific/Decimal/Engineering/Architectural/Metric]: *(Enter an option.)*

Enter the scale factor: *(Specify a distance.)*

Enter the paper width: *(Specify a distance.)*

Enter the paper height: *(Specify a distance.)*

Command prompts in layout mode (paper space):

Enter an option [Align/Create/Scale viewports/Options/Title block/Undo]: *(Enter an option.)*

COMMAND LINE OPTIONS

Align options

Pans the view to align a base point with another viewport.

Enter an option [Angled/Horizontal/Vertical alignment/Rotate view/Undo]: *(Enter an option.)*

Angled specifies the distance and angle from a base point to a second point.

Horizontal aligns views horizontally with a base point in another viewport.

Vertical alignment aligns views vertically with a base point in another viewport.

Rotate view rotates the view about a base point.

Undo undoes the last action.

Create options

Enter option [Delete objects/Create viewports/Undo] <Create>: *(Enter an option.)*

Delete objects erases existing viewports.

Create viewports creates viewports in several configurations. For instance, layout 2 generates the standard engineering layout, and layout 3 arrays viewports along the x and y axes.

Undo undoes the last action.

Scale Viewports options

Select the viewports to scale...

Select objects: *(Pick a viewport.)*

Select objects: *(Press Enter to end object selection.)*

Set the ratio of paper space units to model space units...

Enter the number of paper space units <1.0>: *(Enter a value.)*

Enter the number of model space units <1.0>: *(Enter a value.)*

Select objects selects one or more viewports.

Enter the number of paper space units specifies the paper scale.

Enter the number of model space units specifies the object scale.

Options options

Enter an option [Layer/LImits/Units/Xref] <exit>: *(Enter an option.)*

Layer specifies the layer name for the title block.

Limits specifies whether to reset limits after title block insertion.

Units specifies inch or millimeter paper units.

Xref specifies whether title is inserted as a block or as an external reference.

Title Block options

Enter title block option [Delete objects/Origin/Undo/Insert] <Insert>: *(Enter an option.)*

Delete objects erases an existing title block from the drawing.

Origin relocates the origin.

Undo undoes the last action.

Insert displays the available title blocks.

RELATED SYSTEM VARIABLE

TileMode specifies the current setting of TileMode.

RELATED FILES

mvsetup.dfs is the default settings file for this command.

acadiso.dwg is a template drawing with ISO (international standards) defaults.

All *.dwt* template drawings can also be used as templates.

RELATED COMMANDS

LayoutWizard sets up the viewports via a "wizard."

TIPS

- To create the title block, MvSetup searches the path specified by the AcadPrefix variable. If the appropriate drawing cannot be found, MvSetup creates the default border.
- You can add your own title block with the Add option. Before doing so, create the title block as an AutoCAD drawing.

NavBar

2011 Toggles the display of the navigation bar.

Commands	*Aliases*	*Keyboard Shortcuts*	*Menu*	*Ribbon*
navbar	...	...	**View** ↳**Display** ↳**Navigation Bar**	**View** ↳**Windows** ↳**User Interface** ↳**Navigation Bar**

Command: navbar

Enter an option [ON/OFF] <ON>: *(Enter ON or OFF.)*

Displays the navigation bar interface:

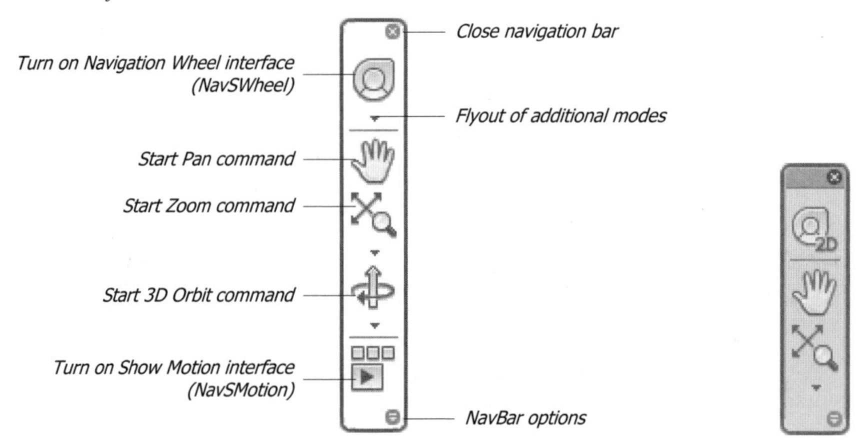

Left: *Navigation bar displayed in model space and block editor*
Right: *Navigation bar displayed in layout tabs.*

UI OPTIONS

displays the Navigation Wheel; see the NavSWheel command. Click the arrow for additional wheels:

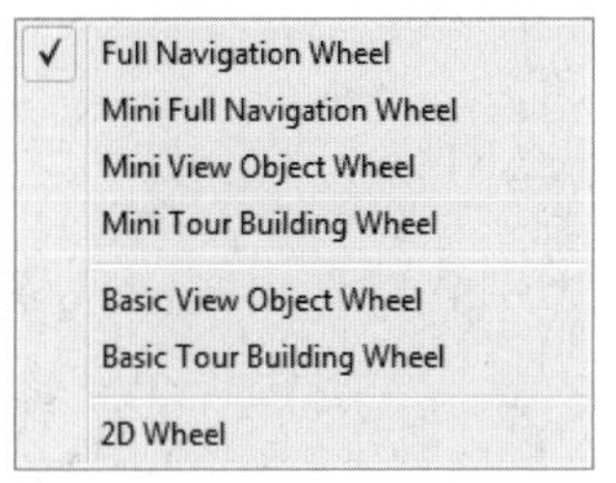

starts the Pan command; press Esc to end.

 starts the Zoom Extents command. Click the arrow for additional zoom modes:

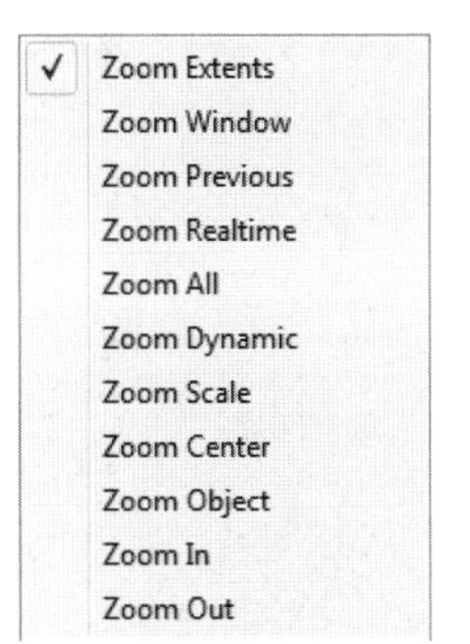

 starts the 3dOrbit command; press Esc to exit. Click the arrow for additional viewing modes:

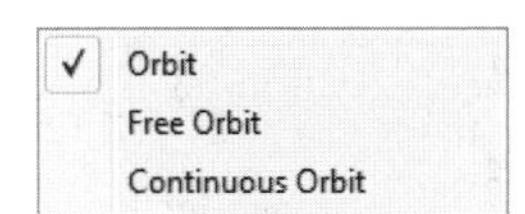

 displays the Show Motion interface; see NavSMotion command.

SHORTCUT MENU OPTIONS

Click button to access this shortcut menu:

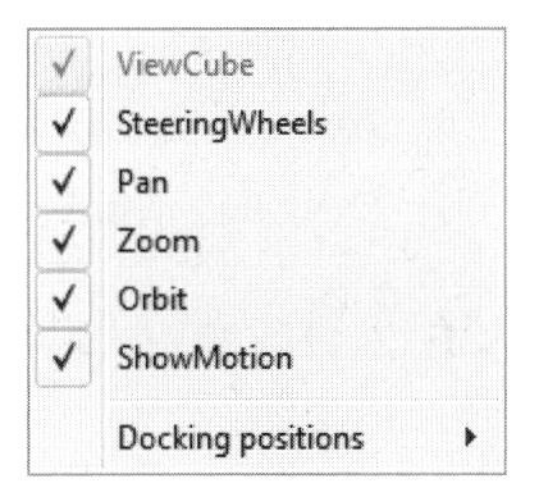

SteeringWheels adds or removes the NavSWheel icon from the navigation bar.

Pan adds or removes the Pan icon from the navigation bar.

Zoom adds or removes the Zoom icon from the navigation bar.

Orbit adds or removes the 3dOrbit icon from the navigation bar.

ShowMotion adds or removes the NavSMotion icon from the navigation bar.

Docking Position docks the navigation bar on the sides of the drawing area or links it to the ViewCube.

RELATED SYSTEM VARIABLE

NavBarDisplay toggles the display of the navigation bar; applies to all viewports and layouts. (The NavBar command applies to the current viewport or space.)

NavSMotion / NavSMotionClose

2009 Opens the shot motion user interface (short for "Navigate Shot MOTION").

Commands	Aliases	Status Bar	Menu	Ribbon
navsmotion	**motion**		**View** ↳**Show Motion**	...
navsmotionclose	**motioncls**	...	...	...

Command: navsmotion

Displays the show motion interface.

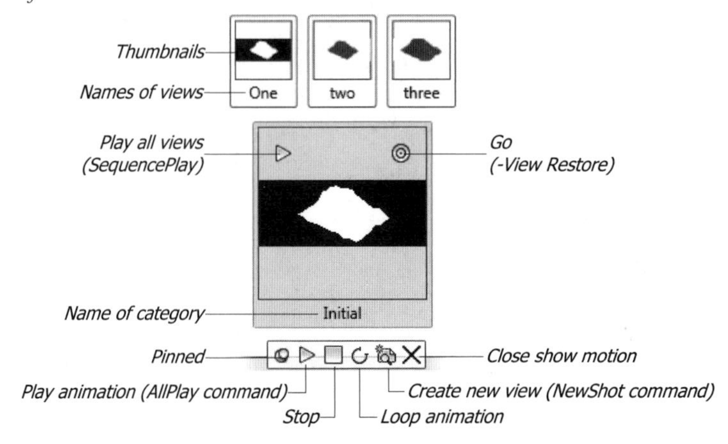

Command: navsmotionclose

Closes the show motion interface.

UI OPTIONS

Play runs the animation associated with the selected view or category; see the SequencePlay command.

Go restores the view; see the View command.

Pinned toggles navigation UI; when pinned, the UI stays in place.

Play All plays all views in all categories; see the AllPlay command.

Stop halts the playback.

Loop repeats the animations until revoked.

Create New View opens the New Shot dialog box; see the New Shot command.

Close closes the show motion interface (NavSMotionClose command).

SHORTCUT MENU OPTIONS

Right-click any thumbnail for this shortcut menu.

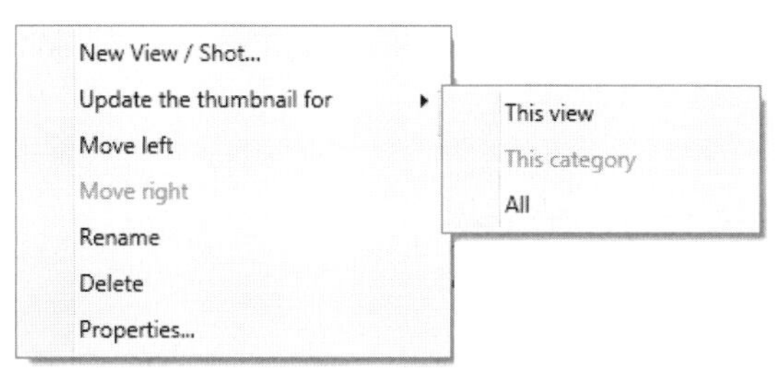

New View/Shot displays the New View/Shot Properties dialog box (NewShot command).

Update The Thumbnail For changes the thumbnail associated with this view or category.

Move Left/Right changes the order by moving the view to the left/right.

Rename changes the name of the view or category.

Delete erases the view or category.

Properties displays the View/Shot Properties dialog box (EditShot command).

RELATED COMMANDS

AllPlay plays back all show motion views in sequence, without prompting.

EditShot prompts for a view name, and then displays the Shot Properties tab of the View / Shot Properties dialog box .

NewShot creates named views with show motion options through the Shot Properties tab of the New View / Shot Properties dialog box.

SequencePlay plays all views in a category.

ViewGo displays the named view as a shortcut to the -View command's Restore option.

ViewPlay plays the animation associated with a named view.

RELATED SYSTEM VARIABLES

ShowMotionPin specifies the default pinning of the Show Motion thumbnail images.

ThumbSize specifies the size of thumbnails.

TIPS

- To create a show motion animation, follow these steps:

  1. Create views and specify animation parameters with the NewShot command.
  2. Run the animation with the undocumented AllShow command.

- Show motion creates animations with views. You set up one or more views in a drawing, and then specify the types of paths, such as zoom or pan (track), and durations. AutoCAD uses show motion to interpolate between the views.
- Show motion only works inside AutoCAD, like the MSlide and VSlide commands. AutoCAD cannot export animations to movie files, but you can use capture software, such as Camtasia or CamStudio, to save the animation.
- View *categories* allow you to have two or more multi-view show motion animations per drawing. Categories were originally designed for sheet sets, but as of AutoCAD 2009 they work with show motion. Categories are defined in the View Categories item of the New View / Shot Properties dialog box.

NavSWheel

2009 Displays one of several steering wheels for 2D and 3D drawing navigation (short for "NAVigation Steering WHEEL").

Command	Aliases	Status Bar	Menu	Ribbon
'navswheel	**wheel** **W** *(Shift+ W)*		**View** **↳SteeringWheels**	**Home** **↳View** **↳Navigate** **↳SteeringWheels**

Command: navswheel

Right-click to display the shortcut menu. Press ESC or ENTER to exit.

Displays large or mini navigation wheel, depending on settings:

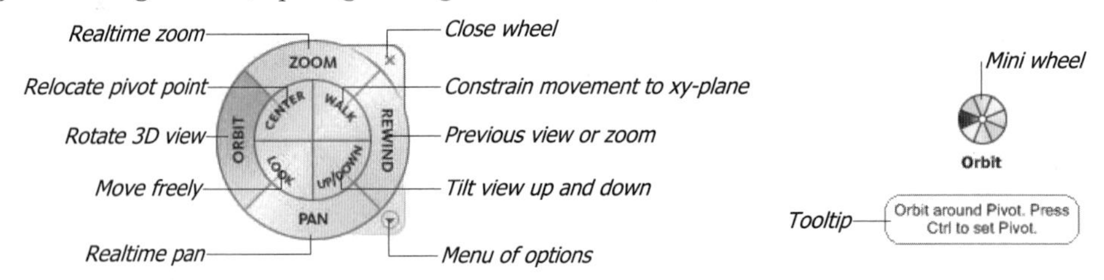

***Left:** Large navigation wheel.*
***Right:** Mini navigation wheel.*

COMMAND LINE OPTION

ESC exits the navigation wheel.

SHORTCUT MENU OPTIONS

Click steering wheel's Menu button, or right-click the wheel for this shortcut menu of options:

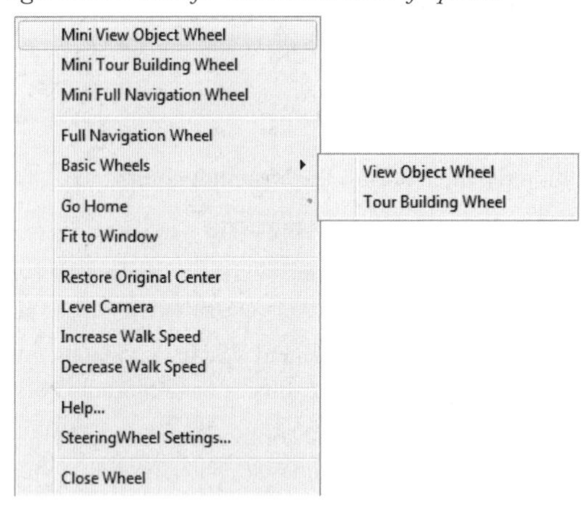

Mini View Object / Tour Building / Full Navigation Wheel switches between mini styles.

Full Navigation / View Object / Tour Building Wheel switches between large styles.

Go Home returns to the original viewpoint.

Fit to Window fits the model to the viewport.

Restore Original Center resets the center point, after being relocated by Center mode.

Level Camera sets z = 0.

Increase / Decrease Walk Speed speeds up and slows down the walk mode.

Help displays online help relevant to this command.

Steering Wheel Settings displays the SteeringWheels Settings dialog box.

Close Wheel exits the command.

STEERINGWHEELS SETTINGS DIALOG BOX

Access this dialog box through the shortcut menu's Settings option:

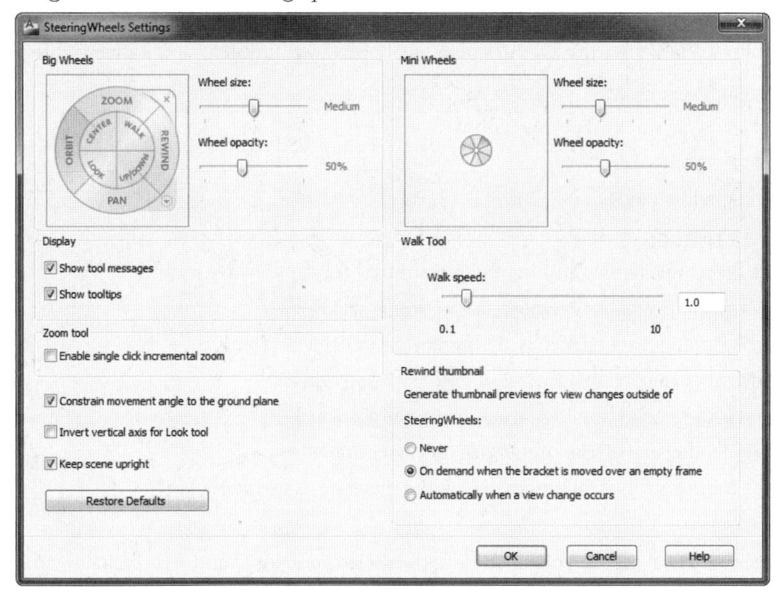

Big Wheels

Wheel Size sets display size of large wheels: small, medium, or large.

Wheel Opacity specifies translucency from 25% (nearly transparent) to 90% (opaque).

Display

☑**Show Tool Messages** toggles display of tools messages.

☑**Show Tooltips** toggles display of wedge and button tooltips.

☐ **Show the Pinned Wheel at Startup** pins the wheel when first activated.

Zoom Tool

☐**Enable Single Click Incremental Zoom** specifies that single clicking the Zoom tool increases magnification by 25%.

☐**Invert Vertical Axis for Look Tool** reverses the mouse drag direction for up and down Look tool motion.

☑**Maintain Up Direction for Orbit Tool** keeps the model upright with the Orbit tool.

☑**Use Selection Sensitivity for Orbit Tool** uses selected objects as pivot points for the Orbit tool.

Restore Defaults resets defaults for this dialog box.

Mini Wheels

Wheel Size sets the display size: small, medium, large, or extra large.

Wheel Opacity specifies translucency from 25% (nearly transparent) to 90% (opaque).

Walk Tool

☐ **Constrain Walk Angle to Ground Plane** maintains the current view along the z direction with the Walk tool.

Walk Speed specifies the Walk tool's speed: 0.1 to 10 units per time.

Rewind Thumbnail

Controls when thumbnails are generated for view changes made without using a wheel. The generated thumbnails are used for the Rewind tool (CaptureThumbnails system variable).

○ **Never** generates no preview thumbnails.

⊙ **On Demand When the Bracket is Moved Over an Empty Frame** generates preview thumbnails when Rewind tool positions the bracket over empty frames.

○ **Automatically When a View Change Occurs** generates preview thumbnails after a view changes.

RELATED COMMANDS

NavVCube toggles the display of the 3D view cube.

VPoint sets static 3D viewpoints.

3dOrbit rotates 3D viewpoints freely.

RELATED SYSTEM VARIABLES

CaptureThumbnails determines how thumbnails are captured for the Rewind tool.

NavSWheelMode specifies the style of navigation steering wheel to display.

NavSWheelOpacityBig changes translucency of big navigation steering wheels.

NavSWheelOpacityMini changes translucency of mini steering wheels.

NavSWheelSizeBig sets the size of the big navigation steering wheel.

NavSWheelSizeMini sets the size of the mini navigation steering wheel.

NavsWheelWalkSpeed controls the walking speed of the navigation steering wheel.

TIPS

- The two kinds of steering wheel are **Regular** (larger with labeled functions) and **Mini** (small without labels).
- The four styles of steering wheel are: View Objects, View Buildings, Full, and 2D.

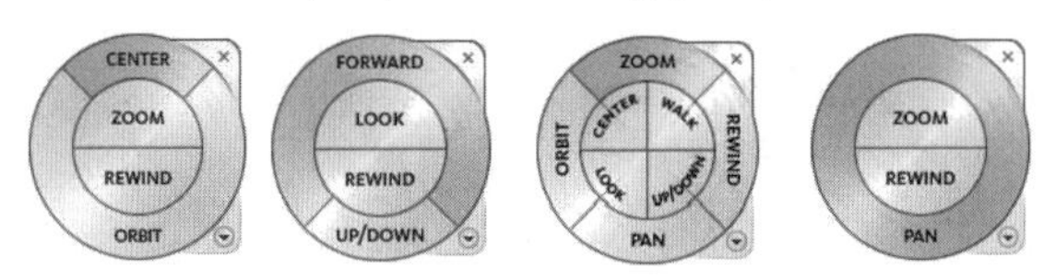

- The Rewind tool shows all previous view changes, and allows you choose the frame of an earlier view point. As you pass the cursor over frames, the underlaying model also changes its viewpoint in real time.

The double-angle frame indicates when you selected a tool from the wheel.

- Press Shift+W to toggle the navigation wheel transparently.
- The undocumented **Antz** command displays all steering wheels at once, after you use AppLoad to load the *Gs_Test.Arx* file.

NavVCube

2009 Displays the 3D navigation cube (short for "Navigation View CUBE").

Command	Alias	Keystroke Shortcuts	Menu	Ribbon
navvcube	**cube**	...	**View**	**View**
			↳**Display**	↳**Windows**
			↳**ViewCube**	↳**User Interface**

Command: navvcube
Enter an option [ON/OFF/Settings] <ON>: *(Enter an option.)*

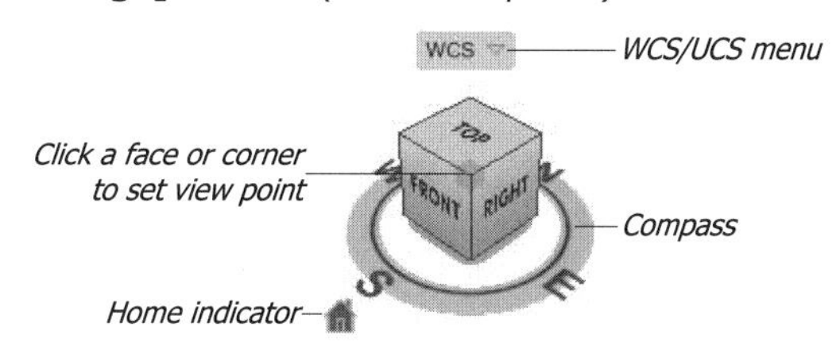

The view cube is not displayed when the visual style is 2D.

COMMAND LINE OPTIONS

ON turns on the viewcube.

OFF turns off the viewcube.

Settings displays the ViewCube Settings dialog box.

SHORTCUT MENU OPTIONS

Access this menu by right-clicking the viewcube:

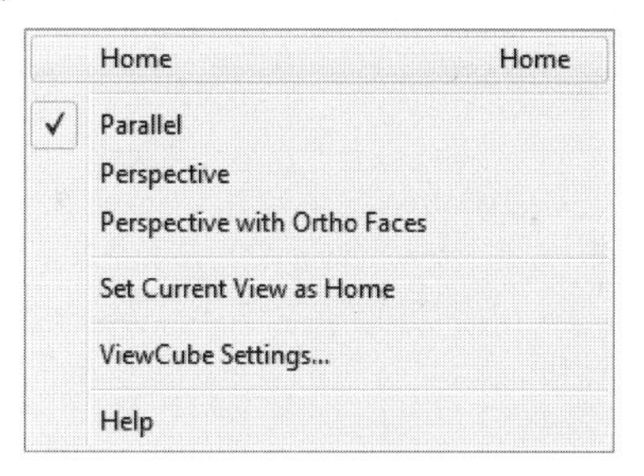

Home sets the viewpoint to the Home view.

Parallel sets view projection to parallel.

Perspective sets view projection to perspective.

Perspective with Ortho Faces keeps faces orthogonal.

Set Current View as Home sets the current 3D view as the home view.

ViewCube Settings displays the ViewCube Settings dialog box.

Help shows online help relative to this command.

VIEWCUBE SETTINGS DIALOG BOX

Access this dialog box through the shortcut menu's ViewCube Settings item:

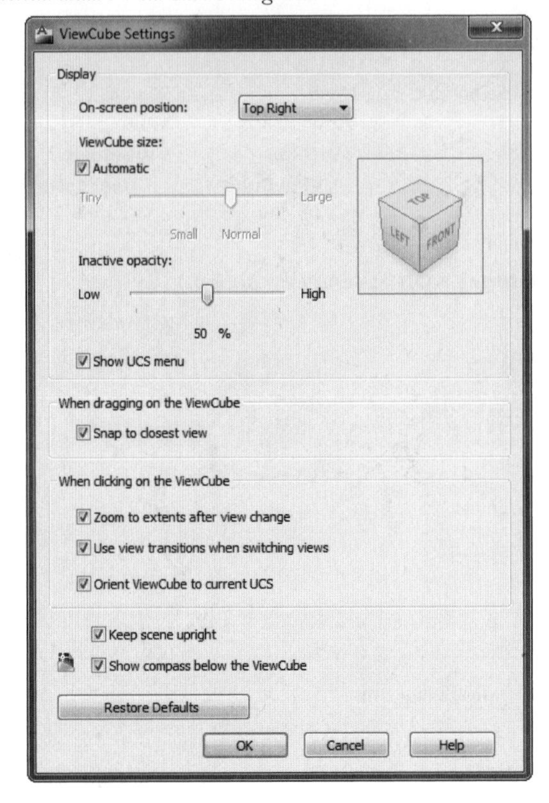

Display

On-screen Position positions the viewcube in the corner of a viewport: bottom or top, right or left.

ViewCube Size changes the size of the viewcube: small, medium, or large.

Inactive Opacity specifies the amount of translucency when the viewcube is inactive, from 0% (invisible) to 100% (opaque).

☑**Show UCS Menu** toggles the display of the viewcube's WCS/UCS menu.

When Dragging on the ViewCube

☑**Snap to Closest View** adjusts the current view to the nearest preset view.

When Clicking on the ViewCube

☑**Zoom to Extents After View Change** fits the model to the viewport after each view change.

☑**Use View Transitions When Switching Views** employs view transitions between view changes.

☑**Orient ViewCube to Current UCS** ensures the viewcube is oriented to the current UCS or WCS (user or world coordinate system).

☑**Keep Scene Upright** maintains a level z-coordinate; prevents upside down views.

☑**Show Compass Below the ViewCube** toggles the compass under the viewcube.

Restore Defaults resets dialog box's options to default values.

RELATED COMMANDS

NavSWheel displays the steering wheel navigator.

NavBar toggles the display of the navigation bar, which is linked to the navigation cube *(new to AutoCAD 2011).*

View sets static 2D and isometric viewpoints.

3dOrbit freely rotates 3D viewpoints.

RELATED SYSTEM VARIABLES

NavVCubeDisplay determines when the navigation cube is displayed.

NavVCubeLocation locates navigation cube in current viewport.

NavVCubeOpacity changes translucency of navigation cube

NavVCubeOrient specifies the navigation cube 's orientation relative to UCS or WCS.

NavVCubeSize specifies the size of the navigation cube.

Perspective toggles between parallel and perspective view modes.

TIPS

- Click corners, edges, or faces to change the 3D viewpoint:

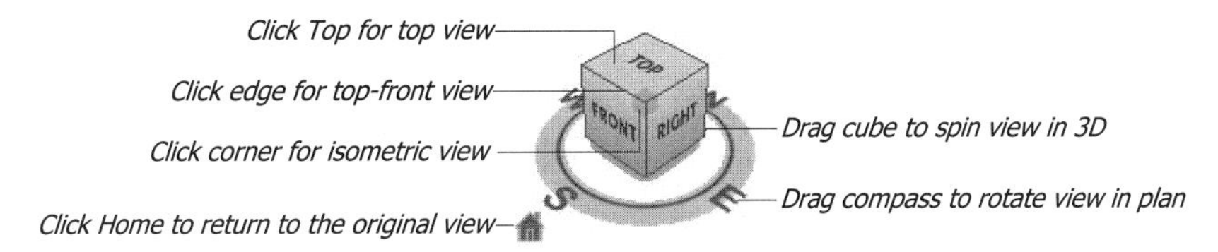

- Drag the cube for a 3dOrbit-like view rotation. Drag the compass to rotate the view in plan view.

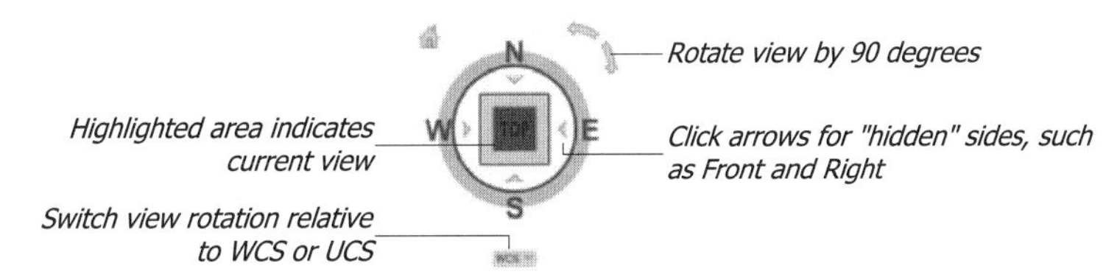

- *Home View* is one of the following:
    - Zoom extents for drawings saved in AutoCAD 2008 and earlier.
    - Isometric view for drawings saved in AutoCAD 2009 and later.
    - View defined by the user: right-click ViewCube, and then choose Set Current View as Home.

- The home view can be used as the saved thumbnail view, instead of the last-saved view. This is controlled by the new Options | Open & Save | Thumbnail View dialog box.
- 3D visual styles default to the perspective viewing mode. Usually, architects want 3D views in perspective, but mechanical designers want 3D views in parallel. Normally, setting the Perspective system variable to 0 returns to parallel viewing mode; clicking Home, however, changes Perspective back to 1. To avoid this, follow these steps:
    1. Select the Home view.
    2. Set Perspective to 0.
    3. Click Set Current View as Home.
- As of AutoCAD 2011, the Navigation Cube also appears in 2D model space.

NetLoad

2005 Loads *.dll* files written with Microsoft's .Net programming interface.

Command	*Aliases*	*Keystroke Shortcuts*	*Menu*	*Ribbon*
netload	...	...	...	...

Command: netload

Displays the Choose .NET Assembly dialog box.

Select a .dll file, and then click Open.

COMMAND LINE OPTIONS

None.

TIP

- Some parts of AutoCAD written in .Net include the Layer dialog box and the migration utility.

New

Rel. 12 Starts new drawings from template drawings, from scratch, or through step-by-step drawing setup "wizards."

Command	*Aliases*	*Keystroke Shortcut*	*Menu Bar*	*Application Menu*
new	**...**	**Ctrl+N**	**File** **↳New**	**New** **↳Drawing**

Command: new

AutoCAD displays one of three interfaces, depending on the settings of the FileDia and Startup variables.

FileDia	***Startup***	***New***
1	1	Displays Startup wizard.
1	0	Displays Select Template dialog box (default).
0	1 *or* 0	Prompts for *.dwt* file at command line.

SELECT TEMPLATE DIALOG BOX

Open creates new drawings based on the selected *.dwt* template file.

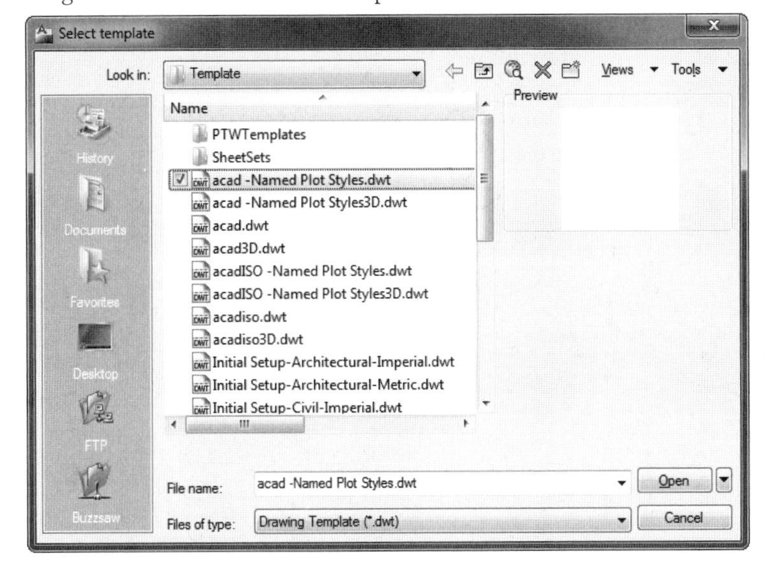

Open with no template - Imperial creates new drawings with default Imperial values.

Open with no template - Metric creates new drawings with default metric values.

STARTUP DIALOG BOX

The Startup wizard is displayed when Startup is set to 1 and when AutoCAD first starts.

The Create New Drawing wizard is similar, and is displayed when subsequent new drawings are opened.

 Open a Drawing page

(This page is not in the Create New Drawing wizard.)

Browse displays the Select File dialog box; see the Open command.

Start from Scratch page

Imperial creates a new drawing with AutoCAD's default values and InsUnits=1 (inches).

Metric creates a new drawing with AutoCAD's default values and InsUnits=4 (millimeters).

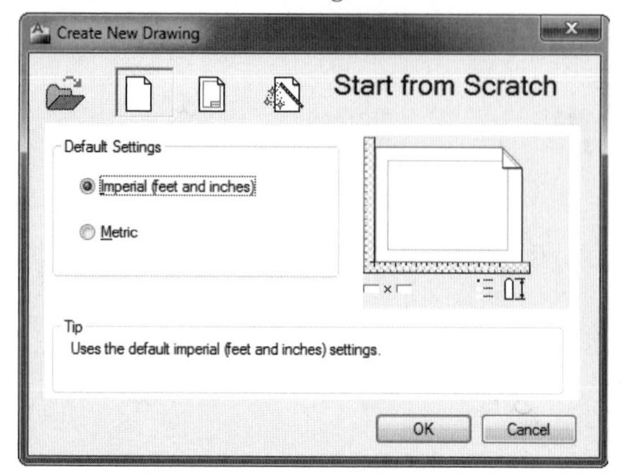

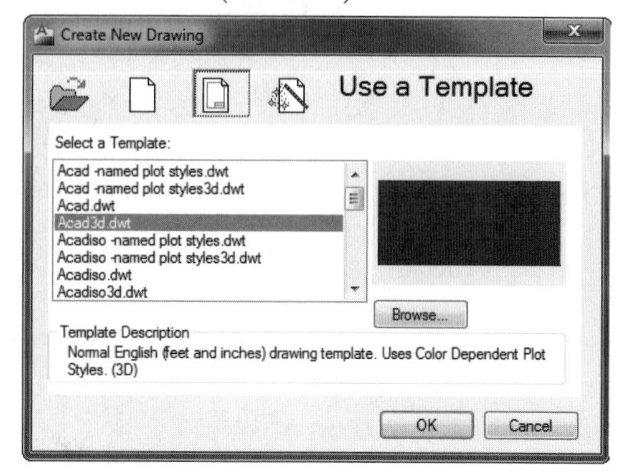

Left to right: *Start from Scratch and Use a Template dialog boxes.*

Use a Template page

Select a Template creates a new drawing based on the selected *.dwt* template file.

Browse displays the Select a Template File dialog box.

Use a Wizard page

Select a Wizard

- **Advanced Setup** sets up a new drawing in several steps.
- **Quick Setup** sets up a new drawing in two steps.

QUICK SETUP WIZARD

This wizard can also be started by the undocumented QkSetupWiz command.

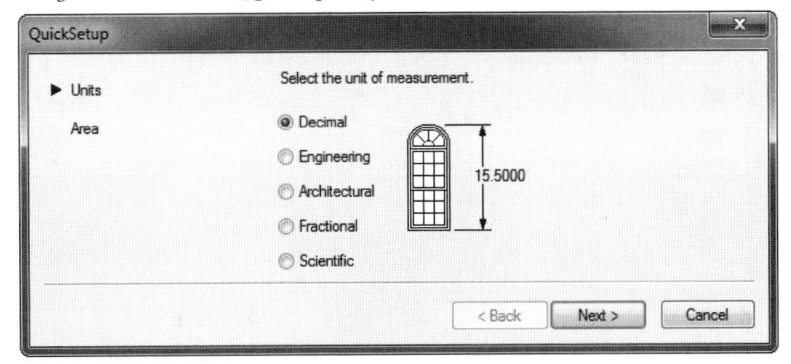

⊙ **Decimal** displays units in decimal (or "metric") notation (default): 123.5000.

○ **Engineering** displays units in feet and decimal inches: 10'-3.5000".

○ **Architectural** displays units in feet, inches and fractional inches: 10'3-1/2".

○ **Fractional** displays units in inches and fractions: 123 1/2.

○ **Scientific** displays units in scientific notation: 1.235E+02.

Cancel cancels the wizard, and returns to the previous drawing.

Back moves back one step.

Next moves forward one step.

Area page

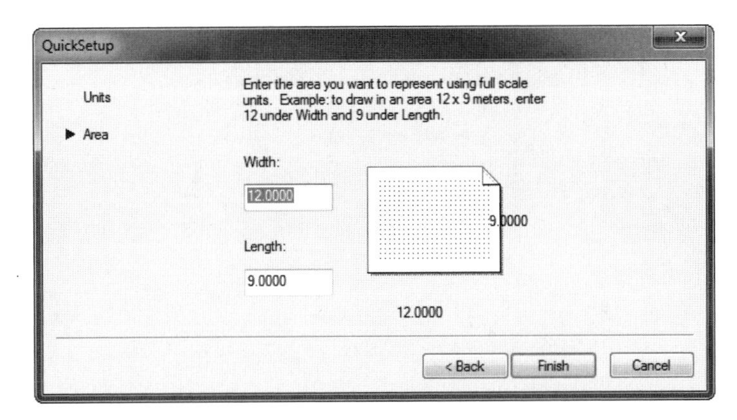

Width specifies the width of the drawing in real-world (not scaled) units; default = 12 units.

Length specifies the length or depth of the drawing in real-world units; default = 9 units.

ADVANCED SETUP WIZARD

This wizard can also be started by the undocumented AdvSetupWiz command.

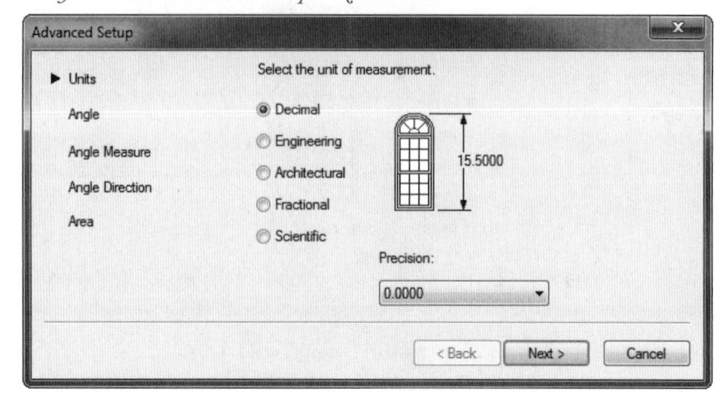

⊙ **Decimal** displays units in decimal (or "metric") notation (default): 123.5000.

○ **Engineering** displays units in feet and decimal inches: 10'-3.5000".

○ **Architectural** displays units in feet, inches, and fractional inches: 10' 3-1/2".

○ **Fractional** displays units in inches and fractions: 123 1/2.

○ **Scientific** displays units in scientific notation: 1.235E+02.

○ **Precision** selects the precision of display up to 8 decimal places or 1/256.

Angle page

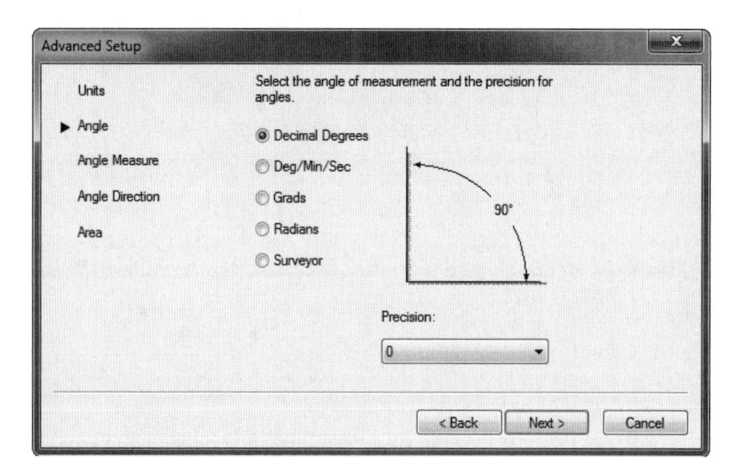

⊙ **Decimal Degrees** displays decimal degrees (default): 22.5000.

○ **Deg/Min/Sec** displays degrees, minutes, and seconds: 22 30.

○ **Grads** displays grads: 25g.

○ **Radians** displays radians: 25r.

○ **Surveyor** displays surveyor units: N 25d0'0" E.

Precision selects a precision up to 8 decimal places.

Angle Measure page

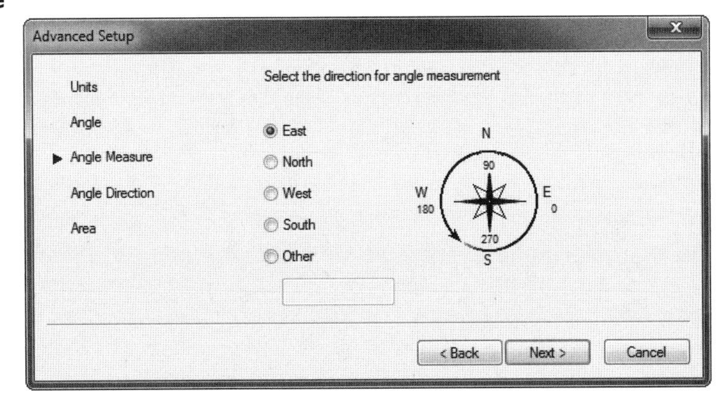

⊙ **East** specifies that zero degrees points East (default).

○ **North** specifies that zero degrees points North.

○ **West** specifies that zero degrees points West.

○ **South** specifies that zero degrees points South.

○ **Other** specifies any of the 360 degrees as zero degrees.

Angle Direction page

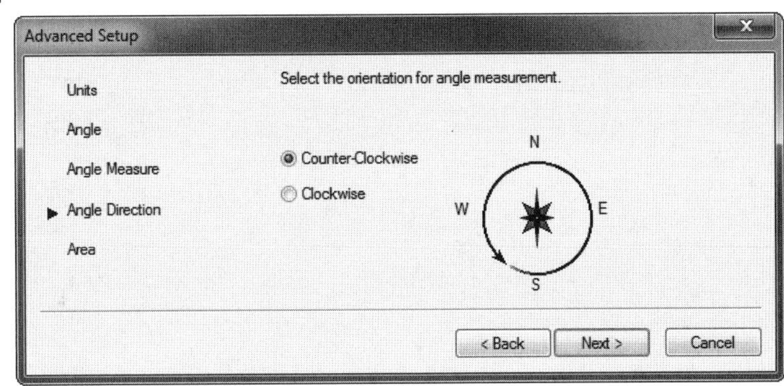

⊙ **Counter-Clockwise** measures positive angles counterclockwise from 0 degrees (default).

○ **Clockwise** measures positive angles clockwise from 0 degrees.

Area page

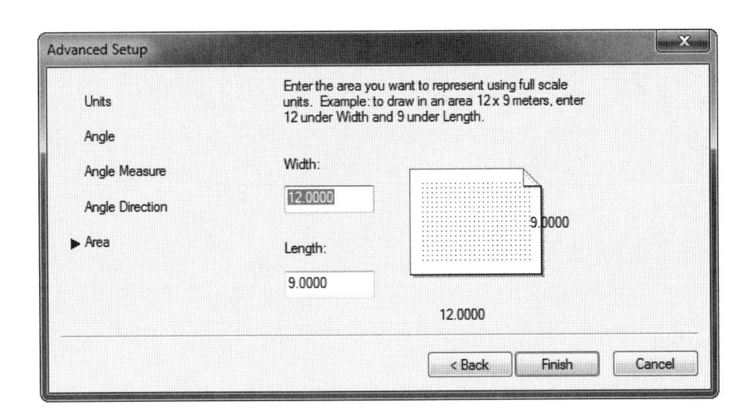

Width specifies the width of the drawing in real-world (not scaled) units; default = 12 units.

Length specifies the length or depth of the drawing in real-world units; default = 9 units.

Command Line Options

When FileDia = 0, AutoCAD prompts you at the command line:

Command: new

Enter template file name or [. (for none)] <default .dwt file path name>: *(Enter the path and name of a .dwt , .dwg, or .dws file.)*

Alternatively, enter the following options:

Enter accepts the default template drawing file.

. (period) eliminates use of a template; AutoCAD uses either *acad.dwt* or *acadiso.dwt*, depending on the setting of the MeasureInit system variable.

~ (tilde) forces the display of the Select Template dialog box.

COMMAND LINE SWITCHES

Switches used by Target field on Shortcut tab in the AutoCAD desktop icon's Properties dialog box.

/b runs a script file after AutoCAD starts; uses the following format:

```
acad.exe "\acad 2011\drawing.dwg" /b "file name.scr"
```

/c specifies the path for alternative hardware configuration file; default = *acad2011.cfg*.

/layout specifies the layout to display (*undocumented*).

/ld loads the specified ARx and DBx applications.

/nohardware prevents AutoCAD from using hardware-accelerated graphics, which can keep AutoCAD from crashing in 3D modeling.

/nologo suppresses the display of the AutoCAD logo screen.

/nossm prevents Sheet Set Manager palette from loading.

/p specifies a user-defined profile to customize AutoCAD's user interface.

/pl publishes a set of drawings defined by a *.dsd* file in the background.

/r restores the default pointing device.

/s specifies additional support folders; maximum is 15 folders, with each folder name separated by a semicolon.

/set specifies the *.dst* sheet set file to load.

/t specifies the *.dwt* template drawing to use.

/v specifies the named view to display upon startup of AutoCAD.

/w specifies the workspace to display first.

RELATED COMMANDS

AdvSetupWiz starts the Advanced Setup wizard (an undocumented command).

QkSetupWiz starts the Quick Setup wizard (an undocumented command).

QNew starts a new drawing based on a predetermined template file.

SaveAs saves the drawing in *.dwg* or *.dwt* formats; creates template files.

RELATED SYSTEM VARIABLES

DbMod indicates whether the drawing has changed since being loaded.

DwgPrefix indicates the path to the drawing.

DwgName indicates the name of the current drawing.

FileDia displays prompts at the 'Command:' prompt.

InsUnits determines whether blocks are inserted in inches, millimeters, or other units.

Startup determines whether the dialog box or the wizard is displayed.

MeasureInit determines whether the units are imperial or metric.

largeObjectSupport determines whether the drawing stores objects larger than 256MB.

RELATED FILES

wizard.ini names and describes template files.

.dwt* are template files stored in *.dwg* format.

TIPS

- Until you name the drawing, AutoCAD names it *drawing1.dwg* (and subsequent drawings created in the same session as *drawing2.dwg*, *drawing3.dwg*, and so on).
- The default template drawing is the last one used, and not necessarily *acad.dwg*.
- Edit and save *.dwt* template drawings to change the defaults for new drawings.
- When you start a new drawing by pressing Ctrl+N, you will find that AutoCAD's behavior differs from that of Microsoft Office programs: these programs display new documents that take on the properties of the current document.
- The Startup system variable can no longer be set in the Options dialog box. The Today window was removed in AutoCAD 2004. It was replaced by the Communications Center.
- As of AutoCAD 2011, this command adds metric or Imperial scale factors to drawings lacking them, based on the setting of the MeasureInit system variable.
- Prior to Release 12, new drawings were started from a text menu that listed numbered options, hence there was no New command — nor Open or Options, for that matter. Users entered 1 to start a new drawing, 2 to open a drawing, and so on.

NewSheetset

2005 Runs the New Sheet Set wizard.

Command	Aliases	Keystroke Shortcuts	Menu Bar	Application Menu
newsheetset	...	...	**Tools** ↳**Wizards** ↳**New Sheet Set**	**New** ↳**Sheet Set**

Command: newsheetset

Displays the Create Sheet Set dialog box.

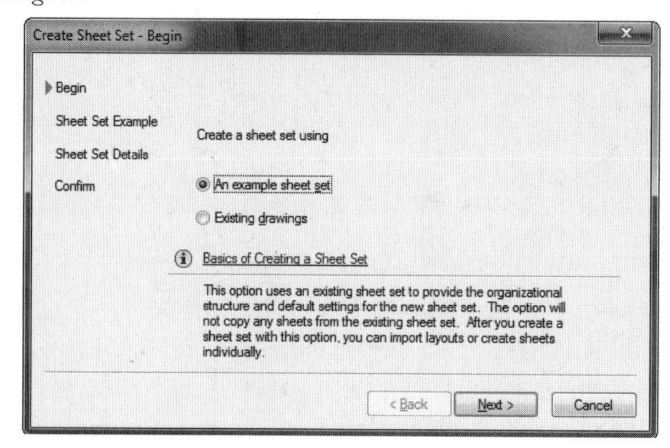

CREATE SHEET SET DIALOG BOX

Create a sheet set using:

⊙**An example sheet set** creates a new sheet set based on templates.

○**Existing drawings** selects one or more folders holding drawings, which are imported into the sheet set.

Click **Next** to continue.

Example Sheet Set

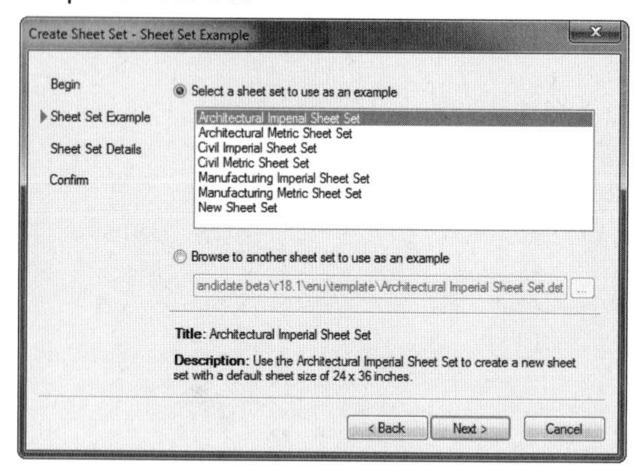

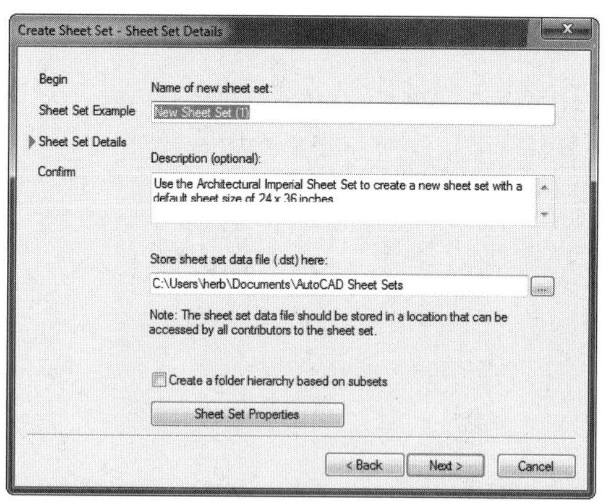

***Left:** Sheet Set Example page.*
***Right:** Sheet Set Details page.*

Sheet Set Example page

⦿ **Select a sheet set to use as an example** selects one of the sheet sets provided by Autodesk.

○ **Browse to another sheet set to use as an example** opens a *.dst* sheet set data file.

Sheet Set Details page

Name of new sheet set names the sheet set.

Description describes the sheet set.

Store sheet set data file here specifies the location of the *.dst* sheet set data file.

☐ **Create a folder hierarchy based on subsets** toggles creation of one or more folders to hold subset files.

Sheet Set Properties displays the Sheet Set Properties dialog box.

Finishes with the Confirm page, illustrated below.

Existing Drawings

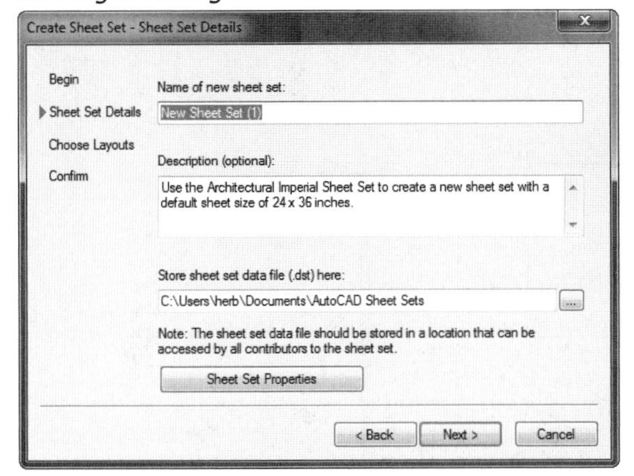

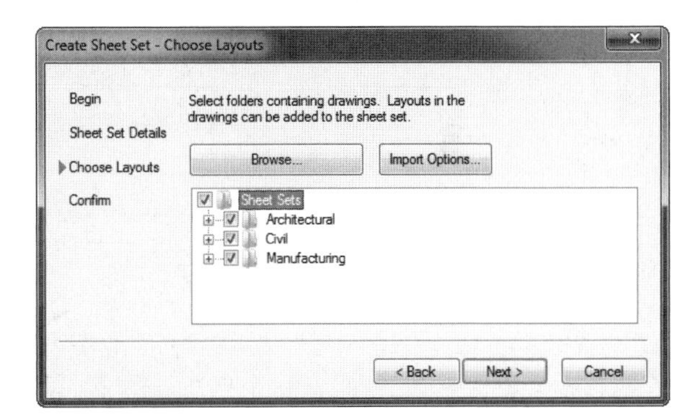

***Left**: Sheet Set Details page.*
***Right**: Choose Layouts page.*

Sheet Set Details page

Name of new sheet set names the sheet set.

Description describes the sheet set.

Store sheet set data file here locates the sheet set data file.

☐ **Create a folder hierarchy based on subsets** toggles creation of one or more folders to hold subset files.

Sheet Set Properties displays the Sheet Set Properties dialog box.

Choose Layouts page

Browse displays Browse for Folder dialog box; select a folder, and then click OK.

Import Options displays the Import Options dialog box.

Confirm page

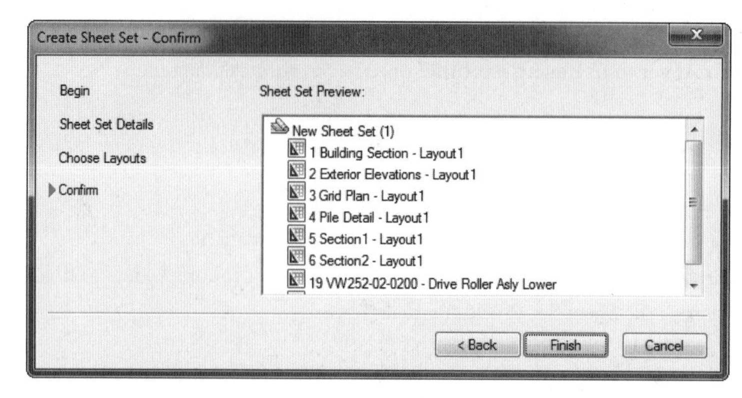

Press Ctrl+C to copy the settings, after selecting all text. (In a text editor, press Ctrl+V to paste the text in a document.)

Click **Back** to change and correct settings; click Finish to create the new sheet set.

SHEET SET PROPERTIES DIALOG BOX

Access this dialog box through the Sheet Set Properties button on the Sheet Set Details page:

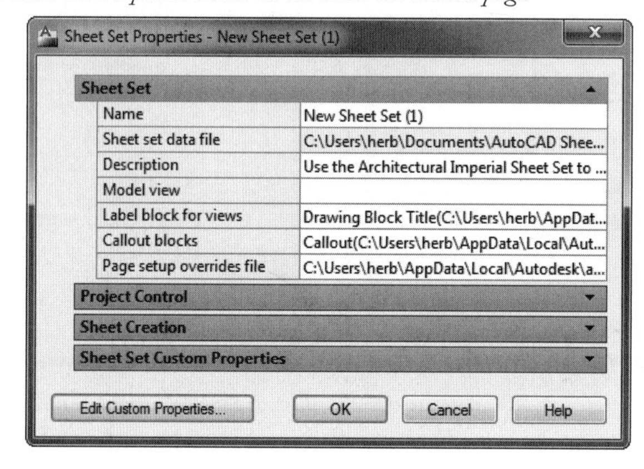

Edit Custom Properties displays the Custom Properties dialog box.

CUSTOM PROPERTIES DIALOG BOX

Access this dialog box through the Edit Custom Properties button on the Sheet Set Properties dialog box:

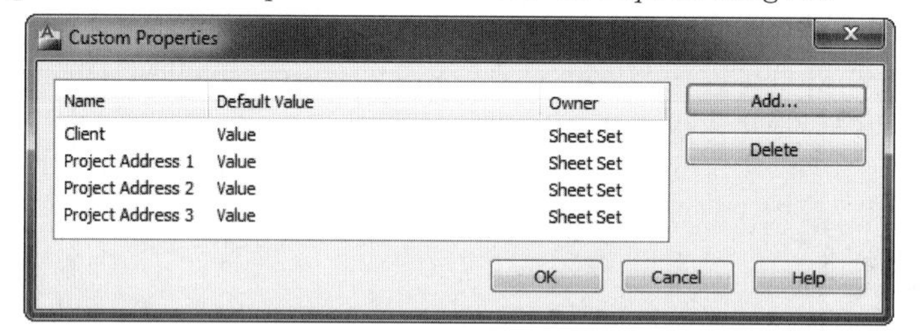

Add displays the Add Custom Property dialog box.

Delete removes the selected custom property without warning; click Cancel to undo erasure.

ADD CUSTOM PROPERTY DIALOG BOX

Access this dialog box through the Add button on the Custom Properties dialog box:

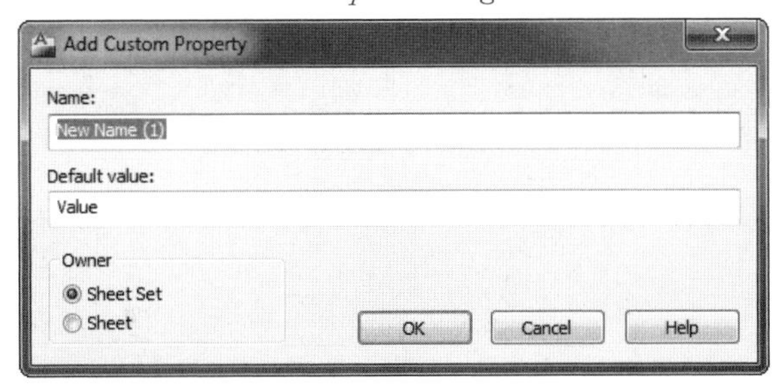

Name names the custom property.

Default value specifies the default value.

⦿ **Sheet set** indicates that the custom property belongs to the sheet set.

○ **Sheet** indicates that the custom property belongs to the sheet.

IMPORT OPTIONS DIALOG BOX

Access this dialog box through the Import Options button on the Choose Layouts page:

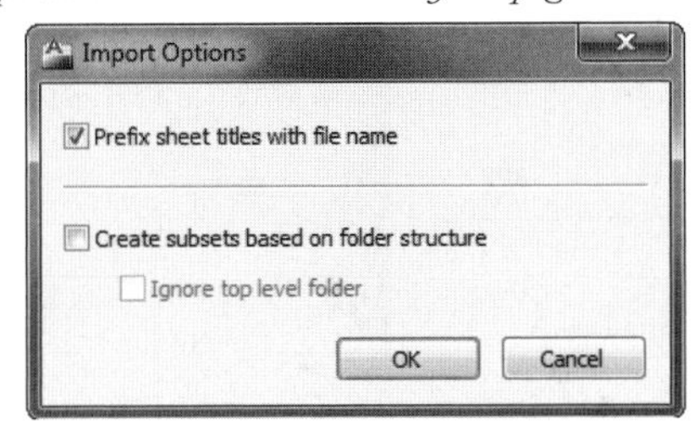

☑ **Prefix sheet titles with file name** tags sheet set names with file names.

☐ **Create subsheets based on folder structure** generates sheets based on folder names.

☐ **Ignore top level folder** ignores the topmost folder in sheet name generation.

RELATED COMMANDS

OpenSheetset opens existing sheet sets.

Sheetset displays the Sheet Set Manager palette.

NewShot

2009 Creates named views and animations with show motion options.

Command	Aliases	Keystroke Shortcuts	Menu	Ribbon
newshot	...	...	...	...

Command: newshot

Displays dialog box.

DIALOG BOX OPTIONS

The content of this dialog box varies with the type of motion shot created and whether the drawing is in model or layout tab.

View Name names the show motion animation.

View Category collects multiple views into a single category name.

View Type toggles between still and animation views:

- **Cinematic** creates moving animations.
- **Still** creates still animations of a given duration.

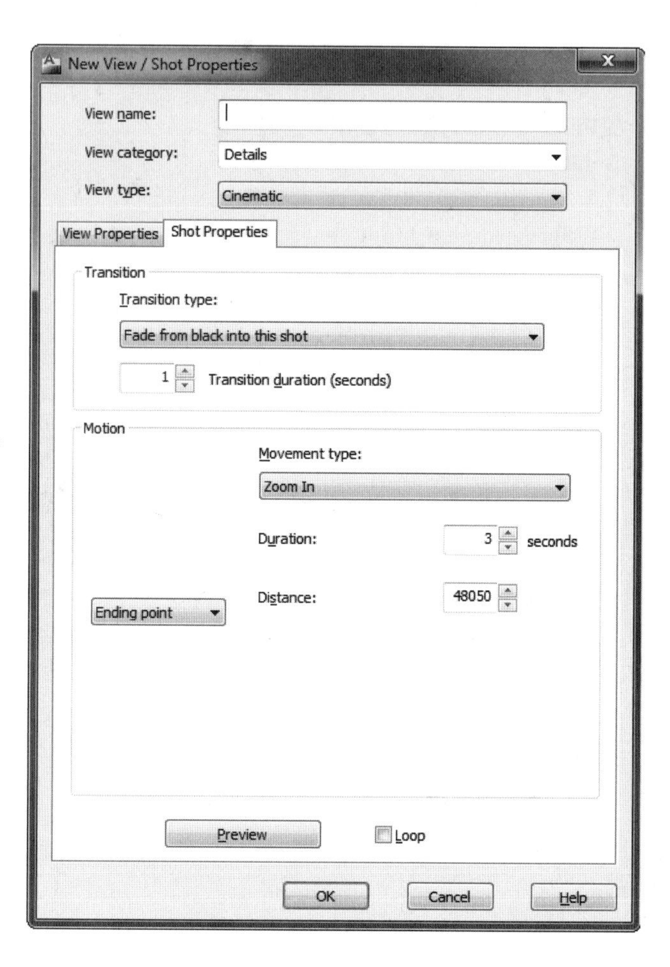

Shot Properties tab

For the View Properties tab, see the View command.

Transition Type specifies the type of transition at the start or end of the animation:

- Cue to shot (no transition); only transition available in layout tabs.
- Fade from white.
- Fade from black.

Transition Duration specifies the length of transition in seconds.

Motion options

Movement specifies the movement during the animation segment.

In model space:

- **Zoom In** or **Zoom Out**.
- **Track Left** or **Track Right** (move).
- **Crane Up** or **Crane Down** (elevate).
- **Look** (rotates objects about you).
- **Orbit** (rotates objects in place).

In paper space:

- **Zoom** or **Pan**.

Playback Duration specifies the length of animation in seconds.

Distance specifies the distance to travel during the duration (sets the zoom speed):

- **Starting Point** specifies that the current view is the start of the animation.
- **Half-way Point** specifies that the current view is at half-way through the animation.

- **Ending Point** specifies that the current view is at the end of the animation.

Crane Up only:

Distance Up specifies the distance the camera moves up.

Distance Back specifies the distance the camera moves back.

Crane Down only:

Distance Forward specifies the distance the camera moves forward.

Distance Down specifies the distance the camera moves down.

☑ **Always Look at Pivot Point** points the camera at the viewpoint.

Look and Orbit only:

Degrees Left / Right specifies the rotation around the z axis.

Degrees Up / Down specifies the rotation around the x,y-plane.

Pan and Zoom only:

Distance (Pan) Left / Right specifies the distance to pan left or right.

Distance (Pan) Up / Down specifies the distance to pan up or down.

Percentage In / Out specifies the percentage to zoom in or out.

Crane Up/Down and Track Left/Right only:

☑ **Always Look at Pivot Point** points the camera at the viewpoint.

Preview dismisses the dialog box to show the transition and motion.

☑ **Loop** repeats the animation until you press Esc.

RELATED COMMANDS

AllPlay plays all show motion views in sequence, without prompting.

EditShot prompts for a view name at the command line, and then displays the Shot Properties tab of the View / Shot Properties dialog box.

NavSMotion command displays the show motion interface.

SequencePlay plays all views in a category.

ViewPlay plays the animation associated with a named view.

TIPS

- Use the ViewPlay or NavSMotion commands to play back the show motion animations created with this command.
- View *categories* were originally designed for sheet sets, but now are adapted for show motion.
- Show motion animation can only be played back inside of AutoCAD; use the AniPath command to create animation that can be saved to external movie files.

Moved Command

NewView is a shortcut to the New option of the View command; it displays the New View dialog box. See the View command. NewView was introduced but undocumented in AutoCAD 2007, and then documented in AutoCAD 2008.

ObjectScale

2008 Adds and removes scale factors from annotatively-scaled objects.

Commands	*Aliases*	*Menu*	*Ribbon*
'objectscale	...	**Modify** ↳**Annotative Object Scale** ↳**Add/Delete Scales**	**Annotate** ↳**Annotation Scaling** ↳**Add/Delete Scales**
'-objectscale	...	...	...

Command: objectscale
Select annotative objects: *(Select one or more objects.)*
Select annotative objects: *(Press Enter.)*

Displays dialog box.

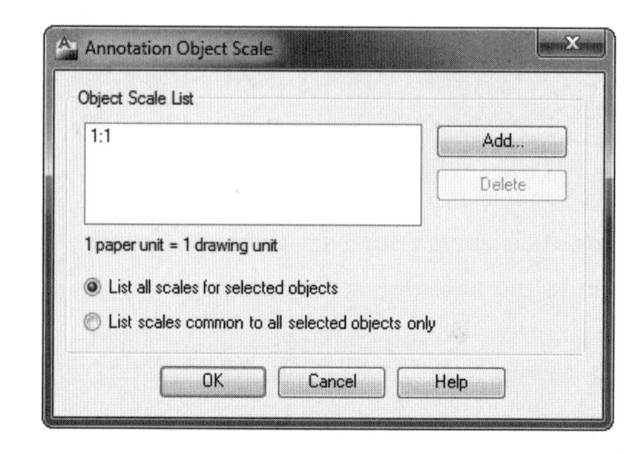

COMMAND LINE OPTION

Select annotative objects chooses annotatively-scaled objects; press Ctrl+A to select everything. (AutoCAD filters out non-annotative objects.)

ANNOTATION OBJECT SCALE DIALOG BOX

Add displays the Add Scales to Object dialog box; detailed later.

Remove removes the selected scale factors.

⊙**List All Scales for Selected Objects** lists scale factors associated with all objects.

○**List Scales Common to All Selected Objects Only** lists scale factors common to all objects.

ADD SCALES TO OBJECT DIALOG BOX

Access this dialog box through the Add button of the Annotation Object Scale dialog box:

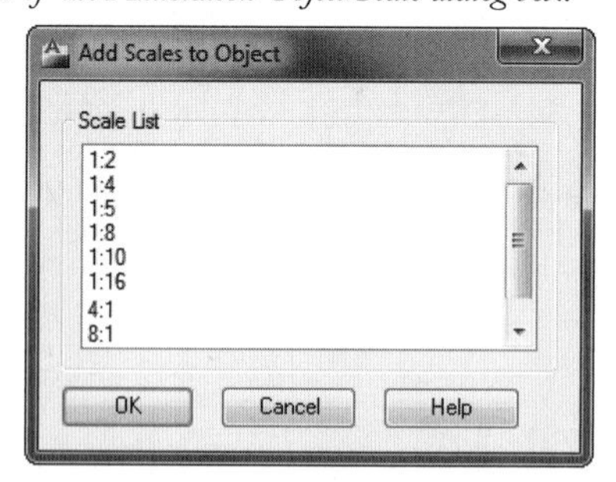

Select one or more scale factors, and then click OK. To select more than one, hold down the Ctrl key.

-OBJECTSCALE command

Command: -objectscale
Select annotative objects: *(Select one or more objects.)*
Select annotative objects: *(Press Enter.)*
Enter an option [Add/Delete] <Add>: *(Type A or D.)*

COMMAND LINE OPTIONS

Add adds scales to the selected objects; AutoCAD prompts, 'Enter named scale to add or [?] '.

Remove removes scale factors; AutoCAD prompts, 'Enter named scale to delete or [?] '.

RELATED COMMANDS

AiObjectScaleAdd adds user-defined scales to annotative objects.

AiObjectScaleRemove removes scales from annotative objects.

AnnoReset resets the scale and location of selected annotative objects to the current setting.

AnnoUpdate updates annotative objects to match the properties of their respective styles.

ScaleListEdit edits the scale factors listed by this command.

RELATED SYSTEM VARIABLES

AnnoAllVisible toggles the display of annotative objects not at the current annotation scale.

AnnoAutoScale adds scale factors automatically to annotative objects when the annotation scale is changed.

CAnnoScale reports the name of the current annotation scale for the current space; a separate scale is used for model space and each paperspace viewport.

CAnnoScaleValue reports the value of the current annotation scale.

HideXrefScales suppresses scale factors imported from xref'ed drawings.

SaveFidelity toggles whether drawings are saved with visual fidelity for earlier releases.

SelectionAnnoDisplay toggles whether alternate scale representations are displayed faded when annotative objects are selected; amount of fading is specified by the XFadeCtl system variable.

TIPS

- Use the drawing bar to change annotation scales on the fly.
- Icons on the drawing bar vary according to the model and layout status.

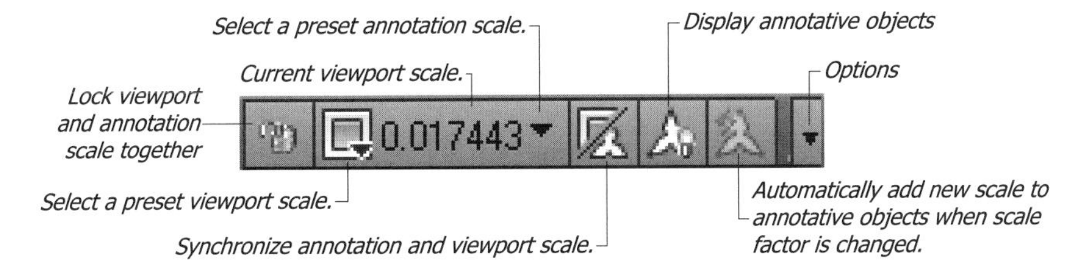

- The ScaleListEdit command allows you to edit the scale factor used by drawings.
- As of AutoCAD 2011, the New command adds metric or Imperial scale factors to template-less drawings based on the setting of the Measure system variable.
- Each drawing contains its own list of scale factors used for annotative objects, viewport scaling, and plotting/publishing.
- The tech editor warns against turn on the Automatically Add New Scale button.

Offset

Ver. 2.5 Draws parallel lines, arcs, circles and polylines; repeats automatically until canceled.

Command	*Aliases*	*Keystroke Shortcuts*	*Menu*	*Ribbon*
offset	**o**	**...**	**Modify** **↳Offset**	**Home** **↳Modify** **↳Offset**

Command: offset
Specify offset distance or [Through/Erase/Layer] <1.0000>: *(Enter a distance or an option.)*
Select object to offset or [Exit/Undo]: *(Select an object.)*
Specify point on side to offset or[Exit/Multiple/Undo]: *(Pick a point.)*
Select object to offset or [Exit/Undo]: *(Select another object, or press Esc to end the command.)*

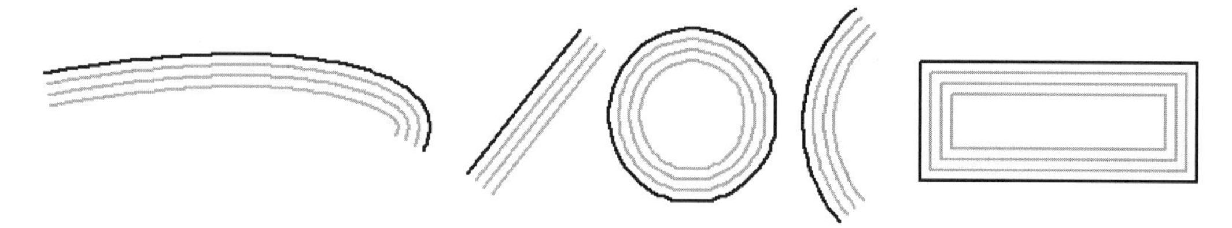

Offset objects shown in gray, with original objects shown black.

COMMAND LINE OPTIONS

Offset distance specifies the perpendicular distance to offset.

Through indicates the offset distance.

Erase erases the original object.

Layer specifies the layer on which to place offset objects.

Exit exits the command, as does Esc.

Undo undoes the last offset operation.

Multiple repeats the offset from the last object.

RELATED COMMANDS

Copy creates one or more copies of a group of objects.

MLine draws up to 16 parallel lines.

RELATED SYSTEM VARIABLES

OffsetDist specifies the current offset distance.

OffsetGapType determines how to close gaps created by offset polylines.

TIPS

- Offsets of curved objects change their radii.
- AutoCAD complains "Unable to offset" when curved objects are smaller than the offset distance.

OleLinks

Rel. 13 Changes, updates, and cancels OLE links between the drawing and other Windows applications (short for Object Linking and Embedding LINKS).

Command	*Aliases*	*Keystroke Shortcuts*	*Menu*	*Ribbon*
olelinks	...	...	**Edit** ↳**OLE Links**	...

Command: olelinks

When no OLE links are in the drawing, the command does nothing.

When at least one OLE object is in the drawing, displays dialog box.

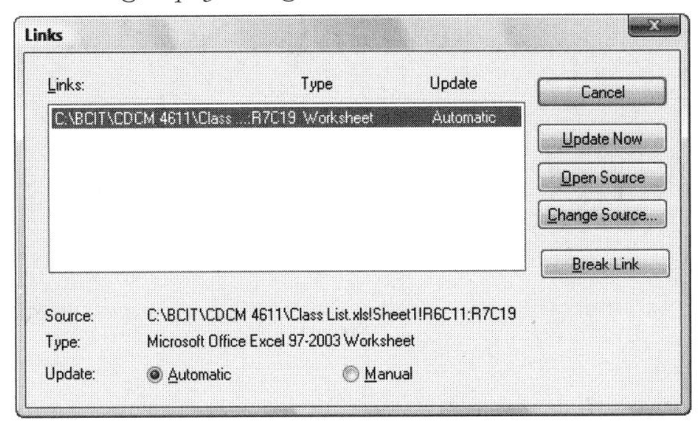

LINKS DIALOG BOX

Links displays a list of linked objects: source file name, type of file, and update mode — automatic or manual.

Update selects either automatic or manual updates.

Update Now updates selected links.

Open Source starts the source application program.

Break Link cancels the OLE link; keeps the object in place.

Change Source displays the Change Source dialog box.

RELATED COMMANDS

InsertObj places an OLE object in the drawing.

PasteSpec pastes objects from the Clipboard as linked objects in the drawing.

OleScale specifies the size, scale, and properties of selected OLE objects.

RELATED SYSTEM VARIABLES

MsOleScale specifies the scale of OLE objects in model space.

OleHide toggles the display of OLE objects.

OleQuality determines the plot quality of OLE objects.

OleStartup loads the source applications for embedded OLE objects before plotting.

RELATED WINDOWS COMMANDS

Edit | Copy copies objects from the source application to the Clipboard.

File | Update updates the linked object in the source application.

OleScale

2000 Modifies the properties of OLE objects.

Command	Aliases	Keystroke Shortcuts	Menu	Ribbon
olescale	...	...	...	...

Select an OLE object before starting this command:

Command: olescale

Displays dialog box.

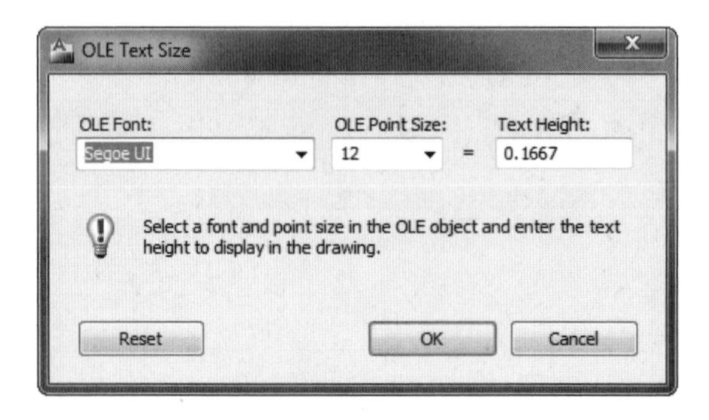

OLE TEXT SIZE DIALOG BOX

The following options apply to text placed as OLE objects in drawings.

OLE Font displays the fonts used by the OLE object, if any.

OLE Point Size displays the text height in point sizes; limited to the point sizes available for the selected font, if any (1 point = $^1/_{72}$ inch).

Text Height specifies the text height in drawing units.

RELATED SYSTEM VARIABLES

MsOleScale specifies the scale of OLE objects in model space.

OleHide toggles the display of OLE objects in the drawing and in plots.

OleQuality specifies the quality of displaying and plotting OLE objects.

OleStartup loads the source application of embedded OLE objects for plots.

RELATED COMMANDS

MsOleScale sets the scale for text in OLE objects.

OleConvert converts OLE objects, if possible (undocumented).

OleOpen opens OLE objects in their source applications (undocumented).

OleReset resets OLE objects (undocumented).

InsertObj inserts an OLE object into the drawing.

OleLinks modifies the link between the object and its source.

PasteSpec allows you to paste an object with a link.

TIPS

- Change OleStartup to 1 to load the OLE source application, which may help improve the plot quality of OLE objects.
- The OLE Plot Quality list box determines the quality of the pasted object when plotted. I recommend the Line Art setting for text, unless the text contains shading and other graphical effects.

Oops

Ver. 1.0 Restores the last-erased group of objects; restores objects removed by the Block and -Block commands.

Command	*Aliases*	*Keystroke Shortcuts*	*Menu*	*Ribbon*
oops	...	...	...	...

Command: oops

COMMAND LINE OPTIONS

None.

RELATED COMMANDS

Block, **Erase**, and **WBlock** can use this command to return erased objects.

U undoes the most recent command.

TIP

- This command restores only the most-recently erased object; use the Undo command to restore earlier objects.

Open

Rel. 12 Loads one or more drawings into AutoCAD.

Command	Alias	Keystroke Shortcuts	Menu	Quick Access Toolbar
open	**openurl**	**Ctrl+O**	**File** ↳**Open**	

Command: open

Displays dialog box:

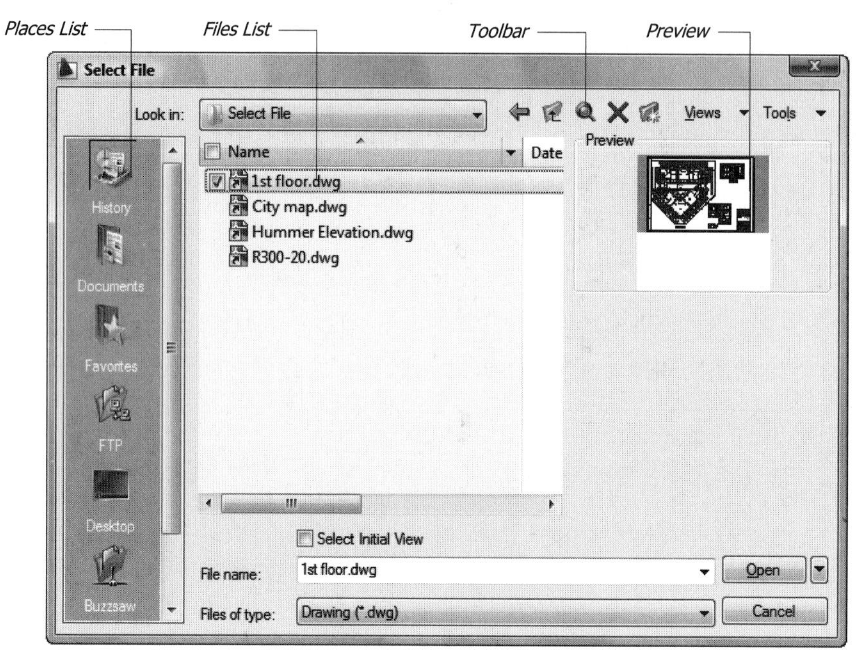

SELECT FILE DIALOG BOX

Look in selects the network drive, hard drive, or folder (subdirectory).

Preview displays a preview image of AutoCAD drawings.

☐ **Select initial view** selects a named view if the drawing has saved views. (After drawing is opened, AutoCAD displays the Select Initial View dialog box listing named views.)

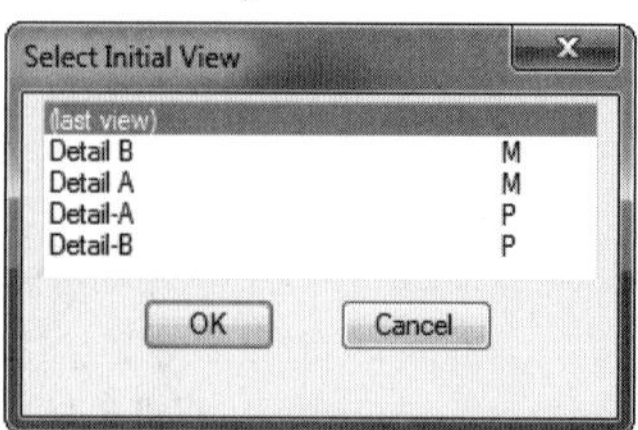

M indicates a view created in model space.

P indicates a view created in paper space (layout mode).

File name names the selected drawing.

Files of type specifies the type of file:

- **Drawing (*.dwg)** AutoCAD drawing file.
- **Standard (*.dws)** drawing standards file; see the Standards command.
- **DXF (*.dxf)** drawing interchange file; see the DxfIn command.
- **Drawing Template File (*.dwt)** template drawing file; see the New command.

Open opens the selected drawing file(s). To open more than one drawing at a time, hold down the Shift key to select a contiguous range of files, and/or hold down the Ctrl key to select two or more non-contiguous files.

Cancel dismisses the dialog box without opening a file.

Open Button

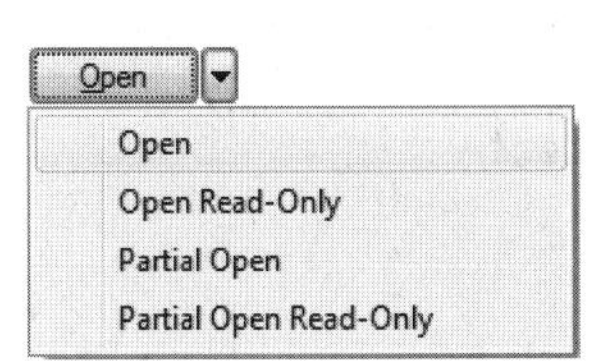

Open opens the drawing.

Open as read-only loads the drawing, but you cannot save changes to the drawing except under another file name. AutoCAD displays "Read Only" on the title bar.

Partial Open loads selected layers or named views; displays the Partial Open dialog box; not available for *.dxf* and template files.

Partial Open Read-Only partially loads the drawing in read-only mode.

PARTIAL OPEN DIALOG BOX

Access this dialog box through the Partial Open option of the Open button:

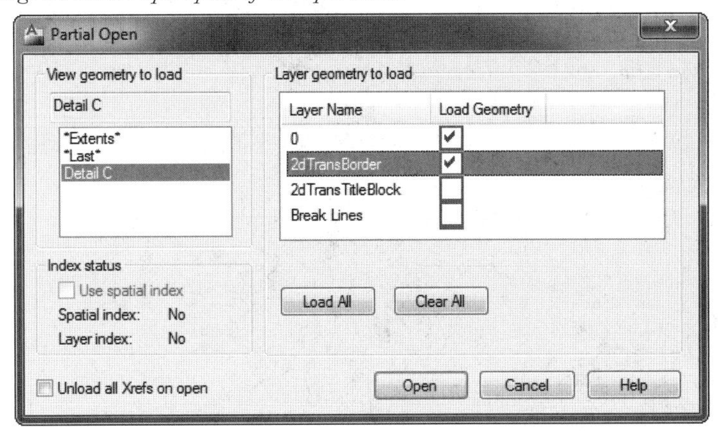

View geometry to load selects the model space views to load; paper space views are not available for partial loading.

Layer Geometry to Load

Load Geometry selects the layers to load.

Load All loads all layers.

Clear All deselects all layer names.

Index Status

Available only when the drawing was saved with spatial indices.

☐ **Use spatial index** determines whether to use the spatial index for loading, if available.

Spatial Index indicates whether the drawing contains the spatial index.

Layer Index indicates whether the drawing contains the layer index.

☐ **Unload all xrefs on open** loads externally-referenced drawings when opening the drawing.

Open opens the drawing, and then partially loads the geometry.

DIALOG BOX TOOLBAR

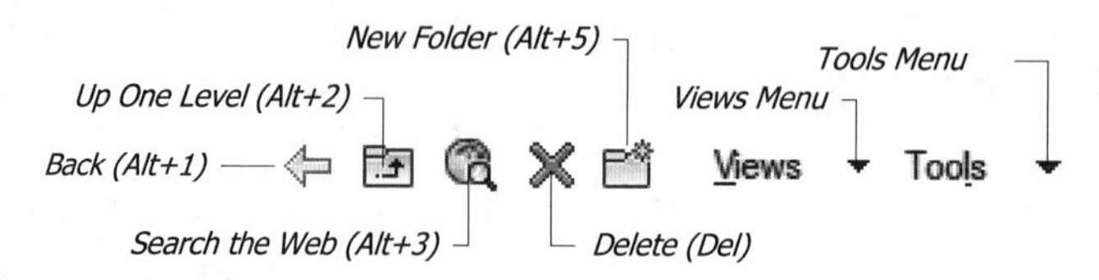

Back returns to the previous folder (keyboard shortcut Alt+1).

Up moves up one level to the next folder or drive (Alt+2).

Search the Web displays the Browse the Web window (Alt+3); see the Browser command.

Delete removes the selected file(s); does not delete folders or drives (Del).

Create New Folder creates new folders (Alt+5).

Views provides display options:

- **List** displays the file and folder names only.
- **Details** displays file and folder names, type, size, and date.
- **Thumbnails** displays thumbnail images of *.dwg* files.
- **Preview** toggles display of the preview window.

Tools provides file-oriented tools:

- **Find** displays the Find dialog box for searching files.
- **Locate** searches for the file along AutoCAD's search paths.
- **Add/Modify FTP Locations** stores the logon names and passwords for FTP (file transfer protocol) sites.
- **Add Current Folder to Places** adds the selected folders to the Places sidebar.
- **Add to Favorites** adds the selected files and folders to the Favorites list.

PLACES LIST

History displays files opened by AutoCAD during the last several weeks.

Documents displays files and folders in the *my documents* folder.

Favorites displays files and folders in the *favorites* folder.

Desktop displays the contents of the *desktop* folder.

FTP displays the FTP Locations list.

Buzzsaw goes to the www.autodesk.com/buzzsaw Web site.

(The Point A and RedSpark destinations were removed from AutoCAD 2004.)

SHORTCUT MENUS

Places List menu

Right-click icons in the Places list.

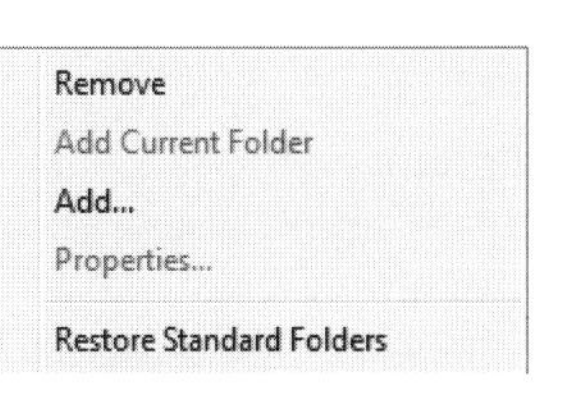

Remove removes a folder from the list.

Add Current Folder adds the selected folder to the list; you can also drag a folder from the file list into the Places list.

Add displays the Add Places Item dialog box. (It's much easier simply to drag a folder onto the Places list.)

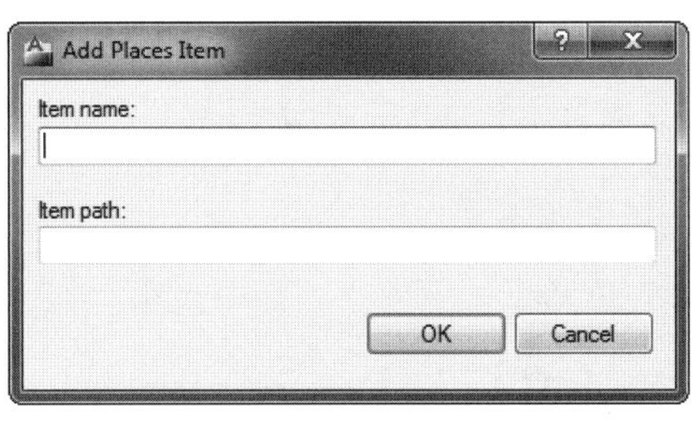

Properties displays the Places Item Properties dialog box, which is identical to the Add Places Item dialog box, and is available only for items you've added.

Restore Standard Folders restores the folders shown above.

File List

Right-click the File list without selecting a file or folder (contents may vary).

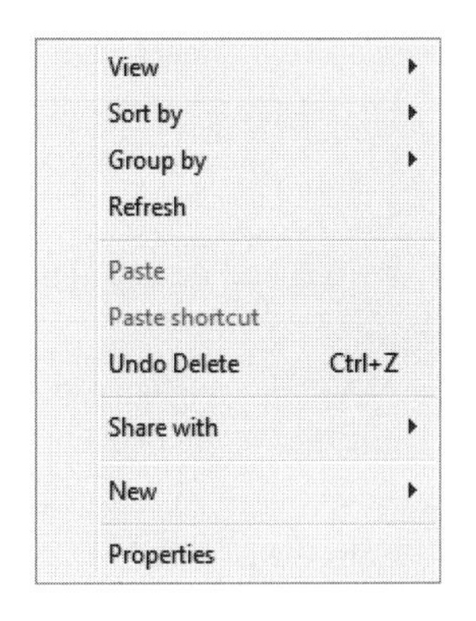

View switches file name views among icons, lists, thumbnails, and details.

Sort /Group By arranges file names using a variety of methods.

Refresh updates the folder listing.

Paste pastes a file from the Clipboard.

Paste Shortcut pastes a file from the Clipboard as a shortcut.

Undo undoes the last operation.

Share With shares the folder over the network.

New creates a new folder (subdirectory) or shortcut.

Properties displays the Properties dialog box of the folder selected by the Look In list.

Right-click a file or folder name.

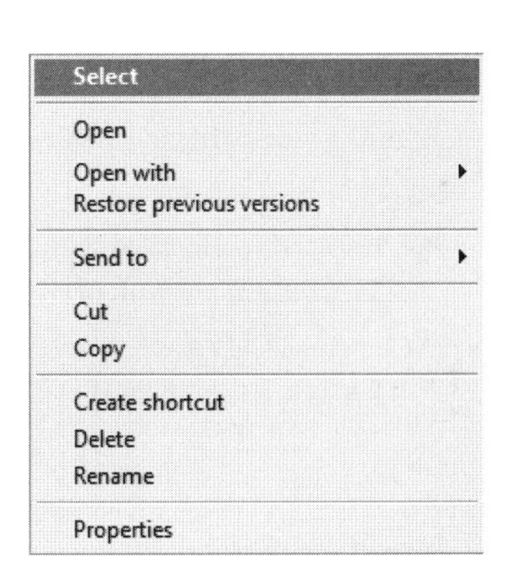

Select opens the drawing in AutoCAD.

Open opens the drawing in AutoCAD also.

Open With chooses another program with which to open this file.

Send To copies the file to another drive; may not work with some software.

Cut cuts the file to the Clipboard.

Copy copies the file to the Clipboard.

Create Shortcut creates a shortcut icon for the selected file.

Delete erases the file; displays a warning dialog box.

Rename renames the file; displays a warning dialog box if you change the extension.

Properties displays the Properties dialog box in read-only mode; use the DwgProps command within AutoCAD to change the settings.

Command Line Options

When FileDia = 0:

Command: open

Enter name of drawing to open <.>: *(Enter name of drawing file.)*

Opening an AutoCAD format file

Enter name specifies the (optional) path and name of the *.dwg*, *.dxf*, or *.dws* file.

If you leave out the extension, AutoCAD assumes a *.dwg* file.

Enter tilde (~) to display the Select File dialog box.

RELATED SYSTEM VARIABLES

TaskBar displays a button on the Windows taskbar for every open drawing.

DbMod indicates whether the drawing has been modified.

DwgCheck checks if the drawing was last edited by AutoCAD.

DwgName contains the drawing's filename.

DwgPrefix contains the drive and folder of the drawing.

DwgTitled indicates whether the drawing has a name other than *drawing1.dwg*.

FileDia toggles the interface between dialog boxes and the command-line.

FullOpen indicates whether the drawing is fully or partially opened.

OpenPartial determines whether drawings can be edited when not yet fully loaded.

RELATED COMMANDS

FileOpen opens drawings without the dialog box.

Import opens drawing files in other formats: 3DS, DGN, SAT, and WMF.

SaveAs saves drawings with new names.

PartiaLoad loads additional portions of partially-opened drawings.

TIPS

- When drawing file names are dragged from Windows Explorer into open drawings, they are inserted as blocks; when dragged to AutoCAD's title bar, they are opened as drawings.
- You can double-click *.dwg* files in Windows Explorer to open them in AutoCAD. If AutoCAD is not yet running, it starts with the last-used profile and vertical application.
- DXF and template files cannot be partially opened.
- After a drawing is partially opened, use PartiaLoad to load additional parts of the drawing.
- When a partially-opened drawing contains a bound xref, only the portion of the xref defined by the selected view is bound to the partially-open drawing.
- You can drag folders into the Places List for quick access to frequently-used files. The same set of folders appears in all of AutoCAD's file-related dialog boxes. The folders in the Places List can be dragged around for a different order. They can be renamed without affecting the folder's actual name.
- As of AutoCAD 2011, this command allows you to ignore all missing SHX fonts and shapes when FontAlt system variable is set to '.' (none).

Aliased Command

OpenUrl was removed from AutoCAD 2000; it is now an alias for the Open command.

OpenDwfMarkup

2005 Opens *.dwf* / *.dwgfx* markup files in AutoCAD, and then loads the Markup Set Manager palette.

Command	*Aliases*	*Keystroke Shortcuts*	*Menu*	*Ribbon*
opendwfmarkup	...	...	**File** **↳Load Markup Set**	...

Command: opendwfmarkup

Displays the Open Markup DWF dialog box. Select a .dwf or .dwfx file, and then click Open.

If the .dwf/.dwfx file contains no markup data, AutoCAD complains:

Filename.dwf does not contain any markup data. Would you like to open this DWF file in the viewer?

Click Yes to open the file in Design Review (nee DWF Composer); click No to cancel.

COMMAND OPTIONS

See the Markup command.

RELATED COMMANDS

DwfAttach attaches DWF files as underlays, including unmarked-up files.

Markup displays the Markup Set Manager, and opens *.dwf* files containing markup data.

MarkupClose closes the Markup Set Manager palette opened by this command

RELATED SYSTEM VARIABLE

MsmState reports whether the Markup Set Manager palette is open.

TIPS

- MSM is short for "markup set manager." DWF is short for "design Web format."
- This command's purpose is to work with marked-up DWF-format files created by the Design Review software, included free with AutoCAD or available from www.autodesk.com/designreview.
- This command replaces the RmlIn command, whose purpose was to import now-obsolete *.rml* redline markup files created by the also now-obsolete Volo View.

OpenSheetset

2005 Loads sheet sets into the current drawing.

Commands	*Aliases*	*Keystroke Shortcuts*	*Menu Bar*	*Application Menu*
opensheetset	...	...	**File** ↳**Open Sheet Set**	**Open** ↳**Sheet Set**
-opensheetset	...	...	...	...

Command: opensheetset

Displays Open Sheet Set dialog box. Select a .dst file, and then click Open.

AutoCAD displays the Sheet Set Manager; see the SheetSet command.

-OPENSHEETSET Command

Command: -opensheetset

Enter name of sheet set to open: *(Enter .dst file name.)*

Enter the path and name of a *.dst* file, or enter tilde (~) to display the file dialog box.

RELATED COMMANDS

NewSheetset creates new sheet sets.

Sheetset opens the Sheet Set Manager palette.

SheetsetHide closes the Sheet Set Manager palette.

RELATED SYSTEM VARIABLES

SsFound records path and file name of the sheet set.

SsLocate toggles whether AutoCAD opens sheet sets associated with the drawing.

SsmAutoOpen toggles whether AutoCAD displays the Sheet Set Manager palette when the drawing is opened.

SmsState reports whether the Sheet Set Manager is active.

FileDia determines whether this command displays a dialog box or command-line prompts.

RELATED COMMAND LINE SWITCHES

These switches are used by the Target field on Shortcut tab in AutoCAD's desktop icon's Properties dialog box:

/nossm prevents Sheet Set Manager palette from loading.

/set specifies the *.dst* sheet set file to load.

TIPS

- The Sheetset command automatically opens the previously-opened sheet set file.
- SSM is short for "sheet set manager."

Options

2000 Sets system and user preferences.

Commands	*Aliases*	*Keystroke Shortcuts*	*Menu Bar*	*Application Menu*
options	**op** **gr, ddgrips** **preferences** **ddselect**	...	**Tools** **↳Options**	**Options**
+options	...	...	...	...

Command: options

Displays dialog box.

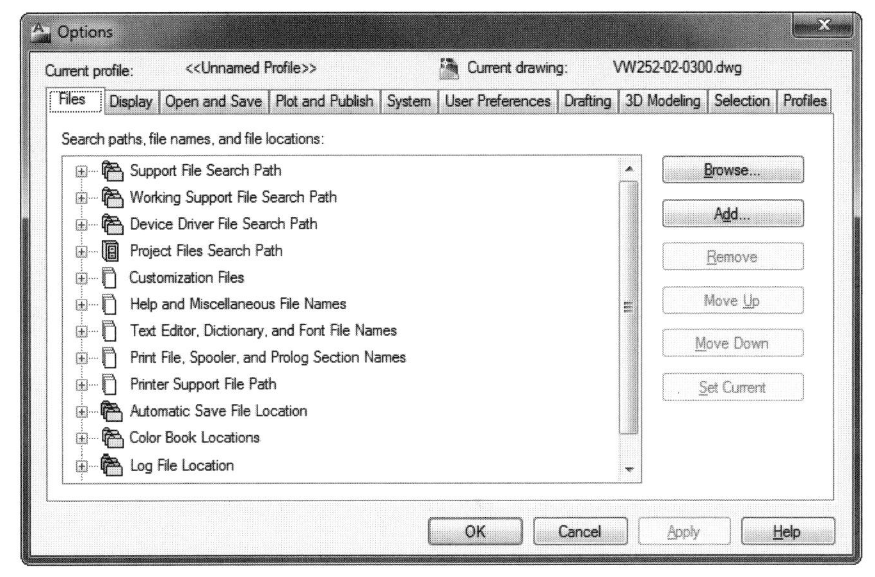

To see more of the Files tab, you can extend the length of the dialog box by dragging down its bottom edge.

— indicates setting is saved to the related system variable in the current drawing.

OPTIONS DIALOG BOX

OK applies the changes, and closes the dialog box.

Apply applies the changes, and keeps the dialog box open.

Files tab

Search paths, file names, and file locations specifies folders and files used by AutoCAD.

Browse displays dialog box for selecting folders or files.

Add adds an item below the selected path or file name.

Remove removes the selected item without warning; click Cancel to undo the removal.

Move Up/Move Down moves the selected item above or below the preceding item; applies to search paths only.

Set Current makes the selected project names and spelling dictionaries current.

Display tab

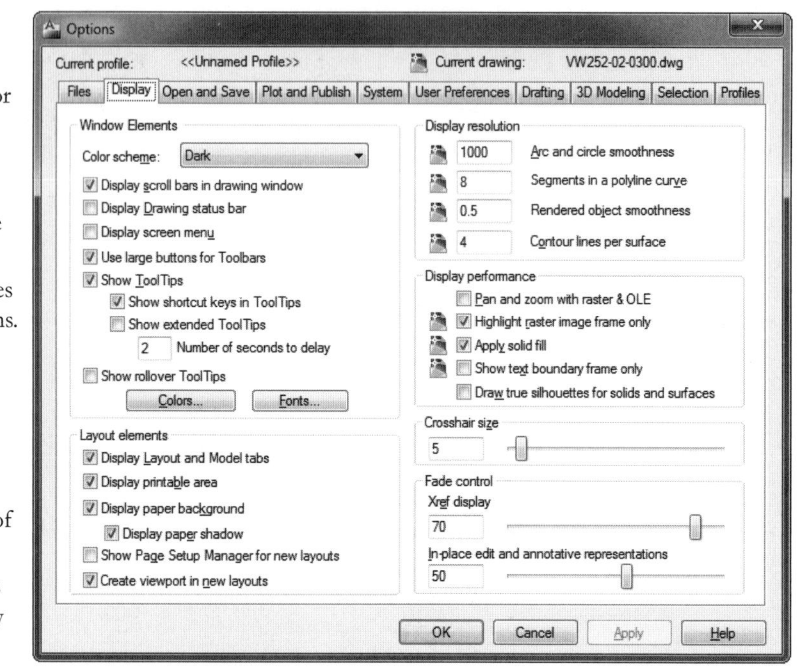

Window Elements options

Color Scheme changes the theme color of AutoCAD's interface between dark and light.

☐ **Display Scroll Bars in Drawing Window** toggles the presence of the horizontal and vertical scroll bars.

☐ **Display Drawing Status Bar** toggles the status bar with annotation buttons.

☐ **Display Screen Menu** toggles the presence of the screen menu.

☐ **Use Large Buttons for Toolbars** switches toolbar buttons between normal and large size.

☑ **Show Tooltips** toggles the display of tooltips.

☑ **Show Shortcut Keys in Tooltips** toggles the display of shortcut key strokes in tooltips.

☑ **Show Extended Tooltips** toggles display of large tooltips (*RolloverTips*).

Number of Seconds to Delay specifies how long AutoCAD waits before displaying an extended tooltip.

☑ **Show Rollover Tooltips** toggles the display of rollover tooltips that report object properties.

Colors specifies colors for user interface; displays Color Options dialog box, detailed later.

Fonts specifies the font for text on the command line; displays the Command Line Window Font dialog box, detailed later.

Layout Elements options

☐ **Display Layout and Model Tabs** toggles the presence of the Model and Layout tabs.

☑ **Display Printable Area** toggles the display of dashed margin lines in layout modes.

☑ **Display Paper Background** toggles the presence of the page in layout modes.

☑ **Display Paper Shadow** toggles the presence of the drop shadow under the page in layout modes.

☐ **Show Page Setup Manager for New Layouts** displays Page Setup dialog box when you create a new layout; this dialog box sets options related to paper and plot settings.

☑ **Create Viewport in New Layouts** toggles automatic creation of a viewport in new layouts.

Display Resolution options

Arc and Circle Smoothness controls the displayed smoothness of circles, arcs, and other curves; range is 1 to 20000 (*ViewRes*); default = 1000.

Segments in a Polyline Curve specifies the number of line segments used to display polyline curves; range is -32767 to 32767 (*SplineSegs*); default = 8.

Rendered Object Smoothness controls the displayed smoothness of shaded and rendered curves; range is 0.01 to 10 (*FacetRes*) default = 0.5.

Contour Lines per Surface specifies the number of contour lines on solid 3D objects; range is 0 to 2047 (*IsoLines*); default = 4.

Display Performance options

☐ **Pan and Zoom with Raster and OLE** toggles the display of raster and OLE images during realtime pan and zoom (*RtDisplay*).

☑ **Highlight Raster Image Frame Only** highlights only the frame and not the entire raster image, when on (*ImageHlt*).

☑ **Apply Solid Fill** toggles the display of solid fills in multilines, traces, solids, solid fills, and wide polylines; this option does not come into effect until you click OK, and then use the Regen command (*FillMode*).

☐ **Show Text Boundary Frame Only** toggles the display of rectangles in place of text; this option does not come into effect until you click OK, and then use the Regen command (*QTextMode*).

☐ **Show Silhouettes in Wireframe** toggles the display of silhouette curves for 3D solid objects; when off, tessellation lines are drawn when hidden-line removal is applied to the 3D object (*DispSilh*).

Crosshair Size specifies the size of the crosshair cursor; range is 1% to 100% of the viewport (*CursorSize*); default = 5%.

Fade Control options

Xref display specifies the amount of fading of attached DWG files; range is 0% to 90%; default = 70%.

In-place Edit and Annotative Representations specifies the amount of fade during in-place reference editing; range is 0% to 90% (*XFadeCtl*); default = 50%.

Open and Save tab

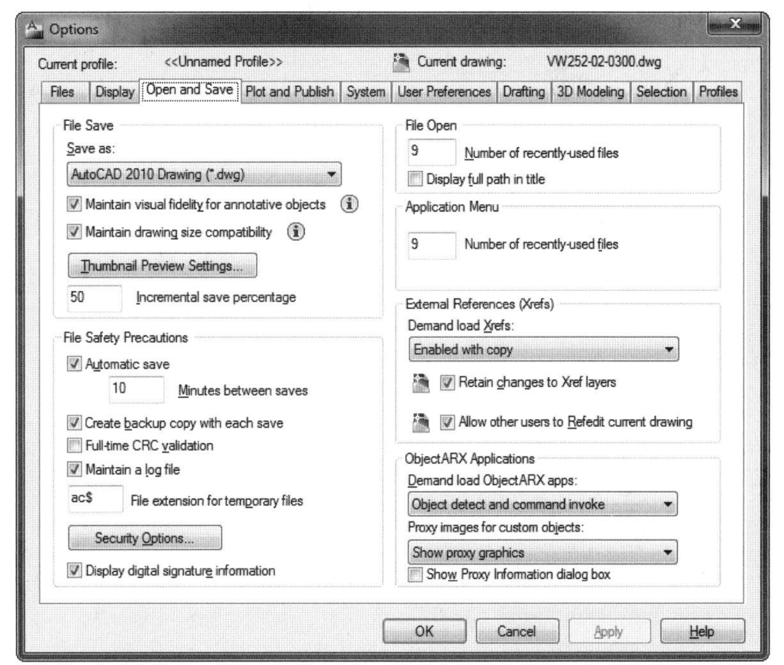

File Save options

Save As specifies the default file format used by the Save and SaveAs commands; default = "AutoCAD 2010 Drawing (*.dwg)."

☑ **Maintain Visual Fidelity for Annotative Objects** places annotatively-scaled objects on their own layers, when exported to AutoCAD 2007 and earlier.

☑ **Maintain Drawing Size Compatibility** toggles use of large objects in drawings; keep turned off to prevent the creation of large objects in drawings, allowing objects to be converted back to AutoCAD 2009 and earlier (*LargeObjectSupport*).

Thumbnail Preview Settings displays Thumbnail Preview Settings dialog box, detailed later.

Incremental Save Percentage indicates the percentage of wasted space allowed in a drawing file before a full save is performed; range is 0% to 100% (*ISavePercent*); default = 50.

File Safety Precautions options

☑ **Automatic Save** saves the drawing automatically at prescribed time intervals (*SaveFile* and *SaveFilePath*).

Minutes Between Saves specifies the duration between automatic saves (*SaveTime*); default = 10 minutes.

☑ **Create Backup Copy with Each Save** creates backup copies when drawings are saved; (*ISavBak*).

☐ **Full-time CRC Validation** performs cyclic redundancy check (CRC) error-checking each time an object is read into the drawing.

☐ **Maintain a Log File** saves the command conversation to a log file (*LogFileMode*); off.

File Extension for Temporary Files specifies the file name extension for temporary files created by AutoCAD (*NodeName*); default = *.ac$*.

Security Options displays the Security Options dialog box; see the SecurityOptions command.

☑ **Display Digital Signature Information** displays digital signature information when opening files with valid digital signatures (*SigWarn*).

File Open options

Number of Recently-used Files to List specifies the number of recently-opened file names to list in the Files menu; minimum = 0; default and maximum = 9.

☐ **Display Full Path in Title** displays the drawing file's path in AutoCAD's titlebar.

Application Menu options

Number of Recently-used Files specifies the number of recently-opened file names to list in the menu browser; minimum = 0; default and maximum = 9.

External References (Xrefs) options

Demand Load Xrefs specifies the style of demand loading of externally-referenced drawings (*XLoadCtl*):

- **Disabled** turns off demand loading.
- **Enabled** turns on demand loading to improve performance, but the drawing cannot be edited by another user; default.
- **Enabled With Copy** turns on demand loading; loads a copy of the drawing so that another user can edit the original.

☑ **Retain Changes to Xref Layers** saves changes to properties for xref-dependent layers (*VisRetain*).

☑ **Allow Other Users to Refedit Current Drawing** allows other users to edit the current drawing when referenced by another drawing (*XEdit*).

ObjectARX Applications options

Demand Load ObjectARX Apps demand-loads an ObjectARx application when the drawing contains proxy objects (*DemandLoad*):

- **Disable Load on Demand** turns off demand loading.
- **Custom Object Detect** demand-loads applications when drawings contain proxy objects.
- **Command Invoke** demand-loads applications when their commands are invoked.
- **Object Detect and Command Invoke** demand-loads the application when the drawing contains proxy objects, or when one of the application's commands is invoked; default.

Proxy Images for Custom Objects specifies how proxy objects are displayed:

- **Do Not Show Proxy Graphics** does not display proxy objects.
- **Show Proxy Graphics** displays proxy objects.
- **Show Proxy Bounding Box** displays a rectangle instead of the proxy object.

☑ **Show Proxy Information Dialog Box** displays a warning dialog box when a drawing contains proxy objects (*ProxyNotice*).

Plot and Publish tab

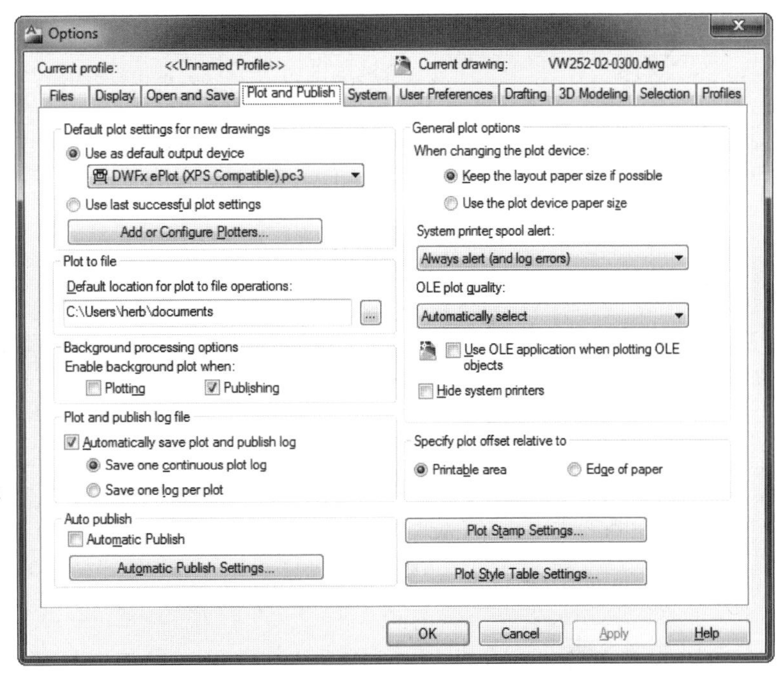

Default Plot Settings for New Drawings options

⦿ **Use As Default Output Device** selects the default output device (printer, plotter, etc).

○ **Use Last Successful Plot Settings** reuses the plot settings from the last successful plot.

Add or Configure Plotters displays Plotter Manager window; see the PlotterManager command.

Plot to File options

Default Location for Plot to File Operations specifies the folder in which to store plot files. (**...** displays dialog box for selecting folder.)

Background Processing options

☐ **Plotting** executes the Plot command in the background (*BackgroundPlot*).

☑ **Publishing** executes the Publish command in the background (*BackgroundPlot*).

Plot and Publish Log File options

☑ **Automatically Save Plot and Publish Log File:**

⦿ **Save One Continuous Plot Log** saves all log data in a single file.

○ **Save One Log Per Plot** saves log data in separate files.

AutoPublish options

☐ **Automatic Publish** exports drawings as DWF, DWFx, or PDF files when saved or closed (*AutoPubFormat*); see the AutoPublish command.

Automatic Publish Settings displays a dialog box of the same name; see the AutoPublish command.

General Plot Options options

When Changing the Plot Device:

⦿ **Keep the Layout Paper Size if Possible** uses the paper size specified by the Page Setup dialog box's Layout Settings tab, if the output device can handle the paper size (*PaperUpdate*); default = on.

○ **Use the Plot Device Paper Size** uses the paper size specified by the *.pc3* plotter configuration file (*PaperUpdate*); default = off.

System Printer Spool Alert displays an alert when a spooled drawing has a conflict:

- **Always Alert (and log errors)** displays alert, and logs the error message.
- **Alert First Time Only (and log errors)** displays the alert once, but logs all error messages.
- **Never Alert (and log first error)** does not display an alert, but log the first error message.
- **Never Alert (do not log errors)** does not display an alert or logs any error messages.

OLE Plot Quality determines the quality of OLE objects when plotted (*OleQuality*); default = Automatically Select; see the OleScale command.

☐ **Use OLE Application When Plotting OLE Objects** launches the application that created the OLE object when plotting a drawing with an OLE object (OleStarup).

☐ **Hide System Printers** hides the names of Windows system printer drivers not specific to AutoCAD.

Specify Plot Offset Relative To options

⊙ **Printable Area** measures offsets to plotter margins.

○ **Edge of Paper** measures offsets to paper edges.

Plot Stamp Settings displays the Plot Stamp dialog box; see the PlotStamp command.

Plot Style Table Settings displays the Plot Style Table Settings dialog box.

System tab

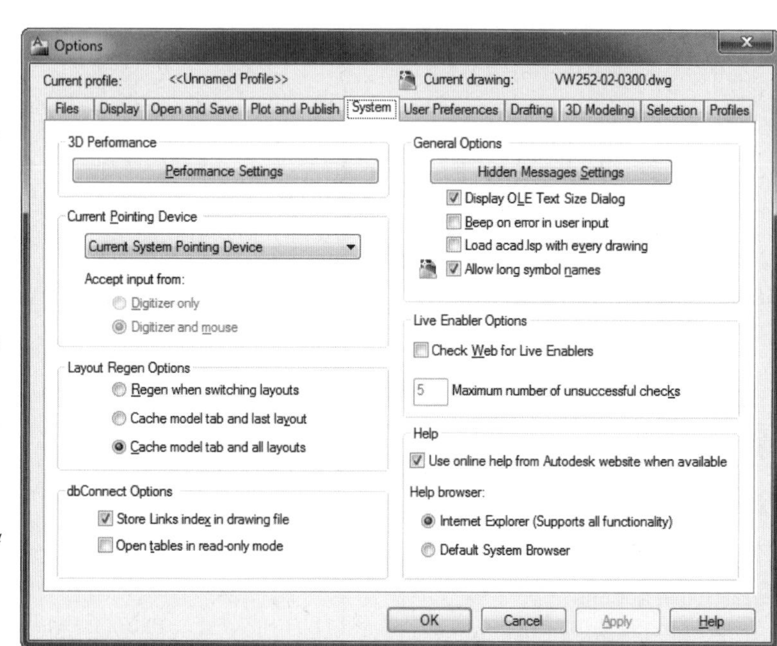

3D Performance options

Performance Settings displays the Adaptive Degradation and Performance Tuning dialog box; see the 3dConfig command.

Current Pointing Device options

Current Pointing Device selects the pointing device driver:

- **Current System Pointing Device** selects the pointing device used by Windows.
- **Wintab Compatible Digitizer** selects a Wintab-compatible digitizer driver.

Accept Input From *(available only when a digitizing tablet is attached):*

○ **Digitizer Only** reads input from the digitizer, and ignores the mouse.

⊙ **Digitizer and Mouse** reads input from the digitizer and the mouse.

Layout Regen options

○ **Regen When Switching Layouts** regenerates the drawing each time layouts are switched.

○ **Cache Model Tab and Last Layout** saves display lists of the model tab and last layout accessed.

⊙ **Cache Model Tab and All Layouts** saves the display list of the model tab and all layouts.

dbConnect Options options

☑ **Store Links Index in Drawing File** ensures faster SQL queries, but creates larger drawing files.

☐ **Open Tables in Read-only Mode** prevents users from writing back to database tables.

General Options options

Hidden Messages Settings lets you turn on warning dialog boxes that have been turned off; displays Show Hidden Message dialog box, detailed later.

☑ **Display OLE Text Size dialog** displays the OLE Properties dialog box after an OLE object is inserted in the drawing; see the OleScale command.

☐ **Beep on error in user input** beeps the computer when AutoCAD detects a user error.

☐ **Load acad.lsp with every drawing** loads the *acad.lsp* file with every drawing (AcadLspAsDoc).

☑ **Allow long symbol names** allows symbol names — such as layers, dimension styles, and blocks — to be up to 255 characters long, and to include letters, numbers, blank spaces, and most punctuation marks; when off, names are limited to 31 characters, and spaces may not be used (*ExtNames*).

The SDI and Startup options were removed from AutoCAD 2008.

Live Enabler Options options

☐ **Check Web for Live Enablers** checks whenever an Internet connection is present.

Maximum number of unsuccessful checks limits how often AutoCAD attempts to call home; default = 5.

Help options *(new to AutoCAD 2011)*

☑ **Use online help from Autodesk website when available** checks for updated help from Autodesk.

Help browser chooses the default browser for displaying help files:

⦿ **Internet Explorer (supports all functionality)**.

○ **Default System Browser** uses another brand of browser, such as Opera, Firefox, or Chrome.

User Preferences tab

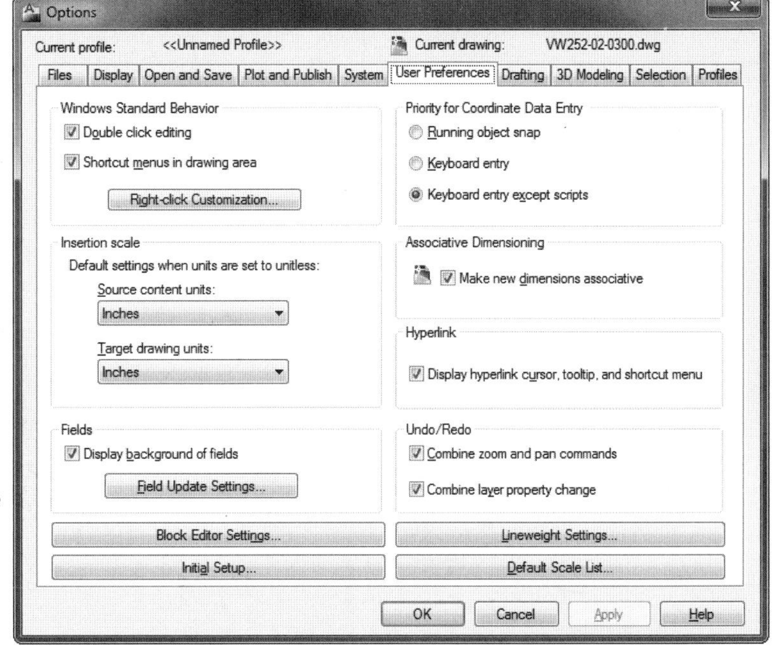

Windows Standard Behavior options

☑ **Double Click Editing** toggles editing objects by double-clicking them; see the Cui command.

☑ **Shortcut Menus in Drawing Area** toggles right-click action in drawing area: when on, displays shortcut menus; when off, equivalent to pressing the Enter key (*ShortCutMenu*).

Right-click Customization displays the Right-Click Customization dialog box, detailed later (*ShortCutMenu*).

Insertion Scale options

Source Content Units specifies the default units when an object is inserted into the drawing from AutoCAD DesignCenter; Unspecified-Unitless means the object is not scaled when inserted (*InsUnitsDefTarget*); default = inches or mm.

Target Drawing Units specifies the default units when "insert units" are not specified by the InsUnits system variable (*InsUnitsDefTarget*); default = inches or mm.

Block Editor Settings specifies initial conditions for the Block Editor; displays the Block Editor Settings dialog box. See the BeSettings command.

Initial Setup specifies basic customization settings for AutoCAD; displays the Initial Setup dialog box. See the InitialSetup command.

Fields options

☑ **Display Background of Fields** displays gray behind field text (*FieldDisplay*).

Field Update Settings displays the Field Update Settings dialog box; see the UpdateField command.

The Hidden Line Settings button was removed from AutoCAD 2007.

Priority for Coordinate Data Entry options

○ **Running Object Snap** means that osnap overrides coordinates entered at the keyboard (*OSnapCoord*).

○ **Keyboard Entry** means that coordinates entered at the keyboard override osnaps .

⊙ **Keyboard Entry Except Scripts** means that coordinates entered at the keyboard override running object snaps, except when coordinates are provided by a script.

Associative Dimensioning options

☑ **Make New Dimension Associative** means that dimensions are associated with objects; when off, dimensions are associated with defpoints (*DimAssoc*).

Hyperlink options

☑ **Display Hyperlink Cursor and Shortcut Menu** displays the hyperlink cursor when the cursor passes over objects containing hyperlinks, and adds the Hyperlink option to shortcut menus when right-clicking objects containing hyperlinks. See the HyperlinkOptions command.

Undo/Reo options

☑ **Combine Zoom and Pan Commands** combines all sequential zooms and pans into a single undo or redo.

☑ **Combine Layer Property Change** combines all changes made to layer properties into a single undo or redo.

Lineweight Settings displays the Lineweight Settings dialog box; see the Lineweight command.

Edit Scale List displays the Edit Scale List dialog box; see the ScaleListEdit command.

Drafting tab

AutoSnap Settings options

☑ **Marker** displays the object snap icon (*AutoSnap*).

☑ **Magnet** turns on the AutoSnap magnet (*AutoSnap*).

☑ **Display AutoSnap Tooltip** displays the AutoSnap tooltip (*AutoSnap*).

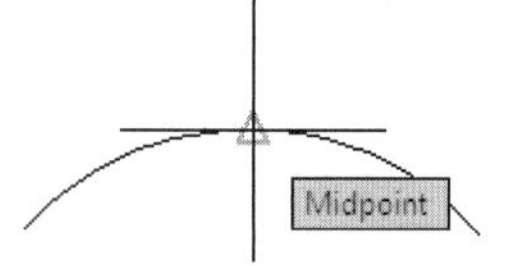

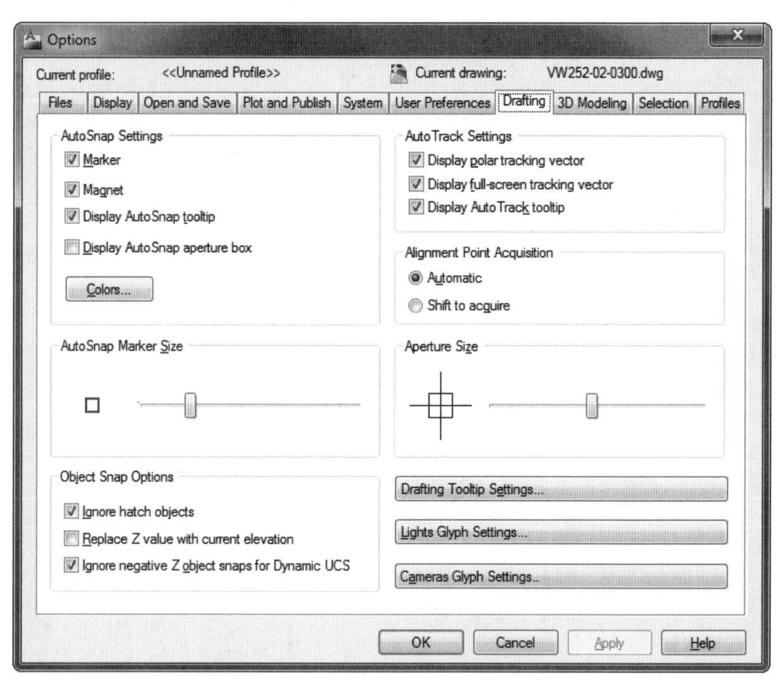

☐ **Display AutoSnap Aperture Box** displays the AutoSnap aperture box (*ApBox*).

Colors specifies the color of the AutoSnap icons; default = color #31. Displays the Drawing Window Colors dialog box.

AutoSnap Marker Size option

AutoSnap Marker Size sets the size for the AutoSnap icon; range is 1 to 20 pixels.

Object Snap Options options

☑ **Ignore Hatch Objects** prevents AutoCAD from snapping to hatch objects.

☐ **Replace Z Value with Current Elevation** makes the z coordinate equal to elevation.

☑ **Ignore Negative Z Object Snaps for Dynamic UCS** does just that (*OsOptions*).

AutoTrack Settings options

☑ **Display Polar Tracking Vector** displays the Polar Tracking vectors at specific angles (*TrackPath*).

☑ **Display Full-screen Tracking Vector** displays the tracking vectors (*TrackPath*).

☑ **Display AutoTrack Tooltip** displays the AutoTrack tooltip (*AutoSnap*).

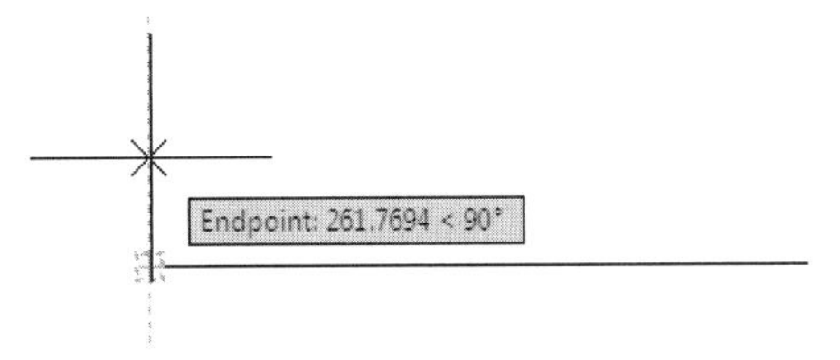

Full-screen tracking vector (dashed line) and AutoTrack tooltip.

Alignment Point Acquisition options

⊙ **Automatic** displays tracking vectors automatically when aperture moves over an object snap.

○ **Shift to acquire** displays tracking vectors when you press Shift and move the aperture over an object snap.

Aperture Size option

Aperture Size sets the size for the aperture; range is 1 to 50 pixels (Aperture); default = 10.

Drafting Tooltip Settings displays the Tooltip Appearance dialog box; see the DSettings command.

Light Glyphs Settings displays the Light Glyphs Appearance dialog box, detailed later.

Camera Glyphs Settings displays the Camera Glyphs Appearance dialog box, detailed later.

3D Modeling tab

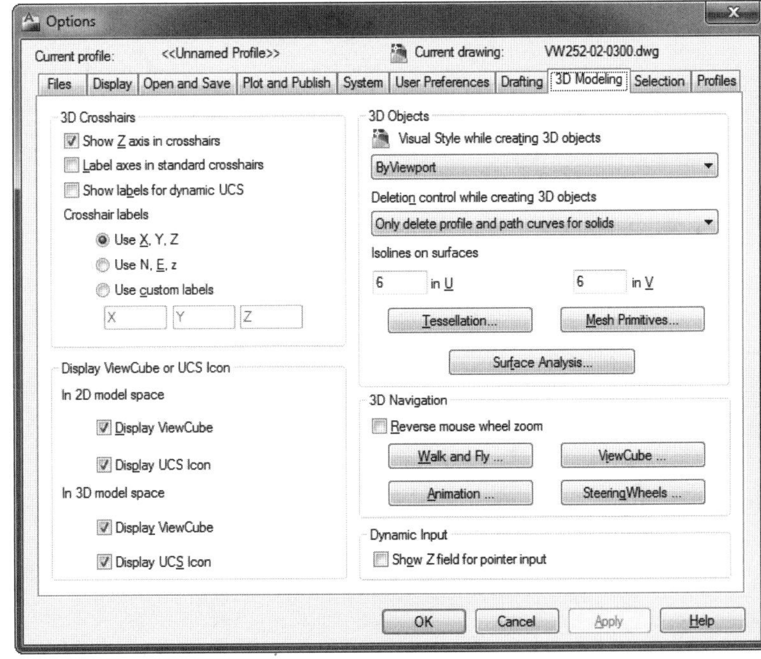

3D Crosshairs options

☑ **Show Z Axis in Crosshairs** toggles display of z axis of the crosshair cursor.

☐ **Label Axes in Standard Crosshairs** toggles whether labels (X, Y, and Z) are displayed with the crosshair cursor.

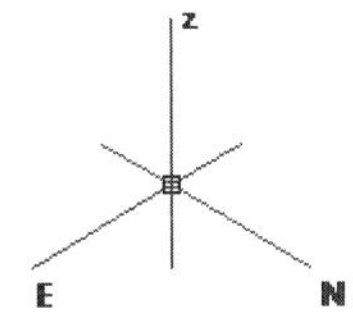

☐ **Show Labels for Dynamic UCS** displays labels during dynamic UCS, even when the axis labels are off.

☐ **Crosshair Labels**:

⊙ **Use X, Y, Z** labels the axes with X, Y, and Z.

○ **Use N, E, z** labels the axes with N (north), E (east), and z.

○ **Use Custom Labels** labels each axis with up to eight user-defined characters.

Display ViewCube or UCS Icon options

In 2D model space *(new to AutoCAD 2011):*

☑ **Display View Cube** toggles display of the ViewCube (*NavVCubeDisplay*).

☑ **Display UCS Icon** toggles display of the UCS icon (*UcsIcon*).

In 3D model space

☑ **Display View Cube** toggles display of the ViewCube (*NavVCubeDisplay*).

☑ **Display UCS Icon** toggles display of the UCS icon (*UcsIcon*).

Dynamic Input option

☐ **Show Z Field for Pointer Input** toggles display of the z coordinate during dynamic input.

3D Objects options

Visual Style While Creating 3D Objects specifies the visual style when you drag 3D primitives into shape (*DragVs*):

- **By Viewport** uses the viewport's setting; see the MView command.
- **2D Wireframe.**
- **3D Hidden.**
- **3D Wireframe.**
- **Conceptual.**
- **Realistic.**

Deletion Control While Creating 3D Objects (*DelObj*):

- **Retain defining geometry.**
- **Delete profile curves.**
- **Delete profile and path curves.**
- **Prompt to delete profile curves.**
- **Prompt to delete profile and path curves.**

Isolines on Surfaces and Meshes specifies the number of isolines in the u and v directions (*SurfU* and *SurfV*).

Tesselation displays the Mesh Tesselation Options dialog box; see the MeshOptions command.

Mesh Primitives displays the Mesh Primitives Options dialog box; see the MeshPrimitiveOptions command.

Surface Analysis displays the Analysis Options dialog box; see the AnalysisOptions command *(new to AutoCAD 2011).*

3D Navigation options

☐ **Reverse Mouse Wheel Zoom** switches the direction of zoom when the mouse wheel is rolled (*ZoomWheel*).

Walk and Fly displays the Walk and Fly Settings dialog box; see the WalkFlySettings command.

Animation displays the Animation Settings dialog box; see the AniPath command.

ViewCube displays the View Cube Settings dialog box; see the NavVCube command.

SteeringWheels displays the Steering Wheels Settings dialog box; see the NavSWheel command .

Selection tab

This tab is also displayed when the old DdGrips command is entered.

Pickbox Size specifies the size of the pickbox; range is 1 to 20 pixels (*PickBox*); default = 3.

Selection Preview options

☑ **When a Command is Active** previews object selection during commands.

☑ **When No Command is Active** previews object selection as cursor passes over objects.

Visual Effect Settings displays the Visual Effects Settings dialog box, detailed later.

Selection Modes options

☑ **Noun/verb Selection** selects an object before executing an editing command (*PickFirst*).

☐ **Use Shift to Add to Selection** requires the Shift key to add and remove objects from selection sets, like Windows (*PickAdd*).

☐ **Press and Drag** creates selection window by dragging (*PickDrag*).

☑ **Implied Windowing** creates a selection window when you pick a point in the drawing that does not pick an object (*PickAuto*).

☑ **Object Grouping** selects the entire group when an object in the group is selected (*PickStyle*).

☐ **Associative Hatch** selects boundary objects, along with the associative hatch patterns.

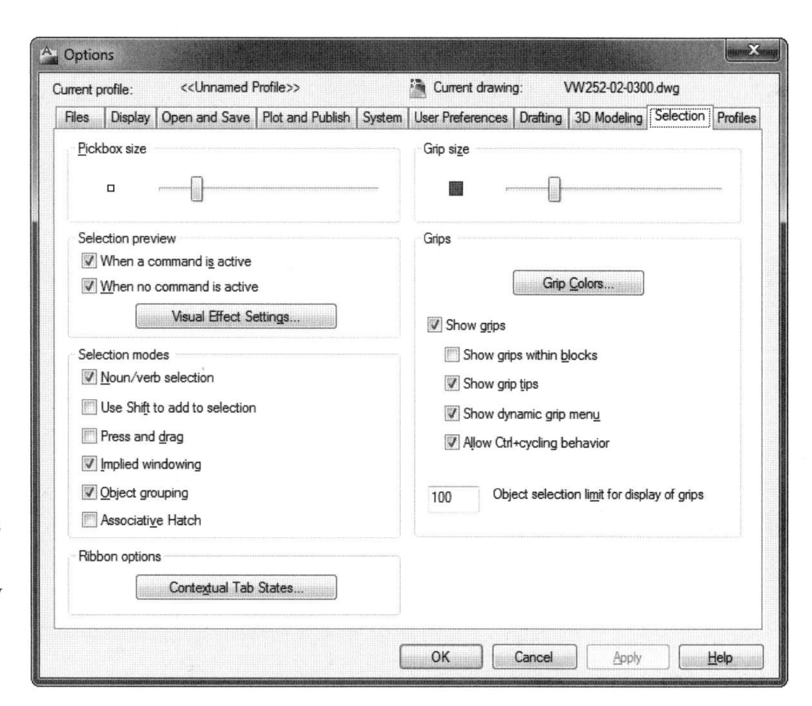

Ribbon Options options

Contextual Tab States controls when context-sensitive tabs appear; displays the Contextual Tab States dialog box, detailed later.

Grip size specifies the size of grips; range is from 1 to 20 pixels (*GripSize*); default = 5.

Grips options

Grip Colors specifies colors of grips; displays the Grip Colors dialog box, detailed later (*new to AutoCAD 2011*).

☑ **Show Grips** toggles all grips universally; replaces the Enable Grips option (*Grips; new to AutoCAD 2011*):

- ☐ **Show Grips in Blocks** displays all grips for every object in the selected block; when off, a single grip at the block's insertion point is displayed (*GripBlock*).
- ☑ **Show Grip Tips** toggles the display of grip tips on custom objects (*GripTips*).
- ☑ **Show Dynamic Grip Menu** toggles the display of ▼ grip menus (*GripMultifunctional; new to AutoCAD 2011*).

Object Selection Limit for Display of Grips limits the number of selected objects that display grips (*GripObjLimit*).

Profiles tab

Set Current sets the selected profile as the current profile.

Add to List displays the Add Profile dialog box; allows you to enter a name and description for the new profile.

Rename displays the Change Profile dialog box; allows you to change the name and the description of the selected profile.

Delete erases the selected profile; the current profile cannot be erased.

Export exports profiles as *.arg* files.

Import imports *.arg* profiles into AutoCAD.

Reset resets the selected profile to AutoCAD's default settings.

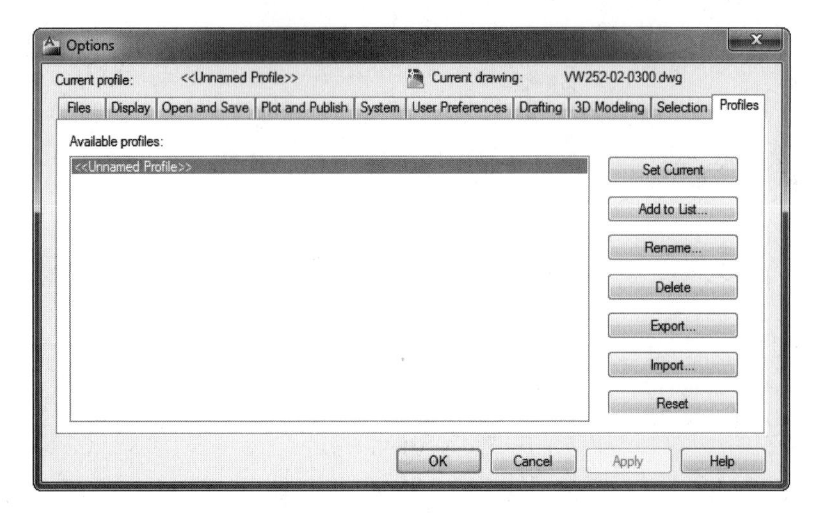

COLOR OPTIONS DIALOG BOX

Access this dialog box through the Colors button in the Display tab:

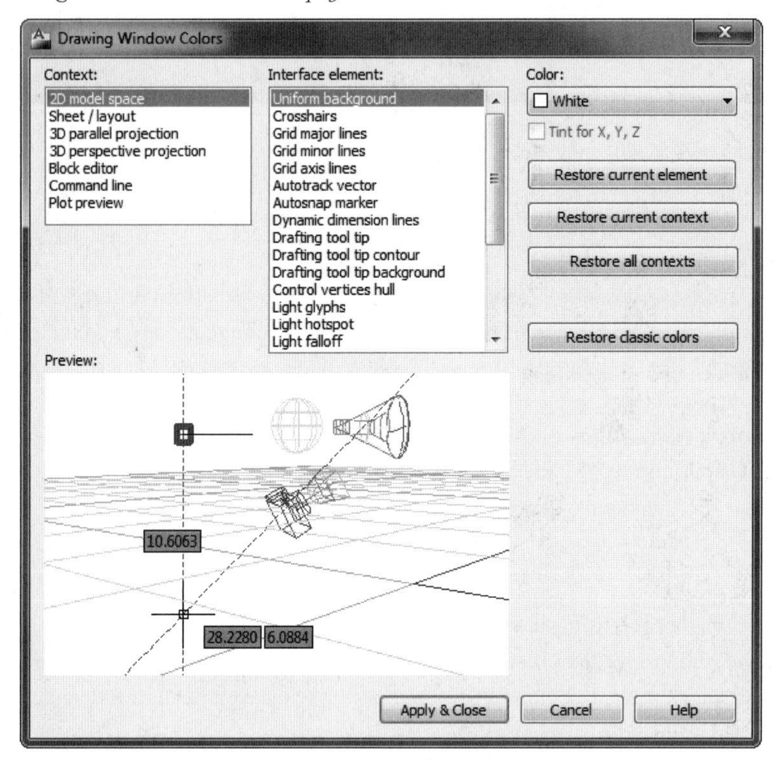

Context selects the user interface.

Interface element selects the individual interface item.

Color selects the color for the interface element.

☐ **Tint for X, Y, Z** selects colors for the crosshair cursor and 3D UCS icon's x, y, z axes.

Restore Current Element sets the selected item back to its default color.

Restore Current Context sets the selected UI element back to its default colors.

Restore All Contexts sets all UIs back to their default colors.

Restore Classic Colors sets all UI colors back to that of AutoCAD 2008 and earlier.

COMMAND LINE WINDOW FONT DIALOG BOX

Access this dialog box through the Font button in the Display tab:

Font selects the font for the command line palette.

Font Style changes the style.

Size specifies the size of the text. The technical editor suggests changing the style and size to bold 14pt to make the text easier to see on classroom screens.

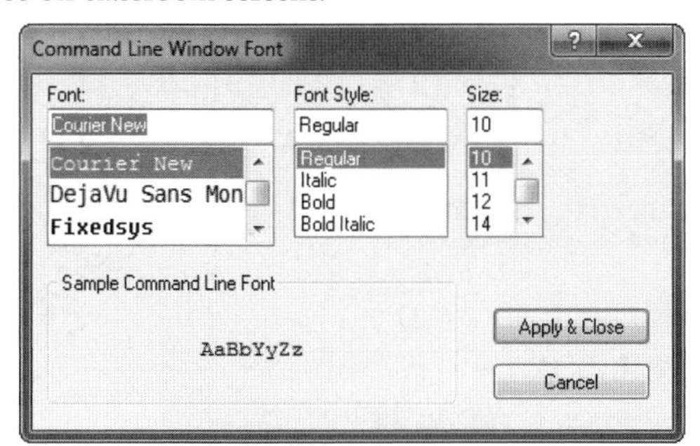

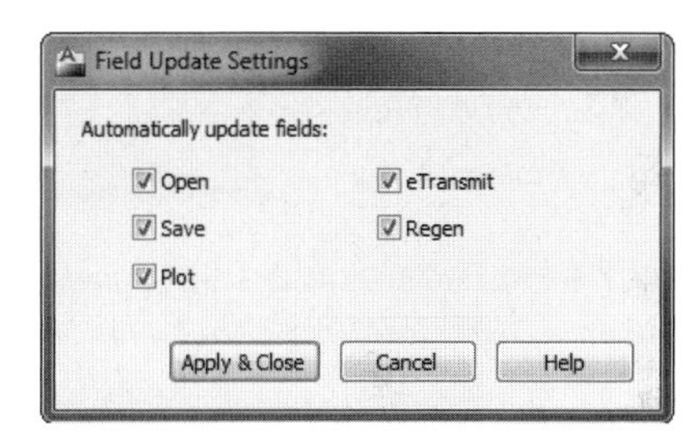

Left to right: *Command Line Window Font and Field Update Settings dialog box.*

FIELD UPDATE SETTINGS DIALOG BOX

Access this dialog box through the Field Update Settings button in the User Preferences tab:

☑ Specifies which commands force AutoCAD to update field text.

PLOT STYLE TABLE SETTINGS DIALOG BOX

Access this dialog box through the Plot Style Table Settings button in the Plot and Publish tab:

Default Plot Style Behavior for New Drawings (*PStylePolicy*):

⊙ **Use Color Dependent Plot Styles** uses color-dependent plot styles in new drawings; plotted colors are defined by a number between 1 and 255.

○ **Use Named Plot Styles** uses named plot styles in new drawings; objects are plotted based on plot style definitions.

Current Plot Style Table Settings options

Default Plot Style Table selects the plot style table to attach to new drawings.

Default Plot Style for Layer 0 specifies the default plot style for Layer 0 in new drawings (*DefLPlStyle*).

Default Plot Style for Objects specifies the default plot style assigned to new objects (*DefPlStyle*). The list displays BYLAYER, BYBLOCK, Normal styles, and any plot styles defined in the currently-loaded plot style table.

Add or Edit Plot Style Tables displays the Plot Style Manager window; see the StylesManager command.

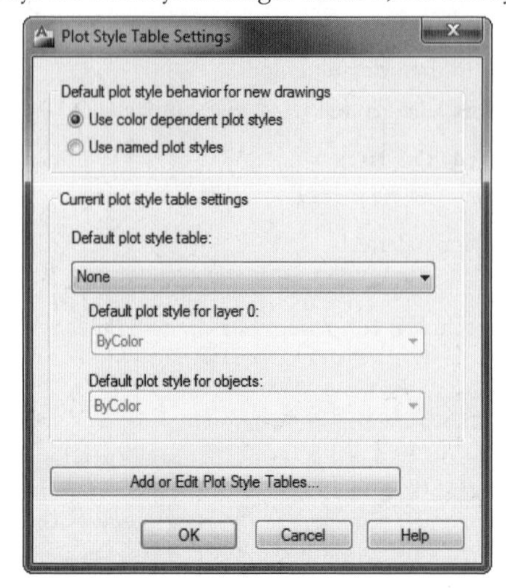

VISUAL EFFECT SETTINGS DIALOG BOX *(redesigned in AutoCAD 2011)*

Access this dialog box through the Visual Effect Settings button of the Selection tab:

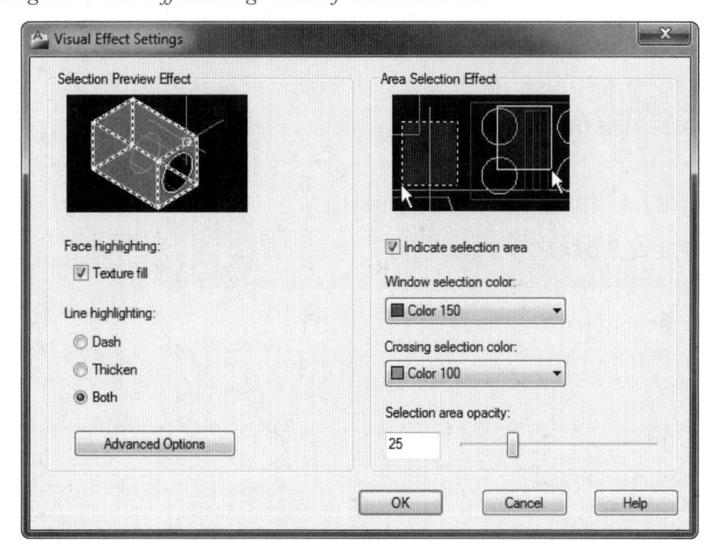

Selection Preview Effect options

Face highlighting

☑ **Texture fill** highlights faces with a texture fill effect *(new to AutoCAD 2011).*

Line highlighting

○ **Dash** changes selected object lines to dashes.

○ **Thicken** thickens selected object lines.

⊙ **Both** changes selected object lines to thick dashes.

Advanced Options displays the Advanced Options dialog box; detailed below.

Area Selection Effect options

☑ **Indicated selection area** toggles filled selection areas.

Window selection color selects the color for Window selection mode.

Crossing selection color selects the color for Crossing selection mode.

Selection area opacity specifies the "see thru ness" of the selection area; 100=opaque.

ADVANCED OPTIONS DIALOG BOX

In the Visual Effect Settings dialog box, click the Advanced Options button:

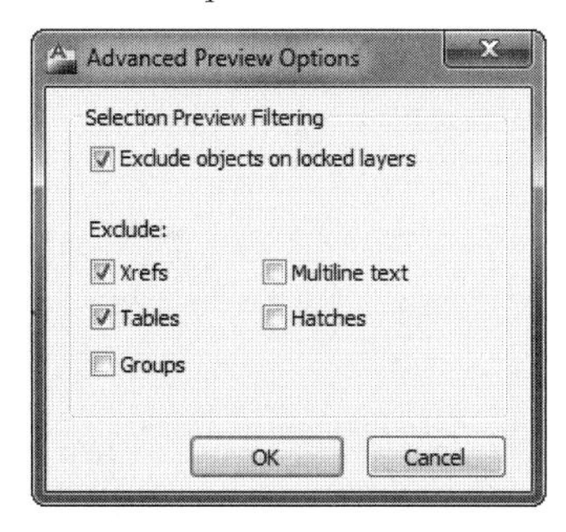

☑**Exclude objects on locked layers** ignores objects on locked layers, which cannot be edited.

☑**Exclude** excludes specific object types.

THUMBNAIL PREVIEW SETTINGS DIALOG BOX

Access this dialog box through the Thumbnail Preview Settings button of the Open and Save tab:

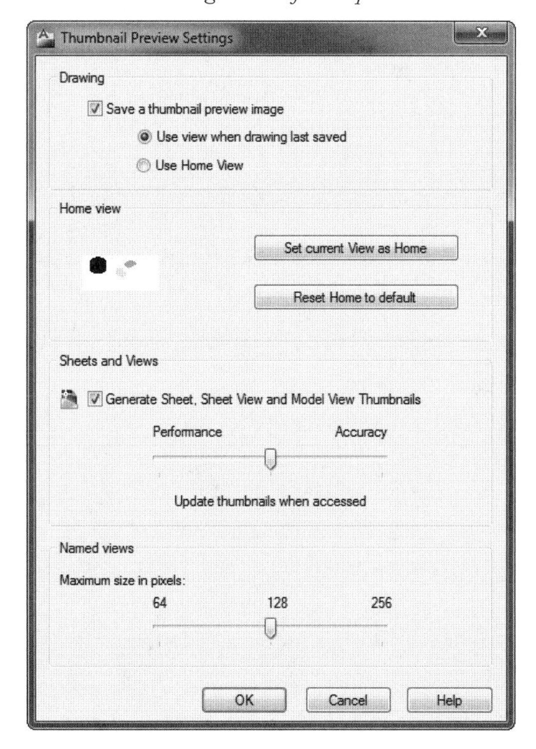

Drawing options

☑ **Save a Thumbnail Preview Image** saves a view of the drawing as a bitmap preview image in the *.dwg* file (*RasterPreview*):

⊙ **Use View when Drawing Last Saved** stores the view when last saved.

○ **Use Home View** stores home view as the thumbnail; see tips.

Home View options

Set Current View as Home View establishes the current view as the drawing's home view.

Reset Home to Default uses the top-left-front isometric view as the home view.

Sheets and Views options

☑ **Generate Sheet, Sheet View and Model View Thumbnails** updates thumbnails.

Update Thumbnails When Accessed:

- **Performance** updates thumbnail previews only when UpdateThumbsNow is executed.
- **Accuracy** updates thumbnail previews with every drawing save.

Named Views option:

Maximum Size in Pixels sets the size of view thumbnail images, ranging from 64 pixels wide to 256.

HIDDEN MESSAGE SETTINGS DIALOG BOX

Access this dialog box through the Hidden Message Settings button of the System tab:

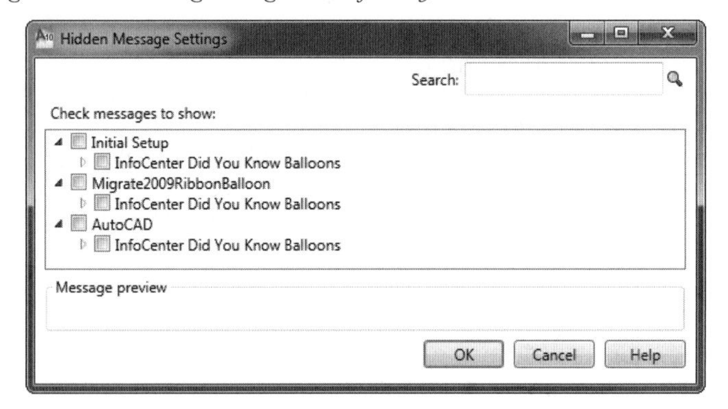

Search searches for message titles.

Check Messages to Show turns on messages hidden by the user.

RIGHT-CLICK CUSTOMIZATION DIALOG BOX

In the User Preferences tab, click the Right Click Customization button:

☐ **Turn on Time-Sensitive Right-Click** toggles Enter-or shortcut menu. The tech editor prefers this option on.

Default Mode:

○ **Repeat Last Command** repeats the previous command, like pressing Enter.

⊙ **Shortcut Menu** displays the appropriate shortcut menu.

Edit Mode:

○ **Repeat Last Command** repeats the previous command, like pressing Enter.

⊙ **Shortcut Menu** displays the appropriate shortcut menu.

Command Mode specifies how shortcut menus operate during commands:

○ **ENTER** disables the shortcut menu, and executes Enter.

○ **Shortcut Menu: Always Enabled** enables the shortcut menu, and disables Enter.

⊙ **Shortcut Menu: Enabled When Command Options Are Present** executes Enter only when no options available.

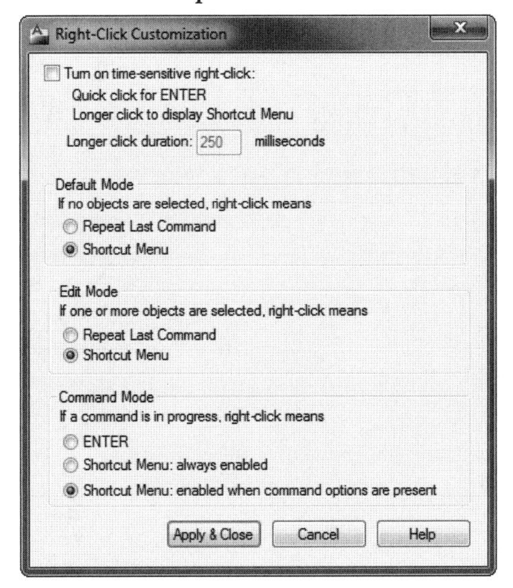

LIGHT AND CAMERA GLYPH APPEARANCE DIALOG BOXES

Access these dialog boxes through the Drafting tab: click the Light Glyph Settings or Camera Glyph Settings button:

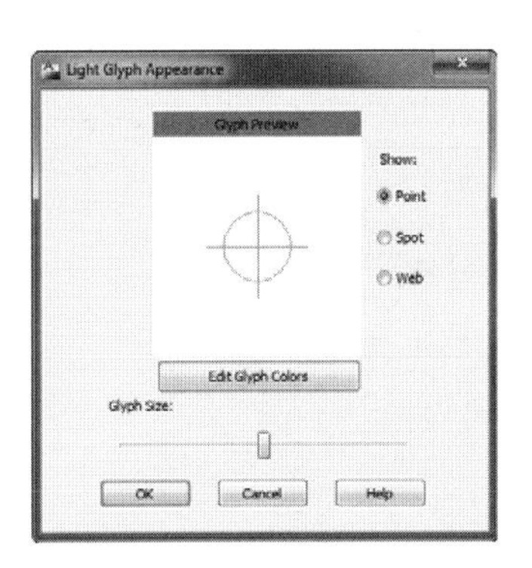

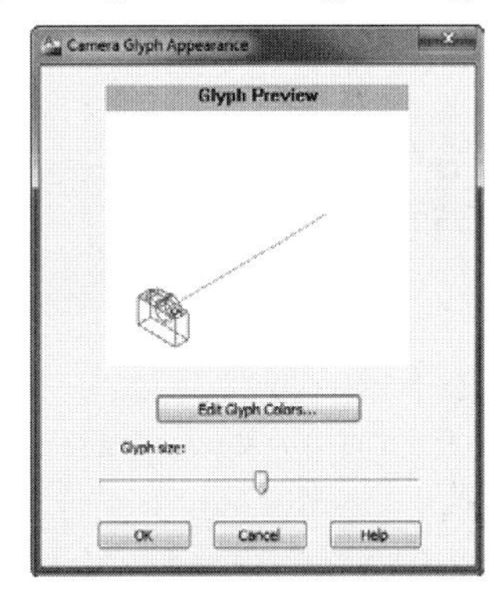

Left to right: *Light Glyph Appearance and Camera Glyph Appearance dialog boxes.*

Show a representation of a glyph for each type of light:

⊙ **Point** glyph for point lights; see the PointLight command.

○ **Spot** glyph for point lights; see the SpotLight command.

○ **Web** glyph for web lights; see the WebLight command.

Edit Glyph Colors changes the color(s) of the glyphs; displays the Drawing Windows Color dialog box.

Glyph Size changes the size of the glyph.

CONTEXTUAL TAB STATES DIALOG BOX

In the Selections tab, click the Contextual Tab States button.

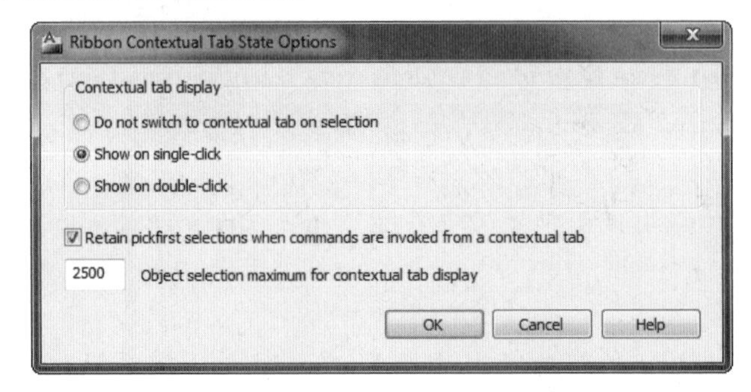

Contextual Tab Display options

Determines when the ribbon displays context-sensitive tabs (*RibbonContextSelect*)::

○ **Do Not Switch To Contextual Tab on Selection** prevents the focus shifting to the ribbon when objects are selected with a single or double-click.

⊙ **Show On Single-click** shifts focus to the ribbon when objects are single-clicked.

○ **Show On Double-click** shifts focus to the ribbon when objects are double-clicked.

Object Selection Options options

Retain Pickfirst Selections When Commands Are Invoked From a Contextual Tab (*RibbonSelectMode*)

☑ A pickfirst selection set remains valid after commands are selected from the ribbon's contextual tab.

☐A pickfisrt selection set is unselected.

Object Selection Maximum for Contextual Tab Display specifies the number of objects that can be selected before the context-sensitive tab no longer appears (*RibbonContextSelLim*).

GRIP COLORS DIALOG BOX *(new to AutoCAD 2011)*

Access this dialog box through the Grip Colors button of the Selection tab:

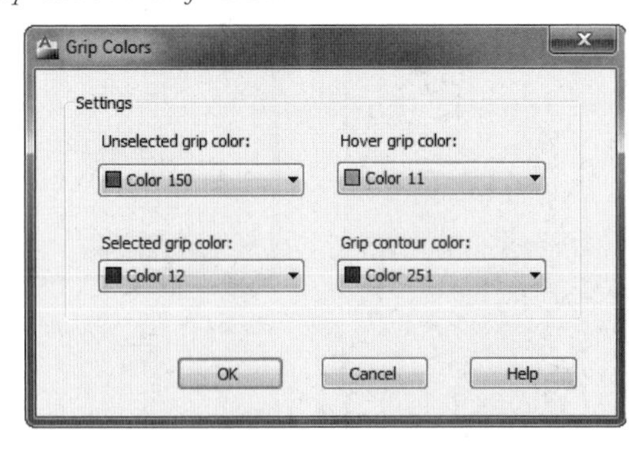

Unselected grip color (cold) specifies the color of unselected grips (*GripColor*).

Hover grip color specifies the color of the grip when the cursor hovers over it (*GripHover*).

Selected grip color (hot) specifies the color of the grip when selected (*GripHot*).

Grip contour color specifies the color of the square that surrounds grips (*GripContour; new to AutoCAD 2011*).

Grip contour (black)

+OPTIONS Command

Displays the tab specified by the index number.

Command: +options
Tab index <0>: *(Enter a digit between 0 and 9.)*

COMMAND LINE OPTION

Tab index specifies the tab to display:

- **0** — Files tab.
- **1** — Display tab.
- **2** — Open and Save tab.
- **3** — Plot and Publish tab.
- **4** — System tab.
- **5** — User Preferences tab.
- **6** — Drafting tab.
- **7** — 3D Modeling tab.
- **8** — Selection tab.
- **9** — Profiles tab.

TIPS

- A larger pickbox makes it easier to select objects generally, but also easier to select objects accidentally.
- The first time you use the DrawOrder command, it turns on all object sort method options.
- *Grips* are small squares that appear on an object when the object is selected at the 'Command' prompt. In other Windows applications, grips are known as *handles*.

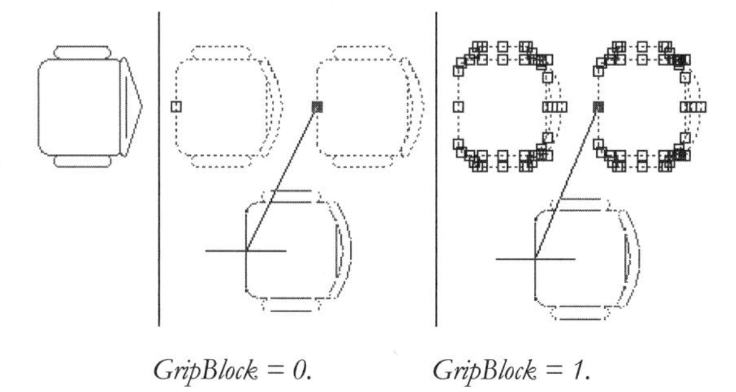

GripBlock = 0. *GripBlock = 1.*

- When an object is first selected, the grips are called *unselected* or *cold* grips. When a grip is selected, it is called a *hot* grip.
- Press Esc to turn off unselected grips; press Esc twice to turn off hot grips.
- AutoCAD 2009 introduced the *home view*:
  - In drawings created by AutoCAD 2008 and earlier, the home view is the extents.
  - In AutoCAD 2009 and later, the home view is the 3/4 isometric (top, left, front) view by default, or a view specified by the user. ViewCube has a home button for returning to this view.
- It's best to choose a crosshair color that is different from your drawing colors.

Ortho

Ver. 1.0 Constrains drawing and editing commands to the vertical and horizontal directions only (short for ORTHOgraphic).

Command	*Aliases*	*Keystroke Shortcuts*		*Status Bar*	*Ribbon*
'ortho	...	**Ctrl+L**	**F8**		...

Command: ortho
Enter mode [ON/OFF] <OFF>: *(Enter ON or OFF.)*

COMMAND LINE OPTIONS

OFF turns off ortho mode.

ON turns on ortho mode.

RELATED COMMANDS

DSettings toggles ortho mode via a dialog box.

Snap rotates the ortho angle.

RELATED SYSTEM VARIABLES

OrthoMode stores the current ortho modes.

PolarMode specifies polar and object snap tracking.

SnapAng specifies the rotation angle of the ortho cursor.

TIPS

- Use ortho mode when you want to constrain your drawing and editing to right angles.
- This command has effectively been replaced by the Polar command, which is not restricted to 90-degree increments.
- Rotate the angle of ortho with the Snap command's Rotate option.
- In isoplane mode, ortho mode constrains the cursor to the current isoplane.
- AutoCAD ignores ortho mode when you enter coordinates from the keyboard and in perspective mode; ortho is also ignored by object snap modes.
- Ortho is not necessarily horizontal or vertical; its orientation is determined by the current UCS and snap alignment.
- Ortho mode can be toggled on and off by clicking ORTHO on the status bar.

-OSnap

Ver. 2.0 Sets and turns on and off object snap modes at the command line (short for Object SNAP.

Command	Alias	Keystroke Shortcuts	Status Bar	Menu	Ribbon
'-osnap	-os	F3		...	Tools ↳Drafting Settings

The OSnap command displays the Drafting Settings dialog box; see the DSettings command.

Command: -osnap
Current osnap modes: Ext
Enter list of object snap modes: *(Enter one or more modes separated by commas.)*

COMMAND LINE OPTIONS

You only need to enter the first three letters as the abbreviation for each option:

APParent snaps to the intersection of two objects that don't physically cross, but appear to intersect on the screen, or would intersect if extended.

CENter snaps to the center point of arcs and circles.

ENDpoint snaps to the endpoint of lines, polylines, traces, and arcs.

EXTension snaps to the extension path of objects.

FROm extends from a point by a given distance.

INSertion snaps to the insertion point of blocks, shapes, and text.

INTersection snaps to the intersection of two objects, to a self-crossing object, or to objects that would intersect if extended.

MIDpoint snaps to the middle of lines and arcs.

NEArest snaps to the object nearest the crosshair cursor.

NODe snaps to a point object.

NONe turns off all object snap modes temporarily.

OFF turns off all object snap modes.

PARallel snaps to a parallel offset.

PERpendicular snaps perpendicularly to objects.

QUAdrant snaps to the quadrant points of circles and arcs.

QUIck snaps to the first object found in the database.

TANgent snaps to the tangent of arcs and circles.

SHORTCUT MENU

Hold down the Ctrl key, and then right-click anywhere in the drawing.

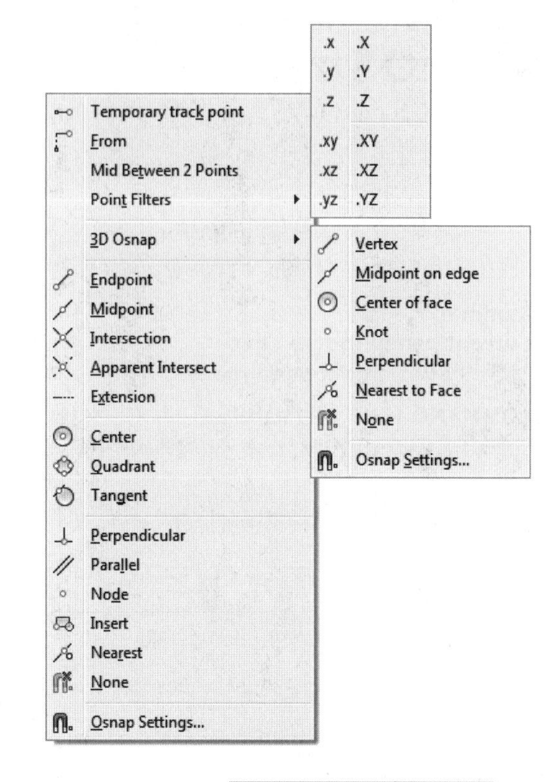

Temporary Track Point invokes Tracking mode.

From locates temporary points during drawing and editing commands; AutoCAD prompts you:

From point: from
Base point: *(Pick a point.)*
<Offset>: *(Pick another point.)*

Mid Between 2 Points locates a point between two pick points; AutoCAD prompts you:

From point: m2p
First point of mid: *(Pick a point.)*
Second point of mid: *(Pick another point.)*

Point filters invokes point filter modes:

.x — **need YZ:** *(Enter value for y and z.)*
.y — **need XZ:** *(Enter value for x and z.)*
.z — **need XY:** *(Enter value for x and y.)*
.xy — **need Z:** *(Enter value for z.)*
.xz — **need Y:** *(Enter value for y.)*
.yz — **need X:** *(Enter value for x.)*

Osnap Settings displays the Object Snap tab of the Drafting Settings dialog box; see the DSettings command.

STATUS BAR OPTIONS

Right-click OSNAP on the status bar:

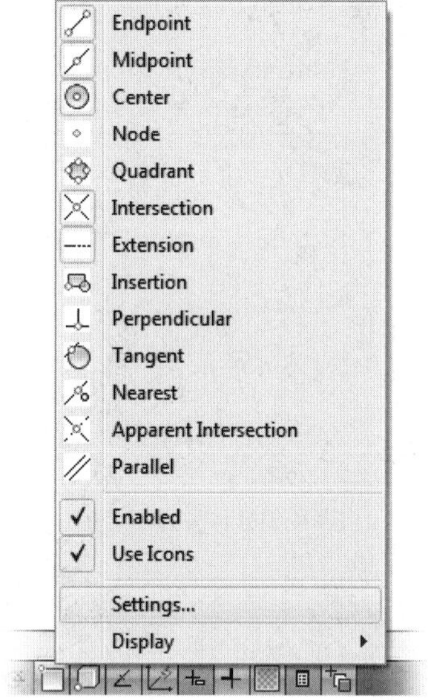

Enabled turns on previously-set object snap modes.

Settings displays the Object Snap tab of the Drafting Settings dialog box.

RELATED COMMANDS

3dOsnap toggles 3D object snap modes *(new to AutoCAD 2011).*

RELATED SYSTEM VARIABLES

Aperture specifies the size of the object snap aperture in pixels.

AutoSnap controls the display of AutoSnap (default = 63).

DgnOsnap toggles snapping to elements in attached MicroStation DGN underlays.

OsnapNodeLegacy determines whether the NODe object snap snaps to text insertion points.

OsOptions toggles between using the z coordinate and the current elevation setting.

OsnapCoord overrides object snaps when you enter coordinates at the 'Command' prompt.

OsnapZ determines whether osnaps uses the z coordinate or the elevation.

OsMode stores the current object snap mode(s).

TempOverrides toggles temporary override keys.

TEMPORARY OVERRIDE KEYS

These keystrokes temporarily override object snaps during commands:

Override	Left Side	Right Side
Disables all osnaps and tracking	SHIFT+D	SHIFT+L
Enables osnap enforcement	SHIFT+S	SHIFT+;
Overrides CENter osnap	SHIFT+C	SHIFT+,
Overrides ENDpoint osnap	SHIFT+E	SHIFT+P
Overrides MIDpoint osnap	SHIFT+V	SHIFT+M
Toggles osnap mode	SHIFT+A	SHIFT+'

TIPS

- The Aperture command controls the drawing area in which AutoCAD searches.
- If AutoCAD finds no snap matching the current modes, then the pick point is selected.
- The m2p modifier can also be entered as mtp.
- The APPint and INT object snap modes should not be used together.
- To turn on more than one object snap at a time, use a comma to separate mode names:

  Enter list of object snap modes: int,end,qua

- The location of object snaps:

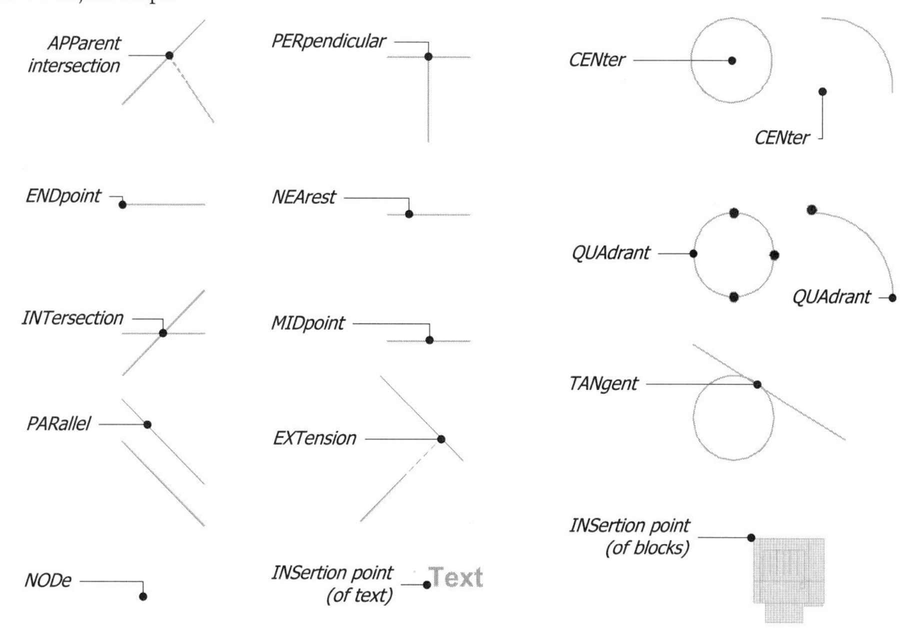

- When InferConstraint is turned on, AutoCAD automatically applies constraints that match current osnap modes when you draw and edit objects.

PageSetup

2000 Sets up model and layout views in preparation for plotting drawings.

Command	*Aliases*	*Shortcut Keystrokes*	*Menu Bar*	*Application Menu*	*Ribbon*
pagesetup	...	...	**File** ↳**Page Setup Manager**	**Print** ↳**Page Setup**	...

Command: pagesetup

Displays dialog box.

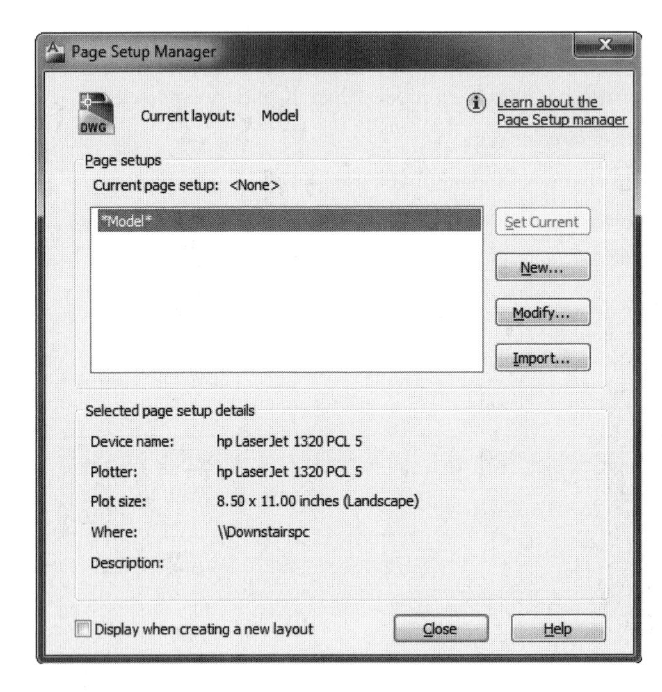

PAGE SETUP MANAGER DIALOG BOX

Set Current selects page setup to be used by the drawing.

New displays the New Page Setup dialog box.

Modify displays the Page Setup dialog box.

Import displays the Select Page Setup From File dialog box; select a *.dwg*, *.dwt*, or *.dxf* file, and then click Open; displays the Import Page Setups dialog box.

☐**Display When Creating a New Layout** displays this dialog box the first time you click a layout tab in new drawings.

NEW PAGE SETUP DIALOG BOX

Access this dialog box through the New button.

New Page Setup Name specifies the name of the setup.

Start With selects the default settings:

- **<None>** — uses no page setups for templates.
- **<Default Output Device>** — selects the plotter specified as the default output device in the Options dialog box.
- **<Previous Plot>** — uses settings from a previous plot job.

- **Layout Name** — selects an existing layout.

OK displays the Page Setup dialog box.

PAGE SETUP DIALOG BOX

Access this dialog box through the Modify button. See the Plot command.

IMPORT PAGE SETUPS DIALOG BOX

Clicking Import displays Page Setup from File dialog box.

Choose a DWG file, and then click Open.

Displays dialog box.

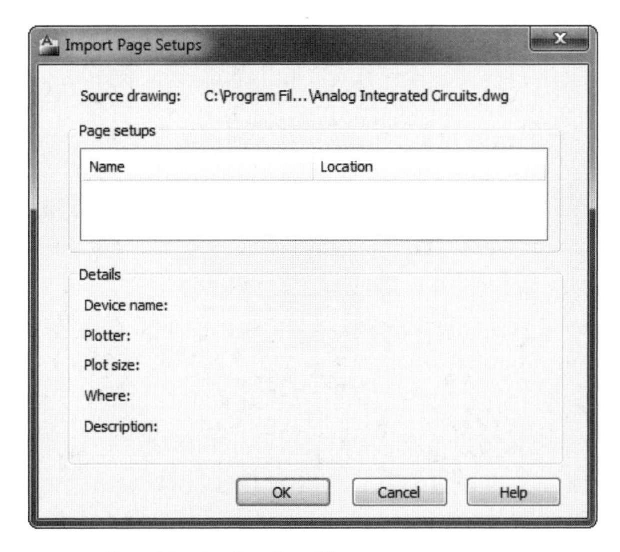

Name lists the names of page setups found in the drawing, if any.

Location indicates the location of the page setups: model or layout.

OK returns to the Page Setup Manager dialog box.

SHORTCUT MENU

Access this menu by right-clicking a page setup name:

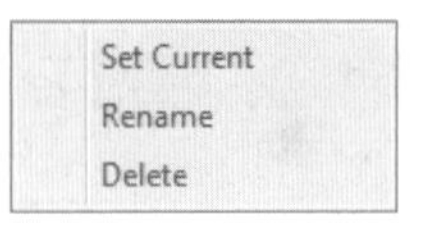

Set Current selects the page setup to use for the drawing.

Rename changes the name of the page setup.

Delete erases the page setup from the drawing, without warning.

RELATED COMMANDS

Layout creates new layouts.

LayoutWizard creates layouts and page setups.

Plot plots drawings based on the settings of the PageSetup command.

TIPS

- The sheet set icon shows when this command is activated from the Sheet Set Manager palette.
- Right-click Model or Layout tabs, and then select Page Setup Manager.

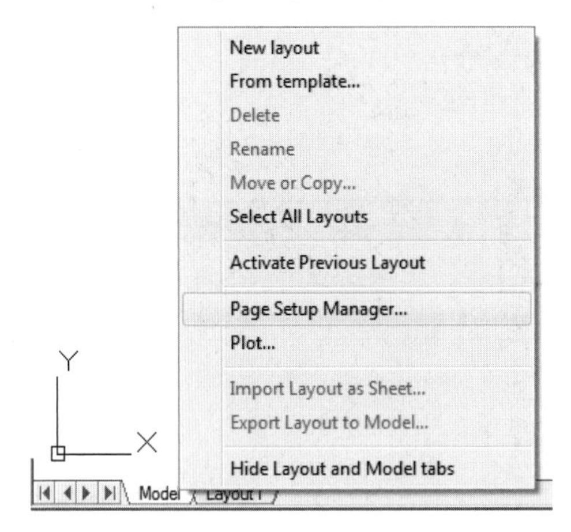

- Multiple page setups can be applied to all or part of a sheet set, such the first ten sheets plotted to an E-size DesignJet plotter, and then the next five sheets on an A-size laser printer.

Pan

Ver. 1.0 Moves the view in the current viewport.

Commands	Aliases	Status Bar	Menu	Ribbon
'pan	**p** **rtpan** **3dpantransparent**		**View** ↳**Pan** ↳**Realtime**	**View** ↳**Navigate** ↳**Pan**
'-pan	**-p**	...	**View** ↳**Pan** ↳**Point**	...

Command: pan

Press Esc or Enter to exit, or right-click to display shortcut menu. *(Move cursor to pan, and then press Esc to exit command.)*

Enters real-time panning mode, and displays hand cursor:

Drag the hand cursor to pan the drawing in the viewport.

Press Enter or Esc to return to the 'Command' prompt.

SHORTCUT MENU OPTIONS

During real-time pan mode, right-click the drawing to display shortcut menu.

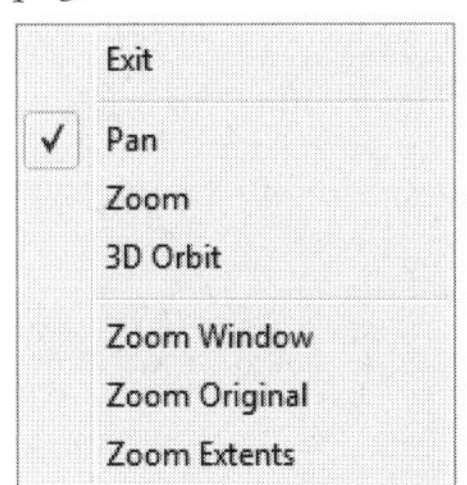

Exit exits real-time pan mode; returns to the 'Command' prompt.

Pan switches to real-time pan mode, when in real-time zoom mode.

Zoom switches to real-time zoom mode; see the Zoom command.

3D Orbit switches to 3D orbit mode; see the 3dOrbit command.

Zoom Window prompts you, 'Press pick button and drag to specify zoom window.'

Zoom Original returns to the view when you first started the Pan command.

Zoom Extents displays the entire drawing.

-PAN Command

Command: -pan
Displacement: *(Pick a point, or enter x, y coordinates.)*
Second point: *(Pick another point, or enter x, y coordinates.)*

COMMAND LINE OPTIONS

Enter or **Esc** exits real-time panning mode.

Displacement specifies the distance and direction to pan the view.

Second point pans to this point.

RELATED COMMANDS

DsViewer displays the Aerial View palette, which pans in an independent palette.

RegenAuto determines how regenerations are handled.

View saves and restores named views.

Zoom pans with the Dynamic option.

NavBar displays the navigation bar with access to Pan and other viewing commands.

3dOrbit pans during perspective mode.

RELATED SYSTEM VARIABLES

MButtonPan determines the action of a mouse's third button or wheel.

RtDisplay toggles the display of raster and OLE images during real-time panning.

VtDuration specifies the duration of view transitions.

VtEnable turns on smooth view transitions.

VtFps specifies the speed of view transitions.

TIPS

- With MButtonPan set to 1 (default), pan by holding down the wheel button and moving the mouse.
- You can pan each viewport independently.
- Use 'Pan to start drawing objects in one area of the drawing, pan over, and then continue working in another area of the drawing.
- You cannot use transparent pan during the DView command, another Pan command, or the View command.
- When the drawing no longer moves during real-time panning, you have reached the panning limit; AutoCAD changes the hand icon to show the limit. (Use Regen to increase the limit.)

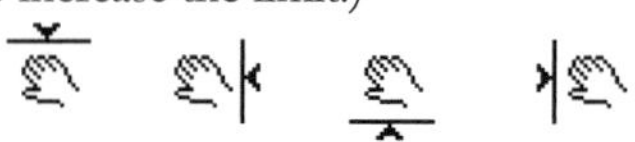

- As an alternative, use the horizontal and vertical scroll bars to pan the drawing.

Parameters / ParametersClose

2010 Opens and closes the Parameters palette.

Commands	Aliases	Shortcut Keystrokes	Menu	Ribbon
parameters	..	...	**Parametric ↳Parameters Manager**	**Parametric ↳Manage ↳Parameters Manager**
parametersclose	...	...	...	...

Command: parameters

Opens the Parameters palette.

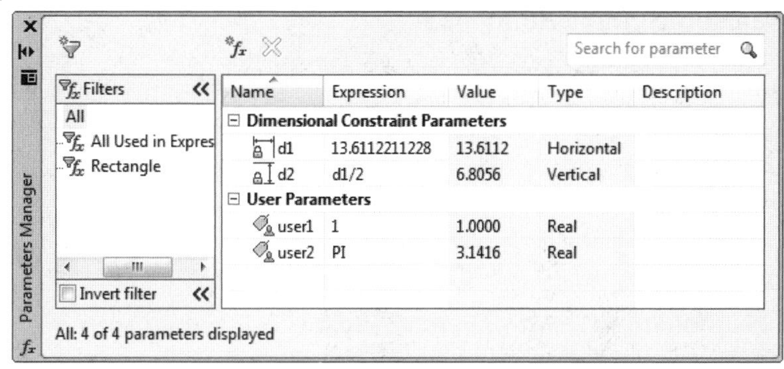

Command: parametersclose

Closes the Parameters palette.

PARAMETERS MANAGER PALETTE

New Group creates a new group by which to filter parameters *(Alt+G; new to AutoCAD 2011).*

Collapse minimizes the filter pane.

☐ **Invert Filter** inverts the effect of the current filter.

Add creates new user expressions.

Remove deletes the selected expression.

Search for parameter uses wildcards to locate parameters amongst long lists of names *(new to AutoCAD 2011).*

Click column headers to sort alphabetically, click again to sort in reverse order:

Name lists the names of variables; AutoCAD assigns names automatically, but you can edit them.

Expression specifies the parametric expression, and consists of constants, names, and expressions (formulae).

Value reports the value of the expression (not user editable).

The following columns are not displayed, by default; right-click header to add and remove columns:

Type reports the type of parameter (not user editable).

Description provides space to describe the parameter.

FILTERS SHORTCUT MENU

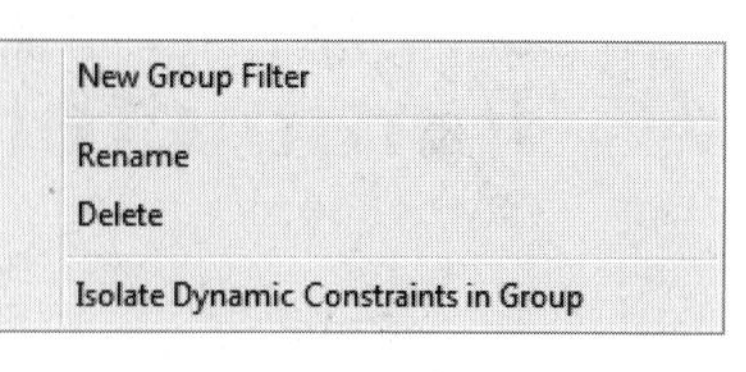

Access this shortcut menu by right-clicking a filter name:

New Group Filter creates a new filter with the generic name 'Group Filter1'.

Rename renames the filter.

Delete removes the filter.

Isolate Dynamic Constraints in Group displays in the drawing only those constraints associated with the group.

PARAMETERS SHORTCUT MENU

Access this shortcut menu by right-clicking a parameter:

Show Filter Tree toggles the filter pane.

Remove from Group Filter removes the parameter from a filter.

Delete Parameter removes the parameter from the drawing.

SELECTED PARAMETER SHORTCUT MENU

Double-click an expression, and then right-click:

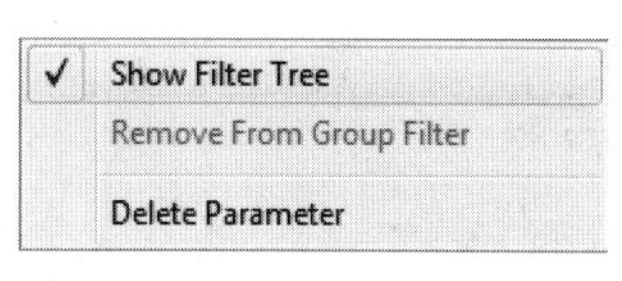

Cut copies the expression to the Clipboard, and then deletes the expression.

Copy copies the expression to the Clipboard.

Paste pastes the contents of the Clipboard.

Delete deletes the expression.

Expression provides a submenu of functions. In addition to those listed by the submenu, you can also use algebraic expressions, including +, -, *, and /.

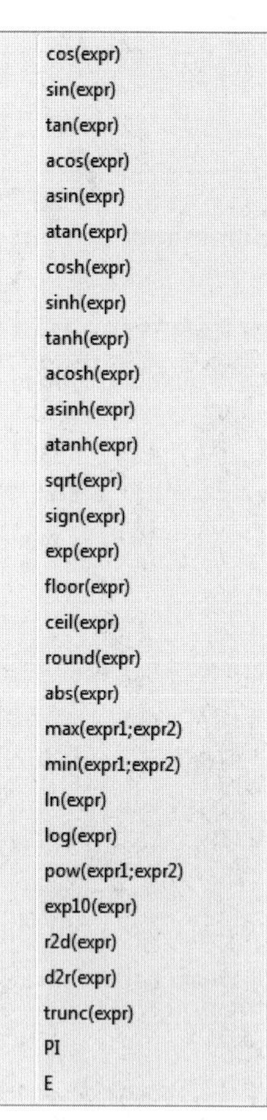

EXPRESSIONS

abs absolute value (makes negative numbers positive)

acos arc cosine

asin arc sine

atan arc tangent

atanh hyperbolic arc tangent

ceil ceiling (rounds up)

cos cosine

cosh hyperbolic cosine

d2r converts degrees to radians

E is the constant 2.718282...

exp exponent

exp10 the exponent to the power of 10

floor floor (rounds down)

ln natural logarithm

log logarithm

max maximum of two expressions

min minimum of two expressions

PI 3.14159265...

pow raises one expression to the power of the second expression

r2d converts radians to degrees

round rounds to the nearest integer

sign changes positive to negative and vice versa

sin sine

sinh hyperbolic sine

sqrt square root of positive numbers

tan tangent

tanh hyperbolic tangent

trunc truncates (removes decimal portion of number)

TIPS

- Parameters are created by the DimConstraint and Dc... group of commands, whichs create constraints that look like dimensions. In the figure below, two dimensional constrains determine the size of the rectangle. The *fx* indicates that *d2* is a formula (expression).
  - Change the value of *d1* to change the size of the rectangle.
  - Change the formula of *d2* to change the aspect ratio of the rectangle.

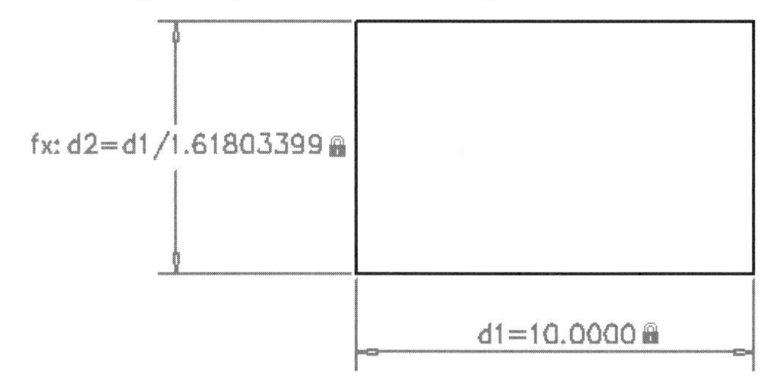

d1 *equals a constant that can be varied by the user.*
d2 *equals d1 divided by 1.61... (the golden mean).*

- See the DimConstraint command for a tutorial on using dimensional constraints and parameters.
- Parameters can be edited directly in the dimension constraints or in the Parameters Manager palette.
- When you change the name of a parameter, AutoCAD automatically changes all other occurrences.
- AutoCAD 2011 added filters to the Parametrics Manager palette.
- You can drag parameters in and out of groups.
- A parameter can belong to more than one group.
- Parameter-like equations can be used by certain commands that include the Expression option, such as ChamferEdge and Extrude.
- Do not use the equals (=) sign in expressions.
- If you isolate a parameter filter group, you can only bring back the others through the ribbon's Parametric | Dimensional | Show All option.

Undocumented Command

PartialCui command loads CUIX files.

PartiaLoad

2006 Loads additional views and layers of a partially-loaded drawing; this command works only with drawings that have been partially loaded.

Commands	Aliases	Shortcut Keystrokes	Menu	Ribbon
partialload	...	...	**File** ↳**Partial Load**	...
-partialload	...	...	...	...

Command: partialload

If the current drawing has not been partially opened through the Open command's Partial Open option, AutoCAD complains, 'Command not allowed unless the drawing has been partially opened.'

Otherwise, displays dialog box.

PARTIAL LOAD DIALOG BOX

See Open command.

-PARTIALOAD Command

Command: -partialload

Specify first corner or [View]: *(Pick a point, or type V.)*

Enter view to load or [?] <*Extents*>: *(Enter a view name, or type ?.)*

Enter layers to load or [?] <none>: *(Enter a layer name, press Enter for no layers, or type ?.)*

COMMAND LINE OPTIONS

Specify first corner specifies a corner to create a new view.

Specify opposite corner specifies the second corner; causes geometry in the new view to be loaded into the drawing.

View options

Enter view to load loads the geometry found in the view into the drawing.

? lists the names of views in the drawing:

Enter view name(s) to list <*>: *(Enter a view name, or type *.)*

Extents loads all geometry into the drawing.

Enter layers to load loads geometry on the layers into the drawing.

? lists the names of layers in the drawing:

Enter layers to list <*>: *(Press Enter.)*

None loads no layers.

RELATED COMMANDS

Open opens a drawing partially via a dialog box.

PartialOpen opens a drawing partially via the command line.

RELATED SYSTEM VARIABLES

FullOpen indicates whether the drawing is fully or partially opened.

OpenPartial determines whether drawings can be edited before they are fully open.

-PartialOpen

2000 Opens a drawing and loads selected layers and views.

Command	*Alias*	*Shortcut Keystrokes*	*Menu*	*Ribbon*
-partialopen	**partialopen**	**...**	**...**	**...**

Command: -partialopen
Enter name of drawing to open <filename.dwg>: *(Enter a file name.)*
Enter view to load or [?]<*Extents*>: *(Enter a view name, or type ?.)*
Enter layers to load or [?]<none>: *(Enter a layer name, or type ?.)*
Unload all Xrefs on open? [Yes/No] <N>: *(Type Y or N.)*

COMMAND LINE OPTIONS

Enter name of drawing to open specifies the name of the DWG file.

Enter view to load loads the geometry found in the view into the drawing.

? lists the names of the views in the drawing.

Extents loads all geometry into the drawing.

Enter layers to load loads geometry on the layers into the drawing.

? lists the names of layers in the drawing.

None loads no layers.

Unload all Xrefs on open:

Yes does not load externally-referenced drawings.

No loads all externally-referenced drawings.

RELATED COMMANDS

Open displays the Select Drawing dialog box; includes the Partial Open option.

PartiaLoad loads additional views or layers of a partially-opened drawing.

RELATED SYSTEM VARIABLES

FullOpen indicates whether the drawing is fully or partially opened.

OpenPartial determines whether drawings can be edited before they are fully open.

TIPS

- PartialOpen is an alias for the -PartialOpen command.
- As an alternative, you can use the Partial Open option of the Open command's Select File dialog box.

PasteAsHyperlink

2000 Pastes object as hyperlinks in drawings; works only if the Clipboard contains appropriate data.

Command	Aliases	Menu	Ribbon
pasteashyperlink	...	**Edit**	**Home**
		↳Paste As Hyperlink	**↳Clipboard**
			↳Paste as Hyperlink

Before using this command, use the Edit | Copy command in AutoCAD or another program to copy drawings, documents, etc. to the Clipboard:

Command: pasteashyperlink

When the Clipboard contains hyperlinkable objects, the following prompt is displayed:

Select objects: *(Select one or more objects.)*

Select objects: *(Press Enter.)*

COMMAND LINE OPTION

Select objects selects the objects to which the hyperlink will be pasted.

RELATED COMMANDS

CopyClip copies a hyperlink to the Clipboard.

Hyperlink adds hyperlinks to selected objects.

SelectURL highlights all objects with hyperlinks.

PasteBlock / PasteOrig

2000 Pastes objects as blocks (or at original coordinates) in drawings.

Command	Aliases	Shortcut Keystroke	Menu	Ribbon
pasteblock	**...**	**Ctrl+Shift+V**	**Edit** **↳Paste as Block**	**Home** **↳Clipboard** **↳Paste as Block**
pasteorig	**...**		**Edit** **↳Paste to Original Coordinates**	**Home** **↳Clipboard** **↳Paste to Original Coordinates**

Command: pasteblock

When the Clipboard contains AutoCAD objects, the following prompt is displayed:

Specify insertion point: *(Pick a point.)*

COMMAND LINE OPTION

Specify insertion point specifies the position for the block.

PASTEORIG Command

Pastes AutoCAD objects from the Clipboard into the current drawing at the objects' original locations (short for PASTE ORIGinal).

Command: pasteorig

COMMAND LINE OPTIONS

None.

RELATED COMMANDS

CopyClip copies a drawing to the Clipboard.

Insert inserts an AutoCAD drawing in the drawing.

PasteBlock pastes AutoCAD objects as a block at a user-specified insertion point.

TIPS

- These commands do not work when the Clipboard contains data that cannot be pasted.
- The PasteBlock command pastes any AutoCAD object as a block, generating a generic block name similar to "A$C65D94228."
- The block is pasted onto the current layer. If the Clipboard contains a block, the block is nested by this command.
- As of AutoCAD 2004, you can use the Ctrl+Shift+V shortcut for the PasteBlock command.
- Use the PasteOrig command to copy objects from one drawing to another.
- The PasteOrig command cannot be used to paste objects into the drawing from which they originate.

PasteClip

Rel. 13 Places objects from the Clipboard in drawings (short for PASTE CLIPboard).

Command	Aliases	Shortcut Keystroke	Menu	Ribbon
pasteclip	...	**Ctrl+V**	**Edit ↳Paste**	**Home ↳Clipboard ↳Paste**

Command: pasteclip

When the Clipboard contains AutoCAD objects, the following prompt is displayed:

Specify insertion point: *(Pick a point.)*

COMMAND LINE OPTION

Specify insertion point specifies the lower-left corner of the pasted object.

RELATED COMMANDS

CopyClip copies selected objects to the Clipboard.

Insert inserts an AutoCAD drawing in the drawing.

InsertObj inserts an OLE object in the drawing.

PasteSpec places the Clipboard object as a pasted or a linked object.

RELATED SYSTEM VARIABLE

OleHide toggles the display of the OLE object (1 = off).

TIPS

- Graphical objects are placed in the drawing as OLE objects.
- Text is usually — but not always — placed in the drawing as Mtext objects.
- Use the PasteBlock command to paste objects as an AutoCAD block.

PasteSpec

2000 Pastes Clipboard objects in the current drawing as embedded, linked, pasted, or AutoCAD objects (short for PASTE SPECial).

Command	*Alias*	*Shortcut Keystrokes*	*Menu*	*Ribbon*
pastespec	**pa**	**...**	**Edit** **↳Paste Special**	**Home** **↳Clipboard** **↳Paste Special**

Command: pastespec

Displays dialog box.

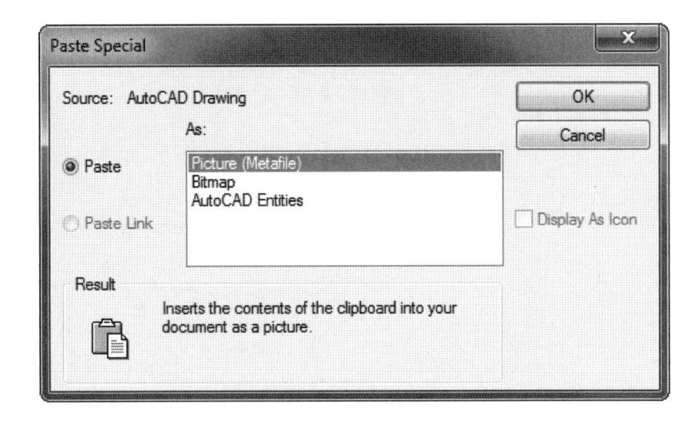

PASTE SPECIAL DIALOG BOX

⦿**Paste** pastes the object as an embedded object.

○**Paste Link** pastes the object as a linked object.

☐**Display as Icon** displays the object as an icon from the originating application.

Change Icon allows you to select the icon.

RELATED COMMANDS

CopyClip copies the drawing to the Clipboard.

InsertObj inserts an OLE object in the drawing.

OleLinks edits the OLE link data.

PasteClip places the Clipboard object as a pasted object.

RELATED SYSTEM VARIABLE

OleHide toggles the display of OLE objects (1 = off).

TIP

- As of AutoCAD 2008, you can paste spreadsheet files directly into the drawing as tables. Follow these steps:
  1. In the spreadsheet program, copy a range of cells to the Clipboard (use Ctrl+C).
  2. In AutoCAD, start the PasteSpec command.
  3. From the As list, select "AutoCAD Entities."
  4. Choose the Paste Link option.
  5. Click OK, and follow the remaining prompts.

PcInWizard

2000 Converts PCP and PC2 plot configurations to PC3 format.

Command	Aliases	Shortcut Keystrokes	Menu	Ribbon
pcinwizard	...	...	**Tools**	...
			↳**Wizards**	
			↳**Import Plot Settings**	

Command: pcinwizard

Displays dialog box.

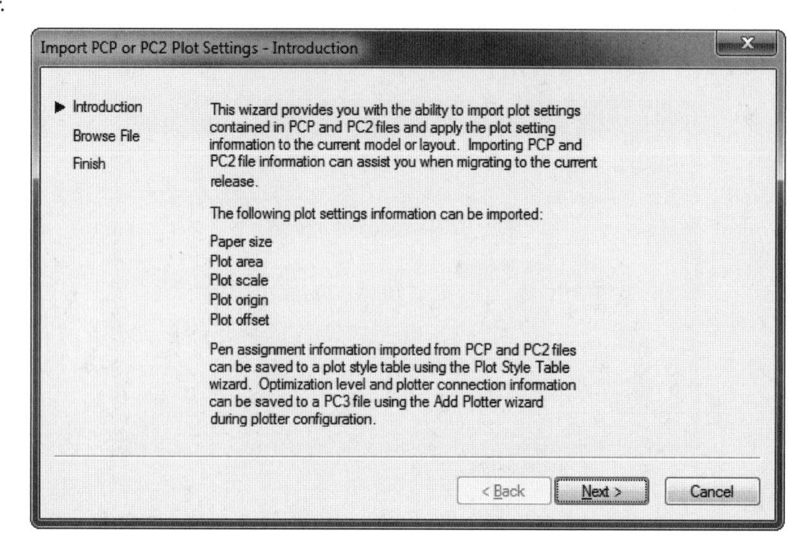

IMPORT PCP OR PC2 PLOT SETTINGS DIALOG BOX

Back returns to the previous step.

Next continues to the next step.

Cancel cancels the wizard.

Browse File page

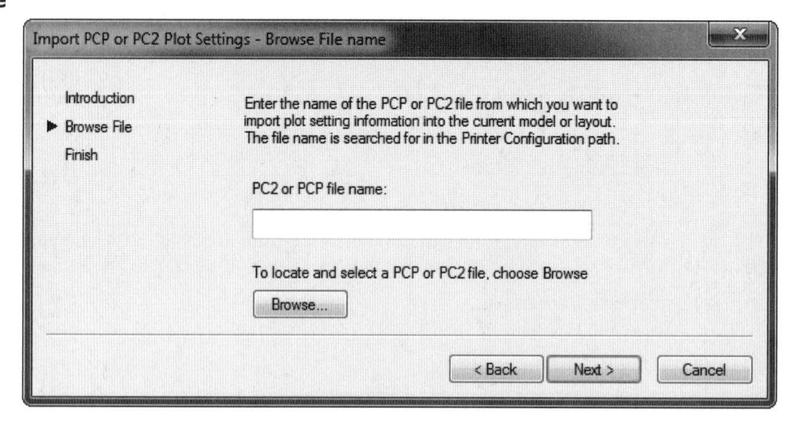

PCP or PC2 filename specifies the name of the *.pcp* or *pc2* file.

Browse displays the Import dialog box.

Finish page

Page Setup displays the Page Setup dialog box; see the PageSetup command.

Finish completes the importation.

RELATED COMMANDS

PageSetup creates and modifies page setups.

Plot uses PC3 files to control the plotter configuration.

TIPS

- This command imports *.pcp* and *.pc2* files, and applies them to the current layout or model tab.
- Use this command only after you create *.pcp* or *.pc2* files with earlier versions of AutoCAD.
- PCP is short for "plotter configuration parameters":
  - *.pcp* files are used by AutoCAD Release 13.
  - *.pc2* files are used by AutoCAD Release 14.
  - *.pc3* files are used by all AutoCAD releases since 2000.
- AutoCAD imports the following information from *.pcp* and *.pc2* files: paper size, plot area, plot scale, plot origin, and plot offset.
- To import color-pen mapping, use the Plot Style Table wizard, run by the *StyShWiz.Exe* program.
- To import the optimization level and plotter connection, use the Add-a-Plotter wizard, run by the *AddPlWiz.Exe* program,

PdfAdjust

2010 Adjusts the contrast, fade, and coloring of PDF underlays.

Command	*Aliases*	*Shortcut Keystrokes*	*Menu*	*Ribbon*
pdfadjust	...	...	**Insert ⇘PDF Underlay**	**Insert ⇘Reference ⇘Adjust**

Command: pdfadjust
Select PDF underlay: *(Choose one or more PDF underlays.)*
Select PDF underlay: *(Choose more underlays, or press Enter to continue.)*
Enter PDF underlay option [Fade/Contrast/Monochrome] <Fade>: *(Enter an option.)*

COMMAND LINE OPTIONS

Select PDF Underlay chooses one or more underlays to adjust; press Ctrl+A to select all underlays in the drawing.

Fade sets the lightness from 0 (no fading) to 100 (invisible); default = 50, somewhat faded.

Contrast changes colors from their primary values, from 0 (minimal contrast) to 100 (maximum contrast); default = 75, normal contrast.

Monochrome changes colors to shades of gray.

RELATED COMMANDS

Adjust adjusts images and other underlay files.

PdfAttach or **Attach** attaches PDF files as underlays.

PdfClip and **Clip** clip PDF underlays.

PdfLayers and **ULayers** toggle the visibility of layers in PDF underlays.

ExportPDF and **Plot** export drawings in PDF format.

RELATED SYSTEM VARIABLES

PdfFrame toggles the visibility of the rectangular frame surrounding PDF underlays.

PdfOsnap toggles osnapping to geometry inside PDF underlays.

TIPS

- This command's default values are affected by the number of PDF underlays selected:
  - One selected — the defaults are displayed for the selected PDF underlay.
  - Two or more selected — the command's generic defaults are displayed.
- To change the background color of the PDF underlay, double-click it to see the Properties palette.
- To make all colors appear black, turn monochrome on and set contrast to 0.
- To hide the PDF underlay, set Fade to 100; to hide the frame also, turn off PdfFame.

PdfAttach

2010 Attaches PDF files as underlays.

Command	Aliases	Shortcut Keystrokes	Menu	Ribbon
pdfattach	...	...	**Insert** **↳PDF Underlay**	**Insert** **↳Reference** **↳Attach**

Command: pdfattach

Displays the Select Reference File dialog box. Select a PDF file, and then click Attach.

Displays the Attach PDF Underlay dialog box.

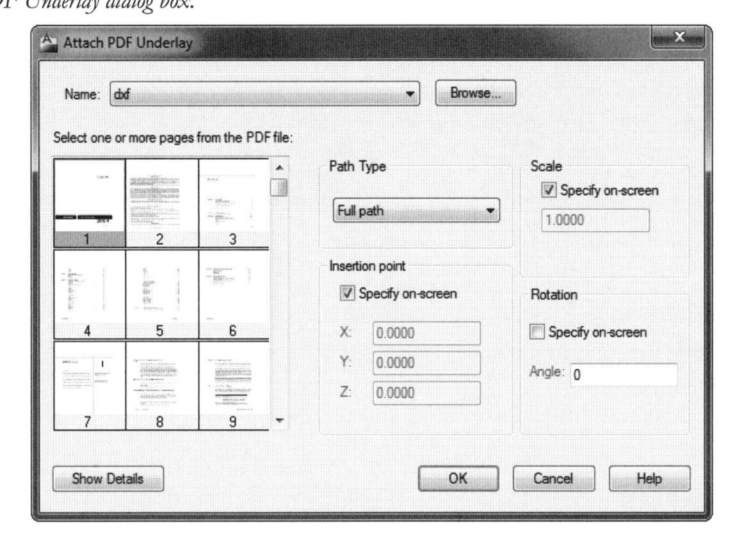

ATTACH PDF UNDERLAY DIALOG BOX

Name specifies the name of the PDF file to be attached as an underlay.

Browse chooses a different PDF file; displays the Select Reference File dialog box.

Select One or More Pages From the PDF File lists the pages found in the PDF file; choose one or more pages to attach. (To select more than one page, hold down the Ctrl key.)

Path Type options

- **Full Path** retains the drive name and path to the folder storing the PDF file.
- **Relative Path** retains the path from the DWG file's folder to the PDF file's folder (starts with \).
- **No Path** ignores the path; the PDF file must reside in the same folder as the DWG file.

Insertion Point options

Specify On-Screen prompts for the insertion point (lower left corner of the PDF underlay) in the drawing:

☑ Displays prompts in the command bar, after the dialog box is closed.

☐ Requires values to be supplied in the dialog box.

X, **Y**, **Z** specify the x, y, and z coordinates of the insertion point.

Scale options

Specify On-Screen prompts for the scale factor (relative to the PDF underlay's insertion point):

☑ Displays prompts in the command bar, after the dialog box is closed.

☐ Requires values to be supplied in the dialog box.

Scale Factor specifies the size relative to 1; negative values mirror the underlay.

Rotation options

Specify On-Screen prompts for the rotation angle (about the PDF underlay's insertion point):

☑ Displays prompts in the command bar, after the dialog box is closed.

☐ Requires values to be supplied in the dialog box.

Angle specifies the rotation angle; positive angles rotate the underlay counterclockwise.

Show Details expands the dialog box to list the found and saved paths. Found In is the full path to the PDF file.

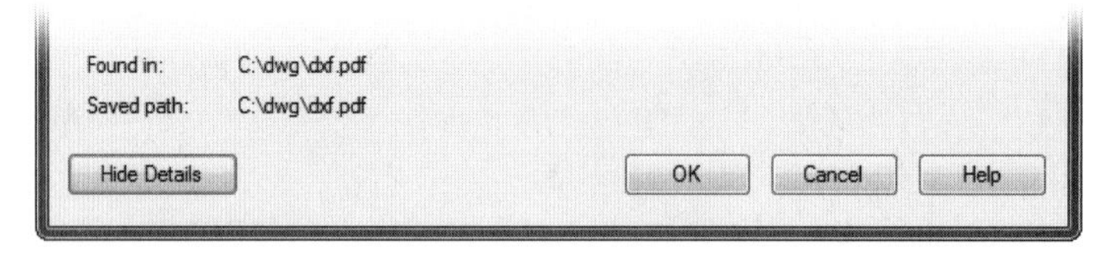

Saved Path specifies the path AutoCAD saves with the drawing; change it with the Path Type droplist.

RELATED COMMANDS

Attach attaches images and other files as underlays.

PdfAdjust or **Adjust** adjusts the contrast, fade, and coloring of PDF underlays.

PdfClip or **Clip** clips PDF underlays.

PdfLayers or **ULayers** toggles the visibility of layers in PDF underlays.

ExportPDF or **Plot** exports drawings in PDF format.

RELATED SYSTEM VARIABLES

PdfFrame toggles the visibility of the rectangular frame surrounding PDF underlays.

PdfOsnap toggles osnapping to geometry inside PDF underlays.

TIPS

- Scale is affected by the InsUnits system variable. *Unitless* applies the scale factor to the underlay's width in AutoCAD's current units; units, such as inches, apply the scale factor to the actual width of the underlay in AutoCAD's current units.
- PDF files are placed with all colors and all layers turned on. Use the PdfAdjust command's Monochrome option to change colors to gray; toggle layer visibility with the PdfLayers command.
- Upon selecting the PDF underlay, the ribbon changes contexts to show PDF-related commands:

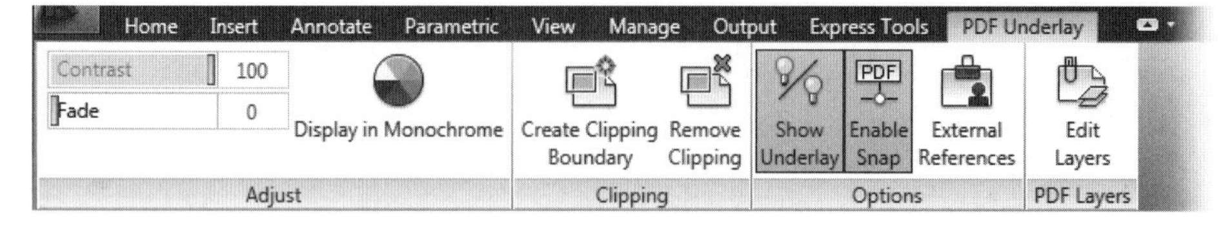

- When you attach a multi-page PDF file, all pages are inserted on top of each other; AutoCAD cannot specify individual insertion points.

PdfClip

2010 Clips PDF underlays to hide portions.

Command	Aliases	Shortcut Keystrokes	Menu	Ribbon
pdfclip	...	...	...	**Insert** ↳**Reference** ↳**Clip**

Command: pdfclip
Select PDF to clip: *(Choose one PDF underlay.)*
Enter PDF clipping option [ON/OFF/Delete/New Boundary] <New Boundary>: *(Enter an option.)*

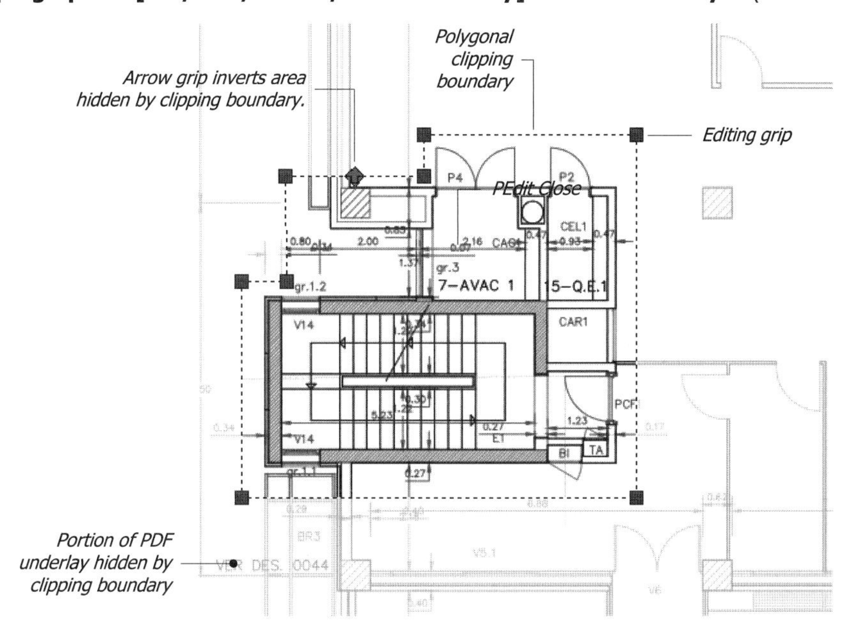

COMMAND LINE OPTIONS

Select PDF to Clip selects a PDF underlay to clip.

ON turns on the previous clipping boundary.

OFF turns off the clipping boundary.

Delete erases the clipping boundary.

New Boundary options
Current mode: Objects inside/outside boundary will be hidden.
Specify clipping boundary:
[Select polyline/Polygonal/Rectangular/Invert clip] <Rectangular>: *(Enter option.)*

Select Polyline prompts you to select a polyline in the drawing, which becomes the clipping boundary.

Polygonal prompts you to pick points that define the clipping boundary; the prompts are similar to those of the PLine command.

Rectangular prompts you to pick two points to define a rectangular boundary.

Invert Clip inverts the display of objects inside or outside the boundary.

Polygonal options

Specify first point: *(Pick a point.)*

Specify next point or [Undo]: *(Pick a point, or type U.)*

Specify next point or [Undo]: *(Pick a point, or type U.)*

Specify next point or [Close/Undo]: *(Pick a point, or enter an option.)*

Specify next point or [Close/Undo]: *(Enter C to close the polygon.)*

Specify first point specifies the start of the first segment of the polygonal clipping path.

Specify next point specifies the next vertex.

Undo undoes the last vertex.

Close closes the polygon clipping path.

Rectangular options

Specify first corner point: *(Pick a point.)*

Specify opposite corner point: *(Pick a point.)*

Specify first corner point specifies one corner of the rectangular clip.

Specify opposite corner point specifies the second corner.

When you select a PDF underlay with a clipped boundary, AutoCAD adds this prompt:

Delete old boundary? [No/Yes] <Yes>: *(Type N or Y.)*

Delete old boundary?

Yes removes the previously-applied clipping path.

No exits the command.

RELATED COMMANDS

Clip clips PDF underlays.

PdfAdjust and **Adjust** changes the look of PDF underlays.

PdfAttach or **Attach** attaches PDF files as underlays.

PdfLayers or **ULayers** toggles the visibility of layers in PDF underlays.

ExportPDF or **Plot** exports drawings in PDF format.

TIPS

- This command adds and removes clipping boundaries.
- Once the boundary is in place, you can edit it interactively: select the boundary, and then drag the grips to change the shape. Click the arrow grip to invert (reverse) the visible area of the underlay. (The hidden portion of the underlay is normally invisible; it is shown above in light gray for illustrative purposes only.)
- Clipping boundaries need not be rectangular or polygonal; they can be any shape. First draw the boundary as a polyline, and then use this command's Select Polyline option to convert the polyline into a clipping boundary.
- To remove the clipping boundary quickly, click the Remove Clipping button on the ribbon's PDF Underlay tab.

PdfLayers

2010 Toggles the visibility of layers in PDF underlays.

Command	*Aliases*	*Shortcut Keystrokes*	*Menu*	*Ribbon*
pdflayers	...	...	...	**Insert** ↳**Reference** ↳**Underlay Layers**

Command: pdflayers

Displays dialog box.

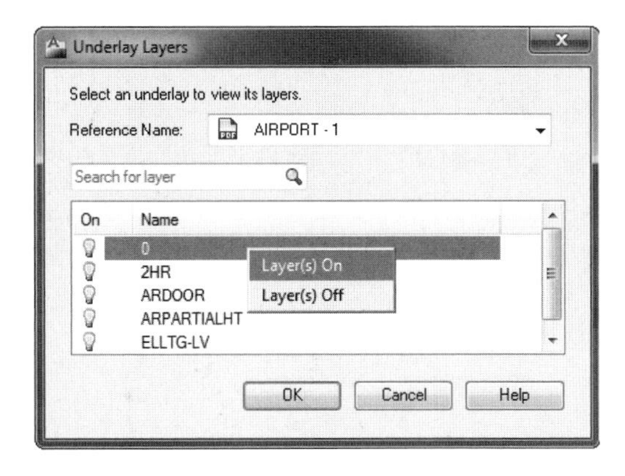

UNDERLAY LAYERS DIALOG BOX

Reference Name selects the PDF underlay; click the droplist, and then choose a name.

Search for Layer filters the list of layer names in real time; enter the first few characters of a layer name. Alternatively, you can choose layers by clicking their names in the dialog box; hold down the Ctrl key to choose two or more.

On sorts layer names by their status, on or off.

Name sorts layer names alphabetically; click Name a second time to sort in reverse order (from Z to A).

SHORTCUT MENU OPTIONS

Access this shortcut menu by right-clicking a layer name:

Layer(s) On turns on the selected layers.

Layer(s) Off turns off the selected layers, and does not plot them.

RELATED COMMANDS

PdfAdjust and **Adjust** adjust PDF underlay files.

PdfAttach or **Attach** attach PDF files as underlays.

PdfClip or **Clip** clip PDF underlays.

PdfLayers or **ULayers** toggle the visibility of layers in PDF underlays.

TIPS

- You can search the layer list using wildcards, such as ? \for any character and * for all characters.
- To hide the entire PDF underlay, click the Show Underlay button on the ribbon's PDF Underlay tab.

PEdit

Ver. 2.1 Edits 2D polylines, 3D polylines, or 3D polygon meshes — depending on the selected object (short for Polyline EDIT).

Command	Alias	Shortcut Keystrokes	Menu	Ribbon
pedit	**pe**	...	**Modify** ↳**Object** ↳**Polyline**	**Home** ↳**Modify** ↳**Edit Polyline**

Command: pedit

Options vary, depending on whether you pick a 2D polyline, 3D polyline, or polymesh.

2D Polyline Operations

The following prompts appear when you select a 2D polyline:

Select polyline or [Multiple]: *(Pick a 2D polyline, or type M.)*

Enter an option [Open *(or* **Close***)***/Join/Width/Edit vertex/Fit/Spline/Decurve/Ltype gen/Reverse/ Undo]:** *(Enter an option.)*

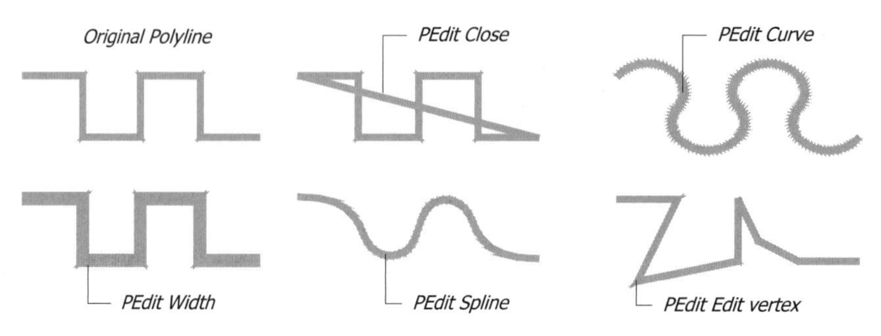

COMMAND LINE OPTIONS

Select polyline chooses a single 2D polyline to edit.

Multiple selects more than one polyline to edit; see options below.

Open opens a closed polyline by removing the last segment.

Close closes an open polyline by joining the two endpoints with a single segment.

Join adds other polylines, arcs, lines, and splines to the current polyline.

Width changes the width of the entire polyline.

Edit vertex options are listed below.

Fit fits a curve to the tangent points of each vertex.

Decurve reverses the effects of Fit and Spline operations.

Spline fits a splined curve along the polyline.

Ltype gen specifies the linetype generation style.

Reverse reverses the direction of the polyline (start point becoming end point, and vice versa).

Undo undoes the most recent PEdit operation.

eXit exits the PEdit command.

Multiple options

When you include any lines, arcs, or splines in the multiple selection set, AutoCAD prompt you:

Convert Lines, Arcs and Splines to polylines [Yes/No]? <Y>: *(Type Y or N.)*

Specify a precision for spline conversion <10>: *(Enter an integer between 0 and 100.)*

Convert checks if you really want to convert the lines, arcs, and splines to part of the polyline.

Specify a precision determines how closely the converted spline adheres to its old path *(new to AutoCAD 2011).*

Edit Vertex options

Enter a vertex editing option

[Next/Previous/Break/Insert/Move/Regen/Straighten/Tangent/Width/eXit] <N>: *(Enter an option.)*

Note: It is much easier to use grips to edit polyline vertices than to use this option:

Break removes a segment or breaks the polyline at a vertex:

- **Next** moves the x marker to the next vertex.
- **Previous** moves the x marker to the previous vertex.
- **Go** performs the break.
- **eXit** exits the Break sub-submenu.

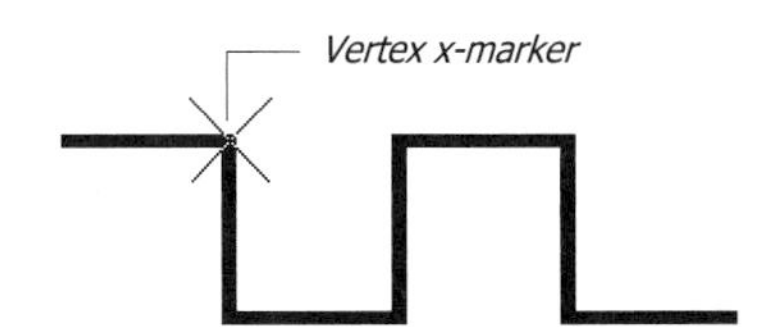

Insert inserts another vertex.

Move relocates a vertex; more easily accomplished with grips editing.

Next moves the x marker to the next vertex.

Previous moves the x marker to the previous vertex.

Regen regenerates the screen to show the effect of the PEdit command.

Straighten draws a straight segment between two vertices:

- **Next** moves the x marker to the next vertex.
- **Previous** moves the x marker to the previous vertex.
- **Go** performs the straightening.
- **eXit** exits the Straighten sub-submenu.

Tangent shows the tangent to the current vertex.

Width changes the width of a segment.

eXit exits the Edit vertex submenu.

3D Polyline Operations

The following prompts appear when a 3D polyline is selected:

Select polyline or [Multiple]: *(Pick a 3D polyline, or type M.)*

Enter an option [Open *(or* **Close)/Join/Edit vertex/Spline curve/Decurve/Reverse/Undo]:** *(Enter an option.)*

COMMAND LINE OPTIONS

Select polyline chooses a single 3D polyline to edit.

Multiple selects more than one polyline to edit; see options listed above.

Open removes the last segment of a closed polyline.

Close closes an open polyline.

Join adds other 3D and 2D polylines, arcs, lines, and splines to the current polyline.

Edit vertex prompts you (see options listed above):

Enter a vertex editing option
[Next/Previous/Break/Insert/Move/Regen/Straighten/eXit] <N>: *(Enter an option)*

Spline curve fits a splined curve along the polyline.

Decurve reverses the effects of the Fit curve and Spline curve operations.

Reverse reverses the direction of the polyline (start points becoming end points, and vice versa).

Undo undoes the most recent PEdit operation.

3D Polygon Mesh Operations

The following prompts appear when a 3D polygon mesh is selected (does not apply to 3D mesh objects):

Select polyline or [Multiple]: *(Pick a 3D polygon mesh, or type M.)*

Enter an option [Edit vertex/Smooth surface/Desmooth *(or* **Smooth surface***)***/Mclose** *(or* **Mopen***)***/ Nclose** *(or* **Nopen***)***/Undo]:** *(Enter an option.)*

COMMAND LINE OPTIONS

Select polyline chooses a single 3D polygon mesh to edit.

Multiple selects more than one 3D polygon mesh to edit.

Edit vertex prompts you (see options listed above for prompts):

Current vertex (0,0).
Enter an option [Next/Previous/Left/Right/Up/Down/Move/REgen/eXit] <N>: *(Enter an option.)*

Desmooth reverses the effect of the Smooth surface options.

Smooth surface smooths the mesh with a Bezier-spline.

Mclose and **Mopen** close and open the mesh in the m-direction.

Mopen opens the mesh in the m-direction.

Nclose and **Nopen** close and open the mesh in the n-direction.

Undo undoes the most recent PEdit operation.

eXit exits the PEdit command.

RELATED COMMANDS

Break breaks 2D polylines at any position.

Chamfer chamfers all vertices of 2D polylines.

Convert converts older polylines to the new lwpolyline format.

Fillet fillets all vertices of 2D polylines.

Join joins like objects, as well as lines, arcs, and polylines into a single polyline.

PLine draws 2D polylines.

3dPoly draws 3D polylines.

RELATED SYSTEM VARIABLES

SplFrame determines the visibility of a polyline spline frame.

SplineSegs specifies the number (-32768 to 32767) of lines or arcs used to draw a splined polyline (default = 8); when the number is negative, arc segments are used; when positive, line segments.

SplineType determines the Bezier-spline smoothing for 2D and 3D polylines.

SurfType determines the smoothing for the Smooth option.

TIPS

- During vertex editing, move the x marker by pressing the spacebar, or the Enter key, or by right-clicking, or pressing mouse button #2. The marker is moved to the next or previous vertex, whichever was used last.
- It's easier to edit the position of vertices with grips, than with this command.
- AutoCAD 2011 added grips to the midpoint of each polyline and polyarc segment; pause the cursor over a grip to access the shortcut menu that converts segments between lines and arcs segments, as well as adds vertices and stretches segments.

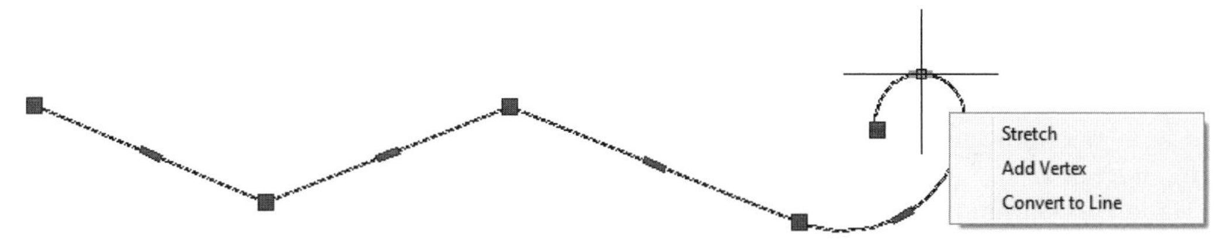

- You can select polyline segments individually by holding down the Ctrl key.

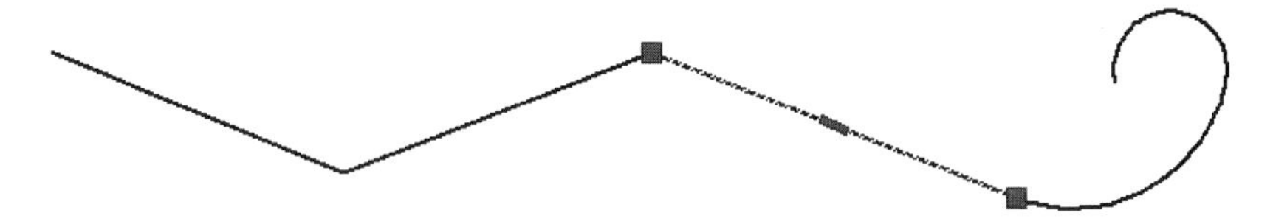

- Cursor icons indicate the editing action being carried out:

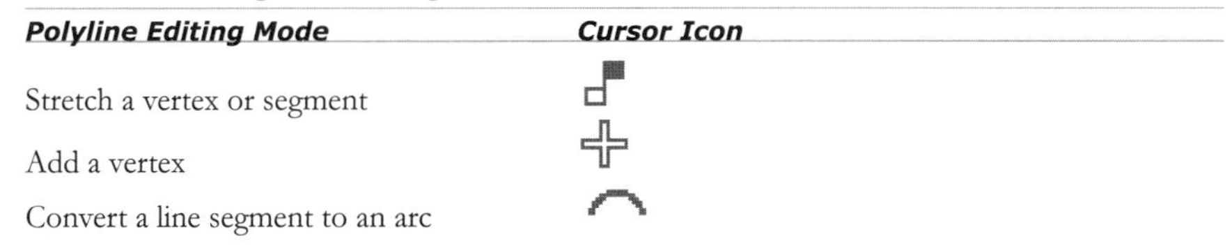

Polyline Editing Mode	***Cursor Icon***
Stretch a vertex or segment	
Add a vertex	
Convert a line segment to an arc	

- Hold down the Ctrl key to add more than one vertex.

PFace

Rel. 11 Draws multisided 3D meshes; intended for use by AutoLISP and ARx programs (short for Poly FACE).

Command	Alias	Shortcut Keystrokes	Menu	Ribbon
pface	...	...	...	...

Command: pface
Specify location for vertex 1: *(Pick point 1.)*
Specify location for vertex 2 or <define faces>: *(Pick point 2, or press Enter.)*
Face 1, vertex 1:
Enter a vertex number or [Color/Layer]: *(Type a number, or enter an option.)*
Face 1, vertex 2:
Enter a vertex number or [Color/Layer] <next face>: *(Type a number, or enter an option.)*
...and so on.

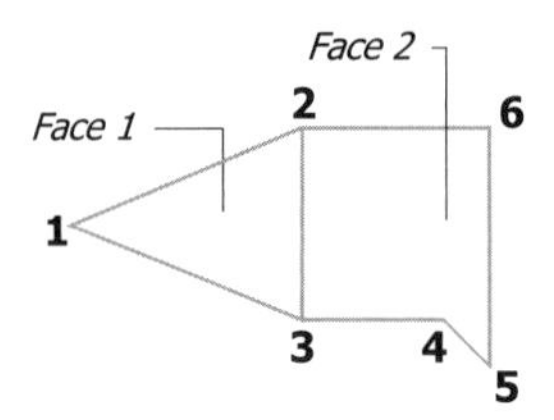

COMMAND LINE OPTIONS

Vertex defines the location of a vertex.

Face defines the faces, based on vertices.

Color gives the face a different color.

Layer places the face on a different layer.

RELATED COMMANDS

3dFace draws three- and four-sided 3D meshes.

3dMesh draws a 3D mesh with polyfaces.

RELATED SYSTEM VARIABLES

PFaceVMax specifies the maximum number of vertices per polyface (default = 4).

SplFrame controls the display of invisible faces (default = 0, not displayed).

TIPS

- The 3dFace command creates 3- and 4-sided meshes, while this command creates meshes of an arbitrary number of sides, and allows you to control the layer and the color of each face.
- The maximum number of vertices in the m- and n-direction is 256 when entered from the keyboard, and 32767 when entered from DXF files or from programs.
- Make an edge invisible by entering a negative number for the beginning vertex of the edge.
- This command is meant for programmers. To draw 3D mesh objects, use these commands instead: 3D, 3dMesh, RevSurf, RuleSurf, EdgeSurf, or TabSurf.

Plan

Rel. 10 Displays the plan view of the WCS, the current UCS, or a named UCS.

Command	*Aliases*	*Shortcut Keystrokes*	*Menu*	*Ribbon*
plan	...	...	**View** **↳3D Views** **↳Plan View**	...

Command: plan

Enter an option [Current ucs/Ucs/World] <Current>: *(Enter an option.)*

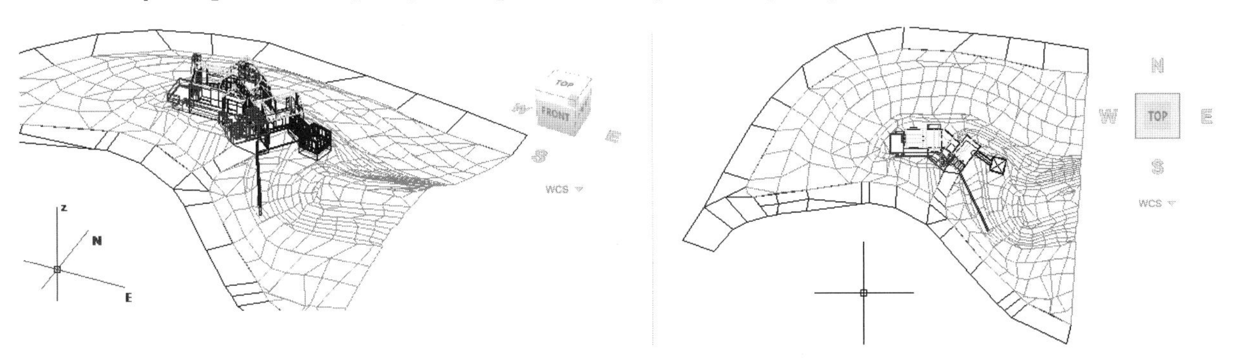

***Left**: Example of a 3D view.*
***Right**: After using the Plan World command.*

COMMAND LINE OPTIONS

Current UCS shows the plan view of the current UCS.

Ucs shows the plan view of the named UCS.

World shows the plan view of the WCS.

RELATED COMMANDS

UCS creates new UCS views.

VPoint changes the viewpoint of 3D drawings.

RELATED SYSTEM VARIABLES

UcsFollow displays the plan view for UCS or WCS automatically.

ViewDir contains the x, y, z coordinates of the current view direction.

TIPS

- Entering VPoint **0,0,0** is an alternative to using this command (or any value for the z coordinate).
- This command turns off perspective mode and clipping planes.
- Plan does not work in paper space; AutoCAD complains, "** Command only valid in Model space **."
- This command is an excellent method for turning off perspective mode.

PlaneSurf

2007 Draws flat rectangular surfaces, or converts 2D closed objects into surfaces (short for PLANar SURFace).

Commands	Aliases	Shortcut Keystrokes	Menu	Ribbon
planesurf	...	...	**Draw** ↳**Modeling** ↳**Surface** ↳**Planar Surface**	**Surface** ↳**Create** ↳**Planar**

Command: planesurf
Specify first corner or [Object] <Object>: *(Pick a point, or type O.)*
Specify other corner: *(Pick another point.)*

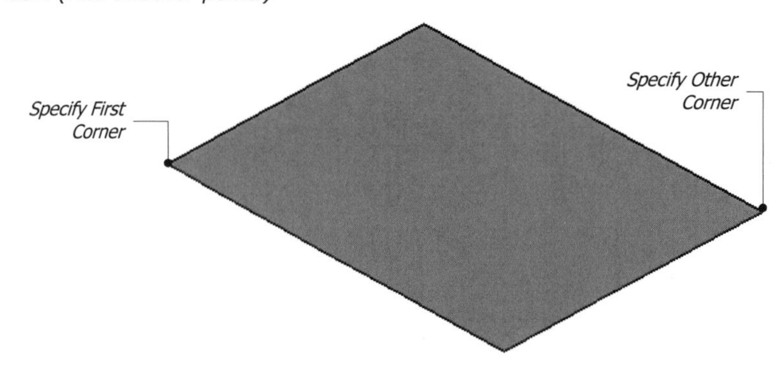

COMMAND LINE OPTIONS

Specify first corner picks the first corner to make a rectangular surface.

Object converts the selected closed 2D object into a surface; choose a line, circle, arc, ellipse, elliptical arc, 2D polyline, planar 3D polyline, or spline.

Specify other corner specifies the other corner of the rectangular surface.

RELATED COMMAND

ConvertToSurface converts non-planar objects to surfaces.

RELATED SYSTEM VARIABLE

Elevation determines the z-height of planar surfaces.

TIPS

- Use this command to create "floors."
- Or ceilings, when Elevation is set to 96".
- Or overcast skies, when Elevation is set to 6,000'.

PLine

Ver. 1.4 Draws complex 2D lines made of straight and curved sections of constant and variable width; treated as a single object (short for Poly LINE).

Command	*Alias*	*Shortcut Keystrokes*	*Menu*	*Ribbon*
pline	**pl**	...	**Draw** ↳**Polyline**	**Home** ↳**Draw** ↳**Polyline**

Command: pline
Specify start point: *(Pick a point.)*
Current line-width is 0.0000
Specify next point or [Arc/Halfwidth/Length/Undo/Width]: *(Pick a point, or enter an option.)*
Specify next point or [Arc/Close/Halfwidth/Length/Undo/Width]: *(Pick a point, or enter an option.)*

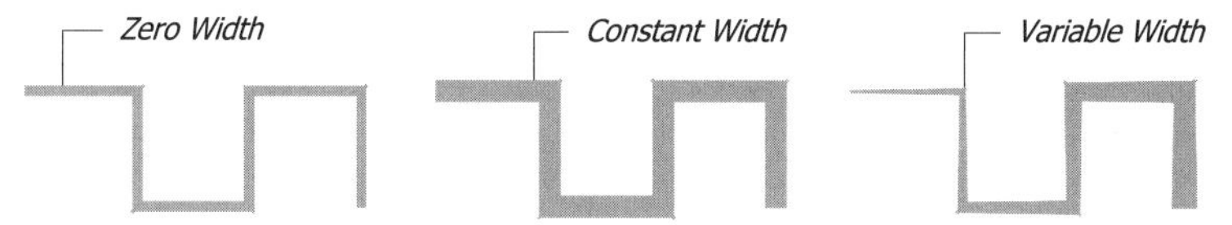

COMMAND LINE OPTIONS

Specify start point indicates the start of the polyline.

Close closes the polyline with a line segment.

Halfwidth indicates the half-width of the polyline.

Length draws a polyline tangent to the last segment.

Undo erases the last-drawn segment.

Width indicates the width of the polyline; start and end widths can be different.

Endpoint of line indicates the endpoint of the polyline.

Arc options

Specify endpoint of arc or
[Angle/CEnter/CLose/Direction/Halfwidth/Line/Radius/Second pt/Undo/Width]: *(Enter an option.)*

Endpoint of arc indicates the arc's endpoint.

Angle indicates the arc's included angle.

CEnter indicates the arc's center point.

CLose uses an arc to close a polyline.

Direction indicates the arc's starting direction.

Halfwidth indicates the arc's half width.

Line switches back to the menu for drawing lines.

Radius indicates the arc's radius.

Second pt draws a three-point arc.

Undo erases the last drawn arc segment.

Width indicates the width of the arc.

RELATED COMMANDS

Boundary draws a polyline boundary.

Donut draws solid-filled circles as polyline arcs.

Ellipse draws ellipses as polyline arcs when PEllipse = 1.

Explode reduces a polyline to lines and arcs with zero width.

Fillet fillets polyline vertices with a radius.

Join joins lines, arcs, and polylines.

PEdit edits the polyline's vertices, widths, and smoothness.

Polygon draws polygons as polylines of up to 1024 sides.

Rectang draws rectangles out of polylines.

Reverse reverses the order of vertices in polylines.

Sketch draws polyline sketches, when SkPoly = 1.

Xplode explodes a group of polylines into lines and arcs of zero width.

3dPoly draws 3D polylines.

RELATED SYSTEM VARIABLES

PLineGen specifies the style of linetype generation.

PLineType controls the conversion of old (pre-Release 14) polylines and the creation of lwpolyline objects by the PLine command.

PLineWid sets the current width of polyline (default = 0.0).

TIPS

- The Boundary command uses a polyline to outline a region automatically; use the List command to find its area.
- If you cannot see a linetype on a polyline, change system variable PlineGen to 1; this regenerates the linetype from one end of the polyline to the other.
- When a polyline is joined to a polyarc, the line is chamfered and never the arc.
- Use the object snap modes INTersection or ENDpoint to snap to the vertices of a polyline.
- You can snap only to the center of wide polylines.
- Polylines can be edited directly using grips; see PEdit command.

Plot

Ver. 1.0 Creates hard copies of drawings on vector, raster, and PostScript plotters; also plots to files on disk.

Commands	Aliases	Shortcut Keystroke	Menu Bar	Application Menu	Quick Access Toolbar
plot	**print** **dwfout**	**Ctrl+P**	**File** ↳**Plot**	**Print** ↳**Print**	
-plot	...	...	...	...	...

Command: plot

Displays dialog box.

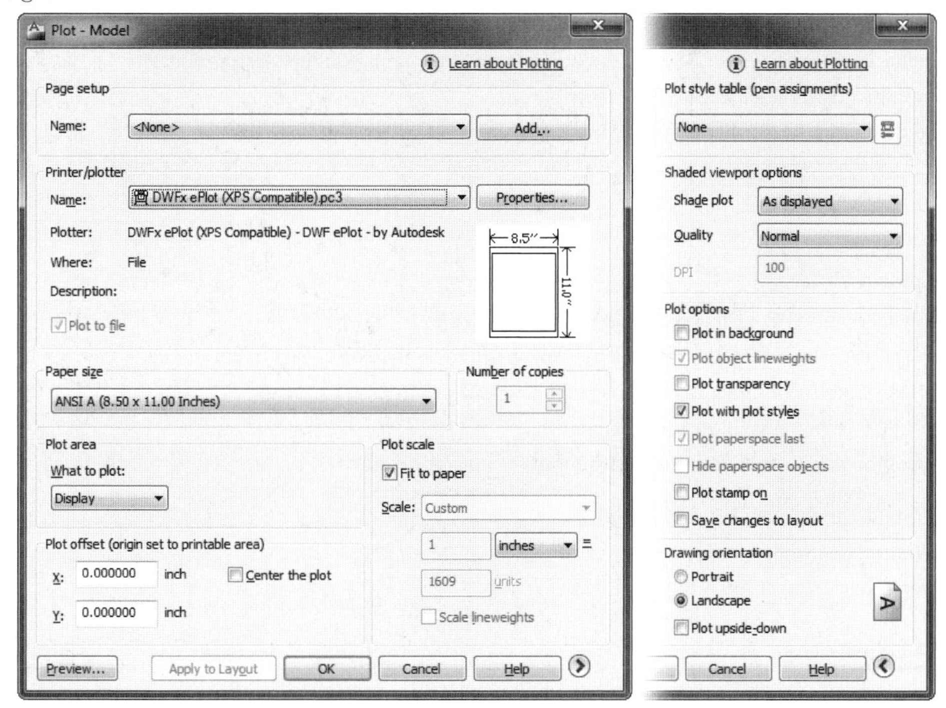

PLOT DIALOG BOX

Page Setup options

Name selects or imports a previous plot setup.

Add prompts for the name of a new plot setup; displays the Add Page Setup dialog box.

Printer/Plotter Configuration options

Name selects a system printer, a *.pc3* named plotter configuration, or previous plot setting.

Properties displays the Plotter Configuration Editor dialog box; see the PlotterManager command.

☐ **Plot to file** plots the drawing to a *.plt* file.

Paper Size option

Paper size displays paper sizes supported by the selected output device.

Plot Area options

What to plot:

- **Layout** plots all parts of the drawing within the margins of the specified paper size; the origin is calculated from 0,0 in the layout (not available in model space).
- **Display** plots all parts of the drawing within the current viewport.
- **Extents** plots the entire drawing.
- **Limits** plots the drawing in the Model tab within a rectangle defined by the Limits command.
- **View** plots a view saved with the View command (not available when no views are named).
- **Window** plots all parts of the drawing within a picked rectangle.

Plot Offset options

Center the plot centers plots on the paper.

X specifies the plot origin offset from x = 0; the origin varies, depending on the plotter model.

Y specifies the plot origin offset from y = 0.

☑**Center the plot** centers plots between page edges or printer margins; unavailable when Layout option is on.

Number of Copies option

Number of copies specifies the number of copies to plot.

Plot Scale options

☑**Fit to paper** scales drawings to fit within the plotter's margins; unavailable when Layout option is on.

Scale specifies the plotting scale; ignored by layouts.

- **Custom** specifies user-defined scale; this list can be edited by the ScaleListEdit command.

Inches / mm specifies the number of inches (or millimeters) on the plotted page.

Unit specifies number of units in the drawing.

☐**Scale lineweights** scales lineweights proportionately to the plot scale.

Preview displays the preview window; see the Preview command.

Apply to layout applies setting to layout.

OK starts the plot.

MORE OPTIONS

Plot Style Table (Pen Assignments) options

Droplist selects a .ctb plot style table file; see the PlotStyle command.

New displays the Add Color-Dependent Plot Style Table wizard; see the StylesManager command.

Edit button displays the Plot Style Table Editor dialog box; see the StylesManager command.

Shaded Viewport Options options

Available only in model space; in layouts, right-click the viewport border, and then select Properties.

Shade Plot plots the viewport as-displayed, in a visual style, or rendered.

Quality sets print quality as draft (wireframe), preview (quarter-resolution, up to 150dpi), normal (half resolution, up to 300dpi), presentation (maximum printer resolution up to 600dpi), maximum (maximum printer resolution), or custom (as specified by DPI text box).

DPI (short for "dots for inch) specifies the resolution for the output device.

Plot Options options

☐ **Plot in background** plots drawings as a background process.

☑ **Plot object lineweights** plots objects with lineweights (available only when plot style table turned off).

☐ **Plot transparency** plots objects with translucency, if applied in the drawing. *Caution*: to simulate object translucency, AutoCAD must rasterize the drawing, which can prolong print times *(new to AutoCAD 2011)*.

☑ **Plot with plot styles** plots objects using the plot styles.

☑ **Plot paperspace last** plots model space objects first, followed by paper space objects (available in layouts only).

☐ **Hide paperspace objects** removes hidden lines before plotting.

☐ **Plot Stamp On** stamps the plot; when on, the Plot Stamp Settings button appears. See PlotStamp command.

☐ **Save changes to layouts** saves changes made in this dialog box to the layout.

Draw Orientation options

⊙ **Portrait** positions the paper vertically.

○ **Landscape** positions the paper horizontally.

☐ **Plot Upside Down** plots the drawing upside down.

-PLOT Command

Command: -plot
Detailed plot configuration? [Yes/No] <No>: *(Type Y or N.)*

Brief Plot Configuration options
Enter a layout name or [?] <Model>: *(Enter a layout name, or type ?.)*
Enter a page setup name <>: *(Enter a page setup name, or press Enter for none.)*
Enter an output device name or [?] <default printer>: *(Enter a printer name, or type* **?** *for a list of printers.)*
Write the plot to a file [Yes/No] <N>: *(Type Y or N.)*
Save changes to page setup [Yes/No]? <N> *(Type Y or N.)*
Proceed with plot [Yes/No] <Y>: *(Type Y or N.)*

Detailed Plot Configuration options
Enter a layout name or [?] <Model>: *(Enter a layout name, or type ?.)*
Enter an output device name or [?] <default printer>: *(Enter a printer name, or type ? for a list of printers.)*
Enter paper size or [?] <Letter 8 ½ x 11 in>: *(Enter a paper size, or type ?.)*
Enter paper units [Inches/Millimeters] <Inches>: *(Type I or M.)*
Enter drawing orientation [Portrait/Landscape] <Landscape>: *(Type P or L.)*
Plot upside down? [Yes/No] <No>: *(Type Y or N.)*
Enter plot area [Display/Extents/Limits/View/Window] <Display>: *(Enter an option.)*
Enter plot scale (Plotted Inches=Drawing Units) or [Fit] <Fit>: *(Enter a scale factor, or type F.)*
Enter plot offset (x,y) or [Center] <0.00,0.00>: *(Enter an offset, or type C.)*
Plot with plot styles? [Yes/No] <Yes>: *(Type Y or N.)*
Enter plot style table name or [?] (enter . for none) <>: *(Enter a name, or enter an option.)*
Plot with lineweights? [Yes/No] <Yes>: *(Type Y or N.)*
Enter shade plot setting [As displayed/legacy Wireframe/legacy Hidden/Visual styles/Rendered] <As displayed>: *(Enter an option.)*

Write the plot to a file [Yes/No] <N>: *(Type Y or N.)*
Save changes to page setup? Or set shade plot quality? [Yes/No/Quality] <N>: *(Enter an option.)*
Proceed with plot [Yes/No] <Y>: *(Type Y or N.)*

COMMAND OPTIONS

Enter a layout name specifies the name of a layout; enter Model for model view or ? for a list of layout names.

Enter an output device name specifies the name of a printer, or enter ? for a list of print devices.

Enter paper size specifies the name of a paper size, or enter ? for a list of paper sizes supported by the print device.

Enter paper units specifies Inches (imperial) or Millimeters (metric) units.

Enter drawing orientation specifies Portrait (vertical) or Landscape (horizontal) orientation.

Plot upside down:

- **Yes** plots the drawing upside down.
- **No** plots the drawing normally.

Enter plot area specifies that the current Display, Extents of the drawing, Limits defined by the Limits command, View name, or window area be plotted.

Enter plot scale specifies the scale using the Plotted Inches=Drawing Units format; or enter F to fit the drawing to the page.

Enter plot offset specifies the distance between the lower left corner of the paper and the drawing; or enter C to center the drawing on the page.

Plot with plot styles:

- **Yes** prompts you to specify the name of a plot style.
- **No** ignores plot styles, and uses color-dependent styles.

Plot with lineweights:

- **Yes** uses lineweight settings to draw thicker lines.
- **No** ignores lineweights.

Enter shade plot setting specifies whether the plot should be rendered As displayed, Wireframe, Hidden lines removed, named Visual style, or Rendered.

Write the plot to a file:

- **Yes** sends the plot to a file.
- **No** sends the plot to the plot device.

Save changes to page setup:

- **Yes** saves the settings you specified during the previous set of questions.
- **No** does not save the settings.
- **Quality** specifies the quality of the plot.

Proceed with plot:

- **Yes** plots the drawing.
- **No** cancels the plot.

Quality options
Enter shade plot quality [Draft/Preview/Normal/pResentation/Maximum/Custom] <Normal>: *(Select an option.)*

Draft, Preview, Normal, pResentation, and Maximum apply preset dpi (dots per inch) settings to the plot. The higher the dpi, the better the quality, but the slower the plot.

Custom allows you to specify the dpi setting; default = 150dpi.

STATUS BAR OPTIONS

Cancel Sheet cancels the current sheet being printed.

Cancel Entire Job cancels the print job.

View Plot and Publish Details displays the Plot and Publish Details dialog box; see the ViewPlotDetails command.

View Plotted File displays PDF and DWF plots in their respective viewers, if installed on your computer.

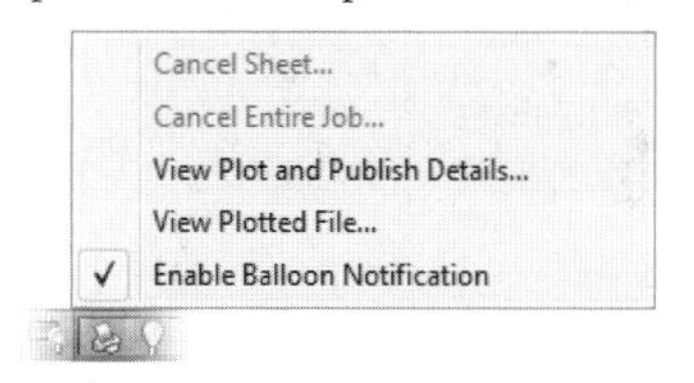

Enable Balloon Notification toggles the display of the balloon reporting the status of print jobs:

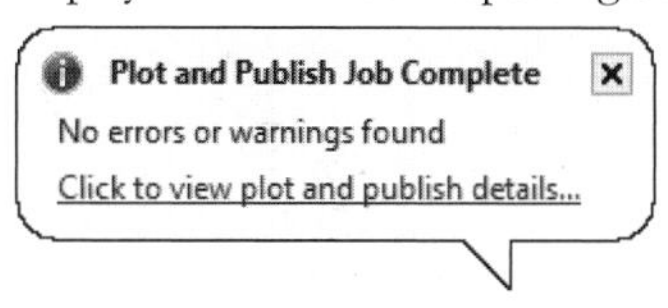

RELATED COMMANDS

PageSetup selects one or more plotter devices.

Preview goes directly to the plot preview screen.

Publish plots multi-page drawings.

PublishToWeb plots drawings to HTML format.

PsOut saves drawings in *.eps* format.

ViewPlotDetails reports on successful and unsuccessful plots.

RELATED SYSTEM VARIABLES

BackgroundPlot toggles background plotting.

BgrdPlotTimeout specifies the time AutoCAD waits for plotter to respond.

FullPlotPath sends either the full path or just the drawing name to the plot spooler software.

RasterPercent specifies the percentage of system memory to allocate to plotting raster images; default = 20%.

RasterDpi specifies the scaling for raster devices.

RasterThreshold specifies the maximum amount of physical memory to assign to plotting raster images; plot fails when memory is insufficient. Default = 20MB.

TextFill toggles the filling of TrueType fonts.

TextQlty specifies the quality of TrueType fonts.

RELATED FILES

.*pc3 holds plotter configuration parameter files.

.*plt holds plot files created with this command.

TIPS

- AutoCAD R12 replaced PrPlot with the Plot command. R13 removed -p freeplot (plotting without using up a network license). AutoCAD 2005 removed the partial preview.
- AutoCAD cannot plot perspective views to-scale, only to-fit.
- You must wait for one background plot to complete before beginning the next one.
- Turn off BackgroundPlot when plotting drawings with scripts.
- As of AutoCAD 2008, the Plot command takes into account these 3dConfig command settings: Emulate Unsupported Hardware Effects, and Texture Compression.
- AutoCAD 2011 adds the transparency option to the Plot dialog box, though not for the -Plot command line.

PlotStamp

2000i Adds information about drawings to hard copy plots.

Commands	*Alias*	*Shortcut Keystrokes*	*Menu*	*Ribbon*
'plotstamp	**ddplotstamp**	...	...	...
'-plotstamp	...	...	...	...

Command: plotstamp

Displays dialog box.

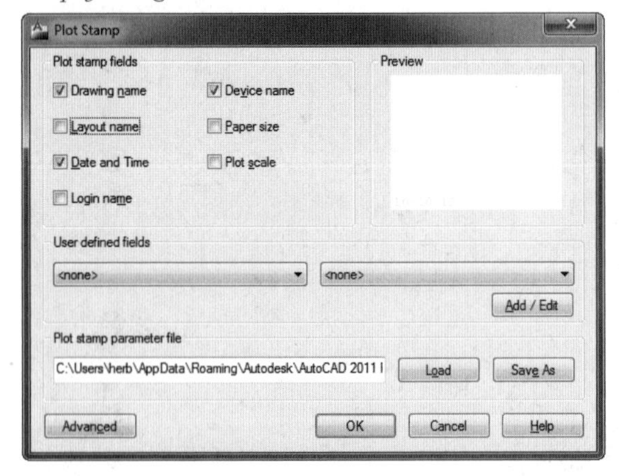

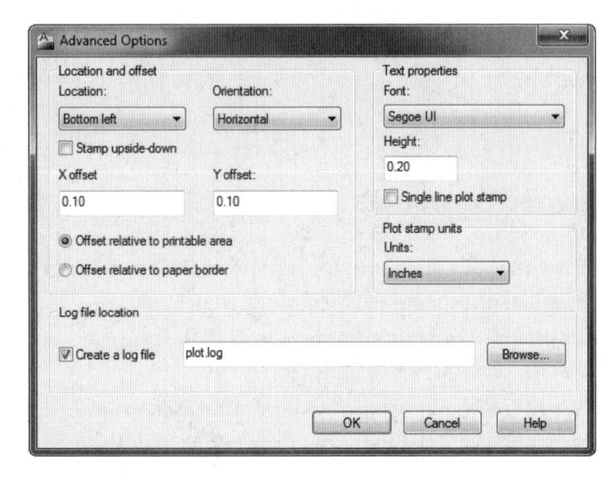

Left: *Plot Stamp dialog box.*
Right: *Advanced Options dialog box.*

PLOT STAMP DIALOG BOX

Plot Stamp Fields options

☑**Drawing name** adds path and name of the drawing.

☐**Layout name** adds layout name.

☑**Date and time** adds short-format date/time of plot.

☐**Login name** adds Windows login name, as stored in the LogInName system variable.

☑**Device name** adds plotting device's name.

☐**Paper size** adds paper size, as currently configured.

☐ **Plot scale** adds the plot scale.

User Defined Fields options

Add/Edit displays the User Defined Fields dialog box.

Plot Stamp Parameter File options

Load/Save displays the Plotstamp Parameter File Name dialog box for opening (or saving) *.pss* files.

Advanced displays the Advanced Options dialog box.

ADVANCED OPTIONS DIALOG BOX

Location and Offset options

Location specifies the location of the plot stamp: Top Left, Bottom Left (default), Bottom Right, or Top Right — relative to the orientation of the drawing on the plotted page.

☑**Stamp upside down** draws the plot stamp upside down; otherwise, right side up.

Orientation rotates the plot stamp relative to the page: Horizontal or Vertical.

X and **Y offset** specify the distance from the corner of the printable area or page; default = 0.1.

⦿**Offset relative to printable area.**

○**Offset relative to paper border.**

Text Properties options

Font selects the font for the plot stamp text.

Height specifies the height of the plot stamp text.

☑**Single line plot stamp** places stamp on a single line; otherwise, on two lines.

Plot Stamp Units option

Units selects either inches, millimeters, or pixels.

Log File Location options

☑ **Create a log file** saves the plot stamp text to a text file.

Browse displays the Log File Name dialog box.

-PLOTSTAMP Command

Command: -plotstamp
Current plot stamp settings:

Displays the current setting of nearly twenty plot stamp settings.

Enter an option [On/OFF/Fields/User Fields/Log file/LOCation/Text properties/UNits]: *(Enter option.)*

COMMAND LINE OPTIONS

On turns on plot stamping.

OFF turns off plot stamping.

Fields specifies plot stamp data: drawing name, layout name, date and time, login name, plot device name, paper size, plot scale, comment, write to log file, log file path, location, orientation, offset, offset relative to, units, font, text height, and stamp on single line.

User fields specifies two user-defined fields.

Log file writes the plotstamp data to a file instead of to the drawing.

LOCation specifies the location and orientation of the plot stamp.

Text properties specifies the font name and height.

UNits specifies the units of measurement: inches, millimeters, or pixels.

RELATED COMMAND

Plot plots the drawing; access the Plot Stamp dialog box via the Plot command's More Options dialog box.

RELATED FILES

.*log is the plot stamp log file; stored in ASCII text format.

.*pss holds the plot stamp parameters; stored in binary format.

TIPS

- This command always uses pen 7 (color 7).
- When options are grayed out in the Plot Stamp dialog box (or -Plotstamp reports 'Current plot stamp file or directory is read only'), this means that the *Inches.pss* or *Mm.pss* file in the *Support* folder is read-only. Use Explorer to change this: (1) right-click each file, (2) select Properties, (3) uncheck Read-only, and (4) click OK.

PlotStyle

2000 Selects and assigns plot styles to objects.

Commands	*Aliases*	*Shortcut Keystrokes*	*Menu Bar*	*Ribbon*	
plotstyle	...	...	**Format** ↳**Plot Style**	...	
-plotstyle	...	...	...	...	...

This command operates only when plot styles are attached to the current drawing:

- *Color-dependent drawings are converted to named plot styles with the ConvertPStyles command.*
- *Future new drawings use named plot styles when PStylePolicy is set to 0.*

Command: plotstyle

When no objects are selected, displays the Current Plot Style dialog box.

When one or more objects are selected, displays the Select Plot Style dialog box.

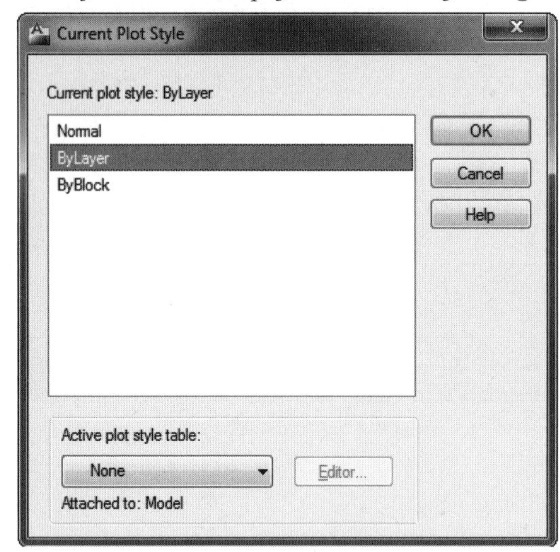

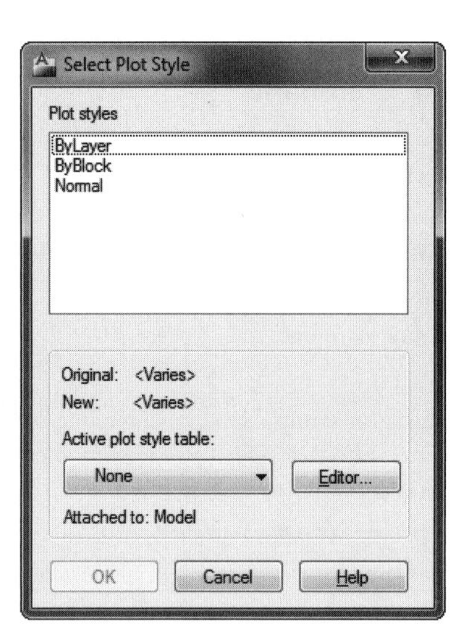

Left: *Current Plot Style dialog box.*
Right: *Select Plot Style dialog box.*

CURRENT PLOT STYLE DIALOG BOX

Displayed when no objects are selected.

Plot styles lists the plot styles available in the drawing.

Active plot style table selects the plot style table to attach to the current drawing.

Editor displays the Plot Style Table Editor dialog box; see the StylesManager command.

SELECT PLOT STYLE DIALOG BOX

Displayed when one or more objects are selected.

Plot styles lists the plot styles available in the drawing.

Active plot style table selects the plot style table to attach to the current drawing.

Editor displays the Plot Style Table Editor dialog box; see the StylesManager command.

-PLOTSTYLE Command

Command: -plotstyle
Current plot style is "Default"
Enter an option [?/Current] : *(Type ? or C.)*
Set current plot style : *(Enter a name.)*
Current plot style is "*plotstylename*"
Enter an option [?/Current] : *(Press Enter.)*

COMMAND LINE OPTIONS

Current prompts you to change plot styles.

Set current plot style specifies the name of the plot style.

Enter exits the command.

? displays plot style names.

RELATED COMMANDS

Plot plots drawings with plot styles.

PageSetup attaches plot style tables to layouts.

StylesManager modifies plot style tables.

Layer specifies plot styles for layers.

RELATED SYSTEM VARIABLES

CPlotStyle specifies the plot style for new objects; defined values include "ByLayer," "ByBlock," "Normal," and "User Defined."

DefLPlStyle specifies the plot style name for layer 0.

DefPlStyle specifies the default plot style for new objects.

PStyleMode indicates whether the drawing is in Color-Dependent or Named Plot Style mode.

PStylePolicy determines whether object color properties are associated with plot style.

TIPS

- This command does not operate until a plot style table has been created for the drawing;.
- A plot style can be assigned to any object and to any layer.
- By default, drawings do not use plot styles.
- A plot style can override the following plot settings: color, dithering, gray scale, pen assignment, screening, linetype, lineweight, end style, join style, and fill style.
- Plot styles are useful for plotting the same layout in different ways, for example, to emphasize objects using different lineweights or colors in each plot.
- Plot style tables can be attached to the Model tab and layout tabs, and can attach different plot style tables to layouts, to create different looks for plots.

PlotterManager

2000 Displays the Plotters window, the Add-A-Plotter wizard, and the PC3 configuration Editor.

Command	*Aliases*	*Shortcut Keystrokes*	*Menu Bar*	*Application Menu*
plottermanager	...	...	**File** **↳Plotter Manager**	**Print** **↳Manage Plotters**

Command: plottermanager

Displays window:

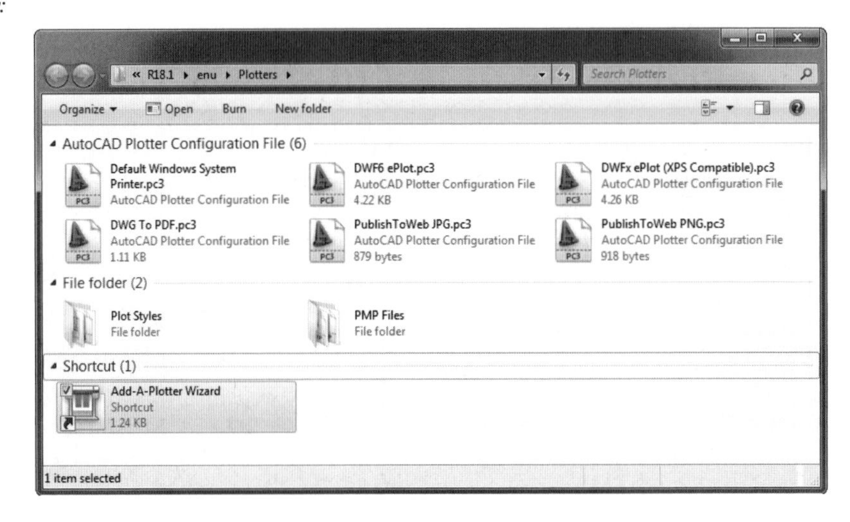

PLOTTERS WINDOW

Add-a-Plotter Wizard adds a plotter configuration; double-click to run the Add Plotter wizard.

*.*pc3* for plotted output parameters; double-click for the Plotter Configuration Editor. See the StylesManager command.

ADD PLOTTER WIZARD

These steps create PC3 plotter configurations for generic HP plotters; similar steps create .pc3 files for other brands of plotters.

Introduction Page

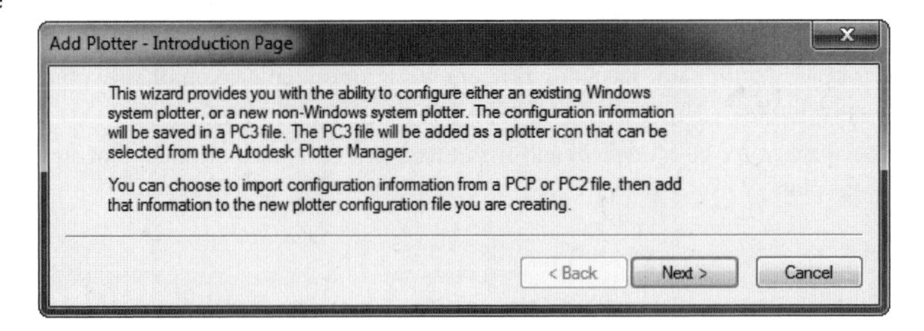

Begin page

My Computer configures printers and plotters connected to your computer.

Network Plotter Server configures printers and plotters connected to other computers on networks.

System Printer configures the default Windows printer.

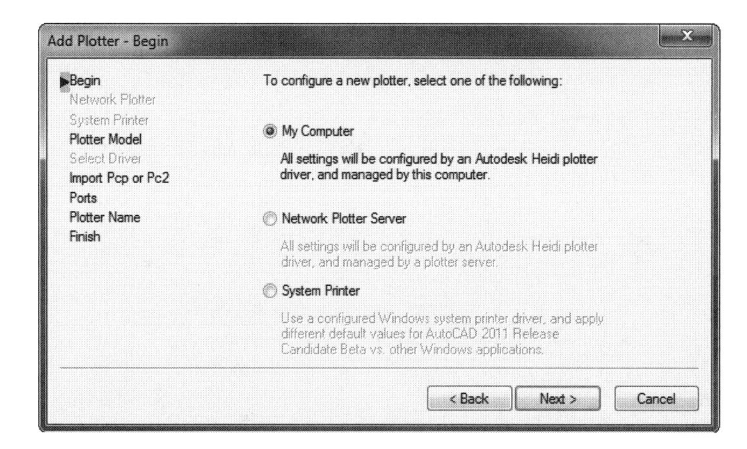

Plotter Model page

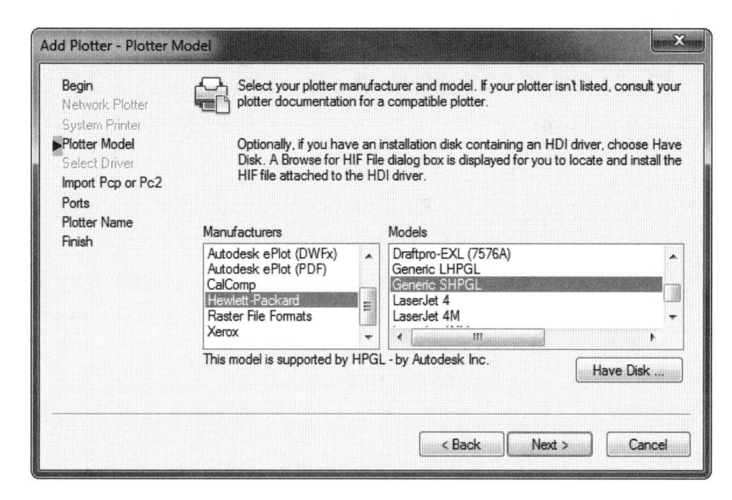

Manufacturer selects a brand name of plotter.

Model selects a specific model number.

Have Disk selects plotter drivers provided by the manufacturer; displays the Open dialog box.

Import Pcp or Pc2 page

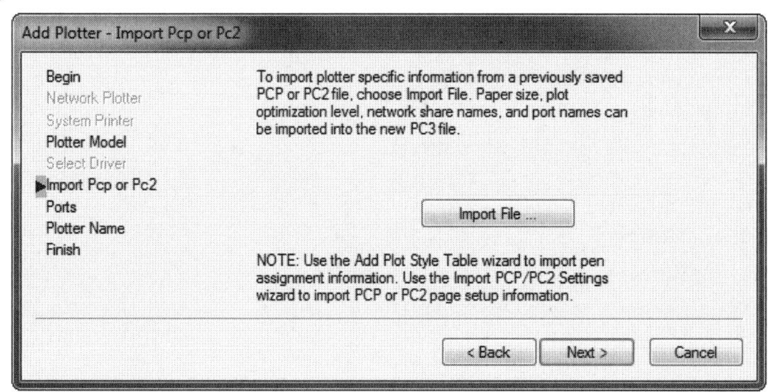

Import File imports *.pcp* and *.pc2* plotter configuration files from earlier versions of AutoCAD.

Ports page

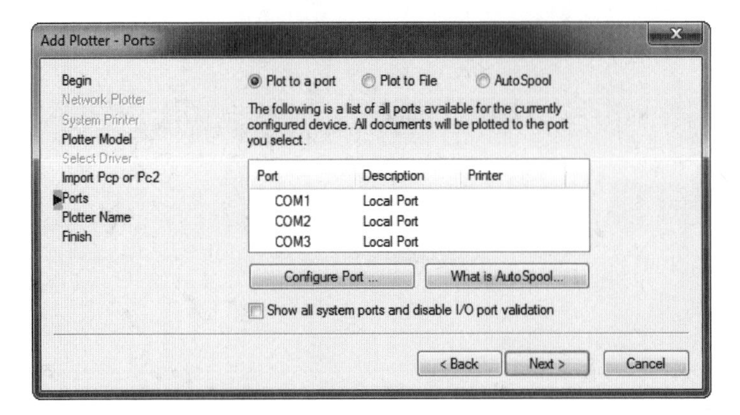

⊙ **Plot to a port** sends the plot to an output port.

○ **Plot to File** plots the drawing to a file with user-defined file names; default file name is the same as the drawing name with the *.plt* extension; PostScript plot files are given the *.eps* extension.

○ **AutoSpool** plots the drawing to files with the file name generated by AutoCAD, and then executes the command specified in the Option dialog box's Files tab. Enter the name of the program that should process the AutoSpool file in Print Spool Executable. You may include these DOS command-line arguments:

Argument	Meaning
%s	Specifies substitute path, spool filename, and extension.
%d	Specifies the path, AutoCAD drawing name, and extension.
%c	Describes the device.
%m	Returns the plotter model.
%n	Specifies the plotter name.
%p	Specifies the plotter number.
%h	Returns the height of the plot area in plot units.
%w	Returns the width of the plot area in plot units.
%i	Specifies the first letter of the plot units.
%l	Specifies the login name (*LogInName system variable*).
%u	Specifies the user name.
%e	Specifies the equal sign (=).
%%	Specifies the percent sign (%).

Port options

Port lists the virtual ports defined by the Windows operating system:

Port	Meaning
USB	Universal serial bus.
COM	Serial port.
LPT	Parallel port.
HDI	Autodesk's Heidi Device Interface.

Description describes the type of port:

Description	Meaning
Local Port	Printer is connected to your computer.
Network Port	Printer is accessible through the network.

Printers describes the brand name of the printer connected to the port.

Configure Port displays Configure Port dialog box for specifying parameters specific to ports, such as time out and protocol.

What is AutoSpool? displays an explanatory dialog box:

☐ **Show All System Ports and Disable I/O Port Validation** prevents AutoCAD from checking whether the port is valid.

Plotter Name page

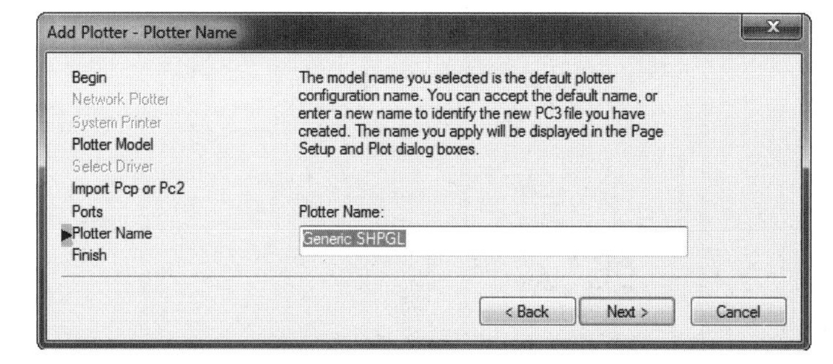

Plotter name names the plotter configuration; you may have many configurations for a single plotter.

Finish Page

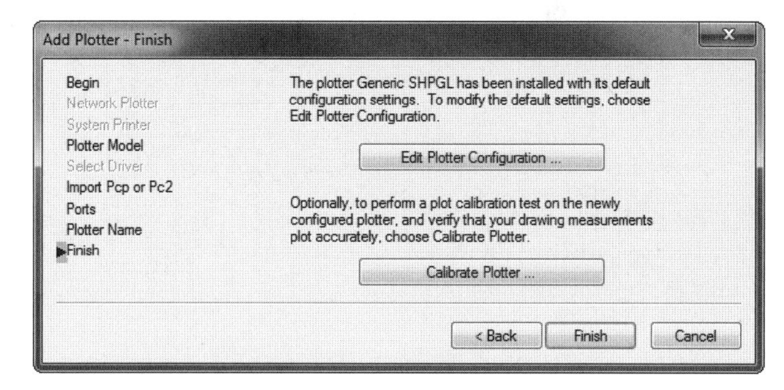

Edit Plotter Configuration displays the Edit Plotter Configuration dialog box.

Calibrate Plotter displays the Calibrate Plotter wizard.

PLOTTER CONFIGURATION EDITOR DIALOG BOX

General tab

Description allows you to provide a detailed description of the plotter configuration.

Ports tab

⊙ **Plot to the following port** sends the plot to an output port.

○ **Plot to File** plots the drawing to a file with a user-defined file name; default file name is the same as the drawing name, with a *.plt* extension; PostScript plot files are given the *.eps* extension.

○ **AutoSpool** plots the drawing to a file with a filename generated by AutoCAD, and then executes the command specified in the Option dialog box's Files tab.

☐ **Show all ports** lists all ports on the computer.

Browse Network displays the Browse for Printer dialog box; selects a printer on the network.

Configure Port displays the Configure Port dialog box; allows you to specify the parameters particular to the port, such as timeout and protocol.

Device and Document Settings tab

Media specifies the paper source, paper size, type, and destination.

Physical Pen Configuration specifies the physical pens in the pen plotter (for pen plotters only).

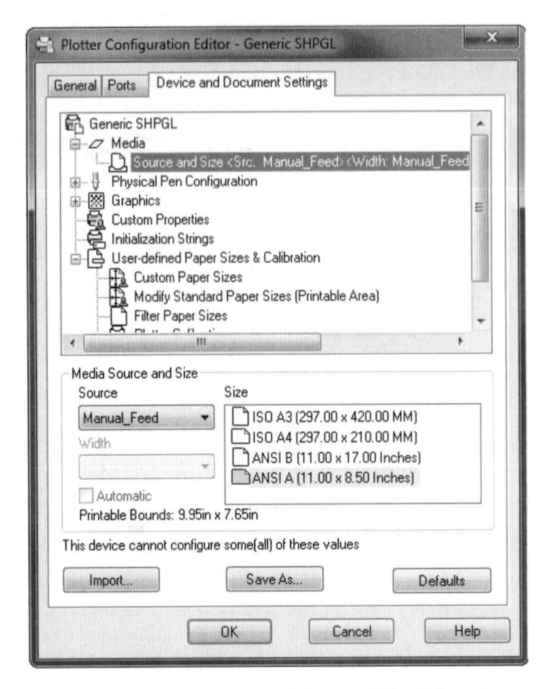

Graphics specifies settings for plotting vector and raster graphics and TrueType fonts.

Custom Properties specifies settings particular to the plotter, printer, or other output device.

Initialization Strings specifies the control codes for pre-initialization, post-initialization, and termination (for non-system plotters only).

Calibration Files and Paper Sizes calibrates the plotter by specifying the *.pmp* file; adds and modifies custom paper sizes; see the Calibrate Plotter wizard.

Import imports *.pcp* and *.pc2* plotter configuration files from earlier versions of AutoCAD.

Save As saves the plotter configuration data to a *.pc3* file.

Defaults resets the plotter configuration settings to the previously-saved values.

RELATED COMMANDS

Plot plots drawings with plot styles.

PageSetup attaches plot style tables to layouts.

StylesManager modifies plot style tables.

AddPlWiz.Exe runs the Add Plotter wizard.

RELATED SYSTEM VARIABLES

PaperUpdate toggles the display of a warning before AutoCAD plots a layout with a paper size different from the size specified by the plotter configuration file.

PlotRotMode controls the orientation of plots.

TIPS

- You can create and edit *.pc3* plotter configuration files without AutoCAD. From the Start button on the Windows toolbar, select Control Panel | Autodesk Plotter Manager.
- The *dwf classic.pc3* file specifies parameters for creating *.dwf* files via the Release 14-compatible DwfOut command.
- The *dwf eplot.pc3* file specifies parameters for creating *.dwf* files via the Plot command.

PngOut

2004 Exports the current view in PNG raster format.

Command	*Aliases*	*Shortcut Keystrokes*	*Menu*	*Ribbon*
pngout	...	...	...	...

Command: pngout

Displays Create Raster File dialog box. Specify a filename, and then click Save.

Select objects or <all objects and viewports>: *(Select objects, or press Enter to select all objects and viewports.)*

COMMAND LINE OPTIONS

Select objects selects specific objects.

All objects and viewports selects all objects and all viewports, whether in model space or in layout mode.

RELATED COMMANDS

BmpOut exports drawings in BMP (bitmap) format.

Image places raster images in the drawing.

JpgOut exports drawings in JPEG (joint photographic expert group) format.

Plot exports drawings in PNG and other raster formats.

TifOut exports drawings in TIFF (tagged image file format) format.

TIPS

- The rendering effects of the VsCurrent command are preserved, but not those of the Render command.
- PNG files are a royalty-free alternative to JPEG files.
- PNG is short for "portable network graphics."

Point

Ver. 1.0 Draws points in 3D space.

Command	*Alias*	*Shortcut Keystrokes*	*Menu*	*Ribbon*
point	**po**	...	**Draw** ↳**Point**	**Home** ↳**Draw** ↳**Multiple Points**

Command: point
Current point modes: PDMODE=0 PDSIZE=0.0000
Specify a point: *(Pick a point.)*

COMMAND LINE OPTION

Point positions a point, or enters a 2D or 3D coordinate.

RELATED COMMANDS

DdPType displays a dialog box for selecting PsMode and PdSize.

Regen regenerates the display to see the new point mode or size.

RELATED SYSTEM VARIABLES

PDMode determines the appearance of a point:

0	*1*	*2*	*3*	*4*
·		+	×	ꜝ
32	*33*	*34*	*35*	*36*
64	*65*	*66*	*67*	*68*
96	*97*	*98*	*99*	*100*

PDSize determines the size of a point:

0 — 5% of height of the ScreenSize system variable (default).
1 — No display.
-10 — *(Negative)* Ten percent of the viewport size.
10 — *(Positive)* Ten pixels in size.

TIPS

- The size and shape of points are determined by the PdSize and PdMode system variables; changing these changes the appearance and size of *all* points in the drawing with the next regeneration (Regen).
- Entering only x, y coordinates places the point at the z coordinate of the current elevation; setting Thickness to a value draws points as lines in 3D space.
- Prefix coordinates with ***** (such as ***1,2,3**) to place points in the WCS, rather than in the UCS. Use the object snap mode NODe to snap to points.
- Points plot according to the setting of the PdMode system variable.

Aliased Commands

PointLight command is accessed through options in the Light command.

PointCloud / PointCloudIndex

2011 Places point cloud data in drawings; converts point cloud files into indexed files.

Command	Aliases	Shortcut Keystrokes	Menu	Ribbon
pointcloud	**pc**	...	...	...
pointcloudindex	**pcindex**	...	**Insert** ↳**Point Cloud** ↳**Index**	**Insert** ↳**Point Cloud** ↳**Index**

Command: pointcloud

Enter an option [Attach/Index/Density/Lock]: *(Enter an option.)*

COMMAND LINE OPTIONS

Attach displays the Select Point Cloud File dialog box; see the PointcloudAttach command, below.

Index converts *.las* indexed laser scan files into point clouds files; see the PointcloudIndex command.

Density limits the number of points displayed at a time *(PointcloudDensity system variable)*; prompts you:

Enter point cloud density <15>: *(Enter an integer between 1 and 100.)*

Lock toggles the locking of point clouds (*PointCloudIndex*); prompts you:

Lock point cloud after attaching? [Yes/No] <No>: *(Type Y or N.)*

POINTCLOUDINDEX Command

Converts point clouds files into index PCG or ISD scan files.

Command: pointcloudindex

Displays the Select Data File dialog box; choose an FLS, FWS, LAS, or XYB file, and then click Open.

Displays the Create Indexed Point Cloud File dialog box; enter a file name, and then click Save.

Converting *filename*.**las to** *filename*.**isd in the background.**

When done, displays balloon on status bar:

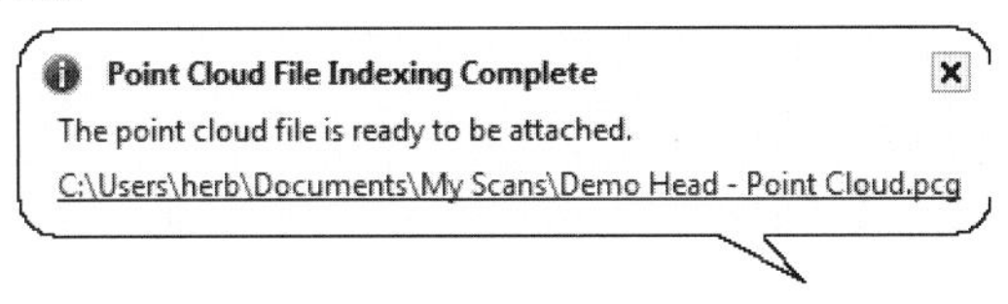

Click the balloon to launch the PointCloudAttach command.

RELATED COMMANDS

PointCloudAttach attaches point cloud data files to drawings.

RELATED SYSTEM VARIABLES

PointCloudAutoUpdate toggles the automatic regeneration of point clouds manipulation and real time panning, zooming, and orbiting.

PointCloudDensity specifies the percentage of points to display simultaneously.

PointCloudLock toggles locking of point clouds.

PointCloudRtDensity specifies the percentage of total points to display during real time zooming, panning, or orbiting.

TIPS

- Point clouds are generated by laser scanners that record the x,y,z coordinates of 3D features of rooms, processing plants, and objects. They are placed in AutoCAD drawings as PointCloud objects.

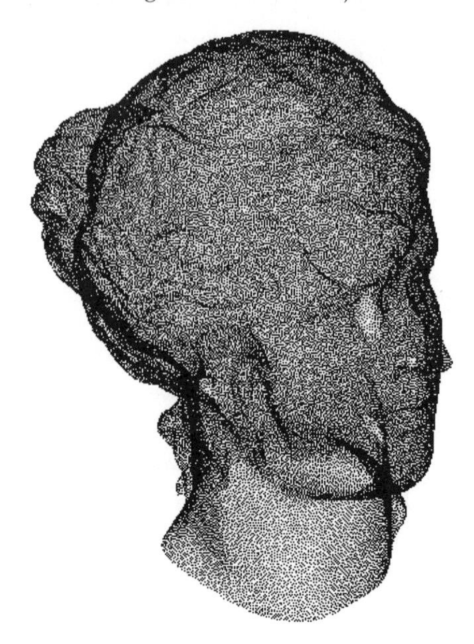

- Always insert point clouds at 0,0,0, since indexed point clouds offset their data from the origin. You can stitch together several clouds by using the same insertion point, scale, and angle for all.
- The following commands can be used with point clouds: Block, Copy, CopyClip, CutClip, Erase, Move, PasteClip, Properties, Rotate, Scale, and Stretch.
- The following object snap modes work with point clouds:
  - NODe snaps to individual points.
  - INSert snaps to the cloud's insertion point.
- The PointCloudDensity and PointCloudRtDensity systems variable are based on the percentage of 1,500,000, which is the maximum number of points per drawing.
  - The default value for PointCloudDensity is 15, which means that 15% or 225,000 points are displayed at a time. This value is distributed among all point clouds in the drawing.
  - The default value of PointCloudRtDensity is 5, which means 5%, or 75,000 points. To improve performance, set this variable to a lower value than PointCloudDensity.
- Locking prevents the point cloud from being moved, rotated and so on

PointCloudAttach

2011 Attach indexed PCG and ISD point cloud files to drawings.

Command	Alias	Shortcut Keystrokes	Menu	Ribbon
pointcloudattach	**pcattach**	...	**Insert** **↳Point Cloud** **↳Attach**	**Insert** **↳Point Cloud** **↳Attach**
-pointcloudattach	...	...	....	...

Command: pointcloudattach

Displays the Select Point Cloud File dialog box. Choose a PCG or ISD file, and then click Open.

Displays the Attach Point Cloud dialog box:

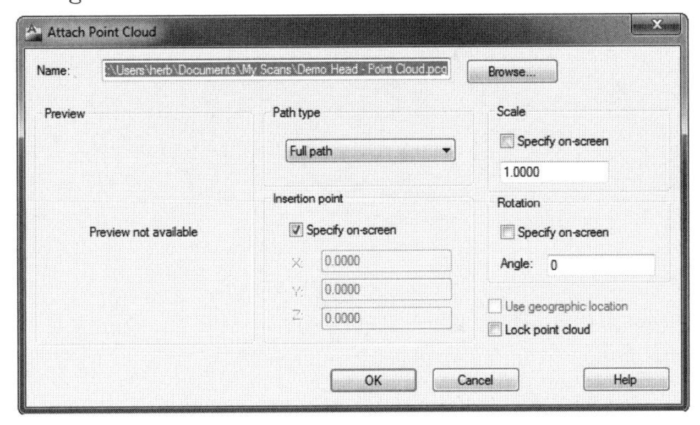

DIALOG BOX OPTIONS

Name specifies the point cloud file.

Path type specifies how much of the file's path AutoCAD remembers.

Insertion point specifies the insertion point of the point cloud; always use 0,0,0.

Scale specifies the scale factor of the point cloud; always use 1.

Rotation specifies the angle of the point cloud; always use 0.

Lock point cloud locks the point cloud so that it cannot be edited.

RELATED COMMAND

PointCloudIndex converts FLS, FWS, LAS, or XYB files to the indexed PCG or ISD files required by this command.

-POINTCLOUDATTACH Command

Command: -pointcloudattach~
Path to point cloud file to attach: *(Enter the file name, or press ~ for the dialog box.)*
Specify insertion point <0,0>: *(Specify a point.)*
Specify scale factor <1>: *(Specify a scale factor.)*
Specify rotation angle <0>: *(Specify a rotation angle.)*
1 point cloud attached

Polygon

Ver. 2.5 Draws 2D polygons of between three and 1024 sides.

Command	Alias	Shortcut Keystrokes	Menu	Ribbon
polygon	**pol**	...	**Draw** **↳Polygon**	**Home** **↳Draw** **↳Polygon**

Command: polygon
Enter number of sides <4>: *(Enter a number.)*
Specify center of polygon or [Edge]: *(Pick a point, or type E.)*
Enter an option [Inscribed in circle/Circumscribed about circle] <I>: *(Type I or C.)*
Specify radius of circle: *(Enter a value, or pick two points.)*

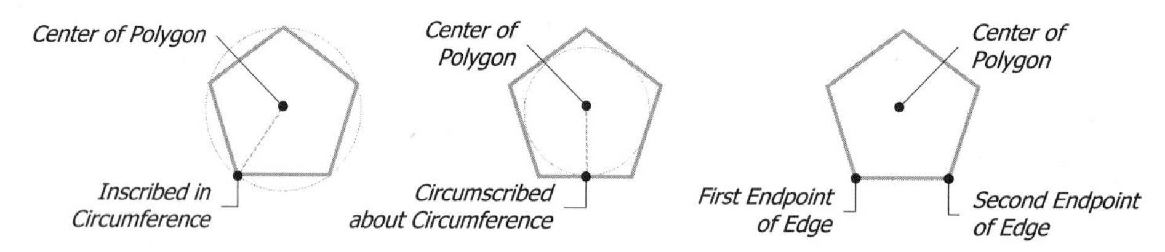

COMMAND LINE OPTIONS

Center of polygon indicates the center point of the polygon; then

- **C** (*Circumscribed*) fits the polygon outside an indicated circumference.
- **I** (*Inscribed*) fits the polygon inside an indicated circumference.

Edge draws the polygon based on the length of one edge.

RELATED COMMANDS

PEdit edits polylines, including polygons.

Rectang draws squares and rectangles.

RELATED SYSTEM VARIABLES

PolySides specifies the most recently-entered number of sides (default = 4).

Reverse reverses the order of vertices in polylines.

PlineConvertMode controls how polylines are converted to splines.

TIPS

- Polygons are drawn as a polyline; use PEdit to edit the polygon, such as its width.
- The pick point determines the polygon's first vertex; polygons are drawn counterclockwise.
- Use the system variable PolySides to preset the default number of polygon sides.
- Use the Snap command to place the polygon precisely; use INTersection or ENDpoint object snap modes to snap to the polygon's vertices.

PolySolid

2007 Draws 3D solid "walls" made of straight and curved sections in a manner similar to polylines.

Command	*Alias*	*Shortcut Keystrokes*	*Menu*	*Ribbon*
polysolid	**psolid**	...	**Draw** ↳**Modeling** ↳**Polysolid**	**Home** ↳**Modeling** ↳**Polysolid**

Command: polysolid

Specify start point or [Object/Height/Width/Justify] <Object>: *(Pick a point, or enter an option.)*

Specify next point or [Arc/Undo]: *(Pick a point, press Enter, or enter an option.)*

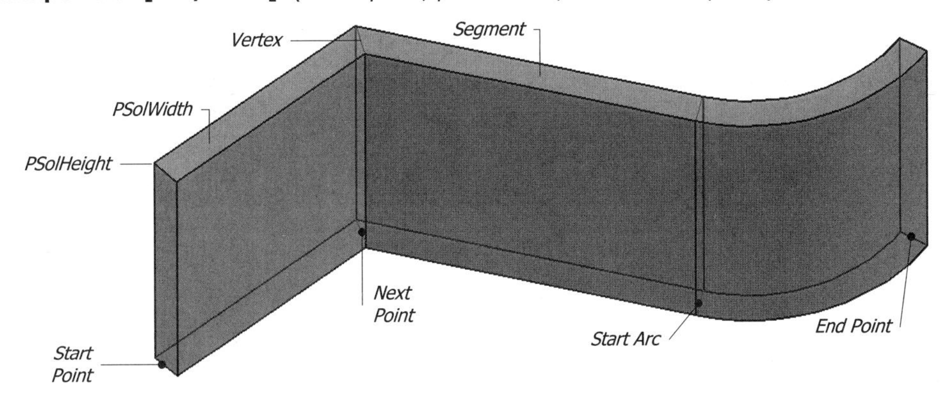

COMMAND LINE OPTIONS

Specify start point indicates the starting point by picking a point or entering x, y coordinates.

Object selects the line, arc, circle, or 2D polyline to convert into a solid.

Height specifies the height of the polysolid; default value is stored in the PSolHeight system variable.

Width specifies the width; default value is found in the PSolWidth system variable.

Justify determines whether the base of the polysolid is drawn to the left, on the center, or to the right of the pick points.

Specify the next point indicates the vertex of the wall; press Enter to stop drawing the polysolid.

Arc switches to arc-drawing mode with prompts similar to that of the PLine command.

Undo undraws the last segment.

Close closes the polysolid by joining the endpoint with the start point.

RELATED SYSTEM VARIABLES

PSolHeight specifies the default height of polysolids.

PSolWidth specifies the default width of polysolids.

Aliased Command

Preferences command was removed from AutoCAD 2000; it is now an alias for the Options command.

PressPull

2007 Extends and shortens the faces of 3D solids.

Command	Aliases	Shortcut Keystrokes	Menu	Ribbon
presspull	...	...	...	**Home** **↳Modeling** **↳Press/Pull**

Command: presspull
Click inside bounded areas to press or pull. *(Pick a face, and then drag.)*

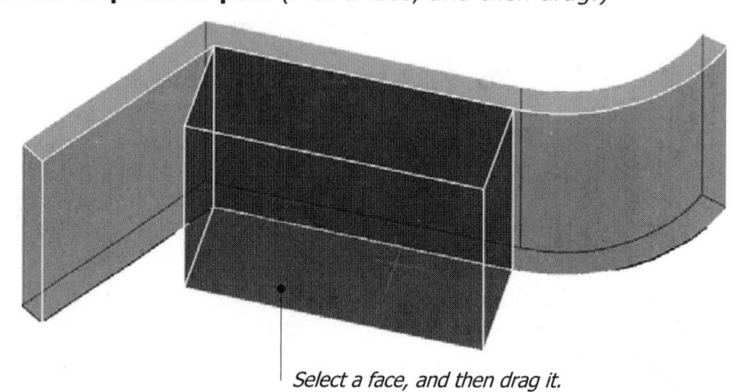

Select a face, and then drag it.

COMMAND LINE OPTION

Click inside bounded areas lets you drag the following faces:

- Areas that can be hatched by picking a point with zero gap tolerance, such as circles.
- Areas enclosed by crossing coplanar linear geometry — includes edges and geometry in blocks.
- Closed polylines, regions, 3D faces, and 2D solids with coplanar vertices.
- Areas created by geometry (including edges on faces) drawn coplanar to any face of 3D solids.

RELATED COMMANDS

Extrude converts 2D areas into 3D solids with optional tapered sides.

Thicken converts surfaces to 3D solids.

TIPS

- This command can be used to lengthen and shorten the faces of 3D solids.
- When used on 2D objects, like circles, the circle is turned into a 3D solid.
- When pushpulling "holes" (cylinders subtracted from solids), the result may be unexpected.
- I recommend using this command only for creating solids from closed boundaries and regions.
- Use Ctrl+pick to select a face, and then use grips editing to press and pull the face.

Preview

Rel. 13 Displays plot preview; bypasses the Plot command.

Command	*Alias*	*Shortcut Keystrokes*	*Menu Bar*	*Application Menu*	*Ribbon*
preview	**pre**	...	**File** **↳Plot Preview**	**Print** **↳Plot Preview**	...

Command: preview

Press ESC or ENTER to exit, or right-click to display shortcut menu.

Displays preview screen:

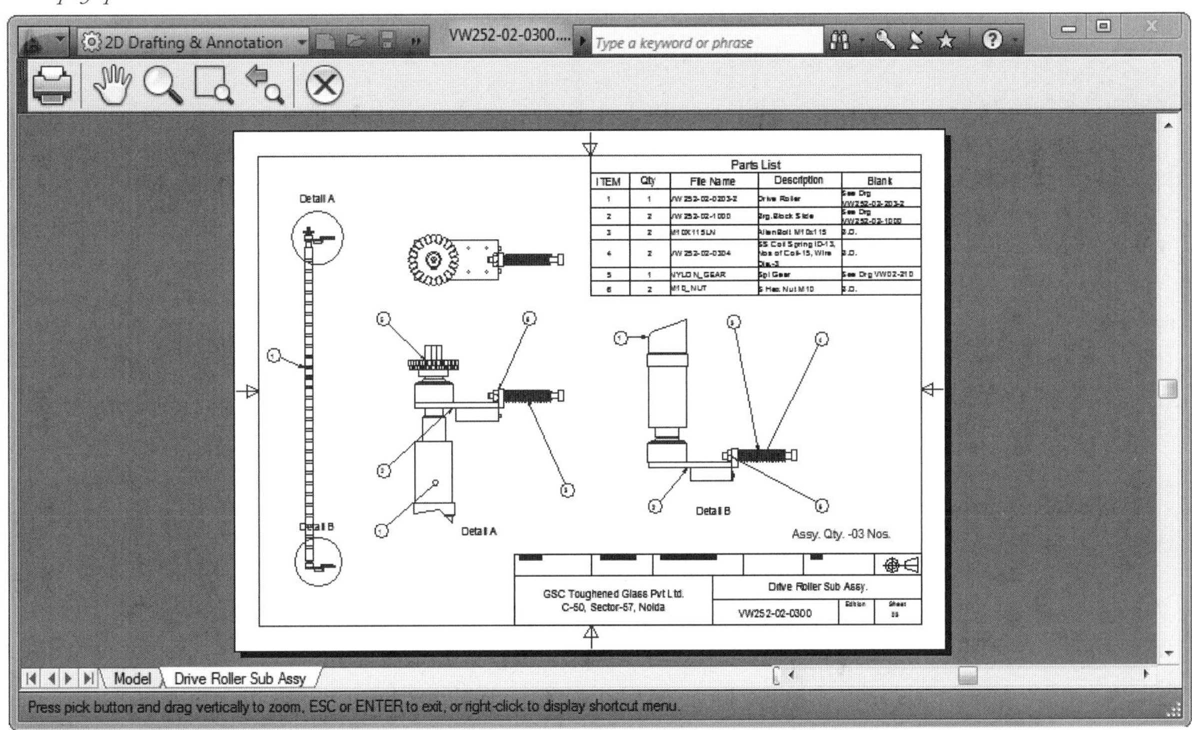

COMMAND LINE OPTION

Esc or **Enter** returns to the drawing window.

Preview toolbar

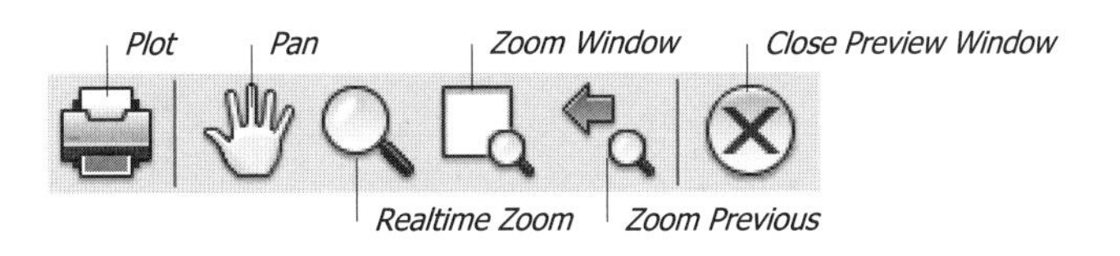

SHORTCUT MENU

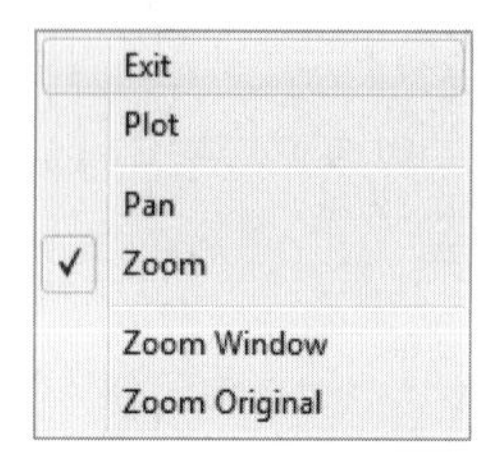

Exit exits preview mode.

Plot plots the drawing; goes immediately to plotting, bypassing the plot dialog box.

Pan pans the preview image in real time.

Zoom enlarges and shrinks the preview image in real time.

Zoom Window zooms into a windowed area.

Zoom Original returns to the original size.

RELATED COMMANDS

Plot plots the drawing.

PageSetup enables plot preview once a plotter is assigned to the layout.

ExportSettings previews exports of drawings in DWF and PDF format.

TIPS

- This command does not operate when no plotter is assigned; use the PageSetup command to assign a plotter to the Model and layout tabs.
- Press Esc to exit preview mode.
- Partial preview mode was removed from AutoCAD 2005.
- To zoom or pan, select the mode, and then hold down the left mouse button. The image zooms or pans as you move the mouse.
- To change the background color of the preview screen, go to Tools | Options | Display | Colors, and then select "Plot Preview Background" from the Window Element droplist.

ProjectGeometry

2011 Projects points, lines, and curves onto 3D solids, surfaces, and regions.

Command	Aliases	Shortcut Keystrokes	Menu	Ribbon
projectgeometry	...	...	...	**Surface** **⇘Project Geometry** **⇘Project to UCS / View / 2 Points**

Command: projectgeometry
SurfaceAutoTrim = 0
Select curves, points to be projected or [PROjection direction]: *(Select one or more objects, or type PRO.)*
Select curves, points to be projected or [PROjection direction]: *(Press Enter to continue.)*

Select a solid, surface, or region for the target of the projection: *(Select an object.)*
Specify the projection direction [View/Ucs/Points] <View>: *(Enter an option.)*
n **object(s) projected successfully.**

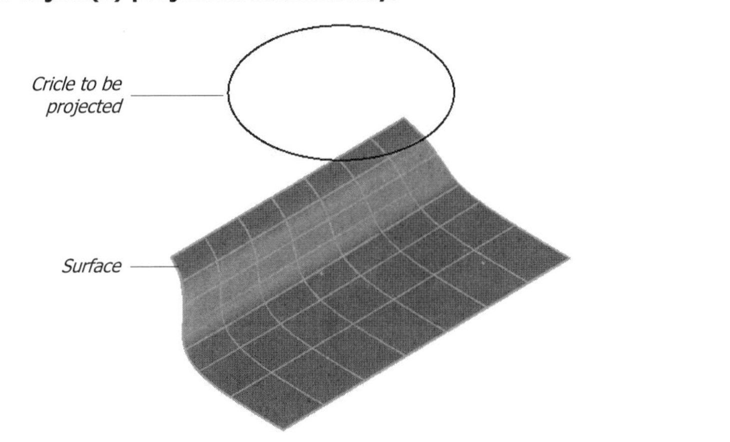

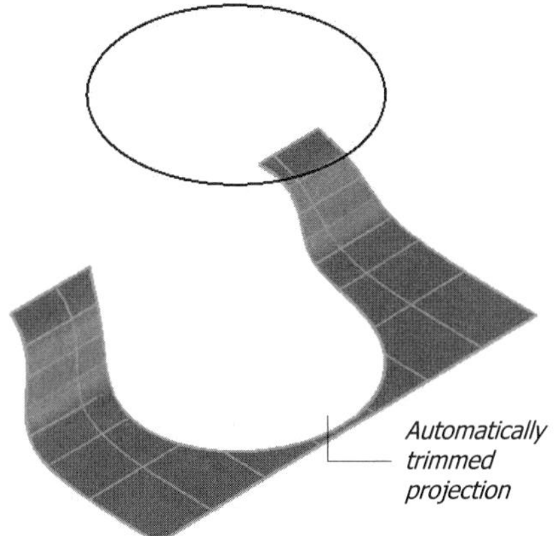

COMMAND LINE OPTIONS

Select curves, points specifies the 2D or 3D objects to project.

PRojection direction repeats the 'Specify the projection direction [View/Ucs/Points]' prompt; see below.

Select a solid, surface, or region specifies the object to project onto.

View projects the geometry along the current view direction.

UCS projects the geometry up or down the z axis of the current UCS.

Points projects the geometry along a path specified by two points; prompts you:

Specify start point of direction: *(Pick a point.)*
Specify end point of direction: *(Pick another point.)*

RELATED COMMANDS

SurfTrim trims surfaces.

Imprint imprints solids.

RELATED SYSTEM VARIABLE

SurfaceAutoTrim determines whether surfaces are trimmed automatically by this command.

TIPS

- This command projects the following objects: points, lines, arcs, circles, ellipses, 2D and 3D polylines (including splined and curve-fit polylines), splines, and helixes.
- After this command projects (imprints) points and curves onto 3D objects, it can trim 3D surfaces (not solids). To do so, change SurfaceAutoTrim to 1 before projecting.
- This command differs from the Imprint command, in that Imprint requires the imprintable object to lie on the solid.
- You can use object snaps to pick points on objects with the Points option.
- Following this command, use the undocumented ImpliedFaceX command to extrude the imprinted object.

Properties / PropertiesClose

2000 Opens and closes the Properties palette for examining and modifying the properties of selected objects.

Commands	Aliases	Shortcut Keystrokes	Menu	Ribbon
'properties	**ch** **mo** **pr, props**	**Ctrl+1**	**Tools** **↳Palettes** **↳Properties**	**View** **↳Palettes** **↳Properties**
	ddchprop **ddmodify**		**Modify** **↳Properties**	
'propertiesclose	**prclose**	**Ctrl+1**	**Tools** **↳Properties**	**...**

You can right-click any object and select Properties from the shortcut menu.

Command: properties

Displays palette with different content, depending on the objects selected. When no objects are selected:

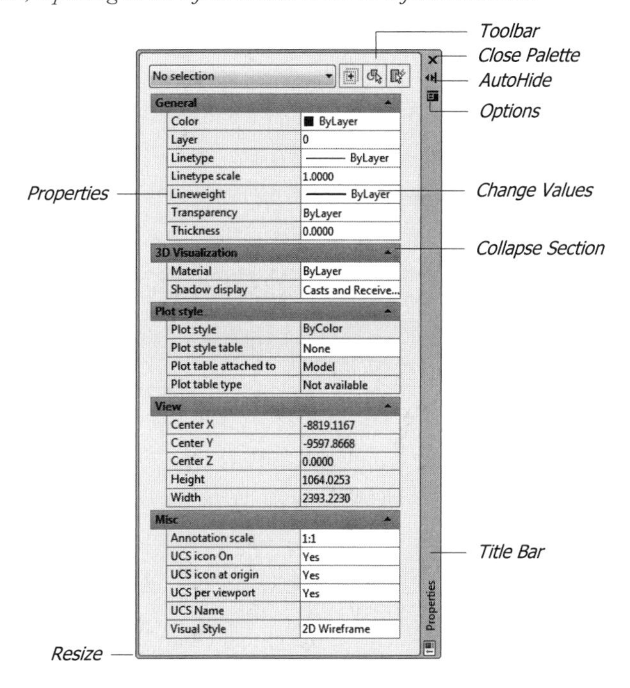

Palette toolbar

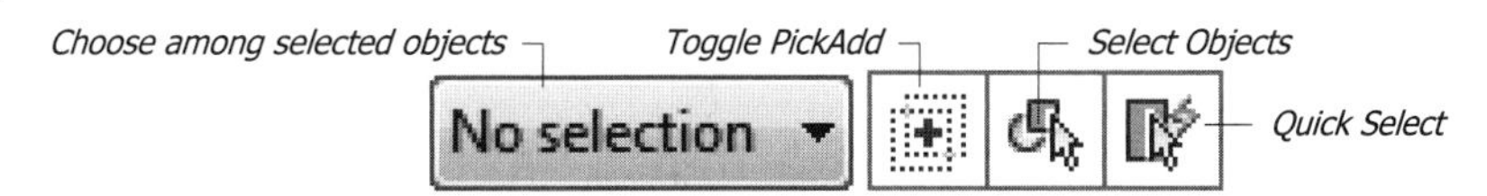

Toggle Value of PickAdd Variable controls selections when the Shift key is held down:

- **Off** adds objects to the selection set.
- **On** replaces objects in the selection set.

Select Objects prompts, 'Select objects:'; you can also just pick objects without this button.

Quick Select displays the Quick Select dialog box; see the QSelect command.

RELATED COMMANDS

QuickProperties displays a floating properties panel near the cursor.

ChProp changes an object's color, layer, linetype, and thickness.

Style creates and changes text styles.

RELATED SYSTEM VARIABLES

OpmState (Object Property Manager) reports whether the Properties palette is open.

CAnnoScale reports the name of the current annotation scale for the current space; a separate scale can be used for model space and paperspace viewports.

CeColor specifies the current color.

CeLtScale specifies the current linetype scale factor.

CeLtype specifies the current linetype.

CeTransparency specifies the current level of translucency *(new to AutoCAD 2011)*.

CLayer specifies the current layer.

CMaterial reports the current material.

CMleaderStyle reports the current multileader style.

Elevation specifies the current elevation in the z-direction.

Thickness specifies the current thickness in the z-direction.

TIPS

- To count objects in the drawing, press Ctrl+A to select all objects. In the figure, the drawing contains 1,861 objects.

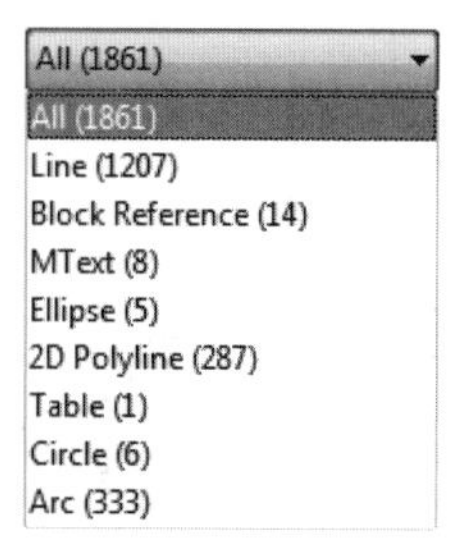

- As an alternative to entering the PropertiesClose command, click the **x** button on the palette's title bar.
- Double-click the title bar to dock and undock the palette within the AutoCAD window.
- The Properties palette can be dragged larger and smaller by its edges.
- When an item is displayed by gray text, it cannot be modified.
- This command can only be used transparently after it has been used at least once non-transparently.

Removed Command

PsDrag command was discontinued with AutoCAD 2000i. It has no replacement.

PSetupIn

2000 Imports user-defined page setups into the current drawing layout (short for Page SETUP IN).

Commands	*Aliases*	*Shortcut Keystrokes*	*Menu*	*Ribbon*
psetupin	...	...	...	...
-psetupin	...	...	...	...

Command: psetupin

Displays the Select File dialog box. Select a .dwg, .dwt, or .dxf file, and then click Open. AutoCAD displays dialog box:

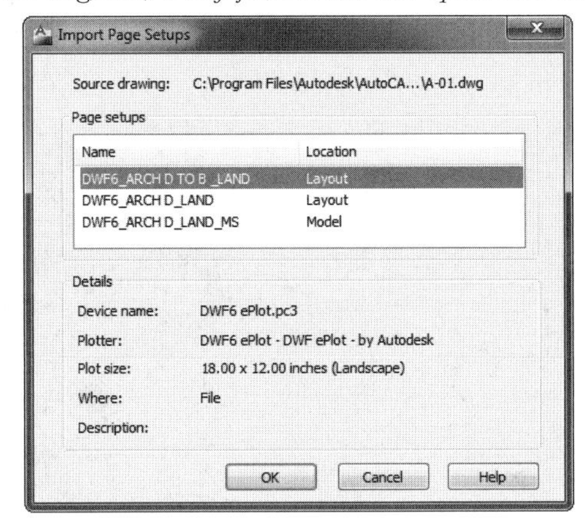

IMPORT PAGE SETTINGS DIALOG BOX

Name lists the names of page setups.

Location lists the location of the setups.

OK closes the dialog box, and loads the page setup.

-PSETUPIN Command

Command: -psetupin
Enter filename: *(Enter .dwg file name.)*
Enter user defined page setup to import or [?]: *(Enter a name, or type ?.)*

COMMAND LINE OPTIONS

Enter filename enters the name of a drawing file.

Enter user defined page setup to import enters the name of a page setup.

? lists the names of page setups in the drawing.

RELATED COMMANDS

PageSetup creates a new page setup configuration.

Plot plots the drawing.

Undocumented Commands

PsFill command fills 2D polyline outlines with raster PostScript patterns (short for PostScript FILL).

PsOut command exports the current drawing as an encapsulated PostScript file.

PSpace

Rel. 11 Switches from model space to paper space in layout mode (short for Paper SPACE).

Command	*Alias*	*Status Bar*	*Menu*	*Ribbon*
pspace	**ps**	PAPER	...	...

This command does not operate in model tab.

Command: pspace

COMMAND LINE OPTIONS

None.

RELATED COMMANDS

MSpace switches from paper space to model space.

MView creates viewports in paper space.

QvLayout shows a thumbnail strip of layouts.

UcsIcon toggles the display of the paper space icon.

Zoom scales paper space relative to model space with the XP option.

RELATED SYSTEM VARIABLES

MaxActVP specifies the maximum number of viewports displaying an image.

PsLtScale specifies the linetype scale relative to paper space.

TileMode allows paper space when set to 0.

TIPS

- Use paper space to lay out multiple views of a single drawing.
- In layout mode, you can switch to paper space by clicking the word MODEL or on the status bar; switch to a layout's model space by clicking the word PAPER or .
- When a drawing is in paper space, AutoCAD displays PAPER or highlights the icon on the status line, and shows the paper space icon:
- In layout mode, switch between paper and model space by double-clicking inside a viewport.
- Paper space is known as "drawing composition" in certain other CAD packages.

Publish

2004 Outputs multiple layout sheets from one or more drawings as a single, multi-page files in DWF, DWFx, 3D DWF, and PDF formats, or hardcopy plot.

Commands	*Aliases*	*Shortcut Keystrokes*	*Menu Bar*	*Application Menu*	*Ribbon*
'publish	...	...	**File** ↳**Publish**	**Print** ↳**Batch Plot**	**Output** ↳**Plot** ↳**Batch Plot**
'-publish	...	...	...	...	

Command: publish

Displays dialog box.

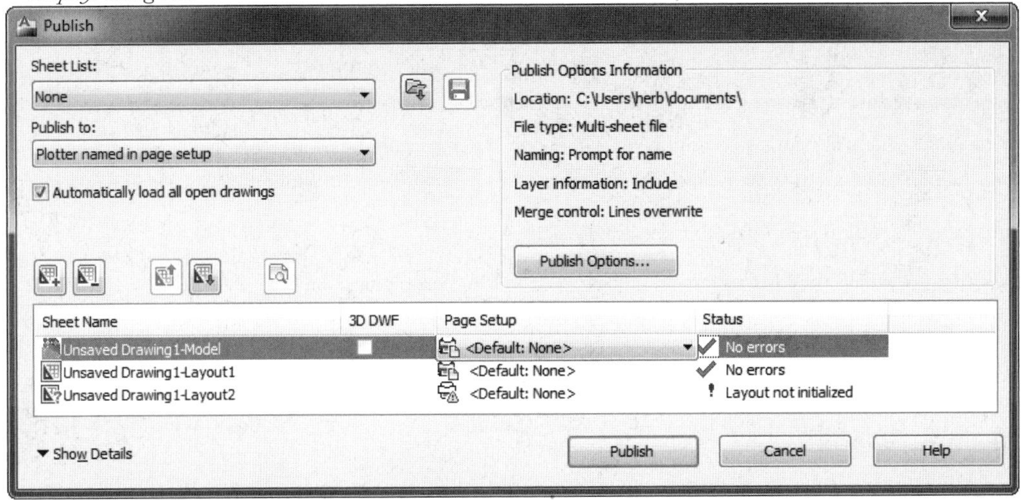

PUBLISH DIALOG BOX

Sheet List names the sheet list, which specifies how to publish a set of drawings. New drawing sets have no sheet lists.

Load Sheet List loads DSD or BP3 files; displays the Load List of Sheets dialog box. Choose a DSD (drawing set description) or BP3 (Batch Plot List) file, and then click **Load**.

Save saves the list of sheets for reuse.

Publish To specifies the plotter or file format: plotter (named in page setup), DWF, DWFx, or PDF.

☑ **Automatically Load All Open Drawings** adds all open drawings and their layouts.

Publish Options displays the Publish Options dialog box.

Add Drawings adds drawings to the sheet list; opens the Select Drawings dialog box.

Remove Drawings removes the selected model/layout tabs from the sheet list.

Move Sheet Up moves the selected sheet up in the list.

Move Sheet Down moves the selected sheet down in the list.

Preview previews the selected sheet; see the Preview command.

Sheet Name concatenates the drawing and layout name with a dash (-); edit sheet names with the Rename option.

3D DWF toggles the output of Model tab between 2D and 3D DWF formats.

Page Setup lists the named page setup for each layout; click to select other setups.

Status displays a message as layouts are published.

Publish generates the *.dwf/ .dwfx* file or plots; displays Now Plotting dialog box.

Show Details expands the dialog box to provide added information about the selected sheet.

Show Details

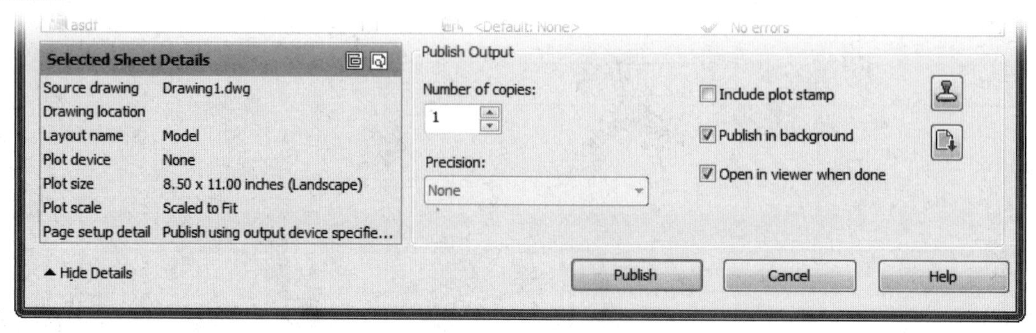

Number of copies allows you make multi-copy prints (available only when plotting).

Precision specifies the precision; see ExportSettings command; available only when publishing to DWF or PDF formats.

☐ **Include plot stamp** adds plot stamp data to the edge of each plot.

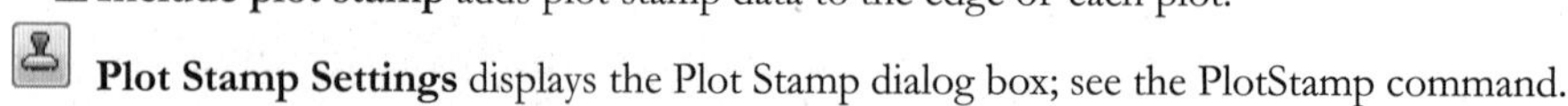

Plot Stamp Settings displays the Plot Stamp dialog box; see the PlotStamp command.

☑ **Publish in Background** returns to the drawing editor more quickly.

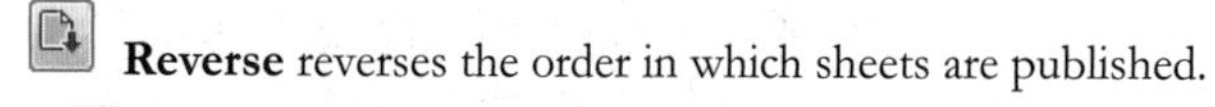

Reverse reverses the order in which sheets are published.

☑ **Open in Viewer When Done** displays DWF files in Design Review or PDF files in Acrobat Viewer v7 or later.

PUBLISH OPTIONS DIALOG BOX

Default Output Location options

Location specifies the folder in which to save the *.dwf* and *.pdf* files.

General DWF/PDF options

Type specifies whether sheets are bundled into a single file:

- **Single-Sheet** generates one file for each sheet.
- **Multi-Sheet DWF** generates a single file for all sheets.

Naming determines how the file is named:

- **Prompt for Name** asks for a name in the file dialog box.
- **Specify Name** names the file in the Name field, below.

Name names the file.

Layer Information toggles inclusion of layer names:

- **Include** includes layer names in exported files.
- **Don't Include** leaves out layer names; merges all graphics into one layer.

Merge Control determines how overlapping objects are handled:

- **Lines Merge** merges overlapping lines.
- **Lines Overwrite** overwrites overlapping lines.

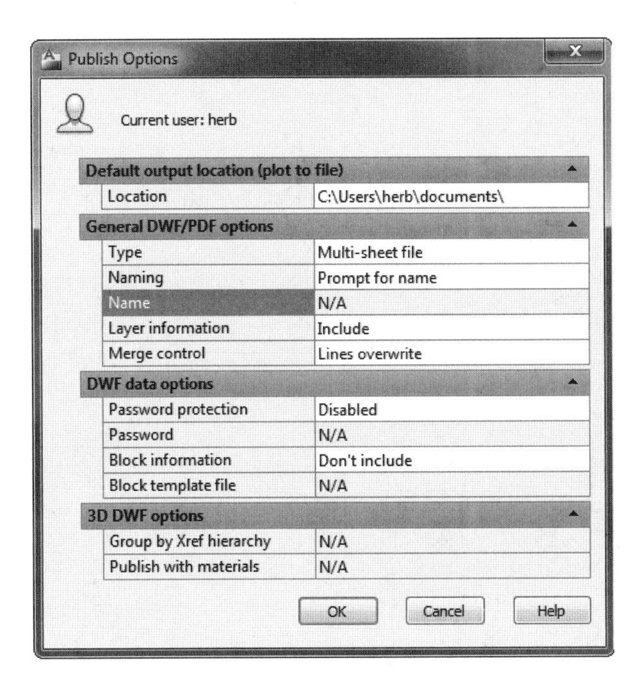

DWF Data Options options

Password Protection toggles use of the passwords:

- **Disabled** leaves out the password.
- **Prompt for Password** prompts you for the password during publishing. Passwords are case-sensitive, and may consist of letters, numbers, punctuation, and non-ASCII characters. *Warning!* Passwords cannot be recovered.
- **Specify Password** prompts the user for a password.

Password specifies the password.

Block information toggles whether blocks are included in *.dwf* files.

Block template file names the *.dxe* or *.blk* files used to format data.

3D DWF options

Grouped by Xref Hierarchy displays objects by xref in the DWF Viewer.

Publish with Materials displays objects with materials in the DWF Viewer.

SHORTCUT MENU

Access this shortcut menu by right-clicking a sheet name in the Sheets to Publish area:

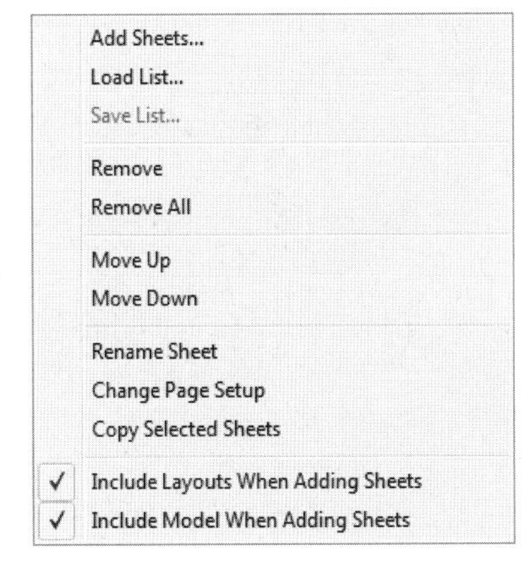

Add Sheets choose drawings from the Select Drawings dialog box.

Load List selects *.dsd* drawing set description or *.bp3* batch plot list file from the displays the Load List of Sheets dialog box.

Save List displays Save List As dialog box; enter a file name, and then click Save.

Remove removes selected sheets from the list, without warning.

Remove All removes all sheets.

Move Up moves the selected sheets up the list.

Move Down moves the selected sheets down the list.

Rename Sheet renames the selected sheet.

Change Page Setup displays the Page Setup droplist.

Copy Selected Sheets copies selected sheets, adding *-Copyn* suffix to sheet name.

☑ **Include Layouts when Adding Sheets** includes all layouts in the sheet set.

☑ **Include Model when Adding Sheets** includes model space with the layouts.

-PUBLISH Command

Command: -publish

If the drawing is not saved, displays Drawing Modified - Save Changes dialog box. Click OK.

Displays Select List of Sheets dialog box. Select a .dsd file, and then click Select.

Immediately plots the drawing set, and generates a log file.

RELATED COMMANDS

AutoPublish automatically publishes drawings as DWF/DWFx files upon saving or closing.

DwfFormat selects DWF or DWFx as the default design Web format.

Plot outputs drawings as *.dwf/.dwfx* files.

PublishToWeb coverts drawings to Web pages.

3dDwf exports model space drawings in 3D DWF/DWFx format.

RELATED SYSTEM VARIABLES

AutomaticPub toggles the AutoPublish feature for DWF/x and PDF files.

PublishCollate toggles whether a set of sheets is published as a single print job.

RELATED STARTUP SWITCH

/pl specifies the *.dsd* file to publish in the background when AutoCAD starts up.

TIPS

- The Model tab is included only when the Include Model When Adding Sheets setting is turned on.
- DWF passwords are case sensitive. If the password is lost, it cannot be recovered, and the *.dwf/.dwfx* file cannot be opened. Erase the file, and then create a new set.
- The -Publish command is good for generating sheets when a *.dsd* (drawing set description) file exists.
- To have AutoCAD publish a set of drawings unattended, follow these steps:
  1. Open the drawings in AutoCAD, and then use the Publish command to create a *.dsd* file. (Click the Save Sheet list button.)
  2. Exit AutoCAD.
  3. Modify the Target of AutoCAD's desktop item's Properties:
     "D:\CAD\AutoCAD 2011\acad.exe" **/pl "C:\path name\file name.dst"**
  4. Close the Properties dialog box.
  5. Double-click the icon, and AutoCAD starts plotting.
- The DWFx format is compatible with Microsoft's XPS format, and can be used with Internet Explorer v7/8 running on Windows Vista and 7.

PublishToWeb

2000i Exports drawings as DWF, JPG, or PNG images embedded in pre-formatted Web pages.

Command	*Aliases*	*Shortcut Keystrokes*	*Menu*	*Ribbon*
'publishtoweb	...	...	**File ↳Publish to Web**	...

Before starting this command, save drawings.

Command: publishtoweb

Displays dialog box.

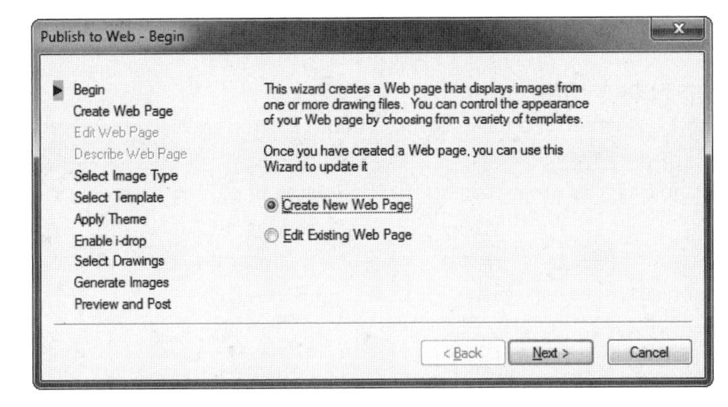

⊙**Create New Web Page** guides you through the steps to create new Web pages from drawings.

○**Edit Existing Web Page** guides you through the steps to edit existing Web pages created earlier by this command.

Create Web Page page

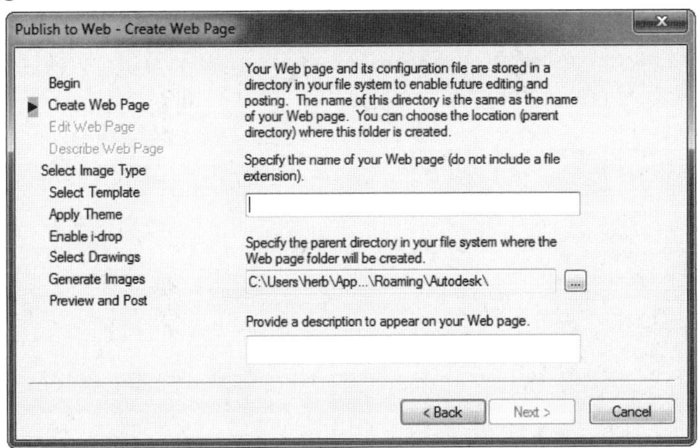

Specify the Name of Your Web Page requires you to enter a file name. (AutoCAD uses the name for the files making up the Web page, which allows you later to edit the Web page.) The name also appears at the top of the Web page.

Provide a Description to Appear On Your Web Page provides a description appearing below the name on the Web page.

Select Image Type page

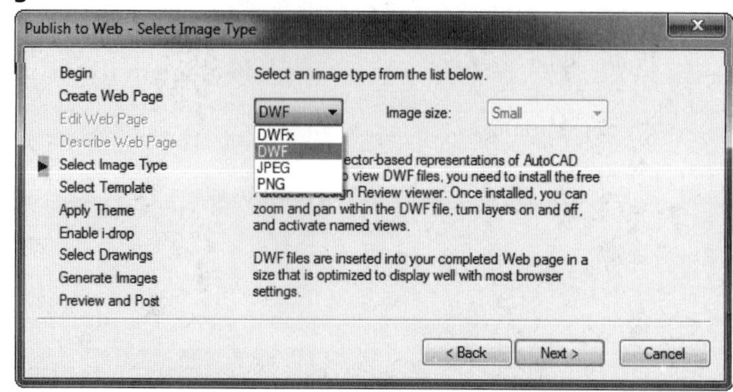

Select an Image Type From the List Below:

- **DWF** or **DWFx** (drawing Web format) vector format displays cleanly, and can be zoomed and panned; a plug-in is required to view DWF files in Web browsers.
- **JPEG** (joint photographic experts group) raster format displayed by all Web browsers; creates artifacts (details that don't exist).
- **PNG** (portable network graphics) raster format does not suffer the artifact problem.

Image size selects a size of raster image (available for JPEG and PNG only):

- **Small**— 789x610 resolution.
- **Medium** — 1009x780 resolution.
- **Large** — 1302x1006 resolution.
- **Extra Large** — 1576x1218 resolution.

Select Template page

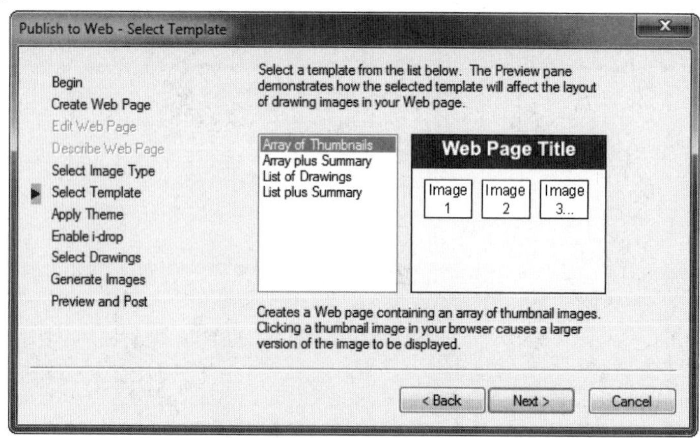

Select a Template From the List Below selects one of the pre-designed formats for the Web page:

Apply Theme page

Select a Theme From the List Below selects one of the pre-designed themes (colors and fonts) for the Web page.

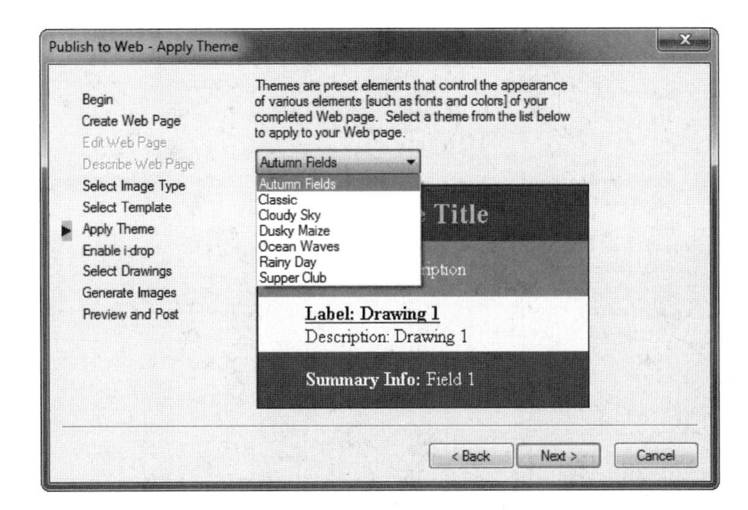

Enable i-drop page

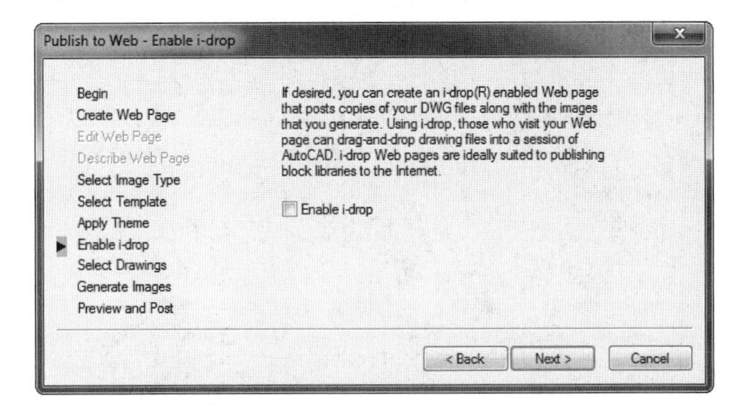

☐ **Enable i-Drop** adds i-drop capability to the Web page. This allows blocks to be dragged from the Web page directly into AutoCAD and other Autodesk software.

Select Drawings page

Image Settings options

Drawing selects the drawing; the current drawing is the default.

Layout selects the name of a layout or Model space.

Label names the image.

Description provides a description that appears with the drawing on the Web page.

Add adds the image setting to the image list.

Update changes the image setting in the image list.

Remove removes the image setting from the image list.

Move Up moves the image setting up the list.

Move Down moves the image setting down list.

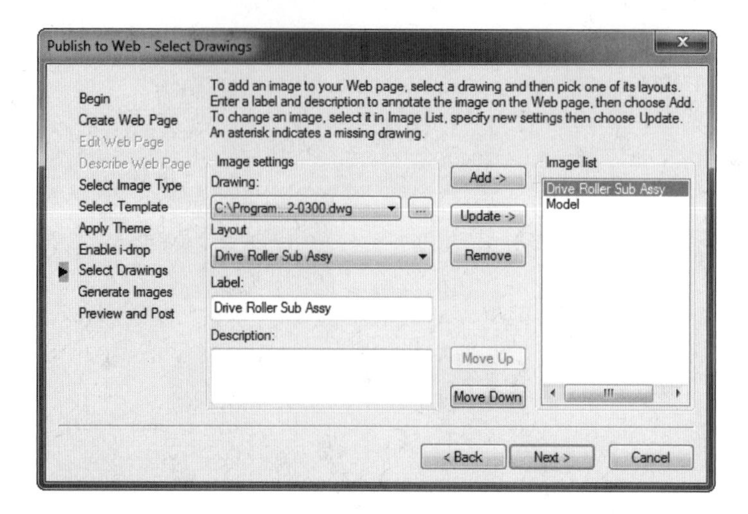

Generate Images page

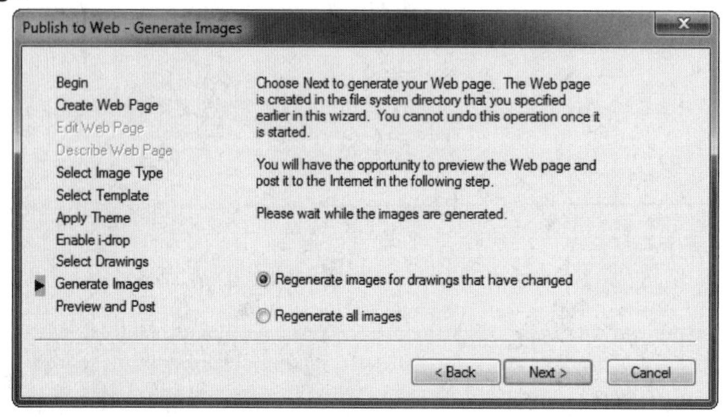

⊙**Regenerate Images For Drawings That Have Changed** updates images for those drawings that have been edited.

○**Regenerate All Images** regenerates all images from all drawings; ensures all are up to date.

Preview and Post page

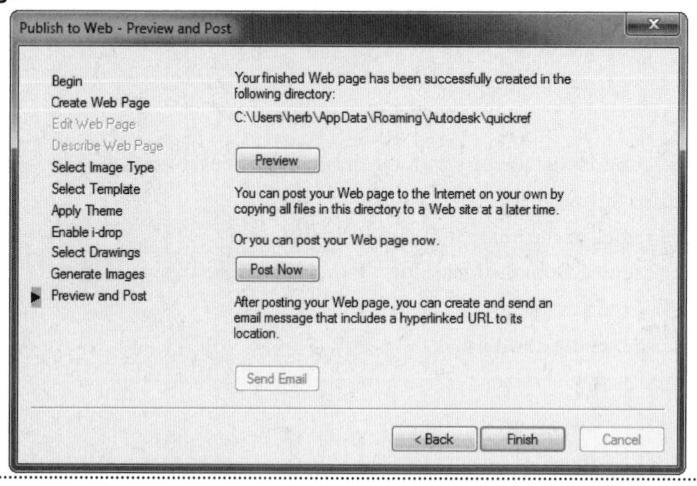

Preview launches the Web browser to preview the resulting Web page.

Post Now uploads the files (HTML, JPEG, PNG, DWF, and so on) to the Web site.

Send Email sends an email to alert others of the posted Web page.

RELATED COMMANDS

Publish exports drawings as multisheet *.dwf* files.

Plot exports drawings as *.dwf* files via the ePlot option.

Hyperlink places hyperlinks in drawings.

RELATED FILES

*.*ptw* are PublishToWeb parameter files, stored in tab-delimited ASCII file.

*.*js* are JavaScript files.

*.*jpg* are joint photographic experts group (raster image) files.

*.*png* are portable network graphics (raster image) files.

*.*dwf* are drawing Web format (vector image) files.

*.*dwfs* are XPS-compatible DXF files.

TIPS

- Use the Regenerate All Images option, unless you have an exceptionally slow computer or a large number of drawings to process. The Generate Images step can take a long time.
- After you click Preview to view the Web page (and after AutoCAD launches the Web browser), click the Back button to make changes, if the result is not to your liking.
- The Post Now option works only if you have correctly set up the FTP (file transfer protocol) parameters. If so, you can have AutoCAD upload the HTML files directly to a Web site. If not, use a separate FTP program to upload the files from the *\windows\applications data\autodesk* folder.
- You can customize the themes and templates by editing the *acwebpublish.css* (themes) and *acwebpublish.xml* (templates) files.
- This command may fail when you use the Options command to redirect templates to a folder other than *\template,* because AutoCAD expects to find the *\ptwtemplate* folder in *\template*. Ensure that the *\ptwtemplate* folder is in the new folder.

Purge

Ver. 2.1 Removes unused, named objects from the drawing: blocks, dimension styles, layers, linetypes, plot styles, shapes, text styles, table styles, application ID tables, multiline styles, zero-length geometry, and empty text objects.

Commands	*Alias*	*Shortcut Keystrokes*	*Menu Bar*	*Application Menu*	*Ribbon*
purge	**pu**	...	**File** ↳**Drawing Utilities** ↳**Purge**	**Drawing Utilities** ↳**Purge**	...
-purge	...	...	...	...	...

Command: purge

Displays dialog box.

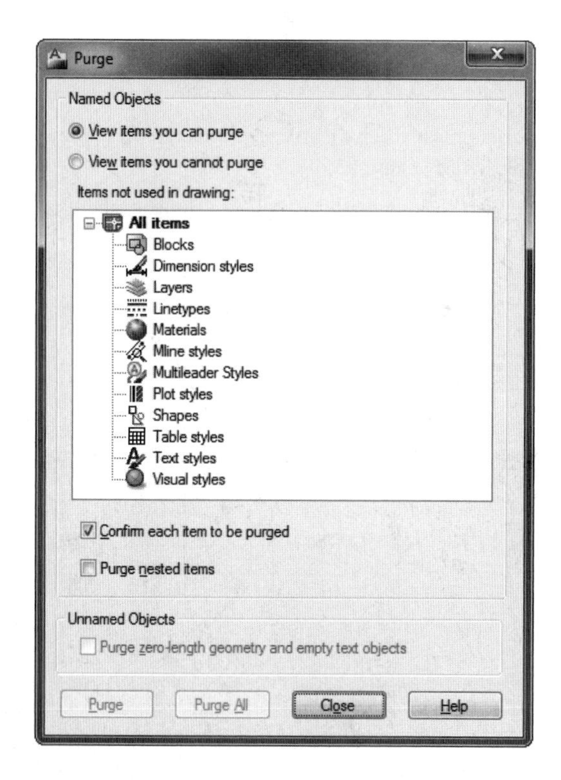

PURGE DIALOG BOX

⊙**View items you can purge** lists objects that can be purged from the drawing.

○**View items you cannot purge** lists objects that cannot be purged from the drawing.

☑**Confirm each item to be purged** displays a confirmation dialog box for each object being purged.

☐**Purge nested items** purges nested objects, such as unused blocks within unused blocks.

Unnamed Objects options

☐**Purge Zero-length Geometry and Empty Text Objects** purges lines and other objects that have no length, and text objects with no text. This option is available only when these invisible objects occur in the drawing.

-PURGE Command

Command: -purge

Enter type of unused objects to purge

[Blocks/Dimstyles/LAyers/LTypes/MAterials/MUltileaderstyles/Plotstyles/SHapes/textSTyles/Mlinestyles/Tablestyles/Visualstyles/Regapps/Zero-length geometry/Empty text objects/All]: *(Enter an option.)*

Sample response:

Purge linetype CENTER? <N> y

Purge linetype CENTER2? <N> y

COMMAND LINE OPTIONS

Blocks purges named but unused (not inserted) blocks.

Dimstyles purges unused dimension styles.

LAyers purges unused layers.

LTypes purges unused linetypes.

MAterials purges unused linetypes.

MUltileaderstyles purges unused mline styles.

Plotstyle purges unused plot styles.

SHapes purges unused shape files.

textSTyles purges unused text styles.

Mlinestyles purges unused multiline styles.

Tablestyles purges unused table styles.

Visualstyles purges unused visual styles.

Regapps purges unused application ID tables of registered applications.

Zero-length geometry purges objects that have no length.

Empty text objects purges blank text objects.

All purges all named objects, if possible.

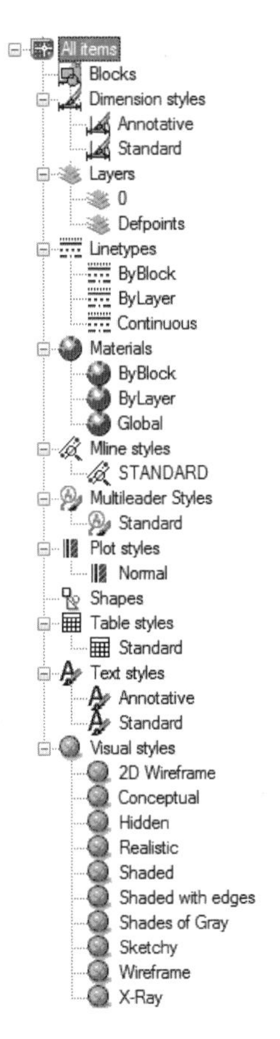

RELATED COMMAND

-WBlock writes the current drawing to disk with the * option, and removes spurious information from the drawing.

TIPS

- On rare occasion, it may be necessary to use this command several times; follow each purge with the Close command, then open the drawing, and purge again. Repeat until the Purge command reports nothing to purge. Typically, however, the Purge Nested Items suffices.
- The View Items You Cannot Purge option lists items being used, as well as system items that can never be purged, as illustrated at right.
- The standard style name "Annotative" was added to AutoCAD 2008.

Pyramid

2007 Draws 3D solid pyramids with pointy or flat tops, and 3 to 32 sides.

Command	*Alias*	*Shortcut Keystrokes*	*Menu*	*Ribbon*
pyramid	**pyr**	**...**	**Draw** ↳**Modeling** ↳**Pyramid**	**Home** ↳ **Modeling** ↳**Pyramid**

Command: pyramid
4 sides Circumscribed
Specify center point of base or [Edge/Sides]: *(Pick a point, or enter an option.)*
Specify base radius or [Inscribed]: *(Pick a point, or type I.)*
Specify height or [2Point/Axis endpoint/Top radius]: *(Pick a point, or enter an option.)*

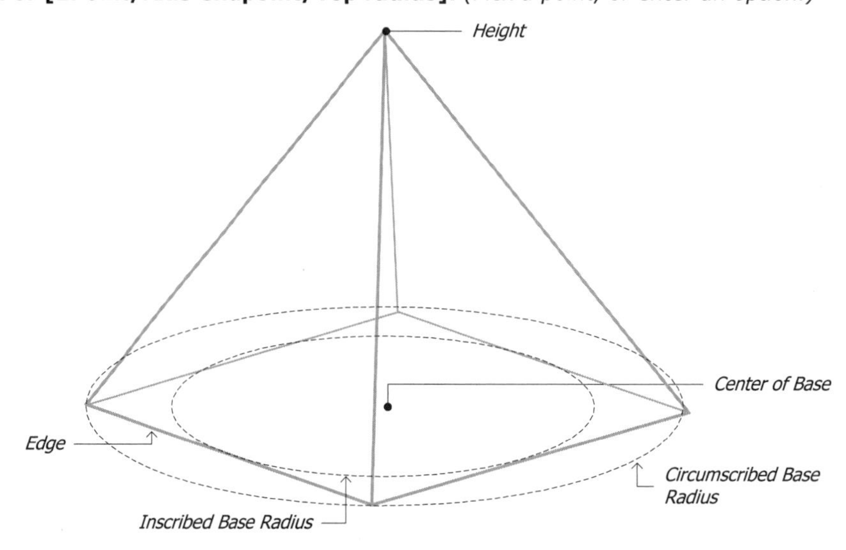

COMMAND LINE OPTIONS

The base is defined by a center point or edge length and the number of sides:

Specify center point of base locates the center of the pyramid's base by defining x, y, and z coordinates or by picking a point in the drawing.

Edge specifies the size of the base by the length of one edge; the base is always made from equilateral sides.

Sides specifies the number of sides; the range is 3 to 32, and the default is 4.

When the base is drawn like a polygon (circumscribed or inscribed):

Specify base radius specifies the radius of a "circle" that circumscribes the base.

Inscribed specifies the radius of a "circle" inscribed within the base.

Specify height indicates the height by specifying x, y, z coordinates or by picking a point in the drawing.

2Point specifies the height by picking two points.

Axis endpoint specifies the location and angle of the top point; this allows the pyramid to be drawn at an angle; the top and bottom are perpendicular to the axis.

Top radius specifies the size of the flat top to the pyramid.

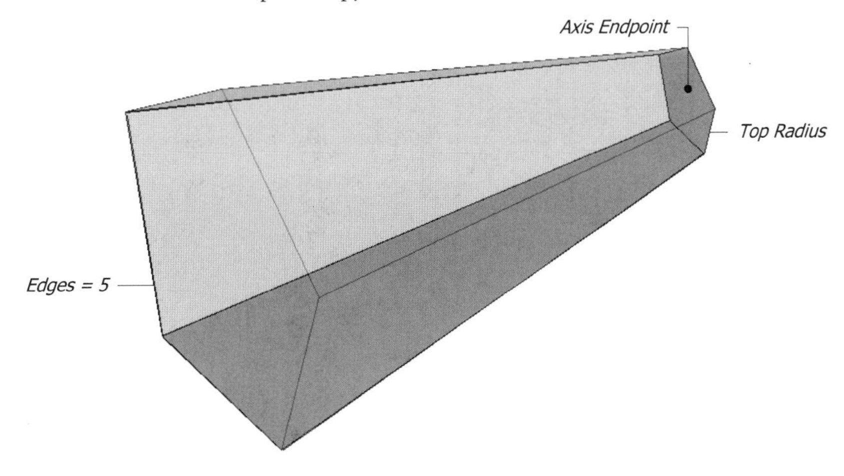

RELATED COMMANDS

Ai_Pyramid draws 3D mesh pyramids.

Cone draws 3D solid cones.

Extrude creates 3D objects with optional tapered sides.

RELATED SYSTEM VARIABLE

DragVs specifies the visual style while creating 3D solids.

TIPS

- To create a pyramid-like object with a non-equilateral base, use the Extrude command.
- After the pyramid is drawn, it can be edited with grips.

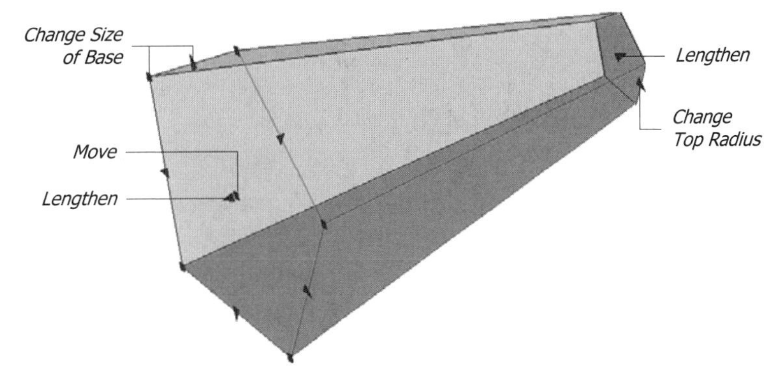

- When the top radius equals the base radius, this command draws a prism.

QDim

2000 Dimensions entire sets of objects (with continuous, baseline, ordinate, radius, diameter, and staggered dimensions) using just three picks (short for Quick DIMensioning).

Command	Aliases	Shortcut Keystrokes	Menu	Ribbon
qdim	...	...	**Dimension** **↳Quick Dimension**	**Annotate** **↳Dimensions** **↳Quick Dimension**

Command: qdim
Select geometry to dimension: *(Select one or more objects; press Ctrl+A to select the entire drawing.)*
Select geometry to dimension: *(Press Enter to end object selection.)*
Specify dimension line position, or
[Continuous/Staggered/Baseline/Ordinate/Radius/Diameter/datumPoint/Edit/seTtings] <Continuous>: *(Enter an option.)*

COMMAND LINE OPTIONS

Select geometry to dimension selects objects to dimension.

Specify dimension line position specifies the location of the dimension line.

Continuous draws continuous dimensions.

Staggered draws staggered dimensions.

Baseline draws baseline dimensions.

Ordinate draws ordinate dimensions relative to the UCS origin.

Radius draws radial dimensions; prompts 'Specify dimension line position:'.

Diameter draws diameter dimensions; prompts 'Specify dimension line position:'.

datamPoint sets a new datum point for ordinate and baseline dimensions; prompts 'Select new datum point:'.

Edit options

Indicate dimension point to remove, or [Add/eXit] <eXit>: *(Select a dimension, point, or enter an option.)*

Indicate dimension to remove selects the dimension point to remove from the dimension.

Add adds a dimension point to the dimension.

eXit returns to dimension drawing mode.

seTtings options

Associative dimension priority [Endpoint/Intersection] <Endpoint>: *(Enter an option.)*

Endpoint applies associative dimensions to endpoints over intersections.

Intersection applies associative dimensions to intersections over endpoints.

RELATED COMMANDS

DimStyle creates dimension styles, which specify the look of a dimension.

Dim*xxx* draws other kinds of dimensions.

QLeader draws leaders.

RELATED SYSTEM VARIABLES

DimStyle specifies the current dimension style.

DimAnno toggles annotative scaling for dimensions.

TIPS

- Example of continuous dimensions:

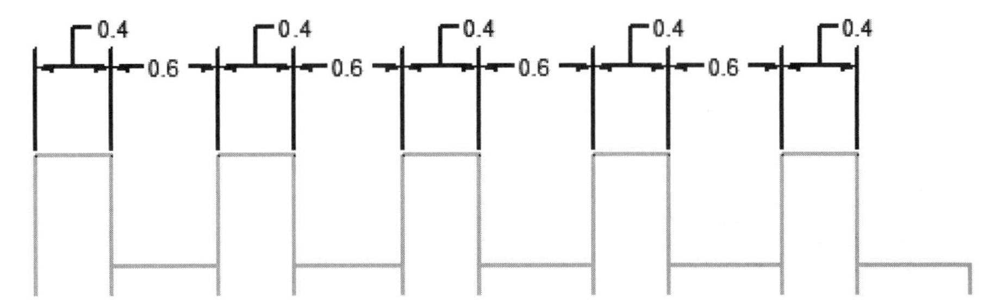

- Example of staggered dimensions:

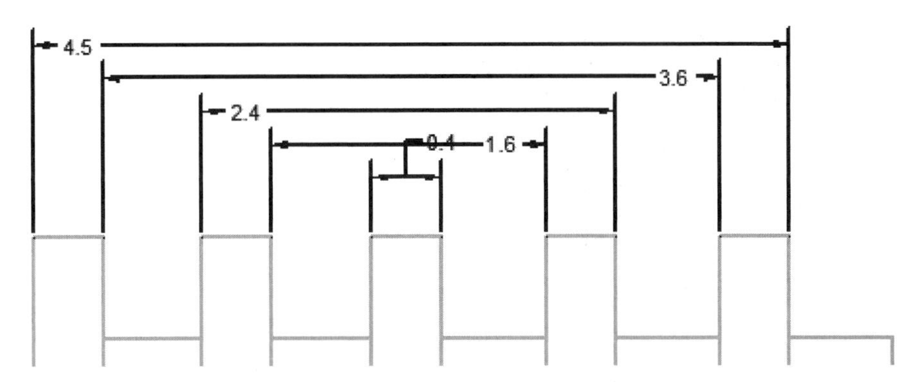

- Example of ordinate dimensions:

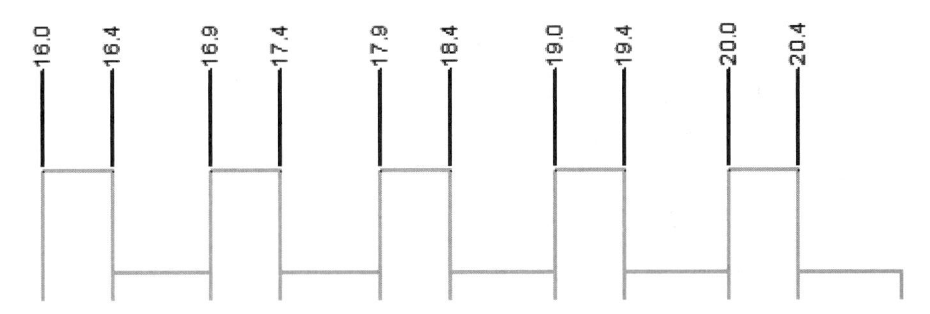

- As of AutoCAD 2004, dimensions created with QDim are fully associative. QDim supports annotative scaling as of AutoCAD 2008.
- This command sometimes fails. AutoCAD complains, "Invalid number of dimension points found."

QLeader

Rel. 14 Draws leaders; dialog box specifies options for custom leaders and annotations (short for Quick LEADER).

Command	Alias	Shortcut Keystrokes	Menu	Ribbon
qleader	**le**	...	...	...

Command: qleader

Specify first leader point, or [Settings] <Settings>: *(Pick point 1 for the arrowhead, or type S.)*

When S is entered, displays dialog box.

Click OK to continue with the command; the prompts vary, depending on the options selected in the dialog box:

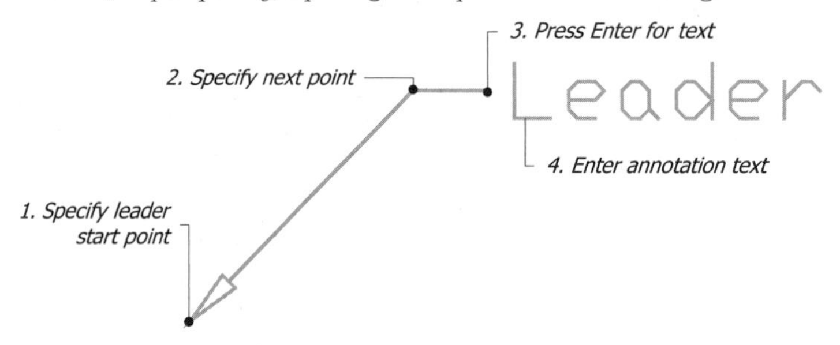

Specify next point: *(Pick point 2 for the leader shoulder.)*
Specify next point: *(Pick point, or press Enter for text options.)*
Specify text width <0.0000>: *(Enter a value, or press Enter.)*
Enter first line of annotation text <Mtext>: *(Enter text, or press Enter for the mtext editor.)*
Enter next line of annotation text: *(Press Enter to end command.)*

COMMAND LINE OPTIONS

Specify first leader point picks the location for the leader's arrowhead; press Enter to display tabbed dialog box.

Specify next point picks the vertices of the leader; press Enter to end the leader line.

Specify text width specifies the width of the bounding box for the leader text.

Enter first line of annotation text specifies the text for leader annotation; press Enter twice to end.

Enter next line of annotation text specifies more text; press Enter once to end.

LEADER SETTINGS DIALOG BOX

Annotation Type options

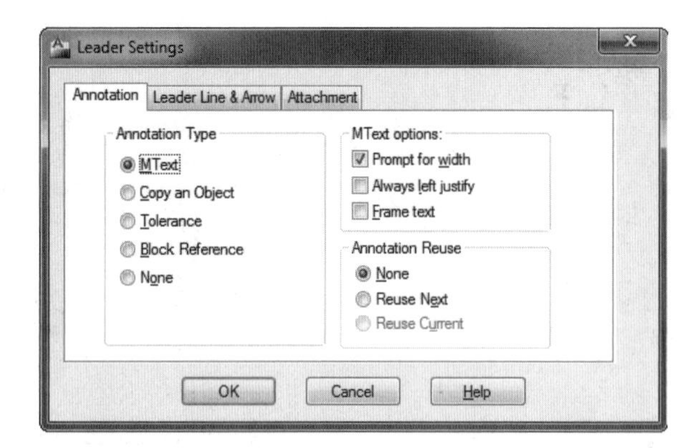

⊙ **MText** prompts you to enter text for the annotation; see MText command.

○ **Copy an Object** attaches any object in the drawing as an annotation.

○ **Tolerance** prompts you to select tolerance symbols for the annotation; see the Tolerance command.

○ **Block Reference** prompts you to select a block for the annotation; see the Block command.

○ **None** attaches no annotation.

MText Options options

☑ **Prompt for width** displays 'Specify text width' prompt.

☐ **Always left justify** left justifies the text, even when the leader is drawn to the right.

☐ **Frame text** places a rectangle around the text.

Annotation Reuse options

⊙ **None** does not retain annotation for next leader.

○ **Reuse Next** remembers the current annotation for the next leader.

○ **Reuse Current** uses the last annotation for the current leader.

Leader Line & Arrow tab

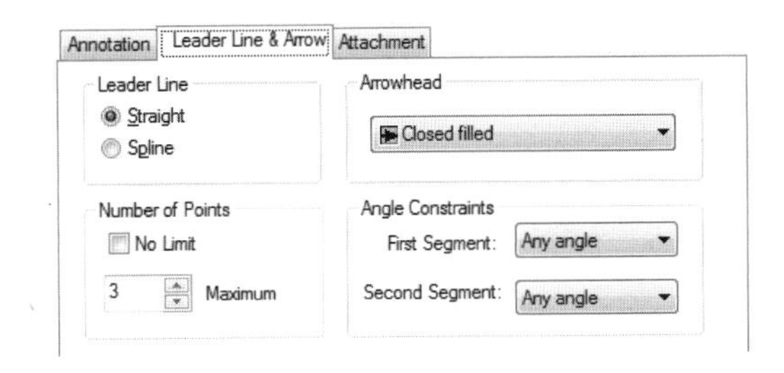

Leader Line options

⊙ **Straight** draws the leader with straight lines.

○ **Spline** draws the leader as a spline curve.

Number of Points options

☐ **No limit** keeps prompting for leader vertex points until you press Enter.

Maximum stops the command prompting for leader vertex points; default=3.

Arrowhead options

Arrowhead selects the type of arrowhead, including Closed filled (default), None, and User Arrow; see DimStyle for a complete list of arrowhead types.

Angle Constraints options

First Segment selects among Any Angle (user-specified), Horizontal (0 degrees), 90, 45, 30, or 15-degree leader line.

Second Segment selects among Any Angle (user-specified), Horizontal, 90, 45, 30, or 15-degree leader line.

Attachment tab

Multi-line Text Attachment options

Text on Left Side positions annotation relative to the last leader segment, when the annotation is located to the left of the leader:

○ Top of top line.

○ Middle of top line.

⊙ Middle of multi-line text.

O Middle of bottom line (default).

O Bottom of bottom line.

Text on Right Side positions annotation relative to the last leader segment, when the annotation is located to the right of the leader.

☐ **Underline bottom line** underlines the last line of leader text.

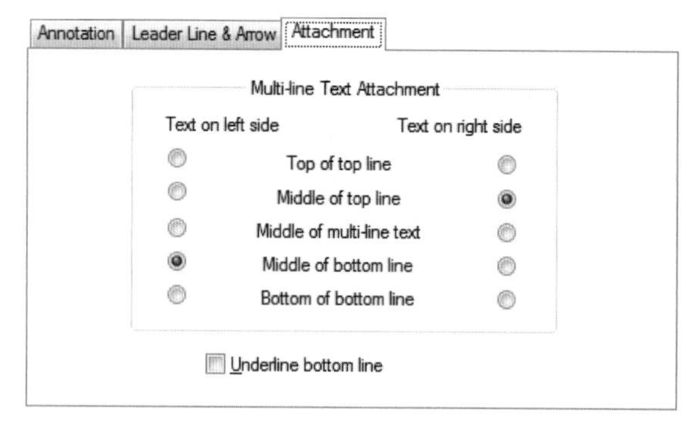

RELATED COMMANDS

TextEdit edits leader text; see the MText command.

DimStyle sets dimension variables, including leaders.

MLeader draws multiline leaders.

Leader draws leaders without dialog boxes.

TIPS

- This command draws leaders, just like the Leader command. The difference is that it brings up a triple-tab dialog box for setting the leader options.
- Some options have interesting possibilities, such as using a mtext, text, block reference, or tolerance in place of the leader text.
- This qleader was drawn with the spline and box text options:

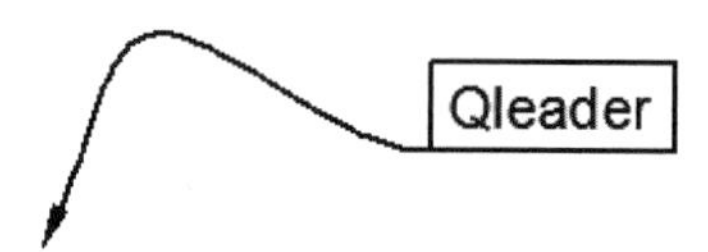

- You may prefer to use the MLeader command due to its greater flexibility.

QNew

2004 Starts new drawings based on a default template file (short for Quick NEW).

Command	Aliases	Keyboard Shorts	Menu	Quick Access Toolbar
qnew	...	...	...	

Command: qnew

Display depends on the value of the Startup system variable. See the New command.

RELATED COMMANDS

New starts new drawings.

SaveAs saves drawings in *.dwg* or *.dwt* format; creates template files.

RELATED SYSTEM VARIABLES

DbMod indicates whether the current drawing has changed since being loaded.

DwgPrefix indicates the path to the current drawing.

DwgName indicates the name of the current drawing.

FileDia displays file prompts at the 'Command' prompt.

RELATED FILES

wizard.ini holds the names and descriptions of template files.

******.dwt* are template files stored in *.dwg* format.

TIPS

- The New and QNew commands operate identically, except when a template drawing is specified by the Option command's Files tab: See the Template Settings nodes.

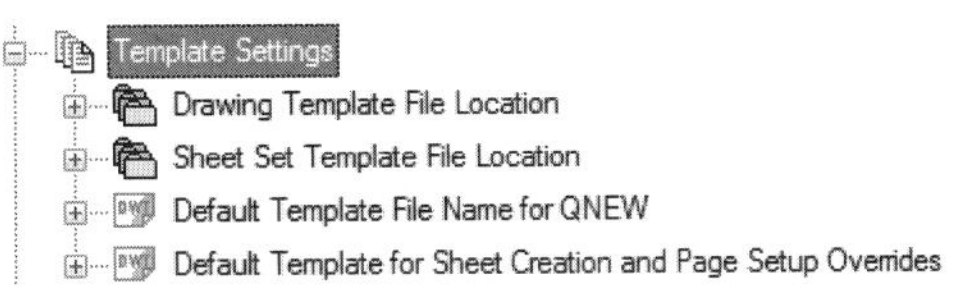

- The Quick Access icon executes the QNew command, while the New | Drawing pick on the application menu executes the New command.

QSave

Rel. 12 Saves the current drawing without requesting a file name (short for Quick SAVE).

Command	*Aliases*	*Shortcut Keystroke*	*Menu Bar*	*Application Menu*	*Quick Start Toolbar*
qsave	...	**Ctrl+S**	**File** ↳**Save**	**Save**	🖫

Command: qsave

If the drawing has never been saved, displays the Save Drawing As dialog box.

COMMAND LINE OPTIONS

None.

RELATED COMMANDS

Quit ends AutoCAD, with or without saving the drawing.

Save saves the drawing, after requesting the filename.

RELATED SYSTEM VARIABLES

DBMod indicates whether the drawing has changed since it was loaded.

DwgName specifies the current drawing filename (default = *drawing1*).

DwgTitled specifies the status of drawing's filename.

AutomaticPub determines whether drawings are also saved in DWF/x or PDF formats.

TIPS

- When the drawing is unnamed, the QSave command displays the Save Drawing As dialog box to request a file name; see the SaveAs command.
- When the drawing file, its folder, or drive (such as a CD-ROM drive) is marked read-only, use the SaveAs command to save the drawing by another file name, or to another folder or drive.
- If the AutomaticPub system variable is turned on, then the drawing is simultaneously exported as a DWF/x or PDF file when being saved. The DWF/x or PDF file is published in the background, if the BackgroundPlot system variable is turned on.
- Use the Options command's Files tab to specify the location of automatic backup files. I recommend that files be saved on another drive, though not a network drive.

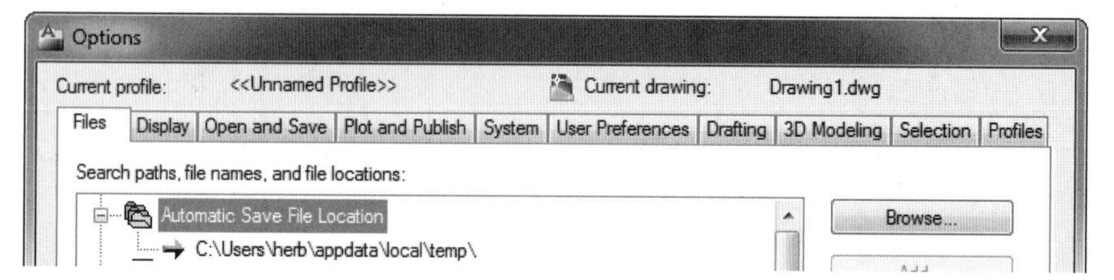

- Auto-saving to network drives can slow down the network; if the network crashes, you can continue to work from the local drive.

QSelect

2000 Creates selection sets of objects based on their properties (short for Quick SELECT).

Command	Aliases	Shortcut Keystrokes	Menu	Ribbon
qselect	...	...	**Tools** ⇘**Quick Select**	**Home** ⇘**Utilities** ⇘**Quick Select**

Command: qselect

Displays dialog box.

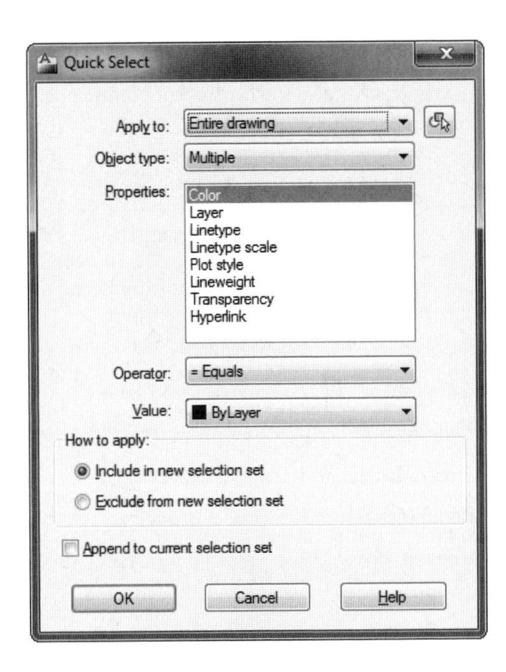

QUICK SELECT DIALOG BOX

Apply to applies the selection criteria to the entire drawing or current selection set; to create a selection set, click the Select Objects button.

Select Objects allows you to select objects; AutoCAD prompts: 'Select objects:'. Right-click or press Enter to return to this dialog box.

Object type lists the objects in the selection set; allows you to narrow the selection criteria to specific types of objects (default = Multiple).

Properties lists the properties valid for the selected object types; when you select more than one object type, only the properties in common are listed.

Operator lists logical operators available for the selected property.

Value specifies the property value for the filter. If known values for the selected property are available, Value becomes a list from which you can choose a value. Otherwise, enter a value.

How to Apply

⊙ **Include in new selection set** creates a new selection set.

○ **Exclude from new selection set** inverts the selection set, excluding all objects that match the selection criteria.

Append to current selection set

☑ adds to the current selection set.

☐ replaces the current selection set.

RELATED COMMANDS

Select selects objects on the command line.

Filter runs a more sophisticated version of the QSelect command.

TIPS

- Use this command to select objects based on their properties; use the Select command to select objects based on their location in the drawing.
- You may select objects before entering the QSelect command, and then add or remove objects from the selection set with the Quick Select dialog box's options.
- Since this command is not transparent, you cannot use it within other commands; instead, use the Properties palette, which has the Quick Select button in the upper right corner.
- Operators include the following:

Operator		Meaning
=	**Equals**	Selects objects equal to the property.
<>	**Not Equal To**	Selects objects different from the property.
>	**Greater Than**	Selects objects greater than the property.
<	**Less Than**	Selects objects less than the property.
*	**Wildcard Match**	Selects objects with matching text.

- As of AutoCAD 2005, this command also selects block insertions and tables.
- As of AutoCAD 2008, this command includes annotative as a property.
- As of AutoCAD 2010, this command includes dynamic dimensions as entities.
- As of AutoCAD 2011, this command includes transparency as a property.
- This command works with the properties of proxy objects created by ObjectARX applications.

QText

Ver. 20 Displays lines of text as rectangular boxes (short for Quick TEXT).

Command	*Aliases*	*Shortcut Keystrokes*	*Menu*	*Ribbon*
'qtext	...	...	...	...

Command: qtext
Enter mode [ON/OFF] <OFF>: *(Type ON or OFF.)*

Illustrated AutoCAD
Quick Reference

Left: *Normally displayed text.*
Right: *Quick text after regeneration.*

COMMAND LINE OPTIONS

ON turns on quick text after the next RegenAll command.

OFF turns off quick text after the next RegenAll command.

RELATED COMMAND

RegenAll regenerates the screen, which makes quick text take effect.

RELATED SYSTEM VARIABLE

QTextMode holds the current state of quick text mode.

TIPS

- A regeneration is required before AutoCAD displays text in quick outline form:

  Command: regenall
  Regenerating model.

- To reduce regen time, use QText to turn lines of text into rectangles, which redraw faster.
- The rectangles displayed by this command are affected by lineweights.
- The length of a qtext box does not necessarily match the actual length of text.
- Turning on QText affects text during plotting; qtext blocks are plotted as rectangles.
- To find invisible text (text made of spaces or empty text), turn on QText, thaw all layers, zoom to extents, and use the RegenAll command. As of AutoCAD 2010, you can use the Purge command to remove empty text objects.

QuickCalc / QcClose

2006 Opens and closes the QuickCalc palette.

Command	Alias	Shortcut Keystrokes	Menu	Ribbon
'quickcalc	**qc**	**Ctrl+8**	**Tools** ↳**Palettes** ↳**Quick Calc**	**View** ↳**Palettes** ↳**Quick Calc**
'qcclose	**...**	**Ctrl+8**	**...**	**...**

Command: quickcalc

Displays palette:

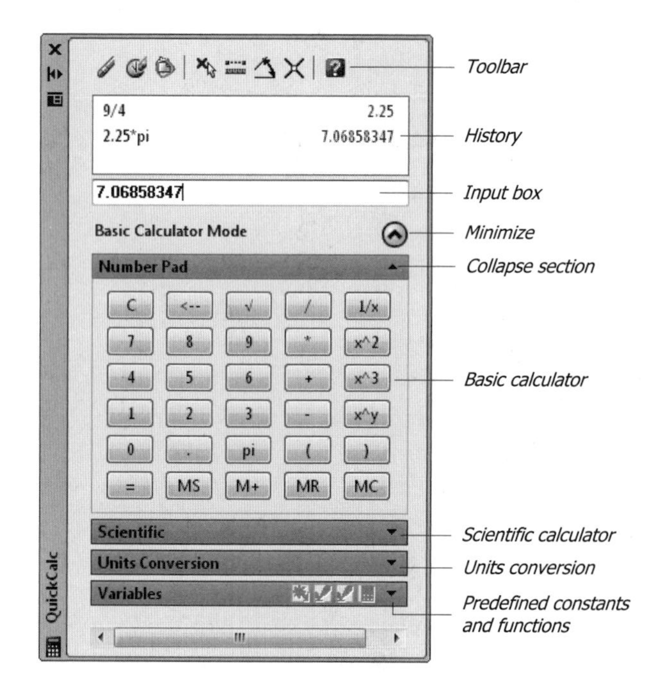

Command: qcclose

Closes the palette.

QUICKCALC PALETTE

Basic Calculator performs arithmetic and algebraic functions.

Scientific Calculator adds logarithmic and trigonometric functions (figure below, left).

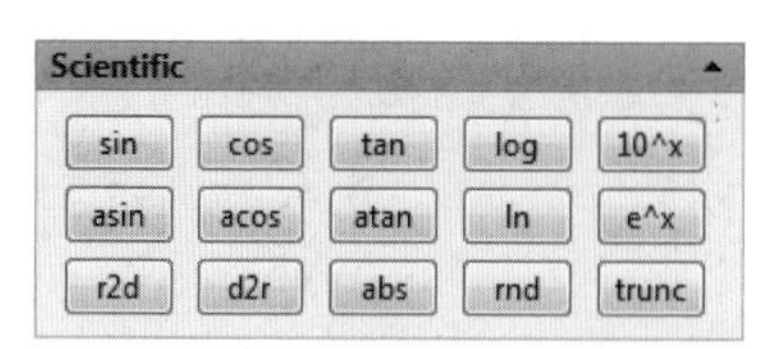

Units Conversion	
Units type	Length
Convert from	Meters
Convert to	Meters
Value to convert	0
Converted value	

Units Conversion converts between units based on the *acad.unt* file (figure above, right). It does not include all conversion factors provided by the file, such as fortnights.

Predefined Constants and Functions defines the values of variables.

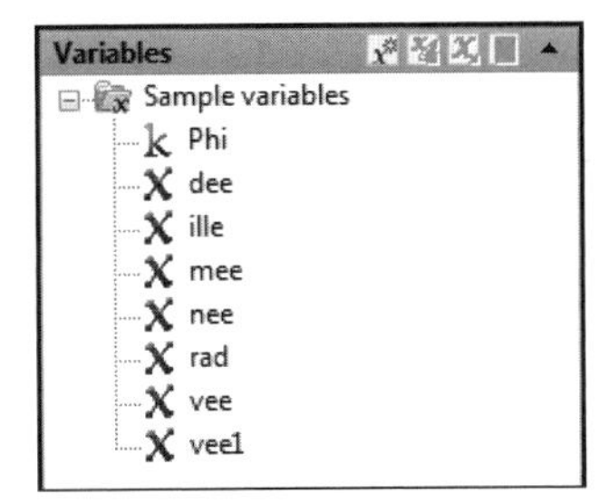

VARIABLE DEFINITION DIALOG BOX

Variable Type options

○ **Constant** stores variables as constant values.

⊙ **Function** stores variables as functions.

Variable Properties options

Name specifies the name of the variable.

Group With stores the variable with a user-defined group.

Value or Expression defines the value of the constant, or the expression for the variable.

Description describes the constant or variable.

PALETTE TOOLBARS

QuickCalc toolbar

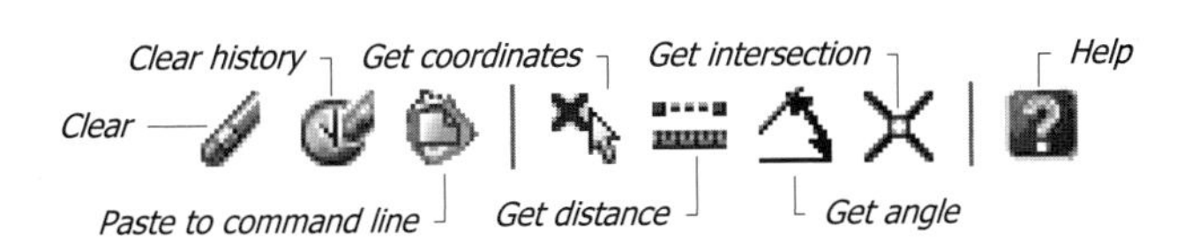

Clear clears the input area.

Clear History clears the history list.

Paste to Command Line places the value on the command line.

Get Coordinates returns the x, y, z coordinates of the pick point.

Get Distance returns the distance between two points picked in the drawing.

Get Angle returns the angle between two points relative to the x axis.

Get Intersection returns the x, y, z coordinates of the intersection of two lines defined by four pick points.

Help displays online help.

Variables toolbar

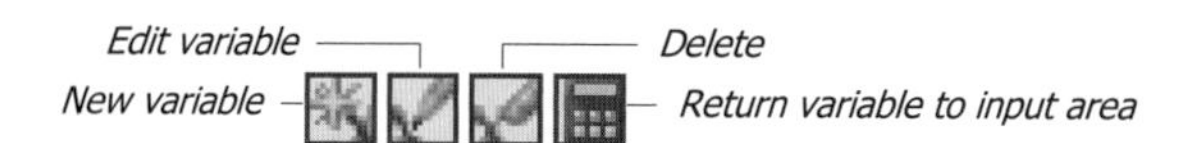

New Variable displays the Variable Definition dialog box.

Edit Variable also opens the Variable Definition dialog box.

Delete erases the selected variable.

Return Variable to Input Area places the variable on the input line.

RELATED COMMAND

Cal displays the geometry calculator function on the command line.

RELATED SYSTEM VARIABLES

CalcInput determines whether mathematical expressions are evaluated in dialog boxes.

QcState reports whether the QuickCalc palette is open or not.

RELATED FILE

acad.unt stores conversion units.

TIPS

- Unlike most calculators, QuickCalc does not calculate answers when you click a function; instead, it builds *expressions*. It expects you to compose expressions: enter the expression, edit it, and then click **=** or press Enter.
- Use the history area to modify expressions and recalculate them.
- You can enter expressions in fields of dialog boxes that expect numbers, such as the one illustrated below:

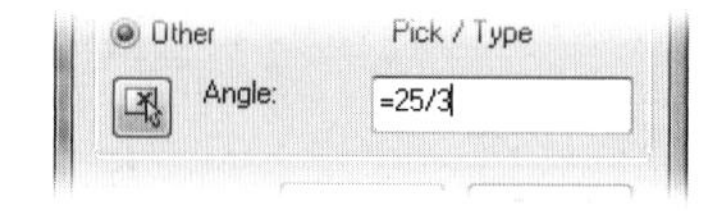

Use the following syntax:

=expression (Press Alt+Enter)

- You can use AutoLISP to perform calculations at the command line:

Command: (/ 25 3)
8.333333

- This command can be used during other commands to provide numeric input. The QuickCalc icon appears in most dialog boxes that accept numeric input.

QuickCui

2008 Opens the Customize User Interface dialog box in a partially-collapsed state.

Command	*Alias*	*Shortcut Keystrokes*	*Menu*	*Ribbon*
quickcui	**qcui**	...	...	...

Command: quickcui

Displays dialog box:

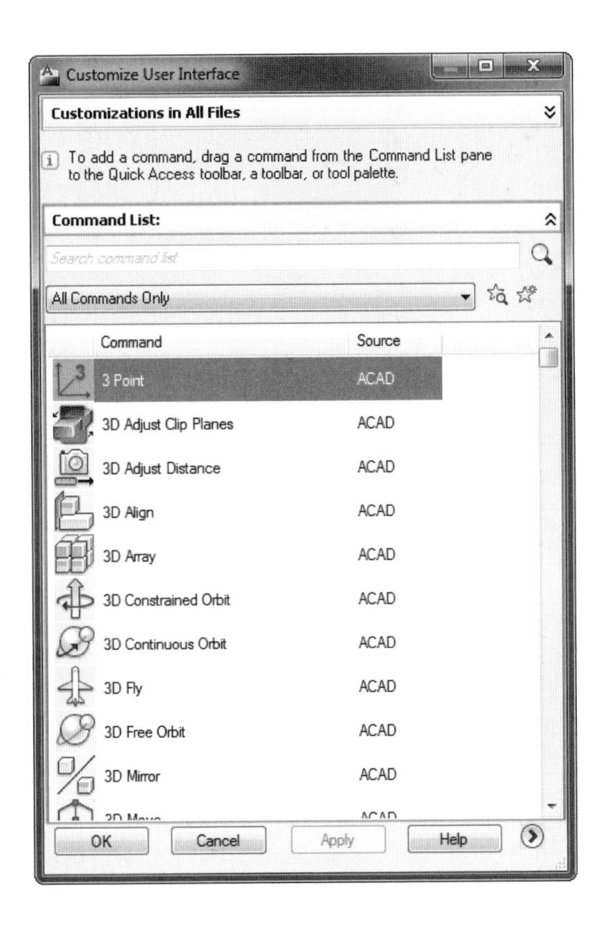

DIALOG BOX OPTION

⊙ opens the rest of the dialog box.

See the CUI command for further details.

RELATED COMMAND

CUI opens the Customize User Interface dialog box.

QuickProperties

2009 Displays a floating panel of object properties near the cursor (an undocumented command).

Command	Aliases	Shortcut Keystroke	Menu	Status Bar
quickproperties	...	**Ctrl+Shift+P**	...	

Command: quickproperties
<Quick Properties panel on>

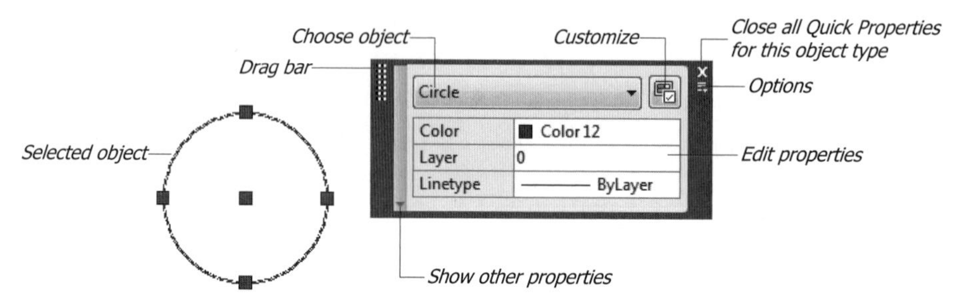

PANEL BUTTONS

Customize displays the Customize User Interface dialog box; see the Cui command.

Options displays a menu of choices.

Show Other Properties expands the panel to list all object properties.

MENU OPTIONS

To access this menu, click the Options button or right-click the edges of the panel.

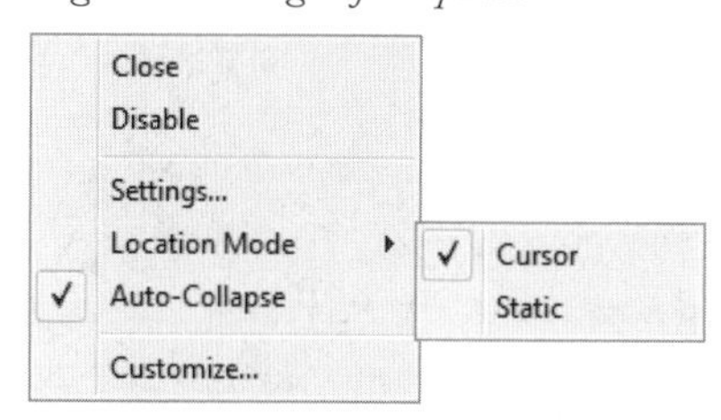

Close closes the panel for the related object type.

Settings displays the Quick Properties tab of the Drafting Settings dialog box; see the DSettings command.

Location Mode determines where the panel floats:

- **Cursor** floats the panel near the cursor.
- **Float** floats the panel in one location.

Auto-Collapse contracts the panel until the cursor passes over it.

Customize displays the Customize User Interface dialog box; see the Cui command.

RELATED COMMANDS

Properties displays the Properties palette.

List reports properties at the command prompt.

RELATED SYSTEM VARIABLES

QpLocation specifies the location of the Quick Properties palette relative to the cursor.

QpMode toggles Quick Properties between open and closed.

TIPS

- To use Quick Properties:
  1. Turn on Quick Properties by clicking its button on the status bar.
  2. Select an object. Notice that the panel appears nearby.
  3. To change a property, choose it, just as in the Properties palette.
  4. To dismiss the panel, don't click the X button, because it prevents the panel from appearing the next time you select the same type of object. Instead, unselect the object, or click the Quick Properties button on the status bar.
- The display properties of this panel can be set in the DSettings dialog box's Quick Properties tab. See the DSettings command.

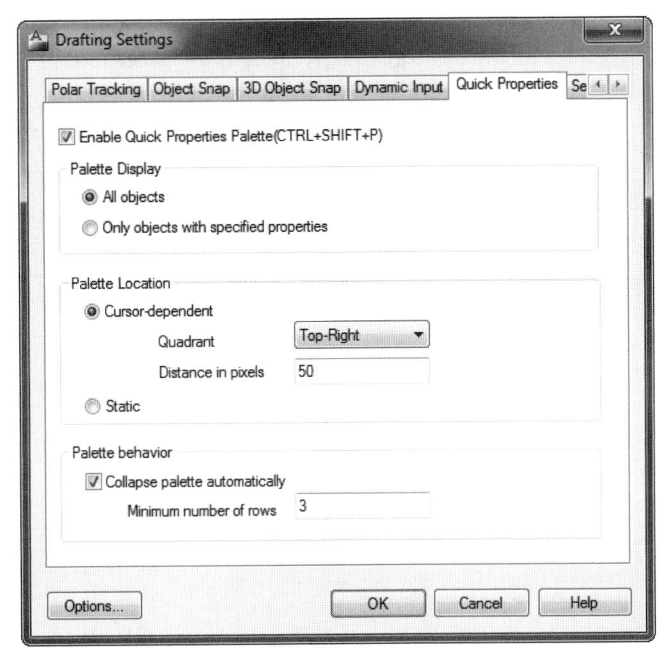

- As an alternative to Quick Properties, use rollover tips, which report properties in a tooltip-like interface. (Use the RolloverTips system variable to toggle the display of rollover tips.) Rollover tips can be customized through the Cui command's Rollover Tips node.

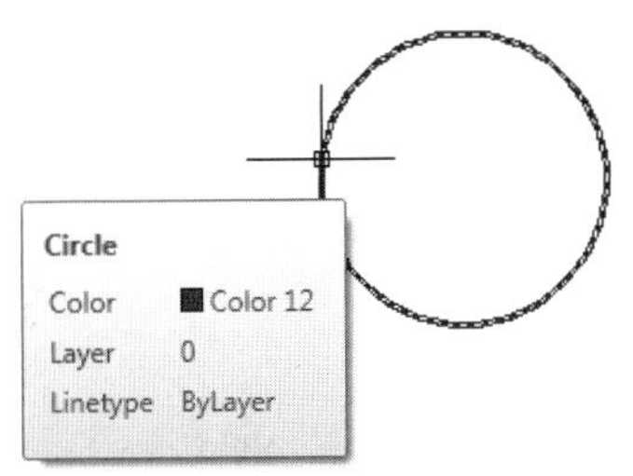

- Quick Properties can be customized with the Cui command's Quick Properties node. You can choose which objects will show Quick Properties, and which properties will be listed for each object.
- As of AutoCAD 2011, Quick Properties are turned off by default. ("Horray!" rejoices the tech editor.)

Quit

Ver. 1.0 Exits AutoCAD without saving changes made to the drawing after the most recent QSave or SaveAs command.

Command	*Alias*	*Shortcut Keystrokes*	*Menu*	*Ribbon*
quit	**exit**	**Ctrl+Q**	**File**	**...**
		Alt+F4	**↳Exit**	

Command: quit

Displays dialog box

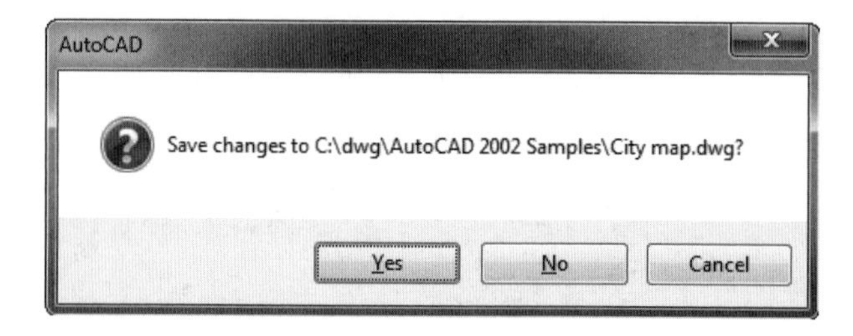

AUTOCAD DIALOG BOX

Yes saves changes before leaving AutoCAD.

No does not save changes.

Cancel does not quit AutoCAD; returns to drawing.

RELATED COMMANDS

Close closes the current drawing.

SaveAs saves the drawing by another name and to another folder or drive.

RELATED SYSTEM VARIABLE

DBMod indicates whether the drawing has changed since it was loaded.

RELATED FILES

.dwg* are AutoCAD drawing files.

.bak* are backup files.

.bkn* are additional backup files.

TIPS

- You can change a drawing, yet preserve its original format:
  1. Use the SaveAs command to save the drawing by another name.
  2. Use the Quit command to preserve the drawing in its most recently-saved state.
- Even if you accidentally save over a drawing, you can recover the previous version — when AutoCAD is set to save backup files automatically. (See the Options command.) Follow these steps:
  1. Use Windows Explorer to rename the drawing file.
  2. Use Windows Explorer to rename the *.bak* (backup) extension to *.dwg*.
- You cannot save changes to a read-only drawing with the Quit command; use the SaveAs command instead.
- As a quick alternative to this command, double click the big red application menu button to close drawings and exit AutoCAD.

QvDrawing / QvDrawingClose

2009 Displays thumbnails of open drawings (short for "Quick View DRAWING").

Commands	Aliases	Status Bar	Menu	Ribbon
qvdrawing	**qvd**		...	...
qvdrawingclose	**qdvc**		...	...

Command: qvdrawing

Displays thumbnail strip of drawing previews:

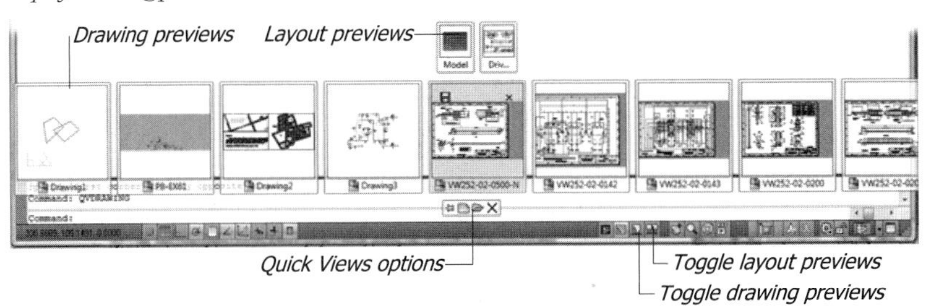

Command: qvdrawingclose

Closes thumbnails strip.

USER INTERFACE OPTIONS

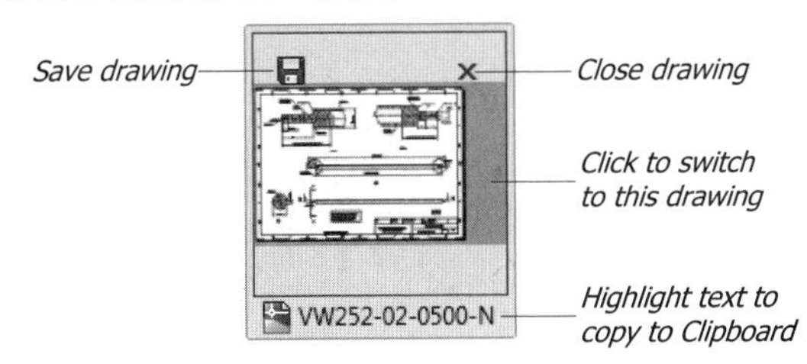

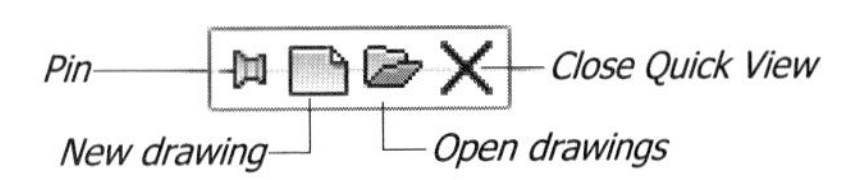

Move cursor over drawing thumbnail to see previews of model and layout tabs.

RELATED COMMANDS

QvLayout displays thumbnails of layouts.

Taskbar displays each drawing's name on a button on the Windows taskbar.

UpdateThumbsNow updates the preview images displayed by this command.

RELATED SYSTEM VARIABLE

QvDrawingPin toggles the pinning of drawing quick views. When pinned, the row of quick view images is always visible.

TIPS

- This command is useful when two or more drawings (and/or two or more layouts) are open in AutoCAD.
- This command is a visual alternative to the Ctrl+Tab shortcut and the Taskbar system variable.

QvLayout / QvLayoutClose

2009 Displays thumbnails of model and layout tabs in open drawings (short for "Quick View LAYOUT").

Commands	Aliases	Status Bar	Menu	Ribbon
qvlayout	**qvl**		...	...
qvlayoutclose	**qvlc**		...	... ...

Command: qvlayout

Displays thumbnail strip of layout previews:

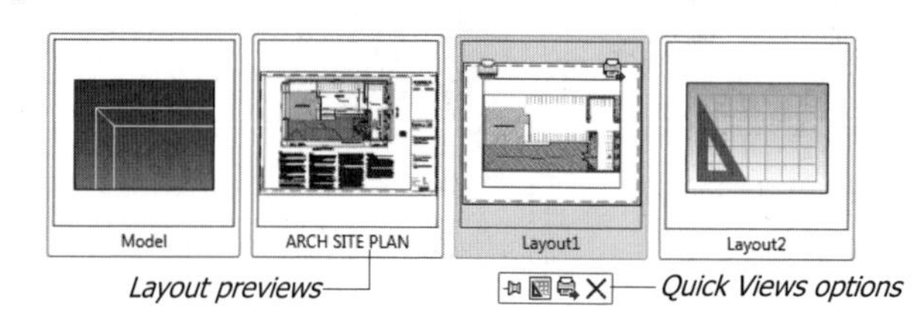

Command: qvlayoutclose

Closes thumbnail strip.

USER INTERFACE OPTIONS

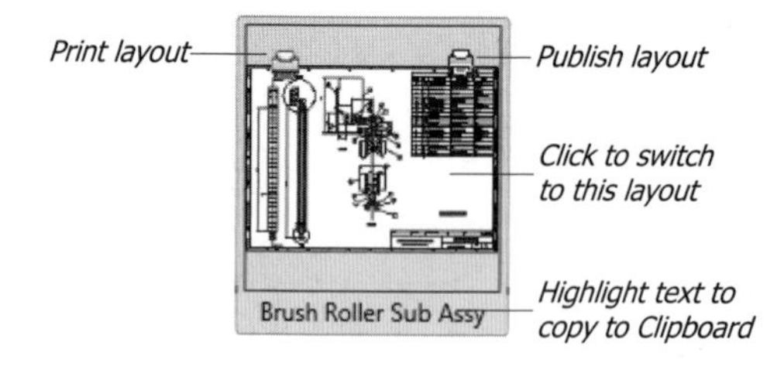

RELATED COMMANDS

QvDrawing displays thumbnails of open drawings.

UpdateThumbsNow updates the preview images displayed by this command.

RELATED SYSTEM VARIABLE

QvLayoutPin toggles the pinning of drawing quick views.

TIPS

- This command is useful for switching between layouts, as well as quickly printing or publishing them.
- Right-click the thumbnails for shortcut menus similar to those displayed by right-clicking layout tabs.

Ray

Rel. 13 Draws semi-infinite construction lines.

Command	Aliases	Shortcut Keystrokes	Menu	Ribbon
ray	...	...	**Draw** ↳**Ray**	**Home** ↳**Draw** ↳**Ray**

Command: ray
Specify start point: *(Pick a point.)*
Specify through point: : *(Pick another point.)*
Specify through point: : *(Press Enter to end the command.)*

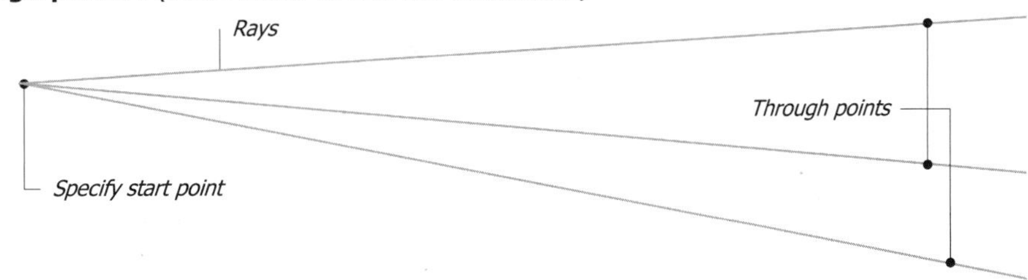

COMMAND LINE OPTIONS

Start point specifies the starting point of the ray.

Through point specifies the point through which the ray passes.

RELATED COMMANDS

Properties modifies rays.

Line draws finite line segments.

XLine draws infinite construction lines.

TIPS

- Ray objects have an endpoint at one end and are "infinite" in length at the other end.
- Rays display and plot, but do not affect the extents of the drawing.
- A ray has all the properties of a line (including color, layer, and linetype), and can be used as a cutting edge for the Trim command.
- The technical editor defines infinite: "You can type in coordinates for the ray's start and through points, keeping in mind these restrictions:
  - Scientific notation allows a maximum of e99.
  - Maximum command line input is 255 characters. Thus, the start point can be -999...999E99, -999...999E99, and the through point can be -999...999E99, -999...999E99 — where ... consists of enough 9s to total 255 characters."

Undocumented and Removed Commands

The undocumented **R14PenWizard** helps create color-dependent plot style tables.

RConfig (render configuration) command was removed from Release 14. It was replaced by Render.

Recover

Re. 12 Attempts to recover damaged drawings without user intervention.

Command	Aliases	Shortcut Keystrokes	Menu Bar	Application Menu	Ribbon
recover	...	...	**File ↳Drawing Utilities ↳Recover**	**Drawing Utilities ↳Recover ↳Recover**	...

Command: recover

Displays the Select File dialog box. Select a .dwg, .dxf, dws, or .dwt file, and then click Open.

Sample output:

```
Drawing recovery.
Drawing recovery log.
Scanning completed.

Validating objects in the handle table.
Valid objects 14526  Invalid objects 0
Validating objects completed.
16 error opening *Model_Space's layout.
Setting layout id to null.
16 error opening *Paper_Space's layout.
Setting layout id to null.

    Salvaged database from drawing.
A vertex was added to a 3D pline (3A18) which had only one vertex.
Auditing Header
Auditing Tables
Auditing Entities Pass 1
Pass 1 14500   objects audited
Auditing Entities Pass 2
Pass 2 14500   objects audited
Auditing Blocks
 10       Blocks audited
Total errors found 2 fixed 2
Erased 0 objects
```

RELATED COMMANDS

Audit checks a drawing for integrity.

DrawingRecovery displays a window listing recoverable drawings.

RecoverAll recovers xrefs as well.

RELATED SYSTEM VARIABLE

RecoverAuto displays recovery notifications of damaged drawing files at the command line or in a dialog box *(new to AutoCAD 2011)*.

TIPS

- The Recover command does not ask permission to repair damaged parts of the drawing file; use the Audit command if you want to control the repair.
- The Quit command discards changes made by the Recover command.
- If the Recover and Audit commands do not fix the problem, try using the DxfOut command, followed by the DxfIn command — assuming the file is not so damaged that it cannot be opened at all.

RecoverAll

2008 Updates and, if necessary, repairs drawings and related xrefs.

Command	Aliases	Menu Bar	Application Menu	Ribbon
recoverall	...	**File** ↳**Drawing Utilities** ↳**Recover Drawings and Xrefs**	**Drawing Utilities** ↳**Recover** ↳**Recover with Xrefs**	...

Command: recoverall

May display an initial warning dialog box.

Displays the Select File dialog box. Select a .dwg, .dws, or .dwt file, and then click Open.

Reports drawing recovery process in the Drawing Recovering Log dialog box:

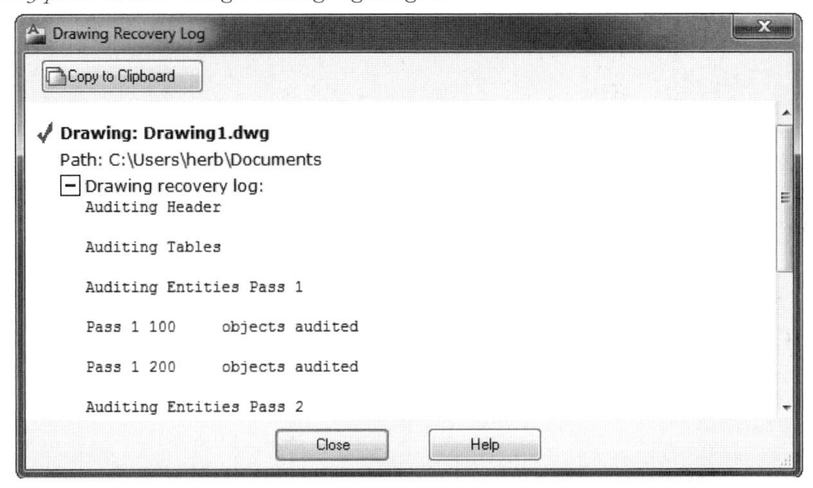

DRAWING RECOVERY LOG DIALOG BOX

Copy to Clipboard copies the report to the Windows Clipboard. In a text editor or word processor, press Ctrl+V to paste the report into the document.

Close closes the dialog box.

RELATED COMMANDS

Recover checks a drawing for integrity.

DrawingRecovery displays a window listing recoverable drawings.

TIP

- The RecoverAll command works like this:

  1. The command opens the selected drawing and related xrefs.
  2. It recovers the files, and it updates custom objects (if an object enabler is available).
  3. Then it saves the files in AutoCAD 2010 format, saving the original files as *.bak* files.
  4. Finally, the command closes the files.

Rectang

Rel. 12 Draws squares and rectangles with a variety of options.

Command	Aliases	Shortcut Keystrokes	Menu	Ribbon
rectang	**rec** **rectangle**	...	**Draw** ↳**Rectangle**	**Home** ↳**Draw** ↳**Rectangle**

Command: rectang

Specify first corner point or [Chamfer/Elevation/Fillet/Thickness/Width]: : *(Pick a point, or enter an option)*

Specify other corner point or [Area/Dimensions/Rotation]: *(Pick another point.)*

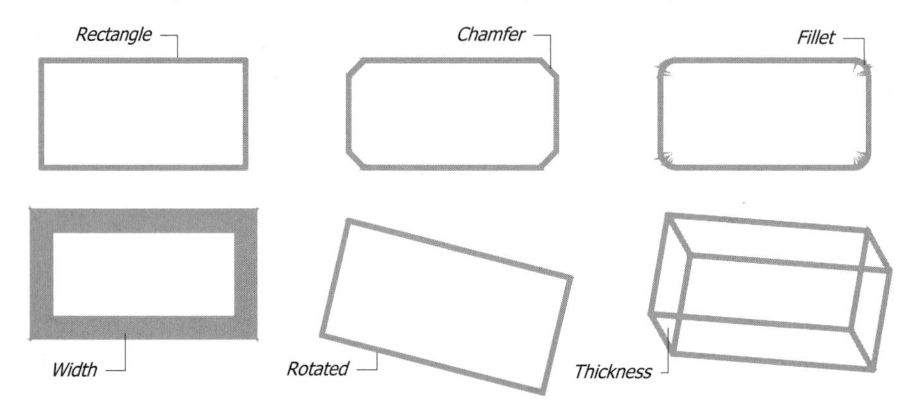

COMMAND LINE OPTIONS

Specify first corner point picks the first corner of the rectangle.

Specify other corner point picks the opposite corner of the rectangle.

Chamfer options

Specify first chamfer distance for rectangles <0.0>: : *(Enter a distance value.)*

Specify second chamfer distance for rectangles <0.0>: *(Enter another distance value.)*

First chamfer distance for rectangles sets the first chamfer distance for all four corners.

Second chamfer distance for rectangles sets the second chamfer distance for all four corners.

Elevation option

Specify the elevation for rectangles <0.0>: *(Enter an elevation value.)*

Elevation for rectangles sets the elevation (*height of the rectangle in the z direction*).

Fillet option

Specify fillet radius for rectangles <1.0>: *(Enter a radius value.)*

Fillet radius for rectangles sets the fillet radius for all four corners of the rectangle.

Thickness option

Specify thickness for rectangles <0.0>: *(Enter a thickness value.)*

Thickness for rectangles sets the thickness of the rectangle's sides in the z direction.

Width option

Specify line width for rectangles <0.0>: *(Enter a width value.)*

Width for rectangles sets the width of all segments of the rectangle's four sides.

Area option

Enter area of rectangle in current units <100.0>: *(Enter an area value.)*

Calculate rectangle dimensions based on [Length/Width] <Length>: *(Type L or W.)*

Enter rectangle length <10.0>:

Length specifies the length of the rectangle; AutoCAD determines the width.

Width specifies the width of the rectangle; AutoCAD determines the length.

Dimensions options

Specify length for rectangles <0.0>: *(Enter a length value.)*

Specify width for rectangles <0.0>: *(Enter a width value.)*

Specify other corner point or [Area/Dimension/Rotation]: *(Pick a point.)*

Specify length specifies the length along the x axis.

Specify width specifies the width along the y axis.

Specify other corner point specifies the orientation of the rectangle.

Rotation options

Specify rotation angle or [Pick points] <0>: *(Enter an angle.)*

Rotation angle specifies the angle from the x axis to rotate the rectangle about its first corner.

Pick points prompts you to pick two points that show the angle.

RELATED COMMANDS

Donut draws solid-filled circles with polylines.

Ellipse draws ellipses with polylines, when PEllipse = 1.

PEdit edits polylines, including rectangles.

PLine draws polylines and polyline arcs.

Polygon draws polygons — 3 to 1024 sides — from polylines.

RELATED SYSTEM VARIABLE

PlineWid saves the current polyline width.

TIPS

- Rectangles are drawn from polylines; use the PEdit command to change the rectangle, such as the width of the polyline.
- The values you set for the Chamfer, Elevation, Fillet, Thickness, and Width options become the default for the next execution of the Rectangle command.
- The pick point determines the location of the rectangle's first vertex; rectangles are drawn counterclockwise.
- Use the Snap command and object snap modes to place the rectangle precisely.
- Use object snap modes ENDpoint or INTersection to snap to the rectangle's vertices.
- This command ignores the settings in the ChamferA, ChamferB, Elevation, FilletRad, PLineWid, and Thickness system variables.

Redefine

Rel. 9 Restores the meaning of AutoCAD commands disabled by the Undefine command.

Command	*Aliases*	*Shortcut Keystrokes*	*Menu*	*Ribbon*
redefine	...	...	...	...

Command: redefine
Enter command name: *(Enter command name.)*

COMMAND LINE OPTION

Enter command name specifies the name of the AutoCAD command to redefine.

RELATED COMMANDS

All AutoCAD commands can be redefined.

Undefine disables the meaning of AutoCAD commands.

TIPS

- Prefix any command with a . (*period*) to redefine the undefinition temporarily, as in

  Command: .line

- Prefix any command with an _ (*underscore*) to make an English-language command work in any linguistic version of AutoCAD, as in

  Command: _line

- You must undefine a command with the Undefine command before using the Redefine command.
- *Warning!* If you undefine the Redefine command, it cannot be redefined with the Redefine command.
- Commands are undefined temporarily; all undefines are cancelled when you exit and restart AutoCAD.

Redo

Ver. 2.5 Reverses the effect of the most-recent U and Undo commands.

Command	Aliases	Shortcut Keystrokes	Menu	Quick Start Toolbar
redo	...	...	**Edit** ↳**Redo**	

Command: redo

COMMAND LINE OPTIONS

None.

RELATED COMMANDS

MRedo redoes more than one undo.

Oops un-erases the most recently-erased objects.

U undoes the most recent AutoCAD command.

Undo undoes the most recent series of AutoCAD commands.

TIPS

- The Redo command is limited to reversing a single undo, while the Undo and U commands can undo operations all the way back to the beginning of the editing session.
- The Redo command must be used immediately following the U or Undo command.
- "Do-do" is what you say after you realized you used Undo once too often and can no longer redo.

Redraw / RedrawAll

Ver. 1.0 Redraws viewports to clean up the screen.

Commands	*Aliases*	*Shortcut Keystrokes*	*Menu*	*Ribbon*
'redraw	**r**	...	...	...
'redrawall	**ra**	...	**View** **⤷Redraw**	...

Command: redraw

Redraws the current viewport.

Command: redrawall

Redraws all viewports.

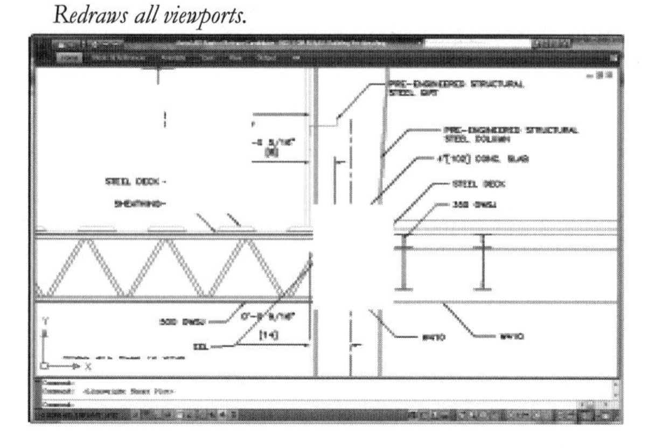

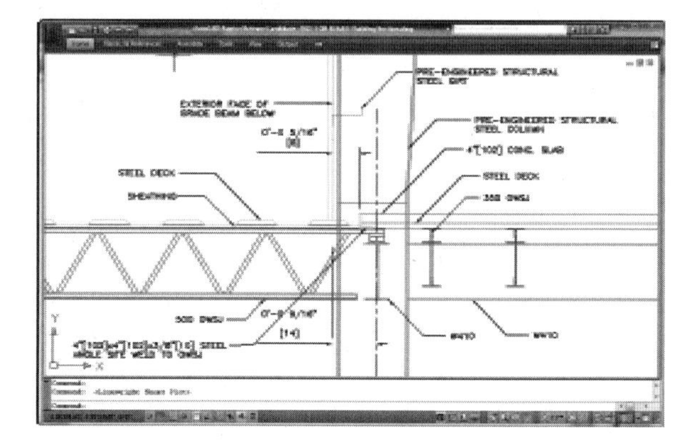

***Left**: Before redraw, portions of the drawing are "missing."*
***Right**: The drawing is clean after the redraw.*

COMMAND LINE OPTION

ESC stops the redraw.

RELATED COMMAND

Regen regenerates the current viewport.

RELATED SYSTEM VARIABLE

SortEnts controls the order of redrawing objects:

0 — Sorts by the order in the drawing database.
1 — Sorts for object selection.
2 — Sorts for object snap.
4 — Sorts for redraw.
8 — Sorts for creating slides.
16 — Sorts for regeneration.
32 — Sorts for plotting.
64 — Sorts for PostScript plotting.

TIPS

- Use this command to clean up the screen after a lot of editing; some commands automatically redraw the screen when they are done.
- Redraw does not affect objects on frozen layers.

RefClose

2000 Saves or discards changes made to reference objects (blocks and xrefs) edited in-place (short for REFerence CLOSE).

Command	Aliases	Menu	Ribbon
refclose	...	**Tools** ↳**Xref and Block In-place Editing** ↳ **Close Reference**	...
		Tools ↳**Xref and Block In-place Editing** ↳**Save Reference Edits**	

Command: refclose
Enter option [Save/Discard reference changes] <Save>: *(Enter an option.)*

COMMAND LINE OPTIONS

Save saves the editing changes made to the block or externally-referenced file.

Discard discards the changes.

RELATED COMMANDS

RefEdit edits blocks and externally-referenced files attached to the current drawing.

Insert inserts a block in the drawing.

XAttach attaches an externally-referenced drawing.

RELATED SYSTEM VARIABLE

RefEditName specifies the file name of the referenced file being edited.

TIPS

- AutoCAD prompts you with a warning dialog box to ensure you really want to discard or save the changes made to the reference:

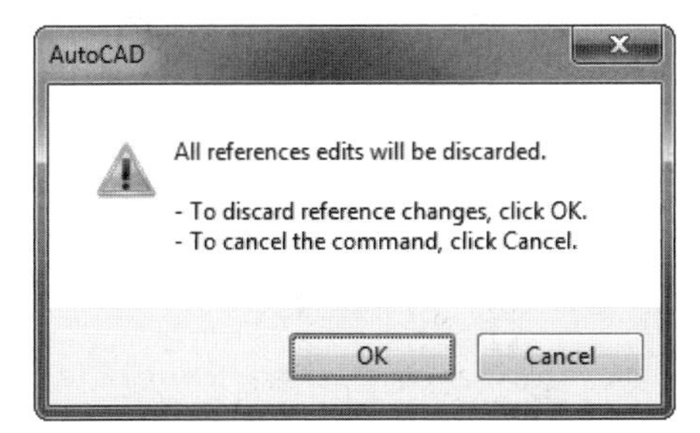

- You can use this command only after the RefEdit command; otherwise AutoCAD reports, "** Command not allowed unless a reference is checked out with RefEdit command **."

RefEdit

2000 Edits-in-place blocks and externally-referenced files attached to the current drawing (short for REFerence EDIT).

Commands	*Aliases*	*Menu*	*Ribbon*
refedit	...	**Tools** **↳Xref and Block In-place Editing** **↳Edit Reference In-Place**	**Insert** **↳Reference** **↳Edit Reference**
-refedit ...	...	...	

Command: refedit

Select reference: *(Pick an externally-referenced drawing or block.)*

Displays dialog box.

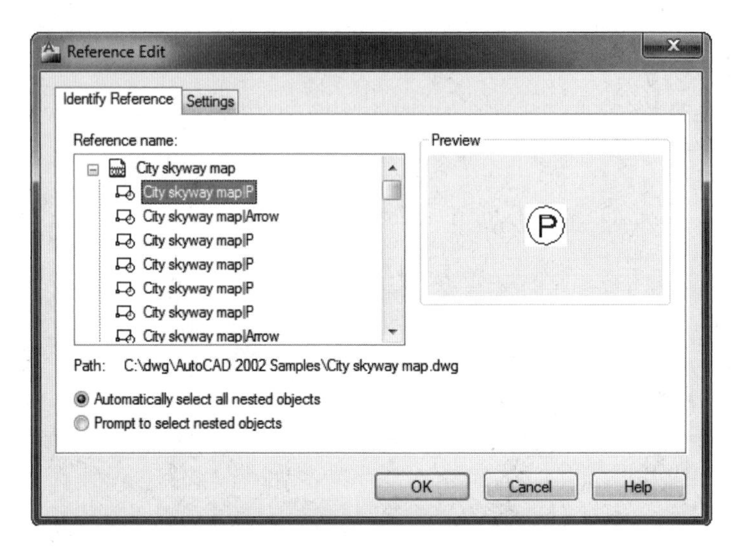

Click OK to continue with the command:

Select nested objects: *(Pick one or more objects.)*

Use REFCLOSE or the Refedit toolbar to end reference editing session.

Displays Edit Reference panel on the ribbon.

COMMAND LINE OPTIONS

Select reference selects an externally-referenced drawing or an inserted block for editing.

Select nested objects selects objects within the reference — this becomes the selection set of objects that you may edit; you may select all nested objects with the All option (with the exception of OLE objects, objects inserted with the MInsert command, and blocks or xrefs with unequal scale factors, which cannot be ref-edited).

REFERENCE EDIT DIALOG BOX

Reference name lists a tree of the selected reference object and its nested references; a single reference can be edited at a time.

Preview displays a preview image of the selected reference.

⦿ **Automatically select all nested objects** selects all nested objects.

○ **Prompt to select nested objects** prompts you: 'Select nested objects.'

Settings Tab

Identify Reference | Settings

☑ Create unique layer, style, and block names
☐ Display attribute definitions for editing
☑ Lock objects not in working set

Create unique layer, style, and block names

☑ Prefixes layer and symbol names of extracted objects with **$*n*$**.

☐ Retains the names of layers and symbols, as in the reference.

Display attribute definitions for editing (available only when an xref contains attributes):

☑ Makes non-constant attributes invisible; attribute definitions can be edited.

☐ Attribute definitions cannot be edited.

Note: When edited attributes are saved back to the block reference, use the AttRedef or BAttMan command to update attributes of the original references.

Lock objects not in working set options:

☑ Locks unselected objects in a manner similar to locking layers.

☐ Does not lock objects.

REFEDIT PANEL

This panel appears in the ribbon automatically after you select reference objects to edit:

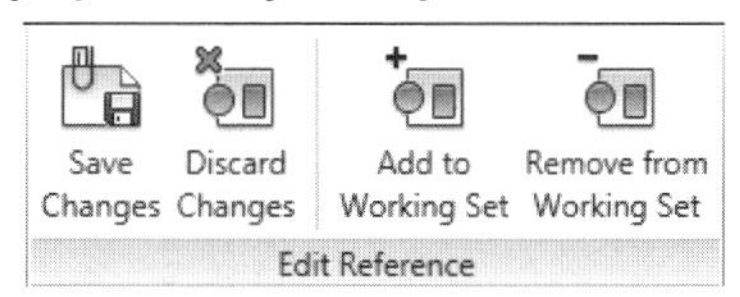

Save Changes executes the RefClose Save command; see the RefClose command.

Discard Changes executes the RefClose Discard command.

Add to Working Set executes the RefSet Add command; see the RefSet command..

Remove from Working Set executes the RefSet Remove command.

-REFEDIT Command

Command: -refedit
Select reference: *(Pick an externally-referenced drawing, or a block.)*
Select nesting level [Ok/Next] <Next>: *(Type O or N.)*
Enter object selection method [All/Nested] <All>: *(Type A or N.)*
Display attribute definitions [Yes/No] <No>: *(Type Y or N.)*
Use REFCLOSE or the Refedit toolbar to end reference editing session.

Displays Refedit toolbar.

COMMAND LINE OPTIONS

Select reference selects an externally-referenced drawing or an inserted block for editing.

Select nesting level selects objects within the reference — this becomes the selection set of objects that you may edit; you may select all nested objects with the All option (with the exception of OLE objects, objects inserted with the MInsert command, and blocks or xrefs with unequal scale factors, which cannot be ref-edited).

Enter object selection method:

All selects all objects.

Nested selects only nested objects.

Display attribute definitions:

Yes makes non-constant attributes invisible; attribute definitions can be edited.

No means attribute definitions cannot be edited.

RELATED COMMANDS

RefSet adds and removes objects from a working set.

RefClose saves or discards editing changes to the reference.

AttRedef redefines blocks and updates attributes.

BAttMan controls all aspects of attributes in blocks.

BEdit edits blocks.

AttIPedit edits attributes in block definitions and insertions directly.

RELATED SYSTEM VARIABLES

RefEditName stores the name of the externally-referenced file or block being edited.

XEdit determines whether the current drawing can be edited while being referenced by another drawing.

XFadeCtl specifies the amount of fading for objects not being edited in place.

TIPS

- You may find it more convenient to use the XEdit or BEdit commands.
- Double-clicking a block launches the Block Editor.
- In layouts, you must be in the proper layout mode to select an xref for editing: paper space for xrefs attached within layouts, model space otherwise.
- OLE objects and objects inserted with the MInsert command cannot be ref-edited.
- AutoCAD identifies the "working set" as those objects that you have selected to edit in-place.
- Objects *not* in a working set cannot be selected.

RefSet

2000 Adds and removes objects from working sets (short for REFerence SET).

Command	*Aliases*	*Menu*	*Ribbon*
refset	...	**Tools** **↳Xref and Block In-place Editing** **↳Add to Working Set**	...
		Tools **↳Xref and Block In-place Editing** **↳Remove from Working Set**	

Command: refset
Transfer objects between the Refedit working set and host drawing...
Enter an option [Add/Remove] <Add>: *(Type A or R.)*
Select objects: *(Pick one or more objects.)*

COMMAND LINE OPTIONS

Add prompts you to select objects to add to the working set.

Remove prompts you to select objects to remove from the working set.

Select objects selects the objects to be added or removed.

RELATED COMMANDS

RefEdit edits reference objects in place.

RefClose saves or discards editing changes to the reference.

RELATED SYSTEM VARIABLES

RefEditName stores the name of the xref or block being edited.

XEdit determines whether the current drawing can be edited while being referenced by another drawing.

XFadeCtl specifies the amount of fading for objects not being edited in place.

TIPS

- The purpose of this command is to add objects to — or remove them from — the "working set" of objects, while you are edit a block or an externally-referenced drawing in-place.
- You cannot add or remove objects on locked layers. AutoCAD complains, '** *n* selected objects are on a locked layer.'

Regen / RegenAll

Ver. 1.0 Regenerates viewports to update the drawing.

Commands	*Aliases*	*Shortcut Keystrokes*	*Menu*	*Ribbon*
regen	**re**	...	**View** **↳Regen**	...
regenall	**rea**	...	**View** **↳Regen All**	...

Command: regen
Regenerating model.

Regenerates the current viewport.

Command: regenall
Regenerating model.

Regenerates all viewports.

COMMAND LINE OPTION

Esc cancels the regeneration.

RELATED COMMANDS

Redraw cleans up the current viewport quickly.

RegenAuto checks with you before doing most regenerations.

ViewRes controls whether zooms and pans are performed at redraw speed.

RELATED SYSTEM VARIABLES

RegenMode toggles automatic regeneration.

WhipArc determines how circles and arcs are displayed:

0 — Circles and arcs displayed as vectors (default).

1 — Circles and arcs displayed as true circles and arcs.

TIPS

- Some commands automatically force a regeneration of the screen; other commands queue the regeneration.
- The Regen command reindexes the drawing database for better display and object selection.
- To save on regeneration time, freeze layers you are not working with, apply QText to turn text into rectangles, and place hatching on its own layer.
- Use the RegenAll command to regenerate all viewports.

RegenAuto

Ver. 1.2 Prompts before performing regenerations, when turned off (short for REGENeration AUTOmatic).

Command	*Aliases*	*Shortcut Keystrokes*	*Menu*	*Ribbon*
'regenauto	...	...	...	...

Command: regenauto
Enter mode [ON/OFF] <ON>: *(Type ON or OFF.)*

COMMAND LINE OPTIONS

OFF turns on "About to regen, proceed?" message.

ON turns off "About to regen, proceed?" message.

RELATED COMMANDS

Regen forces a regeneration in the current viewport.

RegenAll forces a regeneration in all viewports.

RELATED SYSTEM VARIABLES

Expert suppresses "About to regen, proceed?" message when value is greater than 0.

RegenMode specifies the current setting of automatic regeneration:

0 — Off.

1 — On (default).

TIPS

- If a regeneration is caused by a transparent command, AutoCAD delays it and responds with the message, "Regen queued."
- When RegenAuto is set to off, the following prompt is displayed with every command that causes a regeneration:

  Command: regen
  About to regen, proceed? <Y>: *(Press Enter.)*
- AutoCAD Release 12 reduces the number of regenerations by expanding the virtual screen from 16 bits to 32 bits.
- This command was useful when computers were so slow that a regeneration could take several minutes. The technical editor recalls that regenerations were timed to coincide with coffee and lunch breaks.

Region

Rel. 11 Creates 2D regions from closed objects.

Command	*Alias*	*Shortcut Keystrokes*	*Menu*	*Ribbon*
region	**reg**	...	**Draw** ↳**Region**	**Home** ↳**Draw** ↳**Region**

Command: region
Select objects: *(Select one or more closed objects.)*
Select objects: *(Press Enter to end object selection.)*
1 loop extracted.
1 region created.

COMMAND LINE OPTION

Select objects selects objects to convert to a region; AutoCAD discards unsuitable objects.

RELATED COMMANDS

SurfFillet adds surface fillets between pairs of regions *(new to AutoCAD 2011)*.

SurfOffset creates parallel regions at a specified distance from the original *(new to AutoCAD 2011)*.

SurfPatch creates surfaces by fitting caps over regions *(new to AutoCAD 2011)*.

SurfTrim trims at the intersections of regions *(new to AutoCAD 2011)*.

RELATED SYSTEM VARIABLE

DelObj toggles whether objects are deleted during conversion by the Region command.

TIPS

- This command converts closed line sets, closed 2D and planar 3D polylines, and closed curves into region objects.
- This command rejects open objects, intersections, and self-intersecting curves.
- The resulting region is unpredictable when more than two curves share an endpoint.
- Polylines with width lose their width when converted to a region.
- Island are "holes" in regions.

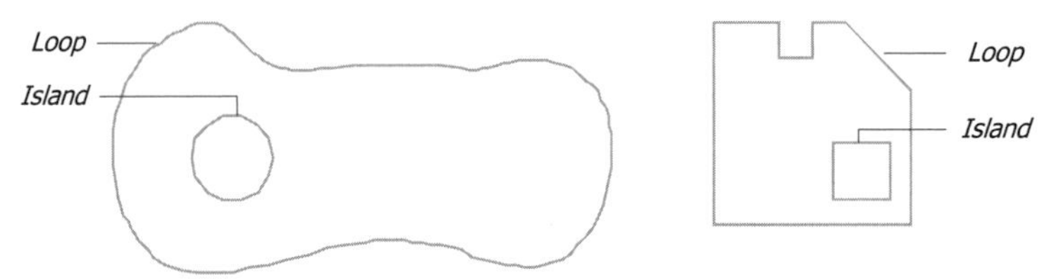

DEFINITIONS

Curve — an object made of circles, ellipses, splines, and joined circular and elliptical arcs.

Island — a closed shape fully within (not touching or intersecting) another closed shape.

Loop — a closed shape made of closed polylines, closed lines, and curves.

Region — a 2D closed area defined as a ShapeManager object.

Reinit

Rel. 12 Reinitializes digitizers and input-output ports, and reloads the *acad.pgp* file (short for REINITialize).

Command	Aliases	Shortcut Keystrokes	Menu	Ribbon
reinit	...	...	...	...

Command: reinit

Displays dialog box.

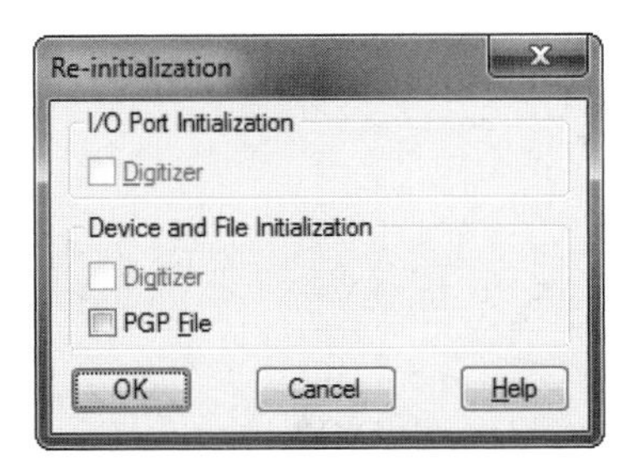

RE-INITIALIZATION DIALOG BOX

I/O Port Initialization option

□ **Digitizer** reinitializes ports connected to the digitizer; grayed out if no digitizer is configured.

Device and File Initialization options

□ **Digitizer** reinitializes the digitizer driver; grayed out if no digitizer is configured.

□ **PGP File** reloads the *acad.pgp* file.

RELATED COMMAND

CuiLoad reloads menu files.

RELATED SYSTEM VARIABLE

Re-Init reinitializes via system variable settings.

RELATED FILES

acad.pgp is the program parameters file in the hidden *\documents and settings\<login>\application data\autodesk\autocad 2011\r18.1\enu\support* folder.

*.*hdi* are device drivers specific to AutoCAD.

TIPS

- This command is almost unnecessary now that most mice, printers, and plotters use USB ports, which do not need to be re-initialized.
- AutoCAD allows you to connect both the digitizer and the plotter to the same port, since you do not need the digitizer during plotting; use this command to reinitialize the digitizer after plotting.
- AutoCAD automatically reinitializes all ports and reloads the *acad.pgp* file each time a drawing is loaded.

Rename

Ver. 2.1 Changes the names of blocks, dimension styles, layers, linetypes, materials, multileader styles, plot styles, text styles, table styles, UCS names, views, and viewports.

Commands	*Aliases*	*Shortcut Keystrokes*	*Menu*	*Ribbon*
rename	**ren**	...	**Format** ↳**Rename**	...
-rename	**-ren**	...	...	...

Command: rename

Displays dialog box.

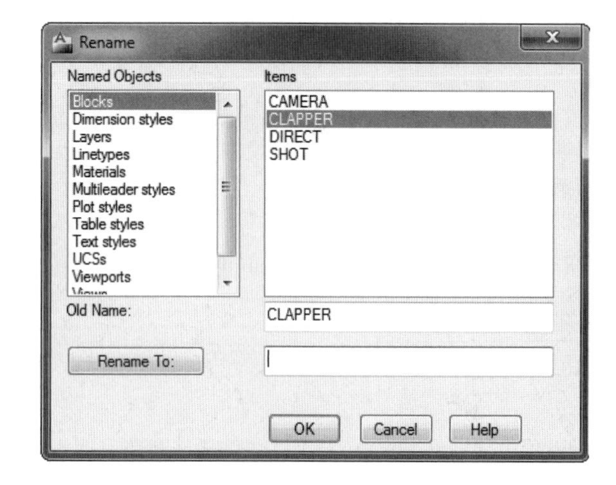

RENAME DIALOG BOX

Named Objects lists named objects in the drawing.

Items lists types of named objects in the current drawing.

Old Name displays the current name of an object to be renamed.

Rename to renames the object.

-RENAME Command

Command: -rename
Enter object type to rename
[Block/Dimstyle/LAyer/LType/Material/multileadeRstyle/Plotstyle/textStyle/Tables style/Ucs/VIew/VPort]: *(Enter an option.)*

Example usage:

[Block/Dimstyle/LAyer/LType/Material/multileadeRstyle/Plotstyle/textStyle/Tablestyle/Ucs/VIew/VPort]: b
Enter old block name: diode-2
Enter new block name: diode-02

COMMAND LINE OPTIONS

Block changes the names of blocks.

Dimstyle changes the names of dimension styles.

LAyer changes the names of layers.

LType changes the names of linetypes.

Material changes the names of materials.

Multileaderstyle changes the names of multileader styles.

Plotstyle changes the names of plot styles.

Tablestyle changes the names of table styles.

textStyle changes the names of text styles.

Ucs changes the names of UCS configurations.

VIew changes the names of view configurations.

VPort changes the names of viewport configurations.

RELATED SYSTEM VARIABLES

CeLType names the current linetype.

CLayer names the current layer.

CMaterial names the current material.

CMLeaderStyle names the current multileader style.

CTableStyle names the current table style.

DimStyle names the current dimension style.

InsName names the current block.

TextStyle names the current text style.

UcsName names the current UCS view.

TIPS

- You cannot rename layer "0", anonymous blocks (such as hatches and dimensions), groups, or linetype "Continuous." You can, however, rename the "Standard" text and dimension style.
- To rename a group of similar names, use * (the wildcard for "all") and ? (the wildcard for a single character).
- You can rename names up to 255 characters in length; some names may be longer, such as those of attached xrefs.
- The Properties command does *not* allow you to rename blocks.
- The PlotStyle option is available only when plot styles are attached to the drawing.
- You cannot use this command during RefEdit.

Render

Rel. 12 Renders 3D objects in an independent window.

Commands	Alias	Shortcut Keystrokes	Menu	Ribbon
render	**rr**	...	**View** ↳**Render** ↳**Render**	**Render** ↳**Render** ↳**Render**
-render	...	...	...	...

Command: render

Displays window (illustrated below); alternatively, it outputs the rendering to the current viewport. Renders a windowed selection or selected objects — depending on the settings in the RPref command.

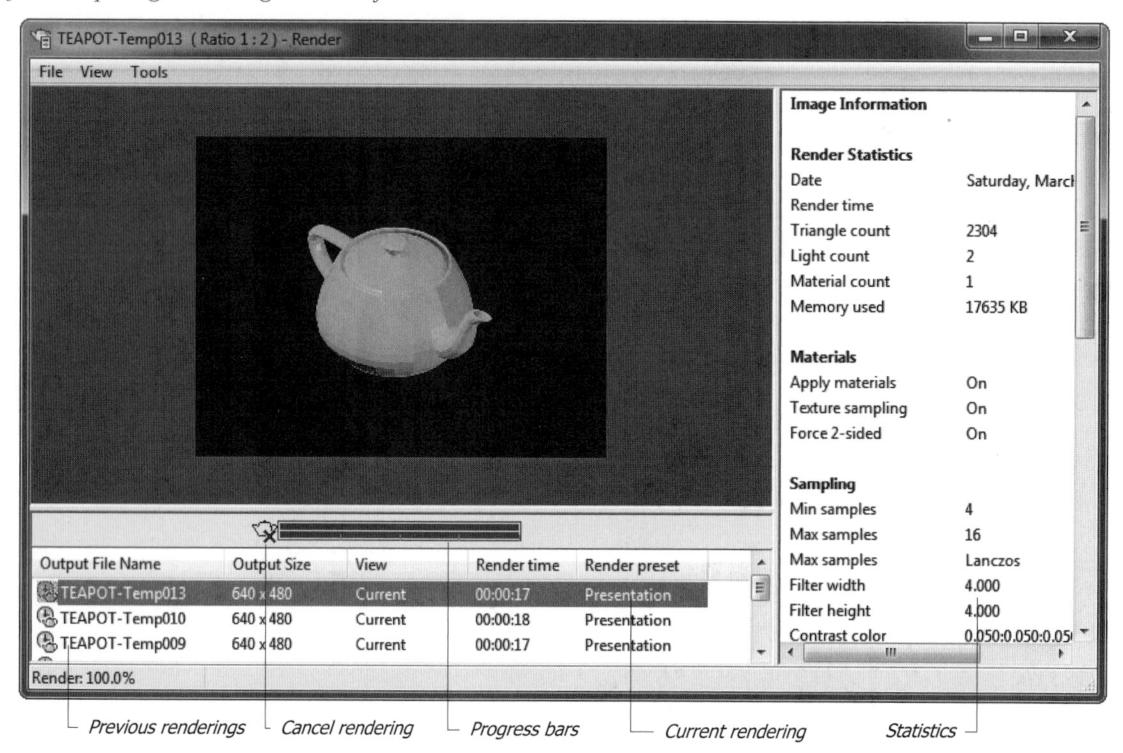

RENDER WINDOW MENU BAR

File options

Save saves the rendered image to a file in BMP, TIFF, PNG, JPEG, Targa, or PCX formats. While the SaveImg command does not apply to this window, it has the same options as this command.

Save Copy saves a copy of the image by another name or to another folder.

Exit closes the Render window; reopen with the RenderWin command.

View options

Status Bar toggles the display of the status bar.

Statistics Pane toggles the display of the statistics pane.

Tools options

Zoom + makes the image up to 64x larger.

Zoom - makes the image up to 64x smaller.

Use the scroll bars or mouse wheel to pan the image. Hold down the Ctrl key and use the mouse wheel to zoom.

PROGRESS METERS

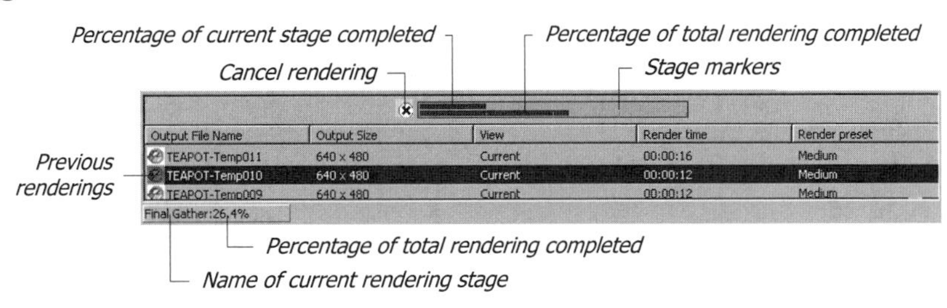

x cancels the rendering, as does the Esc key.

Upper bar shows the progress within the current phase: translation, optional photon emission, optional final gather, and render.

Lower bar shows the progress of the entire rendering; also displayed by the status bar. The history pane reports the time each rendering took.

Status bar reports the name of the current rendering stage, and the current percentage of the total rendering.

SHORTCUT MENU

Right-click the history pane.

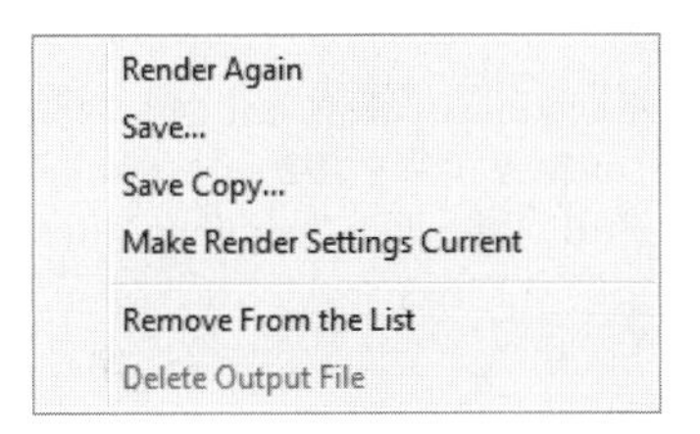

Render Again re-renders the selected history item.

Save saves the rendering to an image file.

Save Copy saves the rendering to a different folder.

Make Render Settings Current loads the rendering style associated with the history item.

Remove From The List removes the history item from the list.

Delete Output File erases the rendering from the image pane

-RENDER Command

Command: -render

Specify render preset [Draft/Low/Medium/High/Presentation/Other] <Medium>: *(Enter a preset name.)*

Specify render destination [Render Window/Viewport] <Render Window>: *(Enter a destination.)*

Draft, **Low**, **Medium**, **High**, **Presentation** specify Autodesk-defined render parameters.

Other allows you to specify a user-defined rendering preset; AutoCAD prompts, 'Specify custom render preset [?]:'.

Render Window outputs the rendering in the Render window.

Viewport renders in the current viewport.

RELATED COMMANDS

RenderEnvironment sets up fog effects.

RenderPresets sets up rendering defaults for use with this command; displays the Render Presets Manager dialog box.

-RenderOutputSize specifies the rendering resolution.

RPref sets up rendering styles; displays the Advanced Render Settings palette.

RenderWin accesses the Render window.

TIPS

- Use the scroll bars to pan the image. Use the mouse wheel to zoom.
- Rendering is performed in four steps: translation, photon emission, final gather, and render. The photon emission and final gather steps occur only if specified in the render preferences.
- Use the -Render command to batch renderings through scripts. It cannot set any rendering parameters, so preset them with the RPref or RenderPresets command by creating named rendering presets.
- The default size of renderings in the Render window is small, just 640 x 480. Use the undocumented -RenderOutputSize command to change the output to any size, such as 1080 x 1920, the resolution for 1080 high-definition screens. Alternatively, select the output size from the droplist below the ribbon's Render tab.

RenderCrop

2007 Renders a windowed area of the current viewport.

Command	*Alias*	*Shortcut Keystrokes*	*Menu*	*Ribbon*
rendercrop	**rc**	...	...	...

Command: rendercrop
Pick crop window to render: *(Pick a point, or enter x, y coordinates.)*
please enter the second point: *(Pick another point.)*

Renders the windowed area:

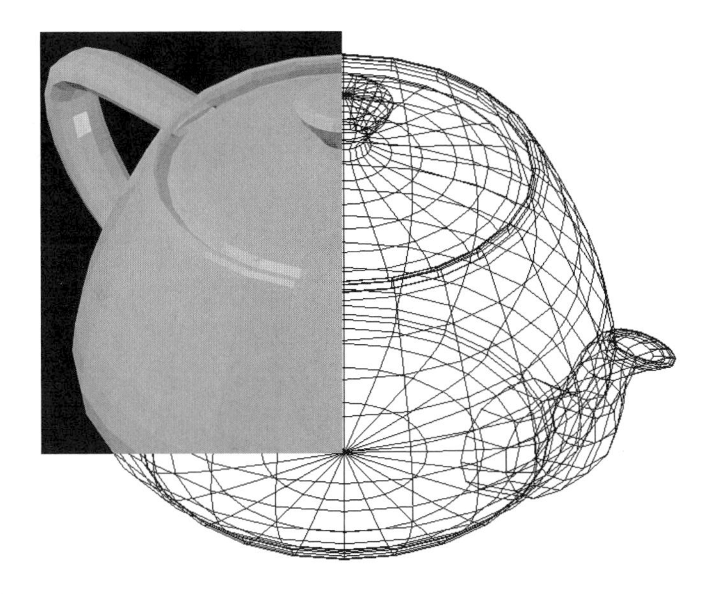

COMMAND OPTIONS

Pick crop window to render picks the first point of the area to be rendered.

please enter the second point picks the other corner of the rectangle. (Note the unusual wording of the prompt.)

RELATED COMMANDS

RenderWin renders the drawing in an independent window.

RPref sets up rendering preferences.

TIPS

- Rendering a windowed area is faster than rendering the entire viewport.
- This command does not allow you to specify rendering preferences; before using this command, run the RPref command to set up preferences.

RenderEnvironment

2007 Sets the parameters for fog effects in renderings.

Command	Alias	Menu	Ribbon
renderenvironment	**fog**	**View**	**Render**
		↳Render	**↳Render**
		↳Render Environment	**↳Environment**

Command: renderenvironment

Displays dialog box.

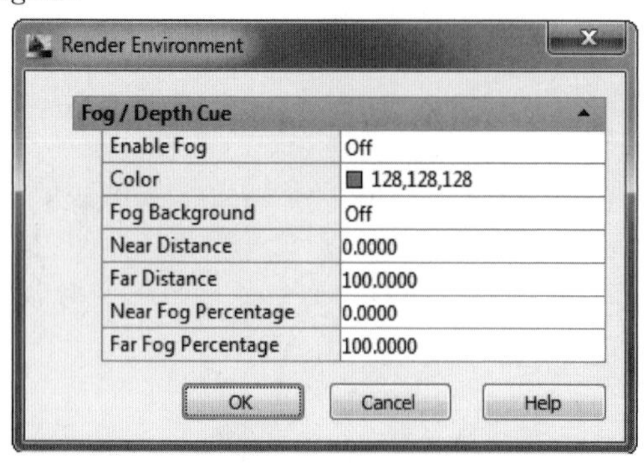

Left: *The Render Environment dialog box.*
Right: *Fog effect applied to a 3D model.*

DIALOG BOX OPTIONS

Enable Fog toggles fog settings.

Color specifies the color of the fog; displays the Select Color dialog box.

Fog Background applies the fog effect to the background; see the View command.

Near/Far Distance indicates the start and end of the fog effect, measured as a decimal (0.0 - 1.0) from the camera.

Near /Far Fog Percentage specifies the near and far intensity of the fog effect.

RELATED COMMAND

Render creates renderings with the fog effect.

TIPS

- First set up the RenderEnvironment command, and then use Render to see the effect.
- The fog can be any color, such as RGB=1,1,1 (HLS-0,1,0) for fog and RGB=0,1,0 (HLS= 0.33,0.5,0) for green alien mist.
- It can be tricky getting the fog effect to work. Use the 3dClip command to set the back clipping plane at the back of the model where you want the fog to have its full effect. Then use RenderEnvironment to set up the following parameters:
    - Enable fog — On
    - Color — White
    - Fog Background — On or Off
    - Near distance — 0.70
    - Far distance — 1.00
    - Near fog percentage — 0.00
    - Far fog percentage — 100

RenderExposure

2008 Modifies the global lighting of scenes rendered with photometric lighting.

Command	Aliases	Shortcut Keystrokes	Menu	Ribbon
renderexposure	...	...	...	**Render** ⇘**Render** ⇘**Adjust Exposure**

Command: renderexposure

- *When LightingUnits = 1 or 2, displays dialog box.*

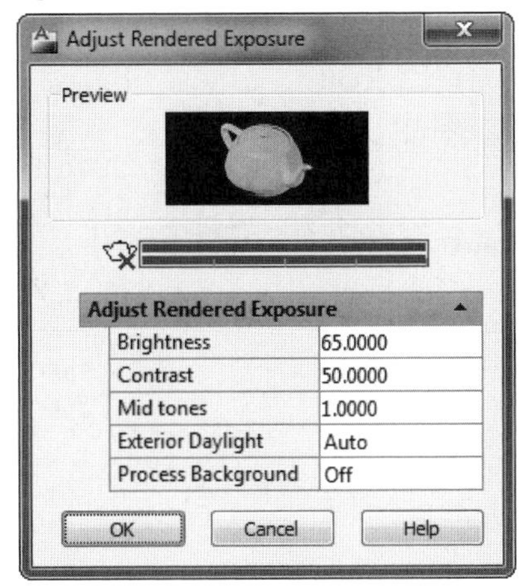

- *When LightingUnits = 0, AutoCAD complains, "Exposure Control is not enabled." To fix the problem, change the value of system variable LightingUnits to 1 or 2.*
- *When InsUnits = 0, this command displays the Units dialog box. Change insertion units to a value other than None.*

ADJUST RENDERED EXPOSURE DIALOG BOX

Brightness ranges from 0 to 200; default = 65.

Contrast ranges from 0 to 100; default = 100.

Mid tones ranges from 0 to 20; default = 1.

Exterior Daylight (a.k.a. sun) changes between auto, off, and on.

Process Background toggles between on and off.

RELATED SYSTEM VARIABLES

LogExpBrightness determines global brightness of photometric lighting.

LogExpContrast determines the global contrast of photometric lighting.

LogExpDaylight toggles global daylight of photometric lighting.

LogExpMidtones determines the global midtones of photometric lighting.

TIP

- Use this command interactively to adjust the lighting of the current 3D drawing, for example, to lighten dark renderings.

-RenderOutputSize

2009 Presets the size (in pixels) of the next rendering (undocumented).

Command	*Aliases*	*Shortcut Keystrokes*	*Menu*	*Ribbon*
-rendereoutputsize	...	...	...	...

Command: -renderoutputsize
Enter render output width: *(Enter a value, such as 1024.)*
Enter render output height: *(Enter a value, such as 800.)*

COMMAND LINE OPTIONS

Enter render output width specifies the width of the rendering, in pixels.

Enter render output height specifies the height, in pixels.

TIPS

- Use this command to reset the output size of the next rendering.
- Standard pixel sizes include the following:

Width	*Height*
Standard aspect ratios:	
640	480
800	600
1024	768
1280	1024
1600	1200
Widescreen aspect ratios:	
1280	800
1440	900
1680	1050
1920	1080
2560	1600

- This command is meant for macros and AutoLISP routines, but can be used by drafters as well.
- You can also set the size of the rendering through the ribbon: Render | Render | Render Output Size:

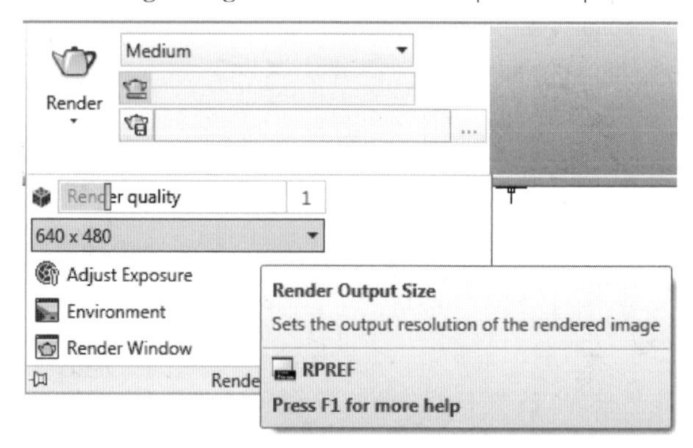

RenderPresets

2007 Creates rendering styles; displays AutoCAD's tallest and widest dialog box.

Command	Aliases	Shortcut Keystrokes	Menu	Ribbon
renderpresets	**rp**	...	...	**Render**
	rfileopt			↳**Render**
				↳**Manage Render Presets**
-renderpresets	...	...	...	...

Command: renderpresets

Displays dialog box.

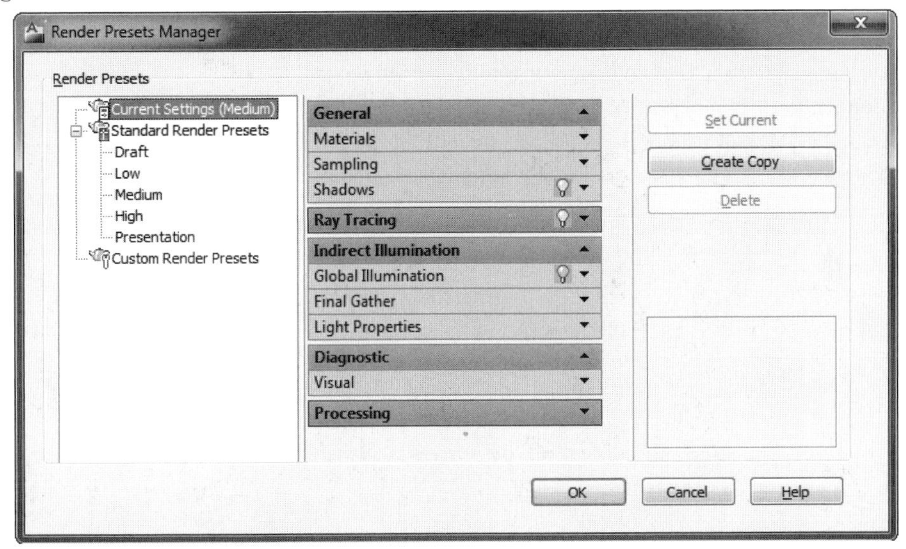

RENDER PRESETS MANAGER DIALOG BOX

See the RPref command for details of the parameters.

Set Current sets the selected rendering style as the default; the Render and RenderCrop commands use this rendering style.

Create Copy copies the parameters so that you can modify them.

Delete removes the named rendering style; you cannot delete the system styles.

-RENDERPRESETS Command

Command: -renderpresets
Enter render presets: *(Enter the name of a rendering preset, such as Medium.)*

COMMAND LINE OPTION

Enter Render Presets specifies the name of a render preset (rendering style), such as Low or Medium. This command is meant for use by macros and AutoLISP routines.

TIPS

- Use this command to create your own rendering styles. AutoCAD comes preset with five: Draft, Low, Medium, High, and Presentation:
  - **Draft** rendering is very fast, but inaccurate and very coarse. Time to render the *teapot.dwg*: 3.5 seconds.
  - **Low** rendering is fast but inaccurate; time to render: 4.5 seconds.
  - **Medium** rendering is slightly slower but much more accurate; time to render: 10 seconds.
  - **High** rendering is slower and more accurate; time to render: 15 seconds.
  - **Presentation** rendering is very slow and very accurate; time to render: 55 seconds.
- The teapot was used in the early days of computer rendering as a complex test to ensure the rendering was completed accurately. Autodesk today uses an icon of the teapot for rendering commands.

- See the RPref command for the content of the Render Presets Manager.

Removed Command

RenderUnload command was removed from Release 14.

RenderWin

2007 Displays the Render window with the most recent rendering; formerly the RendScr command.

Command	Aliases	Shortcut Keystrokes	Menu	Ribbon
renderwin	**rendscr**	...	...	**Render**
	rw			**↳Render**
				↳Render Window

Command: renderwin

Displays the Render window with the most recent rendering, as well as all previous renderings of this AutoCAD session. See the Render command for details.

RELATED COMMANDS

Render generates renderings in the Render window.

RenderCrop generates renderings in a windowed area of the viewport.

RenderPresets sets the rendering parameters.

TIPS

- If this command fails to display the Render window, it's because the window is already open, but covered up by AutoCAD. Look for its button on the taskbar:

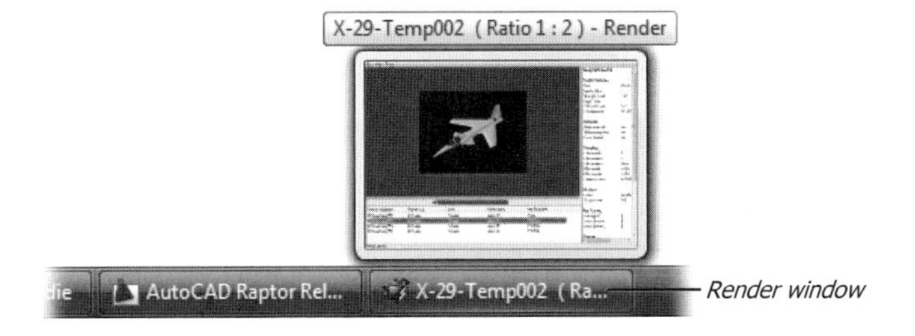

- If no rendering has occurred in this drawing, AutoCAD complains, "RENDERWIN Drawing contains no render history," and does not display the Render window.

Removed & Renamed Commands

Replay command was removed from AutoCAD 2007.

RendScr command is an alias of the RenderWin command as of AutoCAD 2007.

ResetBlock

2006 Resets dynamic block references to their default values.

Command	*Aliases*	*Shortcut Keystrokes*	*Menu*	*Ribbon*
resetblock	...	...	...	...

Command: resetblock
Select objects: *(Select one or more blocks.)*
Select objects: *(Press Enter to end object selection.)*
n **blocks reset.**

COMMAND LINE OPTION

Select objects selects the blocks to be reset.

RELATED COMMAND

BEditor creates and edits dynamic blocks.

TIP

- Dynamic blocks can be changed in many ways, according to the actions defined in the Block Editor. Use this command to reset blocks to their original definition.
- The technical editor comments: "It's interesting to see how many useful commands have access through the menu, toolbar, or ribbon."

Resume

Ver. 2.1 Resumes script files paused by an error, or by pressing the Backspace or ESC keys.

Command	*Aliases*	*Shortcut Keystrokes*	*Menu*	*Ribbon*
'resume	...	...	...	...

Command: resume

COMMAND LINE OPTIONS

BACKSPACE pauses the script file.

ESC stops the script file.

RELATED COMMANDS

RScript reruns the current script file.

Script loads and runs a script file.

RevCloud

2004 Draws revision clouds, and converts objects into revision clouds.

Command	Aliases	Shortcut Keystrokes	Menu	Ribbon
revcloud	...	...	**Draw** ↳**Revision Cloud**	**Home** ↳**Draw** ↳**Revision Cloud**

Command: revcloud
Minimum arc length: 0.5000 Maximum arc length: 0.5000 Style: Normal
Specify start point or [Arc length/Object/Style] <Object>: *(Pick a point, or enter an option.)*
Guide crosshairs along cloud path... *(Move cursor to create cloud.)*
Revision cloud finished.

Left to right: *Normal and calligraphic styles.*

COMMAND LINE OPTIONS

Specify start point specifies the starting point of the cloud.

Guide crosshairs along cloud path outlines the cloud.

Arc Length Options

Specify minimum length of arc <0.5000>: *(Enter a minimum value.)*
Specify maximum length of arc <1.0000>: *(Enter a maximum value.)*

Specify minimum / maximum length of arc specifies the minimum and maximum arc lengths.

Object Options

Select object: *(Select one object to convert to a revision cloud.)*
Reverse direction [Yes/No] <No>: *(Type Y or N.)*

Select object selects the object to convert into a revision cloud.

Reverse direction turns the revision cloud inside-out.

Style Options

Select arc style [Normal/Calligraphy] <Normal>: *(Type N or C.)*

Normal draws clouds with uniform-width arcs.

Calligraphy draws clouds from variable-width polyarcs.

RELATED SYSTEM VARIABLE

DimScale affects the size and width of the arcs.

TIPS

- When the cursor reaches the start point, the revision cloud closes automatically.
- Revision clouds are drawn as polylines; to edit them, select one, and then move its grips.
- The arc length is not available from a system variable, because it is stored in the Windows registry.
- The arc length is multiplied by the value stored in the DimScale system variable.

Reverse

2010 Reverses the direction of lines, polylines, splines, and helices.

Command	Aliases	Shortcut Keystrokes	Menu	Ribbon
reverse	...	...	...	**Home** ↳**Modify** ↳**Reverse**

Command: reverse
Select a line, polyline, spline or helix: *(Choose one or more objects.)*
Select objects: *(Choose more objects, or press Enter to end object selection.)*

COMMAND LINE OPTION

Select a line, polyline, spline or helix chooses the objects.

RELATED COMMANDS

Line creates lines.

PLine creates polylines.

PEdit edits polylines.

Spline creates splines.

SplinEdit edits splines.

Helix creates helixes.

TIP

- The pick order during object creation determines the direction.

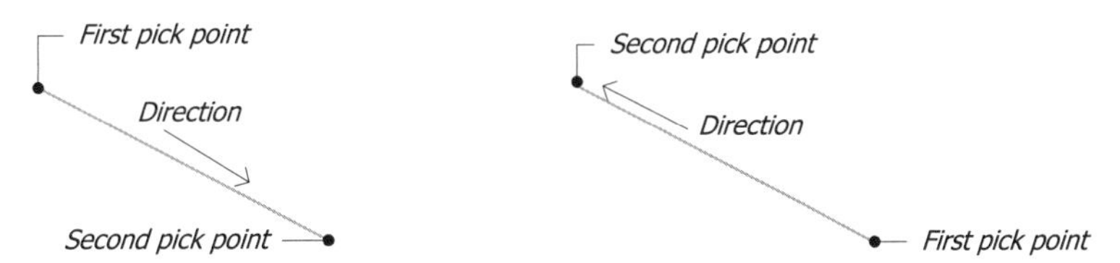

- Use this command to reverse the direction of linetypes.
- The PEdit command includes the Reverse option, which performs the same task as does this command.

Revolve

Rel. 11 Creates 3D solids, 3D mesh, or surface objects by revolving closed 2D objects about an axis.

Command	Aliase	Shortcut Keystrokes	Menu	Ribbon
revolve	**rev**	...	**Draw**	**Home**
			↳**Modeling**	↳ **Modeling**
			↳**Revolve**	↳**Revolve**

Command: revolve

Current wire frame density: ISOLINES=4, Closed profiles creation mode = Solid

Select objects to revolve or [MOde]: *(Select one or more objects, or enter MO.)*

Select objects to revolve or [MOde]: *(Press Enter to end object selection.)*

Specify axis start point or define axis by [Object/X/Y/Z] <Object>: *(Pick a point, or enter an option.)*

Specify angle of revolution or [STart angle/Reverse/EXpression] <360>: *(Enter a value or an option.)*

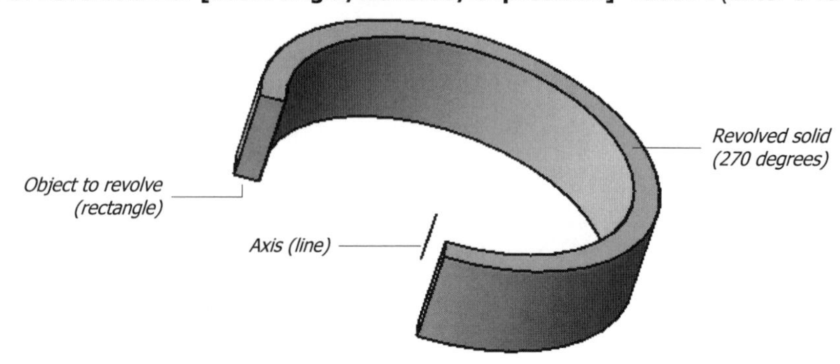

COMMAND LINE OPTIONS

Select objects to revolve selects one or more objects to revolve.

MOde specifies the resulting object as a solid or NURBS surface *(new to AutoCAD 2011).*

Specify axis start point indicates the axis of revolution; you must specify the endpoint.

Object selects the object that determines the axis of revolution.

X uses the positive x axis as the axis of revolution.

Y uses the positive y axis as the axis of revolution.

Specify angle of rotation specifies the amount of rotation; full circle = 360 degrees.

STart angle specifies the starting angle.

Reverse rotates the angle clockwise.

Expression specifies an algebraic or parametric formula that determines the angle *(new to AutoCAD 2011).*

RELATED COMMANDS

Extrude extrudes 2D objects into a 3D solid model.

RevSurf rotates open and closed objects, forming a 3D surface.

TIPS

- Selecting self-intersecting open polylines creates self-intersecting surfaces; closed ones cannot be revolved.
- Selecting more than one object at the 'Select objects to revolve' prompt produces multiple surfaces.

RevSurf

Rel. 10 Generates 3D meshes of revolution defined by a path curve and an axis (short for REVolved SURFace); does not draw sufaces.

Command	*Aliases*	*Shortcut Keystrokes*	*Menu*	*Ribbon*	*Ribbon*
revsurf	**...**	**...**	**Draw** **↳Modeling** **↳Meshes** **↳Revolved Surface**	**Home** **↳3D Modeling** **↳Revolved Surface**	**Mesh Modeling** **↳Primitives** **↳Revolved Surface**

Command: revsurf

Current wire frame density: SURFTAB1=6 SURFTAB2=6

Select object to revolve: *(Select an object.)*

Select object that defines the axis of revolution: *(Select an object.)*

Specify start angle <0>: *(Enter a value.)*

Specify included angle (+=ccw, -=cw) <360>: *(Enter a value.)*

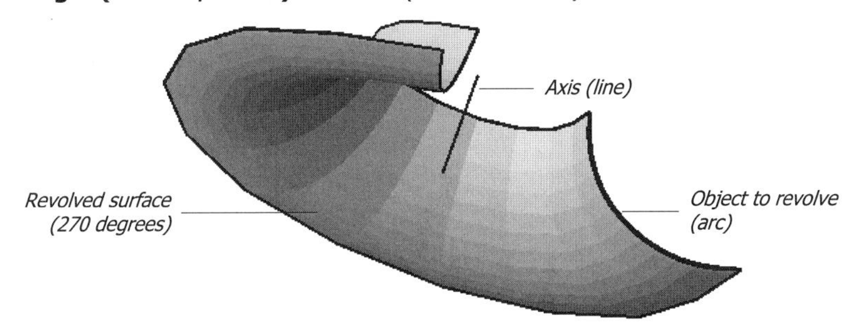

COMMAND LINE OPTIONS

Select object to revolve selects the single object that will be revolved about an axis.

Select object that defines the axis of revolution selects the axis object.

Start angle specifies the starting angle.

Included angle specifies the angle to rotate about the axis.

RELATED COMMANDS

PEdit edits revolved meshes.

Revolve revolves a 2D closed object into a 3D solid.

RELATED SYSTEM VARIABLES

SurfTab1 specifies the mesh density in the m-direction (default = 6).

SurfTab2 specifies the mesh density in the n-direction (default = 6).

MeshType determines if this command creates (old style) polyface mesh or (new style) mesh object.

TIPS

- This command works with open and closed objects.
- If a multi-segment polyline is the axis of revolution, the rotation axis is defined as the vector pointing from the first vertex to the last vertex, ignoring the intermediate vertices.

Ribbon / RibbonClose

2009 Opens and closes the ribbon user interface.

Command	Aliases	Shortcut Keystrokes	Menu	Ribbon
ribbon	**dashboard**	...	**Tools**	...
			↳**Palette**	...
			↳**Ribbon**	
ribbonclose	**dashboardclose**	...	...	...
+ribbon	...	...	...	...

Command: ribbon

Opens the ribbon (redesigned in AutoCAD 2011):

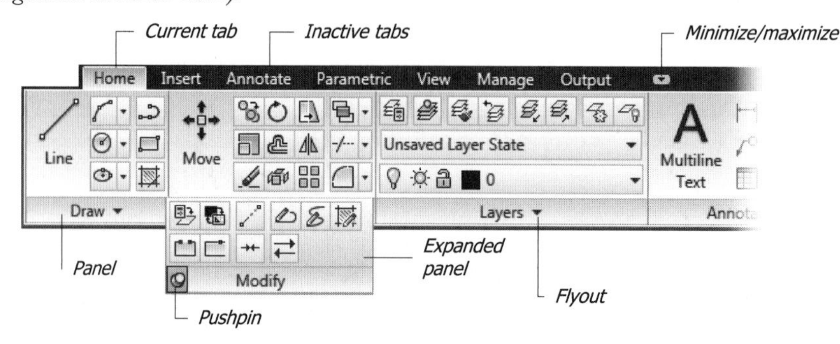

Command: ribbonclose

Closes the ribbon.

USER INTERFACE

Tabs displays groups of panels; every tab contains a different group of panels.

Panels contain groups of buttons and droplists.

Flyouts indicate panels can be expanded.

Pushpin locks flyouts open.

Extended Tooltips display help in two stages: briefly and fully. *(Pause the cursor over a button for at least two seconds.)*

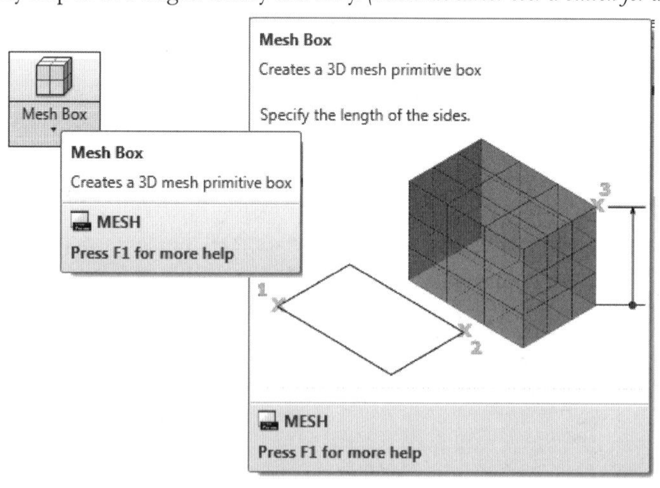

+RIBBON Command

Command: +ribbon

Enter <menu group>,<element id> of the tab to make it active (e.g. ACAD.ID_TabHome): *(Enter the names of the menu group and element id.)*

COMMAND LINE OPTIONS

Enter <menu group>,<element id> specifies the menu group and tab name to display on the ribbon. *Caution*: This command is case-sensitive; names must be entered with correct upper and lower case characters. Follow these steps to obtain the list of element IDs:

1. Start the Cui command.
2. In the Customizations in All Files pane, select the name of a tab in the Ribbon section.

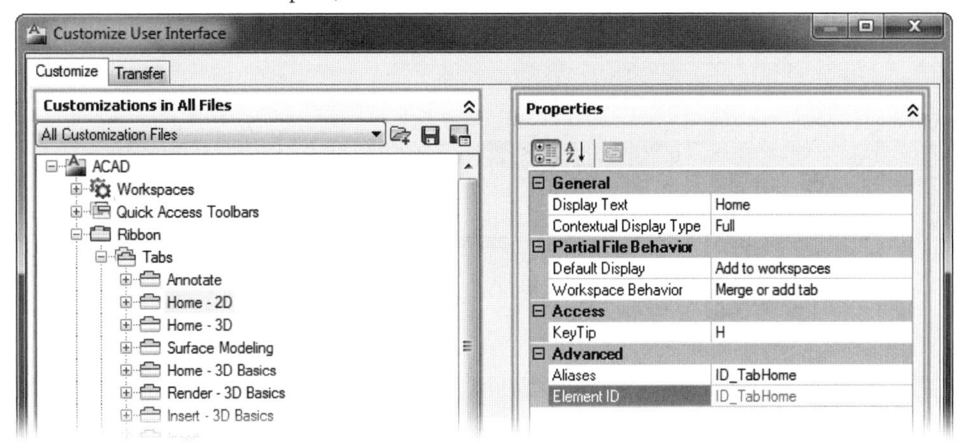

3. In the Properties pane, note the Element ID name, such as ID_TabHome for Home-2D tab.
4. Display the Annotation tab with this command:

   Command: +ribbon

   Enter <menu group>,<element id>: ACAD.ID_TabHome

RELATED COMMANDS

CUI customizes the ribbon.

-Toolbar turns toolbars on and off.

RELATED SYSTEM VARIABLES

RibbonState reports whether the ribbon is displayed (read-only).

RolloverTips toggles the display of rollover tips.

RibbonContextSelect determines when context-sensitive ribbon tabs are displayed.

RibbonContextSelLim determines the number of objects that activate a context-sensitive ribbon tab.

RibbonSelectMode toggles how pickfirst selections interact with context-sensitive ribbon tabs.

MenuBar toggles the display of the menu bar.

StatusBar toggles the display of the status and drawing bars.

TrayIcons toggles the display of the tray.

TIPS

- This command replaces the Dashboard as of AutoCAD 2009, making the Dashboard the shortest-lived user interface in AutoCAD history. When the ribbon is undocked, it resembles the old Dashboard.
- The ribbon changes when you work with mtext, tables, dynamic blocks, block editor, 3D modeling, and other tasks.
- The ribbon can be customized with the CUI command.
- To return to the user interface of AutoCAD 2008 and earlier with the menu bar and toolbars, select the "AutoCAD Classic" workspace with the Workspace command.
- To add the menu bar to the new ribbon-oriented interface, select the down arrow at the right end of the Quick Access toolbar, and then choose the Show Menu Bar item.
- Through the Minimize button, the ribbon can be minimized in three ways:
  - Tabs only:
  - Tabs and panels:
  - Tabs and panel buttons *(new to AutoCAD 2011)*:

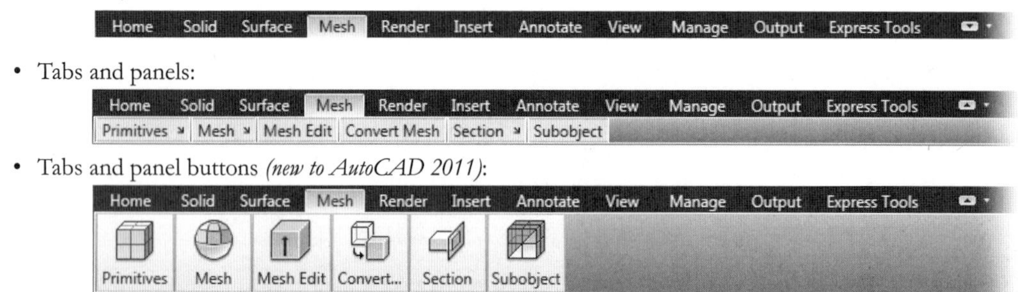

- AutoCAD 2011 adds the simplified ribbon for 3D modeling, accessed through the 3D Basics workspace. Context-sensitive ribbons are added for the Hatch, Gradient, and HatchEdit commands.

Removed Commands

RmlIn command was removed from AutoCAD 2006; its replacement is Markup.

RMat command was removed from AutoCAD 2007; its replacement is MaterialAttach.

Rotate

Ver. 2.5 Rotates objects about an axis perpendicular to the current UCS.

Command	Alias	Shortcut Keystrokes	Menu	Ribbon
rotate	**ro**	...	**Modify** ↳**Rotate**	**Home** ↳**Modify** ↳**Rotate**

Command: rotate
Current positive angle in UCS: ANGDIR=counterclockwise ANGBASE=0
Select objects: *(Select one or more objects.)*
Select objects: *(Press Enter to end object selection.)*
Specify base point: *(Pick a point.)*
Specify rotation angle or [Copy/Reference]: *(Enter a value, or type R.)*

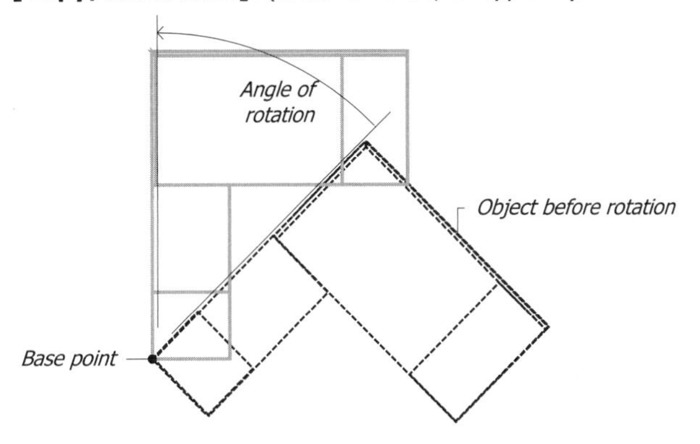

COMMAND LINE OPTIONS

Select objects selects the objects to be rotated.

Specify base point picks the point about which the objects will be rotated.

Specify rotation angle specifies the angle by which the objects will be rotated.

Reference allows you to specify the current rotation angle and new rotation angle.

Copy rotates a copy of the object, leaving the original in place.

RELATED COMMANDS

3dRotate rotates objects in 3D space using a 3D grip tool.

UCS rotates the coordinate system.

Snap command's Rotate option rotates the cursor.

TIPS

- AutoCAD rotates the selected object(s) about the base point.
- At the 'Specify rotation angle' prompt, you can show the rotation by moving the cursor. AutoCAD dynamically displays the new rotated position as you move the cursor.
- Use object snap modes, such as INTersection, to position the base point and rotation angle(s) accurately.

Rotate3D

Rel. 11 Rotates objects about any axis in 3D space with a user interface that's different from the 3dRotate command.

Command	*Aliases*	*Shortcut Keystrokes*	*Menu*	*Ribbon*
rotate3d	...	...	...	...

Command: rotate3d
Current positive angle: ANGDIR=counterclockwise ANGBASE=0
Select objects: *(Select one or more objects.)*
Select objects: *(Press Enter to end object selection.)*

Specify first point on axis or define axis by [Object/Last/View/Xaxis/Yaxis/Zaxis/2points]: *(Pick a point, or enter an option.)*
Specify second point on axis: *(Pick another point.)*
Specify rotation angle or [Reference]: *(Enter a value, or type R.)*

COMMAND LINE OPTIONS

Select objects selects the objects to be rotated.

Specify rotation angle rotates objects by a specified angle: relative rotation.

Reference specifies the starting and ending angle: absolute rotation.

Define Axis By options

Object selects object to specify the rotation axis.

Last selects the previous axis.

View makes the current view direction the rotation axis.

Xaxis makes the x axis the rotation axis.

Yaxis makes the y axis the rotation axis.

Zaxis makes the z axis the rotation axis.

2 points defines two points on the rotation axis.

RELATED COMMANDS

3dRotate rotates objects in 3D space using a 3D grip tool.

Rotate rotates objects in 2D space.

Align rotates, moves, and scales objects in 3D space.

Mirror3d mirrors objects in 3D space.

TIP

- This command is different from 3dRotate:
  - Rotate3D uses the command line.
  - 3dRotate is interactive and is found in the ribbon, toolbars, and on the menu; Rotate3D is not.

RPref / RPrefClose

Rel. 12 Specifies options for the Render and RenderCrop commands (short for Render PREFerences).

Command	*Alias*	*Menu*	*Ribbon*
rpref	**rpr**	**View** **↳Render** **↳Advanced Render Settings**	**Render** **↳Render** **↳Advanced Render Settings**
rprefclose	...	...	...

Command: rpref

Displays palette.

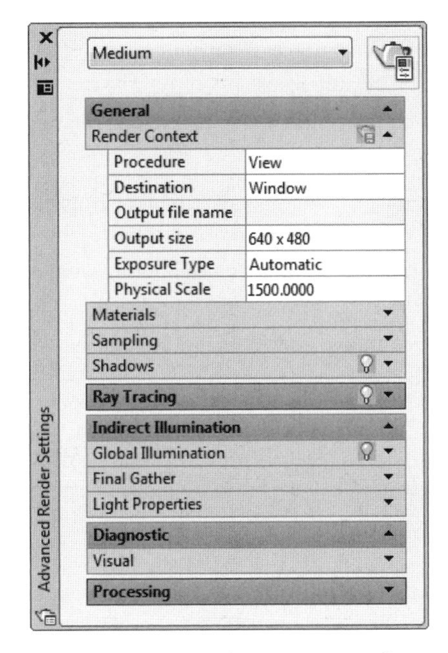

Options are identical to the Render Presets Manager dialog box (RenderPresets command).

Command: rprefclose

Closes the palette.

GENERAL

performs the render operation.

Render Context options

toggles the Output File Name parameter for the Render command's Save option.

Procedure determines which part of the drawing is rendered:

- **View** — renders the entire viewport.
- **Crop** — prompts you to pick two points, like the RenderCrop command.
- **Selected** — prompts you to select the objects to be rendered.

Destination determines where the model will be rendered:

- **Viewport** — model is rendered in the current viewport.

- **Window** — model is rendered in the Render window.

Output file name specifies the name and format of images saved to disc; see SaveImg command.

Output size specifies the size of the rendered image; smaller views render faster.

Materials options

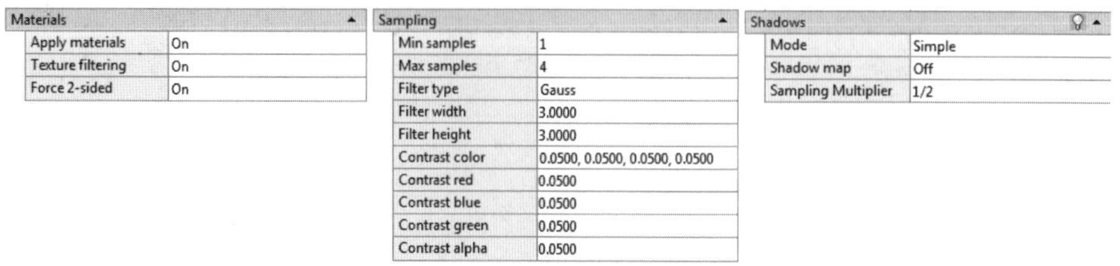

Apply Materials toggles rendering of materials, if applied to the model. If not applied, the model takes on the look defined by the Global material. See the MatEditorOpen command.

Texture Filtering toggles the rendering of texture maps.

Force 2-Sided toggles the rendering of both sides of faces; single-sided is faster.

Sampling options

Min Samples specifies the minimum sample rate (number of samples per pixel). Range is 1/64 (one out of every 64 pixels sampled) to 1024 (1,024 samples per pixel). More samples result in higher quality but slower rendering speed.

Max Samples specifies the maximum sample rate used when neighboring samples exceed the contrast limit (defined below).

Filter Type specifies how multiple samples are combined into one:

- **Box** — sums all samples in the filter area with equal weight (fastest).
- **Gauss** — weights samples using a bell curve.
- **Triangle** — weights samples using a pyramid.
- **Mitchell** — weights samples using a steeper bell curve.
- **Lanczos** — weights samples using a steep bell curve that diminishes effects of samples at edges of the filter area.

Filter Width and **Filter Height** specify the size of the filtered area; range is 0 to 8. Larger values soften the image, but increase rendering time.

Contrast Color selects the threshold RGBA (red, green, blue, alpha) values; displays the Select Color dialog box. Range is from 0 to 1, where 0 is black and 1 is white.

Contrast Red, **Contrast Blue**, and **Contrast Green** indicate threshold values for red, blue, and green components of samples.

Contrast Alpha indicates the threshold value for the alpha component, typically used for transparency.

Shadows options

toggles shadow casting.

Mode specifies the type of shadow casting:

- **Simple** — casts random-order shadows.
- **Sort** — casts shadows from objects to lights.
- **Segments** — casts shadows along the light ray.

Shadow Map Controls toggles shadow mapping:

On — shadow-mapped shadows.

Off — ray-traced shadows.

RAY TRACING

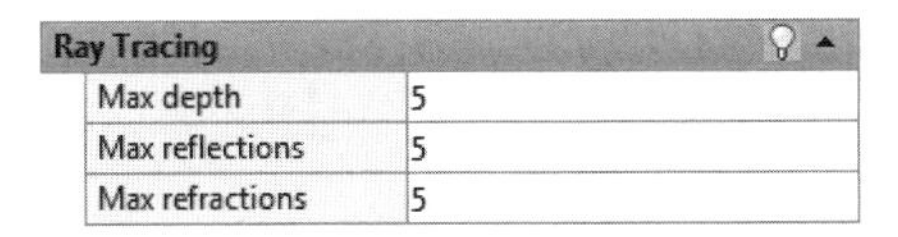

toggles ray tracing.

Max Depth stops ray tracing when the number of reflections reaches this number. If rays refract twice and reflect three times, the ray tracing is stopped when this value is set to 5.

Max Reflections stops ray tracing after this many *reflections* (light bounced off surfaces). Turns off reflections when set to 0.

Max Refractions stops ray tracing after this many *refractions* (light distorted going through an object). Turns off refractions when set to 0.

INDIRECT ILLUMINATION

Indirect Illumination — Global Illumination	
Photons/sample	500
Use radius	Off
Radius	1.0000
Max depth	5
Max reflections	5
Max refractions	5

Final Gather	
Mode	Auto
Rays	200
Radius mode	Off
Max radius	1.0000
Use min	Off
Min radius	0.1000

Light Properties	
Photons/light	10000
Energy multiplier	1.0000

Global Illumination options

toggles global illumination.

Photons/Samples specifies the number of photons used to compute the intensity of global illumination. Range is 1 to 2147483647; more photons make for less noise, more blur, and longer renderings.

Use Radius toggles the use of photon sizing. When off, photons are determined to be 1/10 the radius of the viewport.

Radius specifies the area in which photons are used.

Max Depth, Max Reflections, and **Max Refractions** are the same as for ray tracing.

Final Gather options

toggles gathering (additional global illumination calculations).

Rays specifies the number of rays used to compute indirect illumination. More rays make for less noise but longer renderings.

Radius Mode specifies the radius mode:

- **On** — Max Radius setting is used.
- **Off** — maximum radius is 10 percent of the maximum model radius.
- **View** — Max Radius setting is defined in pixels.

Max Radius specifies maximum radius used for final gathering. Lower values improve quality but increase rendering time.

Use Min Controls toggles the minimum radius setting.

Min Radius specifies minimum radius used for final gathering. Higher values improve quality but increase rendering time.

Light Properties options

Affects how lights behave lights calculate indirect illumination.

Photons/Light specifies the number of photons emitted by each light. Larger values increase accuracy but also increase the rendering time and RAM usage.

Energy Multiplier multiplies the global illumination intensity.

DIAGNOSTIC

Helps track down rendering peculiarities.

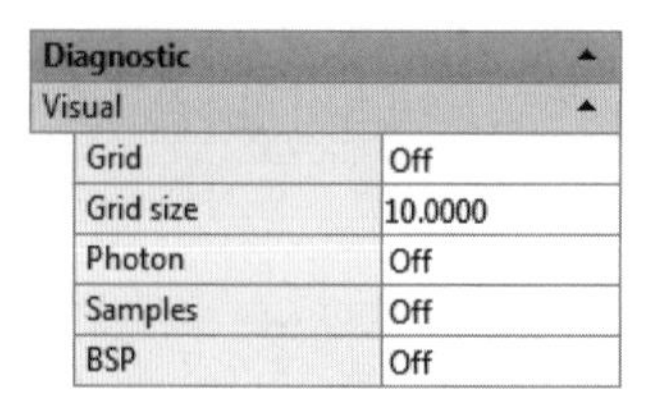

Visual options

Grid shows the coordinate space of:

- **Object** — local u, v, w coordinates for each object.
- **World** — world x, y, z coordinates for all objects.
- **Camera** — imposes a rectangular grid on the view.

Grid Size sets the size of the grid.

Photon renders the effect of a photon map, if present:

- **Density** — renders the photon map; red = high density; blue = low density.
- **Irradiance** — shades photons based on their irradiance; red = maximum irradiance; blue = low irradiance.

Samples toggles the use of sampling.

BSP (binary space partition) helps track down very slow renderings by visualizing the BSP raytrace parameters:

- **Depth** — illustrates the tree depth; red = top faces; blue = deep faces.
- **Size** — colors the leaf size in the tree.

PROCESSING

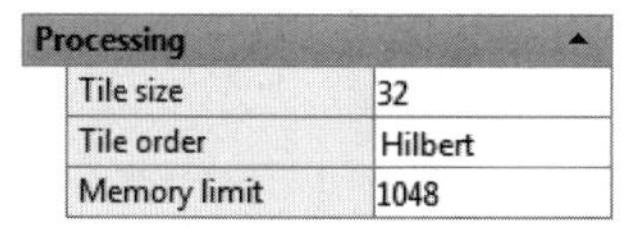

Tile Size specifies the rendering tile size; smaller values result in longer rendering times.

Tile Order specifies the tile render order:

- **Hilbert** — renders the next tile based on the difficulty of switching to the next one.
- **Spiral** — starts at the center and then spirals outward.
- **Left to Right** — bottom to top, left column to right.
- **Right to Left** — bottom to top, right column to left.
- **Top to Bottom** — right to left, top row to bottom.
- **Bottom to Top** — right to left, bottom row to top.

Memory Limit discards geometry to allocate memory for other objects, if this memory limit is reached.

RELATED COMMANDS

Render and **RenderCrop** use these settings to render 3D models.

RenderPresets creates rendering styles.

RScript

Ver. 2.0 Repeats script files (short for Repeat SCRIPT).

Command	*Aliases*	*Shortcut Keystrokes*	*Menu*	*Ribbon*
rscript	...	...	...	...

Command: rscript

COMMAND LINE OPTIONS

None.

RELATED COMMANDS

Resume resumes a script file after being interrupted.

Script loads and runs a script file.

TIPS

- Placed at the end of a script file, this command causes the script file to run repeatedly until canceled with Backspace or Esc.
- If the script repeats itself often, you may want to turn off Undo and log files, which can get large from the repeated commands.

 RuleSurf

Rel. 10 Draws 3D ruled meshes between two objects (short for RULEd SURFace); does not draw surfaces.

Command	*Aliases*	*Shortcut Keystrokes*	*Menu*	*Ribbon*
rulesurf	...	...	**Draw** **↳Modeling** **↳Meshes** **↳Ruled Surf**	**Mesh Modeling** **↳Primitives** **↳Ruled Surfaces**

Command: rulesurf
Select first defining curve: *(Pick an object.)*
Select second defining curve: *(Pick another object.)*

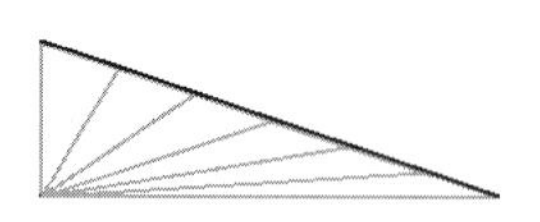
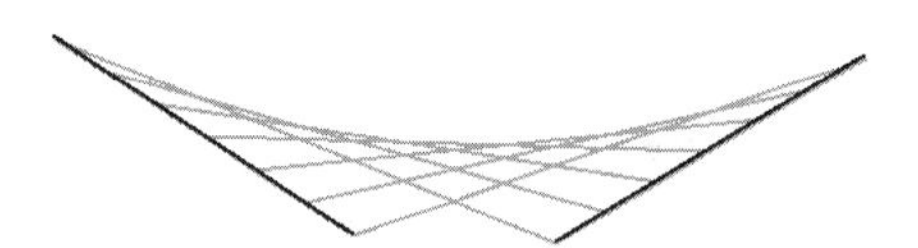
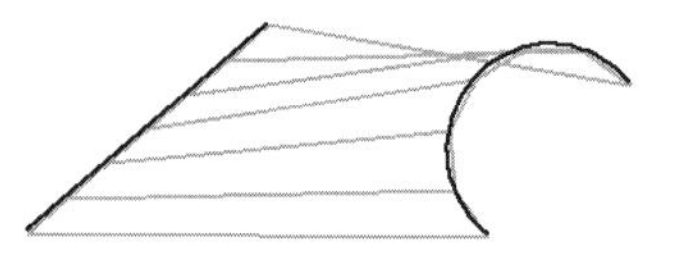

Left to right: *Ruled meshes created between point and line, line and line, and arc and line.*

COMMAND LINE OPTIONS

Select first defining curve selects the first object for the ruled mesh.

Select second defining curve selects the second object for the ruled mesh.

RELATED COMMANDS

EdgeSurf draws meshes bounded by four edges.

RevSurf draws meshes of revolution.

TabSurf draws tabulated meshes.

RELATED SYSTEM VARIABLES

SurfTab1 determines the number of faces drawn.

MeshType specifies whether ruled surfaces are drawn as polyface meshes or 3D mesh objects.

TIPS

- This command uses these objects as the boundary curve: points, lines, arcs, circles, polylines, and 3D polylines.
- Pick order is important: the ruled mesh is drawn starting at the endpoint nearest the pick point. This can result in a twisted surface, as illustrated by the line-line example above.
- The boundary objects must both be either closed or open; the exception is the point object.
- This command begins the drawing of meshes following this procedure:
  - **Open objects** — from the object's endpoint closest to your pick.
  - **Circles** — from the zero-degree quadrant.
  - **Closed polylines** — from the last vertex.
- This command creates meshes with circles in the opposite direction from that of closed polylines.
- This command draws the ruled surfaces as polyface meshes (legacy style) or as 3D mesh objects, depending on the setting of the MeshType system variable.

Save / SaveAs

Saves the current drawing to disk, after prompting for a file name.

Commands	Alias	Shortcut Keystroke	Quick Access	Application Menu
save	**saveurl**	...	...	...
saveas	...	**Ctrl+Shift+S**		**Save As** ↳**Save As**
+saveas	...	...	...	**Save As** ↳**Template, DWT**

Command: save *or* saveas

Displays dialog box.

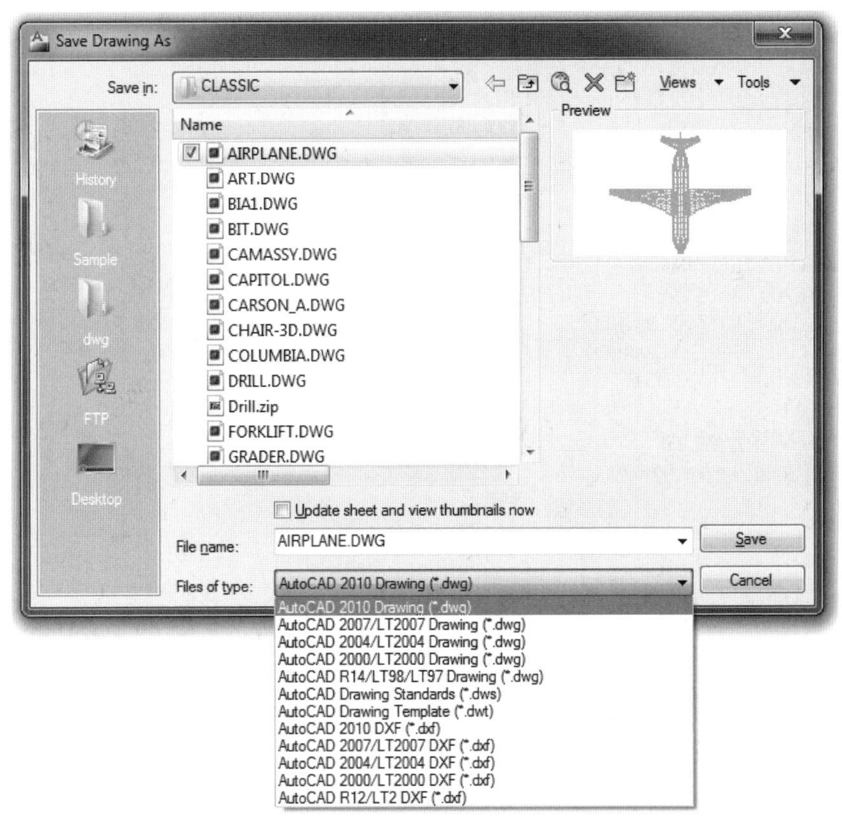

SAVING DRAWING AS DIALOG BOX

Save in selects the folder, hard drive, or network drive in which to save the drawing.

☑**Update sheet and view thumbnails now** updates thumbnail preview images.

File name names the drawing; maximum = 255 characters; default=*drawing1.dwg*.

Save saves the drawing, and then returns to AutoCAD.

Cancel does not save the drawing.

When a drawing of the same name already exists in the same folder, displays dialog box:

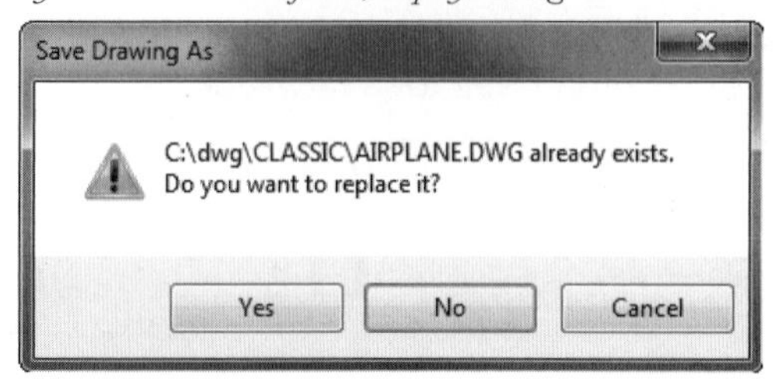

Files of type saves the drawing in a variety of formats:

Save as type	***Saves drawings in format compatible with...***
DWG formats:	
AutoCAD 2010 DWG	AutoCAD 2010/11.
AutoCAD 2007 DWG	AutoCAD 2007/8/9.
AutoCAD 2004/LT 2004 Drawing	AutoCAD 2004/5/6.
AutoCAD 2000/LT 2000 Drawing	AutoCAD 2000/i/2.
AutoCAD R14/LT98/LT97 Drawing	AutoCAD Release 14.
DWS and DWT formats:	
Drawing Standards	AutoCAD standards format.
Drawing Template	Template drawings stored in the *template* folder.
DXF formats:	
AutoCAD 2010 DXF	AC18 DXF, AutoCAD 2010 - 2011.
AutoCAD 2007 DXF	AC17 DXF, AutoCAD 2007 - 2009.
AutoCAD 2004/LT 2004 DXF	AC16 DXF, AutoCAD 2004 - 2006.
AutoCAD 2000/LT 2000 DXF	AC15 DXF, AutoCAD 2000 - 2002.
AutoCAD R12/LT 2 DXF	AC12 DXF format, also compatible with Release 11.

SAVEAS OPTIONS DIALOG BOX

Access this dialog box through Tools | Options:

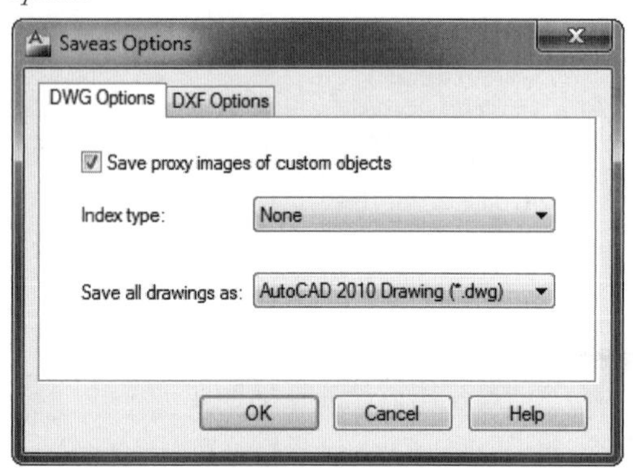

DWG Options tab

Save proxy images of custom objects option:

☑ Saves an image of the custom objects in the drawing file.

☐ Saves a frame around each custom object, instead of an image.

Index type saves indices with drawings; useful only for xrefed and partially-loaded drawings:

- **None** creates no indices (default).
- **Layer** loads layers that are on and thawed.
- **Spatial** loads only the visible portion of clipped xrefs.
- **Layer and Spatial** combines the above two options.

Save all drawings as selects the default format for saving drawings.

DXF Options tab

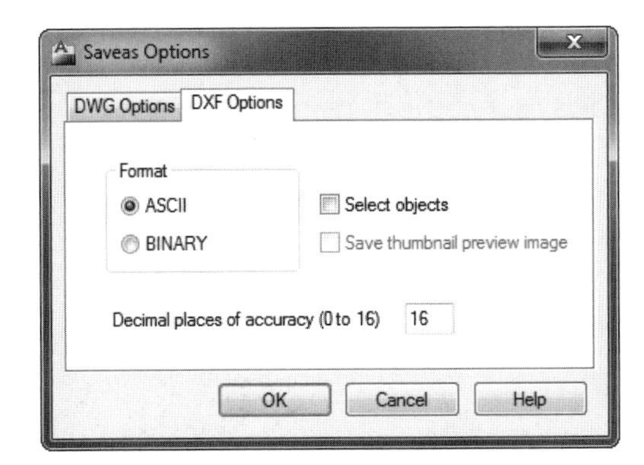

⊙**ASCII** saves the drawing in ASCII format, which can be read by humans – as well as by common text editors – but takes up more disk space.

○**BINARY** saves the drawing in binary format, which takes up less disk space, but cannot be read by some software programs.

☐**Select objects** allows you to save selected objects to the DXF file; AutoCAD prompts you, 'Select objects:'.

☐**Save thumbnail preview image** saves a thumbnail image of the drawing.

Decimal places of accuracy specifies the number of decimal places for real numbers: 0 to 16.

TEMPLATE OPTIONS DIALOG BOX

Displayed when drawing is saved as .dwt template file:

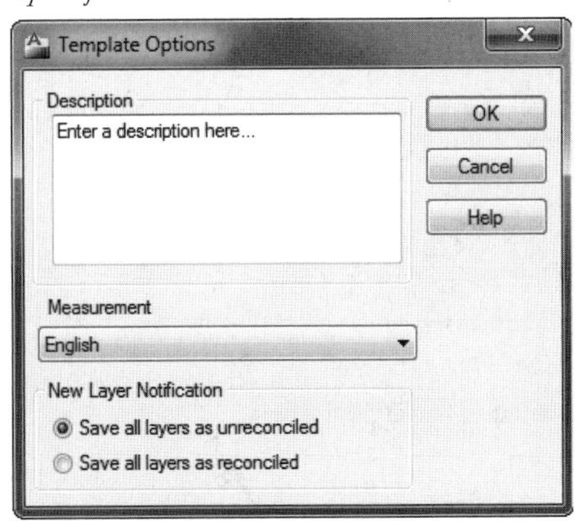

Description describes the template file; stored in *wizard.ini.*

Measurement defines the base measurement system:

- English
- Metric

New Layer Notification determines how to handle new layers added to drawings:

⊙ **Saves All Layers As Unreconciled** alerts the user when the template creates new drawings.

○ **Save All Layers as Reconciled** treats new layers as reconciled (old).

+SAVEAS Command

Specifies the file format at the command prompt.

Command: +saveas

Input save format [dwG/dwT/dwS/dxF/Other]: *(Enter an option.)*

Displays Save Drawing As dialog box.

COMMAND LINE OPTIONS

dwG saves drawings in DWG format.

dwT saves drawings in DWT (template) format.

dwS saves drawings in DWS (drawing standards) format.

dxF saves drawings in DXF (drawing interchange) format.

Other options

Input save format in an integer value: *(Enter a number.)*

Input save format specifies the version or format:

Other	Saves As
0	AutoCAD 2010 DWG
1	AutoCAD 2007 DWG
2	AutoCAD 2004/LT2004 Drawing
3	AutoCAD 2000/LT 2000 Drawing
4	AutoCAD R14/LT98/LT97 Drawing
5	Drawing Standards (*.dws)
6	Drawing Template (*.dwt)
7	AutoCAD 2010 DXF
8	AutoCAD 2007 DXF
9	AutoCAD 2004/LT 2004 DXF
10	AutoCAD 2000/LT 2000 DXF
11	AutoCAD R12/LT2 DXF

RELATED COMMANDS

Export saves the drawing in several vector formats.

Plot saves the drawing in additional raster and vector formats.

Quit exits AutoCAD without saving the drawing.

QSave saves the drawing without prompting for a name.

AutoPublish saves the drawing simultaneously as a DWF file.

RELATED SYSTEM VARIABLES

DBMod indicates that the drawing was modified since being opened.

DwgName specifies the name of the drawing; until then, it is *drawing1.dwg* while unnamed.

AutomaticPub toggles simultaneous-saving as DWF/x or PDF files.

SaveFidelity preserves annotative objects when exported to AutoCAD 2007 or earlier.

LargeObjectSupport toggles support for large objects, such as MText objects with 256KB of text.

TIPS

- The Save and SaveAs commands perform exactly the same function.
- Use these commands to save drawings by other names; for example, saving read-only drawings.
- To control when AutoPublish operates, go to the Option command's Plot and Publish tab, and then click the Automatic DWF Publish Settings button; see the AutoPublish command.
- To save drawings without seeing the Save Drawing As dialog box, use the QSave command.
- All newer releases of AutoCAD can read drawing files produced by older releases, but not the other way around. The Save command translates drawings to make them compatible with earlier releases. The problem is that incompatible objects are modified.
- To save just a part of the drawing, use the WBlock command and then choose the objects to be saved.

Removed Command

SaveAsR12 command was removed from AutoCAD Release 14. Use SaveAs instead, which saves drawings in earlier formats.

SaveImg

Saves a rendering of the current viewport image as a raster file on disk (short for SAVE IMaGe).

Command	Aliases	Shortcut Keystrokes	Menu	Ribbon
saveimg	...	...	**Tools** **↳Display Image** **↳Save**	...

Command: saveimg

Displays the Render Output File dialog box.

Provide a file name, select a format, and then click Save.

One of the following dialog boxes appears:

IMAGE OPTIONS DIALOG BOXES

BMP Image Options dialog box

○ **Monochrome** converts all lines to black.

○ **8 Bits (256 Grayscale)** reduces the image to shades of gray.

○ **8 Bits (256 Colors)** reduces the image to 256 colors.

⊙ **24 Bits (16.7 Million Colors)** maintains all colors in the image.

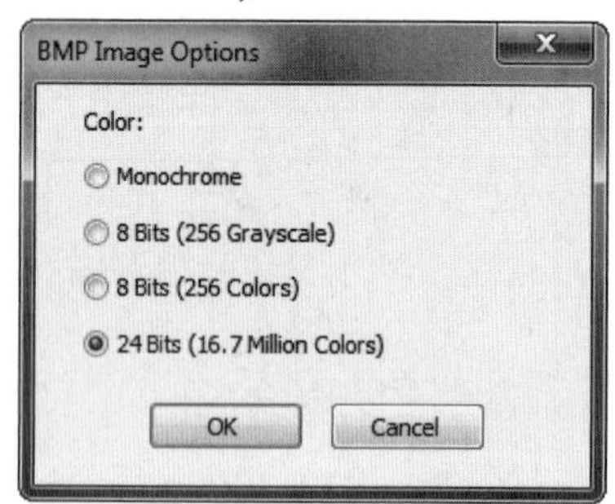

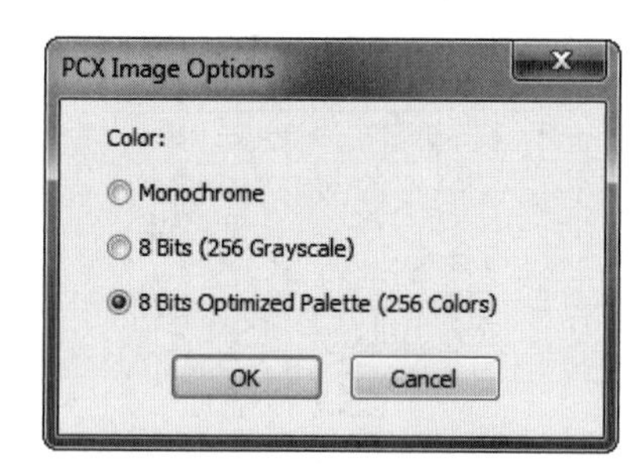

Left: *Options for bitmap images.*
Right: *Options for PCX images.*

PCX Image Options dialog box

○ **Monochrome** converts all lines to black.

○ **8 Bits (256 Grayscale)** reduces the image to shades of gray.

⊙ **8 Bits Optimized (256 Colors)** reduces the image to the best 256 colors possible.

Targa Image Options dialog box

○ **8 Bits (256 Grayscale)** reduces the image to shades of gray.

○ **8 Bits (256 Colors)** reduces the image to 256 colors.

⊙ **24 Bits (16.7 Million Colors)** maintains all colors in the image.

○ **32 Bits (23 Bits + Alpha)** maintains all colors, plus transparency information (alpha).

☐ **Bottom Up** stores data in the file from the bottom to the top of the image.

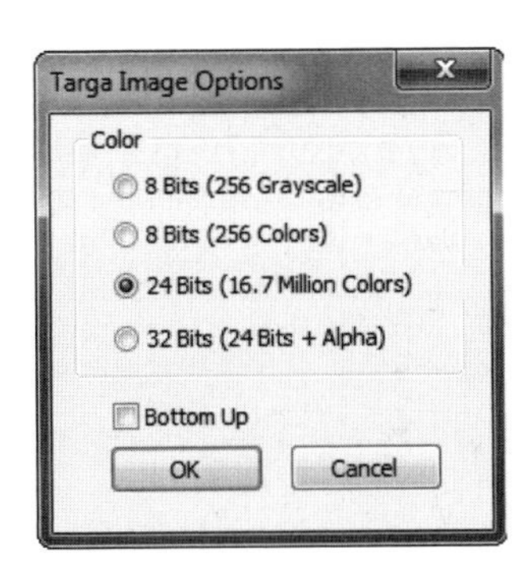

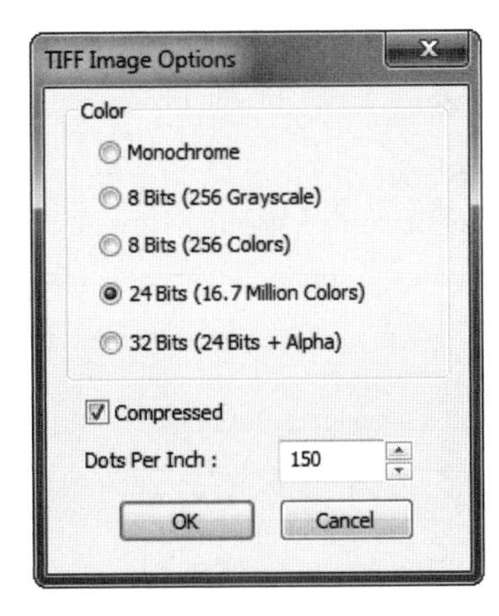

Left: *Options for Targa images.*
Right: *Options for TIFF images.*

TIFF Image Options dialog box

○ **Monochrome** converts all lines to black.

○ **8 Bits (256 Grayscale)** reduces the image to shades of gray.

○ **8 Bits (256 Colors)** reduces the image to 256 colors.

⊙ **24 Bits (16.7 Million Colors)** maintains all colors in the image.

○ **32 Bits (23 Bits + Alpha)** maintains all colors, plus transparency information (alpha).

☑ **Compressed** reduces the size of the file by compressing the data.

Dots Per Inch specifies the resolution of the image; range is 0 to 1000.

PNG Image Options dialog box

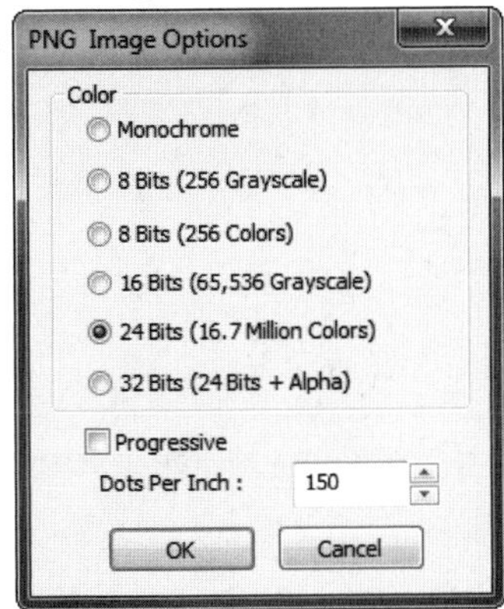

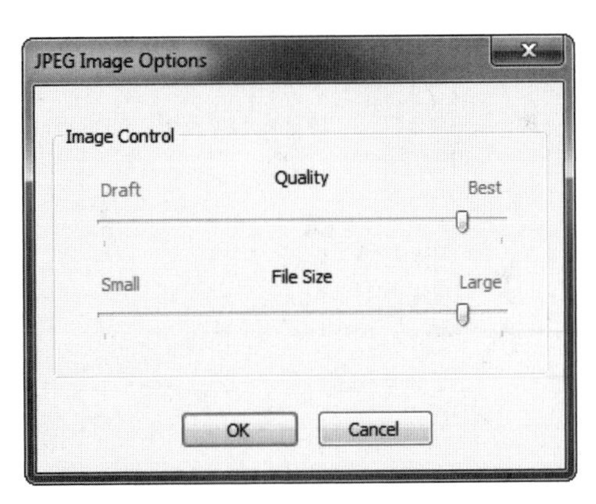

Left: *Options for PNG images.*
Right: *Options for JPEG images.*

○ **Monochrome** converts all lines to black.

○ **8 Bits (256 Grayscale)** reduces the image to shades of gray.

○ **8 Bits (256 Colors)** reduces the image to 256 colors.

⊙ **24 Bits (16.7 Million Colors)** maintains all colors in the image.

○ **32 Bits (23 Bits + Alpha)** maintains all colors, plus transparency information (alpha).

☑**Progressive** displays the image more quickly by showing more of it, until all details are filled in.

Dots Per Inch specifies the resolution of the image; range is 0 to 1000.

JPEG Image Options dialog box

Quality adjusts the amount of compression.

File size adjusts the amount of compression also; increase the compression for smaller file sizes. The sliders are linked, and cannot be set independently.

RELATED COMMANDS

ImageAttach attaches raster images to drawings.

JpgOut saves selected objects in JPEG raster format.

PngOut saves selected objects in PNG raster format.

TifOut saves selected objects in TIFF raster format.

WmfOut saves selected objects in WMF vector format.

CopyClip copies selected objects to the Clipboard.

PRT SCR saves the entire screen to the Clipboard.

TIPS

- This command saves the entire viewport; in contrast, the JpgOut, PngOut, TifOut, and WmfOut commands prompt you to select objects.
- The GIF format, which is one of the two most common formats used to display images on the Internet (and was found in AutoCAD Release 12 and 13), was replaced by the BMP format in AutoCAD Release 14 due to Unisys owning the patent on the compression algorithm employed by GIF.

Aliased Command

SaveUrl was removed from AutoCAD 2000; it now activates the Save command.

Scale

v2.5 Makes selected objects smaller or larger.

Command	*Alias*	*Shortcut Keystrokes*	*Menu*	*Ribbon*
scale	**sc**	...	**Modify** ↳**Scale**	**Home** ↳**Modify** ↳**Scale**

Command: scale
Select objects: *(Select one or more objects.)*
Select objects: *(Press Enter to end object selection.)*
Specify base point: *(Pick a point.)*
Specify a scale factor or [Copy/Reference]: *(Enter scale factor, or type C or R.)*

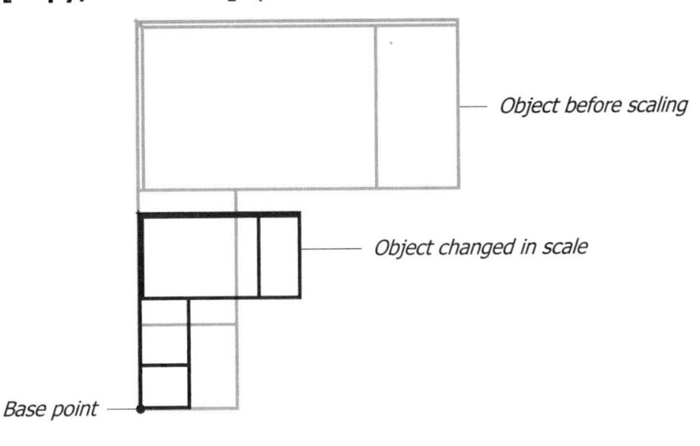

COMMAND LINE OPTIONS

Base point specifies the point from which scaling takes place.

Reference requests a distance, followed by a new distance.

Copy scales a copy of the selected objects, leaving the originals in place.

Scale factor indicates the scale factor, which applies equally in the x and y directions.

Scale Factor	*Meaning*
> **1.0**	Enlarges object(s).
= **1.0**	Makes no change.
> **0.0** *and* < **1.0**	Reduces object(s).
= **0.0** *or negative*	Illegal values.

Reference options
Specify reference length <1>: *(Enter a base length.)*
Specify new length: *(Enter a new length.)*

Specify reference length specifies a baseline length; you can enter a value, or pick two points.

Specify new length specifies a new length that determines the scale factor.

RELATED COMMANDS

Align scales an object in 3D space.

Insert allows a block to be scaled independently in the x, y, and z directions.

Plot allows a drawing to be plotted at any scale.

3dScale scales objects in 3D space.

RELATED SYSTEM VARIABLE

GtDefault determines whether the 3dScale is activated automatically in 3D space when you enter the Scale command.

TIPS

- You can interactively scale the object by moving the cursor to make the object larger and smaller.
- The objects scale from the base point.
- Points are the only objects that cannot be scaled.
- Scale factors larger than 1.0 grow the object; those smaller than 1.0 shrink the object.
- The Reference option is useful for the following situations:
  - Scaling raster images to match the size of a drawing.
  - Making one object the same size as another.
  - Changing the units in a drawing. For example, use 1:12 to change decimal feet into inches, or 2.54:1 to change cm to inches.
- This command changes the size in all dimensions (x, y, z) equally; to change an object in just one dimension, use the Stretch command.
- Raster images cannot be scaled unequally. The workaround is to turn the image into a block, and then insert the block with different x and y scale factors.
- This command has no affect on the drawing's scales.

ScaleListEdit

2006 Edits lists of scale factors for plots, viewports, and annotative objects.

Commands	*Aliases*	*Shortcut Keystrokes*	*Menu*	*Ribbon*
'scalelistedit	...	...	**Format** ↳**Scale List**	**Annotate** ↳**Annotation Scaling** ↳**Scale List**
'-scalelistedit	...	...	...	...

Command: scalelistedit

Displays dialog box.

EDIT SCALE LIST DIALOG BOX

Add adds scale factors to the list; displays the Add Scale Factor dialog box.

Edit changes the selected scale factor; displays the Edit Scale Factor dialog box.

Move Up moves the selected scale factor up the list.

Move Down moves the selected scale factor down the list.

Delete erases the selected scale factors from the list without warning; click Cancel or Restore to recover accidentally-erased scale factors.

Reset restores the list to its original status; AutoCAD warns that custom scale factors will be lost.

ADD / EDIT SCALE FACTOR DIALOG BOX

Access this dialog box by clicking the Add or Edit buttons:

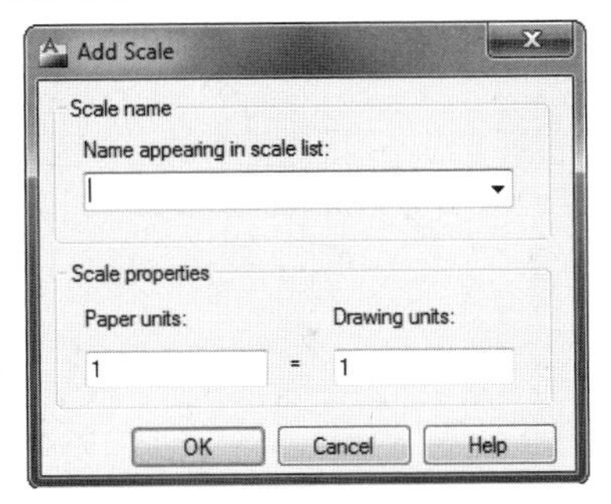

After you click Add, a blank dialog box appears; after you click Edit, the selected scale factors appear in the dialog box.

Scale Name options

Name Appearing in Scale List provides names in place of scale factors. These may be an actual scale factor, such as 1:10.

Scale Properties options

Paper Units specifies the size of objects on paper. This value is usually 1 for drawings larger than the paper.

Drawing Units specifies the size of objects in the drawing. This value is usually 1 for drawings smaller than the paper.

-SCALELISTEDIT Command

Command: -scalelistedit
Enter option [?/Add/Delete/Reset/Exit] <Add>: *(Enter an option.)*
Enter name for new scale: *(Type a name.)*
Enter scale ratio: *(Type the scale ratio, such as 1:1.5.)*

COMMAND LINE OPTIONS

? lists the scale factors in the Text window.

```
    Scale Name                 Paper Units  Drawing Units  Effective Scale
    ----------                 -----------  -------------  ---------------
 1: 1:1                           1.0000        1.0000        1.0000
 2: 1:2                           1.0000        2.0000        0.5000
 3: 1:4                           1.0000        4.0000        0.2500
 4: 1:5                           1.0000        5.0000        0.2000
 5: 1:8                           1.0000        8.0000        0.1250
 6: 1:10                          1.0000       10.0000        0.1000
 7: 1:16                          1.0000       16.0000        0.0625
 8: 1:20                          1.0000       20.0000        0.0500
 9: 1:30                          1.0000       30.0000        0.0333
10: 1:40                          1.0000       40.0000        0.0250
11: 1:50                          1.0000       50.0000        0.0200
12: 1:100                         1.0000      100.0000        0.0100
13: 2:1                           2.0000        1.0000        2.0000
14: 4:1                           4.0000        1.0000        4.0000
15: 8:1                           8.0000        1.0000        8.0000
16: 10:1                         10.0000        1.0000       10.0000
17: 100:1                       100.0000        1.0000      100.0000
18: 1/128" = 1'-0"                0.0078       12.0000        0.0007
19: 1/64" = 1'-0"                 0.0156       12.0000        0.0013
20: 1/32" = 1'-0"                 0.0313       12.0000        0.0026
21: 1/16" = 1'-0"                 0.0625       12.0000        0.0052
22: 3/32" = 1'-0"                 0.0938       12.0000        0.0078
23: 1/8" = 1'-0"                  0.1250       12.0000        0.0104
24: 3/16" = 1'-0"                 0.1875       12.0000        0.0156
25: 1/4" = 1'-0"                  0.2500       12.0000        0.0208
26: 3/8" = 1'-0"                  0.3750       12.0000        0.0313
27: 1/2" = 1'-0"                  0.5000       12.0000        0.0417
28: 3/4" = 1'-0"                  0.7500       12.0000        0.0625
29: 1" = 1'-0"                    1.0000       12.0000        0.0833
30: 1-1/2" = 1'-0"                1.5000       12.0000        0.1250
31: 3" = 1'-0"                    3.0000       12.0000        0.2500
32: 6" = 1'-0"                    6.0000       12.0000        0.5000
33: 1'-0" = 1'-0"                12.0000       12.0000        1.0000
```

Add adds scale factors.

Delete erases scale factors.

Reset restores the list of scale factors to the original factory setting.

Exit exits the command.

RELATED COMMANDS

AiObjectScaleAdd (*undocumented*) adds the current scale factor to annotative objects.

AiObjectScaleRemove (*undocumented*) removes the current scale factor to annotative objects.

RELATED SYSTEM VARIABLES

HideXrefScales hides scale factors imported with xrefs.

MeasureInit specifies Imperial or metric scale factors automatically when new drawings are created without a template file.

TIPS

- This list of scale factors is displayed by any command that needs a scale factor, such as the Plot command, the Properties command for viewports, and the ObjectScale command for annotative objects.
- Use this command to shorten the list of unused scale factors, and to add custom scales. It also affects the scale factors used for annotative scaling.
- As of AutoCAD 2010, this command allows for the creation of custom scale lists.

ScaleText

2002 Changes the height of text objects relative to their insertion points.

Command	Aliases	Shortcut Keystrokes	Menu	Ribbon
scaletext	...	...	**Modify** ↳**Object** ↳**Text** ↳**Scale**	**Annotate** ↳**Text** ↳**Scale**

Command: scaletext
Select objects: *(Select one or more objects.)*
Select objects: *(Press Enter to end object selection.)*
Enter a base point option for scaling
[Existing/Left/Center/Middle/Right/TL/TC/TR/ML/MC/MR/BL/BC/BR] <Existing>: *(Enter an option, or press Enter to keep insertion point as is.)*
Specify new height or [Paper height/Match object/Scale factor] <0.2>: *(Enter a value or option.)*

COMMAND LINE OPTIONS

Enter a base point option for scaling specifies the point from which scaling takes place.

Existing uses the existing insertion point for each text object.

Specify new height specifies the new height of the text; applies only to non-annotative text.

Paper height assigns the annotative property to the text.

Match object matches the height of another text object.

Scale factor scales the text by a factor; cannot use 0 or a negative value.

Scale Factor options
Specify scale factor or [Reference] <2.0000>: *(Enter a value, or type R.)*

Specify scale factor scales the text by a factor; see the Scale command.

Reference supplies a reference value, followed by a new value.

RELATED COMMANDS

Justify changes the justification of text.

Properties changes all aspects of text.

Style creates text styles.

TextEdit edits text.

TIPS

- *Warning!* Scaling text larger may make it overlap nearby text.
- For larger text, use a scale factor of more than 1.0; to reduce the text size, use a scale factor between 0.0 and 1.0. A factor of 1.0 leaves the text unchanged.
- To scale text automatically in model space, use text, dimension, and multileader styles with the annotative property turned on, as noted by the icon.

Removed Command

Scene command was removed from AutoCAD 2007. ("So now we can't make a scene," grumps the technical editor.)

Script

v. 1.4 Runs ASCII files containing sequences of AutoCAD commands and options.

Command	Alias	Shortcut Keystrokes	Menu	Ribbon
'script	**scr**	**...**	**Tools** ↳**Run Script**	**Manage** ↳**Applications** ↳**Run Script**

Command: script

Displays the Select Script File dialog box. Select an .scr file, and then click Open.

Script file begins running as soon as it is loaded.

COMMAND LINE OPTIONS

BACKSPACE interrupts the script.

ESC stops the script.

RELATED COMMANDS

Delay specifies the delay in milliseconds; pauses the script before executing the next command.

Resume resumes a script after it is interrupted.

RScript repeats script files.

STARTUP SWITCH

/b specifies the script file name to run when AutoCAD starts up.

TIPS

- Since the Script command is transparent, it can be used during another command.
- Use the **/s** command-line switch to run a script when AutoCAD starts.
- Prefix the VSlide command with ***** to preload it; this results in a faster slide show:

  ```
  *vslide
  ```

- You can make a script file more flexible — such as pausing for user input, or branching with conditionals — by inserting AutoLISP functions.

Section

Rel. 11 Creates 2D region objects from the intersection of a plane and a 3D solid.

Command	Alias	Shortcut Keystrokes	Menu	Ribbon
section	**sec**	...	...	...

Command: section
Select objects: *(Select one or more objects.)*
Select objects: *(Press Enter to end object selection.)*
Specify first point on section plane by [Object/Zaxis/View/XY/YZ/ZX/3points]: *(Pick a point, or enter an option.)*

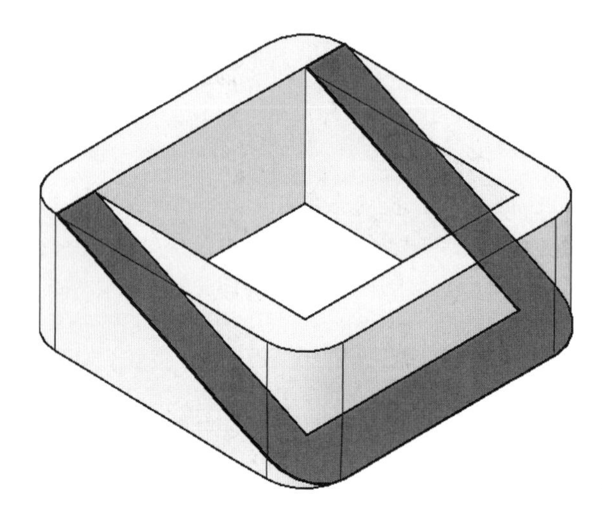

COMMAND LINE OPTIONS

Select objects selects the 3D solid objects to be sectioned.

Object aligns the section plane with a selected object: circle, ellipse, arc, elliptical arc, 2D spline, or 2D polyline.

Zaxis specifies the normal (*z axis*) to the section plane.

View uses the current view plane as the section plane.

XY uses the xy-plane of the current view.

YZ uses the yz-plane of the current view.

ZX uses the zx-plane of the current view.

3 points picks three points to specify the section plane.

RELATED COMMANDS

Slice cuts a slice out of a solid model, creating another 3D solid.

SectionPlane creates "live" (interactive) sections.

TIPS

- Section regions are placed on the current layer, not the object's layer.
- The result of this command is a region object; it can be moved away from the 3D solid.
- One cutting plane can section multiple solids at a time.

SectionPlane

2007 Places "section plane" objects that visually cut 3D objects.

Command	Alias	Shortcut Keystrokes	Menu	Ribbon
sectionplane	**splane**	...	**Draw** ↳**Modeling** ↳**Section Plane**	**Home** ↳**Section** ↳**Section Plane**

Command: sectionplane

Select face or any point to locate section line or [Draw section/Orthographic]: *(Pick a face or a point, or enter an option.)*

Specify through point: *(Pick another point.)*

After the section plane is placed, double-click it to make it "live" (active). See the LiveSection command.

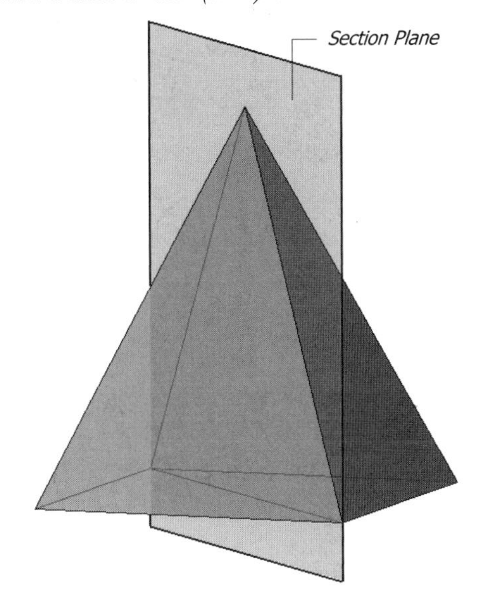

COMMAND LINE OPTIONS

Select face aligns the section plane with the face; you can pick a region or a face on a 3D solid. (There is no need to hold down the Ctrl key to select faces.)

Select any point specifies the starting point of the section plane; if this command insists on aligning the section plane with a face, override it with the Draw Section option.

Draw section places the section plane by picking points.

Orthographic places the section plane parallel to standard views: front, back, top, bottom, left, or right view.

Specify through point positions the section plane.

Draw Section options

It can be helpful to use ENDpoint and QUAdrant object snaps with these prompts.

Specify start point: *(Pick a point, 1.)*

Specify next point: *(Pick another point, 2.)*

Specify next point or ENTER to complete: *(Press Enter to skip to the next prompt, or pick a third point, 3.)*
Specify next point in direction of section view: *(Pick a point to align the section plane, 4.)*

Specify start point specifies the starting point of section plane.

Specify next point draws multi-segment section planes.

Direction of section view specifies the cutaway portion.

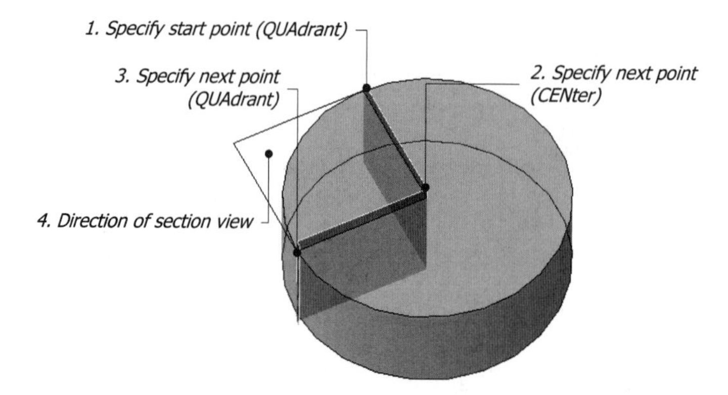

Orthographic option
Align section to: [Front/Back/Top/Bottom/Left/Right]: *(Enter an option.)*

Front, **Back**, **Top**, **Bottom**, **Left**, or **Right** aligns the section plane with current UCS. Live sectioning is turned on automatically (the forward portion of the sectioned objects being removed from view). See the LiveSection command.

RELATED COMMANDS

LiveSection edits section planes.

Section creates region slices from 3D models.

Interfere creates new objects from the interference of two or more 3D solids.

GenerateSection defines the look of generated sections and elevations.

JogSection adds jogs to section planes.

SectionPlaneJob adds jogs with a single click.

SectionPlaneSettings sets properties for this command.

TIPS

- This command is half of a two-part process of working with sections:
    1. Create the section plane with the SectionPlane command.
    2. Edit the section plane with the LiveSection command.
- "Sectionplane" is a distinct object type introduced with AutoCAD 2007.

SectionPlaneJog

2010 Adds jogs to section plane objects with a single click.

Command	*Alias*	*Shortcut Keystrokes*	*Menu*	*Ribbon*
sectionplanejog	**jogsection**	...	...	**Mesh Modeling** **↳Section** **↳Add**

Command: sectionplanejog
Select section object: *(Select one object created by the SectionPlane command.)*
Specify a point on section line to add jog: *(Pick a point on the section plane's line.)*

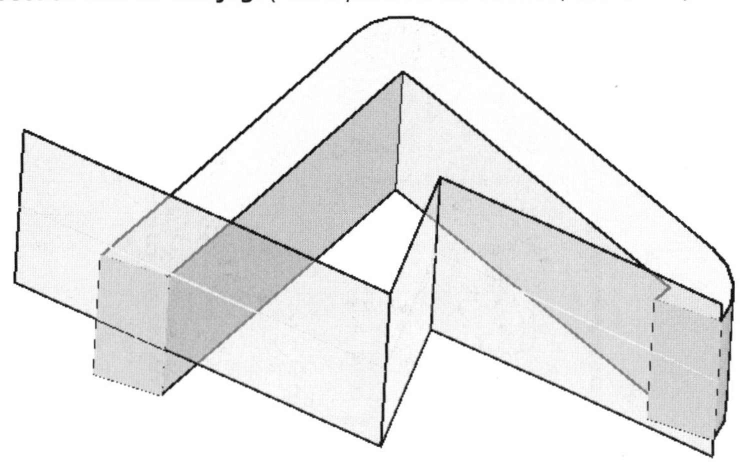

COMMAND LINE OPTIONS

Select objects selects a single section plane object.

Specify a point on section line to add jog specifies the location of the jog.

RELATED COMMANDS

SectionPlane creates section planes.

SectionPlaneSettings sets properties for this command.

TIPS

- Section planes are created by the SectionPlane command; this command does not work with sections created by the Section command.
- Use object snaps to help you pick the point on the section plane's line.

SectionPlaneSettings

Sets the properties of section planes.

Command	Alias	Shortcut Keystrokes	Menu	Ribbon
sectionplanesettings	...	...	...	**Home** **↳Section** **↳Settings**

Command: **sectionplanesettings**

Displays dialog box.

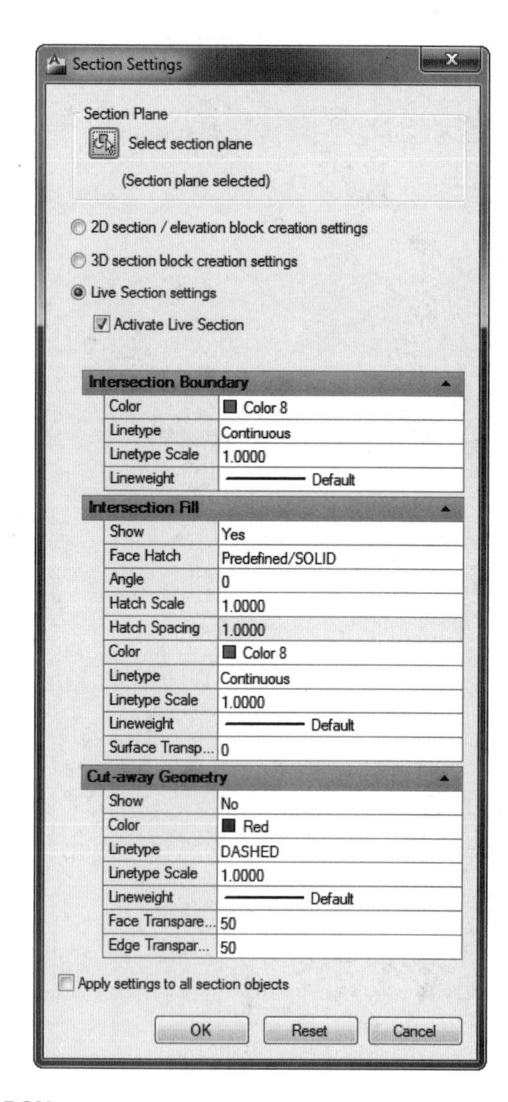

SECTION SETTINGS DIALOG BOX

○ **2D section/elevation block creation settings** displays options in this dialog box for 2D sections and elevations.

○ **3D section block creation settings** displays options in this dialog box for 3D sections.

⊙ **Live Section settings** displays options in this dialog box for live sections.

☑ **Activate Live Section** turns on live sectioning for the selected section.

Intersection Boundary options

Color specifies the color of the intersection boundary.

Linetype specifies the linetype.

Linetype Scale specifies the linetype scale.

Lineweight specifies the lineweight.

Intersection Fill options

Show toggles the display of the intersection fill.

Face Hatch determines if the fill is solid, a hatch pattern, or not present.

Angle specifies the angle of the hatch pattern.

Hatch Scale specifies the hatch pattern scale factor.

Hatch Spacing specifies the spacing between hatch lines.

Color specifies the color of the intersection fill.

Linetype specifies the linetype.

Linetype Scale specifies the linetype scale.

Lineweight specifies the lineweight.

Surface Transparency changes the translucency of the intersection fill; ranges from 0 to 100.

Cutaway Geometry options

Show toggles the display of cutaway geometry (removed by the live section plane).

Color specifies the color of the intersection fill.

Linetype specifies the linetype.

Linetype Scale specifies the linetype scale.

Lineweight specifies the lineweight.

Face Transparency changes the translucency of the cutaway faces.

Edge Transparency changes the translucency of the cutaway edges.

☑ **Apply Settings to All Section Objects** applies settings to future sectionings.

Reset reverts values to default settings.

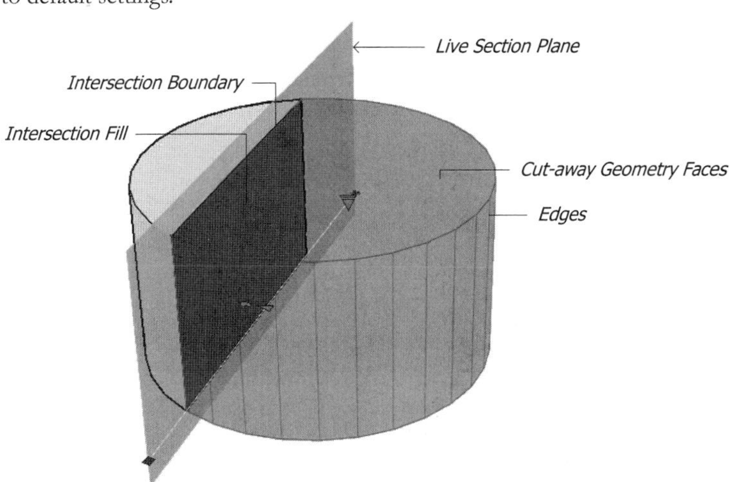

RELATED COMMANDS

SectionPlane creates section planes.

SectionPlaneJog creates jogs in section planes.

TIPS

- Section planes are created by the SectionPlane command; this command does not work with sections created by the Section command.
- Use object snaps to help you pick the point on the section plane's line.

SectionPlaneToBlock

2010 Creates blocks from section planes.

Command	Alias	Shortcut Keystrokes	Menu	Ribbon
sectionplanetoblock	...	...	...	**Mesh Modeling** ↳**Section** ↳**Generate Section**

Command: sectionplanetoblock
Select section object: *(Select an object created by the SectionJog or SectionPlaneJog command.)*

Displays dialog box.

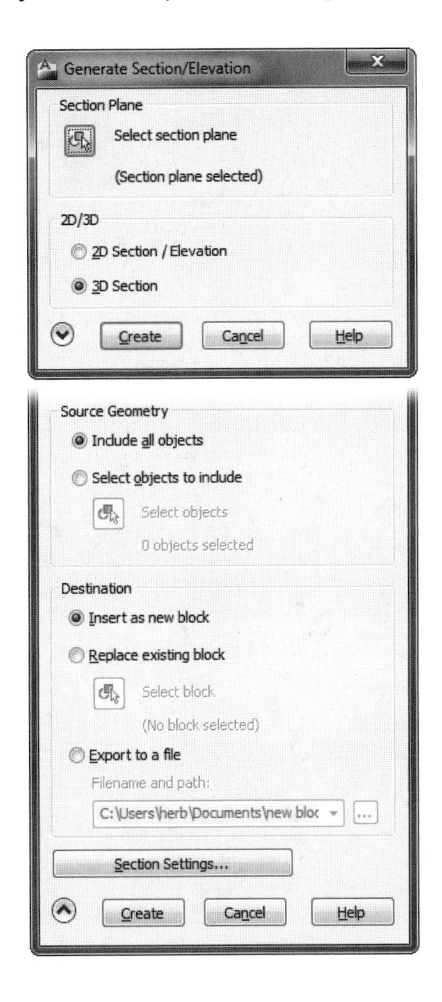

COMMAND LINE OPTION

Select section object selects a section plane object.

GENERATE SECTION/ELEVATION DIALOG BOX

Section Plane option

Select Section Plane selects a section plane (if none is selected yet).

2D/3D options

⊙ **2D Section/Elevation** generates 2D cross sections of section planes.

○ **3D Section** generates 3D cross sections.

More expands the dialog box.

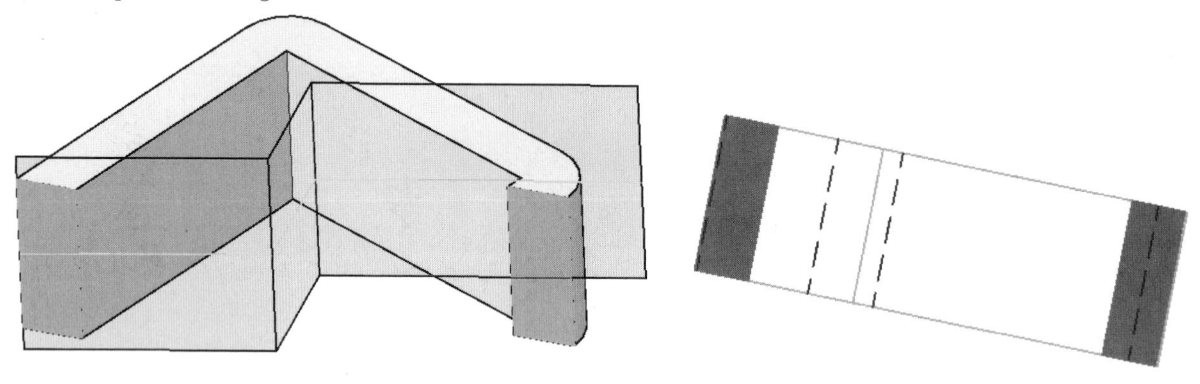

Left: *Section plane.*
Right: *Block created from section plane.*

Source Geometry options

⊙ **Include All Objects** includes all 3D solids, surfaces, meshes, and regions in the drawing, xrefs, and blocks.

○ **Select Objects to Include** selects objects manually from which sections are generated.

Select objects selects the objects to include.

Destination options

⊙ **Insert as New Block** inserts the section as a block in the drawing. After you click Create, AutoCAD presents the standard prompts for inserting blocks; see the -Block command.

○ **Replace Existing Block** replaces an existing block with the new section block.

Select Block selects the block.

○ **Export to a File** saves the section as a DWG file.

Filename and Path specifies the drive, path, and file name for the saved section.

Section Settings opens the Section Settings Dialog Box; see the SectionJogSettings command.

Create closes the dialog box, and then creates the section.

RELATED COMMANDS

SectionPlane places section planes in 3D objects.

GenerateSection displays the Generate Section/Elevation dialog box.

SectionPlaneJog adds jogs to section planes.

TIPS

- Section planes can be edited with dynamic grips.
- Live sectioning works in model space only.

SecurityOptions

2004 Sets passwords and digital signatures for securing drawings against unauthorized access.

Command	*Aliases*	*Shortcut Keystrokes*	*Menu*	*Ribbon*
securityoptions	...	...	...	...

Command: securityoptions

Displays dialog box.

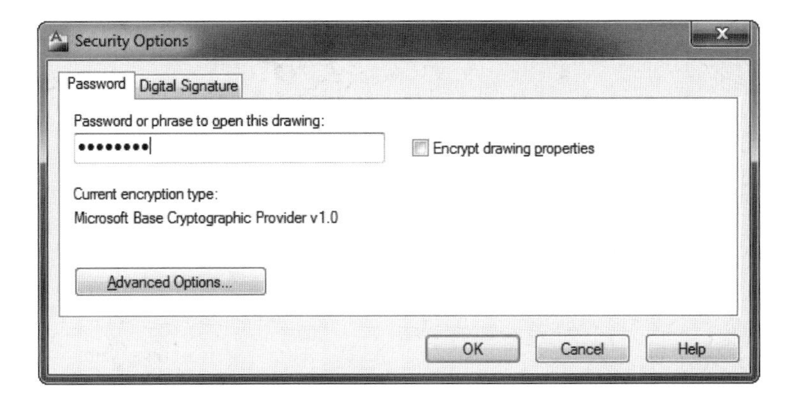

SECURITY OPTIONS DIALOG BOX

Password or phrase to open this drawing assigns a password to lock the drawing.

☐**Encrypt drawing properties** encrypts drawing properties data; see the DwgProps command.

Advanced Options displays the Advanced Options dialog box; available only after a password is entered.

OK displays the Confirm Password dialog box.

Digital Signature tab

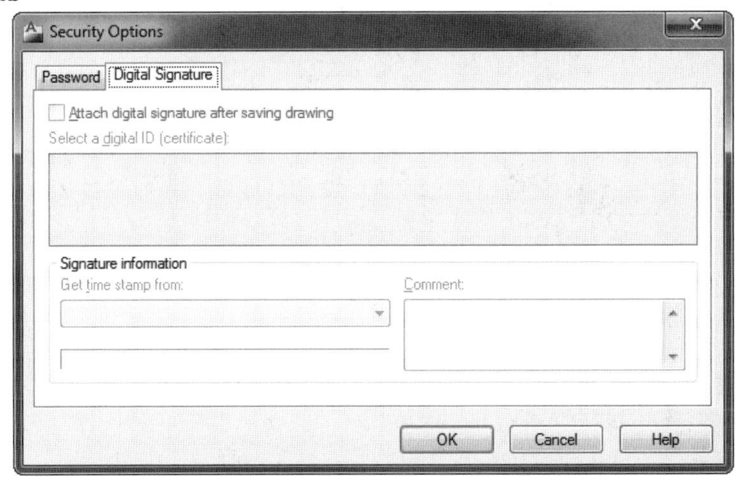

☐**Attach digital signature after saving drawing** attaches the digital signature the next time the drawing is saved.

Select a digital ID (certificate) lists digital signature services.

Signature information

Get time stamp from lists time servers.

Comment provides additional comments to be included with the digital signature.

ADVANCED OPTIONS DIALOG BOX

Choose an encryption provider selects from several encryption types.

Choose a key length selects 40, 48, or 56 bits of encryption.

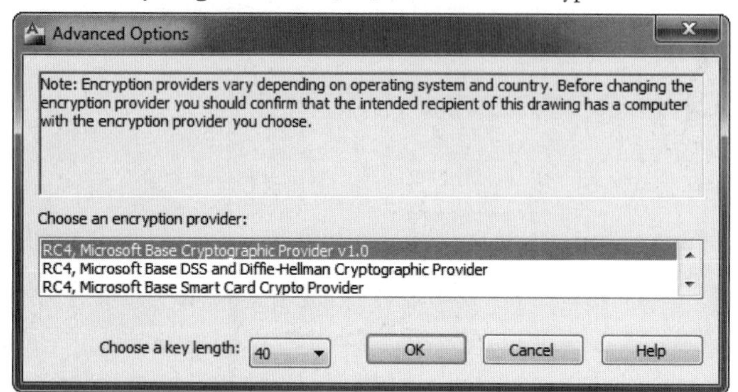

CONFIRM PASSWORD DIALOG BOX

Reenter the password, and then click OK.

RELATED COMMAND

SigValidate displays information about the digital signatures attached to drawings.

RELATED SYSTEM VARIABLE

SigWarn determines whether a warning alerts users that open drawings contain digital signatures.

TIPS

- When opening drawings protected by passwords, AutoCAD displays the Password dialog box for entering the password:

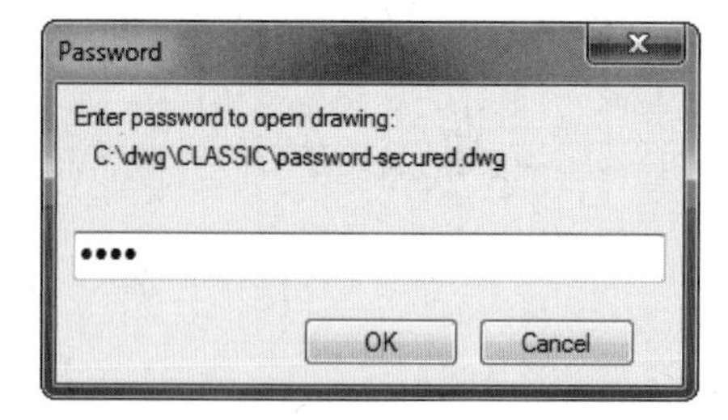

  - When the password entered is correct, AutoCAD opens the drawing.
  - When the password entered is incorrect, AutoCAD gives you a chance to retry.

- To remove the password, open the Security Options dialog box, remove the password, and then click OK.
- *Warning!* If the password is lost, the drawing is not recoverable. Autodesk recommends that you (1) save a backup copy of the drawing not protected with any password, and (2) maintain a list of password names stored in a secure location.
- *Warning!* The encryption provider must be the same on the receiving computers, otherwise the drawings do not open.
- To remove this feature from AutoCAD, install with the Custom option, and then unselect Drawing Encryption. Disgruntled employees have been known to lock out files with their own passwords upon leaving employment.

Seek

2010 Accesses blocks and drawings on Autodesk's Seek web site.

Command	*Alias*	*Shortcut Keystrokes*	*Menu*	*Ribbon*
seek	...	...	...	**Insert** **↳Content** **↳Seek Design Content**

Command: seek

Starts the Web browser, and then opens Autodesk's Seek web site at seek.autodesk.com.

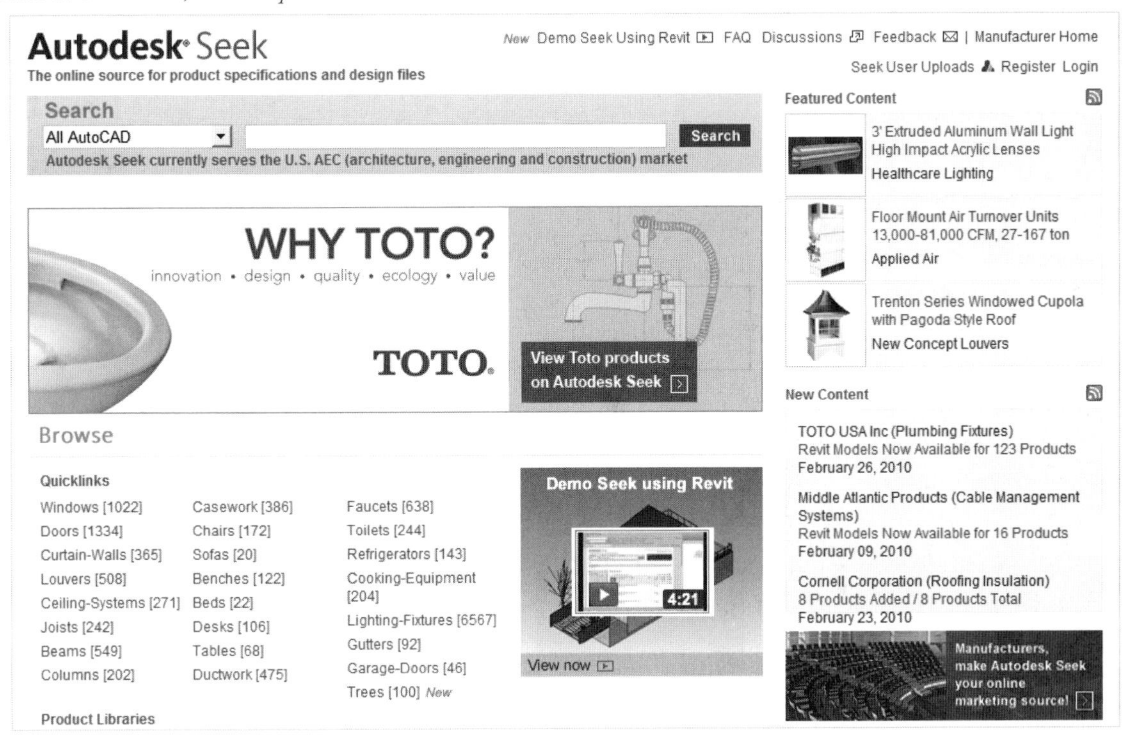

Choose a category, such as Windows, and then select product categories, such as aluminum cladded windows.

Choose the file format, such as DWG, and then click Download Selected.

Save the file to disk, or open it in AutoCAD.

COMMAND LINE OPTIONS

None.

RELATED COMMANDS

ShareWithSeek uploads blocks and drawings to Autodesk's Seek Web site.

Browser opens the default Web browser at the Autodesk Web site.

TIP

- You also access the Seek Web site through the Seek button on the DesignContent palette.

Select

v. 2.5 Creates a selection set of objects based on their location in the drawing.

Command	Aliases	Shortcut Keystrokes	Menu	Ribbon
select	...	...	...	...

Command: select
Select objects: *(Select one or more objects, or enter an option.)*
Select objects: *(Press Enter to end object selection.)*

COMMAND LINE

pick selects a single object.

AU switches from pick mode to C or W mode, depending on whether an object is found at the initial pick point (short for AUtomatic).

ALL selects all objects in the drawing; press Ctrl+A to select all objects.

BOX goes into C or W mode, depending on how the cursor moves.

C selects objects in and crossing the selection box (Crossing).

CL selects application-specific objects (CLass).

CP selects all objects inside and crossing the selection polygon (Crossing Polygon).

F selects all objects crossing a selection polyline (Fence).

G selects objects contained in a named group (Group).

M makes multiple selections before AutoCAD scans the drawing; saves time in a large drawing (Multiple).

SI selects only a single set of objects before terminating the Select command (SIngle).

SU selects vertices, edges, and faces of 3D solids (SUbobject).

W selects all objects inside the selection box (Window).

WP selects objects inside the selection polygon (Windowed Polygon).

A continues to add objects after you use the R option (Add).

L selects the last-drawn object still visible on the screen (Last).

O ends sub-object selection mode (Object).

P selects previously-selected objects (Previous).

R removes objects from the selection set (Remove).

U removes the selected objects most-recently added (Undo).

Enter completes the Select command.

Esc abandons the Select command.

RELATED COMMANDS

All commands that prompt 'Select objects'.

Filter specifies objects to be added to the selection set.

QSelect selects objects based on their properties via a dialog box.

SelectSimilar selects objects based on preset properties *(new to AutoCAD 2011).*

RELATED SYSTEM VARIABLES

PickAdd controls how objects are added to a selection set.

PickAuto controls automatic windowing at the 'Select objects' prompt.

PickDrag controls the method of creating a selection box.

PickFirst controls the command-object selection order.

TIPS

- Objects selected by the Select command become the Previous selection set, and may be selected by any subsequent 'Select objects:' prompt by responding P for "previous."
- To view a list of all selection options, enter any non-valid text at the prompt, such as "asdf":

  Command: select
  Select objects: asdf
  Invalid selection
  Expects a point or
  Window/Last/Crossing/BOX/ALL/Fence/WPolygon/CPolygon/Group/Add/Remove/Multiple/ Previous/Undo/AUto/SIngle/SUbobject/Object
  Select objects:

- Pressing Ctrl+A selects all objects in the drawing, bypassing the Select command.
- All selects all objects in the drawing, except those on frozen and locked layers.
- Crossing selects objects within and crossing the selection rectangle.
- Window selects objects within the selection rectangle.
- Fence selects objects crossing the selection polyline.
- CPolygon selects objects within and crossing the selection polygon.
- WPolygon selects objects within the selection polygon.
- Selection preview highlights objects as the cursor passes over them. This can be useful for determining which objects belong together — for example, lines and arcs vs polylines. However, it slows down the display speed with large, complex objects. In the figure below, the chair block is highlighted by the cursor:

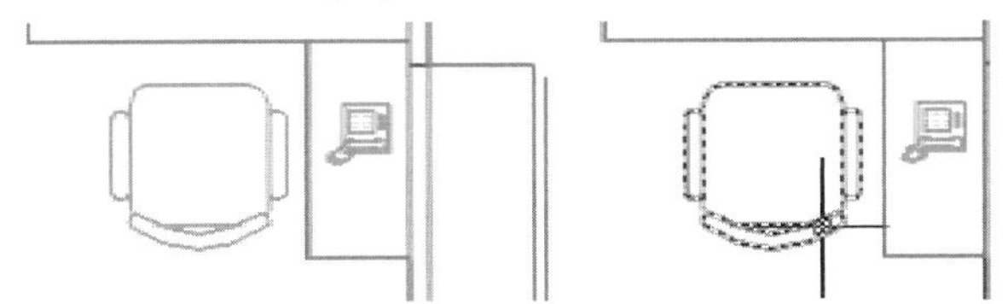

  You can change preview highlighting with the Options command: choose the Selection tab, and then click the Visual Effect Settings button.

- When objects are overlapping, hold down the Ctrl key to cycle through selections.

SelectSimilar

2011 Adds objects with identical properties to the selection set.

Command	*Alias*	*Shortcut Keystrokes*	*Shortcut Menu*	*Ribbon*
selectsimilar	...	...	**Select Similar**	...

Command: selectsimilar

Select objects or [SEttings]: *(Choose one or more objects, or type SE.)*

Select objects or [SEttings]: *(Press Enter to end the command.)*

COMMAND LINE OPTIONS

Select objects chooses one or more objects.

SEttings displays the Settings dialog box.

SETTINGS DIALOG BOX

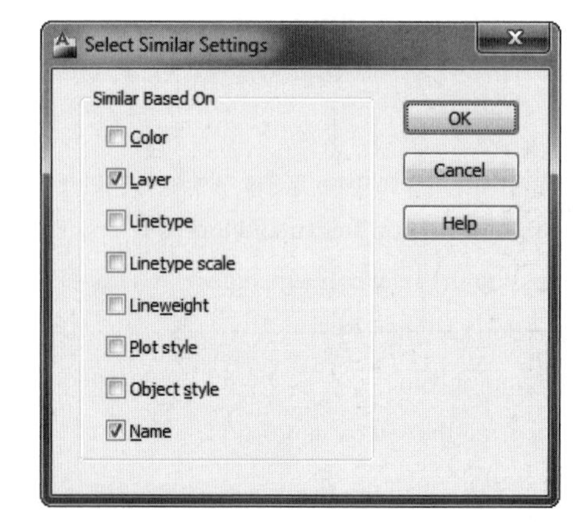

Similar Based On determines the properties to be matched

☑ **Name** matches the names of objects, such as blocks and xrefs.

RELATED COMMANDS

QSelect selects objects based on properties.

AddSelected draws objects whose properties are the same as a selected object *(new to AutoCAD 2011).*

RELATED SYSTEM VARIABLE

SelectSimilarMode sets the default properties for this command.

TIPS

- Use this command to select all blocks that have the same name, or all objects that lie on the same layer.
- To start this command from the shortcut menu, first choose an object, and then select Select Similar from the shortcut menu. AutoCAD immediately chooses all similar objects without prompting you for settings. Use the SelectSimilarMode to preset properties.
- When you select multiple objects, AutoCAD chooses all other objects that share all properties; for example, if two selected objects are on two different layers, all other objects on the two layers are selected.
- This command does not select objects based on type.

SelectURL

Rel. 14 Highlights all objects and areas that have hyperlinks attached to them (undocumented).

Command	*Aliases*	*Shortcut Keystrokes*	*Menu*	*Ribbon*
selecturl	...	...	...	...

Command: selecturl

Highlights hyperlinked objects and areas with grips.

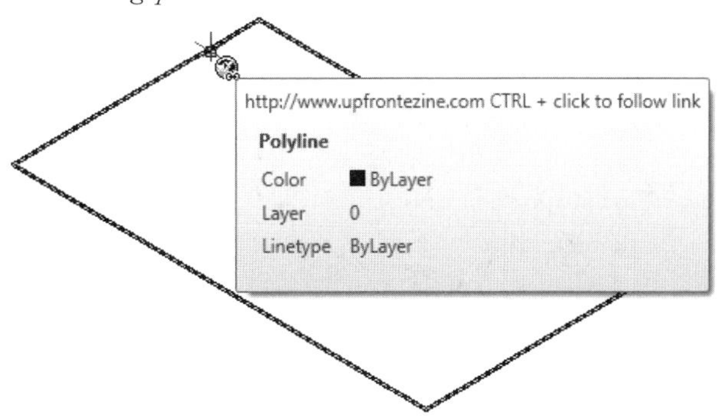

If there are no hyperlinks in the drawing, AutoCAD reports:

No objects with hyperlinks found.

COMMAND LINE OPTIONS

None.

RELATED COMMANDS

AttachURL attaches URLs to objects and areas.

Hyperlink attaches URLs to objects via a dialog box.

TIPS

- Examples of URLs include:

http://www.autodeskpress.com	Autodesk Press Web site.
news://adesknews.autodesk.com	Autodesk news server.
ftp://ftp.autodesk.com	Autodesk FTP server.
http://www.upfrontezine.com	Author Ralph Grabowski's Web site.

- The URL is stored as follows:

Attachment	*URL*
One object	Stored as xdata (extended entity data).
Multiple objects	Stored as xdata in each object.
Area	Stored as xdata in rectangles on layer URLLAYER.

- Do not delete layer URLLAYER. URL is short for "uniform resource locator," the universal file naming convention used by computers and the Internet.

SequencePlay

2009 Plays back ShowMotion animations of all views in a single category (undocumented).

Command	*Alias*	*Shortcut Keystrokes*	*Menu*	*Ribbon*
sequenceplay	**splay**	...	...	...

Command: sequenceplay
Enter view category name to play: *(Enter the name of a category.)*

COMMAND LINE OPTIONS

Enter View Category Name specifies the category name.

RELATED COMMANDS

AllPlay plays all show motion views in sequence, without prompting.

EditShot edits the parameters of ShowMotion animations.

NavSMotion displays the show motion interface.

NewShot creates named views with show motion options.

ViewPlay plays the animation associated with a named view.

3dWalk / **3dPlay** record animations to external video files.

TIPS

- You can have one or more views per category. View categories allow you to have two or more multi-view show motion animations per drawing.
- View *categories* were originally designed for sheet sets, but now are adapted for show motion. They are created in the View Categories item of the New View / Shot Properties dialog box. See the NewShot command.
- This command is meant for use with macros and AutoLISP routines; you would normally use the NavSMotion command to play back animations associated with views and categories..

SetByLayer

2008 Forces overridden properties in viewports to ByLayer.

Command	*Aliases*	*Shortcut Keystrokes*	*Menu*	*Ribbon*
setbylayer	...	...	**Modify**	**Home**
			↳Change to ByLayer	**↳Modify**
				↳Set to ByLayer

Command: setbylayer
Current active settings: Color Linetype Lineweight Material Plot Style
Select objects or [Settings]: *(Select one or more objects, or type S.)*
Change ByBlock to ByLayer? [Yes/No] <Yes>: *(Type Y or N.)*
Include blocks? [Yes/No] <Yes>: *(Type Y or N.)*

COMMAND LINE OPTIONS

Select objects selects one or more objects; enter 'all' for all objects in the drawing.

Settings displays the Settings dialog box, detailed below.

Change ByBlock to ByLayer changes properties from ByBlock to ByLayer.

Include blocks toggles whether blocks should have their properties reset to Bylayer.

SETTINGS DIALOG BOX

Displays dialog box.

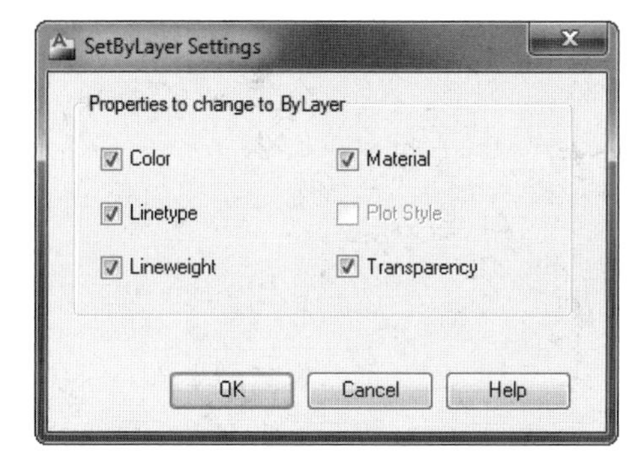

☑**Color** changes the color of selected objects to ByLayer.

☑**Linetype** changes the linetype of selected objects to ByLayer.

☑**Lineweight** changes the lineweight of selected objects to ByLayer.

☑**Material** changes the material of selected objects to ByLayer.

☑**Plot Style** changes the plot style of selected objects to ByLayer.

☑**Transparency** changes the level of transluscency to ByLayer *(new to AutoCAD 2011)*.

RELATED COMMANDS

Change, **Properties**, **Color**, **Linetype**, **Lineweight**, **Materials**, **Plotstyle** override properties.

Layer reports viewport overrides.

RELATED SYSTEM VARIABLES

SetByLayerMode specifies which properties are changed by this command.

VpLayerOverrides reports whether any layer in the current viewport has VP (viewport) property overrides.

SetIDropHandler

2004 Specifies how i-drop objects should be handled in drawings.

Command	Aliases	Shortcut Keystrokes	Menu	Ribbon
'setidrophandler	...	...	...	...

Command: setidrophandler

Displays dialog box:

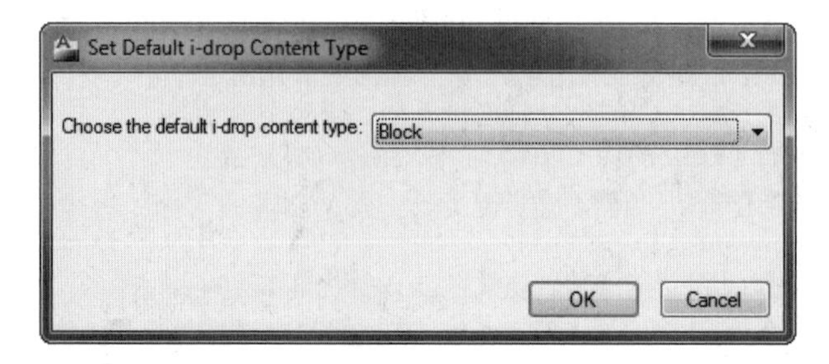

SET DEFAULT I-DROP CONTENT TYPE DIALOG BOX

Choose the default i-drop content type specifies how i-drop objects are handled when placed in drawings.

TIPS

- When i-drop content is dragged from Web pages into drawings, AutoCAD treats the objects as blocks.
- Ever since this command was introduced, its only option has been "Block", so this command serves no purpose.

Removed Command

SetUV command was removed from AutoCAD 2007; its replacement is MaterialMap.

SetVar

v. 2.5 Lists the settings of system variables, and allows you to change variables that are not read-only (short for SET VARiable).

Command	Alias	Shortcut Keystrokes	Menu	Ribbon
'setvar	**set**	...	**Tools** ↳**Inquiry** ↳**Set Variable**	...

Command: setvar
Enter variable name or [?]: *(Enter a name, or type ?.)*

COMMAND LINE OPTIONS

Enter variable name names the system variable you want to access.

? lists the names and settings of system variables.

TIPS

- See Appendix C for a complete list of all system variables found in AutoCAD.
- Example usage of this command:

  Command: setvar
  Enter variable name or [?]: visretain
  Enter new value for VISRETAIN <0>: 1

- Almost all system variables can be entered without the SetVar command. For example:

  Command: visretain
  New value for VISRETAIN <0>: 1

- System variables marked "read only" cannot be changed:

  Command: _pkser
  _PKSER = "341-35000000" (read only)

Replaced Commands

Shade command was replaced in AutoCAD 2000 by the ShadeMode command's Flat+edges option.

ShadeMode command was replaced by the -ShadeMode command in AutoCAD 2007; ShadeMode is now an alias for the VsCurrent command.

-ShadeMode

2000 Generates renderings of 3D models quickly in a variety of modes (replaces the Shade and ShadeMode commands).

Command	*Aliases*	*Shortcut Keystrokes*	*Menu*	*Ribbon*
-shademode	...	...	...	...

Command: -shademode
Current mode: 2D wireframe
Enter option [2D wireframe/3D wireframe/Hidden/Flat/Gouraud/fLat+edges/gOuraud+edges]: *(Enter an option.)*

COMMAND LINE OPTIONS

2D wireframe displays wireframe models in 2D space.

3D wireframe displays wireframe models in 3D space.

Hidden removes hidden faces.

Flat displays flat shaded faces.

fLat+edges displays flat shaded faces, with outlined faces of the background color.

Gouraud displays smooth shaded faces.

gOuraud+edges displays smooth shaded faces, with outlined faces of the background color.

RELATED COMMANDS

VsCurrent displays viewports in a variety of visual styles.

MView specifies hidden-line removal of individual viewports for plotting.

Render performs more realistic renderings than does -ShadeMode.

TIPS

- ShadeMode is now an alias for VsCurrent.
- As an alternative to the -ShadeMode command, the Render command creates high-quality renderings of 3D drawings, but takes somewhat longer to do so.
- The smaller the viewport, the faster the shading.
- The 2D wireframe mode's colors are based on layer and object colors; 3D wireframe colors are based on the color of the rendering materials, if any.
- In Flat mode; each face is filled with shaded grays; in fLat+edges mode, faces are outlined by the background color.
- In Gouraud mode, faces are smoothed; in gOuraud+edges mode, each face is outlined by the background color.

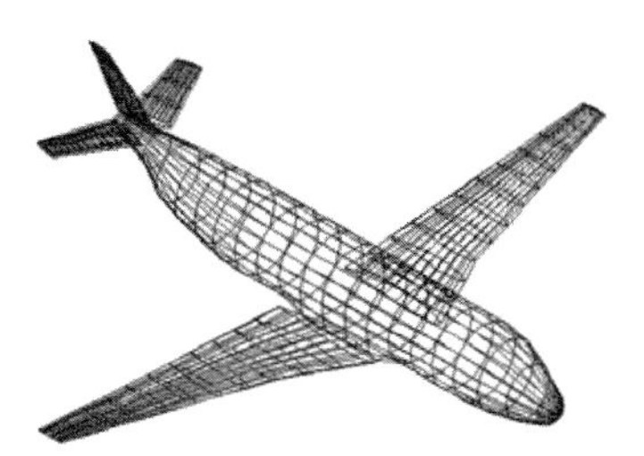
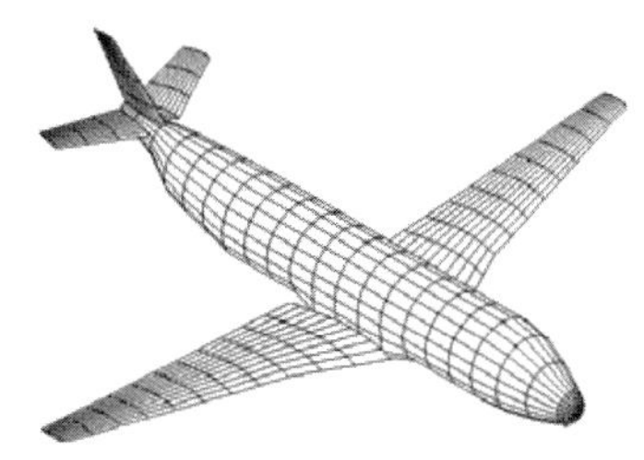
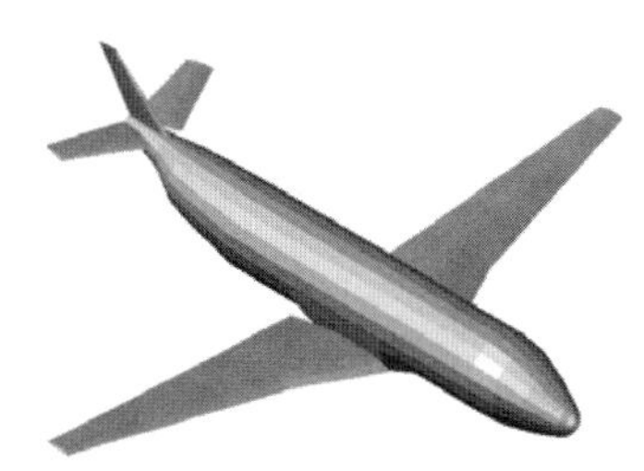

Left: *2D and 3D wireframe modes.*
Center: *Hidden mode.*
Right: *Flat mode.*

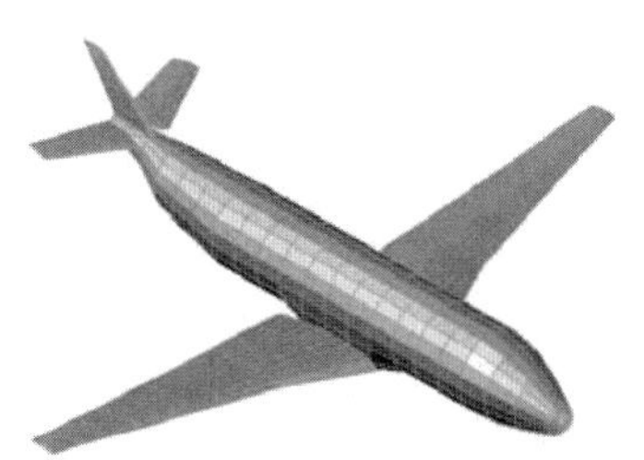
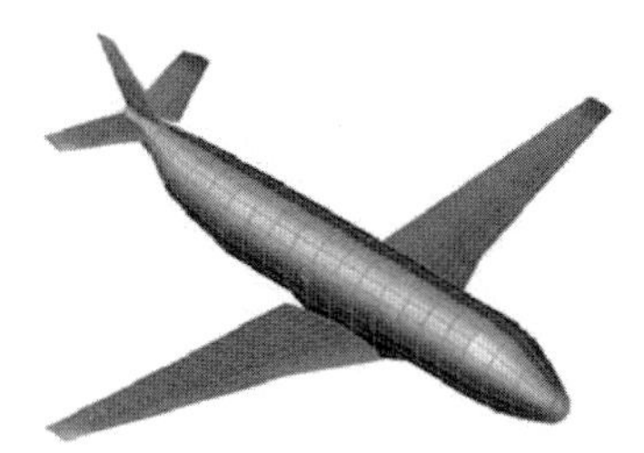

Left: *Gouraud mode.*
Center: *Flat with edges mode.*
Right: *Gouraud with edges mode.*

Shape

v. 1.0 Inserts shapes into drawings.

Command	Aliases	Shortcut Keystrokes	Menu	Ribbon
shape	...	...	...	...

Command: shape
Enter shape name or [?]: *(Enter the name, or type ?.)*
Specify insertion point: *(Pick a point.)*
Specify height <1>: *(Specify the height.)*
Specify rotation angle <0>: *(Specify the angle.)*

COMMAND LINE OPTIONS

Enter shape name indicates the name of the shape to insert.

? lists the names of currently-loaded shapes.

Specify insertion point indicates the insertion point of the shape.

Specify height specifies the height of the shape.

Specify rotation angle specifies the rotation angle of the shape.

RELATED COMMANDS

Load loads SHX-format shape files into the drawings.

Insert inserts blocks into drawings.

Style loads *.shx* font files into drawings.

RELATED SYSTEM VARIABLE

ShpName specifies the default *.shp* file name.

TIPS

- Shapes are used to define the text and symbols found in complex linetypes.

Some electronic shapes.

- Shapes were an early alternative to blocks, but now are used primarily with complex linetypes. They take up extremely small amounts of memory, but are difficult to create.
- Shapes are defined by *.shp* files, which must be compiled into *.shx* files before they can be loaded by the Load command.
- Shapes must be loaded by the Load command before they can be inserted with the Shape command.
- Compile an *.shx* file into an *.shp* file with the Compile command.

ShareWithSeek

2010 Uploads drawings and blocks to Autodesk's Seek web site.

Command	Alias	Menu	Ribbon
sharewithseek	**...**	**File**	**Output**
		↳Share With Autodesk Seek	**↳Autodesk Seek**
			↳Share With Autodesk Seek

Command: sharewithseek

Displays dialog box.

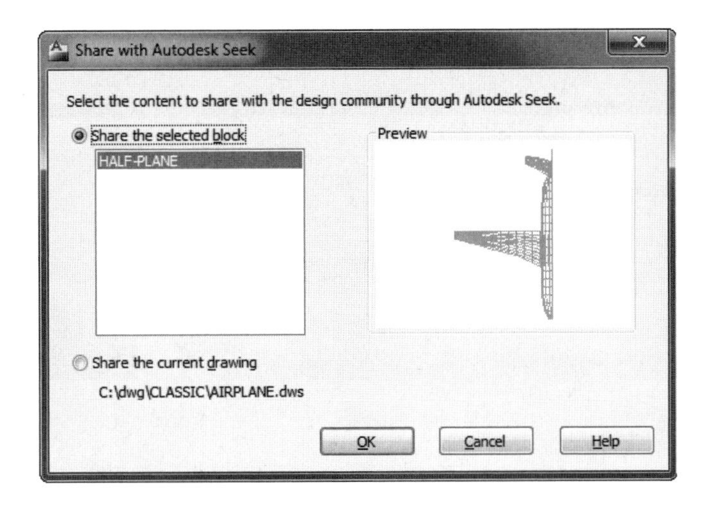

SHARE WITH AUTODESK SEEK DIALOG BOX

Share The Selected Block lists the names of blocks in the drawing, if any; the Preview window shows the selected block.

Share The Current Drawing shows the drawing in the Preview window.

OK uploads the drawing to Autodesk; displays the Progress dialog box, detailed later. When upload is completed, opens the Autodesk Seek Web page in default Web browser; detailed later..

If changes to the drawing have not been saved, prompts you to save them; displays the Save Changes dialog box, detailed later.

SAVE CHANGES DIALOG BOX

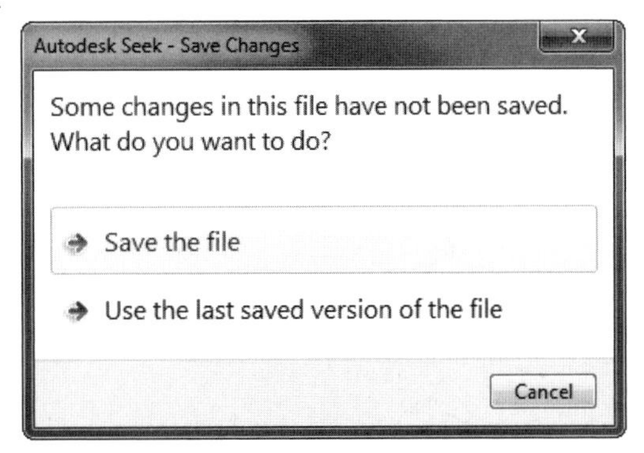

Save the File saves the drawing file before sharing it; opens the Save As dialog box.

Use the Last Saved Version of the File shares the version residing in the current *.dwg* file.

PROGRESS DIALOG BOX

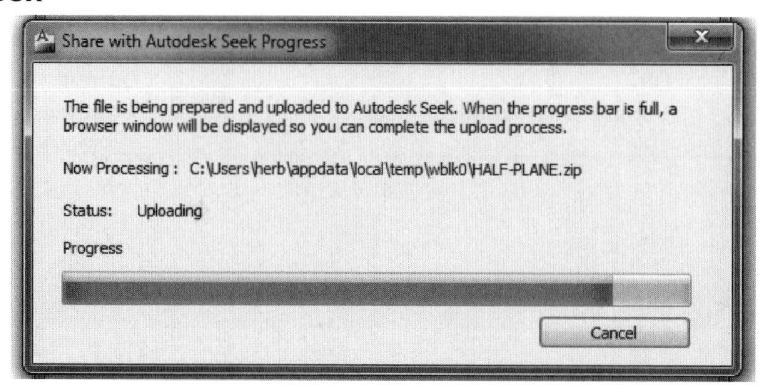

Cancel cancels the upload.

AUTODESK SEEK WEB PAGE

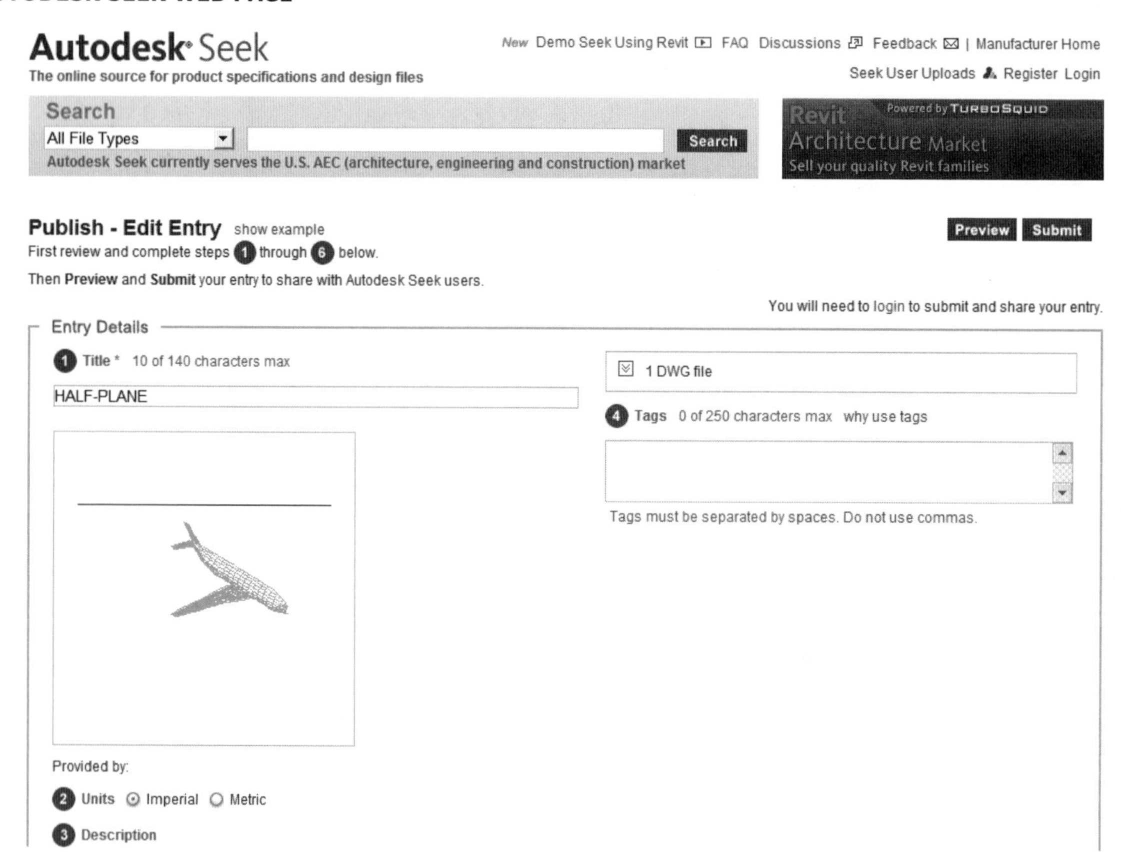

Fill out the information requested:

1. Name the drawing.
2. Specify imperial or metric units.
3. Describe the drawing.
4. Enter *tags*, one-word descriptions of the drawing.
5. Select a category.
6. Specify the type of drawing.

Click **Preview** to ensure the information is correct; incorrect data are highlighted in pink.

And then click **Submit**.

RELATED COMMANDS

Seek downloads blocks and drawings from Autodesk's Seek Web site.

Browser opens the default Web browser at the Autodesk Web site.

Sheetset / SheetsetHide

2005 Displays and hides the Sheetset Manager palette.

Commands	Alias	Shortcut Keystrokes	Menu	Ribbon
sheetset	**ssm**	**Ctrl+4**	**Tools** **↳Palettes** **↳Sheetset Manager**	**View** **↳Palettes** **↳Sheetset Manager**
sheetsethide	**...**	**Ctrl+4**	**Tools** **↳Palettes** **↳✓Sheetset Manager**	**View** **↳Palettes** **↳Sheetset Manager**

Command: sheetset

Displays palette:

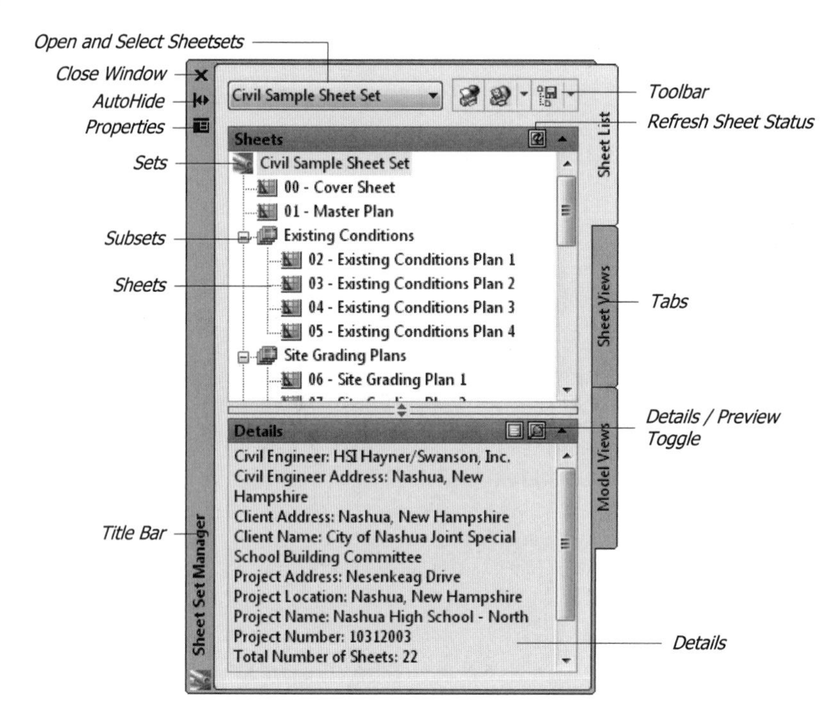

Command: sheetsethide

Closes the palette:

SHEETSET MANAGER PALETTE

x closes the palette; alternatively, use the SheetsetHide command.

Toolbar changes, depending on the tab selected.

Title bar docks against the side of AutoCAD, or can be dragged around.

Tabs change between Sheet List, View List, and Resource Drawings views.

- **Sheet List** displays names of sheets, which are organized into subsets.
- **Sheet Views** displays views for each sheet set; views can be organized into categories.

- **Model Views** displays folders and files used by the sheet set.

Open and select sheet sets displays a droplist of recently-opened sheet sets:

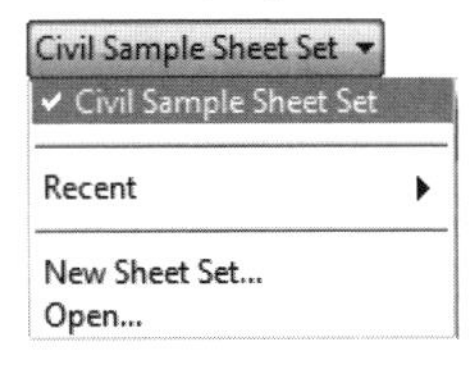

- **Recent** lists recently-opened *.dst* sheet set data files.
- **New Sheet Set** starts the Create Sheet Set wizard; see the NewSheetset command.
- **Open** displays the Open Sheetset dialog box; select a *.dst* file, and then click Open; see the NewSheetset command.

Refresh Sheet Status (**F5**) updates the list of files and information stored in sheet set data *.dst* files, checks resource folders for new and removed drawings, and checks all drawings for new and removed model space views.

Details lists information about selected sheets and sheet sets; see figure below.

Sheet Preview shows preview images of selected sheets; see figure below.

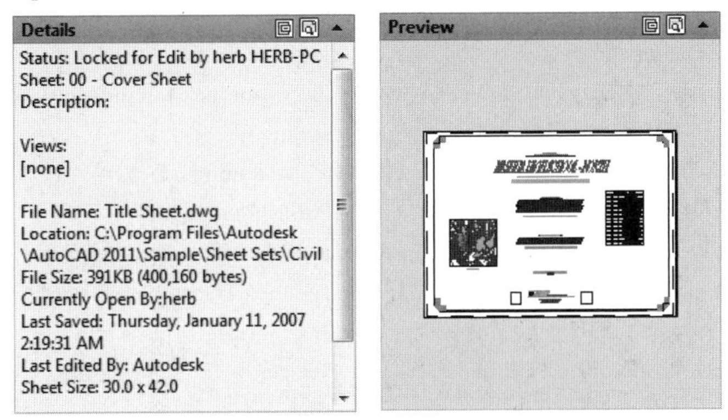

***Left:** Details pane.*
***Right:** Preview pane.*

Stretches the size of the Sheets/Preview panes.

AutoHide reduces palette to title bar when cursor not over the palette.

Properties displays a shortcut menu.

Sheet List toolbar

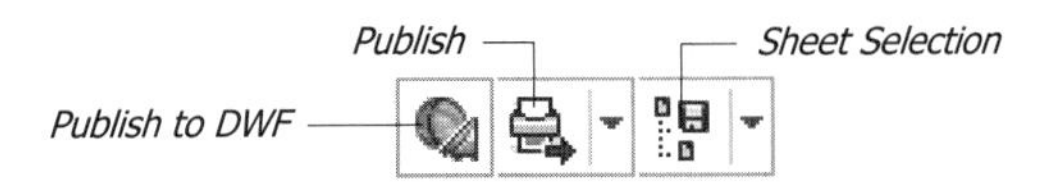

SHORTCUT MENUS

 *Access this menu by right-clicking a sheet set name:*

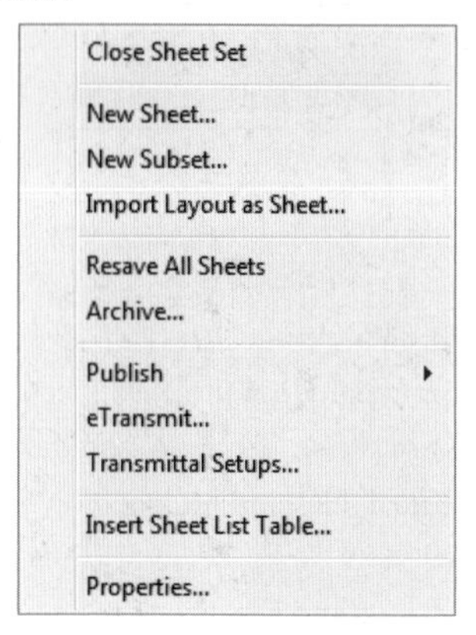

Close Sheet Set closes the entire sheet set; use the Open option to open another sheet set.

New Sheet displays the New Sheet dialog box; enter a number and name, and then click OK.

New Subset displays the Subsheet Properties dialog box; enter a name, and then click OK.

Import Layout as Sheet displays the Import Layout as Sheet dialog box; select a drawing and layouts, and then click OK.

Resave All Sheets saves sheet set information stored in each drawing file.

Archive displays the Archive as Sheet Set dialog box; see the Archive command.

Publish plots the sheet set as a *.dwf* file or to a plotter; see the Publish command.

eTransmit collates the sheet set for transmittal by CD or email; see the eTransmit command.

Transmittal Setups displays the Transmittal Setup dialog box; see the eTransmit command.

Insert Sheet List Table creates a table listing all sheets in the sheet set, and adds it to a sheet.

Properties displays the Sheet Set Properties dialog box; make changes, and then click OK.

Subset

 *Access this menu by right-clicking a subset name:*

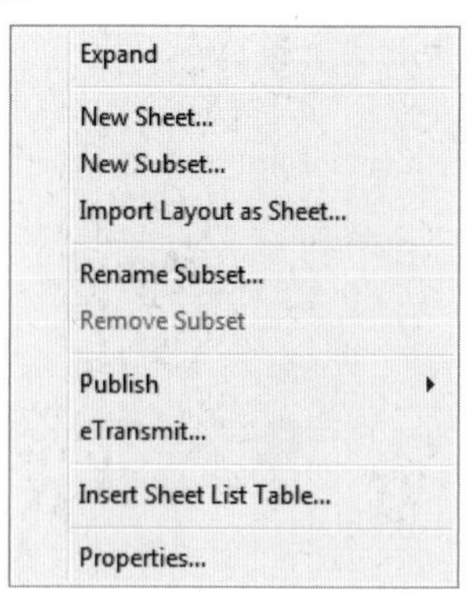

Expand / Collapse shows or hides the names of sheets.

New Sheet creates a new sheet within this subset.

New Subset creates a new subset.

Rename Subset changes the name of the subset.

Remove Subset erases the subset; available only when subset contains no sheets.

Publish plots the sheet set as a *.dwf* file or to a plotter; see the Publish command.

Insert Sheet List Table displays the Sheet List table dialog box, and then inserts a table listing sheet numbers and names into the drawing; prompts you, 'Specify insertion point:'.

Sheet List Table	
Sheet Number	Sheet Title
02	Existing Conditions Plan 1
03	Existing Conditions Plan 2
04	Existing Conditions Plan 3
05	Existing Conditions Plan 4

Properties displays the Subset Properties dialog box; make changes, and then click OK.

Sheet

 *Access this menu by right-clicking a sheet name:*

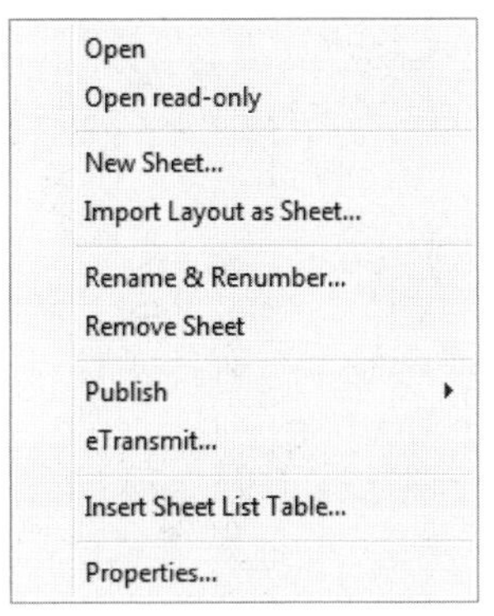

Open opens the drawing in a new window.

Open read-only opens the drawing in read-only mode so that it cannot be edited.

New Sheet creates a new sheet in this subset; displays the New Sheet dialog box.

Import Layout as Sheet imports one or more layouts from any drawing as a sheet; displays Import Layouts as Sheets dialog box.

Rename & Renumber displays the Rename & Renumber Sheet dialog box.

Remove Sheet removes the sheet from the set; does *not* erase the drawing or layout.

Publish plots the sheet set as a *.dwf* file or to a plotter; see the Publish command.

eTransmit collates the sheet set for transmittal by CD or email; see the eTransmit command.

Insert Sheet List Table displays the Sheet List table dialog box, and then inserts a table listing sheet numbers and names into the drawing; prompts you, 'Specify insertion point:'.

Properties displays the Sheet Properties dialog box; make changes, and then click OK.

Sheet Views tab

This tab was formerly called "View List."

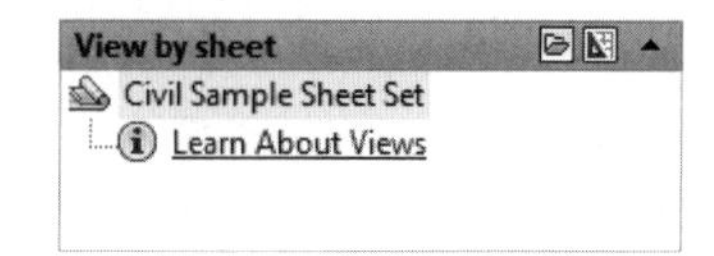

New View Category displays the View Category dialog box.

Category displays sheets by category.

View displays sheets by view name.

Model Views tab

This tab was formerly called "Drawing Resources."

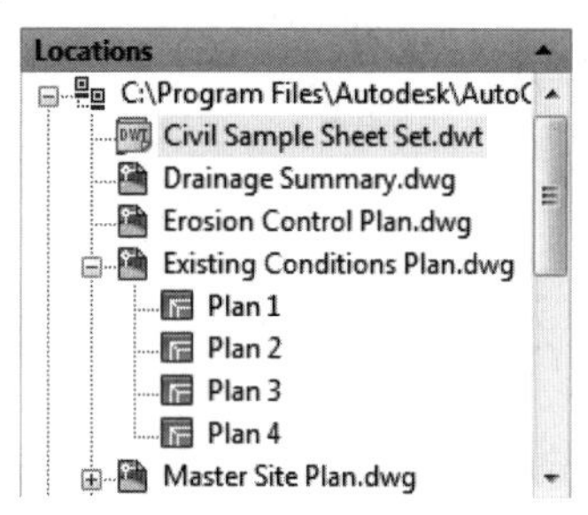

NEW SHEET DIALOG BOX

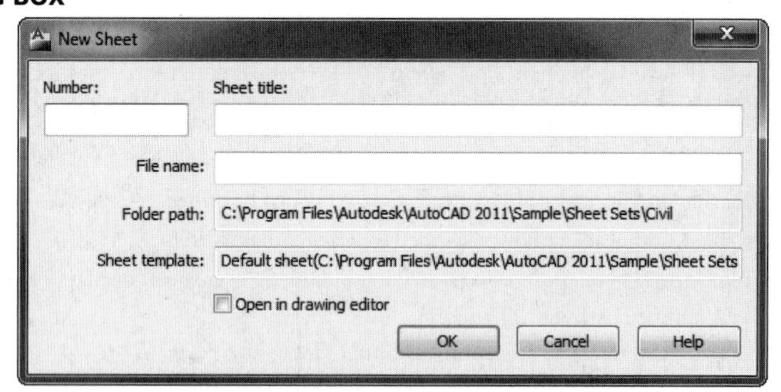

Similar dialog boxes are used for properties of subsets and new sheets.

Number specifies the order number of the sheet; allows sheets to be placed in correct order.

Sheet Title names the sheet, usually with the name of the related layout tab.

File Name names the sheet file, usually consisting of the sheet number and sheet title.

☐**Open in Drawing Editor** opens the new sheet in AutoCAD.

SUBSET PROPERTIES DIALOG BOX

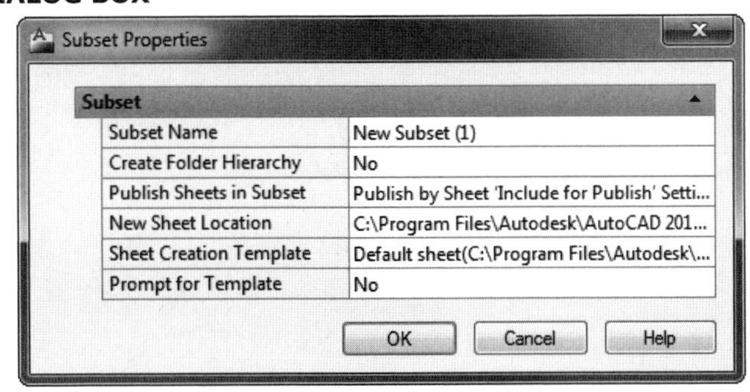

Subset Name specifies the name of the new subset.

Create Folder Hierarchy toggles the creation of folders based on the sheetset name.

Publish Sheets in Subset determines whether sheets in this subset may be published (plotted).

New Sheet Location specifies the path to the folder that will store the sheets.

Sheet Creation Template specifies the DWT sheet template file using this format:

*layoutname [folderpath******]filename***.dwt**

Prompt for Template determines whether the default template is used for new sheets.

IMPORT LAYOUTS AS SHEETS DIALOG BOX

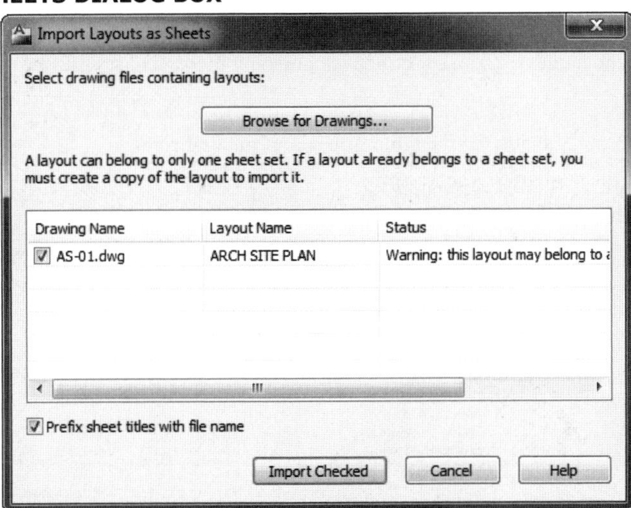

Select a drawing file containing layouts specifies the name of a drawing file.

Browse for Drawings displays the Browse for Folder dialog box; select a folder, and then click Open.

Select layouts to import as sheets lists the names of layouts found in the drawing file.

☑**Prefix sheet titles with file name** adds drawing file names to the beginning of sheet titles.

SHEET LIST TABLE DIALOG BOX

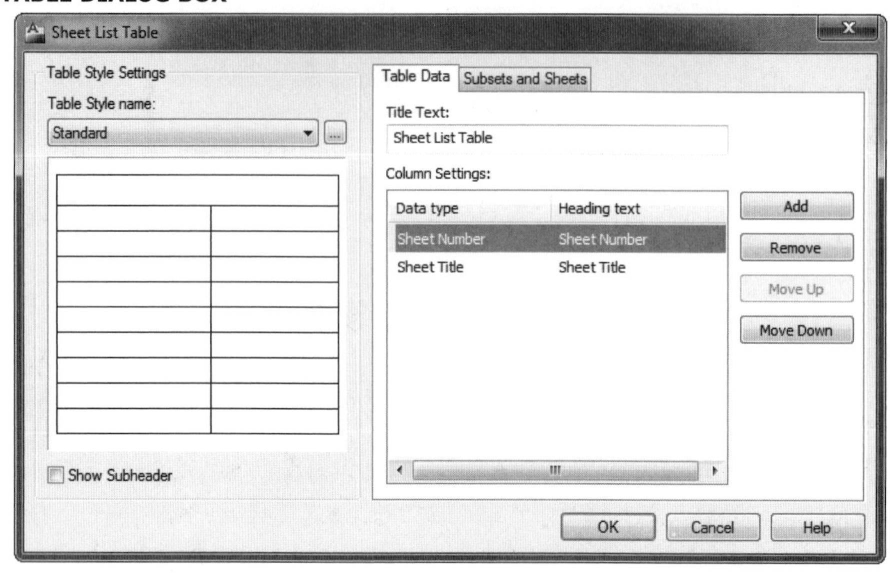

Table Style Name selects a table style defined previously by the TableStyle command.

[...] opens the Table Style dialog box for creating and modifying table styles.

Show Subheader includes the names of subsets, effectively dividing the sheet list table into sections.

Table Data tab

Title Text specifies the name of the title for the table; appears in the first row.

Column Settings specifies the content of each column:

- **Data Type Column** chooses the type of data; click Sheet Number for the droplist of data types.

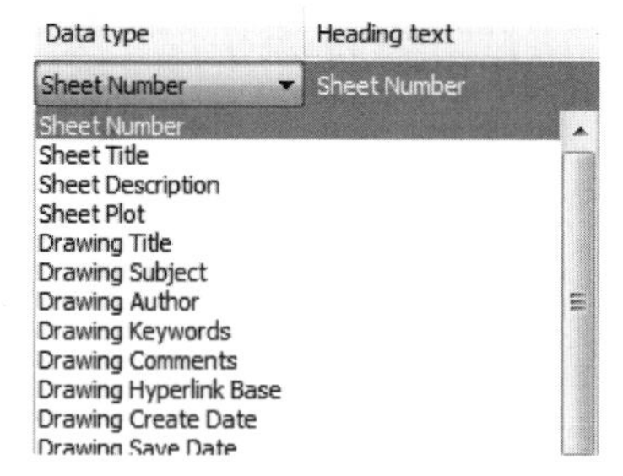

- **Heading Text Column** customizes the heading for each column.

Add / Remove adds and removes sheet number columns.

Move Up / Move Down change the order of the selected item.

Subsets and Sheets tab

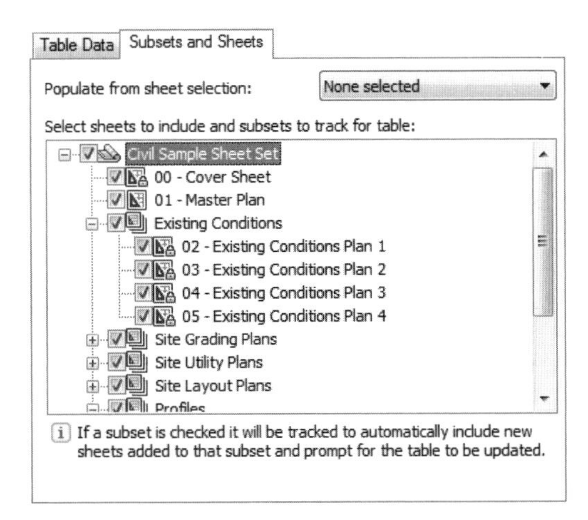

Populate from sheet selection selects a check state from a previously-saved sheet selection.

Select sheets to include and subsets to track for table selects the sheets to include in the table and have updated automatically.

SHEET SET PROPERTIES DIALOG BOX

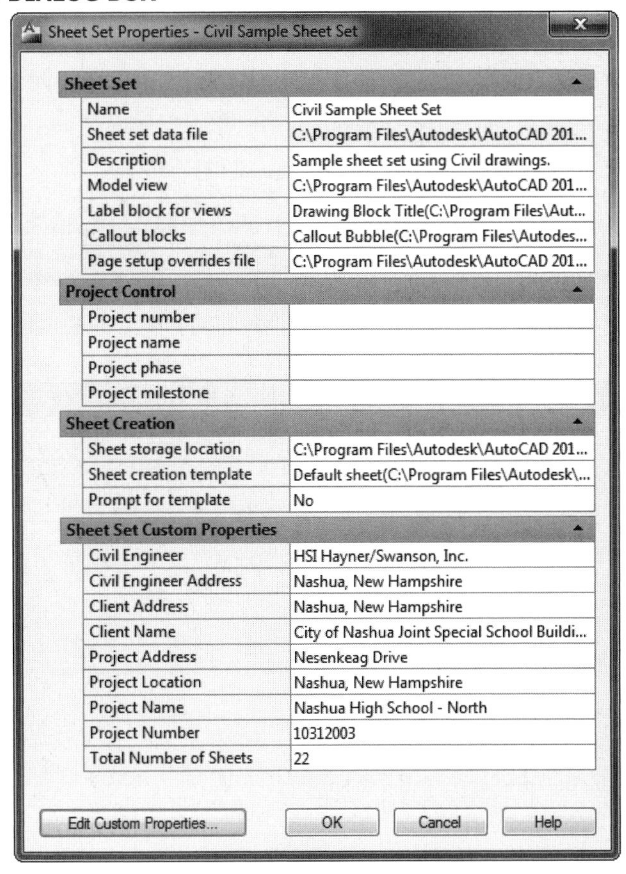

Sheet Set section

Name reports the name of the sheet set.

Sheet Set Data File reports the path to the DST sheet set data file.

Description describes the sheet set.

Model View reports the paths to the folders that contain the DWG drawing files used by the sheet set.

Label Block for Views reports the path to the DWT and DWG files containing label blocks.

Callout Blocks reports the path to the DWT and DWG files containing the callout blocks.

Page Setup Overrides File reports the path to the DWT template file containing page setup overrides.

Other sections

Project Control displays project fields, such as project number, name, and phase.

Sheet Creation displays options for creating sheets, such as folder location and source template.

Sheet Set Custom Properties displays user-defined properties.

Edit Custom Properties displays the Custom Properties dialog box.

CUSTOM PROPERTIES DIALOG BOX

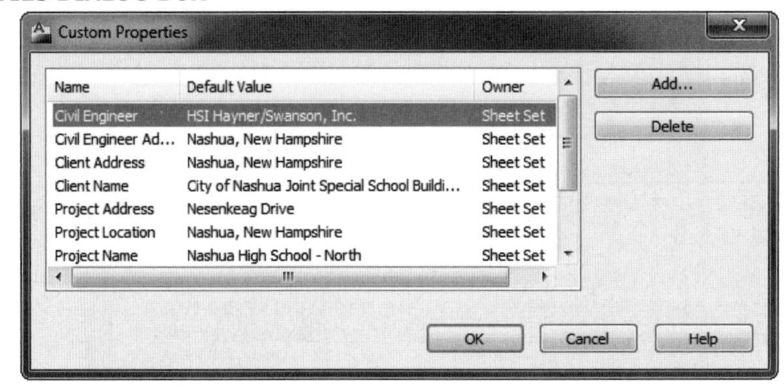

Add / **Remove** adds and removes user-defined properties; see Add Customer Property dialog box.

ADD CUSTOMER PROPERTY DIALOG BOX

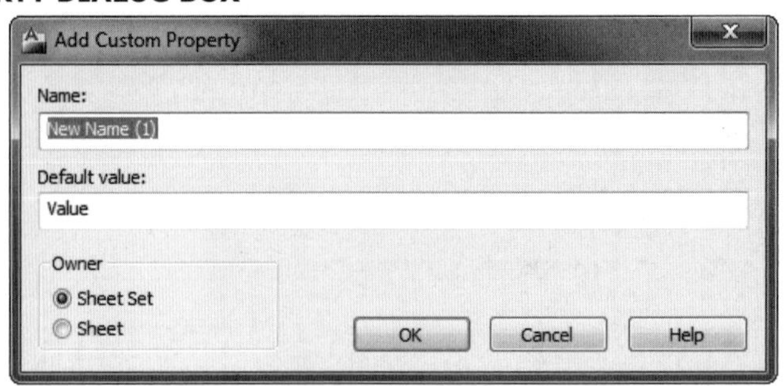

Name names new custom properties.

Default Value specifies the initial value of the custom property.

Owner determines whether the property applies to a sheet set or sheet.

SHEET PROPERTIES DIALOG BOX

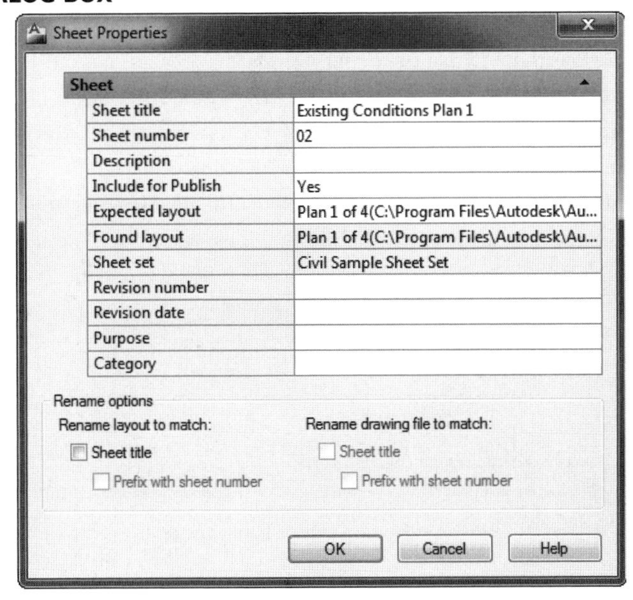

Sheet options

Reports properties and custom properties of the selected sheet.

Rename Options options

See Rename & Renumber Sheet dialog box.

RENAME & RENUMBER SHEET DIALOG BOX

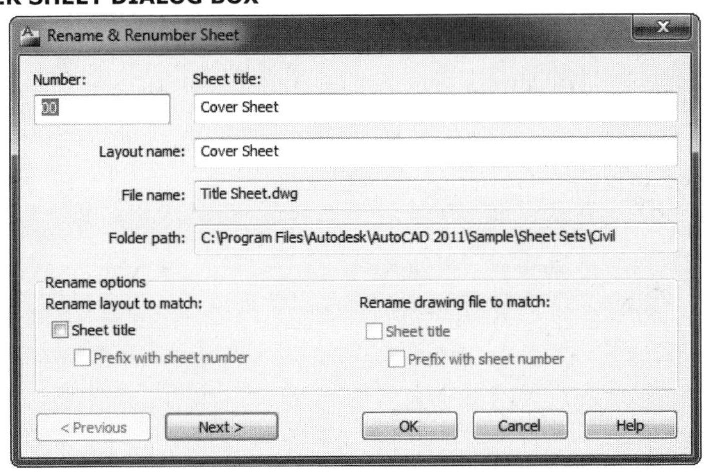

Number numbers the sheet.

Sheet Title names the sheet.

Layout Name names the layout associated with the sheet.

File Name reports the name of the sheet's drawing file.

Folder Path reports the path to the file's folder.

Rename Options options

Rename layout to match:

☐ **Sheet title** matches the layout name to that of the sheet's title.

☐ **Prefix with sheet number** adds the sheet number to the start of the sheet title.

Rename drawing file to match:

☐ **Sheet Title** renames the drawing file to match that of the sheet title.

☐ **Prefix with sheet number** renames the drawing file by adding the sheet number to the start of the file name.

Next / **Previous** loads data for the next (and previous) sheets, allowing you to step through each one.

VIEW CATEGORY DIALOG BOX

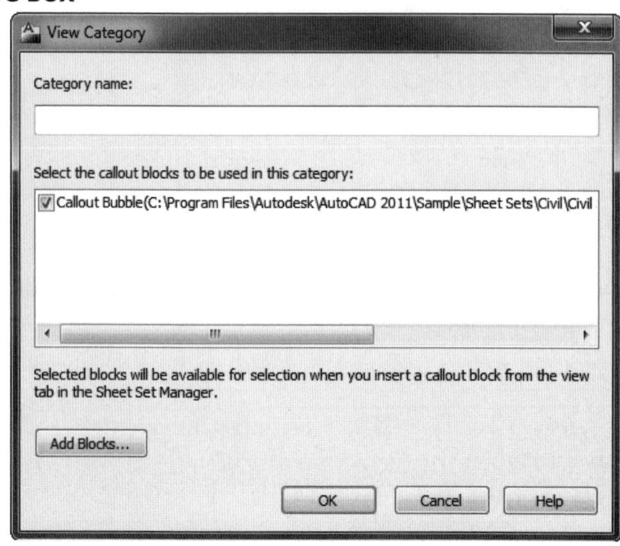

Category Name names the view category.

Select the callout blocks to be used in this category chooses the blocks to use in this view category.

Add Blocks displays the List of Blocks dialog box.

LIST OF BLOCKS DIALOG BOX

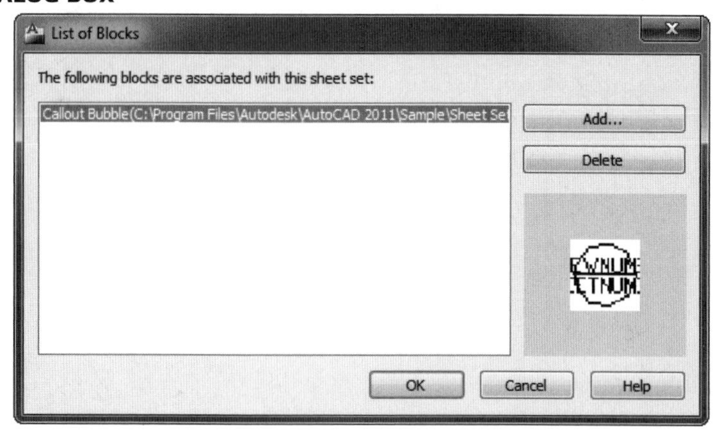

Blocks associated with this sheet set lists the names of available blocks.

Add displays the Select Block dialog box.

Delete removes the selected block from the list.

SELECT BLOCK DIALOG BOX

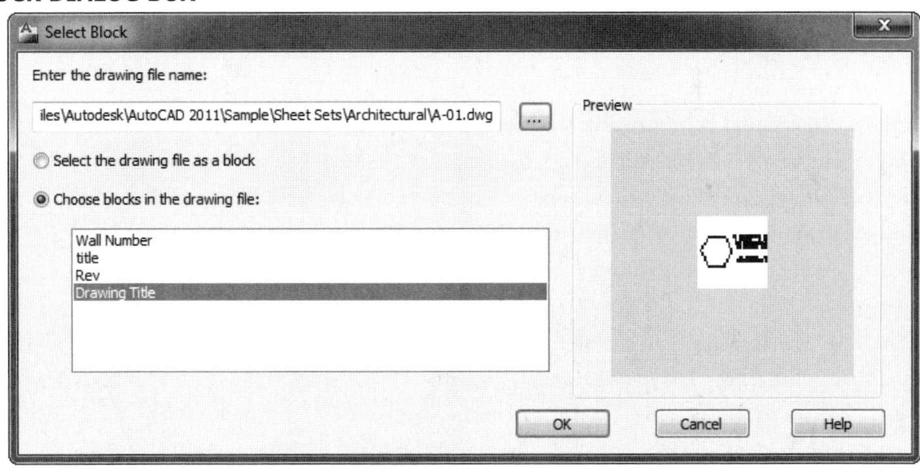

Enter the drawing file name specifies the DWG or DWT file from which to source the block(s).

○ **Select the drawing file as a block** imports the entire drawing as a block definition.

⊙ **Choose blocks in the drawing file** selects a block from the drawing; this option is available only when the drawing contains blocks.

PUBLISH SHEETS DIALOG BOX

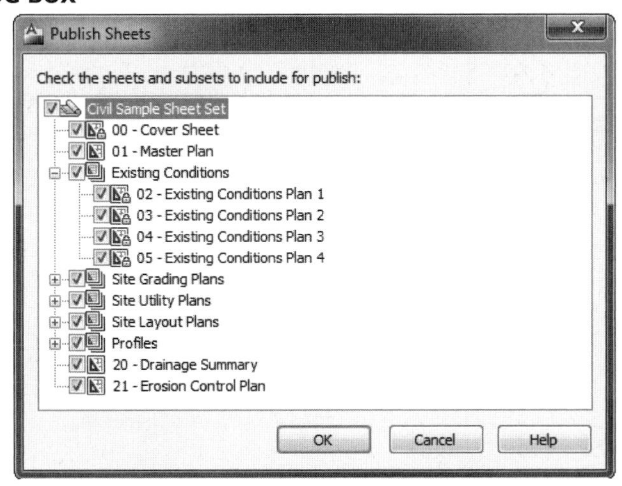

☑ Publishes the sheet with the Publish command.

RELATED COMMANDS

NewSheetset creates new sheet sets.

OpenSheetset opens a specific sheet set.

Field and **MText** embed sheetset data in text.

RELATED SYSTEM VARIABLES

SsFound records the path and file name of current sheet set.

SsLocate toggles whether AutoCAD opens the sheet sets associated with drawings.

SsmAutoOpen toggles whether AutoCAD displays the Sheet Set Manager palette when drawings are opened.

SsmPollTime determines how often AutoCAD updates the Sheet Set Manager.

SsmSheetStatus specifies how data in sheetsets are updated.

SsmState reports whether the Sheet Set Manager palette is open.

STARTUP SWITCHES

/nossm (no Sheet Set Manager) prevents the Sheet Set Manager palette from opening at startup.

/set loads a specified *.dst* sheet set data file at startup.

RELATED FILE

.dst are sheet set data files.

TIPS

- Sheet sets create a hierarchy of drawings, arranged in order; as well, drawings can be organized into subsets (sheets) and categories (views).
- *SSM* is short for "sheet set manager."
- Sheet sets can be created with the NewSheetset command, based either on existing drawings or on another sheet set.
- You can add layouts to sheet sets easily: right-click a layout tab, and then select Import Layout as Sheet from the shortcut menu.

Shell

v. 2.5 Starts a new instance of the Windows command interpreter; runs external commands defined in *acad.pgp*.

Command	*Alias*	*Shortcut Keystrokes*	*Menu*	*Ribbon*
shell	**sh**	...	...	...

Command: shell
OS Command: *(Enter a command, such as **cmd**.)*

Opens window:

```
Administrator: *** AutoCAD Shell Active ***
Microsoft Windows [Version 6.1.7600]
Copyright (c) 2009 Microsoft Corporation.  All rights reserved.

C:\Users\herb\AppData\Local\Autodesk\AutoCAD 2011\R18.1\enu\Template\SheetSets>d
ir
 Volume in drive C is HP
 Volume Serial Number is A2D2-9974

 Directory of C:\Users\herb\AppData\Local\Autodesk\AutoCAD 2011\R18.1\enu\Templa
te\SheetSets

03/18/2010  11:53 AM    <DIR>          .
03/18/2010  11:53 AM    <DIR>          ..
01/20/2010  11:22 AM            80,356 Architectural Imperial.dwt
01/20/2010  11:22 AM            79,431 Architectural Metric.dwt
01/20/2010  11:21 AM            78,147 Civil Imperial.dwt
01/20/2010  11:21 AM            78,532 Civil Metric.dwt
01/20/2010  11:21 AM            82,542 Manufacturing Imperial.dwt
01/20/2010  11:23 AM            78,819 Manufacturing Metric.dwt
               6 File(s)        477,827 bytes
               2 Dir(s)  166,149,369,856 bytes free

C:\Users\herb\AppData\Local\Autodesk\AutoCAD 2011\R18.1\enu\Template\SheetSets>
```

COMMAND LINE OPTIONS

OS Command specifies an operating system command, and then closes the new instance of the Windows command interpreter.

Enter remains in the operating system's command mode for more than one command.

Exit returns to AutoCAD from the OS Command mode.

RELATED COMMAND

Quit exits AutoCAD.

RELATED FILE

acad.pgp is the external command definition file.

TIP

- This command was an early solution to switch tasks in DOS, before Windows. It is rarely used today.
- To view and edit the *acad.pgp* file, use the **Tools | Customize | Edit Custom Files | Program Parameters (acad.pgp)** command. AutoCAD displays the *acad.pgp* file in Notepad.
- Modify the look of the window by clicking its icon (upper left corner) and then choosing Properties.

Aliased Command

ShowMat command was removed from AutoCAD 2007; it now activates the List command.

ShowPalettes

2009 Redisplays all palettes hidden by the HidePalettes command.

Command	*Alias*	*Shortcut Keystroke*	*Menu*	*Ribbon*
showpalettes	...	**Ctrl+Shift+P**	...	...

Command: showpalettes

Displays all hidden palettes:

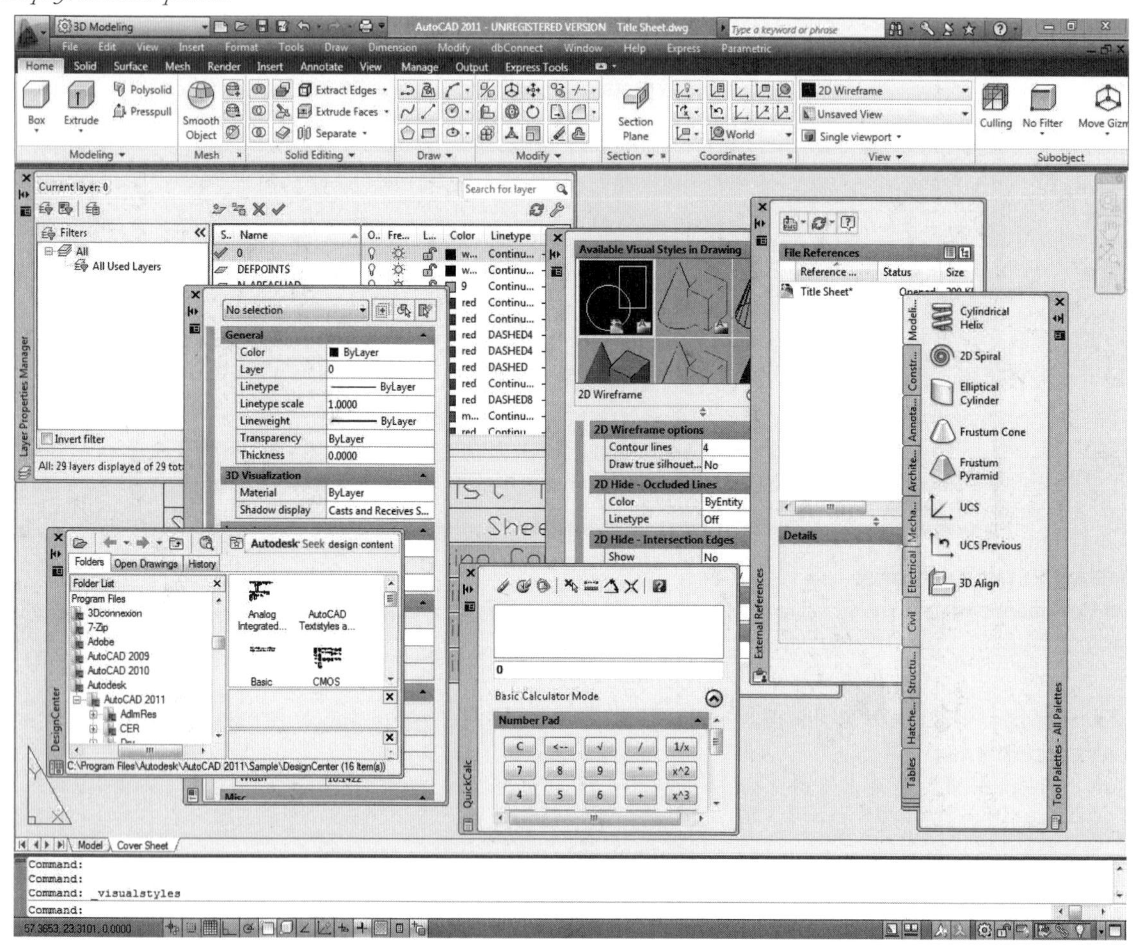

RELATED COMMADS

HidePalettes hides all open palettes.

TIPS

- This command shows all palettes that were previously closed with the HidePalettes command. You can use Ctrl+Shift+P as a keyboard shortcut.
- Palettes are accessed most easily through **Tools | Palettes** on the menu bar.

SigValidate

2004 Displays digital signature information stored in drawings.

Command	*Aliases*	*Shortcut Keystrokes*	*Menu*	*Ribbon*
sigvalidate	...	...	...	...

Command: sigvalidate

Displays dialog box:

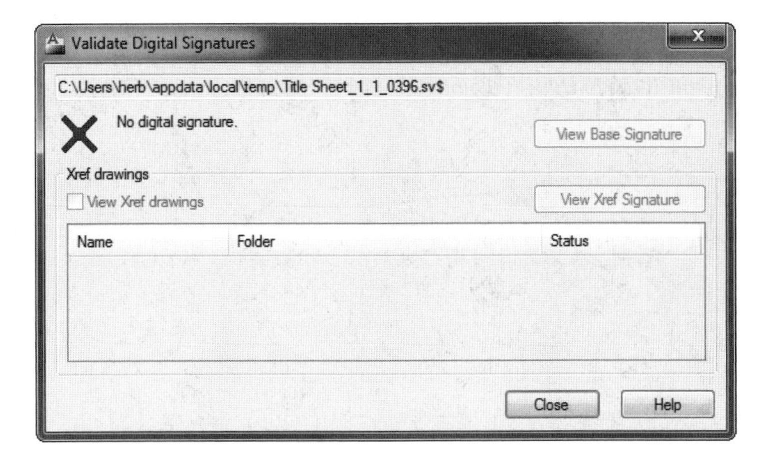

VALIDATE DIGITAL SIGNATURES DIALOG BOX

☐ **Skip Xref Drawings** ignores digital signatures of attached xref drawings.

View Base / Xref Signature displays a dialog box that reports whether the digital signature is valid and if the drawing was modified since last signed.

RELATED COMMAND

SecurityOptions attaches digital signatures to drawings.

TIPS

- Digital signatures validate the authenticity of drawings, and indicate whether the drawing changed since being signed digitally.
- Autodesk notes that digital signatures can become invalid for the following reasons: the drawing was corrupted when the digital signature was attached, the drawing was corrupted in transit, or the digital ID is no longer valid.
- This command requires that the drawing be saved first; if not, AutoCAD reports, "This drawing has been modified. The last saved version will be validated."
- If this command does not work, it was removed from AutoCAD; reinstall with the Custom option, and then select Drawing Encryption.

Sketch

v. 1.4 Allows freehand drawing as a series of lines or polylines.

Command	Aliases	Shortcut Keystrokes	Menu	Ribbon
sketch	...	...	...	...

This command's prompts changed in AutoCAD 2011:

Command: sketch
Type = Lines Increment = 0.1000 Tolerance = 0.5000
Specify sketch or [Type/Increment/toLerance]: *(Enter an option, or drag the cursor.)*
Specify sketch: *(Press Enter to save the sketchings, or Esc to discard them.)*
n **lines recorded.**

COMMAND LINE OPTIONS

Specify sketch draws sketch lines as long as the left moue button is held down; release the button to move to another location in the drawing.

Type determines the object used to record the sketch: line, polyline, or spline *(new to AutoCAD 2011).*

Increment specifies the length of each segment; a segment is drawn each time the cursor moves farther than this distance.

Tolerance (for splines only) specifies how closely the spline follows the sketch *(new to AutoCAD 2011).*

RELATED COMMANDS

Line draws line segments.

PLine draws polyline and polyline arc segments.

Spline draws plines.

RELATED SYSTEM VARIABLES

SketchInc specifies the current recording increment for the Sketch command; default = 0.1 units.

SKPoly controls the type of sketches recorded: (0) lines, (1) polylines, or (3) splines.

SkTolerance specifies how closely the spline matches the sketch; default = 0.5.

TIPS

- The shortcut keystrokes and unique mouse button definitions from earlier releases of AutoCAD are no longer available during this command.
- Menus are unavailable during the Sketch command.

Slice

Rel. 11 Cuts 3D solids and surfaces with planes, creating one or two 3D solids.

Command	*Alias*	*Shortcut Keystrokes*	*Menu*	*Ribbon*
slice	**sl**	...	**Modify**	**Home**
			↳3D Operations	**↳Solid Editing**
			↳Slice	**↳Slice**

Command: slice
Select objects to slice: *(Choose one or more solids, or surfaces.)*
Select objects to slice: *(Press Enter to continue.)*
Specify start point of slicing plane or [planar Object/Surface/Zaxis/View/XY/YZ/ZX/3points] <3points>: *(Enter an option.)*
Specify a point on desired side of the plane or [keep Both sides]: *(Pick a point, or type B.)*

COMMAND LINE OPTIONS

Select objects to slice selects the 3D solids or 3D surfaces to slice.

Slicing Plane options

planar Object aligns the cutting plane with a surface, arc, circle, ellipse or elliptical arc, plines, and 3D polylines.

Zaxis aligns the cutting plane with two normal points.

View aligns the cutting plane with the viewing plane.

XY aligns the cutting plane with the xy-plane of the current UCS.

YZ aligns the cutting plane with the yz-plane of the current UCS.

ZX aligns the cutting plane with the zx-plane of the current UCS.

3 points aligns the cutting plane with three points.

keep Both sides retains both halves of the cut solid model.

Specify a point on the desired side of the plane retains either half of the cut solid model.

TIPS

- This command cannot slice 2D regions, 3D wireframe models, or other 2D shapes.
- It is helpful to use object snaps when specifying points.

Snap

v. 1.0 Sets the drawing "resolution," the origin for the grid and hatches, isometric mode, and the angle of the grid, hatches, and orthographic mode.

Command	*Alias*	*Shortcut Keystrokes*		*Status Bar*	*Ribbon*
'snap	**sn**	**Ctrl+B**	**F9**		**...**

Command: snap

Specify snap spacing or [ON/OFF/Aspect/Style/Type] <0.5>: *(Enter a value, or an option.)*

COMMAND LINE OPTIONS

Snap spacing sets the snap increment.

Aspect sets separate x and y increments.

OFF turns off snap.

ON turns on snap.

Rotate rotates the crosshairs for snap and grid; a hidden option.

Style switches between standard and isometric style.

Type switches between grid or polar snap.

Aspect options

Specify horizontal spacing <0.5>: *(Enter a value.)*

Specify vertical spacing <0.5>: *(Enter a value.)*

Specify horizontal spacing specifies the spacing between snap points along the x-direction.

Specify vertical spacing specifies the spacing between snap points along the y-direction.

Rotate options

Specify base point <0.0,0.0>: *(Enter a value.)*

Specify rotation angle <0>: *(Enter a value.)*

Specify base point specifies the point from which snap increments are determined.

Specify rotation angle rotates the snap about the base point.

Style options

Enter snap grid style [Standard/Isometric] <S>: *(Type S or I.)*

Specify vertical spacing <0.5>: *(Enter a value.)*

Standard specifies rectangular snap.

Isometric specifies isometric snap.

Specify vertical spacing specifies the spacing between snap points along the y-direction.

Type options

Enter snap type [Polar/Grid] <Grid>: *(Type P or G.)*

Polar specifies polar snap.

Grid specifies rectangular snap.

SHORTCUT MENU

On the status bar, click SNAP *to turn snap on and off; right-click for shortcut menu.*

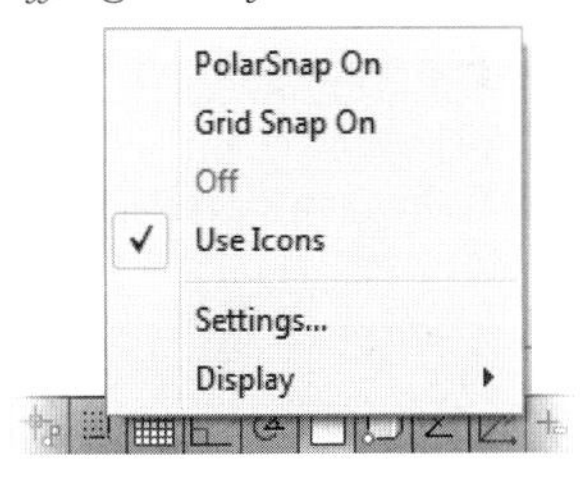

Polar Snap On turns on polar snap.

Grid Snap On turns on snap.

Off turns off snap.

Settings displays the Snap and Grid tab of the Drafting Settings dialog box; see the DSettings command.

RELATED COMMANDS

DSettings sets snap values via a dialog box.

Grid turns on the grid.

Isoplane switches to a different isometric drawing plane.

RELATED SYSTEM VARIABLES

SnapAng specifies the current angle of the snap rotation.

SnapBase specifies the base point of the snap rotation.

SnapIsoPair specifies the current isometric plane setting.

SnapMode determines whether snap is on.

SnapStyl determines the style of snap.

SnapUnit specifies the current snap increment in the x and y directions.

TIPS

- Setting the snap is setting the cursor resolution. For example, setting a snap distance of 0.1 means that when you move the cursor, it jumps in 0.1 increments. You can, however, still type in numerical values of greater resolution, such as 0.1234.
- There is no snap distance in the z-direction.
- The Aspect option is not available when the Style option is set to Isometric; you may, however, rotate the isometric grid.
- As of AutoCAD 2007, the Rotate option is no longer shown, but the option still works.
- The options of the Snap command affect several other commands:

Snap Option	Command	Effect
Style	Ellipse	Adds Isocircle to Ellipse.
SnapBase	Hatch	Relocates the origin of the hatch.
Rotate	Hatch	Rotates the hatching angle.
Rotate	Grid	Rotates the grid display.
Rotate	Ortho	Rotates the ortho angle.

- You can toggle snap mode by clicking the word SNAP on the status bar.
- There are no Crackle or Pop commands to accompany the Snap command, according to the technical editor.

SolDraw

Rel. 13 Creates sections and profiles in viewports created with the SolView command (short for SOLids DRAWing).

Command	Aliases	Shortcut Keystrokes	Menu	Ribbon
soldraw	...	...	**Draw** ↳**Modeling** ↳**Setup** ↳**Drawing**	**Home** ↳**Modeling** ↳**Solid Drawing**

Command: soldraw
Select viewports to draw ..
Select objects: *(Select one or more viewports.)*
Select objects: *(Press Enter to end object selection.)*

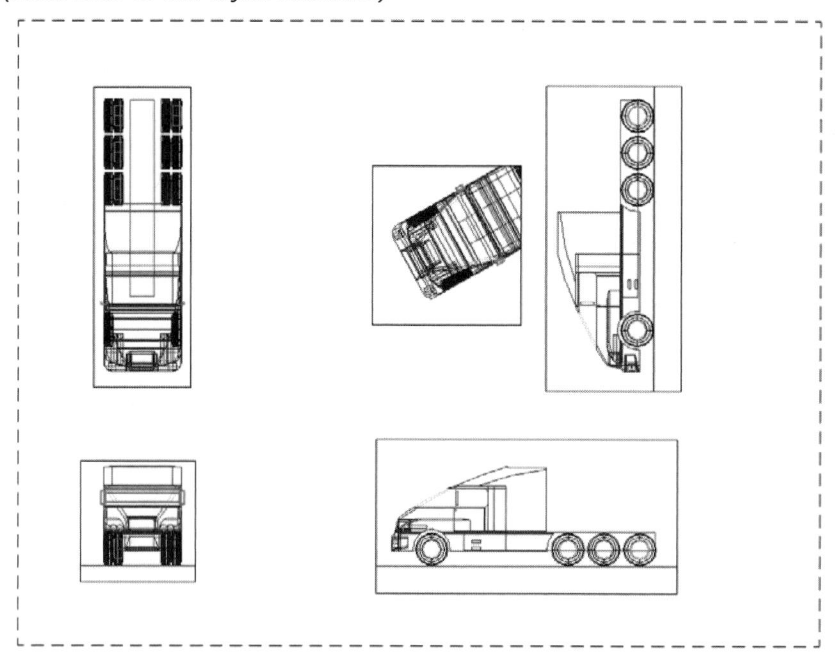

COMMAND LINE OPTION

Select objects selects a viewport; must be a floating viewport in paper space.

RELATED COMMANDS

SolProf creates profile images of 3D solids.

SolView creates floating viewports.

TIPS

- The SolView command must be used before this command.
- This command performs the following actions:

  1. Creates visible and hidden lines representing the silhouette and edges of solids in the viewport.
  2. Projects to a plane perpendicular to the viewing direction.
  3. Generates silhouettes and edges for all 3D solids and portions of solids behind the cutting plane.
  4. Crosshatches sectional views.

- Existing profiles and sections in the selected viewport are erased.
- All layers — except the ones needed to display the profile or section — are frozen in each viewport.
- The following layers are used by SolDraw, SolProf, and SolView:

  viewname-**VIS**
  viewname-**HID**
  viewname-**HAT**
  viewname-**DIM**

- Hatching uses the values set in system variables HpName, HpScale, and HpAng.

Solid

v. 1.0 Draws solid-filled triangles and quadrilaterals; does *not* create 3D solids.

Command	Alias	Shortcut Keystrokes	Menu	Ribbon
solid	**so**	...	...	...

Command: solid
Specify first point: *(Pick a point.)*
Specify second point: *(Pick a point.)*
Specify third point: *(Pick a point.)*
Specify fourth point or <exit>: *(Pick a point, or press Enter to draw triangle.)*

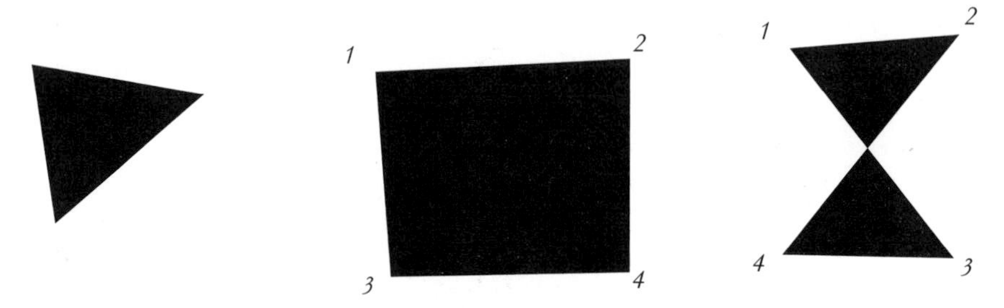

Left: *Three-point solid.*
Right: *Pick order making a difference for four-point solids.*

COMMAND LINE OPTIONS

First point picks the first corner.

Second point picks the second corner.

Third point picks the third corner.

Fourth point picks the fourth corner; alternatively, press Enter to draw a triangle.

Enter draws quadrilaterals; ends the Solid command.

RELATED COMMANDS

Fill turns object fill off and on.

Hatch fills any shape with a solid fill pattern.

Trace draws lines with width.

PLine draws polylines and polyline arcs with width.

RELATED SYSTEM VARIABLE

FillMode determines whether solids are displayed filled or outlined.

SolidEdit

2000 Edits the faces and edges of 3D solids (short for SOLids EDITor).

Command	*Aliases*	*Shortcut Keystrokes*	*Menu*	*Ribbon*
solidedit	...	...	**Modify** ↳**Solid Editing**	**Home** ↳**Solid Editing**

Command: solidedit
Solids editing automatic checking: SOLIDCHECK=1
Enter a solids editing option [Face/Edge/Body/Undo/eXit] <eXit>: *(Enter an option.)*

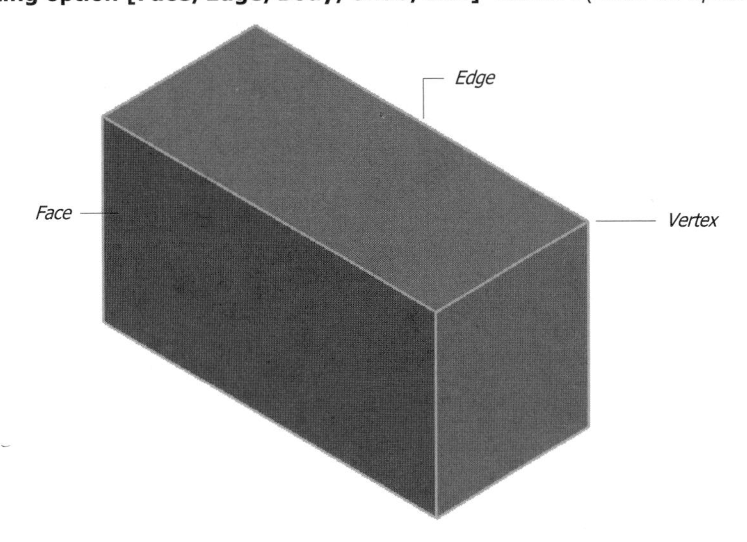

COMMAND LINE OPTIONS

Undo undoes the last editing action, one at a time, up to the start of this command.

eXit exits body mode.

Face options

Enter a face editing option
[Extrude/Move/Rotate/Offset/Taper/Delete/Copy/coLor/Undo/eXit] <eXit>: *(Enter an option.)*

Extrude extrudes one or more faces to the specified distance, or along a path; a positive value extrudes the face in the direction of its *normal* (vector that points from its surface).

Move moves one or more faces the specified distance.

Rotate rotates one or more faces about an axis by a specified angle.

Offset offsets one or more faces by the specified distance, or through a specified point; a positive value increases the size of the solid, while a negative value decreases it.

Taper tapers one or more faces by a specified angle; a positive angle tapers in, while a negative angle tapers out.

Delete removes the selected faces; also removes attached chamfers and fillets.

Copy copies the selected faces as a 3D region or a 3D body object.

Color changes the color of the selected faces.

Edge options

Enter an edge editing option [Copy/coLor/Undo/eXit] <eXit>: *(Enter an option.)*

Copy copies the selected 3D edges as lines, arcs, circles, ellipses, or splines.

coLor changes the color of the selected edges.

Body options

Enter a body editing option [Imprint/seParate solids/Shell/cLean/Check/Undo/eXit] <eXit>: *(Enter an option.)*

Imprint imprints a selection set of arcs, circles, lines, 2D and 3D polylines, ellipses, splines, regions, bodies, and 3D solids on the face of a 3D solid. This function is also available through the Imprint command.

seParate solids separates 3D solids into independent 3D solid objects; the solid objects must have disjointed volumes; this option does *not* separate 3D solids that were joined by a Boolean editing command, such as Intersect, Subtract, and Union.

Shell creates a hollow, thin-walled solid of specified thickness; a positive thickness creates a shell on the inside of the solid, while a negative value creates the shell on the outside of the solid.

cLean removes redundant edges and vertices, imprints, unused geometry, shared edges, and shared vertices.

Check checks whether the object is a 3D solid; duplicates the function of the SolidCheck system variable.

RELATED SYSTEM VARIABLE

SolidCheck toggles solid validation on and off (default = on).

RELATED COMMAND

Imprint imprints the edges of objects onto 3D solid models.

TIPS

- When working with this command, you can select a face, an edge, or an internal point on a face. Alternatively, use the CP (*crossing polygon*), CW (*crossing window*), or F (*fence*) object selection options.
- If you select a closed mesh object, this command asks if you wish to turn it into a solid model.

SolProf

Rel. 13 Creates profile images of 3D solids (short for SOLid PROFile).

Command	*Aliases*	*Shortcut Keystrokes*	*Menu*	*Ribbon*
solprof	...	...	**Draw** ↳**Modeling** ↳**Setup** ↳**Profile**	**Home** ↳**Modeling** ↳**Solid Profile**

Command: solprof
Select objects: *(Select one or more objects.)*
Select objects: *(Press Enter to end object selection.)*
Display hidden profile lines on separate layer? [Yes/No] <Y>: *(Type Y or N.)*
Project profile lines onto a plane? [Yes/No] <Y>: *(Type Y or N.)*
Delete tangential edges? [Yes/No] <Y>: *(Type Y or N.)*
n **solids selected.**

COMMAND LINE OPTIONS

Select objects selects the objects to profile.

Display hidden profile lines on separate layer?

No specifies that all profile lines are visible; a block is created for the profile lines of every selected solid.

Yes generates just two blocks: one for visible lines and one for hidden lines.

Project profile lines onto a plane?

No creates profile lines with 3D objects.

Yes creates profile lines with 2D objects.

Delete tangential edges?

No does not display *tangential edges*, the transition line between two tangent faces.

Yes displays tangential edges.

RELATED COMMANDS

SolDraw creates profiles and sections in viewports.

SolView creates floating viewports.

TIPS

- The SolView command must be used before the SolProf command.
- Solids that share a common volume can produce dangling edges, if you generate profiles with hidden lines. To avoid this, first use the Union command.
- AutoCAD must display a layout, and a model space viewport must be active, before you can use the SolProf command.

SolView

Rel. 13 Creates floating viewports in preparation for the SolDraw and SolProf commands (short for SOLid VIEWs).

Command	Aliases	Shortcut Keystrokes	Menu	Ribbon
solview	...	...	**Draw** ↳**Modeling** ↳**Setup** ↳**View**	**Home** ↳**Modeling** ↳**Solid View**

Command: solview

Enter an option [Ucs/Ortho/Auxiliary/Section]: *(Enter an option.)*

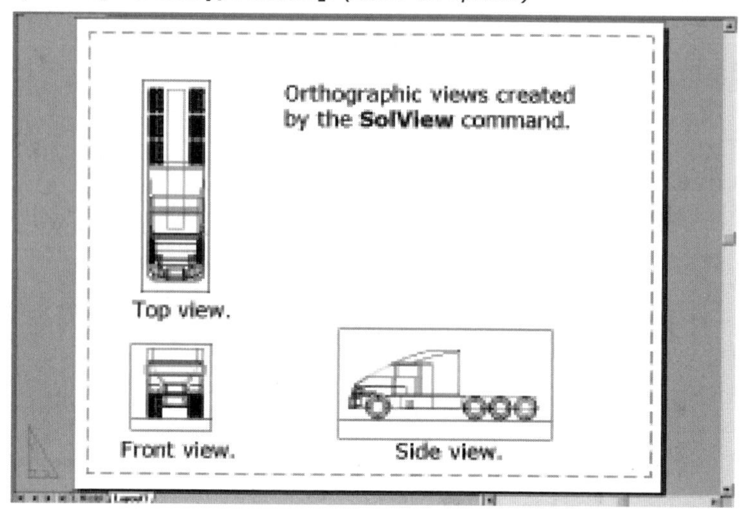

COMMAND LINE OPTIONS

Ucs options

Enter an option [Named/World/?/Current] <Current>: *(Enter an option.)*

Named creates a profile view using the xy-plane of a named UCS.

World creates a profile view using the xy-plane of the WCS.

? lists the names of existing UCSs.

Current creates a profile view using the xy-plane of the current UCS.

Ortho options

Specify side of viewport to project: *(Pick a point.)*

Specify view center: *(Pick a point.)*

Enter view name: *(Enter a name.)*

Pick side of viewport to project selects the edge of one viewport.

View center picks the center of the view.

Clip picks two corners for a clipped view.

View name names the view.

Auxiliary options
Specify first point of inclined plane: *(Pick a point.)*
Specify second point of inclined plane: *(Pick a point.)*
Specify side to view from: *(Pick a point.)*
Specify view center: *(Pick a point.)*
Enter view name: *(Enter a name.)*

Inclined plane's 1st point picks the first point.

Inclined plane's 2nd point picks the second point.

Side to view from determines the view side.

Section options
Specify first point of cutting plane: *(Pick a point.)*
Specify second point of cutting plane: *(Pick a point.)*
Specify side to view from: *(Pick a point.)*
Enter view scale <5.9759>: *(Pick a point.)*
Specify view center <specify viewport>: *(Pick a point.)*
Specify first corner of viewport: *(Pick a point.)*
Specify opposite corner of viewport: *(Pick a point.)*
Enter view name: *(Enter a name.)*

Cutting plane 1st point picks the first point.

Cutting plane 2nd point picks the second point.

Side to view from determines the view side.

Viewscale specifies the scale of the new view.

eXit exits the command.

RELATED COMMANDS

SolDraw creates profiles and sections in viewports.

SolProf creates profile images of 3D solids.

TIPS

- This command creates orthographic, auxiliary, and sectional views.
- This command must be used before the SolDraw command, because it creates the layers required by SolDraw.
- This command is useful for creating layouts for the SolProf command.
- The layers created by this command have the following prefixes:

Layer Name	View
viewname-**VIS**	Visible lines.
viewname-**HID**	Hidden lines.
viewname-**DIM**	Dimensions.
viewname-**HAT**	Hatch patterns for sectional views.

- *Warning!* Autodesk warns that "The information stored on these layers is deleted and updated when you run SolDraw. Do not place permanent drawing information on these layers."
- Lock the viewports immediately after creating them. This prevents subsequent zooming and panning from messing up the viewport scales and alignments.

SpaceTrans

2002 Converts distances between model and space units (short for SPACE TRANSlation).

Commands	*Aliases*	*Shortcut Keystrokes*	*Menu*	*Ribbon*
'spacetrans	...	...	...	...

Command: spacetrans

This command does not operate in Model tab. In model view of a layout tab:

Specify paper space distance <1.000>: *(Enter a value.)*

If the layout contains just one viewport, it is selected automatically; otherwise, you are prompted:

Select a viewport: *(Pick a viewport.)*

Specify model space distance <1.000>: *(Enter a value.)*

COMMAND LINE OPTIONS

Specify paper space distance specifies the paper space length to be converted to its model space equivalent, usually the scale factor.

Select a viewport selects a paper space viewport.

Specify model space distance specifies the model space length to be converted to its paper space equivalent, usually the scale factor.

RELATED COMMANDS

Text places text in the drawing.

SolProf creates profile images of 3D solids.

TIPS

- This command is meant to be used transparently during other commands, and not at the 'Command:' prompt. It does not work in model tab.
- The purpose of this command is to convert lengths between model and paper space.
- Autodesk recommends using this command for converting model space text heights into paper space text heights. Better yet, use annotative text, recommends the technical editor.

Spell

Rel. 13 Checks the spelling of text in the drawing.

Command	Alias	Shortcut Keystrokes	Menu	Ribbon
spell	**sp**	...	**Tools**	**Annotate**
			Annotate	↳**Text**
			↳**Spelling**	↳**Check Spelling**

Command: spell

When all text is recognized, or when spelling check is complete, displays dialog box:

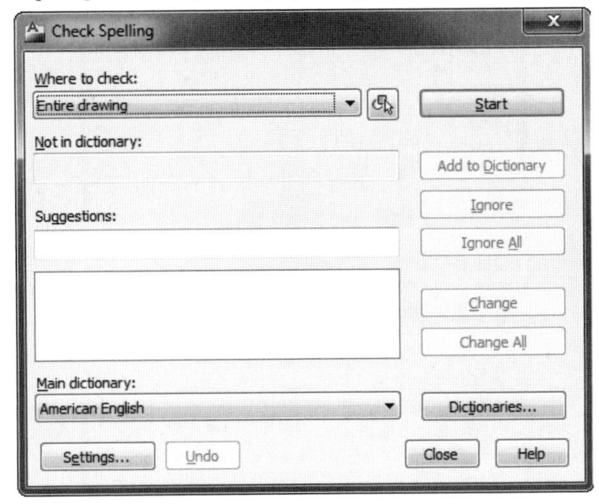

When unrecognized text is found, highlights text and displays dialog box.

CHECK SPELLING DIALOG BOX

Start starts spell checking all words in the current drawing, or those specified by the Settings dialog box.

Add to Dictionary adds the current word to the dictionary file.

Ignore ignores the word, and goes to the next word.

Ignore All ignores all words with this spelling.

Change changes the word to the suggested spelling.

Change All changes all words with this spelling.

Dictionaries selects a different dictionary; displays the Dictionaries dialog box.

Settings displays the Settings dialog box.

DICTIONARIES DIALOG BOX

Main dictionary selects a language for the dictionary.

Custom Dictionaries options

Directory specifies the drive, folder, and filename of the custom dictionary.

Content allows you to add words to the dictionary.

Add adds words (content).

Delete removes custom words from the dictionary.

Import imports custom dictionary CUS files; displays the Add Custom Dictionary File dialog box.

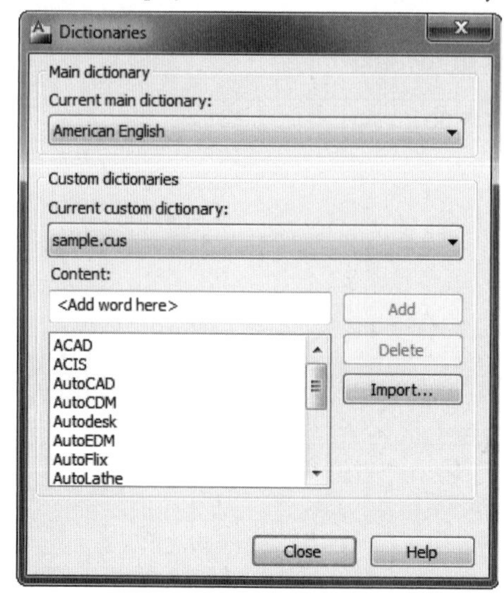

CHECK SPELLING SETTINGS DIALOG BOX

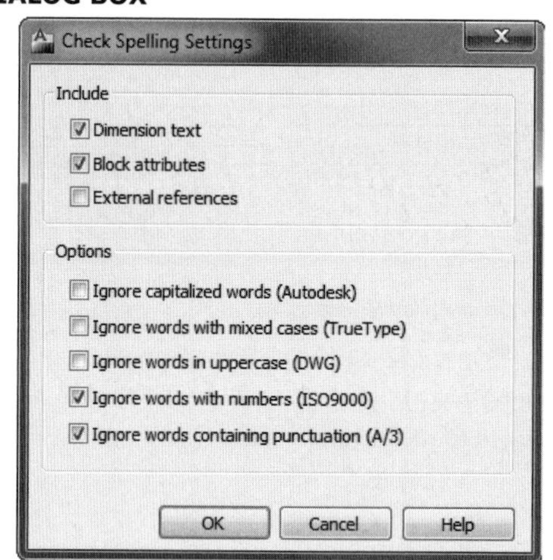

Include options

☑**Dimension Text** spell-checks text in dimensions.

☑**Block Attributes** spell-checks text in attributes.

☐**External References** spell-checks text in externally-referenced drawings.

Options options

☐**Ignore Capitalized Words** ignores words that start with capital letters.

☐**Ignore Words with Mix Cases** ignores words with mixed uppercase and lowercase letters.

☐**Ignore Words in Uppercase** ignores words that are in all uppercase letters.

☑**Ignore Words with Numbers** ignores words that include digits.

☑**Ignore Words Containing Punctuation** ignores words containing punctuation.

RELATED COMMANDS

TextEdit edits text.

MText places paragraph text.

Text places lines of text in the drawing.

AttDef defines attributes.

RELATED SYSTEM VARIABLES

DctCust names the custom spelling dictionary.

DctMain names the main spelling dictionary.

RELATED FILES

enu.dct is the dictionary word file for ENlish Usa; AutoCAD provides 32 dictionary files for 19 languages and variants.

**.cus* are the custom dictionary files.

TIPS

- Spell checkers do not check your spelling, because words that are spelled correctly but used incorrectly (such as *its* and *it's*) are not flagged. Rather, spell checkers look for words that they do not recognize — namely words not in its dictionary file.
- As of AutoCAD 2008, this command zooms into the unrecognized word.
- As of AutoCAD 2009, the Text and MText commands underline misspelled words (real-time spell checking).

- If you accidently add a misspelled word to the custom dictionary, use the Dictionaries button to remove it.
- As of AutoCAD 2010, the U and Undo commands work with Spell, undoing spelling corrections.

Sphere

Rel. 11 Draws 3D spheres as solid models.

Command	Aliases	Shortcut Keystrokes	Menu	Ribbon
sphere	...	...	**Draw** ↳**Modeling** ↳**Sphere**	**Home** ↳**Modeling** ↳**Sphere**

Command: sphere
Specify center point or [3P/2P/Ttr]: *(Pick a point, or enter an option.)*
Specify radius or [Diameter] <1.0>: *(Enter a value, or type D.)*

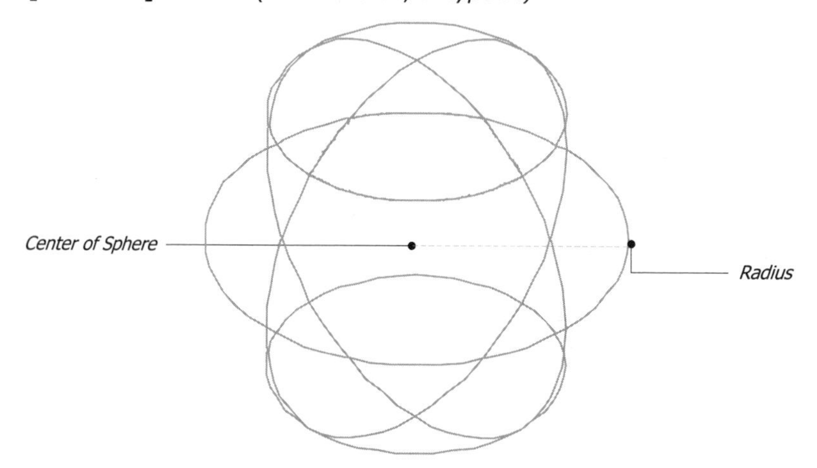

COMMAND LINE OPTIONS

Specify center point locates the center point of the sphere.

3P specifies three points on the outer circumference (equator) of the sphere.

2P specifies two points on the outer diameter of the sphere.

Ttr specifies two tangent points plus the radius of the outer circumference of the sphere.

Radius specifies the radius of the sphere.

Diameter specifies the diameter of the sphere.

RELATED COMMANDS

Cone draws solid cones.

Cylinder draws solid cylinders.

Torus draws solid tori.

Ai_Sphere draws surface meshed spheres.

RELATED SYSTEM VARIABLES

DragVs specifies the visual style during creation of solids.

DispSilh toggles the silhouette display of 3D solids.

IsoLines specifies the number of tessellation lines that define the surface of curved 3D solids; range is from 0 to 2047.

TIPS

- This command places the sphere's central axis parallel to the z axis of the current UCS, with the latitudinal isolines parallel to the xy-plane.
- Notice the effect of the DispSilh system variable, which toggles the silhouette display of 3D solids, after you execute the Hide command:

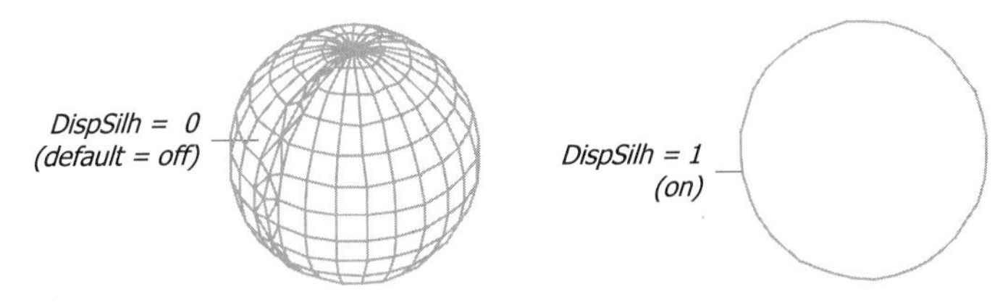

- You must use the Regen command after changing the DispSilh and IsoLines system variables.
- Notice the effect of the IsoLines system variable, which controls the number of tessellation lines used to define the surface of a 3D solid:

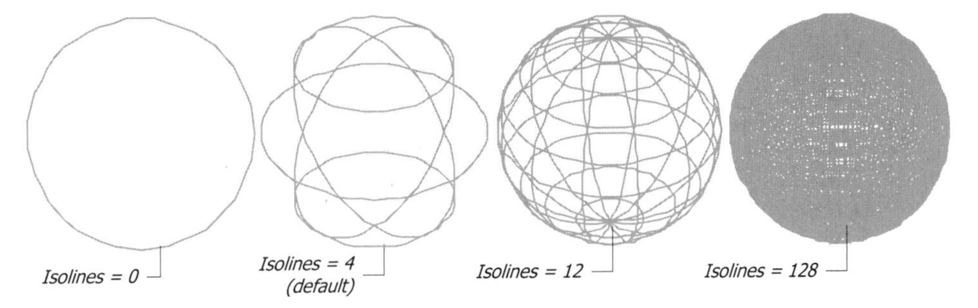

- Spheres can be edited using grips:

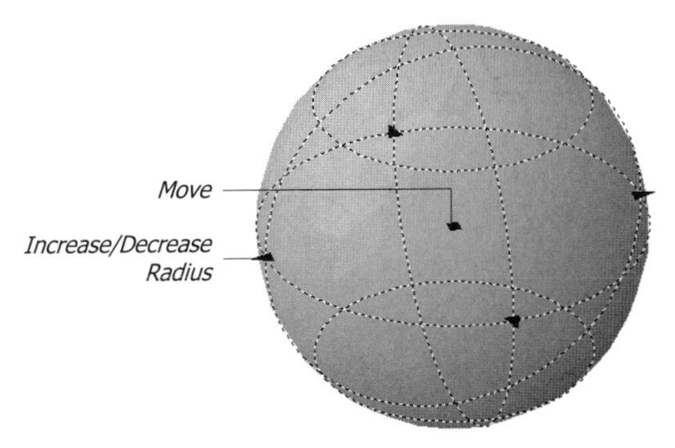

Spline

Rel. 13 Draws NURBS (Non-Uniform Rational Bezier Spline) curves.

Command	*Alias*	*Shortcut Keystrokes*	*Menu*	*Ribbon*
spline	**spl**	...	**Draw** **↳Spline**	**Home** **↳Draw** **↳Spline**

Command: spline
Current settings: Method=Fit   Knots=Chord
Specify first point or [Method/Knots/Object]: *(Pick a point, or enter an option.)*
Enter next point or [start Tangency/toLerance]: *(Pick a point, or enter an option.)*
Enter next point or [end Tangency/toLerance/Undo/Close]:: *(Pick a point, or enter an option, or press Enter to end the command.)*

COMMAND LINE OPTIONS

Specify first point picks the starting point of the spline.
Specify next point picks the next tangent point; press Enter to exit the command.
Close closes the spline at the start point.
Undo undoes to the previous pick point.
start Tangency specifies the tangency of the starting point of the spline.
end Tangency specifies the tangency of the endpoint of the spline.
Object converts 2D and 3D splined polylines into NURBS splines.

Method options *(new to AutoCAD 2011)*

Enter spline creation method [Fit/CV] <Fit>: *(Enter F or C.)*

Fit draws splines with fit points, the method used in earlier releases of AutoCAD.
CV draws splines with control vertices; required when working with NURBS surfaces.

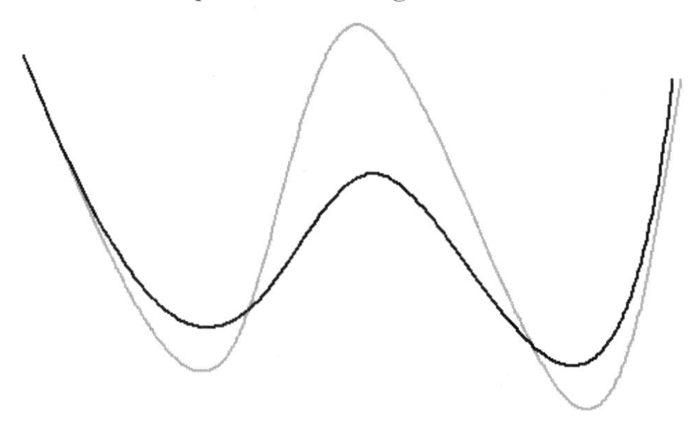

Gray: *Spline drawn with fit points.* ***Black:*** *spline drawn with control vertices.*

Knots options *(new to AutoCAD 2011)*

This option is available for splines drawn with fit points only.

Enter knot parameterization [Chord/Square root/Uniform] <Chord>: *(Enter an option.)*

Chord numbers edit points with real numbers based on the their location.

Square root numbers edit points based on the square root of the chord length between knots.

Uniform numbers edit points with consecutive integers.

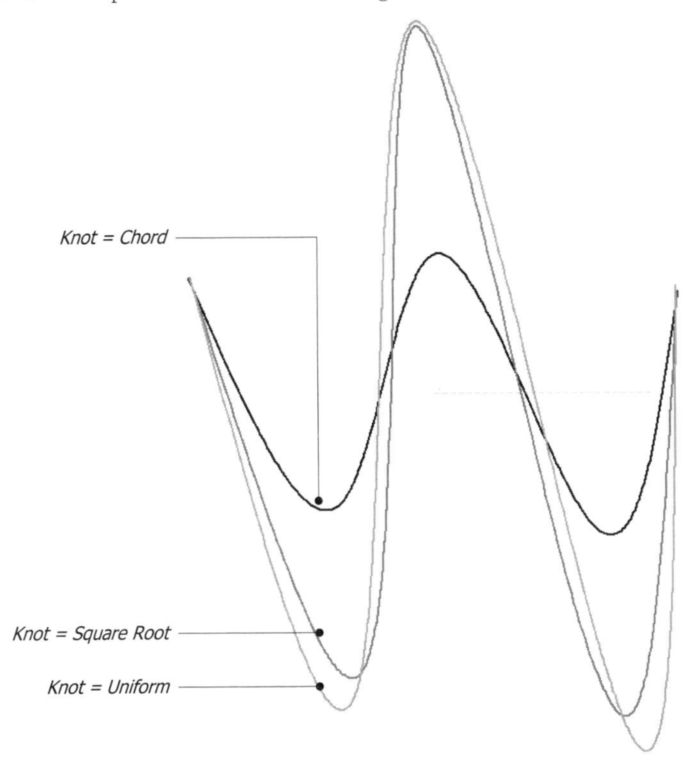

Degree options *(new to AutoCAD 2011)*

This option is available for splines drawn with control vertices only.

Enter degree of spline <3>: *(Enter an integer between 1 and 5.)*

Degree specifies the number of bends between pick points:

Degree	Curve
1	Straight lines
2	Parabolic curves
3	Cubic Bezier curves
4	Quadratic Bezier curves

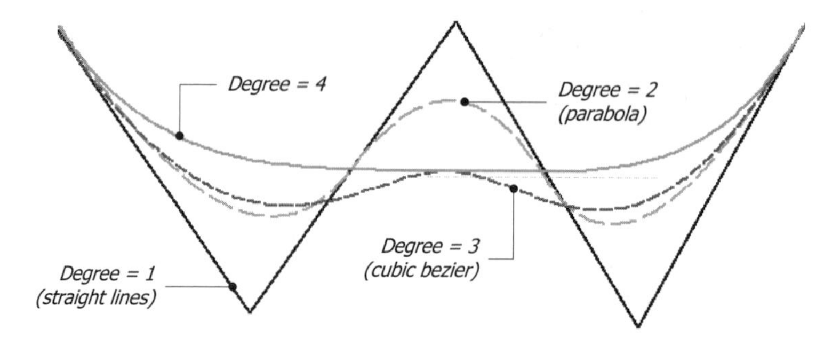

RELATED COMMANDS

PEdit edits splined polylines.

PLine draws splined polylines.

SplinEdit edits NURBS splines.

Sketch sketches with splines.

CvShow and **CvHide** show and hide the control vertices of splines.

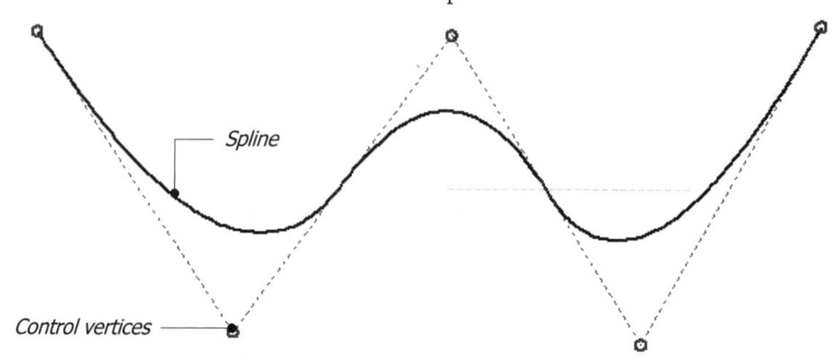

RELATED SYSTEM VARIABLES

DelObj toggles whether the original polyline is deleted with the Object option.

PlineConvertMode determines how polylines are converted to splines.

SplDegree specifies the default degree for new splines created with control vertices

SplKnots specifies the default knot setting when you specify fit points for new splines.

SplMethod toggles the default type of new splines: (0) fit or (1) control vertices *(new to AutoCAD 2011)*.

TIPS

- A closed spline has the same start and end tangent.
- Convert polylines to splines with the PEdit command's Spline option. The PlineConvertMode system variable determines whether the splines are converted to spline segments or arcs.
- The lower the degree, the closer a spline follows its control polyline.
- As of AutoCAD 2011, the SplFrame system variable no longer applies to polylines and splines, but now to helices and smoothed mesh objects; instead, use the CvShow and CvHide commands.

SplinEdit

Rel. 13 Edits NURBS splines.

Command	*Alias*	*Shortcut Keystrokes*	*Menu*	*Ribbon*
splinedit	**spe**	...	**Modify** ↳**Object** ↳**Spline**	**Home** ↳**Modify** ↳**Edit Spline**

Command: splinedit
Select spline: *(Select one spline object.)*
Enter an option [Close/Join/Fit data/Edit vertex/convert to Polyline/Reverse/Undo/eXit] <eXit>:: *(Enter an option.)*

COMMAND LINE OPTIONS

Close / **Open** closes open splines (or opens closed ones).

Join joins the spline to lines and arcs, if endpoints coincide.

convert to Polyline converts the spline to a polyline.

Reverse reverses the spline's direction.

Undo undoes the most-recent edit change.

eXit exits the command.

Fit Data options

Enter a fit data option [Add/Delete/Kink/Move/Purge/Tangents/toLerance/eXit]: *(Enter an option.)*

Add adds fit points.

Delete removes fit points.

Kink adds a kink, a sharp corner *(new to AutoCAD 2011)*.

Move moves fit points.

Purge removes fit point data from the drawing.

Tangents edits the start and end tangents.

toLerance refits the spline with the new tolerance value.

eXit exits suboptions.

Edit Vertex options

Enter a vertex editing option [Add/Delete/Elevate order/Move/Weight/eXit]: *(Enter an option.)*

Add adds a control vertex near your pick point.

Delete reduces the number of control vertices.

Elevate Order increases the number of control vertices; maximum is 26.

Move moves control vertices. Caution: erases all fit points.

Weight changes the weight at control vertices; higher values pull splines closer to control vertices.

RELATED COMMANDS

PEdit edits a splined polyline.

PLine creates polylines, including splined polylines.

Spline draws a NURBS spline.

TIPS

- The spline loses its fit data when you use these SplinEdit options: Refine, Fit Purge, Fit Tolerance followed by Fit Move, and Fit Tolerance followed by Fit Open or Fit Close.
- The maximum order for a spline is 26; once the order has been elevated, it cannot be reduced.
- The larger the weight, the closer the spline is to the control point.
- This command automatically converts spline-fit polylines into splines, even if you do not edit the polyline.
- You may find it easier to edit splines directly using grips and shortcut menus *(new to AutoCAD 2011)*:

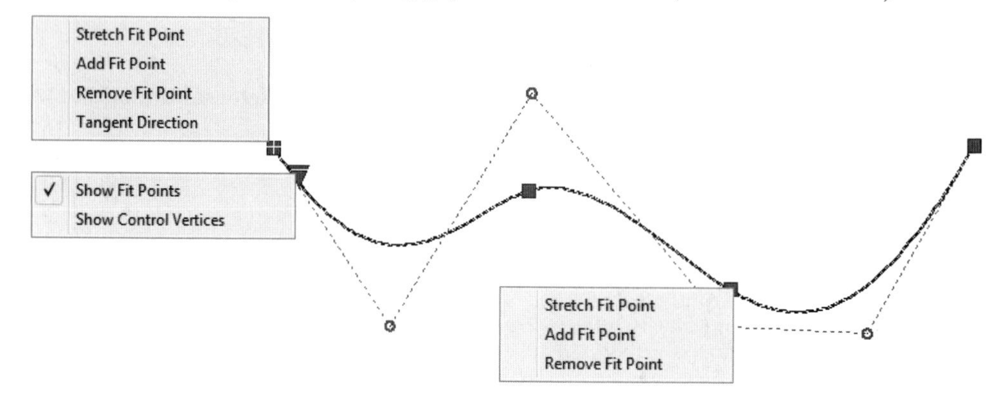

Aliased Command

Spotlight command is an alias for the Light command.

Standards

2002 Loads standards into the current drawing.

Command	*Alias*	*Menu*	*Ribbon*
standards	**sta**	**Tools**	**Manage**
		↳CAD Standards	**↳CAD Standards**
		↳Configure	**↳Configure**

Command: standards

Displays dialog box.

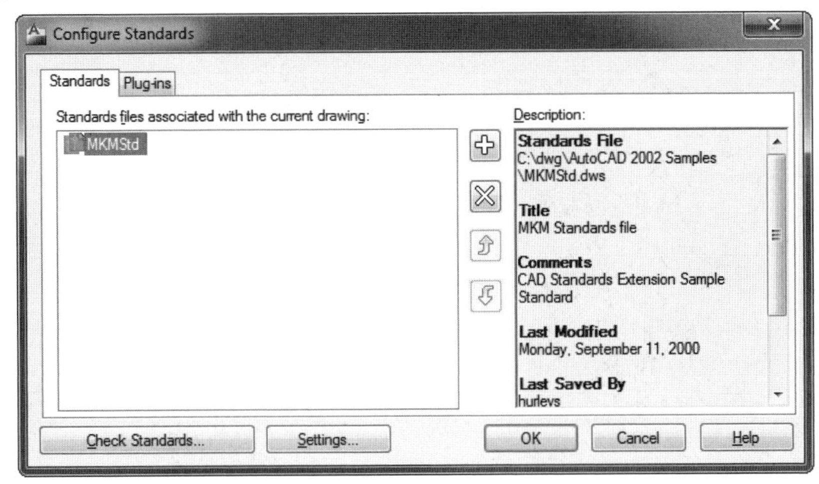

CONFIGURE STANDARDS DIALOG BOX

Add attaches a *.dws* standards file to the drawing displays the Select Standards File dialog box; more than one *.dws* file can be attached (or press the *F3* key);.

Remove Standards removes the selected standards file (*Del*).

Move up moves the selected *.dws* standards file higher in the list; available only when two or more standards files are loaded.

Move down moves the selected standards file lower in the list.

Check Standards displays the Check Standards dialog box; see the CheckStandards command.

Settings displays CAD Standards Settings dialog box; see the CheckStandards command.

Plug-ins tab

Displays information about each item in the CAD standard.

☑ **Dimension Styles** checks if the drawing's dimension styles conform; click to turn off checking.

☑ **Layers** checks if the drawing's layer properties styles conform.

☑ **Linetypes** checks if the drawing's linetype definitions styles conform.

☑ **Text Styles** checks if the drawing's text styles conform.

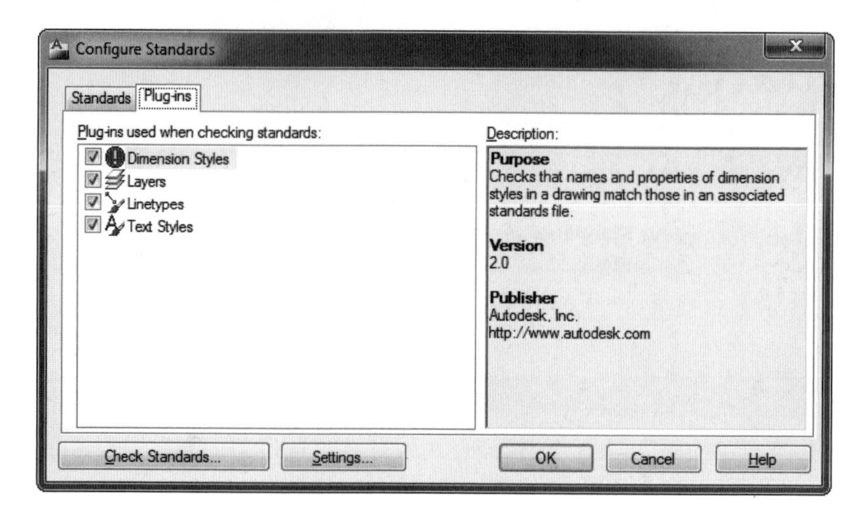

RELATED COMMAND

CheckStandards checks the drawing against standards loaded by this command.

RELATED FILE

*.*dws* are AutoCAD standards files stored in DWG format.

Removed Command

Stats command was removed from AutoCAD 2007.

Status

v. 1.0 Displays information about the current drawing and environment.

Command	Aliases	Shortcut Keystrokes	Menu Bar	Application Menu	Ribbon
'status	...	...	**Tools** ↳**Inquiry** ↳**Status**	**Drawing Utilities** ↳**Status**	...

Command: status

Example output:

```
4913 objects in C:\filename.dwg
Model space limits are X:     0.0000    Y:     0.0000  (Off)
                       X:    12.0000    Y:     9.0000
Model space uses       X:    19.7630    Y:     5.5311
                       X:    90.9009    Y:    73.9771 **Over
Display shows          X:    -3.5363    Y:     5.1582
                       X:   152.0911    Y:    95.6052
Insertion base is      X:     0.0000    Y:     0.0000   Z:     0.0000
Snap resolution is     X:     0.5000    Y:     0.5000
Grid spacing is        X:     0.5000    Y:     0.5000

Current space:         Model space
Current layout:        Model
Current layer:         "0"
Current color:         BYLAYER -- 7 (white)
Current linetype:      BYLAYER -- "Continuous"
Current material:      BYLAYER -- "Global"
Current lineweight:    BYLAYER
Current elevation:     0.0000  thickness:     0.0000
Fill on  Grid off  Ortho off  Qtext off  Snap off  Tablet off
Object snap modes:     Center, Endpoint, Intersection, Extension,
Press ENTER to continue:
Free dwg disk (C:) space: 178763.7 MBytes
Free temp disk (C:) space: 178763.7 MBytes
Free physical memory: 977.8 Mbytes (out of 4094.2M).
Free swap file space: 2817.6 Mbytes (out of 8186.5M).
```

COMMAND LINE OPTIONS

Enter continues the listing.

F2 returns to the graphics screen.

RELATED COMMANDS

DbList lists information about all the objects in the drawing.

List lists information about the selected objects.

Properties lists information about selected objects in a palette.

DEFINITIONS

Model Space limits, Paper Space limits – the x, y coordinates stored in the LimMin and LimMax system variables; 'Off' indicates limits checking is turned off (LimCheck).

Model Space uses, **Paper Space uses** – the x, y coordinates of the lower-left and upper-right extents of objects in the drawing; 'Over' indicates drawing extents exceed the drawing limits.

Display shows – the x, y coordinates of the lower-left and upper-right corners of the current display.

Insertion base is – the x, y, z coordinates, as stored in system variable InsBase.

Snap resolution is – the snap settings, as stored in the SnapUnit system variable.

Grid spacing is – the grid settings, as stored in the GridUnit system variable.

Current space – the indication of whether model space or paper space is current.

Current layout – the name of the current layout.

Current layer, **Current color**, **Current linetype**, **Current lineweight**, **Current plot style**, **Current elevation**, **Material Thickness** – the current values for the layer name, color, linetype name, elevation, and thickness, as stored in system variables CLayer, CeColor, CeLType, CeLweight, CMaterial, CPlotSytle, Elevation, and Thickness.

Fill, **Grid**, **Ortho**, **Qtext**, **Snap**, **Tablet** – the current settings for the fill, grid, ortho, qtext, snap, and tablet modes, as stored in the system variables FillMode, GridMode, OrthoMode, TextMode, SnapMode, and TabMode.

Object Snap modes – the currently-set object modes, as stored in the system variable OsMode.

Free disk (dwg + temp = C) – the amount of free disk space on the drive storing AutoCAD's temporary files, as held by system variable TempPrefix.

Free physical memory – the amount of free RAM.

Free swap file space – the amount of free space in AutoCAD's swap file on disk.

TIPS

- When free disk space becomes close to zero, AutoCAD warns you.
- But when free temp space reaches zero, AutoCAD crashes without warning. With today's terabyte hard drives, this is a rare event, but could happen if AutoCAD were using a smaller-capacity USB thumbdrive. In this case, use the Options command to change the location of AutoCAD's temporary files to a local drive.

StlOut

Rel. 12 Exports 3D solids and bodies in binary or ASCII SLA format (short for STereoLithography OUTput).

Command	*Aliases*	*Shortcut Keystrokes*	*Menu*	*Ribbon*
stlout	...	...	**File ↳Export**	...

Command: stlout
Select a single solid for STL output...
Select objects: *(Select one or more solid objects.)*
Select objects: *(Press Enter to end object selection.)*
Create a binary STL file? [Yes/No] <Y>: *(Type Y or N.)*

Displays the Create STL File dialog box; enter a name, and then click Save.

COMMAND LINE OPTIONS

Select objects selects a single 3D solid object.

Y creates a binary-format *.sla* file.

N creates an ASCII-format *.sla* file.

RELATED COMMANDS

All solid modeling commands.

AcisOut exports 3D solid models to an ASCII SAT-format ACIS file.

AmeConvert converts AME v2.x solid models into ACIS models.

3dPrint converts 3D drawings to STL for delivery to 3D printing service bureaus.

RELATED SYSTEM VARIABLE

FaceTRes determines the resolution of triangulating solid models.

RELATED FILE

***.stl** is the SLA-compatible file with STL extension created by this command.

TIPS

- Solid models not in the positive xyz-octant of the WCS are moved there by this command.
- This command exports 3D solids and watertight meshes.
- Even though this command prompts you twice to 'Select objects', selecting more than one solid causes AutoCAD to union them together into a single solid before exporting.
- The resulting *.stl* file cannot be imported back into AutoCAD, because the file consists of facets.

DEFINITIONS

STL — STereoLithography data file, a faceted representation of the model.

SLA — StereoLithography Apparatus. The technical editor notes this curiosity: SLA is a registered trade mark of 3D Systems, but Stereolithography is not

Stretch

v. 2.5 Lengthens, shortens, or distorts objects.

Command	*Alias*	*Shortcut Keystrokes*	*Menu*	*Ribbon*
stretch	**s**	...	**Modify** ↳**Stretch**	**Home** ↳**Modify** ↳**Stretch**

Command: stretch
Select objects to stretch by crossing-window or crossing-polygon...
Select objects: *(Type C or CP.)*
Specify first corner: *(Pick a point.)*
Specify opposite corner: *(Pick a point.)*
Select objects: *(Press Enter to end object selection.)*
Specify base point or displacement: *(Pick a point.)*
Specify second point of displacement or <use first point as displacement>: *(Pick a point.)*

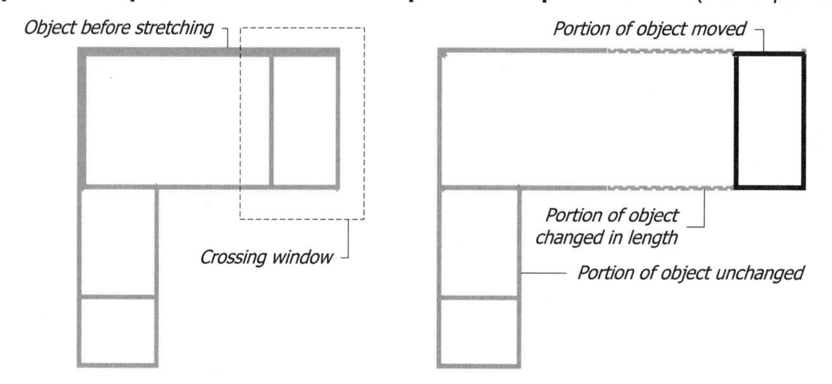

COMMAND LINE OPTIONS

First corner selects object; must be CPolygon or Crossing object selection.

Select objects selects other objects using any selection mode.

Base point indicates the starting point for stretching.

Second point makes the object larger or smaller.

RELATED COMMANDS

Lengthen changes the size of open objects.

Scale increases or decreases the size of objects.

TIPS

- The effect of Stretch on objects is not always obvious; be prepared to use the U command.
- Objects entirely within selection window are moved; objects crossing selection window are stretched.
- This command does not stretch 3D solids, or change the width and tangents of polylines — although it does move them; it does stretch 3D meshes.
- As of AutoCAD 2008, selections made by crossing windows are cumulative, such as when you select multiple crossing windows.

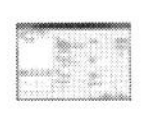

Style

v. 2.0 Creates and modifies text styles, which define the properties of text.

Commands	Aliases	Shortcut Keystrokes	Menu	Ribbon
'style	**st**	...	**Format**	**Home**
	ddstyle		↳**Text Style**	↳**Annotate**
				↳**Text Style**
'-style	...	...	...	...

Command: style

Displays dialog box.

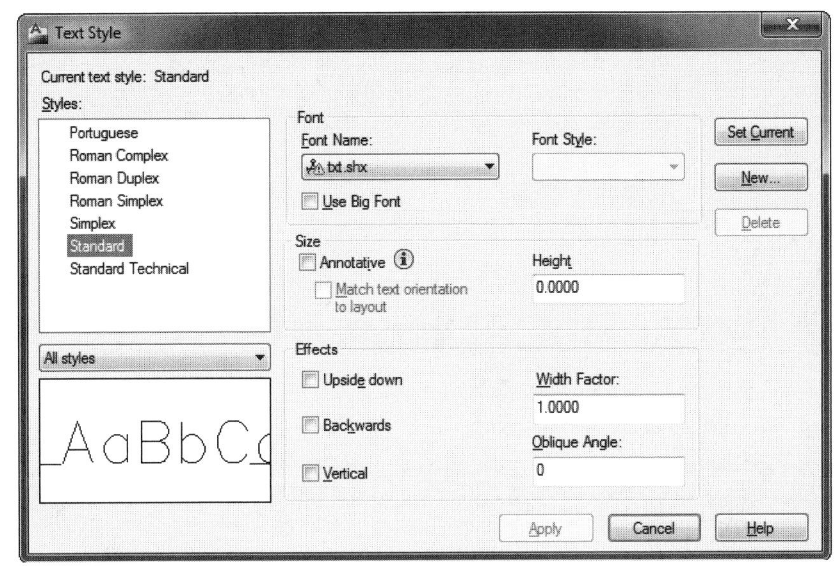

TEXT STYLE DIALOG BOX

To rename a style, right-click its name, and then select Rename from the shortcut menu.

Set Current sets the selected style as current.

New creates new text styles; displays dialog box:

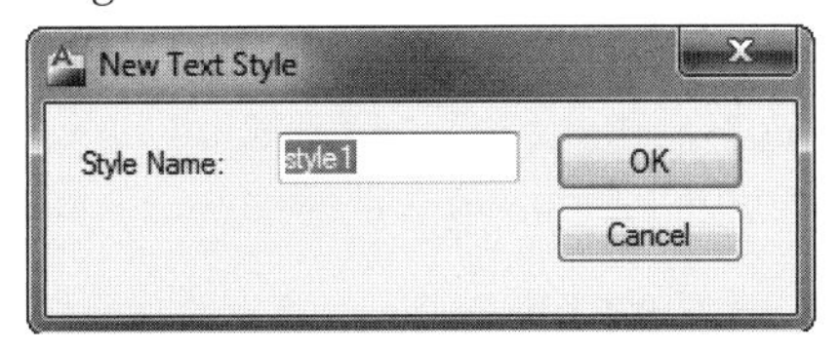

Delete deletes text styles.

Apply applies the changes to the style without exiting the dialog box.

Styles option

- **All styles** lists all styles defined in the drawing.
- **Styles in use** lists those styles used by text in the drawing.

Font options

Font Name lists the names of AutoCAD SHX and TrueType TTF fonts installed in your computer's *windows**fonts* folder. Fonts required by the drawing but missing from your computer show a warning icon *(new to AutoCAD 2011)*.

Font Style selects available TTF font styles: **Bold**, *Italic*, and ***Bold Italic.***

☐ **Use Big Font** specifies the use of Bigfont files, typically for Asian alphabets.

Size options

When Annotative is off:

Height specifies the text height; 0 means users will be prompted for the height during the Text command and other text-related commands.

When Annotative is on:

☐ **Annotative** specifies that the text will be displayed at the correct height, based on the viewport scale factor.

☐ **Match Text Orientation to Layout** specifies that the text orientation in paper space viewports matches the orientation of the layout.

Paper Text Height specifies the height at which text is displayed in paperspace.

Effects options

Not available for all fonts:

☐ **Upside Down** draws text upside down.

☐ **Backwards** draws text backwards.

Width Factor changes the width of characters.

AaBbCcDd AaBbCcDd AaBbCcDd

From left to right: *Width factor angle of 0.5, 1.0 and 1.5 degrees.*

Oblique Angle slants characters forward or backward:

AaBbCcDd AaBbCcDd *AaBbCcDd*

From left to right: *Oblique angle of -15, 0 and 15 degrees.*

☐ **Vertical** draws characters in a vertical column.

-STYLE Command

Command: -style
Enter name of text style or [?] <STANDARD>: *(Enter a name, or type ?.)*
Specify full font name or font filename (TTF or SHX) <txt>: *(Enter a name.)*
Specify height of text or [Annotative] <0.0000>: *(Enter a value or type A.)*
Specify width factor <1.0000>: *(Enter a value.)*
Specify obliquing angle <0>: *(Enter a value.)*
Display text backwards? [Yes/No] <N>: *(Type Y or N.)*
Display text upside-down? [Yes/No] <N>: *(Type Y or N.)*
Vertical? <N> *(Type Y or N.)*
"STANDARD" is now the current text style.

COMMAND LINE OPTIONS

Enter name of text style names the text style; maximum = 31 characters.

? lists the names of styles already defined in the drawing.

Specify full font name or font filename names the font file (SHX or TTF) from which the style is defined.

Specify height of text specifies the height of the text.

Annotative specifies annotative scaling.

Specify width factor specifies the width factor of the text.

Specify obliquing angle specifies the obliquing angle or slant of the text.

Display text backwards toggles mirror writing.

Display text upside-down toggles upside-down writing.

Vertical toggles vertical writing; not available for most fonts.

Annotative options
Create annotative text style [Yes/No] <Yes>: *(Enter Y or N.)*
Match text orientation to layout? [Yes/No] <Yes>: *(Enter Y or N.)*

Annotative specifies that the text will be displayed at the correct height in model space.

Match Text Orientation to Layout makes the text face you, no matter how the viewport display is rotated.

RELATED COMMANDS

Properties changes the style assigned to selected text.

Purge removes unused text style definitions from drawings.

Rename renames text styles.

MText places paragraph text.

Text places a single line of text.

RELATED SYSTEM VARIABLES

TextStyle specifies the current text style.

TextSize specifies the current text height.

TextOutputFileFormat determines the Unicode text format for log files and plotting.

RELATED FILES

*.*shp* is Autodesk's format for vector source fonts.

*.*shx* is Autodesk's format for compiled vector fonts; stored in the *autocad 2011**fonts* folder.

*.*ttf* are TrueType font files; stored in \windows\fonts folder.

TIPS

- The Style command affects the font used with the Text and MText commands, as well as with dimension and table styles.
- A width factor of 0.85 fits in 15% more text without sacrificing legibility, provided the font is legible.
- To use PostScript fonts in drawings, use Compile to convert PostScript *.pfb* files into *.shx* format.
- Text styles used in dimension styles should have a height of 0 to allow DimScale to work correctly.
- While TrueType fonts look better, SHX fonts display faster.
- Avoid using exotic fonts in drawings sent to other computers, which might not have the font. To ensure the fonts are included, use the eTransmit command, which lists all fonts used by drawings.
- The Open command allows you to search for missing fonts, as of AutoCAD 2011.

StylesManager

2000 Displays the Plot Styles window.

Command	Aliases	Shortcut Keystrokes	Menu Bar	Application Menu	Ribbon
stylesmanager	...	...	**File ⇘Plot Style Manager**	**Print ⇘Manage Plot Styles**	...

You can also access this command through the Windows Control Panel: Autodesk Plot Style Manager.

Command: stylesmanager

Displays window, whose content varies:

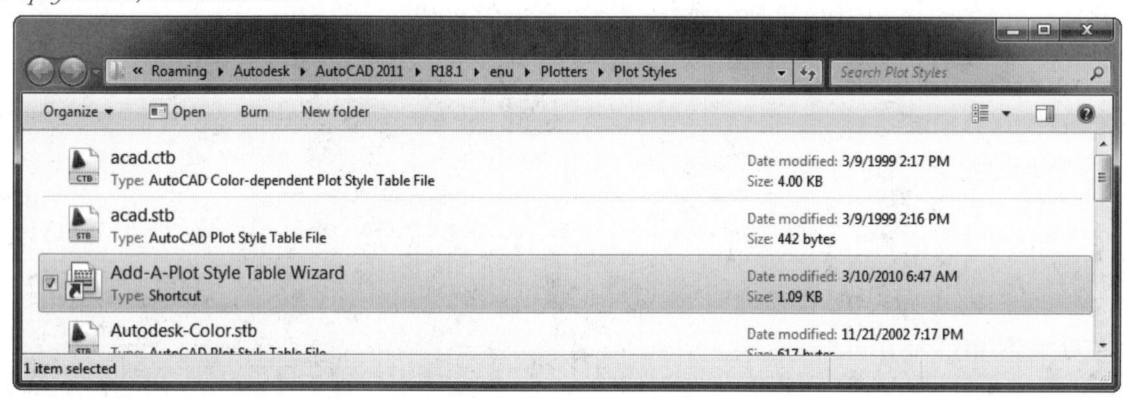

PLOT STYLES WINDOW

.ctb* (color-table based) opens the Plot Style Table Editor dialog box.

***** *.stb* (style-table based) also opens the Plot Style Table Editor dialog box.

Add-A-Plot Style Table Wizard opens Add Plot Style Table wizard; see the R14PenWizard command.

DEFINITIONS

StylesManager — modifies plot styles stored in plot style tables.

Plot styles — when drawings have a plot style table attached to their model and layout tabs, any changes to the plot style change the object using the plot style. Plot styles can be assigned by object or by layer. See the PlotStyle and Layer commands.

Color-table based (.cbt) — assigns colors to objects and layers, as used by older releases of AutoCAD. For example, the color of the object specifies the width of pen. This style of controlling the plot is now called "color dependent."

Style-table based (.stb) — controls every aspect of the plot through "plot styles" in newer releases of AutoCAD. By changing the *.stb* file attached to a layout tab, you immediately change the plot style for all objects and layers in the layout. This, for example, allows a quick change from monochrome to color plotting. A single drawing file can contain multiple plot styles.

PLOT STYLE TABLE EDITOR DIALOG BOX

Access this dialog box by double-clicking a CTB or STB file in the Plot Styles window:

General tab

Description describes the plot style.

☐**Apply global scale factor to non-ISO linetypes** applies the scale factor to all non-ISO linetypes and hatch patterns in the plot.

Scale factor specifies the scale factor.

Save and Close saves the changes, and closes the dialog box.

Cancel cancels the changes, and closes the dialog box.

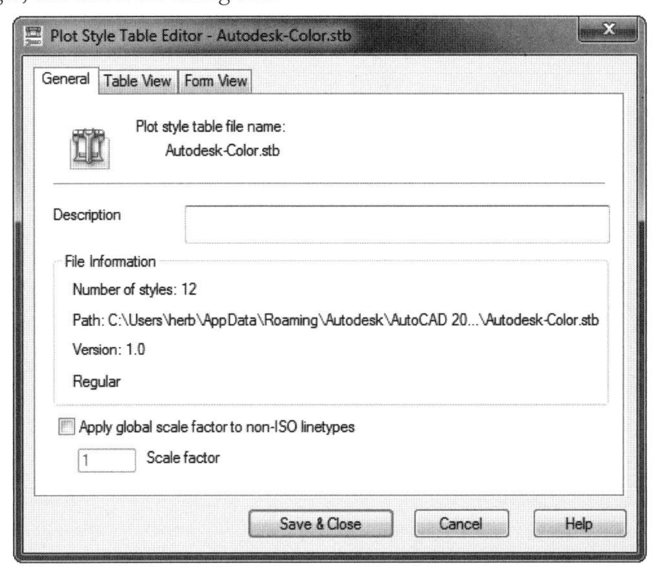

CBT File Table View tab

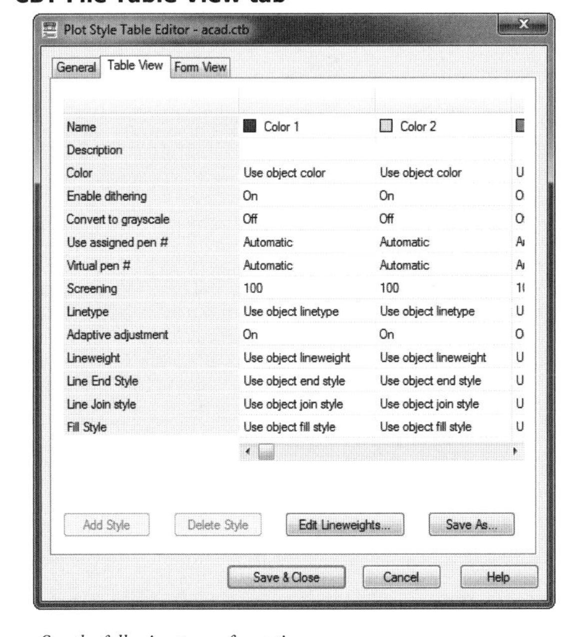

CBT File Form View tab

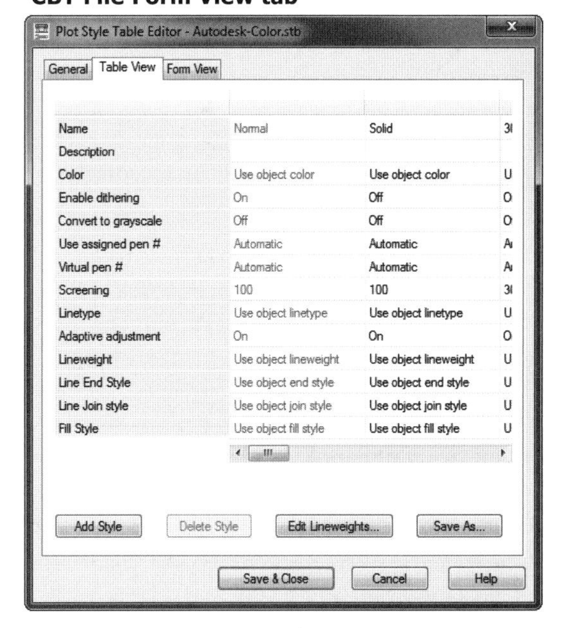

See the following pages for options.

STB File Table View tab

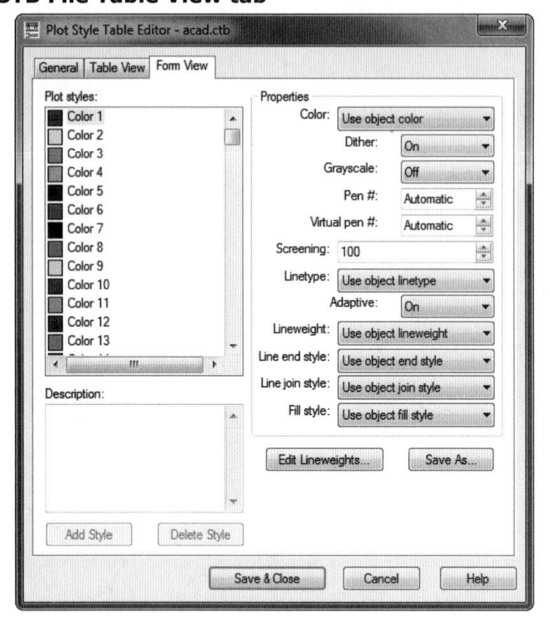

STB File Form View tab

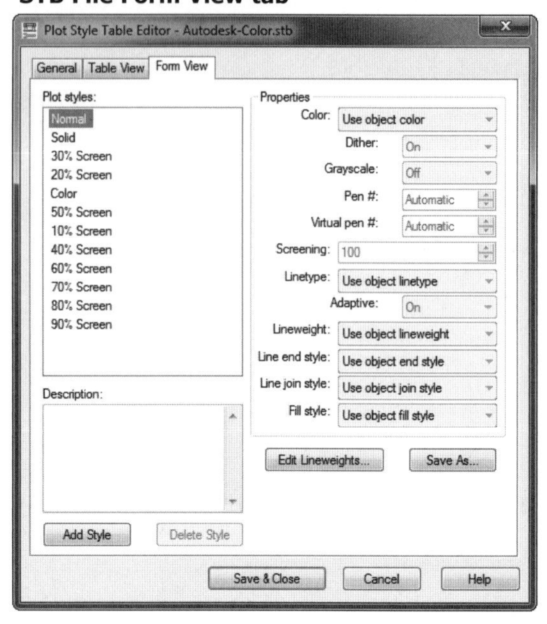

See the following pages for options.

Table View and Form View options are identical

Object color specifies the color of the object.

Description describes the plot style.

Color specifies the plotted color for the objects.

Enable dithering toggles dithering, if the plotter supports dithering, to generate more colors than the plotter is capable of; this setting should be turned off for plotting vectors, and turned on for plotting renderings.

Convert to grayscale converts colors to shades of gray, if the plotter supports gray scale.

Use assigned pen # specifies the pen number of pen plotters; range is 1 to 32.

Virtual pen number specifies the virtual pen number (default = Automatic); range is 1 to 255. A value of 0 (or Automatic) tells AutoCAD to assign virtual pens from ACI (AutoCAD Color Index); this setting is meant for non-pen plotters that can make use of virtual pens.

Screening specifies the intensity of plotted objects; range is 0 (plotted "white") to 100 (full density).

Linetype specifies the linetype with which to plot the objects.

Adaptive adjustment adjusts linetype scale to prevent the linetype from ending in the middle of its pattern; keep off when the plotted linetype scale is crucial.

Lineweight specifies how wide lines are plotted, in millimeters.

Line end style specifies how the ends of lines are plotted.

Line join style specifies how the intersections of lines are plotted.

Fill style specifies how objects are filled.

Add Style adds a plot style.

Delete Style removes a plot style.

Edit Lineweights displays the Edit Lineweights dialog box.

Save As displays the Save As dialog box.

RELATED COMMANDS

ConvertCTB converts legacy CTBs to STBs.

ConvertPStyles switches the current drawing between CTBs and STBs.

TIPS

- To configure a printer or plotter for *virtual pens*, follow these steps:
  1. In the Device and Document Settings tab, open the PC3 Editor dialog box.
  2. In the Vector Graphics section, select 255 Virtual Pens from Color Depth.
- **CTB** is short for "color-dependent based" style table, which is compatible with earlier versions of AutoCAD.
- **STB** is short for "style-table based."
- You can attach a different *.stb* file to each layout and to the model tab in a drawing.
- You can create new *.stb* and *.ctb* files by copying and pasting them; double-click a file to open it for editing.

Subtract

Rel. 11 Removes the volume of one set of 3D models, surfaces, meshes, or 2D regions from another.

Command	*Alias*	*Shortcut Keystrokes*	*Menu*	*Ribbon*
subtract	**su**	**...**	**Modify** **↳Solid Editing** **↳Subtract**	**Home** **↳Solid Editing** **↳Subtract**

Command: subtract
Select objects: *(Select one or more objects.)*
Select objects: *(Press Enter to end object selection.)*
1 solid selected.

Objects to subtract from them...
Select objects: *(Select one or more objects.)*
Select objects: *(Press Enter to end object selection.)*
1 solid selected.

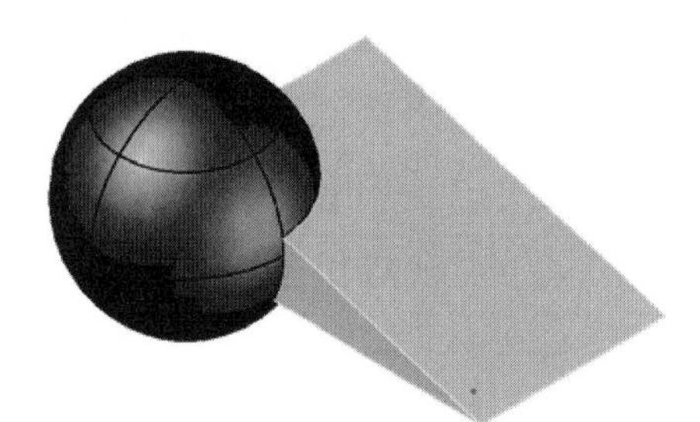

Sphere subtracted from wedge.

COMMAND LINE OPTION

Select objects selects the objects to be subtracted.

RELATED COMMANDS

Interfere finds the volume common to two or more 3D objects.

Intersect removes all but the intersection of two 3D volumes.

Union joins two 3D objects.

RELATED SYSTEM VARIABLE

ShowHist toggles the display of solids' history

TIPS

- AutoCAD subtracts the objects you select *second* from the objects you select *first*.
- You can use this command on 3D solids, meshes, surfaces, and 2D regions. When you select "watertight" meshes, AutoCAD asks if you want the meshes converted to solids or surfaces.
- To subtract one region from another, both must lie in the same plane.
- When an object is fully inside another, AutoCAD performs the subtraction, but reports, "Null solid created — deleted."

SunProperties / SunPropertiesClose

2007 Opens and closes the Sun Properties palette for controlling the sun light.

Commands	Aliases	Shortcut Keystrokes	Menu	Ribbon
sunproperties	...	...	**View** ↳**Render** ↳**Light** ↳**Sun Properties**	**Render** ↳**Sun and Location**
sunpropertiesclose	...	...	...	...

Command: sunproperties

Displays palette.

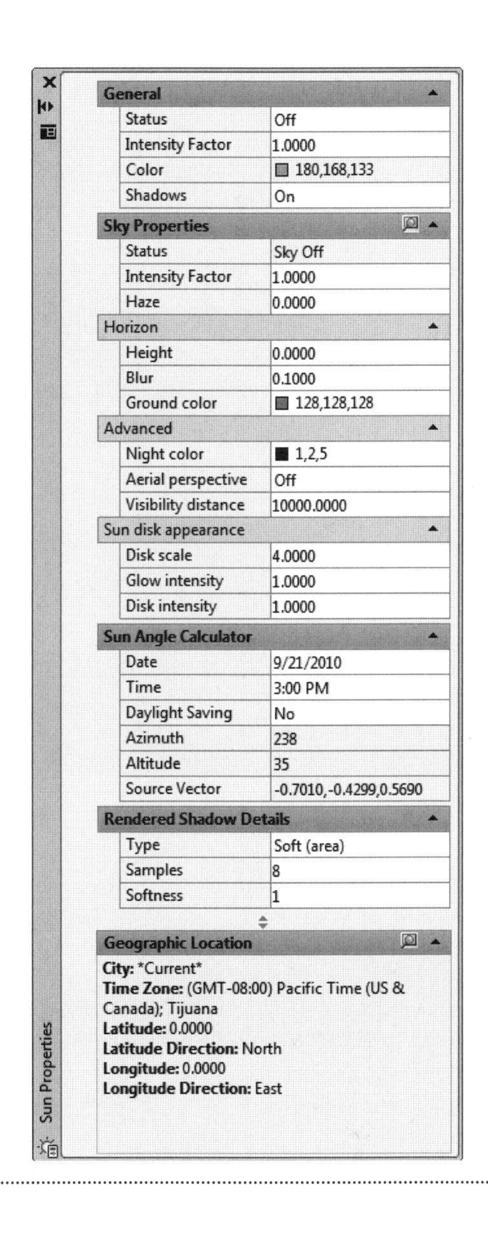

Command: sunpropertiesclose

Closes the palette.

SUN PROPERTIES PALETTE

General options

Status toggles the sun light; inoperative when lighting is not enabled in the drawing.

Intensity Factor specifies the sun's brightness; ranges from 0 (dark) to a very large number, like 9.999E+99 (bright).

Color specifies the color of the light.

Shadows toggles shadows generated by the sun.

Sky Properties options

Status specifies the type of illumination during renderings: none, sky background, or illumination.

Intensity Factor varies the skylight effect.

Haze specifies the magnitude of scattering effects in the atmosphere.

Horizon options

Height specifies the absolute position of the ground plane relative to world zero.

Blur specifies the amount of blur between ground plane and sky.

Ground Color specifies the color of the ground plane.

Rendering of drawing with sun, horizon, haze, and shadows.

Advanced options

Night Color specifies the color of the night sky.

Aerial Perspective toggles aerial perspective.

Visibility Distance specifies the distance at which haze occlusion is 10%.

Sun Disk Appearance options

Disk Scale specifies the scale of the sun disk; leave it at 1 for the correct size.

Glow Intensity specifies the intensity of the sun's glow.

Disk Intensity specifies the intensity of the sun's disk.

Sun Angle Calculator options

Date specifies the day and month.

Time sets the time of day.

Daylight Saving toggles daylight savings time.

Azimuth reports the angle of the sun clockwise from due north; read-only.

Altitude reports the angle of the sun measured vertically from the horizon (90 = overhead); read-only.

Source Vector reports the direction of the sun; read-only.

Rendered Shadow Details options

Type selects the type of shadow, sharp or soft.

Map Size specifies the rendering area of soft shadows; range is 64 to 4096.

Softness specifies the appearance of shadow edges.

Geographic Location option

Edit Geographic Location displays the Geographic Location dialog box; see the GeographicLocation command.

RELATED COMMANDS

GeographicLocation displays the Geographic Location dialog box.

Render renders drawings, showing the effect of the sun light.

Light places lights in drawings.

RELATED SYSTEM VARIABLES

SunPropertiesState reports whether the Sun Properties palette is open.

SunStatus turns the sun on and off.

TIPS

- Since lights can be difficult to place, the easy way to add lighting to 3D models is to turn on the sun light.
- Use the GeographicLocation command to specify the precise location of objects on the surface of the earth.
- The View command's Background option simulates the sun and sky through an option added to AutoCAD 2008.
- To set the date and time, use the Render | Sun Location | Time and Date sliders in the 3D Modeling ribbon.

SurfBlend

2011 Adds continuous blend surfaces between two existing surfaces.

Command	Alias	Shortcut Keystrokes	Menu	Ribbon
surfblend	...	...	**Draw**	**Surface**
			↳**Surfaces**	↳**Create**
			↳**Blend**	↳**Blend**

Command: surfblend
Continuity = G1 - tangent, bulge magnitude = 0.5
Select first surface edges to blend: *(Choose one or more edges of one surface.)*
Select first surface edges to blend: *(Press Enter.)*

Select second surface edges to blend: *(Choose the edges of a second surface.)*
Press Enter to accept the blend surface or [CONtinuity/Bulge magnitude]: *(Press Enter, or enter an option.)*

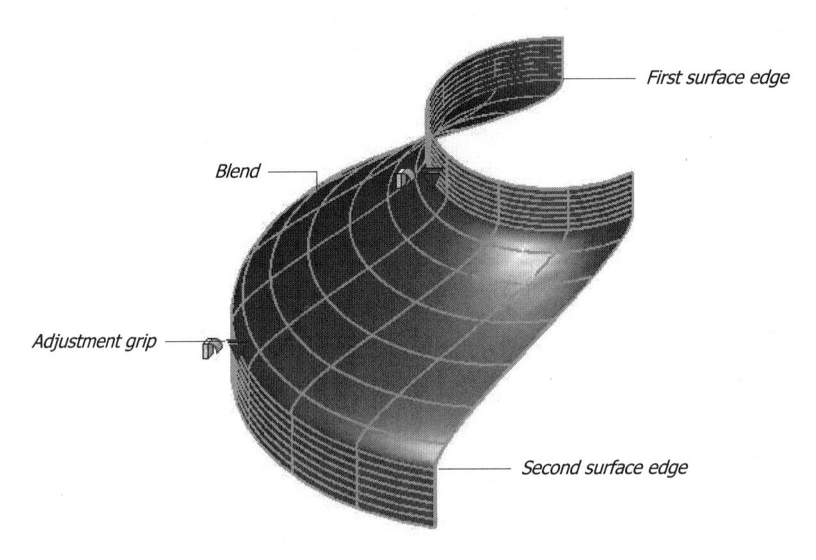

COMMAND LINE OPTIONs

Select first surface edges chooses one or more edges to blend from.

Select second surface edges chooses one or more edges to blend to.

Accept the blend surface places the blend and exits the command.

CONtinuity option

Determines the level of smoothness where the blend surfaces meet the original surfaces.

First edge continuity [G0/G1/G2] <G1>: *(Enter an option.)*
Second edge continuity [G0/G1/G2] <G1>: *(Enter an option.)*

Icon	Level of Smoothness	Meaning
	G0	Position (not tangent or curved)
	G1	Tangent
	G2	Continuous curvature

First / **Second edge** applies the change in continuity to the first (or second) set of edges.

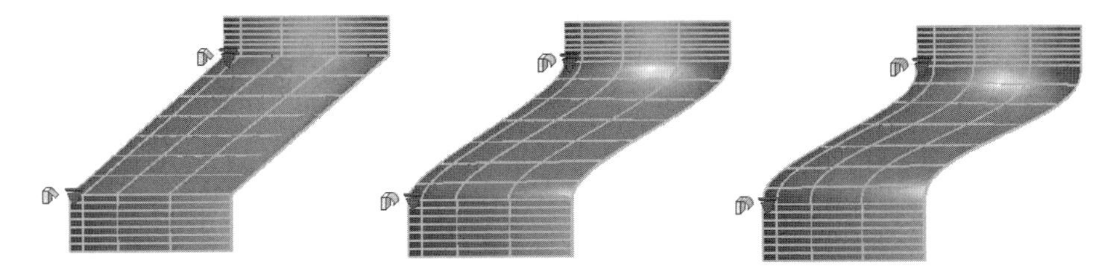

Left to right: *G0, G1, and G2 continuities.*

Bulge magnitude option

Specifies the roundness of where the blend surfaces meet the original surfaces.

First edge bulge magnitude <0.5>: *(Enter a real number between 0 and 1.)*

Second edge bulge magnitude <0.5>: *(Enter a real number between 0 and 1.)*

First / Second edge applies the change in bulge to the first (or second) set of edges.

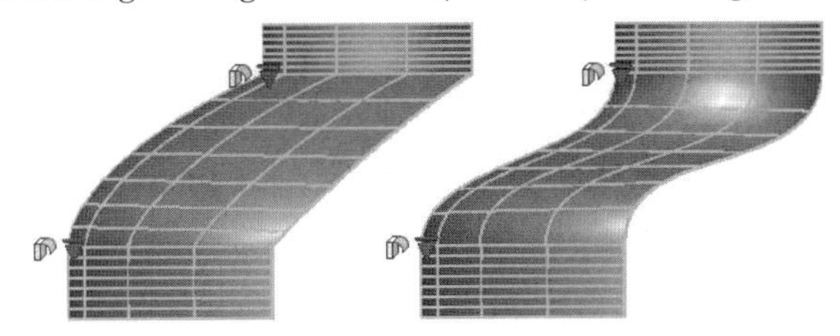

Left to right: *Bulge = 0 and 1.*

RELATED COMMANDS

SurfBlend adds continuous blend surfaces between two existing surfaces.

SurfExtend lengthens surfaces by specified distances.

SurfFillet adds surface fillets between pairs of surfaces or regions.

SurfNetwork creates surfaces between several curves defined by surfaces and edges of solids.

SurfOffset creates parallel surfaces or regions at a specified distance from the original.

SurfPatch creates new surfaces by fitting caps over surface edges that form closed loops.

SurfSculpt creates 3D solids by trimming and combining surfaces, solids, and 3D meshes.

SurfTrim trims at the intersections of surfaces and regions.

SurfUntrim replaces surface areas removed by the SurfTrim command.

RELATED SYSTEM VARIABLES

DelObj determines whether the source objects are erased following this command.

SurfaceAssociativity determines whether the blend maintains its relationship with the source objects.

TIP

- To set the level of curvature, you can enter a G number, or else click the triangular grip, and then choose an option from the context menu.

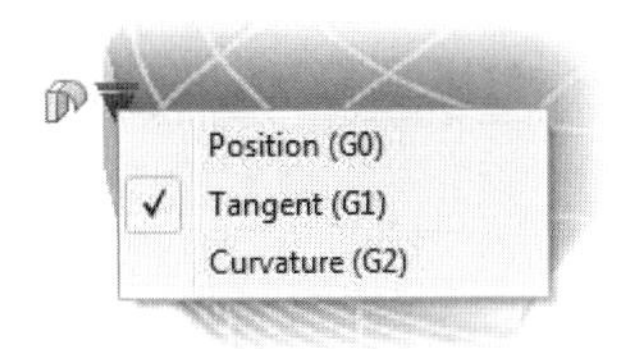

SurfExtend

2011 Lengthens surfaces by specified distances.

Command	Alias	Shortcut Keystrokes	Menu	Ribbon
surfextend	**extendsrf**	**...**	**Modify** **↳Surface Editing** **↳Extend**	**Surface** **↳Edit** **↳Extend**

Command: surfextend
Modes = Extend, Creation = Append
Select surface edges to extend: *(Choose one or more edges.)*
Select surface edges to extend: *(Press Enter to continue.)*
Specify extend distance [Expression/Modes]: *(Specify a distance or an option.)*

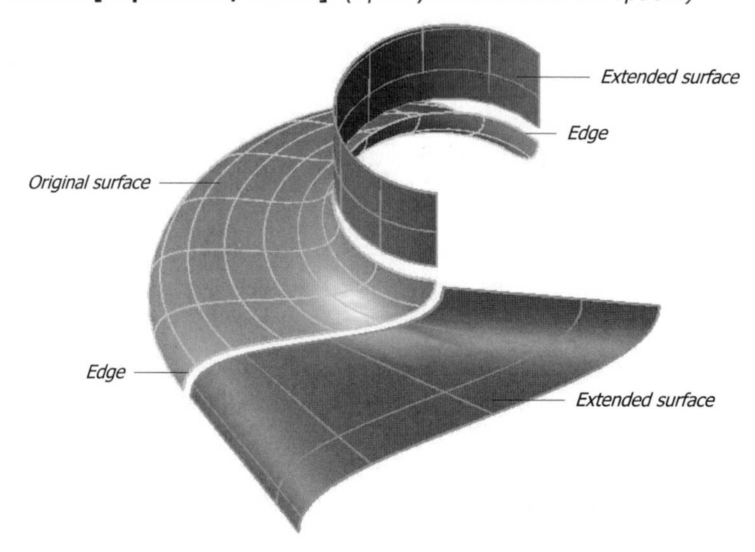

Extended surfaces here moved from the original for clarity.

COMMAND LINE OPTIONS

Select surface edges chooses the edges of surfaces to extend.

Specify extend distance specifies the distance by which to extend the surface.

Expression enters algebraic and parametric formulas to determine the length of the extension.

Modes option

Switches between extension modes and creation types:

Extension mode [Extend/Stretch] <Extend>: *(Type E or S.)*

Extend continues the shape of the surface.

Stretch extrudes the surface.

Creation type [Merge/Append] <Append>: *(Type M or A.)*

Merge extends the surface.

Append creates a new surface adjacent to the original.

RELATED COMMANDS

SurfBlend adds continuous blend surfaces between two existing surfaces.

SurfNetwork creates surfaces between several curves defined by surfaces and edges of solids.

SurfOffset creates parallel surfaces or regions at a specified distance from the original.

SurfPatch creates new surfaces by fitting caps over surface edges that form closed loops.

SurfTrim trims at the intersections of surfaces and regions.

RELATED SYSTEM VARIABLES

SurfaceAssociativity toggles associativity between original and extended surfaces: (0) no associativity or (1) surfaces adjust automatically to modifications made to related ones.

SurfaceAssociativityDrag determines whether associated surfaces are displayed during dragging (move) operations.

TIPS

- The Extend option attempts to mimic the shape and curvature of the source surface.
- The Append option creates a new surface, while the Merge option extends the source surface.

SurfFillet

2011 Adds surface fillets between pairs of intersecting surfaces or regions.

Command	*Alias*	*Shortcut Keystrokes*	*Menu*	*Ribbon*
surffillet	**filletsrf**	**...**	**Draw** **↳Modeling** **↳Surfaces** **↳Fillet**	**Surface** **↳Edit** **↳Fillet**

Command: surffillet
Radius = 0.5000, Trim Surface = yes
Select first surface or region to fillet or [Radius/Trim surface]: *(Pick an object or enter an option.)*
Select second surface or region to fillet or [Radius/Trim surface]: *(Pick an object, or enter an option.)*
Press Enter to accept the fillet surface or [Radius/Trim surfaces]: *(Press Enter, or enter an option.)*

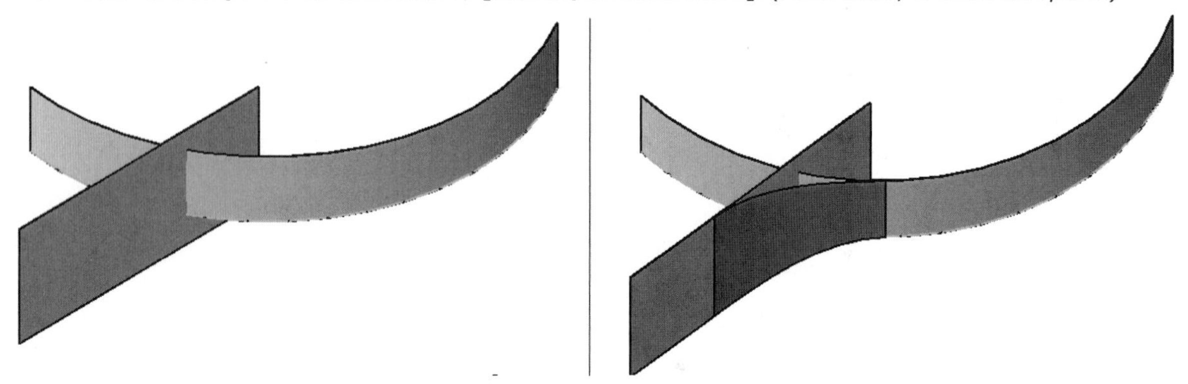

Left: *Two intersecting surfaces.*
Right: *Fillet applied between surfaces.*

COMMAND LINE OPTIONS

Select first surface or region chooses one surface or region.

Select second surface or region chooses the other surface or region.

Accept the fillet surface completes the command by pressing Enter.

Radius specifies the fillet radius:

Specify radius or [Expression] <1.9151>: *(Enter a radius or type E.)*

Expression enters algebraic and parametric formulas to determine the radius of the fillet.

Trim determines whether the two surfaces are trimmed at at the fillet:

Automatically trim surfaces to fillet edge [Yes/No] <Yes>: *(Type Y or N.)*

RELATED COMMANDS

SurfBlend adds continuous blend surfaces between two existing surfaces.

SurfOffset creates parallel surfaces or regions at a specified distance from the original.

SurfPatch creates new surfaces by fitting caps over surface edges that form closed loops.

SurfTrim trims at the intersections of surfaces and regions.

RELATED SYSTEM VARIABLE

FilletRad3D specifies the current fillet radius for 3D objects.

TIPS

- This command create fillets with constant radii.
- At the 'Press Enter to accept the fillet surface' prompt you can interactively adjust the fillet radius by dragging the blue triangle grip.

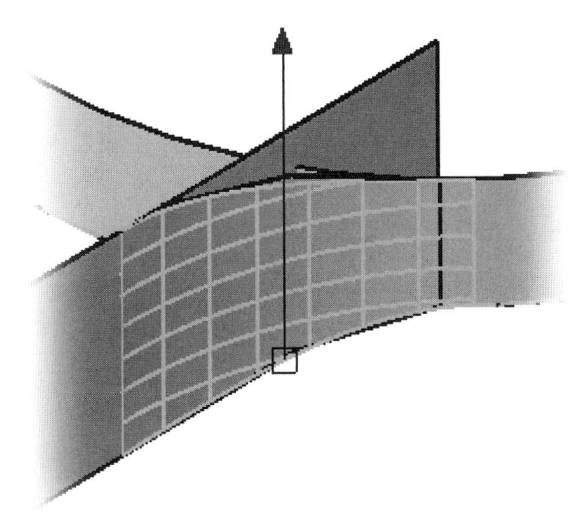

Dragging the arrow grip interactively adjusts the radius of the fillet.

- The result of this command is an independent surface.

SurfNetwork

2011 Creates surfaces between curves and edges of surfaces and solids.

Command	Alias	Shortcut Keystrokes	Menu	Ribbon
surfnetwork	**networksrf**	...	**Draw** ↳**Modeling** ↳**Surfaces** ↳**Network**	**Surface** ↳**Create** ↳**Network**

Command: surfnetwork
Select curves or surface edges in first direction: *(Choose one or more curves.)*
Select curves or surface edges in first direction: *(Press Enter to continue.)*

Select curves or surface edges in second direction: *(Choose one or more curves.)*
Select curves or surface edges in second direction: *(Press Enter to finish.)*

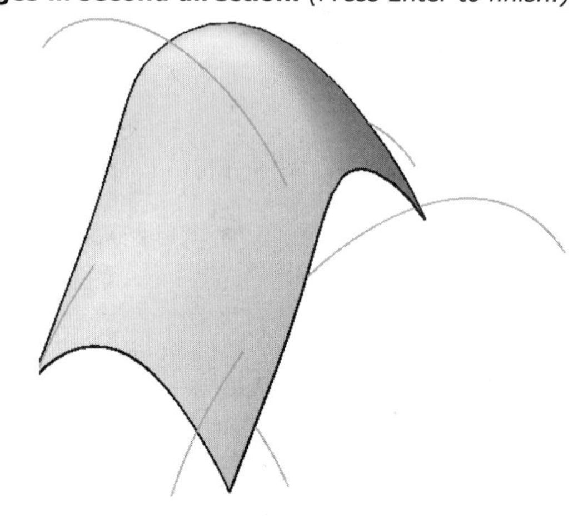

Network surface created between four arcs.

COMMAND LINE OPTION

Select curves or surface edges chooses the curves and edges:

- **First direction** determines the U direction.
- **Second direction** determines V direction.

RELATED COMMANDS

SurfBlend adds continuous blend surfaces between two existing surfaces.

SurfFillet adds surface fillets between pairs of surfaces or regions.

SurfPatch creates new surfaces by fitting caps over surface edges that form closed loops.

SurfSculpt creates 3D solids by trimming and combining surfaces, solids, and 3D meshes.

RELATED SYSTEM VARIABLES

SurfaceAssociativity determines whether the new surface is dependent on the source curves and edge.

DelObj determines whether source curves and edges are retained following this command.

TIP

- *Curves* can be any open object, such as lines and arcs; *edges* must be of open surfaces or edges of regions. While the source objects can be open, they must connected to form a closed area.

SurfOffset

2011 Creates parallel surfaces at a specified distance from surfaces or regions.

Command	Alias	Shortcut Keystrokes	Menu	Ribbon
surfoffset	**offsetscf**	...	**Draw** ↳**Modeling** ↳**Surfaces** ↳**Offset**	**Surface** ↳**Create** ↳**Offset**

Command: surfoffset
Connect adjacent edges = No
Select surfaces or regions to offset: *(Choose one or more surfaces or regions.)*
Select surfaces or regions to offset: *(Press Enter to continue.)*

Specify offset distance or [Flip direction/Both sides/Solid/Connect/Expression] <8.0>: *(Enter a distance or an option.)*

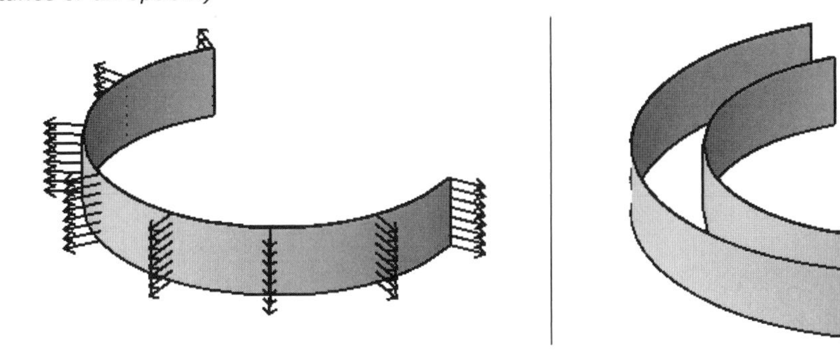

***Left**: Source surface showing offset direction arrows.*
***Right**: Offset surface.*

COMMAND LINE OPTIONS

Select surfaces or regions chooses the surfaces and regions from which to make offsets.

Specify offset distance specifies the distance; the distance can be specified parametrically through the Expression option.

Flip direction flips the direction of the arrows, placing the offset on the opposite side.

Both sides creates two offset surface, one above and one below the original.

Solid creates a 3D solid model of the offset; it thickens the surface by the offset distance, as illustrated below:

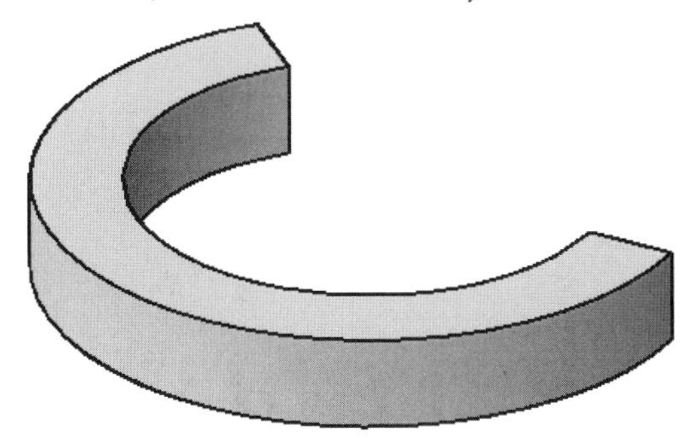

Connect extends offsets to keep them connected:

Keep adjacent edges connected [No/Yes] <Yes>: *(Type Y or N.)*

Expression specifies the offset distance through an equation:

Enter expression: *(Enter a parametric equation.)*

RELATED COMMANDS

SurfBlend adds continuous blend surfaces between two existing surfaces.

SurfExtend lengthens surfaces by specified distances.

SurfFillet adds surface fillets between pairs of surfaces or regions.

SurfNetwork creates surfaces between several curves defined by surfaces and edges of solids.

SurfPatch creates new surfaces by fitting caps over surface edges that form closed loops.

SurfSculpt creates 3D solids by trimming and combining surfaces, solids, and 3D meshes.

SurfTrim trims at the intersections of surfaces and regions.

RELATED SYSTEM VARIABLES

SurfOffsetConnect (*undocumented*) toggles connection between offset surfaces.

TIPS

- Use the Connect option or the SurfOffsetConnect system variable to keep multiple offsets connected.
- This command can convert open surfaces and 2D regions into solids through the Thicken option.

SurfPatch

2011 Creates new surfaces by fitting caps over surface edges that form closed loops.

Command	Alias	Shortcut Keystrokes	Menu	Ribbon
surfpatch	**patch**	...	**Draw** ↳**Modeling** ↳**Surfaces** ↳**Patch**	**Surface** ↳**Create** ↳**Patch**

Command: surfpatch

Continuity = G0 - position, bulge magnitude = 0.5

Select surface edges to patch or <Select Curves>: *(Select one or more edges; press Enter to select curves.)*

Select surface edges to patch or <Select Curves>: *(Press Enter to continue.)*

Press Enter to accept the patch surface or [CONtinuity/Bulge magnitude/CONStrain geometry]: *(Press Enter, or enter an option.)*

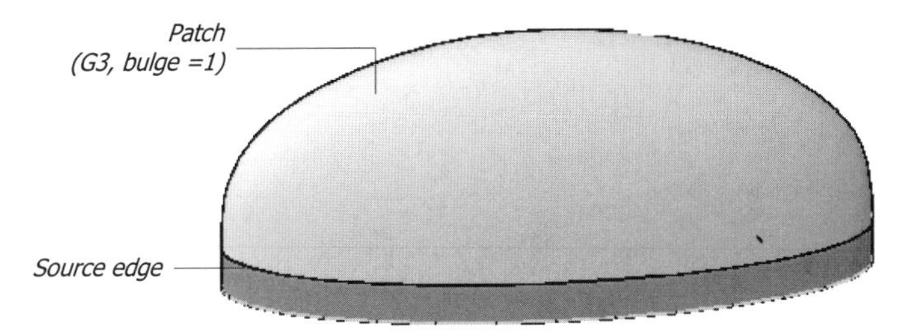

COMMAND LINE OPTIONS

Select surface edges selects one or more edges of surfaces; edges must form a closed loop or rectangle.

Select curves selects one or more curves; either edges or curves can be selected, not both.

Accept the patch surface completes the operations when you press Enter.

CONtinuity specifies the level of smoothness between surfaces. (See the SurfBlend command for details.)

First edge continuity [G0/G1/G2] <G1>: *(Enter an option, or choose a value from the grip.)*

Second edge continuity [G0/G1/G2] <G1>: *(Enter an option, or choose a value from the grip.)*

Bulge magnitude specifies the roundness of the patch. (See the SurfBlend command for details.)

First edge bulge magnitude <0.5>: *(Enter a real number between 0 and 1.)*

Second edge bulge magnitude <0.5>: *(Enter a real number between 0 and 1.)*

CONStrain geometry selects curves to define the shape of the patch:

Select curves or points to constrain patch surface: *(Select one or more objects.)*

RELATED COMMANDS

SurfFillet adds surface fillets between pairs of surfaces or regions.

SurfNetwork creates surfaces between several curves defined by surfaces and edges of solids.

SurfOffset creates parallel surfaces or regions at a specified distance from the original.

SurfSculpt creates 3D solids by trimming and combining surfaces, solids, and 3D meshes.

SurfTrim trims at the intersections of surfaces and regions.

RELATED SYSTEM VARIABLE

SurfaceAssociativity determines whether associativity is maintained between the selected edges (or curves) and the patch surface.

TIPS

- Use this command to close holes in surface models.
- While this command is active, you can adjust the continuity of the patch. Click the triangle grip to access the following shortcut menu:

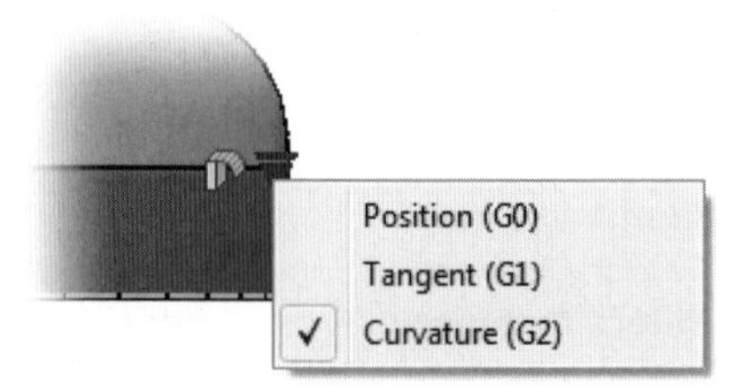

- See the SurfBlend command for examples of how continuity and bulge affect the curvature of patches.

SurfSculpt

2011 Creates 3D solids by trimming and combining surfaces, solids, and 3D meshes automatically.

Command	Alias	Shortcut Keystrokes	Menu	Ribbon
surfsculpt	**createsolid ...**		**Modify** **↳Surface Editing** **↳Sculpt**	**Surface** **↳Edit** **↳Sculpt**

Command: surfsculpt
Mesh conversion set to: Smooth and optimized.
Select surfaces or solids to sculpt into a solid: *(Choose one or more objects.)*
Select surfaces or solids to sculpt into a solid: *(Press Enter.)*

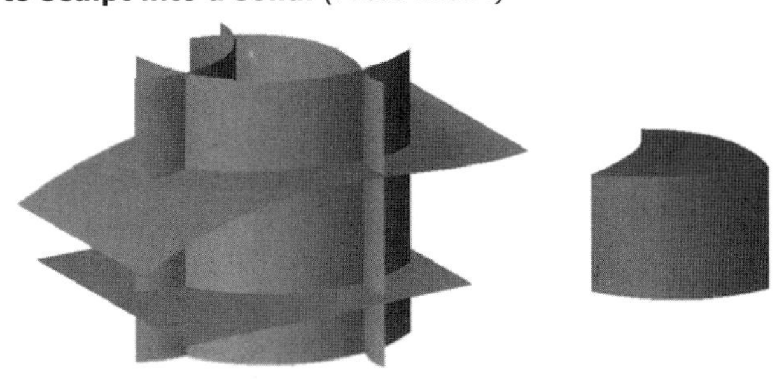

Left to right: *Before and after sculpting intersecting surfaces.*

COMMAND LINE OPTION

Select surface or solids chooses the surfaces, meshes, and solids to combine.

RELATED COMMANDS

SurfBlend adds continuous blend surfaces between two existing surfaces.

SurfExtend lengthens surfaces by specified distances.

SurfFillet adds surface fillets between pairs of surfaces or regions.

SurfNetwork creates surfaces between several curves defined by surfaces and edges of solids.

SurfOffset creates parallel surfaces or regions at a specified distance from the original.

SurfPatch creates new surfaces by fitting caps over surface edges that form closed loops.

SurfSculpt creates 3D solids by trimming and combining surfaces, solids, and 3D meshes.

SurfTrim trims at the intersections of surfaces and regions.

SurfUntrim replaces surface areas removed by the SurfTrim command.

RELATED SYSTEM VARIABLES

SmoothMeshConvert determines the outcome of meshes: smoothed or with flattened faces

TIPS

- Use this command to convert surfaces and 3D meshes into 3D solid models. This command automatically combines (unions) and trims the selected objects.
- This command fails when surfaces have a continuity of G1 or G2, or when surfaces and meshes are not watertight (have gaps).
- The status report "Smooth and optimized" refers to the setting of the SmoothMeshConvert system variable, which applies to 3D meshes only.

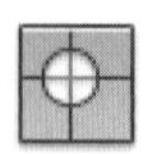

SurfTrim / SurfUntrim

2011 Trims (or replaces) the intersections of surfaces and regions with other geometry.

Command	*Alias*	*Shortcut Keystrokes*	*Menu*	*Ribbon*
surftrim	...	...	**Modify** **↳Surface Editing** **↳Trim**	**Surface** **↳Edit** **↳Trim**
surfuntrim	...	...	**Modify** **↳Surface Editing** **↳Untrim**	**Surface** **↳Edit** **↳Untrim**

Command: surftrim
Extend surfaces = Yes, Projection = Automatic
Select surfaces or regions to trim or [Extend/PROjection direction]: all
Select surfaces or regions to trim or [Extend/PROjection direction]: *(Press Enter to continue.)*

Select cutting curves, surfaces or regions: all
Select cutting curves, surfaces or regions: *(Press Enter to continue.)*

Select area to trim [Undo]: *(Pick surface areas to trim.)*
Select area to trim [Undo]: *(Press Enter to exit the command.)*

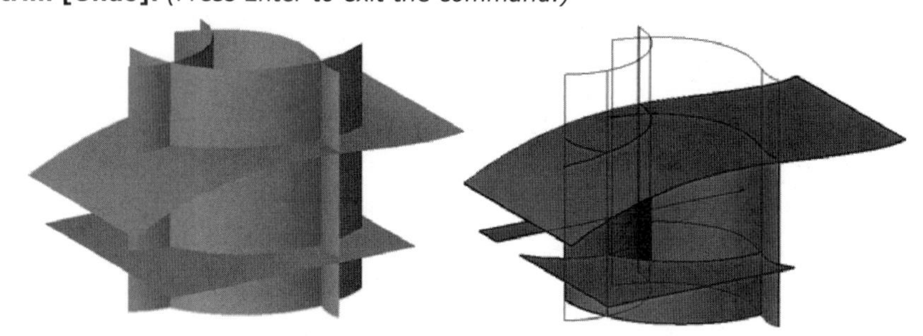

Left: *Intersecting surfaces.*
Right: *Trimmed portions.*

COMMAND LINE OPTIONS

Select surfaces or regions chooses the objects to trim.

Select cutting curves, surfaces, or regions selects the objects that determine the cutting curves: lines, arcs, circles, ellipses, 2D polylines (including curve and spline fit ones), 3D polylines (include spline-fit ones), splines, helices, surfaces, and regions.

Select area to trim chooses the area to trim, inside or outside the cutting curve.

Extend determines whether cutting objects are trimmed at the edges of the trimmed surfaces:

Extend trimming geometry [Yes/No] <Yes>: *(Type Y or N.)*

PROjection direction options

Determines how cutting objects are projected onto the surface:

Specify the projection direction [Automatic/View/Ucs/None] <Automatic>: *(Enter an option.)*

Automatic determines how to project cutting objects based on the current view type.

View projects cutting objects in the direction of the current viewpoint.

UCS projects cutting objects along the z axis of the current UCS.

None does not project cutting objects; trim works only when objects lie on the surfaces.

SURFUNTRIM Command

Replaces surface areas removed by the SurfTrim command.

Command: surfuntrim
Select edges on surface to un-trim or [SURface]: *(Select one or more edges, or type SUR.)*

COMMAND LINE OPTIONS

Select edges chooses the edges to restore to untrimmed state.

SURface selects a surface to replace the trimmed areas:

Select surfaces to un-trim: *(Choose one or more surfaces.)*

RELATED COMMANDS

SurfBlend adds continuous blend surfaces between two existing surfaces.

SurfExtend lengthens surfaces by specified distances.

SurfFillet adds surface fillets between pairs of surfaces or regions.

SurfOffset creates parallel surfaces or regions at a specified distance from the original.

SurfSculpt creates 3D solids by trimming and combining surfaces, solids, and 3D meshes.

Trim trims 2D objects.

RELATED SYSTEM VARIABLES

RebuildOptions determines what happens to original surfaces and trimmed areas on rebuilt surfaces.

SurfaceAssociativity determines whether trimmed surface are updated when trimmed edges are modified.

SurfaceAutoTrim toggles the automatic trimming of surfaces by projected geometry.

SurfTrimAutoExtend (*undocumented*) toggles automatic extension of trim geometry.

SurfTrimProjection (*undocumented*) toggles projection of trim entities onto surfaces.

TIPS

- Autodesk warns that this command does not work when trimmed edges depend on other surface edges that are also trimmed, or when areas were removed by the SurfAutotrim system variable or the ProjectGeometry command.
- To remove trims, you may want to use the U command in place of this command, if still possible.

Sweep

2007 Creates swept 3D solids and surfaces.

Command	Aliases	Shortcut Keystrokes	Menu	Ribbon
sweep	...	...	**Draw**	**Home**
			↳**Modeling**	↳**3D Modeling**
			↳**Sweep**	↳**Sweep**

Command: sweep
Current wire frame density: ISOLINES=4, Closed profiles creation mode = Solid
Select objects to sweep or [MOde]: *(Select one or more objects, or type MO.)*
Select objects to sweep or [MOde]: *(Press Enter.)*
Select sweep path or [Alignment/Base point/Scale/Twist]: *(Pick another object, or enter an option.)*

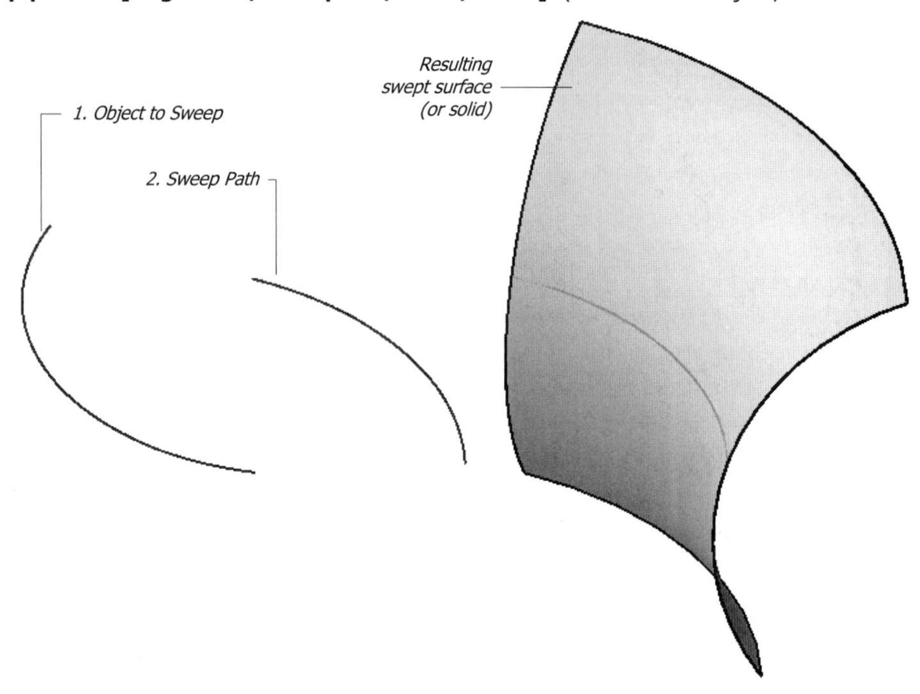

COMMAND OPTIONS

Select objects to sweep selects the objects to be swept.

MOde specifies the resulting object: a solid or a surface *(new to AutoCAD 2011)*.

Select sweep path defines the path along which the objects are swept.

Alignment aligns the path normal (at 90 degrees) to the tangent direction of the path.

Base point relocates the sweep's base point.

Scale resizes the sweep object.

Twist specifies the twist angle.

Twist options

Enter twist angle or allow banking for a non-planar sweep path [Bank]<0.0>: *(Enter an angle, or type B.)*

Twist angle specifies the angle through which the sweep twists, from one end to the other.

Bank causes the sweep to rotate along 3D curved paths: 3D polyline, 3D spline, or helix.

TIPS

- You can use the following objects as the sweep path: lines, arcs, circles, ellipses, elliptical arcs, 2D and 3D polylines, splines, helixes, and edges of surfaces and solids.
- The following objects can be swept: lines, traces, arcs, circles, ellipses, elliptical arcs, 2D polylines, 2D solids, splines, regions, flat 3D faces, flat solid faces, and planar surfaces.
- The figures below illustrate a sweep twisted by 180 degrees:

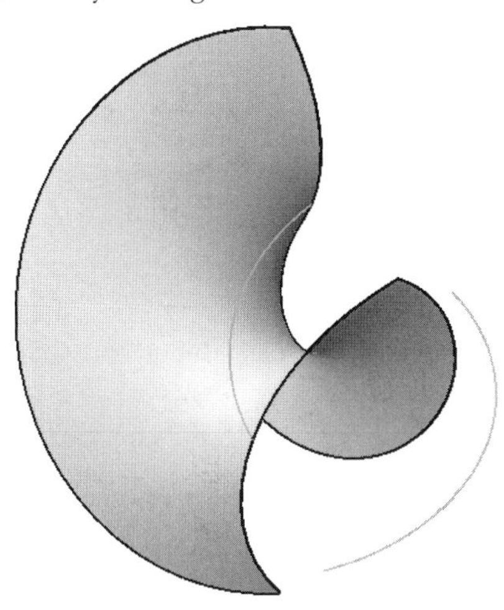

- Apply this command to a circle and a helix to produce a spring:

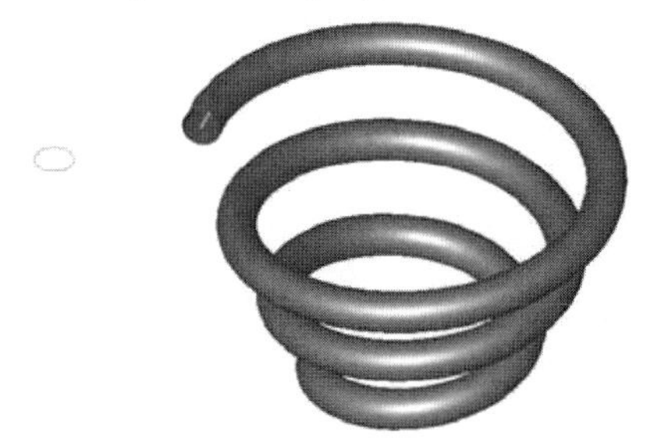

- The Twist option does not work when the path is a closed loop.

SysWindows

Rel. 13 Controls multiple windows (short for SYStem WINDOWS).

Command	*Aliases*	*Shortcut Keystroke*	*Menu*	*Ribbon*
syswindows	...	**F6**	**Window** ↳*varies*	**View** ↳**Window** ↳*varies*

Command: syswindows

Enter an option [Cascade/tile Horizontal/tile Vertical/Arrange icons]: *(Enter an option.)*

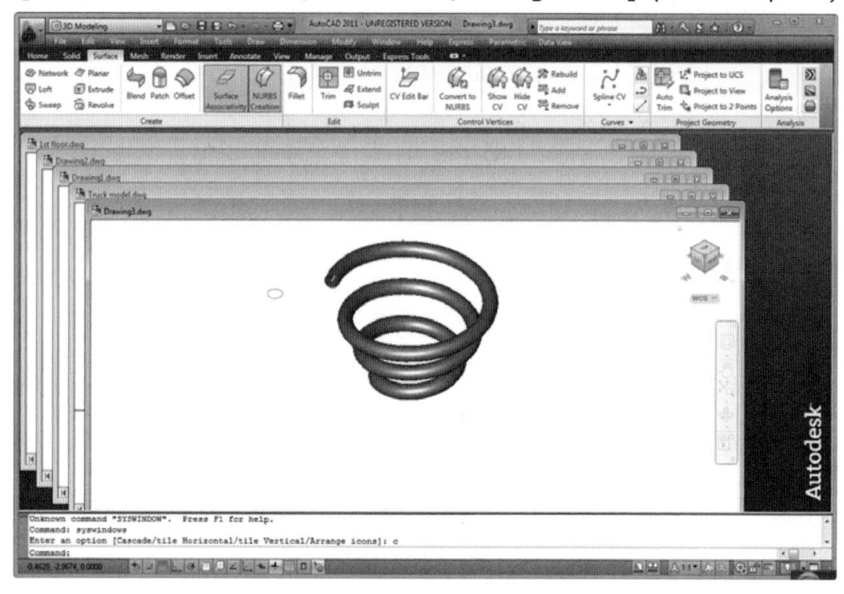

AutoCAD displaying drawings in several tiled windows.

COMMAND LINE OPTIONS

Arrange icons arranges icons in an orderly fashion.

Cascade cascades the window.

tileHorizontal tiles the window horizontally.

tileVertical tiles the window vertically.

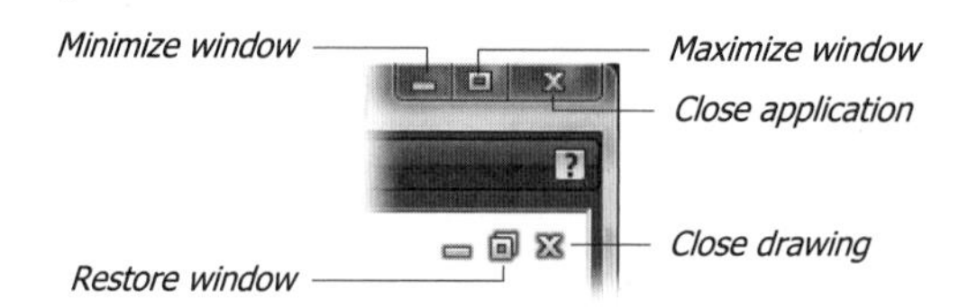

TITLE BAR MENU

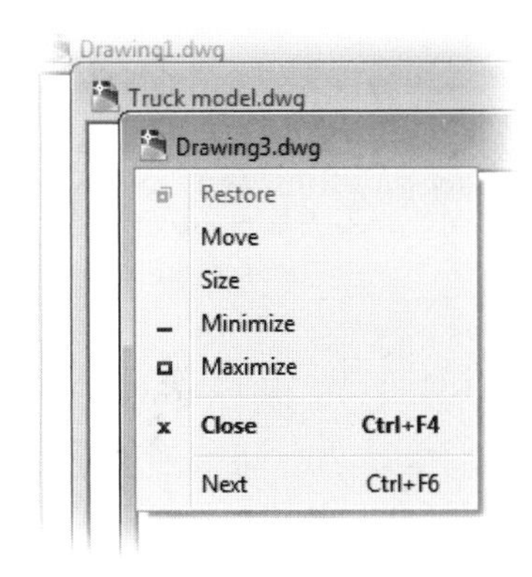

Restore restores the window to its "windowized" size.

Move moves the window.

Size resizes the window.

Minimize minimizes the window.

Maximize maximizes the window.

Close closes the window.

Next switches the focus to the next window.

RELATED COMMANDS

Close closes the current window.

CloseAll closes all windows.

Open opens one or more drawings, each in its own window.

MView creates paper space viewports in a window.

Vports creates model space viewports in a window.

QvDrawing switches between drawings using a thumbnail filmstrip.

RELATED SYSTEM VARIABLES

Taskbar determines whether each drawing opens in a single session of AutoCAD.

SDI forces AutoCAD to load just one drawing at a time.

TIPS

- The SysWindows command had no practical effect until AutoCAD 2000, because AutoCAD Release 13 and 14 supported a single window only.
- Ctrl+F6 quickly switches between currently-loaded drawings.
- The QvDrawings command is faster than Ctrl+F6 when three or more drawings are loaded.
- Ctrl+F4 closes the current drawing

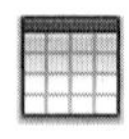

Table

2005 Inserts formatted tables in drawings; links to external data files.

Commands	*Alias*	*Shortcut Keystrokes*	*Menu*	*Ribbon*
table	**tb**	...	**Draw** ↳**Table**	**Home** ↳**Annotation** ↳**Table**
-table	...	....	...	...

Command: table

Displays dialog box:

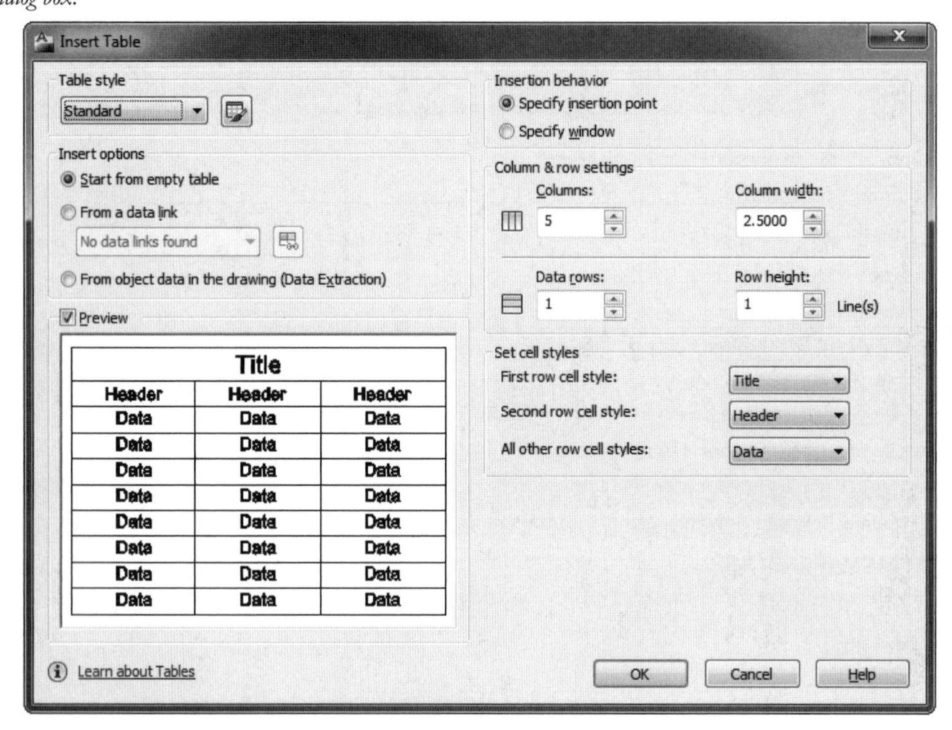

INSERT TABLE DIALOG BOX

Table Style options

Standard ▾ selects from a list of pre-formatted styles; new drawings contain one default style, Standard.

displays the Table Style dialog box to edit or create new styles; see the TableStyle command.

Insert Options options

⊙ **Start from Empty Table** creates a blank table.

○ **From a Data Link** creates a table populated with data from an external spreadsheet file.

No data links found ▾ selects a named data link created previously with the DataLink command.

launches the DataLink command; see the DataLink command.

○ **From Object Data in the Drawing (Data Extraction)** creates a table populated with data from block attributes, object properties, and drawing information; executes the DataExtraction command.

Insertion Behavior options

⊙ **Specify Insertion Point** locates the table by its upper-left corner; after you click OK to exit the dialog box, AutoCAD prompts, 'Specify insertion point:'. Pick a point; AutoCAD inserts the table in the drawing:

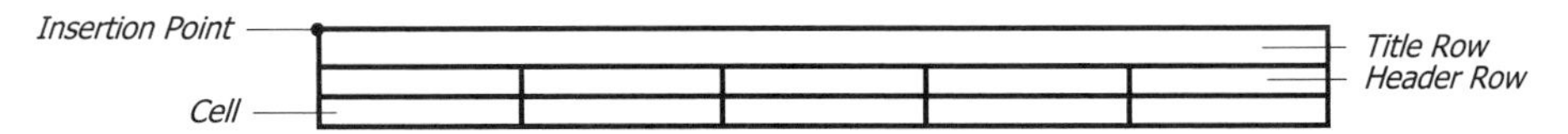

○ **Specify window** fits the table to the window. After you click OK to dismiss the dialog box, AutoCAD prompts you to pick two points: 'Specify first corner:' and 'Specify second corner:'. Pick two points in the drawing; AutoCAD places the table at the first pick point, and fits the table to the second pick point.

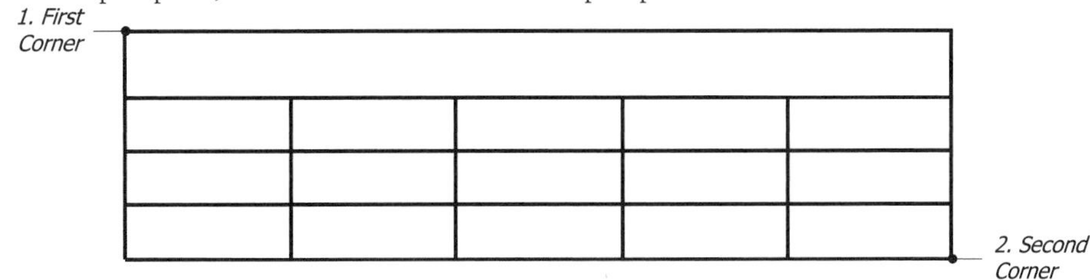

Column & Row Settings options

Default size = 1 row x 5 columns, plus optional title and header rows.

Columns specifies the initial number of columns.

Column width specifies the initial width of columns.

Data Rows specifies the initial number of "data" rows; the table style will determine whether any header and title rows are added above the data rows.

Row height specifies the initial height of rows, in lines.

Set Cell Styles options

First Row Style selects from data, header, or title.

Second Row Style selects from data, header, or title.

All Other Row Cell Styles selects from data, header, or title.

-TABLE Command

Command: -table
Current table style: Standard Cell width: 2.5000 Cell height: 1 line(s)
Enter number of columns or [Auto/from Style/data Link] <5>: *(Enter a number, or enter an option.)*
Enter number of rows or [Auto] <1>: *(Type* A, *or enter a number.)*
Specify insertion point or [Style/Width/Height]: *(Pick a point, or enter an option.)*

COMMAND OPTIONS

Auto creates columns and rows to fit the table.

from Style specifies table formatting, as detailed below.

data Link prompts you for the name of a data link.

Style selects the table style name; see the TableStyle command.

Width specifies the initial width of the table.

Height specifies the initial height of the table.

From Style options

Specify insertion point or [Style/Rows/Columns/Options]: *(Enter O.)*

Enter an insert option [Label text/Data text/Formulas/fIelds/data liNks/Blocks/cell style Overrides] <exit>: *(Enter an option.)*

Label Text inserts rows with the Label cell style, including title and header cell styles.

Data Text inserts rows with the Data cell style, including data cell styles.

Formulas inserts formulas found in the table style.

Fields inserts fields found in the table style.

Data Links inserts data links in the table style.

Blocks inserts blocks found in the table style.

Cell Style Overrides retains cell style overrides found in the table style.

RELATED COMMANDS

MatchCell matches the style of one cell to other cells.

TablEdit edits text in table cells.

TableExport exports tables in CSV format.

TableStyle defines table styles.

DataLink creates named links with external spreadsheet files that can be turned into tables with the Table command.

DataExtraction extracts drawing data to tables.

RELATED SYSTEM VARIABLES

CTableStyle names the current table style.

TableIndicator toggles the display of row numbers and column letters.

TIPS

- To copy tables between drawings, use DesignCenter.
- Whether the table extends up or down from the insertion point depends on its style.
- Tables can be "live": when the spreadsheet data changes, the table can be updated by you. AutoCAD displays a warning balloon in the tray.
- Use the DataExtraction command to create tables of information extracted from the drawing.
- Tables created through the DataLink command display this icon when the cursor pauses over them.
- Linked tables are initially locked to avoid unintentional editing: The lock can be overridden by the TablEdit command's lock options; see the TablEdit command.
- Unlike the Table command, the -Table command cannot set up data links; you first use the DataLink command to create a named link, and then you can use the -Table command to name the data link.
- AutoCAD shows the range of spreadsheet cells with green corner icons: .
- You can autofill cells by dragging an initial value across rows and columns, just as in spreadsheets.
- Tables can flow across multiple columns: enable the Table Breaks option in the Properties palette.
- To create new tables, you can bypass the Table command: copy a range of cells from a spreadsheet, and then paste them into the drawing. This workaround may only work with specific brands and versions of spreadsheet programs.
- Tables cannot be annotative, even if you apply an annotative text style.

TablEdit

2005 Edits cells, rows, and columns of tables.

Command	Aliases	Shortcut Keystrokes	Menu	Ribbon
tabledit	...	...	...	...

Command: tabledit

Pick a table cell: *(Pick inside a cell.)*

When TableToolbar = 0, displays no toolbar; use right-click menu to edit cells (shown below).

When TableToolbar = 1, displays floating toolbar.

When TableToolbar = 2 (default), displays text editor tab on ribbon:

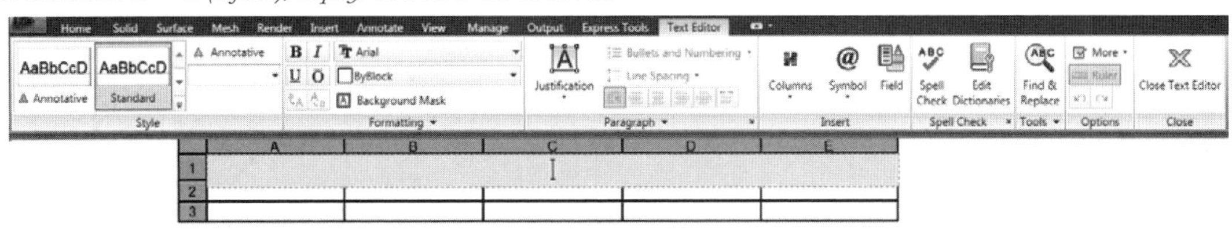

To edit other cells, press the arrow keys. To move to the "next" cell, press Tab. As an alternative to the TablEdit command, double-click cells to edit their content.

TABLE RIBBON

The ribbon, toolbar, and shortcut menu provide similar sets of functions. Described below is shortcut menu:

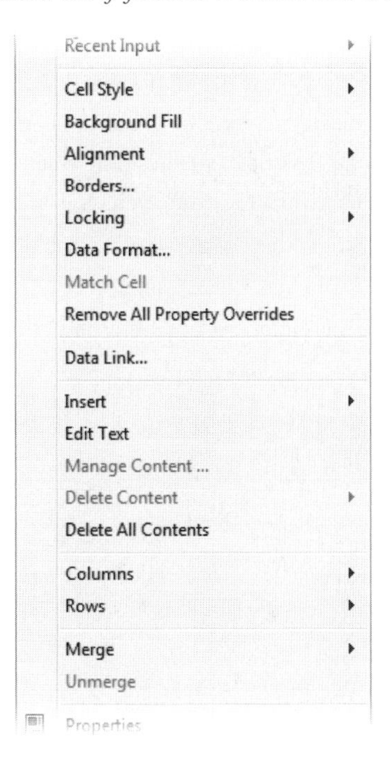

Cell Style selects from preset or user-defined cells; see the TableStyle command.

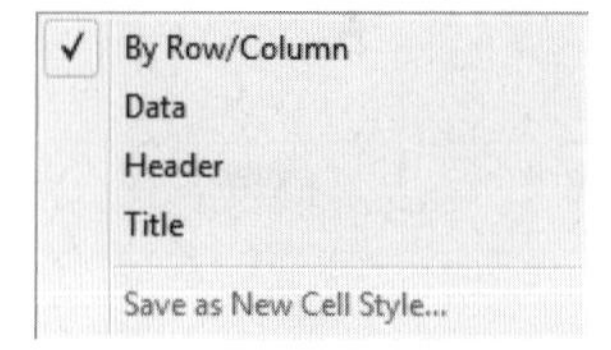

Background Fill changes the color of the cell.

Alignment aligns the content of the cell:

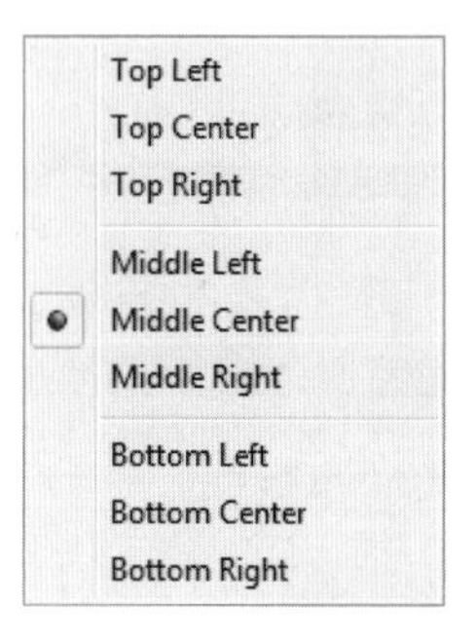

Borders displays the Cell Border Properties dialog box; see the TableStyle command.

Locking specifies unlocked, content locked, format locked, or content and format locked.

Data Format specifies angle, currency, date, percentage, and so on; see the TableStyle command.

Match Cell: copies properties from one cell to others; see the MatchCell command.

Remove All Property Overrides returns the cell to its default format.

Data Link displays the Select A Data Link dialog box; see the DataLink command.

Merge Cells selects among All, By Row, or By Column; operates only when two or more cells are selected.

Insert inserts blocks or fields; see the TInsert and Field commands.

Edit Text edits the text in the cell; see the TextEdit command.

Manage Content displays the Manage Cell Content dialog box.

Delete Content erases the content of the selected cell.

Delete All Contents erases the content of the selected cell.

Columns inserts, deletes, and resizes columns:

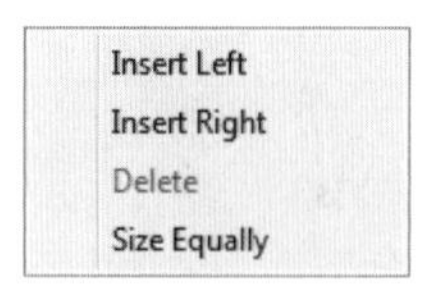

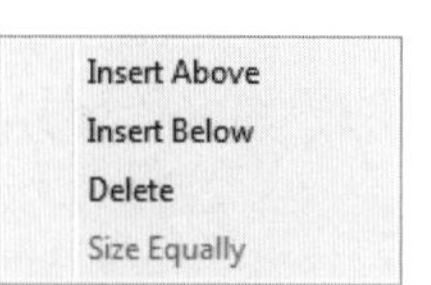

Menu for columns (left) and rows (right)

Rows inserts, deletes, and resizes columns.

Merge merges selected cells into a single cell:

Unmerge reverses the action of the Merge option.

TABLE EDITING WITH GRIPS

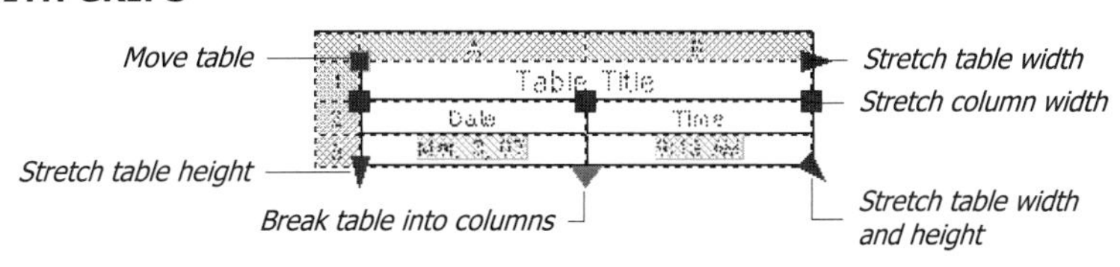

To stretch columns and the table together, hold down the Ctrl key.

RELATED COMMANDS

MatchCell matches the style of one cell to other cells.

EditTableCell edits cells at the command line.

Table creates tables.

TableStyle defines table styles.

Properties changes table style names, and overrides style settings.

RELATED SYSTEM VARIABLES

CTableStyle names the current table style.

TableIndicator toggles the display of row numbers and column letters.

TableToolbar toggles the display of the toolbar when cells are selected in tables.

TIPS

- To pick a cell, you must click its contents.
- If a cell is empty, you can select it by clicking anywhere inside it.
- Alternatively, don't use this command; instead, click a cell wall, and then choose the cell.

TableExport

2005 Exports tables as CSV data files.

Command	*Aliases*	*Shortcut Keystrokes*	*Menu*	*Ribbon*
tableexport	...	...	...	...

Access this command by selecting a table, and then choosing Export from the shortcut menu.

Command: tableexport

Select a table: *(Select one table.)*

Displays the Export Data dialog box.

Enter a file name, select a folder, and then click Save.

TIPS

- *CSV* is short for "comma-separated values." Each table row is one line of data, and each cell is separated by a comma. This format can be read by spreadsheet and database programs.
- To import tables from a spreadsheet, copy and paste it into the drawing. AutoCAD converts the spreadsheet into a table object. Follow these steps:
  1. In the spreadsheet, copy the data to the Clipboard with Ctrl+C.
  2. In AutoCAD, use the Edit | Paste Special command to paste the data:

Format	*Pasted As*
Picture (Metafile)	OLE object.
Device Independent Bitmap	OLE object.
AutoCAD Entities	Table object.
Image Entity	Raster image.
Text	Mtext.
Unicode Text	Mtext.

  To create a live table, ensure that the Paste Link option is selected.

- AutoCAD can import tables from CSV files via the DataLink command.
- To copy tables between drawings, use DesignCenter.
- You can also access this command by selecting the table, right-clicking, and then choosing Export from the shortcut menu.

TableStyle

2005 Creates and edits table styles.

Command	*Alias*	*Shortcut Keystrokes*	*Menu*	*Ribbon*
tablestyle	**ts**	...	**Format**	**Annotate**
			↳Table Style	**↳Tables**
				↳Table Style

Command: tablestyle

Displays dialog box.

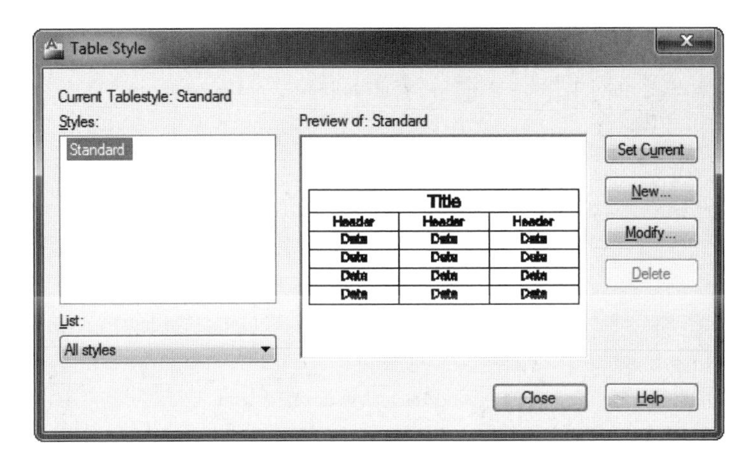

TABLE STYLE DIALOG BOX

Styles lists the styles defined in the drawing; every new drawing has the "Standard" style.

List filters the style names:

- **All Styles** lists all styles.
- **Styles in use** lists the styles used by tables in the drawing.

Set Current sets the selected style as current.

New creates new table styles based on the settings of the current style; displays Create New Table Style dialog box.

Modify changes the settings of the table style; displays the Modify Table Style dialog box.

Delete erases the style from the drawing; unavailable if the "Standard" table style is selected.

NEW/MODIFY TABLE STYLE DIALOG BOX

Access this dialog box through the Modify button:

Starting Table options

Select Table to Start From selects the source for this style:

Selects a table in the drawing; dismisses the dialog box, and prompts you, 'Select a table:'. Pick a table, and the dialog box returns with the table as the source for the new style.

Removes the starting table from the table style; returns to the default style.

General options

Down **Table Direction** determines how the table grows relative to the insertion point:

- **Down** (insertion point is at upper-left corner).

- **Up** (insertion point is at lower-left corner).

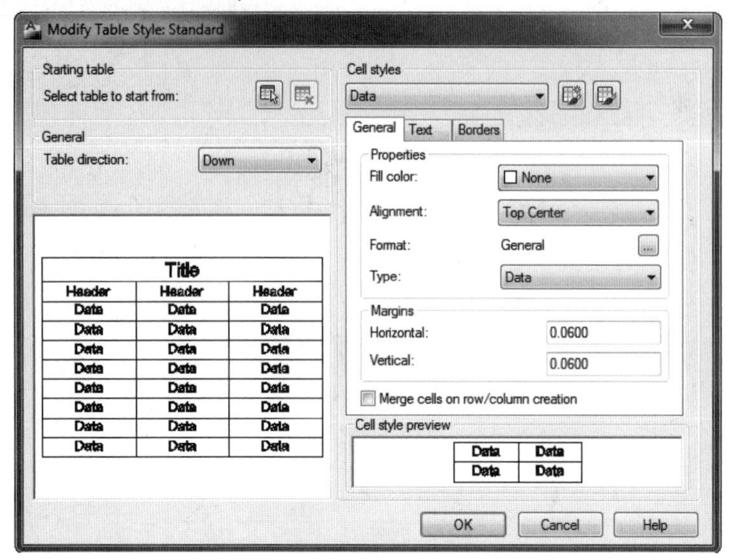

Cell Styles options

Data displays a list of cell format names; defaults are Data, Header, and Title.

creates new table styles through the Create New Cell Style dialog box for naming the new style.

edits table styles through the Manage Cell Styles dialog box, detailed later.

General tab

Properties options

Fill Color selects color for cell backgrounds. Choose Select Color to access the Select Color dialog box; see the Color command.

Alignment specifies the alignment of objects in cells: top left, middle center, bottom right, and so on.

Format displays the Table Cell Format dialog box, identical to that used by the Field command.

Type selects either label or data cells; *label* cells are like titles and headers, while *data* cells are regular cells.

Margins options

Horizontal specifies the horizontal margin, the distance between the cell content and its borders.

Vertical specifies the vertical margin, the distance between the cell content and its borders.

☐ **Merge Cells on Row/column Creation** merges new rows and columns into one cell; this option is meant for creating title rows in tables.

Text tab

Properties options

Text Style selects a predefined text style.

... displays the Text Style dialog box; see the Style command.

Text Height specifies the height of the text; unavailable if the height is defined by text style.

Text Color specifies the color of the text; select Select Color for additional colors.

Text Angle specifies angle of text in cells. Either the text becomes smaller, or the cell expands to accommodate the text.

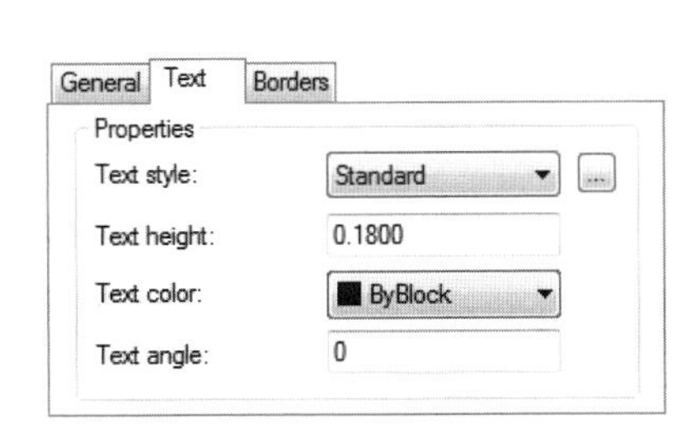

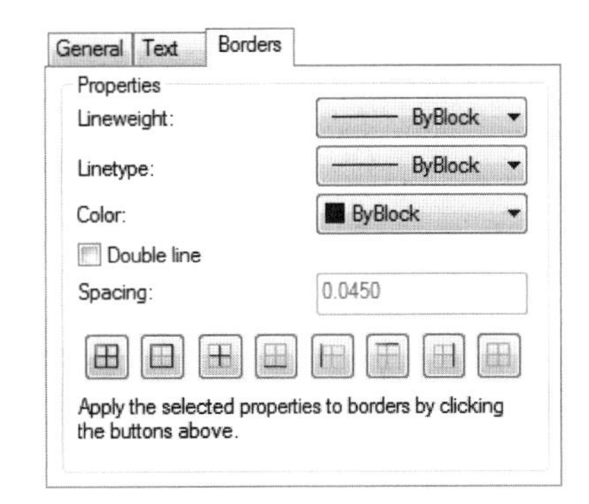

Left to right: *The Text and Borders tabs.*

Borders tab

Lineweight selects a lineweight from the droplist. Lineweights are applied to the center of the border lines, and in some cases thick lineweights may overwrite part of the cell content.

Linetype selects the linetype for the cell borders; click Other to load additional linetypes.

Color chooses the color for the borders.

☐ **Double Line** toggles double-line borders, often used to emphasize cells or to surround the entire table.

Spacing specifies the spacing between double lines.

 specifies the edge to which to apply the properties.

MANAGE CELL STYLES DIALOG BOX

New names new cell styles; displays the Create New Cell Style dialog box.

Rename renames the cell style; standard styles cannot be renamed: Data, Header, and Title.

Delete removes cell styles from the drawing; in-use and standard styles cannot be erased.

RELATED COMMANDS

Table places tables in drawings.

MatchCell matches the style of one cell to other cells.

TableEdit edits text in table cells.

Rename renames table styles.

Purge purges unused table styles from drawings.

DesignCenter shares tables styles between drawings.

Properties changes table style names, and overrides style settings.

RELATED SYSTEM VARIABLE

CTableStyle names the current table style.

TIPS

- To copy table styles from other drawings, copy a table with Ctrl+C, and then paste with Ctrl+V into the other drawing.
- If you erase a table, the style remains in the drawing.
- To revert tables to original styles, select the table, right-click, and select Remove All Property Overrides.

Tablet

v 1.0 Configures and calibrates digitizing tablets, and toggles tablet mode.

Command	*Alias*	*Shortcut Keystrokes*	*Menu*	*Ribbon*
tablet	**ta**	**Ctrl+T F4**	**Tools ↳Tablet**	...

Command: tablet

When no tablet is configured, AutoCAD complains, 'Your pointing device cannot be used as a tablet.'

When a digitizing tablet is configured, AutoCAD continues:

Enter an option [ON/OFF/CAL/CFG]: *(Enter an option.)*

COMMAND LINE OPTIONS

CAL calibrates the coordinates for the tablet.

CFG configures the menu areas on the tablet.

OFF turns off the tablet's digitizing mode.

ON turns on the tablet's digitizing mode.

RELATED SYSTEM VARIABLE

TabMode toggles use of the tablet.

Digitizer (undocumented) reports whether a digitizer is attached to AutoCAD *(new to AutoCAD 2011).*

MaxTouches reports the number of touch points supported by the digitizer *(new to AutoCAD 2011).*

RELATED FILES

acad.cui is the customization source code that defines the functions of tablet menu areas.

tablet.dwg is an AutoCAD drawing of the printed template overlay.

TIPS

- To change the tablet overlay, edit the *tablet.dwg* file (available with AutoCAD Mechanical), and then plot it to fit your digitizer.
- Tablet does not work if a digitizer has not been configured with the Options command.
- AutoCAD supports up to four independent menu areas; macros are specified by the Tablet Menu section of the *acad.cuix* file.
- Menu areas may be skewed, but corners must form a right angle.
- Projective transformation is a limited form of "rubber sheeting": straight lines remain straight, but not necessarily parallel.

DEFINITIONS

Affine Transformation — requires three pick points; sets an arbitrary linear 2D transformation with independent x, y scaling and skewing.

Orthogonal Transformation — requires two pick points; sets the translation; the scaling and rotation angles remain uniform.

Residual Error — proves largest where mapping is least accurate.

Outcome of Fit — reports on the transformation types:

Outcome	*Meaning*
Exact	Enough points to transform data.
Success	More than enough points to transform data.
Impossible	Not enough points to transform data.
Failure	Too many colinear and coincident points.
Canceled	Fitting canceled during projective transformation.

Projective Transformation — maps a perspective projection from one plane to another plane.

RMS Error — specifies root mean square error; smaller is better; measures closeness of fit.

Standard Deviation —when near zero, residual error at each point is roughly the same.

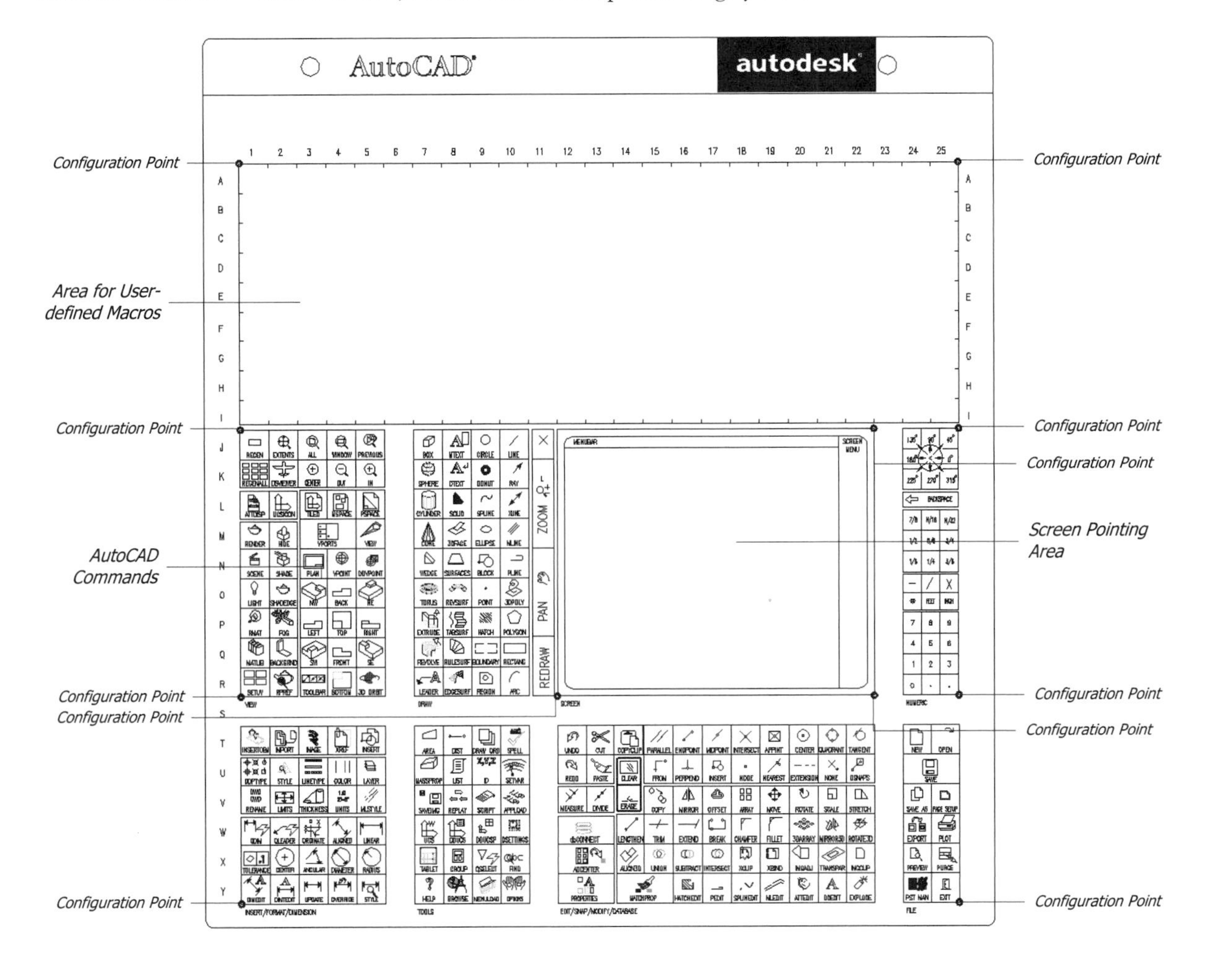

TabSurf

Rel.10 Draws meshes defined by path curves and direction vectors (short for TABulated SURFace).

Command	*Aliases*	*Menu*	*Ribbon*
tabsurf	...	**Draw** **↳Modeling** **↳Meshes** **↳Tabulated Mesh**	**Mesh Modeling** **↳Primitives** **↳Tabulated Surface**

Command: tabsurf
Select object for path curve: *(Select an object.)*
Select object for direction vector: *(Select another object.)*

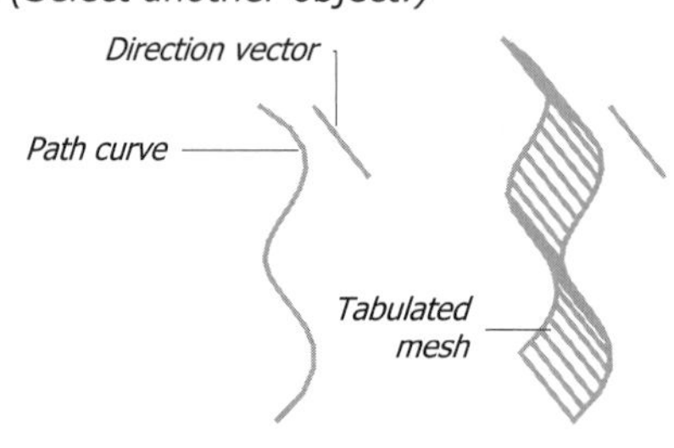

COMMAND LINE OPTIONS

Select object for path curve selects the object that defines the tabulation path.

Select object for direction vector selects the vector that defines the tabulation direction.

RELATED COMMANDS

Edge changes the visibility of 3D face edges.

PEdit edits 3D meshes, such as tabulated ones.

EdgeSurf draws 3D meshes between boundaries.

RevSurf draws revolved 3D meshes around an axis.

RuleSurf draws 3D meshes between open or closed boundaries.

RELATED SYSTEM VARIABLES

SurfTab1 defines the number of tessellations drawn by TabSurf in the *n*-direction.

MeshType determines whether tabulated surfaces are drawn with old polyface meshes or as new mesh objects.

TIPS

- The path curve can be an open or closed object: line, 2D polyline, 3D polyline, arc, circle, or ellipse.
- The direction in which you draw the *direction vector* determines the direction of the extrusion; the length of the vector determines the thickness of the extrusion.
- The number of *m*-direction tabulations is always 1, and lies along the direction vector. The number of *n*-direction tabulations is determined by SurfTab1 along curves only.

Aliased Commands

TargetPoint command is executed by the Light command.

Look for **TaskBar** in Appendix C, the list of system variables.

Text

v. 1.0 Places text, one line at a time, in drawings.

Commands	Aliases	Shortcut Keystrokes	Menu	Ribbon
text	**dtext**	...	**Draw**	**Annotate**
	dt		**↳Text**	**↳Text**
			↳Single Line Text	**↳Single Line Text**
-text	...	...	...	...

Command: text

Current text style: "Standard" Text height: 0.2000 Annotative: No

Specify start point of text or [Justify/Style]: *(Pick a point, or enter an option.)*

Specify height <0.2000>: *(Enter a value.)*

Specify rotation angle of text <0>: *(Enter a value.)*

Displays in-drawing text box.

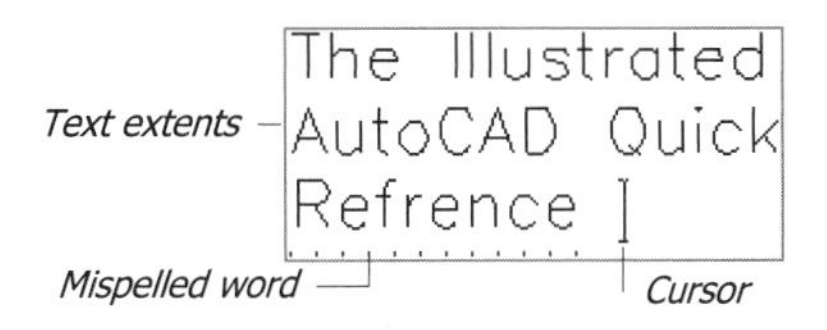

Press Enter once to start a new line of text; press Enter twice to exit the command.

COMMAND LINE OPTIONS

Specify start point of text indicates the starting point of the text.

Justify selects a text justification.

Style specifies the text style; see the Style command.

Specify height indicates the height of the text; this prompt does not appear when the style sets the height to a value other than 0.

Specify paper height indicates the text height in model space; this prompt appears only when an annotative style is used.

Specify rotation angle of text indicates the rotation angle of the text.

Enter continues text one line below the previously-placed text; press Enter twice to exit.

Justify options

Enter an option [Align/Fit/Center/Middle/Right/TL/TC/TR/ML/MC/MR/BL/BC/BR]: *(Enter an option.)*

Align aligns the text between two points with adjusted text height.

Fit fits the text between two points with fixed text height.

Center centers the text along the baseline.

Middle centers the text horizontally and vertically.

Right right-justifies the text.

TL top-left justifies the text.

TC top-center justifies the text.

TR top-right justifies the text.

ML middle-left justifies the text.

MC middle-center justifies the text.

MR middle-right justifies the text.

BL bottom-left justifies the text.

BC bottom-center justifies the text.

BR bottom-right justifies the text.

Style options

Enter style name or [?] <Standard>: *(Enter a name, or type ?.)*

Style name indicates a different style name.

? lists the currently-loaded styles.

SHORTCUT MENU

Access this menu by right-clicking text; see the TextEdit command for option.

Options specific to this command:

SPECIAL SYMBOLS

%%c places the diameter symbol: Ø.

%%d places the degree symbol: °.

%%o starts and stops overlining.

%%p places the plus-minus symbol: ±.

%%u starts and stops underlining.

-TEXT Command

Command: -text

This command is identical to the Text command, except that it ends after you enter one line of text. Also, no bounding box is displayed.

RELATED COMMANDS

TextEdit edits text.

Properties changes all aspects of text.

Style creates new text styles.

MText places paragraph text in drawings.

RELATED SYSTEM VARIABLES

DTextEd toggles between the in-place text editor and the old-style editor.

TextSize is the current height of text.

TextStyle is the current style of text.

TIPS

- You can enter any justification mode at the 'Start point' prompt.
- You can erase text by pressing Backspace, or by selecting text and then pressing the Del key.
- The spacing between lines of text does not necessarily match the snap spacing; the spacing is defined by the font definition file.
- Transparent commands and shortcut keystrokes do not work during the Text command.
- As of AutoCAD 2009, this command performs real-time spell checking.
- At the 'Enter text:' prompt, you can pick any point in the drawing to continue placing text. Press Shift+Tab to move to the previous line of text, or Tab to move to the next text line.

TextScr

v. 2.1 Switches from the AutoCAD window to the Text window (short for TEXT SCReen).

Command	*Aliases*	*Shortcut Keystroke*	*Menu*	*Ribbon*
'textscr	**...**	**F2**	**View** ↳**Display** ↳**Text Window**	**View** ↳**Windows** ↳**Text Window**

Command: textscr

Displays the Text window:

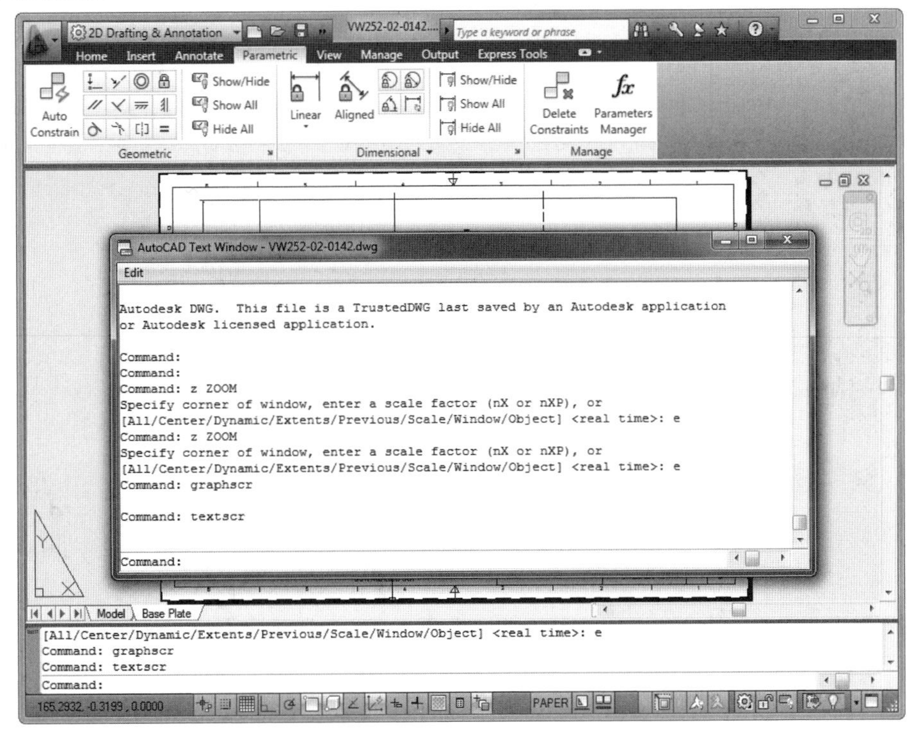

Behind: *Drawing window ("graphics screen").*
In front: *Text window ("screen").*

EDIT MENU

Recent Commands displays the commands most recently input.

Copy copies selected text to the Clipboard.

Copy History copies all text to the Clipboard.

Paste pastes text from the Clipboard into the Text window; available only when the Clipboard contains text.

Paste to CmdLine pastes text from the Clipboard to the command line; available only when the Clipboard contains text.

Options displays the Options dialog box; see the Options command.

SHORTCUT MENU

Access this menu by right-clicking in the Text window:

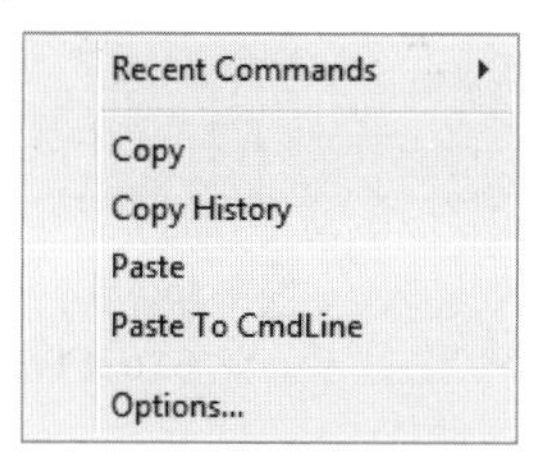

Recent Command displays a list of the last ten used commands.

Copy copies selected text to the Clipboard.

Copy History copies all text to the Clipboard.

Paste pastes text from the Clipboard into the Text window; available only when the Clipboard contains text.

Paste to CmdLine pastes text from the Clipboard to the command line; available only when the Clipboard contains text.

Options displays the Options dialog box; see the Options command.

RELATED COMMAND

GraphScr switches from the Text window to the AutoCAD drawing window.

RELATED SYSTEM VARIABLE

ScreenMode reports whether the screen is in text or graphics mode:

0 — Text screen.

1 — Graphics screen.

2 — Dual screen displaying both text and graphics.

TextEdit

Rel. 11 Edits a single-line text, multiline text, blocks, attribute values, and geometric tolerances (formerly the DdEdit command).

Command	*Aliases*	*Keyboard Shortcuts*	*Menu*	*Ribbon*
textedit	**ed** **ddedit**	...	**Modify** ↳**Object** ↳**Text** ↳**Edit**	...

You can activate this command by double-clicking text.

Command: textedit
Select an annotation object or [Undo]: *(Select a text object, or type U.)*

Select single-line text placed by the Text command, and the direct editing field is displayed.

Select paragraph text placed by the MText command, and the in-place text editor is displayed.

Select an attribute definition (not part of a block definition), and the Edit Attribute Definition dialog is displayed.

Select a block with attributes, and the Enhanced Attribute Editor dialog box is displayed.

Select geometric tolerances, and the Geometric Tolerance dialog box is displayed.

Select a dimension or dimensional constraint, and the in-place text editor is displayed.

Select an annotation object or [Undo]: *(Press Esc to exit command.)*

COMMAND LINE OPTIONS

Undo undoes editing.

Esc ends the command.

SHORTCUT MENU

Access the menu by right-clicking the text; when words are misspelled, the menu shows one or more spelling suggestions.

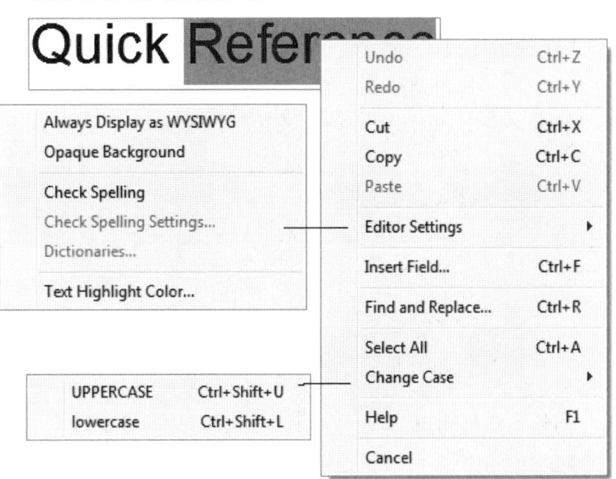

Undo undoes the last action.

Redo redoes the last undo.

Cut copies the selected text to the Clipboard, and then erases it.

Copy copies the selected text to the Clipboard.

Paste pastes text from the Clipboard; unavailable if Clipboard contains no text.

Editor Settings

- **Always Display as WYSIWYG** displays text in-place ("what you see is what you get").
- **Opaque Background** changes the background color to gray.
- **Check Spelling** toggles real-time spell checking.
- **Check Spelling Settings** displays the Check Spelling Settings dialog box; see the Spell command.
- **Dictionaries** displays the Dictionaries dialog box; see the Spell command.
- **Text Highlight Color** displays the Select Color dialog box.

Insert Field displays the Field dialog box; see the Field command.

Find and Replace displays the Find and Replace dialog box; see the Find command.

Select All selects all text.

Change Case changes the selected text between all UPPERCASE and all lowercase.

Help displays online help for this command.

Cancel ignores editing changes for the current annotation object, and dismisses the dialog box.

RELATED COMMANDS

AttIPedit edits attribute values.

EAttEdit edits all text attributes connected with a block.

Field places automatically-updatable field text.

Find searches for text in drawings.

MtEdit edits paragraph text.

Properties edits all text *properties*, including the text itself.

Spell checks the spelling of words in drawings.

RELATED SYSTEM VARIABLES

DTextEd toggles between direct in-place editing and the old Edit Text dialog box.

FontAlt specifies the name of the font to be used when the required font file cannot be located by AutoCAD.

FontMap maps font names.

MirrText determines how the Mirror command affects text.

TextFill toggles the fill of TrueType fonts during plotting and rendering.

TIPS

- TextEdit automatically repeats; press Esc to cancel the command.
- Between AutoCAD 2000 and 2004, this command did edit attribute text.
- As an alternative to this command, double-click text to bring up the appropriate editor.
- AutoCAD 2006 changed this command from a dialog box format to direct editing in drawings.
- Text placed with the Field, Leader, and QLeader commands is mtext; double-click the text to edit.
- Text can also be edited through the Properties window.
- In AutoCAD 2010, Autodesk changed the name of the DdEdit command to TextEdit.

TextToFront

2005 Places all text and dimensions visually on top of overlapping objects.

Commands	*Aliases*	*Menu*	*Ribbon*
texttofront	...	**Tools** **↳Draw Order** **↳Bring Text and Dimensions to Front**	...

Command: texttofront

Bring to front [Text/Dimensions/Both] <Both>: *(Enter an option.)*

nnn **object(s) brought to front.**

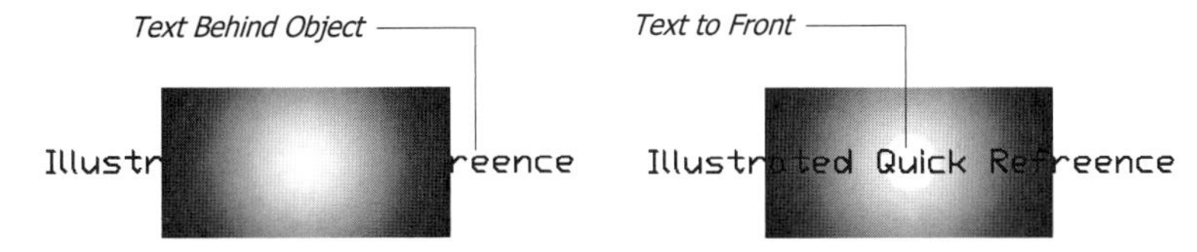

COMMAND LINE OPTIONS

Text brings all text to the front.

Dimensions brings all dimensions to the front.

Both brings both text and dimensions to the front.

RELATED COMMANDS

DrawOrder controls the display order of selected objects.

HatchToBack displays all hatches and gradients behind all other objects *(new to AutoCAD 2011).*

TIPS

- You must apply this command separately to model space and each layout.
- Newer objects are drawn on top of older objects in drawings; to change this, change the DrawOrderCtrl system variable.

Thicken

2007 Creates 3D solids by thickening surfaces.

Command	Aliases	Shortcut Keystrokes	Menu	Ribbon
thicken	...	...	**Modify** ⇘**3D Operations** ⇘**Thicken**	**Home** ⇘**Solid Editing** ⇘**Thicken**

Command: thicken
Select surfaces to thicken: *(Select one or more surfaces.)*
Select surfaces to thicken: *(Press Enter.)*
Specify thickness <0.0>: *(Type a number.)*

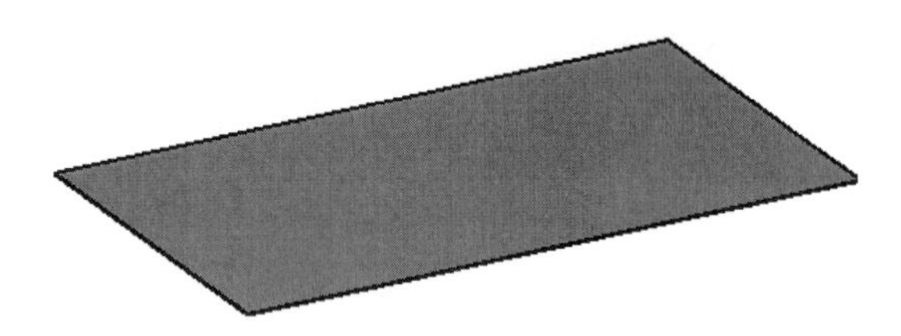

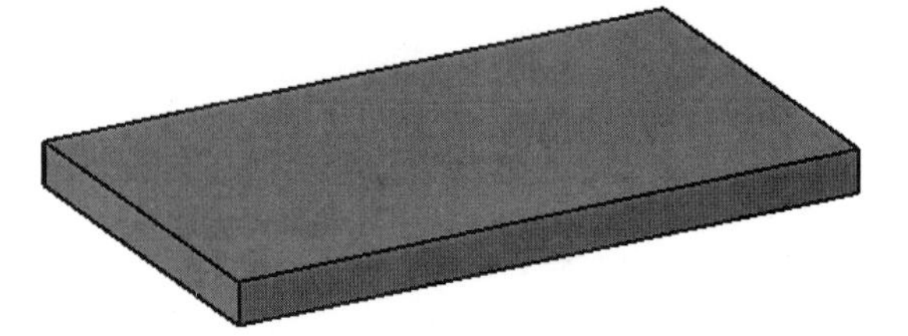

Left: *Surface before thickening.*
Right: *Surface turned into 3D solid.*

COMMAND LINE OPTIONS

Select surfaces to thicken selects one or more surfaces; this command does not work with meshes.

Specify thickness specifies the height of the thickened surface, such as 10. Enter a negative number, such as -10, to thicken the surface downwards.

RELATED COMMANDS

ConvToSolid converts circles and polylines with thickness to 3D solids.

PolySolid converts existing lines, 2D polylines, arcs, and circles to 3D solids.

Extrude thickens 2D objects into solids and surfaces.

RELATED SYSTEM VARIABLE

DelObj determines whether or not the surface is erased following the thickening.

Time

v. 2.5 Displays time-related information about the current drawing.

Command	*Aliases*	*Shortcut Keystrokes*	*Menu*	*Ribbon*
'time	...	...	**Tools** **↳Inquiry** **↳Time**	...

Command: time

Display/ON/OFF/Reset: *(Enter an option.)*

Example output:

```
Current time:             Wednesday, April 19, 2016  4:36:28:562 PM
Times for this drawing:
  Created:                Wednesday, April 19, 2016  3:19:27:359 PM
  Last updated:           Wednesday, April 19, 2016  3:19:27:359 PM
  Total editing time:     0 days 01:17:01:234
  Elapsed timer (on):     0 days 01:17:01:234
  Next automatic save in: <no modifications yet>
```

COMMAND LINE OPTIONS

Display displays the current time and date.

OFF turns off the user timer.

ON turns on the user timer.

Reset resets the user timer.

RELATED COMMAND

Status displays information about the current drawing and environment.

RELATED SYSTEM VARIABLES

CDate is the current date and time.

Date is the current date and time in Julian format.

SaveTime is the interval for automatic drawing saves.

TDCreate is the date and time the drawing was created.

TDInDwg is the time the drawing spent open in AutoCAD.

TDUpdate is the last date and time the drawing was changed.

TDUsrTimer is the current user timer setting.

TIPS

- The time displayed by the Time command is only as accurate as your computer's clock.
- The SaveAs command does not reset timings. The TdCreate system variable retains the original value, while the other timers continue accumulating seconds.
- Starting a new drawing from a template file resets the timers; values in the template do not carry forward.

Removed Command

TiffIn command was removed from AutoCAD R14. Use the ImageAttach command instead.

TifOut

2004 Exports the current viewport in TIFF raster format.

Commands	*Aliases*	*Shortcut Keystrokes*	*Menu*	*Ribbon*
tifout	...	...	...	...

Command: tifout

Displays Create Raster File dialog box. Specify a file name, and then click Save.

Select objects or <all objects and viewports>: *(Select objects, or press Enter to select all objects and viewports.)*

COMMAND LINE OPTIONS

Select objects selects specific objects.

All objects and viewports selects all objects and all viewports, whether in model space or in layout mode.

RELATED COMMANDS

BmpOut exports drawings in BMP (bitmap) format.

Image places raster images in the drawing.

JpgOut exports drawings in JPEG (joint photographic expert group) format.

Plot exports drawings in TIFF and other raster formats.

PngOut exports drawings in PNG (portable network graphics) format.

TIPS

- The rendering effects of the VsCurrent command are preserved, but not those of the Render command.
- TIFF files are commonly used in desktop publishing, but the format produced by AutoCAD cannot be read by some graphics programs.
- TIFF is short for "tagged image file format," and was co-developed by Microsoft and Aldus, the forerunner of Adobe.

TInsert

2005 Inserts a block or a drawing in a table cell (short for Table INSERT).

Commands	*Aliases*	*Shortcut Keystrokes*	*Menu*	*Ribbon*
tinsert	...	...	...	...

Command: tinsert
Pick a table cell: *(Select a single cell in a table.)*

Displays Insert a Block in a Table Cell dialog box.

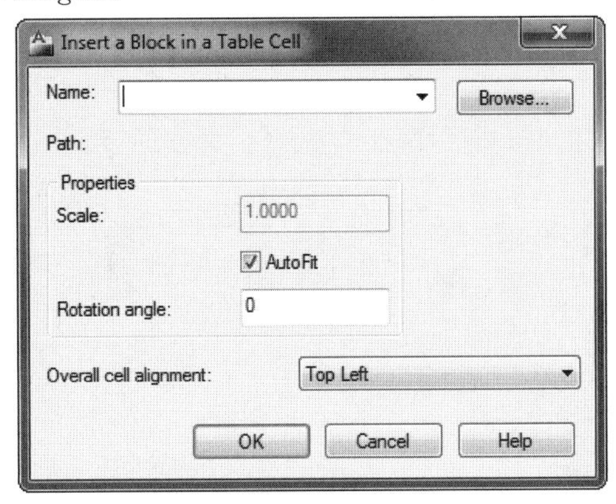

If the block contains attributes, the Enter Attributes dialog box is displayed.

COMMAND LINE OPTION

Pick a table cell specifies the cell in which to place the block.

INSERT A BLOCK IN A TABLE CELL DIALOG BOX

Name specifies the name of the block; the droplist names all blocks found in the current drawing.

Browse displays the Select Drawing File dialog box; select a *.dwg* or *.dxf* file, and click Open.

Properties options

Cell Alignment specifies the position of the block within the cell, from top-left to bottom-right.

Scale sizes the block, and adjusts the cell to fit the block.

☑ **AutoFit** fits the block to the cell.

Rotation Angle specifies the angle at which to place the block.

RELATED COMMANDS

Table creates new tables.

Block creates blocks.

TIP

- Alternatively, select a cell, right-click, and then choose Insert Block from the shortcut menu.

Tolerance

Rel. 13 Places geometric tolerancing symbols and text.

Command	*Alias*	*Shortcut Keystrokes*	*Menu*	*Ribbon*
tolerance	**tol**	**...**	**Dimension** ↳**Tolerance**	**Annotate** ↳**Dimensions** ↳**Tolerance**

Command: tolerance

Displays dialog box.

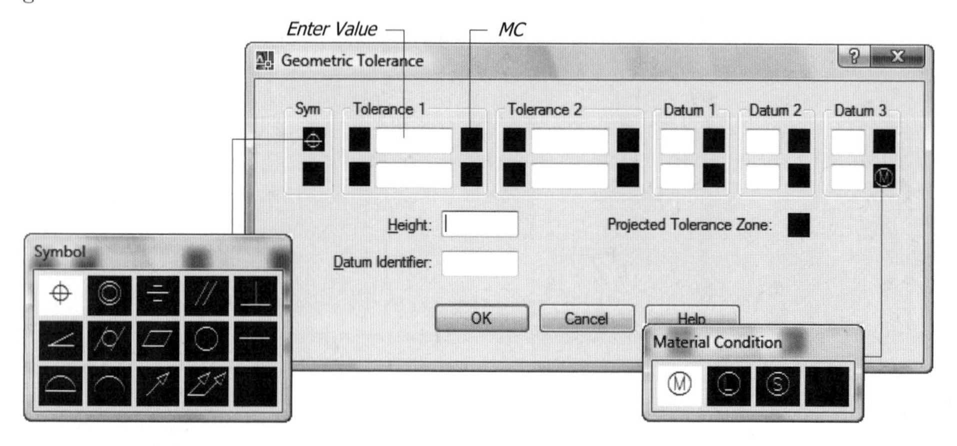

Fill in the dialog box, and then click OK. AutoCAD prompts:

Enter tolerance location: *(Pick a point.)*

GEOMETRIC TOLERANCE DIALOG BOX

Sym displays the Symbol dialog box.

Tolerance 1 and **Tolerance 2** specify the first and second tolerance values:

- **Dia** toggles the optional Ø (diameter) symbol.
- *Value* specifies the tolerance value.
- **MC** displays the Material Condition dialog box.

Datum 1, **Datum 2**, and **Datum 3** specify the datum references:

- **Value** specifies the datum value.
- **MC** displays the Material Condition dialog box.

Height specifies the projected tolerance zone value.

Projected Tolerance Zone toggles the projected tolerance zone symbol.

Datum Identifier creates the datum identifier symbol, such as -A-.

SYMBOL DIALOG BOX

Orientation Symbols

Position.

Concentricity and coaxiality.

Symmetry.

Parallelism.

Perpendicularity.

Angularity.

Form Symbols

Cylindricity.

Flatness.

Circularity and roundness.

Straightness.

Profile Symbols

Profile of the surface.

Profile of the line.

Circular runout.

Total runout.

MATERIAL CONDITION DIALOG BOX

(M) specifies maximum material condition.

(L) specifies least material condition.

(S) specifies regardless of feature size.

RELATED FILES

gdt.shp is the tolerance symbol definition source file.

gdt.shx is the compiled tolerance symbol file.

TIPS

- You can use the DdEdit command to edit tolerance symbols and feature control frames.
- Tolerances can have the annotative property, which ensures that they display at the correct size in model space.
- Tolerances use the current dimension style; if it is annotative, then so is the tolerance text.

DEFINITIONS

Datum — a theoretically-exact geometric reference that establishes the tolerance zone for the feature. These objects can be used as a datum: point, line, plane, cylinder, and other geometry.

Material Condition — symbols that modify the geometric characteristics and tolerance values (modifiers for features that vary in size).

Projected Tolerance Zone — the height of the fixed perpendicular part's extended portion; changes the tolerance to positional tolerance.

Tolerance — the amount of variance from perfect form.

Removed Command

Today command was removed from AutoCAD 2004; it was replaced by the Communication Center.

Torus

Rel. 11 Draws 3D tori as solid models.

Command	Alias	Shortcut Keystrokes	Menu	Ribbon
torus	**tor**	...	**Draw** ↳**Modeling** ↳**Torus**	**Home** ↳**Modeling** ↳**Torus**

Command: torus

Specify center point or [3P/2P/Ttr]: *(Pick point 1.)*

Specify radius or [Diameter] <1.0>: *(Specify the radius, 2, or type D.)*

Specify tube radius or [2Point/Diameter]: *(Specify the radius, 3, or type D.)*

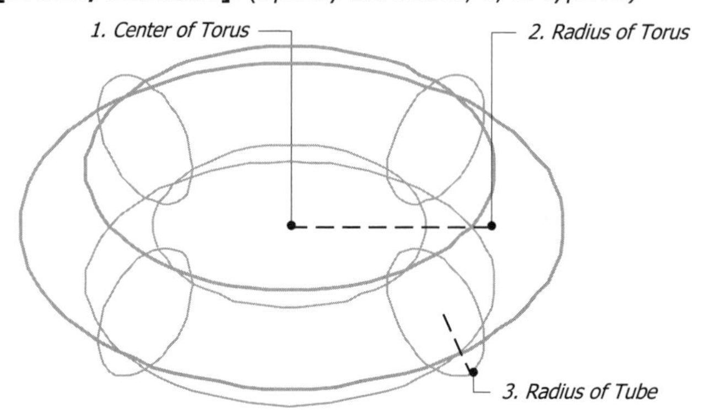

COMMAND LINE OPTIONS

Center point indicates the center of the torus.

3P specifies three points on outer circumference of torus.

2P specifies two points on the outer diameter of the torus.

Ttr specifies two tangent points plus the radius of the outer circumference of the torus.

Specify radius indicates the radius of the torus.

Diameter indicates the diameter of the torus.

Tube radius indicates the radius of the tube.

2Point asks for two points to determine the diameter of the tube.

RELATED COMMANDS

Ai_Torus (undocumented) creates a torus from 3D polyfaces or meshes.

Mesh command's Torus option draws 3D mesh tori.

Cone draws solid cones.

Cylinder draws solid cylinders.

Sphere draws solid spheres.

RELATED SYSTEM VARIABLES

Isolines specifies number of isolines to draw on torus.

DragVs specifies the visual style during torus construction.

TIPS

- When the torus radius is negative, the tube radius must be a larger positive number.
- The radius can no longer be negative number, as in earlier releases of AutoCAD.
- Specify a tube diameter larger than the torus diameter to create a hole-less torus.

- After it is constructed, the torus can be edited by grips.

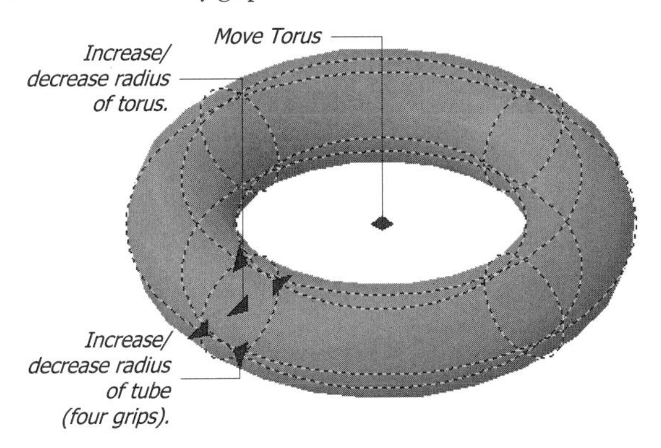

-Toolbar

Rel. 13 Displays and hides toolbars via the command line.

Commands	*Aliases*	*Shortcut Keystrokes*	*Menu*	*Ribbon*
-toolbar	...	...	...	...

The Toolbar and TbConfig commands are aliases for the CUI command; Toolbar and -Toolbar are two different commands.

Command: -toolbar
Enter toolbar name or [ALL]: *(Enter a name, or type ALL.)*
Enter an option [Show/Hide]: *(Type S or H.)*

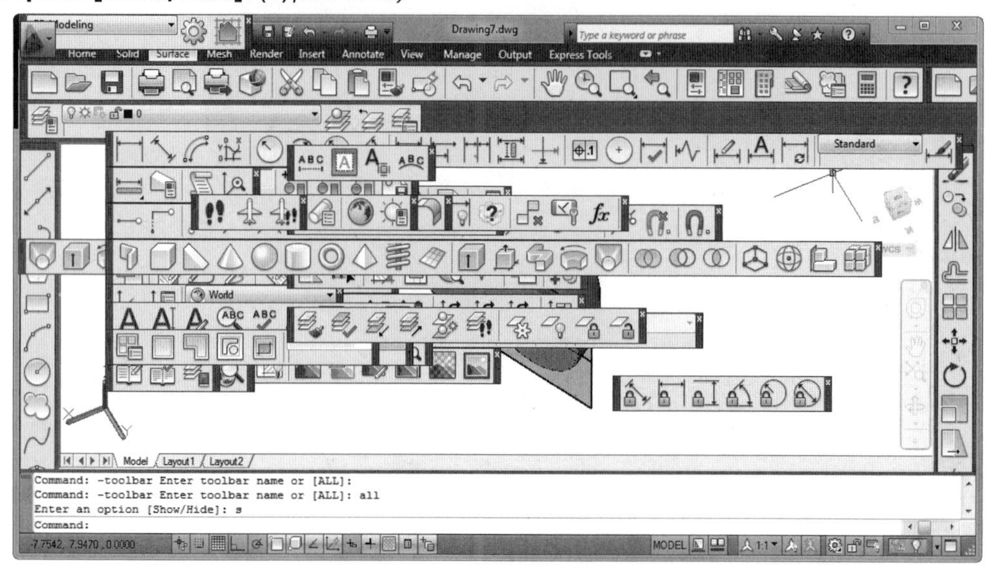

COMMAND LINE OPTIONS

Toolbar name names the toolbar.

ALL applies the command to all toolbars; must be entered in all capital letters.

Show displays the toolbar(s).

Hide dismisses the toolbar(s).

RELATED COMMAND

CUI customizes toolbars via a dialog box.

RELATED SYSTEM VARIABLE

ToolTips toggles the display of tooltips.

RELATED FILE

***.cuix** is a user interface customization file.

TIP

- Right-click any open toolbar for a shortcut menu that accesses toolbars. Use this command when all toolbars are off.

Aliased Command

Toolbar command is now a hardwire alias for the CUI command.

ToolPalettes / ToolPalettesClose

2004 Displays and closes the Tool palette.

Commands	Alias	Shortcut Keystrokes	Menu	Ribbon
'toolpalettes	**tp**	**Ctrl+3**	**Tools** **↳Palettes** **↳Tool Palettes**	**View** **↳Palettes** **↳Tool Palettes**
'toolpalettesclose	**...**	**Ctrl+3**	**...**	**...**

Command: toolpalettes

Opens palette:

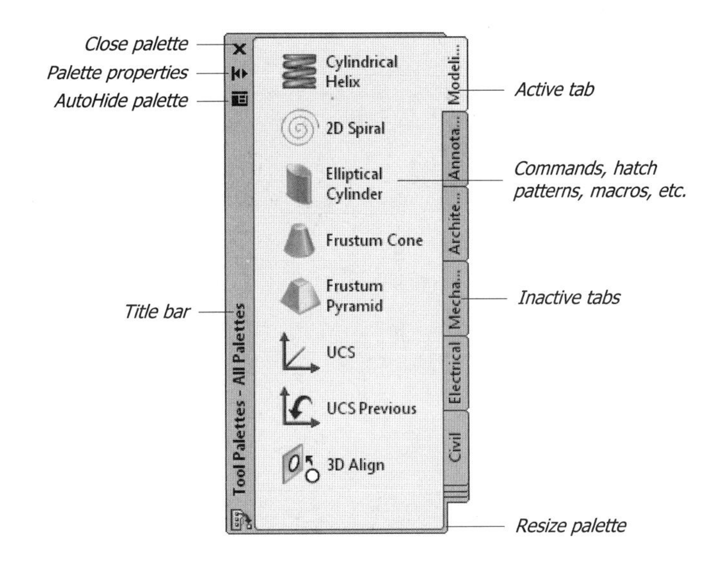

Command: toolpalettesclose

Closes the palette.

SHORTCUT MENUS

Tabs shortcut menu

Right-click the active tab.

Move Up
Move Down
View Options...
Paste
New Palette
Delete Palette
Rename Palette

Move up moves the selected tab up one position.

Move down moves the tab down one position.

View Options displays the View Options dialog box.

Paste pastes a tool previously copied from a palette.

New Palette creates a new tab, and then prompts you for a name.

Delete Palette removes the tab, after prompting you for confirmation.

Rename Palette renames the tab.

Properties shortcut menu

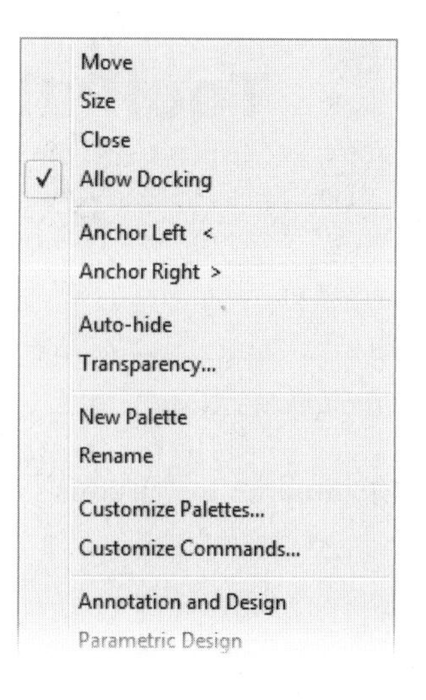

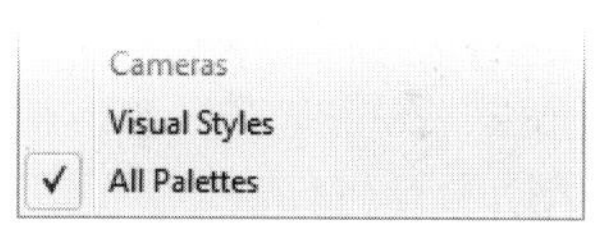

Right-click the palette, or click the Properties button.

Move moves the palette.

Size changes the size of the palette.

Close closes the palette.

Allow Docking toggles whether the palette can be docked.

Anchor Left < and **Right >** *anchors* (docks and minimizes) the palette at the left or right edges of the AutoCAD drawing area.

Auto-hide reduces the palette to the size of its title bar when the cursor is elsewhere.

Transparency displays the Transparency dialog box.

New Palette creates a new tab, and then prompts you for a name.

Rename renames the selected tab.

Customize Palettes displays the Customize dialog box; see the Customize command.

Customize Commands displays the Customize User Interface dialog box; see the Cui command.

All Palettes displays all palettes, as determined by the Customize command.

Palette shortcut menu

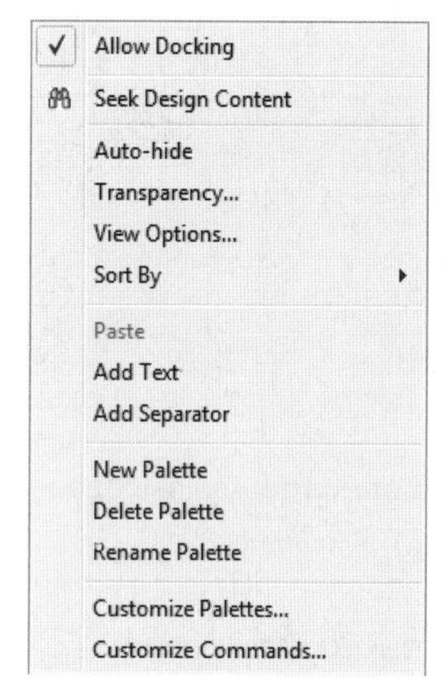

Right-click in the palette, away from any icon. The following options that differ from the above menu:

View Options displays the View Options dialog box.

Sort by sorts the tools:

- **Name** of tool.
- **Type** of tool, such as Command, General, or Pattern.

Paste pastes a tool previously copied from a palette.

Add Text adds text to the palette.

Add Separator adds a horizontal line to the palette.

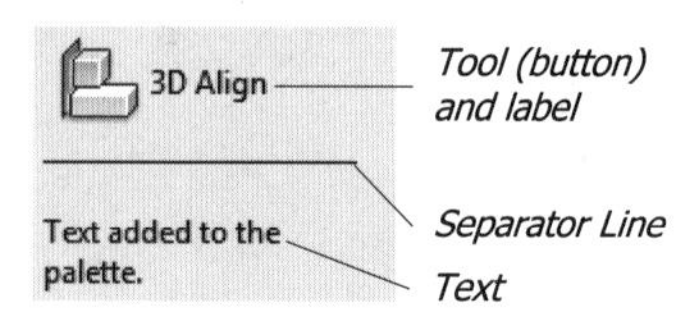

Icon shortcut menu

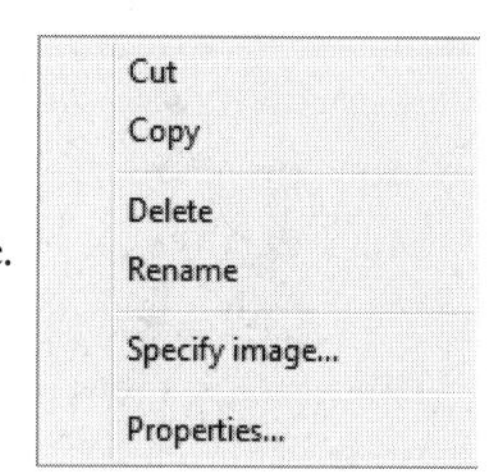

Right-click any icon; contents of menu varies with the type of icon:

Cut copies the icon and its properties to the Clipboard, and then erases the icon.

Copy copies the icon and its properties, which can then be pasted to another palette.

Specify Image displays Select Image File dialog box; choose a BMP, JPG, PNG, GIF, or TIF file.

Properties displays the Properties dialog box.

Blocks also have these options:

Redefine redefines the block.

Block Editor opens the block in the Block Editor.

Update Tool Image updates the icon, should the block change.

DIALOG BOXES

Transparency dialog box

Access this dialog box by right-click the palette's title bar, and then selecting Transparency; transparency is unavailable with hardware acceleration.

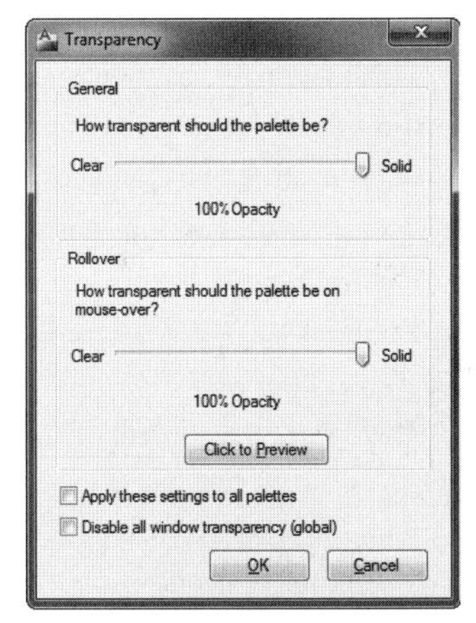

General changes the translucency of the palette from Clear (none) to Solid.

Rollover changes the translucency of rollovers (large tooltips).

Click to Preview previews the translucency effects.

☐ **Apply these settings to all palettes** assigns all of AutoCAD's palettes the same levels of translucency.

☐ **Disable all window transparency (global)** makes all palettes opaque.

View Options dialog box

Access this dialog box by right-clicking the active tab, and then selecting View Options.

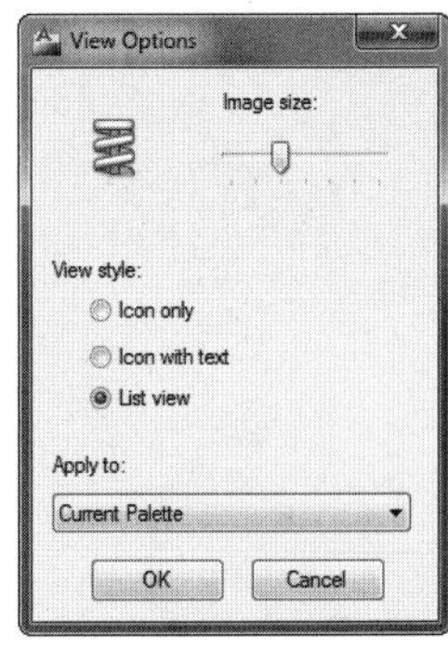

Image Size changes the size of icons from 14 pixels square to 54 pixels.

View style toggles the display of icons and text:

○ **Icon only** displays icons only.

⦿ **Icon with text** displays icons with text below.

○ **List view** displays small icons with text beside.

Apply to determines if changes apply to the current palette or all palettes.

Properties dialog box

Access this dialog box by right-clicking an icon, and then selecting Properties.

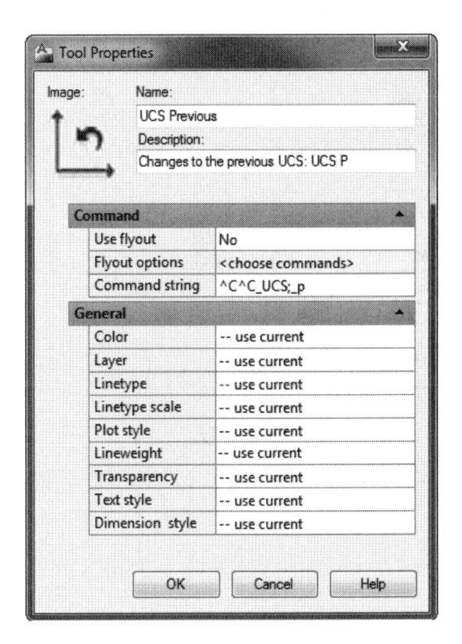

Image displays the icon associated with the tool. To change the icon, right-click it and then select Specify Image.

Name specifies the name that appears on the palette; it can be different from the Command String.

Description is the text that accompanies the icon on the palette; filling it in is optional.

Command and **General** items vary with the type of tool.

RELATED COMMANDS

TpNavigate displays the specified tool palette or palette group.

Customize exports and imports Tool palettes, and creates groups of palettes.

CUI customizes the content of tool palettes.

RELATED SYSTEM VARIABLES

PaletteOpaque determines whether the Tools palette can be transparent.

TpState notes whether the Tools palette is open.

RELATED FILES

***.*xtp* stores the content of Tools palette in an XML-like format.

TIPS

- You can drag objects (while the right mouse button is held down) or toolbar buttons (while the Customize dialog box is open) onto palettes. By editing existing tools, you can store program code.
- To bring content from the DesignCenter into the Tool palette:
  1. In DesignCenter, right-click an item.
  2. From the shortcut menu, select Create Tool Palette.
- The *.xtp* files store content in XML (extended markup language) format.
- The Customize command allows you to create groups of palettes.

TpNavigate

2008 Displays the specified tool palette or palette group (short for "Tool Palettes NAVIGATE").

Command	*Aliases*	*Shortcut Keystrokes*	*Menu*	*Ribbon*
tpnavigate	...	...	...	...

Command: tpnavigate

Specify tool palette to display or [palette Group]: *(Enter the name of a tool palette name, or type G.)*

COMMAND LINE OPTIONS

Specify Tool Palette to Display displays the specified tool palette.

Palette Group prompts, 'Specify palette group to display:'; enter the name of a valid tool palette group.

RELATED COMMAND

ToolPalettes opens the Tools palette.

RELATED SYSTEM VARIABLES

_AuthorPalettepath specifies the path to the folder holding customized Tool palettes (undocumented).

_ToolPalettepath specifies the path to the folder holding Tool palettes (undocumented).

TpState reports whether the Tools palette is open.

TIPS

- You can get the names of groups and palettes from the Customize command.
- This command is meant for use in macros and scripts.
- When you enter an unrecognized name, AutoCAD opens the Tools palette anyway, displaying the last-accessed group.

Trace

v. 1.0 Draws line segments with width.

Command	Aliases	Shortcut Keystrokes	Menu	Ribbon
trace	...	...	...	...

Command: trace
Specify trace width <0.050>: *(Enter a value.)*
Specify start point: *(Pick a point.)*
Specify next point: *(Pick another point.)*
Specify next point: *(Press Enter to end the command.)*

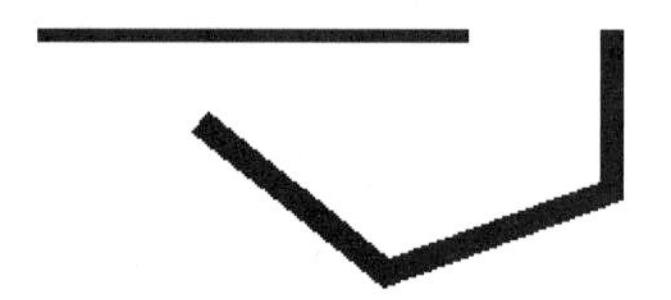

COMMAND LINE OPTIONS

Trace width specifies the width of the trace.

Start point picks the starting point.

Next point picks the next vertex.

Enter exits this command.

RELATED COMMANDS

Line draws lines with zero width.

MLine draws up to 16 parallel lines.

PLine draws polylines and polyline arcs with varying widths.

LWeight gives every object a width.

RELATED SYSTEM VARIABLES

FillMode toggles display of fill or outline traces (default = 1, on).

TraceWid specifies the current width of the trace (default = 0.05).

TIPS

- Trace was meant for drawing traces in printed circuit board designs with the earliest versions of AutoCAD. This command has largely been replaced by the PLine command.
- Traces are drawn along the center line of the pick points. During drawing, the display of a trace segment is delayed by one pick point.
- During the drawing of traces, you cannot back up, because there is no Undo option; if you require this feature, draw wide lines with the PLine command, setting the Width option.
- There is no option for controlling joints (which are always beveled) or end capping (which are always square); if you wish to modify these features, draw wide lines with the MLine command, after setting the solid fill, end cap, and joint options with the MlStyle command.

Tracking

Rel. 14 Locates x and y points visually, relative to other points in the command sequence; *not* a command, but an option modifier.

Command	*Aliases*	*Shortcut Keystrokes*	*Menu*	*Ribbon*
tracking	**tk** **track**	...	...	...

Example usage:

Command: line

Specify first point: *(Pick a point.)*

Specify next point or [Undo]: tk

Enters tracking mode:

First tracking point: *(Pick a point.)*

Next point (Press ENTER to end tracking): *(Pick a point.)*

Next point (Press ENTER to end tracking): *(Press Enter to end tracking.)*

Exits tracking mode:

Specify next point or [Undo]: *(Pick a point.)*

COMMAND LINE OPTIONS

First tracking point picks the first tracking point.

Next point picks the next tracking point.

ENTER exits tracking mode.

RELATED KEYBOARD MODIFIERS

Direct distance entry specifies an angle and relative distance to the next point.

from locates an offset point.

m2p finds a point midway between two picked points.

TIPS

- Tracking is not a command, but an option modifier. It is entered during commands without the ' transparent-command prefix.
- Tracking can be used in conjunction with direct distance entry.
- In tracking mode, AutoCAD automatically turns on ortho mode to constrain the cursor vertically and horizontally.
- If you start tracking in the x direction, the next tracking direction is y, and vice versa.
- You can use tracking as many times as you need to at the 'Specify first point' and 'Specify next point' prompts.

Transparency

Rel. 14 Toggles the transparency of pixels in raster images.

Command	Aliases	Shortcut Keystrokes	Menu	Ribbon
transparency	...	...	**Modify** ↳**Object** ↳**Image** ↳**Transparency**	...

Command: transparency
Select image(s): *(Select one or more images inserted with the Image command.)*
Select image(s): *(Press Enter to end object selection.)*
Enter transparency mode [ON/OFF] <OFF>: *(Type ON or OFF.)*

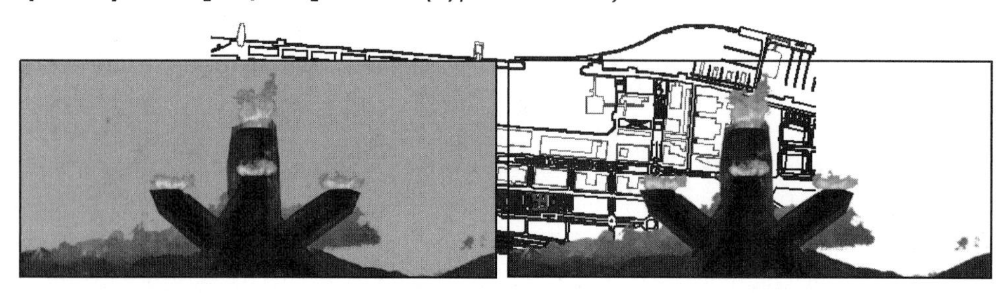

Left: *Transparency off.*
Right: *Transparency on.*

COMMAND LINE OPTIONS

Select image(s) selects the objects whose transparency to change.

ON makes the background pixels transparent.

OFF makes the background pixels opaque.

RELATED COMMANDS

ImageAttach attaches a raster image as an externally-referenced file.

ImageAdjust changes the brightness, contrast, and fading of a raster image.

TIPS

- *Transparent pixels* allow graphics under the image to show through. This command has nothing to do with the transparency property introduced by AutoCAD 2011, which allows you to see through any object.
- This command works only with raster images placed in drawings with the Image command, and is meant for use with raster file formats that allow transparent pixels.
- For this command to work correctly, you must first specify the color of pixel transparent. Use a raster editor, such as PaintShop Pro: when saving the image, designate the color as transparent.

TraySettings

2004 Specifies settings for the icons that appear at the right end of the status bar (known as the "tray").

Command	*Aliases*	*Status Bar*	*Menu*	*Ribbon*
traysettings	...	**Tray Settings**	...	**View** ↳**Windows** ↳**Status Bar**

Command: traysettings

Displays dialog box.

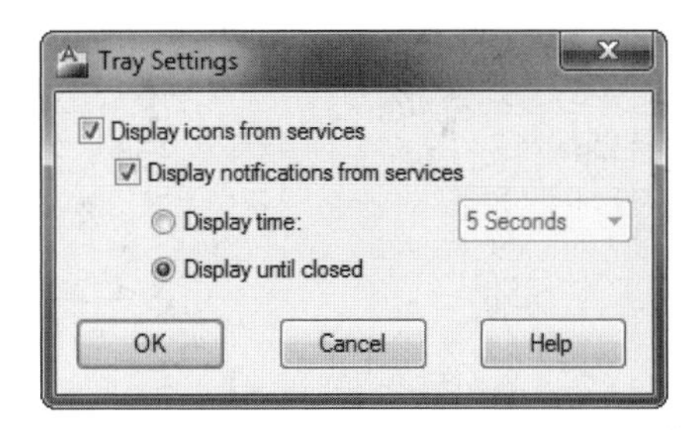

TRAY SETTINGS DIALOG BOX

Display icons from services:

☑Displays "tray" at the right end of the status line; see below.

☐Turns off the tray.

Display notification from services:

☑Displays balloon notifications from a variety of services; see below.

☐Turns off notifications:

○ **Display time** specifies the duration that a notification balloon is displayed.

⊙ **Display until closed** specifies that the notification balloon is displayed until closed by the user.

RELATED SYSTEM VARIABLES

TrayIcons toggles the display of the tray on the status bar.

TrayNotify toggles whether notification balloons are displayed.

TrayTimeOut determines the length of time that notification balloons are displayed.

TIP

- Examples of notification balloons appear below. Click underlined links for more information, or click **X** to dismiss.

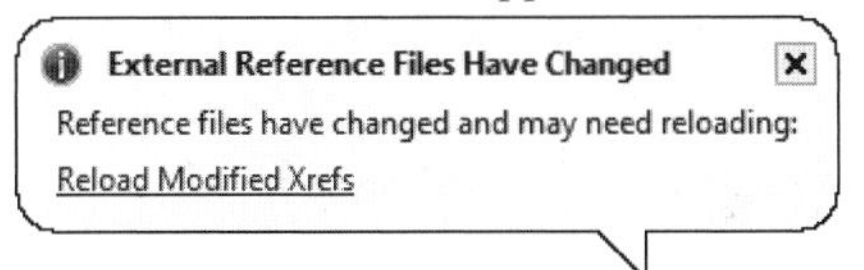

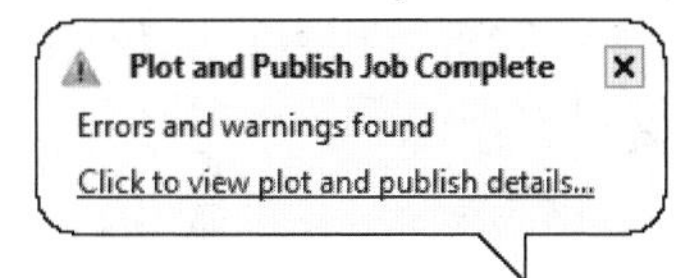

TreeStat

Rel. 12 Displays the status of the drawing's spatial index, including the number and depth of nodes (short for TREE STATistics).

Command	Aliases	Shortcut Keystrokes	Menu	Ribbon
'treestat	...	...	...	...

Command: treestat

Sample output:

```
Deleted objects: 12

Model-space branch
-------------------
Oct-tree, depth limit = 30
Subtree containing objects with defined extents:
    Nodes: 21   Objects: 8   Maximum depth: 23
    Average objects per node: 0.38
    Average node depth: 14.24   Average object depth: 17.00
    Objects at depth 14: 2   17: 5   23: 1
    Nodes with population 0: 18   1: 1   2: 1   5: 1
Total nodes: 24   Total objects: 8

Paper-space branch
-------------------
Quad-tree, depth limit = 20
Subtree containing objects with defined extents:
    Nodes: 1   Objects: 0
    Average objects per node: 0.00
    Average node depth: 5.00
    Nodes with population 0: 1
Total nodes: 4   Total objects: 0
```

RELATED SYSTEM VARIABLES

TreeDepth specifies the size of the tree-structured spatial index in *xxyy* format:

xx — number of model space nodes (default = 30).
yy — number of paper space nodes (default = 20).
-xx — 2D drawing.
+xx — 3D drawing (default).

TreeMax is the maximum number of nodes (default = 10,000,000).

TIPS

- Better performance occurs with fewer objects per oct-tree node.
- When redraws and object selection seem slow, increase the value of TreeDepth.
- Each node consumes 80 bytes of memory.

DEFINITIONS

Oct Tree — the model space branch of the spatial index, where all objects are either 2D or 3D. *Oct* comes from the eight volumes in the xyz coordinate system of 3D space.

Quad Tree — the paper space branch of the spatial index, where all objects are two-dimensional. *Quad* comes from the four areas in the xy coordinate system of 2D space.

Spatial Index — objects indexed by oct-region to record their position in 3D space; it has a tree structure with two primary branches: oct tree and quad tree. Objects are attached to *nodes*; each node is a branch of the *tree*. It's not clear that anyone understands this.

Trim

v. 2.5 Trims lines, arcs, circles, 2D polylines, and hatches to existing and projected cutting lines and views.

Command	*Alias*	*Shortcut Keystrokes*	*Menu*	*Ribbon*
trim	**tr**	**...**	**Modify** **↳Trim**	**Home** **↳Modify** **↳Trim**

Command: trim
Current settings: Projection=UCS Edge=None
Select cutting edges ...
Select objects or <select all>: *(Select one or more objects.)*
Select objects: *(Press Enter to end object selection.)*
Select object to trim or shift-select to extend or [Fence/Crossing/Project/Edge/eRase/Undo] *(Select object, or enter an option.)*
Select object to trim or shift-select to extend or [Fence/Crossing/Project/Edge/eRase/Undo] *(Press Enter to end command.)*

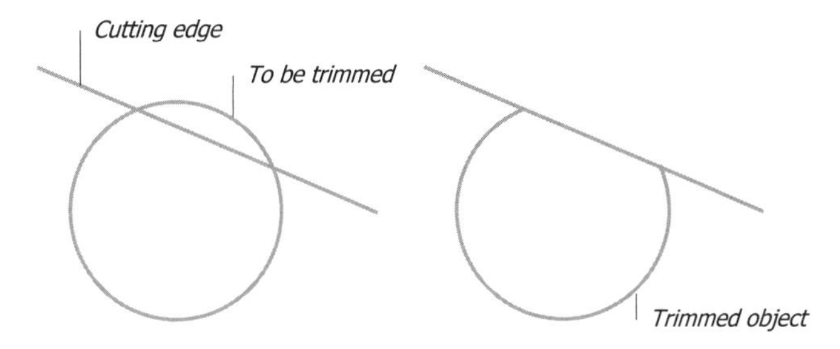

COMMAND LINE OPTIONS

Select objects selects the cutting edges.

Select all selects all objects in the drawing .

Select object to trim trims the object; picked portion is trimmed.

Undo untrims the last trim action.

Fence options

Specify first fence point locates the first point for fence selection mode.

Specify next fence point locates subsequent points; press Enter to exit selection mode.

Crossing options

Specify first corner locates the first corner for crossing selection mode.

Specify opposite corner locates the second corner.

Project options

Enter a projection option [None/Ucs/View] <Ucs>: *(Enter an option.)*

None uses only objects as cutting edges.

Ucs trims at the xy plane of the current UCS.

View trims at the current view plane.

Edge options

Enter an implied edge extension mode [Extend/No extend] <No extend>: *(Type E or N.)*

Extend extends the cutting edge to trim object.

No extend trims only at an actual cutting edge.

Erase option

Select objects to erase or <exit> selects objects to erase; press Enter to exit object selection mode.

RELATED COMMANDS

SurfTrim trims surfaces *(new to AutoCAD 2011)*.

Change changes the size of lines, arcs and circles.

Extend lengthens lines, arcs and polylines.

Lengthen lengthens open objects.

RELATED SYSTEM VARIABLES

EdgeMode determines whether this command projects cutting edges.

ProjMode determines how cutting edges are projected in 3D space.

TIPS

- This command also trims hatches; blocks can be cutting edges, but must be selected with Single, Crossing, Fence, and Select All modes.
- You can select all objects in the drawing by pressing Enter at the 'Select objects' prompt.
- To trim intersecting lines, it can be easier to use the Fillet command with Radius set to 0.
- To have this command extend objects (rather than trim them), hold down the Shift key while selecting objects to extend; the opposite occurs in the Extend command.
- Cutting edges can be trimmed, too.
- When trimming is ambiguous, AutoCAD trims the first object encountered going clockwise inside rectangular crossing windows.

U

V. 2.5 Undoes the most recent AutoCAD command (short for Undo).

Command	*Aliases*	*Shortcut Keystroke*	*Menu*	*Quick Access Toolbar*
u	**...**	**Ctrl+Z**	**Edit** ↳**Undo**	

Command: u

COMMAND LINE OPTIONS

None.

RELATED COMMANDS

Oops unerases the most-recently erased object.

Quit exits the drawing, undoing all changes.

Redo reverses the most-recent undo, if the prior command was U or Undo.

Undo allows more sophisticated control over undo than U.

RELATED SYSTEM VARIABLE

UndoCtl toggles the undo mechanism.

TIPS

- The U command is convenient for stepping back through the design process, undoing one command at a time.
- As of AutoCAD 2006, multiple zooms and pans are grouped into a single undo. This feature can be toggled through the Combine Zooms and Pans option found on the User Preferences tab of the Options dialog box.
- The U command is the same as the Undo 1 command; for greater control over the undo process, use the Undo command.
- The Redo command redoes the most-recent undo only; use MRedo otherwise.
- The Quit command, followed by the Open command, restores the drawing to its original state, if not already saved.
- Because the undo mechanism creates a mirror drawing file on disk, disable the Undo command with system variable UndoCtl (set to 0) when your computer is low on disk space.
- Commands that involve writing to file, plotting, and some display functions (such as Render) are not undone.

Ucs

Rel.10 Defines new coordinate planes, and restores existing UCSs (short for User-defined Coordinate System).

Command	Alias	Shortcut Keystrokes	Menu	Ribbon
ucs	**dducs**	**...**	**Tools** ↳**New UCS**	**View** ↳**Coordinates**

Command: ucs

For compatibility with earlier versions of AutoCAD, you may enter any of the older options at the first prompt.

Current ucs name: *WORLD*

Specify origin of UCS or [Face/NAmed/OBject/Previous/View/World/X/Y/Z/ZAxis] <World>: *(Enter an option.)*

COMMAND LINE OPTIONS

Specify origin of new UCS moves the UCS to a new origin.

Face aligns the UCS with the face of a 3D solid object.

NAmed saves, restores, and deletes named UCSs.

OBject aligns the UCS with a selected object; see tips.

Previous restores the previous UCS orientation.

World aligns the UCS with the WCS.

View aligns the UCS with the current view.

X rotates the UCS about the x axis.

Y rotates the UCS about the y axis.

Z rotates the UCS about the z axis.

ZAxis aligns the UCS with a new origin and z axis.

Face options

Select face of solid object: *(Pick a face.)*

Enter an option [Next/Xflip/Yflip] <accept>: *(Press Enter, or enter an option.)*

Select face selects a face to which the UCS will be applied.

Next selects the next face.

Xflip flips the x axis.

Yflip flips the y axis.

accept accepts the selected face. (Press Enter to accept.)

NAme options

Enter an option [Restore/Save/Delete/?]: *(Enter an option.)*

The following options can be entered directly at the first prompt:

Restore restores a named UCS.

Save saves the current UCS by name.

Del deletes a saved UCS.

? lists the names of saved UCS orientations.

Hidden options

The following options also work with this command:

New defines a new user-defined coordinate system.

Move moves the UCS along the z axis, or specifies a new xy-origin.

orthoGraphic selects a standard orthographic UCS: top, bottom, front, back, left, and right.

Apply applies the UCS setting to a selected viewport, or to all active viewports.

3point aligns the UCS with a point on the positive x-axis and positive xy plane.

STATUS BAR OPTION

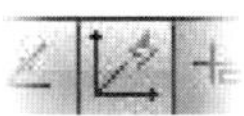

Click Allow/Disallow Dynamic UCS to turn on dynamic UCS mode: the UCS matches the orientation of the selected face temporarily.

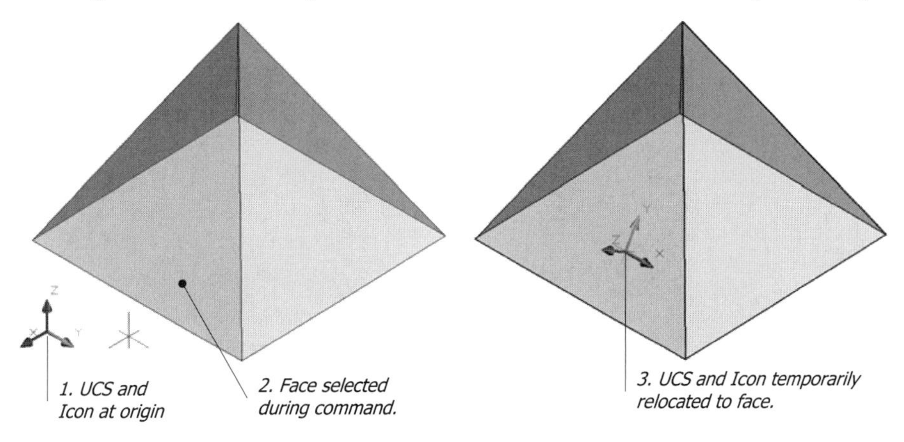

RELATED COMMANDS

UcsMan modifies the UCS via a dialog box.

UcsIcon controls the visibility of the UCS icon.

Plan changes to the plan view of the current UCS.

RELATED SYSTEM VARIABLES

GridDisplay forces the grid to follow the orientation of the dynamic UCS, when set to 8.

UcsAxisAng specifies the default rotation angle when the UCS is rotated around an axis using the X, Y, or Z options of this command.

UcsBase specifies the UCS that defines the origin and orientation of orthographic UCS settings.

UcsDetect toggles dynamic UCS.

UcsFollow aligns the UCS automatically with the view of a newly-activated viewport.

UcsIcon determines the visibility and location of the UCS icon.

UcsOrg specifies the WCS coordinates of UCS icon (default = 0,0,0).

UcsOrtho specifies whether the related UCS is automatically displayed when an orthographic view is restored.

UcsView specifies whether the current UCS is saved when a view is created with the View command.

UcsVp specifies that the UCS reflects the UCS of the currently-active viewport.

UcsXdir specifies the X direction of the current UCS (default = 1,0,0).

UcsYdir specifies the Y direction of the current UCS (default = 0,1,0).

WorldUcs correlates the WCS to the UCS:

0 — Current UCS is WCS.

1 — UCS is the same as WCS (default).

TIPS

- Use the UCS command to draw objects at odd angles in 3D space.
- Although you can create UCSs in paper space, you cannot use 3D viewing commands in paper space.
- Each viewport can have its own UCS.
- UCSs can be aligned with these objects:

Object	*UCS Origin*	*X Axis Passes Through...*
Arc	Center	Endpoint closest to the pick point.
Attributes	Insertion point	Extrusion direction.
Blocks	Insertion point	Extrusion direction.
Circle	Center	Pick point.
Dimension	Midpoint of dimtext	Parallel to the X axis of the dimension.
Line	Endpoint nearest to pick	Parallel to the X axis of the line.
Point	Point	Anywhere.
Polyline	Start point	Start point to next vertex.
2D solid	First point	First two points.
Trace	First point	Centerline
3D Face	First point	First two points.
Shape	Insertion point	Extrusion direction.
MLine	Pick point	Direction of selected segment.
Ray	End of ray	Direction of ray.
Spline	Start of spline	Direction from pick point to start.
Ellipse	Midpoint of axes	Direction from pick point to origin.
Leader	Pick point	Direction of selected segment.
Text	Insertion point	Extrusion direction.

- When the GridDisplay system variable is set to 8, the grid matches the orientation of the dynamic UCS.
- The 3point option is perhaps the most useful, yet it is hidden two levels deep in this command. Fortunately, it features prominently on the ribbon.

DEFINITIONS

UCS — user-defined 2D coordinate system oriented in 3D space; sets a working plane, orients 2D objects, defines the extrusion direction, and the axis of rotation.

WCS — world coordinate system; the default 3D xyz coordinate system.

UcsIcon

Rel.10 Controls the location and display of the UCS icon.

Command	*Aliases*	*Shortcut Keystrokes*	*Menu*	*Ribbon*
ucsicon	...	...	**View**	**View**
			↳**Display**	↳**Coordinates**
			↳**UCS Icon**	↳**Display UCS Icon**

Command: ucsicon

Enter an option [ON/OFF/All/Noorigin/ORigin/Properties] <ON>: *(Enter an option.)*

COMMAND LINE OPTIONS

All applies changes of this command to all viewports.

Noorigin displays the UCS icon in the lower-left corner at all times.

OFF turns off the display of the UCS icon.

ON turns on the display of the UCS icon.

ORigin displays the UCS icon at the current UCS origin, or at lower left corner of the viewport when the origin is off-screen.

Properties displays the UCS Icon dialog box.

UCS ICON DIALOG BOX

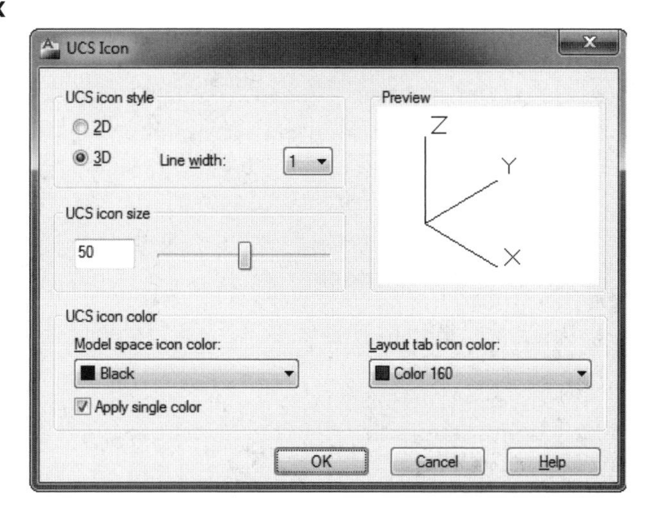

UCS Icon Style options

○**2D** displays flat UCS icon.

⊙**3D** displays tripod UCS icon.

Line width changes the line width from 1 to 2 or 3 pixels.

UCS Icon Size options

Slider bar changes the icon size from 5 to 95 pixels.

UCS Icon Color options

Model space icon color selects the icon's color in model space.

Layout tab icon color selects the icon's color in layout (paper space).

☑ **Apply Single Color** toggles between single and tri-color *(new to AutoCAD 2011)*.

RELATED SYSTEM VARIABLE

UcsIcon determines the visibility and location of the UCS icon.

RELATED COMMAND

UCS creates and controls user-defined coordinate systems.

TIPS

- The UCS icon varies, depending on the current viewpoint relative to the active UCS.
- When AutoCAD switches from 2D wireframe mode to one of the VsCurrent command's 3D options, the UCS icon changes to a rendered 3D icon:

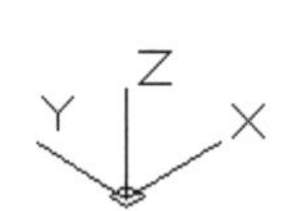

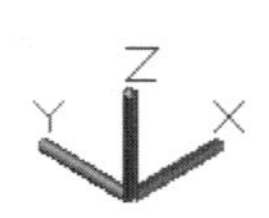

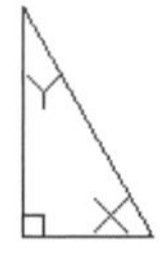

Left to right: *UCS icons in 2D visual style, 3D visual style, and paper space.*

- There is generally no need for the UCS icon in 2D drafting, and it can be safely turned off.
- The Options command's 3D Modeling tab has options for toggling the display of the UCS icon *(redesigned in AutoCAD 2011)*:

Display ViewCube or UCS Icon

In 2D model space

☑ Display ViewCube

☑ Display UCS Icon

In 3D model space

☑ Display ViewCube

☑ Display UCS Icon

- The UCS icon was redesigned in AutoCAD 2011, and takes on the same tri-color scheme as the crosshair cursor: x = red, y = green, and z = blue.

UcsMan

2000

Displays the UCS dialog box (short for "User-defined Coordinate System MANager").

Commands	Aliases	Shortcut Keystrokes	Menu	Ribbon
ucsman	**dducs**	...	**Tools**	**View**
	uc		⤷**Named UCS**	⤷**Coordinates**
	uc			
+ucsman	...	...	...	...

Command: ucsman

Displays dialog box.

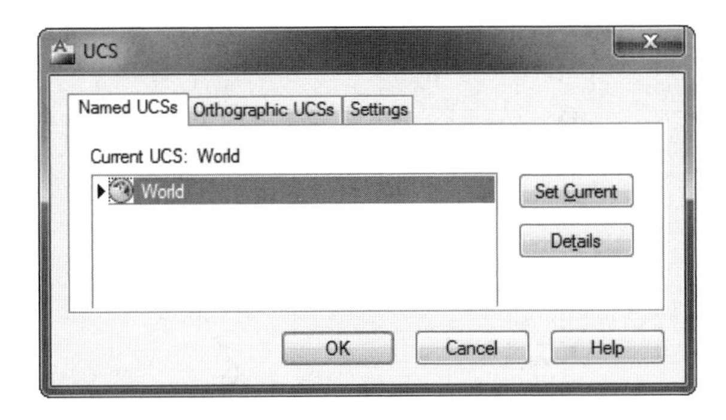

UCS DIALOG BOX

Named UCSs tab

Named UCSs lists the names of the AutoCAD-generated and user-defined coordinate systems of the current viewport in the active drawing; the arrowhead points to the current UCS.

Set Current restores the selected UCS.

Details displays the UCS Details dialog box.

Orthographic UCSs tab

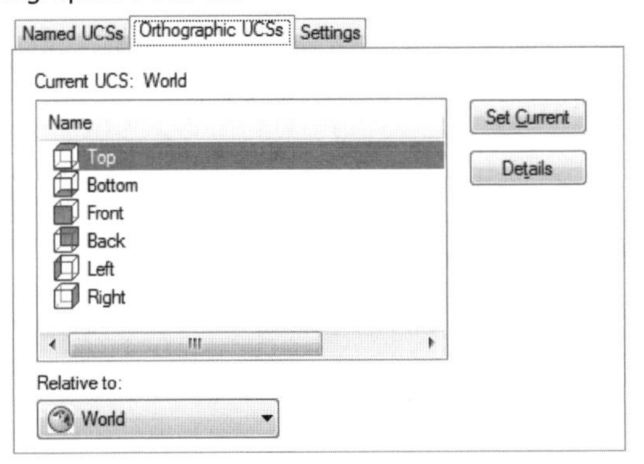

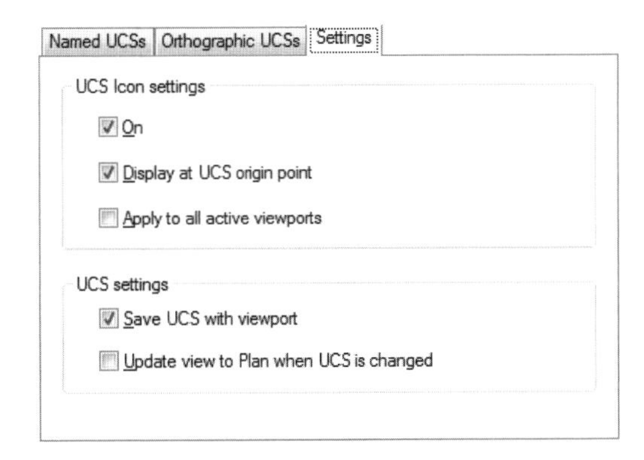

Left to right: *The Orthographic UCSs and Settings tabs.*

Name lists the six standard orthographic UCS views: top, bottom, front, back, left, and right.

Depth specifies the height of the UCS above the x,y plane.

Relative to specifies the orientation of the selected UCS relative to the WCS or to a customized UCS.

Set Current activates the selected UCS.

Details displays the UCS Details dialog box.

Settings tab

UCS icon settings options

☑**On** displays the UCS icon in the current viewport; each viewport can display the UCS icon independently.

Display at UCS origin point:

☑ Displays the UCS icon at the origin of the current UCS.

☐ Displays the UCS icon at the lower-left corner of the viewport.

Apply to all active viewports:

☑Applies these UCS icon settings to all active viewports in the current drawing.

☐Applies to the current viewport only.

UCS settings options

Save UCS with viewport:

☑ Saves the UCS setting with the viewport.

☐ Determines UCS settings from the current viewport.

☑ **Update view to Plan when UCS is changed** restores plan view when the UCS changes.

SHORTCUT MENUS

Access this menu by right-clicking the list in the Named UCSs tab:

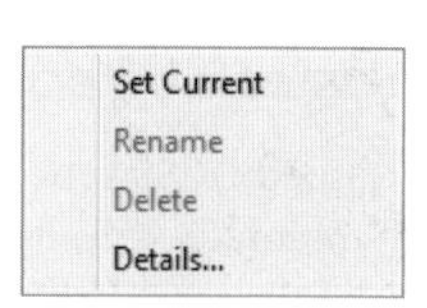

Set Current sets the selected UCS as active.

Rename renames the selected UCS; you cannot rename the World UCS.

Delete erases the selected UCS; you cannot delete the World UCS.

Details displays the UCS Details dialog box.

Access this menu by right-clicking the list in the Orthographic UCS tab:

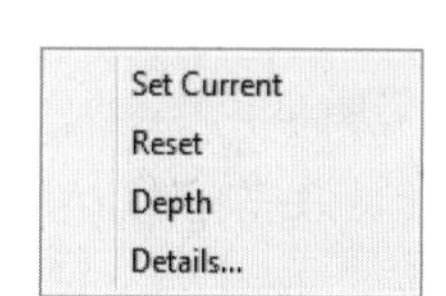

Set Current sets the selected UCS as active.

Reset restores the origin of the selected UCS.

Depth moves the UCS in the z direction; displays dialog box:

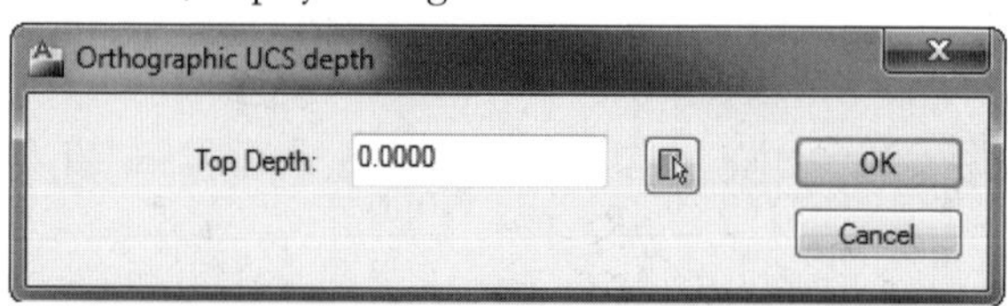

Details displays the UCS Details dialog box.

+UCSMAN Command

Command: +ucsman
Tab index <0>: *(Type 1, 2, or 3.)*

COMMAND LINE OPTION

Tab index displays the tab related to the tab number:

0 — Named UCS tab.
1 — Orthographic UCS tab.
2 — Settings tab.

RELATED COMMANDS

UCS displays the UCS options at the command line.

UcsIcon changes the display of the UCS icon.

Plan displays the plan view of the WCS or a UCS.

TIPS

- Functions of this command were formerly carried out by the DdUcs and DdUcsP commands.
- *Unnamed* is the first entry, when the current UCS is unnamed. *World* is the default for new drawings; it cannot be renamed or deleted. *Previous* is the previous UCS; you can move back through several previous UCSs.

ULayers

2010 Toggles visibility of layers in attached DWF, DWFx, DGN, and PDF underlays (short for "underlay layers").

Command	Aliases	Shortcut Keystroke	Menu	Ribbon
ulayers	...	...	...	**Insert** ↳**Reference** ↳**Underlay Layers**

Command: ulayers

Displays dialog box.

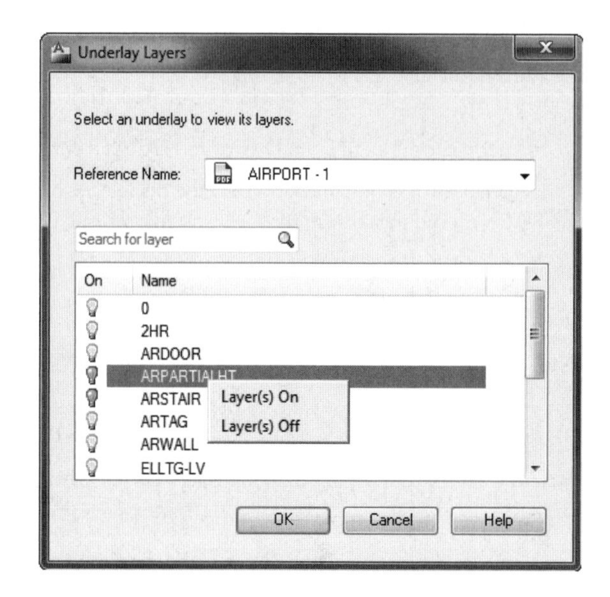

UNDERLAY LAYERS DIALOG BOX

Reference name selects a reference (a.k.a. underlay); the layers of the selected reference are listed below.

Search for layer searches for layer names; useful when the reference contains many layer names.

On shows the status of the layer. A yellow light bulb indicates that the content of the layer is visible.

Names lists the layer names. This list is blank when the attachment contains no layer names.

SHORTCUT MENU

Access this menu by right-clicking a layer name:

Layer(s) On turns on the selected layers.

Layer(s) Off turns off the selected layers. The content of the layers become invisible.

RELATED COMMAND

ExternalReferences attaches DWF/x, DGN, and PDF files as underlays.

TIPS

- To turn more than one layer on (or off) at a time, hold down the **Shift** key and then choose a range of layer names.
- This command does not affect layers in externally referenced drawing files; for them, use the Layer command.

Undefine

Rel. 9 Makes AutoCAD commands unavailable.

Command	*Aliases*	*Shortcut Keystrokes*	*Menu*	*Ribbon*
undefine	...	...	...	...

Command: undefine
Enter command name: *(Enter name.)*

Example usage:
Command: undefine
Enter command name: line
Command: line
Unknown command. Type ? for list of commands.
Command: .line
From point:

COMMAND OPTIONS

Enter command name specifies the name of the command to make unavailable.

. *(period)* is the prefix for undefined commands to redefine them temporarily.

RELATED COMMAND

Redefine redefines an AutoCAD command.

TIPS

- Prefixing undefined commands with a period bypasses the effect of this command. For example:

  Command: .line

- This command allows AutoLISP and ObjectARX to override native AutoCAD commands.
- Commands created by programs cannot be undefined, including the following programming interfaces:

  AutoLISP and Visual LISP.
  ObjectARx.
  Visual Basic for Applications.
  External commands.
  Aliases.

- In menu macros written with international language versions of AutoCAD, precede command names with an underscore character (_) to translate the command name into English automatically.

Undo

V. 2.5 Undoes the effect of the previous command(s).

Command	*Aliases*	*Shortcut Keystrokes*	*Menu*	*Ribbon*
undo	...	...	...	...

Command: undo

Enter the number of operations to undo or [Auto/Control/BEgin/End/Mark/Back] <1>: *(Enter a number, or an option.)*

COMMAND LINE OPTIONS

Auto treats a menu macro as a single command.

Control limits the options of the Undo command.

BEgin groups a sequence of operations (formerly the Group option).

End ends the group option.

Mark sets a marker.

Back undoes back to the marker.

number indicates the number of commands to undo.

Control Options

Enter an UNDO control option [All/None/One] <All>: *(Enter an option.)*

All turns on full undo.

None turns off undo.

One limits the Undo command to a single undo.

RELATED COMMANDS

Oops unerases the most-recently erased object.

Quit leaves the drawing without saving changes.

Redo undoes the most recent undo.

MRedo undoes multiple undoes.

U undoes a single step.

RELATED SYSTEM VARIABLES

UndoCtl determines the state of undo control.

UndoMarks specifies the number of undo marks placed in the Undo control stream.

TIPS

- Since the undo mechanism creates a mirror drawing file on disk, disable the Undo command with system variable UndoCtl (set it to 0) when your computer is low on disk space.
- There are some commands that cannot be undone, such as Save, Plot, and UndoCtl.
- U is not an alias for Undo, but its own command.

Union

2000 Joins two or more solids, regions, or surfaces into a single body.

Command	*Alias*	*Shortcut Keystrokes*	*Menu*	*Ribbon*
union	**uni**	...	**Modify** ↳**Solids Editing** ↳**Union**	**Home** ↳**Solid Editing** ↳**Union**

Command: union
Select objects: *(Select one or more objects.)*
Select objects: *(Select one or more objects.)*
Select objects: *(Press Enter to end command.)*

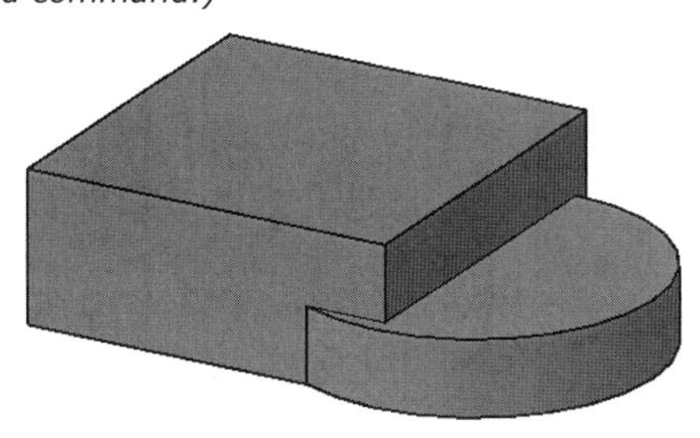

Box and cylinder unioned into a single object.

COMMAND LINE OPTION

Select objects selects the objects to join into a single object; you must select at least two solid, or region, or surface objects.

RELATED COMMANDS

Intersect creates a solid from the intersection of two objects.

Subtract creates a solid by subtracting one object from another.

RELATED SYSTEM VARIABLES

ShowHist toggles display of history in solids.

SolidHist toggles the retention of history in solids.

TIPS

- You must select at least two solid or coplanar region objects. The two objects need not overlap for this command to operate.
- As of AutoCAD 2009, this command works with surface objects.
- As of AutoCAD 2010, this command works with 3D mesh objects, after they are converted to solids or surfaces. When you select two or more mesh objects, AutoCAD asks what you want to do with them:
  - Filter (remove) them from the selection set; they are not unioned.
  - Convert open meshes to surfaces and watertight (closed) meshes to solids.
  - Smoothed solids and surfaces look better; faceted solids and surfaces mimic the faceted look of mesh objects.
- If enabled, history allows you to work with the original objects.

UnisolateObjects Command

See **IsolateObjects** command for details on the UnisolateObjects command.

Units

V. 1.4 Controls the display and format of coordinates and angles, as well as the orientation of angles.

Commands	Aliases	Shortcut Keystrokes	Menu Bar	Application Menu	Ribbon
'units	**un** **ddunits**	...	**Format** **↳Units**	**Drawing Utilities** **↳ Units**	...
'-units	**-un**	...	...	...	

command: units

Displays dialog box.

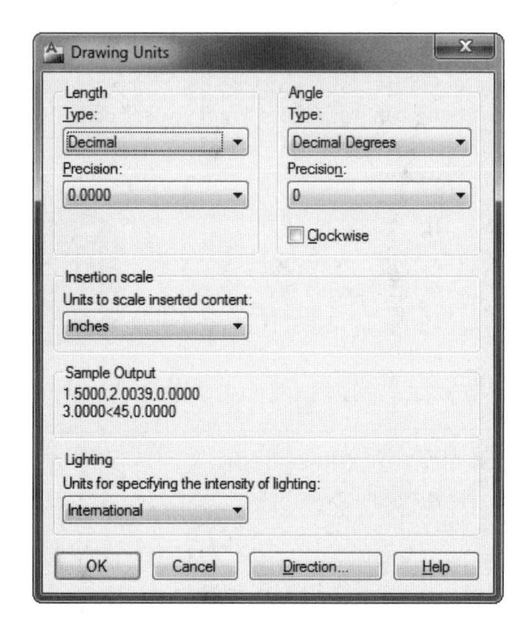

DRAWING UNITS DIALOG BOX

Length options

Type sets the format for units of linear measurement displayed by AutoCAD: Architectural, Decimal, Engineering, Fractional, or Scientific.

Precision specifies the number of decimal places or fractional accuracy.

Angle options

Type sets the current angle format.

Precision sets the precision for the current angle format.

☐ **Clockwise** calculates positive angles in the clockwise direction.

Drawing units for DesignCenter blocks specifies the units when blocks are inserted from the DesignCenter.

Direction displays the Direction Control dialog box.

DIRECTION CONTROL DIALOG BOX

Base Angle options

○ **East** sets the base angle to 0 degrees (default).

○ **North** sets the base angle to 90 degrees.

○ **West** sets the base angle to 180 degrees.

○ **South** sets the base angle to 270 degrees.

⊙ **Other** turns on the Angle option.

Angle sets the base angle to any direction.

Pick an angle dismisses the dialog box temporarily, and allows you to define the base angle by picking two points in the drawing; AutoCAD prompts you 'Pick angle' and 'Specify second point:'.

-UNITS Command

Command: -units

Report formats: (Examples)

1. Scientific	**1.55E+01**
2. Decimal	**15.50**
3. Engineering	**1'-3.50"**
4. Architectural	**1'-3 1/2"**
5. Fractional	**15 1/2**

With the exception of Engineering and Architectural formats, these formats can be used with any basic unit of measurement. For example, Decimal mode is perfect for metric units as well as decimal English units.

Enter choice, 1 to 5 <2>: *(Enter a value.)*

Enter number of digits to right of decimal point (0 to 8) <4>: *(Enter a value.)*

Systems of angle measure: (Examples)

1. Decimal degrees	**45.0000**
2. Degrees/minutes/seconds	**45d0'0"**
3. Grads	**50.0000g**
4. Radians	**0.7854r**
5. Surveyor's units	**N 45d0'0" E**

Enter choice, 1 to 5 <1>: *(Enter a value.)*

Enter number of fractional places for display of angles (0 to 8) <0>: *(Enter a value.)*

Direction for angle 0:
East 3 o'clock = 0
North 12 o'clock = 90
West 9 o'clock = 180
South 6 o'clock = 270
Enter direction for angle 0 <0>: *(Enter a value.)*

Measure angles clockwise? [Yes/No] <N> *(Type Y or N.)*

COMMAND LINE OPTIONS

Report formats selects scientific, decimal, engineering, architectural, or fractional format for length display.

Number of digits to right of decimal point specifies the number of decimal places between 0 and 8.

Systems of angle measure selects decimal degrees, degrees/minutes/seconds, grads, radians, or surveyor's units for angle display.

Denominator of smallest fraction to display specifies the denominator of fraction displays, such as 1/2 or 1/256.

Number of fractional places for display of angles specifies the number of decimal places between 0 and 8.

Direction for angle 0 selects the direction for 0 degrees from east, north, west, or south.

Do you want angles measured clockwise?

- **Yes** measures angles clockwise.
- **No** measures angles counterclockwise.

RELATED SYSTEM VARIABLES

AngBase specifies the direction of zero degrees.

AngDir specifies the direction of angle measurement.

AUnits specifies the units of angles.

AuPrec specifies the displayed precision of angles.

InsUnits specifies the drawing units for blocks dragged from the DesignCenter:

0 — Unitless
1 — Inches
2 — Feet
3 — Miles
4 — Millimeters
5 — Centimeters
6 — Meters
7 — Kilometers
8 — Microinches
9 — Mils
10 — Yards; 3 feet
11 — Angstroms; 0.1 nanometers
12 — Nanometers; 10E-9 meters
13 — Microns; 10E-6 meters
14 — Decimeters; 0.1 meter
15 — Decameters; 10 meters
16 — Hectometers; 100 meters
17 — Gigameters; 10E9 meters
18 — Astronomical Units; 149.597E8 kilometers
19 — Light Years; 9.4605E9 kilometers
20 — Parsecs; 3.26 light years

InsUnitsDefSource specifies source units to be used.

InsUnitsDefTarget specifies target units to be used.

LUnits specifies the units of measurement.

LuPrec specifies the displayed precision of coordinates.

UnitMode toggles the type of display units.

RELATED COMMAND

New sets up drawings with Imperial or metric units.

TIPS

- Because 'Units is a transparent command, you can change units during another command.
- The 'Direction Angle' prompt lets AutoCAD start the angle measurement from any direction.
- AutoCAD accepts the following notations for angle input:

Notation	*Meaning*
<	Specify an angle based on current units setting.
<<	Bypass angle translation set by Units command to use 0-angle-is-east direction and decimal degrees.
<<<	Bypass angle translation; use angle units set by the Units command and 0-angle-is-east direction.

- The system variable UnitMode forces AutoCAD to display units in the same manner that you enter them.
- At one time, the smallest fraction was 1/2048, but now the smallest is 1/256.
- Do not use a suffix — such as 'r' or 'g' — for angles entered as radians or grads; instead, use the Units command to set angle measurement to radians and grads.
- The Drawing units for DesignCenter blocks option is for inserting blocks from the DesignCenter, and especially when the block was created in other units.
- To prevent blocks from being scaled when dragged from the DesignCenter palette, select Unitless.

UpdateField

2005 Forces the update of selected fields.

Commands	*Aliases*	*Shortcut Keystrokes*	*Menu*	*Ribbon*
updatefield	...	...	**Tools** ↳**Update Fields**	**Insert** ↳**Data** ↳**Update Fields**

Command: updatefield
Select objects: *(Select one or more fields.)*
Select objects: *(Press Enter to end field selection.)*
***n* field(s) found**
***n* field(s) updated**

COMMAND LINE OPTION

Select objects selects one or more fields; press Ctrl+A to select all objects in the drawing, including field text.

RELATED COMMANDS

Field places field text, which is automatically updated as its value changes.

Find finds fields.

DdEdit edits the properties of field text.

RELATED SYSTEM VARIABLE

FieldEval determines when fields are updated.

TIP

- This command forces individual fields to update. Depending on the setting of the FieldEval system variable, all fields are updated automatically.

UpdateThumbsNow

2005 Forces the update of preview images in the Sheet Set Manager palette and Quick Views.

Command	*Aliases*	*Shortcut Keystrokes*	*Menu*	*Ribbon*
updatethumbsnow	...	...	...	...

Command: updatethumbsnow

COMMAND LINE OPTIONS

None.

RELATED COMMANDS

SheetSet controls sheet sets.

QVDrawing displays quick views of drawings and layouts.

QVLayout displays quick views of layouts.

RELATED SYSTEM VARIABLE

UpdateThumbnail determines which thumbnails are updated (default = 15):

0 — Thumbnail previews not updated.
1 — Model view thumbnails updated.
2 — Sheet view thumbnails updated.
4 — Sheet thumbnails updated.
8 — Thumbnails updated when sheets and views created, modified, and restored.
16 —Thumbnails updated when drawing saved.

TIPS

- This command controls when preview images are updated; required when the drawing changes.
- Thumbnails are displayed in the Sheet Set Manager and in quick view thumbnails.

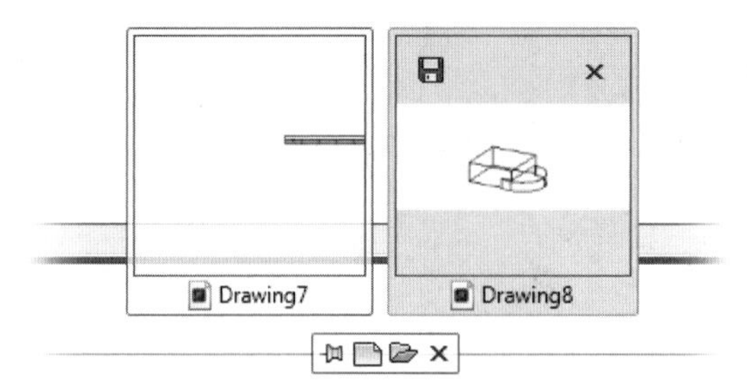

- The SaveAs command's dialog box has the Update Sheet And View Thumbnails Now option.

Deprecated Commands

Autodesk stopped included VBA (Microsoft's Visual Basic for Applications) as of AutoCAD 2010. This affects the **VbaIde**, **VbaLoad** and **VbaUnload**, **VbaMan**, **VbaRun**, and **VbaStmt** commands. VBA is due to removed entirely in AutoCAD 2012.

View

V. 2.0 Saves and displays views by name; creates view categories for sheet sets and ShowMotion animation; sets backgrounds for visual styles; and controls parameters for cameras.

Commands	*Aliases*	*Shortcut Keystrokes*	*Menu*	*Ribbon*
view	**v** **ddview**	...	**View** **↳Named Views**	**View** **↳Named Views**
-view	**-v**	...	**View** **↳3D Views**	...

Command: view

Displays dialog box.

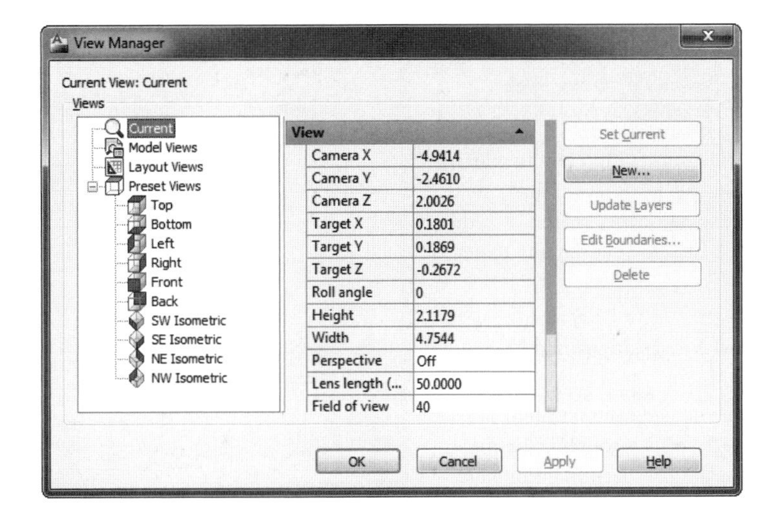

VIEW DIALOG BOX

Set Current restores the named view; alternatively, double-click the view name.

New displays the New View dialog box.

Update Layers updates view to match the layer visibility settings of the current viewport.

Edit Boundaries highlights the view, by graying out the area outside of the view.

Delete removes the named view from the drawing.

NEW VIEW DIALOG BOX

View name names the view — up to 255 characters long.

View Category specifies the default prefix for named views. *Categories* are used by the Sheet Set Manager and by Show Motion animation.

View Type selects Still, Cinematic, or Recorded Walk; affects settings in the Shot Properties tab.

View Properties tab

Boundary options

⊙**Current display** stores the current viewport as the named view.

○**Define window** stores a windowed area as the named view.

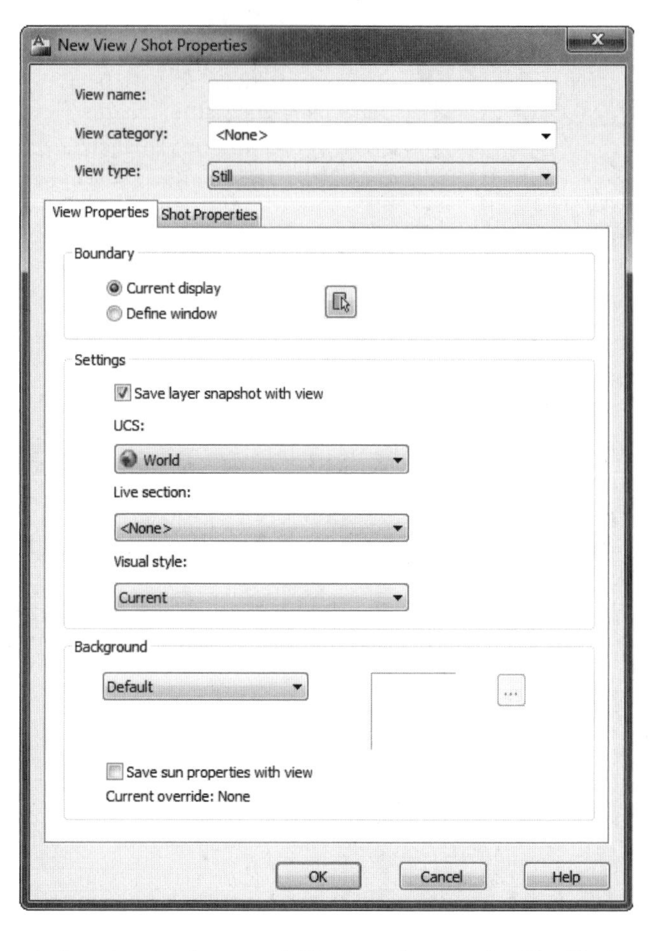

Define View Window dismisses the dialog box temporarily so that you can pick the two corners that define the view. AutoCAD clears the dialog box, and prompts:

Specify first corner: *(Pick a point.)*

Specify opposite corner: *(Pick another point.)*

Specify first corner (or press ENTER to accept): *(Press Enter to return to the dialog box.)*

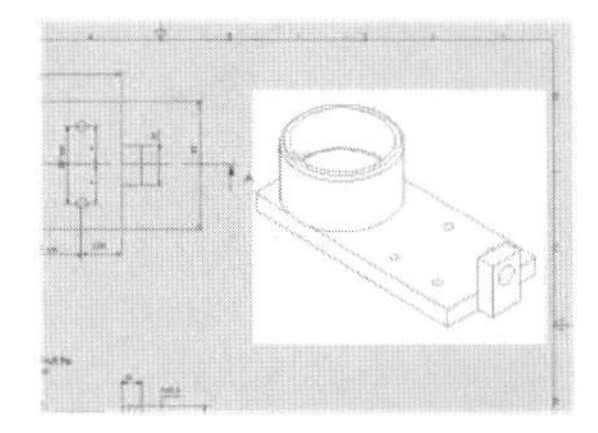

The white rectangle defining the area of the view window.

Settings options

☑ **Save Current Layer Settings with View** toggles whether layer properties (such as freeze/thaw, lock/unlock, plot/no plot, and so on) are stored with the view. You must first use the Layer command to set layer properties.

UCS name specifies the name of a UCS to store with the named view. You must first create named UCSs with the Ucs command.

Live Section specifies the named live section to be displayed when the named view is restored. You must first use the SectionPlane and LiveSection commands to create the named section.

Visual Style specifies the named visual style to be displayed when the view is restored. You must first use the VisualStyles command to create the named visual style.

Background options

selects the background type: default, solid, gradient, image, or sun & sky.

☐ **Save Sun Properties With View** displays simulated sun and sky in the background of the views.

displays the Background dialog box. See the Background command.

Shot Properties tab

See NewShot command.

BACKGROUND DIALOG BOX

The Background command is an alias for this dialog box.

Type selects the type of background:

- **Solid** shows a single color; click the Color bar to display the Select Color dialog box.
- **Gradient** shows two or three colors.
- **Image** shows a picture.

Color displays the Select Color dialog box; choose color, then click OK.

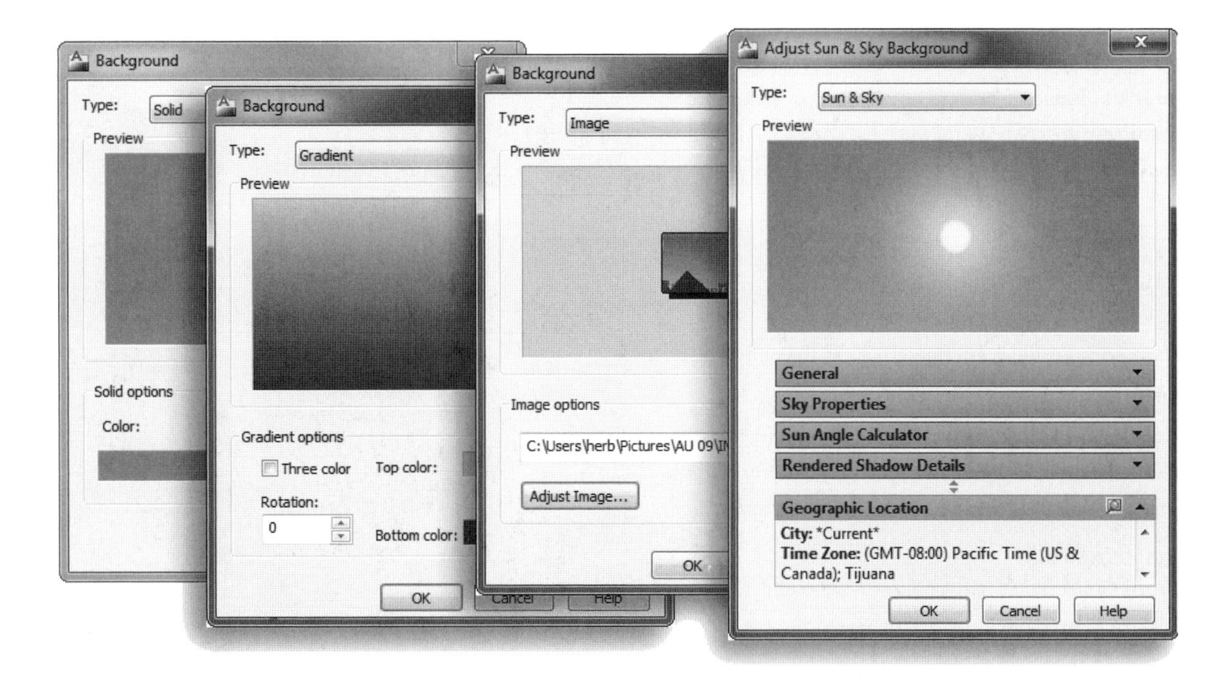

Gradient Options options

Three color toggles gradients between two and three colors.

Rotation displays gradients at an angle; range is from 90 (vertical) to 0 (horizontal) to -90 degrees.

Top color, **Middle color**, and **Bottom color** display the Select Color dialog box.

Image Options options

Browse selects the image from the Select File dialog box. You can choose from Targa, bitmap, TIFF, JPEG, GIF, and PCX formats.

Adjust image displays the Adjust Image dialog box.

Adjust Sun & Sky Background options

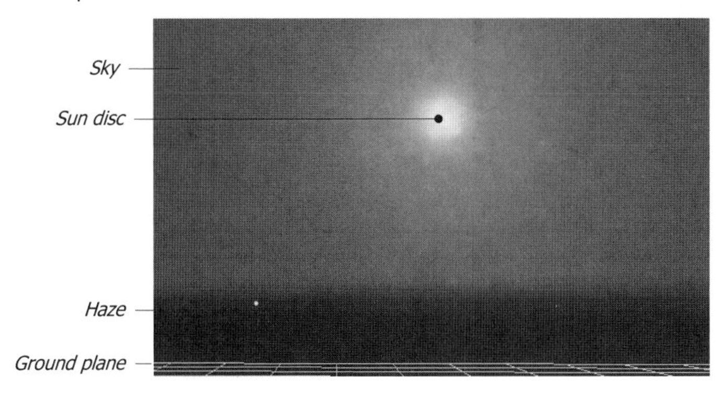

The "Sun & Sky" background displays a sun image placed in the sky background, at a position matching the date, time, and geographic location. As the date and time change, the sun moves through the sky; as it sets, the sky darkens to black. This effect is available only when LightingUnits is set to 1 or 2.

General options

Status toggles the sun effect.

Intensity Factor specifies the brightness of the sun.

Color sets the color of the sun.

Shadows toggles the sun's shadow-casting; leave off for better performance.

Sky Properties options

Status toggles the display of the sky effect.

Intensity Factor specifies the brightness of the sky.

Haze controls the scattering effects in the atmosphere.

Horizon options

Height positions the ground plane relative to world zero.

Blur adjusts the blur between the ground plane and the sky.

Ground Color specifies the color of the ground plane.

Advanced options

Night Color specifies the color of the night sky.

Aerial Perspective toggles aerial perspective.

Visibility Distance sets the distance for 10% haze occlusion.

Sun Disk Appearance options

Disk Scale sizes the sun.

Glow Intensity sets the intensity of the sun's glow.

Disk Intensity sets the intensity of the sun's disk.

Sun Angle Calculator options

Date specifies the date.

Time specifies the time.

Daylight Saving toggles daylight savings time.

Rendered Shadow Details options

Type sets the shadow type: soft is available only with photometric lighting.

Samples specifies the number of samples to take on the solar disk.

Softness specifies the edge of shadow appearance.

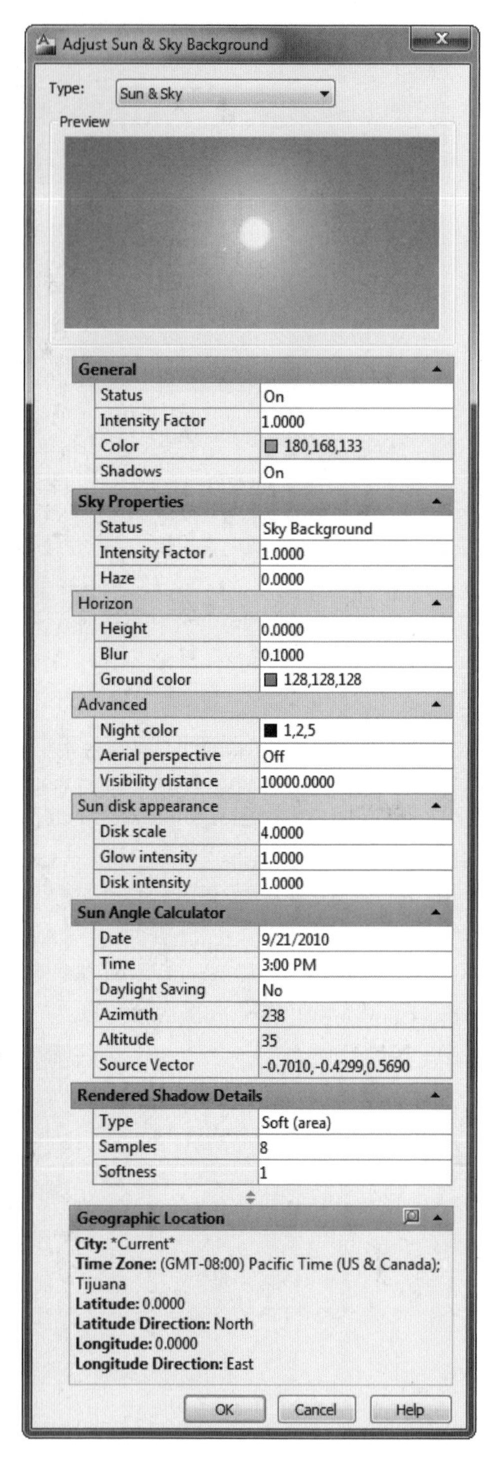

ADJUST IMAGE DIALOG BOX

Known as the SetUV command in AutoCAD 2006 and earlier.

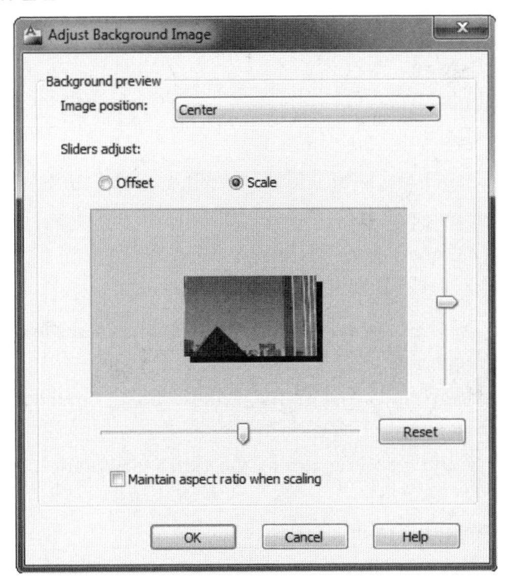

Image position:

- **Center** centers the image in the viewport (figure at left, below).
- **Stretch** fits the image to the viewport (figure at center, below).
- **Tile** repeats the image to fill the viewport (figure at right, below).

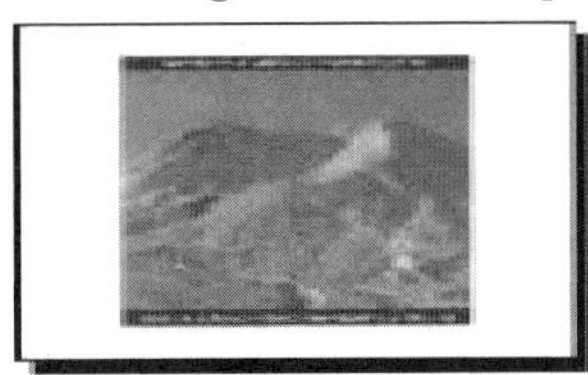

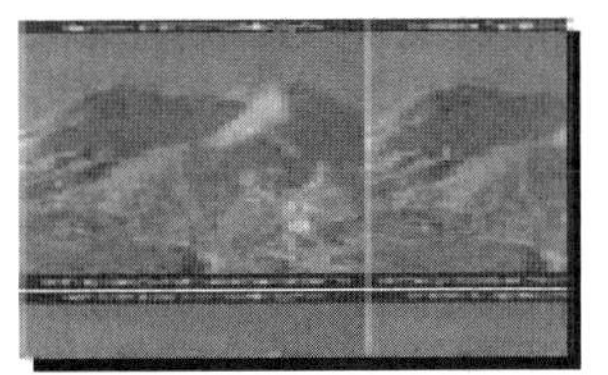

(Images from Mount St. Helens volcano-cam at www.fs.fed.us/gpnf/volcanocams/msh)

Sliders adjust *(neither of the following options works in Stretch mode.)*

⦿ **Offset** moves the image relative to the viewport. When the image goes outside the viewport boundary, its preview appears dimmed.

○ **Scale** changes the size of the image, making it larger or smaller. Negative x scale factors mirror the image; negative y inverts the image.

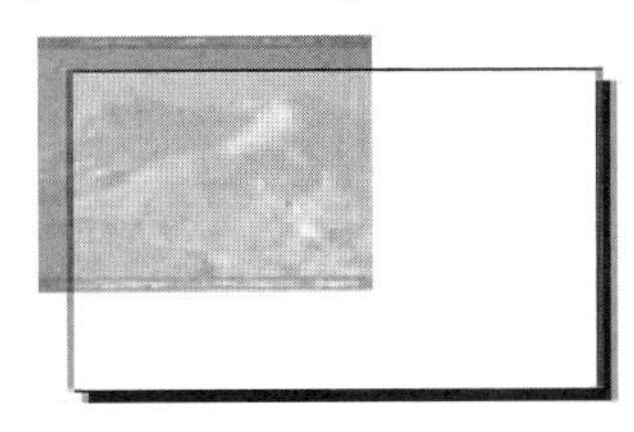

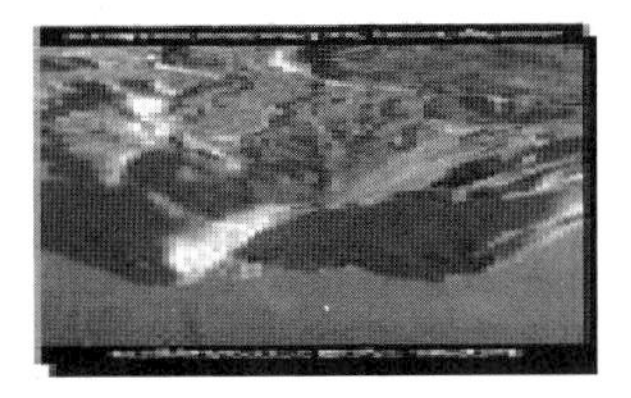

Left to right: *Image moved in viewport, scaled larger than viewport, and upside down (with negative y-scale factor).*

Maintain aspect ratio when scaling prevents the image from distorting due to unequal scaling.

Reset returns the image to its original size and aspect ratio.

SHORTCUT MENU

Access this menu by right-clicking in the Views list.

Set Current sets the selected view as active.

New creates a new view; opens the New View dialog box.

Update Layers updates the view to match the layer visibility settings of the current viewport.

Edit Boundaries highlights the view, by graying out the area outside of the view.

Delete erases the selected view; you cannot delete the Current view.

Set Current
New...
Update Layers
Edit Boundaries...
Delete

VIEW PROPERTIES

View properties views vary with the type of view selected, which must be set in the View dialog box, because the Properties palette cannot select views.) Properties are listed below in alphabetical order:

General	
Name	Overall View
Category	<None>
UCS	World
Layer snapshot	Yes
Annotation scale	1:1
Visual Style	Conceptual
Background overri...	<None>
Live Section	<None>
Animation	
View type	Still
Transition type	Fade from bl...
Transition duration	1.0000
Playback duration	0.5000
View	
Camera X	-4.9414
Camera Y	-2.4610
Camera Z	2.0026
Target X	0.1801
Target Y	0.1869
Target Z	-0.2672
Roll angle	0
Height	2.1179
Width	4.7544
Perspective	Off
Lens length (mm)	50.0000
Field of view	40
Clipping	
Front plane	0.0000
Back Plane	0.0000
Clipping	Off

Background specifies the type of background color or image: none, Solid, Gradient, Image, or edit.

Camera X, **Y**, and **Z** specifies the x, y, and z coordinates of the view's viewpoint (camera); read-only.

Category lists the names of view categories: none or user-defined.

Clipping selects clipping options: Off, Front On, Back On, or Front and Back On.

Field of View indicates the horizontal field of view angle, and is coupled with Lens Length.

Front Plane and **Back Plane** indicate the front (and back) clipping plane's offset distance.

Height indicates the view height above or below the x,y plane; read-only.

Layer Snapshot toggles whether layer parameters are stored with the view: yes or no.

Lens Length (mm) indicates the lens length in millimeters; coupled with Field of View.

Live Section determines whether a live section is displayed when the view is restored.

Location reports the layout associated with the view; read-only.

Name names the view.

Perspective toggles the projection between perspective and parallel: on or off.

Restore Ortho UCS restores the associated UCS when the orthographic view is current.

Roll angle indicates the view's roll angle; read-only.

Set Relative To specifies the base coordinate system for the orthographic view.

Target X, **Y**, and **Z** specifies the x, y, and z coordinates of the view's look-at point (target); read-only.

UCS names the user-defined coordinate system saved with the view.

Viewport Association toggles the view's association with a sheet set's viewport.

Visual Style specifies the visual style to restore with the view: 2D Wireframe, 3D Wireframe, 3D Hidden, Realistic, Conceptual, or user-defined.

Width indicates the view width (field of view); read-only.

-VIEW Command

Command: -view
Enter an option [?/Delete/Orthographic/Restore/Save/sEttings/Window]: *(Enter an option.)*

COMMAND LINE OPTIONS

? lists the names of views saved in the current drawing.

Delete deletes a named view.

Orthographic restores predefined orthographic views.

Restore restores a named view.

Save saves the current view with a name.

sEttings specifies view settings.

Window saves a windowed view with a name.

Orthographic options
Enter an option [Top/Bottom/Front/BAck/Left/Right] <Top>: *(Enter an option.)*
Select Viewport for view: *(Pick a viewport.)*

Enter an option selects a standard orthographic view for current viewport: Top, Bottom, Front, BAck, Left, Right.

Select Viewport for view selects a viewport in Model or Layout tab in which to apply the orthographic view.

sEttings options
Enter an option [Background/Categorize/Layer snapshot/live Section/Ucs/Visual style]: *(Enter option.)*

Background specifies the type of background for the view.

Categorize selects a sheet set category for the view.

Layer snapshot applies the current layer properties to the view.

live Section assigns a section to the view.

Ucs assigns a user-defined coordinate system to the view.

Visual style selects a visual style for the named view.

STARTUP SWITCH

/v specifies the view name to show when AutoCAD starts up.

RELATED COMMANDS

NewView accesses the New View dialog box directly.

NavVCube rotates the viewpoint with the viewcube user interface.

GeographicLocation specifies the location on earth for positioning the sun.

Rename changes the names of views via a dialog box.

UCS creates and displays user-defined coordinate systems.

PartialLoad loads portions of drawings based on view names.

Open opens drawings and optionally starts with a named view.

Plot plots named views.

SheetSet uses named views and categories.

Aliased Command

ViewGo command is a hardwire alias for the -View command's Restore option; added to AutoCAD 2009.

RELATED SYSTEM VARIABLES

DefaultViewCategory specifies the default name for categories.

SkyStatus reports the status of the sun and sky.

ViewCtr specifies the coordinates of the center of the view.

ViewSize specifies the height of the view.

TIPS

- Name views in your drawing to move quickly from one detail to another.
- The Plot command can plot the named views of drawings; the Open command lets you select a view with which to display the drawing.
- Objects outside the window created by the Window option may be displayed, but are not plotted.
- As of AutoCAD 2005, this command creates views for sheets.
- As of AutoCAD 2007, this command controls the background for visual styles. Backgrounds do not appear in 2D wireframe mode.
- As of AutoCAD 2009, this command also handles parameters for Show Motion animation. See the NavSMotion command.
- +View was removed from AutoCAD 2007, because the redesigned dialog box lost its tabs.

ViewPlay

2009 Plays the animation associated with a named view.

Commands	*Aliases*	*Shortcut Keystrokes*	*Menu*	*Ribbon*
viewplay	...	...	...	...

Command: viewplay
Enter view name to play: *(Enter the name of a view.)*

COMMAND LINE OPTION

Enter View Name to Play specifies the view to playback.

RELATED COMMANDS

AllPlay plays all show motion views in sequence, without prompting.

EditShot prompts for a view name at the command line, and then displays the Shot Properties tab of the View / Shot Properties dialog box.

NavSMotion displays the show motion interface.

NewShot creates named views with show motion options; displays the Shot Properties tab of the New View / Shot Properties dialog box.

SequencePlay plays all views in a category; you can have one or more views per category.

ViewGo displays a named view.

TIPS

- This command plays back ShowMotion animations associated with view names; see the NavSMotion and NewShot commands.
- This command is meant for use with macros and AutoLISP routines.

ViewPlotDetails

2005 Reports plotting failures and successes.

Commands	*Aliases*	*Menu Bar*	*Application Menu*	*Ribbon*
viewplotdetails	...	**File** **⇘View Plot and Publish Detail**	**Print** **⇘View Plot and Publish Details**	...

Command: viewplotdetails

Displays dialog box.

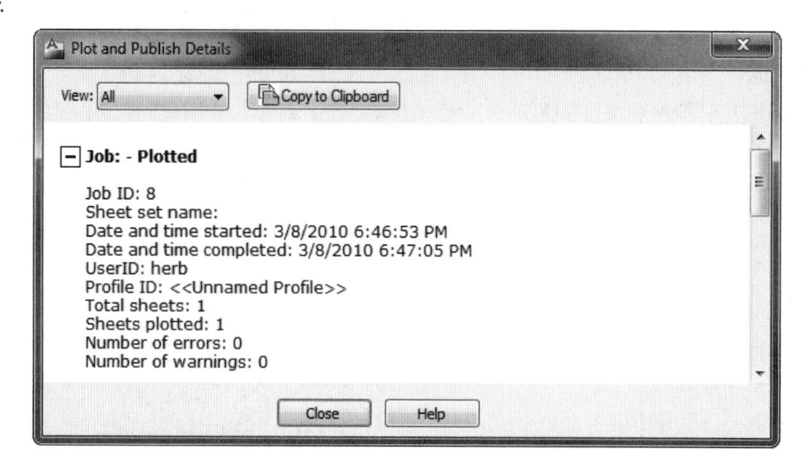

PLOT AND PUBLISH DETAILS DIALOG BOX

Green checkmark indicates a successful plot.

Red X warns of plotting error.

View determines whether all messages are displayed, or just errors:

- All messages.
- Errors only.

Copy to Clipboard copies selected text to the Clipboard.

[−] Collapses text under heading.

[+] Expands text under heading.

RELATED COMMANDS

Plot plots drawings.

TraySettings toggles the display of icons in the tray.

TIPS

- When a plot is completed, AutoCAD displays a balloon. Clicking the blue underlined text displays the dialog box.

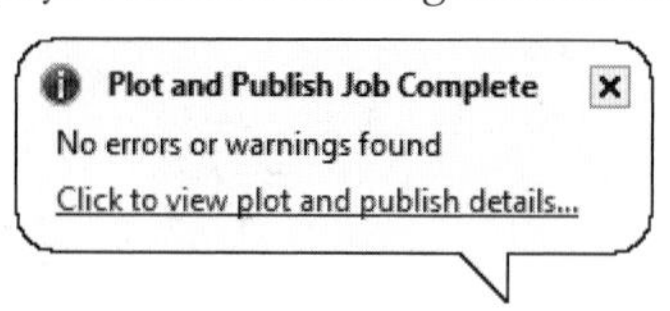

- You can also access this dialog box by right-clicking the Plotting icon in the tray, and then selecting View Plot Details.

ViewRes

V. 2.5 Controls the roundness of curved objects; determines whether zooms and pans are performed as redraws or regens (short for VIEW RESolution).

Commands	*Aliases*	*Shortcut Keystrokes*	*Menu*	*Ribbon*
viewres	...	...	...	...

Command: viewres
Do you want fast zooms? [Yes/No] <Y>: *(Type Y or N.)*
Enter circle zoom percent (1-20000) <1000>: *(Enter a value.)*

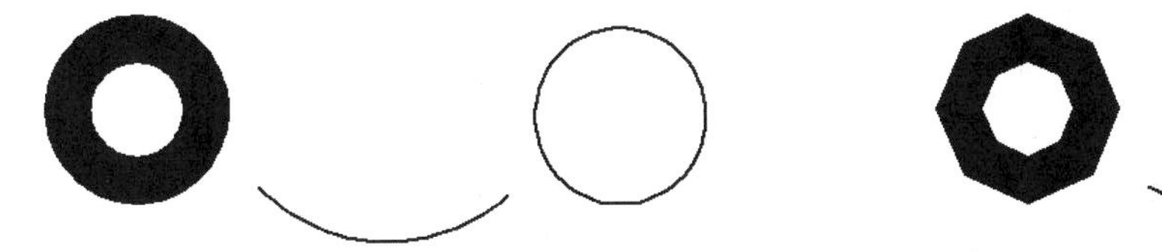

***Left**: Circle zoom percent = 1000.*
***Right**: Circle zoom percent = 1.*

COMMAND LINE OPTIONS

Do you want fast zooms? does not function; retained only for compatibility with macros and scripts.

Enter circle zoom percent specifies that smaller values display faster, but makes circles look less round (above figure); default = 1000.

RELATED SYSTEM VARIABLE

WhipArc toggles the display of circles and arcs as vectors or as true, rounded objects.

RELATED COMMAND

RegenAuto determines whether AutoCAD uses redraws or regens.

TIPS

- Setting WhipArc to 1 is recommended over increasing the value of ViewRes.
- This command was useful for speeding up the display when computers were much slower than they are today.
- Back in the days when "fast zooms" could be toggled, every zoom and pan caused a regeneration when the fast zooms were disabled. Today, AutoCAD tries to make every zoom and pan a redraw.

VisualStyles / VisualStylesClose

2007 Creates and edits visual styles.

Commands	Aliases	Shortcut Keystrokes	Menu	Ribbon
visualstyles	**vsm**	...	**Tools** ↳**Palettes** ↳**Visual Styles**	**Home** ↳**View** ↳**Visual Styles**
visualstylesclose	...	...	...	...
-visualstyles	**-vsm**	...	...	...

Command: visualstyles

Displays palette; AutoCAD 2011 changes the naming and ordering of options.

Command: visualstylesclose

Closes palette.

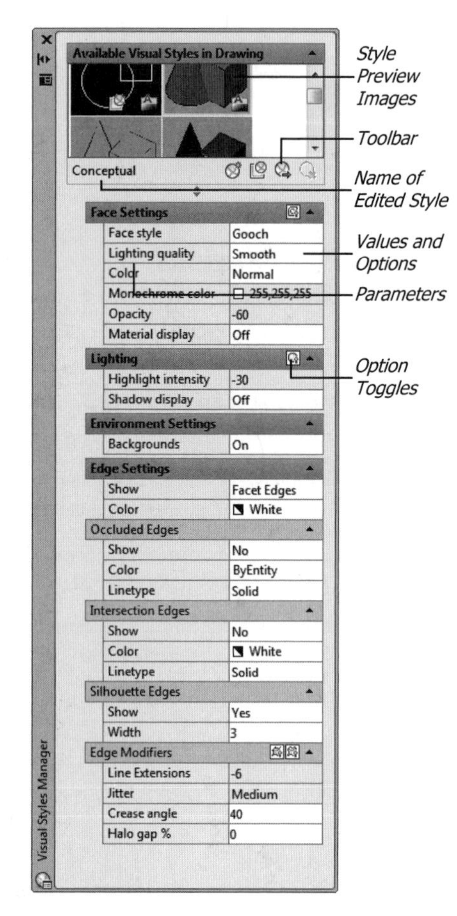

PALETTE OPTIONS

Negative values turn off options, but retain the value. Options found in 2D wirefame mode are prefixed by [2D].

Face Settings options

Face Style selects the style of face shading:

- **None** applies no style to faces, like wireframe mode.
- **Real** applies a rendered style and displays materials, if applied.
- **Gooch** substitutes warm and cool colors for light and dark; allows faces to be seen more easily than with Real.

Lighting Quality switches between faceted, smooth, and smoothest; applies to curved surfaces only.

Color specifies how colors are displayed:

- **Normal** displays face colors normally.
- **Monochrome** displays face colors in the color specified by the Monochrome Color option.
- **Tint** changes the hue and saturation of face colors.
- **Desaturate** reduces the saturation of face colors by 30%.

Monochrome or **Tint Color** specifies the single color for monochrome and tint face colors.

Opacity specifies the transparency of faces; valid values are 0 (opaque), 1 (transparent), 2 - 99 increasing translucency, and 100 (and any negative value) is again opaque.

Material Display determines the level of material realism; when on, turns of Highlight Intensity and Opacity settings.

- **Off** displays no materials for faster display; visual style settings are applied.
- **Materials** displays materials, if available.
- **Materials and Textures** displays textures for maximum realism.

Lighting options

Highlight Intensity specifies the size of highlights on faces without materials. Range is 0 (off) to 100, where larger numbers generate larger areas of intensity.

Shadow Display specifies the type of shadows to display. No lights need be placed in drawings for visual styles to cast shadows. Shadow-casting slows down AutoCAD, so it's best left off unless needed for high-quality images or sun studies.

- **Off** causes objects to cast no shadows.
- **Ground shadows** causes objects to cast shadows on the shadow plane, but not on each other.
- **Mapped shadows** causes objects to cast shadows on the ground and on each other. Graphics board must be capable of full shadow-casting, and hardware acceleration must be turned on; see the 3dConfig command.

Environment Settings option

Backgrounds toggles display of background; backgrounds are created by View command.

Edge Settings options

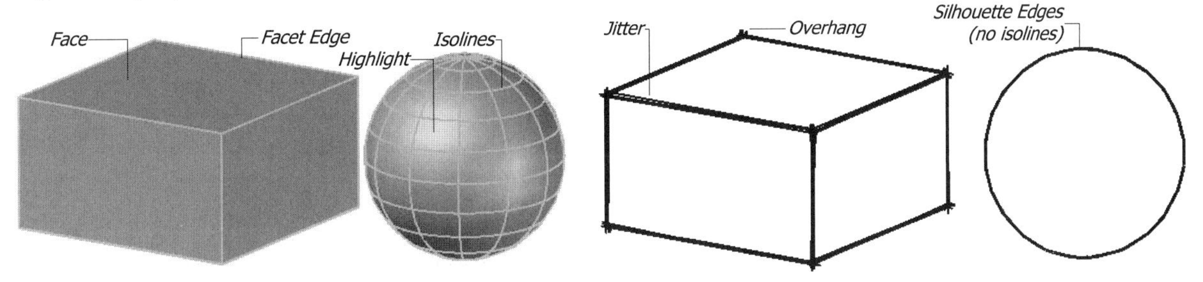

Show specifies which style of edge to display:

- **None** — facets, isolines, and edges are not displayed; this setting cannot be used when the Face Style set to None.
- **Isolines** — isolines are displayed on curved surfaces and on edges on all objects.
- **Facet Edges** — edges are displayed in a color.

Color specifies the color for edges; one color applies to all edges.

The following parameters appear only when Edge Mode = Isolines:

Number of Lines specifies the number of isolines drawn on curved surfaces; range is 0 (none) to 2047 (too many isolines blacken curved objects); a good number is 12. This parameter does not go into effect until after the next Regen command.

Always on Top toggles hidden-line removal of isolines:

- **On** displays all isolines.
- **Off** displays only foreground isolines; isolines located "around the back" are hidden.

Occluded Edges options

[2D] **Show** toggles the visibility of obscured edges and facets:

- **Yes** — hidden lines are shown through the model, although obscured edges are invisible on curved faces when a large crease angle prevents facet lines from showing.
- **No** — hidden lines are not shown.

[2D] **Color** specifies the color for visible obscured edges and facets.

Linetype specifies the line pattern for obscured (hidden) edges and facets. AutoCAD determines the scale factor; linetypes from *acad.lin* cannot be used. The available linetypes are:

- Off
- Short Dash
- Double Long Dash
- Solid
- Long Dash
- Medium Long Dash
- Dashed
- Double Short Dash
- Sparse Dot
- Dotted
- Double Medium Dash

Intersection Edges options

The following parameters appear only when Edge Mode = Facet Edges:

[2D] **Show** toggles the display of intersection edges; best turned off for better performance.

[2D] **Color** and **Linetype** specify the color and linetype of intersection lines, as described above.

Silhouette Edges options

These settings apply to all objects in a viewport equally:

[2D] **Visible** toggles the display of silhouette edges; edges are not displayed when halo gap > 0, opacity > 0 (objects are translucent), or visual styles are wireframe.

Width determines the width of the silhouette edges; range is 1 to 25 pixels.

Edge Modifiers options

These settings apply to isolines and facet edges only:

Line extensions extend edges beyond their boundaries; range is 0 to 100 pixels.

Jitter skews edge lines to create a hand drawn effect: Off (none), Low (few), Medium (medium), or High (high amount of jitter).

The following parameters appear only when Edge Mode = Facet Edges:

Crease Angle specifies the angle beyond which facet edges are not shown on curved surfaces, removing the edge lines; range is from 0 to 180 degrees. A high value, such as 180, turns off the display of facet edges.

Halo Gap % specifies the gap generated between visually overlapping objects; not available in visual styles based on wire-frame modes. Range is 0 to 100 pixels; when greater than 0, silhouette edges are not displayed.

PALETTE TOOLBAR

Create new visual style

Delete the selected visual style

Name of current visual style — Visual Style

Apply the selected visual style to the current viewport

Export the selected visual style to the Tools palette

-VISUALSTYLES Command

Command: -visualstyles

Enter an option [set Current/Saveas/Rename/Delete/?]: *(Enter an option.)*

set Current prompts for the type of visual style to apply: 2dwireframe, 3dwireframe, 3dHidden, Realistic, Conceptual, Shaded, shaded with Edges, shades of Gray, SKetchy, X-ray, or Other. If Other, it prompts for the name.

Saveas saves the current visual style; prompts for a name.

Rename renames a visual style; system styles cannot be renamed.

Delete deletes a visual style; system styles cannot be deleted.

? lists the names of visual styles in the current drawing.

RELATED COMMANDS

VsCurrent sets the visual style.

VsSave saves the current visual style by name.

ToolPalettes stores visual styles.

Rename renames visual styles.

Purge removes unused visual styles.

View applies visual styles to named views.

RELATED SYSTEM VARIABLES

CMaterial specifies the name of the material.

CShadow specifies the type of shadow cast by objects.

DragVs specifies the default visual style for 3D objects.

InterfereObjVs specifies the visual style for interference objects created by Interference.

InterfereVpVs specifies the visual style during interference checking.

Isolines specifies the number of isolines to display on curved solid models.

ShadowPlaneLocation locates the height of the invisible ground plane upon which shadows are cast; can be set to any distance along the z axis, including negative distances.

VsBackgrounds determines whether backgrounds are displayed in visual styles.

VsEdgeColor specifies the edge color; can be any ACI color.

VsEdgeJitter specifies the level of jitter effect.

VsEdgeOverhang extends edge lines beyond intersections.

VsEdges specifies the type of edge to display.

VsEdgeLEx specifies the length of line extensions of edges in visual styles *(new to AutoCAD 2011).*

VsEdgeSmooth specifies the crease angle.

VsFaceHighlight specifies the color of highlights; ignored when VsMaterialMode is on.

VsFaceOpacity controls the transparency/opacity of faces.

VsFaceStyle determines how faces are displayed.

VsHaloGap specifies the "halo" gap (gap between intersecting lines).

VsHidePrecision specifies the accuracy of hides and shades.

VsIntersectionColor specifies the color of intersecting polylines.

VsIntersectionEdges toggles the display of intersecting edges.

VsIntersectionLtype specifies the linetype for intersecting polylines.

VsIsoOntop toggles whether isolines are displayed.

VsLightingQuality toggles the quality of lighting.

VsMaterialMode controls the display of material finishes.

VsMonoColor specifies the monochrome tint.

VObscuredColor specifies the color of obscured lines.

VsObscuredEdges toggles the display of obscured edges.

VsObscuredLtype specifies the linetype of obscured lines.

VsOccludedColor specifies the color of hidden (occluded) lines in visual styles *(new to AutoCAD 2011).*

VsOccludedEdges toggles the display of hidden edges in visual styles *(new to AutoCAD 2011).*

VsOccludedLtype determines the linetype of hidden lines *(new to AutoCAD 2011).*

VsShadows determines the quality of shadows.

VsSilhEdges toggles the display of silhouette edges.

VsSilhWidth specifies the width of silhouette edge lines.

VsState reports whether the Visual Styles window is open.

...whew!

TIPS

- The ShadowPlaneLocation system variable allows you to change the elevation of the "ground" (shadow plane) upon which shadows are cast. The default is 0 units, and is measured in the z direction. This plane is independent of everything else.

AutoCAD 2011 added the following visual styles to the default collection: shaded, shaded with edges, shades of gray, sketchy, and xray.

- The default visual styles look like this:

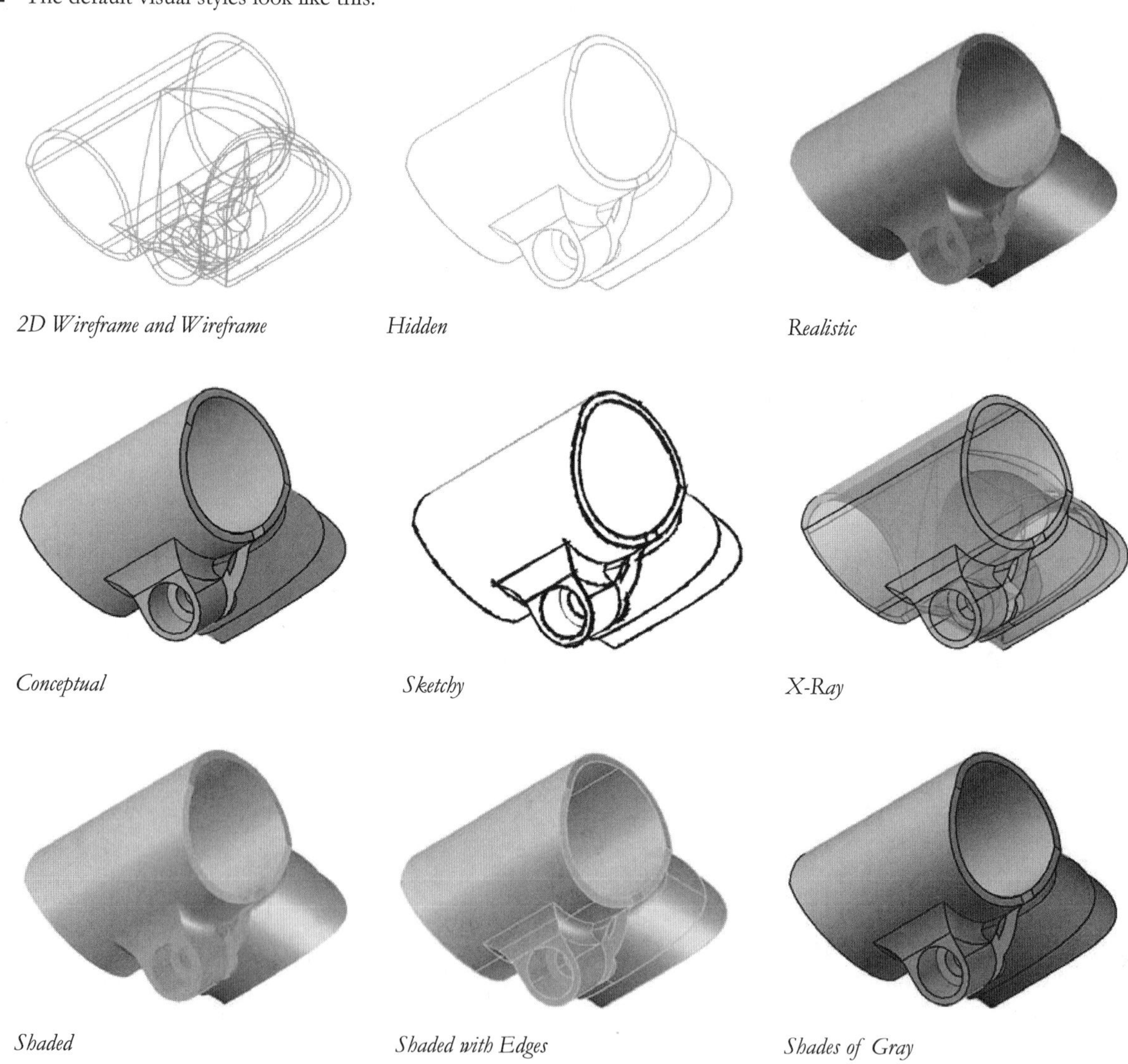

2D Wireframe and Wireframe *Hidden* *Realistic*

Conceptual *Sketchy* *X-Ray*

Shaded *Shaded with Edges* *Shades of Gray*

Removed Command

VlConv command was removed from AutoCAD Release 14; use 3dsIn instead.

VLisp

2000 Opens the VLisp integrated development environment (short for Visual LISP).

Command	*Alias*	*Menu*	*Ribbon*
'vlisp	**vlide**	**Tools**	**Manage**
		↳AutoLISP	**↳Applications**
		↳Visual LISP Editor	**↳Visual LISP Editor**

Command: vlisp

Displays window:

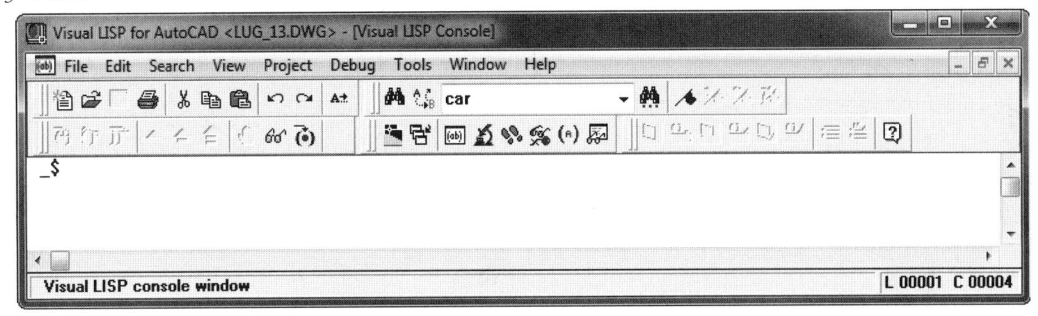

MENU BAR

Select Help | Visual LISP Help Topics for assistance in using this VLISP window.

RELATED COMMAND

AppLoad loads Visual LISP applications, as well as programs written in AutoLISP and other APIs.

TIP

- Sample VLisp code can be found in the *autocad 2011\sample\vlisp* folder.

VpClip

2000 Clips a layout viewport (short for ViewPort CLIPping).

Command	Aliases	Shortcut Keystrokes	Menu	Ribbon
vpclip	...	...	**Modify** ↳**Clip** ↳**Viewport**	**View** ↳**Viewports** ↳**Clip**

Command: vpclip
Select viewport to clip: *(Pick a viewport.)*
Select clipping object or [Polygonal] <Polygonal>: *(Select an object, or type P.)*

The selected viewport disappears, and is replaced by the new clipped viewport.

COMMAND LINE OPTIONS

Select viewport to clip selects the viewport that will be clipped.

Select clipping object selects the object that defines the clipping boundary: closed polyline, circle, ellipse, closed spline, or region.

Polygonal Options
Specify start point: *(Pick a point.)*
Specify next point or [Arc/Close/Length/Undo]: *(Pick a point, or enter an option.)*
Specify next point or [Arc/Close/Length/Undo]: *(Type C to close.)*

Specify start point specifies the starting point for the polygon.

Arc draws an arc segment; see the Arc command.

Close closes the polygon.

Length draws a straight segment of specified length.

Undo undoes the previous polygon segment.

RELATED COMMAND

Mview creates rectangular and polygonal viewports in paper space.

TIPS

- The clipping polyline is placed on the current layer. When the layer is frozen, the viewport is not clipped until the layer is thawed. When the layer is off, the polyline is invisible, but still clips.

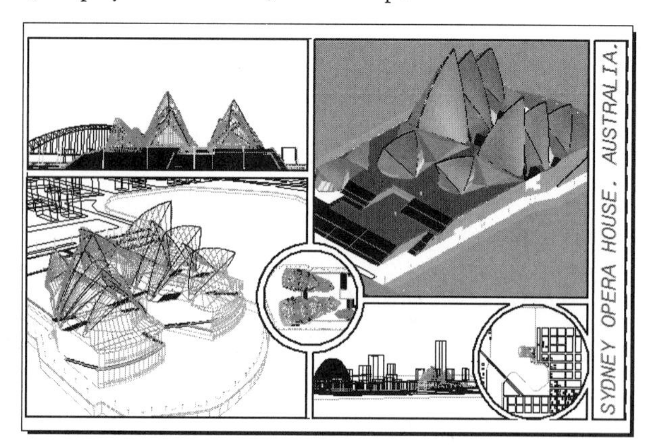

- This command does not operate in Model tab.

VpLayer

Rel. 11 Controls the properties and visibility of layers in viewports, when a layout tab other than Model is selected (short for ViewPort LAYER).

Commands	*Aliases*	*Shortcut Keystrokes*	*Menu*	*Ribbon*
vplayer	...	...	...	...

Command: vplayer
Enter an option [?/Color/LType/LWeight/Freeze/Thaw/Reset/Newfrz/Vpvisdflt]: *(Enter an option.)*
Select a viewport: *(Pick a viewport.)*

COMMAND LINE OPTIONS

Color changes the color associated with a layer; you are prompted to enter the names of layers.

LType changes the linetype associated with a layer; you are prompted to enter the names of layers.

LWeight changes the lineweight associated with a layer; you are prompted to enter the names of layers.

Freeze indicates the names of layers to freeze in this viewport.

Newfrz creates new layers that are frozen in all newly-created viewports (short for NEW FReeZe).

Reset resets the state of layers based on the Vpvisdflt settings.

Thaw indicates the names of layers to thaw in this viewport.

Vpvisdflt determines which layers will be frozen in a newly-created viewport and the default visibility in existing viewports (short for ViewPort VISibility DeFauLT).

? lists the layers frozen in the current viewport.

RELATED COMMANDS

Layer creates and controls layers in all viewports.

MView creates and joins viewports when tilemode is off.

RELATED SYSTEM VARIABLE

TileMode controls whether viewports are tiled (model) or overlapping (layouts).

VpMax / VpMin

2005 Maximizes or minimizes the selected viewport in the AutoCAD window (short for ViewPortMAXimize).

Commands	*Aliases*	*Status Bar*	*Menu*	*Ribbon*
vpmax	...		...	...
vpmin	...		...	...

Command: vpmax

When the layout contains more than one viewport, AutoCAD prompts:

Select a viewport to maximize: *(Select a viewport.)*

AutoCAD maximizes the viewport to fill the entire AutoCAD window, and switches to model space for editing.

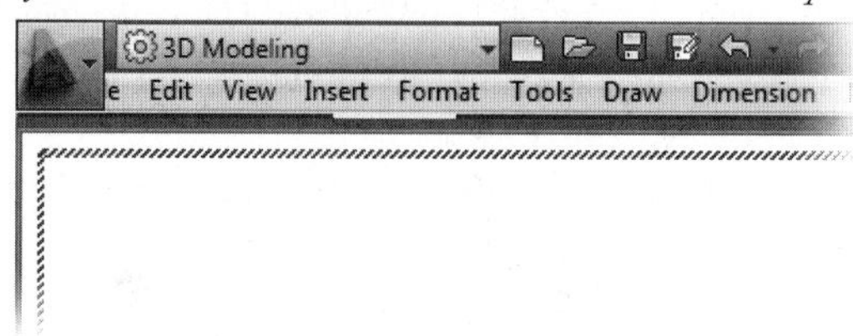

The red dashed border indicates AutoCAD is in VpMax mode.

COMMAND LINE OPTION

Select a viewport to maximize selects the viewport.

RELATED COMMAND

VPorts creates viewports.

RELATED SYSTEM VARIABLE

VpMaximixedState determines whether the viewport is maximized.

TIPS

- Use the VpMin command to return to the layout tab and restore the viewport. Alternatively, you can double-click the red border to restore the viewport.

- When the viewport is maximized, AutoCAD displays the Minimize Viewport icon on the status bar. Click the arrows to move from one viewport to the next.

VPoint

V. 2.1 Changes the viewpoint of 3D drawings (short for ViewPOINT).

Command	*Alias*	*Shortcut Keystrokes*	*Menu*	*Ribbon*
vpoint	**-vp**	...	**View** **↳3D Views** **↳Viewpoint**	...

Command: vpoint
Current view direction: VIEWDIR=0.0000,0.0000,1.0000
Specify a view point or [Rotate] <display compass and tripod>: *(Enter an option, or press Enter for the compass-tripod.)*

COMMAND LINE OPTIONS

Specify a view point indicates the new 3D viewpoint by coordinates.

Rotate indicates the new 3D viewpoint by angle.

Enter brings up visual guides (figure below).

RELATED COMMANDS

DdVpoint adjusts the viewpoint via a dialog box.

NavVCube rotates the viewpoint with the viewcube user interface.

RELATED SYSTEM VARIABLES

VpointX is the x-coordinate of the current 3D view.

VpointY is the y-coordinate of the current 3D view.

VpointZ is the z-coordinate of the current 3D view.

WorldView determines whether VPoint coordinates are in WCS or UCS.

TIPS

- This command works only in model space, and is largely replaced by the 3dOrbit command.
- As the cursor is moved on the compass, the *axis tripod* rotates showing the 3D view direction. To select the view direction, pick a location on the globe and press the pick button.

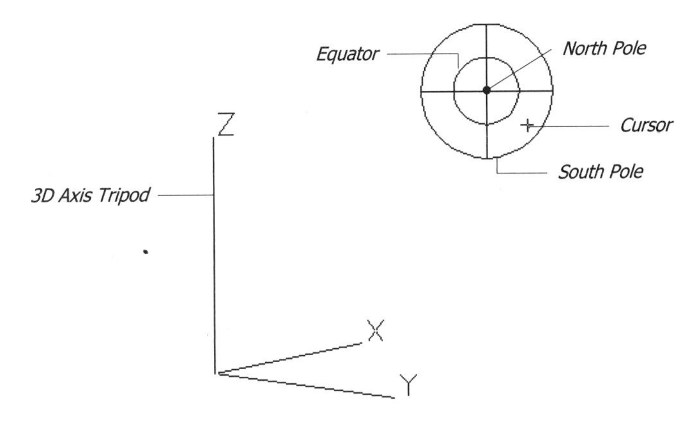

- The *compass* represents the globe, flattened to two dimensions: north pole (0, 0, z) is in the center; equator (x, y, 0), the inner circle; and south pole (0, 0, -z), the outer circle. \

VPorts

Rel. 10 Creates viewports (short for ViewPORTS).

Commands	*Aliases*	*Shortcut Keystroke*	*Menu*	*Ribbon*
vports	**viewports**	**Ctrl+R**	**View** ↳**Viewports**	**View** ↳**Viewports**
+vports	...	...	...	**View** ↳**Viewports** ↳**New**
-vports	...	...	**View** ↳**Viewports** ↳**1 Viewport**	**View** ↳**Viewports** ↳**Join**

Command: vports

Displays tabbed dialog box.

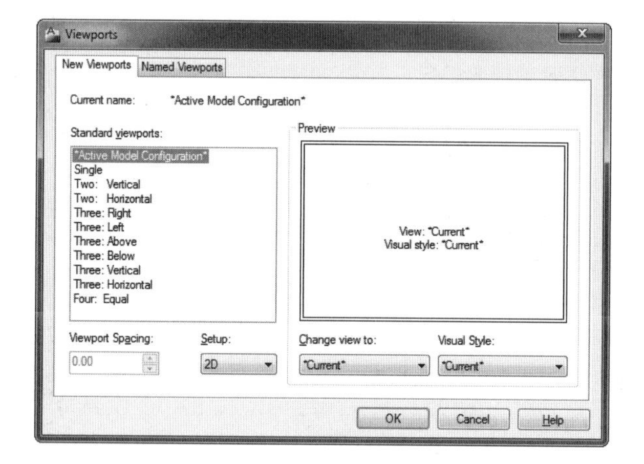

VIEWPORTS DIALOG BOX

In model space:

New name names the viewport configuration; can be up to 255 characters long.

Standard viewports lists the available viewport configurations.

Preview displays a preview of the viewport configuration.

Apply to applies the viewport configuration to one of the following

- Display.
- Current Viewport.

Setup selects 2D or 3D configuration; the 3D option applies orthogonal views, such as top, left, and front.

Change view to selects the type of view; in 3D mode, selects a standard orthogonal view.

In paper space:

Viewport spacing specifies the spacing between the floating viewports; default = 0 units.

Named Viewports tab

Named viewports lists the names of saved viewport configurations.

+VPORTS Command

Command: +vports

Tab index <0>: *(Type 0 or 1.)*

Tab index specifies the tab to display:

0 — New Viewports tab (default).

1 — Named Viewports tab.

-VPORTS Command

Command: -vports

In layout mode, displays MView command prompts. In model space, prompts:

Enter an option [Save/Restore/Delete/Join/SIngle/?/2/3/4] <3>: *(Enter an option.)*

COMMAND LINE OPTIONS

In model space

Save saves the settings of a viewport by name.

Restore restores a viewport definition.

Delete deletes a viewport definition.

Join joins two viewports together as one when they form a rectangle.

SIngle joins all viewports into a single viewport.

? lists the names of saved viewport configurations.

4 divides the current viewport into four.

2 (Two Viewports) options

Horizontal creates one viewport over another.

Vertical creates one viewport beside another (default).

3 (Three Viewports) options

Horizontal creates three viewports over each other.

Vertical creates three viewports beside each other.

Above creates one viewport over two viewports.

Below creates one viewport below two viewports.

Left creates one viewport left of two viewports.

Right creates one viewport right of two viewports (default).

In paper space

Specify corner of viewport or [ON/OFF/Fit/Hideplot/Lock/Object/Polygonal/Restore/LAyer/2/3/4]<Fit>: *(Enter an option.)*

ON turns on the viewport; the objects in the viewport become visible.

OFF turns off the viewport; the objects in the viewport become invisible.

Fit creates one viewport that fills the display area.

Hideplot removes hidden lines when plotting in layout mode.

Lock locks viewports to prevent editing.

Object converts closed polylines, ellipses, splines, regions, or circles into viewports.

Polygonal creates non-rectangular viewports.

LAyer resets layer property overrides to Bylayer in selected viewport; prompts:

Reset viewport layer property overrides back to global properties [Yes/No]?: *(Type Y or N.)*
Select viewports: *(Pick one or more viewports.)*

Other options are identical to those displayed in model tab.

RIGHT CLICK MENU

In paper space, right-click a viewport border for this shortcut menu:

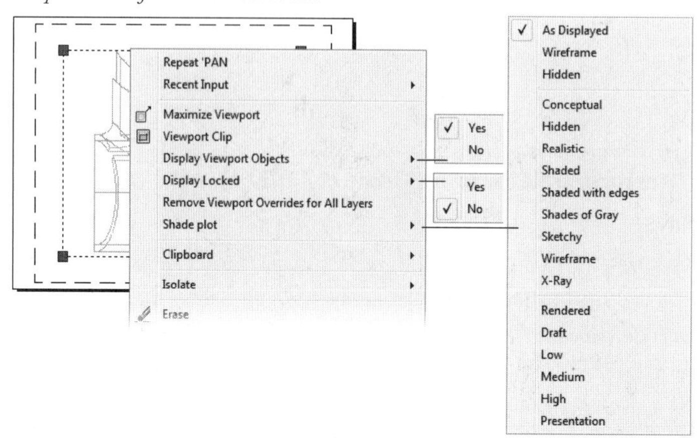

Maximize Viewport maximizes the viewport; see the VpMax command.

Viewport Clip creates, edits, and removes clipping boundaries; see the VpClip command.

Display Viewport Objects toggles visibility of the viewport.

Display Locked toggles locking of the display, which is like locking layers.

Remove Viewport Overrides for All Layers returns properties of objects in this viewport to ByLayer.

Shade Plot selects the type of visual style or rendering to use when plotting this viewport.

RELATED COMMANDS

MView creates viewports in paper space.

RedrawAll redraws all viewports.

RegenAll regenerates all viewports.

VpClip clips a viewport.

RELATED SYSTEM VARIABLES

CvPort identifies the current viewport number.

MaxActVp limits the maximum number of active viewports.

TileMode controls whether viewports can be overlapped or tiled.

VpLayerOverrides reports whether any layer in the current viewport has VP (viewport) property overrides.

VpLayerOverridesMode toggles whether viewport property overrides are displayed or plotted.

VpRotateAssoc toggles whether the viewport contents rotate with the viewport frame.

TIPS

- The Join option joins two viewports only when they form a rectangle in model space.
- You can restore saved viewport arrangements in paper space using the MView command.
- Many display-related commands (such as Redraw and Grid) affect the current viewport only.
- Press Ctrl+R to switch between viewports.
- Use the Layer command to override color, linetype, lineweight, and plot styles on a per-viewport basis.

VsCurrent

2007 Sets the current visual style (short for Visual Style CURRENT).

Command	*Aliases*	*Shortcut Keystrokes*	*Menu*	*Ribbon*
vscurrent	**vs** **shademode**	**...**	**View** **↳Visual Styles**	**...**

Command: vscurrent

Enter an option [2dwireframe/Wireframe/Hidden/Realistic/Conceptual/Shaded/shaded with Edges/shades of Gray/SKetchy/X-ray/Other] <2dwireframe>: *(Enter an option, or type O.)*

COMMAND LINE OPTIONS

2dwireframe displays raster and OLE objects; linetypes, and lineweights are visible; this is AutoCAD's default display mode, and is not a visual style.

3dwireframe looks similar to 2D wireframe, but shows the shaded 3D UCS icon and optionally the compass.

3dHidden removes lines hidden by overlapping objects.

Realistic shades objects, smooths edges, and displays materials, if attached to objects.

Conceptual looks like Realistic, but ranges the colors from "cool to warm" to make the model easier to view.

Shaded shades objects *(new to AutoCAD 2011).*

shaded with Edges shades objects and outlines edges *(new to AutoCAD 2011).*

shades of Gray shades using only gray *(new to AutoCAD 2011).*

SKetchy applies a hand drawn effect *(new to AutoCAD 2011).*

X-ray shades with transparency *(new to AutoCAD 2011).*

Other prompts for the name of a user-defined visual style: "Enter a visual style name or [?]."

? lists the names of existing visual styles.

RELATED COMMANDS

VisualStyles creates and edits visual styles.

VsSave saves the current visual style by name.

ToolPalettes stores visual styles.

TIPS

- Visual styles can be applied through any of the following:
  - Using the VsCurrent command's options.
  - Dragging them from the Visual Styles palette into the drawing; see the VisualStyles command.
  - Selecting them from the Visual Styles toolbar.
  - Dragging them from the Tools palette into the drawing; see the ToolPalettes command.
- Visual styles work only in model space.
- Use the View command's Background option to apply a background image or color to visual styles.

VSlide

V. 2.0 Displays slide files in the current viewport (short for View SLIDE).

Command	*Aliases*	*Shortcut Keystrokes*	*Menu*	*Ribbon*
vslide	...	...	...	...

Command: vslide

Displays Select Slide File dialog box. Select an .sld file, and then click Open.

COMMAND LINE OPTIONS

None.

RELATED COMMANDS

MSlide creates slide files of the current viewport.

Redraw erases slides from the screen.

RELATED AUTODESK PROGRAM

slidelib.exe creates an SLB-format library file from a group of slide files.

RELATED AUTOCAD FILES

*.*sld* stores individual slide files.

*.*slb* stores a library of slide files.

TIP

- The following applies when FileDia is set to 0, or when this command is used in a script:
  - For faster viewing of a series of slides, place an asterisk before the VSlide command to preload the *.sld* slide file, as in:

    Command: *vslide filename

  - Use the following format to display a specific slide stored in an SLB slide library file:

    Command: vslide
    Slide file: acad.slb(slidefilename)

VsSave

2007 Saves the current visual style by name (short for Visual Style SAVE).

Command	Aliases	Shortcut Keystrokes	Menu	Ribbon
vssave	...	...	...	...

Command: vssave

This command operates only in model space.

Save current visual style as or [?]: *(Enter a name, or type ?.)*

COMMAND LINE OPTIONS

Save current visual style as names the current visual style. If the name already exists, AutoCAD asks, "'*name*' already exists. Do you wish to replace the existing visual style? [Yes/No/Try again]."

? lists the names of existing visual styles, such as:

```
2D Wireframe
3D Hidden
3dWireframe
Conceptual
Hidden
Realistic
Shaded
Shaded with edges
Shades of Gray
Sketchy
Wireframe
X-Ray
```

RELATED COMMANDS

VsCurrent sets the current visual style.

VisualStyles creates and edits visual styles.

Rename renames visual styles.

Purge removes unused visual styles from the current drawing.

ToolPalettes stores visual styles and shares them among drawings.

TIPS

- Every drawing contains the visual styles predefined by Autodesk.
- To use a saved visual style in another drawing, drag it from the Visual Styles Manager palette onto the Tools palette. Open the other drawing, and then drag the visual style from the Tool palette into the drawing.
- When the current visual style is 2D Wireframe, the visual style cannot be saved.

VTOptions

2006 Controls view transitions during pans and zooms (short for View Transition OPTIONS).

Command	Aliases	Shortcut Keystrokes	Menu	Ribbon
vtoptions	...	...	...	...

Command: vtoptions

Displays dialog box.

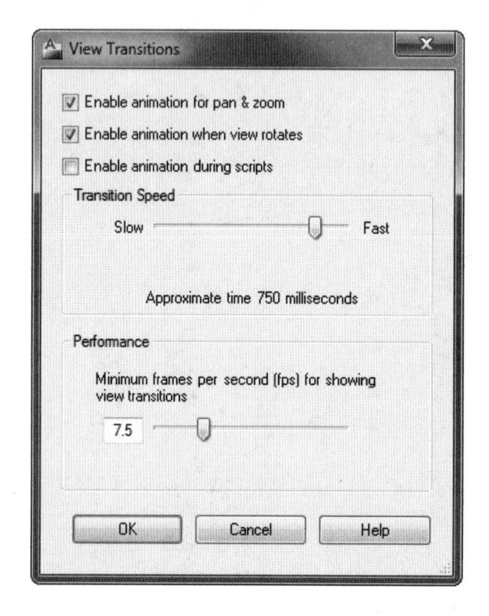

VIEW TRANSITIONS DIALOG BOX

☑**Enable Animation for Pan and Zoom** smooths view transitions during pans and zooms.

☑**Enable Animation When View Rotates** smooths view transitions during view angle changes.

☐**Enable Animation During Scripts** smooths view transitions during scripts.

Transition Speed option

Transition Speed specifies the speed of view transitions in milliseconds.

Performance option

Performance specifies the minimum speed for smooth view transitions in frames per second.

RELATED COMMAND

Zoom utilizes view transitions.

RELATED SYSTEM VARIABLES

VtDuration specifies the duration of view transitions in milliseconds.

VtEnable determines which commands use view transitions.

VtFps specifies the minimum view transition speed in frames per second.

TIPS

- View transitions are particularly useful when viewing in 3D.
- When AutoCAD cannot maintain the transition speed, it switches to an instant transition.
- View transitions make it easier to see where AutoCAD is zooming, but increase the time it takes to complete the zoom.

WalkFlySettings

2007 Presets parameters for the 3dFly and 3dWalk commands.

Command	Aliases	Menu	Ribbon
walkflysettings	...	**View**	**Tools**
		↳**Walk and Fly**	↳**Animation**
		↳**Walk and Fly Settings**	↳**Walk and Fly Settings**

Command: walkflysettings

Displays dialog box:

DIALOG BOX OPTIONS

⊙ **When Entering Walk and Fly Modes** displays window of control keys:

Movement	Left Keyboard	Right Keyboard
Forward	W	Up Arrow
Backward	S	Down Arrow
Left	A	Left Arrow
Right	D	Right Arrow
Turn	*Drag mouse*	*Drag mouse*

○ **Once Per Session** displays the mappings window the first time the 3dWalk and 3dFly commands are entered in an AutoCAD session.

○ **Never** never displays the mappings window.

☑ **Display Position Locator Window** toggles the display of the Position Locator window.

Walk/Fly Step Size specifies the increment of position (step size) in drawing units.

Steps Per Second specifies the number of increments per second.

RELATED COMMANDS

3dFly and **3dWalk** move the viewpoint through 3D scenes.

RELATED SYSTEM VARIABLES

StepsPerSec specifies the number of steps per second.

StepSize specifies the distance per step.

WBlock

V. 1.4 Writes blocks or entire drawings to disk (short for Write BLOCK).

Commands	*Aliases*	*Menu*	*Ribbon*
wblock	**w**	...	...
	acadwblockdialog		
-wblock	**-w**	...	...

Command: wblock

Displays dialog box:

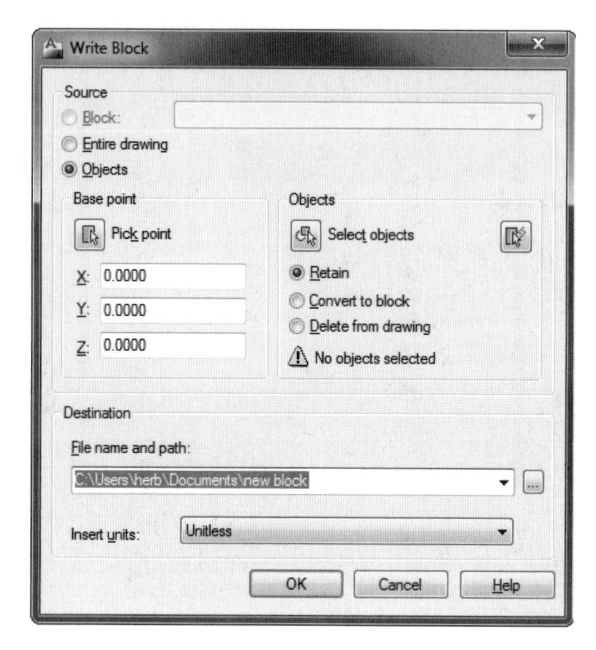

WRITE BLOCK DIALOG BOX

Source options

○ **Block** specifies the name of the block to save as a *.dwg* file.

○ **Entire drawing** selects the current drawing to save as a *.dwg* file.

⊙ **Objects** specifies the objects from the drawing to save as a *.dwg* file.

Base point options

Pick Point dismisses the dialog box temporarily to select the insertion base point.

X specifies the x coordinate of the insertion point.

Y specifies the y coordinate of the insertion point.

Z specifies the z coordinate of the insertion point.

Objects options

Select Objects dismisses the dialog box temporarily to select one or more objects.

Quick Select displays the Quick Select dialog box; see the QSelect command.

⦿ **Retain** retains the selected objects in the current drawing after saving them as a drawing file.

○ **Convert to block** converts the selected objects to a block in the drawing, after saving them as a *.dwg* file; names the block under File Name in the Destination section.

○ **Delete from drawing** deletes selected objects from the drawing, after saving them as a *.dwg* file.

Destination options

File name and path specifies the file name for the block or objects.

displays the Browse for Folder dialog box.

Insert units specifies the units when the *.dwg* file is inserted as a block.

-WBLOCK Command

Command: -wblock

Displays the Create Drawing File dialog box. Name the file, and then click Save.

Enter name of existing block or
[= (block=output file)/* (whole drawing)] <define new drawing>: *(Enter a name, or use = and * options, or press Enter.)*

COMMAND LINE OPTIONS

Enter name of existing block names a current block in the drawing.

= (*equals*) writes block to a *.dwg* file, using the block's name as the file name.

* (*asterisk*) writes the entire drawing to a *.dwg* file.

ENTER creates a *.dwg* drawing file of the selected objects on disk. (The selected objects are erased from the drawing; use the Oops command to bring them back.)

RELATED COMMANDS

Block creates a block from a group of objects.

Insert inserts a block or another drawing into the drawing.

BwBlockAs saves blocks to disk from inside the Block Editor environment.

RELATED SYSTEM VARIABLES

None.

TIPS

- Use the WBlock command to extract blocks from the drawing and store them on a disk drive. This allows the creation of a block library.
- Purge the drawing quickly by using this command on the entire drawing.
- Support for the DesignXML (extended markup language) format was withdrawn from AutoCAD 2004.

Aliased Command

WebLight is an alias for the Light command.

Wedge

Rel.11 Draws 3D wedges as solid models.

Command	Alias	Shortcut Keystrokes	Menu	Ribbon
wedge	**we**	...	**Draw** **↳Modeling** **↳Wedge**	**Home** **↳3D Modeling** **↳Wedge**

Command: wedge
Specify first corner or [Center]: *(Pick a point, or type C.)*
Specify other corner or [Cube/Length]: *(Pick a point, or enter an option.)*
Specify height or [2Point]: *(Pick another point, or enter 2P.)*

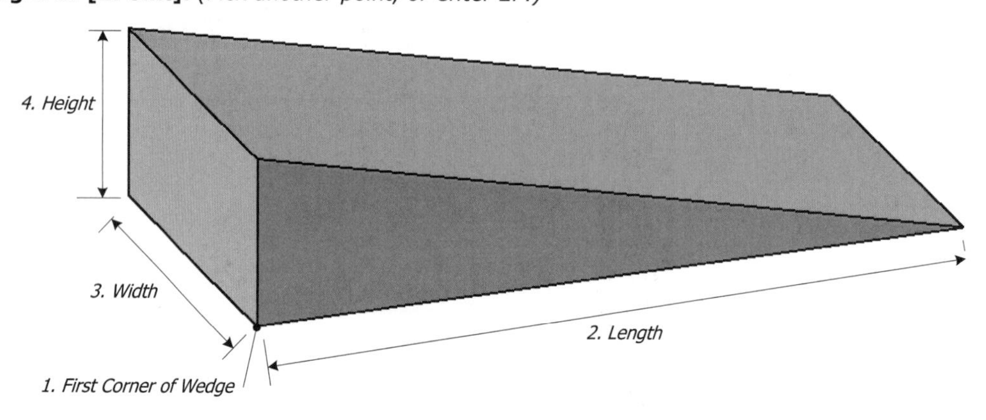

COMMAND LINE OPTIONS

First corner specifies a corner of the wedge.

Center draws the wedge's base about the center of the sloped face.

Cube draws cubic wedges whose width, length, and height are equal.

Length specifies the length, width, and height of the wedge.

2Point picks two points to indicate the height.

Center options
Specify center: *(Pick a point.)*
Specify corner or [Cube/Length]: *(Pick a point, or enter an option.)*
Specify height or [2Point]: *(Pick a point.)*

Specify center of wedge indicates the center of the wedge's inclined face.

Specify opposite corner indicates the distance from the midpoint to one corner.

Cube options
Specify length: *(Specify the length.)*

Specify length indicates the length of all three sides.

Length options
Specify length: *(Specify the length.)*
Specify width: *(Specify the width.)*

Specify height: *(Specify the height.)*

Specify length indicates the length parallel to the x-axis.

Specify width indicates the width parallel to the y-axis.

Specify height indicates the height parallel to the z-axis.

RELATED COMMANDS

Ai_Wedge draws wedges as 3D surface models.

Box draws solid boxes.

Cone draws solid cones.

Cylinder draws solid cylinders.

Sphere draws solid spheres.

Torus draws solid tori.

TIPS

- *Length* means size in the x-direction.
- *Width* means size in the y-direction.
- *Height* means size in the z-direction.
- The IsoLines system variable has no effect on wedges.
- The wedge can be edited using grips:

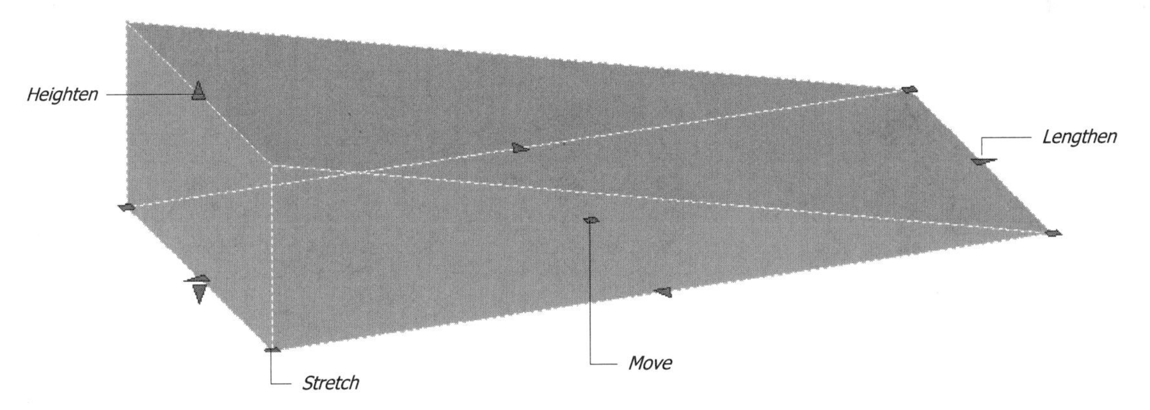

- Use negative values for length, width, and height to draw the wedge in the negative x, y, and z directions.
- The height grows from the first point picked; the second point defines the sharp end of the wedge.

WhoHas

2000 Determines which computer has drawings open.

Command	Aliases	Shortcut Keystrokes	Menu	Ribbon
whohas	...	...	...	...

Command: whohas

Displays the Select Drawing to Query dialog box. Select a drawing file, and then click Open.

When the drawing is open, reports:

Owner: ralphg

Computer's Name : HEATHER

Time Accessed : Monday, March 3, 2011 11:27:28 AM

When the drawing is not open, reports:

User: unknown.

COMMAND LINE OPTIONS

None.

RELATED COMMANDS

Open opens drawings.

XAttach attaches drawings that can be opened by other users.

TIP

- This command is meant for use over networks as a convenient way to find out which users are editing specific drawings.

WipeOut

2004 Places solid fills in areas of the background color to "wipe out" underlaying portions of drawings.

Command	*Aliases*	*Shortcut Keystrokes*	*Menu*	*Ribbon*
wipeout	...	...	**Draw** ↳**Wipeout**	**Home** ↳**Draw** ↳**Wipeout**

Command: wipeout
Specify first point or [Frames/Polyline] <Polyline>: *(Pick a point, or enter option.)*
Specify next point: *(Pick a point.)*
Specify next point or [Undo]: *(Pick a point, or type U.)*
Specify next point or [Close/Undo]: *(Type C to end the command.)*

COMMAND LINE OPTIONS

Specify first/next point specifies the starting and next points of the polygon.

Undo undoes the last segment.

Close closes the polygon.

Frames and Polyline options
Enter mode [ON/OFF] <ON>: *(Type ON or OFF.)*

ON turns on the wipeout boundary polygon.

OFF turns off the boundary polygon.

Select a closed polyline: *(Pick a closed polyline.)*
Erase polyline? [Yes/No] <No>: *(Type Y or N.)*

Select a closed polyline picks a polyline that forms the wipeout boundary.

Erase polyline? determines whether the boundary polyline is erased.

RELATED COMMAND

DrawOrder displays overlapping objects in a different order.

TIPS

- To create a wipeout under text, it's easier to use the MText command's Background Mask.
- Wipeout boundaries can be edited with grips, but cannot be edited when Frames is off.
- The Frames option applies to all wipeouts in the drawing; you cannot turn frames on and off for individual wipeouts.
- To make text appear above the wipeout, use the DrawOrder command to move the text to the Front.
- Freezing or turning off the layer of the wipeout boundary stops the wipeout action.

WmfIn

Rel.12 Imports *.wmf* and *.clp* files (short for Windows MetaFile IN).

Command	*Aliases*	*Menu*	*Ribbon*
wmfin	...	**Insert**	**Insert**
		↳Windows Metafile	**↳Import**
			↳WMF

Command: wmfin

Displays the Import WMF dialog box. Select a file, and then click Open.

Specify insertion point or [Scale/X/Y/Z/Rotate/PScale/PX/PY/PZ/PRotate]: *(Pick a point, or enter an option.)*

Enter X scale factor, specify opposite corner, or [Corner/XYZ] <1>: *(Specify a value, pick a point, or enter an option.)*

Enter Y scale factor <use X scale factor>: *(Specify a value, or press Enter.)*

Specify rotation angle <0>: *(Specify a value, or press Enter.)*

COMMAND LINE OPTIONS

Insertion point picks the insertion point of the lower-left corner of the WMF image.

X scale factor scales the WMF image in the x direction (default = 1).

Corner scales the WMF image in the x and y directions.

XYZ scales the image in the x, y, and z directions.

Y scale factor scales the image in the y direction (default = x scale).

Rotation angle rotates the image (default = 0).

RELATED COMMANDS

WmfOpts specifies options for importing *.wmf* files.

WmfOut exports selected objects in WMF format.

RELATED FILES

**.clp* are Windows Clipboard files.

**.wmf* are Windows Metafiles.

TIPS

- The WMF image is placed as a block with the name WMF0; subsequent placements of *.wmf* files increase the number by one: WMF1, WMF2, and so on.
- The *.clp* files are created by the Windows Clipboard. After using Ctrl+C to copy objects to the Clipboard, you can open the Clipboard Viewer, and then save the image as a *.clp* file.
- Exploding the WMF*n* block results in polylines; even circles, arcs, and text are converted to polylines; solid-filled areas are exploded into solid triangles. The *.clp* file is pasted as a block; when exploded, constituent parts are 2D polylines.
- Use WmfOut and WmfIn to explode splines, text, circles, and arcs to short line segments. (This is useful for translation to CNC machining files.) Do not zoom or pan between WmfOut and WmfIn.
- WmfIn needs a scale factor of 2 to match the size of objects created by WmfOut.

WmfOpts

Rel.12 Specifies options for importing *.wmf* files (short for Windows Meta File OPTionS).

Command	*Aliases*	*Shortcut Keystrokes*	*Menu*	*Ribbon*
wmfopts	...	...	...	...

Command: wmfopts

Displays dialog box:

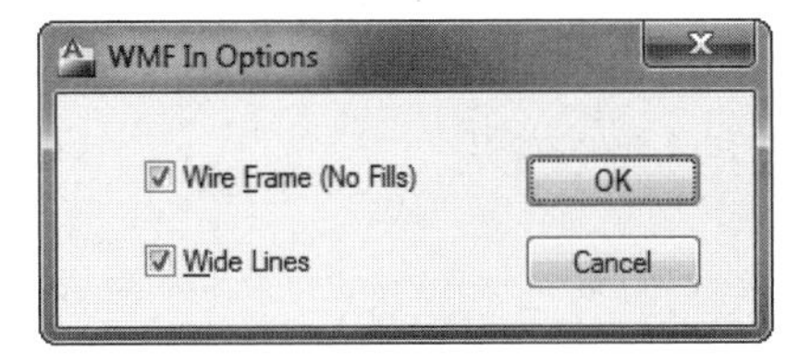

WMF IN OPTIONS DIALOG BOX

Wire Frame

☑ Displays the WMF images with lines only, no filled areas (default).

☐ Displays area fills.

Wide Lines

☑ Displays lines with width (default).

☐ Displays lines with a width of zero.

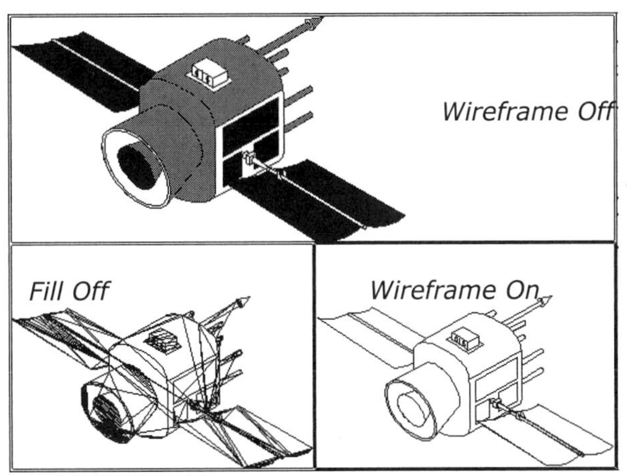

RELATED COMMANDS

WmfIn imports *.wmf* files.

WmfOut exports selected objects in WMF format.

WmfOut

Rel.12 Exports selected objects in WMF format (short for Windows MetaFile OUTput).

Command	*Aliases*	*Shortcut Keystrokes*	*Menu*	*Ribbon*
wmfout	...	...	**File** ↳**Export**	...

Command: wmfout

Displays the Create WMF File dialog box. Enter a file name, and then click Save.

Select objects: *(Select one or more objects.)*

Select objects: *(Press Enter to end object selection.)*

COMMAND LINE OPTION

Select objects selects the objects to export. Press Ctrl+A to select all objects visible in the current viewport.

RELATED SYSTEM VARIABLE

WmfBkgnd toggles the background color of exported *.wmf* files:

0 — Transparent background.

1 — AutoCAD background color.

WmfForegnd switches the foreground and background colors of exported *.wmf* files as required:

0 — Foreground is darker than background color.

1 — Background is darker than foreground color.

RELATED COMMANDS

WmfIn imports files in WMF format.

CopyClip copies selected objects to the Clipboard in several formats, including *.wmf*, also called "picture" format.

TIPS

- The *.wmf* files created by AutoCAD are resolution-dependent; small circles and arcs lose their roundness.
- The All selection does not select all objects in the drawing; instead, the WmfOut command selects all objects *visible* in the current viewport.
- Autodesk needs to update its WMF-related commands to handle EMF (enhanced meta format) files.
- WMF is visually the best format for exporting drawings to office documents, such as those produced by word processors and spreadsheets.
- Because they consist of vectors, WMF images scale correctly in other applications.

WorkSpace

2006 Controls workspaces and settings at the command-line.

Command	*Alias*	*Status Bar*	*Menu*	*Quick Access Toolbar*
workspace	...	⚙	**Tools**	2D Drafting & Annotation
		↳*workspace name*	↳**Workspaces**	
			↳*workspace name*	

Command: workspace
Enter workspace option [setCurrent/SAveas/Edit/Rename/Delete/SEttings/?] <setCurrent>: *(Enter an option.)*

COMMAND LINE OPTIONS

setCurrent makes a workspace current.

SAveas saves the current interface configuration as a named workspace.

Edit allows modifications to workspaces; displays the Customize User Interface dialog box; see the CUI command.

Rename renames workspaces.

Delete erases workspaces.

SEttings displays the Workspace Settings dialog box; see the WsSettings command.

RELATED COMMANDS

Cui creates and modifies workspaces through a dialog box.

WsSave saves the current user interface configuration as a new workspace.

WsSettings specifies options for workspaces.

RELATED SYSTEM VARIABLES

WsCurrent names the current workspace.

STARTUP SWITCH

/w specifies the workspace to show when AutoCAD starts up.

TIPS

- This command is meant for use by scripts and programs.
- The WsCurrent system variable can be used to change workspaces quickly at the command line.
- You can also change workspaces through the Workspaces item on the Windows menu, the droplist on the Workspaces toolbar, the Quick Access toolbar, and the Workspace button on the status bar.

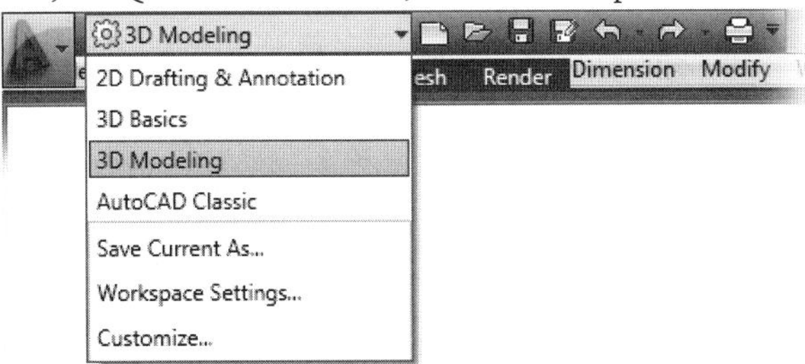

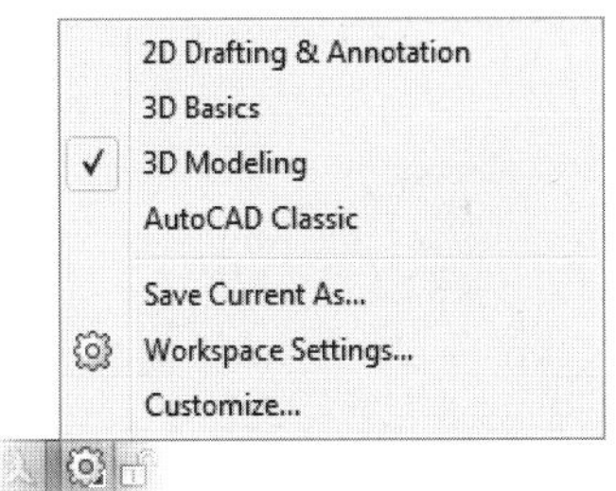

- The default workspace is named "2D Drafting & Annotation." If it has not been edited, you can select this workspace to return AutoCAD to the way it looked when it was first installed on your computer.
- AutoCAD 2011 added the 3D Basics workspace, and installed the Workspace droplist on the Quick Access toolbar.

WsSave

2006 Saves the current user interface as a named workspace (short for WorkSpace Save).

Commands	Aliases	Shortcut Keystrokes	Menu	Ribbon
wssave	...	...	**Tools ↳Workspaces ↳Save Current As**	...
-wssave	...	...	...	...

Command: wssave

Displays dialog box.

SAVE WORKSPACE DIALOG BOX

Name names the workspace. If the name already exists, AutoCAD warns, "Workspace already exists. Do you wish to replace it?" Click Yes or No.

-WSSAVE Command

Command: -wssave

Save Workspace as <AutoCAD Default>: *(Enter a workspace name.)*

OPTIONS

Save Workspace as names the workspace.

RELATED COMMANDS

Cui creates and modifies workspaces through a dialog box.

Workspace creates, saves, and controls workspaces at the command line.

WsSettings specifies options for workspaces.

RELATED SYSTEM VARIABLES

WsCurrent specifies the name of the current workspace.

WsAutoSave saves changes to workspaces automatically *(new to AutoCAD 2011)*.

TIPS

- Workspaces are saved in *.cui* files, and can be edited with the CUI command.
- Set WsAutoSave to 1 so that changes to the user interface are recorded automatically, such as turning on the menu bar.

WsSettings

2006 Specifies which workspaces appear in the Workspace menu.

Command	*Aliases*	*Shortcut Keystrokes*	*Menu*	*Ribbon*
wssettings	...	...	**Tools** **↳Workspaces** **↳Workspace Settings**	...

Command: wssettings

Displays dialog box.

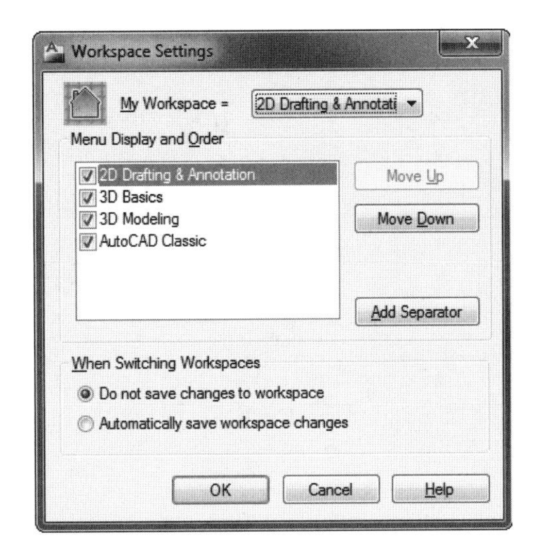

WORKSPACE SETTINGS DIALOG BOX

My Workspace selects the default workspace.

Move Up moves the selected item up the list.

Move Down moves the selected item down the list.

Add Separator adds a gray line above the selected item, which shows up in the Workspaces menu.

⊙ **Do Not Save Changes to Workspace** does not save changes to the user interface in the current workspace before switching to another workspace.

○ **Automatically Save Workspace Changes** saves changes to the user interface before switching to another workspace.

RELATED COMMANDS

Cui creates and modifies workspaces through a dialog box.

Workspace creates, saves, and controls workspaces at the command line.

WsSave saves the current user interface configuration as a new workspace.

RELATED SYSTEM VARIABLE

WsAutoSave saves changes to workspaces automatically *(new to AutoCAD 2011).*

XAttach

Rel.14 Attaches externally-referenced drawings to the current drawing (short for eXternal reference ATTACH).

Command	*Alias*	*Menu*	*Ribbon*
xattach	**xa**	**Insert** ⤷**DWG Reference**	...

Command: xattach

Displays the Select File to Attach dialog box. Select a file, and click Open.

Displays Attach External Reference dialog box.

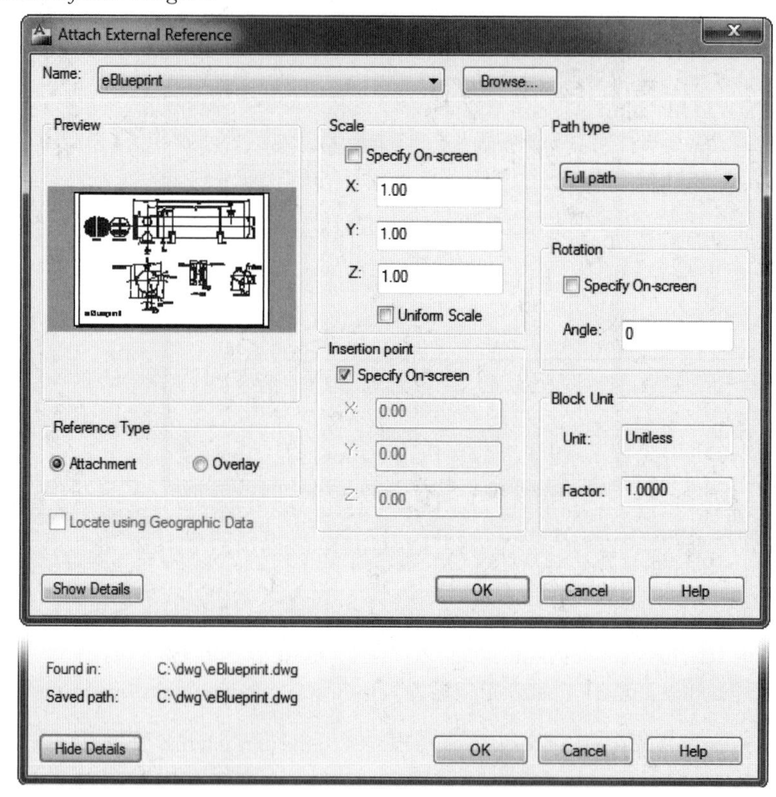

After you click OK, AutoCAD confirms at the command line:

Attach Xref FILENAME: C:\filename.dwg
FILENAME loaded.

EXTERNAL REFERENCE DIALOG BOX

Name specifies the file name of the external *.dwg* file to be attached; the droplist shows the names of currently-attached xrefs (externally-referenced files).

Browse displays the Select File To Attach dialog box.

Reference Type options

⊙ **Attachment** attaches the xref.

○ **Overlay** overlays the xref.

☐ **Locate using Geographic Data** attaches the xref using longitude and latitude; available only when the xref and drawing both contain geographic coordinates.

Show Details expands the dialog box to show the path to the xref. Click **Hide Details** to contract the dialog box.

Scale options

Specify On-screen specifies the scale of the xref in the drawing.

☐ Specify insertion point:

X, Y, Z set the x, y, z scale factors.

☑ AutoCAD prompts you after you click OK to dismiss the dialog box:

Enter X scale factor, specify opposite corner, or [Corner/XYZ] <1>: *(Enter a value, or enter an option.)*

Enter Y scale factor <use X scale factor>: *(Enter a value, or press Enter.)*

(See the -Block command.)

☐ **Uniform scale** sets the y and z scale factors to those of x.

Insertion Point options

Specify On-screen determines if the insertion point is specified in the dialog box or in the command bar.

☐ Specify the insertion point in the dialog box:

X, Y, Z specify the insertion point of the xref in the drawing.

☑ AutoCAD prompts you after you click OK to dismiss the dialog box:

Specify insertion point or [Scale/X/Y/Z/Rotate/PScale/PX/PY/PZ/PRotate]: *(Pick a point, or enter an option.)*

Path Type options

- Full Path saves the xref's file name and full path in the *.dwg* file.
- Relative path saves the xref's file name and the path to it.
- No Path saves only the xref's file name.

Rotation Angle options

Specify On-screen determines if the angle is specified in the dialog box or in the command bar.

☐ Enter the rotation angle in the dialog box:

Angle specifies the rotation of the xref in the drawing.

☑ When on, AutoCAD prompts you after you click OK to dismiss the dialog box:

Specify rotation angle <0>: *(Enter a value, or press Enter.)*

RELATED COMMANDS

Attach, **Xref**, **XAttach**, and **ExternalReferences** attach drawings as xrefs.

XOpen edits xref'ed drawings.

XBind binds portions of xref drawings to the current drawing.

XClip clips the display of xrefs.

ExternalReferences controls xrefs.

RELATED SYSTEM VARIABLES

XRefType determines whether xrefs are attached or overlaid, by default.

XEdit determines whether the drawing may be edited in-place, when being referenced by another drawing.

XLoadPath stores the path of temporary copies of demand-loaded xref drawings.

XRefCtl controls whether *.xlg* external reference log files are written.

ProjectName names the project for the current drawing (default = "").

DemandLoad specifies if and when AutoCAD demand-loads a third-party application when a drawing contains custom objects created by the application.

HideXRefScales hides scale factors imported from xrefs.

IdxCtl controls the creation of layer and spatial indices.

VisRetain specifies how the layer settings — on-off, freeze-thaw, color, and linetype — in xref drawings are defined by the current drawing.

XLoadCtl controls the loading of xref drawings.

TIP

- When AutoCAD cannot find an xref, it searches in the following order:
    - The folder of the current drawing.
    - The project search paths defined in the Options dialog box's Files tab and the ProjectName system variable.
    - The support search paths defined in the Options dialog box's Files tab.
    - The Start In folder specified in the shortcut that launched AutoCAD.

XBind

Rel.11 Binds portions of externally-referenced drawings to the current drawing (short for eXternal BINDing).

Commands	*Aliases*	*Shortcut Keystrokes*	*Menu*	*Ribbon*
xbind	**xb**	...	**Modify** ↳**Object** ↳**External Reference** ↳**Bind**	...
-xbind	**-xb**	...	...	...

Command: xbind

Displays dialog box.

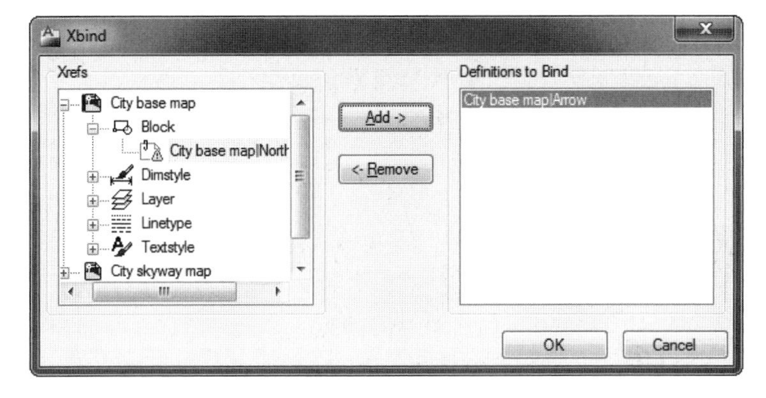

XBIND DIALOG BOX

Xrefs lists externally-referenced drawings, along with their bindable objects: blocks, dimension styles, layer names, linetypes, and text styles.

Definitions to Bind lists definitions that will be bound.

Add adds a definition to the binding list.

Remove removes a definition from the binding list.

-XBIND Command

Command: -xbind

Enter symbol type to bind [Block/Dimstyle/LAyer/LType/Style]: *(Enter an option.)*

Enter dependent name(s): *(Enter one or more names, separated by commas.)*

COMMAND LINE OPTIONS

Block binds blocks to the current drawing.

Dimstyle binds dimension styles to the current drawing.

LAyer binds layer names to the current drawing.

LType binds linetype definitions to the current drawing.

Style binds text styles to the current drawing.

Enter dependent names specifies the named objects to bind.

RELATED COMMANDS

RefEdit edits xref drawings.

ExternalReferences controls xrefs.

TIPS

- This command lets you copy named objects from another drawing to the current drawing.
- Before you can use the XBind command, you must first use the XAttach command to attach an xref to the current drawing.
- Blocks, dimension styles, layer names, linetypes, and text styles are known as "dependent symbols."
- When a dependent symbol is part of an xrefed drawing, AutoCAD uses a vertical bar (|) to separate the xref name from the symbol name, as in *filename|layername*.
- After you use the XBind command, AutoCAD replaces the vertical bar with **0**, as in *filename*0*layername*. The second time you bind that layer from that drawing, XBind increases the digit, as in *filename*1*layername*.
- When the XBind command binds a layer with a linetype (other than Continuous), it automatically binds the linetype.
- When the XBind command binds a block — with a nested block, dimension style, layer, linetype, text style, and/or reference to another xref — it automatically binds those objects as well.

XClip

Rel.12 Clips portions of blocks and externally-referenced drawings (short for eXternal CLIP; formerly the XRefClip command).

Command	*Alias*	*Menu*	*Ribbon*
xclip	**xc**	**Modify** ↳**Clip** ↳**Xref**	**Insert** ↳**Reference** ↳**Clip Xref**

Command: xclip
Select objects: *(Select one or more blocks or xrefs.)*
Select objects: *(Press Enter to end object selection.)*
Enter clipping option
[ON/OFF/Clipdepth/Delete/generate Polyline/New boundary] <New>: *(Enter an option.)*

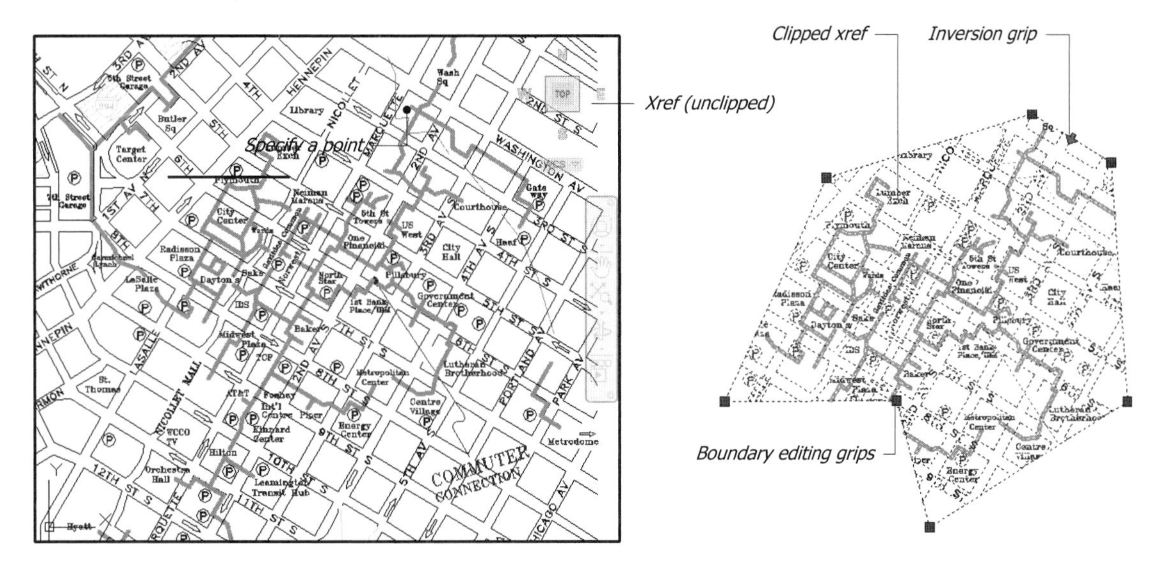

COMMAND LINE OPTIONS

Select objects selects the xref or block, *not* the clipping polyline.

ON turns on clipped display.

OFF turns off clipped display; displays all of the xref or block.

Clipdepth sets front and back clipping planes for 3D xrefs and blocks; see the 3dClip command.

Delete erases the clipping boundary.

generate Polyline extracts the existing boundary as a polyline.

New boundary places new rectangular or polygonal boundaries, inverts the boundary, or converts polylines into boundaries.

New Boundary options

Delete old boundary? [Yes/No] <Yes>: *(Type Y or N.)*

Current mode: Objects inside/outside boundary will be hidden.

Specify clipping boundary:

[Select polyline/Polygonal/Rectangular/Invert clip] <Rectangular>: *(Enter option.)*

Select Polyline prompts you to select a polyline in the drawing, which becomes the clipping boundary.

Polygonal prompts you to pick points that define the clipping boundary; the prompts are similar to those of the PLine command.

Rectangular prompts you to pick two points to define a rectangular boundary.

Invert Clip inverts the display of objects: objects outside the boundary are displayed, or vice versa.

RELATED COMMANDS

XBind binds parts of the xref to the current drawing.

Clip clips attached MicroStation V8 DGN files and DWF/x files.

ExternalReferences controls xrefs.

RELATED SYSTEM VARIABLE

XClipFrame toggles the display of the clipping boundary.

TIPS

- After you draw the clipping boundary and exit this command, the boundary will probably be invisible. Use the XClipFrame system variable to show the clipping boundary.
- The properties of the boundary cannot be changed, other than the layer it is on. All other properties are ignored, such as color, linetype, and so on.
- An xref is limited to a single clipping boundary.
- When you select an existing polyline in the drawing, it must follow these rules:
  - It can be open, but AutoCAD closes it.
  - It must be drawn with straight line segments only (no arcs or splines).
  - It cannot intersect itself.
- A spline-fit polyline results in a curved clip boundary, but a curve-fit polyline does not.
- After you draw a polygonal clipping boundary, you can edit it with the PEdit command.
- Inverted clipped areas are displayed only with the 2D Wireframe visual style. As a workaround, use the Wipeout command.
- Use the Off option to reveal the clipped areas temporarily.
- Use the arrow grip to switch between inside and outside clipping.
- You must use this command to remove clipping boundaries; the Erase command does not work.

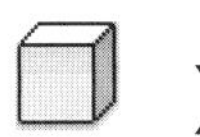

XEdges

2007 Extracts wireframe lines from the edges of 3D solids and surfaces, and from 2D regions (short for eXtract EDGES).

Commands	*Aliases*	*Shortcut Keystrokes*	*Menu*	*Ribbon*
xedges	...	...	**Modify** **↳3D Operations** **↳Extract Edges**	**Home** **↳Solid Editing** **↳Extract Edges**

Command: xedges
Select objects: *(Select one or more solids, surfaces, or regions. Hold down the Ctrl key to select individual edges.)*
Select objects: *(Press Enter.)*

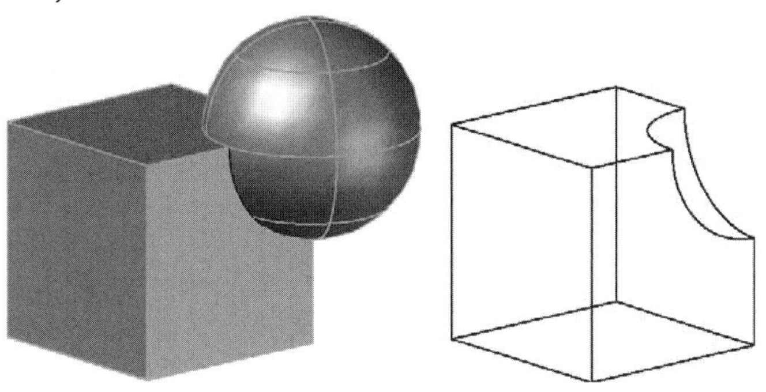

Left: *Original 3D solid.*
Right: *Extracted edges forming 3D wireframe object.*

TIPS

- This command does not work with spheres and tori, because they have no edges. Similarly, only the base of cones can be extracted.
- The command leaves edges "on top" of the original solid. Use grips editing to move the solid out of the way.
- Straight edges become lines; curves become arcs, circles, or splines.
- Xedges are created on the current layer.

XLine

Rel.13 Places infinitely long construction lines in drawings.

Command	Alias	Shortcut Keystrokes	Menu	Ribbon
xline	**xl**	...	**Draw** ↳**Construction Line**	**Home** ↳**Draw** ↳**Construction Line**

Command: xline
Specify a point or [Hor/Ver/Ang/Bisect/Offset]: *(Pick a point, or enter an option.)*
Through point: *(Pick a point.)*
Through point: *(Press Enter to end the command.)*

Draws xlines, as illustrated here:

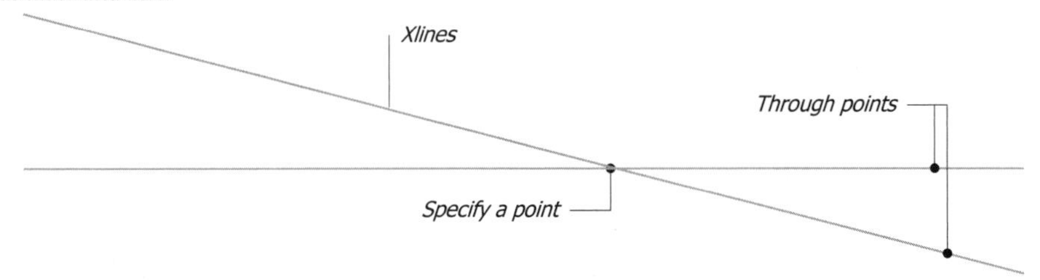

COMMAND LINE OPTIONS

Specify a point picks the midpoint for the xline.

Through point picks another point through which the xline passes.

Ang places the construction line at an angle.

Bisect bisects an angle with the construction line.

From point places the construction line through a point.

Hor places a horizontal construction line.

Offset places the construction line parallel to another object.

Ver places a vertical construction line.

ENTER exits the command.

Angle options
Enter angle of xline <0> or [Reference]: *(Enter an angle, or type R.)*

Enter angle of xline specifies the angle of the xline relative to the x-axis.

Reference specifies the angle relative to two points.

Bisect options
Specify angle vertex point: *(Pick a point.)*
Specify angle start point: *(Pick a point.)*
Specify angle end point: *(Pick a point.)*

Specify angle vertex point specifies the vertex of the angle.

Specify angle start point specifies the angle start point.

Specify angle end point specifies the angle endpoint.

Offset options
Specify offset distance or [Through] <1.0000>: *(Enter a distance, or type T.)*
Select a line object: *(Select a line, xline, ray, or polyline line segment.)*

Specify side to offset: *(Pick a point.)*

Specify offset distance specifies the distance between xlines.

Through picks a point through which the xline should pass.

Select a line object selects the line, xline, ray, or polyline to offset.

Specify side to offset specifies the offset side.

RELATED COMMANDS

Properties modifies characteristics of xline and ray objects.

Ray places semi-infinite construction lines.

RELATED SYSTEM VARIABLE

OffsetDist specifies the current offset distance.

TIPS

- Use xlines to find the bisectors of triangles (using MIDpoint object snap), or to create intersection snap points (using INTersection object snap).

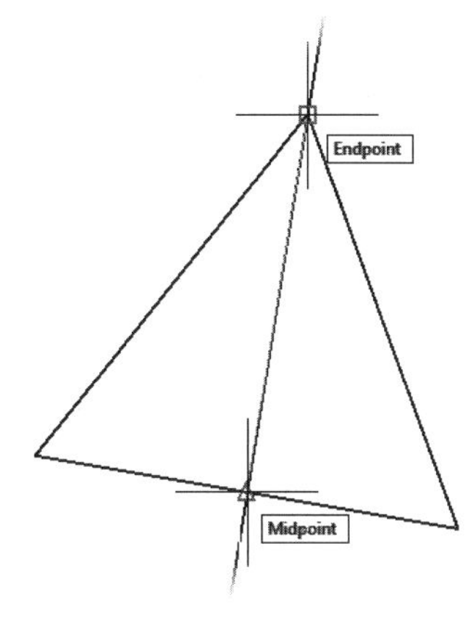

- Ray and xline construction lines are plotted; they do not affect the extents.
- The technical editor asks these philosophical questions:
  - "Which is longer: an xline that is infinitely long in both directions, or a ray that is infinitely long in one direction?"
  - "If you break an xline, it becomes two rays. Now, which is longer?"

XOpen

2004 Opens externally-referenced drawings in new windows (short for eXternal OPEN).

Command	*Aliases*	*Shortcut Keystrokes*	*Menu*	*Ribbon*
xopen	...	...	...	...

Command: xopen
Select xref: *(Select an externally-referenced drawing.)*

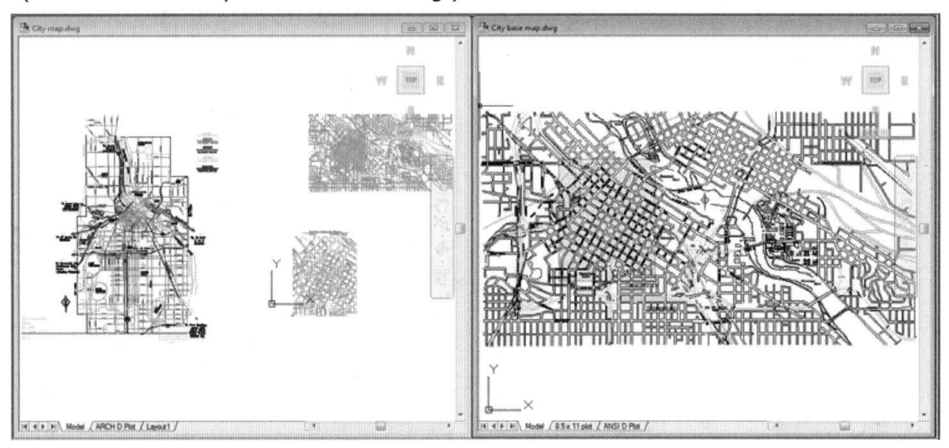

Left: *Original drawing containing xrefs (shown faded).*
Right: *Xref opened in separate window.*

COMMAND LINE OPTION

Select xrefs selects the xref to open; the xref must be inserted in the current drawing.

TIPS

- This command does not work with blocks.
- When you select a non-xref object, AutoCAD complains, 'Object is not an Xref.'

Xplode

Rel.12 Explodes complex objects into simpler objects, with user control (short for "enhanced eXPLODE").

Command	*Alias*	*Shortcut Keystrokes*	*Menu*	*Ribbon*
'xplode	**xp**	...	...	...

Command: xplode
Select objects to XPlode.
Select objects: *(Select one or more objects.)*
Select objects: *(Press Enter to end object selection.)*
Enter an option [Individually/Globally] <Globally>: *(Type I or G.)*
Enter an option [All/Color/LAyer/LType/Inherit from parent block/Explode] <Explode>: *(Enter an option.)*

Left: *Block and polyline.*
Right: *Exploded objects lose attribute data and width .*

COMMAND LINE OPTIONS

Select objects selects objects to be exploded.

Individually allows you to specify options for each selected object.

Globally applies options to all selected objects.

All sets the color, layer, linetype and lineweight of exploded objects.

Color sets the color of objects after they are exploded: red, yellow, green, cyan, blue, magenta, white, bylayer, byblock, or any color number.

LWeight specifies a lineweight.

LAyer sets the layer for the exploded objects.

LType specifies any loaded linetype name for the exploded objects.

Inherit from parent block assigns the color, linetype, lineweight, and layer of the original object to the exploded object.

Explode reduces complex objects into their components.

RELATED COMMANDS

Explode explodes the object without options.

U reverses the explosion.

TIPS

- Examples of complex objects include blocks and polylines; examples of simple objects include lines, circles, and arcs.
- Mirrored blocks can be exploded.
- The LWeight option is not displayed when the lineweight is off.

- The 'Enter an option [Individually/Globally]:' prompt appears only when more than one valid object is selected for explosion.
- Specifying BYLayer for the color or linetype means that the exploded objects take on the color or linetype of the objects' original layer.
- Specifying BYBlock for the color or linetype means that the exploded objects take on the color or linetype of the original object.
- The default layer is the current layer, not the exploded objects' original layer.
- The XPlode command breaks down complex objects as follows:

Object	***Exploded into***
Attribute	Attribute values are deleted; displays attribute definitions.
Block	Component objects.
Helix	Spline.
Leader	Line segments, splines, mtext, and tolerance objects; arrowheads become solids or blocks.
Mesh objects	3D face objects.
Mtext	Text.
Multiline	Line and arc segments.
PlaneSurf	Region.
Polyface mesh	Point, line, or 3D faces.
Polysolid	Region.
Region	Lines, arcs, and splines.
Table	Lines.
2D polyline	Line and arc segments; width and tangency are lost.
3D polyline	Line segments.
3D solid	Planar surfaces become regions; nonplanar surfaces become bodies.
3D body	Single-surface body, regions, or curves.

- The Inherit option works only when the parts were originally drawn with color, linetype, and lineweight set to BYBLOCK, and drawn on layer 0.
- This command cannot explode NURBS surfaces.

Renamed Command

XRef command is now a hardwire alias for the ExternalReferences command.

-XRef

Rel.11 Controls externally-referenced drawings in the current drawing at the command line (short for eXternal REFerence).

Command	Aliases	Shortcut Keystrokes	Menu	Ribbon
-xref	**-xr**	...	...	...

Command: -xref

Enter an option [?/Bind/Detach/Path/Unload/Reload/Overlay/Attach] <Attach>: *(Enter an option.)*

COMMAND LINE OPTIONS

? lists the names of xref files.

Bind makes the xref drawing part of the current drawing.

Detach removes xref files.

Path respecifies paths to xref files.

Unload unloads xref files.

Reload updates the xref files.

Overlay overlays the xref files.

Attach attaches another drawing to the current drawing.

RELATED COMMANDS

ExternalReferences displays a palette of all xrefs, image files, and DWF underlays.

Insert adds another drawing to the current drawing.

RefEdit edits xrefs.

XBind binds parts of xrefs to the current drawing.

XClip clips portions of xrefs.

RELATED SYSTEM VARIABLES

See the XAttach command.

TIPS

- *Warning!* Nested xrefs cannot be unloaded.
- The ExternalReferences command replaces the dialog-based XRef command, as of AutoCAD 2007. Use the undocumented ClassicXref command to access the old dialog box.

Zoom

V. 1.0 Makes the view of drawings larger or smaller in the current viewport.

Command	*Aliases*	*Status Bar*	*Menu*	*Ribbon*
'zoom	**z** **rtzoom**		**View** **↳Zoom**	**View** **↳Navigate** **↳Zoom**

Command: zoom
Specify corner of window, enter a scale factor (nX or nXP), or
[All/Center/Dynamic/Extents/Previous/Scale/Window/Object] <real time>: *(Pick a point, enter an option, or press Enter.)*

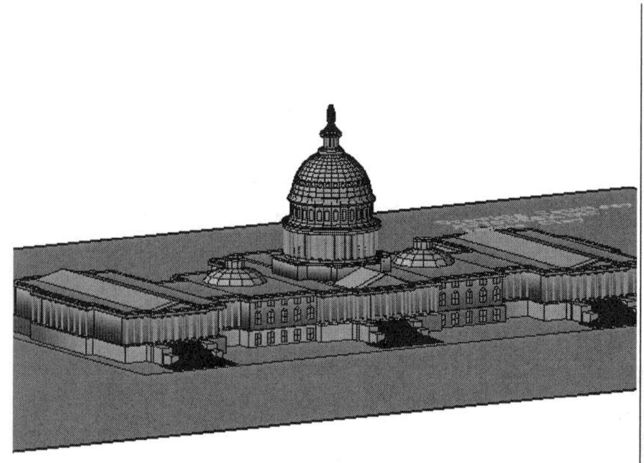

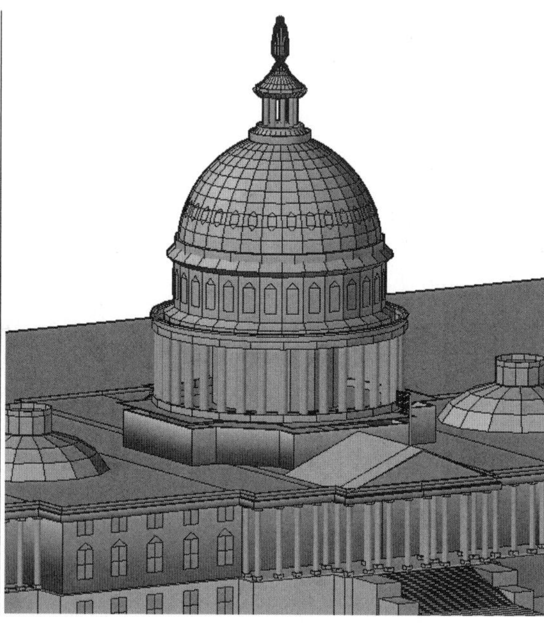

Left to right: *Before and after executing the Zoom command.*

COMMAND LINE OPTIONS

(pick a point) begins the Window option.

realtime starts real-time zoom.

Enter *or* **Esc** ends real-time zoom.

All displays the drawing limits or extents, whichever is greater.

Dynamic brings up the dynamic zoom view.

Extents displays the current drawing extents.

Previous displays the previous view generated by the Pan, View, or Zoom command.

Vmax displays the current virtual screen limits (short for Virtual MAXimum; undocumented).

Window indicates the two corners of the new view.

Object zooms to the extents of selected objects.

Center options
Specify center point: *(Pick a point.)*
Enter magnification or height <>: *(Enter a value.)*

Center point indicates the center point of the new view.

Enter magnification or height indicates a magnification value or height of view.

Left options (undocumented)
Lower left corner point: *(Pick a point.)*
Enter magnification or height <>: *(Enter a value.)*

Lower left corner point indicates the lower-left corner of the new view.

Enter magnification or height indicates a magnification value or height of view.

Scale(X/XP) options

*n***X** displays a new view as a factor of the current view.

*n***XP** displays a paper space view as a factor of model space.

RELATED COMMANDS

DsViewer displays the Aerial View palette, which zooms and pans.

Pan moves the view.

View saves zoomed views by name.

VtOptions controls smooth zoom transitions.

3dZoom performs real-time zooms in perspective viewing mode.

RELATED SYSTEM VARIABLES

ViewCtr reports the coordinates of the current view's center point.

ViewSize reports the height of the current view.

VtDuration specifies the duration of view transitions.

VtEnable turns on smooth view transitions.

VtFps specifies the speed of view transitions.

TIPS

- A magnification of 1x leaves the drawing unchanged; a magnification of 2x makes objects appear larger (zooms in), while 0.5x makes objects appear smaller (zooms out).
- Transparent zoom is *not* possible during the VPoint, Pan, DView, View and 3dOrbit commands.
- During real-time zoom, right-click in the drawing to see the shortcut menu displayed by the 3dOrbit command. But using the wheel mouse is much faster.
- You can use the scroll wheel of the mouse to zoom in and out.

3D

Rel.11 Draws 3D mesh primitives with polygon meshes (short for three Dimensions).

Command	*Aliases*	*Shortcut Keystrokes*	*Menu*	*Ribbon*
3d	...	...	...	...

Command: 3d

Enter an option

[Box/Cone/DIsh/DOme/Mesh/Pyramid/Sphere/Torus/Wedge]: *(Enter an option.)*

See the Ai_ commands for details, such as Ai_Box and Ai_Wedge.

RELATED COMMANDS

Ai_Box draws 3D mesh boxes and cubes.

Ai_Cone draws 3D mesh cones.

Ai_Dish draws 3D mesh dishes.

Ai_Dome draws 3D mesh domes.

Ai_Mesh draws 3D meshes.

Ai_Pyramid draws 3D mesh pyramids.

Ai_Sphere draws 3D mesh spheres.

Ai_Torus draws 3D mesh tori.

Ai_Wedge draws 3D mesh wedges.

RELATED SYSTEM VARIABLES

SurfTab1 controls the mesh density in the M direction of mesh objects.

SurfTab2 controls the mesh density in the N direction of mesh objects.

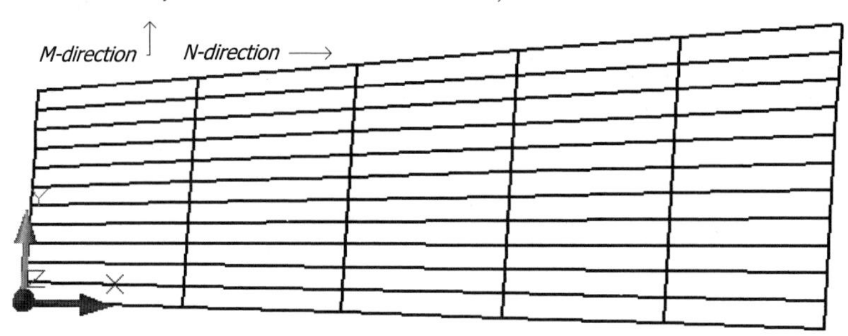

TIPS

- This command creates 3D objects made of 3D polygon meshes, and *not* of 3D solids or surfaces.
- You cannot perform Boolean operations on 3D mesh models.
- To draw cylinders with end caps, apply thickness to circles.
- Use the Ucs or Align command to place 3D mesh models in space; use the VPoint and 3dOrbit commands to view mesh models from different 3D viewpoints.
- You can apply the Hide and Render commands to 3D mesh models.

3dAlign

2007 Moves and rotates objects in 2D and 3D space.

Command	Alias	Shortcut Keystrokes	Menu	Ribbon
3dalign	**3al**	**...**	**Modify** **↳3D Operation** **↳Align**	**Home** **↳Modify** **↳3D Align**

Command: 3dalign
Select objects: *(Select one or more objects.)*
Select objects: *(Press Enter to end object selection.)*

Specify source plane and orientation ...
Specify base point or [Copy]: *(Pick a point, or type C.)*
Specify second point or [Continue] <C>: *(Press Enter, or pick another point.)*
Specify third point or [Continue] <C>: *(Press Enter, or pick a third point.)*

Specify destination plane and orientation ...
Specify first destination point: *(Pick a point.)*
Specify second destination point or [eXit] <X>: *(Press Enter, or pick another point.)*
Specify third destination point or [eXit] <X>: *(Press Enter, or pick a third point.)*

COMMAND LINE OPTIONS

Select objects selects one or more objects to be aligned.

Specify base, **Second**, and **Third point** specify the base, x axis, and y axis, respectively.

Copy copies the objects, instead of moving them.

Continue jumps ahead to the "Specify first destination point" prompt.

Specify First, **Second**, and **Third destination point** specify the basepoint and orientation of the moved/copied/rotated objects.

eXit exits the command.

RELATED COMMANDS

Align moves and rotates objects in 2D space.

3dMove moves objects in 3D space with the assistance of grip tools.

3dRotate rotates objects in 3D space with the assistance of grip tools.

TIP

- Use dynamic UCS to help position the moved/rotated objects more easily.

3dArray

Rel.11 Creates 3D rectangular and polar arrays.

Command	Alias	Shortcut Keystrokes	Menu	Ribbon
3darray	**3a**	...	**Modify** **↳3D Operations** **↳3D Array**	**Home** **↳Modify** **↳3D Array**

Command: 3darray
Select objects: *(Select one or more objects.)*
Select objects: *(Press Enter to end object selection.)*
Enter the type of array [Rectangular/Polar] <R>: *(Type R or P.)*

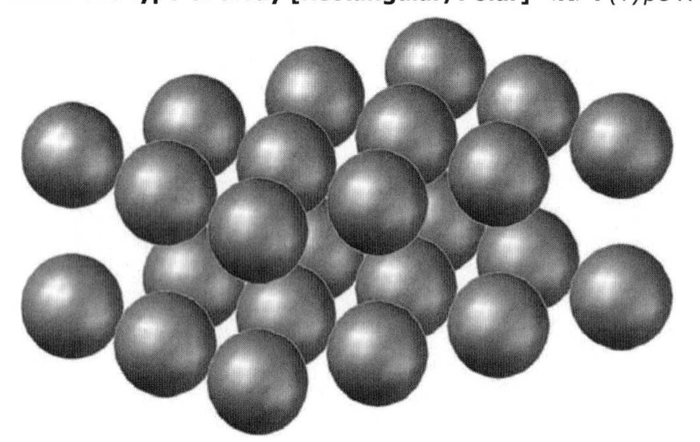

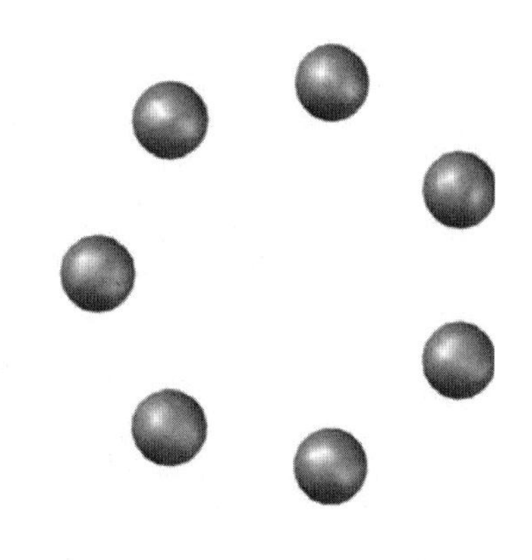

***Left:** 3D rectangular array (4 x 3 x 2)*
***Right:** 3D polar array (7 items).*

COMMAND LINE OPTIONS

Select objects selects the objects to be arrayed.

R creates rectangular 3D arrays.

P creates a polar array in 3D space.

Rectangular Array options
Enter the number of rows (---) <1>: *(Enter a value.)*
Enter the number of columns (|||) <1>: *(Enter a value.)*
Enter the number of levels (...) <1>: *(Enter a value.)*
Specify the distance between rows (---) <1>: *(Enter a value.)*
Specify the distance between columns (|||) <1>: *(Enter a value.)*
Specify the distance between levels (...) <1>: *(Enter a value.)*

Enter the number of rows specifies the number of rows in the x direction.

Enter the number of columns specifies the number of columns in the y direction.

Enter the number of levels specifies the number of levels in the z direction.

Specify the distance between rows specifies the distance between objects in the x direction.

Specify the distance between columns specifies the distance between objects in the y direction.

Specify the distance between levels specifies the distance between objects in the z direction.

Polar Array options

Enter the number of items in the array: *(Enter a number.)*

Specify the angle to fill (+=ccw, -=cw) <360>: *(Enter an angle.)*

Rotate arrayed objects? [Yes/No] <Y>: *(Type Y or N.)*

Specify center point of array: *(Pick a point.)*

Specify second point on axis of rotation: *(Pick a point.)*

Enter the number of items specifies the number of objects to array.

Specify the angle to fill specifies the distance along the circumference that objects are arrayed (default = 360 degrees).

Rotate arrayed objects?

Yes objects rotate so that they face the central axis (default).

No objects do not rotate.

Specify center point of array specifies the center point of the array and one end of the axis.

Specify second point on axis of rotation specifies the other end of the array axis.

ESC interrupts the drawing of arrays.

RELATED COMMANDS

Array creates a rectangular or polar array in 2D space.

Copy creates one or more copies of the selected object.

MInsert creates a rectangular block-array of blocks.

TIP

- This command does not operate in paper space.

3dClip

2000 Performs real-time front and back clipping (short for 3 Dimensional CLIPping).

Command	Alias	Shortcut Keystrokes	Menu	Ribbon
'3dclip	...	...	...	...

Command: 3dclip

Displays window:

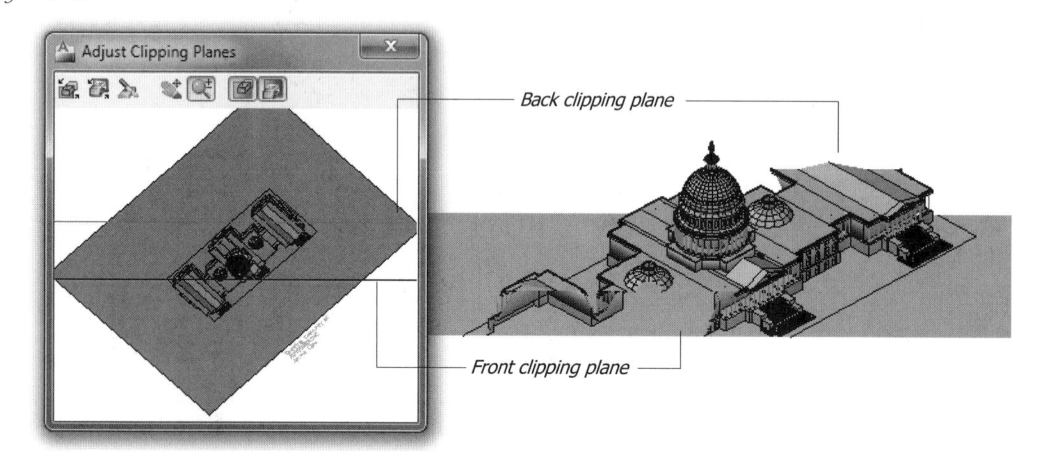

TOOLBAR

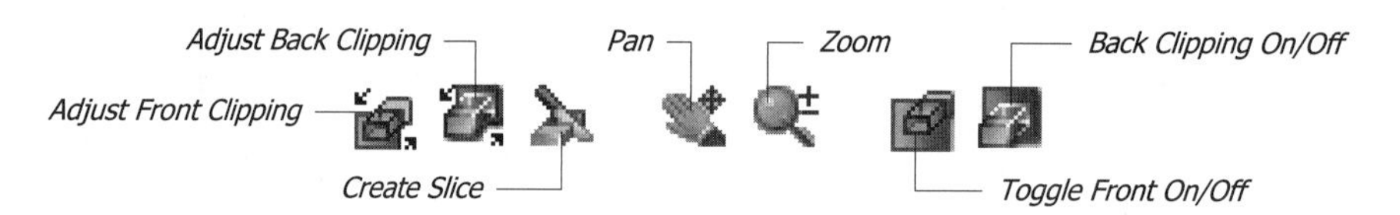

Adjust Front Clipping switches to front clipping mode.

Adjust Back Clipping switches to back clipping mode.

Create Slice switches to slicing — ganged front and back clipping — mode.

Pan moves the model within the window.

Zoom enlarges and reduces the view of the model within the window.

Front Clipping On toggles front clipping on and off.

Back Clipping On toggles back clipping on and off.

RELATED COMMANDS

3dOrbit provides real-time 3D viewing of the drawing.

3dPan performs real-time 3D sideways panning.

3dZoom performs real-time 3D zooming.

RELATED SYSTEM VARIABLES

PerspectiveClip determines the location of eyepoint clipping.

TIPS

- Use this command to hide objects in the front of 3D scenes, or to expose the interior of 3D models.
- Follow these steps to use this command effectively:
  1. Click Toggle Front On/Off to turn on front clipping.
  2. Drag the front clipping line over the model to cut off its front.
  3. Repeat, if necessary, with the back clipping plane.
- When you exit the Adjust Clipping Planes window, AutoCAD remains in 3dOrbit mode, and retains the clipped view. Press Esc to return to the 'Command:' prompt.

3dConfig / GraphicsConfig

2004 Configures display characteristics of your computer's graphics board (short for 3 Dimensional graphics CONFIGiguration).

Commands	*Aliases*	*Shortcut Keystrokes*	*Menu*	*Ribbon*
'3dconfig	...	...	...	**Render** **↳Visual Styles** **↳3D Config**
'-3dconfig	...	...	...	...
graphicsconfig				
-graphicsconfig				

Command: 3dconfig

Displays dialog box.

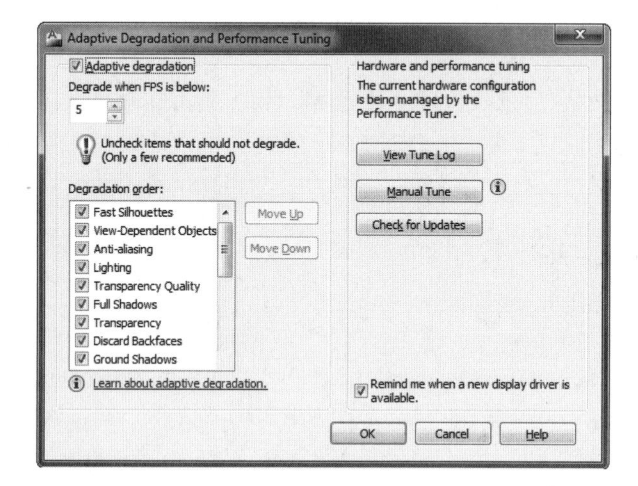

ADAPTIVE DEGRADATION AND PERFORMANCE TUNING DIALOG BOX

Adaptive Degradation options

☑**Degrade When FPS Is Below** determines how AutoCAD degrades the screen image when motion falls below the specified FPS (frames per second), such as during the 3dOrbit command. Default is 5; range is 1 to 60 FPS.

Degradation Order specifies which visual effects are removed from display to prevent further degradation. The first item is degraded or removed first.

Move Up/Down changes the degradation order.

Hardware and Performance Tuning options

View Tune Log shows the results of AutoCAD's automatic evaluation of your computer's graphics board; displays the Performance Tuner Log dialog box.

Manual Tune allows you to set the graphics board's capabilities; displays the Manual Performance Tuning dialog box.

Check for Updates launches Internet Explorer to access Autodesk's certification site.

MANUAL PERFORMANCE TURNING DIALOG BOX

This dialog box is also displayed by the GraphicsConfig command (new to AutoCAD 2011).

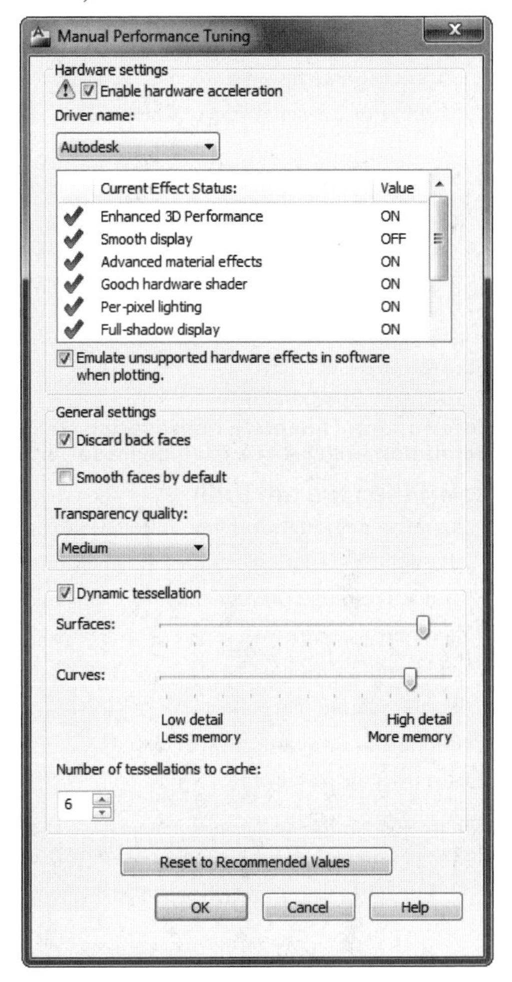

Hardware Settings options

 (green) Graphics board is capable of displaying all graphics required by AutoCAD; hardware acceleration should be on.

 (yellow) Graphics card runs most but not all graphics; turn on hardware acceleration at the risk of AutoCAD locking up.

 (red) Graphics card not capable; hardware acceleration is unavailable.

☑ **Enable Hardware Acceleration** toggles graphics generated by AutoCAD (software) or by the graphics board (hardware). Usually hardware is faster, but if the graphics board lacks the capabilities listed under Current Effect Status, then turn off this option.

Warning! When Geometry Acceleration is turned off, full shadows cannot be displayed.

Driver Name selects the device driver for the graphics board. AutoCAD provides its own drivers; graphics board manufacturers may provide their own.

☐ **Emulate Supported Hardware Effects in Software When Plotting** emulates shadows and other effects not supported by the graphics board when plotting.

General Settings options

☑ **Discard Back Faces** does not draw the backs of faces, which are not seen anyhow.

☐ **Smooth Faces by Default** smooths faces of polymesh objects.

Transparency Quality:

- **Low (Faster)** dithers the image to simulate transparency.
- **Medium** blends the image for improved quality.
- **High (Slower)** specifies highest quality but may result in slower speed (default when hardware acceleration turned on).

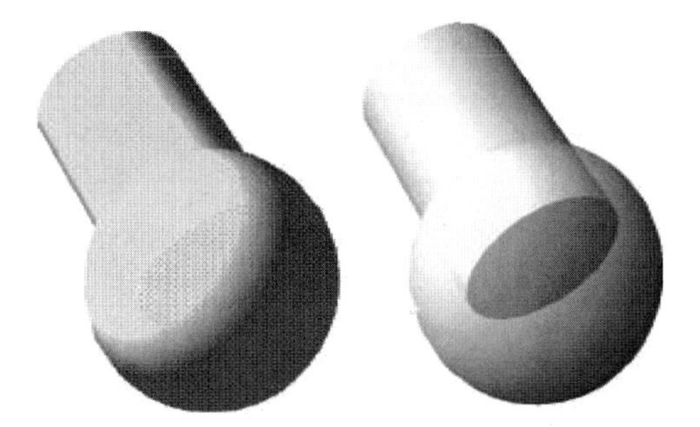

Left: *Transparency simulated by dithering.*
Right: *Hardware-accelerated transparency.*

Dynamic Tessellation options

☑ **Dynamic Tessellation** determines the smoothness of curved 3D objects.

Surfaces specifies the level of surface details; less tessellation uses less memory.

Curves specifies the level of curve details.

Number of Tessellations to Cache:

- **1** specifies the same tessellation for all viewports; some objects may be regenerated during zooming.
- **2** - **15** is required when drawings have two or more viewports with different views.

Reset to Recommended Values returns all options to values selected by Autodesk.

-3DCONFIG Command

Command: -3dconfig
Configure: 3DCONFIG
Enter option [Adaptive degradation/Dynamic tessellation/General options/ acceLeration/Plot emulation/eXit] <Adaptive degradation>: *(Enter an option, or type V for undocumented adVanced options.)*

COMMAND LINE OPTIONS

Adaptive degradation allows AutoCAD to switch to lower-quality rendering to maintain display speed.

Dynamic tessellation specifies the smoothness of faceted objects in 3D drawings.

General options handles options not dependent on the graphics board.

Geometry specifies the display of isolines and backfaces in 3D drawings.

acceLeration specifies hardware or software acceleration.

Plot emulation plots effects not supported by the graphics board, such as shadows.

adVanced lists undocumented options.

eXit exits the command.

Adaptive degradation options
Enter mode [ON/OFF] <ON>: *(Type ON or OFF.)*

ON or **OFF** turns on or off adaptive degradation.

When on, the following options become available:

Enter speed to maintain (0-60fps) <5>: *(Enter a value.)*

Maintain speed fps specifies the speed at which to display, in frames per second; range is 5fps to 60fps.

Dynamic tessellation options
Enter option [Surface tessellation/Curve tessellation/Tessellations to cache/eXit] <Surface tessellations>: *(Enter an option.)*

Surface tessellation specifies the number of tessellation lines to display on surfaces; range is 0 to 100 (default = 92).

Curve tessellation specifies the number of tessellation lines to display on curved surfaces; range is 0 to 100 (default = 87).

Tessellations to cache specifies the number of tessellations to cache; range is 1 to 4 (default = 4).

General options
Enter option [Discard backfaces/Transparency quality/eXit] <Discard backfaces>: *(Enter an option.)*

Discard backfaces toggles how the back sides of 3D objects are processed:

On discards backfaces to enhance performance.

Off displays backfaces.

Configure: Transparency - Enter mode sets the quality of transparency to low, medium, or high.

acceLeration options

Enter option [Hardware/Software/eXit] <Hardware>: *(Enter an option.)*

Hardware configures graphics card to perform 3D display rendering, the faster option.

Software configures 3D display through software, if the graphics board is unable to do so.

Enter option [Driver name/Geometry acceleration/Antialias lines/Shadows enabled/Texture compression/eXit] <Driver name>:

Driver name selects a specific driver for the graphics board.

Geometry acceleration allows a more precise display, if turned on and supported by the graphics board.

Antialias lines uses anti-aliasing to drawn smoother lines.

Shadows enabled enables shadow-casting, if the graphics board is capable. (This option is not displayed when Software is selected.)

Texture compression toggles image compression.

adVanced

Redraw on window expose toggles redrawing windows when they become visible.

-GRAPHCICSCONFIG Command

New to AutoCAD 2011.

Command: -graphicsconfig

Enter option [General options/acceLeration] <General options>: *(Enter an option)*

COMMAND LINE OPTIONS

General options displays the following prompt:

Enter option [Discard backfaces/Smooth faces/Transparency quality/eXit] <Discard backfaces>: *(See the -3dConfig command for options.)*

acceLeration displays the following prompt;

Enter option [Hardware/Software/eXit] <Hardware>: *(See the -3dConfig command for options.)*

RELATED COMMAND LINE SWITCH

/nohardware turns off hardware acceleration when AutoCAD starts up.

TIPS

- Turning on Hardware Acceleration can improve the speed and quality of displays in realistic and conceptual visual styles. However, some effects in AutoCAD no longer work, such as transparent Tools palettes.
- Turning on Hardware Acceleration can cause AutoCAD to crash when starting a 3D drawing.
- When Texture Compression is on, the graphics board uses less memory for storing image-based materials and images; however, opening drawings may take longer and image quality may be reduced.

Relocated Command

For the **3dCOrbit** command, see the 3dOrbit command.

3dDistance / 3dZoom / 3dSwivel / 3dPan

2000 This collection of commands is a subset of 3dOrbit.

Commands	*Aliases*	*Shortcut Keystrokes*	*Menu*
'3ddistance	...	...	**View ⇘Camera ⇘Adjust Distance**
'3dzoom	...	...	...
'3dswivel	...	...	**View ⇘Camera ⇘Swivel**
'3dpan	...	...	...

Command: 3ddistance

Changes the 3D viewing distance interactively, closer to and farther from the target.

Command: 3dzoom

Zooms the 3D view interactively, just like the 3dDistance command.

Command: 3dswivel

Twists the 3D view about the target interactively; the effect is like a rotating pan.

Command: 3dpan

Pans the 3D view side to side interactively; the effect is flat movement.

COMMAND LINE OPTIONS

Enter *or* Esc exits the command.

RELATED COMMANDS

3dClip performs real-time 3D front and back clipping.

3dOrbit performs real-time 3D viewing of the drawing.

3dOrbitCtr specifies the target point for 3D orbiting views.

TIPS

- You can use the 3dPan and 3dZoom commands when the current drawing is in perspective mode; when you try to use the regular Pan and Zoom commands, AutoCAD complains, '** That command may not be invoked in a perspective view **.'
- Autodesk notes that the following commands exist but are not documented:
  - 3dZoomTransparent
  - 3dSwivelTransparent
  - 3dPanTransparent

  These appear to be used by the 3dOrbit command's shortcut menu.

3dDwf

2006 Saves 2D and 3D drawings in 3D DWF and DWFx formats.

Command	Alias	Shortcut Keystrokes	Menu Bar	Application Menu	Ribbon
3ddwf	**3ddwfpublish**	...	...	**Export** ↳**3D DWF**	...

Command: 3ddwf

Displays Export 3D DWF dialog box. Enter a file name, and then click Save.

This command operates only in model tab.

3D DWF PUBLISH DIALOG BOX

Access this dialog box through the Tools | Options menu item in the the Export 3D DWF dialog box.

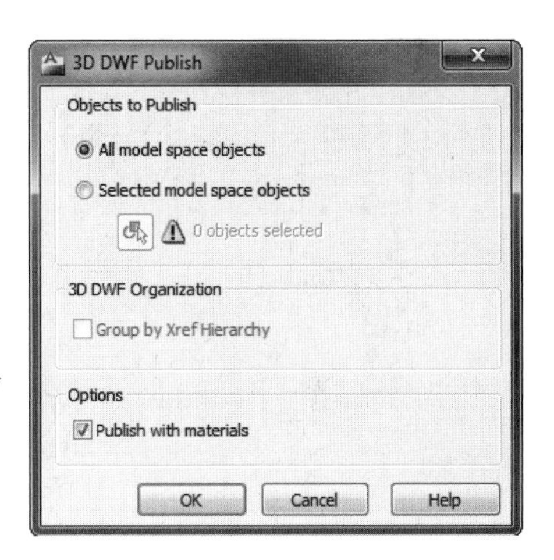

Objects to Publish options

⦿ **All Model Space Objects** saves the entire drawing.

○ **Select Model Space Objects** saves selected objects.

hides dialog box for selecting objects; prompts, "Select objects."

3D DWF Organization options

☐ **Group by Xref Hierarchy** sorts objects by external-reference hierarchy in the DWF Viewer. This option is grayed out when the drawing has links to externally-referenced drawings.

Options options

☑ **Publish with materials** includes material definitions for objects.

RELATED COMMANDS

Plot exports drawings in 2D DWF format.

Publish exports drawing sets in 2D DWF format.

AutoPublish exports drawings in DWF format when saved or closed.

DwfFormat toggles the default format between DWF and DWFx (undocumented).

RELATED SYSTEM VARIABLE

3dDwfPrec specifies the precision of 3D *.dwf* files.

TIPS

- This command works only in model space; it cannot create *.dwf* files of layouts (paper space).
- The level of precision defined by the 3dDwfPrec system variable affects file size and visual accuracy. For example, as precision is increased from level 1 to 6, the *steering.dwf* file size increases 11x, from 45KB to 513KB, while the visual accuracy of curves improves significantly.
- When done saving the drawing in 3D DWF format, AutoCAD asks, "Filename.dwf was published successfully. Would you like to see it now?" Click Yes to view, if Design Review is installed on your computer.

Renamed Command

3dDwfPublish command was renamed 3dDwf in AutoCAD 2007.

3dEditBar

2011 Edits vertices and tangencies of NURBS surfaces.

Command	*Aliases*	*Shortcut Keystrokes*	*Menu Bar*	*Ribbon*
3deditbar	...	...	...	**Surface** **↳Control Vertices** **↳CV Edit Bar**

Command: 3deditbar
Select a NURBS surface to edit: *(Choose one NURBS surface.)*
Select point on NURBS surface. *(Pick a point.)*
Move gizmo to change point location or [Base point/Displacement/Undo/eXit]<exit>: *(Enter an option.)*

COMMAND LINE OPTIONS

Select a NURBS surface chooses the surface to edit.

Select point moves the red UV crosshair cursor along the surface's u and v axes.

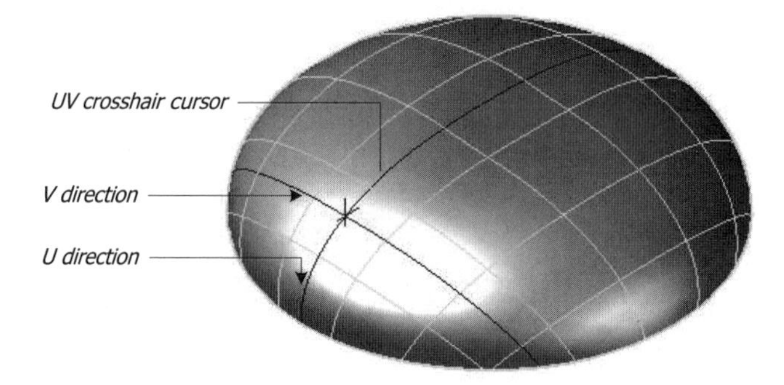

Move gizmo manipulates the surface interactively using the 3DMove gizmo; right-click for additional options.

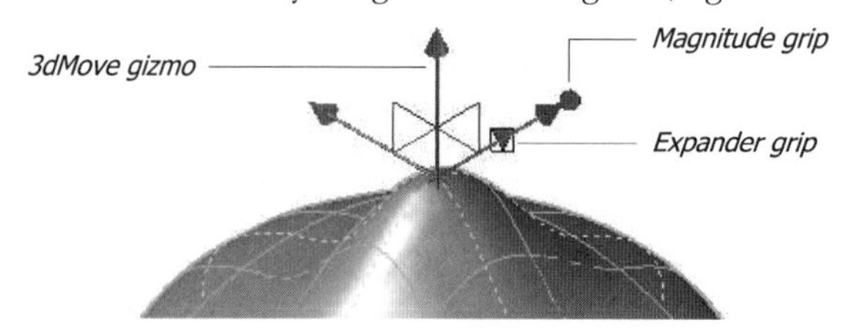

Base point prompt you to specify the edit point: 'Select point on NURBS surface'.

Displacement prompts you to enter a precise distance using x,y,z coordinates entered at the command prompt: 'Specify displacement'.

Undo undoes the last editing action.

Exit Cancels the prompt and returns you to the CV edit bar to continue adding and editing CVs.

Magnitude option

>> Magnitude ** Specify Magnitude or [eXit]:** *(Enter a value, or type X.)*

Specify Magnitude specifies the amount to move the surface; you an also enter the value using dynamic input.

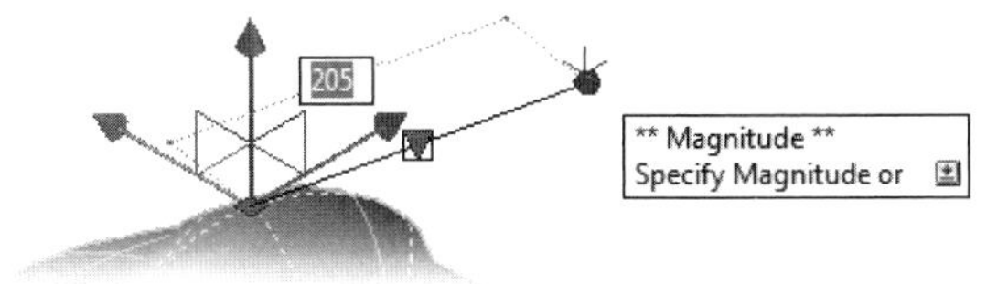

SHORTCUT MENU

Access this menu by right-clicking any grip on the gizmo:

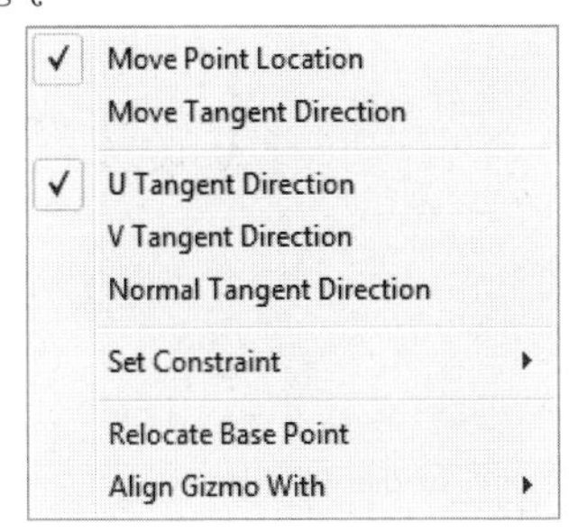

Move Point Location moves the surface to reshape it.

Move Tangent Direction changes tangency to reshape the surface.

U or **V Tangent Direction** constrains tangency in the U or V direction.

Normal Tangent Direction moves the edit bar to edit the tangency of the curve normal to the current UCS.

Set Constraint constrains movement to the X, Y or Z axes or XY, YX, or ZX planes.

Relocate Base Point moves the gizmo to a new location and add a new control vertex.

Align Gizmo With aligns with world UCS, current UCS, or a face.

RELATED COMMANDS

All surface creation and editing commands.

TIPS

- This command allows you interactively to distort NURBS surfaces to create interesting shapes.
- To use this command on other kinds of surfaces, first use the ConvToNurbs command to convert them to NURBS surfaces.
- The U and V directions are equivalent the X and Y axes of the surface, and are independent of the UCS.

3dFace

V. 2.6 Draws 3D faces with three or four corners.

Command	*Alias*	*Shortcut Keystrokes*	*Menu*	*Ribbon*
3dface	**3f**	...	**Draw** ↳**Surfaces** ↳**3D Face**	...

Command: 3dface
Specify first point or [Invisible]: *(Pick a point, or type I followed by the point.)*
Specify second point or [Invisible]: *(Pick a point, or type I followed by the point.)*
Specify third point or [Invisible] <exit>: *(Pick a point, or type I followed by the point.)*
Specify fourth point or [Invisible] <create three-sided face>: *(Pick a point, or type I followed by the point.)*
Specify third point or [Invisible] <exit>: *(Press Enter to exit the command.)*

COMMAND LINE OPTIONS

First point picks the first corner of the face.

Second point picks the second corner of the face.

Third point picks the third corner of the face.

Fourth point picks the fourth corner of the face, or press Enter to create a triangular face.

Invisible makes the edge invisible.

RELATED COMMANDS

3D draws a 3D object: box, cone, dome, dish, pyramid, sphere, torus, or wedge.

Properties modifies the 3d face, including visibility of edges.

Edge changes the visibility of the edges of 3D faces.

EdgeSurf draws 3D surfaces made of 3D meshes.

PEdit edits 3D meshes.

PFace draws generalized 3D meshes.

RELATED SYSTEM VARIABLE

SplFrame controls the visibility of edges.

TIPS

- A 3D face is the same as a 2D solid, except that each corner can have a different z coordinate.
- Unlike the procedure for the Solid command, enter corner coordinates in natural order for 3D faces.
- The i (short for invisible) suffix must be entered before osnap modes, point filters, and corner coordinates.
- *Invisible* 3D faces, where all four edges are invisible, do not appear in wireframe views; they hide objects behind them in hidden-line mode, and are rendered in shaded views.
- You can use the Properties and Edge commands to change the visibility of 3D face edges.
- 3D faces cannot be extruded.

Relocated Command

For the **3dFly** command, see the 3dWalk command.

3dMesh

Rel.10 Draws polygon 3D meshes.

Command	*Aliases*	*Shortcut Keystrokes*	*Menu*	*Ribbon*
3dmesh	...	...	...	...

Command: 3dmesh
Enter size of mesh in M direction: *(Enter a value.)*
Enter size of mesh in N direction: *(Enter a value.)*
Specify location for vertex (0, 0): *(Pick a point.)*
Specify location for vertex (0, 1): *(Pick a point.)*
Specify location for vertex (1, 0): *(Pick a point.)*
Specify location for vertex (1, 1): *(Pick a point.)*

COMMAND LINE OPTIONS

Enter size of mesh in M direction specifies the m-direction mesh size (between 2 and 256).

Enter size of mesh in N direction specifies the n-direction mesh size (between 2 and 256).

Specify location for vertex (*m*,*n*) specifies a 2D or 3D coordinate for each vertex.

RELATED COMMANDS

3D draws a variety of 3D objects.

3dFace draws a 3D face with three or four corners.

Explode explodes a 3D mesh into individual 3D faces.

PEdit edits a 3D mesh.

PFace draws a generalized 3D face.

Xplode explodes a group of 3D meshes.

TIPS

- It is more convenient to use the EdgeSurf, RevSurf, RuleSurf, and TabSurf commands than the 3dMesh command. The 3dMesh command is meant for use by AutoLISP and other programs.
- The range of values for the m- and n-mesh size is 2 to 256.
- The number of vertices = m x n.
- The first vertex is (0,0). The vertices can be at any point in space.
- The coordinates for each vertex in row m must be entered before starting on vertices in row m+1.
- Use the PEdit command to close the mesh, since it is always created open.
- The SurfU and SurfV system variables do not affect the 3D mesh object.

3dMove

2007 Moves objects along the x, y, or z axis.

Command	*Alias*	*Shortcut Keystrokes*	*Menu*	*Ribbon*
3dmove	**3m**	...	**Modify** ↳**3D Operations** ↳**3D Move**	**Home** ↳**Modify** ↳**3D Move**

Command: 3dmove
Current positive angle in UCS: ANGDIR=counterclockwise ANGBASE=0
Select objects: *(Pick one or more objects.)*
Select objects: *(Press Enter.)*
Specify base point or [Displacement] <Displacement>: *(Pick a point, or type D.)*
Specify second point or <use first point as displacement>: *(Pick another point.)*

COMMAND LINE OPTIONS

Select objects selects the objects to be moved.

Specify base point indicates the starting point for the move.

Displacement specifies relative x,y,z-displacement when you press Enter at the next prompt.

Specify second point of displacement indicates the distance to move.

3D Move grip tool

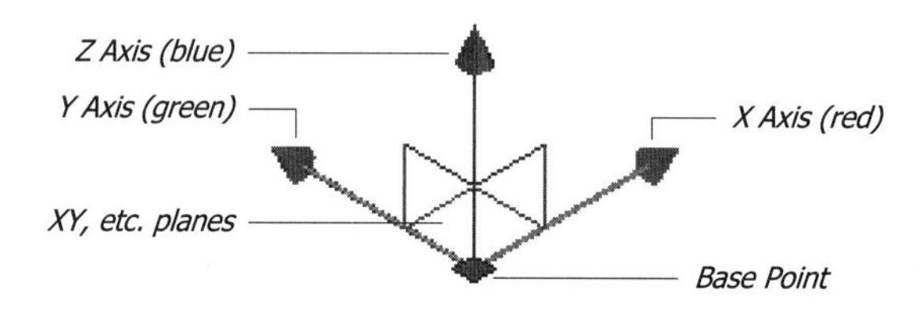

SHORTCUT MENU

Access this menu by right-clicking the gizmo:

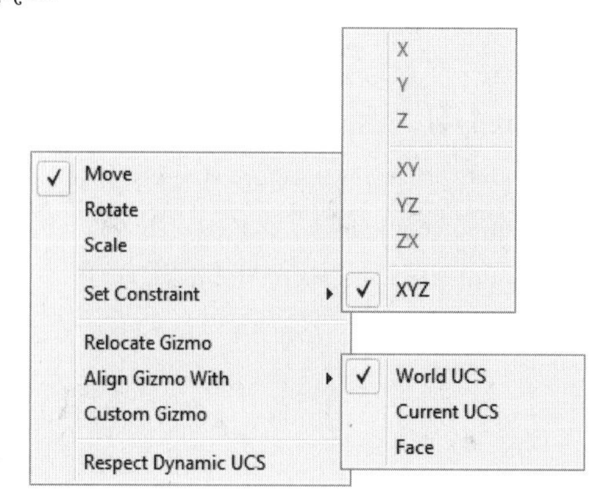

Move / Rotate / Scale switches the gizmo between 3D move, 3D rotate, and 3D scale modes.

Set Constraint locks movement to the X, Y, or Z axis or to the XY, YZ, or ZX planes; choose XYZ to remove the constraint.

Relocate Gizmo moves the gizmo elsewhere in the drawing; alternatively, move the gizmo by dragging its basepoint grip.

Align Gizmo With aligns the gizmo's three axes with the World or current UCS, or to the face of a 3D object.

Custom Gizmo prompts you to locate and orient the gizmo, as follows:

> **Specify Origin of gizmo:** *(Pick a point, or enter x,y,z coordinates.)*
> **Specify point on X-axis or <accept>:** *(Pick a point.)*
> **Specify point on the XY plane or <accept>:** *(Pick a point.)*

Respect Dynamic UCS aligns the gizmo's XY plane with faces or edges temporarily when the gizmo is relocated.

RELATED COMMANDS

Move moves objects in 2D (current UCS) and 3D, but without the grip tools.

3dRotate rotates objects in 3D space.

3dScale resizes objects in 3D space.

RELATED SYSTEM VARIABLES

GtAuto toggles the display of grips tools.

GtDefault determines whether the 3dMove or Move command is activated when the Move command is entered in 3D.

GtLocation locates the grip tool at the UCS icon or on the last selected object.

DefaultGizmo determines which gizmo appears when subobjects are selected: 3dRotate, 3dScale, or 3dMove.

TIPS

- You can drag the base point (in the middle of the grip tool) to relocate the base point.
- When dynamic UCS is turned on, the grip tool aligns itself with the selected face. (Press Ctrl+D or click DUCS to toggle dynamic UCS.)
- This gizmo was redesigned in AutoCAD 2011.

3dOrbit / 3dCOrbit / 3dFOrbit

2000 Provides interactive 3D viewing of drawings.

Commands	Aliases	Shortcut Keystrokes	Menu	Ribbon
'3dorbit	**orbit** **3do**	...	**View** **↳Orbit** **↳Constrained**	**View** **↳Navigate** **↳Orbit**
'3dcorbit	...	...	**View** **↳Orbit** **↳Continuous**	**View** **↳Navigate** **↳Continuous Orbit**
'3dforbit	...	...	**View** **↳Orbit** **↳Free Orbit**	**View** **↳Navigate** **↳Free Orbit**

Command: 3dorbit *or* 3dcorbit *or* 3dforbit

Press ESC or ENTER to exit, or right-click to display shortcut-menu. *(Move the cursor to change the viewpoint, and then press Enter to exit the command.)*

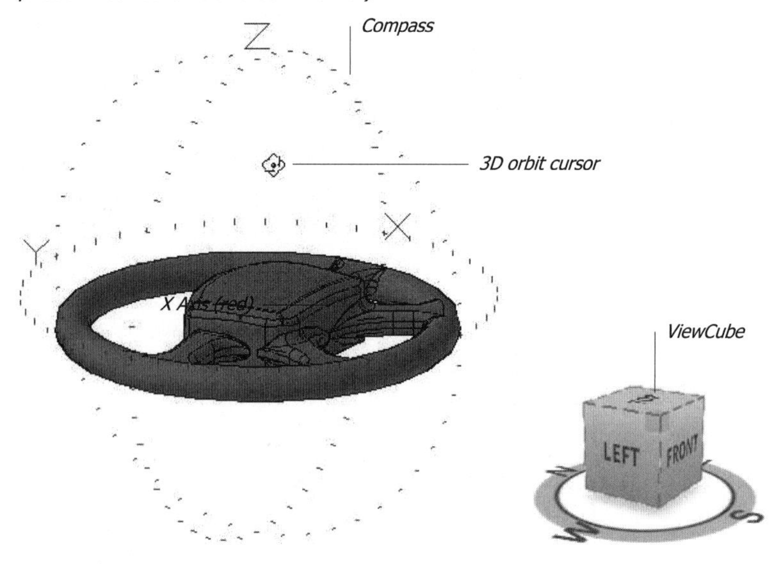

SHORTCUT MENU

Access this menu by right-clicking the drawing during this command:

Exit exits the command.

Current Mode: Constrained Orbit reports the current mode; not a command option.

Other Navigation Modes displays a submenu for accessing other 3D viewing modes.

Enable Orbit Auto Target maintains the view on a target point instead of on the center of the viewport.

Animation Settings displays the Animation Settings dialog box.

Zoom Window zooms into a rectangular area defined by two pick points; see the Zoom command.

Zoom Extents zooms to the extents of the drawing to display all visible objects.

Zoom Previous returns to the prior view.

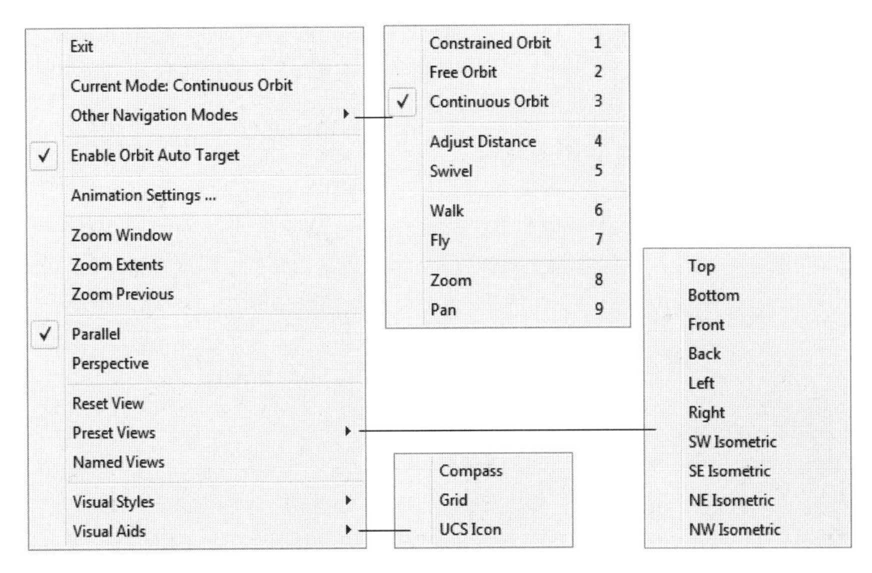

Parallel displays the view in parallel projection; more accurate.

Perspective displays the view in perspective projection; more realistic.

Reset View resets to the view when you first began the command.

Preset Views displays a submenu of standard views, such as top and side.

Named Views displays a submenu of user-named views, if any are defined in the drawing; see the View command.

Visual Styles displays a submenu for selecting the type of visual style (formerly named Shading Modes); see the Visual-Styles command.

Visual Aids displays a submenu of visual aids for navigating in 3D space.

Other Navigation Modes submenu

To switch between navigation modes, press keys 1 - 9 at any time during this command, without having to access the following shortcut submenu.

Constrained Orbit constrains movement to the x,y plane or the z axis; see the 3dOrbit command.

Free Orbit rotates the 3D view freely (3dFOrbit).

Continuous Orbit keeps the view spinning in the direction determined by dragging the mouse across the model (3dCOrbit).

Adjust Distance zooms the viewpoint nearer and farther away (3dDistance).

Swivel pans the viewpoint from side to side (3dSwivelTransparent).

Walk moves through the 3D model at a fixed height (3dWalk).

Fly moves through the 3D model at any height (3dFly).

Zoom zooms in and out (3dZoomTransparent).

Pan pans around (3dPanTransparent).

Visual Aids submenu

Compass toggles display of the compass to help you navigate in 3D space.

Grid toggles the display of the grid as lines, to help you see the x,y plane; see the Grid command.

UCS Icon toggles the display of the 3D UCS icon; see the UcsIcon command.

ANIMATION SETTINGS DIALOG BOX

Access this dialog box by right-clicking the drawing area, and then choosing Animation Settings from the shortcut menu:

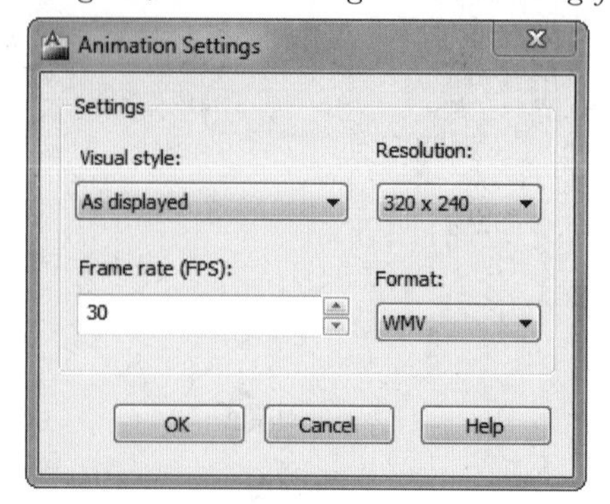

Visual Style selects the visual style by which to save the animation.

Resolution specifies the image quality; ranges from 160x120 (low quality, small files) to 1024x768 (high quality, large file sizes).

Frame rate (FPS) specifies the frames per second, ranging from 1 to 60fps.

Format selects the movie format in which to save the file: Microsoft AVI, QuickTime MOV, generic MPG, or Windows WMV.

RELATED COMMANDS

ViewCube also rotates 3D viewpoints.

3dClip performs real-time 3D front and back clipping.

3dDistance performs real-time 3D forward and backward panning.

3dPan and **3dZoom** perform real-time 3D panning and zooming.

3dWalk and **3dFly** move through 3D views.

AniPath saves flythroughs to movie files.

RELATED SYSTEM VARIABLE

Perspective reports whether the viewport displays a parallel or perspective projection.

TIPS

- Select one or more objects before invoking this command; they will appear while orbiting.
- To record animations during this command, click the Record button in the ribbon's Tools panel:

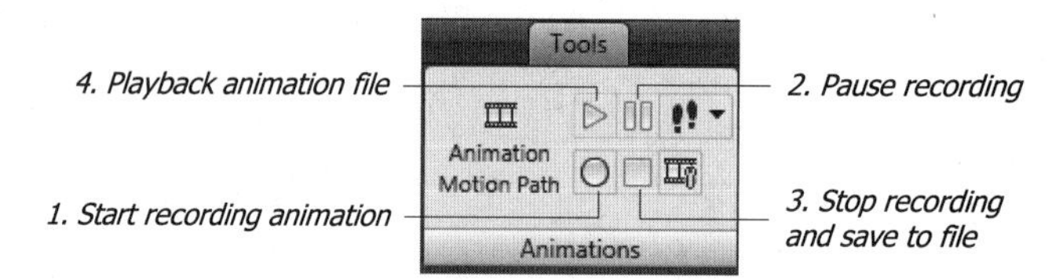

- To activate this command transparently during other commands, hold down the Shift key while pressing the mouse wheel.
- No transparent commands can be entered while this command is active.

3dOrbitCtr

Rel.10 Specifies the target point for 3D orbiting views.

Command	*Aliases*	*Shortcut Keystrokes*	*Menu*	*Ribbon*
'3dorbitctr	...	...	...	...

Command: 3dorbitctr

Specify orbit center point: *(Pick a point.)*

The command then enters 3dOrbit mode; see the 3dOrbit command.

COMMAND LINE OPTION

Specify orbit center point specifies the center of view rotation.

RELATED COMMANDS

3dCOrbit places the drawing in real-time 3D continuous orbit mode.

3dFOrbit performs real-time 3D viewing.

3dOrbit performs real-time 3D viewing fixed on the x,y plane.

TIP

- This command does not place a marker at the center point, oddly enough.

Relocated Command

For the **3dPan** command, see the 3dDistance command.

-3dOsnap

2011 Sets three-dimensional object snap modes.

Command	Aliases	Cursor Menu	Status Bar	Ribbon
'-3dosnap	...	**Ctrl+right** **3D Osnap**		...

3dOsnap command is a hardwire alias for the DSetting command's 3D Osnap tab.

Command: -3dosnap
Current osnap modes: ZVERt,ZCEN
Enter list of object snap modes: *(Enter one or more abbreviations of object snap modes; see list below.)*

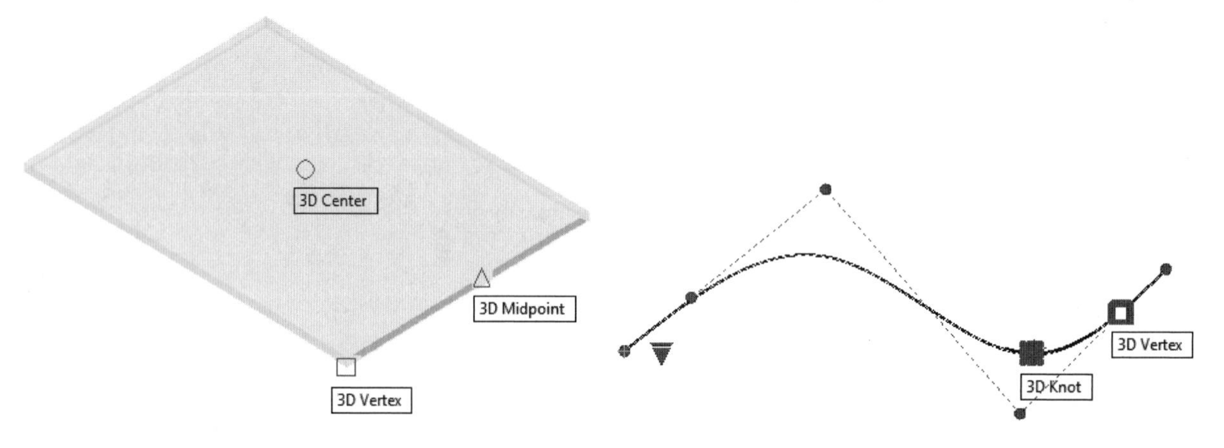

COMMAND LINE OPTIONS

Enter list of object snap modes specifies the 3D osnap modes to turn on. Separate names with commas. Valid modes include the following:

3D Osnap Name	Icon	Abbreviation	Meaning
ZNONe	...	znon	Turns off all 3D osnap modes.
ZCENter	○	zcen	Snaps to the centers of faces (default).
ZKNOt	⊠	zkno	Snaps to spline knots.
ZMIDpoint	△	zmid	Snaps to the midpoints of face edges.
ZNEAr	⧖	znea	Snaps to the nearest face.
ZPERpendicular	⊾	zper	Snaps perpendicular to planar faces.
ZVERtex	□	zver	Snaps to vertices and control vertices (default).

RELATED COMMANDS

-Osnap sets 2D object snap modes.

RELATED SYSTEM VARIABLE

3dOsMode specifies which 3D object snaps are active.

TIPS

- I recommend keeping ZNEAr turned off, since it forces AutoCAD constantly to snap, slowing it down.
- To turn 3D osnaps on and off quickly, right-click the 3D Object Snap button, and then choose the modes to change.

3dPoly

Rel. 10 Draws 3D polylines (short for 3D POLYline).

Command	*Alias*	*Shortcut Keystrokes*	*Menu*	*Ribbon*
3dpoly	**3p**	**...**	**Draw** ↳**3D Polyline**	**Home** ↳**Draw** ↳**3D Polyline**

Command: 3dpoly
Specify start point of polyline: *(Pick a point.)*
Specify endpoint of line or [Undo]: *(Pick a point, or type U.)*
Specify endpoint of line or [Undo]: *(Pick a point, or type U.)*
Specify endpoint of line or [Close/Undo]: *(Pick a point, type U or C, or press Enter to end the command.)*

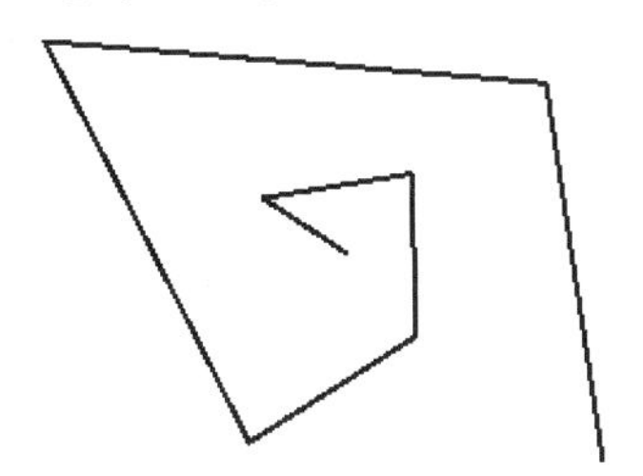

COMMAND LINE OPTIONS

Specify start point indicates the starting point of the 3D polyline.

Close joins the last endpoint with the start point.

Undo erases the last-drawn segment.

Specify endpoint of line indicates the endpoint of the current segment.

Enter ends the command.

RELATED COMMANDS

Explode reduces a 3D polyline to lines and arcs.

PEdit edits 3D polylines.

PLine draws 2D polylines.

TIPS

- Although 3D polylines can only be drawn with straight segments, you can use the PEdit command to turn it into a 3D spline.
- 3D polylines do not support linetypes or widths, although lineweights can fatten up their look.
- Drawing a 3D spline in 3D can be tricky. I find the .xy point filter the easiest assistant:

  Command: 3dpoly
  Specify start point of polyline: *(Pick a point.)*
  Specify endpoint of line or [Undo]: .xy
  of *(Pick another point.)*
  (need Z): (Enter the height, such as 1.)
- As of AutoCAD 2011, 3D polylines can be joined to 2D polylines, lines, arcs, and helices with the Join and PEdit commands.

3dPrint

2010 Packages 3D drawings for delivery to 3D print service bureaus.

Command *Ribbon*	*Alias*	*Shortcut Keystrokes*	*Menu Bar*	*Application Menu*
3dprint ...	...	...	...	**Publish** **↳Send to 3D Print Service**

Command: 3dprint

May display the following dialog box:

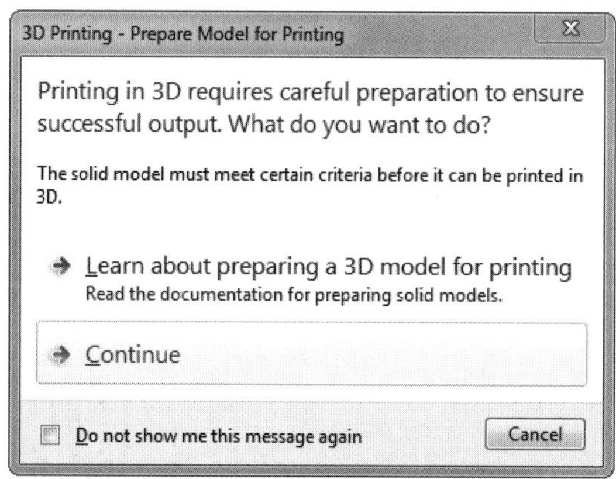

Click Continue.

Select solids or watertight meshes: *(Choose one or more 3D solids or mesh objects.)*

Select solids or watertight meshes: *(Press Enter to continue.)*

Displays Send to 3D Print Service dialog box.

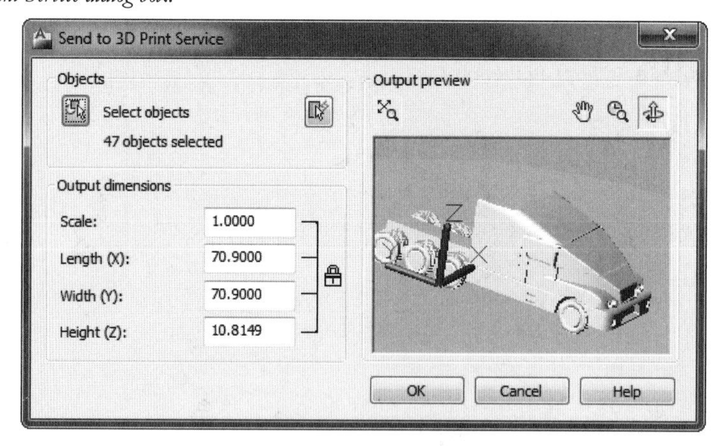

Click OK; AutoCAD displays Create STL File dialog box; see the StlOut command.

Click Save; Internet Explorer opens the "Print an AutoCAD Model Through a 3D Printing Service" Web page.

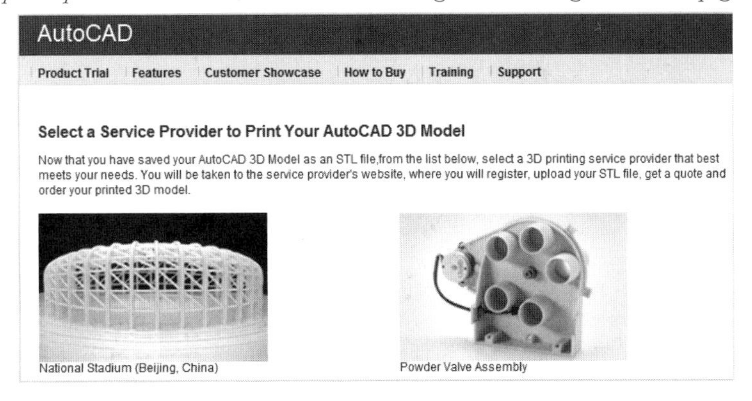

COMMAND LINE OPTION

Select solids or watertight meshes selects the objects to be exported in STL format.

DIALOG BOX OPTIONS

Objects options

Select Objects selects and removes 3D solids and mesh objects; hold down the Shift key to deselect objects.

Quick Select creates a selection set of objects based on their properties; see the QSelect command.

Output Dimensions options

Scale changes the size of the 3D model; enter a value smaller than 1 to make the model smaller. You may need to scale the model small enough to fit the size constraints of the service bureau's 3D printer.

Length (X), **Width (Y)**, **Height (Z)** specify the dimensions along the x, y, and z axes.

Lock locks the x, y, z aspect ratio; click the icon to unlock.

Output Preview options

Zoom Extents displays the model to its extents. Drag the edge of the dialog box to enlarge the preview image.

Realtime Pan moves the model around the preview window; move mouse to pan.

Realtime Zoom enlarges and decreases the size of the model; move mouse to change zoom level.

3dOrbit rotates the model in 3D.

RELATED COMMAND

StlOut exports 3D drawings in STL format (stereolithography).

RELATED SYSTEM VARIABLE

FacetRes sizes the triangular meshes that approximate objects in STL files; 3dPrint sets it to the highest value of 10 automatically.

TIPS

- The selection set can include blocks and xrefs containing solids and watertight meshes. *Watertight* mesh objects have no holes or open gaps. (3D solid models are by nature always watertight.) AutoCAD ignores invalid objects.
- Autodesk's "Print an AutoCAD Model Through a 3D Printing Service" Web page lists the names of several service bureaus that provide 3D printing services. These services are not free, and are very expensive for large 3D models.

3dRotate

2007 Rotates objects about the x, y, or z axis.

Command	*Alias*	*Shortcut Keystrokes*	*Menu*	*Ribbon*
3drotate	**3r**	...	**Modify**	**Home**
			↳**3D Operations**	↳**Modify**
			↳**3D Rotate**	↳**3D Rotate**

Command: 3drotate
Current positive angle in UCS: ANGDIR=counterclockwise ANGBASE=0
Select objects: *(Pick one or more objects.)*
Select objects: *(Press Enter.)*
Specify base point: *(Pick a point; locates the gizmo.)*
Pick a rotation axis: *(Pick one of the three axes.)*
Specify angle start point: *(Pick a point.)*
Specify angle end point: *(Pick another point.)*

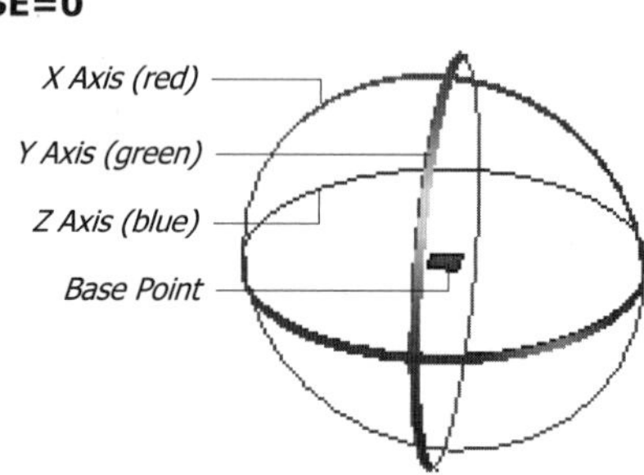

COMMAND LINE OPTIONS

Select objects selects the objects to be rotated.

Specify base point picks the point about which the objects will be rotated.

Pick a rotation axis selects the x (red), y (green), or z (blue) axis about which to rotate; the selected axis turns yellow.

Specify angle start point picks the starting point for rotation.

Specify angle end point picks the ending point for rotation.

RELATED COMMANDS

Rotate rotates objects in 2D space (current UCS).

Rotate3d rotates objects in 3D space without the grip tool.

3dScale resizes objects in 3D space.

3dMove moves objects in 3D space.

RELATED SYSTEM VARIABLES

GtAuto toggles the display of grips tools.

GtDefault determines whether the 3dRotate or Rotate command is activated automatically in 3D space when you enter the Rotate command.

DefaultGizmo determines which gizmo appears when subobjects are selected: 3dRotate, 3dScale, or 3dMove.

GtLocation locates the grip tool at the UCS icon or on the last selected object.

AngDir specifies the direction of positive angles (default =counterclockwise).

AngBase specifies the 0-angle relative to the current UCS (default = 0).

TIPS

- After starting the 3dRotate command, press the Spacebar to switch to the 3dMove command.
- When dynamic UCS is turned on, the grip tool aligns itself with the selected face. (Press Ctrl+D or click DUCS to toggle dynamic UCS.)
- Right-click the gizmo for a shortcut menu of options; see the 3dMove command.

3dScale

2010 Resizes objects in 3D space.

Command	Alias	Shortcut Keystrokes	Menu	Ribbon
3dscale	**3s**	...	**Modify** ⇘**3D Operations** ⇘**3D Scale**	**Home** ⇘**Modify** ⇘**3D scale**

Command: 3dscale
Select objects: *(Pick one or more objects.)*
Select objects: *(Press Enter.)*
Specify base point: *(Pick a point; relocates the gizmo.)*
Pick a scale axis or plane: *(Pick one of the three axes.)*
Specify scale factor or [Copy/Reference] <1.0>: *(Enter a number, pick a point, or enter an option.)*

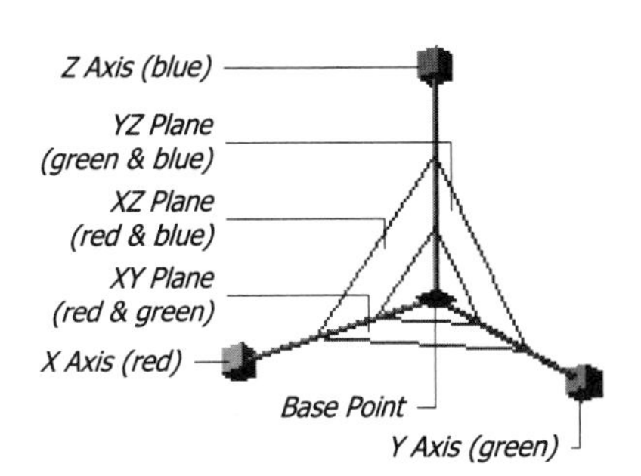

COMMAND LINE OPTIONS

Select objects selects the objects to be scaled.

Specify base point picks the point relative to which the objects will be resized.

Pick a scale axis or plane selects the x (red), y (green), or z (blue) axis about along which to resize; the selected axis turns yellow. Alternatively, choose a plane.

Specify scale factor sizes the selected objects:

<1 decreases the size of the objects.

1 maintains the size of the objects.

>1 increases the size of the objects.

Copy copies the objects; the original maintains its size, the copy is resized.

Reference specifies the scale factor relative to other objects in the drawing.

RELATED COMMANDS

Scale resizes objects in 2D space (current UCS).

Align3d moves and rotates pairs of objects in 3D space.

3dRotate rotates objects in 3D space.

3dMove moves objects in 3D space.

RELATED SYSTEM VARIABLES

GtAuto toggles the display of grips tools.

GtDefault determines whether the 3dScale is activated automatically in 3D space when you enter the Scale command.

GtLocation locates the grip tool at the UCS icon or on the last selected object.

DefaultGizmo determines which gizmo appears when subobjects are selected: 3dRotate, 3dScale, or 3dMove.

TIPS

- Right-click the gizmo for a shortcut menu of options; see the 3dMove command.
- The gizmo was redesigned in AutoCAD 2011.

3dsIn

Rel.13 Imports *.3ds* files created by 3D Studio and other applications (short for 3D Studio IN).

Command	*Aliases*	*Shortcut Keystrokes*	*Menu*	*Ribbon*
3dsin	...	...	**Insert** **↳3D Studio**	...

Command: 3dsin

Displays the 3D Studio File Import dialog box. Select a .3ds file, and then click Open.

AutoCAD displays dialog box.

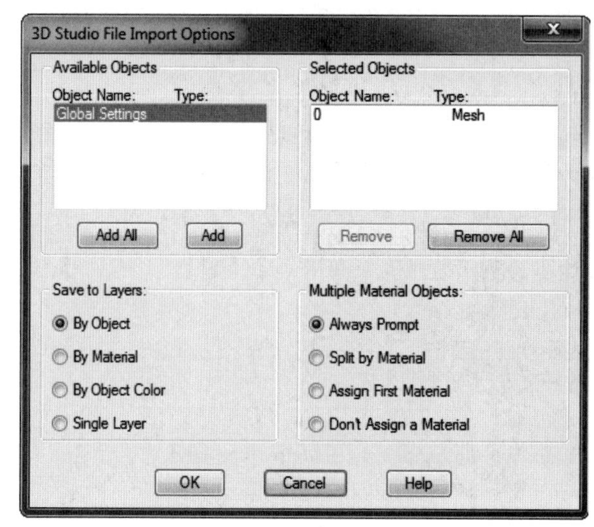

3D STUDIO FILE IMPORT OPTIONS DIALOG BOX

Available and Selected Objects options

Object Name names the object.

Type specifies the type of object.

Add adds the object to the Selected Objects list.

Add All adds all objects to the Selected Objects list.

Remove removes the object from the Selected Objects list.

Remove All removes all objects from the Selected Objects list.

Save to Layers options

⊙**By Object** places each object on its own layer.

○**By Material** places the objects on layers named after materials.

○**By Object Color** places the objects on layers named "Color*nn*."

○**Single Layer** places all objects on layer "AvLayer."

Multiple Material Objects options

⦿ **Always Prompt** prompts for each material.

○ **Split by Material** splits objects with more than one material into multiple objects, each with one material.

○ **Assign First Material** assigns the first material to the entire object.

○ **Don't Assign to a Material** removes all 3D Studio material definitions.

RELATED COMMANDS

FbxImport and **FbxExport** replace the 3dsIn and 3dsOut commands, respectively.

RELATED FILES

*.*3ds* are 3D Studio files.

*.*tga* are converted bitmap and animation files.

TIPS

- You are limited to selecting a maximum of 70 3D Studio objects.
- Conflicting object names are truncated and given a sequence number.
- The By Object option gives the AutoCAD layer the name of the object.
- The By Object Color option places all objects on layer "ColorNone" when no colors are defined in the 3DS file.
- 3D Studio assigns materials to faces, elements, and objects; AutoCAD assigns materials only to objects, colors, and layers.
- 3D Studio bitmaps are converted to *.tga* (Targa format) bitmaps. Converted *.tga* files are saved to the *.3ds* file folder.
- Only the first frame of animation files (CEL, CLI, FLC, and IFL) is converted to Targa bitmap files.
- 3D Studio "ambient lights" lose their color; 3D Studio "omni lights" become point lights in AutoCAD; 3D Studio "cameras" become named views in AutoCAD.

Relocated and Removed Commands

For the **3dSwival** command, see the 3dDistance command.

3dsOut was removed from AutoCAD 2007.

3dWalk / 3dFly

2007 Moves through 3D models interactively.

Commands	Aliases	Shortcut Keystrokes	Menu	Ribbon
'3dwalk	**3dw** **3dnavigate**	...	**View** **↳Walk and Fly** **↳Walk**	...
'3dfly	...	...	**View** **↳Walk and Fly** **↳Fly**	...

Command: 3dwalk *or* 3dfly

Press ESC or ENTER to exit, or right-click to display shortcut-menu. *(Drag mouse to move around, or enter these keystrokes:)*

Movement	Left Side	Right Side of Keyboard
Move forward	W	Up-arrow
Move back	S	Down-arrow
Move left	A	Left-arrow
Move right	D	Right-arrow
Toggle Walk-Fly mode	F	...
Display mappings	Tab	...
Exit walk/fly mode	Esc	Enter

Access the palette by right-clicking the drawing area (both commands displaying the same palette):

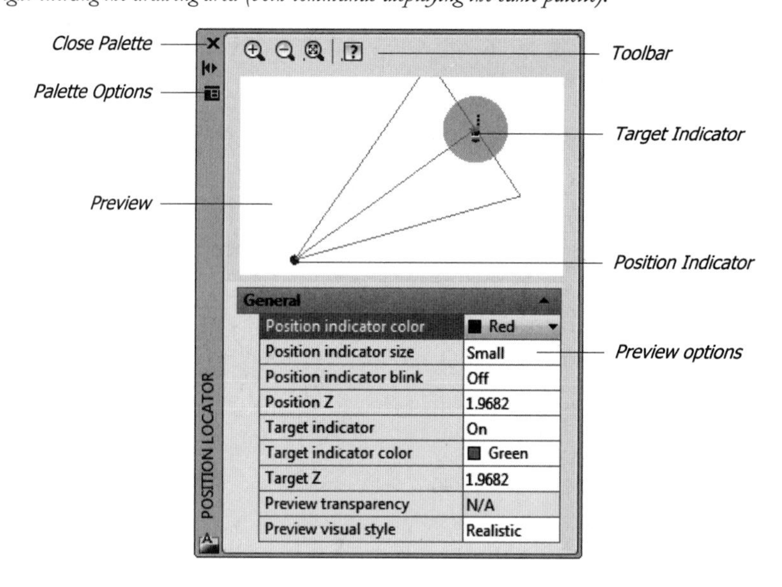

PALETTE OPTIONS

Position Indicator Color specifies the color of the large dot that shows your current position in the 3D model.

Position Indicator Size sets the size of the indicator between Small (the default), Medium, and Large.

Position Indicator Blink toggles blinking of the position indicator.

Position Z specifies the height of the camera or eye.

Target Indicator toggles the display of the target.

Target Indicator Color specifies the color of the target indicator.

Target Z specifies the height of the target.

Preview Transparency changes the translucency of this palette; range is 0 (opaque) to 95 (nearly transparent). N/A reports that the graphics board's settings do not support translucency.

Preview Visual Style specifies the visual style of the preview image; see the VisualStyles command.

PALETTE TOOLBAR

The zoom and pan buttons apply to the preview image in the palette:

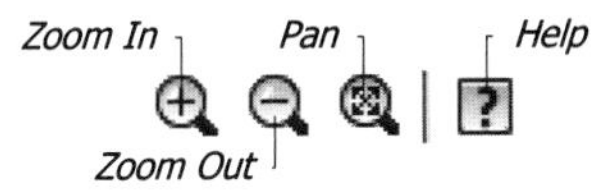

SHORTCUT MENU

The shortcut menu is nearly identical to that of the 3dOrbit command; unique to the 3dWalk and 3dFly commands are these options:

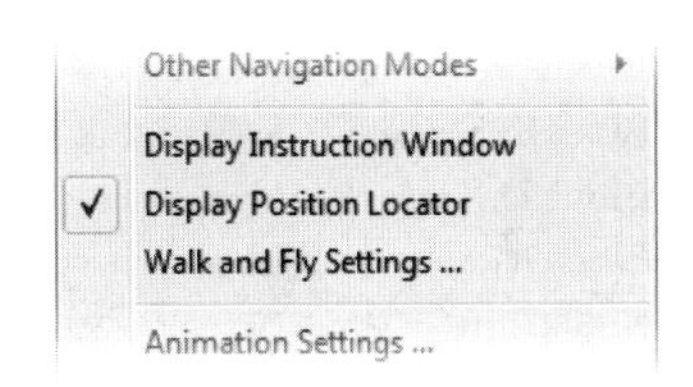

Display Instruction Window lists keyboard shortcuts.

Display Position Locator toggles the display of the Position Locator palette.

Walk and Fly Settings displays the Walk and Fly Settings dialog box; see the WalkFlySettings command.

RELATED COMMANDS

WalkFlySettings presets the step size and rate.

AniPath saves flythroughs to movie files.

3dOrbit circles 3D models.

RELATED SYSTEM VARIABLES

StepsPerSec specifies the number of steps per second.

StepSize specifies the distance per step.

TIPS

- The 3dWalk command keeps the elevation fixed, while 3dFly is independent of elevation.
- If AutoCAD is in parallel mode, this command automatically changes to perspective mode.
- In addition to moving the cursor in the drawing, you can also click-and-drag the view in the Preview area of the Position Locator palette. Notice the hand cursor illustrated below:

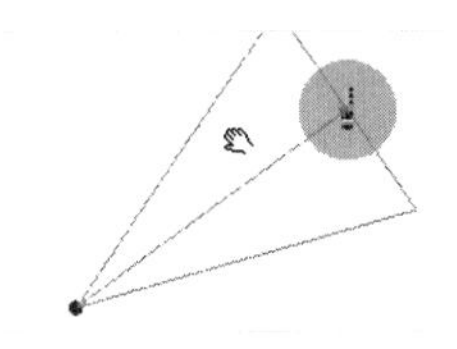

Relocated Command

For the **3dZoom** command, see the 3dDistance command.

APPENDIX A

Express Tools

The following commands are included with the Express Tools package.

A

AcadInfo	Reports on the status of AutoCAD, and stores the data in *acadinfo.txt*.
AliasEdit	Edits the aliases stored in *acad.pgp*.
AlignSpace	Aligns model space objects, whether in different viewports or with objects in paper space.
ArcText	Places text along an arc.
AttIn	Imports attribute data from files in tab-delimited format.
AttOut	Extracts attribute data to files tab-delimited format.

B

BCount	Counts inserted blocks, and then generates a report.
BExtend	Extends objects to those inside blocks and xrefs.
BlockReplace	Replaces all inserts of one block with another.
BlockToXref	Converts blocks to xrefs.
Block?	Lists objects stored in blocks.
BreakLine	Creates the break-line symbol.
BScale	Scales blocks from their insertion points.
BTrim	Trims objects by those inside blocks and xrefs.
Burst	Explodes blocks, and converts attributes to text.

C

CdOrder	Sets the display order by color.
ChUrls	Invokes DdEdit-like editor for editing hyperlinks (URL addresses).
ClipIt	Adds arcs, circles, and polylines to the XClip command.
CopyM	Runs the Copy command with repeat, divide, measure, and array options.

D

DimEx	Exports dimension styles to ASCII files.
DimIm	Imports dimension styles from ASCII files.
DimReassoc	Restores measurements to overridden dimension text.
DumpShx	Decompliles SHX files to source SHP format.
DwgLog	Logs editing actions of drawing files.

E

EditTime	Pauses the timer when you are not editing.
ExpressMenu	Loads the menu for Express Tools.
ExpressTools	Loads the express tools collection and its menu.
ExOffset	Adds options to the Offset command.
ExTrim	Trims all objects at the specified cutting line.
ExPlan	Adds options to the Plan command.

F

Flatten	Reduces 3D drawings to 2D.
FS	(FastSelect) Selects objects that touch the selected object.

G

GAttE	Changes attributes globally.
GetSel	Selects objects based on layer and type.

I

ImageApp	Specifies the name of an external raster image editor.
ImageEdit	Launches the image editor to edit selected images.

J

Julian	Refers to AutoLISP routines for converting calendar dates and times between different formats.

L

LayoutMerge	Combines layouts into a single layout.
Lsp	Provides AutoLISP function searching utilities.
LspSurf	Provides a LISP file viewer.

M

MkLtype	Creates linetypes from selected objects.
MkShape	Creates shapes from selected objects.
MoCoRo	Moves, copies, rotates, and scales objects.
MoveBak	Moves *.bak* files to specified folders.
MPEdit	Acts exactly like the PEdit command.
MStretch	Stretches using multiple selection windows.

N

NCopy	Copies objects nested inside blocks and xrefs.

O

OverKill	Removes overlapping duplicate objects.

P

Plt2Dwg	Imports HPGL files into the drawings.
Propulate	Updates, lists, and clears drawing properties.
PsBScale	Sets and updates the scale of blocks relative to paper space.
PsTScale	Sets text height relative to paper space.

Q

QlAttach	Associates leaders with annotation objects.
QlAttachSet	Associates leaders with annotations.

QlDetachSet	Disassociates leaders from annotations.
QQuit	Closes all drawings, and then exits AutoCAD.

R

ReDirMode	Sets options for the ReDir command.
ReDir	Changes paths for xrefs, images, shapes, and fonts.
RepUrls	Replaces hyperlinks.
Revert	Closes the drawing, and re-opens the original.
RtEdit	Edits remote text objects.
RText	Inserts and edits remote text objects.
RtUcs	Changes UCSs in real time.

S

SaveAll	Saves all drawings.
ShowUrls	Lists URLs in a dialog box.
Shp2blk	Converts a shape definition to a block definition.
Ssx	Creates selection sets.
SuperHatch	Uses images, blocks, external references, or wipeouts as hatch patterns.
SysvDlg	Launches an editor for system variables.

T

TCase	Changes text between Sentence, lower, UPPER, Title, and tOGGLE cASE.
TCircle	Surrounds text and multiline text with circles, slots, and rectangles.
TCount	Prefixes text with sequential numbers.
TextFit	Fits text between points.
TextMask	Places masks behind selected text.
TextUnmask	Removes masks behind selected text.
TFrames	Toggles the frames surrounding images and wipeouts.
TJust	Justifies text created with the MText and AttDef commands.
TOrient	Re-orients text, multiline text, and block attributes.
TScale	Scales text, multiline text, attributes, and attribute definitions.
Txt2Mtxt	Converts single-line to multiline text.
TxtExp	Explodes selected text into polylines.

V

VpScale	Lists the scale of the selected viewports.
VpSynch	Synchronizes viewports with a master viewport.

X

XData	Attaches xdata to objects.
XdList	Lists xdata attached to objects.
XList	Displays properties of objects nested in blocks and xref.

APPENDIX B

Obsolete & Removed Commands

Since AutoCAD v2.5, the following commands have been removed or replaced. Some commands, such as Hide, now operate only in certain modes, such as 2D wireframe visual style.

Command	Introduced	Removed	Replacement	Reaction
AcadBlockDialog	2000	2004	Block	Displays the Block dialog box.
AcadWBlockDialog	2000	2004	WBlock	Displays the Write Block dialog box.
AmeLite	R11	R12	Region	"Unknown command"
AscText	R11	R13	MText	"Unknown command"
Ase...	R12	R13	ASE...	"Unknown command"
	(Most R12 ASE commands were combined into fewer ASE commands with R13.)			
Ase...	R13	2000	dbConnect	"Unknown command"
AseUnload	R12	R14	Arx Unload	"Unknown command"
Assist	2000i	2008	*Infobar*	"Unknown command"
AssistClose	2000i	2008	*none*	"Unknown command"
Axis	v1.4	R12	*none*	"Discontinued command"
B				
Background	R14	2007	View	"Unknown command"
BHatch	R12	2006	Hatch	Displays Hatch & Gradient dialog box.
BMake	R12	2000	Block	Displays Block Definition dialog box.
BPoly	R12	2004	Boundary	Displays the Boundary dialog box.
C				
CConfig	R13	2000	PlotStyle	"Discontinued command"
Config	R12	R14	Options	Displays Options dialog box.
Content	2000	2004	AdCenter	Displays DesignCenter palette.
D				
Dashboard	2007	2009	Ribbon	Displays ribbon.
DashboardClose	2007	2009	RibbonClose	Closes ribbon.
DdAttDef	R12	2000	AttDef	Displays Attribute Def dialog box.
DdAttE	R9	2000	AttEdit	Displays Edit Attributes dialog box.
DdAttExt	R12	2000	AttExt	Displays Attribute Extraction dialog box.
DdChProp	R12	2000	Properties	Displays Properties palette.
DdColor	R13	2000	Color	Displays Select Color dialog box.
DdEModes	R9	R14	Object Properties	Dialog box explains change.
DdGrips	R12	2000	Options	Displays Options dialog box.
DDim	R12	2000	DimStyle	Displays Dimension Style Manager dialog box.

Command	Introduced	Removed	Replacement	Reaction
DdInsert	R12	2000	Insert	Displays Insert dialog box.
DdLModes	R9	R14	Layer	Displays Layer Manager dialog box.
DdLType	R9	R14	Linetype	Displays Linetype Manager dialog box.
DdModify	R12	2000	Properties	Displays Properties palette.
DdOSnap	R12	2000	DSettings	Displays Drafting Settings dialog box.
DdPlotStamp	2000i	2004	PlotStamp	Displays Plot Stamp dialog box.
DdRename	R12	2000	Rename	"Unknown command"
DdRModes	R9	2000	DSettings	Displays Drafting Settings dialog box.
DdSelect	R12	2000	Options	Displays Options dialog. box
DdUcs	R10	2000	UcsMan	Displays UCS dialog box.
DdUcsP	R12	2000	UcsMan	Displays UCS dialog box.
DdUnits	R12	2000	Units	Displays Units dialog box.
DdView	R12	2000	View	Displays View dialog box.
DText	v2.5	2000	Text	Executes Text command.
DLine	R11	R13	MLine	"Unknown command"
DwfOut	R14	2004	Publish	Executes Plot command.
DwfOutD	R14	2000	DwfOut	"Unknown command"
E				
EAttExt	2002	2008	DataExtraction	Runs DataExtraction command.
End	R11	R13	Quit	"Discontinued command"
EndRep	v1.0	v2.5	Minsert	"Discontinued command"
EndSv	v2.0	v2.5	End	"Discontinued command"
EndToday	2000i	2004	none	"Unknown command"
ExpressTools	2000	2002	ExpressTools	Restored in AutoCAD 2004.
F				
Files	v1.4	R14	*Explorer*	"Discontinued command"
FilmRoll	v2.6	R13	none	"Unknown command"
FlatLand	R10	R11	none	Accepts only 0 as a value.
Fog	R14	2007	RenderEnvironment	Executes RenderEnvironment command.
G				
GifIn	R12	R14	ImageAttach	"No longer supported"
H				
HpConfig	R12	2000	PlotStyle	"Discontinued command"
I				
IgesIn	v2.5	R13	none	"Discontinued command"

Command	Introduced	Removed	Replacement	Reaction
IgesOut	v2.5	R13	none	"Discontinued command"
Image	R14	2007	ClassicImage	Displays External References palette.
Impression	2009	2011	none	"Unknown command"
InetCfg	R14	2000	none	"Unknown command"
InetHelp	R14	2000	Help	"Unknown command"
InsertUrl	R14	2000	Insert	Displays Insert dialog.
L				
ListUrl	R14	2000	QSelect	"Unknown command"
M				
MakePreview	R13	R14	RasterPreview	"Discontinued command"
MenuLoad	R13	2006	CuiLoad	Displays Load Customizations dialog box.
MenuUnload	R13	2006	CuiUnload	Displays Unload Customizations dialog box.
MeetNow	2000i	2004	none	"Unknown command"
O				
OceConfig	R13	2000	PlotStyle	"Discontinued command"
OpenUrl	R14	2000	Open	Displays Select File dialog box.
OSnap	v2.0	2000	DSettings	Displays Drafting Settings dialog box.
P				
Painter	R14	2004	MatchProp	Runs MatchProp command.
PcxIn	R12	R14	ImageAttach	"No longer supported"
PsDrag	R12	2000i	none	"Unknown command"
PsIn	R12	2000i	PdfAttach	"Unknown command"
Preferences	R11	2000	Options	Displays Options dialog box.
PrPlot	v2.1	R12	Plot	"Discontinued command"
Q				
QPlot	v1.1	v2.0	SaveImg	"Unknown command"
R				
RConfig	R12	R14	RPref	"Unknown command"
RendScr	R12	2007	RenderWin	Displays Render window.
RenderUnload	R12	R14	Arx Unload	"Unknown command"
Repeat	v1.0	v2.5	Minsert	"Discontinued command"
Replay	R12	2007	none	"Unknown command"
RMat	R13	2007	Materials	Displays Materials palette.
RmlIn	2000i	2006	Markup	"Unknown command"

Command	Introduced	Removed	Replacement	Reaction
S				
SaveAsR12	R13	R14	SaveAs	"Unknown command"
SaveUrl	R14	2000	SaveAs	Displays Save Drawing As dialog box.
Scene	R12	2007	View	"Unknown command"
SetUV	R14	2007	MaterialMap	Runs MaterialMap command.
ShadeMode	2000	2007	-ShadeMode	Runs VsCurrent command.
ShowMat	R13	2007	List	Runs List Command.
Snapshot	v2.0	v2.1	SaveImg	"Unknown command"
Sol...	R11	R13	*(AME commands lost their SOL-prefix)*	
Stats	R12	2007	RenderWin	"Unknown command"
T				
TbConfig	R12	2000i	CUI	Displays CUI dialog box.
TiffIn	R12	R14	ImageAttach	"No longer supported"
Today	2000i	2004	InfoCenter	"Unknown command"
Toolbar	R13	2000i	CUI	Displays CUI dialog box.
V				
+View	...	2007	View	"Unknown command"
VlConv	R13	R14	3dsIn & 3dsOut	"Unknown command"
X				
XRef	R14	2007	ClassicXref	Displays External References palette .
3				
3dDwfPublish	2006	2007	3dDwf	Displays Export 3D DWF dialog box.
3dLine	R9	R11	Line	"Unknown command"
3dsOut	R13	2007	none	"Unknown command"

APPENDIX C

System Variables

AutoCAD stores information about its current state, the drawing, and the operating system in over 800 system variables. These variables help users and programmers determine the state of the AutoCAD system. This appendix lists the names and values of all known system variables.

CONVENTIONS

The following pages list all documented system variables, plus several more not documented by Autodesk. The list uses the following conventions:

 System variable is new to AutoCAD 2011.

Bold System variable is documented by Autodesk.

Italicized System variable is listed neither by the SetVar command nor in Autodesk's documentation.

~~*Strikethru Italic*~~ System variable was removed from this release of AutoCAD or earlier.

System variable must be accessed through the SetVar command. Otherwise, the command of the same name is executed.

Default Indicates the default value, as set in the *acad.dwt* template drawing file. Other template files may have different default values.

TIPS

- The SetVar command lets you change the value of all system variables, except those marked read-only (R/O).
- You can get a list of most system variables with the ? option of the SetVar command, as follows:

Command: setvar
Variable name or ?: ?
Variable(s) to list <*>: *(Press Enter.)*

- In this appendix, "Toggle" means that the system variable has one of two values: 0 (off, closed, disabled) or 1 (on, open, enabled).
- "ACI color" means the system variable takes on one of AutoCAD's 255 basic colors, plus ByLayer (color #256) and ByBlock (color #0). ACI is short for "AutoCAD color index."

Variable	*Default*	*Meaning*
~~_LInfo~~		*Removed from AutoCAD 2004.*
_AuthorPalettePath	*varies*	*Path to the customized Tool palette folder.*
_PkSer (R/O)	*varies*	*Software serial number, such as "117-69999999".*
_Server (R/O)	0	Network authorization code.
_ToolPalettePath	*varies*	*Path to the Tool palette folder.*
_VerNum (R/O)	*varies*	*Internal program build number, such as "Z.54.01".*

A

Variable	*Default*	*Meaning*
AcadLspAsDoc	0	Controls whether *acad.lsp* is loaded into: **0** The first drawing only. **1** Every drawing.
AcadPrefix (R/O)	*varies*	Specifies paths used by AutoCAD search in Options | Files dialog box.
AcadVer (R/O)	"18.0"	Specifies the AutoCAD version number.
AcisOutVer	70	Controls the ACIS version number; values are 15, 16, 17, 18, 20, 21, 30, 40, or 70.
AcGiDumpMode	*0*	*Value of 0 or 1.*
ActPath	""	Specifies paths to action macro files.
ActRecorderState	0	Reports the status of the Action Recorder: **0** Inactive. **1** Actively recording a macro. **2** Actively playing back a macro.
ActRecPath	"C:\users|..."	Specifies current path to recorded action macros.
ActUi	6	Controls state of Active Macro panel on the ribbon: **0** No change. **1** Expands during playback. **2** Expands during recording. **4** Prompts for name and description when recording is complete.
AdcState (R/O)	0	Toggle: reports if DesignCenter is active.
AFlags	0	Controls the default attribute display mode: **0** No mode specified. **1** Invisible. **2** Constant. **4** Verify. **8** Preset. **16** Lock position in block. **32** Multiple-line attributes.
AngBase	0	Controls the direction of zero degrees relative to the UCS.
AngDir	0	Controls the rotation of positive angles: **0** Clockwise. **1** Counterclockwise.
AnnoAllVisible	1	Toggles display of annotative objects at the current scale: **0** Displays only objects matching VP scale. **1** Displays all annotative objects (default).

Variable	*Default*	*Meaning*
AnnoAutoScale	4	Updates annotative objects when the annotation scale is changed: **1** Adds the new annotation scale to annotative objects (except on layers that are off, frozen, locked, or have viewports set to freeze). **2** As above, but includes objects on locked layers (excludes objects on layers that are off, frozen, or have viewports set to freeze). **3** The opposite of 2: applies to annotative objects on all layers, except objects on locked layers. **4** Applies to all objects regardless of layer status. **-1** Same as 1, but turned off. **-2** Same as 2, but turned off. **-3** Same as 3, but turned off. **-4** Same as 4, but turned off.
AnnotativeDwg	0	Toggles whether the drawing acts like an annotative block when inserted into other drawings: **0** Non-annotative block. **1** Annotative block. (This sysvar is read-only when drawing contains annotative text.)
ApBox	0	Toggle: displays the AutoSnap aperture box cursor.
Aperture	10	Controls the object snap aperture in pixels: **1** Minimum size. **50** Maximum size.
AppFrameResources (R/O)	*"pack://application:,,,/AcWindows; etc..."*	*Specifies resource files used by applications.*
ApplyGlobalOpacities		Toggles all palettes between (0) opaque and (1) translucent:
ApState	0	Toggle: reports the state of the ApBox variable.
~~*AssistState* (R/O)~~	~~*0*~~	~~*Removed from AutoCAD 2008.*~~
Area (R/O)	0.0	Reports the area measured by the last Area command.
AttDia	0	Controls the user interface for entering attributes: **0** Command-line prompts. **1** Dialog box.
AttIpe	0	Toggles display of in-place attribute text editor's toolbar: **0** Reduced-function toolbar. **1** Full-function toolbar.
AttMode	1	Controls the display of attributes: **0** Off. **1** Normal. **2** On; displays invisible attributes.
AttMulti	1	Toggles creation of multi-line attributes: **0** Attributes can be single-line only. **1** Attributes can be multi-line.
AttReq	1	Toggles attribute values during insertion: **0** Uses default values. **1** Prompts user for values.

Variable	*Default*	*Meaning*
AuditCtl	0	Toggles creation of *.adt* audit log files: **0** File not created. **1** File created.
AUnits	0	Controls the type of angular units: **0** Decimal degrees. **1** Degrees-minutes-seconds. **2** Grads. **3** Radians. **4** Surveyor's units.
AUPrec	0	Specifies number of decimal places displayed by angles; range is 0 - 8.
AutoDwfPublish	1	Toggles automatic publishing of DWF Files when drawings are saved or closed (to be removed in a future release): **0** DWF are not published. **1** DWF are published.
AutomaticPub	0	Toggles automatic exporting of drawing as DWF and PDF files when drawings are saved or closed: **0** DWF and PDF files are not published. **1** DWF and PDF are published.
AutoSnap	63	Controls the AutoSnap display (sum): **0** Turns off all AutoSnap features. **1** Turns on marker. **2** Turns on SnapTip. **4** Turns on magnetic cursor. **8** Turns on polar tracking. **16** Turns on object snap tracking. **32** Turns on tooltips for polar tracking and object snap tracking.
AuxStat	*0*	*-32768 Minimum value.* *32767 Maximum value.*
~~*AxisMode*~~	*0*	*Removed from AutoCAD 2002.*
AxisUnit	*0.0*	*Obsolete system variable.*

B

Variable	*Default*	*Meaning*
BackgroundPlot	2	Controls background plotting and publishing (ignored during scripts): **0** Plot foreground; publish foreground. **1** Plot background; publish foreground. **2** Plot foreground; publish background. **3** Plot background; publish background.
BackZ	0.0	Controls the location of the back clipping plane offset from the target plane.
BActionBarMode	1	Toggles between action objects and action bars in the Block Editor: **0** Displays legacy action objects (compatible with AutoCAD 2009 and earlier). **1** Displays action bars.
BActionColor	"7"	Specifies ACI text color for actions in Block Editor.
BConStatusMode	0	Toggles the display of the constraint status.

Variable	*Default*	*Meaning*
BDependencyHighlight	1	Toggles highlighting of dependent objects when parameters, actions, or grips selected in Block Editor: **0** Not highlighted. **1** Highlighted.
BGripObjColor	"141"	Specifies ACI color of grips in Block Editor.
BGripObjSize	8	Controls the size of grips in Block Editor; range is 1 to 255.
BgrdPlotTimeout	*20*	*Controls the time out for failed background plots; ranges from 0 to 300 seconds.*
BindType	0	Controls how xref names are converted when being bound or edited: **0** From *xref*|*name* to *xref*0*name*. **1** From *xref*|*name* to *name*.
BlipMode	0	Toggles the display of blip marks.
BlockEditLock	0	Toggles the locking of dynamic blocks being edited: **0** Unlocked. **1** Locked.
BlockEditor (R/O)	0	Reports whether Block Editor is open.
BlockTestWindow (R/O)	0	Reports whether the Block Test window is open.
BParameterColor	"7"	Specifies ACI color of parameters in the Block Editor.
BParameterFont	"Simplex.shx"	Controls the font used for parameter and action text in the Block Editor.
BParameterSize	12	Controls the size of parameter text and features in Block Editor: **1** Minimum. **255** Maximum.
BPTextHorizontal	1	Toggles the alignment of text with dimension lines in Block Editor: **0** Text is aligned with dimension lines. **1** Text is horizontal.
BTMarkDisplay	1	Controls the display of value set markers: **0** Unlocked. **1** Locked.
BVMode	0	Controls the display of invisible objects in the Block Editor: **0** Invisible. **1** Visible and dimmed.

C

Variable	*Default*	*Meaning*
CalcInput	1	Controls how formulas and global constants are evaluated in dialog boxes: **0** Not evaluated. **1** Evaluated after pressing Alt+Enter.
CameraDisplay	0	Toggles the display of camera glyphs.
CameraHeight	0	Specifies the default height of cameras.
CAnnoScale	"1:1"	Names the current annotative scale for the current viewport.
CAnnoScaleValue(R/O)	1.0	Reports the current annotative scale.
CaptureThumbnails	1	Determines how thumbnails are generated by NavSWheel's Restore option: **0** No preview thumbnails generated. **1** Generated when bracket is positioned over empty frames. **2** Generated automatically after every view change.

Variable	Default	Meaning
CBarTransparency	50	Controls the transparency of constraint bars; range is 10 to 90.
CConstraintForm	0	Determines the type of new constraint applied to objects by the DimConstraint command: **0** Applies dynamic constraints. **1** Applies annotative constraints.
CDate (R/O)	*varies*	Specifies the current date and time in the format YyyyMmDd.HhMmSsDd, such as 20080503.18082328
CDynDisplayMode	*0*	*Toggles display of dynamic display.*
CeColor	"BYLAYER"	Controls the current color.
CeLtScale	1.0	Controls the current linetype scaling factor.
CeLType	"BYLAYER"	Controls the current linetype.
CeLWeight	-1	Controls the current lineweight in millimeters; valid values are 0, 5, 9, 13, 15, 18, 20, 25, 30, 35, 40, 50, 53, 60, 70, 80, 90, 100, 106, 120, 140, 158, 200, and 211, plus the following: **-1** BYLAYER. **-2** BYBLOCK. **-3** Default, as defined by LwDefault.
CenterMT	0	Controls how corner grips stretch uncentered multiline text: **0** Center grip moves in same direction; opposite grip stays in place. **1** Center grip stays in place; side grips move in direction of stretch.
CeTransparency	ByLayer	Sets the level of translucency for new objects: **ByLayer** Determined by the object's layer (default). **ByBlock** Determined by the block in which the object resides. **0** Fully opaque (not translucent). **1-90** Range of translucency, as a percentage.
ChamferA	0.0	Specifies the current value of the first chamfer distance.
ChamferB	0.0	Specifies the current value of the second chamfer distance.
ChamferC	0.0	Specifies the current value of the chamfer length.
ChamferD	0	Specifies the current value of the chamfer angle.
ChamMode	0	Toggles the chamfer input mode: **0** Chamfer by two lengths. **1** Chamfer by length and angle.
CipMode	*0*	*Toggles Customer Involvement Program.*
CircleRad	0.0	Specifies the most-recent circle radius.
ClassicKeys	0	Toggles meaning of Ctrl+C: (0) copies objects, or (1) cancels commands.
CLayer	"0"	Specifies name of current layer.
CleanScreenState (R/O)	0	Toggle: reports whether clean screen mode is active.
CliState (R/O)	1	Reports the command line palette.
CMaterial	"ByLayer"	Sets the name of the current material.

Variable	Default	Meaning
CmdActive (R/O)	1	Reports the type of command currently active (used by programs): **1** Regular command. **2** Transparent command. **4** Script file. **8** Dialog box. **16** Dynamic data exchange. **32** AutoLISP command. **64** ARX command.
CmdDia	1	Toggles QLeader's inplace text editor.
CmdEcho	1	Toggles AutoLISP command display: **0** No command echoing. **1** Command echoing.
CmdInputHistoryMax	20	Controls the maximum command input items stored; works with InputHistoryMode.
CmdNames (R/O)	*varies*	Reports name of the command currently active, such as "SETVAR".
CMleaderStyle	"Standard"	Reports name of current mleader style.
CMLJust	0	Controls the multiline justification mode: **0** Top. **1** Middle. **2** Bottom.
CMLScale	1.0	Controls the scale of overall multiline width: **-*n*** Flips offsets of multiline. **0** Collapses to single line. *n* Scales by a factor of *n*.
CMLStyle	"STANDARD"	Specifies the current multiline style name.
Compass	0	Toggles the display of the 3D compass.
ConstraintBarDisplay	1	Toggles the display of constraint bars.
ConstraintBarMode	4031	Determines the geometric constraints displayed by constraint bars: **1** Horizontal **2** Vertical **4** Perpendicular **8** Parallel **16** Tangent **32** Smooth **64** Coincident **128** Concentric **256** Colinear **512** Symmetric **1024** Equal **2048** Fix **4031** All displayed
ConstraintCursorDisplay	*1*	*Toggles the display of the constraint icons that appear next to the cursor: (0) off or (1) on.*
ConstraintInfer	0	Toggles the inference of geometric constraints as objects are drawn and edited: (0) not inferred or (1) inferred.

Variable	Default	Meaning
ConstraintNameFormat	2	Determines how constraint text is displayed: **0** By variable names. **1** By values. **2** By expressions (formulas).
ConstraintRelax	0	Toggles constraints between enforced and relaxed during editing.
ConstraintSolveMode	1	Determines how constraints behave when applied or edited: **0** Geometry size not retained when constraints are applied or modified. **1** Geometry size is retained, even when constraints are applied or modified.
Coords	1	Controls the coordinate display style: **0** Updated by screen picks. **1** Continuous display. **2** Polar display upon request.
CopyMode	0	Toggles whether the Copy command repeats itself: **0** Command repeats copying. **1** Command makes one copy, then exits.
CPlotStyle	"ByColor"	Specifies the current plot style; options for named plot styles are: ByLayer, ByBlock, Normal, and User Defined.
CProfile (R/O)	"<<Unnamed Profile>>"	Specifies the name of the current profile.
CpuTicks (R/O)	*592020023071334.1*	*Reports the number of CPU ticks.*
CrossingAreaColor	100	Specifies the ACI color of crossing rectangle.
CShadow	0	Specifies how shadows are cast by 3D objects (if graphics board is capable; see 3dConfig command): **0** Casts and receives shadows. **1** Casts shadows. **2** Receives shadows. **3** Ignores shadows.
CTab	"Model"	Specifies the name of the current tab.
CTableStyle	"Standard"	Specifies the name of the current table style name.
CullingObj	1	Toggle: subobjects are highlighted when not normal to the current view: (0) not culled or (1) culled.
CullingObjSelection	0	Toggles: hidden objects are selected when a selection window is dragged: (0) not culled so hidden objects are selected, or (1) culled so hidden objects are not selected.
~~*CurrentProfile*~~	"<<Unnamed Profile>>"	*Removed from AutoCAD 2000; replaced by CProfile.*
CursorSize	5	Controls the cursor size as a percent of the viewport size. **1** Minimum size. **100** Full viewport.
CVPort	2	Specifies the current viewport number.

D

Variable	Default	Meaning
~~*DashboardState*~~	*0*	*Replaced by RibbonState.*

Variable	*Default*	*Meaning*
DataLinkNotify	2	Controls reporting on data links: **0** All notifications disabled. **1** Displays data link icon in tray. **2** Displays the icon and a warning balloon in the tray.
Date (R/O)	*varies*	Reports the current date in Julian format, such as 2448860.54043252.
DbcState (R/O)	0	Toggle: specifies whether dbConnect Manager is active.
~~*DBGListAll*~~	*0*	*Removed from AutoCAD 2002.*
DblClkEdit	1	Toggles editing by double-clicking objects.
DBMod (R/O)	0	Reports how the drawing has been modified: **0** No modification since last save. **1** Object database modified. **2** Symbol table modified. **4** Database variable modified. **8** Window modified. **16** View modified. **32** Field modified.
DctCust	"sample.cus"	Specifies the name of custom spelling dictionary.
DctMain	"enu"	Controls the code for spelling dictionary: **enu** American English **eng** British English (ise) **enc** Canadian English **cat** Catalan **csy** Czech **dan** Danish **nld** Dutch (primary) **fin** Finnish **fra** French (accented capitals) **frc** French (unaccented capitals) **deu** German (post-reform) **deo** German (pre-reform) **ita** Italian **nor** Norwegian (Bokmal) **ptb** Portuguese (Brazilian) **ptg** Portuguese (Iberian) **rus** Russian **esp** Spanish **sve** Swedish
DefaultGizmo	0	Determines which gizmo is the default when sub-objects are selected: 3D Move, 3D Rotate, 3D Scale, or none.
DefaultIndex	*0*	*Range is 0 to 255.*
DefaultLighting	1	Toggles distant lighting.
DefaultLightingType	0	Toggles between new (1) and old (0) type of lights.
DefaultViewCategory	*""*	*Specifies the default name for View Category in the View command's New View dialog box.*
DefLPlStyle	"ByColor"	Reports the default plot style for layer 0.
DefPlStyle	"ByColor"	Reports the default plot style for new objects.

Variable	*Default*	*Meaning*
DelObj	1	Toggles the deletion of source objects: **-2** Users are prompted whether to erase all defining objects. **-1** Users are prompted whether to erase profiles and cross sections. **0** Objects retained. **1** Profiles and cross sections erased. **2** All defining objects erased.
DemandLoad	3	Controls application loading when drawing contains proxy objects: **0** Apps not demand-loaded. **1** Apps loaded when drawing opened. **2** Apps loaded at first command. **3** Apps loaded when drawing is opened or at first command.
DgnFrame	0	Controls display of DGN underlay frame: **0** Does not display frame. **1** Displays and plots frame. **2** Displays but does not plot frame.
DgnImportMax	1000000	Limits maximum MicroStation elements to import into AutoCAD; 0 = no limit.
DgnMappingPath (R/O)	"C:\users\..."	Specifies path to folder holding the *dgnsetups.ini* file.
DgnOsnap	1	Toggles object snapping to DGN elements: **0** Does not osnap. **1** Osnaps.
DiaStat (R/O)	0	Reports whether user exited dialog box by clicking: **0** Cancel button. **1** OK button.
Digitizer (R/O)	0	Reports the type of digitizer attached to AutoCAD (read-only): **0** None. **1** Integrated touch. **2** External touch. **4** Integrated pen. **8** External pen. **16** Multiple input. **128** Input devices are ready.

Dimension Variables

Variable	*Default*	*Meaning*
DimADec	0	Controls angular dimension precision: **-1** Use DimDec setting (default). **0** Zero decimal places (minimum). **8** Eight decimal places (maximum).
DimAlt	Off	Toggles alternate units: **On** Enabled. **Off** Disabled.
DimAltD	2	Controls alternate unit decimal places.
DimAltF	25.4	Controls alternate unit scale factor.
DimAltRnd	0.0	Controls rounding factor of alternate units.
DimAltTD	2	Controls decimal places of tolerance alternate units; range is 0 to 8.

Variable	Default	Meaning
DimAltTZ	0	Controls display of zeros in alternate tolerance units: **0** Zeros not suppressed. **1** All zeros suppressed. **2** Includes 0 feet, but suppresses 0 inches. **3** Includes 0 inches, but suppresses 0 feet. **4** Suppresses leading zeros. **8** Suppresses trailing zeros.
DimAltU	2	Controls display of alternate units: **1** Scientific. **2** Decimal. **3** Engineering. **4** Architectural; stacked. **5** Fractional; stacked. **6** Architectural. **7** Fractional. **8** Windows desktop units setting.
DimAltZ	0	Controls the display of zeros in alternate units: **0** Suppresses 0 ft and 0 in. **1** Includes 0 ft and 0 in. **2** Includes 0 ft; suppresses 0 in. **3** Suppresses 0 ft; includes 0 in. **4** Suppresses leading 0 in decimal dimensions. **8** Suppresses trailing 0 in decimal dimensions. **12** Suppresses leading and trailing zeroes.
DimAnno (R/O)	0	Reports whether current dimstyle is: **0** Not annotative. **1** Annotative.
DimAPost	""	Specifies the prefix and suffix for alternate text.
DimArcSym	0	Specifies the location of the arc symbol: **0** Before dimension text. **1** Above the dimension text. **2** Not displayed.
DimAso	*On*	*Toggles associative dimensions:* ***On*** *Dimensions are created associative.* ***Off*** *Dimensions are not associative.*
DimAssoc	2	Controls how dimensions are created: **0** Dimension elements exploded. **1** Single dimension object, attached to defpoints. **2** Single dimension object, attached to geometric objects.
DimASz	0.18	Controls the default arrowhead length.
DimAtFit	3	Controls how text and arrows are fitted when there is insufficient space between extension lines (leader is added when DimTMove = 1): **0** Text and arrows outside extension lines. **1** Arrows first outside, then text. **2** Text first outside, then arrows. **3** Either text or arrows, whichever fits better.

Variable	*Default*	*Meaning*
DimAUnit	0	Controls the format of angular dimensions: **0** Decimal degrees. **1** Degrees.Minutes.Seconds. **2** Grads. **3** Radians. **4** Surveyor units.
DimAZin	0	Controls the display of zeros in angular dimensions: **0** Displays all leading and trailing zeros. **1** Suppresses 0 in front of decimal. **2** Suppresses trailing zeros behind decimal. **3** Suppresses zeros in front and behind the decimal.
DimBlk Closed filled Closed blank Closed Dot Architectural tick Oblique Open Origin indicator Origin indicator 2 Right angle Open 30 Dot small Dot blank Dot small blank Box Box filled Datum triangle Datum triangle filled Integral None	“”	Specifies the name of the arrowhead block: Architectural tick: “Archtick” Box filled: “Boxfilled” Box: “Boxblank” Closed blank: “Closedblank” Closed filled: “” (default) Closed: “Closed” Datum triangle filled: “Datumfilled” Datum triangle: “Datumblank” Dot blanked: “Dotblank” Dot small: “Dotsmall” Dot: “Dot” Integral: “Integral” None: “None” Oblique: “Oblique” Open 30: “Open30” Open: “Open” Origin indication: “Origin” Right-angle: “Open90”
DimBlk1	“”	Specifies the name of first arrowhead’s block; uses same list of names as under DimBlk. . No arrowhead.
DimBlk2	“”	Specifies name of second arrowhead block.
DimCen	0.09	Controls how center marks are drawn: -*n* Draws center lines. **0** No center mark or lines drawn. +*n* Draws center marks of length *n*.
DimClrD	0	ACI color of dimension lines: **0** BYBLOCK (default). **256** BYLAYER.
DimClrE	0	Specifies ACI color of extension lines and leaders.
DimClrT	0	Specifies ACI color of dimension text.
DimConstraintIcon	2	Toggles the display of the lock icon next to dimensional constraints: 0 Does not display the lock icon. 1 Displays icon next to dynamic constraints. 2 Displays the icon next to annotational constraints. 3 Displays the icon next to dynamic and annotational constraints.

Variable	Default	Meaning
DimDec	4	Controls the number of decimal places displayed; range is 0 to 8.
DimDLE	0.0	Controls the length of the dimension line extension.
DimDLI	0.38	Controls the increment of the continued dimension lines.
DimDSep	"."	Specifies the decimal separator (must be a single character).
DimExe	0.18	Controls the extension above the dimension line.
DimExO	0.0625	Specifies the extension line origin offset.
~~*DimFit*~~	*3*	*Obsolete: Autodesk recommends use of DimATfit and DimTMove instead.*
DimFrac	0	Controls the fraction format when DimLUnit set to 4 or 5: **0** Horizontal. **1** Diagonal. **2** Not stacked.
DimFXL	1	Specifies default length of fixed extension lines.
DimFxlOn	0	Toggles fixed extension lines.
DimGap	0.09	Controls the gap between text and the dimension line.
DimJogAngle	45	Specifies default angle for jogged dimension lines.
DimJust	0	Controls the positioning of horizontal text: **0** Center justify. **1** Next to first extension line. **2** Next to second extension line. **3** Above first extension line. **4** Above second extension line.
DimLdrBlk	""	Specifies the name of the block used for leader arrowheads; same as DimBlk. **.** Suppresses display of arrowhead.
DimLFac	1.0	Controls the linear unit scale factor.
DimLim	Off	Toggles the display of dimension limits.
DimLtEx1	""	Specifies linetype for the first extension line.
DimLtEx2	""	Specifies linetype for second extension line.
DimLtype	""	Specifies linetype name for dimension line.
DimLUnit	2	Controls dimension units (except angular); replaces DimUnit: **1** Scientific. **2** Decimal. **3** Engineering. **4** Architectural. **5** Fractional. **6** Windows desktop.
DimLwD	-2	Controls the dimension line lineweight; valid values are BYLAYER, BYBLOCK, or integer multiples of 0.01mm.
DimLwE	-2	Controls the extension lineweight.

Variable	*Default*	*Meaning*
DimPost	""	Specifies the default prefix or suffix for dimension text (maximum 13 characters): **""** No suffix.
DimRnd	0.0	Controls the rounding value for dimension distances.
DimSAh	Off	Toggles separate arrowhead blocks: **Off** Use arrowheads defined by DimBlk. **On** Use arrowheads defined by DimBlk1 and DimBlk2.
DimScale	1.0	Controls the overall dimension scale factor: **0** Value is computed from the scale between current model space viewport and paper space. >**0** Scales text and arrowheads.
DimSD1	Off	Toggles display of the first dimension line: **On** First dimension line is suppressed. **Off** Not suppressed.
DimSD2	Off	Toggles display of the second dimension line: **On** Second dimension line is suppressed. **Off** Not suppressed.
DimSE1	Off	Toggles display of the first extension line: **On** First extension line is suppressed. **Off** Not suppressed.
DimSE2	Off	Toggles display of the second extension line: **On** Second extension line is suppressed. **Off** Not suppressed.
DimSho	*On*	*Toggles dimension updates while dragging:* ***On*** *Dimensions are updated during drag.* ***Off*** *Dimensions are updated after drag.*
DimSOXD	Off	Toggles display of dimension lines outside of extension lines: **On** Dimension lines not drawn outside extension lines. **Off** Are drawn outside extension lines.
DimStyle (R/O)	"STANDARD"	Reports the current dimension style.
DimTAD	0	Controls the vertical position of text: **0** Centers text between extension lines. **1** Places text above dimension line, except when dimension line not horizontal and DimTIH = 1. **2** Places text on side of dimension line farthest from the defining points. **3** Conforms to JIS (Japanese Industry Standard).
DimTDec	4	Specifies number of decimal places for primary tolerances; range 0 - 8.
DimTFac	1.0	Controls the scale factor for tolerance text height.
DimTFill	0	Toggles background fill color for dimension text.
DimTFillClr	0	Specifies background color for dimension text.
DimTIH	On	Toggles alignment of text placed inside extension lines: **Off** Text aligned with dimension line. **On** Text is horizontal.

Variable	*Default*	*Meaning*
DimTIX	Off	Toggles placement of text inside extension lines: **Off** Text placed inside extension lines, if room. **On** Text forced between the extension lines.
DimTM	0.0	Controls the value of the minus tolerance.
DimTMove	0	Controls how dimension text is moved: **0** Moves dimension line with text. **1** Moves text, and adds a leader to the text. **2** Move text anywhere without leader.
DimTOFL	Off	Toggles placement of dimension lines: **Off** Dimension lines not drawn when arrowheads are outside. **On** Dimension lines drawn, even when arrowheads are outside.
DimTOH	On	Toggles text alignment when outside of extension lines: **Off** Text aligned with dimension line. **On** Text is horizontal.
DimTol	Off	Toggles generation of dimension tolerances: **Off** Tolerances not drawn. **On** Tolerances are drawn.
DimTolJ	1	Controls vertical justification of tolerances: **0** Bottom. **1** Middle. **2** Top.
DimTP	0.0	Specifies the value of the plus tolerance.
DimTSz	0.0	Controls the size of oblique tick strokes: **0** Arrowheads. >**0** Oblique strokes.
DimTVP	0.0	Controls the vertical position of text when DimTAD = 0: **1** Turns on DimTAD (=1). >**-0.7** *or* <**0.7** Splits dimension line for text.
DimTxSty	"STANDARD"	Specifies the dimension text style.
DimTxt	0.18	Controls the text height.
DimTxtDirection	Off	Toggles the reading direction of dimension text: 0 Left to right. 1 Right to left.
DimTZin	0	Controls the display of zeros in tolerances: **0** Suppresses 0 ft and 0 in. **1** Includes 0 ft and 0 in. **2** Includes 0 ft; suppresses 0 in. **3** Suppresses 0 ft; includes 0 in. **4** Suppresses leading 0 in decimal dimensions. **8** Suppresses trailing 0 in decimal dimensions. **12** Suppresses leading and trailing zeroes.
~~DimUnit~~	*2*	*Obsolete; replaced by DimLUnit and DimFrac.*
DimUPT	Off	Controls user-positioned text: **Off** Cursor positions dimension line. **On** Cursor also positions text.

Variable	Default	Meaning
DimZIN	0	Controls the display of zero in feet-inches units: **0** Suppresses 0 ft and 0 in. **1** Includes 0 ft and 0 in. **2** Includes 0 ft; suppress 0 in. **3** Suppresses 0 ft; include 0 in. **4** Suppresses leading 0 in decimal dim. **8** Suppresses trailing 0 in decimal dim. **12** Suppresses leading and trailing zeroes.
DispSilh	0	Toggles the silhouette display of 3D solids.
Distance (R/O)	0.0	Reports the distance last measured by the Dist command.
~~*Dither*~~		~~*Removed from Release 14.*~~
DivMeshBoxHeight	3	Specifies the default number of subdivisions along the height of boxes; range is 1 to 256.
DivMeshBoxLength	3	Specifies the default number of subdivisions along the length of boxes; range is 1 to 256.
DivMeshBoxWidth	3	Specifies the default number of subdivisions along the width of boxes; range is 1 to 256.
DivMeshConeAxis	8	Specifies the default number of subdivisions around the base of cones; range is 1 to 256.
DivMeshConeBase	3	Specifies the default number of subdivisions on the base of cones; range is 1 to 256.
DivMeshConeHeight	3	Specifies the default number of subdivisions between the base and the tip of cones; range is 1 to 256.
DivMeshCylAxis	8	Specifies the default number of subdivisions around the trunk of cylinders; range is 1 to 256.
DivMeshCylBase	3	Specifies the default number of subdivisions on the endcaps of cylinders; range is 1 to 256.
DivMeshCylHeight	3	Specifies the default number of subdivisions along the trunk of cylinders.
DivMeshPyrBase	3	Specifies the default number of subdivisions on the base of pyramids; range is 1 to 256.
DivMeshPyrHeight	3	Specifies the default number of subdivisions between the base and the tip of pyramids; range is 1 to 256.
DivMeshPyrLength	3	Specifies the default number of subdivisions along the base of pyramids; range is 1 to 256.
DivMeshSphereAxis	12	Specifies the default number of subdivisions between the two polar ends of spheres; range is 1 to 256.
DivMeshSphereHeight	6	Specifies the default number of subdivisions around spheres; range is 1 to 256.
DivMeshTorusPath	8	Specifies the default number of subdivisions along paths swept by tori profiles; range is 1 to 256.

Variable	Default	Meaning
DivMeshTorusSection	8	Specifies the default number of subdivisions around paths swept by tori profiles; range is 1 to 256.
DivMeshWedgeBase	3	Specifies the default number of subdivisions on the base of wedges; range is 1 to 256.
DivMeshWedgeHeight	3	Specifies the default number of subdivisions along the height of wedges; range is 1 to 256.
DivMeshWedgeLength	4	Specifies the default number of subdivisions along the length of wedges; range is 1 to 256.
DivMeshWedgeSlope	3	Specifies the default number of subdivisions along the slope of wedges; range is 1 to 256.
DivMeshWedgeWidth	3	Specifies the default number of subdivisions along the width of wedges; range is 1 to 256.
DonutId	0.5	Controls the inside diameter of donuts.
DonutOd	1.0	Controls the outside diameter of donuts.
DragMode	2	Controls the drag mode: **0** No drag. **1** On if requested. **2** Automatic.
DragP1	10	Controls the regen drag display; range is 0 to 32767.
DragP2	25	Controls the fast drag display; range is 0 to 32767.
DragVs	""	Specifies the default visual style when 3D solids are created by dragging the cursor; disabled when visual style is 2D wireframe.
DrawOrderCtl	3	Controls the behavior of draw order: **0** Draw order not restored until next regen or drawing reopened. **1** Normal draw order behavior. **2** Draw order inheritance. **3** Combines options 1 and 2.
DrState	0	Toggles the Drawing Recovery palette.
DTextEd	2	Controls the user interface of DText/Text command: **0** Uses in-place text editor. **1** Displays Edit Text dialog box. **2** Click elsewhere in drawing to start new text string.
DwfFrame	2	Display of the frame around DWF overlays: **0** Frame is turned off. **1** Frame is displayed and plotted. **2** Frame is displayed but not plotted.
DwfOsnap	1	Toggles osnapping of the DWF frame.
DwgCheck	0	Toggles checking of whether drawing was edited by software other than AutoCAD: **0** Suppresses dialog box. **1** Displays warning dialog box. **2** Prints warning in command bar.

Variable	Default	Meaning
DwgCodePage (R/O)	*varies*	Same values as SysCodePage.
DwgName (R/O)	*varies*	Reports the current drawing file name, such as "drawing1.dwg".
DwgPrefix (R/O)	*varies*	Reports the drawing's drive and folder, such as "d:\acad 2011\".
DwgTitled (R/O)	0	Reports whether the drawing file name is: **0** "drawing1.dwg". **1** User-assigned name.
~~*DwgWrite*~~		*Removed from AutoCAD Release 14.*
DxEval	12	Controls which commands cause AutoCAD to check for changed external data files (sum): **0** None. **1** Open. **2** Save. **4** Plot. **8** Publish. **12** Plot and Publish. **16** eTransmit and Archive. **32** Save with automatic update. **64** Plot with automatic update. **128** Publish with automatic update. **256** eTransmit/Archive with automatic update.
DynConstraintDisplay	1	Toggles the display of dynamic constraints.
DynConstraintMode	1	Toggles the display of hidden constraints: **0** Constraints remain hidden when objects are selected. **1** Constraints are displayed.
DynDiGrip	31	Controls the dynamic dimensions displayed during grip stretch editing (DynDiVis =2): **0** Displays none. **1** Resulting dimension. **2** Length change dimension. **4** Absolute angle dimension. **8** Angle change dimension. **16** Arc radius dimension. **31** Displays all.
DynDiVis	1	Controls dynamic dimensions displayed during grip stretch editing: **0** First (in the cycle order). **1** First two (in the cycle order). **2** All (as specified by DynDiGrip).
DynMode	3	Controls dynamic input features. (Click DYN on status bar to turn on hidden modes.) **-3** Both on hidden. **-2** Dimensional input on hidden. **-1** Pointer input on hidden. **0** Off. **1** Pointer input on. **2** Dimensional input on. **3** Both on.

Variable	Default	Meaning
DynPiCoords	0	Toggles pointer input coordinates: **0** Relative. **1** Absolute.
DynPiFormat	0	Toggles pointer input coordinates: **0** Polar. **1** Cartesian.
DynPiVis	1	Controls when pointer input is displayed: **0** When user types at prompts for points. **1** When prompted for points. **2** Always.
DynPrompt	1	Toggles display of prompts in Dynamic Input tooltips.
DynToolTips	1	Toggles which tooltips are affected by tooltip appearance settings: **0** Only Dynamic Input value fields. **1** All drafting tooltips.

E

Variable	Default	Meaning
EdgeMode	0	Toggles edge mode for the Trim and Extend commands: **0** Does not extend. **1** Extends cutting edge.
Elevation	0.0	Specifies the current elevation, relative to current UCS.
EnterpriseMenu (R/O)	"."	Reports the path and *.cui* file name.
ErHighlight	1	Toggle: external references are (0) not highlighted or (1) highlighted when selected in the External References dialog box – and vice versa.
EntExts	*1*	*Controls how drawing extents are calculated:* *0 Extents calculated every time; slows down AutoCAD, but uses less memory.* *1 Extents of every object are cached as a two- byte value (default).* *2 Extents of every object are cached as a four- byte value (fastest but uses more memory).*
EntMods (R/O)	*0*	*Increments by one each time an object is modified since the drawing was opened; range 0 to 4.29497E9.*
ErrNo (R/O)	0	Reports error numbers from AutoLISP, ADS, and Arx.
ErState (R/O)	0	Toggle: reports display of the External References palette.
~~ExeDir~~		*Removed from Release 14.*
Expert	0	Controls the display of prompts: **0** Normal prompts. **1** 'About to regen, proceed?' and 'Really want to turn the current layer off?' **2** 'Block already defined. Redefine it?' and 'A drawing with this name already exists. Overwrite it?' **3** Linetype command messages. **4** UCS Save and VPorts Save. **5** DimStyle Save and DimOverride.
ExplMode	1	Toggles whether the Explode and Xplode commands explode non-uniformly scaled blocks: **0** Does not explode. **1** Explodes.

Variable	Default	Meaning
ExportEPlotFormat	2	Specifies the format in which to export drawings: **0** PDF. **1** DWF. **2** DWFx.
ExportModelSpace	0	Specifies which part of model space to export in DWF/x or PDF format: **0** Drawing as currently displayed. **1** Extents of the drawing. **2** Windowed area.
ExportPageSetup	0	Specifies which page setup to use in exporting drawings in DWF/x or PDF format: **0** Default setup. **1** Overridden setup.
ExportPaperSpace	0	Specifies which part of paper space export in DWF/x or PDF format: **0** Current layout only. **1** All layouts.
ExtMax (R/O)	-1.0E+20, -1.0E+20, -1.0E+20	Upper-right coordinate of drawing extents.
ExtMin (R/O)	1.0E+20, 1.0E+20, 1.0E+20	Lower-left coordinate of drawing extents.
ExtNames	1	Controls the format of named objects: **0** Names are limited to 31 characters, and can include A - Z, 0 - 9, dollar ($), underscore (_), and hyphen (-). **1** Names are limited to 255 characters, and can include A - Z, 0 - 9, spaces, and any characters not used by Windows or AutoCAD for special purposes.

F

Variable	Default	Meaning
FaceterDevNormal	40	Specifies the maximum angle between adjacent mesh surfaces and surface normals; range is any positive angle.
FaceterDevSurface	0.0010	Specifies the accuracy of meshes converted from surfaces and solids; range is any positive number: **0** Turns off this option.
FaceterGridRatio	0.0000	Specifies the maximum aspect ratio for mesh subdivisions; range is 0 to 100. Values <1 are inverted: 0.5 = 2.
FaceterMaxEdgeLength	0.0000	Specifies the maximum edge length for meshes converted from surfaces and solids; range is any positive number.
FaceterMaxGrid	4096	Specifies the maximum grid lines in the U and V directions for meshes converted from surfaces and solids; range is 0 to 4096.
FaceTerMeshType	0	Specifies the type of mesh created from solids and surfaces: **0** Optimized mesh type. **1** Quadrilateral type. **2** Triangular type.
FaceterMinUGrid	0	Specifies the minimum number of grid lines in the U direction for meshes converted from solids and surfaces; range is 0 to 1023: **0** Turns off this option.

Variable	*Default*	*Meaning*
FaceTerMinVGrid	0	Specifies the minimum number of grid lines in the V direction for meshes converted from solids and surfaces; range is 0 to 1023: **0** Turns off this option.
FaceterPrimitiveMode	1	Toggles smoothness settings for meshes converted from solids and surfaces: 0 Applies settings from the Mesh Tesselation Options dialog box. 1 Applies settings from the Mesh Primitive Options dialog box.
FaceterSmoothLev	1	Specifies the level of smoothness for meshes converted from solids and surfaces: **<0** Does not smooth objects. **0** Does not smooth objects. **1** Applies smoothness level 1. **2** Applies smoothness level 2. **3** Applies smoothness level 3.
FacetRatio	0	Controls the aspect ratio of facets on rounded 3D bodies: **0** Creates an *n* by 1 mesh. **1** Creates an *n* by *m* mesh.
FaceTRres	0.5000	Controls the smoothness of shaded and hidden-line objects; range is 0.01 to 10.
FbxImportLog	*1*	*Toggles creation of log files that report the status of imported .fbx files: (0) off, or (1) on.*
~~FfLimit~~	...	*Removed from AutoCAD Release 14.*
FieldDisplay	1	Toggles background to field text: **0** No background. **1** Gray background.
FieldEval	31	Controls how fields are updated: **0** Not updated. **1** Updated with Open. **2** Updated with Save. **4** Updated with Plot. **8** Updated with eTransmit. **16** Updated with regeneration.
FileDia	1	Toggles user interface for file-access commands, like Open and Save: **0** Displays command-line prompts. **1** Displays file dialog boxes.
FilletRad	0.0	Specifies the current fillet radius.
FilletRad3d	1.0	Specifies current fillet radius for the FilletEdge and SurfFillet commands:
FillMode	1	Toggles the fill of solid objects, wide polylines, fills, and hatches.
Flatland (R/O)	*0*	*Obsolete system variable.*
FontAlt	"simplex.shx"	Specifies the font used for missing fonts.
FontMap	"acad.fmp"	Specifies the name of the font mapping file.
Force_Paging	*0*	*Ranges from 0 to 4.29497E9.*

Variable	Default	Meaning
Frame	3	Toggles the visibility of frames on all xrefs, images, and DWF, DWFx, PDF, and DGN underlays; overrides the XClipFrame, ImageFrame, DwfFrame, PdfFrame, DgnFrame commands: **0** Not visible; not plotted **1** Visible; plotted. **2** Visible; not plotted. **3** Uses the settings of the XClipFrame, ImageFrame, DwfFrame, PdfFrame, DgnFrame commands.
FrontZ (R/O)	0.0	Reports the front clipping plane offset.
FullOpen (R/O)	1	Reports whether the drawing is: **0** Partially loaded. **1** Fully open.
FullPlotPath	1	Specifies the format of file name sent to plot spooler: **0** Drawing file name only. **1** Full path and drawing file.

G

Variable	Default	Meaning
GeoLatLongFormat	0	Specifies format of latitude and longitude: **0** Decimal degrees. **1** Degrees, minutes, and seconds.
GeoMarkerVisibility	1	Toggles visibility of geographic markers.
GfAng	0	Controls the angle of gradient fill; 0 to 360 degrees.
GfClr1	"RGB 000,000,255"	Specifies the first gradient color in RGB format.
GfClr2	"RGB 255,255,153"	Specifies the second gradient color in RGB format.
GfClrLum	1.0	Controls the level of gray in one-color gradients: 0 Black. 1 White.
GfClrState	1	Specifies the type of gradient fill: 0 Two-color. **1** One-color.
GfName	1	Specifies the style of gradient fill: 1 Linear. 2 Cylindrical. 3 Inverted cylindrical. 4 Spherical. 5 Inverted spherical. 6 Hemispherical. 7 Inverted hemispherical. 8 Curved. 9 Inverted curved.
GfAShift	0	Controls the origin of the gradient fill: 0 Centered. 1 Shifted up and left.

Variable	Default	Meaning
GlobalOpacity	0	Default level of translucency for all palettes when transparency is turned on: **0** Fully transparent. **100** Fully opaque.
GlobCheck	*0*	*Controls reporting on dialog boxes:* **-1** *Turns off local language.* **0** *Turns off.* **1** *Warns if larger than 640x400.* **2** *Reports size in pixels.* **3** *Displays additional information.*
GridDisplay	2	Determines grid display (sum of bit codes): **0** Grid restricted to area specified by the Limits command. **1** Grid is infinite. **2** Adaptive grid, with fewer grid lines when zoomed out. **4** Displays more grid lines when zoomed in. **8** Grid follows the xy plane of the dynamic UCS.
GridMajor	5	Specifies number of minor grid lines per major line.
GridMode	1	Toggles the display of the grid.
GridStyle	0	Switches grid between dots and lines (bit mode): **0** Lined grid is displayed. **1** Dotted grid displayed in 2D model space. **2** Dotted grid displayed in block editor. **4** Dotted grid displayed in sheets and layout tabs.
GridUnit	0.5,0.5	Controls the x, y spacing of the grid.
GripBlock	0	Toggles the display of grips in blocks: **0** At block insertion point. **1** Of all objects within block.
GripColor	150	Specifies ACI color of unselected grips.
GripContour	251	Specifies ACI color of the grip outline.
GripDynColor	140	Specifies ACI color of grips in dynamic blocks.
GripHot	12	Specifies ACI color of selected grips.
GripHover	11	Specifies ACI grip color when cursor hovers.
GripLegacy	*0*	*Toggle of unknown meaning.*
GripMultifunctional	3	Controls access to multifunctional grips: 0 Disabled. 1 Uses Ctrl cycling and the hot grip shortcut menu. 2 Uses the dynamic menu and the hot grip shortcut menu. 3 Uses Ctrl cycling, the dynamic menu, and the hot grip shortcut menu.
GripObjLimit	100	Controls the maximum number of grips displayed; range is 1 to 32767; 0 = grips never suppressed.

Variable	Default	Meaning
Grips	1	Toggles the display of grips: **0** Grips are not displayed. **1** Grips are displayed, except polyline midpoint segment grips. **2** Midpoint grips are displayed on polyline segments.
GripSize	5	Specifies the size of grip; range 1 - 255 pixels.
~~*GripSizeMesh*~~	~~*3*~~	~~*Specifies the size of grips used for mesh objects; range 1 - 255 pixels.*~~
GripSubObjMode	1	Controls the use of Shift to select subobjects: **0** Grips not displayed as hot when subobjects selected. **1** Face, edge, or vertex grips displayed as hot when subobjects of solids, surfaces, and meshes selected (default). **2** Line and arc segment grips display hot on 2D polylines. **3** Combines actions of 1 and 2.
GripTips	1	Toggles display of grip tips when cursor hovers over custom objects: **0** Grip tips and Ctrl-cycling tooltips not displayed. **1** Grip tips and Ctrl-cycling tooltips displayed (default).
GtAuto	1	Toggles display of grip tools.
GtDefault	0	Toggles which commands are the default commands in 3D views: **0** Scale, Move, and Rotate. **1** 3dScale, 3dMove, and 3dRotate.
GtLocation	0	Controls location of grip tools: **0** Aligns grip tool with UCS icon. **1** Aligns with the last selected object.

H

Variable	Default	Meaning
HaloGap	0	Controls the distance by which haloed lines are shortened in 2D wireframe visual style; a percentage of 1 unit.
Handles (R/O)	On	Reports whether object handles (not grips) can be accessed by applications; obsolete.
HatchBoundSet *(R/O)*	*0*	*Reports the contents of the hatch boundary set:* *0 Current viewport (default).* *1 Existing set.*
HatchType *(R/O)*	*0*	*Specifies the type of hatch pattern:* *0 Predefined (default).* *1 User defined.* *2 Custom.*
HelpPrefix	"C:\Program Files\Autodesk\Autocad 2011 \Help\Index.Html"	Specifies path to HTML help file.
HidePrecision	0	Controls the precision of hide calculations in 2D wireframe visual style: **0** Single precision, less accurate, faster. **1** Double precision, more accurate, but slower (recommended).
HideText	On	Controls the display of text during Hide: **On** Text is neither hidden nor hides other objects, unless text object has thickness. **Off** Text is hidden and hides other objects.

Variable	Default	Meaning
HideXrefScales	*1*	*Toggles inclusion of xrefs in the list of scale factors.*
Highlight	1	Toggles object selection highlighting.
HPAng	0.0	Specifies current hatch pattern angle.
HpAnnotative	0	Toggles whether new hatch patterns take on annotative scale factors: 0 Non-annotative. 1 Annotative.
HpAssoc	1	Toggles associativity of hatches: **0** Not associative. **1** Associative.
HpBackgroundColor	"."	Specifies background fill color for hatch patterns: **"None"** *or* **"."** No background color (default). **1** *through* **255** AutoCAD Color Index *Name* Name of the first seven colors, such as "Red." **RGB:** *or* **HSL:** Red-green-blue or hue-saturation-luminosity. *Colorbook* Color book name and number.
HpBound	1	Controls the object created by the Hatch and Boundary commands: **0** Region. **1** Polyline.
HpBoundRetain	0	Toggles whether boundary objects are (0) not created or (1) created for new hatches; the HpBound system variable determines whether the boundaries are region or polylines:
HpColor	"."	Specifies default color for new hatch patterns: **"None"** *or* **"."** No background color (default). **1** *through* **255** AutoCAD Color Index *Name* Name of the first seven colors, such as "Red." **RGB:** *or* **HSL:** Red-green-blue or hue-saturation-luminosity. *Colorbook* Color book name and number.
HpDlgMode	2	Determines when Hatch and Gradient and Hatch Edit dialog boxes are displayed by the Hatch, Gradient, and HatchEdit commands: **0** Dialog boxes are never displayed; enter the seTtings option to display them. **1** Hatch and Gradient dialog boxes are displayed; Hatch Edit is not. **2** Dialog boxes are displayed when the ribbon is not active.
HpDouble	0	Toggles double hatching.
HpDrawOrder	3	Controls draw order of hatches and fills: **0** None. **1** Behind all other objects. **2** In front of all other objects. **3** Behind the hatch boundary. **4** In front of the hatch boundary.
HpGapTol	0	Controls largest gap allowed in hatch boundaries; ranges from 0 to 5000 units.

Variable	Default	Meaning
HpInherit	0	Toggles how MatchProp copies the hatch origin from source object to destination objects: **0** As specified by HpOrigin. **1** As specified by the source hatch object.
HpIslandDetection	1	Determines how islands are hatched: **0** Alternating islands are hatched (Normal mode). **1** Outermost island only is hatched (Outer mode; default). **2** Everything is hatched (Ignore mode).
HpIslandDetectionMode	1	Toggle: type of island detection (controls HpIslandDetection): 0 Legacy island detection. 1 Shiny new island detection method.
HpLastPattern *(R/O)*	*"Ansi31"*	*Reports the name of the last hatch pattern used.*
HpLayer	"Use Current"	Specifies the name of the layer to use for new hatches and fills.
HpMaxlines	1000000	Controls the maximum number of hatch lines; range is 100 - 10,000,000.
HpName	"ANSI31"	Specifies default hatch name.
HpObjWarning	10000	Specifies the maximum number of hatch boundaries that can be selected before AutoCAD flashes warning message; range is 1 to 1073741823.
HpOrigin	0,0	Specifies the default origin for hatch objects.
HpOriginMode	0	Controls the default hatch origin point: **0** Specified by HpOrigin. **1** Bottom-left corner of hatch's rectangular extents. **2** Bottom-right corner of rectangular extents. **3** Top-right corner of rectangular extents. **4** Top-left corner of rectangular extents. **5** Center of hatch's rectangular extents.
HpOriginStoreAsDefault	*0*	*Toggles whether user-defined hatch origin is stored as default value:* ***0*** *Does not store.* ***1*** *Stores specified origin as default.*
HpQuickPreview	On	Toggles whether preview hatches when the cursor is inside boundaries: **Off** Does not preview. **On** Preview.
HpRelativePs *(R/O)*	*Off*	*Toggles whether pattern scaling is relative to the paper space scale factor:* ***Off*** *Does not scale hatch patterns.* ***On*** *Hatch patterns are scaled*
HpScale	1.0	Specifies the current hatch scale factor; cannot be zero.
HpSeparate	0	Controls number of hatch objects made from multiple boundaries: **0** Single hatch object created. **1** Separate hatch object created.
HpSpace	1.0	Controls the default spacing of user-defined hatches; cannot be zero.

Variable	Default	Meaning
HpTransparency	"."	Sets default translucency percentage for new hatches; has no effect on existing patterns: **"Use current"** *or* **"."** Translucency specified by CeTransparency. **"ByLayer"** *or* **"ByBlock"** Translucency specified by the associated layer or block. **0** *to* **90** Range of translucency from 0 (none) to 90 (mostly transparent).
HyperlinkBase	""	Specifies the path for relative hyperlinks.
I		
ImageFrame	1	Toggles the visibility of frames on images: **0** Not visible; not plotted **1** Visible; plotted. **2** Visible; not plotted.
ImageHlt	0	Toggles image frame highlighting when raster images are selected.
ImpliedFace	1	Toggles detection of implied faces.
IndexCtl	0	Controls creation of layer and spatial indices: **0** No indices created. **1** Layer index created. **2** Spatial index created. **3** Both indices created.
InetLocation	"http://www.autodesk.com"	Specifies the default URL for Browser.
InputHistoryMode	15	Controls the content and location of user input history (bitcode sum of): **0** No history displayed. **1** Displayed at the command line, and in dynamic prompt tooltips accessed with Up and Down arrow keys. **2** Current command displayed in the shortcut menu. **4** All commands displayed in the shortcut menu. **8** Blipmark for recent input displayed in the drawing. **15** All on.
InsBase	0.0,0.0,0.0	Controls the default insertion base point relative to the current UCS for Insert and XRef commands.
InsName	""	Specifies the default block name: **.** No default.
InsUnits	1	Specifies drawing units of blocks dragged into drawings: **0** Unitless. **1** Inches. **2** Feet. **3** Miles. **4** Millimeters. **5** Centimeters. **6** Meters. **7** Kilometers. **8** Microinches. **9** Mils. **10** Yards.

Variable	Default	Meaning
		11 Angstroms. **12** Nanometers. **13** Microns. **14** Decimeters. **15** Decameters. **16** Hectometers. **17** Gigameters. **18** Astronomical Units. **19** Light Years. **20** Parsecs.
InsUnitsDefSource	1	Controls source drawing units value; ranges from 0 to 20; see above.
InsUnitsDefTarget	1	Controls target drawing units; see list above.
IntelligentUpdate	20	Controls graphics refresh rate in frames per second; range is 0 (off) to 100 fps.
InterfereColor	"1"	Specifies color of interference objects.
InterfereObjVs	"Realistic"	Specifies visual style of interference objects.
InterfereVpVs	"3d wireframe"	Specifies visual style during interference checking.
IntersectionColor	257	Specifies ACI color of intersection polylines in 2D wireframe visual style (257=ByEntity).
IntersectionDisplay	Off	Toggles display of 3D surface intersections during Hide command in 2D wireframe: **Off** Does not draw intersections. **On** Draws polylines at intersections.
ISaveBak	1	Toggles creation of *.bak* backup files.
ISavePercent	50	Controls the percentage of waste in saved *.dwg* file before cleanup occurs: **0** Slower full saves. **>0** Faster partial saves. **100** Maximum.
IsoLines	4	Controls the number of contour lines on 3D solids; range is 0 - 2047.

L

Variable	Default	Meaning
LargeObjectSupport	0	Toggles support for large-object support in DWG files: **0** Creates objects compatible with AutoCAD 2009 and earlier. **1** Uses large objects in drawings.
LastAngle (R/O)	0	Reports the end angle of last-drawn arc.
LastPoint	0,0,0	Reports the x,y,z coordinates of the last-entered point.
LastPrompt (R/O)	"*varies*"	Reports the last string on the command line.
Latitude	"37.7950"	Specifies last-used angle of latitude.
LayerDlgMode	1	Toggles the command mapped to the Layer command: 0 ClassicLayer command. 1 LayerPalette command.

Variable	*Default*	*Meaning*
LayerEval	1	Controls when newly-added (unreconciled) layers are detected: **0** Never. **1** When xref layers are added to the drawing **2** When new layers are added to drawings and xrefs.
LayerEvalCtl	1	Controls unreconciled layer filters: **0** Disables notification of new layers. **1** Enables notification, as controlled by the LayerEval sysvar.
LayerFilterAlert	2	Controls the deletion of layer filters in excess of 99 filters *and* the number of layers: **0** Off. **1** Deletes all filters without warning, when layer dialog box opened. **2** Recommends deleting all filters when layer dialog box opened. **3** Displays dialog box for selecting filters to erase, upon opening the drawing.
LayerManagerState (R/O)	0	Reports the status of the Layer Properties Manager palette.
LayerNotify	15	Controls when AutoCAD displays alerts about unreconciled layers (bitcode): **0** Off. **1** Plotting. **2** Opening drawings. **4** Loading, reloading, and attaching xrefs. **8** Restoring layer states. **16** Saving drawings. **32** Inserting blocks, etc.
LayLockFadeCtl	90	Controls the fading of locked layers during the LayIso command: **0** Not faded. **>0** Faded up to 90 percent. **<0** Not faded, but the value is saved.
LayoutRegenCtl	2	Controls display list for layouts: **0** Display-list regen'ed with each tab change. **1** Display-list is saved for model tab and last layout tab. **2** Display list is saved for all tabs.
LazyLoad	*0*	*Toggle: unknown purpose.*
LegacyCtrlPick	0	Toggles function of Ctrl+pick: **0** Selects faces, edges, and vertices of 3D solids. **1** Cycles through overlapping objects. **2** Ctrl+click selects sub-objects when SubObjSelectionMode = 0; otherwise, selects sub-objects when Ctrl is not held down.
LensLength (R/O)	50.0	Reports perspective view lens length, in mm.
LightGlyphDisplay	1	Toggles display of light glyph.
LightingUnits	2	Controls the type of lighting used: **0** Generic lighting. **1** International units of photometric lighting. **2** American units of photometric lighting.
LightListState	0	Toggles display of Light List palette.

Variable	Default	Meaning
LightsInBlocks	1	Toggles use of lights in blocks: **0** Off. **1** On.
LimCheck	0	Toggles drawing limits checking.
LimMax	12.0,9.0	Controls the upper right drawing limits.
LimMin	0.0,0.0	Controls the lower left drawing limits.
LinearBrightness	0	Controls the overall brightness of generic lighting in renderings; range is -10 to 10.
LinearContrast	0	Controls the overall contrast of generic lighting in renderings; range is -10 to 10.
LispInit	*1*	*Toggles AutoLISP functions and variables:* *0 Preserved from drawing to drawing.* *1 Valid in current drawing only.*
Locale (R/O)	"enc"	Reports ISO language code; see DctMain.
LocalRootPrefix (R/O)	"d:\docume..."	Reports the path to folder holding local customizable files.
LockUi	0	Controls the position and size of toolbars and palettes; hold down Ctrl key to unlock temporarily (bitcode sum): **0** Toolbars and palettes unlocked. **1** Docked toolbars locked. **2** Docked palettes locked. **4** Floating toolbars locked. **8** Floating palettes locked.
LoftAng1	90	Specifies angle of loft to first cross section; range is 0 to 359.9 degrees.
LoftAng2	90	Specifies angle of loft to second cross section.
LoftMag1	0.0	Specifies magnitude of loft at first cross section; range is 1 to 10.
LoftMag2	0.0	Specifies magnitude of loft at last cross section.
LoftNormals	1	Specifies location of loft normals: **0** Ruled. **1** Smooth. **2** First normal. **3** Last normal. **4** Ends normal. **5** All normal. **6** Use draft angle and magnitude.
LoftParam	7	Specifies the loft shape: **1** Minimizes twists between cross sections. **2** Aligns start-to-end direction of each cross section. **4** Generates simple solids and surfaces, instead of spline solids and surfaces. **8** Closes the surface or solid between the first and last cross-sections.
LogExpBrightness	65.0	Controls the overall brightness of photometric lighting in renderings; range is 0 to 200.

Variable	Default	Meaning
LogExpContrast	50.0	Controls the overall contrast of photometric lighting in renderings; range is 0 to 100.
LogExpDaylight	2	Controls how daylight is displayed photometric renderings: **0** Off. **1** On. **2** Same as sun status.
LogExpMidtones	1.00	Controls the overall midtones of photometric lighting in renderings; range is 0.01 to 20.
LogExpPhysicalScale	1500.000	Scales photometric lights physically.
LogFileMode	0	Toggles writing command prompts to *.log* file.
LogFileName (R/O)	"...\Drawing1.log"	Reports file name and path for *.log* file.
LogFilePath	"d:\acad 2011\"	Specifies path to the *.log* file.
LogInName (R/O)	"*username*"	Reports user's login name; truncated after 30 characters.
~~*LongFName*~~		*Removed from AutoCAD Release 14.*
Longitude	-122.3940	Specifies current angle of longitude (default = San Francisco).
LTScale	1.0	Controls linetype scale factor; cannot be 0.
LUnits	2	Controls linear units display: **1** Scientific. **2** Decimal. **3** Engineering. **4** Architectural. **5** Fractional.
LUPrec	4	Controls decimal places (or inverse of smallest fraction) of linear units; range is 0 to 8.
LwDefault	25	Controls the default lineweight, in millimeters; must be one of the following values: 0, 5, 9, 13, 15, 18, 20, 25, 30, 35, 40, 50, 53, 60, 70, 80, 90, 100, 106, 120, 140, 158, 200, or 211.
LwDisplay	0	Toggles whether lineweights are displayed; setting saved separately for Model space and each layout tab.
LwUnits	1	Toggles units used for lineweights: **0** Inches. **1** Millimeters.

M

Variable	Default	Meaning
MacroTrace	*0*	*Toggles diesel debug mode.*
MatBrowserState (R/O)	0	Reports state of Materials Browser palette: (0) closed or (1) open.
MatEditorState (R/O)	0	Reports state of Materials Editor palette): (0) closed or (1) open.
MaterialsPath (R/O)	""	Reports the path to the material library folder.
MatState	*0*	*No longer operates.*
MaxActVP	64	Controls the maximum number of viewports to display; range is 2 to 64.

Variable	*Default*	*Meaning*
MaxObjMem	*0*	*Controls the maximum number of objects in memory; object pager is turned off when value = 0, <0, or 2,147,483,647.*
MaxSort	1000	Controls the maximum names sorted alphabetically; range 0 - 32767.
MaxTouches (R/O)	0	Reports number of touch points supported by multi-touch digitizers.
MButtonPan	1	Toggles the behavior of the wheel mouse: **0** Behaves as defined by AutoCAD's *.cui* file. **1** Pans when dragging with wheel.
MeasureInit	0	Toggles drawing units for default drawings: **0** English. **1** Metric.
Measurement	0	Toggles current drawing units: **0** English. **1** Metric.
MenuBar	0	Toggles the display of the menu bar.
MenuCtl	1	Toggles the display of submenus in side menu: **0** Only with menu picks. **1** Also with keyboard entry.
MenuEcho	0	Controls menu and prompt echoing (sum): **0** Displays all prompts. **1** Suppresses menu echoing. **2** Suppresses system prompts. **4** Disables ^P toggle. **8** Displays all input-output strings.
MenuName (R/O)	"acad"	Reports path and file name of *.cui* file.
MeshType	1	Toggles the type of mesh used by the RevSurf, TabSurf, RuleSurf, EdgeSurf, and other commands: **0** Polyfaces (compatible with AutoCAD 2009 and earlier). **1** 3D mesh objects (AutoCAD 2010 and later).
Millisecs (R/O)	*248206921*	*Reports number of milliseconds since timing began.*
MirrHatch	0	Toggles mirroring of hatches: **0** Retains hatch angle. **1** Mirrors hatch angle.
MirrText	0	Toggles text handling by Mirror command: **0** Retains text orientation. **1** Mirrors text.
MLeaderScale	1	Scales mleaders based on: **0.0** Makes scale the ratio of model space viewport to paper space. **>0** Specifies scale factor.
ModeMacro	""	Invokes Diesel macros.
MsLtScale	1	Controls how linetypes are scaled: **0** Not scaled by the annotation scale. **1** Scaled by the annotation scale.

Variable	Default	Meaning
MsmState (R/O)	0	Specifies if Markup Set Manager is active: **0** No. **1** Yes.
MsOleScale	1.0	Controls the size of text-containing OLE objects when pasted in model space: **-1** Scales by value of PlotScale. **0** Scales by value of DimScale. **>0** Uses scale factor.
MTextEd	"Internal"	Controls the name of the MText editor: **.** Uses the default editor. **0** Cancels the editing operation. **-1** Uses the secondary editor. **"blank"** Uses MText internal editor. **"Internal"** Uses MText internal editor. **"oldeditor"** Uses the previous internal editor. **"Notepad"** Uses Windows Notepad editor. **":lisped"** Uses Built-in AutoLISP function. *string* Uses another editor with a name fewer than 256 characters long in the form of this syntax: *:AutoLISPtextEditorFunction#TextEditor.*
MTextFixed	2	Controls the mtext editor appearance: **0** Mtext editor is used. **1** Mtext editor remembers its location. **2** Difficult-to-read text is displayed horizontally at a larger size.
MTJigString	"abc"	Specifies the sample text displayed by mtext editor; maximum 10 letters; enter . for no text.
MTextToolbar	1	Toggles toolbar display during MText: **0** Does not display toolbar. **1** Displays Text Formatting toolbar from AutoCAD 2008 and earlier. **2** Displays ribbon's Text Formatting panel.
MyDocumentsPrefix (R/O)	"C:\Documents and Settings*username*\My Documents"	Reports path to *my documents* folder of the logged-in user.

N

Variable	Default	Meaning
NavBarDisplay	1	Toggles display of navigation bar; applies to all viewports and layouts (NavBar command applies to the current viewport or space): **0** Off. **1** On.
NavSWheelMode	2	Specifies style of steering wheel: **0** Big wheel for viewing objects. **1** Big wheel for touring buildings. **2** Big wheel for full navigation. **3** Big wheel for 2D navigation. **4** Mini wheel for viewing objects. **5** Mini wheel for touring buildings. **6** Mini wheel for full navigation.

Variable	Default	Meaning
NavSWheelOpacityBig	50	Changes translucency of big steering wheel. Range is 25% to 90% (almost opaque).
NavSWheelOpacityMini	150	Changes translucency of mini steering wheel. Range is 25% to 90% (almost opaque).
NavSWheelSizeBig	1	Specifies size of the big steering wheel: **0** Small size. **1** Normal size. **2** Large size.
NavSWheelSizeMini	1	Specifies size of the mini steering wheel: **0** Small size. **1** Normal size. **2** Large size. **3** Extra large.
NavSWheelWalkSpeed	*1.0*	*Sets speed of Walk mode; range is 0.1 to 10.*
NavVCubeDisplay	3	Toggles display of the viewcube: **0** Not displayed. **1** Displayed in 3D visual styles, but not 2D **2** Displayed in 2D visual styles, but not 3D.. **3** Displayed in 2D and 3D visual styles .
NavVCubeLocation	0	Locates viewcube in current viewport: **0** Upper-right corner. **1** Upper-left corner. **2** Lower-left corner. **3** Lower-right corner.
NavVCubeOpacity	50	Sets translucency of viewcube; 0 = displayed only when cursor passes over it.
NavVCubeOrient	1	Specifies viewcube's orientation: **0** Relative to WCS. **1** Relative to current UCS.
NavVCubeSize	1	Sizes the viewcube: **0** Small size. **1** Normal size. **2** Large size.
NodeName (R/O)	*"AC$"*	*Reports the name of the network node; range is one to three characters.*
NoMutt	0	Toggles display of messages (a.k.a. muttering) during scripts, LISP, macros: **0** Displays prompt, as normal. **1** Suppresses muttering.
NorthDirection	0	Specifies angle of sun relative to positive y axis, and N direction of viewcube.
NwfState	*1*	*Displays New Features Workshop at AutoCAD start.*

O

Variable	Default	Meaning
ObjectIsolationMode	0	Toggles display of hidden and isolated objects between sessions: **0** Hidden and isolated objects displayed for the current drawing session only. **1** Hidden and isolated settings saved for the next drawing session.

Variable	*Default*	*Meaning*
ObscuredColor	257	Specifies ACI color of objects obscured by Hide in 2D wireframe visual style.
ObscuredLtype	0	Specifies linetype of objects obscured by Hide in 2D wireframe visual mode: **0** Invisible. **1** Solid. **2** Dashed. **3** Dotted. **4** Short dash. **5** Medium dash. **6** Long dash. **7** Double short dash. **8** Double medium dash. **9** Double long dash. **10** Medium long dash. **11** Sparse dot.
OffsetDist	-1.0	Controls current offset distance: <**0** Offsets through a specified point. ≥**0** Uses default offset distance.
OffsetGapType	0	Controls how polylines reconnect when segments are offset: **0** Extends segments to fill gap. **1** Fills gap with fillet (arc segment). **2** Fills gap with chamfer (line segment).
OleFrame	2	Controls visibility of OLE frames: **0** Frame is not displayed and not plotted. **1** Frame is displayed and is plotted. **2** Frame is displayed but is not plotted.
OleHide	0	Controls display and plotting of OLE objects: **0** All OLE objects visible. **1** Visible in paper space only. **2** Visible in model space only. **3** Not visible.
OleQuality	3	Specifies quality of displaying and plotting of embedded OLE objects: **0** Monochrome. **1** Low quality graphics. **2** High quality graphics. **3** Automatically selected mode.
OleStartup	0	Toggles loading of OLE source applications to improve plot quality: **0** Does not load OLE source application. **1** Loads OLE source app when plotting.
OpenPartial	1	Determines whether drawings can be edited when not fully loaded: **0** Drawings must be fully open. **1** Visible portions of drawings can be edited before file is fully open.
OpmState	0	Toggles whether Properties palette is active.
OrthoMode	0	Toggles orthographic mode.

Variable	Default	Meaning
OsMode	4133	Controls current object snap mode (sum): **0** NONe. **1** ENDpoint. **2** MIDpoint. **4** CENter. **8** NODe. **16** QUAdrant. **32** INTersection. **64** INSertion. **128** PERpendicular. **256** TANgent. **512** NEARest. **1024** QUIck. **2048** APPint. **4096** EXTension. **8192** PARallel. **16383** All modes on. **16384** Object snap turned off via OSNAP on the status bar.
OSnapCoord	2	Controls whether keyboard overrides object snap: **0** Object snap overrides keyboard. **1** Keyboard overrides object snap. **2** Keyboard overrides object snap, except in scripts.
OSnapHatch	*0*	*Toggles whether hatches are snapped:* *0 Osnaps ignore hatches.* *1 Hatches are snapped.*
OSnapNodeLegacy	1	Toggles whether osnap snaps to mtext insertion points.
OsnapZ	0	Toggles osnap behavior in z direction: **0** Uses the z-coordinate. **1** Uses the current elevation setting.
OsOptions	3	Determines when objects with negative z values are osnapped: **0** Uses the actual z coordinate. **1** Substitutes z coordinate with the elevation of the current UCS.

P

Variable	Default	Meaning
PaletteOpaque	0	Controls transparency of palettes: **0** Turned off by user. **1** Turned on by user. **2** Unavailable, but turned on by user. **3** Unavailable, and turned off by user.
PaperUpdate	0	Toggles how AutoCAD plots layouts with paper size different from plotter's default: **0** Displays a warning dialog box. **1** Changes paper size to that of the plotter configuration file.

Variable	*Default*	*Meaning*
ParameterCopyMode	1	Determines how constraints and referenced variables are copied: **0** Does not copy constraints. **1** Copies constraints and variables; expressions are constants. **2** Copies constraints, variables, and expressions; references variables; replaces data with constants. **3** Copies constraints, variables, and expressions; references variables; adds variables, if necessary. **4** Copies all; adds variables, if necessary.
ParametersStatus (R/O)	0	Reports whether the Parameters Manager palette is open: **0** Closed. **1** Open.
PdfFrame	1	Toggles the display of the PDF frame for editing and plotting: **0** Not displayed; not plotted. **1** Displayed; plotted. **2** Displayed; not plotted.
PdfOsnap	1	Toggles osnapping between frame and objects in attached PDF files: **0** Geometry not osnapped. **1** Geometry osnapped.
PDMode	0	Controls point display style (sum): **0** Dot. **1** No display. **2** +-symbol. **3** x-symbol. **4** Short line. **32** Circle. **64** Square.
PDSize	0.0	Controls point display size: >**0** Absolute size, in pixels. **0** 5% of drawing area height. <**0** Percentage of viewport size.
PEditAccept	0	Toggles display of the PEdit command's 'Object selected is not a polyline. Do you want to turn it into one? <Y>:' prompt.
PEllipse	0	Toggles object used to create ellipses: **0** True ellipses. **1** Polyline arcs.
Perimeter (R/O)	0.0	Reports perimeter calculated by Area, DbList, and List commands.
Perspective	0	Toggles perspective mode; not available in 2D wireframe visual mode.
PerpectiveClip	5.0	Controls position of eye point clipping, as a percentage; range is 0.01% - 10.0%.
PFaceVMax (R/O)	4	Reports the maximum vertices per 3D face.
PHandle	*0*	*Ranges from 0 to 4.29497E9; unknown usage.*

Variable	Default	Meaning
PickAdd	2	Toggles meaning of Shift key on selection sets: **0** Adds to selection set. **1** Removes from selection set. **2** Persistent; keeps objects selected after the Select command ends (hold down Shift to remove objects from the selection set.
PickAuto	1	Toggles selection set mode: **0** Single pick mode. **1** Automatic windowing and crossing.
PickBox	3	Controls selection pickbox size; range is 0 to 50 pixels.
PickDrag	0	Toggles selection window mode: **0** Pick two corners. **1** Pick a corner; drag to second corner.
PickFirst	1	Toggles command-selection mode: **0** Enter command first. **1** Select objects first.
PickStyle	1	Controls how groups and associative hatches are selected: **0** Includes neither. **1** Includes groups. **2** Includes associative hatches. **3** Includes both.
Platform (R/O)	*"varies"*	Reports the name of the operating system.
PLineConvertMode	0	Determines how polylines are converted to splines: **0** Linear segments. **1** Arc segments.
PLineGen	0	Toggles polyline linetype generation: **0** From vertex to vertex. **1** From end to end.
PLineType	2	Controls automatic conversion and creation of 2D polylines by PLine: **0** Does not convert; creates old-format polylines. **1** Does not convert; creates optimized lwpolylines. **2** Converts polylines in older drawings on open; PLine creates optimized lwpolyline objects.
PLineWid	0.0	Controls current polyline width.
~~*PlotId*~~	""	*Obsolete; has no effect in AutoCAD.*
PlotOffset	0	Toggles the plot offset measurement: **0** Relative to edge of margins. **1** Relative to edge of paper.
PlotRotMode	2	Controls the orientation of plots: **0** Lower left = 0,0. **1** Lower left plotter area = lower left of media. **2** X, y-origin offsets calculated relative to the rotated origin position.
~~*Plotter*~~	*0*	*Obsolete; has no effect in AutoCAD.*

Variable	Default	Meaning
PlotTransparencyOverride	1	Controls when translucent objects are plotted: **0** Does not plot translucency. **1** Uses the settings of the PageSetup and Plot commands. **2** Plots object transparency always; converts drawings to raster, which can take longer to complete.
PlQuiet	0	Toggles display during batch plotting and scripts (replaces CmdDia): **0** Displays plot dialog boxes and nonfatal errors. **1** Logs nonfatal errors; does not display plot dialog boxes.
PointCloudAutoUpdate	1	Toggle: automatic regeneration of point clouds during editing and real time panning, zooming, and orbiting: 0 Manual regeneration required; has no effect on PointCloudRtDensity. 1 Automatically regenerated.
PointCloudDensity	15	Specifies percentage of points to display simultaneously as a percentage of 1,500,000 (the maximum number of points per drawing).
PointCloudLock	0	Toggles locking of point clouds: **0** Not locked. **1** Locked; point clouds cannot be edited, moved, or rotated.
PointCloudRtDensity	5	Specifies percentage of points to display during real time zooming, panning, or orbiting. .
PolarAddAng	""	Holds a list of up to 10 user-defined polar angles; each angle can be up to 25 characters long, each separated with a semicolon (;). For example: 0;15;22.5;45.
PolarAng	90	Controls the increment of polar angle; contrary to Autodesk documentation, you can specify any angle.
PolarDist	0.0	Controls the polar snap increment when SnapStyl is set to 1 (isometric).
PolarMode	0	Controls polar and object snap tracking: **0** Measures polar angles based on current UCS (absolute), track orthogonally; does not use additional polar tracking angles; acquires object tracking points automatically. **1** Measures polar angles from selected objects (relative). **2** Uses polar tracking settings in object snap tracking. **4** Uses additional polar tracking angles (via PolarAng). **8** Acquires object snap tracking points by pressing Shift.
PolySides	4	Controls the default number of polygon sides; range is 3 to 1024.
Popups (R/O)	1	Reports display driver support of AUI: **0** Not available. **1** Available.
PreviewEffect	2	Controls the visual effect for previewing selected objects: **0** Dashed lines. **1** Thick lines. **2** Thick dashed lines.

Variable	Default	Meaning
PreviewFaceEffect	1	Toggles preview selection highlighting of face sub-objects: **0** Faces not highlighted. **1** Faces highlighted with texture fill.
PreviewFilter	7	Controls the exclusion of objects from selection previewing (bitcode sum): **0** No objects excluded. **1** Objects on locked layers. **2** Objects in xrefs. **4** Tables. **7** Objects on locked layers, in xrefs, and tables. **8** Multiline text. **16** Hatch patterns. **32** Groups.
PreviewType	0	Specifies the view for drawing thumbnails: **0** Use last saved view. **1** Use Home view.
Product (R/O)	"AutoCAD"	Reports the name of the software.
Program (R/O)	"acad"	Reports the name of the software's executable file.
ProjectName	""	Controls the project name of the current drawing; searches for xref and image files.
ProjMode	1	Controls the projection mode for Trim and Extend commands: **0** Does not project. **1** Projects to xy plane of current UCS. **2** Projects to view plane.
ProxyGraphics	1	Toggles saving of proxy images in drawings: **0** Not saved; displays bounding box. **1** Image saved with drawing.
ProxyNotice	1	Toggles warning message displayed when drawing contains proxy objects.
ProxyShow	1	Controls the display of proxy objects: **0** Not displayed. **1** All displayed. **2** Bounding box displayed.
ProxyWebSearch	0	Toggles checking for object enablers: **0** Does not check for object enablers. **1** Checks for object enablers when an Internet connection is present.
PsLtScale	1	Toggles paper space linetype scaling: **0** Uses model space scale factor. **1** Uses viewport scale factor.
PSolHeight	4.0	Specifies default height of polysolid objects.
PSolWidth	0.25	Specifies default width of polysolid objects.
PsProlog	*""*	*Specifies the PostScript prologue file name.*
PsQuality	*75*	*Controls resolution of PostScript display, in pixels:* *<0 Display as outlines; no fill.* *0 Displays no fills.* *>0 Displays filled.*

Variable	Default	Meaning
PStyleMode	1	Toggles the plot color matching mode of the drawing: **0** Uses named plot style tables. **1** Uses color-dependent plot style tables.
PStylePolicy (R/O)	1	Reports whether the object color is associated with its plot style: **0** Not associated. **1** Associated.
PsVpScale	0.0	Controls the view scale factor (ratio of units in paper space to units in newly-created model space viewports) 0 = scaled to fit.
PublishAllSheets	1	Determines which sheets (model space and layouts) are loaded automatically into the Publish command's list: **0** Current drawing only. **1** All open drawings.
PublishCollate	1	Controls how sheets are published: **0** Sheet sets are processed one sheet at a time; a separate plot is created for each sheet. **1** Sheets sets are processed as a single job and single plot file.
PublishHatch	1	Determines how hatch patterns are exported in DWF format to Impression: **0** Treated as separate objects. **1** Treated as a single object.
PUcsBase (R/O)	""	Reports name of UCS defining the origin and orientation of orthographic UCS settings; in paper space only.

Q

Variable	Default	Meaning
QAFlags	*0*	*Controls quality assurance flags:* *0 Turned off.* *1 ^C metacharacter cancels grips, just as if user pressed Esc.* *2 Long text screen listings do not pause.* *4 Error and warning messages displayed in command line, instead of in dialog boxes.* *128 Screen picks accepted via the AutoLISP (command) function.*
QaUcsLock	*0*	*Toggle; purpose unknown.*
QcState	0	Toggles whether QuickCalc palette is open.
QpLocation	0	Locates the Quick Properties palette: **0** Near cursor as defined by DSettings command's Quick Properties tab. **1** Floating, in a constant location.
QpMode	-1	Controls Quick Properties palette display: **0** (or negative numbers) Off. **1** On for all objects **2** On only for objects defined in CUI.
QTextMode	0	Toggles quick text mode.
QueuedRegenMax	*2147483647*	*Ranges between very large and very small numbers.*
QvDrawingPin	0	Toggles pinning of drawing quick views.
QvLayoutPin	0	Toggles pinning of layout quick views.

Variable	Default	Meaning
R		
R14RasterPlot	*0*	*Toggle; purpose unknown.*
RasterDpi	300	Controls the conversion of millimeters or inches to pixels, and vice versa; range is 100 to 32767.
RasterPercent	20	Percentage of system memory to allocate to plotting raster images; range is 0 - 100%.
RasterPreview (R/O)	1	Toggles creation of BMP preview image.
RasterThreshold	20	Specifies amount of RAM to allocate to plotting raster images; range is 0 - 2000MB.
Rebuild2dCv	6	Specifies number of control vertices per spline; range is 2 to 32767.
Rebuild2dDegree	3	Specifies degree for splines; range is 1 to 11: **1** Splines are like straight lines (no bends). **2** Splines are parabolic (one bend). **3** Splines are cubic Beziers (two bends).
Rebuild2dOption	1	Toggle: rebuilding of splines: **0** Original curves not deleted. **1** Original curves are deleted.
RebuildDegreeU	3	Specifies U-direction degree for NURBS surfaces; range is 2 to 11.
RebuildDegreeV	3	Specifies V-direction degree for NURBS surfaces; range is 2 to 11.
RebuildOptions	1	Determines fate of original surfaces and trimmed areas on rebuilt surfaces: **0** Keeps original surfaces; trimmed areas are not applied. **1** Deletes original surfaces; trimmed areas are not applied. **2** Keeps original surfaces; trimmed areas applied to rebuilt objects. **3** Deletes original surfaces; trimmed areas are applied.
RebuildU	6	Specifies default number of lines in the U direction for NURBS surfaces; range is 2 to 32767.
RebuildV	6	Specifies the default number of lines in the V direction for NURBS surfaces; range is 2 to 32767.
RecoverAuto	0	Displays recovery notifications of damaged drawing files: **0** Displays a dialog box when opening damaged files; interrupts running scripts. **1** Recovers damaged files automatically; displays dialog box report on recovered file; does not interrupt scripts. **2** Recovers damaged files automatically; displays report on the recovered file at the command prompt.
RecoveryMode	2	Controls recording of drawing recovery information after software failure: **0** Note recorded. **1** Recorded; Drawing Recovery palette does not display automatically. **2** Recorded, and Drawing Recovery palette displays automatically.
RefEditName	""	Specifies the reference file name when in reference-editing mode.

Variable	Default	Meaning
RegenMode	1	Toggles regeneration mode: **0** Regens with each view change. **1** Regens only when required.
Re-Init	0	Controls the reinitialization of I/O devices: **1** Digitizer port. *2 Plotter port; obsolete.* **4** Digitizer. *8 Plotter; obsolete.* **16** Reloads PGP file.
RememberFolders	1	Toggles the path search method: **0** Path specified in AutoCAD desktop properties is default for file dialog boxes. **1** Last path specified by each file dialog box is remembered.
RenderPrefsState	1	Toggles display of the Render Preferences palette.
RenderUserLights	1	Determines which lights are rendered: **0** Default *or* user-defined lights are rendered. **1** Default *and* user-defined lights are rendered.
ReportError	1	Determines if AutoCAD sends an error report to Autodesk: **0** Does not create error reports. **1** Error reports are generated and sent to Autodesk.
RibbonContextSelect	1	Determines when context-sensitive ribbon tabs are displayed: 0 Never. 1 Single-clicking objects or selection sets.. 2 Double-clicking objects or selection sets.
RibbonContextSelLim	2500	Determines the maximum number of objects that activate context-sensitive ribbon tabs; range is 0 to 32767.
RibbonDockedHeight	0	Determines the height of the ribbon when docked: **0** Ribbon sizes itself to height of the selected tab. **1** Minimum height in pixels. **500** Maximum height in pixels.
RibbonSelectMode	1	Determines fate of selections after command is selected from ribbon: **0** Pickfirst selection set is unselected. **1** Pickfirst selection set remains selected.
RibbonState (R/O)	1	Reports whether the ribbon is open.
~~*RIAspect*~~		*Removed from AutoCAD Release 14.*
~~*RIBackG*~~		*Removed from AutoCAD Release 14.*
~~*RIEdge*~~		*Removed from AutoCAD Release 14.*
~~*RIGamit*~~		*Removed from AutoCAD Release 14.*
~~*RIGrey*~~		*Removed from AutoCAD Release 14.*
~~*RIThresh*~~		*Removed from AutoCAD Release 14.*
RoamableRootPrefix (R/O)	"d:\documents and settings*username*\application data\aut..."	Reports the path to the root folder where roamable customized files are located.

Variable	Default	Meaning
RolloverOpacity	0	Specifies the translucency of palettes when cursor moves over them; range is 0 to 100.
RolloverTips	1	Toggles display of rollover tips.
RTDisplay	1	Toggles raster display during real-time zoom and pan: **0** Displays the entire raster image. **1** Displays raster outline only.

S

Variable	Default	Meaning
SaveFidelity	1	Toggles how annotative objects are translated to earlier releases: **0** Makes no changes. **1** Saves each scale representation to a separate layer.
SaveFile (R/O)	""	Reports the automatic save file name.
SaveFilePath	"...\temp\"	Specifies the path for automatic save files.
SaveName (R/O)	""	Reports the drawing's save-as file name.
SaveTime	10	Controls the automatic save interval, in minutes; 0 = disable auto save.
ScreenBoxes (R/O)	0	Reports the maximum number of menu items supported by display; 0 = screen menu turned off.
ScreenMode (R/O)	3	Reports the state of AutoCAD display: **0** Text screen. **1** Graphics screen. **2** Dual-screen display.
ScreenSize (R/O)	*varies*	Reports the current viewport size, in pixels, such as 719.0,381.0.
SDI	*0*	*Controls the multiple-document interface (SDI is "single document interface"):* *0 Turns on MDI.* *1 Turns off MDI. (Only one drawing may be loaded into AutoCAD.)* *2 Disables MDI for apps that cannot support MDI; read-only.* *3 (R/O) Disables MDI for apps that cannot support MDI, even when SDI= 1.*
SelectionAnnoDisplay	1	Toggles how alternate scale representations are displayed when annotative objects selected: **0** Does not display them. **1** Displays them according to the value of XFadeCtl.
SelectionArea	1	Toggles use of colored selection areas.
SelectionAreaOpacity	25	Controls the opacity of color selection areas; range is 0 (transparent) to 100 (opaque).
SelectionCycling	0	Controls selection cycling: **0** Off. **1** On but does not display the list dialog box. **2** On and displays the list dialog box of objects that you be selected.
SelectionPreview	3	Controls selection preview: **0** Off. **1** On when commands are inactive. **2** On when commands prompt for object selection.

Variable	Default	Meaning
SelectSimilarMode	130	Controls matchable properties for SelectSimilar command (bit code): **0** Object type. **1** Color. **2** Layer. **4** Linetype. **8** Linetype scale. **16** Lineweight. **32** Plot style. **64** Text styles, dimension styles, table styles, and so on. **128** Names of referenced objects, such as blocks, xrefs, and images. **130** 2+128 = layer and name of referenced objects.
SetBylayerMode	255	Controls which properties are affected by the SetByLayer command: **0** None. **1** Colors. **2** Linetypes. **4** Lineweights. **8** Materials. **16** Plot styles. **32** ByBlock is changed to Bylayer. **64** Blocks are changed from ByBlock to ByLayer. **128** Includes transparency property. **255** All.
ShadEdge	3	Controls shading by Shade command: **0** Shades only faces. **1** Shades faces, and shows edges in background color. **2** Show edges only in object color. **3** Shows faces in object color, edges in background color.
ShadeDif	70	Controls percentage of diffuse to ambient light; range is 0 to 100%.
ShadowPlaneLocation	0.0	Specifies the default height of the shadow plane.
ShortcutMenu	11	Controls display of shortcut menus (sum): **0** Does not display shortcut menus. **1** Displays default shortcut menus. **2** Displays edit shortcut menus. **4** Displays command shortcut menus when commands are active. **8** Displays command shortcut menus only when options available at the command line. **16** Displays shortcut menus when the right button held down longer.
ShowHist	1	Toggles display of history in solids: **0** Does not display original solids. **1** Displays original solids depending on Show History property settings. **2** Displays all original solids.
ShowLayerUsage	1	Toggles layer-usage icons in Layers dialog box.
ShowMotionPin	1	Specifies pinning of Show Motion images.
ShpName	""	Specifies the default shape name: . Set to no default.

Variable	Default	Meaning
SigWarn	1	Toggles display of dialog box when drawings with digital signatures are opened: **0** Displays only when signature is invalid. **1** Displays always.
SketchInc	0.1	Controls the Sketch command's recording increment.
SkPoly	0	Controls sketch line mode: **0** Records as lines. **1** Records as a polyline. **2** Records as a spline.
SkTolerance	0.5	Specifies how closely spline fits to freehand sketches; range is 0 to 1.
SkyStatus	0	Reports status of sun and sky background: **0** Sky is off. **1** Sky is on. **2** Sky and illumination are on.
SmoothMeshConvert	0	Controls how solids and surfaces are converted to meshes: **0** Converts to smooth models; optimizes or merges coplanar faces. **1** Converts to smooth models; retains original mesh faces. **2** Converts to flattened faces; optimizes or merges coplanar faces. **3** Converts to flattened faces; retains original mesh faces.
SmoothMeshGrid	3	Determines how the facet grid is displayed: **0** Does not display. **1** Displays for smoothing levels 0 and 1. **2** Displays for smoothing levels 0, 1, and 2. **3** Displays levels 0, 1, 2, and 3.
SmoothMeshMaxFace	10000000	Specifies maximum faces on meshes; range is 1 to 16000000.
SmoothMeshMaxLev	4	Specifies maximum smoothness level for meshes; range is 1 - 255.
SmState *(R/O)*	*0*	*Reports state of Show Motion interface: (0) closed or (1) open. (read-only)*
SnapAng	0	Controls rotation angle for snap and grid; when not 0, grid lines are not displayed.
SnapBase	0.0,0.0	Controls current origin for snap and grid.
SnapIsoPair	0	Controls current isometric drawing plane: **0** Left isoplane. **1** Top isoplane. **2** Right isoplane.
SnapMode	0	Toggles snap mode.
SnapStyl	0	Toggles snap style: **0** Normal. **1** Isometric.
SnapType	0	Toggles snap for the current viewport: **0** Standard snap. **1** Polar snap.
SnapUnit	0.5,0.5	Controls x, y spacing for snap distances.
SolidCheck	1	Toggles solid validation.

Variable	Default	Meaning
SolidHist	1	Toggles retention of history in solids.
SortEnts	127	Controls object display sort order: **0** Off. **1** Object selection. **2** Object snap. **4** Redraw. **8** Slide generation. **16** Regeneration. **32** Plot. **64** PostScript output.
SpaceSwitch	*1*	*Either 1 or 9; purpose unknown.*
SplDegree	3	Specifies default degree of new splines created with control vertices; range is 1 to 5.
SplFrame	0	Toggles displays of helix and 3D mesh object controls. (Use CvShow and CvHide for polylines and polyface meshes.) **0** Does not display control polygon of helixes; displays smoothed mesh objects, and does not display invisible edges of 3D faces or polyface meshes. **1** Displays control polygons of helixes; displays unsmoothed mesh objects that are smoothed; displays edges of 3D faces and polyface meshes.
SplineSegs	8	Controls number of line segments that define splined polylines; range is -32768 to 32767. **<0** Draws with fit-curve arcs. **>0** Draws with line segments.
SplineType	6	Controls type of spline curve: **5** Quadratic Bezier spline. **6** Cubic Bezier spline.
SplKnots	0	Specifies default knot setting when using fit points for new splines: 0 Chords. 1 Square root chords. 2 Uniform.
SplMethod	0	Toggles default type of new splines: (0) fit or (1) control vertices.
SsFound	""	Specifies path and file name of sheet sets.
SsLocate	1	Toggles whether sheet set files are opened with drawing: **0** Not opened. **1** Opened automatically.
SsmAutoOpen	1	Toggles whether the Sheet Set Manager is opened with drawing (SsLocate must be 1): **0** Not opened. **1** Opened automatically.
SsmPollTime	60	Controls time interval between automatic refreshes of status data in sheet sets; range is 20 to 600 seconds (SsmSheetStatus = 2).

Variable	Default	Meaning
SsmSheetStatus	2	Controls refresh of status data in sheet sets: **0** Does not refresh automatically. **1** Refreshes when sheet set is loaded or updated. **2** Also refreshes as specified by SsmPollTime.
SsmState (R/O)	0	Toggles whether Sheet Set Manager is open.
StandardsViolation	2	Controls whether alerts are displayed when CAD standards are violated: **0** Displays no alerts. **1** Displays alert when CAD standard violated. **2** Displays icon on status bar when file is opened with CAD standards, and when non-standard objects are created.
Startup	0	Controls which dialog box is displayed by the New and QNew commands: **0** Displays Select Template dialog box. **1** Displays Startup and Create New Drawing dialog box.
~~*StartupToday*~~		~~*Removed from AutoCAD 2004.*~~
StatusBar	1	Controls display of application and drawing status bars: **0** Hides both status bars. **1** Displays the application status bar. **2** Displays both status bars. **3** Displays the drawing status bar.
StepSize	6.0	Specifies length of steps in walk mode; range is 1E-6 to 1E+6.
StepsPerSec	2	Specifies speed of steps in walk mode; range is 1-30.
SubObjSelectionMode	0	Specifies which subobjects are selected by Ctrl+click: **0** All. **1** Only vertices. **2** Only edges. **3** Only faces. **4** Only history subobjects.
SunPropertiesState	0	Toggles display of Sun Properties palette.
SunStatus	0	Toggles display of light by the sun.
SurfaceAssociativity	1	Toggles surface associativity: **0** Surfaces have no associativity to other surfaces. **1** Surfaces adjust automatically to modifications made to related surfaces; DelObj is ignored.
SurfaceAssociativityDrag	1	Specifies how associative surfaces react during dragging (move) operations: **0** No previews; display updated following the drag operations. **1** Reviews first surface only; other surfaces updated following drag operation. **2** Previews all surfaces.
SurfaceAutoTrim	0	Toggles automatic trimming of surfaces by projected geometry: **0** Does not trim. **1** Trims.
SurfaceModelingMode	0	Determines default 3D surface type: **0** Creates procedural surfaces. 1 Creates NURBS surfaces.

Variable	Default	Meaning
SurfOffsetConnect	*0*	*Toggles connection between offset surfaces (0) off or (1) on:*
SurfTrimAutoExtend	*1*	*Toggles automatic extension of trim geometry (0) off or (1) on:*
SurfTrimProjection	*0*	*Toggles projection of trim entities onto surfaces (0) off or (1) on:*
SurfTab1	6	Controls density of m-direction surfaces and meshes; range 5 - 32766.
SurfTab2	6	Specifies density of n-direction surfaces and meshes; range 2 - 32766.
SurfType	6	Controls smoothing of surface by PEdit: **5** Quadratic Bezier spline. **6** Cubic Bezier spline. **8** Bezier surface.
SurfU	6	Controls surface density in m-direction; range is 2 to 200.
SurfV	6	Specifies surface density in n-direction; range is 2 to 200.
SysCodePage (R/O)	"ANSI_1252"	Reports the system code page; set by operating system.

T

Variable	Default	Meaning
TableIndicator	1	Toggles display of column letters and row numbers during table editing.
TableToolbar	1	Toggles display of cell editing toolbar: **0** Does not display. **1** Displays. **2** Displays commands on ribbon.
TabMode	0	Toggles tablet mode.
Target (R/O)	0.0,0.0,0.0	Reports target coordinates in the current viewport.
Taskbar	*1*	*Toggles whether each drawing appears as a button on the Windows taskbar.*
TbCustomize	1	Toggles whether toolbars can be customized.
TDCreate (R/O)	*varies*	Reports the date and time that the drawing was created, such as 2448860.54014699.
TDInDwg (R/O)	*varies*	Reports the duration since the drawing was loaded, such as 0.00040625.
TDuCreate (R/O)	*varies*	Reports the universal date and time when the drawing was created, such as 2451318. 67772165.
TDUpdate (R/O)	*varies*	Reports the date and time of last update, such as 2448860.54014699.
TDUsrTimer (R/O)	*varies*	Reports the decimal time elapsed by user-timer, such as 0.00040694.
TDuUpdate (R/O)	*varies*	Reports the universal date and time of the last save, such as 2451318.67772165.
TempOverrides	1	Toggles temporary overrides.
TempPrefix (R/O)	"d:\temp"	Reports the path for temporary files set by Temp variable.
TextEd (R/O)	0	Reports the status of the text editor: **0** Closed. **1** Open.
TextEval	0	Toggles the interpretation of text input during the -Text command: **0** Literal text. **1** Read **(** and **!** as AutoLISP code.

Variable	Default	Meaning
TextFill	1	Toggles the fill of TrueType fonts when plotted: **0** Outline text. **1** Filled text.
TextOutputFileFormat	0	Controls Unicodes for plot and text window log files: **0** ANSI format. **1** UTF-8 (Unicode). **2** UTF-16LE (Unicode). **3** UTF-16BE (Unicode).
TextQlty	50	Controls the resolution of TrueType fonts when plotted; range is 0 to 100.
TextSize	0.2	Controls the default height of text (2.5 in metric units).
TextStyle	"Standard"	Specifies the default name of text style.
Thickness	0.0	Controls the default object thickness.
ThumbSize	1	Specifies the size of thumbnail images: **0** 64x64 pixels. **1** 128x128 pixels. **2** 256x256 pixels.
TileMode	1	Toggles the view mode: **0** Displays layout tab. **1** Displays model tab.
TimeZone	-80000	Specifies current time zone.
ToolTipMerge	0	Toggles the merging of tooltips during dynamic display.
ToolTips	1	Toggles the display of tooltips.
TpState (R/O)	0	Reports if Tool Palettes palette is open.
TraceWid	0.0500	Specifies current width of traces.
TrackPath	0	Controls display of polar and object snap tracking alignment paths: **0** Displays object snap tracking path across the entire viewport. **1** Displays object snap tracking path between the alignment point and "From point" to cursor location. **2** Turns off polar tracking path. **3** Turns off polar and object snap tracking paths.
TransparencyDisplay	1	Toggles display of translucency in objects: (0) off or (1) on.
TrayIcons	1	Toggles the display of the tray on status bar.
TrayNotify	1	Toggles service notifications displayed by the tray.
TrayTimeout	0	Controls length of time that tray notifications are displayed; range is 0 to 10 seconds.
TreeDepth	3020	Controls the maximum branch depth (in *xxyy* format): *xx* Model-space nodes. *yy* Paper-space nodes. *>0* 3D drawing. *<0* 2D drawing.
TreeMax	10000000	Controls the memory consumption during drawing regeneration.

Variable	Default	Meaning
TrimMode	1	Toggles trims during Chamfer and Fillet: **0** Leaves selected edges in place. **1** Trims selected edges.
TSpaceFac	1.0	Controls the mtext line spacing distance measured as a factor of "normal" text spacing; ranges from 0.25 to 4.0.
TSpaceType	1	Controls the type of mtext line spacing: **1** At Least adjusts line spacing based on the height of the tallest character in a line of mtext. **2** Exactly uses the specified line spacing; ignores character height.
TStackAlign	1	Controls vertical alignment of stacked text: **0** Bottom aligned. **1** Center aligned. **2** Top aligned.
TStackSize	70	Controls size of stacked text as a percentage of the current text height; range is 25 to 125%.

U

Variable	Default	Meaning
UcsAxisAng	90	Controls the default angle for rotating the UCS around an axis (via the UCS command using the X, Y, or Z options); valid values limited to: 5, 10, 15, 18, 22.5, 30, 45, 90, or 180.
UcsBase	""	Specifies name of UCS that defines the origin and orientation of orthographic UCS settings.
UcsDetect	1	Toggles dynamic UCS mode.
UcsFollow	0	Toggles view displayed with new UCSs: **0** Does not change. **1** Aligns UCS with new view automatically.
UcsIcon	3	Controls display of the UCS icon: **0** Off. **1** On at lower-left corner. **2** On at UCS origin, if possible.
UcsName (R/O)	""	Reports the name of current UCS view: **""** Current UCS is unnamed.
UcsOrg (R/O)	0.0,0.0,0.0	Reports origin of current UCS relative to WCS.
UcsOrtho	1	Controls whether the related orthographic UCS settings are restored automatically: **0** Does not change UCS setting when orthographic view is restored. **1** Restores related ortho UCS automatically when an ortho view is restored.
UcsView	1	Toggles whether the current UCS is saved with a named view.
UcsVp	1	Toggles whether the UCS in active viewports remains fixed (locked) or changes (unlocked) to match the UCS of the current viewport.
UcsXDir (R/O)	1.0,0.0,0.0	Reports the x-direction of current UCS relative to WCS.
UcsYDir (R/O)	0.0,1.0,0.0	Reports the y-direction of current UCS relative to WCS.

Variable	Default	Meaning
UndoCtl (R/O)	21	Reports the status of undo (bitsum): **0** Disabled. **1** Enabled. **2** Limited to one command. **4** Set to auto-group mode. **8** Set to active group mode. **16** Set to combined zooms and pans.
UndoMarks (R/O)	0	Reports the number of undo marks.
UnitMode	0	Toggles the type of units display: **0** As set by Units command. **1** As entered by user.
UOSnap	1	Toggles object snapping for geometry in DWF, DWFx, PDF, and DGN underlays; overrides the DwfOsnap, PdfOsnap, and DgnOsnap system variables: **0** Disabled. **1** Enabled. **2** Setting dependent on DwfOsnap, PdfOsnap, and DgnOsnap.
UpdateThumbnail	15	Controls how thumbnails are updated (sum): **0** Thumbnail previews not updated. **1** Sheet views updated. **2** Model views updated. **4** Sheets updated. **8** Updated when sheets or views are created, modified, or restored. **16** Updated when the drawing is saved.
UseAcis	*0*	*Toggle involving ACIS.*
UserI1 *thru* **UserI5**	0	Stores user-definable integer variables.
UserR1 *thru* **UserR5**	0.0	Stores user-definable real variables.
UserS1 *thru* **UserS5**	""	Stores user-definable string variables; values are not saved.

V

Variable	Default	Meaning
ViewCtr (R/O)	*varies*	Reports x, y, z coordinates of center of current view, such as 15,9,56.
ViewDir (R/O)	*varies*	Reports current view direction relative to UCS: 0,0,1=plan view.
ViewMode (R/O)	0	Reports the current view mode: **0** Normal view. **1** Perspective mode on. **2** Front clipping on. **4** Back clipping on. **8** UCS-follow on. **16** Front clip not at eye.
ViewSize (R/O)	*varies*	Reports the height of current view in drawing units.
ViewTwist (R/O)	0	Reports the twist angle of current view.
VisRetain	1	Controls xref drawing's layer settings: **0** Does not save xref-dependent layer settings in the current drawing. **1** Saves xref-dependent layer settings in the current drawing, and take precedence over settings in the xref'ed drawing the next time the current drawing is loaded.

Variable	Default	Meaning
VpLayerOverrides (R/O)	0	Reports whether VP layer properties are overridden in the current viewport: **0** None. **1** At least one layer.
VpLayerOverridesMode	1	Controls display and plot of VP layer property overrides: **0** Not displayed or plotted. **1** Displayed and plotted.
VpMaximizedState (R/O)	0	Reports whether viewport is maximized by VpMax command.
VpRotateAssoc	1	Toggles whether the contents are rotated with viewports: **0** Views are not rotated by viewports. **1** Views are rotated by viewports.
VsaCurvatureHigh	1.0	Specifies value above which surface curvature is displayed green; range is any real number.
VsaCurvatureLow	-1.0	Specifies value below which surface curvature is displayed blue; range is any real number.
VsaCurvatureType	0	Controls type of curvature analysis used by AnalysisCurvature command: **0** Gaussian evaluates areas of high and low curvature. **1** Mean evaluates mean curvature of u and v surface values. **2** Maximum evaluates maximum curvature of u and v surface values. **3** Minimum evaluates minimum curvature of u and v surface values.
VsaDraftangleHigh	3	Specifies draft angle above which portions of model are displayed green during draft analysis: **-90** Surface is parallel to UCS with surface normal facing opposite direction from construction plane. **0** Surface is perpendicular to the construction plane. **90** Surface is parallel to construction plane with surface normal facing same direction as UCS.
VsaDraftangleLow	-3	Specifies draft angle below which portions of model are displayed with blue during draft analysis: **-90** Surface is parallel to UCS with surface normal facing opposite direction from construction plane. **0** Surface is perpendicular to construction plane. **90** Surface is parallel to construction plane with surface normal facing same direction as UCS.
VsaZebraColor1	"Rgb:255,255,255"	Specifies first color of zebra analysis stripes; uses ACI, RGB, HSL, or ColorBook values.
VsaZebraColor2	"Rgb:0,0,0"	Specifies second color of zebra analysis stripes.
VsaZzebraDirection	90	Specifies angle of zebra stripes; range is 0 to 90 degrees.
VsaZebraSize	45	Specifies width of stripes in zebra analysis displays; range is 1 to 100 pixels.
VsaZebraType	1	Controls type of zebra analysis display: **0** Chrome ball effect. **1** Cylinder effect.

Variable	Default	Meaning
VsBackgrounds	1	Toggles display of backgrounds in visual styles.
VsEdgeColor	"ByEntity"	Specifies the edge color.
VsEdgeJitter	-2	Specifies the level of pencil effect: **0** *or* **-*n*** None. **1** Low. **2** Medium. **3** High.
VsEdgeOverhang	-6	Specifies extension of pencil lines beyond edges; range is 1 to 100 pixels; -*n* = none.
VsEdges	1	Specifies types of edges to display: **0** Displays no edges. **1** Display isolines. **2** Displays facets and edges.
VsEdgeSmooth	1	Specifies crease angle; range is 0 to 180 degrees.
VsEdgeLEx	-6	Specifies length of line extensions of edges in visual styles. Negative number turns off extensions; range is 0 to 100 pixels.
VsFaceColorMode	0	Determines color of faces.
VsFaceHighlight	-30	Specifies the size of highlights; range is -100 to 100. Ignored when VsMaterialMode is 1 or 2 and objects have materials attached.
VsFaceOpacity	-60	Controls the transparency/opacity of faces; range is -100 to 100 (fully opaque).
VsFaceStyle	0	Determines how faces are displayed: **0** No rendering. **1** Real. **2** Gooch.
VsHaloGap	0	Specifies halo gap; range is 0 to 100 pixels.
VsHidePrecision	0	Toggles accuracy of hides and shades.
VsIntersectionColor	"7 (white)"	Specifies the color of intersecting polylines.
VsIntersectionEdges	0	Toggles the display of intersecting edges.
VsIntersectionLtype	1	Specifies the linetype for intersecting lines: **0** Off. **1** Solid. **2** Dashed. **3** Dotted. **4** Short dash. **5** Medium dash. **6** Long dash. **7** Double-short dash. **8** Double-medium dash. **9** Double-long dash. **10** Medium-long dash. **11** Sparse dot.
VsIsoOntop	0	Toggles whether isolines are displayed.

Variable	*Default*	*Meaning*
VsLightingQuality	1	Toggles the quality of lighting: **0** Displays facets. **1** Smooths facets. **2** Turns on per-pixel lighting.
VsMaterialMode	0	Controls the display of material finishes: **0** Does not display materials or texturees. **1** Displays materials only. **2** Displays materials and textures.
VSMax (R/O)	*varies*	Reports the upper-right corner of virtual screen, such as 37.46,27.00,0.00.
VSMin (R/O)	*varies*	Reports the lower-left corner of virtual screen, such as -24.97,-18.00,0.0.
VsMonoColor	"RGB:255,255,255"	Specifies the monochrome tint.
VsObscuredColor	"ByEntity"	Specifies color of obscured lines.
VsObscuredEdges	1	Toggles display of obscured edges.
VsObscuredLtype	1	Specifies linetype of obscured lines; see VsIntersectionLtype.
VsOccludedColor	"Byentity"	Specifies color of hidden (occluded) lines in visual styles; uses ACI, RGB, HSL, or ColorBook values.
VsOccludedEdges	1	Toggles display of hidden (occluded) edges in visual styles: (0) off or (1) on.
VsOccludedLtype	1	Controls linetype of hidden (occluded) lines in visual styles. Changing this system variable creates a new unsaved visual style. Range is 1 to 11: **1** Solid lines (default for most visual styles). **2** Dashed lines (default for hidden and shaded with edges). **3** Dotted lines. **4** Short dashes. **5** Medium dashes. **6** Long dashes. **7** Double short dashes. **8** Double medium dashes. **9** Double long dashes. **10** Medium long dashes. **11** Sparse dots.
VsShadows	0	Determines the quality of shadows: **0** Does not display shadows. **1** Displays ground shadows. **2** Displays full shadows.
VsSilhEdges	0	Toggles the display of silhouette edges.
VsSilhWidth	5	Specifies the width of silhouette edge lines; range is 1 to 25 pixels.
VsState	0	Toggles the Visual Styles palette.
VtDuration	750	Controls the duration of smooth view transition; range is 0 to 5000 seconds.

Variable	Default	Meaning
VtEnable	3	Controls smooth view transitions for pans, zooms, view rotations, and scripts: **0** Turned off. **1** Pan and zoom. **2** View rotation. **3** Pan, zoom, and view rotation. **4** During scripts only. **5** Pan and zoom during scripts. **6** View rotation during scripts. **7** Pan, zoom, and view rotation during scripts.
VtFps	7	Controls minimum speed for smooth view transitions; range is 1 to 30 frames per second.

W

Variable	Default	Meaning
WbDefaultBrowser (R/O)	*0*	*Specifies which Web browser is used for the new Web-based help system:* 0 Default system browser. 1 Internet Explorer.
WbHelpOnline (R/O)	*1*	*Toggles Web-based help is accessed from Autodesk when online: (0) local access only or (1) local and Autodesk online access.*
WbHelpType (R/O)	*1*	*Toggles the source of help files:* ***0*** *CHM (traditional compiled help files).* ***1*** *HTML (hypertext markup language).*
WhipArc	0	Toggles display of circular objects: **0** Displays as connected vectors. **1** Displays as true circles and arcs.
WhipThread	1	Controls multithreaded processing on two CPUs (if present) during redraws and regens: **0** Performs single-threaded calculations. **1** Makes regenerations multi-threaded. **2** Makes redraws multi-threaded. **3** Makes regens and redraws multi-threaded.
WindowAreaColor	150	Specifies ACI color of windowed selection area.
WmfBkgnd	Off	Toggles background of *.wmf* files: **Off** Background is transparent. **On** Background is same as AutoCAD's background color.
WmfForegnd	Off	Toggles foreground colors of exported WMF images: **Off** Foreground is darker than background. **On** Foreground is lighter than background.
WorldUcs (R/O)	1	Toggles matching of WCS with UCS: **0** Current UCS does not match WCS. **1** UCS matches WCS.
WorldView	1	Toggles view during 3dOrbit, DView, and VPoint commands: **0** Current UCS. **1** WCS.

Variable	*Default*	*Meaning*
WriteStat (R/O)	1	Toggle: reports whether *.dwg* file is read-only: **0** Drawing file cannot be written to. **1** Drawing file can be written to.
WsAutosave	0	Toggles automatic saving of changes to workspaces: (0) off or (1) on.
WsCurrent	"*varies*"	Controls name of current workspace.

X

Variable	*Default*	*Meaning*
XClipFrame	2	Toggles visibility of xref clipping boundary: **0** Hides and does not plot the xclip frame. **1** Displays and plots xclip frames. **2** Displays but does not plot xclip frames.
XDwgFadeCtl	70	Specifies the amount attached xrefs are shown faded: **0** Not faded. **>0** Increasingly faded, to a maximum of 90%. **<0** Not faded, but fade value is retained.
XEdit	1	Toggles editing of xrefs: **0** Cannot in-place refedit. **1** Can in-place refedit.
XFadeCtl	50	Controls faded display of objects not being edited in-place: **0** No fading; minimum value. **90** 90% fading; maximum value.
XLoadCtl	2	Controls demand loading: **0** Demand loading turned off; entire drawing is loaded. **1** Demand loading turned on; xref file opened. **2** Demand loading turned on; a *copy* of the xref file is opened.
XLoadPath	"...\temp"	Specifies path for storing temporary copies of demand-loaded xref files.
XRefCtl	0	Toggles creation of *.xlg* xref log files.
XrefNotify	2	Controls notification of updated and missing xrefs: **0** Displays no alert. **1** Displays icon indicating xrefs are attached; yellow alert indicates missing xrefs. **2** Also displays balloon messages when an xref is modified.
XrefType	0	Toggles xrefs: **0** Attached. **1** Overlaid.

Z

Variable	*Default*	*Meaning*
ZoomFactor	60	Controls the zoom level via mouse wheel; range from 3 to 100.
ZoomWheel	0	Switches the zoom direction when mouse wheel is rotated forward: **0** Zooms in (opposite of most other software, including Inventor). **1** Zooms out.

3

Variable	Default	Meaning
3dConversionMode	1	Controls how material and light definitions are converted when pre-AutoCAD 2008 drawings are opened: **0** None converted. **1** Converted automatically. **2** Converted after prompted.
3dDwfPrc	2	Level of precision in drawings exported as *.dwf* files: **1** 1 **2** 0.5 **3** 0.2 **4** 0.1 **5** 0.01 **6** 0.001
3dOsMode	11	Specifies the current 3D object snaps modes (bit code): **0** Disables all 3D osnap modes. **1** Temporarily disables all 3D osnap modes (ZNON). **2** Snaps to vertices (ZVER). **4** Snaps to midpoints of edges (ZMID). **8** Snaps to centers of faces (ZCEN). **16** Snaps to spline knots (ZKNO). **32** Snaps to the perpendiculars of faces (ZPER). **64** Snaps to objects nearest to faces (ZNEA). **128** Turns on all 3D osnap modes. **11** 1 + 2 + 8 = center of a face and vertex.
3dSelectionMode	1	Toggles how visually overlapping objects are selected (other than in 2D and 3D wireframe visual modes; this system variable is slated for removal in a future release of AutoCAD): **0** Uses traditional 3D selection. 1 **Uses line-of-sight selection.**

APPENDIX D

Command Aliases

The following commands have aliases, as defined by the *acad.pgp* file.

Command	Alias(es)
A	
actrecord	arr
actstop	ars
-actstop	-ars
actuserinput	aru
actusermessage	arm
-actusermessage	-arm
adcenter	adc, adcenter, content, dc, dcenter
align	al
allplay	aplay
analysiscurvature	curvatureanalysis
analysisdraftangle	draftangleanalysis
analysiszebra	zebra
appload	ap
arc	a
area	aa
array	ar
-array	-ar
attdef	att
-attdef	-att
attdef	ddattdef
attedit	ate
-attedit	-ate, atte
attedit	ddatte
attext	ddattext
attipedit	ati
B	
baction	ac
bclose	bc
bcparameter	cparam
bedit	be
block	acadblockdialog, b, bmake, bmod
-block	-b
boundary	bo, bpoly
-boundary	-bo
bparameter	param
break	br
bsave	bs
bvstate	bvs
C	
camera	cam
chamfer	cha
change	-ch
checkstandards	chk
circle	c
color	col, colour, ddcolor
commandline	cli
constraintbar	cbar
constraintsettings	csettings
copy	co, cp
ctablestyle	ct
cvadd	insertcontrolpoint
cvhide	pointoff
cvrebuild	rebuild
cvremove	removecontrolpoint
cvshow	pointon
cylinder	cyl
D	
dataextraction	dx
datalink	dl
datalinkupdate	dlu
dbconnect	aad, aex, ali, aro, ase, asq, dbc
ddedit	ed
ddgrips	gr
ddvpoint	vp
delconstraint	delcon
dist	di
divide	div
donut	do, doughnut
drawingrecovery	drm
draworder	dr
dsettings	ddrmodes, ds, se
dsviewer	av

Command	Alias(es)
dview	dv

Dimensions

Command	Alias(es)
dimaligned	dal, dimali
dimangular	dan, dimang
dimarc	dar
dimbaseline	dba, dimbase
dimcenter	dce
dimconstraint	dcon
dimcontinue	dco, dimcont
dimdiameter	ddi, dimdia
dimdisassociate	dda
dimedit	ded, dimed
dimjogged	djo, jog
dimjogline	djl
dimlinear	dimhorizontal, dimlin, dimrotated, dimvertical
dimlinear	dli
dimordinate	dimord
dimordinate	dor, dimover, dov
dimradius	dimrad, dra
dimreassociate	dre
dimstyle	d, ddim, dimsty, dst
dimtedit	dimted

E

Command	Alias(es)
editshot	eshot
ellipse	el
erase	e
explode	x
export	exp
-export	-qpub
exportdwf	edwf
exportdwfx	edwfx
exportpdf	epdf
-exporttoautocad	aectoacad
extend	ex
externalreferences	er
extrude	ext

F

Command	Alias(es)
fillet	f
filter	fi
flatshot	fshot

G

Command	Alias(es)
geographiclocation	geo, north, northdir
geomconstraint	gcon
gradient	gd
group	g
-group	-g

H

Command	Alias(es)
hatch	bh, h
-hatch	-h
hatchedit	he
hatchtoback	hb
hide	hi
hidepalettes	poff

I

Command	Alias(es)
image	im
-image	-im
imageadjust	iad
imageattach	iat
imageclip	icl
import	imp
insert	ddinsert, i, inserturl
-insert	-i
insertobj	io
interfere	inf
intersect	in
isolateobjects	isolate

J

Command	Alias(es)
join	j

L

Command	Alias(es)
layer	ddlmodes, la
-layer	-la
layerstate	las, lman
-layout	lo
leader	lead
lengthen	len
line	l

Command	Alias(es)
linetype	ddltype, lt, ltype
-linetype	-lt, -ltype
list	li, ls, showmat
ltscale	lts
lweight	lineweight, lw

M

Command	Alias(es)
markup	msm
matbrowseropen	mat, rmat
matchprop	ma, painter
materialmap	setuv
materials	finish, rmat, mat
measure	me
measuregeom	mea
meshcrease	crease
meshrefine	refine
meshsmooth	convtomesh, smooth
meshsmoothless	less
meshsmoothmore	more
meshsplit	split
meshuncrease	uncrease
mirror	mi
mirror3d	3dmirror
mleader	mld
mleaderalign	mla
mleadercollect	mlc
mleaderedit	mle
mleaderstyle	mls
mline	ml
move	m
mspace	ms
mtext	mt, t
-mtext	-t
mview	mv

N

Command	Alias(es)
navsmotion	motion
navsmotionclose	motioncls
navswheel	wheel
navvcube	cube
newshot	nshot
newview	nview

Command	Alias(es)

O

Command	Alias(es)
offset	o
open	dxfin, openurl
options	op, preferences
osnap	ddosnap, os
-osnap	-os

P

Command	Alias(es)
pan	p
-pan	-p
parameters	par
-parameters	-par
-partialopen	partialopen
pastespec	pa
pedit	pe
pline	pl
plot	dwfout, print
plotstamp	ddplotstamp
point	po
pointcloud	pc
pointcloudattach	pcattach
pointcloudindex	pcindex
pointlight	freepoint
polygon	pol
polysolid	psolid
preview	pre
properties	ch, ddchprop, ddmodify, mo, pr, props
propertiesclose	prclose
pspace	ps
publishtoweb	ptw
purge	pu
-purge	-pu
pyramid	pyr

Q

Command	Alias(es)
qleader	le
quickcalc	qc
quickcui	qcui
quit	exit
qvdrawing	qvd
qvdrawingclose	qvdc
qvlayout	qvl
qvlayoutclose	qvlc

Command	Alias(es)

R

Command	Alias(es)
rectang	rec
rectang	rectangle
redraw	r
redrawall	ra
regen	re
regenall	rea
region	reg
rename	ren
-rename	-ren
render	rr
rendercrop	rc
renderenvironment	fog
renderpresets	rfileopt, rp
renderwin	rendscr, rw
revolve	rev
ribbon	dashboard
ribbonclose	dashboardclose
rotate	ro
rpref	rpr

S

Command	Alias(es)
save	saveurl
saveas	dxfout
scale	sc
script	scr
section	sec
sectionplane	splane
sectionplanejog	jogsection
sectionplanetoblock	generatesection
sequenceplay	splay
setvar	set
shademode	sha, shade
sheetset	ssm
showpalettes	pon
slice	sl
snap	sn
solid	so
spell	sp
spline	spl
splinedit	spe
standards	sta
stretch	s
style	ddstyle, st
subtract	su
surfblend	blendsrf
surfextend	extendsrf
surffillet	filletsrf
surfnetwork	networksrf
surfoffset	offsetsrf
surfpatch	patch
surfsculpt	createsolid

T

Command	Alias(es)
table	tb
tablestyle	ts
tablet	ta
text	dt, dtext
textedit	tedit
thickness	th
tilemode	ti, tm
tolerance	tol
toolbar	to
toolpalettes	tp
torus	tor
trim	tr

U

Command	Alias(es)
ucs	dducs
ucsman	dducs, dducsp, uc
union	uni
unisolateobjects	unhide, unisolate
units	ddunits, un
-units	-un

V

Command	Alias(es)
view	ddview, v
-view	-v
viewgo	vgo
viewplay	vplay
visualstyles	vsm
-visualstyles	-vsm
vpoint	-vp
vports	viewports
vscurrent	vs

Command	Alias(es)
W	
wblock	acadwblockdialog, w
-wblock	-w
wedge	we
X	
xattach	xa
xbind	xb
-xbind	-xb
xclip	xc
xline	xl
xref	xr
-xref	-xr
Z	
zoom	z
3	
3dalign	3al
3darray	3a
3dface	3f
3dmove	3m
3dorbit	3do, orbit
3dpoly	3p
3dprint	3dp, 3dplot, rapidprototype
3drotate	3r
3dscale	3s
3dwalk	3dnavigate, 3dw

APPENDIX E

Undocumented Commands

The following commands operate in AutoCAD, but are not documented by Autodesk. Some of these commands are documented in this book. And some may require an ARX file to be loaded.

A

AdcCustomNavigate	Points to folder containing registered custom objects for DesignCenter's Custom tab.
Ai_Box	Draws square and rectangular 3D polyface box objects.
Ai_Cone	Draws 3D polyface cone objects with optional truncated tops.
Ai_Dish	Draws 3D polyface dish objects (bottom half of spheres).
Ai_Dome	Draws 3D polyface dome objects (top half of spheres).
Ai_Mesh	Draws 3D polymesh objects.
Ai_Pyramid	Draws 3D polyface pyramid objects with three- or four-sided bases and optional truncated tops.
Ai_Sphere	Draws 3D polyface sphere objects.
Ai_Torus	Draws 3D polyface donut objects.
Ai_Wedge	Draws 3D polyface wedge objects.
Ai_CircTan	Draws circles tangent to three points.
Ai_Custom_Safe	Accesses Autodesk support web pages.
Ai_Product_Support_ Safe	Accesses Autodesk support web pages.
Ai_Training_Safe	Accesses Autodesk support web pages.
Ai_Access_SubCtr	Accesses Autodesk support web pages.
AiDimFlipArrow	Reverses the direction of selected arrowheads.
AiDimPrec	Changes the displayed precision of existing dimensions.
AiDimStyle	Saves and applies preset dimension styles
Ai_Dim_TextAbove	Moves dimension text above dimension lines to match the JIS standard.
Ai_Dim_TextCenter	Centers text vertically on the dimension line, but not horizontally.
Ai_Dim_TextHome	Centers text horizontally on the dimension line, but not vertically.
AiDimTextMove	Moves dimension text (0) with the dimension line, (1) with a leader, or (2) independently of the dimension line.
Ai_EditCustFile	Edits custom files.
Ai_Fms	Switches drawings to layout mode, and then goes to floating model space (short for floating model space).
Ai_Pspace	Switches drawings to layout mode.
Ai_SelAll	Selects all objects in drawings in the current space, including those off-screen.
AiObjectScaleAdd	Adds the current annotative scale to annotative objects.
AiObjectScaleRemove	Removes the current annotative scale from annotative objects.
AllPlay	Plays all show-motion views in sequence, without prompts.
Antz	Controls the ViewCube orientation from the command line; displays all SteeringWheels at once.

B

Browser2	Prompts for a web address, and then launches the web browser.

C

CaptureTransients	Toggles the status of transient capturing.
Clip-Tahoe+1	Acts as a placeholder for the unified Clip command from older releases of AutoCAD.
CustomerInvolvementProgram	Asks if you want to provide information to Autodesk about your use of AutoCAD.

D

DbConnect / DbClose	Open and close the dbConnect Manager palette that connects objects with rows in external database tables.
DimHorizontal	Draws horizontal linear dimensions, like DimLinear with the H option.
DimRotated	Draws rotated linear dimensions, like DimLinear with the R option.
DimVertical	Draws vertical linear dimensions, like DimLinear with the V option.
DumpMemAlloc	Controls memory dumps at the command line.
Dxfin	Imports DXF files; displays the Select File dialog box.
Dxfout	Exports DXF files; to change export options, choose Tools	Options.

E

EditTableCell	Performs all cell editing operations at the command prompt.
ExAcReload	Reloads xrefs, and then reports whether they have changed.
-Export	Specifies the default output format, DWF, DWFx, or PDF.

F

FbxListMaterials	Lists the names of materials in the current drawing.
FbxSetMaterial	Sets the default material for the current drawing.

G

GenerateSection	Generates sections of 3D solid models.
GlClipBackOn	Positions the back clipping plane.
GlClipFront	Positions the front clipping plane.
GsAutoOrbit	Rotates the contents of the current viewport around the x and y axes.
GsAutoZoomPan	Zooms and pans the contents of the current viewport.
Gsb1	Rotates, pans, and zooms the current drawing, and then displays the result.
Gsb2	Runs the wireframe and gouraud-shaded benchmarks at different speeds.
Gsb3	Runs the wireframe and gouraud-shaded benchmarks at different speeds.
Gsb4	Runs the wireframe and gouraud-shaded benchmarks at different speeds.
Gsb5	Runs the benchmark in wireframe, hidden-line, flat-shaded, and gouraud-shaded modes.
GsbXyAutomated	Spins the content of the current viewport, and then reports the time and speed results.
GsbXy	Rotates 3D model in all rendering modes about the x and y axes, as listed below.
GsbXyFlat	Runs benchmark in flat-shaded mode about the x and y axes.
GsbXyG3	Runs benchmark in a faster gouraud-shaded mode about the x and y axes.
GsbXyGouraud	Runs benchmark in gouraud-shaded mode about the x and y axes.
GsbXyHidden	Runs benchmark in hidden-line removal mode about the x and y axes.
GsbXyWireframe	Runs benchmark in wireframe mode about the x and y axes.
GsClipBack	Positions the back clipping plane.
GsClipFrontOn	Positions the front clipping plane.
GsDolly	Moves the viewpoint along an axis.

GsDebug	Selects a debugging mode: Bias, CloudFailure, nVerts, deVice (OpenGL, DirectX 9 or 10), GrabImage (save as PNG), HUD (heads-up display), and Multisample rate.
GsOrbit	Rotates the display in 3D.
GsPan	Pans the display around.
GsVsTest	Displays arrays of boxes, spheres, teapots, and other shapes made of VsTestEntity objects.
GsZoom	Zooms the display in and out.

I

ImpledFaceX	Extracts loops from 3D solids to create regions.
InkBorderDisplay	Toggles the display of a border around the ink objects. (You first must load the asdktpctest.arx application into AutoCAD with the AppLoad command.)
Inkcolor	Specifies the color of the ink.
Inkgesture	Enables ink gesture mode in AutoCAD.
Inkhilite	Toggles highlight mode.
Inkolecreate	Creates OLE (object linking and embedding) objects of ink.
Inkpenwidth	Specifies the width of ink.
Inkreco	Converts ink to text by attempting to recognize the handwriting.
Inkrline	Toggles redline mode.
Inktransparency	Toggles redline mode.

M

MaterialAssign	Assigns materials to objects.
MtEdit	Acts as a hardwire alias for the TextEdit command as of AutoCAD 2010.
Mtprop	Changes the properties of multiline text.

O

OleConvert	Converts OLE objects, if possible, although it rarely is.
OleOpen	Opens OLE objects in their source applications.
OleReset	Resets OLE objects.
OptChProp	Changes the styles of objects at the command line: text, dimension, table, or multiline style.

P

PartialCui	Loads CUIX files.
PCloud	Displays a point cloud made of RenderBuffer objects.
Psfill	Fills 2D polyline outlines with raster PostScript patterns.
PsOut	Exports the current drawing as an encapsulated PostScript file.

Q

QuickProperties	Displays a floating panel of object properties near the cursor.

R

R14PenWizard	Create color-dependent plot style tables.
-RenderOutputSize	Presets the size (in pixels) of the next rendering.
Resize	Resizes the AutoCAD window to 942 x 534.

S

SequencePlay Plays back showmotion animations of all views in a single category.

T

TCloud Displays a point cloud made of triangles.

3

3dPan2 Pans drawings in real time.

3dZoom2 Zooms drawings in real time.

AutoCAD 2011 Icons & Shortcuts

Command Prefixes

' Specifies transparent command:
`From point: 'zoom`

'? Provides context-sensitive help:
`Command: line '?`

~ Forces display of file dialog box:
`Command: -insert ~`

- Forces display on command line:
`Command: -mtext`

+ Prompts for tab number:
`Command: +options`

. Forces use of undefined command:
`Command: .line`

_ Forces English cmd in int'l version:
`Command: _line`

multiple Automatically repeats command:
`Command: multiple circle`

(Begins AutoLISP functions

) Ends AutoLISP function:
`Radius: (/ 3.2 2.0)`

$(Begins Diesel macro

Coordinate Input Symbols

Default directions:
X positive = Right
Y positive = Up
Z positive = Right-hand rule

x,y 2D Cartesian coordinates:
`From point: 2,3`

x,y,z 3D Cartesian coordinates:
`From point: 2,3,4`

@x,y Relative coordinates:
`From point: @2,3`

#x,y Absolute coordinates:
`From point: #2,3`

d<a 2D polar coordinates:
`From point: 2<45`

d<a,z 3D cylindrical coordinates:
`From point: 2<45,5`

@d<a,z Relative cylindrical coordinates:
`From point: @2<45,5`

d<a<a 3D spherical coordinates:
`From point: 5<45<30`

New in AutoCAD 2011

Angle Modifiers

Angle defaults:
0 degrees = East (3 o'clock)
Positive direction = Counterclockwise

'	Minutes
"	Seconds
.	Decimal of a degree or seconds
d	Degrees
r	Radian
g	Grad
<	Angle
<<	Force 0 degrees = east, and decimal degrees
<<<	Force 0=east, but uses current format
+	Counterclockwise
–	Clockwise

Selection Set Modes

Ctrl+*pick*	Selects faces
Pick	Selects one object
ALL	Selects all objects
AU	AUtomatic: *(pick)* or BOX
BOX	Left to right = Crossing; right to left = Window
C	Crossing
CP	Crossing polygon
F	Fence
G	Group
L	Last
M	Multiple (no highlighting)
P	Previous
SI	Single selection
SU	Selects faces, edges, or vertices
W	Window
WP	Window polygon

Modify selections:

A	Adds to selection set (default)
R	Removes from selection set
Shift	Removes from selection set
U	Undoes changes to selection set

Option Modifiers

tt	Tracking
m2p	Midpoint between two points
from	Offsets from temp ref point
(mouse icon)	Direct distance entry: move mouse, and enter a distance